Calcium Signaling

A subject collection from *Cold Spring Harbor Perspectives in Biology*

OTHER SUBJECT COLLECTIONS FROM COLD SPRING HARBOR PERSPECTIVES IN BIOLOGY

The Golgi
Germ Cells
The Mammary Gland as an Experimental Model
The Biology of Lipids: Trafficking, Regulation, and Function
Auxin Signaling: From Synthesis to Systems Biology
The Nucleus
Neuronal Guidance: The Biology of Brain Wiring
Cell Biology of Bacteria
Cell–Cell Junctions
Generation and Interpretation of Morphogen Gradients
Immunoreceptor Signaling
NF-κB: A Network Hub Controlling Immunity, Inflammation, and Cancer
Symmetry Breaking in Biology
The Origins of Life
The p53 Family

Calcium Signaling

A subject collection from *Cold Spring Harbor Perspectives in Biology*

EDITED BY

Martin D. Bootman
The Babraham Institute

Michael J. Berridge
The Babraham Institute

James W. Putney
National Institutes of Health

H. Llewelyn Roderick
*The University of Cambridge
and The Babraham Institute*

COLD SPRING HARBOR LABORATORY PRESS
Cold Spring Harbor, New York • www.cshlpress.com

Calcium Signaling

A Subject Collection from *Cold Spring Harbor Perspectives in Biology*
Articles online at www.cshperspectives.org

All rights reserved
© 2012 by Cold Spring Harbor Laboratory Press, Cold Spring Harbor, New York
Printed in the United States of America

Executive Editor	Richard Sever
Managing Editor	Maria Smit
Project Manager	Barbara Acosta
Permissions Administrator	Carol Brown
Cover Designer	Denise Weiss
Publisher	John Inglis

Front cover artwork: The image depicts the complex regulation of adenylyl cyclase (AC) by Ca^{2+}. Ca^{2+} and cyclic AMP are key intracellular messengers with many bidirectional links to control each other's activity. In particular, Ca^{2+} controls the production of cyclic AMP in a variety of ways. Such regulation can be direct or via an intermediary Ca^{2+}-sensitive enzyme. Some ACs specifically respond to Ca^{2+} derived from store-operated Ca^{2+} channels (SOCCs). Ca^{2+} can also regulate AC by binding to calmodulin (CaM), and the Ca^{2+}/CaM complex can then affect AC activity. Ca^{2+}-bound CaM can also activate Ca^{2+}/CaM-activated kinase and calcineurin, both of which may regulate AC. Indirectly, $G\beta\gamma$ subunits from $G\alpha q$-linked receptors can also regulate AC activity. In addition, $G\alpha q$ can activate phospholipase C, which converts phosphatidylinositol 4,5-bisphosphate (PIP_2) to diacylglycerol and inositol trisphosphate ($InsP_3$). DAG activates protein kinase C (PKC), which can also modulate the activity of AC. $InsP_3$ binds to, and activates, its receptors ($InsP_3R$) on the endoplasmic reticulum (ER), thereby releasing Ca^{2+} from the ER stores into the cytoplasm. This emptying of the ER Ca^{2+} stores triggers extracellular Ca^{2+} entry via SOCC. Further details can be found in the article by Halls and Cooper.

Library of Congress Cataloging-in-Publication Data

Calcium signaling / edited by Martin D. Bootman ... [et al.].
    p. cm.
 Includes index.
 ISBN 978-0-87969-903-1 (hardcover : alk. paper)
 1. Calcium--Physiological effect. 2. Cellular signal transduction.
 I. Bootman, Martin D.
   QP535.C2C266319 2011
   612.3'924--dc23
                          2011024271

10 9 8 7 6 5 4 3 2 1

All World Wide Web addresses are accurate to the best of our knowledge at the time of printing.

Authorization to photocopy items for internal or personal use, or the internal or personal use of specific clients, is granted by Cold Spring Harbor Laboratory Press, provided that the appropriate fee is paid directly to the Copyright Clearance Center (CCC). Write or call CCC at 222 Rosewood Drive, Danvers, MA 01923 (978-750-8400) for information about fees and regulations. Prior to photocopying items for educational classroom use, contact CCC at the above address. Additional information on CCC can be obtained at CCC Online at http://www.copyright.com/.

All Cold Spring Harbor Laboratory Press publications may be ordered directly from Cold Spring Harbor Laboratory Press, 500 Sunnyside Blvd., Woodbury, New York 11797-2924. Phone: 1-800-843-4388 in Continental U.S. and Canada. All other locations: (516) 422-4100. FAX: (516) 422-4097. E-mail: cshpress@cshl.edu. For a complete catalog of all Cold Spring Harbor Laboratory Press publications, visit our website at http://www.cshlpress.com/.

Contents

Preface, vii

CALCIUM SIGNAL GENERATION/MODULATION AND HOMEOSTASIS

Calcium Entry

Voltage-Gated Calcium Channels, 1
William A. Catterall

The Role of Transient Receptor Potential Cation Channels in Ca^{2+} Signaling, 25
Maarten Gees, Barbara Colsoul, and Bernd Nilius

Store-Operated Calcium Channels: New Perspectives on Mechanism and Function, 57
Richard S. Lewis

mGluR1/TRPC3-Mediated Synaptic Transmission and Calcium Signaling in Mammalian Central Neurons, 81
Jana Hartmann, Horst A. Henning, and Arthur Konnerth

Calcium Release

Ryanodine Receptors: Structure, Expression, Molecular Details, and Function in Calcium Release, 97
Johanna T. Lanner, Dimitra K. Georgiou, Aditya D. Joshi, and Susan L. Hamilton

IP_3 Receptors: Toward Understanding Their Activation, 119
Colin W. Taylor and Stephen C. Tovey

NAADP Receptors, 141
Antony Galione

Calcium Oscillations, 159
Geneviève Dupont, Laurent Combettes, Gary S. Bird, and James W. Putney

Calcium Buffers and Pumps

Cytosolic Ca^{2+} Buffers, 177
Beat Schwaller

Organellar Calcium Buffers, 197
Daniel Prins and Marek Michalak

Contents

The Plasma Membrane Ca^{2+} ATPase and the Plasma Membrane Sodium Calcium Exchanger Cooperate in the Regulation of Cell Calcium, 213
Marisa Brini and Ernesto Carafoli

The Ca^{2+} Pumps of the Endoplasmic Reticulum and Golgi Apparatus, 229
Ilse Vandecaetsbeek, Peter Vangheluwe, Luc Raeymaekers, Frank Wuytack, and Jo Vanoevelen

FUNCTIONAL ASPECTS OF CALCIUM SIGNALING

Proteins Mediating Effects of Calcium

The Diversity of Calcium Sensor Proteins in the Regulation of Neuronal Function, 253
Hannah V. McCue, Lee P. Haynes, and Robert D. Burgoyne

Regulation by Ca^{2+}-Signaling Pathways of Adenylyl Cyclases, 273
Michelle L. Halls and Dermot M.F. Cooper

Protein Kinase C: The "Masters" of Calcium and Lipid, 295
Peter Lipp and Gregor Reither

Physiological Processes Regulated by Calcium

Ca^{2+} Signaling During Mammalian Fertilization: Requirements, Players, and Adaptations, 313
Takuya Wakai, Veerle Vanderheyden, and Rafael A. Fissore

Calcium Signaling in Neuronal Development, 337
Sheila S. Rosenberg and Nicholas C. Spitzer

Visualization of Ca^{2+} Signaling During Embryonic Skeletal Muscle Formation in Vertebrates, 351
Sarah E. Webb and Andrew L. Miller

Calcium Signaling in Synapse-to-Nucleus Communication, 371
Anna M. Hagenston and Hilmar Bading

Calcium Signaling in Cardiac Myocytes, 403
Claire J. Fearnley, H. Llewelyn Roderick, and Martin D. Bootman

Calcium Signaling in Smooth Muscle, 423
David C. Hill-Eubanks, Matthias E. Werner, Thomas J. Heppner, and Mark T. Nelson

Calcium in Cell Death and Disease

Apoptosis and Autophagy: Decoding Calcium Signals that Mediate Life or Death, 443
Michael W. Harr and Clark W. Distelhorst

Endoplasmic-Reticulum Calcium Depletion and Disease, 461
Djalila Mekahli, Geert Bultynck, Jan B. Parys, Humbert De Smedt, and Ludwig Missiaen

Index, 491

Preface

TWENTY YEARS AGO, IT WAS JUST ABOUT POSSIBLE TO READ and digest most of the papers on calcium signaling that appeared in such disparate areas as muscle function, neurobiology, and development. Scanning lists of publications contained in the *Current Contents* booklet would reveal that only very few pieces of a colossal jigsaw puzzle had been unearthed, but with each publication additional pieces were added, giving some clarity to the overall picture. The situation is rather different now. Libraries are virtual, and publications can be immediately searched and accessed. However, the growth of the calcium signaling literature over the past decades has been such that it is now impossible to read extensively as a generalist. Indeed, a simple search for "calcium signaling" returns more than 35,000 hits in PubMed. One thus feels genuinely sorry for students and postdocs who join this fast-moving area. They must read so much just to catch up. The published literature is no less daunting for established scientists trying to explore the wider relevance of the biology they focus on.

That is where this volume comes in: It provides a detailed expert snapshot of the calcium signaling field as it stands right now and gives some insight into the history of the discoveries too. The chapters illustrate the considerable breadth of calcium signaling mechanisms used by cells. Further, they describe the impact of calcium signals on a diverse range of cellular processes. Reading through them provides insight into both the generic nature of calcium signaling and also its unique tissue- and function-specific characteristics in some settings. The grand challenges are to understand how cellular calcium signaling proteomes determine the physiology of cells and how subtle changes in those proteomes lead to disease. This volume goes some way to explaining both issues in many different situations. It can be loosely grouped into chapters dealing with calcium signal generation/modulation and chapters exploring downstream consequences of calcium signals. However, the considerable overlap between these themes emphasizes the fact that cellular calcium signaling proteomes are plastic and can both determine and be determined by the specific characteristics of calcium signals.

We would like to express our sincere thanks to Richard Sever at Cold Spring Harbor Laboratory Press, who provided the motivation to compile this volume. Grateful thanks also go to Joan Ebert and Barbara Acosta at Cold Spring Harbor Laboratory Press, who tracked the progress of all the manuscripts and gave timely reminders of things that needed to be done. Finally, we must thank all the authors, who have taken time out of their busy schedules to write such excellent chapters.

MARTIN D. BOOTMAN
MICHAEL J. BERRIDGE
JAMES W. PUTNEY
H. LLEWELYN RODERICK
April 15, 2011

Calcium Signaling

A subject collection from *Cold Spring Harbor Perspectives in Biology*

Voltage-Gated Calcium Channels

William A. Catterall

Department of Pharmacology, University of Washington, Seattle, Washington 98195-7280
Correspondence: wcatt@uw.edu

Voltage-gated calcium (Ca^{2+}) channels are key transducers of membrane potential changes into intracellular Ca^{2+} transients that initiate many physiological events. There are ten members of the voltage-gated Ca^{2+} channel family in mammals, and they serve distinct roles in cellular signal transduction. The Ca_V1 subfamily initiates contraction, secretion, regulation of gene expression, integration of synaptic input in neurons, and synaptic transmission at ribbon synapses in specialized sensory cells. The Ca_V2 subfamily is primarily responsible for initiation of synaptic transmission at fast synapses. The Ca_V3 subfamily is important for repetitive firing of action potentials in rhythmically firing cells such as cardiac myocytes and thalamic neurons. This article presents the molecular relationships and physiological functions of these Ca^{2+} channel proteins and provides information on their molecular, genetic, physiological, and pharmacological properties.

PHYSIOLOGICAL ROLES OF VOLTAGE-GATED Ca^{2+} CHANNELS

Ca^{2+} channels in many different cell types activate on membrane depolarization and mediate Ca^{2+} influx in response to action potentials and subthreshold depolarizing signals. Ca^{2+} entering the cell through voltage-gated Ca^{2+} channels serves as the second messenger of electrical signaling, initiating many different cellular events (Fig. 1). In cardiac and smooth muscle cells, activation of Ca^{2+} channels initiates contraction directly by increasing cytosolic Ca^{2+} concentration and indirectly by activating calcium-dependent calcium release by ryanodine-sensitive Ca^{2+} release channels in the sarcoplasmic reticulum (Reuter 1979; Tsien 1983; Bers 2002). In skeletal muscle cells, voltage-gated Ca^{2+} channels in the transverse tubule membranes interact directly with ryanodine-sensitive Ca^{2+} release channels in the sarcoplasmic reticulum and activate them to initiate rapid contraction (Catterall 1991; Tanabe et al. 1993). The same Ca^{2+} channels in the transverse tubules also mediate a slow Ca^{2+} conductance that increases cytosolic concentration and thereby regulates the force of contraction in response to high-frequency trains of nerve impulses (Catterall 1991). In endocrine cells, voltage-gated Ca^{2+} channels mediate Ca^{2+} entry that initiates secretion of hormones (Yang and Berggren 2006). In neurons, voltage-gated Ca^{2+} channels initiate synaptic transmission (Tsien et al. 1988; Dunlap et al. 1995; Catterall and Few 2008). In many different cell types, Ca^{2+} entering the cytosol via voltage-gated Ca^{2+} channels regulates enzyme activity, gene expression, and other biochemical processes (Flavell and Greenberg 2008). Thus, voltage-gated Ca^{2+} channels are

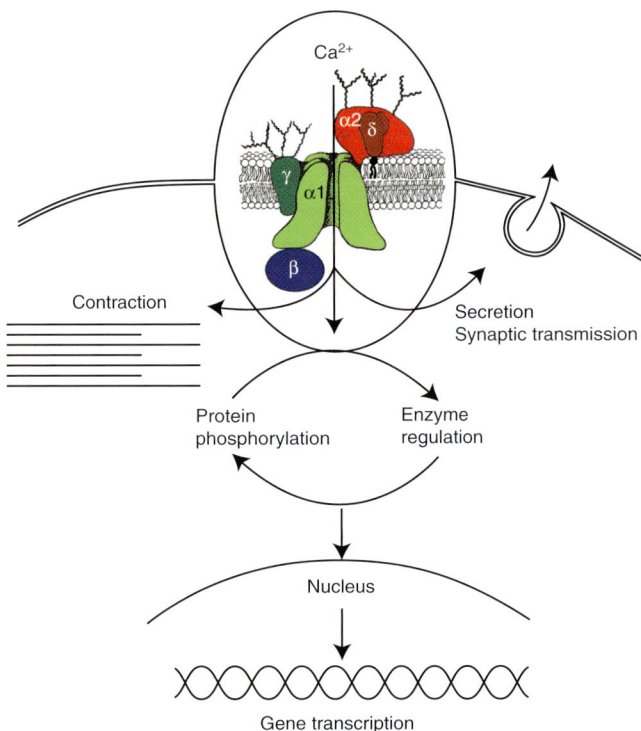

Figure 1. Signal transduction by voltage-gated Ca^{2+} channels. Ca^{2+} entering cells initiates numerous intracellular events, including contraction, secretion, synaptic transmission, enzyme regulation, protein phosphorylation/dephosphorylation, and gene transcription. (*Inset*) Subunit structure of voltage-gated Ca^{2+} channels. The five-subunit complex that forms high-voltage-activated Ca^{2+} channels is illustrated with a central pore-forming α1 subunit, a disulfide-linked glycoprotein dimer of α2 and δ subunits, an intracellular β subunit, and a transmembrane glycoprotein γ subunit (in some Ca^{2+} channel subtypes). As described in the text, this model is updated from the original description of the subunit structure of skeletal muscle Ca^{2+} channels. (Adapted from Takahashi et al. 1987).

the key signal transducers of electrical excitability, converting the electrical signal of the action potential in the cell surface membrane to an intracellular Ca^{2+} transient. Signal transduction in different cell types involves different molecular subtypes of voltage-gated Ca^{2+} channels, which mediate voltage-gated Ca^{2+} currents with different physiological, pharmacological, and regulatory properties.

Ca^{2+} CURRENT TYPES DEFINED BY PHYSIOLOGICAL AND PHARMACOLOGICAL PROPERTIES

Since the first recordings of Ca^{2+} currents in cardiac myocytes (reviewed in Reuter 1979), it has become apparent that there are multiple types of Ca^{2+} currents as defined by physiological and pharmacological criteria (Tsien et al. 1988; Bean 1989a; Llinás et al. 1992). In cardiac, smooth, and skeletal muscle, the major Ca^{2+} currents are distinguished by high voltage of activation, large single channel conductance, slow voltage-dependent inactivation, marked up-regulation by cAMP-dependent protein phosphorylation pathways, and specific inhibition by Ca^{2+} antagonist drugs including dihydropyridines, phenylalkylamines, and benzothiazepines (Table 1) (Reuter 1979; Tsien et al. 1988). These Ca^{2+} currents have been designated L-type, as they have slow voltage-dependent inactivation and therefore are long lasting when Ba^{2+} is the current carrier and there is no Ca^{2+}-dependent inactivation (Tsien et al.

Table 1. Subunit composition and function of Ca^{2+} channel types

Ca^{2+} current type	α1 Subunits	Specific blocker	Principal physiological functions	Inherited diseases
L	$Ca_v1.1$	DHPs	Excitation-contraction coupling in skeletal muscle, regulation of transcription	Hypokalemic periodic paralysis
	$Ca_v1.2$	DHPs	Excitation-contraction coupling in cardiac and smooth muscle, endocrine secretion, neuronal Ca^{2+} transients in cell bodies and dendrites, regulation of enzyme activity, regulation of transcription	Timothy syndrome: cardiac arrhythmia with developmental abnormalites and autism spectrum disorders
	$Ca_v1.3$	DHPs	Endocrine secretion, cardiac pacemaking, neuronal Ca^{2+} transients in cell bodies and dendrites, auditory transduction	
	$Ca_v1.4$	DHPs	Visual transduction	Stationary night blindness
N	$Ca_v2.1$	ω-CTx-GVIA	Neurotransmitter release, Dendritic Ca^{2+} transients	
P/Q	$Ca_v2.2$	ω-Agatoxin	Neurotransmitter release, Dendritic Ca^{2+} transients	Familial hemiplegic migraine, cerebellar ataxia
R	$Ca_v2.3$	SNX-482	Neurotransmitter release, Dendritic Ca^{2+} transients	
T	$Ca_v3.1$	None	Pacemaking and repetitive firing	
	$Ca_v3.2$		Pacemaking and repetitive firing	Absence seizures
	$Ca_v3.3$			

Abbreviations: DHP, dihydropyridine; ω-CTx-GVIA, ω-conotoxin GVIA from the cone snail *Conus geographus*; SNX-482, a synthetic version of a peptide toxin from the tarantula *Hysterocrates gigas*.

1988). L-type Ca^{2+} currents are also recorded in endocrine cells where they initiate release of hormones (Yang and Berggren 2006) and in neurons where they are important in regulation of gene expression, integration of synaptic input, and initiation of neurotransmitter release at specialized ribbon synapses in sensory cells (Tsien et al. 1988; Bean 1989a; Flavell and Greenberg 2008). L-type Ca^{2+} currents are subject to regulation by second messenger–activated protein phosphorylation in several cell types as discussed below.

Electrophysiological studies of Ca^{2+} currents in starfish eggs (Hagiwara et al. 1975) first revealed Ca^{2+} currents with different properties from L-type, and these were subsequently characterized in detail in voltage-clamped dorsal root ganglion neurons (Carbone and Lux 1984; Fedulova et al. 1985; Nowycky et al. 1985). In comparison to L-type, these novel Ca^{2+} currents activated at much more negative membrane potentials, inactivated rapidly, deactivated slowly, had small single channel conductance, and were insensitive to conventional Ca^{2+} antagonist drugs available at that time (Table 1). They were designated low-voltage-activated Ca^{2+} currents for their negative voltage dependence (Carbone and Lux 1984) or T-type Ca^{2+} currents for their transient openings (Nowycky et al. 1985).

Whole-cell voltage clamp and single-channel recording from dissociated dorsal root ganglion neurons revealed an additional Ca^{2+} current, N-type (Table 1) (Nowycky et al. 1985). N-type Ca^{2+} currents were initially distinguished by their intermediate voltage dependence and rate of inactivation—more negative and faster than L-type but more positive and slower than T-type (Nowycky et al. 1985). They are insensitive to organic L-type Ca^{2+} channel blockers but blocked by the cone snail peptide ω-conotoxin GVIA and related peptide toxins (Tsien et al. 1988; Olivera et al. 1994). This pharmacological profile has become the primary method to distinguish N-type Ca^{2+} currents, because the voltage dependence and kinetics of N-type Ca^{2+} currents in different neurons vary considerably.

Analysis of the effects of other peptide toxins revealed three additional Ca^{2+} current types (Table 1). P-type Ca^{2+} currents, first recorded in Purkinje neurons (Llinás and Yarom 1981; Llinás et al. 1989), are distinguished by high sensitivity to the spider toxin ω-agatoxin IVA (Mintz et al. 1992). Q-type Ca^{2+} currents, first recorded in cerebellar granule neurons (Randall and Tsien 1995), are blocked by ω-agatoxin IVA with lower affinity. R-type Ca^{2+} currents in cerebellar granule neurons are resistant to most subtype-specific organic and peptide Ca^{2+} channel blockers (Randall and Tsien 1995) and may include multiple channel subtypes (Tottene et al. 1996). They can be blocked selectively in some cell types by the peptide SNX-482 derived from the tarantula *Hysterocrates gigas* (Newcomb et al. 1998). Although L-type and T-type Ca^{2+} currents are recorded in a wide range of cell types, N-, P-, Q-, and R-type Ca^{2+} currents are most prominent in neurons.

MOLECULAR PROPERTIES OF Ca^{2+} CHANNELS

Subunit Structure

Ca^{2+} channels purified from skeletal muscle transverse tubules are complexes of α1, α2, β, γ, and δ subunits (Fig. 1) (Curtis and Catterall 1984, 1986; Flockerzi et al. 1986; Hosey et al. 1987; Leung et al. 1987; Striessnig et al. 1987; Takahashi et al. 1987). Analysis of the biochemical properties, glycosylation, and hydrophobicity of these five subunits led to a model comprising a principal transmembrane α1 subunit of 190 kDa in association with a disulfide-linked α2δ dimer of 170 kDa, an intracellular phosphorylated β subunit of 55 kDa, and a transmembrane γ subunit of 33 kDa (Fig. 1) (Takahashi et al. 1987).

The α1 subunit is a protein of about 2000 amino acid residues in length with an amino acid sequence and predicted transmembrane structure like the previously characterized, pore-forming α subunit of voltage-gated sodium channels (Fig. 2) (Tanabe et al. 1987). The amino acid sequence is organized in four repeated domains (I–IV), which each contains six transmembrane segments (S1–S6) and a

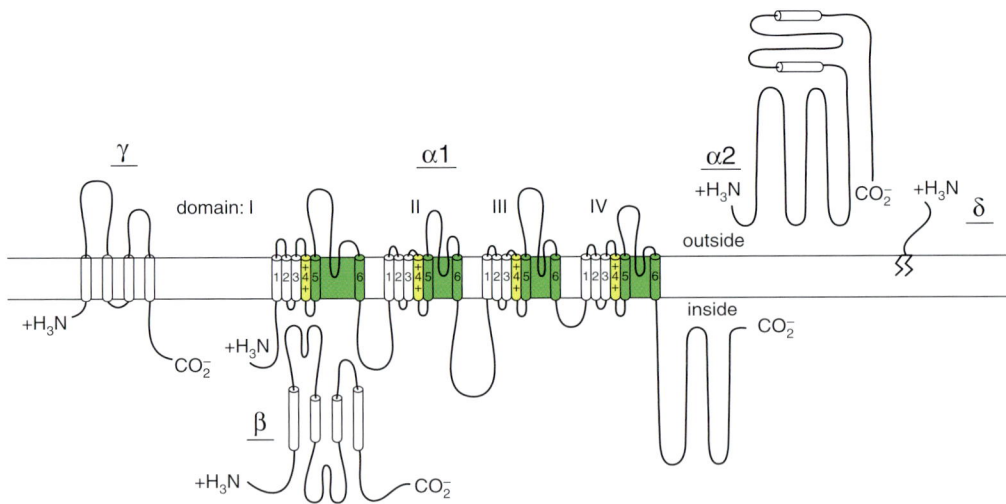

Figure 2. Subunit structure of Ca^{2+} channels. The structures of Ca^{2+} channel subunits are illustrated as transmembrane folding models; predicted α helices are depicted as cylinders; the lengths of lines correlate approximately to the lengths of the polypeptide segments represented; and the zigzag line on the δ subunit illustrates its glycophosphatidylinositol anchor.

membrane-associated loop between transmembrane segments S5 and S6. As expected from biochemical analysis (Takahashi et al. 1987a), the intracellular β subunit has predicted α helices but no transmembrane segments (Fig. 2) (Ruth et al. 1989), whereas the γ subunit is a glycoprotein with four transmembrane segments (Fig. 2) (Jay et al. 1990). The cloned α2 subunit has many glycosylation sites and several hydrophobic sequences (Ellis et al. 1988), but biosynthesis studies indicate that it is an extracellular, extrinsic membrane glycoprotein, attached to the membrane through disulfide linkage to the δ subunit (Fig. 2) (Gurnett et al. 1996). The δ subunit is encoded by the 3′ end of the coding sequence of the same gene as the α2 subunit, and the mature forms of these two subunits are produced by posttranslational proteolytic processing and disulfide linkage (Fig. 2) (De Jongh et al. 1990). Although it was initially assumed that the δ subunit was anchored to the membrane via a single membrane segment, recent work argues persuasively that further posttranslational processing actually cleaves the predicted transmembrane segment and replaces it with a glycophosphatidylinositol membrane anchor (Fig. 2) (Davies et al. 2010).

Purification of cardiac Ca^{2+} channels labeled by dihydropyridine Ca^{2+} antagonists identified subunits of the sizes of the α1, α2δ, β, and γ subunits of skeletal muscle Ca^{2+} channels (Chang and Hosey 1988; Schneider and Hofmann 1988; Kuniyasu et al. 1992), whereas immunoprecipitation of Ca^{2+} channels from neurons labeled by dihydropyridine Ca^{2+} antagonists revealed α1, α2δ, and β subunits but no γ subunit (Ahlijanian et al. 1990). Purification and immunoprecipitation of N-type and P/Q-type Ca^{2+} channels labeled by ω-conotoxin GVIA and ω-agatoxin IVA, respectively, from brain membrane preparations also revealed α1, α2δ, and β subunits but not γ subunits (McEnery et al. 1991; Martin-Moutot et al. 1995; Witcher et al. 1995a; Liu et al. 1996). More recent experiments have unexpectedly revealed a novel γ subunit (stargazin), which is the target of the *stargazer* mutation in mice (Letts et al. 1998), and a related series of seven γ subunits is expressed in brain and other tissues (Klugbauer et al. 2000). These γ-subunit-like proteins can modulate the voltage dependence of $Ca_V2.1$ channels expressed in nonneuronal cells, so they may be associated with these Ca^{2+} channels in vivo. However, the stargazin-like γ subunits (also called transmembrane AMPA receptor modulators [TARPs]) are the primary modulators of glutamate receptors in the postsynaptic membranes of brain neurons (Nicoll et al. 2006), and it remains to be determined whether they are also associated with voltage-gated Ca^{2+} channels in brain neurons in vivo.

Three-Dimensional Structure of Ca^{2+} Channels

The three-dimensional structure of Ca^{2+} channels is not known at high resolution. Low-resolution structural models have been developed from image reconstruction analysis of $Ca_V1.1$ channels purified from skeletal muscle membranes (Serysheva et al. 2002; Wang et al. 2002; Wolf et al. 2003), and some of the structural features have been associated with the α1, β, and α2δ subunits (Fig. 3A). Further high-resolution structural analysis will be required to confirm these initial structural models. The three-dimensional structure of the $Ca_V\beta$ subunits has been determined at high resolution by X-ray crystallography (Fig. 3B) (Chen et al. 2004; Van Petegem et al. 2004). These subunits contain conserved SH3 and guanylate kinase domains like the MAGUK family of scaffolding proteins. These two domains are arrayed side-by-side in the $Ca_V\beta$ subunit (Fig. 3B). The $Ca_V\beta$ subunits bind to a single site in the α1 subunits (the α interaction domain, AID) (Pragnell et al. 1994), which is located in the first half of the intracellular loop connecting domains I and II. The AID forms an α helix that is bound tightly to a groove in the guanylate kinase domain of the $Ca_V\beta$ subunit. This tight, multipoint binding interaction likely sustains the association between Ca^{2+} channel α1 and β subunits throughout the lifetime of the Ca^{2+} channel complex at the cell surface membrane. MAGUK proteins often bind more than one protein partner, so $Ca_V\beta$ subunits may

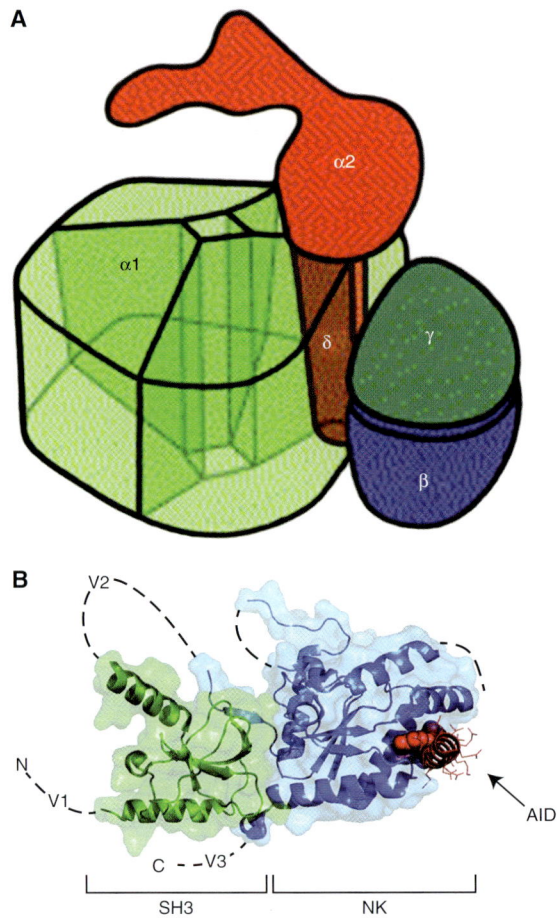

Figure 3. Three-dimensional architecture of Ca^{2+} channels. (*A*) Illustration of the skeletal muscle $Ca_V1.1$ channel based on cryo-electronmicroscopy. This drawing assumes pseudo-fourfold symmetry of the α1 subunit. The view shows the extracellular side with the α2 subunit. The α1, γ, and δ subunits are embedded into the lipid membrane (not shown), which separates the extracellular α2 subunit from the cytosol. α2 is anchored via the disulfide-linked δ subunit within the α1 subunit. The proposed model allows for a tight interaction between α1 and δ as well as α1 and γ. (*B*) Structure of the $Ca_V\beta$ subunit with the α interaction domain (AID). Coordinates are for the $Ca_V\beta2a$–$Ca_V1.2$ AID complex with SH3 (green) and NK (blue) domains are indicated. V1, V2, and V3 show the locations of the three variable domains that are absent from the structure. The AID (red) binds to a deep groove in the NK domain. AID residues tyrosine, tryptophan, and isoleucine are shown as CPK. The remaining residues are shown as lines.

also interact with other intracellular proteins, and several potential binding partners are under active investigation.

Functions of Ca^{2+} Channel Subunits

Expression of the α1 subunit is sufficient to produce functional skeletal muscle Ca^{2+} channels, but with low expression level and abnormal kinetics and voltage dependence of the Ca^{2+} current (Perez-Reyes et al. 1989). Coexpression of the α2δ subunit and especially the β subunit enhanced the level of expression and conferred more normal gating properties (Lacerda et al. 1991; Singer et al. 1991). As for skeletal muscle Ca^{2+} channels, coexpression of β subunits has a large effect on the level of expression and the voltage dependence and kinetics of gating of cardiac and neuronal Ca^{2+} channels (reviewed in Hofmann et al. 1994; Dolphin 2003). In general, the level of expression is increased and the voltage dependence of activation and inactivation is shifted to more negative membrane potentials, and the rate of inactivation is increased. However, these effects are different for the individual β subunit isoforms. For example, the β2a subunit slows channel inactivation in most subunit combinations. Coexpression of α2δ subunits also increases expression and enhances function of Ca^{2+} channels, but to a lesser extent and in a more channel-specific way than do β subunits (Arikkath and Campbell 2003; Davies et al. 2007). In general, γ subunits have smaller effects.

Ca^{2+} Channel Diversity

The different types of Ca^{2+} currents are primarily defined by different α1 subunits, and ten different ones have been characterized by cDNA cloning and functional expression in mammalian cells or *Xenopus* oocytes (Table 1). These subunits can be divided into three structurally and functionally related families (Ca_V1, Ca_V2, and Ca_V3) (Snutch and Reiner 1992; Ertel et al. 2000). L-type Ca^{2+} currents are mediated by the Ca_V1 type of α1 subunits, which have about 75% amino acid sequence identity among them. The Ca_V2 type Ca^{2+} channels form a distinct subfamily with <40% amino acid sequence identity with Ca_V1 α1 subunits but >70% amino acid sequence identity among themselves. Cloned $Ca_V2.1$ subunits (Mori et al. 1991; Starr et al. 1991) conduct P- or Q-type Ca^{2+} currents, which are inhibited by ω-agatoxin IVA. $Ca_V2.2$ subunits conduct N-type Ca^{2+} currents blocked with high affinity by ω-conotoxin GVIA (Dubel et al. 1992; Williams et al. 1992). Cloned $Ca_V2.3$ subunits form R-type Ca^{2+} channels, which are resistant to both organic Ca^{2+} antagonists specific for L-type Ca^{2+} currents and the peptide toxins specific for N-type or P/Q-type Ca^{2+} currents (Soong et al. 1994). T-type Ca^{2+} currents are mediated by the Ca_V3 Ca^{2+} channels (Perez-Reyes et al. 1998). These α1 subunits are only distantly related to the other known homologs, with <25% amino acid sequence identity. These results reveal a surprising structural dichotomy between the T-type, low-voltage-activated Ca^{2+} channels and the high-voltage-activated Ca^{2+} channels. Evidently, these two lineages of Ca^{2+} channels diverged very early in evolution of multicellular organisms. Single representatives of the Ca_V1, Ca_V2, and Ca_V3 subfamilies are present in invertebrate genomes, including the worm *Caenorhabditis elegans* and the fruit fly *Drosophila*.

The diversity of Ca^{2+} channel structure and function is substantially enhanced by multiple β subunits. Four β subunit genes have been identified, and each is subject to alternative splicing to yield additional isoforms (reviewed in Hofmann et al. 1994; Dolphin 2003). In Ca^{2+} channel preparations isolated from brain, individual Ca^{2+} channel α1 subunit types are associated with multiple types of β subunits, although there is a different rank order in each case (Pichler et al. 1997; Witcher et al. 1995b). The different β subunit isoforms cause different shifts in the kinetics and voltage dependence of gating, so association with different β subunits can substantially alter the physiological function of an α1 subunit. Genes encoding four α2δ subunits have been described (Klugbauer et al. 1999), and the α2δ isoforms produced by these different genes have selective effects on the level of functional expression and the voltage dependence of different α1 subunits (Davies et al. 2007).

Molecular Basis for Ca^{2+} Channel Function

Intensive studies of the structure and function of the related pore-forming subunits of Na^+, Ca^{2+}, and K^+ channels have led to identification of their principal functional domains (reviewed in Catterall 2000a,b; Yi and Jan 2000; Bichet et al. 2003; Yu et al. 2005). Each domain of the principal subunits consists of six transmembrane α helices (S1–S6) and a membrane-associated loop between S5 and S6 (Fig. 2). The S4 segments of each homologous domain serve as the voltage sensors for activation, moving outward and rotating under the influence of the electric field and initiating a conformational change that opens the pore. The S5 and S6 segments and the membrane-associated pore loop between them form the pore lining of the voltage-gated ion channels. The narrow external end of the pore is lined by the pore loop, which contains a pair of glutamate residues in each domain that are required for Ca^{2+} selectivity, a structural feature that is unique to Ca^{2+} channels (Heinemann et al. 1992). Remarkably, substitutions that add only three glutamate residues in the pore loops between the S5 and S6 segments in domains II, III, and IV of sodium channels are sufficient to confer Ca^{2+} selectivity (Heinemann et al. 1992; Sather and McCleskey 2003). The inner pore is lined by the S6 segments, which form the receptor sites for the pore-blocking Ca^{2+} antagonist drugs specific for L-type Ca^{2+} channels (Hockerman et al. 1997a,b). All Ca^{2+} channels share these general structural features, but the amino acid residues that confer high affinity for the organic Ca^{2+} antagonists used in therapy of cardiovascular diseases are present only in the Ca_V1 family of Ca^{2+} channels, which conduct L-type Ca^{2+} currents.

Ca_V1 CHANNELS AND EXCITATION-RESPONSE COUPLING

Ca_V1 channels serve to couple depolarization of the plasma membrane to a wide range of cellular responses (Fig. 1). Three widely studied examples are excitation-contraction coupling in muscle, excitation-transcription coupling in nerve and muscle, and excitation-secretion coupling in endocrine cells and at specialized ribbon synapses.

Ca_V1 CHANNELS AND EXCITATION-CONTRACTION COUPLING

Mechanisms of Excitation-Contraction Coupling

Ca_V1 channels initiate excitation-contraction coupling in skeletal, cardiac, and smooth muscle. There are striking mechanistic differences between excitation-contraction coupling in skeletal muscle and cardiac muscle. In skeletal muscle, entry of external Ca^{2+} is not required for initiation of contraction (Armstrong et al. 1972). $Ca_V1.1$ channels in the transverse tubules are thought to interact directly with the ryanodine-sensitive Ca^{2+} release channels (RyR1) of the sarcoplasmic reticulum (Numa et al. 1990), as observed in high-resolution electron microscopy (Block et al. 1988), and the voltage-driven conformational changes in their voltage-sensing domains are thought to directly induce activation of RyR1 (Numa et al. 1990). Reconstitution of excitation-contraction coupling in myocytes from mutant mice requires both $Ca_V1.1$ and RyR1 proteins and their relevant sites of protein–protein interaction (Tanabe et al. 1990; Nakai et al. 1998), and functional expression of the $Ca_V1.1$ channel in skeletal muscle requires its RyR1 binding partner (Nakai et al. 1996).

In contrast to skeletal muscle, entry of Ca^{2+} is required for excitation-contraction coupling in cardiac myocytes, and Ca^{2+} entry via $Ca_V1.2$ channels triggers activation of the RyR2 and initiates Ca^{2+}-induced Ca^{2+}-release, activation of actomyosin, and contraction (Fabiato 1983; Bers 2002). Release of Ca^{2+} from the sarcoplasmic reticulum via RyR2 greatly amplifies the cellular Ca^{2+} transient and is required for effective initiation of contraction. All three steps in the cascade of Ca^{2+} transport processes—Ca^{2+} entry via $Ca_V1.2$ channels, Ca^{2+} release via RyR, and Ca^{2+} uptake into the sarcoplasmic reticulum by SERCA Ca^{2+} pumps—are tightly regulated by second messenger signaling networks (Bers 2002). The section below considers the regulation of Ca_V1 channels in excitation-contraction coupling.

Regulation of Excitation-Contraction Coupling via Ca_V1 Channels

As part of the flight-or-flight response, the rate and force of contraction of both skeletal and cardiac muscle are increased through the activity of the sympathetic nervous system. Release of catecholamines stimulates β-adrenergic receptors (β-ARs), which increases the force of skeletal and cardiac muscle contraction and the heart rate (Reuter 1983; Tsien et al. 1986). In cardiac muscle, Ca^{2+} influx through $Ca_V1.2$ channels is responsible for initiating excitation-contraction coupling, and increased Ca^{2+} channel activity via the PKA pathway is primarily responsible for the increase in contractility. $Ca_V1.2$ channels are modulated by the β-adrenergic receptor/cAMP signaling. Activation of β-adrenergic receptors increases L-type Ca^{2+} currents through PKA-mediated phosphorylation of the $Ca_V1.2$ channel protein and/or associated proteins (Tsien 1973; Reuter and Scholz 1977; Osterrieder et al. 1982; McDonald et al. 1994).

The pore-forming α1 subunit and the auxiliary β subunits of skeletal muscle $Ca_V1.1$ channels (Curtis and Catterall 1985; Flockerzi et al. 1986; Takahashi et al. 1987) and cardiac $Ca_V1.2$ channels (Hell et al. 1993b; De Jongh et al. 1996; Haase et al. 1996; Puri et al. 1997) are phosphorylated by PKA. These α1 subunits are also truncated by proteolytic processing of the carboxy-terminal domain (Fig. 4) (De Jongh et al. 1989, 1991, 1996; Hulme et al. 2005). Voltage-dependent potentiation of $Ca_V1.1$ channels on the 50-msec time scale requires PKA phosphorylation (Sculptoreanu et al. 1993) as well as PKA anchoring via an A kinase anchoring protein (AKAP) (Johnson et al. 1994, 1997), suggesting close association of PKA and Ca^{2+} channels. A novel, plasma membrane–targeted AKAP (AKAP15) is associated with both $Ca_V1.1$ channels (Gray et al. 1997, 1998) and $Ca_V1.2$ channels (Hulme et al. 2003), and may mediate their regulation by PKA. This AKAP (also known as AKAP18 [Fraser et al. 1998]) binds to the carboxy-terminal domain of $Ca_V1.1$ channels (Hulme et al. 2002) and $Ca_V1.2$ channels (Hulme et al. 2003) via a novel modified leucine zipper interaction near the primary sites of PKA phosphorylation. Block of this interaction by competing peptides prevents PKA regulation of Ca^{2+} currents in intact skeletal and cardiac myocytes (Hulme et al. 2002, 2003, 2006b). These physiological results suggest that a Ca^{2+} channel signaling complex containing AKAP15 and PKA is formed in both skeletal and cardiac muscle, and this conclusion is supported by specific colocalization of these proteins in both skeletal and cardiac myocytes and specific coimmunoprecipitation of this complex from both tissues (Hulme et al. 2002, 2003, 2006a). Remarkably, block of kinase anchoring is as effective as block of kinase activity in preventing $Ca_V1.1$ and $Ca_V1.2$ channel regulation, consistent with the conclusion that PKA targeting via leucine zipper interactions is absolutely required for regulation of Ca_V1 channels in intact skeletal and cardiac myocytes.

Proteolytic Processing and Regulation via the Carboxy-Terminal Domain

The distal carboxy-terminal domains of skeletal muscle and cardiac Ca^{2+} channels are proteolytically processed in vivo (Fig. 4B) (De Jongh et al. 1991, 1996). Nevertheless, the most prominent in vitro PKA phosphorylation sites of both proteins are located beyond the site of proteolytic truncation (Rotman et al. 1992, 1995; De Jongh et al. 1996; Mitterdorfer et al. 1996), and interaction of AKAP15 and PKA with the distal carboxy-terminal domain through a leucine zipper motif is required for regulation of cardiac Ca^{2+} channels in intact myocytes (Hulme et al. 2003). These results imply that the distal carboxy-terminal domain remains associated with the proteolytically processed cardiac $Ca_V1.2$ channel, and this is supported by evidence that the distal carboxyl-terminus can bind to the truncated $Ca_V1.1$ and $Ca_V1.2$ channels in vitro (Gerhardstein et al. 2000; Gao et al. 2001; Hulme et al. 2005) and in transfected cells (Hulme et al. 2002; Hulme et al. 2006b). Moreover, formation of this complex dramatically inhibits cardiac Ca^{2+} channel function in intact mammalian cells (Hulme et al. 2006b). Deletion of the distal carboxy-terminal near the site of proteolytic processing increases Ca^{2+} channel activity (Wei et al. 1994; Hulme et al. 2006b). However, noncovalent association of the cleaved distal carboxy-terminal reduces channel activity more than 10-fold, to a level much below that of channels with an intact carboxyl-terminus (Hulme et al. 2006b). Thus, proteolytic processing produces an autoinhibited Ca^{2+} channel complex containing noncovalently bound distal

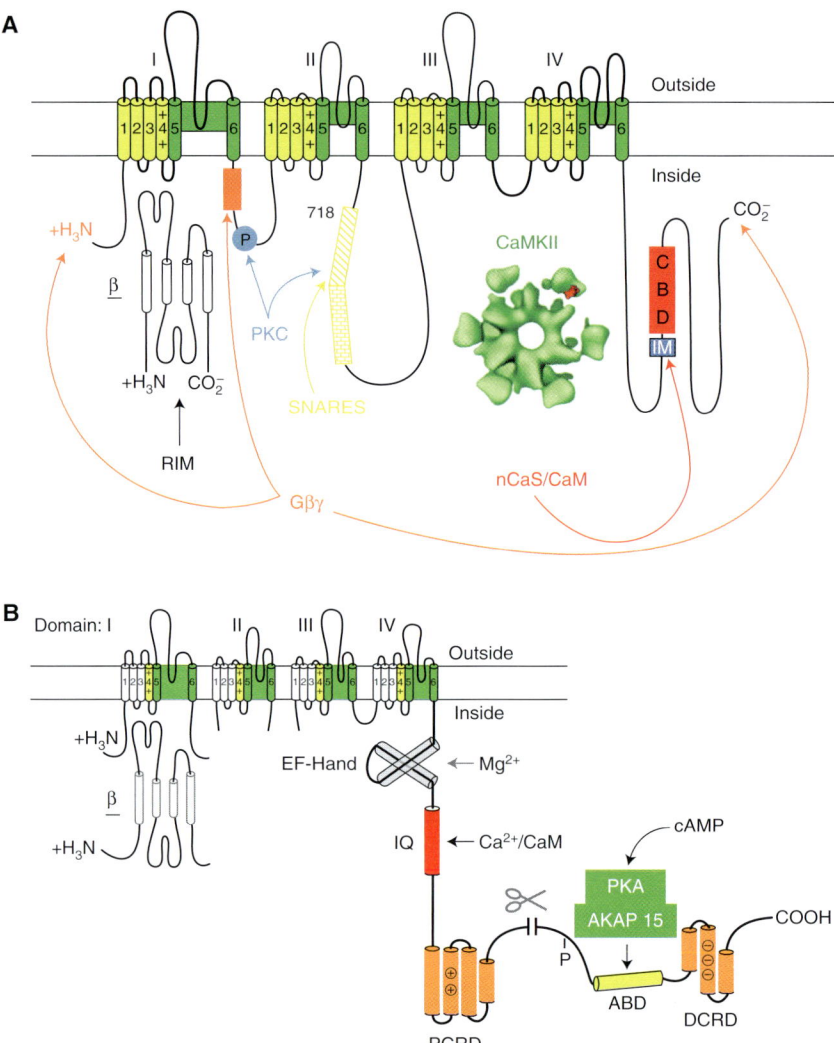

Figure 4. Ca^{2+} channel signaling complexes. (*A*) The presynaptic Ca^{2+} channel signaling complex. A presynaptic Ca^{2+} channel α1 subunit is illustrated as a transmembrane folding diagram as in Figure 2. Sites of interaction of SNARE proteins (the synprint site), Gβγ subunits, protein kinase C (PKC), CaMKII, and CaM and CaS proteins are illustrated. IM, IQ-like motif; CBD, CaM binding domain. (*B*) The cardiac Ca^{2+} channel signaling complex. The carboxy-terminal domain of the cardiac Ca^{2+} channels is shown in expanded presentation to illustrate the regulatory interactions clearly. ABD, AKAP15 binding domain; DCRD, distal carboxy-terminal regulatory domain; PCRD, proximal carboxy-terminal regulatory domain; scissors, site of proteolytic processing. The DCRD binds to the PCRD through a modified leucine zipper interaction.

carboxyl-terminus with AKAP15 and PKA associated through a modified leucine zipper interaction (Fig. 4B). This autoinhibited complex appears to be the primary substrate for regulation of cardiac Ca^{2+} channels by the β-adrenergic receptor/PKA pathway in vivo, and PKA up-regulation results from phosphorylation of a single site near the end of the proximal carboxy-terminal domain at the interface with the distal carboxy-terminal domain (Fig. 4B) (Hulme et al. 2006b; Emrick et al. 2010; Fuller et al. 2010).

Ca^{2+} Binding Proteins

In addition to their regulation by the PKA/AKAP15 signaling complex, cardiac Ca^{2+} channels have calmodulin bound to their carboxy-terminal domain through an IQ motif (Fig. 4B), and Ca^{2+} binding to calmodulin causes Ca^{2+}-dependent inactivation (Peterson et al. 1999; Qin et al. 1999; Zühlke et al. 1999). Activation of $Ca_V1.2$ channels in the presence of Ba^{2+} as the permeant ion results in inward Ba^{2+} currents that activate rapidly and inactivate slowly via a voltage-dependent inactivation process. In contrast, in the presence of Ca^{2+} as the permeant ion, Ca^{2+} currents are rapidly inactivated via Ca^{2+}/calmodulin-dependent inactivation. The Ca^{2+}-dependent inactivation process is crucial for limiting Ca^{2+} entry during long cardiac action potentials. In light of these results, it is evident that both the cAMP and Ca^{2+} second messenger pathways regulate $Ca_V1.2$ channels locally, dependent on associated regulatory proteins in Ca^{2+} channel signaling complexes.

Ca_V1 CHANNELS IN EXCITATION-TRANSCRIPTION COUPLING

Ca^{2+} entering neurons through L-type Ca^{2+} currents conducted by Ca_V1 channels has a privileged role in regulation of gene transcription, compared to similar amounts of Ca^{2+} entering via other voltage-gated or ligand-gated ion channels (Flavell and Greenberg 2008). This unique access of Ca_V1 channels to regulation of transcription might arise from preferential localization, which could provide Ca^{2+} in the vicinity of transcriptional regulators, preferential interaction with binding partners, which could be activated by local Ca^{2+} entry and carry the regulatory signal to the nucleus, or nuclear targeting of a subunit or domain of the Ca_V1 channel itself, which would serve to regulate transcription directly. It is likely that all three of these mechanisms are involved based on recent experiments.

Ca_V1 channels are localized in higher density in the cell bodies and proximal dendrites of neurons compared to Ca_V2 and Ca_V3 channels, which are more prevalent in nerve terminals and dendrites, respectively (Westenbroek et al. 1990; Hell et al. 1993a). This preferential localization would favor Ca^{2+} entry through these channels in control of transcription in the nucleus. However, this effect seems insufficient to fully account for the dominance of this Ca^{2+} entry pathway.

Studies with selective Ca^{2+} buffers indicate that only a local increase in Ca^{2+} is required for up-regulation of transcription in neurons (Wheeler et al. 2008). These findings suggest that specifically bound Ca^{2+}-dependent regulatory proteins may respond to local Ca^{2+} entering via Ca_V1 channels and regulate transcription. Calmodulin is a resident Ca^{2+}-dependent regulator of Ca_V1 channels (Pitt et al. 2001), and calmodulin binding to the proximal carboxy-terminal domain of $Ca_V1.2$ channels is required for regulation of transcription in neurons (Bito et al. 1996; Dolmetsch et al. 2001). Thus, calmodulin itself might serve as a regulator by binding local Ca^{2+}, changing conformation to the active form, and moving to the nucleus (Bito et al. 1996; Deisseroth et al. 1998). However, there are large pools of free and Ca^{2+}-bound calmodulin throughout the cell, so additional mechanisms must be engaged to specifically move Ca^{2+}/calmodulin complexes from the Ca_V1 channels to the nucleus in the context of this mode of regulation. Calcineurin bound to the distal carboxy-terminal domain of Ca_V1 channels also is a potential transcriptional regulator through dephosphorylation of regulatory proteins (Oliveria et al. 2007). In cultured hippocampal neurons, dephosphorylation of the nuclear factor of activated T cells (NFAT) by calcineurin bound to $Ca_V1.2$ channels induces its dissociation, movement to the nucleus, and regulation of transcription (Oliveria et al. 2007). This pathway appears to have all of the necessary elements for selective regulation of gene transcription by Ca^{2+} entering neurons via $Ca_V1.2$ channels and has the precedent that it is a crucial element in gene regulation in lymphocytes by a similar mechanism.

The distal carboxy-terminal domain of the Ca_V1 channel itself has also been proposed as a transcriptional regulator (Gomez-Ospina et al. 2006). The large carboxy-terminal domain of $Ca_V1.1$ and $Ca_V1.2$ channels is proteolytically processed in vivo near its center (De Jongh et al. 1991, 1996), leaving a noncovalently associated distal carboxy-terminal domain of more than 300 residues intact to regulate channel activity (Fig. 4B) (Hulme et al. 2006b). In neurons, this proteolytic cleavage process is regulated by Ca^{2+} and blocked by calpain inhibitors (Hell et al. 1996). The distal carboxy-terminal domain can be detected in the nuclei of a subset of neurons in the developing brain and in neurons in cell culture (Gomez-Ospina et al. 2006), opening the possibility of direct effects on transcription in the nucleus. Indeed, the distal carboxy-terminal domain can regulate the transcription of a substantial set of other genes in neurons (Gomez-Ospina et al. 2006), as well as the transcription of the gene encoding the $Ca_V1.2$ channel itself in cardiac myocytes (Schroder et al. 2009). This regulatory mechanism also has all of the necessary elements to give selective regulation of gene expression by $Ca_V1.2$ channels, but it remains unknown how the parallel effects of the distal carboxyl-terminus on regulation of channel activity versus migration to the nucleus and regulation of transcription are controlled. At least in neurons, it seems that only a small fraction of the distal carboxy-terminal is located in the nucleus (Gomez-Ospina et al. 2006), so it may be that most of the proteolytically processed distal carboxy-terminal domain remains associated with $Ca_V1.2$ channels as an autoinhibitory regulator of channel activity while a small fraction dissociates and moves to the nucleus to regulate transcription.

Ca_V1 CHANNELS IN EXCITATION-SECRETION COUPLING

Ca^{2+} entry via Ca_V1 channels initiates secretion of hormones from endocrine cells (Artalejo et al. 1994; Yang and Berggren 2006) and release of neurotransmitters at specialized ribbon synapses in sensory-transduction neurons (Table 1) (Kollmar et al. 1997; Barnes and Kelly 2002). The relative role of individual Ca_V1 channel subtypes in secretion, as well as the contribution of Ca_V2 channels, differs among cell types and species. In the pancreas, the requirement for L-type Ca^{2+} currents for insulin secretion is greater in mouse than in human β cells (Eliasson et al. 2008; Braun et al. 2009). In adrenal chromaffin cells, L-type Ca^{2+} currents conducted by $Ca_V1.2$ and $Ca_V1.3$ channels trigger secretion of catecholamines, and their activity is strongly regulated by second messenger signaling pathways, including cAMP (Marcantoni et al. 2007).

Neurotransmitter release at specialized ribbon synapses is continuous, similar to hormone secretion in some physiological circumstances, and Ca_V1 channels are specifically required for this mode of synaptic transmission. In photoreceptors, $Ca_V1.4$ channels are primarily responsible for Ca^{2+} entry that triggers exocytosis of neurotransmitters (Table 1) (Barnes and Kelly 2002). Mutations in the $Ca_V1.4$ channel in humans lead to stationary night blindness (Bech-Hansen et al. 1998; Striessnig et al. 2010). In auditory hair cells, $Ca_V1.3$ channels conduct the L-type Ca^{2+} currents that trigger neurotransmitter release (Kollmar et al. 1997). Deletion of the gene encoding $Ca_V1.3$ channels causes deafness in mice (Platzer et al. 2000). The distal carboxy-terminal domain plays an autoregulatory role in both $Ca_V1.3$ and $Ca_V1.4$ channels (Singh et al. 2006, 2008), but it is not known whether it is subject to proteolytic processing in vivo. $Ca_V1.3$ channels are regulated by multiple interacting proteins (Cui et al. 2007; Jenkins et al. 2010), which may be important in tuning their activity to fit the specific requirements of hair cells transmitting auditory information at different frequencies.

Ca_V2 CHANNELS IN SYNAPTIC TRANSMISSION

Presynaptic Ca^{2+} channels conduct P/Q-, N-, and R-type Ca^{2+} currents, which initiate synaptic transmission (Table 1). The efficiency of neurotransmitter release depends on the third or fourth power of the entering Ca^{2+}. This steep dependence of neurotransmission on Ca^{2+} entry makes the presynaptic Ca^{2+} channel an exceptionally sensitive and important target of regulation. In the nervous system, $Ca_V2.1$ channels conducting P/Q-type Ca^{2+} currents and $Ca_V2.2$ channels conducting N-type Ca^{2+} currents are the predominant pathways for Ca^{2+} entry initiating fast release of classical neurotransmitters like glutamate, acetylcholine, and GABA. Extensive studies indicate that they are controlled by many different protein interactions with their intracellular domains, which serve as a platform for Ca^{2+}-dependent signal transduction (Fig. 4A).

SNARE Proteins

Ca^{2+} entry through voltage-gated Ca^{2+} channels initiates exocytosis by triggering the fusion of secretory vesicle membranes with the plasma membrane through actions on the SNARE protein complex of syntaxin, SNAP-25, and VAMP/synaptobrevin (reviewed in Bajjalieh and Scheller 1995; Sudhof 1995, 2004). The function of the SNARE protein complex is regulated by interactions with numerous proteins, including the synaptic vesicle Ca^{2+}-binding protein synaptotagmin. Presynaptic $Ca_V2.1$ and $Ca_V2.2$ channels interact directly with the SNARE proteins through a specific synaptic protein interaction (synprint) site in the large intracellular loop connecting domains II and III (Fig. 4A) (Sheng et al. 1994; Rettig et al. 1996). This interaction is regulated by Ca^{2+} and protein phosphorylation (Sheng et al. 1996; Yokoyama et al. 1997, 2005). Synaptotagmin also binds to the synprint site of Ca_V2 channels (Charvin et al. 1997; Sheng et al. 1997; Wiser et al. 1997). Injection into presynaptic neurons of peptides that block SNARE protein interactions with Ca_V2 channels inhibits synaptic transmission, consistent with the conclusion that interaction with SNARE proteins is required to position docked synaptic vesicles near Ca^{2+} channels for fast exocytosis (Mochida et al. 1996; Rettig et al. 1997). These results define a second functional activity of the presynaptic Ca^{2+} channel–targeting docked synaptic vesicles to a source of Ca^{2+} for effective transmitter release.

In addition to this functional role of interaction between Ca^{2+} channels and SNARE proteins in the anterograde process of synaptic transmission, these interactions also have retrograde regulatory effects on Ca^{2+} channel function. Coexpression of the plasma membrane SNARE proteins syntaxin or SNAP-25 with $Ca_V2.1$ or $Ca_V2.2$ channels reduces the level of channel expression and inhibits Ca^{2+} channel activity by shifting the voltage dependence of steady-state inactivation during long depolarizing prepulses toward more negative membrane potentials (Bezprozvanny et al. 1995; Wiser et al. 1996; Zhong et al. 1999). The inhibitory effects of syntaxin are relieved by coexpression of SNAP-25 and synaptotagmin to form a complete SNARE complex (Wiser et al. 1997; Tobi et al. 1999; Zhong et al. 1999), which has the effect of enhancing activation of Ca_V2 channels with nearby docked synaptic vesicles that have formed complete SNARE complexes and are ready for release. These processes fine-tune the efficiency of neurotransmitter release at frog neuromuscular junctions, where peptide and cDNA reagents can be used to modify synaptic function in vivo (Keith et al. 2007).

G Protein Modulation

N-type and P/Q-type Ca^{2+} currents are regulated through multiple G protein coupled pathways (Hille 1994; Jones et al. 1997; Ikeda and Dunlap 1999). Although there are several G protein signaling pathways that regulate these channels, one common pathway that has been best studied at both cellular and molecular levels is voltage dependent and membrane delimited (i.e., a pathway without soluble intracellular messengers whose effects can be reversed by strong depolarization). Inhibition of Ca^{2+} channel activity is typically caused by a positive shift in the voltage dependence and a slowing of channel activation (Bean 1989b). These effects are relieved by strong depolarization resulting in facilitation of Ca^{2+} currents (Marchetti et al. 1986; Bean 1989b). Synaptic transmission is inhibited by neurotransmitters through this mechanism. G-protein α subunits are thought to confer specificity in receptor coupling, but Gβγ subunits are responsible for modulation of Ca^{2+} channels. Cotransfection of cells with the Ca^{2+} channel α1 and β subunits plus Gβγ causes a shift in the voltage dependence of Ca^{2+} channel activation to more positive membrane potentials and reduces the steepness of voltage-dependent activation, effects that closely mimic the actions of neurotransmitters and guanyl nucleotides on N-type and P/Q-type Ca^{2+} channels in neurons and neuroendocrine cells (Herlitze et al. 1996). In contrast, transfection with a range of Gα subunits does not have this effect. This voltage shift can be reversed by strong positive prepulses resulting in voltage-dependent facilitation of the Ca^{2+} current in the presence of Gβγ, again closely mimicking the effects of neurotransmitters and guanyl nucleotides on Ca^{2+} channels. Similarly, injection or expression of Gβγ subunits in sympathetic ganglion neurons induces facilitation and occludes modulation of N-type channels by norepinephrine, but Gα subunits do not (Herlitze et al. 1996; Ikeda 1996). These results point to the Gβγ subunits as the primary regulators of presynaptic Ca^{2+} channels via this voltage-dependent pathway through direct protein–protein interactions (Fig. 4A).

Possible sites of G protein βγ subunit interaction with Ca^{2+} channels have been extensively investigated by construction and analysis of channel chimeras, by G protein binding experiments, and by site-directed mutagenesis and expression (Fig. 4A). Evidence from G protein binding and site-directed mutagenesis experiments points to the intracellular loop between domains I and II (L_{I-II}) as a crucial site of G protein regulation, and peptides from this region of Ca$_V$2.2 prevent inhibition of channel activity by Gβγ, presumably by binding to Gβγ and competitively inhibiting its access to Ca^{2+} channels (De Waard et al. 1997; Herlitze et al. 1997; Zamponi et al. 1997). This region of the channel binds Gβγ in vitro as well as in vivo in the yeast two-hybrid assay (De Waard et al. 1997; Zamponi et al. 1997; Garcia et al. 1998). Increasing evidence also points to segments in the amino- and carboxy-terminal domains of Ca^{2+} channels that are also required for G protein regulation (Zhang et al. 1996; Page et al. 1997, 1998; Qin et al. 1997; Canti et al. 1999; Li et al. 2004). As the amino- and carboxy-terminal domains are likely to interact with each other in the folded channel protein, a second site of interaction for G proteins may be formed at their intersection.

Ca^{2+} Binding Proteins

Ca^{2+}-dependent facilitation and inactivation of presynaptic Ca^{2+} channels was observed in patch clamp recordings of presynaptic nerve terminals in the rat neurohypophysis (Branchaw et al. 1997) and the calyx of Held synapse in the rat brainstem (Forsythe et al. 1998b). During tetanic stimulation at this synapse, Ca$_V$2.1 channel currents show both Ca^{2+}-dependent facilitation and inactivation (Borst and Sakmann 1998; Cuttle et al. 1998; Forsythe et al. 1998a), which results in facilitation and depression of excitatory postsynaptic responses (Borst and Sakmann 1998; Cuttle et al. 1998; Forsythe et al. 1998b). Ca^{2+}-dependent facilitation and inactivation are also observed for cloned and expressed Ca$_V$2.1 channels expressed in mammalian cells (Lee et al. 1999, 2000). A novel CaM-binding site was identified by yeast two-hybrid screening in the carboxy-terminal domain of the pore-forming α$_1$2.1 subunit of Ca$_V$2.1 channels (Lee et al. 1999). This CaM-binding domain (CBD) (Fig. 4A) is located on the carboxy-terminal side of the sequence in α$_1$2.1 that corresponds to the IQ-domain that is required for CaM modulation of cardiac Ca$_V$1.2 channels (Peterson et al. 1999; Qin et al. 1999; Zühlke et al. 1999). The modified IQ domain of α$_1$2.1 begins with the amino acid sequence IM instead of IQ and has other changes that would be predicted to substantially reduce its affinity for CaM. CaM binding to the CBD is Ca^{2+}-dependent. Both Ca^{2+}-dependent facilitation and inactivation are blocked by coexpression of a CaM inhibitor peptide (Lee et al. 1999), suggesting that Ca^{2+}-dependent modulation of Ca$_V$2.1 channels in neurons is caused by two sequential interactions with CaM or a related Ca^{2+}-binding protein.

The mechanism for Ca^{2+}-dependent facilitation and inactivation of Ca$_V$2.1 channels involves CaM binding to two adjacent subsites—the CBD and the upstream IQ-like motif

(Lee et al. 2003). The IQ-like motif is required for facilitation, whereas the CBD is required for inactivation. In addition, the two lobes of CaM are also differentially involved in these two processes. Mutation of the two EF hands in the carboxy-terminal lobe primarily prevents facilitation, whereas mutation of the EF hands in the amino-terminal lobe primarily prevents inactivation (DeMaria et al. 2001; Erickson et al. 2001; Lee et al. 2003). FRET studies indicate that apo-calmodulin can bind to $Ca_V2.1$ channels in intact cells and binding is enhanced by Ca^{2+} binding to calmodulin (Erickson et al. 2001). Altogether, these results support a model in which the two lobes of CaM interact differentially with the modified IQ domain and the CBD to effect bi-directional regulation, with the high-affinity carboxy-terminal lobe primarily controlling facilitation through interactions with the IQ-like domain and the lower-affinity amino-terminal lobe primarily controlling inactivation through interactions with the CBD. This biphasic regulation of $Ca_V2.1$ channels causes synaptic facilitation and depression in transfected sympathetic ganglion neuron synapses in which neurotransmission is initiated by transfected $Ca_V2.1$ channels (Mochida et al. 2008).

CaM is the most well-characterized member of a superfamily of Ca^{2+} sensor (CaS) proteins, many of which differ from CaM in having neuron-specific localization, amino-terminal myristoylation, and amino acid substitutions that prevent Ca^{2+} binding to one or two of the EF hands (Haeseleer and Palczewski 2002). The CaS protein CaBP1 binds to the CBD, but not the IQ-like domain, of $\alpha_1 2.1$ and its binding is Ca^{2+}-independent (Lee et al. 2002). CaBP1 causes a strong enhancement of the rate of inactivation, a positive shift in the voltage-dependence of activation, and a loss of Ca^{2+}-dependent facilitation of $Ca_V2.1$ channels, which would combine to reduce the activity of these channels. Because it coimmunoprecipitates and colocalizes with $Ca_V2.1$ channels in the brain (Lee et al. 2002), CaBP1 may be an important determinant of $Ca_V2.1$ channel function in neurons and may contribute to the diversity of function of these channels in the nervous system. Visinin-like protein 2 (VILIP-2) is a neuronal Ca^{2+}-binding protein that is distantly related to CaBP-1 (Haeseleer and Palczewski 2002). Consistent with these structural differences, VILIP-2 has opposite effects on $Ca_V2.1$ channels than CaBP-1 (Lautermilch et al. 2005). Coexpression of VILIP-2 causes slowed inactivation and enhanced facilitation, but its binding and effects are Ca^{2+}-independent like CaBP-1. VILIP-2 may serve as a positive modulator of synaptic transmission, prolonging Ca^{2+} channel opening, and enhancing facilitation. Differential expression of CaBP1 and VILIP-2 at synapses would lead to opposite modulation of synaptic transmission in response to trains of action potentials and opposing input–output functions at the synapse.

Ca_V3 CHANNELS AND FREQUENCY MODULATION

Molecular Properties of Ca_V3 Channels

Ca^{2+} channels of the Ca_V3 subfamily conduct T-type Ca^{2+} currents (Catterall et al. 2005). These Ca^{2+} currents are activated at comparatively negative membrane potentials, in the same range as Na^+ currents in most cells, and they have fast voltage-dependent inactivation compared to other Ca^{2+} currents (Nowycky et al. 1985). These Ca^{2+} currents are therefore well-suited for rhythmic firing of action potentials. They are also well-suited for generation of large Ca^{2+} transients because they are activated at negative membrane potentials where the driving force for Ca^{2+} entry is large. A family of three Ca_V3 channel $\alpha1$ subunits have been characterized by cDNA cloning and sequencing (Catterall et al. 2005). Remarkably, these Ca^{2+} channel subunits have the same molecular organization as Ca_V1 and Ca_V2 channels but are only $\sim$25% identical in amino acid sequence (Catterall et al. 2005). This is a similar level of amino acid sequence identity as Ca^{2+} channels have with Na^+ channels, indicating that these subfamilies of Ca^{2+} channels separated from each other at the same point of evolution as Na^+ channels separated from Ca^{2+} channels. Although Ca_V3 channels are similar in structure to Ca_V1 and Ca_V2 channels, there is no clear evidence at present that they interact with the same set of auxiliary subunits. In fact, the prevailing view is that the α1 subunits function independently of other subunits. This would be unique among the families of Na^+ and Ca^{2+} channels.

Functional Roles of Ca_V3 Channels

As expected from their functional properties, Ca_V3 channels are important in repetitively firing tissues. In the sino-atrial node of the heart, they conduct an important component of the pacemaker current that generates the heartbeat (Mangoni et al. 2006). In the relay neurons of the thalamus, they are crucial for generation of the rhythmic bursts of action potentials that drive sleep spindles and control sleep (Lee et al. 2004). Moreover, mutations in Ca_V3 channels cause absence epilepsy, in which the affected individuals transiently enter a sleep-like state that interrupts their normal activities (Kim et al. 2001; Song et al. 2004). In the adrenal cortex, they are important in regulation of synthesis and secretion of aldosterone (Welsby et al. 2003).

Regulation of Ca_V3 Channels

In neurons, dopamine and other neurotransmitters inhibit T-type Ca^{2+} currents via a pathway that is specific for the Gβ2 subunit (Wolfe et al. 2003). As for Ca_V2 channels, G protein βγ subunits bind directly to Ca_V3 channels and regulate them (DePuy et al. 2006). The site of interaction is in the intracellular loop connecting domains II and III (DePuy et al. 2006). In addition, in adrenal glomerulosa cells, angiotensin II regulates aldosterone secretion via enhanced activation of $Ca_V3.2$ channels (Welsby et al. 2003). This regulation is mediated by a signaling complex of CaMKII bound to the intracellular loop connecting domains II and III (Yao et al. 2006). Phosphorylation of a single serine residue in this intracellular loop negatively shifts the voltage dependence of activation and thereby substantially increases Ca^{2+} current at negative membrane potentials (Yao et al. 2006). It is unknown at this stage whether binding of CaMKII is required for physiological regulation or whether binding of the kinase per se has any regulatory effect.

THE EFFECTOR CHECKPOINT MODEL OF Ca^{2+} CHANNEL REGULATION

In closing this article on Ca^{2+} signaling via voltage-gated Ca^{2+} channels, it is interesting to introduce an emerging theme that unites several aspects of the localized regulation of these proteins. Ca^{2+} channel signaling complexes are formed when the effectors and regulators of the Ca^{2+} signal bind to the intracellular domains of Ca^{2+} channels to effectively receive and respond to the local Ca^{2+} signal. In four cases, binding of the effectors of the Ca^{2+} signal has been shown to enhance the activity of the Ca_V1 and Ca_V2 channels. First, in skeletal muscle, interactions of the plasma membrane $Ca_V1.1$ channel with the ryanodine-sensitive Ca^{2+} release channel in the sarcoplasmic reticulum, which serves as the effector of excitation-contraction coupling, greatly increase the functional activity of the $Ca_V1.1$ channels (Nakai et al. 1996a). Second, as described above, interaction with individual plasma membrane SNARE proteins inhibits the activity of Ca_V2 channels, but formation of complete SNARE complex containing synaptotagmin, the effector of exocytosis, relieves this inhibition and enhances Ca^{2+} channel activity (Bezprozvanny et al. 1995; Wiser et al. 1996, 1997; Zhong et al. 1999). Third, binding of Ca^{2+}/CaM-dependent protein kinase II, an effector of Ca^{2+}-dependent regulatory events, to a site in the carboxy-terminal domain of $Ca_V2.1$ channels substantially increases their activity (Jiang et al. 2008). Finally, binding of RIM, a regulator of SNARE protein function, to the $Ca_Vβ$ subunits substantially increases Ca_V2 channel activity (Kiyonaka et al. 2007). The common thread in all of these diverse examples of Ca^{2+} channel regulation by interacting proteins is that binding of an effector ready to respond to the Ca^{2+} signal enhances the activity of the Ca^{2+} channel. Thus, this mechanism provides a functional checkpoint of the fitness of a

Ca^{2+} channel to carry out its physiological role, and enhances its activity if it passes this checkpoint criterion. This "effector checkpoint" mechanism would serve to focus Ca^{2+} entry on the Ca_V channels that are ready to use the resulting Ca^{2+} signal to initiate a physiological intracellular signaling process. It seems likely that further studies will reveal more examples of this form of regulation and that it may be a unifying theme in the regulation of Ca^{2+} signaling by Ca_V channels.

REFERENCES

Ahlijanian MK, Westenbroek RE, Catterall WA. 1990. Subunit structure and localization of dihydropyridine-sensitive calcium channels in mammalian brain, spinal cord, and retina. *Neuron* **4:** 819–832.

Arikkath J, Campbell KP. 2003. Auxiliary subunits: Essential components of the voltage-gated calcium channel complex. *Curr Opin Neurobiol* **13:** 298–307.

Armstrong CM, Bezanilla FM, Horowicz P. 1972. Twitches in the presence of ethylene glycol bis (-aminoethyl ether)-N,N′-tetracetic acid. *Biochim Biophys Acta* **267:** 605–608.

Artalejo CR, Adams ME, Fox AP. 1994. Three types of calcium channel trigger secretion with different efficacies in chromaffin cells. *Nature* **367:** 72–76.

Bajjalieh SM, Scheller RH. 1995. The biochemistry of neurotransmitter secretion. *J Biol Chem* **270:** 1971–1974.

Barnes S, Kelly ME. 2002. Calcium channels at the photoreceptor synapse. *Adv Exp Med Biol* **514:** 465–476.

Bean BP. 1989a. Classes of calcium channels in vertebrate cells. *Annu Rev Physiol* **51:** 367–384.

Bean BP. 1989b. Neurotransmitter inhibition of neuronal calcium currents by changes in channel voltage dependence. *Nature* **340:** 153–156.

Bech-Hansen NT, Naylor MJ, Maybaum TA, Pearce WG, Koop B, Fishman GA, Mets M, Musarella MA, Boycott KM. 1998. Loss-of-function mutations in a calcium-channel α1 subunit gene in Xp11.23 cause incomplete X-linked congenital stationary night blindness. *Nat Genet* **19:** 264–267.

Bers DM. 2002. Cardiac excitation-contraction coupling. *Nature* **415:** 198–205.

Bezprozvanny I, Scheller RH, Tsien RW. 1995. Functional impact of syntaxin on gating of N-type and Q-type calcium channels. *Nature* **378:** 623–626.

Bichet D, Haass FA, Jan LY. 2003. Merging functional studies with structures of inward-rectifier potassium channels. *Nat Rev Neurosci* **4:** 957–967.

Bito H, Deisseroth K, Tsien RW. 1996. CREB phosphorylation and dephosphorylation: A Ca^{2+}- and stimulus duration-dependent switch for hippocampal gene expression. *Cell* **87:** 1203–1214.

Block BA, Imagawa T, Campbell KP, Franzini-Armstrong C. 1988. Structural evidence for direct interaction between the molecular components of the transverse tubule/sarcoplasmic reticulum junction in skeletal muscle. *J Cell Biol* **107:** 2587–2600.

Borst JG, Sakmann B. 1998. Facilitation of presynaptic calcium currents in the rat brainstem. *J Physiol* **513:** 149–155.

Branchaw JL, Banks MI, Jackson MB. 1997. Ca^{2+}- and voltage-dependent inactivation of Ca^{2+} channels in nerve terminals of the neurohypophysis. *J Neurosci* **17:** 5772–5781.

Braun M, Ramracheya R, Johnson PR, Rorsman P. 2009. Exocytotic properties of human pancreatic β-cells. *Ann NY Acad Sci* **1152:** 187–193.

Canti C, Page KM, Stephens GJ, Dolphin AC. 1999. Identification of residues in the N terminus of α1B critical for inhibition of the voltage-dependent calcium channel by Gβγ. *J Neurosci* **19:** 6855–6864.

Carbone E, Lux HD. 1984. A low voltage-activated, fully inactivating Ca channel in vertebrate sensory neurones. *Nature* **310:** 501–502.

Catterall WA. 1991. Excitation-contraction coupling in vertebrate skeletal muscle: A tale of two calcium channels. *Cell* **64:** 871–874.

Catterall WA. 2000a. From ionic currents to molecular mechanisms: The structure and function of voltage-gated sodium channels. *Neuron* **26:** 13–25.

Catterall WA. 2000b. Structure and regulation of voltage-gated calcium channels. *Annu Rev Cell Dev Biol* **16:** 521–555.

Catterall WA, Few AP. 2008. Calcium channel regulation and presynaptic plasticity. *Neuron* **59:** 882–901.

Catterall WA, Perez-Reyes E, Snutch TP, Striessnig J. 2005. International Union of Pharmacology. XLVIII. Nomenclature and structure-function relationships of voltage-gated calcium channels. *Pharmacol Rev* **57:** 411–425.

Chang FC, Hosey MM. 1988. Dihydropyridine and phenylalkylamine receptors associated with cardiac and skeletal muscle calcium channels are structurally different. *J Biol Chem* **263:** 18929–18937.

Charvin N, Lévêque C, Walker D, Berton F, Raymond C, Kataoka M, Shoji-Kasai Y, Takahashi M, De Waard M, Seagar MJ. 1997. Direct interaction of the calcium sensor protein synaptotagmin I with a cytoplasmic domain of the α1A subunit of the P/Q-type calcium channel. *EMBO J* **16:** 4591–4596.

Chen YH, Li MH, Zhang Y, He LL, Yamada Y, Fitzmaurice A, Shen Y, Zhang H, Tong L, Yang J. 2004. Structural basis of the α1β subunit interaction of voltage-gated Ca^{2+} channels. *Nature* **429:** 675–680.

Cui G, Meyer AC, Calin-Jageman I, Neef J, Haeseleer F, Moser T, Lee A. 2007. Ca^{2+}-binding proteins tune Ca^{2+}-feedback to $Ca_V1.3$ channels in mouse auditory hair cells. *J Physiol* **585:** 791–803.

Curtis BM, Catterall WA. 1984. Purification of the calcium antagonist receptor of the voltage-sensitive calcium channel from skeletal muscle transverse tubules. *Biochem* **23:** 2113–2118.

Curtis BM, Catterall WA. 1985. Phosphorylation of the calcium antagonist receptor of the voltage-sensitive calcium channel by cAMP-dependent protein kinase. *Proc Natl Acad Sci* **82:** 2528–2532.

Curtis BM, Catterall WA. 1986. Reconstitution of the voltage-sensitive calcium channel purified from skeletal muscle transverse tubules. *Biochemistry* **25:** 3077–3083.

Cuttle MF, Tsujimoto T, Forsythe ID, Takahashi T. 1998. Facilitation of the presynaptic calcium current at an auditory synapse in rat brainstem. *J Physiol* **512:** 723–729.

Davies A, Hendrich J, Van Minh AT, Wratten J, Douglas L, Dolphin AC. 2007. Functional biology of the α2δ subunits of voltage-gated calcium channels. *Trends Pharmacol Sci* **28:** 220–228.

Davies A, Kadurin I, Alvarez-Laviada A, Douglas L, Nieto-Rostro M, Bauer CS, Pratt WS, Dolphin AC. 2010. The α2δ subunits of voltage-gated calcium channels form GPI-anchored proteins, a posttranslational modification essential for function. *Proc Natl Acad Sci* **107:** 1654–1659.

De Jongh KS, Merrick DK, Catterall WA. 1989. Subunits of purified calcium channels: A 212-kDa form of α1 and partial amino acid sequence of a phosphorylation site of an independent β subunit. *Proc Natl Acad Sci* **86:** 8585–8589.

De Jongh KS, Warner C, Catterall WA. 1990. Subunits of purified calcium channels. α_2 and δ are encoded by the same gene. *J Biol Chem* **265:** 14738–14741.

De Jongh KS, Warner C, Colvin AA, Catterall WA. 1991. Characterization of the two size forms of the α1 subunit of skeletal muscle L-type calcium channels. *Proc Natl Acad Sci* **88:** 10778–10782.

De Jongh KS, Murphy BJ, Colvin AA, Hell JW, Takahashi M, Catterall WA. 1996. Specific phosphorylation of a site in the full-length form of the α1 subunit of the cardiac L-type calcium channel by cAMP-dependent protein kinase. *Biochemistry* **35:** 10392–10402.

De Waard M, Liu HY, Walker D, Scott VES, Gurnett CA, Campbell KP. 1997. Direct binding of G-protein βγ complex to voltage-dependent calcium channels. *Nature* **385:** 446–450.

Deisseroth K, Heist EK, Tsien RW. 1998. Translocation of calmodulin to the nucleus supports CREB phosphorylation in hippocampal neurons. *Nature* **392:** 198–202.

DeMaria CD, Soong TW, Alseikhan BA, Alvania RS, Yue DT. 2001. Calmodulin bifurcates the local Ca^{2+} signal that modulates P/Q-type Ca^{2+} channels. *Nature* **411:** 484–489.

DePuy SD, Yao J, Hu C, McIntire W, Bidaud I, Lory P, Rastinejad F, Gonzalez C, Garrison JC, Barrett PQ. 2006. The molecular basis for T-type Ca^{2+} channel inhibition by G protein β2γ2 subunits. *Proc Natl Acad Sci* **103:** 14590–14595.

Dolmetsch RE, Pajvani U, Fife K, Spotts JM, Greenberg ME. 2001. Signaling to the nucleus by an L-type calcium channel-calmodulin complex through the MAP kinase pathway. *Science* **294:** 333–339.

Dolphin AC. 2003. β subunits of voltage-gated calcium channels. *J Bioenerg Biomembr* **35:** 599–620.

Dubel SJ, Starr TVB, Hell J, Ahlijanian MK, Enyeart JJ, Catterall WA, Snutch TP. 1992. Molecular cloning of the α1 subunit of an omega-conotoxin-sensitive calcium channel. *Proc Natl Acad Sci* **89:** 5058–5062.

Dunlap K, Luebke JI, Turner TJ. 1995. Exocytotic calcium channels in mammalian central neurons. *Trends Neurosci* **18:** 89–98.

Eliasson L, Abdulkader F, Braun M, Galvanovskis J, Hoppa MB, Rorsman P. 2008. Novel aspects of the molecular mechanisms controlling insulin secretion. *J Physiol* **586:** 3313–3324.

Ellis SB, Williams ME, Ways NR, Brenner R, Sharp AH, Leung AT, Campbell KP, McKenna E, Koch WJ, Hui A, et al. 1988. Sequence and expression of mRNAs encoding the α1 and α2 subunits of a DHP-sensitive calcium channel. *Science* **241:** 1661–1664.

Emrick MA, Sadilek M, Konoki K, Catterall WA. 2010. β-adrenergic-regulated phosphorylation of the skeletal muscle $Ca_V1.1$ channel in the fight-or-flight response. *Proc Natl Acad Sci* **107:** 18712–18717.

Erickson MG, Alseikhan BA, Peterson BZ, Yue DT. 2001. Preassociation of calmodulin with voltage-gated Ca^{2+} channels revealed by FRET in single living cells. *Neuron* **31:** 973–985.

Ertel EA, Campbell KP, Harpold MM, Hofmann F, Mori Y, Perez-Reyes E Schwartz A, Snutch TP, Tanabe T, Birnbaumer L, et al. 2000. Nomenclature of voltage-gated calcium channels. *Neuron* **25:** 533–535.

Fabiato A. 1983. Calcium-induced release of calcium from the cardiac sarcoplasmic reticulum. *Am J Physiol* **245:** C1–C14.

Fedulova SA, Kostyuk PG, Veselovsky NS. 1985. Two types of calcium channels in the somatic membrane of newborn rat dorsal root ganglion neurones. *J Physiol* **359:** 431–446.

Flavell SW, Greenberg ME. 2008. Signaling mechanisms linking neuronal activity to gene expression and plasticity of the nervous system. *Annu Rev Neurosci* **31:** 563–590.

Flockerzi V, Oeken HJ, Hofmann F, Pelzer D, Cavalie A, Trautwein W. 1986. Purified dihydropyridine-binding site from skeletal muscle t-tubules is a functional calcium channel. *Nature* **323:** 66–68.

Forsythe ID, Tsujimoto T, Barnes-Davies M, Cuttle MF, Takahashi T. 1998. Inactivation of presynaptic calcium current contributes to synaptic depression at a fast central synapse. *Neuron* **20:** 797–807.

Fraser IDC, Tavalin SJ, Lester LB, Langeberg LK, Westphal AM, Dean RA, Marrion NV, Scott JD. 1998. A novel lipid-anchored A-kinase anchoring protein facilitates cAMP-responsive membrane events. *EMBO J* **17:** 2261–2272.

Fuller MD, Emrick MA, Sadilek M, Scheuer T, Catterall WA. 2010. Molecular mechanism of calcium channel regulation in the fight-or-flight response. *Sci Signal* **3:** ra70.

Gao T, Cuadra AE, Ma H, Bunemann M, Gerhardstein BL, Cheng T, Eick RT, Hosey MM. 2001. C-terminal fragments of the α1C ($Ca_V1.2$) subunit associate with and regulate L-type calcium channels containing C-terminal-truncated α1C subunits. *J Biol Chem* **276:** 21089–21097.

Garcia DE, Li B, Garcia-Ferreiro RE, Hernández-Ochoa EO, Yan K, Gautam N, Catterall WA, Mackie K, Hille B. 1998. G-protein β subunit specificity in the fast membrane-delimited inhibition of Ca^{2+} channels. *J Neurosci* **18:** 9163–9170.

Gerhardstein BL, Gao T, Bunemann M, Puri TS, Adair A, Ma H, Hosey MM. 2000. Proteolytic processing of the C terminus of the α1C subunit of L-type calcium channels and the role of a proline-rich domain in membrane tethering of proteolytic fragments. *J Biol Chem* **275**: 8556–8563.

Gomez-Ospina N, Tsuruta F, Barreto-Chang O, Hu L, Dolmetsch R. 2006. The C terminus of the L-type voltage-gated calcium channel $Ca_V 1.2$ encodes a transcription factor. *Cell* **127**: 591–606.

Gray PC, Tibbs VC, Catterall WA, Murphy BJ. 1997. Identification of a 15-kDa cAMP-dependent protein kinase-anchoring protein associated with skeletal muscle L-type calcium channels. *J Biol Chem* **272**: 6297–6302.

Gray PC, Johnson BD, Westenbroek RE, Hays LG, Yates JR3rd, Scheuer T, Catterall WA, Murphy BJ. 1998. Primary structure and function of an A kinase anchoring protein associated with calcium channels. *Neuron* **20**: 1017–1026.

Gurnett CA, De Waard M, Campbell KP. 1996. Dual function of the voltage-dependent Ca^{2+} channel α2δ subunit in current stimulation and subunit interaction. *Neuron* **16**: 431–440.

Haase H, Bartel S, Karczewski P, Morano I, Krause EG. 1996. In-vivo phosphorylation of the cardiac L-type calcium channel β subunit in response to catecholamines. *Mol Cell Biochem* **163–164**: 99–106.

Haeseleer F, Palczewski K. 2002. Calmodulin and calcium-binding proteins: Variations on a theme. *Adv Exp Med Biol* **514**: 303–317.

Hagiwara S, Ozawa S, Sand O. 1975. Voltage clamp analysis of two inward current mechanisms in the egg cell membrane of a starfish. *J Gen Physiol* **65**: 617–644.

Heinemann SH, Terlau H, Stühmer W, Imoto K, Numa S. 1992. Calcium channel characteristics conferred on the sodium channel by single mutations. *Nature* **356**: 441–443.

Hell JW, Westenbroek RE, Warner C, Ahlijanian MK, Prystay W, Gilbert MM, Snutch TP, Catterall WA. 1993a. Identification and differential subcellular localization of the neuronal class C and class D L-type calcium channel α1 subunits. *J Cell Biol* **123**: 949–962.

Hell JW, Yokoyama CT, Wong ST, Warner C, Snutch TP, Catterall WA. 1993b. Differential phosphorylation of two size forms of the neuronal class C L-type calcium channel α1 subunit. *J Biol Chem* **268**: 19451–19457.

Hell JW, Westenbroek RE, Breeze LJ, Wang KKW, Chavkin C, Catterall WA. 1996. N-methyl-D-aspartate receptor-induced proteolytic conversion of postsynaptic class C L-type calcium channels in hippocampal neurons. *Proc Natl Acad Sci* **93**: 3362–3367.

Herlitze S, Garcia DE, Mackie K, Hille B, Scheuer T, Catterall WA. 1996. Modulation of calcium channels by G protein βγ subunits. *Nature* **380**: 258–262.

Herlitze S, Hockerman GH, Scheuer T, Catterall WA. 1997. Molecular determinants of inactivation and G protein modulation in the intracellular loop connecting domains I and II of the calcium channel α1A subunit. *Proc Natl Acad Sci* **94**: 1512–1516.

Hille B. 1994. Modulation of ion-channel function by G-protein-coupled receptors. *Trends Neurosci* **17**: 531–536.

Hockerman GH, Peterson BZ, Johnson BD, Catterall WA. 1997a. Molecular determinants of drug binding and action on L-type calcium channels. *Annu Rev Pharmacol Toxicol* **37**: 361–396.

Hockerman GH, Peterson BZ, Sharp E, Tanada TN, Scheuer T, Catterall WA. 1997b. Construction of a high-affinity receptor site for dihydropyridine agonists and antagonists by single amino acid substitutions in a non-L-type calcium channel. *Proc Natl Acad Sci* **94**: 14906–14911.

Hofmann F, Biel M, Flockerzi V. 1994. Molecular basis for calcium channel diversity. *Annu Rev Neurosci* **17**: 399–418.

Hosey MM, Barhanin J, Schmid A, Vandaele S, Ptasienski J, O'Callahan C, Cooper C, Lazdunski M. 1987. Photoaffinity labelling and phosphorylation of a 165 kilodalton peptide associated with dihydropyridine and phenylalkylamine-sensitive calcium channels. *Biochem Biophys Res Commun* **147**: 1137–1145.

Hulme JT, Ahn M, Hauschka SD, Scheuer T, Catterall WA. 2002. A novel leucine zipper targets AKAP15 and cyclic AMP-dependent protein kinase to the C terminus of the skeletal muscle calcium channel and modulates its function. *J Biol Chem* **277**: 4079–4087.

Hulme JT, Lin TW, Westenbroek RE, Scheuer T, Catterall WA. 2003. β-adrenergic regulation requires direct anchoring of PKA to cardiac $Ca_V 1.2$ channels via a leucine zipper interaction with A kinase-anchoring protein 15. *Proc Natl Acad Sci* **100**: 13093–13098.

Hulme JT, Konoki K, Lin TW, Gritsenko MA, Camp DG 2nd, Bigelow DJ, Catterall WA. 2005. Sites of proteolytic processing and noncovalent association of the distal C-terminal domain of $Ca_V 1.1$ channels in skeletal muscle. *Proc Natl Acad Sci* **102**: 5274–5279.

Hulme JT, Westenbroek RE, Scheuer T, Catterall WA. 2006a. Phosphorylation of serine 1928 in the distal C-terminal of cardiac $Ca_V 1.2$ channels during β-adrenergic regulation. *Proc Natl Acad Sci* **103**: 16574–16579.

Hulme JT, Yarov-Yarovoy V, Lin TW-C, Scheuer T, Catterall WA. 2006b. Autoinhibitory control of the $Ca_V 1.2$ channel by its proteolytically cleaved distal C-terminal domain. *J Physiol (Lond)* **576**: 87–102.

Ikeda SR. 1996. Voltage-dependent modulation of N-type calcium channels by G-protein β/γ subunits. *Nature* **380**: 255–258.

Ikeda SR, Dunlap K. 1999. Voltage-dependent modulation of N-type calcium channels: Role of G protein subunits. *Adv Second Messenger Phosphoprotein Res* **33**: 131–151.

Jay SD, Ellis SB, McCue AF, Williams ME, Vedvick TS, Harpold MM, Campbell KP. 1990. Primary structure of the γ subunit of the DHP-sensitive calcium channel from skeletal muscle. *Science* **248**: 490–492.

Jenkins MA, Christel CJ, Jiao Y, Abiria S, Kim KY, Usachev YM, Obermair GJ, Colbran RJ, Lee A. 2010. Ca^{2+}-dependent facilitation of $Ca_V 1.3$ Ca^{2+} channels by densin and Ca^{2+}/calmodulin-dependent protein kinase II. *J Neurosci* **30**: 5125–5135.

Jiang X, Lautermilch NJ, Watari H, Westenbroek RE, Scheuer T, Catterall WA. 2008. Modulation of $Ca_V 2.1$ channels by Ca^{2+}/calmodulin-dependent protein kinase II bound to the C-terminal domain. *Proc Natl Acad Sci* **105**: 341–346.

Johnson BD, Scheuer T, Catterall WA. 1994. Voltage-dependent potentiation of L-type Ca^{2+} channels in skeletal muscle cells requires anchored cAMP-dependent protein kinase. *Proc Natl Acad Sci* **91:** 11492–11496.

Johnson BD, Brousal JP, Peterson BZ, Gallombardo PA, Hockerman GH, Lai Y, Scheuer T, Catterall WA. 1997. Modulation of the cloned skeletal muscle L-type Ca^{2+} channel by anchored cAMP-dependent protein kinase. *J Neurosci* **17:** 1243–1255.

Jones LP, Patil PG, Snutch TP, Yue DT. 1997. G-protein modulation of N-type calcium channel gating current in human embryonic kidney cells (HEK 293). *J Physiol (Lond)* **498:** 601–610.

Keith RK, Poage RE, Yokoyama CT, Catterall WA, Meriney SD. 2007. Bidirectional modulation of transmitter release by calcium channel/syntaxin interactions in vivo. *J Neurosci* **27:** 265–269.

Kim D, Song I, Keum S, Lee T, Jeong MJ, Kim SS, McEnery MW, Shin HS. 2001. Lack of the burst firing of thalamocortical relay neurons and resistance to absence seizures in mice lacking α1G T-type Ca^{2+} channels. *Neuron* **31:** 35–45.

Kiyonaka S, Wakamori M, Miki T, Uriu Y, Nonaka M, Bito H, Beedle AM, Mori E, Hara Y, De Waard M, et al. 2007. RIM1 confers sustained activity and neurotransmitter vesicle anchoring to presynaptic calcium channels. *Nat Neurosci* **10:** 691–701.

Klugbauer N, Lacinová L, Marais E, Hobom M, Hofmann F. 1999. Molecular diversity of the calcium channel α2δ subunit. *J Neurosci* **19:** 684–691.

Klugbauer N, Dai S, Specht V, Lacinova L, Marais E, Bohn G, Hofmann F. 2000. A family of γ-like calcium channel subunits. *FEBS Lett* **470:** 189–197.

Kollmar R, Montgomery LG, Fak J, Henry LJ, Hudspeth AJ. 1997. Predominance of the α1D subunit in L-type voltage-gated Ca^{2+} channels of hair cells in the chicken's cochlea. *Proc Natl Acad Sci* **94:** 14883–14888.

Kuniyasu A, Oka K, Ide-Yamada T, Hatanaka Y, Abe T, Nakayama H, Kanaoka Y. 1992. Structural characterization of the dihydropyridine receptor-linked calcium channel from porcine heart. *J Biochem (Tokyo)* **112:** 235–242.

Lacerda AE, Kim HS, Ruth P, Perez-Reyes E, Flockerzi V, Hofmann F, Birnbaumer L, Brown AM. 1991. Normalization of current kinetics by interaction between the α1 and β subunits of the skeletal muscle dihydropyridine-sensitive calcium channel. *Nature* **352:** 527–530.

Lautermilch NJ, Few AP, Scheuer T, Catterall WA. 2005. Modulation of $Ca_V2.1$ channels by the neuronal calcium-binding protein visinin-like protein-2. *J Neurosci* **25:** 7062–7070.

Lee A, Wong ST, Gallagher D, Li B, Storm DR, Scheuer T, Catterall WA. 1999. Ca^{2+}/calmodulin binds to and modulates P/Q-type calcium channels. *Nature* **399:** 155–159.

Lee A, Scheuer T, Catterall WA. 2000. Calcium/calmodulin dependent inactivation and facilitation of P/Q-type calcium channels. *J Neurosci* **20:** 6830–6838.

Lee A, Westenbroek RE, Haeseleer F, Palczewski K, Scheuer T, Catterall WA. 2002. Differential modulation of $Ca_V2.1$ channels by calmodulin and Ca^{2+}-binding protein 1. *Nat Neurosci* **5:** 210–217.

Lee A, Zhou H, Scheuer T, Catterall WA. 2003. Molecular determinants of Ca^{2+}/calmodulin-dependent regulation of $Ca_V2.1$ channels. *Proc Natl Acad Sci* **100:** 16059–16064.

Lee J, Kim D, Shin HS. 2004. Lack of δ waves and sleep disturbances during non-rapid eye movement sleep in mice lacking α1G subunit of T-type calcium channels. *Proc Natl Acad Sci* **101:** 18195–18199.

Letts VA, Felix R, Biddlecome GH, Arikkath J, Mahaffey CL, Valenzuela A, Bartlett IFS, Mori Y, Campbell KP, Frankel WN. 1998. The mouse stargazer gene encodes a neuronal calcium-channel γ subunit. *Nature Genet* **19:** 340–347.

Leung AT, Imagawa T, Campbell KP. 1987. Structural characterization of the 1,4-dihydropyridine receptor of the voltage-dependent calcium channel from rabbit skeletal muscle. Evidence for two distinct high molecular weight subunits. *J Biol Chem* **262:** 7943–7946.

Li B, Zhong H, Scheuer T, Catterall WA. 2004. Functional role of a C-terminal Gβγ-binding domain of $Ca_V2.2$ channels. *Mol Pharmacol* **66:** 761–769.

Liu H, De Waard M, Scott VES, Gurnett CA, Lennon VA, Campbell KP. 1996. Identification of three subunits of the high affinity omega-conotoxin MVIIC-sensitive calcium channel. *J Biol Chem* **271:** 13804–13810.

Llinás R, Yarom Y. 1981. Electrophysiology of mammalian inferior olivary neurones in vitro. Different types of voltage-dependent ionic conductances. *J Physiol (Lond)* **315:** 569–584.

Llinás RR, Sugimori M, Cherksey B. 1989. Voltage-dependent calcium conductances in mammalian neurons. The P channel. *Ann NY Acad Sci* **560:** 103–111.

Llinás R, Sugimori M, Hillman DE, Cherksey B. 1992. Distribution and functional significance of the P-type, voltage-dependent calcium channels in the mammalian central nervous system. *Trends Neurosci* **15:** 351–355.

Mangoni ME, Traboulsie A, Leoni AL, Couette B, Marger L, Le Quang K, Kupfer E, Cohen-Solal A, Vilar J, Shin HS, et al. 2006. Bradycardia and slowing of the atrioventricular conduction in mice lacking $Ca_V3.1$/α1G T-type calcium channels. *Circ Res* **98:** 1422–1430.

Marcantoni A, Baldelli P, Hernandez-Guijo JM, Comunanza V, Carabelli V, Carbone E. 2007. L-type calcium channels in adrenal chromaffin cells: Role in pace-making and secretion. *Cell Calcium* **42:** 397–408.

Marchetti C, Carbone E, Lux HD. 1986. Effects of dopamine and noradrenaline on Ca channels of cultured sensory and sympathetic neurons of chick. *Pflugers Arch* **406:** 104–111.

Martin-Moutot N, Leveque C, Sato K, Kato R, Takahashi M, Seagar M. 1995. Properties of omega conotoxin MVIIC receptors associated with α1A calcium channel subunits in rat brain. *FEBS Lett* **366:** 21–25.

McDonald TF, Pelzer S, Trautwein W, Pelzer DJ. 1994. Regulation and modulation of calcium channels in cardiac, skeletal, and smooth muscle cells. *Physiol Rev* **74:** 365–507.

McEnery MW, Snowman AM, Sharp AH, Adams ME, Snyder SH. 1991. Purified omega-conotoxin GVIA receptor of rat brain resembles a dihydropyridine-sensitive L-type calcium channel. *Proc Natl Acad Sci* **88:** 11095–11099.

Mintz IM, Adams ME, Bean BP. 1992. P-type calcium channels in rat central and peripheral neurons. *Neuron* **9**: 85–95.

Mitterdorfer J, Froschmayr M, Grabner M, Moebius FF, Glossmann H, Striessnig J. 1996. Identification of PKA phosphorylation sites in the carboxyl terminus of L-type calcium channel α-1 subunits. *Biochemistry* **35**: 9400–9406.

Mochida S, Sheng ZH, Baker C, Kobayashi H, Catterall WA. 1996. Inhibition of neurotransmission by peptides containing the synaptic protein interaction site of N-type Ca^{2+} channels. *Neuron* **17**: 781–788.

Mochida S, Few AP, Scheuer T, Catterall WA. 2008. Regulation of presynaptic $Ca_V2.1$ channels by Ca^{2+} sensor proteins mediates short-term synaptic plasticity. *Neuron* **57**: 210–216.

Mori Y, Friedrich T, Kim MS, Mikami A, Nakai J, Ruth P, Bosse E, Hofmann F, Flockerzi V, Furuichi T, et al. 1991. Primary structure and functional expression from complementary DNA of a brain calcium channel. *Nature* **350**: 398–402.

Nakai J, Dirksen RT, Nguyen HT, Pessah IN, Beam KG, Allen PD. 1996. Enhanced dihydropyridine receptor channel activity in the presence of ryanodine receptor. *Nature* **380**: 72–75.

Nakai J, Sekiguchi N, Rando TA, Allen PD, Beam KG. 1998. Two regions of the ryanodine receptor involved in coupling with L-type Ca^{2+} channels. *J Biol Chem* **273**: 13403–13406.

Newcomb R, Szoke B, Palma A, Wang G, Chen XH, Hopkins W, Cong R, Miller J, Urge L, Tarczy-Hornoch K, et al. 1998. Selective peptide antagonist of the class E calcium channel from the venom of the tarantula *Hysterocrates gigas*. *Biochemistry* **37**: 15353–15362.

Nicoll RA, Tomita S, Bredt DS. 2006. Auxiliary subunits assist AMPA-type glutamate receptors. *Science* **311**: 1253–1256.

Nowycky MC, Fox AP, Tsien RW. 1985. Three types of neuronal calcium channel with different calcium agonist sensitivity. *Nature* **316**: 440–443.

Numa S, Tanabe T, Takeshima H, Mikami A, Niidome T, Nishimura S, Adams BA, Beam KG. 1990. Molecular insights into excitation-contraction coupling. *Cold Spring Harb Symp Quant Biol* **55**: 1–7.

Olivera BM, Miljanich GP, Ramachandran J, Adams ME. 1994. Calcium channel diversity and neurotransmitter release: The omega-conotoxins and omega-agatoxins. *Annu Rev Biochem* **63**: 823–867.

Oliveria SF, Dell'Acqua ML, Sather WA. 2007. AKAP79/150 anchoring of calcineurin controls neuronal L-type Ca^{2+} channel activity and nuclear signaling. *Neuron* **55**: 261–275.

Osterrieder W, Brum G, Hescheler J, Trautwein W, Flockerzi V, Hofmann F. 1982. Injection of subunits of cyclic AMP-dependent protein kinase into cardiac myocytes modulates Ca^{2+} current. *Nature* **298**: 576–578.

Page KM, Stephens GJ, Berrow NS, Dolphin AC. 1997. The intracellular loop between domains I and II of the B-type calcium channel confers aspects of G-protein sensitivity to the E-type calcium channel. *J Neurosci* **17**: 1330–1338.

Page KM, Cantí C, Stephens GJ, Berrow NS, Dolphin AC. 1998. Identification of the amino terminus of neuronal Ca^{2+} channel α1 subunits α1B and α1E as an essential determinant of G-protein modulation. *J Neurosci* **18**: 4815–4824.

Perez-Reyes E, Kim HS, Lacerda AE, Horne W, Wei XY, Rampe D, Campbell KP, Brown AM, Birnbaumer L. 1989. Induction of calcium currents by the expression of the α1 subunit of the dihydropyridine receptor from skeletal muscle. *Nature* **340**: 233–236.

Perez-Reyes E, Cribbs LL, Daud A, Lacerda AE, Barclay J, Williamson MP, Fox M, Rees M, Lee JH. 1998. Molecular characterization of a neuronal low-voltage-activated T-type calcium channel. *Nature* **391**: 896–900.

Peterson BZ, DeMaria CD, Yue DT. 1999. Calmodulin is the Ca^{2+} sensor for Ca^{2+}-dependent inactivation of L-type calcium channels. *Neuron* **22**: 549–558.

Pichler M, Cassidy TN, Reimer D, Haase H, Krause R, Ostler D, Striessnig J. 1997. β subunit heterogeneity in neuronal L-type calcium channels. *J Biol Chem* **272**: 13877–13882.

Pitt GS, Zuhlke RD, Hudmon A, Schulman H, Reuter H, Tsien RW. 2001. Molecular basis of calmodulin tethering and calcium-dependent inactivation of L-type calcium channels. *J Biol Chem* **276**: 30794–30802.

Platzer J, Engel J, Schrott-Fischer A, Stephan K, Bova S, Chen H, Zheng H, Striessnig J. 2000. Congenital deafness and sinoatrial node dysfunction in mice lacking class D L-type Ca^{2+} channels. *Cell* **102**: 89–97.

Pragnell M, De Waard M, Mori Y, Tanabe T, Snutch TP, Campbell KP. 1994. Calcium channel β subunit binds to a conserved motif in the I-II cytoplasmic linker of the α1 subunit. *Nature* **368**: 67–70.

Puri TS, Gerhardstein BL, Zhao XL, Ladner MB, Hosey MM. 1997. Differential effects of subunit interactions on protein kinase A- and C-mediated phosphorylation of L-type calcium channels. *Biochemistry* **36**: 9605–9615.

Qin N, Platano D, Olcese R, Stefani E, Birnbaumer L. 1997. Direct interaction of Gβγ with a C-terminal Gβγ-binding domain of the Ca^{2+} channel α1 subunit is responsible for channel inhibition by G protein-coupled receptors. *Proc Natl Acad Sci* **94**: 8866–8871.

Qin N, Olcese R, Bransby M, Lin T, Birnbaumer L. 1999. Ca^{2+} induced inhibition of the cardiac Ca^{2+} channel depends on calmodulin. *Proc Natl Acad Sci* **96**: 2435–2438.

Randall A, Tsien RW. 1995. Pharmacological dissection of multiple types of calcium channel currents in rat cerebellar granule neurons. *J Neurosci* **15**: 2995–3012.

Rettig J, Sheng Z-H, Kim DK, Hodson CD, Snutch TP, Catterall WA. 1996. Isoform-specific interaction of the α1A subunits of brain Ca^{2+} channels with the presynaptic proteins syntaxin and SNAP-25. *Proc Natl Acad Sci* **93**: 7363–7368.

Rettig J, Heinemann C, Ashery U, Sheng ZH, Yokoyama CT, Catterall WA, Neher E. 1997. Alteration of Ca^{2+} dependence of neurotransmitter release by disruption of Ca^{2+} channel/syntaxin interaction. *J Neurosci* **17**: 6647–6656.

Reuter H. 1979. Properties of two inward membrane currents in the heart. *Annu Rev Physiol* **41**: 413–424.

Reuter H. 1983. Calcium channel modulation by neurotransmitters, enzymes and drugs. *Nature* **301**: 569–574.

Reuter H, Scholz H. 1977. The regulation of calcium conductance of cardiac muscle by adrenaline. *J Physiol* **264**: 49–62.

Rotman EI, De Jongh KS, Florio V, Lai Y, Catterall WA. 1992. Specific phosphorylation of a COOH-terminal site on the full-length form of the α1 subunit of the skeletal muscle calcium channel by cAMP-dependent protein kinase. *J Biol Chem* **267**: 16100–16105.

Rotman EI, Murphy BJ, Catterall WA. 1995. Sites of selective cAMP-dependent phosphorylation of the L-type calcium channel α1 subunit from intact rabbit skeletal muscle myotubes. *J Biol Chem* **270**: 16371–16377.

Ruth P, Röhrkasten A, Biel M, Bosse E, Regulla S, Meyer HE, Flockerzi V, Hofmann F. 1989. Primary structure of the β subunit of the DHP-sensitive calcium channel from skeletal muscle. *Science* **245**: 1115–1118.

Sather WA, McCleskey EW. 2003. Permeation and selectivity in calcium channels. *Annu Rev Physiol* **65**: 133–159.

Schneider T, Hofmann F. 1988. The bovine cardiac receptor for calcium channel blockers is a 195-kDa protein. *Eur J Biochem* **174**: 369–375.

Schroder E, Byse M, Satin J. 2009. L-type calcium channel C terminus autoregulates transcription. *Circ Res* **104**: 1373–1381.

Sculptoreanu A, Scheuer T, Catterall WA. 1993. Voltage-dependent potentiation of L-type Ca^{2+} channels due to phosphorylation by cAMP-dependent protein kinase. *Nature* **364**: 240–243.

Serysheva II, Ludtke SJ, Baker MR, Chiu W, Hamilton SL. 2002. Structure of the voltage-gated L-type Ca^{2+} channel by electron cryomicroscopy. *Proc Natl Acad Sci* **99**: 10370–10375.

Sheng Z-H, Rettig J, Takahashi M, Catterall WA. 1994. Identification of a syntaxin-binding site on N-type calcium channels. *Neuron* **13**: 1303–1313.

Sheng Z-H, Rettig J, Cook T, Catterall WA. 1996. Calcium-dependent interaction of N-type calcium channels with the synaptic core-complex. *Nature* **379**: 451–454.

Sheng Z-H, Yokoyama C, Catterall WA. 1997. Interaction of the synprint site of N-type Ca^{2+} channels with the C2B domain of synaptotagmin I. *Proc Natl Acad Sci* **94**: 5405–5410.

Singer D, Biel M, Lotan I, Flockerzi V, Hofmann F, Dascal N. 1991. The roles of the subunits in the function of the calcium channel. *Science* **253**: 1553–1557.

Singh A, Gebhart M, Fritsch R, Sinnegger-Brauns MJ, Poggiani C, Hoda JC, Engel J, Romanin C, Striessnig J, Koschak A. 2008. Modulation of voltage- and Ca^{2+}-dependent gating of $Ca_V1.3$ L-type calcium channels by alternative splicing of a C-terminal regulatory domain. *J Biol Chem* **283**: 20733–20744.

Snutch TP, Reiner PB. 1992. Calcium channels: Diversity of form and function. *Curr Opin Neurobiol* **2**: 247–253.

Song I, Kim D, Choi S, Sun M, Kim Y, Shin HS. 2004. Role of the α1G T-type calcium channel in spontaneous absence seizures in mutant mice. *J Neurosci* **24**: 5249–5257.

Soong TW, Stea A, Hodson CD, Dubel SJ, Vincent SR, Snutch TP. 1994. Structure and functional expression of a member of the low voltage-activated calcium channel family. *Science* **260**: 1133–1136.

Starr TVB, Prystay W, Snutch TP. 1991. Primary structure of a calcium channel that is highly expressed in the rat cerebellum. *Proc Natl Acad Sci* **88**: 5621–5625.

Striessnig J, Knaus HG, Grabner M, Moosburger K, Seitz W, Lietz H, Glossmann H. 1987. Photoaffinity labelling of the phenylalkylamine receptor of the skeletal muscle transverse-tubule calcium channel. *FEBS Lett* **212**: 247–253.

Striessnig J, Bolz HJ, Koschak A. 2010. Channelopathies in $Ca_V1.1$, $Ca_V1.3$, and $Ca_V1.4$ voltage-gated L-type Ca^{2+} channels. *Pflugers Arch* **460**: 361–374.

Sudhof TC. 1995. The synaptic vesicle cycle: A cascade of protein-protein interactions. *Nature* **375**: 645–653.

Sudhof TC. 2004. The synaptic vesicle cycle. *Annu Rev Neurosci* **27**: 509–547.

Takahashi M, Seagar MJ, Jones JF, Reber BF, Catterall WA. 1987. Subunit structure of dihydropyridine-sensitive calcium channels from skeletal muscle. *Proc Natl Acad Sci* **84**: 5478–5482.

Tanabe T, Takeshima H, Mikami A, Flockerzi V, Takahashi H, Kangawa K, Kojima M, Matsuo H, Hirose T, Numa S. 1987. Primary structure of the receptor for calcium channel blockers from skeletal muscle. *Nature* **328**: 313–318.

Tanabe T, Beam KG, Adams BA, Niidome T, Numa S. 1990. Regions of the skeletal muscle dihydropyridine receptor critical for excitation-contraction coupling. *Nature* **346**: 567–569.

Tanabe T, Mikami A, Niidome T, Numa S, Adams BA, Beam KG. 1993. Structure and function of voltage-dependent calcium channels from muscle. *Ann NY Acad Sci* **707**: 81–86.

Tobi D, Wiser O, Trus M, Atlas D. 1999. N-type voltage-sensitive calcium channel interacts with syntaxin, synaptotagmin and SNAP-25 in a multiprotein complex. *Receptor Channel* **6**: 89–98.

Tottene A, Moretti A, Pietrobon D. 1996. Functional diversity of P-type and R-type calcium channels in rat cerebellar neurons. *J Neurosci* **16**: 6353–6363.

Tsien RW. 1973. Adrenaline-like effects of intracellular iontophoresis of cyclic AMP in cardiac Purkinje fibres. *Nat New Biol* **245**: 120–122.

Tsien RW. 1983. Calcium channels in excitable cell membranes. *Annu Rev Physiol* **45**: 341–358.

Tsien RW, Bean BP, Hess P, Lansman JB, Nilius B, Nowycky MC. 1986. Mechanisms of calcium channel modulation by β-adrenergic agents and dihydropyridine calcium agonists. *J Mol Cell Cardiol* **18**: 691–710.

Tsien RW, Lipscombe D, Madison DV, Bley KR, Fox AP. 1988. Multiple types of neuronal calcium channels and their selective modulation. *Trends Neurosci* **11**: 431–438.

Van Petegem F, Clark KA, Chatelain FC, Minor DL Jr. 2004. Structure of a complex between a voltage-gated calcium channel β subunit and an α subunit domain. *Nature* **429**: 671–675.

Wang MC, Velarde G, Ford RC, Berrow NS, Dolphin AC, Kitmitto A. 2002. 3D structure of the skeletal muscle dihydropyridine receptor. *J Mol Biol* **323:** 85–98.

Wei XNA, Lacerda AE, Olcese R, Stefani E, Perez-Reyes E, Birnbaumer L. 1994. Modification of Ca^{2+} channel activity by deletions at the carboxyl terminus of the cardiac α1 subunit. *J Biol Chem* **269:** 1635–1640.

Welsby PJ, Wang H, Wolfe JT, Colbran RJ, Johnson ML, Barrett PQ. 2003. A mechanism for the direct regulation of T-type calcium channels by Ca^{2+}/calmodulin-dependent kinase II. *J Neurosci* **23:** 10116–10121.

Westenbroek RE, Ahlijanian MK, Catterall WA. 1990. Clustering of L-type Ca^{2+} channels at the base of major dendrites in hippocampal pyramidal neurons. *Nature* **347:** 281–284.

Wheeler DG, Barrett CF, Groth RD, Safa P, Tsien RW. 2008. CaMKII locally encodes L-type channel activity to signal to nuclear CREB in excitation-transcription coupling. *J Cell Biol* **183:** 849–863.

Williams ME, Brust PF, Feldman DH, Patthi S, Simerson S, Maroufi A, McCue AF, Velicelebi G, Ellis SB, Harpold MM. 1992. Structure and functional expression of an omega-conotoxin-sensitive human N-type calcium channel. *Science* **257:** 389–395.

Wiser O, Bennett MK, Atlas D. 1996. Functional interaction of syntaxin and SNAP-25 with voltage-sensitive L- and N-type Ca^{2+} channels. *EMBO J* **15:** 4100–4110.

Wiser O, Tobi D, Trus M, Atlas D. 1997. Synaptotagmin restores kinetic properties of a syntaxin-associated N-type voltage sensitive calcium channel. *FEBS Lett* **404:** 203–207.

Witcher DR, De Waard M, Kahl SD, Campbell KP. 1995a. Purification and reconstitution of N-type calcium channel complex from rabbit brain. *Method Enzymol* **238:** 335–348.

Witcher DR, De Waard M, Liu H, Pragnell M, Campbell KP. 1995b. Association of native calcium channel β subunits with the α1 subunit interaction domain. *J Biol Chem* **270:** 18088–18093.

Wolf M, Eberhart A, Glossmann H, Striessnig J, Grigorieff N. 2003. Visualization of the domain structure of an L-type Ca^{2+} channel using electron cryo-microscopy. *J Mol Biol* **332:** 171–182.

Wolfe JT, Wang H, Howard J, Garrison JC, Barrett PQ. 2003. T-type calcium channel regulation by specific G-protein βγ subunits. *Nature* **424:** 209–213.

Yang SN, Berggren PO. 2006. The role of voltage-gated calcium channels in pancreatic β-cell physiology and pathophysiology. *Endocr Rev* **27:** 621–676.

Yao J, Davies LA, Howard JD, Adney SK, Welsby PJ, Howell N, Carey RM, Colbran RJ, Barrett PQ. 2006. Molecular basis for the modulation of native T-type Ca^{2+} channels in vivo by Ca^{2+}/calmodulin-dependent protein kinase II. *J Clin Invest* **116:** 2403–2412.

Yi BA, Jan LY. 2000. Taking apart the gating of voltage-gated potassium channels. *Neuron* **27:** 423–425.

Yokoyama CT, Sheng Z-H, Catterall WA. 1997. Phosphorylation of the synaptic protein interaction site on N-type calcium channels inhibits interactions with SNARE proteins. *J Neurosci* **17:** 6929–6938.

Yokoyama CT, Myers SJ, Fu J, Mockus SM, Scheuer T, Catterall WA. 2005. Mechanism of SNARE protein binding and regulation of Ca_v2 channels by phosphorylation of the synaptic protein interaction site. *Mol Cell Neurosci* **28:** 1–17.

Yu FH, Yarov-Yarovoy V, Gutman GA, Catterall WA. 2005. Overview of molecular relationships in the voltage-gated ion channel superfamily. *Pharmacol Rev* **57:** 387–395.

Zamponi GW, Bourinet E, Nelson D, Nargeot J, Snutch TP. 1997. Crosstalk between G proteins and protein kinase C mediated by the calcium channel α1 subunit. *Nature* **385:** 442–446.

Zhang JF, Ellinor PT, Aldrich RW, Tsien RW. 1996. Multiple structural elements in voltage-dependent Ca^{2+} channels support their inhibition by G proteins. *Neuron* **17:** 991–1003.

Zhong H, Yokoyama C, Scheuer T, Catterall WA. 1999. Reciprocal regulation of P/Q-type calcium channels by SNAP-25, syntaxin and synaptotagmin. *Nat Neurosci* **2:** 939–941.

Zühlke RD, Pitt GS, Deisseroth K, Tsien RW, Reuter H. 1999. Calmodulin supports both inactivation and facilitation of L-type calcium channels. *Nature* **399:** 159–162.

The Role of Transient Receptor Potential Cation Channels in Ca^{2+} Signaling

Maarten Gees, Barbara Colsoul, and Bernd Nilius

KU Leuven, Department of Molecular Cell Biology, Laboratory Ion Channel Research, Campus Gasthuisberg, Leuven, Belgium

Correspondence: Bernd.Nilius@med.kuleuven.be

The 28 mammalian members of the super-family of transient receptor potential (TRP) channels are cation channels, mostly permeable to both monovalent and divalent cations, and can be subdivided into six main subfamilies: the TRPC (canonical), TRPV (vanilloid), TRPM (melastatin), TRPP (polycystin), TRPML (mucolipin), and the TRPA (ankyrin) groups. TRP channels are widely expressed in a large number of different tissues and cell types, and their biological roles appear to be equally diverse. In general, considered as polymodal cell sensors, they play a much more diverse role than anticipated. Functionally, TRP channels, when activated, cause cell depolarization, which may trigger a plethora of voltage-dependent ion channels. Upon stimulation, Ca^{2+} permeable TRP channels generate changes in the intracellular Ca^{2+} concentration, [Ca^{2+}]$_i$, by Ca^{2+} entry via the plasma membrane. However, more and more evidence is arising that TRP channels are also located in intracellular organelles and serve as intracellular Ca^{2+} release channels. This review focuses on three major tasks of TRP channels: (1) the function of TRP channels as Ca^{2+} entry channels; (2) the electrogenic actions of TRPs; and (3) TRPs as Ca^{2+} release channels in intracellular organelles.

Transient receptor potential (TRP) channels constitute a large and functionally versatile family of cation-conducting channel proteins, which have been mainly considered as polymodal unique cell sensors. The first TRP channel gene was discovered in *Drosophila melanogaster* (Montell and Rubin 1989) in the analysis of a mutant fly whose photoreceptors failed to retain a sustained response to maintained light stimuli. So far, more than 50 TRP channels have been identified with representative members in many species. The evolutionary first TRP channels in protists, chlorophyte algae, choanoflagellates, yeast, and fungi are primary chemo-, thermo-, or mechanosensors (Cai 2008; Wheeler and Brownlee 2008; Chang et al. 2010; Matsuura et al. 2009). Many of these functions are remarkably conserved from protists, worms, and flies to humans (Montell 2005; Pedersen et al. 2005; Nilius et al. 2007; Damann et al. 2008). More than 50 *trp* genes have been cloned so far that comprise approximately 20% of the known genes encoding ion channels. In mammals, 28 TRP channels were found and classified according to homology into 6 subfamilies: TRPC (canonical), TRPV (vanilloid),

TRPM (melastatin), TRPA (ankyrin), TRPML (mucolipin), and TRPP (polycystin) (Fig. 1). TRPs are expressed in numerous excitable and nonexcitable tissues, if not in all cell types. They are involved in manifold physiological functions, ranging from pure sensory functions, such as pheromone signaling, taste transduction, nociception, and temperature sensation, over homeostatic functions, such as Ca^{2+} and Mg^{2+} reabsorption and osmoregulation, to many other motile functions, such as muscle contraction and vaso-motor control. We are still at the very beginning of identifying all the diverse physiological functions of this intriguing ion channel family, and our knowledge about TRP channel expression and functioning in various tissues of mammals is limited. Accumulating evidence, however, suggests that TRP channels play prominent roles in the regulation of the intracellular calcium level in both excitable and nonexcitable cells.

The molecular architecture of TRP channels is reminiscent of voltage-gated channels and comprises six putative transmembrane segments (S1–S6), intracellular N- and C-termini, and a pore-forming reentrant loop between S5 and S6 (Gaudet 2008b). The length of the cytosolic tails varies greatly between TRP channel subfamilies, as do their structural and functional domains (for detailed reviews see Owsianik et al. 2006a). The TRPC and TRPM family members all contain a 25-amino-acid motif (the TRP domain) containing a TRP box C-terminal to S6, but this domain is not present in the other families. Although TRPC and TRPV family members contain 3-4 ankyrin repeats in their N-terminal cytoplasmic tail, TRPA1 contains 14 ankyrin repeats, and they are not present in the other families. Lastly, TRPC and TRPM family members contain protein-rich sequences in the region C-terminal of the TRP domain (known as the TRP box 2).

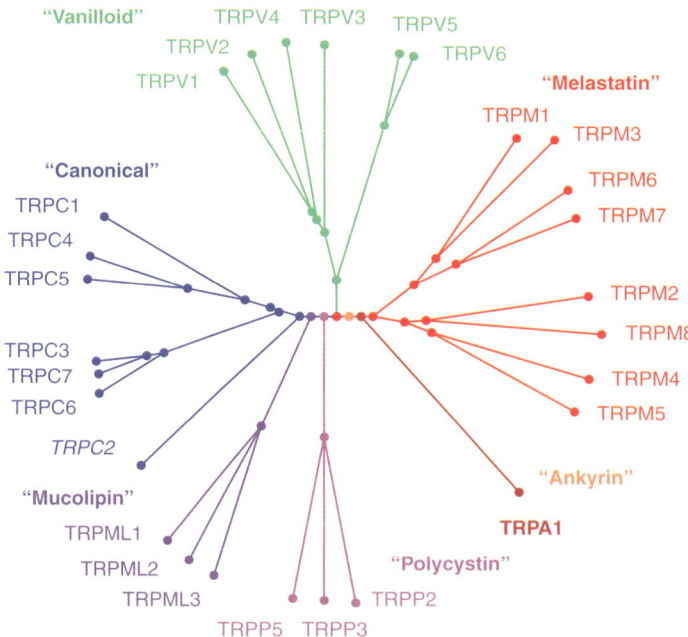

Figure 1. Phylogenetic tree of the mammalian TRP-channel superfamily. TRPC (canonical), TRPM (melastatin), TRPV (vanilloid), TRPA (ankyrin), TRPP (polycystin), and TRPML (mucolipin) are the only identified subfamilies in mammals.

The TRP box is most likely important for binding of phosphatidylinositol phosphates, such as PI(4,5)P$_2$ (Rohacs 2007). So far, our knowledge of the three-dimensional structure is limited, as only parts of TRP proteins have been crystallized. Most of the TRP channels probably form tetramers, in which the capacity to function as homo- or heteromers is still a matter of debate. However, increasing evidence suggests heteromultimeric channel assembly within one subfamily, creating a variety of different channels with unique properties, as compared to homomers (Strubing et al. 2001; Smith et al. 2002). Topics to be explored further include the association with accessory proteins (e.g., beta subunits) and the forming of signalplexes (Montell 2003; Peng et al. 2007; Redondo et al. 2008), the various mechanisms of insertion and retrieval in and from the plasma membrane, and the general processes for the regulation of mainly intracellular location or their trafficking to the plasma membrane.

Importantly, most, if not all, TRP channels are modulated by Ca^{2+} itself, which generates positive or negative feedback loops. Thus, regarding the modulation of Ca^{2+} signaling, TRP channels provide a huge plasticity to the overall control of the intracellular Ca^{2+} concentration [Ca^{2+}]$_i$.

This review focuses on the functional role of TRP channels as modulators of intracellular Ca^{2+} signaling. Changes in the concentration of free cytosolic Ca^{2+} ([Ca^{2+}]$_i$) are of fundamental importance in different stages of the cell cycle, starting from the fertilization and embryonic pattern formation, to cell differentiation and proliferation, and cell death. Furthermore, [Ca^{2+}]$_i$ plays a role in different cellular processes including transmitter release, muscle contraction, and gene transcription (Berridge et al. 2000). TRP channels can contribute to changes in [Ca^{2+}]$_i$, either by acting as Ca^{2+}-entry pathways in the plasma membrane or by changing the membrane polarization; in this way modulating the driving force for Ca^{2+} entry mediated by alternative pathways. Alternatively, [Ca^{2+}]$_i$ can be elevated by the release from intracellular stores (Bootman et al. 2001). In addition, TRP channels are functionally linked with voltage-dependent Ca^{2+}-entry channels that are activated by depolarization, for example, due to TRP gating. By changing the membrane potential and local Ca^{2+} gradients, TRP channels contribute to modulating the driving force for Ca^{2+} entry and provide intracellular pathways for Ca^{2+} release from cellular organelles.

For more detailed information on TRP channels regarding structure, gating, and special functional aspects, we refer a wealth of excellent reviews (Desai and Clapham 2005; Montell 2005; Ramsey et al. 2006; Nilius et al. 2007; Vennekens et al. 2008; Latorre et al. 2009; Vriens et al. 2009). For more detailed information, we direct the interested reader to databases such as http://www.ensembl.org/index.html and http://www.iuphar-db.org/DATABASE/FamilyMenuForward?familyId=78 (see also Clapham et al. 2009).

TRPs AS Ca^{2+} ENTRY CHANNELS

Ca^{2+} Permeable TRP Pores

Although most TRPs are Ca^{2+} permeable, the selectivity varies greatly between the different members with P$_{Ca}$/P$_{Na}$ ratios ranging from <1 for TRPM1 to >100 for TRPV5 and TRPV6 (Fig. 2). This variance reflects different pore structures and obviously also differences in the dynamic pore behavior; for example, pore dilation by activations with various agonists (Chung et al. 2008; Karashima et al. 2010). In general, there is no high homology in the primary structure of the putative selectivity filter regions throughout all TRP subfamilies (Owsianik et al. 2006b). For TRPV5 and TRPV6, it is shown that the Ca^{2+}-permeability depends on D^{542} in TRPV5 and the corresponding D^{541} in TRPV6 (Nilius et al. 2001). As TRPV5 and TRPV6 form homo- and heteromultimers, it appears that the Ca^{2+} selectivity in these channels depends on a ring of four aspartate residues in the pore of the channel, corresponding to the ring of four negatively-charged residues (aspartates and/or glutamates) in the pore of voltage-gated Ca^{2+} channels (Ellinor et al. 1995; Hoenderop et al. 2003). It is shown that neutralization of the D^{546} in

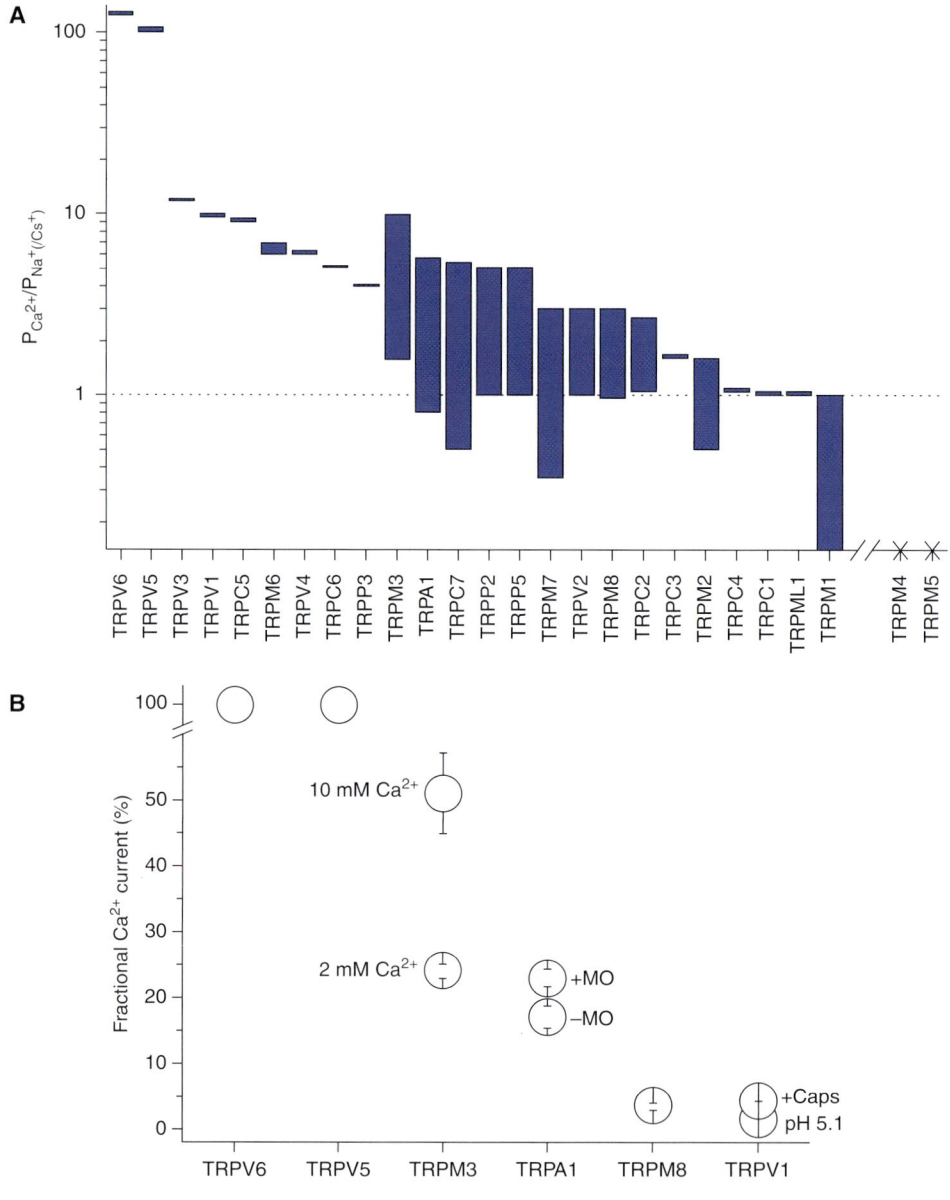

Figure 2. Ca^{2+} selectivity of TRP channels. (*A*) Ca^{2+} selectivity of TRP channels expressed as $P_{Ca^{2+}}/P_{Na^+(/Cs^+)}$ (depending on availability). Bars indicate maximal and minimal values found in literature; TRPM4 and TRPM5 are impermeable to Ca^{2+} (for TRPM1, values are deduced from Oancea et al. 2009). (*B*) Fractional Ca^{2+} current of TRP channels as found in literature (for TRPM3 Oberwinkler personal communication, see Drews et al. 2010).

TRPV1 and the corresponding D^{682} in TRPV4 also reduce the Ca^{2+} permeability. For TRPV4, additional neutralization of D^{672} reduces the divalent selectivity even further, while introducing a negative charge, instead of M^{680} abolishing the Ca^{2+} permeability completely (Garcia-Martinez et al. 2000; Voets et al. 2002). More recently, it is shown that also in the *Drosophila* TRP an aspartate (D^{621}) plays an important role in determining the Ca^{2+} permeability (Liu et al. 2007a), whereas neutralization of the negative charges in the loops between TM5 and

TM6 of TRPC5 (E^{543}, E^{595}, E^{598}) and TRPC1 (all seven, D to N and E to Q) resulted in a decreased La^{2+} and Ca^{2+} permeability (Jung et al. 2003; Liu et al. 2003). This suggests that the negatively-charged residues determining the pore properties of TRPC1 and TRPC5 are located in the distal parts of the putative pore entrance. A pore study of TRPM4, which is impermeable to Ca^{2+}, showed that substitution of residues E^{981} to A^{986} with the selectivity filter of TRPV6 yielded a functional channel with the gating hallmarks of TRPM4 (activation by Ca^{2+}, voltage dependence) and with a TRPV6-like sensitivity to block by extracellular Ca^{2+}, Mg^{2+}, and Ca^{2+} permeation (Nilius et al. 2005a). Two recent studies showed that E^{1024} and D^{1031} play an important role in the Ca^{2+} permeation through TRPM6, and that restoration of only two residues in the human TRPM2 (Q981E/P983Y) to the evolutionary, more ancient Nudix-linked channel residues, significantly increased the Ca^{2+} permeability, whereas introducing the newer sequence in TRPM7 (E1047Q/Y1049P) resulted in the loss of Ca^{2+} permeability (Topala et al. 2007; Mederos y Schnitzler et al. 2008). Another study showed that neutralizing the charges of residues E^{1052} and E^{1047} of TRPM7 and the corresponding residues in TRPM6 (E^{1024} and E^{1029}) reduced the Ca^{2+} permeation (Li et al. 2007).

Although pore structures have been considered as relatively stable, evidence is accumulating that for some TRP channels the pore diameter and also Ca^{2+} permeation depend on the mode of activation; for example, the activating agonist. Such dynamic pore behavior was first observed for P2X channels (Egan and Khakh 2004). Activation of TRPV1 leads to time-dependent and agonist-concentration-dependent increases in relative permeability to large cations and changes in Ca^{2+} permeability that parallels a pore dilation. TRPV1 agonists showed different capabilities for evoking ionic selectivity changes. Very likely, for TRPV1, protein kinase C-dependent phosphorylation of S^{800} in the TRPV1 C-terminus increases agonist-evoked ionic selectivity changes. Thus, the qualitative signaling properties of TRPV1 are dynamically modulated during channel activation, a process that probably shapes TRPV1 (Chung et al. 2008). A similar situation holds for TRPA1. From the relative permeability of the nonstimulated TRPA1 channel to cations of different sizes, a diameter of ~ 11 Å is calculated for the channel pore, which fits with the pore size of TRPM6, TRPV1, and TRPP2, but is significantly larger for the highly Ca^{2+} permeable channels TRPV5 and TRPV6 (~ 7.5 Å and 5.4 Å). Under conditions of activation by electrophilic compounds such as AITC (allyl isothiocyanate, or mustard oil, MO), the TRPA1 channel undergoes a pore dilation by ~ 3 Å. This dynamic pore behavior was coupled with an increased Ca^{2+} permeation and an increased fraction of Ca^{2+} contributing to the total current. Upon MO stimulation, P_{Ca}/P_{Na} changed from ~ 5.7 to 7.9, and the fractional Ca^{2+} current from 17.0% to 23.3%. This pore dilation is probably not present for nonelectrophilic agonists and also disappeared when a negatively-charged residue in the pore, D^{918}, was changed to noncharged residues. Again, this negative charge in the TRPA1 pore determines the Ca^{2+} entry (Chen et al. 2009; Karashima et al. 2010). Also, for TRPV5, a pH-dependent pore dilation has been described (Yeh et al. 2005). This dynamic pore behavior adds a new regulation mechanism to agonist-induced Ca^{2+} entry. Pore dilation, modulation of the fractional Ca^{2+} current, and changes of the Ca^{2+} permeation are modulator properties for Ca^{2+} signaling in likely many TRP channels.

Thus, depending on the pore structure, TRP channels vary in their pore size and in the permeation for Ca^{2+}, which is indicated by the P_{Ca}/P_{Na} (or P_{Cs}) ratios. Obviously, the Ca^{2+} permeation also varies under different experimental conditions, might be influenced by associated proteins, and even shows a dynamic behavior for some TRP channels. Figure 2 gives an overview of Ca^{2+} permeation properties and their variability.

TRPCs

The mammalian members of the TRPC family can by divided into 4 subfamilies on the basis of functional similarities and sequence alignment: TRPC1, TRPC2, TRPC3/6/7, and TRPC4/5. TRPC channels in general are nonselective Ca^{2+} permeable cation channels, but the selectivity ratio P_{Ca}/P_{Na} varies significantly between the different family members (see Fig. 2). With a few exceptions, expression of TRPC family members is broad; thus, generally, most cell types contain multiple TRPCs (for examples, see Montell et al. 2002; Montell 2005). The characterization is further complicated by different heterotetramers. It is shown that TRPC1 can form heteromers with TRPC4 and 5, and the TRPC subfamilies TRPC4/5, and TRPC3/6/7 can form heteromers among themselves, with properties that can differ significantly from those of the homotetramers (Strubing et al. 2001; Goel et al. 2002; Hofmann et al. 2002; Strubing et al. 2003; Schilling and Goel 2004). Activation of TRPC channels occurs mainly via different isoforms of phospholipase C (PLC) (Venkatachalam et al. 2002).

The relation between the constituents, STIM1 and ORAI1, of the best characterized store-operated Ca^{2+} channel (SOC) carrying I_{CRAC} and the contribution of TRPC channels is still a matter of dispute. It was first shown that TRPC1/STIM1 and ORAI1 very likely form ternary complexes to contribute to a SOC channel (Ambudkar et al. 2007). ORAI1 proteins may interact with TRPCs and act as regulatory subunits that confer STIM1-mediated store depletion sensitivity to these channels (Lu et al. 2010; Liao et al. 2007). STIM1 has been shown to bind to TRPC1, TRPC4, and TRPC5 and is therefore involved in store-operated Ca^{2+} entry (SOCE) (Yuan et al. 2007; Sours-Brothers et al. 2009). The involvement of TRPC channels in SOCE may depend on special membrane structures, such as lipid rafts (Pani et al. 2008). In general, a majority of reports described TRPC1 as a store-operated channel whose gating mechanism is still to be elucidated (Worley et al. 2007; Kim et al. 2009c; Ng et al. 2009). So far, as concluded from a plethora of experimental evidence, TRPC channels might, under certain circumstances, act as SOCs, but are to be clearly distinguished from the calcium-release-activated calcium channels, as they show distinct properties: high Ca^{2+} selectivity; very small single channel conductance; distinct Ca^{2+} dependent modulation, e.g., fast and slow inactivation; and slow decay in divalent free solutions (for reviews and more detailed descriptions, see Vaca 2010; Bolotina 2008; Birnbaumer 2009; Kiselyov and Patterson 2009; Yuan et al. 2009).

TRPC1 (ENSG00000144935; TRPC1) is activated by the neuronal metabotropic glutamate receptor mGluR1 and thus contributes to the slow excitatory postsynaptic potential (EPSP) (Kim et al. 2003). Furthermore, TRPC1 provides an important route for Ca^{2+} entry after agonist, growth factor, and PKC induction in different cell types such as endothelial cells (Kamouchi et al. 1999; Nilius and Droogmans 2001; Tiruppathi et al. 2006), platelets (Authi 2007), smooth muscle cells (Dietrich et al. 2006), and B-lymphocytes (Mori et al. 2002). One report identifies TRPC1 as the mechanosensitive cation channel, responsible for transducing membrane stretch in cationic currents (Maroto et al. 2005), but this is disputed (Gottlieb et al. 2008). In *Trpc1* KO mice, it is shown that the salivary gland fluid secretion regulated by neurotransmitters is severely reduced (Liu et al. 2007b). TRPC1 is also activated by orexin A, a peptide hormone associated to the regulation of sleep/wakefulness states, alertness, and appetite (Larsson et al. 2005). More recently, it is shown that knockdown of *Trpc1* in zebrafish impaired angiogenesis, an effect that could be rescued by reintroducing TRPC1, which is reminiscent of the role of TRPCs in axon guidance (Yu et al. 2010).

Trpc2 (ENSMUSG00000058020; TRPC2) in humans is a pseudogene (Yildirim and Birnbaumer 2007), but in rodents it plays an important role in pheromone detection via the vomeronasal sensory neurons (VSN) (Yildirim and Birnbaumer 2007). TRPC2 is also shown to be important for the Ca^{2+} signaling in spermatozoa after egg ZP3 stimulation (Yildirim and Birnbaumer 2007). Lastly, TRPC2 is shown to be involved in Ca^{2+} release from the intracellular stores (Gailly and Colson-Van Schoor 2001; Tong et al. 2004; Yildirim and Birnbaumer 2007).

Expression of TRPC3 (ENSG00000138741; TRPC3) is highest in brain, smooth, and cardiac muscle cells (Riccio et al. 2002b; Clapham 2003). TRPC3 is a constitutively active receptor-operated channel that can be further stimulated by DAG (Lemonnier et al. 2008). As all TRPCs, TRPC3 can interact directly via a CIRB region with both IP_3R and Calmodulin (CaM). TRPC3 channel activation by IP_3 can lead to the constriction of cerebral arteries (Wedel et al. 2003; Xi et al. 2008) and is involved in synaptogenesis and growth-cone guidance (Amaral and Pozzo-Miller 2007b). The *Trpc3* gene was found to be damaged in human T-cell mutants defective in Ca^{2+} influx; introduction of the complete human *TRPC3* cDNA into those mutants rescued the Ca^{2+} currents, as well as TCR-dependent Ca^{2+} signals (Philipp et al. 2003). TRPC3 activation by purinergic receptors results in both Ca^{2+} influx and depolarization of endothelial cells and vasoconstriction in smooth muscle cells (Ahmmed and Malik 2005; Kwan et al. 2007).

Expression of TRPC4 (ENSG00000100991; TRPC4) is found in endothelium and smooth muscle cells (Beech 2005; Tiruppathi et al. 2006), intestinal pacemaker cells (ICC) (Kim et al. 2006), in many brain regions (Zechel et al. 2007), adrenal glands (Philipp et al. 2000), and in kidneys (Freichel et al. 2005). It is suggested that TRPC4 is an essential component of the nonselective cation channel involved in neuromodulation of stomach smooth muscle after muscarinic stimulation (Lee et al. 2005). Furthermore, a reduced agonist-induced Ca^{2+} entry and vasorelaxation is shown in the vascular endothelium of TRPC4-deficient mice (Freichel et al. 2001).

Similar to TRPC4, TRPC5 (ENSG00000072315; TRPC5) is expressed in multiple tissues, including brain tissue (Hofmann et al. 2000). In the CNS, it is shown that TRPC5 can form heteromeric cation channels with TRPC1, and these heteromultimers may play an important role during brain development (Strubing et al. 2001; Strubing et al. 2003). In neurons, TRPC5 is loaded to vesicular packages for neuronal transport via association with synaptotagmin and stathmin-2 (Greka et al. 2003). TRPC5 shows a striking voltage dependence, shifting between outwardly rectifying and doubly rectifying shapes (called phases) depending on the time in the activation-deactivation cycle. These phase transitions can be modulated by external factors such as La^{3+} and the scaffolding protein EBP50 (Obukhov and Nowycky 2004; Obukhov and Nowycky 2008). Vesicular insertion of TRPC5 from a subplasmalemmal reserve pool is shown to be regulated by EGF-RTK, in a manner depending on PI3K, Rac, and phosphatidylinositol 4-phosphate 5-kinase (PIP(5)K) (Bezzerides et al. 2004). This process is shown to be important for the regulation of hippocampal neurite length and growth-cone morphology (Bezzerides et al. 2004). Noticeably, TRPC5 is required for muscarinic persistent responses involved in establishing a transient working memory in the entorhinal cortex (Zhang et al. 2010). More recently, it was shown that TRPC5 is important for amygdala function and fear-related behavior (Riccio et al. 2009). TRPC5 is also activated by nitric oxide (NO), which was shown to be achieved by nitrosylation of residues C^{553} and C^{558} (Yoshida et al. 2006).

TRPC6 (ENSG00000137672; TRPC6) and TRPC7 (ENSG00000069018; TRPC7) are closely related, but whereas expression of TRPC6 is highest in the lung and brain, TRPC7 is mainly expressed in the kidney and pituitary gland (Hofmann et al. 2000; Riccio et al. 2002a; Montell 2005). It is shown that thrombin activation of TRPC6 can induce Ca^{2+} entry in platelets (Hassock et al. 2002). The channel is also shown to be an important part of the vascular α_1-activated Ca^{2+}-permeable cation channel in smooth muscle (Inoue et al. 2001; Jung et al. 2002). Furthermore, *Trpc6* KO mice showed an elevated blood pressure and increased vascular smooth muscle contractility that was only partly recovered by the constitutively-active TRPC3-type channels, which are up-regulated in the smooth muscle cells of *Trpc6* KO mice (Dietrich et al. 2005). Lastly, TRPC6 channel activity at the slit diaphragm is shown to be essential for proper regulation of podocyte structure and function (Reiser et al. 2005; Graham et al. 2007). The functional role of TRPC7 is still unclear, but it is suggested that TRPC7 conducts Ca^{2+} in AT1-induced myocardial apoptosis via a calcineurin-dependent pathway and can thereby contribute to the process of heart failure (Satoh et al. 2007).

TRPVs

Similar to the TRPC family, the TRPV (vanilloid) family can be divided into four subfamilies on the basis of structure and function, namely TRPV1/TRPV2, TRPV3, TRPV4, and TRPV5/6 (Vennekens et al. 2008). As mentioned above, TRPV5 and 6 are the only highly Ca^{2+}-selective channels in the TRP channel family (Nilius et al. 2000; Vennekens et al. 2000; Nilius et al. 2001), whereas TRPV1–4 are nonselective cation channels (permeability ratio P_{Ca}/P_{Na} between $\sim$1 and $\sim$15; see Fig. 2) that are activated by temperature and by numerous other stimuli (Nilius et al. 2003; Nilius et al. 2004; Vennekens et al. 2008; Vriens et al. 2009). All channels of the TRPV family contain 3–6 NH_2-terminal ankyrin repeats (for details, see Gaudet 2008a; Gaudet 2008b; Gaudet 2009).

TRPV1 (ENSG00000196689; TRPV1) was the first mammalian TRPV family member to be discovered and has been studied most extensively. Expression of TRPV1 was first identified in the pain-sensitive neurons of the dorsal root ganglion (DRG) and trigeminal ganglion (TG) neurons, but is also present in the terminals of spinal and peripheral nerves. TRPV1 expression is also shown in multiple non-neuronal cell types (Hayes et al. 2000). Activation of TRPV1 is voltage-dependent and can be induced by capsaicin and temperature ($>42°C$) (Voets et al. 2004a). *Trpv1* KO mice showed the importance of the channel in the detection and integration of different painful chemical and thermal stimuli (Caterina et al. 1997). In the pancreas, it is shown that TRPV1 is involved in the release of substance P (Nathan et al. 2001). In the bladder, TRPV1 is an important target for the treatment of cystitis-induced bladder overactivity, but the exact localization is still under debate (De Ridder and Baert 2000; Charrua et al. 2007; Wang et al. 2008; Everaerts et al. 2009).

TRPV2 (ENSG00000187688; TRPV2) has 50% sequence identity to TRPV1 and is also expressed in DRG neurons, different brain regions, and non-neuronal tissues, including GI tract and smooth muscle cells (Vennekens et al. 2008). Similar to TRPV1, TRPV2 is also activated by heat but only at higher, noxious temperatures ($>52°C$) compared to TRPV1 (Caterina et al. 1999). Growth factors, such as insulin-like growth factor (IGF-1), can activate TRPV2 by vesicular insertion in the membrane, a process that can be associated to myocyte degeneration caused by the disruption of dystrophin-glycoprotein complexes (Kanzaki et al. 1999; Iwata et al. 2003). Furthermore, in a dystrophin-deficient (*mdx*) mouse, a model for muscular dystrophy, it was shown that expression of a dominant-negative TRPV2 reduced the muscle damage (Iwata et al. 2009). TRPV2 has also been described as a mechano-sensor in vascular smooth muscle cells, as it can function as a stretch-activated channel (Muraki et al. 2003; Beech et al. 2004). More recently, it is shown that TRPV2 is of fundamental importance in innate immunity, as early phagocytosis was impaired in macrophages lacking the cation channel (Link et al. 2010).

TRPV3 (ENSG00000167723; TRPV3) is expressed in DRG and TG neurons, the brain, the tongue, and the testis (Smith et al. 2002; Xu et al. 2002; Chung et al. 2003; Chung et al. 2005). Expression is also high in the skin, keratinocytes, and in the cells surrounding hair follicles (Peier et al. 2002b; Gopinath et al. 2005; Moqrich et al. 2005; Asakawa et al. 2006; Xu et al. 2006; Mandadi et al. 2009). TRPV3 is activated by innocuous warm temperatures ($>30–33°C$), and the natural compounds camphor, thymol carvacrol, and eugenol, which are also potent sensitizers for temperature activation of the channel (Peier et al. 2002b; Xu et al. 2002; Moqrich et al. 2005; Xu et al. 2006). The importance of TRPV3 as a temperature sensor is shown in *Trpv3* KO mice, in which the responses to innocuous and noxious heat are dramatically diminished, whereas responses to other sensory modalities remained unaltered (Moqrich et al. 2005). More recently, it is shown that heating of keratinocytes causes release of ATP that can consequently activate termini of neighboring DRG neurons, a process that is compromised in keratinocytes from TRPV3-deficient mice (Mandadi et al. 2009). TRPV3 is required for forming the skin-barrier function and keratinocytes cornification and forms a signalplex with TGF-α/EGFR (Cheng et al. 2010a).

TRPV4 (ENSG00000111199; TRPV4) is a channel that is widely expressed in the brain, DRG neurons, and multiple non-neuronal tissues including bone, chondrocytes, insulin-secreting β-cells, keratinocytes, smooth muscle cells, hair cells of the inner ear, and different epithelial cell types (Vennekens et al. 2008; Everaerts et al. 2010). TRPV4 can be activated by moderate temperatures ($>24°C$) and is, as such, constitutively active at normal body temperatures. Other activating stimuli include shear stress, cell swelling, anandamide, arachidonic acid, and 4α-phorbol 12,13-didecanoate (4α-PDD) (Watanabe et al. 2002a; Watanabe et al. 2002b; Nilius et al. 2003; Watanabe et al. 2003; Nilius et al. 2004). It is reported that TRPV4 plays a role in thermoregulation via epidermal keratinocytes (Chung et al. 2003; Chung et al. 2004). As a mechanical and osmotic stimulus-induced nociceptor, TRPV4 seems important in DRGs and TGs (Alessandri-Haber et al. 2003; Liedtke and Friedman 2003; Suzuki et al. 2003). Lastly, it is shown that TRPV4 expressed in osteoblasts and osteoclasts may play a role in bone formation and remodelling (Masuyama et al. 2008; Mizoguchi et al. 2008) whereas in chondrocytes, it is shown to regulate the Sox9 pathway involved in the regulation of chondrocyte polarity and differentiation and in endochondral ossification (Muramatsu et al. 2007).

TRPV5 (ENSG00000127412; TRPV5) and TRPV6 (ENSG00000165125; TRPV6) are close homologs and the only members of the TRP family that are highly Ca^{2+}-selective. TRPV5 and TRPV6 can function as homomers, but TRPV5/6 heterotetramers are also formed (Hoenderop et al. 2003). Both channels are constitutively active and are essential for Ca^{2+} reabsorption in the kidney (TRPV5) and in the intestine (TRPV6) (den Dekker et al. 2003; Nijenhuis et al. 2003a; Nijenhuis et al. 2003b). Both channels are tightly regulated by extracellular and intracellular Ca^{2+} concentrations, although their kinetics differ (Voets et al. 2001; Nilius et al. 2002; Voets et al. 2003; Hoenderop et al. 2005). The striking Ca^{2+}-dependent inactivation probably reflects a Ca^{2+}-induced PI(4,5)P$_2$ depletion (Thyagarajan et al. 2008; Thyagarajan et al. 2009).

TRPMs

The members of the TRPM (melastatin) family are divided into 4 groups on the basis of sequence homology: TRPM1/3, TRPM2/8, TRPM4/5, and TRPM 6/7. Ca^{2+} permeability in the TRPM family ranges from impermeable to Ca^{2+} (TRPM4 and 5) to highly Ca^{2+} permeable (TRPM3, 6 and 7; see Fig. 2). Unlike the previously-discussed TRP channels, TRPM channels lack the N-terminal ankyrin repeats.

Lower expression levels of TRPM1 (ENSG00000134160; TRPM1) in malignant melanoma cell lines suggested that TRPM1 had a tumor suppressor function, but this was debated in further research (Duncan et al. 1998; Duncan et al. 2001; Miller et al. 2004). It is thought that TRPM1 is a constitutively open, nonselective cation channel, but little is known about the functional properties and cellular functions of this channel, partly because of the huge number of different splice variants. More recently, it was shown that a TRPM1 long-form (TRPM1-L) plays an important role in the ON pathway of retinal bipolar cells, and this might explain the complete congenital stationary night blindness seen in patients with mutations in *Trpm1* (Koike et al. 2010; Audo et al. 2009; Li et al. 2009; Shen et al. 2009; van Genderen et al. 2009).

TRPM2 (ENSG00000142185; TRPM2) is a chanzyme, forming a nonselective cation channel fused C-terminally to an enzymatic ADP-ribose pyrophosphatase domain (Perraud et al. 2001; Perraud et al. 2003). Expression of TRPM2 is found highest in the brain, but is also found in different peripheral cell types (Kraft and Harteneck 2005).

It is shown that activation of TRPM2 causes predisposition to apoptosis and cell death and that inhibition of TRPM2 is neuroprotective, probably because TRPM2 is activated by H_2O_2 and functions as a sensor for the cellular redox status (Kuhn et al. 2005; McNulty and Fonfria 2005; Zhang et al. 2006). This activation is prevented by a truncated TRPM2 isoform (TRPM2-S), generated by alternative splicing of the full-length protein (TRPM2-L) (Zhang et al. 2003a).

Transcription of TRPM3 (ENSG000000-83067; *TRPM3*), similar to TRPM1, results in a number of different mRNA species, and this variability is further enhanced by the presence of different starting positions and C-terminal ends (Grimm et al. 2003; Lee et al. 2003). Expression of TRPM3 has been shown in the human brain and kidney, although it was undetectable in a mouse kidney (Grimm et al. 2003; Lee et al. 2003). TRPM3 forms a channel permeable to divalent cations and is activated by D-erythro-sphingosine, pregnenolone sulfate, activation of an endogenous muscarinic receptor, and by a decreased extracellular osmolarity. This activation by hypotonicity argues for a role for TRPM3 in the renal osmo-homeostasis (Grimm et al. 2003; Lee et al. 2003; Grimm et al. 2005).

TRPM4 (ENSG00000130529; TRPM4) and TRPM5 (ENSG00000070985; TRPM5) are two closely-related cation channels that are ubiquitously expressed; whereas TRPM4 expression is highest in the heart, pancreas, and placenta, TRPM5 expression is found mainly in the intestine, taste buds, and pancreas but also in the stomach, lung, testis, and brain (Ullrich et al. 2005; Fonfria et al. 2006; Zhang et al. 2003b; Kokrashvili et al. 2009; Colsoul et al. 2010). As these channels are impermeable to Ca^{2+}, they do not function as Ca^{2+}-entry channels. They do, however, play a role in $[Ca^{2+}]_i$ modulation, which is discussed later.

TRPM6 (ENSG00000119121; TRPM6) and TRPM7 (ENSG00000092439; TRPM7) are highly homologous channel kinases, with expression in the kidney and intestine for TRPM6 and ubiquitously for TRPM7 (Runnels et al. 2001; Monteilh-Zoller et al. 2003; Voets et al. 2004b). TRPM6 is highly permeable to Mg^{2+}, activated by low $[Mg^{2+}]_I$, and is shown to be important in the Mg^{2+} homeostasis and reabsorption in the kidney and intestine (Schlingmann et al. 2002; Voets et al. 2004b; Schlingmann et al. 2005). Apart from Mg^{2+} and Ca^{2+}, TRPM7 is also permeable to divalent cations and is responsible for the uptake of these trace metal ions (Monteilh-Zoller et al. 2003). Furthermore, TRPM7 is shown to be involved in the regulation of the cell cycle and in neurotoxic death (Wolf and Cittadini 1999; Aarts et al. 2003; Aarts and Tymianski 2005). Zebrafish with a TRPM7 mutation have a defective skeletogenesis with kidney-stone formation and have an increased cell death of the melanophores (Elizondo et al. 2005; McNeill et al. 2007).

TRPM8 (ENSG000000144481; TRPM8) cDNA was isolated from prostate cancer cells, but was later shown to be widely expressed, with high expression in a subset of pain- and temperature-sensitive neurons (Tsavaler et al. 2001; McKemy et al. 2002; Peier et al. 2002a). TRPM8 is activated by cold temperatures (8–28°C) and by chemicals such as icilin and menthol, known to produce a cooling sensation (Voets et al. 2004a; Dhaka et al. 2006). Although the role for TRPM8 in the progression of cancer cells is highly debated, it is clear that it acts as a cold thermosensor in sensory neurons.

TRPA

TRPA1 (ENSG00000104321; TRPA1) is the only member of the TRPA (ankyrin) family characterized by the 14 NH_2 terminal ankyrin repeats (Story et al. 2003). It is expressed in hair cells and in the sensory DRG and TG neurons (Story et al. 2003; Corey et al. 2004). TRPA1 is activated by noxious cold and different chemicals including allyl isothiocyanate (the pungent compound in mustard oil), allicin (from garlic), cinnamaldehyde (from cinnamon), mentol (from mint), tetra-hydrocannabinoid (from marijuana), nicotine (from tobacco) and bradykinin (Patapoutian et al. 2003; Story et al. 2003; Macpherson et al. 2005; Karashima et al. 2007; Macpherson et al. 2007; Karashima et al. 2009; Talavera et al. 2009). TRPA1 plays an important role in cold temperature and chemical-induced nociception and in the transduction mechanism through which these irritants and other endogenous proalgesics elicit inflammatory pain (Story et al. 2003; Bautista et al. 2006; Kwan et al. 2006; Karashima et al. 2009). TRPA1 is also suggested as an interesting target for the treatment of cough in humans, as TRPA1 agonists can evoke coughing, an effect that is reduced in the presence of TRPA1 antagonists (Andre et al. 2009; Birrell et al. 2009).

TRPPs

The TRPP (polycystin) family comprises eight members, from which only the polycystic kidney disease 2 (PKD2 or TRPP2) and the PKD2-like (TRPP3 and TRPP5, or, according to Clapham et al. 2009, named TRPP1, TRPP2, and TRPP3) are shown to be channels. The cation-permeable TRPP channels have a P_{Ca}/P_{Na} between 1 and 5 and do not contain a TRP domain nor ankyrin repeats (see Fig. 2) (Delmas 2004; Delmas et al. 2004).

TRPP2 (ENSG00000118762; TRPP2) is widely expressed but most present in the kidney. It is localized to both motile and nonmotile cilia, in which it seems to be a mechanosensor involved in the nodal ciliary movement (Delmas et al. 2004). TRPP2's role in this crucial process for correct organ localization during development is consistent with the left to right asymmetry defects found in animal models lacking TRPP2 (Pennekamp et al. 2002; Bisgrove et al. 2005). Via a coiled-coil domain, TRPP2 and TRPP1 can form a functional polycystin complex, which appears to be essential for pressure-sensing in the kidney, a process altered in autosomal-dominant polycystic kidney disease (Sharif-Naeini et al. 2009).

TRPP3 (ENSG00000107593; TRPP3) and TRPP5 (ENSG00000078795; TRPP5) are less studied, and although TRPP3 is widely expressed, expression of TRPP5 is mainly shown in the testes (Keller et al. 1994; Nomura et al. 1998; Guo et al. 2000). The Ca^{2+}-permeable, nonselective TRPP3 channel is shown to be activated by alkalization and, as such, is thought to be involved in acid-sensing of sour tastes and in the cerebrospinal fluid (Huang et al. 2006; Ishimaru et al. 2006; Shimizu et al. 2009). TRPP5 might play a role in calcium homeostasis and, as such, may contribute to cell proliferation, apoptosis, (Xiao et al. 2009) and to spermatogenesis (Guo et al. 2000; Chen et al. 2008; Xiao et al. 2009).

TRPMLs

The TRPML (mucolipin) family contains three mammalian members: TRPML1 (ENSG000-00090674; TRPML1), TRPML2 (ENSG000-00153898; TRPML2), and TRPML3 (ENSG0-0000055732; TRPML3). The TRPML proteins show only low homology with the other TRP channels and are comparatively shorter. As TRPML ion channels are mainly localized in endosomes and lysosomes, their characteristics will be further elucidated below (Cheng et al. 2010b).

TRP CHANNELS DEPOLARIZE EXCITABLE CELLS AND CHANGE INWARDLY-DRIVING FORCES FOR Ca^{2+} ENTRY

Activation of all TRP channels as nonselective cation channels causes a cell depolarization. This depolarizing action of TRP channels is often underestimated. As outlined, many TRP channels have a relatively small fractional Ca^{2+} current. TRPV1, the classical example for TRP-channel-activated Ca^{2+} entry, has only a fractional Ca^{2+} current of less than 5% (Zeilhofer et al. 1997). Also, TRPM8 has a small fractional Ca^{2+} current of $\sim$3%. Only TRPV5, TRPV6, and, in addition, TRPA1 and TRPM3 are characterized by a high fractional Ca^{2+} current. For TRPM3, this fraction is around 24% of the total current and even increases to $\sim$51% in 10 mM Ca^{2+} (Drews et al. 2010). Therefore, next to their role as Ca^{2+}-entry channels, it has to be considered that TRPs may have an important function as depolarizing ion channels.

First, TRP function concerns excitable cells expressing voltage-operated Ca^{2+} channels (CaVs, VOCCs) (Fig. 3A). Activation of TRP channels would trigger gating of those channels

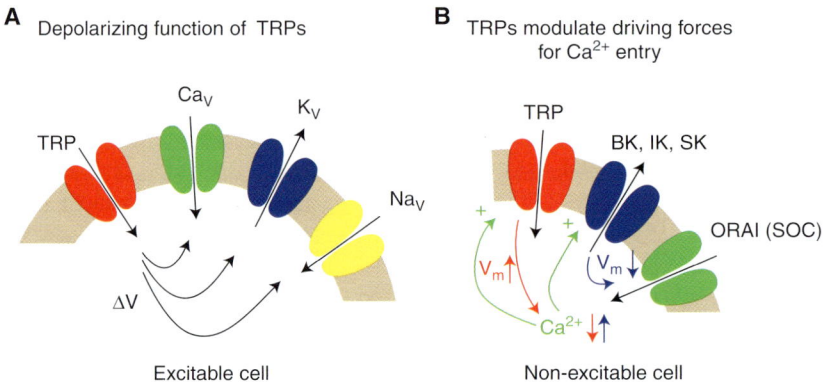

Figure 3. TRP channels depolarize excitable cells and modulate the driving force for Ca^{2+} entry. (*A*) Depolarization of excitable cells upon opening of TRP channels regulates voltage-dependent Ca^{2+}, K^+, and Na^+ channels. (*B*) Membrane depolarization by TRP channels results in a reduced Ca^{2+} entry via ORAI, whereas hyperpolarization of the membrane by BK, IK, or SK channels results in an increased Ca^{2+} influx. This Ca^{2+} then modulates TRP and BK, IK, and SK function to fine-tune the $[Ca^{2+}]_i$ content.

or, if maintained depolarizations are generated, may induce Ca^{2+}-channel inactivation. In addition, the firing pattern of neuronal cells will be modulated by conductance changes via TRP channel activation of inhibition. Examples have been described in detail, indicating that electrogenic effects of TRP channels might be even more important than a direct contribution to $[Ca^{2+}]_i$ changes by mediating Ca^{2+} entry. Only some are mentioned here. All TRPCs are expressed in the brain and obviously have a significant role in regulation of the firing pattern of neurons (Ramsey et al. 2006; Huang et al. 2007; Boisseau et al. 2008; Gokce et al. 2009). They are involved in brain development, synaptogenesis, growth-cone guidance, dendritic growth, spine forming, and many other functions coupled to electrogenesis (Amaral and Pozzo-Miller 2007b; Wen et al. 2007; Tai et al. 2008; for a review, see Talavera et al. 2008; Shim et al. 2009). Persistent neuronal activity lasting seconds to minutes allows the transient storage of memory traces in entorhinal cortex and, thus, could play a major role in working memory. This firing property involves intrinsic properties in cortical neurons by the recruitment of a nonselective cation conductance, probably via heteromeric TRPC channels (TRPC4/5) (Zhang et al. 2010). TRPC3 and TRPC7 channels are expressed in rhythmically active ventral respiratory group islands in the Pre-Bötzinger complex. TRPC3/7 mediate inward currents underlying the pacemaker activity and enhance respiratory rhythm activity (Ben-Mabrouk and Tryba 2010). TRPC3 in the hippocampus is involved in theta-burst stimulation of mossy fibers (MF). Brief theta-burst stimulation of mossy fibers induced a long-lasting depression in the amplitude of EPSCs mediated by both AMPA and NMDA receptors and a reduction in neurotransmitter release. This depression depends on BDNF-mediated activation of TRPC3, and probably has an electrical component (Amaral and Pozzo-Miller 2007a; Li et al. 2010). It has also been shown that TRPV1 mediates long-term depression, including a Ca^{2+}-independent mechanism (Gibson et al. 2008). In general, TRPV1 seems to be involved in the regulation of LTP and LTD (Li et al. 2008). TRPV1 is involved in the activity-dependent facilitation of glutamatergic transmission from solitary tract (ST) afferents. Afferent activation triggered long-lasting asynchronous glutamate release only from TRPV1-expressing synapses, resulting in postsynaptic EPSCs. This release depends on presynaptic TRPV1s and depolarization-dependent activation of

VOCCs. This interaction provides a new form of synaptic plasticity and brings a new integrative feature to the CNS and autonomic regulation (Peters et al. 2010). Another striking interaction between TRPV1 and VOCC has been shown for the action of TNF-α on nociception, which depends on a TRPV1-induced shift in the membrane potential and a modulation of VOCCs thereupon (mainly N-type, less L- and P-/Q-type) (Hagenacker et al. 2009). Temperature can dynamically influence the hippocampal neural activities. TRPV4 seems to be an important player in this signaling cascade: Activation of TRPV4 depolarizes the resting membrane potential in hippocampal neurons by allowing cation influx and potentiates neuronal firing. This effect is absent in TRPV4-deficient mice and to evoke firing, larger depolarizations are required in these mice. Thus, TRPV4 is a key regulator for hippocampal neural excitabilities, also a Ca^{2+}-independent mechanism (Shibasaki et al. 2007).

Several functions of TRP channels involved in the generation of pacemaking inward currents are known. In cardiac muscle, TRPC-mediated inward currents might participate in pacemaking (Ju and Allen 2007). In ileal smooth muscle cell, TRPC4 underlies the muscarinic inward current, which triggers depolarization and the contractile response of these intestinal muscles (Tsvilovskyy et al. 2009). Many other examples for a close interaction between TRP channels and VOCC can be found in the literature.

Second, TRP channels can regulate the driving forces for Ca^{2+} entry, mainly in nonexcitable cells, via depolarization (negative-feedback regulation) or hyperpolarization (positive-feedback regulation) via Ca^{2+}-dependent activation of other ion channels, such as K^+ channels (Fig. 3B). Again, many examples are described in detail. For instance, TRPM4 and TRPM5 are both activated by an increase in $[Ca^{2+}]_i$ but are impermeable for Ca^{2+}. They influence, however, Ca^{2+} entry through Ca^{2+}-permeable channels such as the ORAI/STIM complex (CRAC), by decreasing Ca^{2+} entry due to depolarization (for a striking negative-feedback example, see mast cells; Nilius and Vennekens 2006; Vennekens et al. 2007). Inhibition of TRPM4 in mast cells causes cell hyperpolarization following antigen activation of these cells. In turn, the CRAC-mediated Ca^{2+} influx will be increased, resulting in an elevated release of histamine and interleukins and consequently in an aggravated allergic response (Vennekens et al. 2007; Vennekens et al. 2008). Conversely, activation of TRPM4 might be a means to weaken allergic responses. This interaction between depolarizing TRPM4/5 and the hyperpolarization by Ca^{2+}-activated K^+ channels, such as BK_{Ca} (Slo1 or $K_{Ca}1.1$, $K_{Ca}4.1-4.2$), IKs ($K_{Ca}3.1$), and SKs ($K_{Ca}2.1-2.3$) generates a fine-tuning of Ca^{2+} entry in many nonexcitable cells, such as several blood cell types and endothelial cells. Another striking example is the close interaction and physical association of TRPC1 with BK_{Ca} in vascular smooth muscle cells. Activation of TRPC1 causes hyperpolarization, which in turn could serve to reduce agonist-induced membrane depolarization, thereby preventing excessive contraction of VSMCs to contractile agonists (Kwan et al. 2009). A similar BKCa-TRPC6 association is considered for the dynamic regulation of the filter slit function in podocytes of the renal glomeruli (Kim et al. 2009a).

Third, TRP channels are targets of changes of $[Ca^{2+}]_i$ themselves. They can be activated or inhibited by Ca^{2+}. Probably the earliest Ca^{2+}-activated channel in phylogenies is the yeast mechanosensor TRPY1, in which Ca^{2+} binds to clusters of negative charges in the C-terminus and greatly enhances the force-induced activation (Su et al. 2009). Other TRP channels that are activated by Ca^{2+} include TRPC1, TRPC4, TRPC5, TRPC6, TRPV4, TRPM2, TRPA1, TRPM4, and TRPM5. Many, if not all, TRP channels are modulated by Ca^{2+}, often via complex signaling cascades including Ca^{2+}/CaM binding, Ca^{2+}-dependent PLC modulation, and Ca^{2+}-dependent PKC activation (for TRPM4, see Nilius et al. 2005b). Some TRPs have more Ca^{2+}-binding sites, often overlapping with PIP2- and especially CaM-binding sites, for which binding domains have been identified in detail for TRPC1 through 7. CaM may also interrupt binding sites for many proteins such as MARCKS, GAP43, GRK5, EGFR, and the ErbB family (for a review, see Gordon-Shaag et al. 2008). In addition to Ca^{2+}-dependent activation, all TRP channels show an activity-dependent inactivation mediated by Ca^{2+}, mostly at higher concentrations than needed for activation. Possible mechanisms include Ca^{2+}-dependent kinases; Ca^{2+}-dependent phosphatases; Ca^{2+}-regulated PLCs, which modulate the important TRP-channel modulator $PI(4,5)P_2$ (Nilius et al. 2006); and direct interaction with Ca^{2+}/CaM (Nilius et al. 2005b; Gordon-Shaag et al. 2008; Nilius et al. 2008). This interplay between Ca^{2+}-dependent activation and inactivation provides a huge diversity of TRP-channel modulation in native cells.

TRP CHANNELS AS INTRACELLULAR CALCIUM-RELEASE CHANNELS

The main dogma so far has been that most of the TRP channels exert their functional effects by their strategic localization in the plasma membrane, where they act as ion channels. Their role as scaffolding proteins or intracellular proteins serving different cell functions has not yet been considered systematically. In fact, most, if not all, TRP channels are also located in intracellular organelles, in which the sarco/endoplasmic reticulum and endosomes, lysosomes, and autophagosomes are only the best-studied compartments. The role of mitochondrial-, Golgi-, nuclear- and peroxisomal-TRP channels is not yet understood. TRP channels can be found in intracellular membranes that form part of the biosynthetic or secretory pathway, mainly, when they are on the way to the plasma membrane. Alternatively, TRP channels may play a role in intracellular organelles, participating in maintaining/establishing vesicular ion homeostasis or regulating membrane trafficking. Following the division of intracellular organelles into two groups (group 1: endocytotic, secretory, and autophagic with ER, Golgi apparatus, secretory vesicles/granules, endosomes, autophagosomes, and lysosomes; and group 2: mitochondria, peroxisomes, and nucleus), all have intraluminal-Ca^{2+} concentrations much higher than the cytosolic-Ca^{2+} concentrations, ranging from μM to mM (Dong et al. 2010). Therefore, Ca^{2+} homeostasis in these organelles will be a crucial cell function. Increasing evidence points to a role of several TRP channels as intracellular, calcium-release channels, which is the focus of this review. An important problem is that many TRP channels, as discussed above, require $PI(4,5)P_2$ for activation (see Rohacs 2007; Rohacs and Nilius 2007; Nilius et al. 2008; Rohacs 2009). Because intracellular membranes lack this phospholipid, it is intriguing to speculate that other substituents may compensate for the requirement of $PI(4,5)P_2$ for TRP functioning (Fig. 4).

TRP Channels in Endoplasmic and Sarcoplasmic Reticulum

First, evidence for a role for TRPV1 as an intracellular, calcium-release channel comes from experiments showing that activation of heterologously expressed TRPV1 in COS-7 cells and native TRPV1 in dorsal root ganglion cells gives rise to an increase in intracellular calcium in the absence of extracellular calcium (Olah et al. 2001). It was also already known that a weak agonist of TRPV1, anandamide, induces intracellular calcium release via a PLC-independent mechanism (Felder et al. 1993). Further experiments revealed that TRPV1 is localized to the ER and Golgi compartments. Activation of TRPV1 by capsaicin induces calcium release from an IP_3-sensitive but thapsigargin-insensitive store (Turner et al. 2003). Calcium gradients in the Golgi are maintained by SERCA and by the thapsigargin-insensitive secretory pathway (SPCA). Intracellular stores can be dissected based on their release and/or refilling capacities; within the IP_3-sensitive store, there exists a compartment that also contains functional TRPV1 molecules that mediate release. The TRPV1/IP_3R-containing store is apparently thapsigargin-insensitive. Depletion of this calcium store can occur without I_{CRAC} activation. Due to the thapsigargin-insensitivity of TRPV1-induced calcium release, the most likely location for TRPV1 is in the SPCA-positive

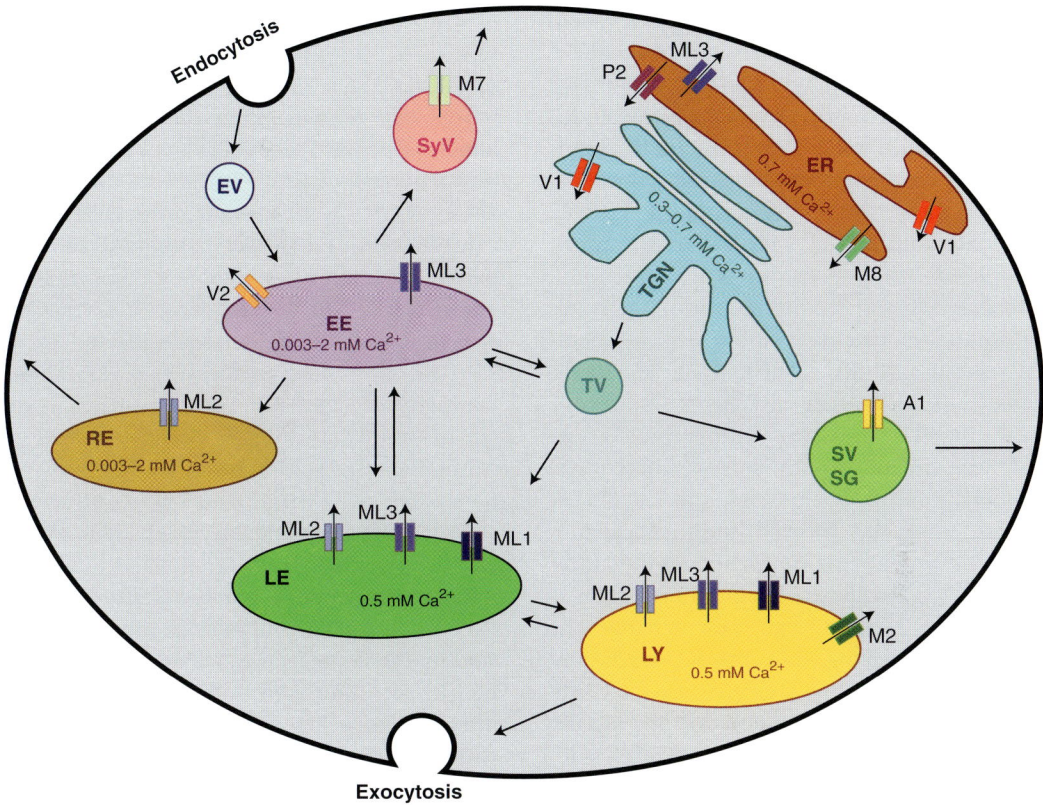

Figure 4. Expression of TRP channels in intracellular compartments. Early endosomes (EE) are derived from the plasma membrane via endocytotic vesicles (EV). The cargo from these early endosomes can either go back to the plasma membrane via the recycling endosomes (RE) or follow the late endocytotic pathway via late endosomes (LE) to the lysosomes (LY). Intermediate transport vesicles (TV) are derived from the ER and/or the trans-Golgi-network (TGN). The content in these transport vesicles can either be delivered to early endosomes, late endosomes, or be transported to the plasma membrane via secretory vesicles (SV) or secretory granules (SG). Synaptic vesicles (SyV) are derived from early endosomes and release neurotransmitters in the extracellular space. Only intracellular locations of TRP channels are indicated.

Golgi compartments (Turner et al. 2003). However, other studies suggest that TRPV1- and thapsigargin-sensitive internal calcium pools substantially overlap, and that TRPV1 immunoreactivity co-localizes with a marker in the ER. Still, although TRPV1 forms agonist-sensitive channels in the ER, which upon activation release calcium from internal stores, it fails to activate endogenous store-operated Ca^{2+} entry (Wisnoskey et al. 2003).

TRPM8 is amongst others expressed in LNCaP (lymph node carcinoma of the prostate) cells and sensory neurons from dorsal root and trigeminal ganglia (McKemy et al. 2002; Zhang and Barritt 2004; Thebault et al. 2005; Ramsey et al. 2006). Several lines of evidence suggest that TRPM8 is involved in menthol-induced calcium release from intracellular stores. TRPM8 is shown to be expressed in both the endoplasmic reticulum and the plasma membrane of the androgen-responsive prostate cancer LNCaP cells and to mediate, respectively, Ca^{2+} inflow and Ca^{2+} release from intracellular stores in these cells (Zhang and Barritt 2004). These cells show a current in response to cold and menthol with biophysical properties (strong inward rectification and high calcium selectivity) different than those described for

TRPM8 in a heterologous expression system and in sensory neurons (Thebault et al. 2005). This current could nevertheless be suppressed by experimental maneuvers that decrease endogenous TRPM8 mRNA or protein, which is explained by the extraplasmalemmal localization of TRPM8 in these cells. Indeed, TRPM8 in the ER from LNCaP cells is able to support Ca^{2+} release in response to cold or menthol, and this ER Ca^{2+} store depletion activates Ca^{2+} entry via plasma membrane store-operated channels (SOC) (Thebault et al. 2005). However, the fact that TRPM8 is responsible for the menthol-induced calcium release is controversial. Indeed, experiments in different cell lines (HEK293, LNCaP, CHO, and COS cells) indicate that the menthol-induced calcium release is potentiated at higher temperatures (Mahieu et al. 2007). Furthermore, overexpression of TRPM8 does not enhance the menthol-induced calcium release; icilin and eucalyptol, 2 more potent agonists of TRPM8 than menthol, do not induce calcium release, unlike geraniol and linalool, which are structurally related to menthol. These data indicate that menthol induces intracellular calcium release in a TRPM8-independent manner. Clearly, more experiments are needed to unravel the molecular identity of this calcium-release channel.

Trpp2 and *Pkd1*, or polycystin-1, are genes mutated in autosomal-dominant polycystic kidney disease (ADPKD). Both proteins are expressed in many tissues, including kidney, heart, liver, pancreas, brain, and muscles (Hughes et al. 1995; Geng et al. 1997; Luo et al. 2003). PKD1 is a large glycoprotein localized to the plasma membrane and primary cilia (Hanaoka et al. 2000; Newby et al. 2002; Yoder et al. 2002). TRPP2 has a dynamic expression pattern with localization reported in mitotic spindles, the ER/SR, basolateral plasma membrane, and together with PKD1 in the primary cilium (Cai et al. 1999; Foggensteiner et al. 2000; Yoder et al. 2002; Rundle et al. 2004). Expression of TRPP2 in the ER is caused by ER-retention signals in the cytoplasmic tail, which will prevent trafficking of TRPP2 toward the plasma membrane (Cai et al. 1999; Hanaoka et al. 2000). Masking or deleting this ER-retention motif allows plasma membrane expression of TRPP2 (Cai et al. 1999). Co-immunoprecipitation experiments show that polycystin-1, which is present at the plasma membrane, is physically associating with polycystin-2 (Tsiokas et al. 1997). This might mean that TRPP2 can function as a surface membrane calcium-permeable channel or signaling complex, chaperoned by TRPP1 (Hanaoka et al. 2000). On the other side, TRPP2 may be located in the ER, tethering close to the surface membrane via the physical association with TRPP1 (Koulen et al. 2002). Single channel recordings from ER microsomes fused to lipid bilayers reveal that polycystin-2 is a high conductance channel that is permeable to divalent cations (Koulen et al. 2002). It functions as a calcium-release channel in a porcine kidney cell line (LLC-PK1) to augment calcium transients initiated by the IP_3 receptor after receptor stimulation by, for example, vasopressin. This calcium release does not seem to require coassembly with polycystin-1. Cells expressing a mutant channel protein, which retains the protein interactions with wild-type polycystin-2, do not show this calcium release, strongly indicating that TRPP2 itself forms the channel (Koulen et al. 2002), in contrast with the polycystin-1-dependent channel activity of TRPP2 at the plasma membrane (Hanaoka et al. 2000). TRPP2 can be activated by intracellular calcium; it shows a bell-shaped response curve to Ca^{2+} that is dependent on phosphorylation at the C-terminus of the protein (Cai et al. 2004). These data suggest that TRPP2 functions as a new type of calcium-release channel, with properties (ER location, calcium-dependent calcium-channel activity) allowing it to mediate calcium-induced calcium release. Recent data suggest that the function of TRPP2 as a calcium-induced calcium-release channel is dependent on interaction with the IP_3 receptor, and that TRPP2 is activated by a local rise in cytosolic Ca^{2+} generated by IP_3-induced calcium release (Sammels et al. 2010). The pathomechanism by which the TRPP2 channel, together with PKD1, contributes to ADPKD is still not completely understood. In the primary cilium, the TRPP2/PKD1 complex acts as a mechanosensor that provides flow-activated Ca^{2+} entry and plays a role in Ca^{2+} signaling in tubular renal cells. This mechanosensitive property of the complex is deficient in PKD1-deficient cells. Additionally, the PKD1/TRPP2 complex regulates transfer of the helix-loop-helix (HLH) protein Id2, a crucial regulator of cell proliferation and differentiation, into the nucleus. An enhanced nuclear localization of Id2 in renal epithelial cells from AKPKD patients constitutes a mechanism for the hyperproliferative phenotype and may cause cyst formation (Benezra 2005; Li et al. 2005). Furthermore, TRPP2 may play a role in the increased apoptotic rate reported in ADPKD patients and several models of polycystic kidney diseases. Indeed, the Ca^{2+} content of the ER determines the sensitivity of the cell to apoptotic stress, and TRPP2 is suggested to function as an antiapoptotic ion channel that regulates the Ca^{2+} concentration in the ER (Wegierski et al. 2009).

TRP Channels in Endosomes and Lysosomes

The endolysosome system is comprised of early endosomes, recycling endosomes, late endosomes, and lysosomes (Luzio et al. 2007b; Dong et al. 2010). Early endosomes are derived from the plasma membrane via endocytosis. Components in these early endosomes can either be recycled back to the plasma membrane via the recycling endosomes or follow the late endosome pathway. Lysosomes are derived from late endosomes and are filled with hydrolytic enzymes. Some common features of endolysosomes are an acidic pH (established and maintained by the vacuolar (V)-ATPase H^+ pump and essential for the degradative function of the hydrolases) and a positive membrane potential (assumed to be in the range of $+30$ to $+110$ mV) that provides a driving force for calcium release into the cytosol. Indeed, endolysosomes also serve as storage for intracellular calcium, containing a luminal calcium concentration of ~ 0.5 mM that is maintained by an unidentified H^+-Ca^{2+} exchanger. Next to other factors, such as small GTPase Rabs and phosphoinositides (PIP), membrane trafficking is regulated by the release of calcium from endolysosomes. In fact, the membrane fusion process that drives the movement of endocytosed material and enzymes within the endocytic pathway depends on calcium ions, presumably released from the endocytic organelles (Pryor et al. 2000). However, the ion channels responsible for endolysosomal calcium release are still undefined. Candidates include several members of the TRP family, such as TRPMLs, TRPV2, and TRPM2.

TRPMLs, or mucolipins, constitute a family of inwardly-rectifying, calcium-permeant, but probably proton-impermeant, cation channels and consist of three members: TRPML1, TRPML2, and TRPML3. All TRPMLs colocalize with lysosome-associated membrane protein (Lamp-1) or Rab7, indicating localization to late endosomes and lysosomes (Karacsonyi et al. 2007; Thompson et al. 2007; Kim et al. 2009b; Martina et al. 2009). In addition, TRPML2 is found in recycling endosomes (Karacsonyi et al. 2007), whereas TRPML3 also localizes to early endosomes and the plasma membrane (Kim et al. 2009b). However, it has been shown that TRPMLs interact to form homo- and heteromultimers, and that the presence of either TRPML1 or TRPML2 specifically influences the spatial distribution of TRPML3. Indeed, TRPML1 and TRPML2 homomultimers form lysosomal proteins, whereas TRPML3 homomultimers are probably channels in the endoplasmic reticulum. However, TRPML3 localizes to lysosomes when coexpressed with either TRPML1 or TRPML2, and is comparably mislocalized when lysosomal targeting of TRPML1 and TRPML2 is disrupted. Conversely, TRPML3 does not cause retention of TRPML1 or TRPML2 in the endoplasmic reticulum. These data demonstrate that there is a hierarchy controlling the subcellular distributions of the TRPMLs, such that TRPML1 and TRPML2 dictate the localization of TRPML3 and not vice versa (Venkatachalam et al. 2006).

TRPML1 (or Mucolipin 1) is the best described member of the TRPML subfamily and is expressed in almost every tissue with the highest level of expression in the brain, kidney, spleen, liver, and heart (Sun et al. 2000). TRPML1 follows a direct or indirect pathway to reach lysosomes. Newly synthesized proteins, together with TRPML1, are routed directly from the Golgi to lysosomes by recognition of the N-terminal E^{11}TERLLL motif of TRPML1 (direct pathway). The indirect pathway comprises trafficking to the plasma membrane, followed by internalization to early endosome that is guided by the C-terminal E^{573}EHSLL motif (Curcio-Morelli et al. 2009; Samie et al. 2009). TRPML1 is a TRP-channel-related protein that shows strong topological homology with the polycystin-2 channel and contains an internal Ca^{2+}- and Na^+-channel-pore region (Sun et al. 2000). Several studies suggest that TRPML1 is a nonselective cation channel with a large conductance and permeability to Na^+, K^+, and Ca^{2+} (LaPlante et al. 2002; Raychowdhury et al. 2004). However, its localization in the endolysosomal membrane makes it difficult to characterize. TRPML3 is the only TRPML channel found in the plasma membrane, making it easier to characterize the channel properties via whole-cell current measurements. This revealed that TRPML3 is an inwardly rectifying, Ca^{2+}-permeable cation channel (Cuajungco and Samie 2008). The current is inhibited by an acid extracytosolic (analogous to the luminal side) pH (Kim et al. 2008). Notably, elimination of TRPML3 regulation by extracytosolic pH has the same functional and cellular phenotype as the A419P (Va) mutation, a gain-of-function mutation that causes the varitint-waddler phenotype (a disease characterized by deafness, circling behavior, and pigmentation defects) (Di Palma et al. 2002). This mutation is likely to disrupt channel-gating by locking the channel in an open state, making the channel constitutively active and yielding much larger currents (Cuajungco and Samie 2008); although basic properties, such as I-V characteristics, single channel conductance, and ion selectivity have not changed (Kim et al. 2007; Nagata et al. 2008). The same mutation has been made in TRPML1 and TRPML2 channels in order to effectively characterize the pore properties of these channels using whole-cell recordings. This revealed, for both channels, inwardly rectifying, Ca^{2+}-permeable cation currents, in accordance with TRPML3 channel properties (Dong et al. 2009; Samie et al. 2009). At least for TRPML1, several properties of this mutated channel, such as IV relationship, kinetics, and voltage dependence, are similar to the wild-type channel measured in the lysosomal membrane (Dong et al. 2008), indicating that the activating Va mutation is a useful approach for characterizing the pore properties of TRPML1. Obviously, these channels provide a novel class of TRP channels that are functionally still much less characterized than other TRPs (for an excellent review, see Cheng et al. 2010b; Dong et al. 2010).

The physiological functions of the TRPML channels are still under intensive investigation. Mucolipins could regulate the ionic conditions within, and acidity of, the endocytic organelles, in this way regulating the activity and delivery of the digestive enzymes within the endocytic pathway (Puertollano and Kiselyov 2009). However, the calcium permeability of TRPML channels allows a role in different intracellular processes regulated by calcium. Mutations in the *Trpml1* gene cause mucolipidosis type IV, a lysosomal storage disease characterized by severe psychomotor retardation and ophthalmologic abnormalities. Indeed, all cell types show an abnormal accumulation of phospholipids, sphingolipids, acidic mucopolysaccharides, and cholesterol in swollen and enlarged LEL-like vacuoles (Puertollano and Kiselyov 2009). In agreement with this, cells from *Trpml1* KO mice show abnormal lipid accumulation and enlarged vacuoles (Venugopal et al. 2007). These enlarged LELs seem to be the late endosome-lysosome hybrid organelles from which lysosomes are formed in the normal situation. Enlarged endolysosomes can result from either uncontrolled and excessive fusion, defective membrane fission, or impaired organellar osmoregulation (Luzio et al. 2007a). The defect observed in *Trpml1*-deficient cells is most likely related to the sorting and trafficking processes of the endocytic pathway, more specifically the formation of transport vesicles from the LEL compartment to the trans-Golgi-network and the reformation of lysosomes from the late endosome-lysosome hybrid organelles (Thompson et al. 2007). The Ca^{2+} permeability of TRPML1 seems to be required for the membrane fission from LEL compartments or late endosome-lysosome hybrids and the biogenesis of both late retrograde transport vesicles and lysosomes. Indeed, membrane fission and stabilization of transport vesicles are dependent on luminal calcium and intraluminal calcium release (Luzio et al. 2007a). The major mucolipidosis type IV disease mutations of TRPML1 are expected to produce no protein (Altarescu et al. 2002). However, two mutations (V446L and ΔF408) do retain TRPML1 channel function. WT TRPML1 is inhibited by a reduction of pH, and this pH-dependent regulation seems to be lost in these two mutations (Raychowdhury et al. 2004). The recent generation of the *Trpml1* KO mouse, a murine model for MLIV, will help to further elucidate the function of TRPML1 (Venugopal et al. 2007). Interestingly, a *Drosophila* model for mucolipidosis type IV has been developed that mimicks some of the key disease features, such as abnormal intracellular accumulation of macromolecules, motor defects, and neurodegeneration. Here, the basis for macromolecule retention is a defective autophagy, which results in oxidative stress and impaired synaptic transmission. Late-apoptotic cells accumulate in the brains of *trpml-* (the only TRPML member in the fly) deficient flies. The accumulation of those cells, and also the degree of motor deficits, were suppressed by expression of TRPML in neurons, glia, and hematopoietic cells. Because hematopoietic cells in humans are involved in clearance of apoptotic cells, it seems possible that bone marrow transplantation might have a potential as a new therapeutic strategy for MLIV (Venkatachalam et al. 2008).

TRPML1 is also thought to mediate a NAADP-activated intracellular calcium release channel. Indeed, nicotinic acid adenine dinucleotide phosphate (NAADP) is an attractive candidate for mediating endolysosomal calcium release (Churchill et al. 2002; Calcraft et al. 2009). In fact, a lysosomal NAADP-sensitive Ca^{2+}-release channel, different from sarcoplasmic reticulum Ca^{2+}-release channels, could be measured in lysosomes from native rat liver cells and bovine coronary arterial myocytes (Zhang and Li 2007; Zhang et al. 2009). This current could be blocked by using an anti-TRPML1 antibody or a TRPML1-specific siRNA, suggesting involvement of TRPML1. However, the molecular identity of the NAADP receptor remains controversial (Galione and Churchill 2002). Another attractive family of ion channels comprises the two-pore channel family (TPCs) (Patel et al. 2010). Indeed, these channels provide a family of NAADP receptors, with TPC1 and TPC3 being expressed on endolysosomal and TPC2 on lysosomal membranes (Calcraft et al. 2009). Both TPC1 and TPC2 have been shown to be an important component of NAADP-elicited calcium release (Brailoiu et al. 2009; Zong et al. 2009). Thus, the nature of the NAADP-receptor and the contribution of different ion channels to NAADP action remains a matter of debate.

Much less is known about the function of TRPML2 and TRPML3. TRPML2 is expressed in B-lymphocytes at different cell stages. Overexpression of TRPML2 in these cells induces accumulation of enlarged lysosomal structures, indicating that TRPML2 might participate in the regulation of the specialized lysosomal compartment of B-lymphocytes, and in this way may be critical for normal immune response (Song et al. 2006). Furthermore, TRPML2 has been reported to have a regulatory role in the trafficking of proteins along the Arf6-regulated pathway. Indeed, TRPML2 colocalizes with glycosylphosphatidylinositol-anchored proteins (GPI-APs), such as CD59, and overexpression of a TRPML2 inactive mutant decreases recycling of CD59 to the plasma membrane, indicating that TRPML2 localizes to the Arf6-regulated pathway and regulates sorting of GPI-APs (Karacsonyi et al. 2007). Overexpression of TRPML3 leads to reduced constitutive and regulated endocytosis, increased autophagy, and marked exacerbation of autophagy evoked by various cell stressors with nearly complete recruitment of TRPML3 into the autophagosomes (Kim et al. 2009b). These data indicate that TRPML3 is a prominent regulator of endocytosis, membrane trafficking, and autophagy, perhaps by controlling the Ca^{2+} in the vicinity of cellular organelles that is necessary to regulate these cellular events (Kim et al. 2009b). Furthermore, calcium release from endolysosomes by TRPML3 may be important for efficient endosomal acidification. Indeed, after internalization from the plasma membrane, the endosomes contain a high concentration of calcium that is rapidly released in order to allow acidification of the compartment. TRPML3 is a good candidate to mediate this efflux, since the channel is inhibited by low pH and would still be active at the characteristic pH of the early endosomes, but will become inactive once the acidification has taken place, in this way preventing further Ca^{2+} efflux and acidification (Martina et al. 2009). Given the clinical importance of TRPMLs, these channels obviously define a new group of pharmacological targets as introduced recently by small molecule activators of TRPML3 (Grimm et al. 2010).

TRPV2 localizes to early endosomes, and activation of TRPV2 in these intracellular vesicles is suggested to cause calcium-dependent fusion between endosomal membranes (Saito et al. 2007). Endogenous ionic currents measured in an isolated enlarged endosome showed pharmacological similarities (inhibition by Ruthenium Red and activation by 2-aminoethyldiphenyl borate) with the TRPV2 channel. This current was inhibited by a decrease in the luminal pH and an increase in the luminal chloride concentration, two features known to occur after endocytosis (Saito et al. 2007). The current hypothesis is that activation of TRPV2 leads to Ca^{2+}-dependent fusion between endosomal membranes. Further experiments are needed to confirm this theory.

TRPM2 serves a dual role as a plasma membrane Ca^{2+}-influx channel and as an intracellular, calcium-release channel in pancreatic beta cells (Lange et al. 2009). Indeed, intracellular ADPR elicits intracellular calcium release in these cells, which is dependent on TRPM2 channels and comes from lysosomes. The lysosomal calcium release through TRPM2 contributes to H_2O_2-mediated beta cell death (Lange et al. 2009).

TRP Channels in Secretory Vesicles, Secretory Granules, and Synaptic Vesicle

Much less is known about TRP channels in these compartments. TRPA1 is expressed in secretory vesicles and granules (Dong et al. 2010). Icilin, a TRPA1 agonist, elicits intracellular Ca^{2+} release from secretory vesicles in HEK cells heterologously expressing TRPA1 (Prasad et al. 2008). TRPA1 is a Ca^{2+}-activated channel (Zurborg et al. 2007) and may therefore contribute to Ca^{2+} release from these organelles. TRPM7 is localized in synaptic vesicles of sympathetic neurons and bound to several proteins of the synaptic vesicle fusion apparatus, including synapsin I and synaptotagmin I. Furthermore, ion conductance via TRPM7 is critical to neurotransmitter release upon an increase in $[Ca^{2+}]_i$ and might in this way play a role in forming the exocytotic machinery (Krapivinsky et al. 2006). Finally, the role of TRP channels in intracellular organelles is a just-appearing and extremely challenging topic.

CONCLUSIONS

TRP channels have been extensively reviewed in the last 10 years. It is obvious now that they function not only as polymodal cell sensors in sensory processes, but that they are unique channels involved in many cell functions. We want to stress in this review that TRP channels are not only important for Ca^{2+} entry via the plasma membrane. This function might even be overestimated, given the often very small fractional Ca^{2+} currents through these channels. They obviously play an important role in electrogenesis, regulating the activity for voltage-dependent ion channels including Ca^{2+} channels (Ca_V's, VOCCs). They are also important regulators of the driving forces for Ca^{2+} entry via other Ca^{2+}-permeable channels such as ORAIs, CNG, and NMDA receptors. TRPs must also be considered as targets of changes in $[Ca^{2+}]_I$, which again creates a huge diversity and versatility of these channels in the whole process of Ca^{2+} signaling. Finally, TRPs are intracellular channels, and may even act in some cases as scaffolding proteins forming signaling complexes. Their role as intracellular channels is just emerging and will undoubtedly be the focus of future research.

ACKNOWLEDGMENTS

We are grateful to all members of the Laboratory for Ion Channel Research, KU Leuven, for helpful comments and discussion. M.G. was supported by a Doctoral Fellowship of the Research Foundation-Flanders. This work was supported by grants from Interuniversity Attraction Poles Programme - Belgian State - Belgian Science Policy, P6/28, the Research Foundation-Flanders (G.0172.03 and G.0565.07), the Research Council of the KU Leuven (GOA 2004/07), and the Flemish Government (Excellentiefinanciering, EF/95/010). MG and BC contributed equally to this work.

REFERENCES

Aarts M, Iihara K, Wei WL, Xiong ZG, Arundine M, Cerwinski W, MacDonald JF, Tymianski M. 2003. A key role for TRPM7 channels in anoxic neuronal death. *Cell* **115:** 863–877.

Aarts MM, Tymianski M. 2005. TRPM7 and ischemic CNS injury. *Neuroscientist* **11:** 116–123.

Ahmmed GU, Malik AB. 2005. Functional role of TRPC channels in the regulation of endothelial permeability. *Pflugers Arch* **451:** 131–142.

Alessandri-Haber N, Yeh JJ, Boyd AE, Parada CA, Chen X, Reichling DB, Levine JD. 2003. Hypotonicity induces TRPV4-mediated nociception in rat. *Neuron* **39:** 497–511.

Altarescu G, Sun M, Moore DF, Smith JA, Wiggs EA, Solomon BI, Patronas NJ, Frei KP, Gupta S, Kaneski CR, et al. 2002. The neurogenetics of mucolipidosis type IV. *Neurology* **59:** 306–313.

Amaral MD, Pozzo-Miller L. 2007a. BDNF Induces Calcium Elevations Associated with IBDNF, a Non-Selective Cationic Current Mediated by TRPC Channels. *J Neurophysiol* **98:** 2476–2482.

Amaral MD, Pozzo-Miller L. 2007b. TRPC3 channels are necessary for brain-derived neurotrophic factor to activate a nonselective cationic current and to induce dendritic spine formation. *J Neurosci* **27:** 5179–5189.

Ambudkar IS, Ong HL, Liu X, Bandyopadhyay BC, Cheng KT. 2007. TRPC1: the link between functionally distinct store-operated calcium channels. *Cell Calcium* **42:** 213–223.

Andre E, Gatti R, Trevisani M, Preti D, Baraldi PG, Patacchini R, Geppetti P. 2009. Transient receptor potential ankyrin receptor 1 is a novel target for pro-tussive agents. *Br J Pharmacol* **158:** 1621–1628.

Asakawa M, Yoshioka T, Matsutani T, Hikita I, Suzuki M, Oshima I, Tsukahara K, Arimura A, Horikawa T, Hirasawa T, et al. 2006. Association of a mutation in TRPV3 with defective hair growth in rodents. *J Invest Dermatol* **126:** 2664–2672.

Audo I, Kohl S, Leroy BP, Munier FL, Guillonneau X, Mohand-Said S, Bujakowska K, Nandrot EF, Lorenz B, Preising M, et al. 2009. TRPM1 is mutated in patients with autosomal-recessive complete congenital stationary night blindness. *Am J Hum Genet* **85:** 720–729.

Authi KS. 2007. TRP channels in platelet function. *Handb Exp Pharmacol* **179:** 425–443.

Bautista DM, Jordt SE, Nikai T, Tsuruda PR, Read AJ, Poblete J, Yamoah EN, Basbaum AI, Julius D. 2006. TRPA1 mediates the inflammatory actions of environmental irritants and proalgesic agents. *Cell* **124:** 1269–1282.

Beech DJ. 2005. Emerging functions of 10 types of TRP cationic channel in vascular smooth muscle. *Clin Exp Pharmacol Physiol* **32:** 597–603.

Beech DJ, Muraki K, Flemming R. 2004. Non-selective cationic channels of smooth muscle and the mammalian homologues of Drosophila TRP. *J Physiol* **559:** 685–706.

Benezra R. 2005. Polycystins: inhibiting the inhibitors. *Nat Cell Biol* **7:** 1064–1065.

Ben-Mabrouk F, Tryba AK. 2010. Substance P modulation of TRPC3/7 channels improves respiratory rhythm regularity and ICAN-dependent pacemaker activity. *Eur J Neurosci* **31:** 1219–1232.

Berridge MJ, Lipp P, Bootman MD. 2000. The versatility and universality of calcium signalling. *Nat Rev Mol Cell Biol* **1:** 11–21.

Bezzerides VJ, Ramsey IS, Kotecha S, Greka A, Clapham DE. 2004. Rapid vesicular translocation and insertion of TRP channels. *Nat Cell Biol* **6:** 709–720.

Birnbaumer L. 2009. The TRPC class of ion channels: a critical review of their roles in slow, sustained increases in intracellular Ca(2+) concentrations. *Annu Rev Pharmacol Toxicol* **49:** 395–426.

Birrell MA, Belvisi MG, Grace M, Sadofsky L, Faruqi S, Hele DJ, Maher SA, Freund-Michel V, Morice AH. 2009. TRPA1 agonists evoke coughing in guinea pig and human volunteers. *Am J Respir Crit Care Med* **180:** 1042–1047.

Bisgrove BW, Snarr BS, Emrazian A, Yost HJ. 2005. Polaris and Polycystin-2 in dorsal forerunner cells and Kupffer's vesicle are required for specification of the zebrafish left-right axis. *Dev Biol* **287:** 274–288.

Boisseau S, Kunert-Keil C, Lucke S, Bouron A. 2008. Heterogeneous distribution of TRPC proteins in the embryonic cortex. *Histochem Cell Biol* **131:** 355–363.

Bolotina VM. 2008. Orai, STIM1 and iPLA2beta: a view from a different perspective. *J Physiol* **586:** 3035–3042.

Bootman MD, Collins TJ, Peppiatt CM, Prothero LS, MacKenzie L, De Smet P, Travers M, Tovey SC, Seo JT, Berridge MJ, et al. 2001. Calcium signalling-an overview. *Semin Cell Dev Biol* **12:** 3–10.

Brailoiu E, Churamani D, Cai X, Schrlau MG, Brailoiu GC, Gao X, Hooper R, Boulware MJ, Dun NJ, Marchant JS, et al. 2009. Essential requirement for two-pore channel 1 in NAADP-mediated calcium signaling. *J Cell Biol* **186:** 201–209.

Cai X. 2008. Unicellular Ca^{2+} Signaling 'Toolkit' at the Origin of Metazoa. *Mol Biol Evol* **25:** 1357–1361.

Cai Y, Anyatonwu G, Okuhara D, Lee KB, Yu Z, Onoe T, Mei CL, Qian Q, Geng L, Wiztgall R, et al. 2004. Calcium dependence of polycystin-2 channel activity is modulated by phosphorylation at Ser812. *J Biol Chem* **279:** 19987–19995.

Cai Y, Maeda Y, Cedzich A, Torres VE, Wu G, Hayashi T, Mochizuki T, Park JH, Witzgall R, Somlo S. 1999. Identification and characterization of polycystin-2, the PKD2 gene product. *J Biol Chem* **274:** 28557–28565.

Calcraft PJ, Ruas M, Pan Z, Cheng X, Arredouani A, Hao X, Tang J, Rietdorf K, Teboul L, Chuang KT, et al. 2009. NAADP mobilizes calcium from acidic organelles through two-pore channels. *Nature* **459:** 596–600.

Caterina MJ, Rosen TA, Tominaga M, Brake AJ, Julius D. 1999. A capsaicin-receptor homologue with a high threshold for noxious heat. *Nature* **398:** 436–441.

Caterina MJ, Schumacher MA, Tominaga M, Rosen TA, Levine JD, Julius D. 1997. The capsaicin receptor: a heat-activated ion channel in the pain pathway. *Nature* **389:** 816–824.

Chang Y, Schlenstedt G, Flockerzi V, Beck A. 2010. Properties of the intracellular transient receptor potential (TRP) channel in yeast, Yvc1. *FEBS Lett* **584:** 2028–2032.

Charrua A, Cruz CD, Cruz F, Avelino A. 2007. Transient receptor potential vanilloid subfamily 1 is essential for the generation of noxious bladder input and bladder overactivity in cystitis. *J Urol* **177:** 1537–1541.

Chen J, Kim D, Bianchi BR, Cavanaugh EJ, Faltynek CR, Kym PR, Reilly RM. 2009. Pore dilation occurs in TRPA1 but not in TRPM8 channels. *Mol Pain* **5:** 3.

Chen Y, Zhang Z, Lv XY, Wang YD, Hu ZG, Sun H, Tan RZ, Liu YH, Bian GH, Xiao Y, et al. 2008. Expression of Pkd2l2 in testis is implicated in spermatogenesis. *Biol Pharm Bull* **31:** 1496–1500.

Cheng X, Jin J, Hu L, Shen D, Dong X-P, Samie MA, Knoff J, Eisinger B, Liu M-L, Huang SM, et al. 2010a. TRP Channel Regulates EGFR Signaling in Hair Morphogenesis and Skin Barrier Formation. *Cell* **141:** 331–343.

Cheng X, Shen D, Samie M, Xu H. 2010b. Mucolipins: Intracellular TRPML1-3 channels. *FEBS Lett* **584:** 2013–2021.

Chung MK, Guler AD, Caterina MJ. 2005. Biphasic currents evoked by chemical or thermal activation of the heat-gated ion channel, TRPV3. *J Biol Chem* **280:** 15928–15941.

Chung MK, Guler AD, Caterina MJ. 2008. TRPV1 shows dynamic ionic selectivity during agonist stimulation. *Nat Neurosci* **11:** 555–564.

Chung MK, Lee H, Caterina MJ. 2003. Warm temperatures activate TRPV4 in mouse 308 keratinocytes. *J Biol Chem* **278:** 32037–32046.

Chung MK, Lee H, Mizuno A, Suzuki M, Caterina MJ. 2004. 2-aminoethoxydiphenyl borate activates and sensitizes the heat-gated ion channel TRPV3. *J Neurosci* **24:** 5177–5182.

Churchill GC, Okada Y, Thomas JM, Genazzani AA, Patel S, Galione A. 2002. NAADP mobilizes Ca(2+) from reserve granules, lysosome-related organelles, in sea urchin eggs. *Cell* **111:** 703–708.

Clapham DE. 2003. TRP channels as cellular sensors. *Nature* **426:** 517–524.

Clapham DE, Nilius B, Owsianik G. 2009. Transient Receptor Potential Channels. Last modified on 2009-10-14. (IUPHAR-DB). http://www.iuphar-db.org/DATABASE/FamilyMenuForward?familyId=78.

Colsoul B, Schraenen A, Lemaire K, Quintens R, Van Lommel L, Segal A, Owsianik G, Talavera K, Voets T, Margolskee RF, et al. 2010. Loss of high-frequency glucose-induced Ca^{2+} oscillations in pancreatic islets correlates with impaired glucose tolerance in Trpm5-/- mice. *Proc Natl Acad Sci U S A* **107:** 5208–5213.

Corey DP, Garcia-Anoveros J, Holt JR, Kwan KY, Lin SY, Vollrath MA, Amalfitano A, Cheung EL, Derfler BH, Duggan A, et al. 2004. TRPA1 is a candidate for the mechanosensitive transduction channel of vertebrate hair cells. *Nature* **432:** 723–730.

Cuajungco MP, Samie MA. 2008. The varitint-waddler mouse phenotypes and the TRPML3 ion channel mutation: cause and consequence. *Pflugers Arch* **457:** 463–473.

Curcio-Morelli C, Zhang P, Venugopal B, Charles FA, Browning MF, Cantiello HF, Slaugenhaupt SA. 2010. Functional multimerization of mucolipin channel proteins. *J Cell Physiol* **222:** 328–335.

Damann N, Voets T, Nilius B. 2008. TRPs in Our Senses. *Curr Biol* **18:** R880–R889.

De Ridder D, Baert L. 2000. Vanilloids and the overactive bladder. *BJU Int* **86:** 172–180.

Delmas P. 2004. Polycystins: from mechanosensation to gene regulation. *Cell* **118:** 145–148.

Delmas P, Padilla F, Osorio N, Coste B, Raoux M, Crest M. 2004. Polycystins, calcium signaling, and human diseases. *Biochem Biophys Res Commun* **322:** 1374–1383.

den Dekker E, Hoenderop JG, Nilius B, Bindels RJ. 2003. The epithelial calcium channels, TRPV5 & TRPV6: from identification towards regulation. *Cell Calcium* **33:** 497–507.

Desai BN, Clapham DE. 2005. TRP channels and mice deficient in TRP channels. *Pflugers Arch* **451:** 11–18.

Dhaka A, Viswanath V, Patapoutian A. 2006. Trp ion channels and temperature sensation. *Annu Rev Neurosci* **29:** 135–161.

Di Palma F, Belyantseva IA, Kim HJ, Vogt TF, Kachar B, Noben-Trauth K. 2002. Mutations in Mcoln3 associated with deafness and pigmentation defects in varitint-waddler (Va) mice. *Proc Natl Acad Sci U S A* **99:** 14994–14999.

Dietrich A, Chubanov V, Kalwa H, Rost BR, Gudermann T. 2006. Cation channels of the transient receptor potential superfamily: their role in physiological and pathophysiological processes of smooth muscle cells. *Pharmacol Ther* **112:** 744–760.

Dietrich A, Mederos YSM, Gollasch M, Gross V, Storch U, Dubrovska G, Obst M, Yildirim E, Salanova B, Kalwa H, et al. 2005. Increased vascular smooth muscle contractility in TRPC6-/- mice. *Mol Cell Biol* **25:** 6980–6989.

Dong XP, Cheng X, Mills E, Delling M, Wang F, Kurz T, Xu H. 2008. The type IV mucolipidosis-associated protein TRPML1 is an endolysosomal iron release channel. *Nature* **455:** 992–996.

Dong XP, Wang X, Shen D, Chen S, Liu M, Wang Y, Mills E, Cheng X, Delling M, Xu H. 2009. Activating mutations of the TRPML1 channel revealed by proline-scanning mutagenesis. *J Biol Chem* **284:** 32040–32052.

Dong XP, Wang X, Xu H. 2010. TRP Channels of Intracellular Membranes. *J Neurochem* **113:** 313–328.

Drews A, Loch S, Mohr F, Rizun O, Lambert S, Oberwinkler J. 2010. The fractional calcium current through fast ligand-gated TRPM channels. *Acta Physiologica* **198:** 227.

Duncan LM, Deeds J, Cronin FE, Donovan M, Sober AJ, Kauffman M, McCarthy JJ. 2001. Melastatin expression and prognosis in cutaneous malignant melanoma. *J Clin Oncol* **19:** 568–576.

Duncan LM, Deeds J, Hunter J, Shao J, Holmgren LM, Woolf EA, Tepper RI, Shyjan AW. 1998. Down-regulation of the novel gene melastatin correlates with potential for melanoma metastasis. *Cancer Res* **58:** 1515–1520.

Egan TM, Khakh BS. 2004. Contribution of calcium ions to P2X channel responses. *J Neurosci* **24:** 3413–3420.

Elizondo MR, Arduini BL, Paulsen J, MacDonald EL, Sabel JL, Henion PD, Cornell RA, Parichy DM. 2005. Defective skeletogenesis with kidney stone formation in dwarf zebrafish mutant for trpm7. *Curr Biol* **15:** 667–671.

Ellinor PT, Yang J, Sather WA, Zhang JF, Tsien RW. 1995. Ca^{2+} channel selectivity at a single locus for high-affinity Ca^{2+} interactions. *Neuron* **15:** 1121–1132.

Everaerts W, Nilius B, Owsianik G. 2010. The vanilloid transient receptor potential channel Trpv4: From structure to disease. *Prog Biophys Mol Biol* **103:** 2–17.

Everaerts W, Sepulveda MR, Gevaert T, Roskams T, Nilius B, De Ridder D. 2009. Where is TRPV1 expressed in the bladder, do we see the real channel? *Naunyn Schmiedebergs Arch Pharmacol* **379:** 421–425.

Felder CC, Briley EM, Axelrod J, Simpson JT, Mackie K, Devane WA. 1993. Anandamide, an endogenous cannabimimetic eicosanoid, binds to the cloned human cannabinoid receptor and stimulates receptor-mediated signal transduction. *Proc Natl Acad Sci U S A* **90:** 7656–7660.

Foggensteiner L, Bevan AP, Thomas R, Coleman N, Boulter C, Bradley J, Ibraghimov-Beskrovnaya O, Klinger K, Sandford R. 2000. Cellular and subcellular distribution of polycystin-2, the protein product of the PKD2 gene. *J Am Soc Nephrol* **11:** 814–827.

Fonfria E, Murdock PR, Cusdin FS, Benham CD, Kelsell RE, McNulty S. 2006. Tissue distribution profiles of the human TRPM cation channel family. *J Recept Signal Transduct Res* **26:** 159–178.

Freichel M, Suh SH, Pfeifer A, Schweig U, Trost C, Weissgerber P, Biel M, Philipp S, Freise D, Droogmans G, et al. 2001. Lack of an endothelial store-operated Ca^{2+} current impairs agonist-dependent vasorelaxation in TRP4-/- mice. *Nat Cell Biol* **3:** 121–127.

Freichel M, Vennekens R, Olausson J, Stolz S, Philipp SE, Weissgerber P, Flockerzi V. 2005. Functional role of TRPC proteins in native systems: implications from knockout and knock-down studies. *J Physiol* **567:** 59–66.

Gailly P, Colson-Van Schoor M. 2001. Involvement of trp-2 protein in store-operated influx of calcium in fibroblasts. *Cell Calcium* **30:** 157–165.

Galione A, Churchill GC. 2002. Interactions between calcium release pathways: multiple messengers and multiple stores. *Cell Calcium* **32:** 343–354.

Garcia-Martinez C, Morenilla-Palao C, Planells-Cases R, Merino JM, Ferrer-Montiel A. 2000. Identification of an aspartic residue in the P-loop of the vanilloid receptor that modulates pore properties. *J Biol Chem* **275:** 32552–32558.

Gaudet R. 2008a. A primer on ankyrin repeat function in TRP channels and beyond. *Mol Biosyst* **4:** 372–379.

Gaudet R. 2008b. TRP channels entering the structural era. *J Physiol* **586:** 3565–3575.

Gaudet R. 2009. Divide and Conquer: High Resolution Structural Information on TRP Channel Fragments. *J Gen Physiol* **133:** 231–237.

Geng L, Segal Y, Pavlova A, Barros EJ, Lohning C, Lu W, Nigam SK, Frischauf AM, Reeders ST, Zhou J. 1997. Distribution and developmentally regulated expression of murine polycystin. *Am J Physiol* **272:** F451–F459.

Gibson HE, Edwards JG, Page RS, Van Hook MJ, Kauer JA. 2008. TRPV1 channels mediate long-term depression at synapses on hippocampal interneurons. *Neuron* **57:** 746–759.

Goel M, Sinkins WG, Schilling WP. 2002. Selective association of TRPC channel subunits in rat brain synaptosomes. *J Biol Chem* **277:** 48303–48310.

Gokce O, Runne H, Kuhn A, Luthi-Carter R. 2009. Short-term striatal gene expression responses to brain-derived neurotrophic factor are dependent on MEK and ERK activation. *PLoS ONE* **4:** e5292.

Gopinath P, Wan E, Holdcroft A, Facer P, Davis JB, Smith GD, Bountra C, Anand P. 2005. Increased capsaicin receptor TRPV1 in skin nerve fibres and related vanilloid receptors TRPV3 and TRPV4 in keratinocytes in human breast pain. *BMC Womens Health* **5:** 2.

Gordon-Shaag A, Zagotta WN, Gordon SE. 2008. Mechanism of Ca^{2+}-dependent desensitization of TRP channels. *Channels* **2:** 125–129.

Gottlieb P, Folgering J, Maroto R, Raso A, Wood TG, Kurosky A, Bowman C, Bichet D, Patel A, Sachs F, et al. 2008. Revisiting TRPC1 and TRPC6 mechanosensitivity. *Pflugers Arch* **455:** 1097–1103.

Graham S, Ding M, Sours-Brothers S, Yorio T, Ma JX, Ma R. 2007. Downregulation of TRPC6 protein expression by high glucose, a possible mechanism for the impaired Ca^{2+} signaling in glomerular mesangial cells in diabetes. *Am J Physiol Renal Physiol* **293:** F1381–F1390.

Greka A, Navarro B, Oancea E, Duggan A, Clapham DE. 2003. TRPC5 is a regulator of hippocampal neurite length and growth cone morphology. *Nat Neurosci* **6:** 837–845.

Grimm C, Jors S, Saldanha SA, Obukhov AG, Pan B, Oshima K, Cuajungco MP, Chase P, Hodder P, Heller S. 2010. Small Molecule Activators of TRPML3. *Chem Biol* **17:** 135–148.

Grimm C, Kraft R, Sauerbruch S, Schultz G, Harteneck C. 2003. Molecular and functional characterization of the melastatin-related cation channel TRPM3. *J Biol Chem* **278:** 21493–21501.

Grimm C, Kraft R, Schultz G, Harteneck C. 2005. Activation of the melastatin-related cation channel TRPM3 by

D-erythro-sphingosine [corrected]. *Mol Pharmacol* **67**: 798–805.

Guo L, Schreiber TH, Weremowicz S, Morton CC, Lee C, Zhou J. 2000. Identification and characterization of a novel polycystin family member, polycystin-L2, in mouse and human: sequence, expression, alternative splicing, and chromosomal localization. *Genomics* **64**: 241–251.

Hagenacker T, Czeschik JC, Schafers M, Busselberg D. 2010. Sensitization of voltage activated calcium channel currents for capsaicin in nociceptive neurons by tumor-necrosis-factor-alpha. *Brain Res Bull* **81**: 157–163.

Hanaoka K, Qian F, Boletta A, Bhunia AK, Piontek K, Tsiokas L, Sukhatme VP, Guggino WB, Germino GG. 2000. Co-assembly of polycystin-1 and -2 produces unique cation-permeable currents. *Nature* **408**: 990–994.

Hassock SR, Zhu MX, Trost C, Flockerzi V, Authi KS. 2002. Expression and role of TRPC proteins in human platelets: evidence that TRPC6 forms the store-independent calcium entry channel. *Blood* **100**: 2801–2811.

Hayes P, Meadows HJ, Gunthorpe MJ, Harries MH, Duckworth DM, Cairns W, Harrison DC, Clarke CE, Ellington K, Prinjha RK, et al. 2000. Cloning and functional expression of a human orthologue of rat vanilloid receptor-1. *Pain* **88**: 205–215.

Hoenderop JG, Nilius B, Bindels RJ. 2005. Calcium absorption across epithelia. *Physiol Rev* **85**: 373–422.

Hoenderop JG, Voets T, Hoefs S, Weidema F, Prenen J, Nilius B, Bindels RJ. 2003. Homo- and heterotetrameric architecture of the epithelial Ca^{2+} channels TRPV5 and TRPV6. *Embo J* **22**: 776–785.

Hofmann T, Schaefer M, Schultz G, Gudermann T. 2000. Transient receptor potential channels as molecular substrates of receptor-mediated cation entry. *J Mol Med* **78**: 14–25.

Hofmann T, Schaefer M, Schultz G, Gudermann T. 2002. Subunit composition of mammalian transient receptor potential channels in living cells. *Proc Natl Acad Sci U S A* **99**: 7461–7466.

Huang AL, Chen X, Hoon MA, Chandrashekar J, Guo W, Trankner D, Ryba NJ, Zuker CS. 2006. The cells and logic for mammalian sour taste detection. *Nature* **442**: 934–938.

Huang WC, Young JS, Glitsch MD. 2007. Changes in TRPC channel expression during postnatal development of cerebellar neurons. *Cell Calcium* **42**: 1–10.

Hughes J, Ward CJ, Peral B, Aspinwall R, Clark K, San Millan JL, Gamble V, Harris PC. 1995. The polycystic kidney disease 1 (PKD1) gene encodes a novel protein with multiple cell recognition domains. *Nat Genet* **10**: 151–160.

Inoue R, Okada T, Onoue H, Hara Y, Shimizu S, Naitoh S, Ito Y, Mori Y. 2001. The transient receptor potential protein homologue TRP6 is the essential component of vascular alpha(1)-adrenoceptor-activated $Ca(2+)$-permeable cation channel. *Circ Res* **88**: 325–332.

Ishimaru Y, Inada H, Kubota M, Zhuang H, Tominaga M, Matsunami H. 2006. Transient receptor potential family members PKD1L3 and PKD2L1 form a candidate sour taste receptor. *Proc Natl Acad Sci U S A* **103**: 12569–12574.

Iwata Y, Katanosaka Y, Arai Y, Komamura K, Miyatake K, Shigekawa M. 2003. A novel mechanism of myocyte degeneration involving the Ca^{2+}-permeable growth factor-regulated channel. *J Cell Biol* **161**: 957–967.

Iwata Y, Katanosaka Y, Arai Y, Shigekawa M, Wakabayashi S. 2009. Dominant-negative inhibition of Ca^{2+} influx via TRPV2 ameliorates muscular dystrophy in animal models. *Hum Mol Genet* **18**: 824–834.

Ju YK, Allen DG. 2007. Store-operated Ca^{2+} entry and TRPC expression; possible roles in cardiac pacemaker tissue. *Heart Lung Circ* **16**: 349–355.

Jung S, Muhle A, Schaefer M, Strotmann R, Schultz G, Plant TD. 2003. Lanthanides potentiate TRPC5 currents by an action at extracellular sites close to the pore mouth. *J Biol Chem* **278**: 3562–3571.

Jung S, Strotmann R, Schultz G, Plant TD. 2002. TRPC6 is a candidate channel involved in receptor-stimulated cation currents in A7r5 smooth muscle cells. *Am J Physiol Cell Physiol* **282**: C347–C359.

Kamouchi M, Philipp S, Flockerzi V, Wissenbach U, Mamin A, Raeymaekers L, Eggermont J, Droogmans G, Nilius B. 1999. Properties of heterologously expressed hTRP3 channels in bovine pulmonary artery endothelial cells. *J Physiol (Lond)* **518**: 345–358.

Kanzaki M, Zhang YQ, Mashima H, Li L, Shibata H, Kojima I. 1999. Translocation of a calcium-permeable cation channel induced by insulin-like growth factor-I. *Nat Cell Biol* **1**: 165–170.

Karacsonyi C, Miguel AS, Puertollano R. 2007. Mucolipin-2 localizes to the Arf6-associated pathway and regulates recycling of GPI-APs. *Traffic* **8**: 1404–1414.

Karashima Y, Damann N, Prenen J, Talavera K, Segal A, Voets T, Nilius B. 2007. Bimodal action of menthol on the transient receptor potential channel TRPA1. *J Neurosci* **27**: 9874–9884.

Karashima Y, Prenen J, Talavera K, Janssens A, Voets T, Nilius B. 2010. Agonist-Induced Changes in $Ca(2+)$ Permeation through the Nociceptor Cation Channel TRPA1. *Biophys J* **98**: 773–783.

Karashima Y, Talavera K, Everaerts W, Janssens A, Kwan KY, Vennekens R, Nilius B, Voets T. 2009. TRPA1 acts as a cold sensor in vitro and in vivo. *Proc Natl Acad Sci U S A* **106**: 1273–1278.

Keller SA, Jones JM, Boyle A, Barrow LL, Killen PD, Green DG, Kapousta NV, Hitchcock PF, Swank RT, Meisler MH. 1994. Kidney and retinal defects (Krd), a transgene-induced mutation with a deletion of mouse chromosome 19 that includes the Pax2 locus. *Genomics* **23**: 309–320.

Kim BJ, So I, Kim KW. 2006. The relationship of TRP channels to the pacemaker activity of interstitial cells of Cajal in the gastrointestinal tract. *J Smooth Muscle Res* **42**: 1–7.

Kim EY, Alvarez-Baron CP, Dryer SE. 2009a. TRPC3 and TRPC6 associate with BK_{Ca} channels: Role in BK_{Ca} trafficking to the surface of cultured podocytes. *Mol Pharmacol* **75**: 466–477.

Kim HJ, Li Q, Tjon-Kon-Sang S, So I, Kiselyov K, Muallem S. 2007. Gain-of-function mutation in TRPML3 causes the mouse Varitint-Waddler phenotype. *J Biol Chem* **282**: 36138–36142.

Kim HJ, Li Q, Tjon-Kon-Sang S, So I, Kiselyov K, Soyombo AA, Muallem S. 2008. A novel mode of TRPML3 regulation by extracytosolic pH absent in the varitint-waddler phenotype. *Embo J* **27**: 1197–1205.

Kim HJ, Soyombo AA, Tjon-Kon-Sang S, So I, Muallem S. 2009b. The Ca(2+) channel TRPML3 regulates membrane trafficking and autophagy. *Traffic* **10**: 1157–1167.

Kim MS, Zeng W, Yuan JP, Shin DM, Worley PF, Muallem S. 2009c. Native Store-operated Ca^{2+} Influx Requires the Channel Function of Orai1 and TRPC1. *J Biol Chem* **284**: 9733–9741.

Kim SJ, Kim YS, Yuan JP, Petralia RS, Worley PF, Linden DJ. 2003. Activation of the TRPC1 cation channel by metabotropic glutamate receptor mGluR1. *Nature* **426**: 285–291.

Kiselyov K, Patterson RL. 2009. The integrative function of TRPC channels. *Front Biosci* **14**: 45–58.

Koike C, Obara T, Uriu Y, Numata T, Sanuki R, Miyata K, Koyasu T, Ueno S, Funabiki K, Tani A, et al. 2010. TRPM1 is a component of the retinal ON bipolar cell transduction channel in the mGluR6 cascade. *Proc Natl Acad Sci U S A* **107**: 332–337.

Kokrashvili Z, Rodriguez D, Yevshayeva V, Zhou H, Margolskee RF, Mosinger B. 2009. Release of endogenous opioids from duodenal enteroendocrine cells requires Trpm5. *Gastroenterology* **137**: 598–606, 606 e1–2.

Koulen P, Cai Y, Geng L, Maeda Y, Nishimura S, Witzgall R, Ehrlich BE, Somlo S. 2002. Polycystin-2 is an intracellular calcium release channel. *Nat Cell Biol* **4**: 191–197.

Kraft R, Harteneck C. 2005. The mammalian melastatin-related transient receptor potential cation channels: an overview. *Pflugers Arch* **451**: 204–211.

Krapivinsky G, Mochida S, Krapivinsky L, Cibulsky SM, Clapham DE. 2006. The TRPM7 Ion Channel Functions in Cholinergic Synaptic Vesicles and Affects Transmitter Release. *Neuron* **52**: 485–496.

Kuhn FJ, Heiner I, Luckhoff A. 2005. TRPM2: a calcium influx pathway regulated by oxidative stress and the novel second messenger ADP-ribose. *Pflugers Arch* **451**: 212–219.

Kwan HY, Huang Y, Yao X. 2007. TRP channels in endothelial function and dysfunction. *Biochim Biophys Acta* **1772**: 907–914.

Kwan HY, Shen B, Ma X, Kwok YC, Huang Y, Man YB, Yu S, Yao X. 2009. TRPC1 Associates With BK_{Ca} Channel to Form a Signal Complex in Vascular Smooth Muscle Cells. *Circ Res* **104**: 670–U207.

Kwan KY, Allchorne AJ, Vollrath MA, Christensen AP, Zhang DS, Woolf CJ, Corey DP. 2006. TRPA1 contributes to cold, mechanical, and chemical nociception but is not essential for hair-cell transduction. *Neuron* **50**: 277–289.

Lange I, Yamamoto S, Partida-Sanchez S, Mori Y, Fleig A, Penner R. 2009. TRPM2 functions as a lysosomal Ca^{2+}-release channel in beta cells. *Sci Signal* **2**: ra23.

LaPlante JM, Falardeau J, Sun M, Kanazirska M, Brown EM, Slaugenhaupt SA, Vassilev PM. 2002. Identification and characterization of the single channel function of human mucolipin-1 implicated in mucolipidosis type IV, a disorder affecting the lysosomal pathway. *FEBS Lett* **532**: 183–187.

Larsson KP, Peltonen HM, Bart G, Louhivuori LM, Penttonen A, Antikainen M, Kukkonen JP, Akerman KE. 2005. Orexin-A-induced Ca^{2+} entry: evidence for involvement of trpc channels and protein kinase C regulation. *J Biol Chem* **280**: 1771–1781.

Latorre R, Zaelzer C, Brauchi S. 2009. Structure-functional intimacies of transient receptor potential channels. *Q Rev Biophys* **42**: 201–246.

Lee KP, Jun JY, Chang IY, Suh SH, So I, Kim KW. 2005. TRPC4 is an essential component of the nonselective cation channel activated by muscarinic stimulation in mouse visceral smooth muscle cells. *Mol Cells* **20**: 435–441.

Lee N, Chen J, Sun L, Wu S, Gray KR, Rich A, Huang M, Lin JH, Feder JN, Janovitz EB, et al. 2003. Expression and characterization of human transient receptor potential melastatin 3 (hTRPM3). *J Biol Chem* **278**: 20890–20897.

Lemonnier L, Trebak M, Putney JW Jr. 2008. Complex regulation of the TRPC3, 6 and 7 channel subfamily by diacylglycerol and phosphatidylinositol-4,5-bisphosphate. *Cell Calcium* **43**: 506–514.

Li HB, Mao RR, Zhang JC, Yang Y, Cao J, Xu L. 2008. Anti-stress Effect of TRPV1 Channel on Synaptic Plasticity and Spatial Memory. *Biol Psychiatry* **64**: 286–292.

Li M, Du J, Jiang J, Ratzan W, Su LT, Runnels LW, Yue L. 2007. Molecular determinants of Mg^{2+} and Ca^{2+} permeability and pH sensitivity in TRPM6 and TRPM7. *J Biol Chem* **282**: 25817–25830.

Li X, Luo Y, Starremans PG, McNamara CA, Pei Y, Zhou J. 2005. Polycystin-1 and polycystin-2 regulate the cell cycle through the helix-loop-helix inhibitor Id2. *Nat Cell Biol* **7**: 1102–1112.

Li Y, Calfa G, Inoue T, Amaral MD, Pozzo-Miller L. 2010. Activity-Dependent release of endogenous BDNF from mossy fibers evokes a TRPC3 current and Ca^{2+} elevations in Ca3 pyramidal neurons. *J Neurophysiol* **103**: 2846–2856.

Li Z, Sergouniotis PI, Michaelides M, Mackay DS, Wright GA, Devery S, Moore AT, Holder GE, Robson AG, Webster AR. 2009. Recessive mutations of the gene TRPM1 abrogate ON bipolar cell function and cause complete congenital stationary night blindness in humans. *Am J Hum Genet* **85**: 711–719.

Liao Y, Erxleben C, Yildirim E, Abramowitz J, Armstrong DL, Birnbaumer L. 2007. Orai proteins interact with TRPC channels and confer responsiveness to store depletion. *Proc Natl Acad Sci U S A* **104**: 4682–4687.

Liedtke W, Friedman JM. 2003. Abnormal osmotic regulation in trpv4-/- mice. *Proc Natl Acad Sci U S A* **100**: 13698–13703.

Link TM, Park U, Vonakis BM, Raben DM, Soloski MJ, Caterina MJ. 2010. TRPV2 has a pivotal role in macrophage particle binding and phagocytosis. *Nat Immunol* **11**: 232–239.

Liu CH, Wang T, Postma M, Obukhov AG, Montell C, Hardie RC. 2007a. In vivo identification and manipulation of the Ca^{2+} selectivity filter in the Drosophila transient receptor potential channel. *J Neurosci* **27**: 604–615.

Liu X, Cheng KT, Bandyopadhyay BC, Pani B, Dietrich A, Paria BC, Swaim WD, Beech D, Yildirim E, Singh BB, et al. 2007b. Attenuation of store-operated Ca^{2+} current impairs salivary gland fluid secretion in TRPC1(-/-) mice. *Proc Natl Acad Sci U S A* **104**: 17542–17547.

Liu X, Singh BB, Ambudkar IS. 2003. TRPC1 is required for functional store-operated Ca^{2+} channels. Role of acidic amino acid residues in the S5-S6 region. *J Biol Chem* **278:** 11337–11343.

Lu M, Branstrom R, Berglund E, Hoog A, Bjorklund P, Westin G, Larsson C, Farnebo LO, Forsberg L. 2010. Expression and association of TRPC subtypes with Orai1 and STIM1 in human parathyroid. *J Mol Endocrinol* **44:** 285–294.

Luo Y, Vassilev PM, Li X, Kawanabe Y, Zhou J. 2003. Native polycystin 2 functions as a plasma membrane Ca^{2+}-permeable cation channel in renal epithelia. *Mol Cell Biol* **23:** 2600–2607.

Luzio JP, Bright NA, Pryor PR. 2007a. The role of calcium and other ions in sorting and delivery in the late endocytic pathway. *Biochem Soc Trans* **35:** 1088–1091.

Luzio JP, Pryor PR, Bright NA. 2007b. Lysosomes: fusion and function. *Nat Rev Mol Cell Biol* **8:** 622–632.

Macpherson LJ, Dubin AE, Evans MJ, Marr F, Schultz PG, Cravatt BF, Patapoutian A. 2007. Noxious compounds activate TRPA1 ion channels through covalent modification of cysteines. *Nature* **445:** 541–545.

Macpherson LJ, Geierstanger BH, Viswanath V, Bandell M, Eid SR, Hwang S, Patapoutian A. 2005. The pungency of garlic: activation of TRPA1 and TRPV1 in response to allicin. *Curr Biol* **15:** 929–934.

Mahieu F, Owsianik G, Verbert L, Janssens A, De Smedt H, Nilius B, Voets T. 2007. TRPM8-independent menthol-induced Ca^{2+} release from endoplasmic reticulum and Golgi. *J Biol Chem* **282:** 3325–3336.

Mandadi S, Sokabe T, Shibasaki K, Katanosaka K, Mizuno A, Moqrich A, Patapoutian A, Fukumi-Tominaga T, Mizumura K, Tominaga M. 2009. TRPV3 in keratinocytes transmits temperature information to sensory neurons via ATP. *Pflugers Arch* **458:** 1093–1102.

Maroto R, Raso A, Wood TG, Kurosky A, Martinac B, Hamill OP. 2005. TRPC1 forms the stretch-activated cation channel in vertebrate cells. *Nat Cell Biol* **7:** 179–185.

Martina JA, Lelouvier B, Puertollano R. 2009. The calcium channel mucolipin-3 is a novel regulator of trafficking along the endosomal pathway. *Traffic* **10:** 1143–1156.

Masuyama R, Vriens J, Voets T, Karashima Y, Owsianik G, Vennekens R, Lieben L, Torrekens S, Moermans K, Vanden Bosch A, et al. 2008. TRPV4-mediated calcium influx regulates terminal differentiation of osteoclasts. *Cell Metab* **8:** 257–265.

Matsuura H, Sokabe T, Kohno K, Tominaga M, Kadowaki T. 2009. Evolutionary conservation and changes in insect TRP channels. *BMC Evol Biol* **9:** 228.

McKemy DD, Neuhausser WM, Julius D. 2002. Identification of a cold receptor reveals a general role for TRP channels in thermosensation. *Nature* **416:** 52–58.

McNeill MS, Paulsen J, Bonde G, Burnight E, Hsu MY, Cornell RA. 2007. Cell death of melanophores in zebrafish trpm7 mutant embryos depends on melanin synthesis. *J Invest Dermatol* **127:** 2020–2030.

McNulty S, Fonfria E. 2005. The role of TRPM channels in cell death. *Pflugers Arch* **451:** 235–242.

Mederos y Schnitzler M, Waring J, Gudermann T, Chubanov V. 2008. Evolutionary determinants of divergent calcium selectivity of TRPM channels. *Faseb J* **22:** 1540–1551.

Miller AJ, Du J, Rowan S, Hershey CL, Widlund HR, Fisher DE. 2004. Transcriptional regulation of the melanoma prognostic marker melastatin (TRPM1) by MITF in melanocytes and melanoma. *Cancer Res* **64:** 509–516.

Mizoguchi F, Mizuno A, Hayata T, Nakashima K, Heller S, Ushida T, Sokabe M, Miyasaka N, Suzuki M, Ezura Y, et al. 2008. Transient receptor potential vanilloid 4 deficiency suppresses unloading-induced bone loss. *J Cell Physiol* **216:** 47–53.

Monteilh-Zoller MK, Hermosura MC, Nadler MJ, Scharenberg AM, Penner R, Fleig A. 2003. TRPM7 provides an ion channel mechanism for cellular entry of trace metal ions. *J Gen Physiol* **121:** 49–60.

Montell C. 2003. The venerable inveterate invertebrate TRP channels. *Cell Calcium* **33:** 409–417.

Montell C. 2005. The TRP superfamily of cation channels. *Sci STKE* **2005:** re3.

Montell C, Birnbaumer L, Flockerzi V. 2002. The TRP channels, a remarkably functional family. *Cell* **108:** 595–598.

Montell C, Rubin GM. 1989. Molecular characterization of the Drosophila trp locus: a putative integral membrane protein required for phototransduction. *Neuron* **2:** 1313–1323.

Moqrich A, Hwang SW, Earley TJ, Petrus MJ, Murray AN, Spencer KS, Andahazy M, Story GM, Patapoutian A. 2005. Impaired thermosensation in mice lacking TRPV3, a heat and camphor sensor in the skin. *Science* **307:** 1468–1472.

Mori Y, Wakamori M, Miyakawa T, Hermosura M, Hara Y, Nishida M, Hirose K, Mizushima A, Kurosaki M, Mori E, et al. 2002. Transient receptor potential 1 regulates capacitative Ca(2+) entry and Ca(2+) release from endoplasmic reticulum in B lymphocytes. *J Exp Med* **195:** 673–681.

Muraki K, Iwata Y, Katanosaka Y, Ito T, Ohya S, Shigekawa M, Imaizumi Y. 2003. TRPV2 is a component of osmotically sensitive cation channels in murine aortic myocytes. *Circ Res* **93:** 829–838.

Muramatsu S, Wakabayashi M, Ohno T, Amano K, Ooishi R, Sugahara T, Shiojiri S, Tashiro K, Suzuki Y, Nishimura R, et al. 2007. Functional gene screening system identified TRPV4 as a regulator of chondrogenic differentiation. *J Biol Chem* **282:** 32158–32167.

Nagata K, Zheng L, Madathany T, Castiglioni AJ, Bartles JR, Garcia-Anoveros J. 2008. The varitint-waddler (Va) deafness mutation in TRPML3 generates constitutive, inward rectifying currents and causes cell degeneration. *Proc Natl Acad Sci U S A* **105:** 353–358.

Nathan JD, Patel AA, McVey DC, Thomas JE, Prpic V, Vigna SR, Liddle RA. 2001. Capsaicin vanilloid receptor-1 mediates substance P release in experimental pancreatitis. *Am J Physiol Gastrointest Liver Physiol* **281:** G1322–1328.

Newby LJ, Streets AJ, Zhao Y, Harris PC, Ward CJ, Ong AC. 2002. Identification, characterization, and localization of a novel kidney polycystin-1-polycystin-2 complex. *J Biol Chem* **277:** 20763–20773.

Ng LC, McCormack MD, Airey JA, Singer CA, Keller PS, Shen XM, Hume JR. 2009. TRPC1 and STIM1 mediate capacitative Ca^{2+} entry in mouse pulmonary arterial smooth muscle cells. *J Physiol* **587:** 2429–2442.

Nijenhuis T, Hoenderop JG, Nilius B, Bindels RJ. 2003a. (Patho)physiological implications of the novel epithelial Ca^{2+} channels TRPV5 and TRPV6. *Pflugers Arch* **446:** 401–409.

Nijenhuis T, Hoenderop JG, van der Kemp AW, Bindels RJ. 2003b. Localization and regulation of the epithelial Ca^{2+} channel TRPV6 in the kidney. *J Am Soc Nephrol* **14:** 2731–2740.

Nilius B, Droogmans G. 2001. Ion channels and their functional role in vascular endothelium. *Physiol Rev* **81:** 1415–1459.

Nilius B, Mahieu F, Prenen J, Janssens A, Owsianik G, Vennekens R, Voets T. 2006. The Ca^{2+}-activated cation channel TRPM4 is regulated by phosphatidylinositol 4,5-biphosphate. *EMBO Journal* **25:** 467–478.

Nilius B, Owsianik G, Voets T. 2008. Transient receptor potential channels meet phosphoinositides. *Embo J* **27:** 2809–2816.

Nilius B, Owsianik G, Voets T, Peters JA. 2007. Transient Receptor Potential Channels in Disease. *Physiol Rev* **87:** 165–217.

Nilius B, Prenen J, Hoenderop JG, Vennekens R, Hoefs S, Weidema AF, Droogmans G, Bindels RJ. 2002. Fast and slow inactivation kinetics of the Ca^{2+} channels ECaC1 and ECaC2 (TRPV5 and TRPV6). Role of the intracellular loop located between transmembrane segments 2 and 3. *J Biol Chem* **277:** 30852–30858.

Nilius B, Prenen J, Janssens A, Owsianik G, Wang C, Zhu MX, Voets T. 2005a. The selectivity filter of the cation channel TRPM4. *J Biol Chem* **280:** 22899–22906.

Nilius B, Prenen J, Tang J, Wang C, Owsianik G, Janssens A, Voets T, Zhu MX. 2005b. Regulation of the Ca^{2+} Sensitivity of the Nonselective Cation Channel TRPM4. *J Biol Chem* **280:** 6423–6433.

Nilius B, Vennekens R. 2006. From cardiac cation channels to the molecular dissection of the transient receptor potential channel TRPM4. *Pflugers Arch Europ J Physiol* **453:** 313–321.

Nilius B, Vennekens R, Prenen J, Hoenderop JG, Bindels RJ, Droogmans G. 2000. Whole-cell and single channel monovalent cation currents through the novel rabbit epithelial Ca^{2+} channel ECaC. *J Physiol* **527:** 239–248.

Nilius B, Vennekens R, Prenen J, Hoenderop JG, Droogmans G, Bindels RJ. 2001. The single pore residue Asp542 determines Ca^{2+} permeation and Mg^{2+} block of the epithelial Ca^{2+} channel. *J Biol Chem* **276:** 1020–1025.

Nilius B, Vriens J, Prenen J, Droogmans G, Voets T. 2004. TRPV4 calcium entry channel: a paradigm for gating diversity. *Am J Physiol Cell Physiol* **286:** C195–C205.

Nilius B, Watanabe H, Vriens J. 2003. The TRPV4 channel: structure-function relationship and promiscuous gating behaviour. *Pflugers Arch* **446:** 298–303.

Nomura H, Turco AE, Pei Y, Kalaydjieva L, Schiavello T, Weremowicz S, Ji W, Morton CC, Meisler M, Reeders ST, et al. 1998. Identification of PKDL, a novel polycystic kidney disease 2-like gene whose murine homologue is deleted in mice with kidney and retinal defects. *J Biol Chem* **273:** 25967–25973.

Oancea E, Vriens J, Brauchi S, Jun J, Splawski I, Clapham DE. 2009. TRPM1 forms ion channels associated with melanin content in melanocytes. *Sci Signal* **2:** ra21.

Obukhov AG, Nowycky MC. 2004. TRPC5 activation kinetics are modulated by the scaffolding protein ezrin/radixin/moesin-binding phosphoprotein-50 (EBP50). *J Cell Physiol* **201:** 227–235.

Obukhov AG, Nowycky MC. 2008. TRPC5 channels undergo changes in gating properties during the activation-deactivation cycle. *J Cell Physiol* **216:** 162–171.

Olah Z, Szabo T, Karai L, Hough C, Fields RD, Caudle RM, Blumberg PM, Iadarola MJ. 2001. Ligand-induced dynamic membrane changes and cell deletion conferred by vanilloid receptor 1. *J Biol Chem* **276:** 11021–11030.

Owsianik G, D'Hoedt D, Voets T, Nilius B. 2006a. Structure-function relationship of the TRP channel superfamily. *Rev Physiol Biochem Pharmacol* **156:** 61–90.

Owsianik G, Talavera K, Voets T, Nilius B. 2006b. Permeation and selectivity of trp channels. *Annu Rev Physiol* **68:** 685–717.

Pani B, Ong HL, Liu X, Rauser K, Ambudkar IS, Singh BB. 2008. Lipid rafts determine clustering of STIM1 in endoplasmic reticulum-plasma membrane junctions and regulation of store-operated Ca^{2+} entry (SOCE). *J Biol Chem* **283:** 17333–17340.

Patapoutian A, Peier AM, Story GM, Viswanath V. 2003. ThermoTRP channels and beyond: mechanisms of temperature sensation. *Nat Rev Neurosci* **4:** 529–539.

Patel S, Marchant JS, Brailoiu E. 2010. Two-pore channels: Regulation by NAADP and customized roles in triggering calcium signals. *Cell Calcium* **47:** 480–490.

Pedersen SF, Owsianik G, Nilius B. 2005. TRP channels: an overview. *Cell Calcium* **38:** 233–252.

Peier AM, Moqrich A, Hergarden AC, Reeve AJ, Andersson DA, Story GM, Earley TJ, Dragoni I, McIntyre P, Bevan S, et al. 2002a. A TRP channel that senses cold stimuli and menthol. *Cell* **108:** 705–715.

Peier AM, Reeve AJ, Andersson DA, Moqrich A, Earley TJ, Hergarden AC, Story GM, Colley S, Hogenesch JB, McIntyre P, et al. 2002b. A heat-sensitive TRP channel expressed in keratinocytes. *Science* **296:** 2046–2049.

Peng L, Popescu DC, Wang N, Shieh BH. 2007. Anchoring TRP to the INAD macromolecular complex requires the last 14 residues in its carboxyl terminus. *J Neurochem* **104:** 1526–1535.

Pennekamp P, Karcher C, Fischer A, Schweickert A, Skryabin B, Horst J, Blum M, Dworniczak B. 2002. The ion channel polycystin-2 is required for left-right axis determination in mice. *Curr Biol* **12:** 938–943.

Perraud AL, Fleig A, Dunn CA, Bagley LA, Launay P, Schmitz C, Stokes AJ, Zhu Q, Bessman MJ, Penner R, et al. 2001. ADP-ribose gating of the calcium-permeable LTRPC2 channel revealed by Nudix motif homology. *Nature* **411:** 595–599.

Perraud AL, Schmitz C, Scharenberg AM. 2003. TRPM2 Ca^{2+} permeable cation channels: from gene to biological function. *Cell Calcium* **33:** 519–531.

Peters JH, McDougall SJ, Fawley JA, Smith SM, Andresen MC. 2010. Primary Afferent Activation of Thermosensitive TRPV1 Triggers Asynchronous Glutamate Release at Central Neurons. *Neuron* **65:** 657–669.

Philipp S, Strauss B, Hirnet D, Wissenbach U, Mery L, Flockerzi V, Hoth M. 2003. TRPC3 mediates T-cell receptor-dependent calcium entry in human T-lymphocytes. *J Biol Chem* **278:** 26629–26638.

Philipp S, Trost C, Warnat J, Rautmann J, Himmerkus N, Schroth G, Kretz O, Nastainczyk W, Cavalie A, Hoth M, et al. 2000. TRP4 (CCE1) protein is part of native calcium release-activated Ca^{2+}-like channels in adrenal cells. *J Biol Chem* **275:** 23965–23972.

Prasad P, Yanagihara AA, Small-Howard AL, Turner H, Stokes AJ. 2008. Secretogranin III directs secretory vesicle biogenesis in mast cells in a manner dependent upon interaction with chromogranin A. *J Immunol* **181:** 5024–5034.

Pryor PR, Mullock BM, Bright NA, Gray SR, Luzio JP. 2000. The role of intraorganellar Ca(2+) in late endosome-lysosome heterotypic fusion and in the reformation of lysosomes from hybrid organelles. *J Cell Biol* **149:** 1053–1062.

Puertollano R, Kiselyov K. 2009. TRPMLs: in sickness and in health. *Am J Physiol Renal Physiol* **296:** F1245–F1254.

Ramsey IS, Delling M, Clapham DE. 2006. An introduction to TRP channels. *Annu Rev Physiol* **68:** 619–647.

Raychowdhury MK, Gonzalez-Perrett S, Montalbetti N, Timpanaro GA, Chasan B, Goldmann WH, Stahl S, Cooney A, Goldin E, Cantiello HF. 2004. Molecular pathophysiology of mucolipidosis type IV: pH dysregulation of the mucolipin-1 cation channel. *Hum Mol Genet* **13:** 617–627.

Redondo PC, Jardin I, Lopez JJ, Salido GM, Rosado JA. 2008. Intracellular Ca(2+) store depletion induces the formation of macromolecular complexes involving hTRPC1, hTRPC6, the type II IP(3) receptor and SERCA3 in human platelets. *Biochim Biophys Acta* **1783:** 1163–1176.

Reiser J, Polu KR, Moller CC, Kenlan P, Altintas MM, Wei C, Faul C, Herbert S, Villegas I, Avila-Casado C, et al. 2005. TRPC6 is a glomerular slit diaphragm-associated channel required for normal renal function. *Nat Genet* **37:** 739–744.

Riccio A, Li Y, Moon J, Kim KS, Smith KS, Rudolph U, Gapon S, Yao GL, Tsvetkov E, Rodig SJ, et al. 2009. Essential role for TRPC5 in amygdala function and fear-related behavior. *Cell* **137:** 761–772.

Riccio A, Mattei C, Kelsell RE, Medhurst AD, Calver AR, Randall AD, Davis JB, Benham CD, Pangalos MN. 2002a. Cloning and functional expression of human short TRP7, a candidate protein for store-operated Ca^{2+} influx. *J Biol Chem* **277:** 12302–12309.

Riccio A, Medhurst AD, Mattei C, Kelsell RE, Calver AR, Randall AD, Benham CD, Pangalos MN. 2002b. mRNA distribution analysis of human TRPC family in CNS and peripheral tissues. *Brain Res Mol Brain Res* **109:** 95–104.

Rohacs T. 2007. Regulation of TRP channels by PIP2. *Pflugers Arch Europ J Physiol* **453:** 753–762.

Rohacs T. 2009. Phosphoinositide regulation of non-canonical transient receptor potential channels. *Cell Calcium* **45:** 554–565.

Rohacs T, Nilius B. 2007. Regulation of transient receptor potential (trp) channels by phosphoinositides. *Pflugers Arch* **455:** 157–168.

Rundle DR, Gorbsky G, Tsiokas L. 2004. PKD2 interacts and co-localizes with mDia1 to mitotic spindles of dividing cells: role of mDia1 IN PKD2 localization to mitotic spindles. *J Biol Chem* **279:** 29728–29739.

Runnels LW, Yue L, Clapham DE. 2001. TRP-PLIK, a bifunctional protein with kinase and ion channel activities. *Science* **291:** 1043–1047.

Saito M, Hanson PI, Schlesinger P. 2007. Luminal chloride-dependent activation of endosome calcium channels: patch clamp study of enlarged endosomes. *J Biol Chem* **282:** 27327–27333.

Samie MA, Grimm C, Evans JA, Curcio-Morelli C, Heller S, Slaugenhaupt SA, Cuajungco MP. 2009. The tissue-specific expression of TRPML2 (MCOLN-2) gene is influenced by the presence of TRPML1. *Pflugers Arch* **459:** 79–91.

Sammels E, Devogelaere B, Mekahli D, Bultynck G, Missiaen L, Parys JB, Cai Y, Somlo S, De Smedt H. 2010. Polycystin-2 activation by inositol 1,4,5-trisphosphate-induced Ca^{2+} release requires its direct association with the inositol 1,4,5-trisphosphate receptor in a signaling microdomain. *J Biol Chem* **285:** 18794–18805.

Satoh S, Tanaka H, Ueda Y, Oyama J, Sugano M, Sumimoto H, Mori Y, Makino N. 2007. Transient receptor potential (TRP) protein 7 acts as a G protein-activated Ca^{2+} channel mediating angiotensin II-induced myocardial apoptosis. *Mol Cell Biochem* **294:** 205–215.

Schilling WP, Goel M. 2004. Mammalian TRPC channel subunit assembly. *Novartis Found Symp* **258:** 18–30.

Schlingmann KP, Sassen MC, Weber S, Pechmann U, Kusch K, Pelken L, Lotan D, Syrrou M, Prebble JJ, Cole DE, et al. 2005. Novel TRPM6 mutations in 21 families with primary hypomagnesemia and secondary hypocalcemia. *J Am Soc Nephrol* **16:** 3061–3069.

Schlingmann KP, Weber S, Peters M, Niemann Nejsum L, Vitzthum H, Klingel K, Kratz M, Haddad E, Ristoff E, Dinour D, et al. 2002. Hypomagnesemia with secondary hypocalcemia is caused by mutations in TRPM6, a new member of the TRPM gene family. *Nat Genet* **31:** 166–170.

Sharif-Naeini R, Folgering JH, Bichet D, Duprat F, Lauritzen I, Arhatte M, Jodar M, Dedman A, Chatelain FC, Schulte U, et al. 2009. Polycystin-1 and -2 dosage regulates pressure sensing. *Cell* **139:** 587–596.

Shen Y, Heimel JA, Kamermans M, Peachey NS, Gregg RG, Nawy S. 2009. A transient receptor potential-like channel mediates synaptic transmission in rod bipolar cells. *J Neurosci* **29:** 6088–6093.

Shibasaki K, Suzuki M, Mizuno A, Tominaga M. 2007. Effects of body temperature on neural activity in the hippocampus: regulation of resting membrane potentials by transient receptor potential vanilloid 4. *J Neurosci* **27:** 1566–1575.

Shim S, Yuan JP, Kim JY, Zeng W, Huang G, Milshteyn A, Kern D, Muallem S, Ming G-L, Worley PF. 2009. Peptidyl-Prolyl Isomerase FKBP52 Controls Chemotropic Guidance of Neuronal Growth Cones via Regulation of TRPC1 Channel Opening. *Neuron* **64:** 471–483.

Shimizu T, Janssens A, Voets T, Nilius B. 2009. Regulation of the murine TRPP3 channel by voltage, pH, and changes in cell volume. *Pflugers Arch* **457:** 795–807.

Smith GD, Gunthorpe MJ, Kelsell RE, Hayes PD, Reilly P, Facer P, Wright JE, Jerman JC, Walhin JP, Ooi L, et al. 2002. TRPV3 is a temperature-sensitive vanilloid receptor-like protein. *Nature* **418:** 186–190.

Song Y, Dayalu R, Matthews SA, Scharenberg AM. 2006. TRPML cation channels regulate the specialized lysosomal compartment of vertebrate B-lymphocytes. *Eur J Cell Biol* **85:** 1253–1264.

Sours-Brothers S, Ding M, Graham S, Ma R. 2009. Interaction between TRPC1/TRPC4 assembly and STIM1 contributes to store-operated Ca^{2+} entry in mesangial cells. *Exp Biol Med* **234:** 673–682.

Story GM, Peier AM, Reeve AJ, Eid SR, Mosbacher J, Hricik TR, Earley TJ, Hergarden AC, Andersson DA, Hwang SW, et al. 2003. ANKTM1, a TRP-like channel expressed in nociceptive neurons, is activated by cold temperatures. *Cell* **112:** 819–829.

Strubing C, Krapivinsky G, Krapivinsky L, Clapham DE. 2001. TRPC1 and TRPC5 form a novel cation channel in mammalian brain. *Neuron* **29:** 645–655.

Strubing C, Krapivinsky G, Krapivinsky L, Clapham DE. 2003. Formation of novel TRPC channels by complex subunit interactions in embryonic brain. *J Biol Chem* **278:** 39014–39019.

Su Z, Zhou X, Loukin SH, Haynes WJ, Saimi Y, Kung C. 2009. The use of yeast to understand TRP-channel mechanosensitivity. *Pflugers Arch* **458:** 861–867.

Sun M, Goldin E, Stahl S, Falardeau JL, Kennedy JC, Acierno JS, Bove C, Kaneski CR, Nagle J, Bromley MC, et al. 2000. Mucolipidosis type IV is caused by mutations in a gene encoding a novel transient receptor potential channel. *Hum Mol Genet* **9:** 2471–2478.

Suzuki M, Mizuno A, Kodaira K, Imai M. 2003. Impaired pressure sensation in mice lacking TRPV4. *J Biol Chem* **278:** 22664–22668.

Tai Y, Feng S, Ge R, Du W, Zhang X, He Z, Wang Y. 2008. TRPC6 channels promote dendritic growth via the CaMKIV-CREB pathway. *J Cell Sci* **121:** 2301–2307.

Talavera K, Gees M, Karashima Y, Meseguer VM, Vanoirbeek JA, Damann N, Everaerts W, Benoit M, Janssens A, Vennekens R, et al. 2009. Nicotine activates the chemosensory cation channel TRPA1. *Nat Neurosci* **12:** 1293–1299.

Talavera K, Nilius B, Voets T. 2008. Neuronal TRP channels: thermometers, pathfinders and life-savers. *Trends Neurosci* **31:** 287–295.

Thebault S, Lemonnier L, Bidaux G, Flourakis M, Bavencoffe A, Gordienko D, Roudbaraki M, Delcourt P, Panchin Y, Shuba Y, et al. 2005. Novel role of cold/menthol-sensitive transient receptor potential melastatine family member 8 (TRPM8) in the activation of store-operated channels in LNCaP human prostate cancer epithelial cells. *J Biol Chem* **280:** 39423–39435.

Thompson EG, Schaheen L, Dang H, Fares H. 2007. Lysosomal trafficking functions of mucolipin-1 in murine macrophages. *BMC Cell Biol* **8:** 54.

Thyagarajan B, Benn B, Lukacs V, Christakos S, Rohacs T. 2009. Phospholpase C mediated regulation of TRPV6 channels: implications in active intestinal Ca^{2+} transport. *Mol Pharmacology* **75:** 608–616.

Thyagarajan B, Lukacs V, Rohacs T. 2008. Hydrolysis of phosphatidylinositol 4,5-bisphosphate mediates calcium induced inactivation of TRPV6 channels. *J Biol Chem* **283:** 14980–14987.

Tiruppathi C, Ahmmed GU, Vogel SM, Malik AB. 2006. Ca^{2+} signaling, TRP channels, and endothelial permeability. *Microcirculation* **13:** 693–708.

Tong Q, Chu X, Cheung JY, Conrad K, Stahl R, Barber DL, Mignery G, Miller BA. 2004. Erythropoietin-modulated calcium influx through TRPC2 is mediated by phospholipase Cgamma and IP_3R. *Am J Physiol Cell Physiol* **287:** C1667–C1678.

Topala CN, Groenestege WT, Thebault S, van den Berg D, Nilius B, Hoenderop JG, Bindels RJ. 2007. Molecular determinants of permeation through the cation channel TRPM6. *Cell Calcium* **41:** 513–523.

Tsavaler L, Shapero MH, Morkowski S, Laus R. 2001. Trp-p8, a novel prostate-specific gene, is up-regulated in prostate cancer and other malignancies and shares high homology with transient receptor potential calcium channel proteins. *Cancer Res* **61:** 3760–3769.

Tsiokas L, Kim E, Arnould T, Sukhatme VP, Walz G. 1997. Homo- and heterodimeric interactions between the gene products of PKD1 and PKD2. *Proc Natl Acad Sci U S A* **94:** 6965–6970.

Tsvilovskyy VV, Zholos AV, Aberle T, Philipp SE, Dietrich A, Zhu MX, Birnbaumer L, Freichel M, Flockerzi V. 2009. Deletion of TRPC4 and TRPC6 in mice impairs smooth muscle contraction and intestinal motility in vivo. *Gastroenterology* **137:** 1415–1424.

Turner H, Fleig A, Stokes A, Kinet JP, Penner R. 2003. Discrimination of intracellular calcium store subcompartments using TRPV1 (transient receptor potential channel, vanilloid subfamily member 1) release channel activity. *Biochem J* **371:** 341–350.

Ullrich ND, Voets T, Prenen J, Vennekens R, Talavera K, Droogmans G, Nilius B. 2005. Comparison of functional properties of the Ca^{2+}-activated cation channels TRPM4 and TRPM5 from mice. *Cell Calcium* **37:** 267–278.

Vaca L. 2010. SOCIC: the store-operated calcium influx complex. *Cell Calcium* **47:** 199–209.

van Genderen MM, Bijveld MM, Claassen YB, Florijn RJ, Pearring JN, Meire FM, McCall MA, Riemslag FC, Gregg RG, Bergen AA, et al. 2009. Mutations in TRPM1 are a common cause of complete congenital stationary night blindness. *Am J Hum Genet* **85:** 730–736.

Venkatachalam K, Hofmann T, Montell C. 2006. Lysosomal Localization of TRPML3 Depends on TRPML2 and the Mucolipidosis-associated Protein TRPML1. *J Biol Chem* **281:** 17517–17527.

Venkatachalam K, Long AA, Elsaesser R, Nikolaeva D, Broadie K, Montell C. 2008. Motor deficit in a Drosophila model of mucolipidosis type IV due to defective clearance of apoptotic cells. *Cell* **135:** 838–851.

Venkatachalam K, van Rossum DB, Patterson RL, Ma HT, Gill DL. 2002. The cellular and molecular basis of store-operated calcium entry. *Nat Cell Biol* **4:** E263–E272.

Vennekens R, Hoenderop JG, Prenen J, Stuiver M, Willems PH, Droogmans G, Nilius B, Bindels RJ. 2000. Permeation and gating properties of the novel epithelial $Ca(2+)$ channel. *J Biol Chem* **275:** 3963–3969.

Vennekens R, Olausson J, Meissner M, Bloch W, Mathar I, Philipp SE, Schmitz F, Weissgerber P, Nilius B, Flockerzi

V, et al. 2007. Increased IgE-dependent mast cell activation and anaphylactic responses in mice lacking the calcium-activated nonselective cation channel TRPM4. *Nat Immunol* **8:** 312–320.

Vennekens R, Owsianik G, Nilius B. 2008. Vanilloid transient receptor potential cation channels: an overview. *Curr Pharm Des* **14:** 18–31.

Venugopal B, Browning MF, Curcio-Morelli C, Varro A, Michaud N, Nanthakumar N, Walkley SU, Pickel J, Slaugenhaupt SA. 2007. Neurologic, gastric, and opthalmologic pathologies in a murine model of mucolipidosis type IV. *Am J Hum Genet* **81:** 1070–1083.

Voets T, Droogmans G, Wissenbach U, Janssens A, Flockerzi V, Nilius B. 2004a. The principle of temperature-dependent gating in cold- and heat-sensitive TRP channels. *Nature* **430:** 748–754.

Voets T, Janssens A, Prenen J, Droogmans G, Nilius B. 2003. Mg^{2+}-dependent gating and strong inward rectification of the cation channel TRPV6. *J Gen Physiol* **121:** 245–260.

Voets T, Nilius B, Hoefs S, van der Kemp AW, Droogmans G, Bindels RJ, Hoenderop JG. 2004b. TRPM6 forms the Mg^{2+} influx channel involved in intestinal and renal Mg^{2+} absorption. *J Biol Chem* **279:** 19–25.

Voets T, Prenen J, Fleig A, Vennekens R, Watanabe H, Hoenderop JG, Bindels RJ, Droogmans G, Penner R, Nilius B. 2001. CaT1 and the calcium release-activated calcium channel manifest distinct pore properties. *J Biol Chem* **276:** 47767–47770.

Voets T, Prenen J, Vriens J, Watanabe H, Janssens A, Wissenbach U, Bodding M, Droogmans G, Nilius B. 2002. Molecular determinants of permeation through the cation channel TRPV4. *J Biol Chem* **277:** 33704–33710.

Vriens J, Appendino G, Nilius B. 2009. Pharmacology of vanilloid transient receptor potential cation channels. *Mol Pharmacol* **75:** 1262–1279.

Wang ZY, Wang P, Merriam FV, Bjorling DE. 2008. Lack of TRPV1 inhibits cystitis-induced increased mechanical sensitivity in mice. *Pain* **139:** 158–167.

Watanabe H, Davis JB, Smart D, Jerman JC, Smith GD, Hayes P, Vriens J, Cairns W, Wissenbach U, Prenen J, et al. 2002a. Activation of TRPV4 channels (hVRL-2/mTRP12) by phorbol derivatives. *J Biol Chem* **277:** 13569–13577.

Watanabe H, Vriens J, Prenen J, Droogmans G, Voets T, Nilius B. 2003. Anandamide and arachidonic acid use epoxyeicosatrienoic acids to activate TRPV4 channels. *Nature* **424:** 434–438.

Watanabe H, Vriens J, Suh SH, Benham CD, Droogmans G, Nilius B. 2002b. Heat-evoked activation of TRPV4 channels in a HEK293 cell expression system and in native mouse aorta endothelial cells. *J Biol Chem* **277:** 47044–47051.

Wedel BJ, Vazquez G, McKay RR, St JBG, Putney JW Jr. 2003. A calmodulin/inositol 1,4,5-trisphosphate (IP_3) receptor-binding region targets TRPC3 to the plasma membrane in a calmodulin/IP_3 receptor-independent process. *J Biol Chem* **278:** 25758–25765.

Wegierski T, Steffl D, Kopp C, Tauber R, Buchholz B, Nitschke R, Kuehn EW, Walz G, Kottgen M. 2009. TRPP2 channels regulate apoptosis through the Ca^{2+} concentration in the endoplasmic reticulum. *Embo J* **28:** 490–499.

Wen Z, Han L, Bamburg JR, Shim S, Ming GL, Zheng JQ. 2007. BMP gradients steer nerve growth cones by a balancing act of LIM kinase and Slingshot phosphatase on ADF/cofilin. *J Cell Biol* **178:** 107–119.

Wheeler GL, Brownlee C. 2008. Ca(2+) signalling in plants and green algae - changing channels. *Trends Plant Sci* **13:** 506–514.

Wisnoskey BJ, Sinkins WG, Schilling WP. 2003. Activation of vanilloid receptor type I in the endoplasmic reticulum fails to activate store-operated Ca^{2+} entry. *Biochem J* **372:** 517–528.

Wolf FI, Cittadini A. 1999. Magnesium in cell proliferation and differentiation. *Front Biosci* **4:** D607–D617.

Worley PF, Zeng W, Huang GN, Yuan JP, Kim JY, Lee MG, Muallem S. 2007. TRPC channels as STIM1-regulated store-operated channels. *Cell Calcium* **42:** 205–211.

Xi Q, Adebiyi A, Zhao G, Chapman KE, Waters CM, Hassid A, Jaggar JH. 2008. IP_3 constricts cerebral arteries via IP_3 receptor-mediated TRPC3 channel activation and independently of sarcoplasmic reticulum Ca^{2+} release. *Circ Res* **102:** 1118–1126.

Xiao Y, Lv X, Cao G, Bian G, Duan J, Ai J, Sun H, Li Q, Yang Q, Chen T, et al. 2010. Overexpression of Trpp5 contributes to cell proliferation and apoptosis probably through involving calcium homeostasis. *Mol Cell Biochem* **339:** 155–161.

Xu H, Delling M, Jun JC, Clapham DE. 2006. Oregano, thyme and clove-derived flavors and skin sensitizers activate specific TRP channels. *Nat Neurosci* **9:** 628–635.

Xu H, Ramsey IS, Kotecha SA, Moran MM, Chong JA, Lawson D, Ge P, Lilly J, Silos-Santiago I, Xie Y, et al. 2002. TRPV3 is a calcium-permeable temperature-sensitive cation channel. *Nature* **418:** 181–186.

Yeh BI, Kim YK, Jabbar W, Huang CL. 2005. Conformational changes of pore helix coupled to gating of TRPV5 by protons. *Embo J* **24:** 3224–3234.

Yildirim E, Birnbaumer L. 2007. TRPC2: molecular biology and functional importance. *Handb Exp Pharmacol*: 53–75.

Yoder BK, Hou X, Guay-Woodford LM. 2002. The polycystic kidney disease proteins, polycystin-1, polycystin-2, polaris, and cystin, are co-localized in renal cilia. *J Am Soc Nephrol* **13:** 2508–2516.

Yoshida T, Inoue R, Morii T, Takahashi N, Yamamoto S, Hara Y, Tominaga M, Shimizu S, Sato Y, Mori Y. 2006. Nitric oxide activates TRP channels by cysteine S-nitrosylation. *Nat Chem Biol* **2:** 596–607.

Yu PC, Gu SY, Bu JW, Du JL. 2010. TRPC1 Is Essential for In Vivo Angiogenesis in Zebrafish. *Circ Res* **106:** 1221–1232.

Yuan JP, Kim MS, Zeng W, Shin DM, Huang G, Worley PF, Muallem S. 2009. TRPC channels as STIM1-regulated SOCs. *Channels (Austin)* **3:** 221–225.

Yuan JP, Zeng W, Huang GN, Worley PF, Muallem S. 2007. STIM1 heteromultimerizes TRPC channels to determine their function as store-operated channels. *Nat Cell Biol* **9:** 636–645.

Zechel S, Werner S, von Bohlen Und Halbach O. 2007. Distribution of TRPC4 in developing and adult murine brain. *Cell Tissue Res* **328:** 651–656.

Zeilhofer HU, Kress M, Swandulla D. 1997. Fractional Ca^{2+} currents through capsaicin- and proton-activated ion channels in rat dorsal root ganglion neurones. *J Physiol* **503:** 67–78.

Zhang F, Jin S, Yi F, Li PL. 2009. TRP-ML1 functions as a lysosomal NAADP-sensitive Ca^{2+} release channel in coronary arterial myocytes. *J Cell Mol Med* **13:** 3174–3185.

Zhang F, Li PL. 2007. Reconstitution and characterization of a nicotinic acid adenine dinucleotide phosphate (NAADP)-sensitive Ca^{2+} release channel from liver lysosomes of rats. *J Biol Chem* **282:** 25259–25269.

Zhang L, Barritt GJ. 2004. Evidence that TRPM8 is an androgen-dependent Ca^{2+} channel required for the survival of prostate cancer cells. *Cancer Res* **64:** 8365–8373.

Zhang W, Chu X, Tong Q, Cheung JY, Conrad K, Masker K, Miller BA. 2003a. A novel TRPM2 isoform inhibits calcium influx and susceptibility to cell death. *J Biol Chem* **278:** 16222–16229.

Zhang W, Hirschler-Laszkiewicz I, Tong Q, Conrad K, Sun SC, Penn L, Barber DL, Stahl R, Carey DJ, Cheung JY, et al. 2006. TRPM2 is an ion channel that modulates hematopoietic cell death through activation of caspases and PARP cleavage. *Am J Physiol Cell Physiol* **290:** C1146–C1159.

Zhang Y, Hoon MA, Chandrashekar J, Mueller KL, Cook B, Wu D, Zuker CS, Ryba NJ. 2003b. Coding of sweet, bitter, and umami tastes: different receptor cells sharing similar signaling pathways. *Cell* **112:** 293–301.

Zhang Z, Reboreda A, Alonso A, Barker PA, Seguela P. 2010. TRPC channels underlie cholinergic plateau potentials and persistent activity in entorhinal cortex. *Hippocampus* (in press).

Zong X, Schieder M, Cuny H, Fenske S, Gruner C, Rotzer K, Griesbeck O, Harz H, Biel M, Wahl-Schott C. 2009. The two-pore channel TPCN2 mediates NAADP-dependent Ca(2+)-release from lysosomal stores. *Pflugers Arch* **458:** 891–899.

Zurborg S, Yurgionas B, Jira JA, Caspani O, Heppenstall PA. 2007. Direct activation of the ion channel TRPA1 by Ca(2+). *Nat Neurosci* **10:** 277–279.

Store-Operated Calcium Channels: New Perspectives on Mechanism and Function

Richard S. Lewis

Department of Molecular and Cellular Physiology, Stanford University School of Medicine, Stanford, California 94305

Correspondence: rslewis@stanford.edu

Store-operated calcium channels (SOCs) are a nearly ubiquitous Ca^{2+} entry pathway stimulated by numerous cell surface receptors via the reduction of Ca^{2+} concentration in the ER. The discovery of STIM proteins as ER Ca^{2+} sensors and Orai proteins as structural components of the Ca^{2+} release-activated Ca^{2+} (CRAC) channel, a prototypic SOC, opened the floodgates for exploring the molecular mechanism of this pathway and its functions. This review focuses on recent advances made possible by the use of STIM and Orai as molecular tools. I will describe our current understanding of the store-operated Ca^{2+} entry mechanism and its emerging roles in physiology and disease, areas of uncertainty in which further progress is needed, and recent findings that are opening new directions for research in this rapidly growing field.

Calcium is a remarkably multifunctional signaling ion, at the heart of diverse biological processes that direct the birth, development, function, and eventual death of cells, tissues, and organisms. Cells use a diverse array of transporters and channels to regulate intracellular Ca^{2+} concentration ($[Ca^{2+}]_i$). A major pathway present in nearly all metazoan cells is the store-operated Ca^{2+} channel (SOC). The defining feature of SOCs, the one that distinguishes them from all other classes of Ca^{2+} channels discussed in this volume, is their activation by the reduction of Ca^{2+} concentration in the lumen of the ER ($[Ca^{2+}]_{ER}$). Though they were originally described in nonexcitable cells (cells lacking the ability to fire action potentials), they are now known to be present in virtually all cells, including excitable cells like skeletal muscle and neurons (Parekh and Putney 2005).

Physiologically, SOCs are most commonly activated by stimuli that release Ca^{2+} from the ER. This generally involves receptors that activate phospholipase C to produce inositol 1,4,5-trisphosphate (IP_3) and activate IP_3 receptors in the ER, but can also result from Ca^{2+}-induced Ca^{2+} release through ER/SR ryanodine receptors. The notion that ER Ca^{2+} depletion can control Ca^{2+} entry was first formulated by Jim Putney 25 years ago as the "capacitative calcium entry" hypothesis based on observations that Ca^{2+} entry triggered by muscarinic agonists was more closely linked to the emptiness of the ER store than to IP_3 elevation or occupation of the muscarinic receptor (Putney 1986). The introduction of thapsigargin

(TG) (Thastrup et al. 1989), a sarcoendoplasmic reticulum Ca^{2+}-ATPase (SERCA) inhibitor that depletes ER Ca^{2+} independently of receptors and IP_3, and methods for measuring $[Ca^{2+}]_i$ in single cells (fura-2, indo-1, etc.) (Grynkiewicz et al. 1985) provided powerful tools that greatly accelerated progress in establishing store-operated Ca^{2+} entry (SOCE, as it was later renamed) as a ubiquitous Ca^{2+} entry pathway. TG-induced Ca^{2+} entry was soon shown to occur in dozens of cell types (Putney and Bird 1993; Parekh and Penner 1997), though nothing was known about the diversity of pathways that might be involved, let alone their molecular basis.

A major step forward was the identification of store-operated Ca^{2+} currents in mast cells and T cells. This achievement arose initially from attempts to identify Ca^{2+} conductances triggered by secretory agonists in mast cells and antigen receptors in T cells (Penner et al. 1988; Lewis and Cahalan 1989). In both cases, extremely small currents were detected and linked to large $[Ca^{2+}]_i$ rises, suggesting a high Ca^{2+} selectivity. In T cells, the current was shown to activate spontaneously during whole-cell recordings and in response to the T-cell mitogen phytohemagglutinin in perforated-patch recordings (Lewis and Cahalan 1989). Soon after, fluorescence-based studies showed that Ca^{2+} influx triggered through T cell mitogens shared several features with TG-induced influx, suggesting that T cell receptor agonists activated the store-dependent pathway (Mason et al. 1991; Sarkadi et al. 1991). These two paths of research converged when Hoth and Penner described a highly Ca^{2+}-selective current in mast cells that was activated in whole-cell recordings spontaneously (by Ca^{2+} chelators), by IP_3, and by ionomycin, and called it the Ca^{2+} release-activated Ca^{2+} (CRAC) current (Hoth and Penner 1992). In Jurkat T cells, Zweifach and Lewis showed that TG activated a similar current, which appeared to be identical to the mitogen-stimulated current described earlier, and made the first estimate of its characteristically tiny conductance (~20 femtosiemens, far too small to resolve single-channel currents) (Zweifach and Lewis 1993). These initial studies defined a membrane conductance that over the next decade would be described biophysically and pharmacologically in detail, providing a characteristic "fingerprint" culled from its ion selectivity, unitary conductance, and regulation by intra- and extracellular Ca^{2+} and pharmacological inhibitors (Parekh and Penner 1997; Prakriya et al. 2004).

Among several currents that were described as store-operated in different cells, the CRAC current emerged as the prototype because of its extensive characterization and the weight of evidence showing that it could be activated by ER Ca^{2+} depletion independently of surface receptors or changes in cytosolic $[Ca^{2+}]$. This included activation by intracellular Ca^{2+} chelators, SERCA inhibitors or ionomycin at constant intracellular $[Ca^{2+}]_i$, and by the Ca^{2+} chelator TPEN loaded into the ER (Prakriya et al. 2004; Parekh and Putney 2005). In fact, the CRAC channel is the only store-operated channel whose input–output relation is known. This relation, first examined by Hofer et al. (1998) and later quantified by Luik et al. (2008) using an ER-targeted cameleon protein, shows that I_{CRAC} is a highly nonlinear function of $[Ca^{2+}]_{ER}$, with a Hill coefficient of ~4 and a $K_{1/2}$ of 170 μM. Given a resting $[Ca^{2+}]_{ER}$ of ~400 μM, these results suggest that the ER must be depleted by ~25% before I_{CRAC} begins to activate significantly.

Over the two decades after Putney formalized the capacitative Ca^{2+} entry hypothesis, many mechanisms were proposed as the link between Ca^{2+} store depletion and SOCE. Among these, diffusible messengers released from the ER, insertion of CRAC channels into the plasma membrane, and conformational coupling of CRAC channels with IP_3 receptors in the ER were the most extensively studied, but in the absence of molecular substrates were difficult to establish (Prakriya et al. 2004; Parekh and Putney 2005). The CRAC channel fingerprint proved useful in ruling out a number of candidate genes for the CRAC channel itself, particularly members of the transient receptor potential (TRP) protein family, but its identity remained a mystery (Prakriya et al. 2004; Parekh and Putney 2005). The discoveries of the ER

Ca^{2+} sensor STIM1 in 2005 and the CRAC channel protein Orai1 a year later marked an unmistakable turning point in the field, as they provided the first and most essential molecular tools with which to dissect the SOCE mechanism. The history of these discoveries and the early revelations they afforded have been reviewed extensively (Cahalan et al. 2007; Wu et al. 2007; Fahrner et al. 2009; Putney 2009; Várnai et al. 2009; Hogan et al. 2010). In this review, I will summarize our current understanding of how Ca^{2+} store depletion leads to Ca^{2+} entry at a molecular level, and the role of STIM oligomerization and additional proteins in this process. I will also describe how these discoveries and the ensuing studies have increased awareness of SOCE roles in physiology and disease, and have created entirely new directions for research. Throughout I will emphasize work on STIM1 and Orai1 mainly because they have been the most extensively studied isoforms, but will discuss other STIM and Orai isoforms to highlight important functional differences. For more information on these other isoforms, the reader is referred to the reviews cited above.

STIM AND ORAI: A MOLECULAR BASIS FOR STORE-OPERATED CALCIUM ENTRY

RNAi screens for inhibitors of SOCE led to identification of the STIM protein family. Using small-scale RNAi screens Roos et al. (2005) identified STIM in *Drosophila* S2 cells and by homology the mammalian homologs STIM1 and STIM2, while Liou et al. (2005) identified STIM1 and STIM2 in a HeLa cell screen. Shortly thereafter, the Rao, Kinet, and Cahalan groups identified the Orai/CRACM protein family (Orai/CRACM1, 2, and 3) by genome-wide RNAi screens in S2 cells for inhibition of SOCE (Vig et al. 2006b; Zhang et al. 2006) or of NFAT translocation combined with linkage analysis of a family in which a loss of CRAC channel function led to severe combined immunodeficiency (SCID) (Feske et al. 2006).

STIM1 was quickly identified as a Ca^{2+} sensor for SOCE because it displayed an EF-hand-like sequence on the predicted luminal side, and neutralization of acidic residues to reduce Ca^{2+} binding by the EF hand elicited constitutive Ca^{2+} entry independent of store depletion, in effect mimicking the store-depleted state (Liou et al. 2005; Zhang et al. 2005). Of the three Orai isoforms in mammals, Orai1 was most closely connected to the CRAC channel because a point mutation (R91W) in Orai1 led to a complete loss of I_{CRAC} in human T cells (Feske et al. 2005, 2006), Orai1 is the predominant isoform in these cells, and its properties (fast Ca^{2+}-dependent inactivation, sensitivity to 2-aminoethyldiphenyl borate (2-APB), and predominant expression in human T cells) most closely match those of CRAC (Lis et al. 2007). Orai1 was shown to be the pore-forming subunit of the CRAC channel by changes in I_{CRAC} calcium selectivity caused by mutating highly conserved glutamates (particularly E106D) in the *trans*-membrane domains (Prakriya et al. 2006; Vig et al. 2006a; Yeromin et al. 2006).

The Molecular Choreography of SOCE

Depletion of ER Ca^{2+} stores activates SOCE through a distinctive rearrangement of STIM1 and Orai1 in the cell. At resting levels of $[Ca^{2+}]_{ER}$ (~400–500 μM), STIM1 and Orai1 are both mobile proteins that are relatively diffusely localized throughout the ER and plasma membranes, respectively (Fig. 1A). Within seconds, store depletion causes STIM1 to accumulate at the cell periphery, forming "puncta" in confocal or total internal reflection fluorescence (TIRF) microscopic images of cells expressing fluorescently labeled STIM1 (Liou et al. 2005; Zhang et al. 2005). These peripheral puncta correspond at the ultrastructural level to specialized regions of smooth ER positioned within ~10–20 nm of the plasma membrane, known as ER-PM junctions (Fig. 1B) (Wu et al. 2006; Lur et al. 2009; Orci et al. 2009). In Jurkat T cells expressing ER-targeted horseradish peroxidase, electron microscopic studies show that native ER–PM junctions cover <5% of the cell surface (Wu et al. 2006). Over roughly the same time period, Orai1 accumulates at PM sites directly opposite STIM1 (Luik et al. 2006; Xu et al. 2006), where STIM1 binds to Orai1 and opens the CRAC channel, generating highly localized

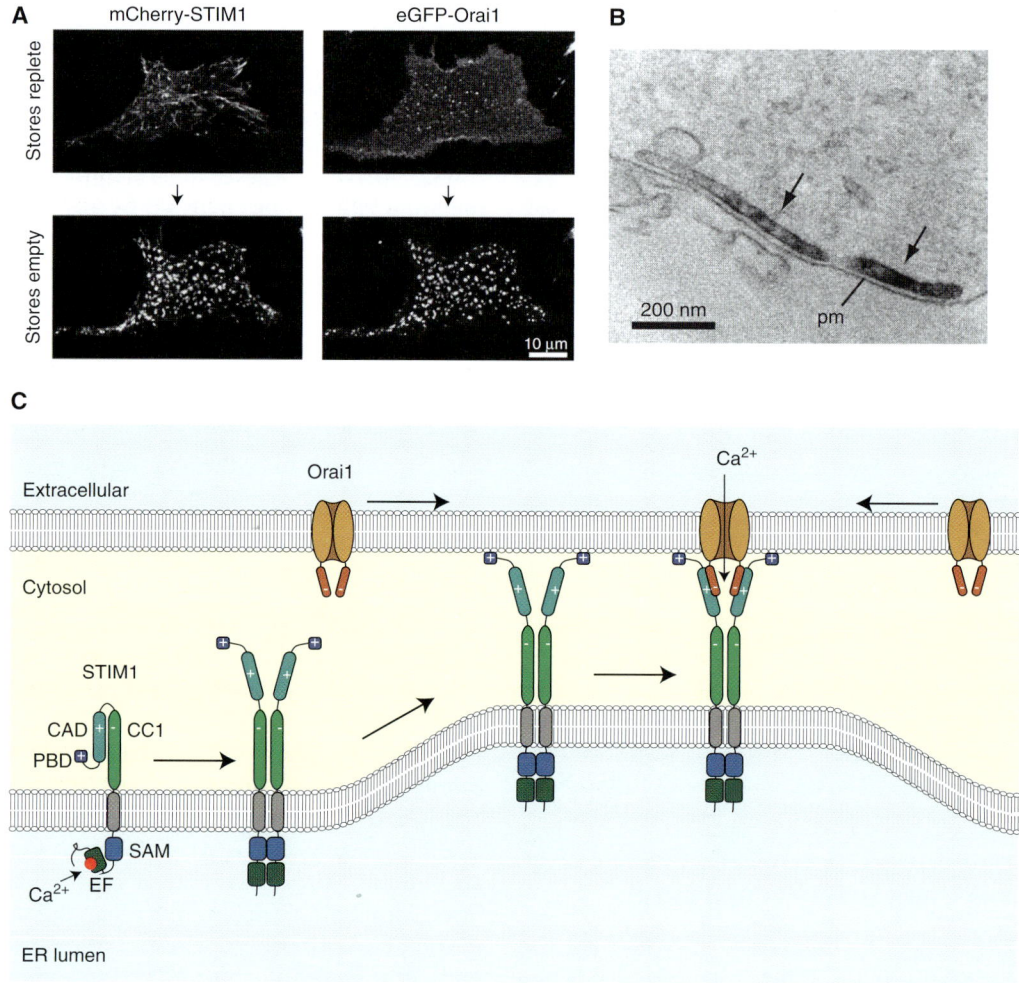

Figure 1. The molecular choreography of store-operated calcium entry. (A) In this resting HEK 293 cell, mCherry-STIM1 is localized throughout the ER and eGFP-Orai1 is dispersed throughout the PM of the cell footprint. Following store depletion, both proteins redistribute into colocalized puncta. (Panel A is from Park et al. (2009) and reprinted with permission from Elsevier © 2009.) (B) Electron micrograph showing accumulation of HRP-STIM1 in the ER (arrows) at ER-PM junctions in a Jurkat T cell after store depletion. (Panel B is modified from Luik et al. (2006) and reprinted with permission from The Rockefeller University Press © 2006.) (C) A current model for SOCE, divided into four phases from left to right. For simplicity and to emphasize the interactions of STIM1 and Orai1, full stoichiometries are not shown (STIM1 is likely to be a dimer at rest and after store depletion at least a tetramer, and 8 STIM1s probably interact with each CRAC channel for full activation). At far left, STIM1 is pictured in its resting state when Ca^{2+} stores are replete. The Ca^{2+}-bound EF hand interacts with the SAM domain, and electrostatic interactions between acidic residues in CC1 (−) and basic residues in CAD (+) prevent CAD from interacting with Orai. On store depletion, Ca^{2+} is released from the EF hand, allowing STIM1 to oligomerize (shown here schematically as a dimer) and assume an extended conformation that exposes CAD and the polybasic domain. Oligomers move to ER-PM junctions by diffusion in the ER membrane, and accumulate there through interaction of the polybasic domain with phosphoinositides in the PM. At the junction, CRAC channels diffusing in the PM bind to the STIM1 CAD via electrostatic interaction of a coiled-coil region of the Orai1 carboxyl terminus (−) with basic CAD residues (+). This interaction traps CRAC channels and combined with CAD interactions with the Orai1 amino terminus (not shown) opens them at the ER-PM junction. EF, canonical EF hand; SAM, sterile α-motif; CC1, coiled-coil 1; CAD, CRAC activation domain; PBD, polybasic domain.

Ca^{2+} "hot spots" visible by TIRF microscopy (Luik et al. 2006). Puncta formation is reversible; store refilling causes both proteins to slowly disperse from the junctions and revert to their diffuse localization (Liou et al. 2005; Várnai et al. 2007; Malli et al. 2008; Smyth et al. 2008).

The assembly of these elementary units of SOCE in response to store depletion is in most cells quite slow, occurring over seconds to tens of seconds, which largely reflects the time required for STIM1 and Orai1 to migrate to ER-PM junctions. The timing of these events may have physiologic consequence; for example, in T cells, the slow speed of CRAC complex assembly and disassembly may generate Ca^{2+} oscillations by creating lags between changes in [Ca^{2+}]$_{ER}$ and CRAC channel activity (Dolmetsch and Lewis 1994). However, activation of SOCE is not always so slow. Recent studies indicate that STIM1 in skeletal muscle is prelocalized to the triads where the sarcoplasmic reticulum comes to within 10 nm of the plasma membrane transverse tubule (Stiber et al. 2008). In these cells, SOCE can occur in <1 sec of Ca^{2+} release from the SR, and is thought to help replenish SR Ca^{2+} during high-frequency stimulation (Edwards et al. 2010). Thus, differential targeting of STIM1 and Orai1 in the resting state may adjust the activation kinetics of SOCE over orders of magnitude as needed in different cells.

Studies of the SOCE mechanism have reached a point where many of the underlying events can be understood in terms of the interactions of specific domains in STIM1 and Orai1 (Figs. 2 and 3). Below I discuss the functional roles of these domains in carrying out the major steps of the SOCE mechanism, diagrammed in Fig. 1C.

Oligomerization of STIM Is a Regulatory Switch for SOCE

The oligomerization of STIM is key to the regulation of its function. Early work on the isolated EF hand/sterile-α motif (EF-SAM) domain from STIM1 showed that removal of Ca^{2+} caused it to transition from a monomer to dimers and oligomers, prompting Ikura and colleagues to suggest that oligomerization of the EF-SAM domains may be the trigger for SOCE (Stathopulos et al. 2006). Supporting

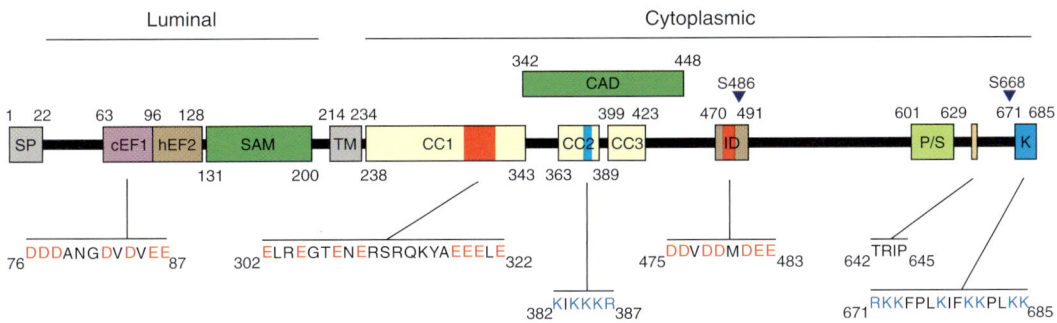

Figure 2. Functional organization of STIM1. The major functional domains of STIM1 are indicated by residue numbers relative to the translation initiation site. SP, signal peptide; cEF1, canonical EF hand; hEF2, hidden (noncanonical) EF hand; SAM, sterile α-motif; TM, transmembrane domain; CC1–3, coiled-coil domains 1–3; CAD, CRAC activation domain; ID, inactivation domain; P/S, proline-serine-rich domain; K, polybasic domain. CAD is highly similar to SOAR (aa 344–442) and Ccb9 (aa 339–444), not shown. Sequences are displayed for regions with established functions: the Ca^{2+} sensing domain in cEF1, the basic region of CC2 involved in autoinhibitory binding to the acidic region of CC1 and stimulatory binding to Orai1, the acidic residues in the ID of STIM1 required for Ca^{2+}-dependent inactivation, the EB1 binding sequence (TRIP) that links STIM1 to the tips of growing microtubules, and the carboxy-terminal polybasic domain that targets STIM1 to the ER-PM junction. Phosphorylation sites S486 and S668 help suppress STIM1 activity and SOCE during mitosis. Acidic and basic residues are shown in red and blue, respectively.

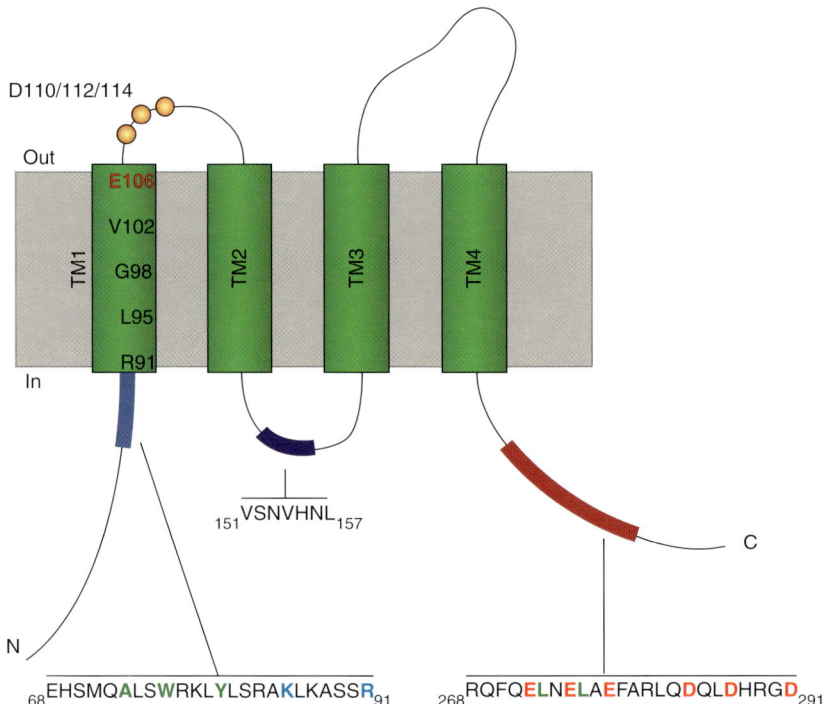

Figure 3. Functional organization of Orai1. This schematic layout of a single Orai1 subunit includes the four transmembrane domains (TM1–TM4), amino and carboxyl termini, and the connecting loops. Functionally significant residues and sequences are shown. The indicated residues in TM1 line the aqueous pore, E106 (red) forms the ion selectivity filter, and R91 inhibits channel function when substituted by bulky hydrophobic residues. D110/112/114 help determine sensitivity to block by lanthanides. In the amino terminus proximal to the PM, residues 68–91 form a Ca^{2+}/CaM binding domain, with residues essential for CaM binding and CDI indicated in bold green (A73, W76, Y80). This region also binds CAD and contains residues required for channel opening in bold blue (K85, R91). A domain in the intracellular loop is required for CDI, and a probable coiled-coil domain in the carboxyl terminus contains hydrophobic (bold green) and acidic (bold red) residues critical for binding STIM1 CAD.

this idea, its Ca^{2+} affinity of 0.2–0.6 mM (Stathopulos et al. 2006) is similar to the $[Ca^{2+}]_{ER}$ dependence of STIM1 translocation into puncta ($K_{1/2}$ ~200 μM) (Brandman et al. 2007; Luik et al. 2008). Importantly, increased Förster resonance energy transfer (FRET) between CFP- and YFP-labeled STIM1 after store depletion showed that STIM1 oligomerizes in cells after store depletion but prior to accumulation in puncta (Fig. 1C) (Liou et al. 2007; Malli et al. 2008; Muik et al. 2008). Subsequent NMR studies revealed a structural basis for this process. In the resting state, a single Ca^{2+} ion bound to the canonical EF hand exposes a hydrophobic cleft in STIM1 that interacts with hydrophobic anchor side chains from the SAM domain (Stathopulos et al. 2008). In this way, high $[Ca^{2+}]_{ER}$ prevents EF-SAM from engaging in intermolecular interactions, essentially providing a "brake" on spontaneous activation of SOCE. When Ca^{2+} is released from the EF hand (as during store depletion), hydrophobic regions are exposed, leading to oligomerization of the EF-SAM domains.

Although STIM1 and STIM2 are closely related in domain organization and show >65% sequence similarity, STIM2 appears to have a lower affinity for Ca^{2+}, such that it is partially Ca^{2+}-free and localized to ER-PM junctions in resting cells (Brandman et al. 2007).

The lower Ca^{2+} affinity of STIM2 may enable it to respond to smaller fluctuations of $[Ca^{2+}]_{ER}$, and by finely modulating SOCE act as a homeostatic regulator of cytosolic and ER $[Ca^{2+}]$ (Brandman et al. 2007). A reduced Ca^{2+} affinity may also enable STIM2 to play an active role in signaling, by sustaining a low level of Ca^{2+} entry through CRAC channels under conditions where stores have refilled enough to deactivate STIM1 (Oh-Hora et al. 2008). The full range of STIM2 functions is not yet understood.

The causal link between STIM1 oligomerization and SOCE was demonstrated by Luik et al. (2008), who replaced the EF-SAM domain of STIM1s with FRB or FKBP protein modules, and artificially cross-linked the modified STIM1 with a rapamycin analog. These conditions were sufficient to cause the modified STIM1s to accumulate in puncta and activate I_{CRAC} independently of store depletion. Thus, the reorganization of STIM1 and Orai1 into colocalized complexes appears to be an autonomous process that is held in check simply by the state of oligomerization of STIM1.

Interestingly, in addition to structural changes in the EF-SAM domain, interactions of the cytoplasmic domains are also critical for the oligomerization of full-length STIM1 following store depletion. Although a truncated STIM1 containing only the EF-SAM domain attached to the TM domain dimerizes in the ER on store depletion, these dimers appear to be unstable (Covington et al. 2010). Addition of the cytosolic coiled-coil 1 (CC1) and CRAC activation domain (CAD) (Fig. 2, and described below) is necessary for the protein to form stable higher-order oligomers after store depletion. Knowing at a structural level how these domains interact will be central to understanding how structural changes in the EF-SAM domains are transmitted across the ER membrane to trigger the conformational changes in the cytosolic region of STIM1 that lead to its activation.

The conformational changes in STIM1 that occur in response to oligomerization have recently begun to come to light. Several complementary studies suggest that oligomerization causes the STIM1 cytoplasmic domain to unfold and unmasks a hidden activation domain. Balla and colleagues first noted that an acidic sequence in the Orai1 carboxyl terminus, thought to interact with a basic region in coiled-coil 2 (CC2) of STIM1 (K384-386) (Calloway et al. 2009, 2010), also resembles a region in CC1 of STIM1 (Korzeniowski et al. 2010). They hypothesized that the CC2 basic residues may engage this CC1 acidic domain in the resting state, creating an autoinhibitory clamp when stores are full (Figs. 1C and 2). Neutralizing the CC1 acidic domain led to spontaneous activation, whereas neutralizing the CC2 basic domain prevented interaction with Orai1, supporting a model in which oligomerization destabilizes the intramolecular interaction, exposing CAD so that it can interact instead with the Orai1 carboxyl terminus. Intramolecular binding of the acidic and basic domains in STIM1 remains to be shown directly; however, concurrent studies of soluble STIM1 cytoplasmic fragments support the model. From a combination of FRET measurements and biophysical analysis, Muik et al. (2011) suggest that the cytoplasmic domain of STIM1 adopts a folded conformation at rest but elongates on binding to Orai1. Mutations of hydrophobic residues in CC1, CC2, and coiled-coil 3 (CC3) appeared to elongate the STIM1 fragment in the absence of Orai1, suggesting that coiled-coil formation in all three regions is involved in maintaining the resting folded conformation.

Accumulation of STIM at ER-PM Junctions

Two main mechanisms have been proposed to explain how STIM1 moves to ER-PM junctions. FRAP studies have suggested that STIM1 moves passively by diffusion with an estimated diffusion coefficient (D) of 0.1–0.15 $\mu m^2/s$ in the ER membrane (Liou et al. 2007; Covington et al. 2010). Thus, a simple but attractive model is that STIM1 oligomers diffuse passively to junctions where they are trapped by interaction with the PM or Orai (Fig. 1C and see below). However, STIM1 may also be transported within cells by an active microtubule-dependent mechanism. When growing microtubules cross the

ER, the microtubule tip-binding protein EB1 binds to STIM1 and pulls out a new ER tubule through a "tip attachment complex" mechanism (Grigoriev et al. 2008; Honnappa et al. 2009). Although this interaction is weakened by store depletion and does not appear necessary for puncta formation (Grigoriev et al. 2008), one could imagine that STIM1-laden ER tubules are propelled in this way toward the cell periphery by microtubules, and when they encounter local sites of elevated IP$_3$, local ER depletion releases the tubule near the PM with STIM1 primed to interact with the PM and establish or stabilize a junction. Interestingly, in two EM studies, store depletion caused the number of ER-PM junctions to increase, and overexpression of STIM1 further amplified this effect (Wu et al. 2006; Orci et al. 2009). Microtubules have also been observed aligned with cortical ER, perhaps as a memory trace of the ER's journey to the PM (Orci et al. 2009).

STIM1 movement to the cell periphery after store depletion may be regulated by the energy state of the cell. Mitochondrial depolarization impedes the movement of STIM1 oligomers to ER-PM junctions and correspondingly inhibits SOCE (Singaravelu et al. 2011). This effect appears to depend on mitofusin-2, a mitochondrial membrane protein that helps form short tethers between mitochondria and ER. One idea is that this intereference with SOCE may provide protection from Ca^{2+} overload in damaged cells with de-energized mitochondria (Singaravelu et al. 2011).

The formation of STIM1 puncta itself is a highly nonlinear function of $[Ca^{2+}]_{ER}$ ($K_{1/2}$ ~200 μM, Hill coefficient of 4-8), consistent with the idea that multiple STIM1s, each bearing a single Ca^{2+}-binding EF hand, must oligomerize to be able to accumulate at ER-PM junctions (Brandman et al. 2007; Luik et al. 2008). To form puncta in the nominal absence of Orai, STIM1 requires an intact carboxy-terminal polybasic domain, which has been proposed to form an amphipathic α-helix that interacts with phosphoinositides in the plasma membrane, in effect creating a diffusion trap (Huang et al. 2006; Liou et al. 2007; Park et al. 2009). In line with this hypothesis, conditions that deplete phosphatidylinositol 4,5-bisphosphate (PIP$_2$) and phosphatidylinositol 3,4,5-trisphosphate (PIP$_3$) together strongly inhibit STIM1 puncta formation, and STIM1/2 polybasic domains bind to liposomes containing these lipids in vitro (Ercan et al. 2009; Walsh et al. 2010a). Depletion of PIP$_2$ alone by recruitment of an active 5-phosphatase to the PM only partially inhibits STIM1 puncta formation (Korzeniowski et al. 2009; Walsh et al. 2010a). However, the phosphatidylinositol 3-kinase and phosphatidylinositol 4-kinase inhibitor LY294002 largely eliminates SOCE without affecting STIM1 puncta, revealing effects of phosphatidylinositol 4-phosphate (PI 4-P) more directly on Orai1 (Korzeniowski et al. 2009). Together, these studies reveal potentially complex functions of lipids in SOCE, which may involve effects of PIP$_2$ and PIP$_3$ on STIM1 localization and of PI 4-P on Orai1 opening, and further work will be needed to sort out these various roles.

Deletion of the C-terminal polybasic domain reveals a second mechanism for STIM1 accumulation at junctions, mediated by binding to Orai1 through the CAD domain. While STIM1 lacking the polybasic domain forms puncta if coexpressed with Orai1 (Park et al. 2009), I$_{CRAC}$ activation is delayed under these conditions (Li et al. 2007). These results suggest that the polybasic domain-PM interaction may act to accelerate I$_{CRAC}$ activation by creating clusters of STIM1 that more efficiently trap CRAC channels as they diffuse through ER-PM junctions.

Accumulation and Activation of CRAC Channels at ER-PM Junctions

Orai1 is highly mobile in the plasma membrane with a diffusion coefficient of ~0.1 μm^2/s (Park et al. 2009; Madl et al. 2010), and most evidence supports a diffusion-trap mechanism in which Orai1 randomly encounters STIM1 at the ER–PM junctions and is trapped there by binding directly to STIM1 (Fig. 1C and see below). A recent study suggests that binding of only 1–2 STIM1s is sufficient for trapping a

CRAC channel (Hoover and Lewis 2011). Lateral movement of Orai1 into the junctions is consistent with a visible loss of Orai1 in the spaces between STIM1 puncta after store depletion. However, store depletion may also contribute to SOCE by regulating Orai1 levels in the PM. In resting oocytes Orai1 appears to recycle constantly between the PM and an endosomal compartment, with $\sim$80% of the total Orai1 pool residing in the PM (Yu et al. 2010). Interestingly, the endosomal pool of Orai1 disappears on store depletion, prompting the suggestion that Orai1 binding to STIM1 at ER-PM junctions prevents its reuptake into endosomes. In platelets there is evidence for Orai1 insertion into the PM, but in this case in response to elevated $[Ca^{2+}]_i$ following activation of SOCE (Woodard et al. 2008). A powerful example of how regulation of Orai1 trafficking can modulate SOCE occurs during meiosis, when internalization of Orai1 helps inactivate SOCE (Yu et al. 2009).

Several groups have identified a minimal domain of STIM1 that binds to and powerfully activates CRAC channels, called CAD [CRAC Activation Domain; aa 342–448 (Park et al. 2009)], SOAR [STIM1-Orai Activating Region; aa 344–442 (Yuan et al. 2009)], or Ccb9 [aa 339–444 (Fig. 2) (Kawasaki et al. 2009)]. Park et al. showed that the binding is direct, based on GST pulldowns and copurification of Orai1 and CAD. Constitutive activation of I_{CRAC} by CAD binding to Orai1 offers the strongest evidence to date in favor of a conformational coupling mechanism involving direct protein-protein interactions, and argues against a diffusible messenger model (though it does not rule out the possibility that a diffusible messenger could modulate CRAC or provide a parallel activation pathway).

CAD and longer fragments of the STIM1 cytosolic domain appear to interact more strongly with the carboxyl terminus of Orai1 than to the amino terminus in vitro (Park et al. 2009; Zhou et al. 2010a). Although the carboxyl terminus is necessary and sufficient for Orai1 to colocalize with STIM1 in puncta, the amino terminus is neither necessary nor sufficient (Li et al. 2007; Muik et al. 2008).

The carboxyl terminus is weakly predicted to form a coiled-coil, and interactions with STIM1 and colocalization of Orai1 with STIM1 are inhibited by mutations predicted to destabilize the coiled-coil, such as L273S/D and L276D (Fig. 3) (Navarro-Borelly et al. 2008; Muik et al. 2008; Frischauf et al. 2009). The structural basis of the STIM-Orai carboxyl terminal interaction is not known, but may involve the formation of heteromeric STIM-Orai coiled-coils (Frischauf et al. 2009) or more restricted electrostatic interactions between acidic residues in the Orai1 carboxyl terminus (aa 272–291) and a basic region in the CC2 domain within CAD (K384–386) (Calloway et al. 2009, 2010; Korzeniowski et al. 2010).

Binding of STIM1 to the Orai1 amino terminus, though weaker than to the carboxyl terminus, does appear to be critical for channel activation. Because channels with a deletion of residues 1–73 from Orai1 still open (though at a reduced level), most attention has been focused on the highly conserved aa 73–91 (the approximate beginning of TM1) (Li et al. 2007). Deletion of aa 73–84 prevents CAD binding to the amino terminus and channel opening (Park et al. 2009). Mutations such as K85E (Lis et al. 2010) and R91W, the mutation found originally in T cells from SCID patients (Feske et al. 2006), completely prevent channel opening, apparently without preventing STIM1-Orai1 binding (Navarro-Borelly et al. 2008; Derler et al. 2009a), although R91W may interact with STIM1 somewhat less strongly than wild-type Orai1 (Fig. 3) (Muik et al. 2011). Interestingly, K85E does not prevent CAD binding to the amino terminus, suggesting that this residue is required for allosteric changes leading to gate opening. R91 may serve a similar function (Derler et al. 2009a). Interestingly, this membrane-proximal amino-terminal region overlaps with a calmodulin (CaM) binding site (aa 68–91), and several residues in this region (A73, Y80, W76) appear to be essential for CaM binding and Ca^{2+}-dependent inactivation (Fig. 3) (Mullins et al. 2009). Thus, the membrane-proximal region of the amino terminus is intimately tied to both activation and inactivation of CRAC channels and may provide important clues about the coupling of STIM1 binding to channel gating.

Questions of Stoichiometry: STIM1, Orai1, and the CRAC Channel Complex

A number of persistent questions apply to the native stoichiometries of STIM1 and Orai1 that are central to understanding the functional interactions of these proteins. In resting cells inactive STIM1 is often referred to as a dimer, based on coimmunoprecipitation of orthogonally labeled STIM1s and the high resting FRET between STIM1s labeled with CFP and YFP (Baba et al. 2006; Covington et al. 2010). However, the possibility that protein overexpression in these studies may have favored formation of dimers from endogenous monomers through mass action cannot be excluded. It is important to note that even if resting dimers are an artifact of overexpression, they do not seem to perturb either the inactivity of STIM1 in resting cells or the activation of SOCE following depletion. Unfortunately, measuring the resting STIM1 stoichiometry in vivo is not a simple task. Dimeric interactions may be too labile to survive solubilization, and measuring single-molecule photobleaching of GFP-tagged STIM1 is problematic because of its mobility and by the fact that fixation depletes Ca^{2+} stores, possibly promoting higher-order oligomerization (E. Covington and R. Lewis, unpubl.).

More importantly, the stoichiometry of "active" STIM1 following store depletion is unknown. The CAD fragment forms a tetramer in solution (Park et al. 2009), whereas the longer STIM1 cytosolic fragments generally appear as dimers (Ji et al. 2008; Penna et al. 2008; Muik et al. 2009; Yuan et al. 2009; Zhou et al. 2010a) that may need to unfold to expose the CAD and activate Orai1 (Korzeniowski et al. 2010; Muik et al. 2011). These findings raise the intriguing possibility that on store depletion, STIM1 undergoes a conformational change that results in formation of a tetramer through interactions among multiple CADs (Covington et al. 2010), an idea that remains to be tested.

The subunit stoichiometry of the CRAC channel has also been the subject of debate. A tetramer of Orai1 subunits was first proposed by Mignen et al. based mainly on the ability of a dominant-negative pore mutant (E106Q) to inhibit currents produced by dimeric or trimeric but not tetrameric Orai1 concatemers (Mignen et al. 2008). In cells transfected with monomeric Orai1-GFP and subsequently fixed, quantitation of photobleaching steps of single Orai1-GFP particles also supported a tetrameric stoichiometry (Ji et al. 2008), consistent with population brightness analysis of mobile Orai1-GFP after photobleaching in unfixed cells (Madl et al. 2010). On the other hand, biochemical analysis and single-molecule photobleaching of *Drosophila* Orai-GFP (dOrai-GFP) in the presence or absence of dSTIM cytosolic fragments suggested instead that Orai in resting cells is dimeric, and that STIM binding brings Orai dimers together to form the tetrameric functional channel (Penna et al. 2008). These studies raise fundamental questions about how the subunit stoichiometry of the CRAC channel is regulated, the role of STIM1 in this process, and whether changes in stoichiometry are reversible. New approaches, including structural studies will likely be necessary to resolve these questions.

Finally, recent studies have begun to show how the STIM1:Orai1 binding stoichiometry determines the extent of CRAC channel activation. Li et al. (2011) knocked variable numbers of STIM1 binding sites out of tetrameric Orai1 concatemers, revealing a graded relationship between the number of STIM1s bound and activity. Occupation of four sites was needed to evoke maximal activity, with each site thought to bind two STIM1s, leading to the conclusion that eight STIM1s must bind to fully activate the channel. Hoover and Lewis (2011) came to similar conclusions using native STIM1 and Orai1 coupled to mCherry and GFP. In cells with the same STIM1 expression, current magnitude increased with increasing amounts of Orai1, but dropped abruptly to near zero when the STIM1:Orai1 ratio in puncta fell below ∼2 (or 8/channel). Both studies may explain the enigma that overexpression of Orai1 alone often suppresses endogenous I_{CRAC} (Mercer et al. 2006; Peinelt et al. 2006; Soboloff et al. 2006; Li et al. 2007); by depleting a limited supply of STIM1, excess Orai1 may reduce the STIM1:Orai1 binding stoichiometry to a level that cannot open the channel. The two studies differ in the nonlinearity of CRAC channel activation, which appeared to be a much steeper function of the STIM1:Orai1 ratio with monomeric Orai1 than with tetrameric concatemers, raising the question of whether activation of SOCE at the single-channel level is "all-or-none" or graded with the number of STIM1s bound. A previous study using noise analysis concluded that store depletion, Ca^{2+}-dependent potentiation, and low doses of 2-APB through the stepwise recruitment of silent channels to a highly active state (Po = 0.8) rather than a smooth increase in open probability of all the channels (Prakriya and Lewis 2006). Thus, even if channel activation is a graded function of occupancy by STIM1, when a diffusing CRAC channel encounters a cluster of STIM1 at the ER-PM junction, it may quickly become fully occupied and activate in an "all-or-none" manner.

Are STIM and Orai the Only Players in SOCE?

Overexpression of STIM1 and Orai1 in HEK cells generates massive CRAC-like currents after store depletion, up to 100 times the size of endogenous currents (Mercer et al. 2006; Peinelt et al. 2006; Zhang et al. 2006). These findings initially suggested that these two proteins may be sufficient on their own to account for SOCE. The ability of cytosolic STIM1 fragments to activate CRAC channels outside of ER-PM junctions and independently of store depletion lends further support to this view. Zhou et al. tested the idea more rigorously using a mutant yeast system in which Orai1 was expressed and trapped in secretory vesicles (Zhou et al. 2010a). Because yeast does not express any homologs of Orai or STIM, it was considered unlikely they would express any possible cofactors of SOCE. Purified cytosolic fragments of STIM1 were able to bind to the vesicles and elicit Ca^{2+} flux, strongly suggesting that the two proteins can in fact function autonomously.

Although these findings support the idea that interactions of STIM1 and Orai1 alone are sufficient to elicit SOCE, they do not rule out important modulatory roles for additional proteins in vivo. An apt analogy can be made to the mechanism of exocytosis, in which purified SNARE proteins are capable of reconstituting membrane fusion to some degree by themselves in vitro but auxiliary proteins are needed to attain the high efficiency, speed, and control of secretion observed in living cells (Südhof and Rothman 2009). Orai1 in resting cells was first proposed to associate with another protein based on its exclusion from ER-PM junctions engineered by chemical crossbridging to have a reduced gap of ~9 nm (Várnai et al. 2007). One candidate for such a partner could be CRACR2A, a widely expressed cytosolic protein that copurifies with Orai1 in HeLa cells (Srikanth et al. 2010). CRACR2A binds to both Orai1 and STIM1 to form a ternary complex, and contains two EF hands that trigger its dissociation at elevated Ca^{2+}. Knockdown of CRACR2A in Jurkat cells reduced SOCE by about half, consistent with a role of enhancing the coupling between STIM1 and Orai1. Interestingly, CRACR2A's binding site on the Orai1 amino terminus overlaps with that of Ca^{2+}-CaM, whose binding is linked to Ca^{2+}-dependent inactivation (CDI) of I_{CRAC} (Mullins et al. 2009). Increased local $[Ca^{2+}]_i$ would be expected to cause CRACR2A to unbind, but it is not yet known whether this is necessary to allow Ca^{2+}-CaM to bind and trigger CDI.

STIM1 may also associate with proteins other than Orai. Although a truncated STIM1 containing only the luminal and transmembrane domains diffuses at the expected rate for a single transmembrane ER protein, addition of the cytosolic CC1 domain slows it by about twofold (Covington et al. 2010), and oligomerization in response to store depletion slows it by another factor of two (Liou et al. 2007; Covington et al. 2010). These changes in diffusion rate are too great to be easily explained by multimerization of the transmembrane domains, suggesting instead that STIM1 interacts with cytoskeletal, ER, or other cytoplasmic proteins through its CC1 domain (Covington et al. 2010). Slowing could reflect the steric hindrance of a long coiled-coil structure protruding up to ~15 nm from the ER membrane, weak binding of STIM1 along the length of microtubules (Smyth et al. 2007; Honnappa et al. 2009), or interactions with proteins like CRACR2A (Srikanth et al. 2010) or protein complexes associated with mitochondria-ER tethers (Singaravelu et al. 2011). Golli, a widely expressed, alternatively spliced form of myelin basic protein, is another recent candidate for a STIM1-binding protein. Knockdown and overexpression experiments suggest that it inhibits SOCE in T cells when localized to the PM via a myristoylation site (Feng et al. 2006). Golli appears to interact with the isolated carboxyl terminus of STIM1 in a Ca^{2+}-dependent manner, but thus far evidence for interaction with full-length STIM1 in cells is somewhat weak (Walsh et al. 2010b). Given the large number of hits in the original RNAi screens for SOCE suppression, continuing efforts to pull out STIM1- and Orai-interacting proteins are likely to reveal additional mechanisms that fine-tune SOCE under physiological conditions.

Toward a Molecular Basis for CRAC Channel Properties

Ion Selectivity and Permeation

The CRAC channel is unusual among Ca^{2+} channels in its pore properties. It has among the highest selectivity for Ca^{2+} over monovalent cations, comparable to that of Ca_V channels (Hoth 1995), but has a narrow pore diameter (3.9 Å) and a conductance that is among the smallest of any ion channel, being estimated from noise analysis at ~20 fS for Ca^{2+} (Zweifach and Lewis 1993; Prakriya and Lewis 2006). Initial studies identified mutations in several highly conserved residues that reduced the divalent ion selectivity or the sensitivity to block by trivalent cations. E106 and E190 (and their equivalent positions in dOrai) emerged as being critical for divalent selectivity, with D110/112/114 participating in block by trivalent cations (Fig. 3) (Prakriya et al. 2006; Vig et al. 2006; Yeromin et al. 2006). To attain a more precise picture of the pore, subsequent studies applied scanning cysteine mutagenesis and probed for disulfide crosslinking (Zhou et al. 2010b) or accessibility to thiol-reactive reagents (McNally et al. 2009) as indicators of pore-lining positions. The shared conclusion was that TM1 residues line the ion-conductive pore, but those in TM3 do not, indicating that the effects of E190 mutations on ion selectivity were most likely the results of allosteric changes in protein structure (Fig. 3). The ability of Cd^{2+} to block the cysteine-substituted pore are consistent with a narrow pore diameter along its length (McNally et al. 2009). The emerging model of the CRAC channel pore is that of a rather flexible vestibule at the extracellular surface, a selectivity filter formed by a quartet of glutamates near the extracellular end of the pore, followed by a long narrow channel that may explain the low overall conductance. By comparing cysteine accessibility in closed vs. open channels, these methods should also be applicable to identifying the location of channel gates.

Ca^{2+}-dependent Inactivation (CDI)

Like many Ca^{2+}-permeable channels, CRAC channels are regulated by negative feedback from the Ca^{2+} that enters the cell. Rapid Ca^{2+}-dependent inactivation (CDI) occurs over tens of ms and is thought to result from binding of Ca^{2+} to sites several nm from the mouth of the channel (Hoth and Penner 1992, 1993; Zweifach and Lewis 1995). A role for STIM1 in this process was first suggested by the observation that CAD peptide activates Orai1 without allowing CDI (Park et al. 2009). Studies of truncated STIM1 subsequently led to the identification of a region carboxy-terminal to CAD that fully restored CDI (Derler et al. 2009b; Lee et al. 2009; Mullins et al. 2009). The inactivation domain (ID_{STIM}; aa 470–491) contains a stretch of seven acidic residues that exert a powerful influence over the extent of CDI (Fig. 2). Neutralization of pairs of residues either reduced or enhanced CDI (Mullins et al. 2009), whereas neutralization of 5–7 residues completely eliminated CDI (Derler et al. 2009b; Lee et al. 2009; Mullins et al. 2009). Preliminary evidence from ^{45}Ca overlays suggested that this domain may bind Ca^{2+} weakly, but the effects of neutralizing mutations on Ca^{2+} binding were not in every case consistent with their effects on CDI (Mullins et al. 2009). Thus, its potential role as a Ca^{2+} binding site for CDI is at present undetermined.

Evidence is stronger for CaM as the Ca^{2+} sensor for CDI. Early work by Rychkov et al. showed evidence for a role of CaM, based on partial suppression of CDI by overexpressing a mutant Ca^{2+}-nonbinding CaM or a CaM inhibitory peptide (Litjens et al. 2004). Subsequent work revealed that CaM binds in a Ca^{2+}-dependent manner to the Orai1 amino terminus in a region adjacent to the start of TM1 (aa 68–90; Fig. 3) (Mullins et al. 2009). Mutations that prevent CaM binding (A73E, W76E/A/S, and Y80E) completely eliminate CDI, whereas mutations that preserve binding (Y80A/S) display CDI with accelerated kinetics. Much remains to be understood about how CaM operates to control CDI, such as the location of the apo-CaM binding site, and the role of the CaM binding domain in controlling CDI kinetics.

In addition to the amino terminus, other regions of Orai1 also participate in CDI. Alanine substitutions in the intracellular II-III loop at aa 151–154 eliminate CDI, whereas application of a soluble wild-type peptide containing this region reduces current unless it bears the alanine mutations (Fig. 3) (Srikanth et al. 2010). Srikanth et al. suggest that the II-III loop of Orai1 may function as a blocking particle in CDI to inhibit conduction. Another study has shown that mutations that increase Orai1 pore size (E106D, E190Q) greatly reduce or eliminate CDI (Yamashita et al. 2007). These findings raise the possibility that the inactivation gate may involve elements of the selectivity filter, as has been proposed for Ca_V and K_V channels.

One complication in designing and interpreting CDI experiments is that the STIM1:Orai1 binding stoichiometry has a powerful effect on the extent of CDI. Scrimgeour et al. (2009) initially reported that CDI was normal in cells expressing a high ratio of STIM1:Orai1 but was lost in cells with low expression ratios. In an extension of this work, we found that CDI fell in parallel with channel activation as the STIM1:Orai1 ratio fell below ~ 2, suggesting that full inactivation requires binding of eight STIM1s to the channel (Hoover and Lewis 2011). Thus, the number of STIM1s bound to a CRAC channel determines not only its ability to open, but its propensity to inactivate in response to Ca^{2+} entry.

One important concept that has emerged from the work on CDI described above is that the CRAC channel itself is not merely a multimer of Orai1, but is more accurately considered as a multiprotein complex. It binds CaM, which acts as a subunit to promote CDI. In addition, the requirement for STIM1 in enabling the channel to gate in response to an extracellular ligand (Ca^{2+}) implies that STIM1 also acts as a channel subunit and not merely as an activating ligand.

FUNCTIONAL ROLES OF STIM AND ORAI: INSIGHTS FROM GENETICS

Determining the precise functions of STIM and Orai in intact organisms has long been frustrated by a lack of highly specific pharmacological inhibitors. However, new insights and unexpected roles have been revealed by the phenotypes of genetically engineered STIM1/2 and Orai1 mutant mice as well as human patients with naturally occurring loss-of-function mutations in STIM1 or Orai1. Thus far, these studies have established important functions for STIM and Orai in the immune system, skeletal muscle, and platelets, among other cells and tissues.

Three distinct mutations causing loss of Orai1 function have been described in unrelated human families, as well as one frameshift mutation in STIM1 leading to a loss of protein expression; in mice, Orai1-, STIM1-, and STIM2-deficient mice have been engineered by homologous recombination or insertional mutagenesis (for reviews see Feske 2010). It is noteworthy that the clinical pathophysiologies caused by a loss of STIM1 or Orai1 function mostly overlap, suggesting that STIM1's critical functions in humans are performed primarily through Orai1 rather than other targets (Feske 2010). There are differences in phenotypes between humans and mice that may be caused by environment (the real world vs. animal care facilities or pathogen-free conditions), genetic background (e.g., highly inbred mouse strains), or the expression patterns of different Orai isoforms. Here I will attempt to identify the generalities that apply to both humans and mice and describe some of the unexpected results; a fuller discussion of these genetic studies is available in several excellent reviews (Baba and Kurosaki 2009; Feske 2009, 2010; Vig and Kinet 2009; Hogan et al. 2010).

Immune Cells

Loss of STIM1 or Orai1 function in humans causes severe combined immunodeficiency (SCID), a lethal condition marked by recurrent opportunistic infections, the only therapy being a functional hematopoietic stem cell transplant. Although normal numbers of T cells are present in these patients, they appear largely nonfunctional because of an inability of antigen receptor engagement to generate Ca^{2+} signals that regulate gene expression during T cell activation (Feske et al. 2001). These findings have triggered intense interest in developing specific pharmacologic inhibitors of the CRAC channel to treat a variety of autoimmune disorders. In mice, the absence of Orai1 or STIM1 function causes multiple defects of cytokine expression in T cells (interleukin (IL)- 2, interferon-γ, IL-4, IL-10) (Gwack et al. 2008; Oh-Hora et al. 2008), and in mast cells, defective synthesis or release of inflammatory mediators (tumor necrosis factor-α, IL-6, serotonin, and leukotriene C_4) leads to a reduced anaphylaxis response in vivo (Baba et al. 2008; Vig et al. 2008). B-cell targeted deletion of STIM1 and STIM2 revealed a novel role for SOCE in enabling B cells to limit autoimmunity; in these animals reduced secretion of the anti-inflammatory cytokine IL-10 was linked to increased severity of experimental autoimmune encephalomyelitis, a model for multiple sclerosis (Matsumoto et al. 2011).

Interestingly, loss of STIM or Orai1 function does not eliminate all immune activity. T cell-dependent antibody responses and graft vs. host disease response appear relatively unaffected in $STIM1^{-/-}$ mice (Beyersdorf et al. 2009). Partial function of B cells is also evident in STIM1- or Orai1-deficient patients who develop lymphoproliferative disease and an autoimmune attack on their own neutrophils and thrombocytes (McCarl et al. 2009; Picard et al. 2009). A related autoimmune phenotype was observed in STIM1/STIM2 deficient mice (Oh-Hora et al. 2008; Oh-hora 2009). The autoimmunity is likely caused by defective generation of regulatory T (T_{reg}) cells that normally act to suppress T and B cell function. Finally, while there is much evidence supporting a role of sustained TCR-mediated Ca^{2+} signals in positive and negative selection of T cells during development in the thymus, development appears to proceed normally in the absence of STIM1 or Orai1; the same is true for development of mast cells and platelets. These findings raise a number of intriguing questions. Can other Ca^{2+} channels substitute for STIM1 and Orai1 (e.g., Orai2 or 3, Ca_V, or TRPC channels, P2X receptors, or something else?), or are the Ca^{2+} requirements for some cell functions low enough that ER Ca^{2+} release alone is sufficient? Finally, what makes the development of T_{reg} cells relatively sensitive to the loss of STIM or Orai1, while that of the other T cell subsets is not affected?

Platelets

Ca^{2+} is an essential signal for platelet activation events leading to clotting, hemostasis and thrombosis. Platelets express several parallel Ca^{2+} entry pathways, which made it difficult to define the role of SOCE in specific platelet functions until the generation of genetically engineered mice. In mice that are STIM1- or Orai1-deficient or homozygous for the nonfunctional Orai1 R93W gene, platelets show deficient Ca^{2+} responses to collagen and thrombin with defects in activation and thrombus formation in vitro (Varga-Szabo et al. 2008; Bergmeier et al. 2009; Braun et al. 2009).

Interestingly, these mice are protected in experimental models of arterial thrombosis and ischemic brain infarction, while displaying only a mild prolongation of bleeding time. Thus, STIM1 and Orai1 may not be absolutely essential for most of the rapid responses of platelets to activating factors (e.g., shape change, integrin activation, and secretion), but are perhaps more important for late responses such as stabilization of the thrombus, clot retraction, coagulant activity, and recruitment of cells to sites of injury (Varga-Szabo et al. 2009). These findings suggest the potential therapeutic use of CRAC channel blockers to treat ischemic cardiovascular or cerebrovascular disease.

Sweat Glands and Teeth

STIM1 and Orai1 appear to be required for the normal development and function of some ectodermally-derived tissues, such as sweat glands and teeth. In humans lacking Orai1 function, eccrine sweat glands do not function properly (whether this is a developmental or a purely functional defect is unclear), and this leads to dry skin and defects in thermoregulation manifest as recurrent fever (McCarl et al. 2009). The dental enamel in patients lacking STIM1 or Orai1 expression also fails to develop properly, implicating SOCE in the transepithelial transport of Ca^{2+} during formation of enamel (Picard et al. 2009; McCarl et al. 2009). STIM1- and Orai1-knockout mice may provide a useful experimental model for this poorly understood process.

Skeletal Muscle

More unexpected roles of STIM1 and Orai1 have emerged from genetic studies of skeletal muscle. The first hints for a function came from observations that CRAC-deficient SCID patients display a mild muscle myopathy with reduced strength and endurance (Partiseti et al. 1994; McCarl et al. 2009). STIM1 and Orai1 were later shown to be highly expressed in both human and mouse skeletal muscle (Williams et al. 2001; Gwack et al. 2008; Stiber et al. 2008; Vig et al. 2008; McCarl et al. 2009), and STIM1-deficient mice exhibit reduced muscle mass and susceptibility to fatigue (Stiber et al. 2008). The reduced mass is presumably caused by a defect in muscle development, as knockdown of STIM1, STIM2, or Orai1 or 3 inhibits the differentiation of myoblasts into myotubes and underlying gene activation events (Darbellay et al. 2008, 2010). Early fatigue is consistent with a role for STIM1 and Orai1 in maintaining SR Ca^{2+} content during periods of high frequency muscle activity, when slow depletion of Ca^{2+} from the SR would normally activate SOCE to replenish the SR Ca^{2+} deficit and thus maintain Ca^{2+} release transients and the force of contraction (Pan et al. 2002; Stiber et al. 2008). An intriguing feature of SOCE in muscle is that it is orders of magnitude faster than in nonexcitable cells (Launikonis and Ríos 2007), commencing in some cases within milliseconds of Ca^{2+} release from the SR (Edwards et al. 2010). Several models have been proposed to explain how prelocalization of STIM1 and Orai1 at the triadic junction may confer these extraordinarily rapid activation kinetics (Dirksen 2009; Launikonis et al. 2010).

FUTURE DIRECTIONS

Formation and Function of the ER-PM Junctions

Because SOCE depends absolutely on the direct contact of STIM and Orai at ER-PM junctions, the number, size, and intermembrane dimensions of these structures will exert profound effects on SOCE. However, little is known about how ER-PM junctions form and disassemble. Depletion of stores and STIM1 overexpression increases the number and extent of ER-PM junctions (Wu et al. 2006; Orci et al. 2009), suggesting a role for STIM1 in their formation or stabilization. Similarly, overexpression or knockdown of the widely expressed SR/ER protein junctate increases or decreases the abundance of junctions, respectively, although this was store-independent (Treves et al. 2004, 2010). Overexpression of TRPC3 also increased junction formation, suggesting that multiple proteins may participate in bridging the two membranes. Periodic electron densities have been reported between the ER and PM at junctate-induced junctions, but thus far their molecular basis is unknown, and the relationship to junctions in which STIM and Orai accumulate is unclear (Treves et al. 2010).

The highly restricted region of Ca^{2+} entry at ER-PM junctions offers attractive opportunities for local signaling within Ca^{2+} microdomains to effector proteins (Luik et al. 2006). SOCE has been shown to activate or repress particular enzymes with high specificity, including adenylate cyclases (Willoughby and Cooper 2007), the plasma membrane Ca^{2+} pump (Bautista and Lewis 2004), and phospholipase A_2 (Chang et al. 2008). The evidence for local signaling is mostly based on two approaches. One is to show that BAPTA, a fast Ca^{2+} chelator, can interrupt the activation of the target by SOCE, whereas the slower buffer EGTA cannot, suggesting that the target is located somewhere between the calculated capture distances of the two chelators (tens of nanometers from the channels). The other approach is to show that when SOCE and non-SOCE pathways are engaged to produce the same global $[Ca^{2+}]_i$ increase, only SOCE activates the target. Using these approaches, recent studies suggest that transcriptional pathways may be activated preferentially near CRAC channels (di Capite et al. 2009; Ng et al. 2009). Ca^{2+} entry through CRAC channels has long been known to drive gene expression through NFAT and other factors, and the efficiency and specificity of gene activation depends on the amplitude and duration of sustained Ca^{2+} signals and the frequency of Ca^{2+} oscillations triggered through CRAC channels (Dolmetsch et al. 1997, 1998). Recently, Parekh and colleagues presented evidence that NFAT activation, assayed by nuclear accumulation, is triggered preferentially in the close vicinity of CRAC channels rather than in the cytosol (Kar et al. 2011). This finding is remarkable, given that the NFAT-activating phosphatase calcineurin (CN) is present in the cytoplasm and its Ca^{2+} sensitivity is sufficiently high to dephosphorylate NFAT at submicromolar levels of global cytosolic Ca^{2+} (Stemmer and Klee 1994). One possible explanation is that cytoplasmic CN activity is weak enough to be subverted by cytoplasmic NFAT kinases, whereas junctional CN activity more completely dephosphorylates NFAT, enabling it to run the cytoplasmic kinase gauntlet and reach the nucleus. Further studies may shed light on the details of this mechanism.

Regulation of SOCE in a Physiological Context

One expansive goal for further research is to determine the many levels at which SOCE can be regulated under physiological conditions. The great majority of studies to date have been conducted using TG, high intracellular IP_3, ionomycin, or chelators, all of which deplete the ER completely throughout the cell. Physiological responses are likely to be much more localized, particularly when stimuli are spatially restricted. A particularly interesting example occurs during immune synapse formation between T cells and antigen-presenting cells (APCs), in which the T cell receptors engage peptide-MHC complexes on the APC. Shortly after cell–cell contact, STIM1 and Orai1 change their localization, accumulating initially at the synapse and then appearing to move in tandem to the distal end of the cell (Barr et al. 2008; Lioudyno et al. 2008). Ca^{2+} imaging shows a domain of elevated $[Ca^{2+}]_i$ at the synapse, where it could potentially affect synapse stability, cytoskeletal remodeling, secretion, and other Ca^{2+}-dependent events (Lioudyno et al. 2008). These studies raise many interesting questions about the mechanisms involved in polarizing the SOCE machinery and the potentially different consequences of localizing STIM1-Orai1 complexes to the two poles of the cell.

STIM1 is known to be phosphorylated in situ (Manji et al. 2000), but functional effects are only beginning to be explored. Among multiple sites of phosphorylation, several (S575, S608, S621) are phosphorylated by ERK1/2 and this increases after store depletion (Pozo-Guisado et al. 2010). Alanine mutations at these sites significantly repressed SOCE, suggesting that phosphorylation may serve as a powerful positive modulator. Conversely, Smyth et al. linked phosphorylation in the cytoplasmic region to the suppression of SOCE during mitosis (Smyth et al. 2009). Among multiple sites, phosphorylation at S486 and S668 were found to be most potent in preventing STIM1 puncta formation and suppressing SOCE (Fig. 2). The physiological role of SOCE suppression is not yet clear, although a slight slowing of cell proliferation was seen in cells overexpressing a truncated STIM1 that functions during mitosis. It will be interesting to see if phosphorylation affects cell function independently of suppressing Ca^{2+} entry. From these early studies it is already clear that phosphorylation has the potential to modulate CRAC channel activity in response to multiple environmental cues.

Oxidation may also affect STIM1 and Orai1 function independently of ER Ca^{2+}, possibly modulating SOCE under conditions of cellular stress or inflammation. Oxidation by H_2O_2 or buthionine sulfoximine can cause STIM1 to form puncta and open Orai1 channels (Hawkins et al. 2010). One proposed mechanism involves S-glutathionylation of C56, which then reduces Ca^{2+} binding by the adjacent EF hand. In this way STIM1 may act as a cellular sensor for oxidant stress, although an indirect effect of oxidants on SOCE through depletion of Ca^{2+} stores has not been excluded. On the other hand, oxidation also inhibits SOCE by reducing Orai1 activity (Bogeski et al. 2010). This effect has been traced primarily to C195, and occurs with Orai1 but not Orai3. Interestingly, SOCE in effector T cells is more resistant to inhibition by ROS than in naïve T cells, which may result from increased expression of Orai3 relative to Orai1 in effector cells. Bogeski et al. propose that the increased Orai3:Orai1 ratio is an adaptation that allows effector T cells to proliferate and optimize cytokine production in the oxidizing environment of injured or inflamed tissue. Understanding how the stimulatory and inhibitory effects of oxidation are integrated during responses to physiological stimuli is an important goal.

Temperature is another factor that has been largely overlooked in studies of STIM1-Orai1 function; even though the majority of studies use mammalian proteins, room temperature has been the norm. Interestingly, Xiao et al. (2011) showed that STIM1 is in fact highly temperature-dependent, and it begins to form puncta without store depletion at temperatures $>35°C$. Interestingly (and fortunately for us), this does not by itself activate Ca^{2+} entry, but Orai1 channels open during subsequent cooling. It is not known whether binding to Orai1 is completely inhibited at the high temperature, or some more subtle change is involved. One proposal is that this thermosensitivity may prime immune cells for greater activity during fever as they circulate between the body core and the cooler periphery, or play a role in skeletal muscle physiology during intense exercise (Xiao et al. 2011). These findings also raise the possibility that at mammalian physiological temperatures the $[Ca^{2+}]_{ER}$ threshold for STIM1 oligomerization may be shifted to higher $[Ca^{2+}]_{ER}$ than has been measured at room temperature (Brandman et al. 2007; Luik et al. 2008), making STIM1 more responsive to small changes in $[Ca^{2+}]_{ER}$ than currently thought.

Multiple Roles for STIM1 in Ca^{2+} Channel Regulation

TRPC Channels

Mammalian TRPC channels are weakly Ca^{2+} permeable channels with a long history of association with SOCE, but whether they are truly store-dependent is still being debated (DeHaven et al. 2009; Wang et al. 2010). The strongest evidence of store-dependent activation comes from charge-swapping experiments suggesting that a di-lysine motif in STIM1 interacts with a conserved di-aspartate domain in TRPC1 and other TRPCs to activate these channels following store depletion (Zeng et al. 2008). One complication in assessing whether TRPCs are store-operated has been that PLC-coupled agonists have been used in most cases to activate the channels, stimulating second-messenger (diacylglycerol, PIP_2, IP_3) and store-depletion pathways in parallel. The definitive tests for store-operated function (i.e., channel activation by store depletion independent of receptor activation or changes in $[Ca^{2+}]_i$) have in most cases not been performed, or have given ambiguous results. For example, TRPC7, once thought to be store-operated because it is activated by TG, was later found not to be activated by IP_3 or ionomycin even though they also empty the ER (DeHaven et al. 2009). One step toward settling this debate would be to test whether a variety of receptor-independent store-depletion stimuli (SERCA blockers, cytosolic and ER Ca^{2+} chelators, lipophilic Ca^{2+} ionophores, TPEN) all activate TRPC channels under conditions of heavy cytosolic buffering at resting $[Ca^{2+}]_i$ levels. Unfortunately, these tests also complicated by potential crosstalk between Orai1 channels and TRPC channels mediated through local increases in $[Ca^{2+}]_i$ entering via Orai1. In fact, the relation between STIM1, Orai1, and TRPCs could be quite complex, given evidence that TRPC channel activation or insertion into the PM may be affected by local changes in $[Ca^{2+}]_i$ through other channels (Gross et al. 2009; Cheng et al. 2011). A critical challenge for future studies is to understand this relation and how TRPC channels act in concert with STIM1 and Orai1 to shape store-dependent Ca^{2+} signals.

ARC Channels

Arachidonate-regulated Ca^{2+} (ARC) channels are similar in terms of Ca^{2+} selectivity and conductance to the CRAC channel, but differ in lacking CDI and sensitivity to 2-APB, and most importantly are activated by arachidonic acid rather than store depletion (Shuttleworth 2009). STIM1 is necessary for their activation, but only the fraction of STIM1 that is constitutively present in the PM (comprising about 10%–20% of the total STIM1 pool in most cells) (Mignen et al. 2007). Concatemer studies suggest that ARC channels form as a pentamer of Orai subunits (3 × Orai1 + 2 × Orai3), and glutamine substitutions for glutamates at E106 in Orai1 or the corresponding location in Orai3 (E81) both inhibit conductance, suggesting the formation of a pentameric glutamatergic selectivity filter (Mignen et al. 2009). Ion selectivity studies with less extreme aspartate substitutions and sizing the wt pentameric pore under divalent-free conditions will help validate this model and help relate its ion selectivity mechanism to that of the tetrameric CRAC channel. Tail swapping experiments have indicated that arachidonic acid selectivity requires two amino-terminal Orai3 regions in the pentamer (Thompson et al. 2010). Much remains to be learned about how STIM1 regulates the ARC channels, and how ARC and CRAC channel synthesis are controlled in cells that express both Orai1 and Orai3.

Ca_V Channels

STIM1 was recently and unexpectedly discovered to exert reciprocal actions on store-operated and voltage-gated Ca^{2+} ($Ca_V1.2$) channels (Park et al. 2010; Wang et al. 2010). STIM1 inhibits $Ca_V1.2$ channel function in two ways. First, store-depletion acts through STIM1 to acutely inhibit voltage-gated opening of $Ca_V1.2$; this effect ranges from a modest 15% (Park et al. 2010) to nearly total inhibition (Wang et al. 2010). In addition, STIM1 expression can nearly completely suppress Cav1.2 by stimulating its internalization (Park et al. 2010). These effects appear to be caused by binding of the CAD domain of STIM1 to the carboxyl terminus of $Ca_V1.2$. Many of the details of the mechanisms remain to be worked out, and further studies will be needed to understand how general these phenomena are and whether the crosstalk extends to other members of the Ca_V family. The effects of STIM1 on $Ca_V1.2$ channel function may help explain how PLC-coupled receptors inhibit voltage-gated Ca^{2+} signaling, and in cells that express both $Ca_V1.2$ and Orai1, the reciprocal regulation of these channels provides a new mechanism to enable the selective activation of channel-specific signaling pathways.

CONCLUSION

The identification of the STIM and Orai genes has led to a remarkable wave of progress in understanding the mechanism of SOCE, to the point where the major steps are now known and have been described in at least a qualitative way. There clearly remains much that is not understood at a mechanistic level, and application of increasingly sophisticated physiological, cell biological, biochemical and structural approaches will be needed to fill in these gaps. It is hoped that with a more detailed mechanistic picture of CRAC channel function and regulation, we may better understand the full diversity of their functions in vivo, and ultimately develop new strategies to control them in the treatment of human disease.

ACKNOWLEDGMENTS

Work in the Lewis laboratory is supported by the National Institutes of Health (NIH GM45374).

REFERENCES

Baba Y, Kurosaki T. 2009. Physiological function and molecular basis of STIM1-mediated calcium entry in immune cells. *Immunol Rev* **231:** 174–188.

Baba Y, Hayashi K, Fujii Y, Mizushima A, Watarai H, Wakamori M, Numaga T, Mori Y, Iino M, Hikida M, et al. 2006. Coupling of STIM1 to store-operated Ca^{2+} entry through its constitutive and inducible movement in the endoplasmic reticulum. *Proc Natl Acad Sci* **103:** 16704–16709.

Baba Y, Nishida K, Fujii Y, Hirano T, Hikida M, Kurosaki T. 2008. Essential function for the calcium sensor STIM1 in mast cell activation and anaphylactic responses. *Nat Immunol* **9:** 81–88.

Barr VA, Bernot KM, Srikanth S, Gwack Y, Balagopalan L, Regan CK, Helman DJ, Sommers CL, Oh-Hora M, Rao A, et al. 2008. Dynamic movement of the calcium sensor STIM1 and the calcium channel Orai1 in activated T-cells: Puncta and distal caps. *Mol Biol Cell* **19:** 2802–2817.

Bautista DM, Lewis RS. 2004. Modulation of plasma membrane calcium-ATPase activity by local calcium microdomains near CRAC channels in human T cells. *J Physiol* **556:** 805–817.

Bergmeier W, Oh-Hora M, McCarl CA, Roden RC, Bray PF, Feske S. 2009. R93W mutation in Orai1 causes impaired calcium influx in platelets. *Blood* **113:** 675–678.

Beyersdorf N, Braun A, Vögtle T, Varga-Szabo D, Galdos RR, Kissler S, Kerkau T, Nieswandt B. 2009. STIM1-independent T cell development and effector function in vivo. *J Immunol* **182:** 3390–3397.

Bogeski I, Kummerow C, Al-Ansary D, Schwarz EC, Koehler R, Kozai D, Takahashi N, Peinelt C, Griesemer D, Bozem M, et al. 2010. Differential redox regulation of ORAI ion channels: A mechanism to tune cellular calcium signaling. *Sci Signal* **3:** ra24.

Brandman O, Liou J, Park W, Meyer T. 2007. STIM2 is a feedback regulator that stabilizes basal cytosolic and endoplasmic reticulum Ca^{2+} levels. *Cell* **131:** 1327–1339.

Braun A, Varga-Szabo D, Kleinschnitz C, Pleines I, Bender M, Austinat M, Bösl M, Stoll G, Nieswandt B. 2009. Orai1 (CRACM1) is the platelet SOC channel and essential for pathological thrombus formation. *Blood* **113:** 2056–2063.

Cahalan MD, Zhang SL, Yeromin AV, Ohlsen K, Roos J, Stauderman KA. 2007. Molecular basis of the CRAC channel. *Cell Calcium* **42:** 133–144.

Calloway N, Vig M, Kinet J-P, Holowka D, Baird B. 2009. Molecular clustering of STIM1 with Orai1/CRACM1 at the plasma membrane depends dynamically on depletion of Ca^{2+} stores and on electrostatic interactions. *Mol Biol Cell* **20:** 389–399.

Calloway NT, Holowka DA, Baird BA. 2010. A Basic Sequence In STIM1 Promotes Ca^{2+} Influx By Interacting With The C-Terminal Acidic Coiled-Coil Of Orai1. *Biochemistry* **49:** 1067–1071.

Chang W-C, di Capite J, Singaravelu K, Nelson C, Halse V, Parekh AB. 2008. Local Ca^{2+} influx through Ca^{2+} release-activated Ca^{2+} (CRAC) channels stimulates production of an intracellular messenger and an intercellular pro-inflammatory signal. *J Biol Chem* **283:** 4622–4631.

Cheng KT, Liu X, Ong HL, Swaim W, Ambudkar IS. 2011. Local Ca^{2+} entry via Orai1 regulates plasma membrane recruitment of TRPC1 and controls cytosolic Ca^{2+} signals required for specific cell functions. *PLoS Biol* **9:** e1001025.

Covington ED, Wu MM, Lewis RS. 2010. Essential role for the CRAC activation domain in store-dependent oligomerization of STIM1. *Mol Biol Cell* **21:** 1897–1907.

Darbellay B, Arnaudeau S, König S, Jousset H, Bader C, Demaurex N, Bernheim L. 2008. STIM1 and Orai1-dependent store-operated calcium entry regulates human myoblast differentiation. *J Biol Chem* **284:** 5370–5380.

Darbellay B, Arnaudeau S, Ceroni D, Bader CR, Konig S, Bernheim L. 2010. Human muscle economy myoblast differentiation and excitation-contraction coupling use the same molecular partners, STIM1 and STIM2. *J Biol Chem* **285:** 22437–22447.

DeHaven WI, Jones BF, Petranka JG, Smyth JT, Tomita T, Bird GS, Putney JW. 2009. TRPC channels function independently of STIM1 and Orai1. *J Physiol* **587:** 2275–2298.

Derler I, Fahrner M, Carugo O, Muik M, Bergsmann J, Schindl R, Frischauf I, Eshaghi S, Romanin C. 2009a. Increased hydrophobicity at the N terminus/membrane interface impairs gating of the severe combined immunodeficiency-related ORAI1 mutant. *J Biol Chem* **284:** 15903–15915.

Derler I, Fahrner M, Muik M, Lackner B, Schindl R, Groschner K, Romanin C. 2009b. A Ca^{2+} release-activated Ca^{2+} (CRAC) modulatory domain (CMD) within STIM1 mediates fast Ca^{2+}-dependent inactivation of ORAI1 channels. *J Biol Chem* **284:** 24933–24938.

di Capite J, Ng SW, Parekh AB. 2009. Decoding of cytoplasmic Ca^{2+} oscillations through the spatial signature drives gene expression. *Curr Biol* **19:** 853–858.

Dirksen RT. 2009. Checking your SOCCs and feet: The molecular mechanisms of Ca^{2+} entry in skeletal muscle. *J Physiol* **587:** 3139–3147.

Dolmetsch RE, Lewis RS. 1994. Signaling between intracellular Ca^{2+} stores and depletion-activated Ca^{2+} channels generates $[Ca^{2+}]_i$ oscillations in T lymphocytes. *J Gen Physiol* **103:** 365–388.

Dolmetsch RE, Lewis RS, Goodnow CC, Healy JI. 1997. Differential activation of transcription factors induced by Ca^{2+} response amplitude and duration. *Nature* **386:** 855–858.

Dolmetsch RE, Xu K, Lewis RS. 1998. Calcium oscillations increase the efficiency and specificity of gene expression. *Nature* **392:** 933–936.

Edwards JN, Murphy RM, Cully TR, von Wegner F, Friedrich O, Launikonis BS. 2010. Ultra-rapid activation and deactivation of store-operated Ca^{2+} entry in skeletal muscle. *Cell Calcium* **47:** 458–467.

Ercan E, Momburg F, Engel U, Temmerman K, Nickel W, Seedorf M. 2009. A conserved, lipid-mediated sorting mechanism of yeast Ist2 and mammalian STIM proteins to the peripheral ER. *Traffic* **10:** 1802–1818.

Fahrner M, Muik M, Derler I, Schindl R, Fritsch R, Frischauf I, Romanin C. 2009. Mechanistic view on domains mediating STIM1-Orai coupling. *Immunol Rev* **231:** 99–112.

Feng J-M, Hu YK, Xie L-H, Colwell CS, Shao XM, Sun X-P, Chen B, Tang H, Campagnoni AT. 2006. Golli protein negatively regulates store depletion-induced calcium influx in T cells. *Immunity* **24:** 717–727.

Feske S. 2009. ORAI1 and STIM1 deficiency in human and mice: Roles of store-operated Ca^{2+} entry in the immune system and beyond. *Immunol Rev* **231:** 189–209.

Feske S. 2010. CRAC channelopathies. *Pflugers Arch* **460:** 417–435.

Feske S, Giltnane J, Dolmetsch R, Staudt LM, Rao A. 2001. Gene regulation mediated by calcium signals in T lymphocytes. *Nat Immunol* **2:** 316–324.

Feske S, Prakriya M, Rao A, Lewis R. 2005. A severe defect in CRAC Ca^{2+} channel activation and altered K^+ channel gating in T cells from immunodeficient patients. *J Exp Med* **202:** 651–662.

Feske S, Gwack Y, Prakriya M, Srikanth S, Puppel S, Tanasa B, Hogan P, Lewis R, Daly M, Rao A. 2006. A mutation in Orai1 causes immune deficiency by abrogating CRAC channel function. *Nature* **441:** 179–185.

Frischauf I, Muik M, Derler I, Bergsmann J, Fahrner M, Schindl R, Groschner K, Romanin C. 2009. Molecular determinants of the coupling between STIM1 and Orai channels: Differential activation of Orai1-3 channels by a STIM1 coiled-coil mutant. *J Biol Chem* **284:** 21696–21706.

Grigoriev I, Gouveia SM, van der Vaart B, Demmers J, Smyth JT, Honnappa S, Splinter D, Steinmetz MO, Putney JW, Hoogenraad CC, Akhmanova A. 2008. STIM1 is a MT-plus-end-tracking protein involved in remodeling of the ER. *Curr Biol* **18:** 177–182.

Gross S, Guzman G, Wissenbach U, Philipp S, Zhu M, Bruns D, Cavalie A. 2009. TRPC5 is a Ca^{2+}-activated channel functionally coupled to Ca^{2+}-selective ion channels. *J Biol Chem* **284:** 34423–34432.

Grynkiewicz G, Poenie M, Tsien RY. 1985. A new generation of Ca^{2+} indicators with greatly improved fluorescence properties. *J Biol Chem* **260:** 3440–3450.

Gwack Y, Srikanth S, Oh-Hora M, Hogan PG, Lamperti ED, Yamashita M, Gelinas C, Neems DS, Sasaki Y, Feske S, Rao A, et al. 2008. Hair loss and defective T- and B-cell function in mice lacking ORAI1. *Mol Cell Biol* **28:** 5209–5222.

Hawkins BJ, Irrinki KM, Mallilankaraman K, Wang Y, Bhanumathy CD, Subbiah R, Ritchie MF, Soboloff J, Baba Y, Kurosaki T, Madesh M, et al. 2010. S-glutathionylation activates STIM1 and alters mitochondrial homeostasis. *J Cell Biol* **190:** 391–405.

Hofer AM, Fasolato C, Pozzan T. 1998. Capacitative Ca^{2+} entry is closely linked to the filling state of internal Ca^{2+} stores: A study using simultaneous measurements of I_{CRAC} and intraluminal $[Ca^{2+}]$. *J Cell Biol* **140:** 325–334.

Hogan PG, Lewis RS, Rao A. 2010. Molecular basis of calcium signaling in lymphocytes: STIM and ORAI. *Annu Rev Immunol* **28:** 491–533.

Honnappa S, Gouveia SM, Weisbrich A, Damberger FF, Bhavesh NS, Jawhari H, Grigoriev I, van Rijssel, Buey RM, Lawera A, Steinmetz MO, et al. 2009. An EB1-binding motif acts as a microtubule tip localization signal. *Cell* **138:** 366–376.

Hoover PJ, Lewis RS. 2011. Stoichiometric requirements for trapping and gating of CRAC channels by stromal interaction molecule 1 (STIM1). *Proc Natl Acad Sci* doi: 10.1073/pnas.1101664108.

Hoth M. 1995. Calcium and barium permeation through calcium release-activated calcium (CRAC) channels. *Pflugers Archiv Eur J Physiol* **430:** 315–322.

Hoth M, Penner R. 1992. Depletion of intracellular calcium stores activates a calcium current in mast cells. *Nature* **355:** 353–356.

Hoth M, Penner R. 1993. Calcium release-activated calcium current in rat mast cells. *J Physiol* **465:** 359–386.

Huang G, Zeng W, Kim J, Yuan J, Han L, Muallem S, Worley P. 2006. STIM1 carboxyl-terminus activates native SOC, I_{crac} and TRPC1 channels. *Nat Cell Biol* **8:** 1003–1010.

Ji W, Xu P, Li Z, Lu J, Liu L, Zhan Y, Chen Y, Hille B, Xu T, Chen L. 2008. Functional stoichiometry of the unitary calcium-release-activated calcium channel. *Proc Natl Acad Sci* **105:** 13668–13673.

Kar P, Nelson C, Parekh AB. 2011. Selective activation of the transcription factor NFAT1 by calcium microdomains near CRAC channels. *J Biol Chem* **286:** 14795–14803.

Kawasaki T, Lange I, Feske S. 2009. A minimal regulatory domain in the C terminus of STIM1 binds to and activates ORAI1 CRAC channels. *Biochem Biophys Res Commun* **385:** 49–54.

Korzeniowski M, Popovic M, Szentpetery Z, Varnai P, Stojilkovic S, Balla T. 2009. Dependence of stim1/orai1 mediated calcium entry on plasma membrane phosphoinositides. *J Biol Chem* **284:** 21027–21035.

Korzeniowski MK, Manjarrés IM, Varnai P, Balla T. 2010. Activation of STIM1-Orai1 involves an intramolecular switching mechanism. *Sci Signal* **3:** ra82.

Launikonis BS, Ríos E. 2007. Store-operated Ca^{2+} entry during intracellular Ca^{2+} release in mammalian skeletal muscle. *J Physiol* **583:** 81–97.

Launikonis BS, Murphy RM, Edwards JN. 2010. Toward the roles of store-operated Ca^{2+} entry in skeletal muscle. *Pflugers Arch* **460**: 813–823.

Lee KP, Yuan JP, Zeng W, So I, Worley PF, Muallem S. 2009. Molecular determinants of fast Ca^{2+}-dependent inactivation and gating of the Orai channels. *Proc Natl Acad Sci* **106**: 14687–14692.

Lewis RS, Cahalan MD. 1989. Mitogen-induced oscillations of cytosolic Ca^{2+} and transmembrane Ca^{2+} current in human leukemic T cells. *Cell Regul* **1**: 99–112.

Li Z, Lu J, Xu P, Xie X, Chen L, Xu T. 2007. Mapping the interacting domains of STIM1 and Orai1 in Ca^{2+} release-activated Ca^{2+} channel activation. *J Biol Chem* **282**: 29448–29456.

Li Z, Liu L, Deng Y, Ji W, Du W, Xu P, Chen L, Xu T. 2011. Graded activation of CRAC channel by binding of different numbers of STIM1 to Orai1 subunits. *Cell Res* **21**: 305–315.

Liou J, Kim M, Heo W, Jones J, Myers J, Ferrell J, Meyer T. 2005. STIM is a Ca^{2+} sensor essential for Ca^{2+}-store-depletion-triggered Ca^{2+} influx. *Curr Biol* **15**: 1235–1241.

Liou J, Fivaz M, Inoue T, Meyer T. 2007. Live-cell imaging reveals sequential oligomerization and local plasma membrane targeting of stromal interaction molecule 1 after Ca^{2+} store depletion. *Proc Natl Acad Sci* **104**: 9301–9306.

Lioudyno M, Kozak J, Penna A, Safrina O, Zhang S, Sen D, Roos J, Stauderman K, Cahalan M. 2008. Orai1 and STIM1 move to the immunological synapse and are up-regulated during T cell activation. *Proc Natl Acad Sci* **105**: 2011–2016.

Lis A, Peinelt C, Beck A, Parvez S, Monteilh-Zoller M, Fleig A, Penner R. 2007. CRACM1, CRACM2, and CRACM3 are store-operated Ca^{2+} channels with distinct functional properties. *Curr Biol* **17**: 794–800.

Lis A, Zierler S, Peinelt C, Fleig A, Penner R. 2010. A single lysine in the N-terminal region of store-operated channels is critical for STIM1-mediated gating. *J Gen Physiol* **136**: 673–686.

Litjens T, Harland M, Roberts M, Barritt G, Rychkov G. 2004. Fast Ca^{2+}-dependent inactivation of the store-operated Ca^{2+} current (I_{SOC}) in liver cells: A role for calmodulin. *J Physiol* **558**: 85–97.

Luik R, Wu M, Buchanan J, Lewis R. 2006. The elementary unit of store-operated Ca^{2+} entry: Local activation of CRAC channels by STIM1 at ER-plasma membrane junctions. *J Cell Biol* **174**: 815–825.

Luik R, Wang B, Prakriya M, Wu M, Lewis R. 2008. Oligomerization of STIM1 couples ER calcium depletion to CRAC channel activation. *Nature* **454**: 538–542.

Lur G, Haynes L, Prior I, Gerasimenko O, Feske S, Petersen O, Burgoyne R, Tepikin A. 2009. Ribosome-free Terminals of Rough ER Allow Formation of STIM1 Puncta and Segregation of STIM1 from IP_3 Receptors. *Curr Biol* **19**: 1648–1653.

Madl J, Weghuber J, Fritsch R, Derler I, Fahrner M, Frischauf I, Lackner B, Romanin C, Schütz GJ. 2010. Resting-state Orai1 diffuses as homotetramer in the plasma membrane of live mammalian cells. *J Biol Chem* **285**: 41135–41142.

Malli R, Naghdi S, Romanin C, Graier WF. 2008. Cytosolic Ca^{2+} prevents the subplasmalemmal clustering of STIM1: An intrinsic mechanism to avoid Ca^{2+} overload. *J Cell Sci* **121**: 3133–3139.

Manji S, Parker N, Williams R, van Stekelenburg L, Pearson R, Dziadek M, Smith P. 2000. STIM1: A novel phosphoprotein located at the cell surface. *Biochim Biophys Acta* **1481**: 147–155.

Mason MJ, Mahaut-Smith MP, Grinstein S. 1991. The role of intracellular Ca^{2+} in the regulation of the plasma membrane Ca^{2+} permeability of unstimulated rat lymphocytes. *J Biol Chem* **266**: 10872–10879.

Matsumoto M, Fujii Y, Baba A, Hikida M, Kurosaki T, Baba Y. 2011. The calcium sensors STIM1 and STIM2 control B cell regulatory function through interleukin-10 production. *Immunity* **34**: 703–714.

McCarl C-A, Picard C, Khalil S, Kawasaki T, Röther J, Papolos A, Kutok J, Hivroz C, LeDeist F, Plogmann K, Ehl S, Feske S, et al. 2009. ORAI1 deficiency and lack of store-operated Ca^{2+} entry cause immunodeficiency, myopathy, and ectodermal dysplasia. *J Allergy Clin Immunol* **124**: 1311–1318.e7.

McNally BA, Yamashita M, Engh A, Prakriya M. 2009. Structural determinants of ion permeation in CRAC channels. *Proc Natl Acad Sci* **106**: 22516–22521.

Mercer J, Dehaven W, Smyth J, Wedel B, Boyles R, Bird G, Putney J. 2006. Large store-operated calcium selective currents due to co-expression of Orai1 or Orai2 with the intracellular calcium sensor, Stim1. *J Biol Chem* **281**: 24979–24990.

Mignen O, Thompson JL, Shuttleworth TJ. 2007. STIM1 regulates Ca^{2+} entry via arachidonate-regulated Ca^{2+}-selective (ARC) channels without store depletion or translocation to the plasma membrane. *J Physiol (Lond)* **579**: 703–715.

Mignen O, Thompson JL, Shuttleworth TJ. 2008. Orai1 subunit stoichiometry of the mammalian CRAC channel pore. *J Physiol (Lond)* **586**: 419–425.

Mignen O, Thompson JL, Shuttleworth TJ. 2009. The molecular architecture of the arachidonate-regulated Ca^{2+}-selective ARC channel is a pentameric assembly of Orai1 and Orai3 subunits. *J Physiol* **587**: 4181–4197.

Muik M, Frischauf I, Derler I, Fahrner M, Bergsmann J, Eder P, Schindl R, Hesch C, Polzinger B, Fritsch R, Romanin C, et al. 2008. Dynamic coupling of the putative coiled-coil domain of ORAI1 with STIM1 mediates ORAI1 channel activation. *J Biol Chem* **283**: 8014–8022.

Muik M, Fahrner M, Derler I, Schindl R, Bergsmann J, Frischauf I, Groschner K, Romanin C. 2009. A cytosolic homomerization and a modulatory domain within STIM1 C terminus determine coupling to ORAI1 channels. *J Biol Chem* **284**: 8421–8426.

Muik M, Fahrner M, Schindl R, Stathopulos P, Frischauf I, Derler I, Plenk P, Lackner B, Groschner K, Ikura M, Romanin C. 2011. STIM1 couples to ORAI1 via an intramolecular transition into an extended conformation. *EMBO J* **30**: 1678–1689.

Mullins F, Park C, Dolmetsch R, Lewis R. 2009. STIM1 and calmodulin interact with Orai1 to induce Ca^{2+}-dependent inactivation of CRAC channels. *Proc Natl Acad Sci* **106**: 15495–15500.

Navarro-Borelly L, Somasundaram A, Yamashita M, Ren D, Miller RJ, Prakriya M. 2008. STIM1-Orai1 interactions and Orai1 conformational changes revealed by live-cell FRET microscopy. *J Physiol* **586:** 5383–5401.

Ng S-W, Nelson C, Parekh AB. 2009. Coupling of Ca^{2+} microdomains to spatially and temporally distinct cellular responses by the tyrosine kinase Syk. *J Biol Chem* **284:** 24767–24772.

Oh-Hora M. 2009. Calcium signaling in the development and function of T-lineage cells. *Immunol Rev* **231:** 210–224.

Oh-Hora M, Yamashita M, Hogan P, Sharma S, Lamperti E, Chung W, Prakriya M, Feske S, Rao A. 2008. Dual functions for the endoplasmic reticulum calcium sensors STIM1 and STIM2 in T cell activation and tolerance. *Nat Immunol* **9:** 432–443.

Orci L, Ravazzola M, Le Coadic M, Shen W-W, Demaurex N, Cosson P. 2009. STIM1-induced precortical and cortical subdomains of the endoplasmic reticulum. *Proc Natl Acad Sci* **106:** 19358–19362.

Pan Z, Yang D, Nagaraj RY, Nosek TA, Nishi M, Takeshima H, Cheng H, Ma J. 2002. Dysfunction of store-operated calcium channel in muscle cells lacking mg29. *Nat Cell Biol* **4:** 379–383.

Parekh AB, Penner R. 1997. Store depletion and calcium influx. *Physiol Rev* **77:** 901–930.

Parekh AB, Putney JW. 2005. Store-operated calcium channels. *Physiol Rev* **85:** 757–810.

Park CY, Hoover PJ, Mullins FM, Bachhawat P, Covington ED, Raunser S, Walz T, Garcia KC, Dolmetsch RE, Lewis RS. 2009. STIM1 clusters and activates CRAC channels via direct binding of a cytosolic domain to Orai1. *Cell* **136:** 876–890.

Park CY, Shcheglovitov A, Dolmetsch R. 2010. The CRAC channel activator STIM1 binds and inhibits L-type voltage-gated calcium channels. *Science* **330:** 101–105.

Partiseti M, Le Deist F, Hivroz C, Fischer A, Korn H, Choquet D. 1994. The calcium current activated by T cell receptor and store depletion in human lymphocytes is absent in a primary immunodeficiency. *J Biol Chem* **269:** 32327–32335.

Peinelt C, Vig M, Koomoa D, Beck A, Nadler M, Koblan-Huberson M, Lis A, Fleig A, Penner R, Kinet J. 2006. Amplification of CRAC current by STIM1 and CRACM1 (Orai1). *Nat Cell Biol* **8:** 771–773.

Penna A, Demuro A, Yeromin AV, Zhang SL, Safrina O, Parker I, Cahalan MD. 2008. The CRAC channel consists of a tetramer formed by Stim-induced dimerization of Orai dimers. *Nature* **456:** 116–120.

Penner R, Matthews G, Neher E. 1988. Regulation of calcium influx by second messengers in rat mast cells. *Nature* **334:** 499–504.

Picard C, McCarl C-A, Papolos A, Khalil S, Lüthy K, Hivroz C, LeDeist F, Rieux-Laucat F, Rechavi G, Rao A, Fischer A, Feske S. 2009. STIM1 mutation associated with a syndrome of immunodeficiency and autoimmunity. *N Engl J Med* **360:** 1971–1980.

Pozo-Guisado E, Campbell DG, Deak M, Alvarez-Barrientos A, Morrice NA, Alvarez IS, Alessi DR, Martín-Romero FJ. 2010. Phosphorylation of STIM1 at ERK1/2 target sites modulates store-operated calcium entry. *J Cell Sci* **123:** 3084–3093.

Prakriya M, Lewis R. 2006. Regulation of CRAC channel activity by recruitment of silent channels to a high open-probability gating mode. *J Gen Physiol* **128:** 373–386.

Prakriya M, Lewis R, Maue R, Bittar E. 2004. Store-operated calcium channels: Properties, functions and the search for a molecular mechanism. *Adv Molec Cell Biol* **32:** 121–140.

Prakriya M, Feske S, Gwack Y, Srikanth S, Rao A, Hogan P. 2006. Orai1 is an essential pore subunit of the CRAC channel. *Nature* **443:** 230–233.

Putney JW. 1986. A model for receptor-regulated calcium entry. *Cell Calcium* **7:** 1–12.

Putney JW. 2009. Capacitative calcium entry: From concept to molecules. *Immunol Rev* **231:** 10–22.

Putney JW, Bird GS. 1993. The inositol phosphate-calcium signaling system in nonexcitable cells. *Endocr Rev* **14:** 610–631.

Roos J, DiGregorio P, Yeromin A, Ohlsen K, Lioudyno M, Zhang S, Safrina O, Kozak J, Wagner S, Cahalan M, Stauderman K, et al. 2005. STIM1, an essential and conserved component of store-operated Ca^{2+} channel function. *J Cell Biol* **169:** 435–445.

Sarkadi B, Tordai A, Homolya L, Scharff O, Gárdos G. 1991. Calcium influx and intracellular calcium release in anti-CD3 antibody-stimulated and thapsigargin-treated human T lymphoblasts. *J Membr Biol* **123:** 9–21.

Scrimgeour N, Litjens T, Ma L, Barritt GJ, Rychkov GY. 2009. Properties of Orai1 mediated store-operated current depend on the expression levels of STIM1 and Orai1 proteins. *J Physiol* **587:** 2903–2918.

Shuttleworth TJ. 2009. Arachidonic acid, ARC channels, and Orai proteins. *Cell Calcium* **45:** 602–610.

Singaravelu K, Nelson C, Bakowski D, Martins de Brito O, Ng SW, Di Capite J, Powell T, Scorrano L, Parekh AB. 2011. Mitofusin 2 regulates STIM1 migration from the Ca^{2+} store to the plasma membrane in cells with depolarised mitochondria. *J Biol Chem* **286:** 12189–12201.

Smyth J, Dehaven W, Bird G, Putney J. 2007. Role of the microtubule cytoskeleton in the function of the store-operated Ca^{2+} channel activator STIM1. *J Cell Sci* **120:** 3762–3771.

Smyth J, DeHaven W, Bird G, Putney J Jr. 2008. Ca^{2+}-store-dependent and -independent reversal of Stim1 localization and function. *J Cell Sci* **121:** 762–772.

Smyth JT, Petranka JG, Boyles RR, DeHaven WI, Fukushima M, Johnson KL, Williams JG, Putney JW. 2009. Phosphorylation of STIM1 underlies suppression of store-operated calcium entry during mitosis. *Nat Cell Biol* **11:** 1465–1472.

Soboloff J, Spassova M, Tang X, Hewavitharana T, Xu W, Gill D. 2006. Orai1 and STIM reconstitute store-operated calcium channel function. *J Biol Chem* **281:** 20661–20665.

Srikanth S, Jung H-J, Kim K-D, Souda P, Whitelegge J, Gwack Y. 2010. A novel EF-hand protein, CRACR2A, is a cytosolic Ca^{2+} sensor that stabilizes CRAC channels in T cells. *Nat Cell Biol* **12:** 436–446.

Srikanth S, Jung HJ, Ribalet B, Gwack Y. 2010. The intracellular loop of Orai1 plays a central role in fast inactivation of Ca^{2+} release-activated Ca^{2+} channels. *J Biol Chem* **285:** 5066–5075.

Stathopulos P, Li G, Plevin M, Ames J, Ikura M. 2006. Stored Ca^{2+} depletion-induced oligomerization of STIM1 via the EF-SAM region: An initiation mechanism for capacitive Ca^{2+} entry. *J Biol Chem* **281:** 35855–35862.

Stathopulos PB, Zheng L, Li G-Y, Plevin MJ, Ikura M. 2008. Structural and mechanistic insights into STIM1-mediated initiation of store-operated calcium entry. *Cell* **135:** 110–122.

Stemmer, Paul M, Klee, Claude B. 1994. Dual calcium ion regulation of calcineurin by calmodulin and calcineurin B. *Biochemistry* **33:** 6859–6866.

Stiber J, Hawkins A, Wang S, Burch J, Graham V, Ward CC, Seth M, Finch E, Malouf N, Williams RS, Rosenberg P, et al. 2008. STIM1 signalling controls store-operated calcium entry required for development and contractile function in skeletal muscle. *Nat Cell Biol* **10:** 688–697.

Südhof TC, Rothman JE. 2009. Membrane fusion: Grappling with SNARE and SM proteins. *Science* **323:** 474–477.

Thastrup O, Dawson AP, Scharff O, Foder B, Cullen PJ, Drøbak BK, Bjerrum PJ, Christensen SB, Hanley MR. 1989. Thapsigargin, a novel molecular probe for studying intracellular calcium release and storage. *Agents Actions* **27:** 17–23.

Thompson J, Mignen O, Shuttleworth TJ. 2010. The N-terminal domain of Orai3 determines selectivity for activation of the store-independent ARC channel by arachidonic acid. *Channels (Austin)* **4:** 398–410.

Treves S, Franzini-Armstrong C, Moccagatta L, Arnoult C, Grasso C, Schrum A, Ducreux S, Zhu MX, Mikoshiba K, Girard T, Zorzato F, et al. 2004. Junctate is a key element in calcium entry induced by activation of $InsP_3$ receptors and/or calcium store depletion. *J Cell Biol* **166:** 537–548.

Treves S, Vukcevic M, Griesser J, Armstrong CF, Zhu MX, Zorzato F. 2010. Agonist-activated Ca^{2+} influx occurs at stable plasma membrane and endoplasmic reticulum junctions. *J Cell Sci* **123:** 4170–4181.

Varga-Szabo D, Braun A, Kleinschnitz C, Bender M, Pleines I, Pham M, Renné T, Stoll G, Nieswandt B. 2008. The calcium sensor STIM1 is an essential mediator of arterial thrombosis and ischemic brain infarction. *J Exp Med* **205:** 1583–1591.

Varga-Szabo D, Braun A, Nieswandt B. 2009. Calcium signaling in platelets. *J Thromb Haemost* **7:** 1057–1066.

Várnai P, Tóth B, Tóth DJ, Hunyady L, Balla T. 2007. Visualization and manipulation of plasma membrane-endoplasmic reticulum contact sites indicates the presence of additional molecular components within the STIM1-Orai1 Complex. *J Biol Chem* **282:** 29678–29690.

Várnai P, Hunyady L, Balla T. 2009. STIM and Orai: The long-awaited constituents of store-operated calcium entry. *Trends Pharmacol Sci* **30:** 118–128.

Vig M, Kinet J-P. 2009. Calcium signaling in immune cells. *Nat Immunol* **10:** 21–27.

Vig M, Beck A, Billingsley J, Lis A, Parvez S, Peinelt C, Koomoa D, Soboloff J, Gill D, Fleig A, Penner R, et al. 2006a. CRACM1 multimers form the ion-selective pore of the CRAC channel. *Curr Biol* **16:** 2073–2079.

Vig M, Peinelt C, Beck A, Koomoa DL, Rabah D, Koblan-Huberson M, Kraft S, Turner H, Fleig A, Penner R, Kinet J-P. 2006b. CRACM1 is a plasma membrane protein essential for store-operated Ca^{2+} entry. *Science* **312:** 1220–1223.

Vig M, DeHaven W, Bird G, Billingsley J, Wang H, Rao P, Hutchings A, Jouvin M, Putney J, Kinet J. 2008. Defective mast cell effector functions in mice lacking the CRACM1 pore subunit of store-operated calcium release-activated calcium channels. *Nat Immunol* **9:** 89–96.

Walsh CM, Chvanov M, Haynes LP, Petersen OH, Tepikin AV, Burgoyne RD. 2010a. Role of phosphoinositides in STIM1 dynamics and store-operated calcium entry. *Biochem J* **425:** 159–168.

Walsh CM, Doherty MK, Tepikin AV, Burgoyne RD. 2010b. Evidence for an interaction between Golli and STIM1 in store-operated calcium entry. *Biochem J* **430:** 453–460.

Wang Y, Deng X, Mancarella S, Hendron E, Eguchi S, Soboloff J, Tang XD, Gill DL. 2010. The calcium store sensor, STIM1, reciprocally controls Orai and $Ca_V1.2$ channels. *Science* **330:** 105–109.

Williams R, Manji S, Parker N, Hancock M, Van Stekelenburg L, Eid J, Senior P, Kazenwadel J, Shandala T, Saint R. 2001. Identification and characterization of the STIM (stromal interaction molecule) gene family: Coding for a novel class of transmembrane proteins. *Biochem J* **357:** 673–685.

Willoughby D, Cooper DMF. 2007. Organization and Ca^{2+} regulation of adenylyl cyclases in cAMP microdomains. *Physiol Rev* **87:** 965–1010.

Woodard GE, Salido GM, Rosado JA. 2008. Enhanced exocytotic-like insertion of Orai1 into the plasma membrane upon intracellular Ca^{2+} store depletion. *Am J Physiol, Cell Physiol* **294:** C1323–C1331.

Wu M, Buchanan J, Luik R, Lewis R. 2006. Ca^{2+} store depletion causes STIM1 to accumulate in ER regions closely associated with the plasma membrane. *J Cell Biol* **174:** 803–813.

Wu M, Luik R, Lewis R. 2007. Some assembly required: Constructing the elementary units of store-operated Ca^{2+} entry. *Cell Calcium* **42:** 163–172.

Xiao B, Coste B, Mathur J, Patapoutian A. 2011. Temperature-dependent STIM1 activation induces Ca^{2+} influx and modulates gene expression. *Nat Chem Biol* **7:** 351–358.

Xu P, Lu J, Li Z, Yu X, Chen L, Xu T. 2006. Aggregation of STIM1 underneath the plasma membrane induces clustering of Orai1. *Biochem Biophys Res Commun* **350:** 969–976.

Yamashita M, Navarro-Borelly L, McNally B, Prakriya M. 2007. Orai1 mutations alter ion permeation and Ca^{2+}-dependent fast inactivation of CRAC channels: Evidence for coupling of permeation and gating. *J Gen Physiol* **130:** 525–540.

Yeromin A, Zhang S, Jiang W, Yu Y, Safrina O, Cahalan M. 2006. Molecular identification of the CRAC channel by altered ion selectivity in a mutant of Orai. *Nature* **443:** 226–229.

Yu F, Sun L, Machaca K. 2009. Orai1 internalization and STIM1 clustering inhibition modulate SOCE inactivation during meiosis. *Proc Natl Acad Sci* **106:** 17401–17406.

Yu F, Sun L, Machaca K. 2010. Constitutive recycling of the store-operated Ca^{2+} channel Orai1 and its internalization during meiosis. *J Cell Biol* **191**: 523–535.

Yuan J, Zeng W, Dorwart M, Choi Y, Worley P, Muallem S. 2009. SOAR and the polybasic STIM1 domains gate and regulate Orai channels. *Nat Cell Biol* **11**: 337–343.

Zeng W, Yuan JP, Kim MS, Choi YJ, Huang GN, Worley PF, Muallem S. 2008. STIM1 gates TRPC channels, but not Orai1, by electrostatic interaction. *Mol Cell* **32**: 439–448.

Zhang S, Yu Y, Roos J, Kozak J, Deerinck T, Ellisman M, Stauderman K, Cahalan M. 2005. STIM1 is a Ca^{2+} sensor that activates CRAC channels and migrates from the Ca^{2+} store to the plasma membrane. *Nature* **437**: 902–905.

Zhang SL, Yeromin AV, Zhang XH-F, Yu Y, Safrina O, Penna A, Roos J, Stauderman KA, Cahalan MD. 2006. Genome-wide RNAi screen of Ca^{2+} influx identifies genes that regulate Ca^{2+} release-activated Ca^{2+} channel activity. *Proc Natl Acad Sci* **103**: 9357–9362.

Zhou Y, Meraner P, Kwon HT, Machnes D, Masatsugu O-H, Zimmer J, Huang Y, Stura A, Rao A, Hogan PG. 2010a. STIM1 gates the store-operated calcium channel ORAI1 in vitro. *Nat Struct Mol Biol* **17**: 112–116.

Zhou Y, Ramachandran S, Oh-hora M, Rao A, Hogan PG. 2010b. Pore architecture of the ORAI1 store-operated calcium channel. *Proc Natl Acad Sci* **107**: 4896–4901.

Zweifach A, Lewis RS. 1993. Mitogen-regulated Ca^{2+} current of T lymphocytes is activated by depletion of intracellular Ca^{2+} stores. *Proc Natl Acad Sci* **90**: 6295–6299.

Zweifach A, Lewis RS. 1995. Rapid inactivation of depletion-activated calcium current (I_{CRAC}) due to local calcium feedback. *J Gen Physiol* **105**: 209–226.

mGluR1/TRPC3-Mediated Synaptic Transmission and Calcium Signaling in Mammalian Central Neurons

Jana Hartmann, Horst A. Henning, and Arthur Konnerth

Institute of Neuroscience and Center for Integrated Protein Science, Technical University of Munich, Munich, Germany

Correspondence: jana.hartmann@lrz.tum.de

Metabotropic glutamate receptors type 1 (mGluR1s) are required for a normal function of the mammalian brain. They are particularly important for synaptic signaling and plasticity in the cerebellum. Unlike ionotropic glutamate receptors that mediate rapid synaptic transmission, mGluR1s produce in cerebellar Purkinje cells a complex postsynaptic response consisting of two distinct signal components, namely a local dendritic calcium signal and a slow excitatory postsynaptic potential. The basic mechanisms underlying these synaptic responses were clarified in recent years. First, the work of several groups established that the dendritic calcium signal results from IP_3 receptor-mediated calcium release from internal stores. Second, it was recently found that mGluR1-mediated slow excitatory postsynaptic potentials are mediated by the transient receptor potential channel TRPC3. This surprising finding established TRPC3 as a novel postsynaptic channel for glutamatergic synaptic transmission.

Glutamate is the predominant neurotransmitter used by excitatory synapses in the mammalian brain (Hayashi 1952; Curtis et al. 1959). At postsynaptic sites, glutamate binds to two different classes of receptors, namely the ionotropic glutamate receptors (iGluRs) and the metabotropic glutamate receptors (mGluRs) (Sladeczek et al. 1985; Nicoletti et al. 1986; Sugiyama et al. 1987). The iGluRs represent ligand-gated nonselective cation channels that underlie excitatory postsynaptic currents (EPSCs). Based on their subunit composition, gating, and permeability properties, they are subdivided into three groups named after specific agonists: AMPA- (α-amino-3-hydroxy-5-methyl-4-isoxazolepropionic acid), NMDA receptors (*N*-methyl D-aspartate receptors) and kainate receptors (Alexander et al. 2009). The other class of glutamate receptors, the mGluRs, consists of receptors that are coupled to G proteins and act through distinct downstream signaling cascades. They are structurally different from iGluRs and characterized by the presence of seven transmembrane domains (Houamed et al. 1991; Masu et al. 1991). The mGluRs exist as homodimers that do not by themselves form an ion-permeable pore in the membrane (Ozawa et al. 1998). To date, eight different genes (and more splice variants) encoding mGluRs have been identified and form the mGluR1 through mGluR8 subtypes (Alexander et al. 2009). Based on the amino acid sequence homology, downstream signal transduction pathways, and pharmacological properties, each of the subtypes was assigned to one of three groups. Group I receptors consist of mGluR1 and mGluR5 that positively couple to the phospholipase C (PLC). The receptors mGluR2 and mGluR3 constitute group II, whereas the remaining mGluRs, namely mGluR4, mGluR6, mGluR7, and mGluR8, belong to group III. Both groups II and III inhibit the adenylyl cyclase and thereby reduce the concentration of cAMP in the cytosol.

Of all different subtypes, mGluR1 is the most abundantly expressed mGluR in the mammalian central nervous system. In the brain, mGluR1 is highly expressed in the olfactory bulb, dentate gyrus, and cerebellum (Lein et al. 2007). The highest expression level of mGluR1 in the brain is found in Purkinje cells, the principal neurons of the cerebellar cortex (Shigemoto et al. 1992; Lein et al. 2007). Together with the AMPA receptors, mGluR1s are part of the excitatory synapses formed between parallel fibers and Purkinje cells (Fig. 1A). Each Purkinje cell is innervated by 100,000–200,000 parallel fibers (Ito 2006) that are axons of the cerebellar granule cells, the most abundant type of neuron in the brain. A second type of excitatory input to Purkinje cells is represented by the climbing fibers that originate in the inferior olive in the brain stem (Ito 2006). The two excitatory synaptic inputs to Purkinje cells are important determinants for the main functions of the cerebellum, including the real-time control of movement precision, error-correction, and control of posture as well as the procedural learning of complex movement sequences and conditioned responses.

It is expected that mGluR1 is involved in many of these cerebellar functions. This view is supported by the observation that mGluR1-deficient knockout mice show severe impairments in motor coordination. In particular, the gait of these mice is strongly affected as well as their ability for motor learning and general coordination (Aiba et al. 1994). The phenotype of the general mGluR1-knockout mice is rescued by the insertion of the gene encoding mGluR1 exclusively into cerebellar Purkinje cells (Ichise et al. 2000) and blockade of mGluR1 expression only in Purkinje cells of adult mice leads to impaired motor coordination (Nakao et al. 2007). These findings established mGluR1 in Purkinje cell as synaptic receptors that are indispensable for a normal cerebellar function.

Synaptic transmission involving mGluR1s is found at both parallel fiber-Purkinje cell synapses (Batchelor and Garthwaite 1993; Batchelor et al. 1994) as well as at climbing fiber-Purkinje cell synapses (Dzubay and Otis 2002). Most of our knowledge on the mGluR1 was gained from the analysis of the parallel fiber synapses. The parallel fiber synapse is quite unique in the central nervous system regarding its endowment with neurotransmitter receptors. In contrast to most other glutamatergic synapses in the mammalian brain, it lacks functional NMDA receptors (Shin and Linden 2005). The entire synaptic transmission at these synapses relies on AMPA receptors and on mGluR1 (Takechi et al. 1998). Although AMPA receptors are effectively activated even with single shock stimuli (Konnerth et al. 1990; Llano et al. 1991b), activation of mGluRs requires repetitive stimulation (Batchelor and Garthwaite 1993; Batchelor et al. 1994; Batchelor and Garthwaite 1997; Takechi et al. 1998). A possible explanation for the need of repetitive stimulation may relate to the observation that mGluR1s are found mostly at the periphery of the subsynaptic region (Nusser et al. 1994). At these sites outside the synaptic cleft, glutamate levels that are sufficiently high for receptor activation may be reached only with repetitive stimulation.

At parallel fiber-Purkinje cell synapses, repetitive stimulation produces an initial AMPA receptor postsynaptic signal component, followed by a more prolonged mGluR1 component (Fig. 1). Figure 1B shows a current clamp recording of this response consisting of an early burst of action potentials, followed by a prolonged depolarization known as a "slow excitatory postsynaptic potential" (slow EPSP)

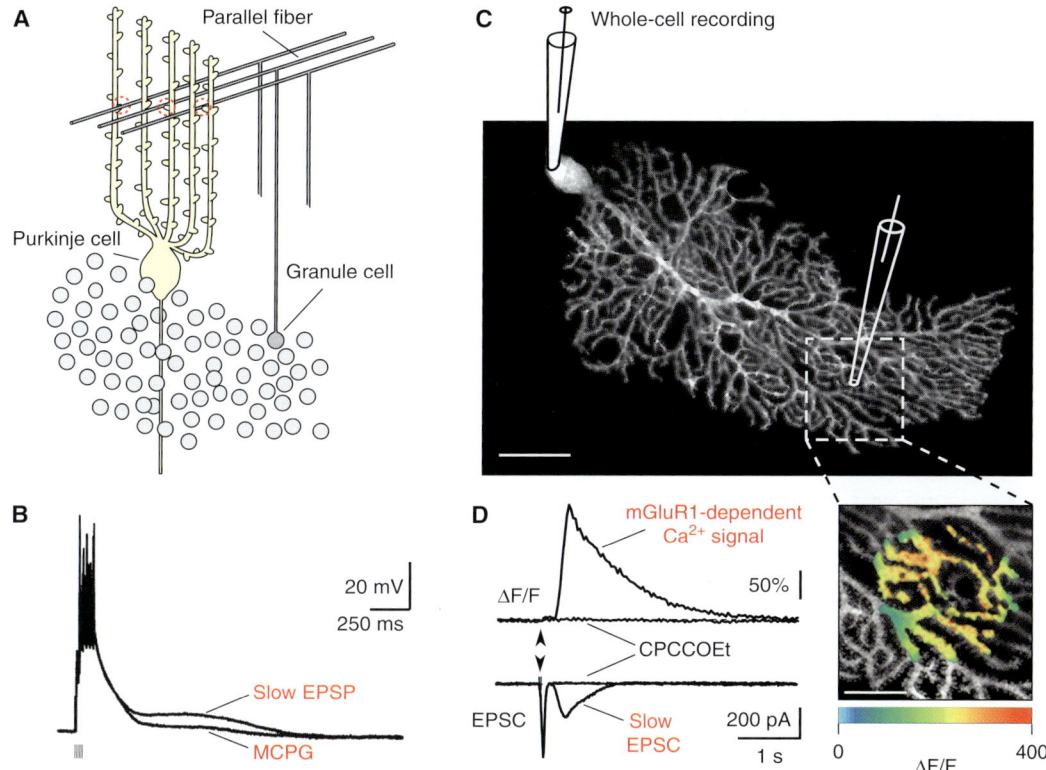

Figure 1. Parallel fiber-evoked mGluR1-dependent signals. (*A*) Diagram showing the parallel fiber synaptic input to Purkinje cell dendrites. (*B*) Microelectrode recording of glutamatergic postsynaptic potentials from a Purkinje cell in an acute slice of adult rat cerebellum. Short trains of stimuli to the parallel fibers (5–6 at 50 Hz) caused summation of the early AMPA receptor-dependent EPSPs (leading to spike firing) and a slow, delayed, depolarizing potential (slow EPSP), which was reversibly inhibited by antagonist of mGluRs (+)-MCPG (1mM). (*C*) Confocal image of a patch-clamped Purkinje cell in a cerebellar slice of an adult mouse. The patch-clamp pipette and the glass capillary used for electrical stimulation of parallel fibers are depicted schematically. The site of stimulation is shown at higher magnification in *D*. (*D*) *Left*: Parallel fiber-evoked (five pulses at 200 Hz, in 10 mM CNQX) synaptic responses consisting of a dendritic mGluR1-dependent Ca^{2+} transient ($\Delta F/F$, *top*) and an early rapid and a slow excitatory postsynaptic current (EPSC, *bottom*). Block of the mGluR1-dependent components by the group I-specific mGluR-antagonist CPCCOEt (200 μM) is shown as indicated. *Right*: Pseudocolor image of the synaptic Ca^{2+} signal. (*B*, Reprinted with modifications, with permission, from Batchelor and Gaithwaite 1997 [Nature Publishing Group].)

(Batchelor and Garthwaite 1993; Batchelor et al. 1994; Batchelor and Garthwaite 1997). Voltage-clamp recordings allow a clear separation of the initial rapid, AMPA receptor mediated excitatory postsynaptic current (EPSC) and the mGluR1-mediated slow EPSC (Fig. 1D) (Takechi et al. 1998; Hartmann et al. 2008). In addition of inducing the slow EPSPs, mGluR1s mediate a large and highly localized dendritic calcium transient in cerebellar Purkinje cells (Fig. 1D) (Llano et al. 1991a; Finch and Augustine 1998; Takechi et al. 1998).

mGluR1-DEPENDENT POSTSYNAPTIC Ca^{2+} RELEASE FROM INTERNAL STORES

The mGluR1-mediated synaptic Ca^{2+} transients constitute a distinct class of postsynaptic response because they may occur independently from changes in membrane potential (Takechi et al. 1998). The generation of these Ca^{2+} transients is the result of a cascade of signaling events. During the initial step, binding of glutamate to the mGluR1 activates the phospholipase C (PLC). This is followed by inositoltrisphosphate (IP_3) production and accumulation in the cytosol. IP_3 binds to its own receptor channels that are located in the endoplasmic reticulum (ER) membrane and are permeable for Ca^{2+} (Verkhratsky 2005). At parallel fiber-Purkinje cell synapses, the release of Ca^{2+} from ER stores forms a characteristic local signaling response (Finch and Augustine 1998; Takechi et al. 1998). An important feature of the mGluR1-dependent Ca^{2+} signal is its spatial restriction ranging from dendritic regions (Fig. 1C) to small dendritic terminal branchlets or even to single spines (Fig. 2). Focal synaptic stimulation involving the activation of just a few neighboring parallel fibers causes mGluR1-dependent Ca^{2+} signals that occur at tiny spino-dendritic regions (Fig. 2, spine 2-dendrite 2) or in individual spines (Fig. 2, spine 1) (Fig. 2B) (Takechi et al. 1998). However, there is evidence that nominally similar parallel fiber synaptic inputs may cause variable postsynaptic Ca^{2+} transients, depending apparently on the dendritic location (Hartmann et al. 2004). Possibly, this variability may relate to a heterogeneity of the organization of the smooth ER network in the Purkinje cell dendrites and spines (Harris and Stevens 1988; Martone et al. 1993; Terasaki et al. 1994). Alternatively, there may be an uneven distribution of IP_3 receptors, ryanodine receptors (RyRs), Ca^{2+}-ATPases and luminal Ca^{2+}-binding proteins, like calsequestrin, calnexin and calreticulin, in the dendritic tree of Purkinje cells (Villa et al.

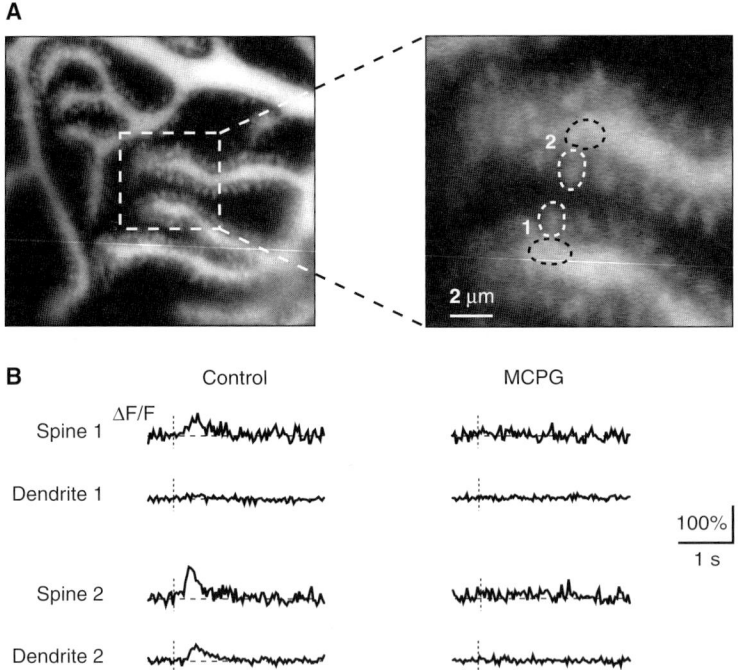

Figure 2. Identification of synaptic Ca^{2+}-release signals in spines and dendritic microdomains (Takechi et al. 1998). (A) Images of spiny dendrites of a Purkinje cell, indicating the sites of fluorescence measurements. (B) Fluorescence measurements taken from the dendrites shown in A. In spine 1, the synaptic Ca^{2+} transient was restricted to an individual spine. Note the complete absence of any Ca^{2+} signal in the immediate dendritic vicinity. A large synaptic Ca^{2+} transient was detected in spine 2 and a smaller signal occurred in the subspine region. MCPG (1 mM) completely blocked all synaptic Ca^{2+} transients.

1991; Takei et al. 1992; Villa et al. 1992; Nori et al. 1993).

A more detailed dissection of the intracellular signaling cascade shows that mGluR1s couple to their downstream effectors through members of the G_q protein subclass of heterotrimeric G proteins (Masu et al. 1991). Only two members of this G protein family, $G\alpha_q$ and $G\alpha_{11}$, are expressed in the brain (Nakamura et al. 1991; Wilkie et al. 1991). Both are present also in Purkinje cells (Tanaka et al. 2000) and it has been shown that the $G\alpha_q/G\alpha_{11}$ immunoreactivity colocalizes with mGluR1 in Purkinje cell spines (Tanaka et al. 2000). However, measurements of postsynaptic Ca^{2+} signals in Purkinje cells in mice deficient for either $G\alpha_q$ or $G\alpha_{11}$ revealed that only $G\alpha_q$, but not $G\alpha_{11}$, is required for mGluR-mediated Ca^{2+} transients (Hartmann et al. 2004).

G_q proteins activate the PLCβ of which four subtypes (PLCβ1-4) have been identified (Blank et al. 1991; Berstein et al. 1992; Rebecchi and Pentyala 2000). Purkinje cells express all subtypes with the exception of the PLCβ2 (Tanaka and Kondo 1994; Roustan et al. 1995; Watanabe et al. 1998; Lein et al. 2007). PLCβ1 has the lowest and PLCβ4 the highest expression level in Purkinje cells (Lein et al. 2007). PLCβ3 prevails in the caudal cerebellum (Hirono et al. 2001), whereas PLCβ4 is reciprocally expressed in the rostral lobuli of the cerebellum (Kano et al. 1998). PLCβ1 is present primarily in the somata of Purkinje cells and is, therefore, unlikely to play a major role in synaptic transmission at parallel fiber synapses that are all located in the spiny dendrites (Kano et al. 1998).

Of all three IP_3 receptor subtypes, IP_3R1 is expressed in Purkinje cells with extraordinary high density (Sharp et al. 1999; Lein et al. 2007). IP_3R2 is present in low amounts, whereas IP_3R3 is not found in Purkinje cells (Lein et al. 2007). Most remarkably, the signaling properties of the IP_3R1 in its native environment in the Purkinje cell cytosol differ largely from those detected in isolated preparations or other cell types (Khodakhah and Ogden 1993; Fujiwara et al. 2001). With an EC50 of 25.8 μM (Fujiwara et al. 2001) its sensitivity for IP_3 is exceptionally low, indicating that 10- to 20-fold higher concentrations of IP_3 are required to evoke Ca^{2+}-release from stores as compared with astrocytes, hepatocytes, exocrine cells and vascular endothelium (Khodakhah and Ogden 1993). Intriguingly, IP_3Rs isolated from Purkinje cells have a sensitivity for IP_3 that is similar to that found in other cell types (Fujiwara et al. 2001). In general, the IP_3R forms a "macro" signal complex in which it operates as an integrator and regulator for signaling cascades. The most important regulators of the IP_3R are Ca^{2+} ions, calmodulin, the immunophilin FKBP12, ATP, and protein kinases (reviewed in Patterson et al. [2004]; Mikoshiba [2007]). Notably, calmodulin specifically inhibits IP_3-binding to IP_3R1 receptors and IP_3-evoked Ca^{2+} mobilization in a Ca^{2+}-independent manner (Patel et al. 1997; Cardy and Taylor 1998). Because of its high expression in Purkinje cells, calmodulin may be responsible for the low affinity of the IP_3R1 for IP_3 in the Purkinje cell cytosol. A Ca^{2+}-independent binding to the IP_3R1 has been reported also for CaBP1, a member of the neuronal Ca^{2+} sensor family of Ca^{2+} binding proteins. The interaction of CaBP1 with the amino terminus of the IP_3R1 results in a weakened IP_3 binding to the receptor, similarly to the effect of calmodulin (Kasri et al. 2004). However, because of its very low expression in Purkinje cells (Lein et al. 2007), CaBP1 does not seem to be a major cause for the specific features of IP_3R-dependent Ca^{2+}-signaling in this cell type. Instead, it is assumed that, in addition to calmodulin, in Purkinje cells the IP_3R1 is specifically regulated by proteins like the IP_3 inhibitor IRI (Watras et al. 2000) or the carbonic anhydrase-related protein CARP (Nogradi et al. 1997). Interestingly, the expression of CARP is restricted to Purkinje cells (Nogradi et al. 1997) in which it reduces the affinity of IP_3R1 for IP_3 by binding to the modulatory domain of IP_3R1 (Hirota et al. 2003). On the behavioral level, the relevance of the carbonic anhydrase activity for IP_3R-dependent signaling in Purkinje cells is emphasized by the fact that a mutation in the carbonic anhydrase related protein 8 (Car8) is associated with a pronounced ataxic gait (Jiao et al. 2005), whereas on the ultrastructural level,

Purkinje cells show various synaptic anomalies (Hirasawa et al. 2007). Even though the affinity of the IP$_3$R1 for its agonist in Purkinje cells is low, the magnitude of Ca^{2+} release signal from the ER Ca^{2+} store is much larger than in other cell types (Khodakhah and Ogden 1993; Ogden and Capiod 1997). Furthermore, in comparison to cells in other tissues, in Purkinje cells the kinetics of activation and Ca^{2+}-dependent inactivation (Bezprozvanny et al. 1991) of the IP$_3$R1 are particularly fast (Khodakhah and Ogden 1995; Ogden and Capiod 1997). For a cytosolic IP$_3$ concentration of 38 μM following photorelease of "caged" IP$_3$, the efflux of Ca^{2+} ions from the ER was estimated to be $\approx$1 μM/ms in Purkinje cells and only $\approx$0.03 μM/ms in cells from peripheral tissues (Khodakhah and Ogden 1993). Together, these observations suggest that IP$_3$-dependent Ca^{2+}-release in Purkinje cells is particularly fine-tuned for signaling on the time scale of synaptic transmission in the millisecond range, which takes place in a localized space adjacent to the site of synaptic inputs (Takechi et al. 1998).

An increasing amount of evidence indicates that the mGluR1-mediated synaptic Ca^{2+} release signal from stores plays an important role in activity dependent synaptic plasticity. Thus, parallel fiber synapse-mediated IP$_3$R1-dependent Ca^{2+} transients are required for the induction of long-term depression (LTD) at parallel fiber-Purkinje cell synapses (Kasono and Hirano 1995; Khodakhah and Armstrong 1997; Finch and Augustine 1998; Inoue et al. 1998; Daniel et al. 1999). LTD, the presumed cellular basis of motor learning in the cerebellum (Ito 2000) requires the conjunctive stimulation of parallel and climbing fiber inputs (Gao et al. 2003). During LTD induction, parallel fiber firing activates mGluR1-dependent pathways that include IP$_3$R1-mediated release of Ca^{2+} ions from ER Ca^{2+} stores, whereas climbing fiber activity strongly depolarizes the Purkinje cells and induces influx of Ca^{2+} ions through voltage-gated Ca^{2+} channels (Knöpfel et al. 1991). The increased cytosolic concentrations of Ca^{2+} and IP$_3$ are, thus, the result of the concerted parallel and climbing fiber activity.

Because of their sensitivity to both IP$_3$ and Ca^{2+} (Bezprozvanny et al. 1991), IP$_3$R1s act as coincidence detectors for the two excitatory inputs onto Purkinje cells and mediate an amplification of the postsynaptic Ca^{2+} release signal. This facilitates the induction of LTD (Wang et al. 2000; Doi et al. 2005). Experimental evidence indicates that IP$_3$R1-mediated Ca^{2+} release in the spines themselves is of outstanding importance for the induction of LTD. Thus, LTD was absent in a myosin-Va mutant mouse line, in which the ER does not extend into dendritic spines (Miyata et al. 2000). Furthermore, IP$_3$R-dependent Ca^{2+} release in single spines elicited by sparse parallel fiber stimulation is able to induce LTD specifically at the activated input (Wang et al. 2000).

MECHANISMS UNDERLYING mGluR1-DEPENDENT SYNAPTIC DEPOLARIZATION

As mentioned above, repetitive stimulation of parallel fiber inputs evokes a slow EPSC that is sensitive to mGluR-specific antagonists (Batchelor and Garthwaite 1993; Batchelor et al. 1994). Remarkably, the frequencies that are most effective in brain slices (Batchelor and Garthwaite 1997) resemble those that are encountered with the stimulation of parallel (Isope et al. 2004) or mossy fibers in vivo (Chadderton et al. 2004). The slow EPSC is characterized by a characteristic time course: it starts with a latency of 100–200 ms after the stimulation of afferent parallel fibers and lasts for about 1 second (Fig. 1D)(Batchelor et al. 1994). The long search for the mechanisms underlying the slow EPSC involved pharmacological experiments that indicated that the mGluR1-mediated slow EPSC is not mediated by hyperpolarization-activated cation channels, purinergic receptors (Canepari et al. 2001), Na$^+$/Ca^{2+}-exchangers (Hirono et al. 1998) or voltage-gated Ca^{2+} channels (Tempia et al. 2001). Instead, the ionic properties of the slow EPSCs are reminiscent of currents that permeate through the canonical transient receptor potential (TRPC) channels. Indeed, a few years ago (Kim et al. 2003) reported experiments that suggested that the mGluR1-dependent slow EPSC is mediated by TRPC1.

Later studies investigated in more detail TRPC channels in Purkinje cells (Huang et al. 2007; Lein et al. 2007; Hartmann et al. 2008). Although the general notion of an involvement of TRPCs was confirmed, the new experiments did not provide evidence for a specific role of TRPC1. Instead, several lines of evidence pointed to TRPC3 as the postsynaptic channel that mediates the slow EPSC. Thus, the TRPC3 protein was found to be abundant in the somatodendritic compartment of Purkinje cells in the adult mouse brain (Fig. 3A,B) (Hartmann and Konnerth 2008). The number of mRNA copies for TRPC3 is 8–10 times higher in single Purkinje cells than that of TRPC1 whereas mRNA of the other subunits is present in low amounts. Furthermore, of all TRPC subunits the expression of TRPC3 is most tightly coupled to a developmental time window in which the outgrowth of the dendritic tree of Purkinje cells (Hendelman and Aggerwal 1980), the formation of parallel fiber synapses and the elimination of supernumerary climbing fibers (Scelfo and Strata 2005) take place (Fig. 3C) (Huang et al. 2007). The outstanding role of TRPC3 channels for mGluR1-dependent synaptic transmission in Purkinje cells was unambiguously shown by a recent study that analyzed mGluR1-mediated transmission in specifically designed TRPC-deficient mice (Hartmann et al. 2008) (Figs. 4A–F). The importance of TRPC3 and the mGluR1-dependent slow EPSC for cerebellar function was further emphasized by the behavioral impairments that were observed in TRPC3 null mutant mice. These mice show a movement deficit of their hindpaws that leads to an ataxic wide-based gait and poor performance when they walk on a horizontal ladder or an elevated beam (Hartmann et al. 2008).

The role of TRPC3 for mGluR1-mediated signaling and sensorimotor integration was further corroborated in a study that used mice with a gain-of-function mutation in the *Trpc3* gene. A single amino acid exchange in the TRPC3 protein (T635A) results in loss of a phosphorylation site of TRPC3 and altered gating of the channel. Therefore, mGluR1-dependent inward currents are increased and motor coordination is heavily impaired. These so-called *moonwalker* (*Mwk*) mice display massive symptoms of ataxia. Remarkably, and in distinct contrast to the TRPC3 knockout mice (Hartmann et al. 2008), increased TRPC3-mediated signaling in *Mwk* mice is accompanied by a reduced outgrowth of the dendritic arborization of Purkinje cells and even Purkinje cell loss after 4 months of age (Becker et al. 2009). It is remarkable that the absence of TRPC3 is less deleterious for Purkinje cell development than the excess of TRPC3-mediated signaling. The high impact of TRPC3 overexpression may relate to the fact that the developmental increase in TRPC3 expression in Purkinje cells during the first two postnatal weeks (Huang et al. 2007) coincides with the most intensive phase of dendritic growth in these neurons (Hendelman and Aggerwal 1980).

The signaling cascade that links mGluR1 and TRPC3 is not yet fully elucidated. There is evidence that both mGluR1-evoked signal components, the Ca^{2+} release signal and the slow EPSC, are mediated by G_q proteins, but subsequently follow divergent pathways (Fig. 5). It has been shown that both $G\alpha_q$ and $G\alpha_{11}$ contribute to the generation of the slow EPSC (Hartmann et al. 2004). However, pharmacological manipulations that block the mGluR1-dependent Ca^{2+} release from internal stores do not affect the mGluR1-dependent depolarization (Finch and Augustine 1998; Takechi et al. 1998). Importantly, the two signal components of mGluR1-mediated transmission, the Ca^{2+} signal and the slow EPSC, may occur independently of each other (Hirono et al. 1998; Takechi et al. 1998; Tempia et al. 1998; Hartmann et al. 2008), as expected from processes that involve distinct intracellular signaling pathways.

The role of PLCβ in the generation of the mGluR1-dependent slow EPSC is controversial. An involvement of PLCβ was indicated by the observation that slow EPSC were absent in PLCβ4 deficient mutant mice (Sugiyama et al. 1999). However, slow EPSCs seem to be insensitive to the PLCβ antagonist U73122 (Hirono et al. 1998; Canepari et al. 2001; Glitsch 2010). It has been reported that U73122 abolished

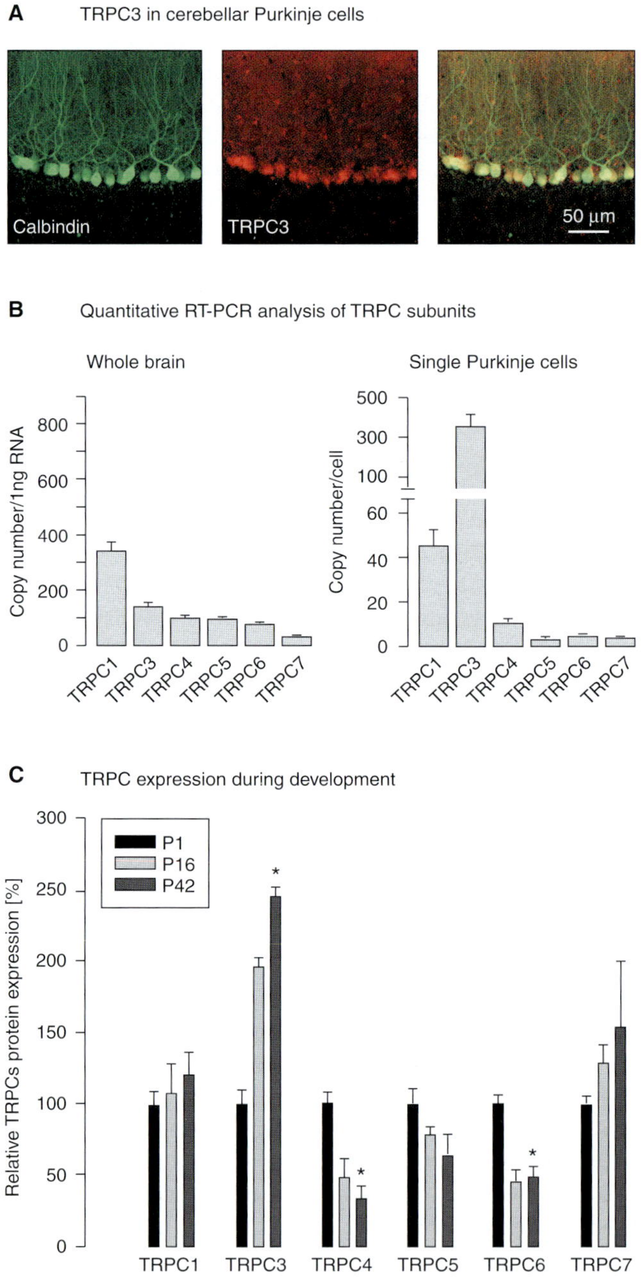

Figure 3. Expression of TRPC channel subunits in the murine brain and single Purkinje cells. (*A*) A dual-channel confocal scan of an immunohistochemical staining in an acute cerebellar slice. Calbindin D28k immunoreactivity is shown in green (*left*) and that for TRPC3 in red (*middle*). *Right*: Merged images (Hartmann et al. 2008). (*B*) Copy numbers of TRPC subunit mRNA detected in 1ng total RNA of mouse whole brain (*left*) and in single Purkinje cells (*right*). (*C*) Relative TRPC expression at postnatal days 1 (P1), 16 (P16) and 42 (P42), normalized to average expression value obtained at P1 for a given TRPC subunit. For the quantification, relative Western blot band intensities were analyzed. Data represent averages from three to four independent Western blots. There is a significant increase in TRPC3 expression and a significant decrease in both TRPC4 and TRPC6 expression (unpaired t-tests). (Reprinted with modifications, with permission, from Huang et al. 2007 [Elsevier].)

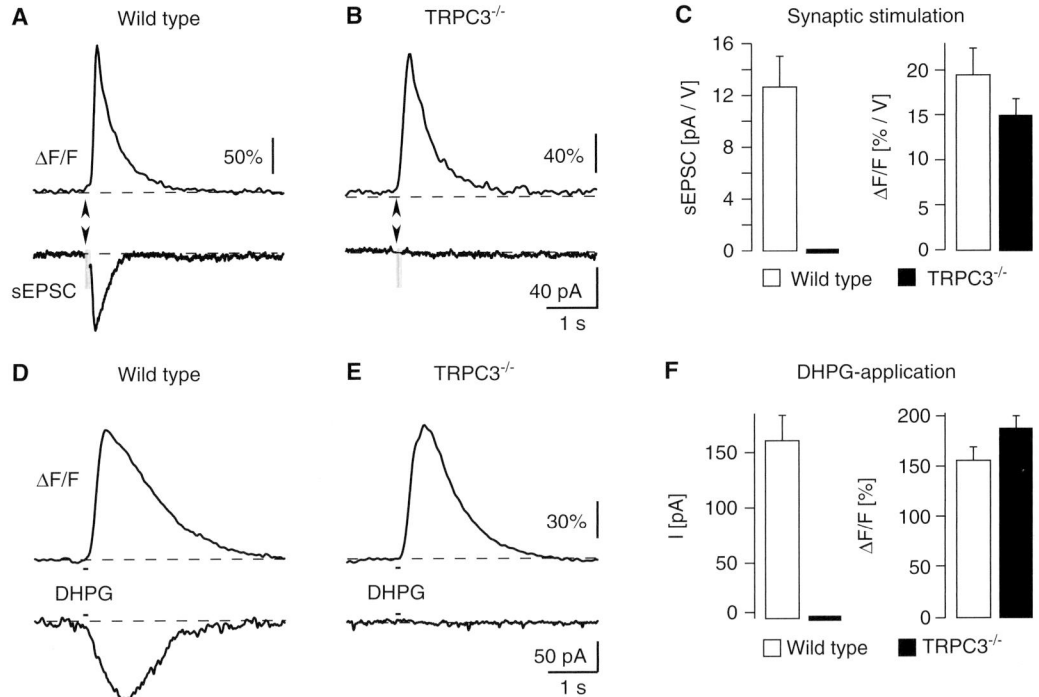

Figure 4. Analysis of TRPC3-deficient mice (Hartmann et al. 2008). (*A*) Slow EPSC in a wild-type mouse (*lower trace*) and the corresponding local dendritic Ca^{2+} response (*upper trace*). (*B*) Similar recording in a TRPC3$^{-/-}$ mouse. (*C*) Summary graphs for normalized (to stimulation strength) slow EPSCs and Ca^{2+} transients. (*D*) DHPG evoked a slow inward current (*lower trace*) and a local Ca^{2+} transient (*upper trace*) in a wild-type mouse. (*E*) Similar recording in a TRPC3$^{-/-}$ mouse. (*F*) Summary of DHPG-evoked current (*right*) and Ca^{2+} (*left*) responses.

the mGluR1-dependent Ca^{2+} release signal in rat cerebellar slices (Takechi et al. 1998). A possible explanation for these discrepancies may be that the genetic removal of PLCβ disrupted the tight assembly of the molecular components of the mGluR1-dependent signaling cascade at the postsynaptic density (Nakamura et al. 2004) and that PLCβ activity is generally not needed for the signal transduction between mGluR1 and TRPC3. A recent study indicated that the activation of mGluR1 initiates two independent signaling pathways downstream of the G_q proteins: one dependent on the PLCβ and another one dependent on phospholipase D1 (Fig. 5) (Glitsch 2010). It was shown in that paper that activation of mGluR1 is followed by the translocation of the phospholipase D to the plasma membrane and that the TRPC3-dependent slow EPSCs in Purkinje cell require the activity of the phospholipase D (PLD). However, PLD most likely does not gate the TRPC3 directly, as suggested by experiments that were performed in a cell line transfected with human TRPC3 (Glitsch 2010). In Purkinje cells, DHPG-evoked inward currents are sensitive to antagonists of the small G proteins of the rho family. From other studies it is known that PLD1 is one of the effectors of rho GTPases (Jenkins and Frohman 2005; Weernink et al. 2007). Thus, it is possible that the activation of TRPC3 through mGluR1 involves a rho GTPase-dependent pathway that depends on PLD1 (Glitsch 2010).

Another subtype of the PLC, namely PLCγ, which is also found in Purkinje cells (Lein et al. 2007) has been shown to activate TRPC3 in other cell types (van Rossum et al. 2005; Lockwich et al. 2008; Tong et al. 2008) including

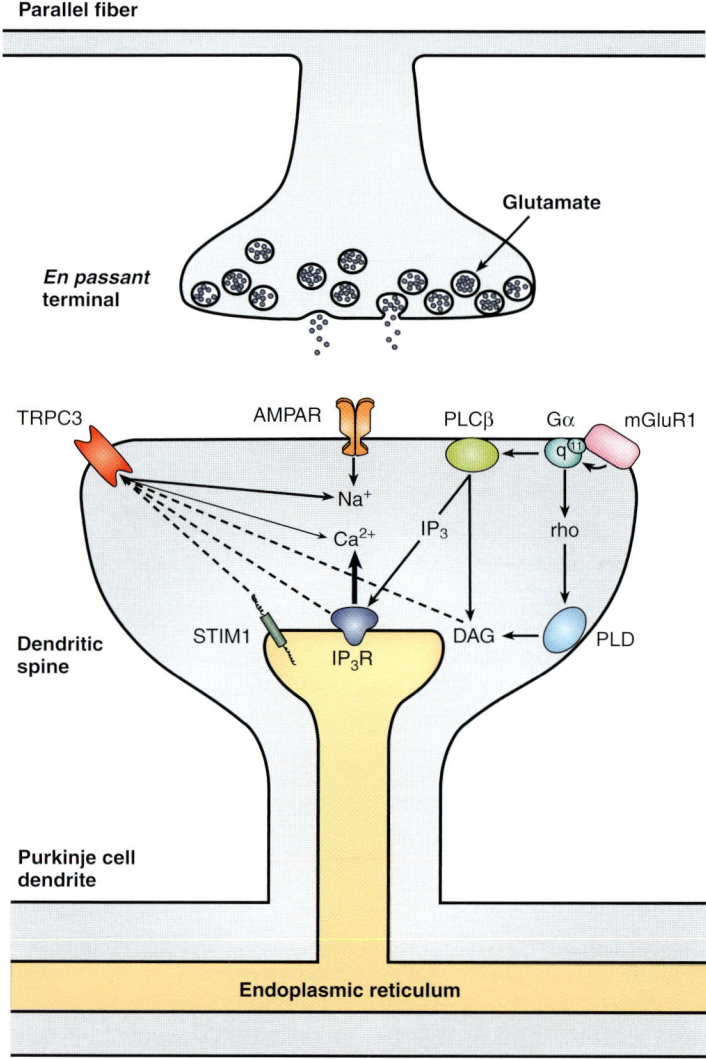

Figure 5. Model of glutamatergic synaptic signaling at parallel fiber-Purkinje cell synapses. Glutamate released from presynaptic terminals binds to AMPA receptors and the mGluR1. Influx of sodium ions through AMPA receptor channels leads to a fast postsynaptic depolarization. The mGluR1 couples predominantly to $G\alpha_q$ but also to $G\alpha_{11}$. This is followed by the activation of the phospholipase Cβ that cleaves phosphatidyl-4,5-bisphosphate (PIP$_2$) into inositoltrisphosphate (IP$_3$) and diacylglycerol (DAG). By binding to its receptor (IP$_3$R) in the endoplasmic reticulum membrane, IP$_3$ releases Ca^{2+}-ions from this intracellular Ca^{2+}-store. TRPC3 channels in Purkinje cells can be activated following stimulation of the PLD and require activity of a small G protein of the Rho family. Possible other candidates for TRPC3 gating are DAG, the stromal interaction molecule 1 (STIM1), the amino terminus of the IP$_3$R1 a.o.

neurons (Li et al. 1999). However, the internal effector molecule that activates the TRPC3 remains to be determined. It is well established that, when activated, the PLCs, in addition to the soluble factor IP$_3$, produce the lipophilic compound diacylglycerol (DAG) (Weernink et al. 2006). In heterologous expression systems, TRPC3 was shown to be directly regulated by diacylglycerol (DAG) (Fig. 5) (Hofmann et al. 1999; Lemonnier et al. 2008) and by the substrate of PLCs, namely phosphatidylinositol-4,5-bisphosphate (Lemonnier et al. 2008). In transfected HEK293 cells, the opening of TRPC3 channels is induced by their binding to the amino-terminal domain of the activated IP_3 receptor (Kiselyov et al. 1998; Kiselyov et al. 1999). In the same cell type, there is also an obligatory role of tyrosine kinase Src (Vazquez et al. 2004). In pancreatic acinar cells, the scaffolding protein Homer 1b/c is critical for the regulation of the activity of TRPC3 by the IP_3 receptor (Kim et al. 2006). Finally, another interesting component involved in TRPC3 signaling is the stromal interacting molecule I (STIM1) that has an essential role for the Ca^{2+} homeostasis in the ER in many cell types (Roos et al. 2005; Zhang et al. 2005). STIM1 is a Ca^{2+} sensor that, following depletion of ER Ca^{2+} stores, activates a Ca^{2+} influx through Orai channels in the plasma membrane of nonneuronal cells (Zhang et al. 2005). This so-called store-operated Ca^{2+} entry (SOCE) ensures the replenishment of ER Ca^{2+} stores. Because of their opening downstream of the PLC and their permeability for Ca^{2+} ions, TRPC channels were proposed to function as store-operated channels (Zhu et al. 1996; Zitt et al. 1997). In HEK293 cells, it has been shown that STIM1 exerts a gating action on TRPC3 by intermolecular electrostatic interaction (Zeng et al. 2008). TRPC3 opens when two positively charged lysine-residues in the carboxyl terminus of STIM1 (684 and 685) get in contact with two negatively charged residues of aspartate (697 and 698) in the intracellular carboxy-terminal domain of TRPC3. Purkinje cells express all the factors mentioned above (Lein et al. 2007) and it is important to note that TRPC3 is part of a multi-protein signaling complex that is located postsynaptically and, in addition to the mGluR1, contains also the IP_3R1, PLCβ4 (Nakamura et al. 2004), Homer proteins (Tu et al. 1999; Yuan et al. 2003; Kim et al. 2006) and STIM1 and other TRPC subunits (Yuan et al. 2007). However, whether SOCE requires TRPC channels and whether they are part of the STIM1-Orai-complex is controversially discussed (Liao et al. 2007; Liao et al. 2008; DeHaven et al. 2009).

As most other TRP channels, TRPC3 channels are permeable not only for Ca^{2+} but also for Na^+ ions (Zhu et al. 1996; Alexander et al. 2009). It is, therefore, not surprising that TRPC3-mediated slow EPSCs are associated with dendritic Na^+ transients in Purkinje cells (Knöpfel et al. 2000). These synaptic Na^+ signals are restricted to dendritic regions that are innervated by the afferent parallel fibers. In addition to Na^+ transients, mGluR1-dependent slow EPSCs are also associated with dendritic Ca^{2+} transients that may result in part from Ca^{2+} entry through TRPC3 channels (Tempia et al. 2001). However, it remains unclear how Ca^{2+} entry through TRPC3 channels compares to the Ca^{2+} release signal from internal stores. The answer to this question is important for a better understanding of the induction of LTD, a process known to require a transient elevation in intradendritic Ca^{2+} concentration (Konnerth et al. 1992).

Another open question concerns the possible molecular partners of TRPC3 channels in Purkinje cells. In view of the fact that Purkinje cells express various TRPCs (Fig. 3), TRPC3 may form heteromers with one or several of those channels (Zagranichnaya et al. 2005). The specific arrangement may determine the relative composition of the ion fluxes (Nilius et al. 2007). Furthermore, the existence of heteromeric complexes of TRPC- and Orai-channels, which may be regulated by STIM1, has been suggested and is controversially discussed at present (Liao et al. 2007; Ong et al. 2007; Cheng et al. 2008; Liao et al. 2008; Zeng et al. 2008; DeHaven et al. 2009). Because all three subtypes of Orai-proteins are expressed in Purkinje cells (Hartmann et al. 2010), such complexes represent possible candidates for the ion channels underlying the mGluR1-dependent slow EPSC.

CONCLUSIONS

In the mammalian central nervous system, mGluR1 is involved in a variety of physiological functions including memory, cognition, the sensation of pain and fear (Gravius et al. 2010). An extensively studied function of the mGluR1 is its role in cerebellar sensorimotor integration. The absence of mGluR1 activity in the cerebellar cortex has deleterious consequences for synaptic plasticity (for review, see Rose and Konnerth [2001]) as well as for motor coordination (Aiba et al. 1994; Ichise et al. 2000; Coesmans et al. 2003; Nakao et al. 2007). Even a mild interference with mGluR1-dependent signaling cascades invariably leads to ataxia (Hartmann et al. 2004; Hartmann et al. 2008). At parallel fiber-to-Purkinje cell synapses, important aspects of mGluR1-dependent signaling became clear in recent years (Fig. 5). Thus, it was found that mGluR1 induces via PLCβ the release of Ca^{2+} ions from internal Ca^{2+} stores in dendrites and spines (Takechi et al. 1998). This process was shown to be important for synaptic plasticity and for motor learning (Miyata et al. 2000). In addition to mediating the release of Ca^{2+} ions from internal stores, mGluR1 was shown to activate the cation channels TRPC3 (Hartmann et al. 2008). This surprising observation established TRPC3 as a novel postsynaptic channel for glutamatergic synaptic transmission. The gating of TRPC3 downstream of the synaptic activation of mGluR1 has not been entirely clarified, but seems to involve various signaling events, one of which is the rho GTPase-dependent activation of the PLD1 (Glitsch 2010). Other candidate signaling molecules include the IP_3R itself (Kiselyov et al. 1999), diacylglycerol (Hofmann et al. 1999) and STIM1 (Zeng et al. 2008). A promising experimental strategy for the elucidation of Purkinje cell-specific intracellular signaling events downstream of mGluR1 may involve the functional analysis of cell type-selective mutant mice in vitro and in vivo.

REFERENCES

Aiba A, Kano M, Chen C, Stanton ME, Fox GD, Herrup K, Zwingman TA, Tonegawa S. 1994. Deficient cerebellar long-term depression and impaired motor learning in mGluR1 mutant mice. *Cell* **79:** 377–388.

Alexander SPH, Mathie A, Peters JA. 2009. Guide to Receptors and Channels (GRAC), 4th Edition. *Br J Pharmacol* **158:** S1-254.

Batchelor AM, Garthwaite J. 1993. Novel synaptic potentials in cerebellar Purkinje cells: probable mediation by metabotropic glutamate receptors. *Neuropharmacology* **32:** 11–20.

Batchelor AM, Garthwaite J. 1997. Frequency detection and temporally dispersed synaptic signal association through a metabotropic glutamate receptor pathway. *Nature* **385:** 74–77.

Batchelor AM, Madge DJ, Garthwaite J. 1994. Synaptic activation of metabotropic glutamate receptors in the parallel fibre-Purkinje cell pathway in rat cerebellar slices. *Neuroscience* **63:** 911–915.

Becker EB, Oliver PL, Glitsch MD, Banks GT, Achilli F, Hardy A, Nolan PM, Fisher EM, Davies KE. 2009. A point mutation in TRPC3 causes abnormal Purkinje cell development and cerebellar ataxia in moonwalker mice. *Proc Natl Acad Sci* **106:** 6706–6711.

Berstein G, Blank JL, Smrcka AV, Higashijima T, Sternweis PC, Exton JH, Ross EM. 1992. Reconstitution of agonist-stimulated phosphatidylinositol 4,5-bisphosphate hydrolysis using purified m1 muscarinic receptor, Gq/11, and phospholipase C-b 1. *J Biol Chem* **267:** 8081–8088.

Bezprozvanny I, Watras J, Ehrlich BE. 1991. Bell-shaped calcium-response curves of Ins(1,4,5)P3- and calcium-gated channels from endoplasmic reticulum of cerebellum. *Nature* **351:** 751–754.

Blank JL, Ross AH, Exton JH. 1991. Purification and characterization of two G-proteins that activate the b1 isozyme of phosphoinositide-specific phospholipase C. Identification as members of the Gq class. *J Biol Chem* **266:** 18206–18216.

Canepari M, Papageorgiou G, Corrie JE, Watkins C, Ogden D. 2001. The conductance underlying the parallel fibre slow EPSP in rat cerebellar Purkinje neurones studied with photolytic release of L-glutamate. *J Physiol* **533:** 765–772.

Cardy TJ, Taylor CW. 1998. A novel role for calmodulin: Ca^{2+}-independent inhibition of type-1 inositol trisphosphate receptors. *Biochem J* **334:** 447–455.

Chadderton P, Margrie TW, Häusser M. 2004. Integration of quanta in cerebellar granule cells during sensory processing. *Nature* **428:** 856–860.

Cheng KT, Liu X, Ong HL, Ambudkar IS. 2008. Functional requirement for Orai1 in store-operated TRPC1-STIM1 channels. *J Biol Chem* **283:** 12935–12940.

Coesmans M, Smitt PA, Linden DJ, Shigemoto R, Hirano T, Yamakawa Y, van Alphen AM, Luo C, van der Geest JN, Kros JM, et al. 2003. Mechanisms underlying cerebellar motor deficits due to mGluR1-autoantibodies. *Ann Neurol* **53:** 325–336.

Curtis DR, Phillis JW, Watkins JC. 1959. Chemical excitation of spinal neurones. *Nature* **183:** 611–612.

Daniel H, Levenes C, Fagni L, Conquet F, Bockaert J, Crepel F. 1999. Inositol-1,4,5-trisphosphate-mediated rescue of cerebellar long-term depression in subtype 1 metabotropic glutamate receptor mutant mouse. *Neuroscience* **92:** 1–6.

DeHaven WI, Jones BF, Petranka JG, Smyth JT, Tomita T, Bird GS, Putney JW Jr. 2009. TRPC channels function independently of STIM1 and Orai1. *J Physiol* **587:** 2275–2298.

Doi T, Kuroda S, Michikawa T, Kawato M. 2005. Inositol 1,4,5-trisphosphate-dependent Ca^{2+} threshold dynamics detect spike timing in cerebellar Purkinje cells. *J Neurosci* **25**: 950–961.

Dzubay JA, Otis TS. 2002. Climbing fiber activation of metabotropic glutamate receptors on cerebellar Purkinje neurons. *Neuron* **36**: 1159–1167.

Finch EA, Augustine GJ. 1998. Local calcium signalling by inositol-1,4,5-trisphosphate in Purkinje cell dendrites. *Nature* **396**: 753–756.

Fujiwara A, Hirose K, Yamazawa T, Iino M. 2001. Reduced IP_3 sensitivity of IP_3 receptor in Purkinje neurons. *Neuroreport* **12**: 2647–2651.

Gao W, Dunbar RL, Chen G, Reinert KC, Oberdick J, Ebner TJ. 2003. Optical imaging of long-term depression in the mouse cerebellar cortex in vivo. *J Neurosci* **23**: 1859–1866.

Glitsch MD. 2010. Activation of native TRPC3 cation channels by phospholipase D. *FASEB Journal* **24**: 318–325.

Gravius A, Pietraszek M, Dekundy A, Danysz W. 2010. Metabotropic glutamate receptors as therapeutic targets for cognitive disorders. *Curr Top Med Chem* **10**: 187–206.

Harris KM, Stevens JK. 1988. Dendritic spines of rat cerebellar Purkinje cells: serial electron microscopy with reference to their biophysical characteristics. *J Neurosci* **8**: 4455–4469.

Hartmann J, Konnerth A. 2008. Mechanisms of metabotropic glutamate receptor-mediated synaptic signaling in cerebellar Purkinje cells. *Acta Physiol (Oxf)* **195**: 79–90.

Hartmann J, Blum R, Kovalchuk Y, Adelsberger H, Kuner R, Durand GM, Miyata M, Kano M, Offermanns S, Konnerth A. 2004. Distinct roles of Ga_q and Ga_{11} for Purkinje cell signaling and motor behavior. *J Neurosci* **24**: 5119–5130.

Hartmann J, Dragicevic E, Adelsberger H, Henning HA, Sumser M, Abramowitz J, Blum R, Dietrich A, Freichel M, Flockerzi V, et al. 2008. TRPC3 channels are required for synaptic transmission and motor coordination. *Neuron* **59**: 392–398.

Hartmann J, Karl R, Henning AH, Sakimura K, Baba Y, Kurosaki T, Konnerth A. 2010. A role of STIM and Orai proteins in synaptic signaling of cerebellar Purkinje cells. *Acta Physiol* **198**: P-TUE-112.

Hayashi T. 1952. A physiological study of epileptic seizures following cortical stimulation in animals and its application to human clinics. *Jpn J Physiol* **3**: 46–64.

Hendelman WJ, Aggerwal AS. 1980. The Purkinje neuron: I. A Golgi study of its development in the mouse and in culture. *J Comp Neurol* **193**: 1063–1079.

Hirasawa M, Xu X, Trask RB, Maddatu TP, Johnson BA, Naggert JK, Nishina PM, Ikeda A. 2007. Carbonic anhydrase related protein 8 mutation results in aberrant synaptic morphology and excitatory synaptic function in the cerebellum. *Mol Cell Neurosci* **35**: 161–170.

Hirono M, Konishi S, Yoshioka T. 1998. Phospholipase C-independent group I metabotropic glutamate receptor- mediated inward current in mouse Purkinje cells. *Biochem Biophys Res Commun* **251**: 753–758.

Hirono M, Sugiyama T, Kishimoto Y, Sakai I, Miyazawa T, Kishio M, Inoue H, Nakao K, Ikeda M, Kawahara S, et al. 2001. Phospholipase Cb4 and Protein Kinase Ca and/or Protein Kinase CbI Are Involved in the Induction of Long Term Depression in Cerebellar Purkinje Cells. *J Biol Chem* **276**: 45236–45242.

Hirota J, Ando H, Hamada K, Mikoshiba K. 2003. Carbonic anhydrase-related protein is a novel binding protein for inositol 1,4,5-trisphosphate receptor type 1. *Biochem J* **372**: 435–441.

Hofmann T, Obukhov AG, Schaefer M, Harteneck C, Gudermann T, Schultz G. 1999. Direct activation of human TRPC6 and TRPC3 channels by diacylglycerol. *Nature* **397**: 259–263.

Houamed KM, Kuijper JL, Gilbert TL, Haldeman BA, O'Hara PJ, Mulvihill ER, Almers W, Hagen FS. 1991. Cloning, expression, and gene structure of a G protein-coupled glutamate receptor from rat brain. *Science* **252**: 1318–1321.

Huang WC, Young JS, Glitsch MD. 2007. Changes in TRPC channel expression during postnatal development of cerebellar neurons. *Cell Calcium* **42**: 1–10.

Ichise T, Kano M, Hashimoto K, Yanagihara D, Nakao K, Shigemoto R, Katsuki M, Aiba A. 2000. mGluR1 in cerebellar Purkinje cells essential for long-term depression, synapse elimination, and motor coordination. *Science* **288**: 1832–1835.

Inoue T, Kato K, Kohda K, Mikoshiba K. 1998. Type 1 inositol 1,4,5-trisphosphate receptor is required for induction of long-term depression in cerebellar Purkinje neurons. *J Neurosci* **18**: 5366–5373.

Isope P, Franconville R, Barbour B, Ascher P. 2004. Repetitive firing of rat cerebellar parallel fibres after a single stimulation. *J Physiol* **554**: 829–839.

Ito M. 2000. Mechanisms of motor learning in the cerebellum. *Brain Res* **886**: 237–245.

Ito M. 2006. Cerebellar circuitry as a neuronal machine. *Prog Neurobiol* **78**: 272–303.

Jenkins J, Frohman M. 2005. Phospholipase D: a lipid centric review. *Cell Mol Life Sci* **62**: 2305–2360.

Jiao Y, Yan J, Zhao Y, Donahue LR, Beamer WG, Li X, Roe BA, Ledoux MS, Gu W. 2005. Carbonic anhydrase-related protein VIII deficiency is associated with a distinctive lifelong gait disorder in waddles mice. *Genetics* **171**: 1239–1246.

Kano M, Hashimoto K, Watanabe M, Kurihara H, Offermanns S, Jiang H, Wu Y, Jun K, Shin HS, Inoue Y, et al. 1998. Phospholipase Cb4 is specifically involved in climbing fiber synapse elimination in the developing cerebellum. *Proc Natl Acad Sci* **95**: 15724–15729.

Kasono K, Hirano T. 1995. Involvement of inositol trisphosphate in cerebellar long-term depression. *NeuroReport* **6**: 569–572.

Kasri NN, Holmes AM, Bultynck G, Parys JB, Bootman MD, Rietdorf K, Missiaen L, McDonald F, De Smedt H, Conway SJ, et al. 2004. Regulation of $InsP_3$ receptor activity by neuronal Ca^{2+}-binding proteins. *Embo J* **23**: 312–321.

Khodakhah K, Armstrong CM. 1997. Induction of long-term depression and rebound potentiation by inositol trisphosphate in cerebellar Purkinje neurons. *Proc Natl Acad Sci* **94**: 14009–14014.

Khodakhah K, Ogden D. 1993. Functional heterogeneity of calcium release by inositol trisphosphate in single

Purkinje neurons, cultured cerebellar astrocytes, and peripheral tissues. *Proc Natl Acad Sci* **90:** 4976–4980.

Khodakhah K, Ogden D. 1995. Fast activation and inactivation of inositol trisphosphate-evoked Ca^{2+} release in rat cerebellar Purkinje neurons. *J Physiol (Lond)* **487:** 343–358.

Kim JY, Zeng W, Kiselyov K, Yuan JP, Dehoff MH, Mikoshiba K, Worley PF, Muallem S. 2006. Homer 1 mediates store- and inositol 1,4,5-trisphosphate receptor-dependent translocation and retrieval of TRPC3 to the plasma membrane. *J Biol Chem* **281:** 32540–32549.

Kim SJ, Kim YS, Yuan JP, Petralia RS, Worley PF, Linden DJ. 2003. Activation of the TRPC1 cation channel by metabotropic glutamate receptor mGluR1. *Nature* **426:** 285–291.

Kiselyov K, Mignery GA, Zhu MX, Muallem S. 1999. The N-terminal domain of the IP_3 receptor gates store-operated hTrp3 channels. *Mol Cell* **4:** 423–429.

Kiselyov K, Xu X, Mozhayeva G, Kuo T, Pessah I, Mignery G, Zhu X, Birnbaumer L, Muallem S. 1998. Functional interaction between $InsP_3$ receptors and store-operated Htrp3 channels. *Nature* **396:** 478–482.

Knöpfel T, Anchisi D, Alojado ME, Tempia F, Strata P. 2000. Elevation of intradendritic sodium concentration mediated by synaptic activation of metabotropic glutamate receptors in cerebellar Purkinje cells. *Eur J Neurosci* **12:** 2199–2204.

Knöpfel T, Vranesic I, Staub C, Gahwiler BH. 1991. Climbing Fibre Responses in Olivo-cerebellar Slice Cultures. II. Dynamics of Cytosolic Calcium in Purkinje Cells. *Eur J Neurosci* **3:** 343–348.

Konnerth A, Dreessen J, Augustine GJ. 1992. Brief dendritic calcium signals initiate long-lasting synaptic depression in cerebellar Purkinje cells. *Proc Natl Acad Sci* **89:** 7051–7055.

Konnerth A, Llano I, Armstrong CM. 1990. Synaptic currents in cerebellar Purkinje cells. *Proc Natl Acad Sci* **87:** 2662–2665.

Lein ES, Hawrylycz MJ, Ao N, Ayres M, Bensinger A, Bernard A, Boe AF, Boguski MS, Brockway KS, Byrnes EJ, et al. 2007. Genome-wide atlas of gene expression in the adult mouse brain. *Nature* **445:** 168–176.

Lemonnier L, Trebak M, Putney JW Jr. 2008. Complex regulation of the TRPC3, 6 and 7 channel subfamily by diacylglycerol and phosphatidylinositol-4,5-bisphosphate. *Cell Calcium* **43:** 506–514.

Li HS, Xu XZ, Montell C. 1999. Activation of a TRPC3-dependent cation current through the neurotrophin BDNF. *Neuron* **24:** 261–273.

Liao Y, Erxleben C, Abramowitz J, Flockerzi V, Zhu MX, Armstrong DL, Birnbaumer L. 2008. Functional interactions among Orai1, TRPCs, and STIM1 suggest a STIM-regulated heteromeric Orai/TRPC model for $SOCE/I_{crac}$ channels. *Proc Natl Acad Sci* **105:** 2895–2900.

Liao Y, Erxleben C, Yildirim E, Abramowitz J, Armstrong DL, Birnbaumer L. 2007. Orai proteins interact with TRPC channels and confer responsiveness to store depletion. *Proc Natl Acad Sci* **104:** 4682–4687.

Llano I, Dreessen J, Kano M, Konnerth A. 1991a. Intradendritic release of calcium induced by glutamate in cerebellar Purkinje cells. *Neuron* **7:** 577–583.

Llano I, Marty A, Armstrong CM, Konnerth A. 1991b. Synaptic- and agonist-induced excitatory currents of Purkinje cells in rat cerebellar slices. *J Physiol (Lond)* **434:** 183–213.

Lockwich T, Pant J, Makusky A, Jankowska-Stephens E, Kowalak JA, Markey SP, Ambudkar IS. 2008. Analysis of TRPC3-interacting proteins by tandem mass spectrometry. *J Proteome Res* **7:** 979–989.

Martone ME, Zhang Y, Simpliciano VM, Carragher BO, Ellisman MH. 1993. Three-dimensional visualization of the smooth endoplasmic reticulum in Purkinje cell dendrites. *J Neurosci* **13:** 4636–4646.

Masu M, Tanabe Y, Tsuchida K, Shigemoto R, Nakanishi S. 1991. Sequence and expression of a metabotropic glutamate receptor. *Nature* **349:** 760–765.

Mikoshiba K. 2007. IP_3 receptor/Ca^{2+} channel: from discovery to new signaling concepts. *J Neurochem* **102:** 1426–1446.

Miyata M, Finch EA, Khiroug L, Hashimoto K, Hayasaka S, Oda SI, Inouye M, Takagishi Y, Augustine GJ, Kano M. 2000. Local calcium release in dendritic spines required for long-term synaptic depression. *Neuron* **28:** 233–244.

Nakamura F, Ogata K, Shiozaki K, Kameyama K, Ohara K, Haga T, Nukada T. 1991. Identification of two novel GTP-binding protein alpha-subunits that lack apparent ADP-ribosylation sites for pertussis toxin. *J Biol Chem* **266:** 12676–12681.

Nakamura M, Sato K, Fukaya M, Araishi K, Aiba A, Kano M, Watanabe M. 2004. Signaling complex formation of phospholipase Cb4 with metabotropic glutamate receptor type 1alpha and 1,4,5-trisphosphate receptor at the perisynapse and endoplasmic reticulum in the mouse brain. *Eur J Neurosci* **20:** 2929–2944.

Nakao H, Nakao K, Kano M, Aiba A. 2007. Metabotropic glutamate receptor subtype-1 is essential for motor coordination in the adult cerebellum. *Neurosci Res* **57:** 538–543.

Nicoletti F, Iadarola MJ, Wroblewski JT, Costa E. 1986. Excitatory amino acid recognition sites coupled with inositol phospholipid metabolism: developmental changes and interaction with a1-adrenoceptors. *Proc Natl Acad Sci* **83:** 1931–1935.

Nilius B, Owsianik G, Voets T, Peters JA. 2007. Transient receptor potential cation channels in disease. *Physiol Rev* **87:** 165–217.

Nogradi A, Jonsson N, Walker R, Caddy K, Carter N, Kelly C. 1997. Carbonic anhydrase II and carbonic anhydrase-related protein in the cerebellar cortex of normal and lurcher mice. *Brain Res Dev Brain Res* **98:** 91–101.

Nori A, Villa A, Podini P, Witcher DR, Volpe P. 1993. Intracellular Ca^{2+} stores of rat cerebellum: heterogeneity within and distinction from endoplasmic reticulum. *Biochem J* **291:** 199–204.

Nusser Z, Mulvihill E, Streit P, Somogyi P. 1994. Subsynaptic segregation of metabotropic and ionotropic glutamate receptors as revealed by immunogold localization. *Neuroscience* **61:** 421–427.

Ogden D, Capiod T. 1997. Regulation of Ca^{2+} release by $InsP_3$ in single guinea pig hepatocytes and rat Purkinje neurons. *J Gen Physiol* **109:** 741–756.

Ong HL, Cheng KT, Liu X, Bandyopadhyay BC, Paria BC, Soboloff J, Pani B, Gwack Y, Srikanth S, Singh BB, et al. 2007. Dynamic assembly of TRPC1/STIM1/Orai1 ternary complex is involved in store operated calcium influx: Evidence for similarities in SOC and CRAC channel components. *J Biol Chem* **282:** 9105–9116.

Ozawa S, Kamiya H, Tsuzuki K. 1998. Glutamate receptors in the mammalian central nervous system. *Prog Neurobiol* **54:** 581–618.

Patel S, Morris SA, Adkins CE, O'Beirne G, Taylor CW. 1997. Ca^{2+}-independent inhibition of inositol trisphosphate receptors by calmodulin: redistribution of calmodulin as a possible means of regulating Ca^{2+} mobilization. *Proc Natl Acad Sci* **94:** 11627–11632.

Patterson RL, Boehning D, Snyder SH. 2004. Inositol 1,4,5-trisphosphate receptors as signal integrators. *Annu Rev Biochem* **73:** 437–465.

Rebecchi MJ, Pentyala SN. 2000. Structure, function, and control of phosphoinositide-specific phospholipase C. *Physiol Rev* **80:** 1291–1335.

Roos J, DiGregorio PJ, Yeromin AV, Ohlsen K, Lioudyno M, Zhang S, Safrina O, Kozak JA, Wagner SL, Cahalan MD, et al. 2005. STIM1, an essential and conserved component of store-operated Ca^{2+} channel function. *J Cell Biol* **169:** 435–444.

Rose CR, Konnerth A. 2001. Stores not just for storage. intracellular calcium release and synaptic plasticity. *Neuron* **31:** 519–522.

Roustan P, Abitbol M, Menini C, Ribeaudeau F, Gerard M, Vekemans M, Mallet J, Dufier JL. 1995. The rat phospholipase Cb4 gene is expressed at high abundance in cerebellar Purkinje cells. *Neuroreport* **6:** 1837–1841.

Scelfo B, Strata P. 2005. Correlation between multiple climbing fibre regression and parallel fibre response development in the postnatal mouse cerebellum. *Eur J Neurosci* **21:** 971–978.

Sharp AH, Nucifora FC Jr, Blondel O, Sheppard CA, Zhang C, Snyder SH, Russell JT, Ryugo DK, Ross CA. 1999. Differential cellular expression of isoforms of inositol 1,4,5-triphosphate receptors in neurons and glia in brain. *J Comp Neurol* **406:** 207–220.

Shigemoto R, Nakanishi S, Mizuno N. 1992. Distribution of the mRNA for a metabotropic glutamate receptor (mGluR1) in the central nervous system: an in situ hybridization study in adult and developing rat. *J Comp Neurol* **322:** 121–135.

Shin JH, Linden DJ. 2005. An NMDA receptor/nitric oxide cascade is involved in cerebellar LTD but is not localized to the parallel fiber terminal. *J Neurophysiol* **94:** 4281–4289.

Sladeczek F, Pin JP, Recasens M, Bockaert J, Weiss S. 1985. Glutamate stimulates inositol phosphate formation in striatal neurones. *Nature* **317:** 717–719.

Sugiyama H, Ito I, Hirono C. 1987. A new type of glutamate receptor linked to inositol phospholipid metabolism. *Nature* **325:** 531–533.

Sugiyama T, Hirono M, Suzuki K, Nakamura Y, Aiba A, Nakamura K, Nakao K, Katsuki M, Yoshioka T. 1999. Localization of phospholipase Cb isozymes in the mouse cerebellum. *Biochem Biophys Res Commun* **265:** 473–478.

Takechi H, Eilers J, Konnerth A. 1998. A new class of synaptic response involving calcium release in dendritic spines. *Nature* **396:** 757–760.

Takei K, Stukenbrok H, Metcalf A, Mignery GA, Südhof TC, Volpe P, De Camilli P. 1992. Ca^{2+} stores in Purkinje neurons: endoplasmic reticulum subcompartments demonstrated by the heterogeneous distribution of the $InsP_3$ receptor, Ca^{2+}-ATPase, and calsequestrin. *J Neurosci* **12:** 489–505.

Tanaka O, Kondo H. 1994. Localization of mRNAs for three novel members (b 3, b 4 and g 2) of phospholipase C family in mature rat brain. *Neurosci Lett* **182:** 17–20.

Tanaka J, Nakagawa S, Kushiya E, Yamasaki M, Fukaya M, Iwanaga T, Simon MI, Sakimura K, Kano M, Watanabe M. 2000. Gq protein alpha subunits Ga_q and Ga_{11} are localized at postsynaptic extra-junctional membrane of cerebellar Purkinje cells and hippocampal pyramidal cells. *Eur J Neurosci* **12:** 781–792.

Tempia F, Alojado ME, Strata P, Knöpfel T. 2001. Characterization of the mGluR(1)-mediated electrical and calcium signaling in Purkinje cells of mouse cerebellar slices. *J Neurophysiol* **86:** 1389–1397.

Tempia F, Miniaci MC, Anchisi D, Strata P. 1998. Postsynaptic current mediated by metabotropic glutamate receptors in cerebellar Purkinje cells. *J Neurophysiol* **80:** 520–528.

Terasaki M, Slater NT, Fein A, Schmidek A, Reese TS. 1994. Continuous network of endoplasmic reticulum in cerebellar Purkinje neurons. *Proc Natl Acad Sci* **91:** 7510–7514.

Tong Q, Hirschler-Laszkiewicz I, Zhang W, Conrad K, Neagley DW, Barber DL, Cheung JY, Miller BA. 2008. TRPC3 is the erythropoietin-regulated calcium channel in human erythroid cells. *J Biol Chem* **283:** 10385–10395.

Tu JC, Xiao B, Naisbitt S, Yuan JP, Petralia RS, Brakeman P, Doan A, Aakalu VK, Lanahan AA, Sheng M, et al. 1999. Coupling of mGluR/Homer and PSD-95 complexes by the Shank family of postsynaptic density proteins. *Neuron* **23:** 583–592.

van Rossum DB, Patterson RL, Sharma S, Barrow RK, Kornberg M, Gill DL, Snyder SH. 2005. Phospholipase Cgamma1 controls surface expression of TRPC3 through an intermolecular PH domain. *Nature* **434:** 99–104.

Vazquez G, Wedel BJ, Kawasaki BT, Bird GS, Putney JW Jr. 2004. Obligatory role of Src kinase in the signaling mechanism for TRPC3 cation channels. *J Biol Chem* **279:** 40521–40528.

Verkhratsky A. 2005. Physiology and pathophysiology of the calcium store in the endoplasmic reticulum of neurons. *Physiol Rev* **85:** 201–279.

Villa A, Podini P, Clegg DO, Pozzan T, Meldolesi J. 1991. Intracellular Ca^{2+} stores in chicken Purkinje neurons: differential distribution of the low affinity-high capacity Ca^{2+} binding protein, calsequestrin, of Ca^{2+} ATPase and of the ER lumenal protein, Bip. *J Cell Biol* **113:** 779–791.

Villa A, Sharp AH, Racchetti G, Podini P, Bole DG, Dunn WA, Pozzan T, Snyder SH, Meldolesi J. 1992. The endoplasmic reticulum of Purkinje neuron body and dendrites: molecular identity and specializations for Ca^{2+} transport. *Neuroscience* **49:** 467–477.

Wang SS, Denk W, Hausser M. 2000. Coincidence detection in single dendritic spines mediated by calcium release. *Nat Neurosci* **3:** 1266–1273.

Watanabe M, Nakamura M, Sato K, Kano M, Simon MI, Inoue Y. 1998. Patterns of expression for the mRNA corresponding to the four isoforms of phospholipase Cb in mouse brain. *Eur J Neurosci* **10:** 2016–2025.

Watras J, Orlando R, Moraru II. 2000. An endogenous sulfated inhibitor of neuronal inositol trisphosphate receptors. *Biochemistry* **39:** 3452–3460.

Weernink O, Lopéz de Jesus PA, Schmidt M. 2007. Phospholipase D signaling: orchestration by PIP2 and small GTPases. *Naunyn Schmiedeberg Arch Pharmacol* **374:** 399–411.

Weernink PAO, Han L, Jakobs KH, Schmidt M. 2006. Dynamic phospholipid signaling by G protein-coupled receptors. *Biochimica et Biophysica acta* **1768:** 888–900.

Wilkie TM, Scherle PA, Strathmann MP, Slepak VZ, Simon MI. 1991. Characterization of G-protein a subunits in the G_q class: expression in murine tissues and in stromal and hematopoietic cell lines. *Proc Natl Acad Sci* **88:** 10049–10053.

Yuan JP, Kiselyov K, Shin DM, Chen J, Shcheynikov N, Kang SH, Dehoff MH, Schwarz MK, Seeburg PH, Muallem S, et al. 2003. Homer binds TRPC family channels and is required for gating of TRPC1 by IP_3 receptors. *Cell* **114:** 777–789.

Yuan JP, Zeng W, Huang GN, Worley PF, Muallem S. 2007. STIM1 heteromultimerizes TRPC channels to determine their function as store-operated channels. *Nat Cell Biol* **9:** 636–645.

Zagranichnaya TK, Wu X, Villereal ML. 2005. Endogenous TRPC1, TRPC3, and TRPC7 proteins combine to form native store-operated channels in HEK-293 cells. *J Biol Chem* **280:** 29559–29569.

Zeng W, Yuan JP, Kim MS, Choi YJ, Huang GN, Worley PF, Muallem S. 2008. STIM1 Gates TRPC Channels, but Not Orai1, by Electrostatic Interaction. *Mol Cell* **32:** 439–448.

Zhang SL, Yu Y, Roos J, Kozak JA, Deerinck TJ, Ellisman MH, Stauderman KA, Cahalan MD. 2005. STIM1 is a Ca^{2+} sensor that activates CRAC channels and migrates from the Ca^{2+} store to the plasma membrane. *Nature* **437:** 902–905.

Zhu X, Jiang M, Peyton M, Boulay G, Hurst R, Stefani E, Birnbaumer L. 1996. trp, a novel mammalian gene family essential for agonist-activated capacitative Ca^{2+} entry. *Cell* **85:** 661–671.

Zitt C, Obukhov AG, Strubing C, Zobel A, Kalkbrenner F, Luckhoff A, Schultz G. 1997. Expression of TRPC3 in Chinese hamster ovary cells results in calcium-activated cation currents not related to store depletion. *J Cell Biol* **138:** 1333–1341.

Ryanodine Receptors: Structure, Expression, Molecular Details, and Function in Calcium Release

Johanna T. Lanner, Dimitra K. Georgiou, Aditya D. Joshi, and Susan L. Hamilton

Baylor College of Medicine, Department of Molecular Physiology and Biophysics, Houston, Texas 77030

Correspondence: susanh@bcm.edu

Ryanodine receptors (RyRs) are located in the sarcoplasmic/endoplasmic reticulum membrane and are responsible for the release of Ca^{2+} from intracellular stores during excitation-contraction coupling in both cardiac and skeletal muscle. RyRs are the largest known ion channels (>2MDa) and exist as three mammalian isoforms (RyR 1–3), all of which are homotetrameric proteins that interact with and are regulated by phosphorylation, redox modifications, and a variety of small proteins and ions. Most RyR channel modulators interact with the large cytoplasmic domain whereas the carboxy-terminal portion of the protein forms the ion-conducting pore. Mutations in RyR2 are associated with human disorders such as catecholaminergic polymorphic ventricular tachycardia whereas mutations in RyR1 underlie diseases such as central core disease and malignant hyperthermia. This chapter examines the current concepts of the structure, function and regulation of RyRs and assesses the current state of understanding of their roles in associated disorders.

Intracellular Ca^{2+} is an important secondary messenger for signal transduction and is essential for cellular processes such as excitation-contraction coupling (E-C coupling). The major source of intracellular Ca^{2+} is the sarcoplasmic reticulum (SR) in striated muscle and the endoplasmic reticulum (ER) in other cell types. There are two major Ca^{2+} release channels localized in the SR/ER, the ryanodine receptors (RyRs) (Otsu et al. 1990) and inositol 1,4,5-triphosphate receptors (IP$_3$Rs) (Nixon et al. 1994). The present article reviews the structure, regulation, expression, and function of the RyRs. RyRs exist in three isoforms (RyR 1-3) and are named after the plant alkaloid ryanodine, which binds to RyRs with high affinity and specificity and displays preferential interactions with the open state of the channel allowing its usage to evaluate the functional state of the channel (Imagawa et al. 1987; Inui et al. 1987; Lai et al. 1988; Chu et al. 1990). Ryanodine at nanomole concentrations locks the channel in an open subconductance state and inhibits the channel at high concentrations (>100 μM) (Meissner et al. 1986; Lai et al. 1989; McGrew et al. 1989). RyRs are homotetamers with a total molecular mass of >2 MDa (each subunit is >550 kDa) (Inui et al. 1987; Lai et al. 1988). RyRs are modulated (see Fig. 1) directly or indirectly by the dihydropyridine receptor

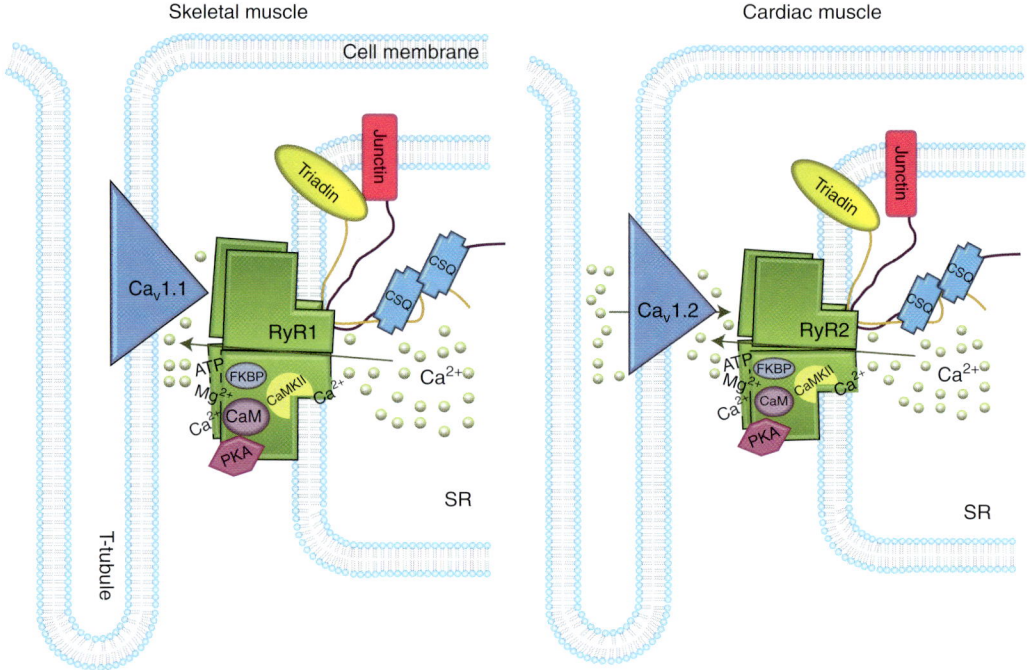

Figure 1. Schematic figure of the interaction between RyR and various modulators. *Left* panel illustrates skeletal muscle and *right* panel shows cardiac muscle. Modulators bind to the RyR tetramer but are for simplicity only depicted on one monomer.

(DHPR; also known as L-type Ca^{2+} channel, $Ca_V1.1/1.2$) and by various ions, small molecules and proteins, e.g., Ca^{2+}, Mg^{2+}, protein kinase A (PKA), FK506 binding proteins (FKBP12 and 12.6), calmodulin (CaM), Ca^{2+}/calmodulin-dependent protein kinase II (CaMKII), calsequestrin (CSQ), triadin, junctin (Smith 1986; Tanabe et al. 1990; Ikemoto et al. 1991; Sabbadini et al. 1992; Wang and Best 1992; Brillantes et al. 1994; Chen and MacLennan 1994; Yang et al. 1994; Ma et al. 1995; Mayrleitner et al. 1995; Tripathy et al. 1995; Timerman et al. 1996; Nakai et al. 1998; Moore et al. 1999b; Rodney et al. 2000). RyR1 and RyR2 are crucial for E-C coupling in both skeletal and cardiac muscle, respectively. The action potential travels to transverse tubules (t-tubules) resulting in SR Ca^{2+} release, either by mechanical coupling to DHPR in skeletal muscle (Rios and Brum 1987) or by Ca^{2+} induced Ca^{2+} release in cardiac muscle (Endo 1977). Increases in cytoplasmic Ca^{2+} initiate muscle contraction. Ca^{2+} is then pumped back to the SR by sarcoplasmic reticulum Ca^{2+} ATPase (SERCA) leading to relaxation (Nakai et al. 1998; Fill and Copello 2002). RyRs also play important roles in signal transduction in the nervous system and in osteoclasts where they contribute to secretion, synaptic plasticity, learning, and apoptosis (Zaidi et al. 1992; Chavis et al. 1996; Schwab et al. 2001).

RYANODINE RECEPTOR GENES AND ISOFORMS

There are three known mammalian isoforms of RyR: RyR1, RyR2, and RyR3. RyR1 was first detected in skeletal muscle (Takeshima et al. 1989; Zorzato et al. 1990), RyR2 was first found in cardiac muscles (Nakai et al. 1990; Otsu et al. 1990), and RyR3, previously referred to as the brain isoform, was found in brain (Hakamata et al. 1992). RyR1 is the most thoroughly examined isoform because of its high expression levels and ease of purification from skeletal muscle. In humans, the gene encoding RyR1 is located on chromosome 19q13.2 and spans 104 exons.

The gene encoding RyR2 is located on chromosome 1q43 and spans 102 exons, whereas the RyR3 gene with 103 exons is on chromosome 15q13.3-14. RyR1, 2, and 3 are located in chromosomes 7A3, 13A2, and 2E4 in mice (Mattei et al. 1994). In nonmammalian vertebrates RyRα and RyRβ are highly homologous to the three mammalian isoforms (Oyamada et al. 1994; Ottini et al. 1996). RyRs have been identified in Drosophila (D) melanigaster, Caenorhabditis (C) elegans, and Homarus americanus (Takeshima et al. 1994; Maryon et al. 1996; Quinn et al. 1998). The three mammalian isoforms are 65% identical in sequence (Hakamata et al. 1992) with three major regions of diversity: D1, between residues 4254 and 4631 in skeletal sequence and 4210 and 4562 in cardiac sequence; D2 between residues 1342 and 1403 in skeletal sequence and residues 1353 and 1397 in the cardiac sequence; D3, between residues 1872 and 1923 in skeletal sequence and 1852 and 1890 in the cardiac sequence. Region D2 is critical for the mechanical interaction between RyR1 and $Ca_V1.1$ (Perez et al. 2003) and mutations in D1 alter Ca^{2+} and caffeine sensitivity of RyR1 (Du et al. 2000). D3 may contain Ca^{2+} dependent inactivation sites (Hayek et al. 1999). In addition to these diverse regions, the large cytoplasmic domain is the site of both interaction with a large number of the modulators of channel activity, and many of the mutations that underlie the RyR channelopathies. Four-fifths of the RyR protein is cytoplasmic with ~one-fifth luminal and membrane spanning domains.

EXPRESSION OF RYANODINE RECEPTORS

The RyR1 isoform is primarily expressed in skeletal muscles (Takeshima et al. 1989; Zorzato et al. 1990) and is located in the junctional region of the terminal SR (Franzini-Armstrong and Nunzi 1983). RyR1 also appears to be expressed at low levels in cardiac muscle, smooth muscle (Neylon et al. 1995), stomach, kidney, thymus (Nakai et al. 1990; Giannini et al. 1995), cerebellum, Purkinje cells, adrenal glands, ovaries, and testis (Marks et al. 1989; Takeshima et al. 1989; Furuichi et al. 1994; Ottini et al. 1996). Recently it has been shown that RyR1 is also expressed in B-lymphocytes (Vukcevic et al. 2010).

The predominant form of RyR in cardiac muscle is RyR2 (Nakai et al. 1990; Otsu et al. 1990). Recently a splice variant of RyR2 was identified in the heart that increases susceptibility to apoptosis (Valdivia 2007). RyR2 is also expressed at high levels in Purkinje cells of cerebellum and cerebral cortex (Lai et al. 1992; Nakanishi et al. 1992; Sharp et al. 1993; Furuichi et al. 1994) and in low levels in stomach, kidney, adrenal glands, ovaries, thymus, and lungs (Kuwajima et al. 1992; Giannini et al. 1995).

RyR3 is expressed in hippocampal neurons, thalamus, Purkinje cells, corpus striatum (Hakamata et al. 1992; Lai et al. 1992; Furuichi et al. 1994), skeletal muscles (highest expression in the diaphragm) (Neylon et al. 1995; Marks et al. 1989), the smooth muscle cells of the coronary vasculature, lung, kidney, ileum, jejunum, spleen, stomach of mouse and aorta, uterus, ureter, urinary bladder, and esophagus of rabbit (Giannini et al. 1992; Hakamata et al. 1992; Giannini et al. 1995; Ottini et al. 1996).

Nonmammalian vertebrates, such as birds, fish, and chickens express isoforms RyRα and β (O'Brien et al. 1993). RyRα is abundant in skeletal muscles and its expression is lower in brain (cerebellum) (Oyamada et al. 1994). RyRβ is expressed in various tissues including skeletal and cardiac muscles, cerebellum, lungs, and stomach (Oyamada et al. 1994). In C. elegans RyRs are found in body wall, vulval, anal, and pharyngeal muscles (Hamada et al. 2002). In D. melanogaster a single isoform of RyR is expressed in digestive tract and nervous system (Vazquez-Martinez et al. 2003).

ROLE OF RYANODINE RECEPTORS IN HUMAN DISEASES

Mutations in both RyR1 and RyR2 are associated with a number of human diseases. Mutations in the RYR1 gene underlie several debilitating and/or life-threatening muscle diseases including malignant hyperthermia (MH) (MacLennan et al. 1990), heat/exercise induced exertional rhabdomyolysis (Capacchione et al. 2010), central core disease (CCD) (Zhang et al.

1993), multiminicore disease (MmD) (Ferreiro et al. 2002), and atypical periodic paralyses (APP) (Zhou et al. 2010). Mutations in RyR2 cause/are associated with catecholaminergic polymorphic ventricular tachycardia (CPVT) and arrhythmogenic right ventricular dysplasia type 2 (ARVD2) (Phillips et al. 1994, Zhang et al. 1993, Magee et al. 1956). Today around 300 mutations have been identified and linked to diseases associated with RyR (Fig. 2).

MH is an autosomal dominant disease in which genetically susceptible individuals respond to inhalation anesthetics (e.g., halothane) and muscle relaxants (e.g., succinylcholine) with sustained muscle contractions (Mickelson and Louis 1996). More than 150 different point mutations in the RYR1 gene have been identified and linked to MH (Fig. 2). The majority of RyR1 mutations linked to MH cluster in the cytoplasmic domains of RyR1 (amino acids 35 to 614 and 2129 to 2458). Another cluster of mutations is found near the carboxyl terminus (4637 to 4973) (Phillips et al. 1994; Quane et al. 1994; Lynch et al. 1999; Monnier et al. 2000; Scacheri et al. 2000; Tilgen et al. 2001). MH is often a silent disorder that goes undetected until the patient undergoes surgery or is exposed to high ambient temperatures ($\sim 37^\circ$C) (Jurkat-Roth et al. 2000). The underlying physiological consequence of MH is abnormal calcium homeostasis with increase sensitivity of channel opening in response to activators (Tong et al. 1999).

An MH episode is characterized by elevations in body temperature, metabolic acidosis, hypoxia, tachycardia, skeletal muscle rigidity, and rhabdomyolysis (Denborough et al. 1962; Ellis et al. 1988; Pamukcoglu 1988; Britt et al. 1991; Ryan and Tedeschi 1997) and is life threatening if not immediately treated with dantrolene, currently the only clinically approved treatment for MH (Ward et al. 1986; Zhao et al. 2001; Paul-Pletzer et al. 2002). The incidence of MH is ~ 1 in 15,000 anesthetized children and ~ 1 in 50,000 to 100,000 anesthetized adults (MacLennan 1992; Strazis and Fox 1993; Rosenberg et al. 2007). Another disorder related

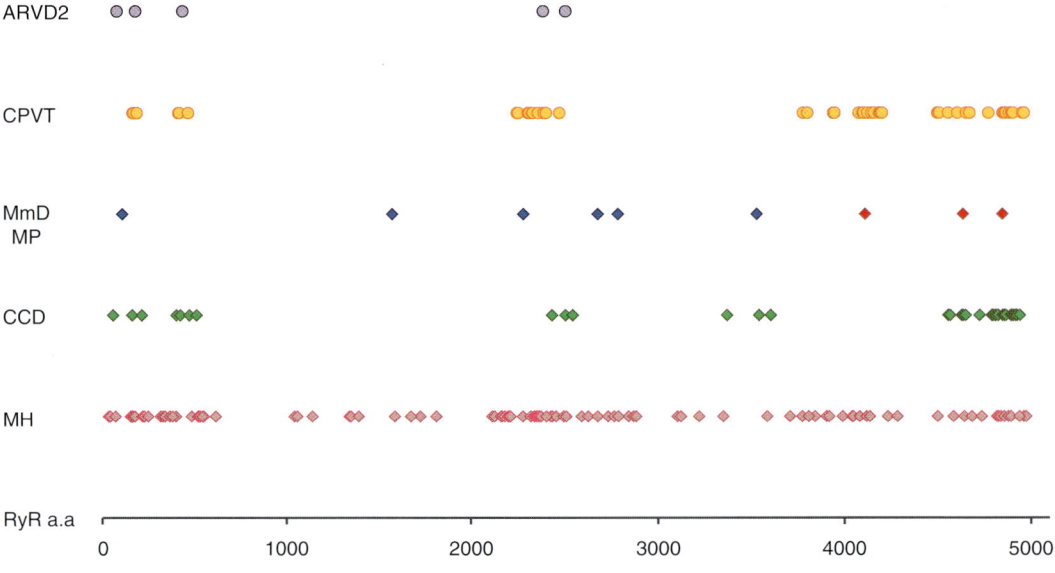

Figure 2. Linear presentation of RyR primary sequence outlining known mutations associated with skeletal and cardiac muscle diseases. Diamond shaped dots represent RyR1 mutations associated with: MH (151 mutations), CCD (63 mutations), and MmD (blue, 6 mutations) or other myopathies (MP, red, 4 mutations). Circles represent RyR2 mutations associated with: CPVT (53 mutations) and ARVD2 (5 mutations). Data collected from HGMD® (database of human gene mutation data) until 2006, UniProt 2007–2009, and publication by Vukcevic et al. 2010).

to MH is heat/exercise-induced exertional rhabdomyolysis; a clinical syndrome where heat/exercise-induced triggers breakdown of striated muscles that results in renal failure, hyperkalemia, and multi-organ failure. Approximately 26,000 cases are identified per year in United States (Capacchione et al. 2010).

CCD is a congenital myopathy in humans, and is characterized by hypotonia and muscle weakness of lower extremities leading to delayed attainment of motor skills (Dirksen and Avila 2002; Lueck and Dirksen 2004; Robinson et al. 2006). Slow-twitch (type I) skeletal muscles (e.g., soleus) of CCD of patients exhibit amorphous areas (central cores) that lack mitochondria and oxidative enzyme activity (Magee and Shy 1956; Shuaib et al. 1987). In some cases SR and t-tubules also degenerate and resting Ca^{2+} concentrations elevate with or without luminal store depletion (Tong et al. 1999). The majority of CCD causing mutations are in the pore-forming domain of RyR1 (Lynch et al. 1999; Monnier et al. 2000; Scacheri et al. 2000; Tilgen et al. 2001). Incidence of CCD is ∼1 in 16,000 of total congenital myopathies and ∼1 in 100,000 live births (Jungbluth 2007a).

MmD is an autosomal recessive myopathy characterized by weakness in axial and proximal limb muscles, hypoxia, and muscle biopsies showing characteristic mini cores due to lack of oxidative enzyme activity (Jungbluth 2007b; Sharma et al. 2007).

APP are dominant and genetically heterogeneous conditions characterized by muscle weakness and are divided into hypokalemic periodic paralysis and hyperkalemic periodic paralysis. Mutations on SCN4A and CASNAS1S gene that codes for α1s subunit of DHPR have been identified as general cause of hypokalemic- and hyperkalemic-periodic paralysis, respectively. Recently in a patient suffering from MmD the RyR1 mutation Arg2939Lys has been identified and clinical features of the patient are reminiscent of hyperkalemic periodic paralysis, suggesting a new RyR1-related form of periodic paralysis with additional myopathy features (Zhou et al. 2010).

Mutations in RyR2 produce altered Ca^{2+} homeostasis leading to ARVD2 (Dalla Volta et al. 1961; Marcus et al. 1982; Fontaine et al. 1984) and CPVT (Marks et al. 2002; Priori et al. 2002; Laitinen et al. 2003). ARVD2 an autosomal dominant cardiac disease characterized by replacement of myocytes with fibrofatty tissue leading to ventricular arrhythmias (Corrado et al. 2000). Mutations in RyR2 are detected at three regions that are homologous to the mutations on RyR1 associated with MH and CCD. Studies in ARVD2 suggest that Ca^{2+} leakage from myocardial SR via dysfunctional RyR2 is associated with development of ventricular arrhythmias (Tiso et al. 2001). The incidence of ARVD2 is ∼1 in 10,000 adults in United States (Fontaine et al. 2001). CPVT is characterized by stress-induced ventricular tachycardia (Marks et al. 2002; Priori et al. 2002; Laitinen et al. 2003). Roles for protein kinase A (PKA) and Ca^{2+}/calmodulin dependent protein kinase II (CaMKII) phosphorylation and enhancement of RyR2 open probability in these arrhythmias has been suggested (Valdivia et al. 1995; Marx et al. 2000; Wehrens et al. 2004). In single channel recordings it has been shown that CPVT RyR2 mutation Arg4496Cys increases open probability at low Ca^{2+} concentrations (∼5 nM) but not at higher concentrations (∼150 nM) (Jiang et al. 2002; Wehrens and Marks 2003). In patients with CPVT increased PKA phosphorylation and leaky RyR2 channels was observed during β-adrenergic stress and exercise. Binding studies in vitro suggested that the mutant RyR2 associated with CPVT have lower affinity for FKBP12.6 (Wehrens et al. 2003). Later studies suggested that CPVT RyR2 expressing cells are more sensitive to β-adrenergic receptor stimulation (by either isoproterenol or forskolin) and have prolonged Ca^{2+} transients under these conditions. This sensitivity does not appear to be caused by differences either in RyR2 phosphorylation or loss of FKBP12.6 (George et al. 2003). Nonsense or missense mutations in the calsequestrin 2 gene have also been associated with autosomal recessive form of CPVT (Lahat et al. 2001; Postma et al. 2002).

RyR3 is the least studied ryanodine receptor, and consequently little is known of its function. Recently, RyR3 was suggested to play a role in Alzheimer's disease, and up-regulation of RyR3 in cortical neurons is neuroprotective in TgCRND8 mouse model of Alzheimer's disease (Supnet et al. 2009).

ULTRASTRUCTURAL STUDIES ON RYANODINE RECEPTOR

The Size Challenge

RyRs, the largest known ion channels (Takeshima et al. 1989; Nakai et al. 1990; Otsu et al. 1990; Zorzato et al. 1990; Hakamata et al. 1992), are large conductance channels (Smith et al. 1985; Smith et al. 1986b) capable of creating rapid transient increases of cytosolic Ca^{2+}. Analysis of the primary structure of RyRs reveals several functional motifs seen in other proteins; but the role of these motifs in RyRs function has not yet been elucidated (see review Hamilton and Serysheva 2009). The importance of RyRs in mammalian physiology and disease drives the need for high resolution structural information. The massive size, multiple modulators, and the dynamic nature of RyRs make their structural analysis a challenge. Advances in single-particle electron cryomicroscopy (cryo-EM) and crystal structures of small fragments ($\sim$200 amino acids) of the protein (Amador 2009; Lobo and Van Petegem 2009) are beginning to elucidate many important structural features.

Structural Studies on RyRs

Most cryo-EM studies on RyRs, (Radermacher et al. 1992; Radermacher et al. 1994; Serysheva et al. 1995; Orlova et al. 1996; Sharma et al. 1998; Serysheva et al. 1999; Benacquista et al. 2000; Sharma et al. 2000; Ludtke et al. 2005; Samsó et al. 2005; Serysheva et al. 2005; Serysheva et al. 2008; Samsó et al. 2009) and all the subnanometer resolution analysis (Serysheva et al. 2008; Samsó et al. 2009) have focused on the RyR1, however, some progress has been made with RyR2 (Sharma et al. 1998; Liu et al. 2002) and RyR3 (Sharma et al. 2000; Liu et al. 2001). Overall, the structures of all three isoforms are similar, consistent with the high sequence homology ($\sim$65%). However, the small differences seen are important, because they reflect variations in the primary sequence and are likely to be related to the specialized functions of each isoform.

RyRs form homotetramers of square prism shape and are arrayed in the SR where they control the release of Ca^{2+}. The cytoplasmic area of the channel (280 Å $\times$ 280 Å $\times$ 120 Å) is connected with the transmembrane region (120 Å $\times$ 120 Å $\times$ 60 Å) (Fig. 3). The membrane region constitutes approximately one-fifth of the channel and is localized to the carboxy terminal of the protein and forms the ion-conducting pore. The cytoplasmic/sarcoplasmic area that is also called the "foot" is a huge area with cavities and micro-structures that facilitate interactions with solvent, small molecules, and protein modulators. The corners of the cytoplasmic area, also called "clamps" are connected through the "handle" domain that surrounds the "central rim" domain of the cytoplasmic area. This area is connected to the membrane region through the "column." These structural domains have been divided in to 15 subdomains. The clamps (Fig. 3, subdomains 5, 6, 7, 8, 9, 10) undergo major conformational changes during the opening and closing of the channel (Serysheva et al. 2008; Samsó et al. 2009), are likely to participate in intermolecular interactions with neighboring RyRs, and are the sites of interactions with modulators (Wagenknecht et al. 1994; Wagenknecht et al. 1996; Wagenknecht et al. 1997; Samsó and Wagenknecht 2002; Samsó et al. 2006; Sharma et al. 2006; Meng et al. 2009). Two of the areas of high divergence in the primary sequence of the RyR isoforms were mapped in the clamps (Zhang et al. 2003; Liu et al. 2004). At subnanometer resolution seven α-helixes and three β-sheets have been localized to the clamp domain (Serysheva et al. 2008).

The handle domain that is formed by subdomains 3 and 4 (Fig. 3) has been found to contain an expanded region of divergence (Liu et al. 2002), and a β-sheet mapped on the subdomain 4 (Serysheva et al. 2008). In total seven β-sheets and 36 α-helixes at various orientations have been mapped on the cytoplasmic region

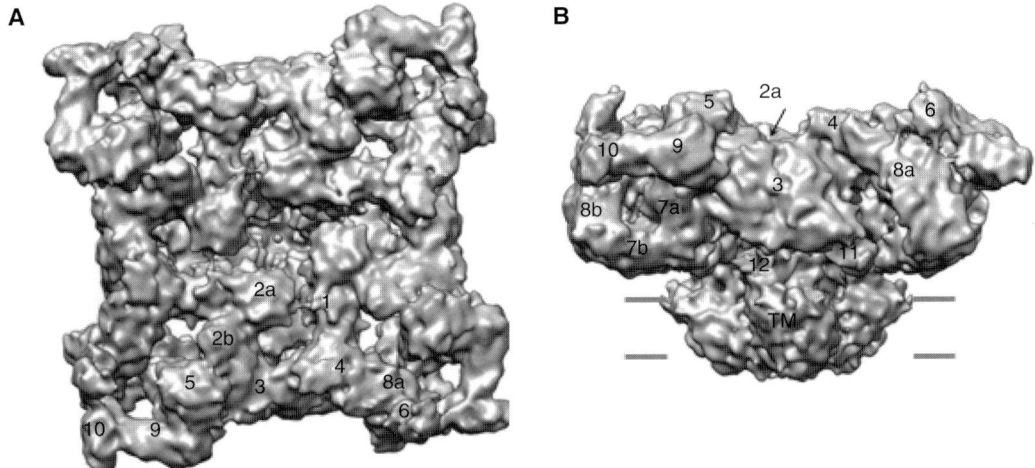

Figure 3. Cryo-EM reconstruction of RyR1 at 9.6 Å resolution. Cytoplasmic domain (A) and side view (B) of RyR1 with the different subdomains mapped by Irina Serysheva (See section "structural studies" for detailed information).

of the channel (Serysheva et al. 2008). These structures appear interconnected and merge toward the center of the molecule (Samsó et al. 2005). Two more β-sheets have been found in central rim (Fig. 3, subdomains 1,2) and one in the column (Fig. 3, subdomains 11,12), which also has eight α-helixes (Serysheva et al. 2008) that maintain the connection of the cytoplasmic and transmembrane regions (Samsó et al. 2005).

Current three-dimensional reconstructions at 8–10 Å resolution suggest five (Ludtke et al. 2005; Serysheva et al. 2008) or six α-helixes (Samsó et al. 2005) with different orientation in the transmembrane region of the closed channel. Although these studies (Ludtke et al. 2005; Samsó et al. 2005) were performed at similar conditions, they differ in interpretations, one study suggesting conformational similarity to the open K^+ channel (Ludtke et al. 2005) and the other suggesting a structure more similar to the closed K^+ channel (Samsó et al. 2005; Samsó et al. 2009). The conformation of the open RyR has so far only been proposed to resemble the conformation of the open K^+ channel (Samsó et al. 2009). The pore region has been predicted to consist of between 4 and 12 transmembrane segments (Takeshima et al. 1989; Zorzato et al. 1990; Tunwell et al. 1996;

Du et al. 2002). Most of these models place both amino- and carboxy-termini in the cytoplasm. The ion conducting pore has been proposed to be located in the lumenal region and to include the GVRAGGGIGD amino acid sequence (4891–4900 human RyR1) (Zhao et al. 1999; Du et al. 2001) which is conserved between RyRs. The sequence GGIG has been proposed as the selectivity filter (Balshaw et al. 1999; Gao et al. 2000) based on the similarity with the consensus selectivity filter of K^+ channels. During channel opening, massive movements of cytoplasmic and transmembrane masses take place and result in a 4 Å increase of the ion gate (Samsó et al. 2009).

Crystal Structure of the Amino-Terminal Domain

The first crystal structures of RyRs are from the amino-terminal domain. The first 210 amino acids of the RyR1 structure (rabbit) was at 2.5 Å resolution (PDB ID code 3HSM) (Amador 2009), and a similar fragment (amino acids 9–205 of rabbit RyR1) was resolved at 2.9 Å resolution (PDB ID code 3ILA) (Lobo and Van Petegem 2009). The structure of the first 217 amino acids of mouse RyR2 was at 2.5 Å resolution (PDB ID code 3IM5) (Lobo and

Van Petegem 2009). These domains of RyR1 and RyR2 display the same overall fold, and consist of a β-trefoil domain flanked by a rigid α-helix. Furthermore, minor differences are seen in the loops connecting the various β-strands in the two isoforms (Amador 2009; Lobo and Van Petegem 2009).

Effects of Disease-Causing Mutations on the Structure of RyR1

An area rich in disease-associated mutations has been identified in the amino-terminal domain of RyR1 where mutations associated with MH and CCD have been located. Six are in a short loop between Gln156 and Asp167, and the rest either in different β-strands (Cys35, Arg178, and Tyr179) or loops (Leu14 and Arg45) (Amador 2009). Similar clustering of disease-causing mutations is seen in RyR2. Mutations found in RyR2 associated with CPVT and ARVD2, Arg169, and Arg176 (which correspond to Arg156 and Arg163 MH mutations of RyR1) are in a short loop with Pro164. Two more disease-causing mutations, Ala77 and Val186, are located in the rigid α-helix and in a small loop close to a β-strand, respectively (Lobo and Van Petegem 2009). These mutations in the amino-terminal of RyRs have been proposed to affect the interaction of RyR1 with modulators because they appear to cause only local changes in the structure (Amador 2009; Lobo and Van Petegem 2009). Another area of the amino-terminal region, amino acids 414–466 of RyR2 contains more than half of the amino-terminal mutations associated with cardiac arrhythmias and sudden death (Wang et al. 2007) and was mapped to a location between subdomains 5 and 9 using green fluorescent protein insertion into the primary structure and difference mapping of cryo-EM reconstructed structures (Wang et al. 2007). Notably, mutations at positions Glu161, Arg164, Arg402, and Ile404 of RyR1 and the mutations Arg169, Ile417, and Arg418 of RyR2 are located in the suggested FKBP binding pocket (Serysheva et al. 2008); whereas the Ile4898Thr CCD mutation appears to be located in the proposed selectivity filter in the pore region.

RYR REGULATION

RyRs, together with $Ca_v1.1/Ca_v1.2$, PKA, FKBP12 and 12.6, CaM, CaMKII, triadin, junction, and calsequestrin form the core of the macromolecular complex that regulates SR Ca^{2+} release. Thus, RyR structure, function and regulation are likely to be defined within this complex macromolecular set of interactions. Despite the ~65% sequence homology, the different RyR isoforms respond differently to some modulators. Most of the RyRs modulators interact with the cytoplasmic region of the channel, suggesting that they allosterically regulate channel gating. The locations of the binding sites for some of the modulators have been predicted from the primary structure, interactions with RyR1 fragments, and from difference mapping in the cryo-EM structures.

$Ca_v1.1$ and 1.2

The voltage dependent Ca^{2+} channels ($Ca_V1.1$ and $Ca_V1.2$, skeletal and cardiac isoforms respectively), also known as dihydropyridine receptors (DHPRs), are composed of multiple subunits: α_1, α_2/δ, β and γ for $Ca_V1.1$. $Ca_V1.2$ has α_1, α_2/δ, and β but the γ subunit has not been identified. The α_1 subunit is both the voltage sensing and the pore forming subunit. DHPRs and RyRs are targeted to either side of the narrow junctional gap that separates the external and internal membrane systems in striated muscle. They are arranged so that bidirectional structural and functional coupling can occur between the proteins. The molecular mechanism of E-C coupling is fundamentally different between skeletal and cardiac muscle (Rios and Brum 1987; Bers and Stiffel 1993; Garcia et al. 1994; Lamb 2000). In skeletal muscle a physical interaction between $Ca_v1.1$ and RyR1 is required for E-C coupling and SR Ca^{2+} release; referred to as voltage-induced Ca^{2+} release (Lamb 2000). In contrast, RyR2 Ca^{2+} release in cardiac muscle is initiated by Ca^{2+} influx via $Ca_V1.2$, designated Ca^{2+} induced Ca^{2+}-release (see review Bers 2002). Thus, cardiac E-C coupling is dependent on extracellular Ca^{2+} and functional coupling between $Ca_v1.2$

and RyR2, which is assumed to rely on spatial proximity between the proteins rather than physical interaction. Because of the role of the direct physical interaction between $Ca_V1.1$ and RyR1 in skeletal muscle, E-C coupling can proceed for long periods in the absence of extracellular Ca^{2+} (Armstrong CM 1972; Dulhunty and Gage 1988). The $Ca_V1.1$ and RyR1 interaction in skeletal muscle is dependent on the strict geometrical alignments between the two proteins, which has been shown with electron microscopy in different muscle preparations (Takekura et al. 1994; Protasi et al. 1998) A critical determinant of E-C coupling in skeletal muscle is the α_{1S} II-III loop of $Ca_v1.1$, with a minor involvement of the α_{1S} I-II loop (Tanabe et al. 1990; Nakai et al. 1998; Kugler et al. 2004). Other parts in addition to the α_{1S} II-III loop are also suggested to be able to interact with RyR, e.g., the carboxyl terminus and the α_{1S} III-IV loop (Slavik et al. 1997; Leong and MacLennan 1998). The β_{1a} carboxyl terminus is also important for E-C coupling (Beurg et al. 1999; Sheridan et al. 2003). Multiple regions of RyR1 are likely to be involved in the interactions with $Ca_v1.1$ (Nakai et al. 1998; Proenza et al. 2002; Protasi et al. 2002; Perez et al. 2003; Sheridan et al. 2006). The sites of interaction appear to be distributed over a large part of the cytoplasmic region, among them the divergent region 2 (Sheridan et al. 2006), which has been mapped to the clamp domains (Liu et al. 2004). Freeze-fracture studies suggest subdomains 4 and 6 as the most likely locations for contact with $Ca_V1.1$ (Paolini et al. 2004).

Mechanical E-C coupling may also play a role in other tissues that express Ca_v isoforms and RyR isoforms; for instance there is evidence for both cardiac-like Ca^{2+}-induced Ca^{2+}-release and skeletal-like protein–protein interactions in neurons (Chavis et al. 1996; Mouton et al. 2001).

Ca^{2+}, Mg^{2+}, and ATP

Ca^{2+}, Mg^{2+}, and ATP are important small molecule regulators of RyRs. Mg^{2+} and ATP modulate RyRs in the cytoplasm and Ca^{2+} regulates RyRs both in the cytoplasm and in the lumen of SR. The cardiac and skeletal isoform of RyR show some difference in the regulation by these ligands, which has been linked to the different characteristics of E-C coupling in the two isoforms (Lamb 2000; Laver et al. 2001). Ca^{2+} has both direct effects on RyRs and also regulates RyRs via CaM and CaMKII. RyR1 activity shows a bell-shaped dependence on Ca^{2+} concentration and is activated by low Ca^{2+} concentration (~ 1 μM) by binding to specific high-affinity Ca^{2+} sites, and inhibited by high Ca^{2+} concentration (~ 1 mM) by binding to less selective low-affinity Ca^{2+} sites (Meissner et al. 1986; Meissner 1994; Meissner et al. 1997). Luminal Ca^{2+}, i.e., SR Ca^{2+} is also suggested to stimulate Ca^{2+} release in skeletal and cardiac muscle. Single channel measurements with increased store Ca^{2+} load have shown an increase in the sensitivity of Ca^{2+} release in the presence of cytosolic agonists such as ATP and caffeine (Smith et al. 1986; Sitsapesan 1995; Györke and Györke 1998; Xu et al. 1998; Laver et al. 2004). However, there have been conflicting results regarding luminal Ca^{2+} regulation and the source of these discrepancies is unclear. These might reflect altered mechanisms because of different experimental conditions, for example, differences in membrane preparation that results in variation of the protein complexes formed between RyR and calsequestrin, triadin, and junctin that are known to regulate RyR via luminal Ca^{2+} (Györke et al. 2004). Mg^{2+} is believed to inhibit RyRs by two mechanisms: reducing RyR open probability by competing with high-affinity Ca^{2+} activation sites, or by binding to less selective low-affinity Ca^{2+} sites that also mediate Ca^{2+} inhibition (Meissner et al. 1986; Laver et al. 1997). Noteworthy, there is a predicted difference in Mg^{2+} inhibition among RyR1, RyR2, and RyR3. RyR2 and RyR3 are activated by Ca^{2+} to greater extent than RyR1 and require higher Ca^{2+} concentration for inactivation. Hence, at elevated cytoplasmic Ca^{2+} concentrations Mg^{2+} causes potent RyR1 inhibition and relatively little inhibition of RyR2 and RyR3 (Lamb 2000; Meissner 2002).

ATP is an activator of RyRs. Various other adenine nucleotides (ADP, AMP, cAMP, adenosine, and adenine) also potentiate SR Ca^{2+} release but are less efficacious than ATP (Meissner 1984). In vitro studies have shown that skeletal muscle RyR1 can be activated by ATP in the absence of Ca^{2+}, but Ca^{2+} needs to be present for maximal activation (Meissner 1984; Meissner et al. 1986; Laver et al. 2001). Cardiac RyR is not activated by ATP in the absence of Ca^{2+}. However, ATP augments the Ca^{2+} induced activation of RyR2, but the effects are more modest than those seen with RyR1 (Xu et al. 1996; Kermode et al. 1998). In cells, most ATP is in complex with Mg^{2+}. Therefore, it is probable that under physiological conditions the MgATP complex rather than free ATP regulates Ca^{2+} release. The presence of high concentrations of free Mg^{2+} in cells and its inhibitory effects on RyR makes it difficult to determine the different effects of ATP and MgATP.

Calmodulin

CaM is a ubiquitously expressed 17-kDa Ca^{2+}-binding protein that regulates RyRs by direct binding. CaM is also known to bind to and regulate $Ca_v1.1$ and $Ca_v1.2$ (Tang et al. 2002; Ohrtman et al. 2008; Halling et al. 2009). CaM contains four EF-hand Ca^{2+} binding pockets (two in the carboxy-terminal domain and two in the amino-terminal domain of the protein) and binds to one site per RyR subunit (four per tetramer) (Moore et al. 1999a). All three RyR isoforms bind and are regulated by CaM both in its Ca^{2+}-free (apoCaM) and Ca^{2+}-bound (CaCaM) states (Tripathy et al. 1995; Yamaguchi et al. 2005). ApoCaM is a partial agonist whereas CaCaM is an inhibitor of RyR1 and SR Ca^{2+} release (Rodney et al. 2000). CaM binding site involves amino acids 3614–3643 of the RyR1 rabbit sequence (Takeshima et al. 1989; Moore et al. 1999b; Yamaguchi et al. 2003; Zhang et al. 2003). Cryo-EM difference mapping of the three-dimensional structures of RyR1 with and without added CaM has suggested that the CaCaM binding site is located in subdomain 3. The site seems to be displaced to ∼33 Å in the presence of Ca^{2+} with respect to its position for apoCaM (Wagenknecht et al. 1994; Wagenknecht et al. 1997; Samsó and Wagenknecht 2002). This displacement could be caused by a movement of the CaM upon binding calcium and/or a movement of the CaM binding site when RyR1 binds Ca^{2+}. The structure of CaM bound to a RyR1 peptide (3614–3643) has been visualized by NMR residual dipolar coupling (Maximciuc et al. 2006). Amino acids 3615–3628 contact the carboxy lobe of CaM, whereas amino acids 3628–3637 bind the amino lobe of CaM. In cardiac muscle, CaM shifts the Ca^{2+}-dependence of RyR2 activation to higher Ca^{2+} concentrations and hence decreases the RyR2 opening at all Ca^{2+} concentrations (Balshaw et al. 2001; Yamaguchi et al. 2003). Recently reduced affinity for CaM binding to RyR2 with PKA phosphorylation was found in a CPVT-associated mouse model (Arg2474Ser), resulting in spontaneous local Ca^{2+} release events leading to lethal arrythmias (Xu et al. 2010). In addition to CaM, a number of other EF-hand containing proteins have been recognized to interact with and regulate RyR, including calumenin and S100A1 (Jung et al. 2006; Wright et al. 2008). S100A1 has been found to compete with CaM for the RyR binding site (Wright et al. 2008). The questions of which of these EF-hand proteins actually regulate RyRs in vivo have yet to be answered.

Calsequestrin

Calsequestrin (CSQ) is the major intra-SR Ca^{2+} buffer. There are two genes encoding CSQ; type 1 CSQ (CSQ1) expressed in skeletal muscle and type 2 CSQ (CSQ2) expressed in cardiac and low levels in slow-twitch skeletal muscle. In addition to functioning as a Ca^{2+} buffer, CSQ forms oligomers in the lumen and interacts with the RyR anchoring proteins junctin and triadin embedded in the SR membrane. Together these three proteins appear to regulate RyR activity. The molecular details underlying these interactions have not been elucidated in either skeletal or cardiac muscle (Beard et al. 2009; Györke et al. 2009).

CSQ1 and CSQ2 appear to have unique isoform-specific properties in skeletal and cardiac muscle. CSQ1 reduces the activity of

RyR1 whereas CSQ2 increases the open probability of RyR1 and RyR2 (Wei et al. 2009). CSQ1-mediated inhibition of Ca^{2+} release during a single action potential may tune RyR1 activation to stimulation frequency and maintain Ca^{2+} release with repeated stimulation. In cardiac muscle CSQ2 may facilitate high rates of Ca^{2+} efflux through RyR2 during systole resulting in fast activation of contraction.

Recently, a mutation in the CSQ2 gene was linked to exercise-induced cardiac death caused by CPVT, although under basal conditions the cardiac contractility is apparently normal in subjects lacking functional CSQ2 (Postma et al. 2002). Knollmann and coworkers showed that *Casq2*-null mice are viable and display normal SR Ca^{2+} release and contractile function under basal conditions. However, exposure to catecholamines in *Casq2*-null myocytes caused increased diastolic SR Ca^{2+} leak, resulting in premature spontaneous SR Ca^{2+} releases that triggered beats indicating that these mice are susceptible to catecholaminergic ventricular arrhythmias (Knollmann et al. 2006).

FK506-Binding Protein 12 and 12.6 (FKBP12 and FKBP12.6)

FKBPs are named according to their molecular mass and belong to the immunophilins, a family of highly conserved proteins that bind immunosuppressive drugs such as FK506 and rapamycin. FKBPs are expressed in most tissues and are involved in a number of biochemical processes such as protein folding, receptor signaling, protein trafficking, and transcription. FKBP12 and FKBP12.6 (also known as calstabin 1 and 2, respectively) physically interact with all three isoforms of RyR but have different expression levels and binding affinity in different tissues (Chelu et al. 2004). FKBP12 copurifies with RyR1 (Jayaraman et al. 1992; Brillantes et al. 1994) and FKBP12.6 copurifies with RyR2 (Timerman et al. 1995; Timerman et al. 1996; Barg et al. 1997; Jeyakumar et al. 2001; Masumiya et al. 2003). Although somewhat controversial, a component of the FKBP12 binding site appears to be located between amino acids 2458 and 2468 of RyR1 (Rabbit sequence, SwissProt accession #P11716). Mutation of the amino acid Val2461 abolishes the FKBP12 binding (Gaburjakova et al. 2001; Avila et al. 2003). The amino-terminal and the carboxy-terminal regions of RyR2 have also been suggested to interact with FKBP12.6 (Masumiya et al. 2003; Xiao et al. 2004; Zissimopoulos and Lai 2005). Difference mapping of three-dimensional reconstructions of RyR with and without FKBP12 or 12.6 places the FKBPs binding site between subdomains 3, 5, and 9 (Wagenknecht et al. 1996; Wagenknecht et al. 1997; Samsó et al. 2006; Sharma et al. 2006). In agreement with this localization, FRET studies have localized the FKBP12 (Cornea et al. 2009) and the FKBP12.6 (Cornea et al. 2010) binding site to the same area as the model from Samsó et al 2006. Furthermore, both FKBP12 and 12.6 bind RyR1 and RyR2 in the same orientation (Cornea et al. 2010). Comparison of this location with the docking of the IP3 homology model, which includes the suppressor domain and the IP3-binding core region both with high sequence similarity to RyR1 amino-terminus, suggests a binding pocket for FKBP12 formed by Glu161, Arg164, Arg402, and Ile404 (Serysheva et al. 2008).

In mammals FKBP12 and 12.6 bind to RyRs with a stoichiometry of four FKBPs per RyR homotetramer (Jayaraman et al. 1992; Timerman et al. 1993; Qi et al. 1998). Under physiological conditions (i.e., the absence of immunosuppressive drugs), FKBPs are though to bind to RyRs with high affinity and stabilize the closed state of the channel (Ahern et al. 1994; Brillantes et al. 1994; McCall et al. 1996; Ahern et al. 1997; Marx et al. 1998; Marx et al. 2001). Removal of FKBP12, by preventing rebinding with an immunosuppressive drug or as the result of a genetic FKBP deficiency leads to greater open probability of the channel and longer mean open times (Ahern et al. 1997; Marx et al. 1998; Shou et al. 1998). Furthermore, FKBP12 displacement in skeletal muscle alters the coupling between RyR1 and $Ca_V1.1$. The consequences of these changes are dependent on muscle type and activity (Tang et al. 2004). In cardiac muscle, FKBP12 deficiency results in cardiomyopathy and ventricle septal defects that mimic human congenital heart disorder (Shou et al. 1998).

PKA and CaMKII Phosphorylation

The importance of RyR phosphorylation in modulation of Ca^{2+} release from SR was first established in the heart (Takasago et al. 1989). The functional consequences of phosphorylation on RyR function and the identity of the enzymes involved have been the focus of considerable debate. RyRs have several potential phosphorylation sites in their cytoplasmic domains. PKA, CaMKII, and cGMP-dependent kinase (PKG) have all been shown to phosphorylate RyR isoforms (Rodriguez et al. 2003; Wehrens et al. 2004; Xiao et al. 2006; Huke and Bers 2007).

The "fight or flight" response is a classic physiological stress pathway that involves activation of the sympathetic nervous system (SNS) that among other effects results in larger and faster Ca^{2+} transients and subsequently stronger and faster muscle contractions (Bers 2002). SNS activation causes β-adrenergic stimulation of the muscle, which via an intracellular signaling cascade results in activation of PKA. SNS-activated PKA phosphorylates RyR, altering its gating properties, but also phosphorylates several other key proteins involved in Ca^{2+} handling such as troponin I and phospholamban (Valdivia et al. 1995; Li et al. 2000; Kentish et al. 2001; Reiken et al. 2003). Modified RyR function is associated with increased SR Ca^{2+} leak in heart, which could contribute to reduced contractile function and increased propensity to arrhythmias. Altered phosphorylation of RyR2 has been suggested as one possible explanation for RyR dysfunction. Marks and colleagues propose that a hyper-adrenergic state that occurs in heart failure or during extreme stress, including exercise, leads to hyperphosphorylation of RyR serine residues (Ser2030, Ser2809 in RyR2 and Ser2843 in RyR1). They also suggest that hyperphosphorylation causes FKBPs to dissociate from RyRs, producing "leaky channels" (i.e., channels prone to open at rest). Such leaky channels could underlie increased risk for arrhythmias in heart failure and contribute to decreased muscle force production by reducing SR Ca^{2+} store content (Marx et al. 2000; Reiken et al. 2003). However, other groups have not found PKA-dependent hyperphosphorylation in failing hearts (Xiao et al. 2005). In addition, other laboratories suggest that CamKII-dependent phosphorylation of RyR2 is involved in enhanced SR Ca^{2+} leak and reduced SR Ca^{2+} load in heart failure and may contribute to arrhythmias and contractile dysfunction (Ai et al. 2005; Chelu et al. 2009; Curran et al. 2010; Neef et al. 2010). Phosphorylation of other targets of these kinases (troponin I, sarcolemmal Ca^{2+} channels, and phospholamban) could also alter the Ca^{2+} handling in cardiac and skeletal muscle.

CaMKII is modulated by changes in intracellular Ca^{2+} ($[Ca^{2+}]_i$), although little is still known quantitatively about the role of dynamic $[Ca^{2+}]_i$ fluctuations in the activation of CaMKII (Huke and Bers 2007; Aydin et al. 2007). CaMKII phosphorylates the same residues on RyR1 as PKA and also phosphorylates several other proteins such as troponin I, sarcolemmal Ca^{2+} channels, and phospholamban. Ser2808 on RyR2 was first described as a CaMKII phosphorylation site, but it was later shown that both PKA and PKG also phosphorylate this site (Witcher et al. 1991; Rodriguez et al. 2003; Wehrens et al. 2004; Xiao et al. 2006; Huke and Bers 2007). Ser2814, however, appears to only be phosphorylated by CaMKII whereas Ser2030 is only phosphorylated by PKA (Wehrens et al. 2004; Xiao et al. 2006).

Reactive Oxygen Species and Reactive Nitrogen Species

Sulfhydryl groups (SH, also called thiol) of cysteine (Cys) residues are potential targets for reduction/oxidation (redox) modifications of proteins. Alteration in the redox state of SH groups of two neighboring cysteine residues can lead to formation or breaking of disulfide bonds, which can modify both the structure and function of proteins. Low concentrations of redox active molecules (reactive oxygen species/reactive nitrogen species; ROS/RNS) constitute a basal endogenous redox buffering system that reversibly interacts with proteins.

Both RyR1 and RyR2 have nitric oxide (NO) covalently bound to cysteines (i.e., S-nitrosylation) and this posttranslational modification is reversible (Xu et al. 1998; Eu et al. 2000; Sun et al. 2008). High levels of ROS/RNS are able to irreversibly modify and even damage proteins in cardiac ischemia-reperfusion injury (Ferdinandy and Schulz 2003).

RyR is an established redox-sensitive channel and alterations in its redox state can result in either activation (Stoyanovsky et al. 1997; Eager and Dulhunty 1998) or inactivation (Boraso and Williams 1994; Marengo et al. 1998). Other key components of Ca^{2+} regulation and E-C coupling, e.g., SERCA and Ca_v's, are also redox modulated. RyR has $\sim$100 cysteines per subunit and $\sim$20 of them have been estimated to be free for redox modifications by oxidation, nitrosylation, or alkylation by the redox active molecule glutathione (Zable et al. 1997; Xu et al. 1998). A number of redox-sensitive cysteines have been identified in both the open and closed state of the channel and appear to be distributed across the primary structure of cytoplasmic region (Voss et al. 2004; Aracena et al. 2006). Several of these sites have been mapped to the clamp domains like Cys36 and Cys315 (Liu et al. 2005; Amador 2009; Hamilton and Serysheva 2009; Lobo and Van Petegem 2009), whereas the Cys3635 is located in the subdomain 3 in the CaM binding site (Moore et al. 1999b; Sun et al. 2001). S-nitrosylation of Cys3635 has been shown to reverse the CaM inhibition on RyR1 and to activate the channel (Moore et al. 1999b). The S-nitrosylation of Cys3635 appears to occur only at physiological tissue O_2 tension (pO_2; $\sim$10 mm Hg) and facilitates muscle contraction (Eu et al. 2003). Moreover, increased S-nitrosylation-induced RyR1 activity is suggested to sensitize RyR1 to environmental heat stress and MH crises (Durham et al. 2008). Increased RyR1 nitrosylation has also been observed in muscle dystrophy and is thought to contribute to muscle weakness by increased SR Ca^{2+} leak (Bellinger et al. 2009). In comparison to RyR1, no specific redox sensitive cysteine residues have yet been identified for RyR2. RyR2 is also pO_2-responsive but is not activated or S-nitrosylated directly by NO; instead activation and S-nitrosylation of RyR2 requires S-nitrosoglutathione (Sun et al. 2008).

CONCLUDING REMARKS

Primary sequence and location of several mutations are identified for RyR, but unanswered questions and debated topics still remain regarding the tertiary structure, the macromolecular interactions, and the regulation of RyR. The large size of RyR makes it more challenging to study, but new insights into the detailed structure of RyR are emerging with the continuous improvement and refinement of technologies such as cryo-EM and FRET-based assays. Resolution of the structure of RyR is progressing steadily, and ultimately we will have a map including carbon backbones, side-chains, membrane spanning regions and binding sites of interacting molecules. Along with our understanding of RyRs structure, it is likely that the number of known modulators that interact with RyR will also increase. Although the basic role and function of RyR in E-C coupling in skeletal and cardiac muscle is well established, further refinement of our understanding of the many modulators of RyR will be important in the development of therapeutics for treatment of cardiac and skeletal muscle diseases.

ACKNOWLEDGMENT

Special thanks to Irina Serysheva for providing the Cryo-EM figure. JT Lanner is supported by a fellowship from the Swedish Research Council.

REFERENCES

Ahern GP, Junankara PR, Dulhunty AF. 1994. Single channel activity of the ryanodine receptor calcium release channel is modulated by FK-506. *FEBS Lett* **352**: 369–374.

Ahern GP, Junankar PR, Dulhunty AF. 1997. Subconductance states in single-channel activity of skeletal muscle ryanodine receptors after removal of FKBP12. *Biophys J* **72**: 146–162.

Ai X, Curran JW, Shannon TR, Bers DM, Pogwizd SM. 2005. Ca^{2+}/Calmodulin-Dependent Protein Kinase Modulates Cardiac Ryanodine Receptor Phosphorylation and Sarcoplasmic Reticulum Ca^{2+} Leak in Heart Failure. *Circ Res* **97**: 1314–1322.

Amador FJ, Liu S, Ishiyama N, Plevin MJ, Wilson A, Maclennan DH, Ikura M. 2009. Crystal structure of type I ryanodine receptor amino-terminal -trefoil domain reveals a disease-associated mutation "hot spot" loop. *Proc Natl Acad Sci* **106:** 11040–11044.

Aracena P, Aguirre P, Munoz P, Nunez MT. 2006. Iron and glutathione at the crossroad of redox metabolism in neurons. *Biol Res* **39:** 157–165.

Armstrong CM, BF, Horowicz P. 1972. Twitches in the presence of ethylene glycol bis(-aminoethyl ether)-N,N'-tetracetic acid. *Biochim Biophys Acta* **267:** 605–608.

Avila G, Lee EH, Perez CF, Allen PD, Dirksen RT. 2003. FKBP12 Binding to RyR1 Modulates Excitation-Contraction Coupling in Mouse Skeletal Myotubes. *J Biol Chem* **278:** 22600–22608.

Aydin J, Korhonen T, Tavi P, Allen DG, Westerblad H, Bruton JD. 2007. Activation of Ca^{2+}-dependent protein kinase II during repeated contractions in single muscle fibres from mouse is dependent on the frequency of sarcoplasmic reticulum Ca^{2+} release. *Acta Physiolog* **191:** 131–137.

Balshaw D, Gao L, Meissner G. 1999. Luminal loop of the ryanodine receptor: A pore-forming segment? *Proc Natl Acad Sci* **96:** 3345–3347.

Balshaw DM, Xu L, Yamaguchi N, Pasek DA, Meissner G. 2001. Calmodulin Binding and Inhibition of Cardiac Muscle Calcium Release Channel (Ryanodine Receptor). *J Biol Chem* **276:** 20144–20153.

Barg S, Copello JA, Fleischer S. 1997. Different interactions of cardiac and skeletal muscle ryanodine receptors with FK-506 binding protein isoforms. *Am J Physiol Cell Physiol* **272:** C1726–C1733.

Beard N, Wei L, Dulhunty A. 2009. Ca^{2+} signaling in striated muscle: the elusive roles of triadin, junctin, and calsequestrin. *Eur Biophys J* **39:** 27–36.

Bellinger AM, Reiken S, Carlson C, Mongillo M, Liu X, Rothman L, Matecki S, Lacampagne A, Marks AR. 2009. Hypernitrosylated ryanodine receptor calcium release channels are leaky in dystrophic muscle. *Nat Med* **15:** 325–330.

Benacquista BL, Sharma MR, Samsó M, Zorzato F, Treves S, Wagenknecht T. 2000. Amino Acid Residues 4425 4621 Localized on the Three-Dimensional Structure of the Skeletal Muscle Ryanodine Receptor. *Biophys J* **78:** 1349–1358.

Bers DM. 2002. Cardiac excitation-contraction coupling. *Nature* **415:** 198–205.

Bers DM, Stiffel VM. 1993. Ratio of ryanodine to dihydropyridine receptors in cardiac and skeletal muscle and implications for E-C coupling. *Am J Physiol Cell Physiol* **264:** C1587–C1593.

Beurg M, Ahern CA, Vallejo P, Conklin MW, Powers PA, Gregg RG, Coronado R. 1999. Involvement of the Carboxy-Terminus Region of the Dihydropyridine Receptor β1a Subunit in Excitation-Contraction Coupling of Skeletal Muscle. *Biophys J* **77:** 2953–2967.

Boraso A, Williams AJ. 1994. Modification of the gating of the cardiac sarcoplasmic reticulum Ca^{2+}-release channel by H2O2 and dithiothreitol. *Am J Physiol Heart Circ Physiol* **267:** H1010–H1016.

Brillantes AB, Ondrias K, Scott A, Kobrinsky E, Ondriasova E, Moschella MC, Jayaraman T, Landers M, Ehrlich BE, Marks AR. 1994. Stabilization of calcium release channel (ryanodine receptor) function by FK506-binding protein. *Cell* **77:** 513–523.

Britt LD, Dascombe WH, Rodriguez A. 1991. New horizons in management of hypothermia and frostbite injury. *Surg Clin North Am* **71:** 345–370.

Capacchione JF, Sambuughin N, Bina S, Mulligan LP, Lawson TD, Muldoon SM. 2010. Exertional rhabdomyolysis and malignant hyperthermia in a patient with ryanodine receptor type 1 gene, L-type calcium channel alpha-1 subunit gene, and calsequestrin-1 gene polymorphisms. *Anesthesiology* **112:** 239–244.

Chavis P, Fagni L, Lansman JB, Bockaert J. 1996. Functional coupling between ryanodine receptors and L-type calcium channels in neurons. *Nature* **382:** 719–722.

Chelu MG, Danila CI, Gilman CP, Hamilton SL. 2004. Regulation of Ryanodine Receptors by FK506 Binding Proteins. *Trends Cardiovasc Med* **14:** 227–234.

Chelu MG, Sarma S, Sood S, Wang S, van Oort RJ, Skapura DG, Li N, Santonastasi M, Muller FU, Schmitz W, et al. 2009. Calmodulin kinase II–mediated sarcoplasmic reticulum Ca^{2+} leak promotes atrial fibrillation in mice. *J Clin Invest* **119:** 1940-1951.

Chen SR, MacLennan DH. 1994. Identification of calmodulin-, Ca^{2+}-, and ruthenium red-binding domains in the Ca^{2+} release channel (ryanodine receptor) of rabbit skeletal muscle sarcoplasmic reticulum. *J Biol Chem* **269:** 22698–22704.

Chu A, Diaz-Muñoz M, Hawkes MJ, Brush K, Hamilton SL. 1990. Ryanodine as a probe for the functional state of the skeletal muscle sarcoplasmic reticulum calcium release channel. *Mol Pharmacol* **37:** 735–741.

Cornea RL, Nitu F, Gruber S, Kohler K, Satzer M, Thomas DD, Fruen BR. 2009. FRET-based mapping of calmodulin bound to the RyR1 Ca^{2+} release channel. *Proc Natl Acad Sci* **106:** 6128–6133.

Cornea RL, Nitu FR, Samsó M, Thomas DD, Fruen BR. 2010. Mapping the ryanodine receptor (RyR) FK506-binding protein (FKBP) subunit using fluorescence resonance energy transfer (FRET). *J Biol Chem* **285:** 19219–19226.

Corrado D, Basso C, Thiene G. 2000. Arrhythmogenic right ventricular cardiomyopathy: diagnosis, prognosis, and treatment. *Heart* **83:** 588–595.

Curran J, Brown KH, Santiago DJ, Pogwizd S, Bers DM, Shannon TR. 2010. Spontaneous Ca waves in ventricular myocytes from failing hearts depend on Ca^{2+}-calmodulin-dependent protein kinase II. *J Mol Cell Cardiol* **49:** 25–32.

Dalla Volta S, Battaglia G, Zerbini E. 1961. "Auricularization" of right ventricular pressure curve. *Am Heart J* **61:** 25–33.

Denborough MA, Forster JFA, Lovell RRH, Maplestone PA, Villiers JD. 1962. Anaesthetic Deaths in a Family. *Br J Anaesth* **34:** 395–396.

Dirksen RT, Avila G. 2002. Altered ryanodine receptor function in central core disease: leaky or uncoupled Ca^{2+} release channels? *Trends Cardiovasc Med* **12:** 189–197.

Du GG, Khanna VK, MacLennan DH. 2000. Mutation of divergent region 1 alters caffeine and Ca^{2+} sensitivity of the skeletal muscle Ca^{2+} release channel (ryanodine receptor). *J Biol Chem* **275:** 11778–11783.

Du GG, Guo X, Khanna VK, MacLennan DH. 2001. Functional Characterization of Mutants in the Predicted Pore Region of the Rabbit Cardiac Muscle Ca^{2+} Release Channel (Ryanodine Receptor Isoform 2). *J Biol Chem* **276:** 31760–31771.

Du GG, Sandhu B, Khanna VK, Guo XH, MacLennan DH. 2002. Topology of the Ca^{2+} release channel of skeletal muscle sarcoplasmic reticulum (RyR1). *Proc Natl Acad Sci* **99:** 16725–16730.

Dulhunty AF, Gage PW. 1988. Effects of extracellular calcium concentration and dihydropyridines on contraction in mammalian skeletal muscle. *J Physiol* **399:** 63–80.

Durham WJ, Aracena-Parks P, Long C, Rossi AE, Goonasekera SA, Boncompagni S, Galvan DL, Gilman CP, Baker MR, Shirokova N, et al. 2008. RyR1 S-Nitrosylation Underlies Environmental Heat Stroke and Sudden Death in Y522S RyR1 Knockin Mice. *Cell* **133:** 53–65.

Eager KR, Dulhunty AF. 1998. Activation of the Cardiac Ryanodine Receptor by Sulfhydryl Oxidation is Modified by Mg^{2+} and ATP. *J Memb Biol* **163:** 9–18.

Ellis FR, Halsall PJ, Harriman DG. 1988. Malignant hyperpyrexia and sudden infant death syndrome. *Br J Anaesth* **60:** 28–30.

Endo M. 1977. Calcium release from the sarcoplasmatic reticulum. *Physiol Rev* **57:** 71–108.

Eu JP, Hare JM, Hess DT, Skaf M, Sun J, Cardenas-Navina I, Sun QA, Dewhirst M, Meissner G, Stamler JS. 2003. Concerted regulation of skeletal muscle contractility by oxygen tension and endogenous nitric oxide. *Proc Natl Acad Sci* **100:** 15229–15234.

Eu JP, Sun J, Xu L, Stamler JS, Meissner G. 2000. The Skeletal Muscle Calcium Release Channel: Coupled O2 Sensor and NO Signaling Functions. *Cell* **102:** 499–509.

Ferdinandy P, Schulz R. 2003. Nitric oxide, superoxide, and peroxynitrite in myocardial ischaemia-reperfusion injury and preconditioning. *Br J Pharmacol* **138:** 532–543.

Ferreiro A, Monnier N, Romero NB, Leroy JP, Bonnemann C, Haenggeli CA, Straub V, Voss WD, Nivoche Y, Jungbluth H, et al. 2002. A recessive form of central core disease, transiently presenting as multi-minicore disease, is associated with a homozygous mutation in the ryanodine receptor type 1 gene. *Ann Neurol* **51:** 750–759.

Fill M, Copello JA. 2002. Ryanodine receptor calcium release channels. *Physiol Rev* **82:** 893–922.

Fontaine G, Frank R, Guiraudon G, Pavie A, Tereau Y, Chomette G, Grosgogeat Y. 1984. Significance of intraventricular conduction disorders observed in arrhythmogenic right ventricular dysplasia. *Arch Mal Coeur Vaiss* **77:** 872–879.

Fontaine G, Gallais Y, Fornes P, Hebert JL, Frank R. 2001. Arrhythmogenic right ventricular dysplasia/cardiomyopathy. *Anesthesiology* **95:** 250–254.

Franzini-Armstrong C, Nunzi G. 1983. Junctional feet and particles in the triads of a fast-twitch muscle fibre. *J Muscle Res Cell Mot* **4:** 233–252.

Furuichi T, Furutama D, Hakamata Y, Nakai J, Takeshima H, Mikoshiba K. 1994. Multiple types of ryanodine receptor/Ca^{2+} release channels are differentially expressed in rabbit brain. *J Neurosci* **14:** 4794–4805.

Gaburjakova M, Gaburjakova J, Reiken S, Huang F, Marx SO, Rosemblit N, Marks AR. 2001. FKBP12 Binding Modulates Ryanodine Receptor Channel Gating. *J Biol Chem* **276:** 16931–16935.

Gao L, Balshaw D, Xu L, Tripathy A, Xin C, Meissner G. 2000. Evidence for a Role of the Lumenal M3-M4 Loop in Skeletal Muscle Ca^{2+} Release Channel (Ryanodine Receptor) Activity and Conductance. *Biophys J* **79:** 828–840.

Garcia J, Tanabe T, Beam KG. 1994. Relationship of calcium transients to calcium currents and charge movements in myotubes expressing skeletal and cardiac dihydropyridine receptors. *J Cell Biol* **103:** 125–147.

George CH, Higgs GV, Lai FA. 2003. Ryanodine receptor mutations associated with stress-induced ventricular tachycardia mediate increased calcium release in stimulated cardiomyocytes. *Circ Res* **93:** 531–540.

Giannini G, Clementi E, Ceci R, Marziali G, Sorrentino V. 1992. Expression of a ryanodine receptor-Ca^{2+} channel that is regulated by TGF-beta. *Science* **257:** 91–94.

Giannini G, Conti A, Mammarella S, Scrobogna M, Sorrentino V. 1995. The ryanodine receptor/calcium channel genes are widely and differentially expressed in murine brain and peripheral tissues. *J Cell Biol* **128:** 893–904.

Györke I, Györke S. 1998. Regulation of the Cardiac Ryanodine Receptor Channel by Luminal Ca^{2+} Involves Luminal Ca^{2+} Sensing Sites. *Biophys J* **75:** 2801–2810.

Györke S, Stevens SCW, Terentyev D. 2009. Cardiac calsequestrin: quest inside the SR. *J Physiol* **587:** 3091–3094.

Györke I, Hester N, Jones LR, Györke S. 2004. The Role of Calsequestrin, Triadin, and Junctin in Conferring Cardiac Ryanodine Receptor Responsiveness to Luminal Calcium. *Biophys J* **86:** 2121–2128.

Hakamata Y, Nakai J, Takeshima H, Imoto K. 1992. Primary structure and distribution of a novel ryanodine receptor/calcium release channel from rabbit brain. *FEBS Lett* **312:** 229–235.

Halling DB, Georgiou DK, Black DJ, Yang G, Fallon JL, Quiocho FA, Pedersen SE, Hamilton SL. 2009. Determinants in Ca_V1 Channels That Regulate the Ca^{2+} Sensitivity of Bound Calmodulin. *J Biol Chem* **284:** 20041–20051.

Hamada T, Sakube Y, Ahnn J, Kim DH, Kagawa H. 2002. Molecular dissection, tissue localization and Ca^{2+} binding of the ryanodine receptor of *Caenorhabditis elegans*. *J Mol Biol* **324:** 123–135.

Hamilton SL, Serysheva II. 2009. Ryanodine Receptor Structure: Progress and Challenges. *J Biol Chem* **284:** 4047–4051.

Hayek SM, Zhao J, Bhat M, Xu X, Nagaraj R, Pan Z, Takeshima H, Ma J. 1999. A negatively charged region of the skeletal muscle ryanodine receptor is involved in Ca^{2+}-dependent regulation of the Ca^{2+} release channel. *FEBS Lett* **461:** 157–164.

Huke S, Bers DM. 2007. Temporal dissociation of frequency-dependent acceleration of relaxation and protein phosphorylation by CaMKII. *J Mol Cell Cardiol* **42:** 590–599.

Ikemoto N, Antoniu B, Kang JJ, Meszaros LG, Ronjat M. 1991. Intravesicular calcium transient during calcium release from sarcoplasmic reticulum. *Biochemistry* **30:** 5230–5237.

Imagawa T, Smith JS, Coronado R, Campbell KP. 1987. Purified ryanodine receptor from skeletal muscle sarcoplasmic reticulum is the Ca^{2+}-permeable pore of the calcium release channel. *J Biol Chem* **262**: 16636–16643.

Inui M, Saito A, Fleischer S. 1987. Purification of the ryanodine receptor and identity with feet structures of junctional terminal cisternae of sarcoplasmic reticulum from fast skeletal muscle. *J Biol Chem* **262**: 1740–1747.

Jayaraman T, Brillantes AM, Timerman AP, Fleischer S, Erdjument-Bromage H, Tempst P, Marks AR. 1992. FK506 binding protein associated with the calcium release channel (ryanodine receptor). *J Biol Chem* **267**: 9474–9477.

Jeyakumar LH, Ballester L, Cheng DS, McIntyre JO, Chang P, Olivey HE, Rollins-Smith L, Barnett JV, Murray K, Xin H-B, et al. 2001. FKBP Binding Characteristics of Cardiac Microsomes from Diverse Vertebrates. *Biochem Biophys Res Comm* **281**: 979–986.

Jiang D, Xiao B, Zhang L, Chen SR. 2002. Enhanced basal activity of a cardiac Ca^{2+} release channel (ryanodine receptor) mutant associated with ventricular tachycardia and sudden death. *Circ Res* **91**: 218–225.

Jung DH, Mo SH, Kim DH. 2006. Calumenin, a multiple EF-hands Ca^{2+}-binding protein, interacts with ryanodine receptor-1 in rabbit skeletal sarcoplasmic reticulum. *Biochem Biophys Res Comm* **343**: 34–42.

Jungbluth H. 2007a. Central core disease. *Orphanet J Rare Dis* **2**: 25.

Jungbluth H. 2007b. Multi-minicore Disease. *Orphanet J Rare Dis* **2**: 31.

Jurkat-Rott K. McCarthy T. Lehmann-Horn F. 2000. Genetics and pathogenesis of malignant hyperthermia. *Muscle Nerve* **23**: 4–17.

Kentish JC, McCloskey DT, Layland J, Palmer S, Leiden JM, Martin AF, Solaro RJ. 2001. Phosphorylation of Troponin I by Protein Kinase A Accelerates Relaxation and Crossbridge Cycle Kinetics in Mouse Ventricular Muscle. *Circ Res* **88**: 1059–1065.

Kermode H, Williams AJ, Sitsapesan R. 1998. The Interactions of ATP, ADP, and Inorganic Phosphate with the Sheep Cardiac Ryanodine Receptor. *Biophys J* **74**: 1296–1304.

Knollmann BRC, Chopra N, Hlaing T, Akin B, Yang T, Ettensohn K, Knollmann BEC, Horton KD, Weissman NJ, Holinstat I, et al. 2006. Casq2 deletion causes sarcoplasmic reticulum volume increase, premature Ca^{2+} release, and catecholaminergic polymorphic ventricular tachycardia. *J Clin Invest* **116**: 2510–2520.

Kugler G, Weiss RG, Flucher BE, Grabner M. 2004. Structural Requirements of the Dihydropyridine Receptor α1S II-III Loop for Skeletal-type Excitation-Contraction Coupling. *J Biol Chem* **279**: 4721–4728.

Kuwajima G, Futatsugi A, Niinobe M, Nakanishi S, Mikoshiba K. 1992. Two types of ryanodine receptors in mouse brain: skeletal muscle type exclusively in Purkinje cells and cardiac muscle type in various neurons. *Neuron* **9**: 1133–1142.

Lahat H, Pras E, Olender T, Avidan N, Ben-Asher E, Man O, Levy-Nissenbaum E, Khoury A, Lorber A, Goldman B, et al. 2001. A missense mutation in a highly conserved region of CASQ2 is associated with autosomal recessive catecholamine-induced polymorphic ventricular tachycardia in Bedouin families from Israel. *Am J Hum Genet* **69**: 1378–1384.

Lai FA, Dent M, Wickenden C, Xu L, Kumari G, Misra M, Lee HB, Sar M, Meissner G. 1992. Expression of a cardiac Ca^{2+}-release channel isoform in mammalian brain. *Biochem J* **288**: 553–564.

Lai FA, Erickson HP, Rousseau E, Liu QY, Meissner G. 1988. Purification and reconstitution of the calcium release channel from skeletal muscle. *Nature* **331**: 315–319.

Lai FA, Misra M, Xu L, Smith HA, Meissner G. 1989. The ryanodine receptor Ca^{2+}-release channel complex of skeletal muscle sarcoplasmic reticulum. Evidence for a cooperatively coupled, negatively charged homotetramer. *J Biol Chem* **264**: 16776–16785.

Laitinen PJ, Swan H, Kontula K. 2003. Molecular genetics of exercise-induced polymorphic ventricular tachycardia: identification of three novel cardiac ryanodine receptor mutations and two common calsequestrin 2 amino-acid polymorphisms. *Eur J Hum Genet* **11**: 888–891.

Lamb G. 2000. Excitation & Contraction Coupling In Skeletal Muscle: Comparisons with Cardiac Muscle. *Clin Exp Pharmacol Physiol* **27**: 216–224.

Laver DR, Baynes TM, Dulhunty AF. 1997. Magnesium Inhibition of Ryanodine-Receptor Calcium Channels: Evidence for Two Independent Mechanisms. *J Memb Biol* **156**: 213–229.

Laver D, Lenz G, Lamb G. 2001. Regulation of the calcium release channel from rabbit skeletal muscle by the nucleotides ATP, AMP, IMP and adenosine. *J Physiol* **537**: 763–778.

Laver D, O'Neill E, Lamb G. 2004. Luminal Ca^{2+}-regulated Mg^{2+} inhibition of skeletal RyRs reconstituted as isolated channels or coupled clusters. *J Physiol* **124**: 741–758.

Leong P, MacLennan DH. 1998. The Cytoplasmic Loops between Domains II and III and Domains III and IV in the Skeletal Muscle Dihydropyridine Receptor Bind to a Contiguous Site in the Skeletal Muscle Ryanodine Receptor. *J Biol Chem* **273**: 29958–29964.

Li L, Desantiago J, Chu G, Kranias EG, Bers DM. 2000. Phosphorylation of phospholamban and troponin I in beta -adrenergic-induced acceleration of cardiac relaxation. *Am J Physiol Heart Circ Physiol* **278**: H769–H779.

Liu Z, Wang R, Zhang J, Chen SRW, Wagenknecht T. 2005. Localization of a Disease-associated Mutation Site in the Three-dimensional Structure of the Cardiac Muscle Ryanodine Receptor. *J Biol Chem* **280**: 37941–37947.

Liu Z, Zhang J, Li P, Chen SRW, Wagenknecht T. 2002. Three-dimensional Reconstruction of the Recombinant Type 2 Ryanodine Receptor and Localization of Its Divergent Region 1. *J Biol Chem* **277**: 46712–46719.

Liu Z, Zhang J, Sharma MR, Li P, Chen SRW, Wagenknecht T. 2001. Three-dimensional reconstruction of the recombinant type 3 ryanodine receptor and localization of its amino terminus. *Proc Natl Acad Sci* **98**: 6104–6109.

Liu Z, Zhang J, Wang R, Wayne Chen SR, Wagenknecht T. 2004. Location of Divergent Region 2 on the Three-dimensional Structure of Cardiac Muscle Ryanodine Receptor/Calcium Release Channel. *J Mol Biol* **338**: 533–545.

Lobo PA, Van Petegem F. 2009. Crystal Structures of the N-Terminal Domains of Cardiac and Skeletal Muscle

Ryanodine Receptors: Insights into Disease Mutations. *Structure* **17**: 1505–1514.

Ludtke SJ, Serysheva II, Hamilton SL, Chiu W. 2005. The Pore Structure of the Closed RyR1 Channel. *Structure* **13**: 1203–1211.

Lueck JD, Dirksen RT. 2004. Ryanodinopathies: muscle disorders linked to mutations in ryanodine receptors. *Basic Appl Myol* **14**: 339–352.

Lynch PJ, Tong J, Lehane M, Mallet A, Giblin L, Heffron JJ, Vaughan P, Zafra G, MacLennan DH, McCarthy TV. 1999. A mutation in the transmembrane/luminal domain of the ryanodine receptor is associated with abnormal Ca^{2+} release channel function and severe central core disease. *Proc Natl Acad Sci* **96**: 4164–4169.

Ma J, Bhat MB, Zhao J. 1995. Rectification of skeletal muscle ryanodine receptor mediated by FK506 binding protein. *Biophys J* **69**: 2398–2404.

MacLennan DH. 1992. The genetic basis of malignant hyperthermia. *Trends Pharmacol Sci* **13**: 330–334.

MacLennan DH, Duff C, Zorzato F, Fujii J, Phillips M, Korneluk RG, Frodis W, Britt BA, Worton RG. 1990. Ryanodine receptor gene is a candidate for predisposition to malignant hyperthermia. *Nature* **343**: 559–561.

Magee KR, Shy GM. 1956. A new congenital nonprogressive myopathy. *Brain* **79**: 610–621.

Marcus FI, Fontaine GH, Guiraudon G, Frank R, Laurenceau JL, Malergue C, Grosgogeat Y. 1982. Right ventricular dysplasia: a report of 24 adult cases. *Circulation* **65**: 384–398.

Marengo JJ, Hidalgo C, Bull R. 1998. Sulfhydryl oxidation modifies the calcium dependence of ryanodine-sensitive calcium channels of excitable cells. *Biophy J* **74**: 1263–1277.

Marks AR, Priori S, Memmi M, Kontula K, Laitinen PJ. 2002. Involvement of the cardiac ryanodine receptor/calcium release channel in catecholaminergic polymorphic ventricular tachycardia. *J Cell Physiol* **190**: 1–6.

Marks AR, Tempst P, Hwang KS, Taubman MB, Inui M, Chadwick C, Fleischer S, Nadal-Ginard B. 1989. Molecular cloning and characterization of the ryanodine receptor/junctional channel complex cDNA from skeletal muscle sarcoplasmic reticulum. *Proc Natl Acad Sci* **86**: 8683–8687.

Marx SO, Ondrias K, Marks AR. 1998. Coupled Gating Between Individual Skeletal Muscle Ca^{2+} Release Channels (Ryanodine Receptors). *Science* **281**: 818–821.

Marx SO, Reiken S, Hisamatsu Y, Gaburjakova M, Gaburjakova J, Yang YM, Rosemblit N, Marks AR. 2001. Phosphorylation-dependent regulation of ryanodine receptors: a novel role for leucine/isoleucine zippers. *J Cell Biol* **153**: 699–708.

Marx SO, Reiken S, Hisamatsu Y, Jayaraman T, Burkhoff D, Rosemblit N, Marks AR. 2000. PKA Phosphorylation Dissociates FKBP12.6 from the Calcium Release Channel (Ryanodine Receptor): Defective Regulation in Failing Hearts. *Cell* **101**: 365–376.

Maryon EB, Coronado R, Anderson P. 1996. Unc-68 encodes a ryanodine receptor involved in regulating *C. elegans* body-wall muscle contraction. *J Cell Biol* **134**: 885–893.

Masumiya H, Wang R, Zhang J, Xiao B, Chen SRW. 2003. Localization of the 12.6-kDa FK506-binding Protein (FKBP12.6) Binding Site to the NH2-terminal Domain of the Cardiac Ca^{2+} Release Channel (Ryanodine Receptor). *J Biol Chem* **278**: 3786–3792.

Mattei MG, Giannini G, Moscatelli F, Sorrentino V. 1994. Chromossomal localization of murine ryanodine receptor genes RYR1, RYR2, and RYR3 by in situ hybridization. *Genomics* **22**: 202–204.

Maximciuc AA, Putkey JA, Shamoo Y, MacKenzie KR. 2006. Complex of Calmodulin with a Ryanodine Receptor Target Reveals a Novel, Flexible Binding Mode. *Structure* **14**: 1547–1556.

Mayrleitner M, Chandler R, Schindler H, Fleischer S. 1995. Phosphorylation with protein kinases modulates calcium loading of terminal cisternae of sarcoplasmic reticulum from skeletal muscle. *Cell Calcium* **18**: 197–206.

McCall E, Li L, Satoh H, Shannon TR, Blatter LA, Bers DM. 1996. Effects of FK-506 on Contraction and Ca^{2+} Transients in Rat Cardiac Myocytes. *Circ Res* **79**: 1110–1121.

McGrew SG, Wolleben C, Siegl P, Inui M, Fleischer S. 1989. Positive cooperativity of ryanodine binding to the calcium release channel of sarcoplasmic reticulum from heart and skeletal muscle. *Biochemistry* **28**: 1686–1691.

Meissner G. 1984. Adenine nucleotide stimulation of Ca^{2+}-induced Ca^{2+}-release in sarcoplasmic reticulum. *J Biol Chem* **259**: 2365–2374.

Meissner G. 1994. Ryanodine Receptor/Ca^{2+} Release Channels and Their Regulation by Endogenous Effectors. *Ann Rev Physiol* **56**: 485–508.

Meissner G. 2002. Regulation of mammalian ryanodine receptors. *Frontiers in Biosci* **7**: d2072–2080.

Meissner G, Darling E, Eveleth J. 1986. Kinetics of rapid calcium release by sarcoplasmic reticulum. Effects of calcium, magnesium, and adenine nucleotides. *Biochemistry* **25**: 236–244.

Meissner G, Rios E, Tripathy A, Pasek DA. 1997. Regulation of Skeletal Muscle Ca^{2+} Release Channel (Ryanodine Receptor) by Ca^{2+} and Monovalent Cations and Anions. *J Biol Chem* **272**: 1628–1638.

Meng X, Wang G, Viero C, Wang Q, Mi W, Su X-D, Wagenknecht T, Williams AJ, Liu Z, Yin C-C. 2009. CLIC2-RyR1 Interaction and Structural Characterization by Cryo-electron Microscopy. *J Mol Biol* **387**: 320–334.

Mickelson JR, Louis CF. 1996. Malignant hyperthermia: excitation-contraction coupling, Ca^{2+} release channel, and cell Ca^{2+} regulation defects. *Physiol Rev* **76**: 537–592.

Monnier N, Romero NB, Lerale J, Nivoche Y, Qi D, MacLennan DH, Fardeau M, Lunardi J. 2000. An autosomal dominant congenital myopathy with cores and rods is associated with a neomutation in the RYR1 gene encoding the skeletal muscle ryanodine receptor. *Hum Mol Genet* **9**: 2599–2608.

Moore CP, Zhang J-Z, Hamilton SL. 1999b. A Role for Cysteine 3635 of RYR1 in Redox Modulation and Calmodulin Binding. *J Biol Chem* **274**: 36831–36834.

Moore CP, Rodney G, Zhang J-Z, Santacruz-Toloza L, Strasburg G, Hamilton SL. 1999a. Apocalmodulin and Ca^{2+} Calmodulin Bind to the Same Region on the Skeletal Muscle Ca^{2+} Release Channel. *Biochemistry* **38**: 8532–8537.

Mouton J, Marty I, Villaz M, Feltz A, Maulet Y. 2001. Molecular interaction of dihydropyridine receptors with type-1 ryanodine receptors in rat brain. *Biochem J* **354:** 597–603.

Nakai J, Imagawa T, Hakamata Y, Shigekawa M, Takeshima H, Numa S. 1990. Primary structure and functional expression from cDNA of the cardiac ryanodine receptor/calcium release channel. *FEBS letters* **271:** 169–177.

Nakai J, Sekiguchi N, Rando TA, Allen PD, Beam KG. 1998. Two regions of the ryanodine receptor involved in coupling with L-type Ca^{2+} channels. *J Biol Chem* **273:** 13403–13406.

Nakanishi S, Kuwajima G, Mikoshiba K. 1992. Immunohistochemical localization of ryanodine receptors in mouse central nervous system. *Neurosci Res* **15:** 130–142.

Neef S, Dybkova N, Sossalla S, Ort KR, Fluschnik N, Neumann K, Seipelt R, Schondube FA, Hasenfuss G, Maier LS. 2010. CaMKII-Dependent Diastolic SR Ca^{2+} Leak and Elevated Diastolic Ca^{2+} Levels in Right Atrial Myocardium of Patients with Atrial Fibrillation. *Circ Res* **106:** 1134–1144.

Neylon CB, Richards SM, Larsen MA, Agrotis A, Bobik A. 1995. Multiple types of ryanodine receptor/Ca^{2+} release channels are expressed in vascular smooth muscle. *Biochem Biophys Res Commun* **215:** 814–821.

Nixon GF, Mignery GA, Somlyo AV. 1994. Immunogold localization of inositol 1,4,5-trisphosphate receptors and characterization of ultrastructural features of the sarcoplasmic reticulum in phasic and tonic smooth muscle. *J Muscle Res Cell Motil* **15:** 682–700.

O'Brien J, Meissner G, Block BA. 1993. The fastest contracting muscles of nonmammalian vertebrates express only one isoform of the ryanodine receptor. *Biophys J* **65:** 2418–2427.

Ohrtman J, Ritter B, Polster A, Beam KG, Papadopoulos S. 2008. Sequence Differences in the IQ Motifs of $Ca_V1.1$ and $Ca_V1.2$ Strongly Impact Calmodulin Binding and Calcium-dependent Inactivation. *J Biol Chem* **283:** 29301–29311.

Orlova EV, Serysheva II, van Heel M, Hamilton SL, Chiu W. 1996. Two structural configurations of the skeletal muscle calcium release channel. *Nat Struct Mol Biol* **3:** 547–552.

Otsu K, Willard HF, Khanna VK, Zorzato F, Green NM, MacLennan DH. 1990. Molecular cloning of cDNA encoding the Ca^{2+} release channel (ryanodine receptor) of rabbit cardiac muscle sarcoplasmic reticulum. *J Biol Chem* **265:** 13472–13483.

Ottini L, Marziali G, Conti A, Charlesworth A, Sorrentino V. 1996. Alpha and beta isoforms of ryanodine receptor from chicken skeletal muscle are the homologues of mammalian RyR1 and RyR3. *Biochem J* **315:** 207–216.

Oyamada H, Murayama T, Takagi T, Iino M, Iwabe N, Miyata T, Ogawa Y, Endo M. 1994. Primary structure and distribution of ryanodine-binding protein isoforms of the bullfrog skeletal muscle. *J Biol Chem* **269:** 17206–17214.

Pamukcoglu T. 1988. Sudden death due to malignant hyperthermia. *Am J Forensic Med Pathol* **9:** 161–162.

Paolini C, Fessenden JD, Pessah IN, Franzini-Armstrong C. 2004. Evidence for conformational coupling between two calcium channels. *Proc Natl Acad Sci* **101:** 12748–12752.

Paul-Pletzer K, Yamamoto T, Bhat MB, Ma J, Ikemoto N, Jimenez LS, Morimoto H, Williams PG, Parness J. 2002. Identification of a Dantrolene-binding Sequence on the Skeletal Muscle Ryanodine Receptor. *J Biol Chem* **277:** 34918–34923.

Perez CF, Mukherjee S, Allen PD. 2003. Amino acids 1-1,680 of ryanodine receptor type 1 hold critical determinants of skeletal type for excitation-contraction coupling. Role of divergence domain D2. *J Biol Chem* **278:** 39644–39652.

Phillips MS, Khanna VK, De Leon S, Frodis W, Britt BA, MacLennan DH. 1994. The substitution of Arg for Gly2433 in the human skeletal muscle ryanodine receptor is associated with malignant hyperthermia. *Hum Mol Genet* **3:** 2181–2186.

Postma AV, Denjoy I, Hoorntje TM, Lupoglazoff JM, Da Costa A, Sebillon P, Mannens MM, Wilde AA, Guicheney P. 2002. Absence of calsequestrin 2 causes severe forms of catecholaminergic polymorphic ventricular tachycardia. *Circ Res* **91:** e21–e26.

Priori SG, Napolitano C, Memmi M, Colombi B, Drago F, Gasparini M, DeSimone L, Coltorti F, Bloise R, Keegan R, et al. 2002. Clinical and molecular characterization of patients with catecholaminergic polymorphic ventricular tachycardia. *Circulation* **106:** 69–74.

Proenza C, O'Brien J, Nakai J, Mukherjee S, Allen PD, Beam KG. 2002. Identification of a Region of RyR1 That Participates in Allosteric Coupling with the $\alpha 1S$ (CaV1.1) II-III Loop. *J Biol Chem* **277:** 6530–6535.

Protasi F, Franzini-Armstrong C, Allen PD. 1998. Role of Ryanodine Receptors in the Assembly of Calcium Release Units in Skeletal Muscle. *J Cell Biol* **140:** 831–842.

Protasi F, Paolini C, Nakai J, Beam KG, Franzini-Armstrong C, Allen PD. 2002. Multiple Regions of RyR1 Mediate Functional and Structural Interactions with $\alpha 1S$-Dihydropyridine Receptors in Skeletal Muscle. *Biophys J* **83:** 3230–3244.

Qi Y, Ogunbunmi EM, Freund EA, Timerman AP, Fleischer S. 1998. FK-binding Protein Is Associated with the Ryanodine Receptor of Skeletal Muscle in Vertebrate Animals. *J Biol Chem* **273:** 34813–34819.

Quane KA, Keating KE, Healy JM, Manning BM, Krivosic-Horber R, Krivosic I, Monnier N, Lunardi J, McCarthy TV. 1994. Mutation screening of the RYR1 gene in malignant hyperthermia: detection of a novel Tyr to Ser mutation in a pedigree with associated central cores. *Genomics* **23:** 236–239.

Quinn KE, Castellani L, Ondrias K, Ehrlich BE. 1998. Characterization of the ryanodine receptor/channel of invertebrate muscle. *Am J Physiol* **274:** R494–R502.

Radermacher M, Rao V, Grassucci R, Frank J, Timerman AP, Fleischer S, Wagenknecht T. 1994. Cryo-electron microscopy and three-dimensional reconstruction of the calcium release channel/ryanodine receptor from skeletal muscle. *J Cell Biol* **127:** 411–423.

Radermacher M, Wagenknecht T, Grassucci R, Frank J, Inui M, Chadwick C, Fleischer S. 1992. Cryo-EM of the native structure of the calcium release channel/ryanodine receptor from sarcoplasmic reticulum. *Biophys J* **61:** 936–940.

Reiken S, Lacampagne A, Zhou H, Kherani A, Lehnart SE, Ward C, Huang F, Gaburjakova M, Gaburjakova J, Rosemblit N, et al. 2003. PKA phosphorylation activates the calcium release channel (ryanodine receptor) in skeletal muscle: defective regulation in heart failure. *J Cell Biol* **160:** 919–928.

Rios E, Brum G. 1987. Involvement of dihydropyridine receptors in excitation-contraction coupling in skeletal muscle. *Nature* **325:** 717–720.

Robinson R, Carpenter D, Shaw MA, Halsall J, Hopkins P. 2006. Mutations in RYR1 in malignant hyperthermia and central core disease. *Hum Mutat* **27:** 977–989.

Rodney GG, Williams BY, Strasburg GM, Beckingham K, Hamilton SL. 2000. Regulation of RYR1 activity by Ca^{2+} and calmodulin. *Biochem* **39:** 7807–7812.

Rodriguez P, Bhogal MS, Colyer J. 2003. Stoichiometric Phosphorylation of Cardiac Ryanodine Receptor on Serine 2809 by Calmodulin-dependent Kinase II and Protein Kinase A. *J Biol Chem* **278:** 38593–38600.

Rosenberg H, Davis M, James D, Pollock N, Stowell K. 2007. Malignant hyperthermia. *Orphanet J Rare Dis* **2:** 21.

Ryan JF, Tedeschi LG. 1997. Sudden unexplained death in a patient with a family history of malignant hyperthermia. *J Clin Anesth* **9:** 66–68.

Sabbadini RA, Betto R, Teresi A, Fachechi-Cassano G, Salviati G. 1992. The effects of sphingosine on sarcoplasmic reticulum membrane calcium release. *J Biol Chem* **267:** 15475–15484.

Samsó M, Wagenknecht T. 2002. Apocalmodulin and Ca^{2+}-Calmodulin Bind to Neighboring Locations on the Ryanodine Receptor. *J Biol Chem* **277:** 1349–1353.

Samsó M, Shen X, Allen PD. 2006. Structural Characterization of the RyR1-FKBP12 Interaction. *J Mol Biol* **356:** 917–927.

Samsó M, Wagenknecht T, Allen PD. 2005. Internal structure and visualization of transmembrane domains of the RyR1 calcium release channel by cryo-EM. *Nat Struct Mol Biol* **12:** 539–544.

Samsó M, Feng W, Pessah IN, Allen PD. 2009. Coordinated Movement of Cytoplasmic and Transmembrane Domains of RyR1 upon Gating. *PLoS Biol* **7:** e1000085.

Scacheri PC, Hoffman EP, Fratkin JD, Semino-Mora C, Senchak A, Davis MR, Laing NG, Vedanarayanan V, Subramony SH. 2000. A novel ryanodine receptor gene mutation causing both cores and rods in congenital myopathy. *Neurology* **55:** 1689–1696.

Schwab Y, Mouton J, Chasserot-Golaz S, Marty I, Maulet Y, Jover E. 2001. Calcium-dependent translocation of synaptotagmin to the plasma membrane in the dendrites of developing neurones. *Brain Res Mol Brain Res* **96:** 1–13.

Serysheva I, Hamilton S, Chiu W, Ludtke S. 2005. Structure of Ca^{2+} release channel at 14Å resolution. *J Mol Biol* **345:** 427–431.

Serysheva II, Ludtke SJ, Baker ML, Cong Y, Topf M, Eramian D, Sali A, Hamilton SL, Chiu W. 2008. Subnanometer-resolution electron cryomicroscopy-based domain models for the cytoplasmic region of skeletal muscle RyR channel. *Proc Natl Acad Sci* **105:** 9610–9615.

Serysheva II, Orlova EV, Chiu W, Sherman MB, Hamilton SL, Heel Mv. 1995. Electron cryomicroscopy and angular reconstitution used to visualize the skeletal muscle calcium release channel. *Nat Struct Mol Biol* **2:** 18–24.

Serysheva II, Schatz M, van Heel M, Chiu W, Hamilton SL. 1999. Structure of the Skeletal Muscle Calcium Release Channel Activated with Ca^{2+} and AMP-PCP. *Biophys J* **77:** 1936–1944.

Sharma MR. 2006. Three-Dimensional Visualization of FKBP12.6 Binding to an Open Conformation of Cardiac Ryanodine Receptor. *Biophys J* **90:** 164–172.

Sharma MC, Gulati S, Sarkar C, Jain D, Kalra V, Suri V. 2007. Multi-minicore disease: a rare form of myopathy. *Neurol India* **55:** 50–53.

Sharma MR, Jeyakumar LH, Fleischer S, Wagenknecht T. 2000. Three-dimensional Structure of Ryanodine Receptor Isoform Three in Two Conformational States as Visualized by Cryo-electron Microscopy. *J Biol Chem* **275:** 9485–9491.

Sharma MR, Penczek P, Grassucci R, Xin H-B, Fleischer S, Wagenknecht T. 1998. Cryoelectron Microscopy and Image Analysis of the Cardiac Ryanodine Receptor. *J Biol Chem* **273:** 18429–18434.

Sharp AH, McPherson PS, Dawson TM, Aoki C, Campbell KP, Snyder SH. 1993. Differential immunohistochemical localization of inositol 1,4,5-trisphosphate- and ryanodine-sensitive Ca^{2+} release channels in rat brain. *J Neurosci* **13:** 3051–3063.

Sheridan DC, Cheng W, Ahern CA, Mortenson L, Alsammarae D, Vallejo P, Coronado R. 2003. Truncation of the Carboxyl Terminus of the Dihydropyridine Receptor β1a Subunit Promotes Ca^{2+} Dependent Excitation-Contraction Coupling in Skeletal Myotubes. *Biophys J* **84:** 220–237.

Sheridan DC, Takekura H, Franzini-Armstrong C, Beam KG, Allen PD, Perez CF. 2006. Bidirectional signaling between calcium channels of skeletal muscle requires multiple direct and indirect interactions. *Proc Natl Acad Sci* **103:** 19760–19765.

Shou W, Aghdasi B, Armstrong DL, Guo Q, Bao S, Charng M-J, Mathews LM, Schneider MD, Hamilton SL, Matzuk MM. 1998. Cardiac defects and altered ryanodine receptor function in mice lacking FKBP12. *Nature* **391:** 489–492.

Shuaib A, Paasuke RT, Brownell KW. 1987. Central core disease. Clinical features in 13 patients. *Medicine (Baltimore)* **66:** 389–396.

Sitsapesan R, Williams AJ. 1995. The gating of the sheep skeletal sarcoplasmic reticulum Ca^{2+}-release channel is regulated by luminal Ca^{2+}. *J Membr Biol* **146:** 133–144.

Slavik KJ, Wang JP, Aghdasi B, Zhang JZ, Mandel F, Malouf N, Hamilton SL. 1997. A carboxy-terminal peptide of the alpha 1-subunit of the dihydropyridine receptor inhibits Ca^{2+}-release channels. *Am J Physiol Cell Physiol* **272:** C1475–C1481.

Smith JS. 1986. Single-channel calcium and barium currents of large and small conductance from sarcoplasmic reticulum. *Biophys J* **50:** 921–928.

Smith JS, Coronado R, Meissner G. 1985. Sarcoplasmic reticulum contains adenine nucleotide-activated calcium channels. *Nature* **316:** 446–449.

Smith J, Coronado R, Meissner G. 1986. Single channel measurements of the calcium release channel from skeletal muscle sarcoplasmic reticulum. Activation by Ca^{2+} and ATP and modulation by Mg^{2+}. *J Gen Physiol* **88:** 573–588.

Stoyanovsky D, Murphy T, Anno PR, Kim YM, Salama G. 1997. Nitric oxide activates skeletal and cardiac ryanodine receptors. *Cell Calcium* **21**: 19–29.

Strazis KP, Fox AW. 1993. Malignant hyperthermia: a review of published cases. *Anesthesia Analgesia* **77**: 297–304.

Sun J, Xin C, Eu JP, Stamler JS, Meissner G. 2001. Cysteine-3635 is responsible for skeletal muscle ryanodine receptor modulation by NO. *Proc Natl Acad Sci* **98**: 11158–11162.

Sun J, Yamaguchi N, Xu L, Eu JP, Stamler JS, Meissner G. 2008. Regulation of the Cardiac Muscle Ryanodine Receptor by O_2 Tension and S-Nitrosoglutathion. *Biochemistry* **47**: 13985–13990.

Supnet C, Noonan C, Richard K, Bradley J, Mayne M. 2009. Up-regulation of the type 3 ryanodine receptor is neuroprotective in the TgCRND8 mouse model of Alzheimer's disease. *J Neurochem* **112**: 356–265.

Takasago T, Imagawa T, Shigekawa M. 1989. Phosphorylation of the Cardiac Ryanodine Receptor by cAMP-Dependent Protein Kinase. *J Biochem* **106**: 872–877.

Takekura H, Sun X, Franzini-Armstrong C. 1994. Development of the excitation-contraction coupling apparatus in skeletal muscle: peripheral and internal calcium release units are formed sequentially. *J Muscle Res Cell Motility* **15**: 102–118.

Takeshima H, Nishi M, Iwabe N, Miyata T, Hosoya T, Masai I, Hotta Y. 1994. Isolation and characterization of a gene for a ryanodine receptor/calcium release channel in Drosophila melanogaster. *FEBS Lett* **337**: 81–87.

Takeshima H, Nishimura S, Matsumoto T, Ishida H, Kangawa K, Minamino N, Matsuo H, Ueda M, Hanaoka M, Hirose T, et al. 1989. Primary structure and expression from complementary DNA of skeletal muscle ryanodine receptor. *Nature* **339**: 439–445.

Tanabe T, Beam KG, Adams BA, Niidome T, Numa S. 1990. Regions of the skeletal muscle dihydropyridine receptor critical for excitation-contraction coupling. *Nature* **346**: 567–569.

Tang W, Sencer S, Hamilton SL. 2002. Calmodulin modulation of proteins involved in excitation-contraction coupling. *Frontiers Biosci* **7**: d1583–1589.

Tang W, Ingalls CP, Durham WJ, Snider J, Reid MB, Wu G, Matzuk MM, Hamilton SL. 2004. Altered excitation-contraction coupling with skeletal muscle specific FKBP12 deficiency. *FASEB J*: 04–1587fje.

Tilgen N, Zorzato F, Halliger-Keller B, Muntoni F, Sewry C, Palmucci LM, Schneider C, Hauser E, Lehmann-Horn F, Muller CR, et al. 2001. Identification of four novel mutations in the C-terminal membrane spanning domain of the ryanodine receptor 1: association with central core disease and alteration of calcium homeostasis. *Hum Mol Genet* **10**: 2879–2887.

Timerman AP, Ogunbumni E, Freund E, Wiederrecht G, Marks AR, Fleischer S. 1993. The calcium release channel of sarcoplasmic reticulum is modulated by FK-506-binding protein. Dissociation and reconstitution of FKBP-12 to the calcium release channel of skeletal muscle sarcoplasmic reticulum. *J Biol Chem* **268**: 22992–22999.

Timerman AP, Onoue H, Xin H-B, Barg S, Copello J, Wiederrecht G, Fleischer S. 1996. Selective Binding of FKBP12.6 by the Cardiac Ryanodine Receptor. *J Biol Chem* **271**: 20385–20391.

Timerman AP, Wiederrecht G, Marcy A, Fleischer S. 1995. Characterization of an Exchange Reaction between Soluble FKBP-12 and the FKBP Ryanodine Receptor Complex. *J Biol Chem* **270**: 2451–2459.

Tiso N, Stephan DA, Nava A, Bagattin A, Devaney JM, Stanchi F, Larderet G, Brahmbhatt B, Brown K, Bauce B, et al. 2001. Identification of mutations in the cardiac ryanodine receptor gene in families affected with arrhythmogenic right ventricular cardiomyopathy type 2 (ARVD2). *Hum Mol Genet* **10**: 189–194.

Tong J, McCarthy TV, MacLennan DH. 1999. Measurement of resting cytosolic Ca^{2+} concentrations and Ca^{2+} store size in HEK-293 cells transfected with malignant hyperthermia or central core disease mutant Ca^{2+} release channels. *J Biol Chem* **274**: 693–702.

Tripathy A, Xu L, Mann G, Meissner G. 1995. Calmodulin activation and inhibition of skeletal muscle Ca^{2+} release channel (ryanodine receptor). *Biophy J* **69**: 106–119.

Tunwell RE, Wickenden C, Bertrand BM, Shevchenko VI, Walsh MB, Allen PD, Lai FA. 1996. The human cardiac muscle ryanodine receptor-calcium release channel: identification, primary structure and topological analysis. *Biochem J* **318**: 477–487.

Valdivia HH. 2007. One gene, many proteins: alternative splicing of the ryanodine receptor gene adds novel functions to an already complex channel protein. *Circ Res* **100**: 761–763.

Valdivia HH, Kaplan JH, Ellis-Davies GCR, Lederer WJ. 1995. Rapid Adaptation of Cardiac Ryanodine Receptors: Modulation by Mg^{2+} and Phosphorylation. *Science* **267**: 1997–2000.

Vazquez-Martinez O, Canedo-Merino R, Diaz-Muñoz M, Riesgo-Escovar JR. 2003. Biochemical characterization, distribution and phylogenetic analysis of Drosophila melanogaster ryanodine and IP3 receptors, and thapsigargin-sensitive Ca^{2+} ATPase. *Journal of Cell Science* **116**: 2483–2494.

Voss AA, Lango J, Ernst-Russell M, Morin D, Pessah IN. 2004. Identification of Hyperreactive Cysteines within Ryanodine Receptor Type 1 by Mass Spectrometry. *J Biol Chem* **279**: 34514–34520.

Vukcevic M, Broman M, Islander G, Bodelsson M, Ranklev-Twetman E, Müller CR, Treves S. 2010. Functional Properties of RYR1 Mutations Identified in Swedish Patients with Malignant Hyperthermia and Central Core Disease. *Anesthesia & Analgesia* **111**: 185–190.

Wagenknecht T, Berkowitz J, Grassucci R, Timerman AP, Fleischer S. 1994. Localization of calmodulin binding sites on the ryanodine receptor from skeletal muscle by electron microscopy. *Biophys J* **67**: 2286–2295.

Wagenknecht T, Grassucci R, Berkowitz J, Wiederrecht GJ, Xin HB, Fleischer S. 1996. Cryoelectron microscopy resolves FK506-binding protein sites on the skeletal muscle ryanodine receptor. *Biophys J* **70**: 1709–1715.

Wagenknecht T, Radermacher M, Grassucci R, Berkowitz J, Xin H-B, Fleischer S. 1997. Locations of Calmodulin and FK506-binding Protein on the Three-dimensional Architecture of the Skeletal Muscle Ryanodine Receptor. *J Biol Chem* **272**: 32463–32471.

Wang J, Best PM. 1992. Inactivation of the sarcoplasmic reticulum calcium channel by protein kinase. *Nature* **359:** 739–741.

Wang R, Chen W, Cai S, Zhang J, Bolstad J, Wagenknecht T, Liu Z, Chen SRW. 2007. Localization of an NH2-terminal Disease-causing Mutation Hot Spot to the "Clamp"? Region in the Three-dimensional Structure of the Cardiac Ryanodine Receptor. *J Biol Chem* **282:** 17785–17793.

Ward A, Chaffman M, Sorkin E. 1986. Dantrolene. A review of its pharmacodynamic and pharmacokinetic properties and therapeutic use in malignant hyperthermia, the neuroleptic malignant syndrome and an update of its use in muscle spasticity. *Drugs* **32:** 130–168.

Wehrens XH, Marks AR. 2003. Altered function and regulation of cardiac ryanodine receptors in cardiac disease. *Trends Biochem Sci* **28:** 671–678.

Wehrens XH, Lehnart SE, Huang F, Vest JA, Reiken SR, Mohler PJ, Sun J, Guatimosim S, Song LS, Rosemblit N, et al. 2003. FKBP12.6 deficiency and defective calcium release channel (ryanodine receptor) function linked to exercise-induced sudden cardiac death. *Cell* **113:** 829–840.

Wehrens XHT, Lehnart SE, Reiken SR, Marks AR. 2004. Ca^{2+}/Calmodulin-Dependent Protein Kinase II Phosphorylation Regulates the Cardiac Ryanodine Receptor. *Circ Res* **94:** e61–e70.

Wei L, Hanna AD, Beard NA, Dulhunty AF. 2009. Unique isoform-specific properties of calsequestrin in the heart and skeletal muscle. *Cell Calcium* **45:** 474–484.

Witcher DR, Kovacs RJ, Schulman H, Cefali DC, Jones LR. 1991. Unique phosphorylation site on the cardiac ryanodine receptor regulates calcium channel activity. *J Biol Chem* **266:** 11144–11152.

Wright NT, Prosser BL, Varney KM, Zimmer DB, Schneider MF, Weber DJ. 2008. S100A1 and Calmodulin Compete for the Same Binding Site on Ryanodine Receptor. *J Biol Chem* **283:** 26676–26683.

Xiao B, Jiang MT, Zhao M, Yang D, Sutherland C, Lai FA, Walsh MP, Warltier DC, Cheng H, Chen SRW. 2005. Characterization of a Novel PKA Phosphorylation Site, Serine-2030, Reveals No PKA Hyperphosphorylation of the Cardiac Ryanodine Receptor in Canine Heart Failure. *Circ Res* **96:** 847–855.

Xiao B, Sutherland C, Walsh MP, Chen SRW. 2004. Protein Kinase A Phosphorylation at Serine-2808 of the Cardiac Ca^{2+}-Release Channel (Ryanodine Receptor) Does Not Dissociate 12.6-kDa FK506-Binding Protein (FKBP12.6). *Circ Res* **94:** 487–495.

Xiao B, Zhong G, Obayashi M, Yang D, Chen K, Walsh MP, Shimoni Y, Cheng H, Ter Keurs H, Chen SRW. 2006. Ser-2030, but not Ser-2808, is the major phosphorylation site in cardiac ryanodine receptors responding to protein kinase A activation upon β-adrenergic stimulation in normal and failing hearts. *Biochem J* **396:** 7–16.

Xu L, Mann G, Meissner G. 1996. Regulation of Cardiac Ca^{2+} Release Channel (Ryanodine Receptor) by Ca^{2+}, H^+, Mg^{2+}, and Adenine Nucleotides Under Normal and Simulated Ischemic Conditions. *Circ Res* **79:** 1100–1109.

Xu L, Eu JP, Meissner G, Stamler JS. 1998. Activation of the cardiac calcium release channel (ryanodine receptor) by poly-S-nitrosylation. *Science* **279:** 234–237.

Xu X, Yano M, Uchinoumi H, Hino A, Suetomi T, Ono M, Tateishi H, Oda T, Okuda S, Doi M, et al. 2010. Defective calmodulin binding to the cardiac ryanodine receptor plays a key role in CPVT-associated channel dysfunction. *Biochemical and Biophysical Research Communications* **394:** 660–666.

Yamaguchi N, Xu L, Pasek DA, Evans KE, Chen SRW, Meissner G. 2005. Calmodulin Regulation and Identification of Calmodulin Binding Region of Type-3 Ryanodine Receptor Calcium Release Channel. *Biochemistry* **44:** 15074–15081.

Yamaguchi N, Xu L, Pasek DA, Evans KE, Meissner G. 2003. Molecular Basis of Calmodulin Binding to Cardiac Muscle Ca^{2+} Release Channel (Ryanodine Receptor). *J Biol Chem*: 23480–23486.

Yang HC, Reedy MM, Burke CL, Strasburg GM. 1994. Calmodulin interaction with the skeletal muscle sarcoplasmic reticulum calcium channel protein. *Biochemistry* **33:** 518–525.

Zable AC, Favero TG, Abramson JJ. 1997. Glutathione modulates ryanodine receptor from skeletal muscle sarcoplasmic reticulum. Evidence for redox regulation of the Ca^{2+} release mechanism. *J Biol Chem* **272:** 7069–7077.

Zaidi M, Shankar VS, Towhidul Alam AS, Moonga BS, Pazianas M, Huang CL. 1992. Evidence that a ryanodine receptor triggers signal transduction in the osteoclast. *Biochemical and Biophysical Research Communications* **188:** 1332–1336.

Zhang Y, Chen HS, Khanna VK, De Leon S, Phillips MS, Schappert K, Britt BA, Browell AK, MacLennan DH. 1993. A mutation in the human ryanodine receptor gene associated with central core disease. *Nat Genet* **5:** 46–50.

Zhang J, Liu Z, Masumiya H, Wang R, Jiang D, Li F, Wagenknecht T, Chen SRW. 2003. Three-dimensional Localization of Divergent Region 3 of the Ryanodine Receptor to the Clamp-shaped Structures Adjacent to the FKBP Binding Sites. *J Biol Chem* **278:** 14211–14218.

Zhao F, Li P, Chen SRW, Louis CF, Fruen BR. 2001. Dantrolene Inhibition of Ryanodine Receptor Ca^{2+} Release Channels. *J Biol Chem* **276:** 13810–13816.

Zhao M, Li P, Li X, Zhang L, Winkfein RJ, Chen SRW. 1999. Molecular Identification of the Ryanodine Receptor Pore-forming Segment. *J Biol Chem* **274:** 25971–25974.

Zhou H, Lillis S, Loy RE, Ghassemi F, Rose MR, Norwood F, Mills K, Al-Sarraj S, Lane RJ, Feng L, et al. 2010. Multi-minicore disease and atypical periodic paralysis associated with novel mutations in the skeletal muscle ryanodine receptor (RYR1) gene. *Neuromuscul Disord* **20:** 166–173.

Zissimopoulos S, Lai FA. 2005. Interaction of FKBP12.6 with the Cardiac Ryanodine Receptor C-terminal Domain. *J Biol Chem* **280:** 5475–5485.

Zorzato F, Fujii J, Otsu K, Phillips M, Green NM, Lai FA, Meissner G, MacLennan DH. 1990. Molecular cloning of cDNA encoding human and rabbit forms of the Ca^{2+} release channel (ryanodine receptor) of skeletal muscle sarcoplasmic reticulum. *J Biol Chem* **265:** 2244–2256.

IP$_3$ Receptors: Toward Understanding Their Activation

Colin W. Taylor and Stephen C. Tovey

Department of Pharmacology, University of Cambridge, Cambridge CB2 1PD, United Kingdom

Correspondence: cwt1000@cam.ac.uk

Inositol 1,4,5-trisphosphate receptors (IP$_3$R) and their relatives, ryanodine receptors, are the channels that most often mediate Ca^{2+} release from intracellular stores. Their regulation by Ca^{2+} allows them also to propagate cytosolic Ca^{2+} signals regeneratively. This brief review addresses the structural basis of IP$_3$R activation by IP$_3$ and Ca^{2+}. IP$_3$ initiates IP$_3$R activation by promoting Ca^{2+} binding to a stimulatory Ca^{2+}-binding site, the identity of which is unresolved. We suggest that interactions of critical phosphate groups in IP$_3$ with opposite sides of the clam-like IP$_3$-binding core cause it to close and propagate a conformational change toward the pore via the adjacent N-terminal suppressor domain. The pore, assembled from the last pair of transmembrane domains and the intervening pore loop from each of the four IP$_3$R subunits, forms a structure in which a luminal selectivity filter and a gate at the cytosolic end of the pore control cation fluxes through the IP$_3$R.

A BRIEF HISTORY OF IP$_3$ RECEPTORS

Sidney Ringer, in his famous correction to an earlier paper, showed that Ca^{2+} entry can evoke a physiological response by demonstrating that beating of the frog heart requires extracellular Ca^{2+} (Ringer 1883). Almost a century passed before it became clear that this Ca^{2+} entry, via voltage-gated Ca^{2+} channels, was not directly responsible for contraction, but instead provided the trigger for a much larger release of Ca^{2+} from stores within the sarcoplasmic reticulum (SR). The latter is mediated by type-2 ryanodine receptors (RyR) (Fabiato 1983; Cheng et al. 1993), which like many Ca^{2+} channels, are able both to transport Ca^{2+} through an open pore and respond to it. These observations highlight two general points. First, cells call upon two sources of Ca^{2+} to evoke increases in cytosolic Ca^{2+} concentration; second, interactions between these Ca^{2+} fluxes across the plasma membrane and the membranes of intracellular stores are important determinants of the physiological response. The same points apply to the Ca^{2+} signals evoked by receptors that stimulate phospholipase C (PLC) and, thereby, formation of inositol 1,4,5-trisphosphate (IP$_3$).

The biochemical sequence linking these receptors to formation of IP$_3$ emerged in the 1980s (Michell et al. 1989; Berridge 2005), but work in the decade before had established that many receptors regulate many different responses by increasing the cytosolic Ca^{2+} concentration (Rasmussen 1970; Berridge 1975). In his influential review, Bob Michell (Michell 1975), building on work showing that many of these receptors also stimulate phospholipid turnover (Hokin and Hokin 1953), had suggested a causal link between phosphoinositide hydrolysis and Ca^{2+} signals. Here, as in many studies, the emphasis was on Ca^{2+} entry, with a consensus only slowly emerging that Ca^{2+} fluxes across both the plasma membrane and the membranes of intracellular stores contribute to cytosolic Ca^{2+} signals (Rasmussen 1970; Berridge 1975; Williams 1980; Putney et al. 1981). In the years following Michell's review, decisive evidence, much of it coming from Mike Berridge's elegant studies of blowfly salivary gland, established that phosphoinositide hydrolysis is, as predicted by Michell, required for PLC-linked receptors to evoke Ca^{2+} signals (Berridge and Fain 1979). The same preparation was used to show that IP_3 is the first water-soluble product of the signaling pathway (Berridge 1983). IP_3, thus, emerged as a prime candidate for the cytosolic messenger linking events at the plasma membrane to release of Ca^{2+} from intracellular stores. Paradoxically, it was to be many years before the links between receptors that stimulate PLC and Ca^{2+} entry were resolved. These came with elaboration of the pathways linking empty Ca^{2+} stores to Ca^{2+} entry, the so-called store-operated Ca^{2+} entry pathway (Putney 1997; Park et al. 2009), and recognition that many trp channels are regulated by products of PLC activity (Nilius et al. 2007). IP_3 receptors (IP_3R) also contribute more directly to Ca^{2+} entry across the plasma membrane either because, at least in some cells, IP_3R are functionally expressed in the plasma membrane (Dellis et al. 2006; Dellis et al. 2008), or perhaps through their direct interactions with other plasma membrane Ca^{2+} channels (Kiselyov et al. 1999). Here, we focus solely on Ca^{2+} release from the endoplasmic reticulum (ER) by IP_3R. Some of the key steps in the evolution of our current understanding of IP_3R are listed in Table 1.

The role of the SR as the intracellular source of Ca^{2+} signals in striated muscle was long-established (Endo et al. 1970), but there was no such agreement on the identity of the organelle from which Ca^{2+} was released in other cells. Competing claims suggested roles for mitochondria or the ER. Evidence that in resting hepatocytes only the ER contains appreciable amounts of Ca^{2+} (Burgess et al. 1983) was quickly followed by the demonstration that IP_3 evoked Ca^{2+} release from a non-mitochondrial Ca^{2+} store in permeabilized pancreatic acinar cells (Streb et al. 1983). Countless groups quickly replicated these findings in many cells, and within months it was universally accepted that the ER is the major Ca^{2+} store from which IP_3 stimulates Ca^{2+} release in most animal cells (Berridge and Irvine 1984; Berridge and Irvine 1989). Subsequent work has suggested that IP_3 may also stimulate Ca^{2+} release from the Golgi apparatus (Pinton et al. 1998), from within the nucleus (Gerasimenko et al. 1995; Echevarria et al. 2003; Marchenko et al. 2005), and perhaps also from secretory vesicles (Gerasimenko et al. 1996), but ER remains the major IP_3-sensitive Ca^{2+} store. Evidence that IP_3 stimulates Ca^{2+} efflux from the ER (rather than inhibiting Ca^{2+} uptake) and the first single channel recordings (Ehrlich and Watras 1988) established that the IP_3R is an IP_3-gated, Ca^{2+}-permeable channel. The first studies of ^{32}P-IP_3 binding (Spät et al. 1986) were followed by purification of IP_3R from cerebellum (Maeda et al. 1988; Supattapone et al. 1988) and then cloning of the first IP_3R subtype (IP_3R1) (Furuichi et al. 1989; Mignery et al. 1989). Subsequent studies identified two additional genes encoding vertebrate IP_3R (IP_3R2 and IP_3R3) and a single gene in invertebrates (Taylor et al. 1999). It remains far from clear whether plants express related IP_3R (Krinke et al. 2007). These studies established that IP_3R are unusually large proteins, comprising tetramers of closely-related subunits, each with about 2700 amino acid residues. RyR are even larger: they, too, are tetramers, but the subunits are almost twice the size of IP_3R ($\sim$5000 residues). This progress with identifying IP_3R together with single channel recordings of IP_3R, initially in artificial lipid bilayers and later in native membranes (Foskett et al. 2007; Rahman et al. 2009), provided the foundations from which to explore the structural determinants of IP_3R behavior. The advances toward understanding the molecular mechanisms

Table 1. Landmarks en route to a structural analysis of IP$_3$ receptor behavior.

	RyR	IP$_3$R
1883	Ca^{2+} entry required for heart contraction[1]	
1953		Acetylcholine stimulates turnover of phospholipids[2]
1975		Phosphoinositide hydrolysis proposed to cause Ca^{2+} signals[3]
1977	Ca^{2+} waves occur at fertilization[4]	
1977	Ca^{2+}-induced Ca^{2+} release in SR[5]	
1979		Phosphoinositide hydrolysis required for receptor-stimulated Ca^{2+} signals[6]
1980	Introduction of Quin 2[7] and facile loading methods[8]	
1983		IP$_3$ is first water-soluble product of PLC[9]
1983		IP$_3$ stimulates Ca^{2+} release from a non-mitochondrial store[10]
1985	Ryanodine, selective RyR ligand[11]	
1985	Single channel records of RyR[12]	
1986		Frequency-coded Ca^{2+} spikes[13]
1987		Ca^{2+} regulates IP$_3$R[14,15]
1987	RyR1 purified[16]	IP$_3$R1 purified[17]
1988		Single channel records of IP$_3$R[18]
1989	Cloning of RyR1[19]	Cloning of IP$_3$R1[20,21]
1990		Elementary Ca^{2+}-release events[22]
1993	Elementary Ca^{2+}-release events[23]	
2002		Atomic structure of IBC[24]
2005		Atomic structure of SD[25]
2009	Atomic structure of N-terminal of RyR[26,27]	

[1]Ringer (1883).
[2]Hokin & Hokin (1953).
[3]Michell (1975).
[4]Ridgeway et al. (1977).
[5]Endo (1977).
[6]Berridge & Fain (1979).
[7]Tsien (1980).
[8]Tsien (1981).
[9]Berridge (1983).
[10]Streb et al. (1983).
[11]Sutko et al. (1985).
[12]Smith et al. (1985).
[13]Woods et al. (1986).
[14]Iino (1987).
[15]Iino (1990).
[16]Imagawa et al. (1987).
[17]Supattapone et al. (1988).
[18]Ehrlich & Watras (1988).
[19]Takeshima et al. (1989).
[20]Mignery et al. (1989).
[21]Furuichi et al. (1989).
[22]Parker & Ivorra (1990).
[23]Cheng et al. (1993).
[24]Bosanac et al. (2002).
[25]Bosanac et al. (2005).
[26]Amador et al. (2009).
[27]Lobo & Van Petegem (2009).

of IP$_3$R behavior were accompanied by similar progress with RyR (Table 1). Recurrent themes, to which we return, are the similarities between RyR and IP$_3$R, and the many instances where observations of one channel family have informed further analysis of the other. Very recently, a third family of intracellular Ca^{2+} channels, unrelated to RyR and IP$_3$R, has been implicated in Ca^{2+} signaling. These are the two-pore channels (TPC) that are activated by NAADP and release Ca^{2+} from acidic Ca^{2+} stores, including lysosomes and endosomes (Patel et al. 2010; Zhu et al. 2010). Several trp (transient receptor protein) channels, in addition to their roles in the plasma membrane, may also mediate release of Ca^{2+} from intracellular stores (Gees et al. 2010).

Parallel to work addressing the workings of IP$_3$R, there was growing interest in the spatiotemporal complexity of cytosolic Ca^{2+} signals. Ca^{2+} waves were first observed during fertilization. These waves were proposed to result from Ca^{2+}-induced Ca^{2+} release (CICR) and were followed by smaller repetitive Ca^{2+} transients (Ridgway et al. 1977; Gilkey 1983). It was, however, the work of Peter Cobbold that focused most attention on the complexity of intracellular Ca^{2+} signals (Woods et al. 1986). Just as the activity of a nerve is conveyed by the frequency of its action potentials, Cobbold demonstrated that in hepatocytes the concentration of the extracellular stimulus determined the frequency of the cytosolic Ca^{2+} transients. As these ideas gathered momentum (Berridge 1995), evidence accumulated in support of cells using the information provided by frequency-encoded Ca^{2+} spikes as an efficient means of regulating cellular activity (Dolmetsch et al. 1997; Li et al. 1998; Berridge et al. 2000; Dupont et al. 2003). The single greatest contributor to progress in understanding the genesis of these intracellular Ca^{2+} signals was the introduction, by Roger Tsien in 1980, of simple, minimally disruptive methods for measuring the free cytosolic Ca^{2+} concentration in intact cells (Tsien 1980; Tsien 1981). These methods, in combination with improved optical microscopy, allowed Ian Parker to begin to resolve the subcellular organization of IP$_3$-evoked Ca^{2+} signals (Parker and Ivorra 1990;
Parker et al. 1996). He showed that as the IP$_3$ concentration increases, it triggers a hierarchy of elementary Ca^{2+} release events, beginning with the openings of single IP$_3$R (Ca^{2+} blips), progressing to the coordinated openings of a cluster of several IP$_3$R (Ca^{2+} puffs) and finally, with sufficient IP$_3$, culminating in a regenerative Ca^{2+} wave invading the entire cell (Bootman et al. 1997; Demuro and Parker 2007). The demonstration, in 1987 by Masamitsu Iino, that IP$_3$R are stimulated by cytosolic Ca^{2+} (Iino 1987), and the later widespread recognition that all IP$_3$R are biphasically regulated by cytosolic Ca^{2+} (Iino 1990; Taylor and Laude 2002), provided what has become the most widely accepted explanation for the recruitment of elementary Ca^{2+}-release events. Namely, that CICR, already an established feature of RyR (Endo et al. 1970), allows an active IP$_3$R to propagate its activity to neighboring IP$_3$R.

These observations and accumulating evidence that local Ca^{2+} signals can selectively regulate local events (Rizzuto et al. 1993; Berridge et al. 2000; Dyer et al. 2005; Willoughby and Cooper 2007) prompted a re-assessment of the ways in which Ca^{2+} signals convey information. It became untenable to think of responses to graded changes in the intensity of the extracellular stimulus as being simply encoded in graded changes in global cytosolic Ca^{2+} concentration. Ca^{2+} entering the cytosol via one channel can regulate different proteins to Ca^{2+} entering via another (Berridge et al. 2000; Dyer et al. 2005; Willoughby and Cooper 2007). Hence, the spatial organization of the changes in cytosolic Ca^{2+} concentration profoundly affects the physiological response, and that presents many opportunities for delivering different Ca^{2+} signals in response to different stimuli or different stimulus intensities. The duration of each Ca^{2+} increase, whether local or global, is also important in determining not only the amplitude of the response, but also its nature, because Ca^{2+}-binding proteins differ in their responses to transient and sustained signals. Finally, the frequency with which Ca^{2+} signals are delivered can determine both the nature and amplitude of the cellular response. The key point is that the versatility of Ca^{2+} as an intracellular messenger capable of regulating diverse cellular events depends largely on the spatiotemporal complexity of cytosolic Ca^{2+} signals (Berridge et al. 2000). If we are to understand how Ca^{2+} functions as a ubiquitous intracellular messenger, we must explain how IP_3-evoked Ca^{2+} signals grow from the opening of a single IP_3R to much larger events. That explanation depends, ultimately, on putting IP_3R into appropriate places within the cell, and on the interactions between IP_3 and Ca^{2+} in regulating the opening of IP_3R. In recent reviews (Taylor et al. 2009a; Taylor et al. 2009b) and original reports, we have described how IP_3R are co-translationally targeted to the ER and then retained there by sequences within their transmembrane domains (TMD) (Parker et al. 2004; Pantazaka and Taylor 2010). We have also suggested that within the ER, IP_3 causes IP_3R to assemble into small clusters within which their regulation by both IP_3 and Ca^{2+} is retuned to facilitate the Ca^{2+}-mediated recruitment of IP_3R activity by an active neighbor (Rahman and Taylor 2009; Rahman et al. 2009). Here, we focus entirely on the interactions between Ca^{2+} and IP_3 in regulating IP_3R activity, and the extent to which we can explain those interactions at the structural level.

REGULATION OF IP_3 RECEPTORS BY Ca^{2+} AND IP_3

Activation of IP_3R requires both IP_3 and its permeating ion, Ca^{2+} (Finch et al. 1991; Marchant and Taylor 1997; Adkins and Taylor 1999; Taylor and Laude 2002; Foskett et al. 2007). There are reports of IP_3-independent activation of IP_3R by CaBP1 (Yang et al. 2002), a member of the neuronal Ca^{2+}-sensor family, and by $G\beta\gamma$ subunits (Zeng et al. 2003), but the physiological relevance is unclear (Haynes et al. 2004; Nadif Kasri et al. 2004). The current consensus is that binding of IP_3 to the IP_3R is essential for its activation, but whether all four IP_3-binding sites of the tetrameric IP_3R must be occupied is unresolved. Positively cooperative responses to IP_3 in some (Dufour et al. 1997; Marchant and Taylor 1997; Tu et al. 2005a), though not all, studies (Finch et al. 1991; Watras et al. 1991;

Laude et al. 2005), and delays before the first response to IP_3 that decrease with increasing IP_3 concentration (Marchant and Taylor 1997), indicate that channel opening requires occupancy of more than one IP_3-binding site. However, gating by IP_3 of heteromeric IP_3R in which at least one subunit is mutated to prevent IP_3 binding suggests that occupancy of fewer than four IP_3-binding sites may be sufficient to cause some channel opening (Boehning and Joseph 2000a). IP_3R subtypes differ in their affinities for IP_3, with the general consensus being that IP_3R2 is more sensitive than IP_3R1, and both are considerably more sensitive than IP_3R3 (Tu et al. 2005b; Iwai et al. 2007). In the cellular context, however, differences in expression level (Dellis et al. 2006; Tovey et al. 2010), subcellular distribution (Petersen et al. 1999), post-transcriptional and post-translational modifications, and association of IP_3R with accessory proteins (Patterson et al. 2004) may be more important determinants of sensitivity.

Soon after the first report of IP_3-evoked Ca^{2+} release, cytosolic Ca^{2+} was shown also to regulate IP_3R (Suematsu et al. 1984; Jean and Klee 1986); thereafter, it emerged that the effects of Ca^{2+} were biphasic, with modest increases in cytosolic Ca^{2+} concentration enhancing responses to IP_3, while higher concentrations were inhibitory (Iino 1987; Iino 1990; Finch et al. 1991; Parys et al. 1992; Marshall and Taylor 1993). This provided yet another parallel with RyR, which are also biphasically regulated by Ca^{2+} (Hamilton 2005). The coregulation of IP_3R by IP_3 and Ca^{2+} in permeabilized cells was confirmed by single-channel recordings of IP_3R1 reconstituted into lipid bilayers (Bezprozvanny et al. 1991; Striggow and Ehrlich 1996; Kaftan et al. 1997; Ramos-Franco et al. 1998a; Ramos-Franco et al. 1998b; Tu et al. 2002; Tu et al. 2005b) and in native nuclear membranes (Stehno-Bittel et al. 1995; Mak et al. 1998; Boehning et al. 2001a; Marchenko et al. 2005). In each case, the single-channel open probability (P_o) of IP_3-activated channels displayed a bell-shaped dependence on cytosolic Ca^{2+} concentration. Evidence that purified IP_3R1 could be stimulated, but not inhibited, by cytosolic Ca^{2+} (Thrower et al. 1998; Michikawa et al.

1999) raised the possibility that Ca^{2+} inhibition might be mediated by an accessory protein, although it has yet to be identified. The same explanation perhaps accounts for some reports, often derived from bilayer recordings, in which Ca^{2+} was suggested not to inhibit IP_3R2 or IP_3R3 (Horne and Meyer 1995; Hagar et al. 1998; Miyakawa et al. 1999; Ramos-Franco et al. 2000). The balance of opinion, supported by numerous studies of all three IP_3R subtypes and using both single-channel and Ca^{2+}-efflux studies, is that all three IP_3R subtypes are biphasically regulated by cytosolic Ca^{2+} (Marshall and Taylor 1993; Oancea and Meyer 1996; Dufour et al. 1997; Missiaen et al. 1998; Miyakawa et al. 1999; Swatton et al. 1999; Boehning and Joseph 2000b; Mak et al. 2000; Mak et al. 2001; Tu et al. 2005a). Two independent Ca^{2+}-binding sites, which differ in their interactions with different bivalent cations and in their affinities for Ca^{2+}, mediate the stimulatory and inhibitory effects of cytosolic Ca^{2+} (Marshall and Taylor 1994; Striggow and Ehrlich 1996; Hajnóczky and Thomas 1997). Both sites are essential elements of many models proposed to explain regenerative Ca^{2+} signals (Lechleiter et al. 1991; Berridge 1997). This core biphasic pattern of regulation by cytosolic Ca^{2+} may be modulated by other intracellular signals (and these, too, may have contributed to some of the disparate findings) and by processing of IP_3R. Ca^{2+}-dependent inhibition of IP_3R3, for example, is very sensitive to cytoplasmic ATP (Tu et al. 2005b), and the neuronal $S2^+$ splice variant of IP_3R1 has a broader Ca^{2+}-dependence than the peripheral $S2^-$ form (Tu et al. 2002). However, IP_3 is the major influence on what Ca^{2+} does to IP_3R: The two ligands are essential co-agonists of IP_3R (Finch et al. 1991). Activation of IP_3R1 by Ca^{2+} is positively cooperative, enabling P_o to reach its maximum value over a narrow range of Ca^{2+} concentrations, suggesting that IP_3R1 may be well suited to mediating CICR and regenerative Ca^{2+} signals. Activation of IP_3R3 is less cooperative, occurs over a broader range of Ca^{2+} concentrations, and requires lesser activation, making it well suited as a trigger for Ca^{2+} release as the level of IP_3 increases (Mak et al. 2001; Foskett et al. 2007).

Foskett and colleagues have argued, from their analyses of patch-clamp recordings of nuclear IP_3R, that IP_3 decreases the sensitivity of the IP_3R to inhibition by cytosolic Ca^{2+}, and that this alone is the means whereby IP_3 stimulates channel opening (Mak et al. 1998; Mak et al. 2001; Ionescu et al. 2006). This simple explanation, where IP_3 serves only to relieve tonic inhibition by resting Ca^{2+} concentrations, is impossible to reconcile with their observation that pretreatment of cells with Ca^{2+}-free media abolishes Ca^{2+} inhibition without preventing IP_3 from activating IP_3R (Mak et al. 2003). This simple model was later elaborated to include at least three different Ca^{2+} sensors (Mak et al. 2003), but at the core of this revised scheme is a single Ca^{2+}-binding site that switches from being inhibitory in the absence of IP_3 to stimulatory in its presence (Mak et al. 2003). The essential feature of this scheme is consistent with our initial model, derived from rapid superfusion analysis, which suggests that IP_3 both relieves Ca^{2+} inhibition and promotes binding of Ca^{2+} to a stimulatory site (Marchant and Taylor 1997; Adkins and Taylor 1999). The latter is essential for the channel to open. We, however, argue that the stimulatory and inhibitory Ca^{2+}-binding sites are distinct (Marshall and Taylor 1994). We suggest, therefore, that the essential role of IP_3 is to promote Ca^{2+} binding to a stimulatory Ca^{2+}-binding site. IP_3, by priming this site, allows Ca^{2+} to provide instantaneous control over whether the channel opens (Fig. 1A).

The structural basis for Ca^{2+}-regulation of IP_3R is unresolved: it may be either direct, via Ca^{2+} binding to a site intrinsic to the IP_3R or via an accessory Ca^{2+}-binding protein (Taylor et al. 2004). Stimulation of IP_3R by cytosolic Ca^{2+} is universally observed even with purified IP_3R reconstituted into lipid bilayers (Ferris et al. 1989; Hirota et al. 1995; Michikawa et al. 1999), suggesting that this essential Ca^{2+}-binding site probably resides within the primary sequence of the IP_3R. At least seven cytosolic Ca^{2+}-binding sites have been identified within IP_3R1 (Sienaert et al. 1996; Sienaert et al. 1997), but the physiological relevance of these sites is unresolved. Two of the sites (residues 304-381 and 378-450) are within the IP_3-binding

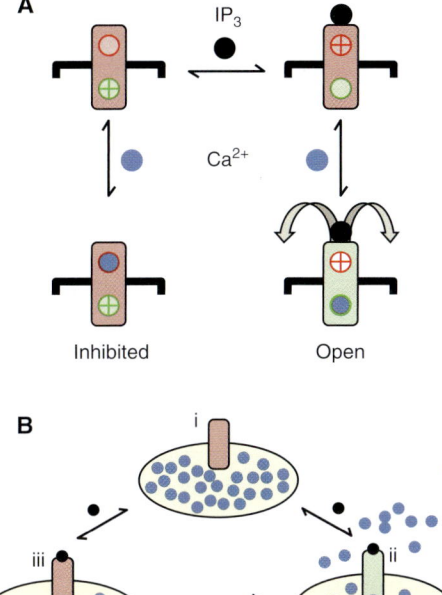

Figure 1. Regulation of IP$_3$R by cytosolic and luminal Ca^{2+}. (*A*) Binding of IP$_3$ (black circle) to the IP$_3$R determines whether a stimulatory (green) or inhibitory (red) Ca^{2+}-binding site is available (Adkins and Taylor 1999). IP$_3$ binding causes the stimulatory site to become accessible and the inhibitory site to be concealed; binding of Ca^{2+} (blue circle) to the former then triggers opening of the channel. (*B*) Luminal Ca^{2+} is proposed to tune the sensitivity of the IP$_3$R to cytosolic IP$_3$ and Ca^{2+} such that full stores (i) are most sensitive to IP$_3$. As the IP$_3$R opens (ii) and the stores lose Ca^{2+}, they are proposed to lose sensitivity to IP$_3$ until eventually the IP$_3$R closes, despite the continued presence of the cytosolic stimuli, trapping Ca^{2+} within the ER (iii). Conversely, stores regain their sensitivity to IP$_3$ as the stores refill, perhaps thereby determining the interval between Ca^{2+} spikes in stimulated cells (Berridge 2007).

core, for which there is a high-resolution structure (Bosanac et al. 2002). This structure shows two surface-exposed clusters of acidic residues that overlap with residues in the second N-terminal Ca^{2+}-binding region. However, point mutations of several of these acidic residues had no effect on Ca^{2+}-regulation of IP$_3$R (Joseph et al. 2005). The remaining Ca^{2+}-binding sites fall within the central region of the IP$_3$R (Sienaert et al. 1996; Sienaert et al. 1997). The site between residues 1347–1426 is interesting because its proximity to a calmodulin (CaM)-binding region is reminiscent of RyR, which have two CaM-binding regions within ~200 residues of high-affinity Ca^{2+}-binding sites, and a third flanked by two high-affinity Ca^{2+}-binding sites (Chen and MacLennan 1994). Interactions between these sites have been proposed to contribute to regulation of RyR by Ca^{2+} and CaM (Chen and MacLennan 1994). None of the Ca^{2+}-binding sites within IP$_3$R contain EF-hands or any other known Ca^{2+}-binding motif, and none have obvious sequence similarity with similar regions in RyR. However, each site has clusters of negatively charged residues that may coordinate Ca^{2+} (Sienaert et al. 1997). There is presently no evidence to link any of these sites directly to Ca^{2+} regulation of IP$_3$R. The only tangible link between specific residues and Ca^{2+} regulation comes from mutagenesis of a glutamate residue that is conserved in all IP$_3$R and RyR. Mutation of this residue in RyR massively reduced the Ca^{2+} sensitivity of the channel (Chen et al. 1998; Li and Chen 2001). Mutation of the same residue (Glu-2100) to another acidic residue (Asp) caused a ~5- to 10-fold decrease in the Ca^{2+}-sensitivity of the IP$_3$R to both stimulation and inhibition, abolished oscillatory Ca^{2+} transients in response to agonist stimulation, and reduced the Ca^{2+}-binding affinity of a large fragment that includes the residue (Miyakawa et al. 2001; Tu et al. 2003). A rather puzzling aspect of these results is the observation that mutation of a single residue similarly attenuates both stimulation and inhibition by Ca^{2+}, when other evidence suggests that the two effects are mediated by distinct sites. This, together with the lack of direct evidence that Ca^{2+} is coordinated by the conserved glutamate, leaves open the possibility that rather than itself contributing to an essential Ca^{2+}-binding site, this residue may be allosterically coupled to the site.

Ca^{2+}-mediated inhibition of IP$_3$R is widely assumed to contribute to termination of local cytosolic Ca^{2+} signals, but it remains far from clear whether such inhibition is mediated by Ca^{2+} binding directly to IP$_3$R or to an associated protein (Taylor and Laude 2002). The effects of

Ca^{2+} on IP_3 binding differ between subtypes: It inhibits binding to IP_3R1 (Worley et al. 1987; Supattapone et al. 1988; Joseph et al. 1989; Varney et al. 1990; Richardson and Taylor 1993; Benevolensky et al. 1994; Cardy et al. 1997; Yoneshima et al. 1997), but the effects of Ca^{2+} on IP_3 binding to IP_3R from cells expressing predominantly IP_3R2 or IP_3R3 are confused (Pietri et al. 1990; Mohr et al. 1993; Marshall and Taylor 1994; Cardy et al. 1997; Yoneshima et al. 1997; Lin et al. 2000; Swatton and Taylor 2002). These conflicting results, and evidence that purified IP_3R1 is not inhibited by Ca^{2+} (Danoff et al. 1988; Richardson and Taylor 1993; Benevolensky et al. 1994; Lin et al. 2000), lend some support to the idea that Ca^{2+} inhibition may be mediated by an accessory protein. It is, however, noteworthy that deletion of the suppressor domain (SD, residues 1-223) of IP_3R1, which appears not to include a Ca^{2+}-binding site, abolishes inhibition of IP_3 binding by Ca^{2+} (Sienaert et al. 2002). This suggests that effective regulation by an accessory protein might require the SD.

Calmodulin (CaM) is one candidate for the accessory protein through which Ca^{2+} inhibition is exercised (Nadif Kasri et al. 2002; Taylor and Laude 2002). CaM is a ubiquitously expressed, EF-hand containing, Ca^{2+}-binding protein that serves as the Ca^{2+}-sensor for many cellular events (Gnegy 1993). All IP_3R subtypes are inhibited by Ca^{2+}-CaM (Hirota et al. 1999; Michikawa et al. 1999; Missiaen et al. 1999; Adkins et al. 2000; Missiaen et al. 2000), and CaM has been shown to restore Ca^{2+} inhibition to purified IP_3R (Hirota et al. 1999; Michikawa et al. 1999; Nosyreva et al. 2002). Yet, it has proven difficult to relate these functional effects of CaM to either its effects on IP_3 binding or to identified CaM-binding sites within IP_3R. CaM inhibits IP_3 binding to IP_3R1 in a Ca^{2+}-independent manner (Patel et al. 1997; Cardy and Taylor 1998), through a site that probably lies within the SD (Adkins et al. 2000; Sienaert et al. 2002). Its properties are clearly inconsistent with the ability of CaM to inhibit IP_3R function only in the presence of Ca^{2+}. There is a high-affinity Ca^{2+}-CaM-binding site within the central region of IP_3R1 and IP_3R2, but not IP_3R3 (Yamada et al. 1995; Lin et al. 2000). However, mutations that prevented Ca^{2+}-CaM binding to this site had no affect on Ca^{2+}-dependent inhibition of IP_3R (Zhang and Joseph 2001; Nosyreva et al. 2002). This evidence and the absence of the site from IP_3R3 suggest that the central Ca^{2+}-CaM-binding site cannot be responsible for Ca^{2+} inhibition of IP_3R. An additional high-affinity Ca^{2+}-CaM-binding site is created in IP_3R1 after removal of the S2 splice region: While this may increase the Ca^{2+}-CaM sensitivity of peripheral $S2^-$ IP_3R1, it is not a universal candidate for mediating Ca^{2+} inhibition of IP_3R (Islam et al. 1996; Lin et al. 2000). Recently, it was suggested that bound CaM is essential for IP_3R function because a peptide antagonist of CaM inhibited IP_3-evoked Ca^{2+} release (Nadif Kasri et al. 2006). It is now clear that this peptide acts directly on IP_3R, with no requirement for CaM (Sun and Taylor 2008). While this eliminates an essential role for tethered CaM, it raises the intriguing possibility that an endogenous CaM-like structure might be essential for IP_3R activation (Sun and Taylor 2008). In summary, all IP_3R subtypes are inhibited by Ca^{2+}-CaM, but the molecular basis of this inhibition has not been established. It seems, on balance, that CaM is unlikely to be the accessory protein through which Ca^{2+} universally inhibits IP_3R. That need not preclude a role for CaM in modulating IP_3R function (Taylor and Laude 2002), just as it does for RyR (Chen et al. 1997; Fruen et al. 2000; Rodney et al. 2001), but we must look elsewhere for the site through which Ca^{2+} inhibits IP_3R.

We turn now to the luminal surface of the IP_3R, where, and again drawing parallels with RyR, we consider regulation by luminal Ca^{2+}. Persuasive evidence suggests that Ca^{2+} release by RyR may be terminated before Ca^{2+} stores are entirely depleted because luminal Ca^{2+} is required to maintain RyR activity (Györke and Györke 1998; Launikonis et al. 2006; Jiang et al. 2008), possibly via its interaction with calsequestrin, a luminal high-capacity Ca^{2+}-binding protein (Launikonis et al. 2006; Terentyev et al. 2006). A similar scheme has been proposed to account for two features of IP_3-evoked Ca^{2+} release: the initiation of Ca^{2+} release after the quiescent interspike interval during repetitive Ca^{2+} spikes (Berridge 2007) and quantal Ca^{2+} release via IP_3R. The latter describes the situation wherein unidirectional Ca^{2+} efflux from intracellular stores terminates before the stores have fully emptied after stimulation with submaximally effective concentrations of IP_3 without loss of their ability to respond to a further increase in IP_3 concentration (Muallem et al. 1989; Meyer and Stryer 1990; Taylor and Potter 1990; Oldershaw et al. 1991; Bootman et al. 1992; Brown et al. 1992; Combettes et al. 1992; Ferris et al. 1992; Hirota et al. 1995). The proposal is that luminal Ca^{2+} sets the gain on the regulation by cytosolic IP_3 and Ca^{2+}, so that as the luminal free Ca^{2+} concentration falls, it causes the sensitivity of the IP_3R to IP_3 to fall until, as Ca^{2+} leaks from the ER, the IP_3R closes despite the continued presence of cytosolic IP_3 and residual Ca^{2+} within the ER (Irvine 1990). Conversely, as stores refill between Ca^{2+} spikes in an intact cell, the model predicts that the sensitivity of the IP_3R increases until it exceeds the threshold at which prevailing cytosolic IP_3 and Ca^{2+} concentrations become sufficient to trigger opening (Fig. 1B). Despite the enduring appeal of the model, evidence that luminal Ca^{2+} directly regulates IP_3R is not yet entirely convincing.

Stores loaded with Ca^{2+} have been shown to become more sensitive to IP_3 in some studies (Missiaen et al. 1992; Nunn and Taylor 1992; Oldershaw and Taylor 1993; Parys et al. 1993; Missiaen et al. 1994; Horne and Meyer 1995; Combettes et al. 1996; Tanimura and Turner 1996), but not in others (Combettes et al. 1992; Shuttleworth 1992; Combettes et al. 1993; van de Put et al. 1994). However, even the supportive results do not eliminate the possibility that the increased sensitivity to IP_3 arises from having Ca^{2+} pass through active IP_3R and increase their sensitivity from the cytosolic surface. Similar difficulties have plagued analyses of the effects of luminal Ca^{2+} on RyR (Tripathy and Meissner 1996; Laver 2007; Laver 2009). In bilayer recordings of IP_3R1, where essential accessory proteins may be lost, luminal Ca^{2+} failed to potentiate the Ca^{2+} release evoked by IP_3 (Bezprozvanny and Ehrlich 1994). Despite the caveats, regulation of IP_3R by luminal Ca^{2+} deserves serious consideration. A high-affinity Ca^{2+}-binding site within the luminal loop linking TMD 5 and 6 (Sienaert et al. 1996) contains conserved acidic residues that could mediate luminal Ca^{2+} regulation, although the sub-μM affinity of this site for Ca^{2+} would be ill-suited to detecting likely changes in luminal Ca^{2+} concentration. Luminal accessory proteins, akin to those that regulate RyR, are another possibility, with ERp44 being one candidate. ERp44 belongs to the thioredoxin protein family and regulates IP_3R in a pH- and luminal Ca^{2+}-dependent manner (Higo et al. 2005). Binding of ERp44 to the TMD5-6 loop of IP_3R inhibits channel activity, and the interaction is disrupted by high concentrations of Ca^{2+} consistent with the suggestion that luminal Ca^{2+} might enhance IP_3R activity.

To summarize, IP_3 works by tuning the Ca^{2+} sensitivity of the IP_3R: It stimulates Ca^{2+} binding to a stimulatory site and inhibits Ca^{2+} binding to an inhibitory site (Fig. 1A). Binding to the stimulatory site is the trigger for opening of the pore. The identity of neither Ca^{2+}-binding site is known: The stimulatory site probably resides within the IP_3R itself, but the inhibitory site may require an accessory protein, though this is unlikely to be CaM. Luminal Ca^{2+} may further tune the sensitivity of the IP_3R to regulation by its cytosolic ligands, but this remains unproven.

STRUCTURAL DETERMINANTS OF IP_3R ACTIVATION

Judged by their primary amino acid sequences, all known IP_3R subunits are assumed to have a similar architecture. Each subunit, of about 2700 residues, comprises three major regions: the N-terminal to which IP_3 binds, the C-terminal region with its six transmembrane regions (TMD) (Galvan et al. 1999), and a large intervening sequence (Fig. 2A). Functional IP_3Rs are tetrameric, assembled either from identical subunits or from mixtures of the three subtypes and their many splice variants (Taylor et al. 1999; Foskett et al. 2007). Several structures of the entire IP_3R1 have been published, each derived from single particle analysis of images from electron microscopy (Hamada and Mikoshiba

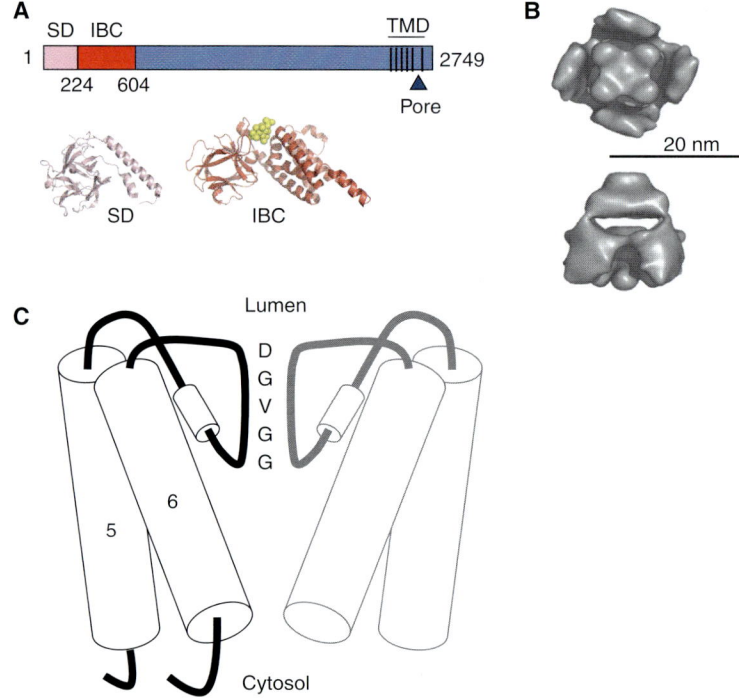

Figure 2. Major structural domains of IP$_3$R. (*A*) The three key regions defined by the primary sequence of a single IP$_3$R subunit are highlighted: the N-terminal with its SD and IBC, the C-terminal region with its six TMD and pore, and the large central region. Atomic structures of the SD (Bosanac et al. 2005) and IBC with IP$_3$ bound (Bosanac et al. 2002) are also shown. (*B*) Two views of the IP$_3$R derived from single particle analysis (da Fonseca et al. 2003) (*top*, from the cytosol; *bottom*, across the ER membrane with the ER lumen at the top). (*C*) A possible structure of the IP$_3$R pore, with a luminal selectivity filter and a constriction formed by the tepee-like structure of TMD6. Only two of the four IP$_3$R subunits are shown.

2002; Jiang et al. 2002a; da Fonseca et al. 2003; Hamada et al. 2003; Serysheva et al. 2003; Sato et al. 2004). These studies confirm the tetrameric state of IP$_3$R, but variability between the structures and their relatively low resolution (~30 Å) have, so far, limited any realistic interpretation of the structural basis of IP$_3$R activation (Taylor et al. 2004) (Fig. 2B). Whether structures of recombinant IP$_3$R will contribute to resolving this impasse remains to be seen (Wolfram et al. 2010).

There has been more progress with RyR, although only recently has the resolution of these structures (~30 Å) improved on that obtained for IP$_3$R. These structures of native RyR, and all three subtypes of recombinant RyR reveal a shape like a square mushroom with a very large, open cytoplasmic structure tethered to a much smaller TMD region (the stalk). At ~30 Å resolution, the structures of the three RyR subtypes are almost indistinguishable, and because they, like the three subtypes of IP$_3$R, share about 65% sequence identity, it seems reasonable to suppose that the 3D structures of all IP$_3$R are also likely to be similar to each other. These studies of RyR have identified positions of critical residues within the 3D structure, the sites to which accessory proteins bind, and conformational changes associated with opening of the pore (Orlova et al. 1996; Serysheva et al. 2005; Wang et al. 2007; Jones et al. 2008). Activation of RyR is associated with considerable changes in both the pore and cytoplasmic regions: The four corners of the latter dip down toward the SR, while the central region lifts away from it (Samso et al. 2009). It is noteworthy, in the context of schemes for activation of IP_3R (see below), that large movements of some cytoplasmic domains of RyR1 appear to occur around hinges that link them to relatively immobile domains.

The highest resolution maps (~10Å), although still insufficient to map 3D structure to primary sequence, have come close to defining the likely secondary structure of the pore of RyR1 (Ludtke et al. 2005; Samso et al. 2005; Samso et al. 2009). This region appears to have six α-helices (Samso et al. 2009), consistent with models of RyR that suggest six TMD (Meur et al. 2007). Along the central axis, it has a luminal constriction (probably the selectivity filter, see below) and a tepee-like assembly of four inner helices (likely to be TMD6), with the apex pointing into the cytoplasmic structure. By analogy with MthK channels, this constriction may form the gate of the RyR. Kinking of the inner helix around a central Gly residue causes splitting of the tepee and thereby opening of the channel for MthK (Jiang et al. 2002b). One structure (Samso et al. 2009) is consistent with a similar mechanism operating for RyR1, but another structure (Ludtke et al. 2005) and mutagenesis of the critical Gly (G4863 in RyR1) (Wang et al. 2003) contradict it. These insights into the possible workings of the RyR pore are significant for IP_3R, because it is within the pore region (TMD5-6) that RyR and IP_3R share the greatest sequence similarity. We turn, therefore, to the pore of the IP_3R to explore its properties and structure.

All IP_3R (like all RyR) are cation channels with extremely large conductance, but only modest selectivity for Ca^{2+} over monovalent cations (permeability ratio, $P_{Ca}/P_K \sim 6$) (Williams et al. 2001; Foskett et al. 2007). The voltage-gated and store-operated Ca^{2+} channels that mediate Ca^{2+} entry across the plasma membrane are vastly more selective ($P_{Ca}/P_K > 1000$). In the ER, where most IP_3Rs are located, this lack of selectivity is unlikely to be a problem because Ca^{2+} is probably the only cation with an appreciable electrochemical gradient across the ER membrane. In effect, the ER Ca^{2+} pump (SERCA), by creating a steep Ca^{2+} concentration gradient across the ER membrane, assumes responsibility for determining which cations flow through an open IP_3R. Indeed, the K^+ permeability of IP_3R and RyR may facilitate rapid Ca^{2+} release by allowing K^+ to move into the ER to electrically compensate the efflux of Ca^{2+} (Gillespie and Fill 2008). The pore of the IP_3R, like that of RyR, is formed by the final pair of TMD (TMD5-6) and the luminal loop that links them from each of the four subunits (Ramos-Franco et al. 1999; Williams et al. 2001) (Fig. 2C). The loop includes a sequence (GGVGD in IP_3R) similar to that of the selectivity filter of K^+ channels (Balshaw et al. 1999), consistent with the idea that the overall architecture of the pore region may be broadly similar to that of K^+ channels (MacKinnon 2004). For both IP_3R and RyR, however, the pore must be larger and less-selective than for K^+ channels, and probably able to accommodate only one cation at a time (Williams et al. 2001). This model for the IP_3R pore, where TMD5 (the outer helix) and TMD6 (inner helix) cradle a short pore helix and selectivity filter (Fig. 2C), is consistent with mutagenesis of residues within this region affecting ion permeation (Boehning et al. 2001b; Dellis et al. 2006; Dellis et al. 2008; Schug et al. 2008), with biophysical evidence that the narrowest region of the pore lies close to the luminal entrance of the RyR (Williams et al. 2001) and with the intermediate resolution structures of the pore region of RyR1 (Samso et al. 2009). A conserved acidic residue (D2550 in IP_3R1) at the luminal end of the selectivity filter (Fig. 2C) contributes to the modest Ca^{2+} selectivity of IP_3R (Boehning et al. 2001b; Dellis et al. 2008) and RyR (Gao et al. 2000; Wang et al. 2005; Gillespie 2008), but the structural determinants of ion selectivity and permeation by IP_3R are otherwise poorly understood. The changes in pore structure that allow it to open are also minimally understood. Indeed, mutation of the conserved Gly within TMD6 of IP_3R (G2586 in IP_3R1), which might have been thought to provide the gating hinge (Samso et al. 2009), appears not to prevent IP_3 from opening IP_3R (Schug et al. 2008). In short, aside from knowing the regions of primary sequence that form the IP_3R pore (TMD5-6) and a rather vague notion that its structure perhaps resembles that of K^+ channels, we have only the most rudimentary knowledge of the structural determinants of how the IP$_3$R pore opens and selects between ions.

The conformational changes in the IP$_3$R that lead to opening of its pore are initiated by IP$_3$ binding to the IP$_3$-binding core (IBC, residues 224–604 in IP$_3$R1) (Fig. 3A). Although IP$_3$ is the only endogenous ligand of the IBC, there are many synthetic agonists, all of which have structures equivalent to the equatorial 6-hydroxyl and the 4- and 5-phosphate groups of IP$_3$ (Fig. 3A) (Rossi et al. 2010). It is noteworthy that neither of the immediate products of IP$_3$ metabolism, IP$_2$ and IP$_4$, binds to the IBC; both metabolic pathways are therefore effective means of terminating activation of IP$_3$R by IP$_3$. An atomic structure of the IBC with IP$_3$ bound (Bosanac et al. 2002) shows IP$_3$ held within a clam-like structure in which the phosphate groups of IP$_3$ are coordinated by basic residues (Fig. 3A). The two sides of the clam, the α- and β-domains, form a network of interactions with the essential groups of IP$_3$. The 4-phosphate is hydrogen-bonded with residues in the β-domain, the 5-phosphate forms hydrogen bonds with residues predominantly in the α-domain, and the 6-hydroxyl interacts with the backbone of a residue within the α-domain. It is easy to imagine how these interactions with IP$_3$ might pull the α- and β-domains together, causing the clam to close in a manner similar to glutamate binding to ionotropic glutamate receptors (Mayer 2006). Structures of the IBC without IP$_3$ bound are urgently needed to assess this proposal, but two lines of evidence lend circumstantial support. First, the IBC adopts a more constrained structure when it binds IP$_3$ (Chan et al. 2007). Second, adenophostins, which are high-affinity agonists of IP$_3$R (Rossi et al. 2010), retain some activity after loss of the 3-phosphate (analogous to the 5-phosphate of IP$_3$), probably because their adenine moiety

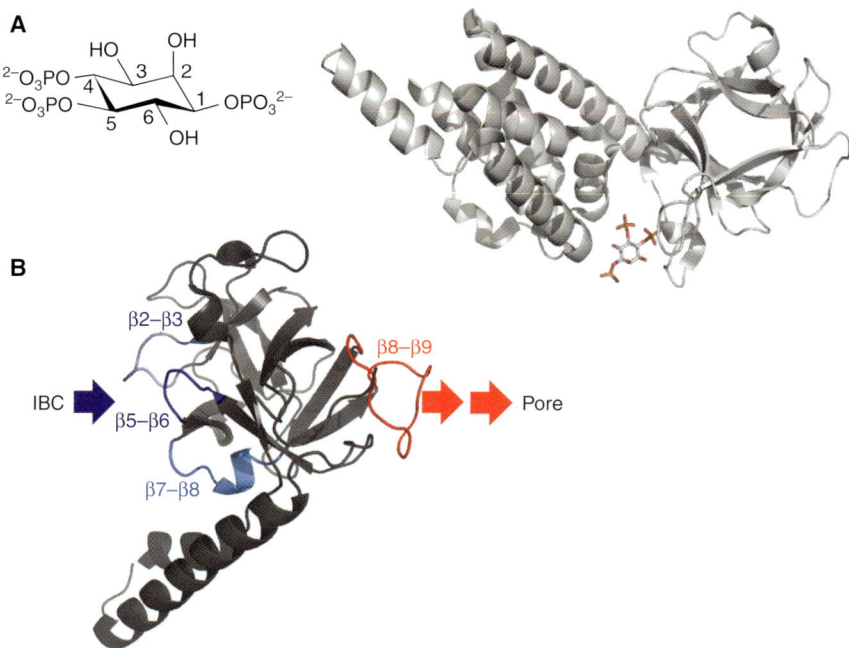

Figure 3. Initiation of IP$_3$R activation by IP$_3$. (*A*) The structure of IP$_3$, with its critical vicinal 4,5-bisphosphate and 6-hydroxyl groups, is shown alongside the structure of the IBC with IP$_3$ bound. The latter shows the 4- and 5-phosphates contacting the β- and α-domains, respectively (Bosanac et al. 2002), and thereby pulling the clam into a more closed state. (*B*) Structure of the SD (Bosanac et al. 2005) showing possible sites of interaction with the IBC and downstream domains through which it signals to the pore. See text for further details.

interacts strongly with a residue in the α-domain and thereby partially mimics the clam-closure that would otherwise require the 5-phosphate to bind to the α-domain (Sureshan et al. 2009). We envisage, therefore, that when IP$_3$ binds to the IBC, the essential vicinal phosphate groups through their contacts with the α- and β-domains effectively cross-bridge the two sides of the clam-like structure, causing it to close, and thereby initiate the processes that will culminate in opening of the pore.

It is worth commenting briefly on available antagonists of IP$_3$R because of their obvious value as experimental tools. There are no specific antagonists of IP$_3$R, although with appropriate caution some antagonists can yield useful insight (Michelangeli et al. 1995). Heparin is a competitive antagonist of IP$_3$ (Worley et al. 1987), although it is not membrane-permeant and, among many additional effects, it uncouples G-protein-coupled receptors from their G proteins (Dasso and Taylor 1991) and activates RyR (Ehrlich et al. 1994). 2-aminoethyl diphenylboronate (2-APB) is membrane-permeant and inhibits IP$_3$-evoked Ca^{2+} release without affecting IP$_3$ binding (Maruyama et al. 1997); its mechanism of action is unresolved. However, 2-APB also inhibits Ca^{2+} uptake and many other Ca^{2+} channels. It has recently aroused interest as a modulator of STIM and, therefore, store-operated Ca^{2+} entry (Goto et al. 2010). A screen of 2-APB analogues with selectivity for store-operated Ca^{2+} entry may yet also provide IP$_3$R-selective antagonists (Goto et al. 2010). Xestospongins, isolated from an Australian sponge, are high-affinity membrane-permeant inhibitors of IP$_3$-evoked Ca^{2+} release that do not affect IP$_3$ binding (Gafni et al. 1997), but they, too, have side effects (Solovyova et al. 2002). High concentrations of caffeine inhibit IP$_3$-evoked Ca^{2+} release (Parker and Ivorra 1991) without affecting IP$_3$ binding (Worley et al. 1987), but caffeine also stimulates RyR, inhibits cyclic nucleotide phosphodiesterases, and interferes with many Ca^{2+} indicators. Membrane-permeant peptide antagonists of IP$_3$R may provide another potential source of selective antagonists (Sun and Taylor 2008).

How IP$_3$ binding to the IBC leads to binding of Ca^{2+} to the IP$_3$R, and thereby opening of the pore, remains largely unknown, but it is clear that the suppressor domain (SD, residues 1-223 of IP$_3$R1), which is connected to the IBC by a flexible linkage (Chan et al. 2007), plays an essential role. The clearest evidence is that IP$_3$ binds to IP$_3$R without an SD, but it fails to open the pore (Uchida et al. 2003; Szlufcik et al. 2006). The name of this region derives from the observation that, although the SD itself is unlikely to make any direct contacts with IP$_3$, its presence decreases the affinity of IP$_3$R for IP$_3$ (Uchida et al. 2003). We have interpreted this effect to reflect the use of binding energy from the binding of IP$_3$ to the IBC to cause a conformational change within the SD. This interpretation gains considerable support from our analysis of partial agonists of the IP$_3$R (Rossi et al. 2009). The crux of our argument is that the energy provided by agonist binding drives both the conformational changes that lead to receptor activation and tighter binding of the ligand to its receptor. There is, therefore, a play-off between these two claims on the binding energy. Partial agonists, because they less effectively activate the receptor, divert more binding energy into stabilizing the binding, while full agonists evoke more substantial conformational changes; therefore, less binding energy remains to stabilize binding. Our results show that although full and partial agonists bind with similar affinities to the IBC, the SD causes the affinity of full agonists to decrease more than for partial agonists (Rossi et al. 2009). Quantitative analyses of these results lead to the conclusion that the most energetically costly conformational change in the IP$_3$R evoked by IP$_3$ occurs within its N-terminal (residues 1-604), and that these conformational changes pass entirely via the SD to the pore region (Rossi et al. 2009). We suggest, therefore, that the SD is the essential link between IP$_3$ binding to the IBC and the subsequent conformational changes that lead to opening of the pore. Without a structure of the entire N-terminal of the IP$_3$R, we can only speculate on the physical relationship between the IBC and SD, but our results with partial agonists and mutagenesis are consistent with three exposed loops of the SD (β2–β3, β5–β6, and β7–β8, blue in Fig. 3B) being the most likely sites of interaction with the IBC (Rossi et al. 2009).

Remarkably, and despite their rather low sequence identities (~30%), the crystal structures of the SD from IP$_3$R1 (Bosanac et al. 2005) and of the analogous N-terminal regions from RyR1 and RyR2 (Amador et al. 2009; Lobo and Van Petegem 2009) are extremely similar. Several mutations associated with malignant hyperthermia and central core disease (RyR1) or catecholaminergic polymorphic ventricular tachycardia (RyR2), all of which impair the normal regulation of gating, are clustered in an exposed loop (β8–β9) of the N-terminal of RyR (Amador et al. 2009). Furthermore, and consistent with the N-terminal of the RyR mediating essential interdomain interactions, a peptide derived from this region causes RyR2 to open spontaneously, apparently by uncoupling an interaction between the endogenous loop and a central region of the RyR that includes residues 2460–2495 (Oda et al. 2005; Tateishi et al. 2009). In light of the conservation of structure between IP$_3$R and RyR, it is tempting to speculate that the same loop in the SD of the IP$_3$R (β8–β9, red in Fig. 3B) may mediate transfer of conformational changes onward toward the pore. Co-immunoprecipitation studies have suggested an interaction between the N-terminal of IP$_3$R1 (most likely the SD) and the pore region of an adjacent subunit (Boehning and Joseph 2000a), perhaps mediated by the cytosolic loop linking TMD4 to TMD5 (Schug and Joseph 2006). An attractive possibility, therefore, is that the SD (perhaps its β8-β9 loop) interacts directly with the short cytosolic helix linking TMD4 and TMD5, and thereby gates the pore (Schug and Joseph 2006; Rossi et al. 2009). Such an interaction would require that the SD comes very close to the pore, but the exact location of the SD within the 3D structure of either the IP$_3$R or RyR is unknown. The N-terminal of the RyR probably lies within the clamp region at the periphery of the large square cytoplasmic structure (Wang et al. 2007), and it does change shape during RyR activation (Samso et al. 2009). Yet, in this location the N-terminal is too far from the pore to interact directly with the TMD4-5 loop, consistent perhaps with evidence that in RyR the N-terminal may interact directly with a neighboring domain that includes residues from the central part of the primary sequence (Wang et al. 2007). These observations and the evidence that the uncoupling peptide derived from the N-terminal of RyR2 appears to interact with residues remote from the pore (Oda et al. 2005; Tateishi et al. 2009), suggest that the links between the SD and pore may, at least for RyR, be indirect.

In summary, we suggest that IP$_3$R activation is initiated when IP$_3$ binds to the IBC, and perhaps thereby causes closure of its clam-like structure. That conformational change, which must also initiate the events that allow Ca^{2+} to bind to a stimulatory site, is passed to the rest of the IP$_3$R entirely via the SD. The location of that Ca^{2+}-binding site and, therefore, the structural links between it and the SD, are unresolved. We speculate that one face of the SD interacts directly with the IBC, and the opposite face interacts with the structure through which conformational changes pass to the pore. The pore is a relatively nonselective, large-conductance cation channel formed by the tetrameric assembly of the TMD5-6 regions of each subunit. Its structure is unresolved but likely to be broadly similar to K$^+$ channels with a selectivity filter and gate at opposite ends of its membrane-spanning structure.

ACKNOWLEDGMENTS

Work from the authors' laboratory is supported by the Wellcome Trust.

REFERENCES

Adkins CE, Morris SA, De Smedt H, Török K, Taylor CW. 2000. Ca^{2+}-calmodulin inhibits Ca^{2+} release mediated by type-1, -2 and -3 inositol trisphosphate receptors. *Biochem J* **345:** 357–363.

Adkins CE, Taylor CW. 1999. Lateral inhibition of inositol 1,4,5-trisphosphate receptors by cytosolic Ca^{2+}. *Curr Biol* **9:** 1115–1118.

Amador FJ, Liu S, Ishiyama N, Plevin MJ, Wilson A, Maclennan DH, Ikura M. 2009. Crystal structure of type I ryanodine receptor amino-terminal β-trefoil domain reveals a disease-associated mutation "hot spot" loop. *Proc Natl Acad Sci USA* **106:** 11040–11044.

Balshaw D, Gao L, Meissner G. 1999. Luminal loop of the ryanodine receptor: A pore-forming segment? *Proc Natl Acad Sci USA* **96:** 3345–3347.

Benevolensky D, Moraru II, Watras J. 1994. Micromolar calcium decreases the affinity of inositol trisphosphate receptor in vascular smooth muscle. *Biochem J* **299:** 631–636.

Berridge MJ. 1975. The interaction of cyclic nucleotides and calcium in the control of cellular activity. In *Advances in Cyclic Nucleotide and Protein Phosphorylation Research*, (ed. P Greengard, GA Robison), pp. 1–98. Raven Press, New York.

Berridge MJ. 1983. Rapid accumulation of inositol trisphosphate reveals that agonists hydrolyse polyphosphoinositides instead of phosphatidylinositol. *Biochem J* **212:** 849–858.

Berridge MJ, (ed.) 1995. *CIBA Foundation Symposium Calcium waves, gradients and oscillations.* John Wiley and Sons, Chichester.

Berridge MJ. 1997. Elementary and global aspects of calcium signalling. *J Physiol* **499:** 291–306.

Berridge MJ. 2005. Unlocking the secrets of cell signaling. *Annu Rev Physiol* **67:** 1–21.

Berridge MJ. 2007. Inositol trisphosphate and calcium oscillations. *Biochem Soc Symp* **74:** 1–7.

Berridge MJ, Fain JN. 1979. Inhibition of phosphatidylinositol synthesis and the inactivation of calcium entry after prolonged exposure of the blowfly salivary gland to 5-hydroxytryptamine. *Biochem J* **178:** 59–69.

Berridge MJ, Irvine RF. 1984. Inositol trisphosphate, a novel second messenger in cellular signal transduction. *Nature* **312:** 315–321.

Berridge MJ, Irvine RF. 1989. Inositol phosphates and cell signalling. *Nature* **341:** 197–205.

Berridge MJ, Lipp P, Bootman MD. 2000. The versatility and universality of calcium signalling. *Nat Rev Mol Cell Biol* **1:** 11–21.

Bezprozvanny I, Ehrlich BE. 1994. Inositol (1,4,5)-trisphosphate (InsP$_3$)-gated Ca channels from cerebellum: conduction properties for divalent cations and regulation by intraluminal calcium. *J Gen Physiol* **104:** 821–856.

Bezprozvanny I, Watras J, Ehrlich BE. 1991. Bell-shaped calcium-response curves for Ins(1,4,5)P$_3$- and calcium-gated channels from endoplasmic reticulum of cerebellum. *Nature* **351:** 751–754.

Boehning D, Joseph SK. 2000a. Direct association of ligand-binding and pore domains in homo- and heterotetrameric inositol 1,4,5-trisphosphate receptors. *EMBO J* **19:** 5450–5459.

Boehning D, Joseph SK. 2000b. Functional properties of recombinant type I and type III inositol 1,4,5-trisphosphate receptor isoforms expressed in COS-7 cells. *J Biol Chem* **275:** 21492–21499.

Boehning D, Joseph SK, Mak D-OD, Foskett JK. 2001a. Single-channel recordings of recombinant inositol trisphosphate receptors in mammalian nuclear envelope. *Biophys J* **81:** 117–124.

Boehning D, Mak D-OD, Foskett JK, Joseph SK. 2001b. Molecular determinants of ion permeation and selectivity in inositol 1,4,5-trisphosphate receptor Ca^{2+} channels. *J Biol Chem* **276:** 13509–13512.

Bootman MD, Berridge MJ, Lipp P. 1997. Cooking with calcium: the recipes for composing global signals from elementary events. *Cell* **91:** 367–373.

Bootman MD, Berridge MJ, Taylor CW. 1992. All-or-nothing Ca^{2+} mobilization from the intracellular stores of single histamine-stimulated HeLa cells. *J Physiol* **450:** 163–178.

Bosanac I, Alattia J-R, Mal TK, Chan J, Talarico S, Tong FK, Tong KI, Yoshikawa F, Furuichi T, Iwai M, et al. 2002. Structure of the inositol 1,4,5-trisphosphate receptor binding core in complex with its ligand. *Nature* **420:** 696–700.

Bosanac I, Yamazaki H, Matsu-ura T, Michikawa M, Mikoshiba K, Ikura M. 2005. Crystal structure of the ligand binding suppressor domain of type 1 inositol 1,4,5-trisphosphate receptor. *Mol Cell* **17:** 193–203.

Brown GR, Sayers LG, Kirk CJ, Michell RH, Michelangeli F. 1992. The opening of the inositol 1,4,5-trisphosphate-sensitive Ca^{2+} channel in rat cerebellum is inhibited by caffeine. *Biochem J* **282:** 309–312.

Burgess GM, McKinney JS, Fabiato A, Leslie BA, Putney JW Jr. 1983. Calcium pools in saponin-permeabilized guinea pig hepatocytes. *J Biol Chem* **258:** 15336–15345.

Cardy TJA, Taylor CW. 1998. A novel role for calmodulin: Ca^{2+}-independent inhibition of type-1 inositol trisphosphate receptors. *Biochem J* **334:** 447–455.

Cardy TJA, Traynor D, Taylor CW. 1997. Differential regulation of types 1 and 3 inositol trisphosphate receptors by cytosolic Ca^{2+}. *Biochem J* **328:** 785–793.

Chan J, Whitten AE, Jeffries CM, Bosanac I, Mal TK, Ito J, Porumb H, Michikawa T, Mikoshiba K, Trewhella J, et al. 2007. Ligand-induced conformational changes via flexible linkers in the amino-terminal region of the inositol 1,4,5-trisphosphate receptor. *J Mol Biol* **373:** 1269–1280.

Chen SRW, Ebisawa K, Li X, Zhang L. 1998. Molecular identification of the ryanodine receptor Ca^{2+} sensor. *J Biol Chem* **273:** 14675–14678.

Chen SRW, Li X, Ebisawa K, Zhang L. 1997. Functional characterization of the recombinant type 3 Ca^{2+} release channel (ryanodine receptor) expressed in HEK293 cells. *J Biol Chem* **272:** 24234–24236.

Chen SRW, MacLennan DH. 1994. Identification of calmodulin-, Ca^{2+}- and ruthenium red-binding domains in the Ca^{2+} release channel (ryanodine receptor) of rabbit skeletal muscle sarcoplasmic reticulum. *J Biol Chem* **269:** 22698–22704.

Cheng H, Lederer WJ, Cannell MB. 1993. Calcium sparks. Elementary events underlying excitation-contraction coupling in heart muscle. *Science* **262:** 740–744.

Combettes L, Cheek TR, Taylor CW. 1996. Regulation of inositol trisphosphate receptors by luminal Ca^{2+} contributes to quantal Ca^{2+} mobilization. *EMBO J* **15:** 2086–2093.

Combettes L, Claret M, Champeil P. 1992. Do submaximal InsP$_3$ concentrations only induce partial release discharge of permeabilized hepatocyte calcium pools because of the concomitant reduction of intraluminal Ca^{2+} concentration? *FEBS Lett* **301:** 287–290.

Combettes L, Claret M, Champeil P. 1993. Calcium control of InsP$_3$-induced discharge of calcium from permeabilised hepatocyte pools. *Cell Calcium* **14:** 279–292.

da Fonseca PCA, Morris SA, Nerou EP, Taylor CW, Morris EP. 2003. Domain organisation of the type 1 inositol 1,4,5-trisphosphate receptor as revealed by single-particle analysis. *Proc Natl Acad Sci USA* **100:** 3936–3941.

Danoff SK, Supattapone S, Snyder SH. 1988. Characterization of a membrane protein from brain mediating the inhibition of inositol 1,4,5-trisphosphate receptor binding by calcium. *Biochem J* **254:** 701–705.

Dasso LLT, Taylor CW. 1991. Heparin and other polyanions uncouple a_1-adrenoceptors from G-proteins. *Biochem J* **280:** 791–795.

Dellis O, Dedos S, Tovey SC, Rahman T-U-, Dubel SJ, Taylor CW. 2006. Ca^{2+} entry through plasma membrane IP_3 receptors. *Science* **313:** 229–233.

Dellis O, Rossi AM, Dedos SG, Taylor CW. 2008. Counting functional IP_3 receptors into the plasma membrane. *J Biol Chem* **283:** 751–755.

Demuro A, Parker I. 2007. Multi-dimensional resolution of elementary Ca^{2+} signals by simultaneous multi-focal imaging. *Cell Calcium* **43:** 367–374.

Dolmetsch RE, Lewis RS, Goodnow CC, Healy JI. 1997. Differential activation of transcription factors induced by Ca^{2+} response amplitude and duration. *Nature* **386:** 855–858.

Dufour J-F, Arias IM, Turner TJ. 1997. Inositol 1,4,5-trisphosphate and calcium regulate the calcium channel function of the hepatic inositol 1,4,5-trisphosphate receptor. *J Biol Chem* **272:** 2675–2681.

Dupont G, Houart G, De Koninck P. 2003. Sensitivity of CaM kinase II to the frequency of Ca^{2+} oscillations: a simple model. *Cell Calcium* **34:** 485–497.

Dyer JL, Liu Y, Pino de la Huerga I, Taylor CW. 2005. Long-lasting inhibition of adenylyl cyclase selectively mediated by inositol 1,4,5-trisphosphate-evoked calcium release. *J Biol Chem* **280:** 8936–8944.

Echevarria W, Leite MF, Guerra MT, Zipfel WR, Nathanson MH. 2003. Regulation of calcium signals in the nucleus by a nucleoplasmic reticulum. *Nat Cell Biol* **5:** 440–446.

Ehrlich BE, Kaftan E, Bezprozvannaya S, Bezprozvanny I. 1994. The pharmacology of intracellular Ca^{2+}-release channels. *Trends Pharmacol Sci* **15:** 145–149.

Ehrlich BE, Watras J. 1988. Inositol 1,4,5-trisphosphate activates a channel from smooth muscle sarcoplasmic reticulum. *Nature* **336:** 583–586.

Endo M. 1977. Calcium release from the sarcoplasmic reticulum. *Physiol Rev* **57:** 71–108.

Endo M, Tanaka M, Ogawa Y. 1970. Calcium induced release of calcium from the sarcoplasmic reticulum of skinned skeletal muscle fibres. *Nature* **228:** 34–36.

Fabiato A. 1983. Calcium-induced release of calcium from cardiac sarcoplasmic reticulum. *Am J Physiol* **245:** C1–C14.

Ferris CD, Cameron AM, Huganir RL, Snyder SH. 1992. Quantal calcium release by purified reconstituted inositol 1,4,5-trisphosphate receptors. *Nature* **356:** 350–352.

Ferris CD, Huganir RL, Supattapone S, Snyder SH. 1989. Purified inositol 1,4,5-trisphosphate receptor mediates calcium flux in reconstituted lipid vesicles. *Nature* **342:** 87–89.

Finch EA, Turner TJ, Goldin SM. 1991. Calcium as a coagonist of inositol 1,4,5-trisphosphate-induced calcium release. *Science* **252:** 443–446.

Foskett JK, White C, Cheung KH, Mak DO. 2007. Inositol trisphosphate receptor Ca^{2+} release channels. *Physiol Rev* **87:** 593–658.

Fruen BR, Bardy JM, Byrem TM, Strasburg GM, Louis CF. 2000. Differential Ca^{2+} sensitivity of skeletal and cardiac muscle ryanodine receptors in the presence of calmodulin. *Am J Physiol* **279:** C724–C733.

Furuichi T, Yoshikawa S, Miyawaki A, Wada K, Maeda M, Mikoshiba K. 1989. Primary structure and functional expression of the inositol 1,4,5-trisphosphate-binding protein P_{400}. *Nature* **342:** 32–38.

Gafni J, Munsch JA, Lam TH, Catlin MC, Costa LG, Molinski TF, Pessah IN. 1997. Xestospongins: Potent membrane permeable blockers of the inositol 1,4,5-trisphosphate receptor. *Neuron* **19:** 723–733.

Galvan DL, Borrego-Diaz E, Perez PJ, Mignery GA. 1999. Subunit oligomerization, and topology of the inositol 1,4,5-trisphosphate receptor. *J Biol Chem* **274:** 29483–29492.

Gao L, Balshaw D, Xu L, Tripathy A, Xin C, Meissner G. 2000. Evidence for a role of the lumenal M3-M4 loop in skeletal muscle Ca^{2+} release channel (ryanodine receptor) activity and conductance. *Biophys J* **79:** 828–840.

Gees M, Colsoul B, Nilius B. 2010. The Role of Transient Receptor Potential Cation Channels in Ca^{2+} Signaling. *Cold Spring Harb Perspect Biol* doi: 10.1101/cshperspect. a003962.

Gerasimenko OV, Gerasimenko JV, Belan PV, Petersen OH. 1996. Inositol trisphosphate and cyclic ADP-ribose-mediated release of Ca^{2+} from single isolated pancreatic zymogen granules. *Cell* **84:** 473–480.

Gerasimenko OV, Gerasimenko JV, Tepikin AV, Petersen OH. 1995. ATP-dependent accumulation and inositol trisphosphate- or cyclic ADP-ribose-mediated release of Ca^{2+} from the nuclear envelope. *Cell* **80:** 439–444.

Gilkey JC. 1983. Roles of calcium and pH in activation of eggs of the medaka fish, Oryzias latipes. *J Cell Biol* **97:** 669–678.

Gillespie D. 2008. Energetics of divalent selectivity in a calcium channel: the ryanodine receptor case study. *Biophys J* **94:** 1169–1184.

Gillespie D, Fill M. 2008. Intracellular calcium release channels mediate their own countercurrent: the ryanodine receptor case study. *Biophys J* **95:** 3706–3714.

Gnegy ME. 1993. Calmodulin in neurotransmitter and hormone action. *Annu Rev Pharmacol Toxicol* **33:** 45–70.

Goto J, Suzuki AZ, Ozaki S, Matsumoto N, Nakamura T, Ebisui E, Fleig A, Penner R, Mikoshiba K. 2010. Two novel 2-aminoethyl diphenylborinate (2-APB) analogues differentially activate and inhibit store-operated Ca^{2+} entry via STIM proteins. *Cell Calcium* **47:** 1–10.

Györke I, Györke S. 1998. Regulation of the cardiac ryanodine receptor channel by luminal Ca^{2+} involves luminal Ca^{2+} sensing sites. *Biophys J* **75:** 2801–2810.

Hagar RE, Burgstahler AD, Nathanson MH, Ehrlich BE. 1998. Type III $InsP_3$ receptor channel stays open in the presence of increased calcium. *Nature* **296:** 81–84.

Hajnóczky G, Thomas AP. 1997. Minimal requirements for calcium oscillations driven by the IP$_3$ receptor. *EMBO J* **16:** 3533–3543.

Hamada K, Mikoshiba K. 2002. Two-state conformational changes in inositol 1,4,5-trisphosphate receptor regulated by calcium. *J Biol Chem* **277:** 21115–21118.

Hamada K, Terauchi A, Mikoshiba K. 2003. Three-dimensional rearrangements with inositol 1,4,5-trisphosphate receptor by calcium. *J Biol Chem* **278:** 52881–52889.

Hamilton SL. 2005. Ryanodine receptors. *Cell Calcium* **38:** 253–260.

Haynes LP, Tepikin AV, Burgoyne RD. 2004. Calcium-binding protein 1 is an inhibitor of agonist-evoked, inositol 1,4,5-trisphosphate-mediated calcium signaling. *J Biol Chem* **279:** 547–555.

Higo T, Hattori M, Nakamura T, Natsume T, Michikawa T, Mikoshiba K. 2005. Subtype-specific and ER lumenal environment-dependent regulation of inositol 1,4,5-trisphosphate receptor type 1 by ERp44. *Cell* **120:** 85–98.

Hirota J, Michikawa T, Miyawaki A, Furuichi T, Okura I, Mikoshiba K. 1995. Kinetics of calcium release by immunoaffinity-purified inositol 1,4,5-trisphosphate receptor in reconstituted lipid vesicles. *J Biol Chem* **270:** 19046–19051.

Hirota J, Michikawa T, Natsume T, Furuichi T, Mikoshiba K. 1999. Calmodulin inhibits inositol 1,4,5-trisphosphate-induced calcium release through the purified and reconstituted inositol 1,4,5-trisphosphate receptor type 1. *FEBS Lett* **456:** 322–326.

Hokin MR, Hokin LE. 1953. Enzyme secretion and the incorporation of P^{32} into phospholipides of pancreas slices. *J Biol Chem* **203:** 967–977.

Horne JH, Meyer T. 1995. Luminal calcium regulates the inositol trisphosphate receptor of rat basophilic leukemia cells at the cytosolic side. *Biochemistry* **34:** 12738–12746.

Iino M. 1987. Calcium dependent inositol trisphosphate-induced calcium release in the guinea-pig taenia caeci. *Biochem Biophys Res Commun* **142:** 47–52.

Iino M. 1990. Biphasic Ca^{2+} dependence of inositol 1,4,5-trisphosphate-induced Ca^{2+} release in smooth muscle cells of the guinea pig taenia caeci. *J Gen Physiol* **95:** 1103–1122.

Imagawa T, Smith JS, Coronado R, Campbell KP. 1987. Purified ryanodine receptor from skeletal muscle sarcoplasmic reticulum is the Ca^{2+}-permeable pore of the calcium release channel. *J Biol Chem* **262:** 16636–16643.

Ionescu L, Cheung KH, Vais H, Mak DO, White C, Foskett JK. 2006. Graded recruitment and inactivation of single InsP$_3$ receptor Ca^{2+}-release channels: implications for quantal Ca^{2+} release. *J Physiol* **573:** 645–662.

Irvine RF. 1990. "Quantal" Ca^{2+} release and the control of Ca^{2+} entry by inositol phosphates - a possible mechanism. *FEBS Lett* **262:** 5–9.

Islam MO, Yoshida Y, Koga T, Kojima M, Kanagawa K, Imai S. 1996. Isolation and characterization of vascular smooth muscle inositol 1,4,5-trisphosphate receptor. *Biochem J* **316:** 295–302.

Iwai M, Michikawa T, Bosanac I, Ikura M, Mikoshiba K. 2007. Molecular basis of the isoform-specific ligand-binding affinity of inositol 1,4,5-trisphosphate receptors. *J Biol Chem* **282:** 12755–12764.

Jean T, Klee CB. 1986. Calcium modulation of inositol 1,4,5-trisphosphate-induced calcium release from neuroblastoma x glioma hybrid (NG108-15) microsomes. *J Biol Chem* **261:** 16414–16420.

Jiang D, Chen W, Xiao J, Wang R, Kong H, Jones PP, Zhang L, Fruen B, Chen SR. 2008. Reduced threshold for luminal Ca^{2+} activation of RyR1 underlies a causal mechanism of porcine malignant hyperthermia. *J Biol Chem* **283:** 20813–20820.

Jiang Q-X, Thrower EC, Chester DW, Ehrlich BE, Sigworth FJ. 2002a. Three-dimensional structure of the type 1 inositol 1,4,5-trisphosphate receptor at 24 Å resolution. *EMBO J* **21:** 3575–3581.

Jiang Y, Lee A, Chen J, Cadene M, Chait BT, MacKinnon R. 2002b. The open pore conformation of potassium channels. *Nature* **417:** 523–526.

Jones PP, Meng X, Xiao B, Cai S, Bolstad J, Wagenknecht T, Liu Z, Chen SR. 2008. Localization of PKA phosphorylation site, Ser^{2030}, in the three-dimensional structure of cardiac ryanodine receptor. *Biochem J* **410:** 261–270.

Joseph SK, Brownell S, Khan MT. 2005. Calcium regulation of inositol 1,4,5-trisphosphate receptors. *Cell Calcium* **38:** 539–546.

Joseph SK, Rice HL, Williamson JR. 1989. The effect of external calcium and pH on inositol trisphosphate-mediated calcium release from cerebellum microsomal fractions. *Biochem J* **258:** 261–265.

Kaftan EJ, Ehrlich BE, Watras J. 1997. Inositol 1,4,5-trisphosphate (InsP$_3$) and calcium interact to increase the dynamic range of InsP$_3$ receptor-dependent calcium signaling. *J Gen Physiol* **110:** 529–538.

Kiselyov K, Mignery GA, Zhu MX, Muallem S. 1999. The N-terminal domain of the IP$_3$ receptor gates store-operated hTrp3 channels. *Mol Cell* **4:** 423–429.

Krinke O, Novotna Z, Valentova O, Martinec J. 2007. Inositol trisphosphate receptor in higher plants: is it real? *J Exp Bot* **58:** 361–376.

Laude AJ, Tovey SC, Dedos S, Potter BVL, Lummis SCR, Taylor CW. 2005. Rapid functional assays of recombinant IP$_3$ receptors. *Cell Calcium* **38:** 45–51.

Launikonis BS, Zhou J, Royer L, Shannon TR, Brum G, Rios E. 2006. Depletion "skraps" and dynamic buffering inside the cellular calcium store. *Proc Natl Acad Sci USA* **103:** 2982–2987.

Laver DR. 2007. Ca^{2+} stores regulate ryanodine receptor Ca^{2+} release channels via luminal and cytosolic Ca^{2+} sites. *Biophys J* **92:** 3541–3555.

Laver DR. 2009. Luminal Ca^{2+} activation of cardiac ryanodine receptors by luminal and cytoplasmic domains. *Eur J Biophys* **39:** 19–26.

Lechleiter J, Girard S, Peralta E, Clapham D. 1991. Spiral calcium wave propagation and annihilation in *Xenopus laevis* oocytes. *Science* **252:** 123–126.

Li P, Chen SR. 2001. Molecular basis of Ca^{2+} activation of the mouse cardiac Ca^{2+} release channel (ryanodine receptor). *J Gen Physiol* **118:** 33–44.

Li W-H, Llopis J, Whitney M, Zlokarnik G, Tsien RY. 1998. Cell-permeant caged InsP$_3$ ester shows that Ca^{2+} spike frequency can optimize gene expression. *Nature* **392:** 936–941.

Lin C, Widjaja J, Joseph SK. 2000. The interaction of calmodulin with alternatively spliced isoforms of the type-I inositol trisphosphate receptor. *J Biol Chem* **275**: 2305–2311.

Lobo PA, Van Petegem F. 2009. Crystal structures of the N-terminal domains of cardiac and skeletal muscle ryanodine receptors: insights into disease mutations. *Structure* **17**: 1505–1514.

Ludtke SJ, Serysheva II, Hamilton SL, Chiu W. 2005. The pore structure of the closed RYR1 channel. *Structure* **13**: 1203–1211.

MacKinnon R. 2004. Potassium channels and the atomic basis of selective ion conduction (Nobel Lecture). *Angew Chem Int Edn Engl* **43**: 4265–4277.

Maeda N, Niinobe M, Nakahira K, Mikoshiba K. 1988. Purification and characterization of P_{400} protein, a glycoprotein characteristic of purkinje cell from mouse cerebellum. *J Neurochem* **51**: 1724–1730.

Mak D-OD, McBride S, Foskett JK. 1998. Inositol 1,4,5-tris-phosphate activation of inositol tris-phosphate receptor Ca^{2+} channel by ligand tuning of Ca^{2+} inhibition. *Proc Natl Acad Sci USA* **95**: 15821–15825.

Mak D-O, McBride S, Foskett JK. 2001. Regulation by Ca^{2+} and inositol 1,4,5-trisphosphate ($InsP_3$) of single recombinant type 3 $InsP_3$ receptor channels: Ca^{2+} activation uniquely distinguishes types 1 and 3 $InsP_3$ receptors. *J Gen Physiol* **117**: 435–446.

Mak D-O, McBride SMJ, Petrenko NB, Foskett JK. 2003. Novel regulation of calcium inhibition of the inositol 1,4,5-trisphosphate receptor calcium-release channel. *J Gen Physiol* **122**: 569–581.

Mak DO, McBride S, Raghuram V, Yue Y, Joseph SK, Foskett JK. 2000. Single-channel properties in endoplasmic reticulum membrane of recombinant type 3 inositol trisphosphate receptor. *J Gen Physiol* **115**: 241–256.

Marchant JS, Taylor CW. 1997. Cooperative activation of IP_3 receptors by sequential binding of IP_3 and Ca^{2+} safeguards against spontaneous activity. *Curr Biol* **7**: 510–518.

Marchenko SM, Yarotskyy VV, Kovalenko TN, Kostyuk PG, Thomas RC. 2005. Spontaneously active and $InsP_3$-activated ion channels in cell nuclei from rat cerebellar Purkinje and granule neurones. *J Physiol* **565**: 897–910.

Marshall ICB, Taylor CW. 1993. Biphasic effects of cytosolic calcium on $Ins(1,4,5)P_3$-stimulated Ca^{2+} mobilization in hepatocytes. *J Biol Chem* **268**: 13214–13220.

Marshall ICB, Taylor CW. 1994. Two calcium-binding sites mediate the interconversion of liver inositol 1,4,5-trisphosphate receptors between three conformational states. *Biochem J* **301**: 591–598.

Maruyama T, Kanaji T, Nakade S, Kanno T, Mikoshiba K. 1997. 2APB, 2-aminoethoxydiphenyl borate, a membrane-penetrable modulator of $Ins(1,4,5)P_3$-induced Ca^{2+} release. *J Biochem* **122**: 498–505.

Mayer ML. 2006. Glutamate receptors at atomic resolution. *Nature* **440**: 456–462.

Meur G, Parker AKT, Gergely FV, Taylor CW. 2007. Targeting and retention of type 1 ryanodine receptors to the endoplasmic reticulum. *J Biol Chem* **282**: 23096–23103.

Meyer T, Stryer L. 1990. Transient calcium release induced by successive increments of inositol 1,4,5-trisphosphate. *Proc Natl Acad Sci USA* **87**: 3841–3845.

Michelangeli F, Mezna M, Tovey S, Sayers LG. 1995. Pharmacological modulators of the inositol 1,4,5-trisphosphate receptor. *Neuropharmacol* **34**: 1111–1122.

Michell RH. 1975. Inositol phospholipids and cell surface receptor function. *Biochim Biophys Acta* **415**: 81–147.

Michell RH, Drummond AH, Downes CP. 1989. *Inositol Lipids in Cell Signalling*. Academic Press Ltd, London.

Michikawa T, Hirota J, Kawano S, Hiraoka M, Yamada M, Furuichi T, Mikoshiba K. 1999. Calmodulin mediates calcium-dependent inactivation of the cerebellar type 1 inositol 1,4,5-trisphosphate receptor. *Neuron* **23**: 799–808.

Mignery GA, Südhof TC, Takei K, De Camilli P. 1989. Putative receptor for inositol 1,4,5-trisphosphate similar to ryanodine receptor. *Nature* **342**: 192–195.

Missiaen L, De Smedt H, Bultynck G, Vanlingen S, De Smet P, Callewaert G, Parys J. 2000. Calmodulin increases the sensitivity of type 3 inositol 1,4,5-trisphosphate receptors to Ca^{2+} inhibition in human bronchial mucosal cells. *Mol Pharmacol* **57**: 564–567.

Missiaen L, De Smedt H, Parys JB, Casteels R. 1994. Co-activation of inositol trisphosphate-induced Ca^{2+} release by cytosolic Ca^{2+} is loading-dependent. *J Biol Chem* **269**: 7238–7242.

Missiaen L, Parys JB, Sienaert I, Maes K, Kunzelmann K, Takahashi M, Tanzawa K, De Smedt H. 1998. Functional properties of the type-3 $InsP_3$ receptor in 16HBE14o-bronchial mucosal cells. *J Biol Chem* **273**: 8983–8986.

Missiaen L, Parys JB, Weidema AF, Sipma H, Vanlingen S, De Smet P, Callewaert G, De Smedt H. 1999. The bell-shaped Ca^{2+}-dependence of the inositol 1,4,5-trisphosphate induced Ca^{2+} release is modulated by $Ca^{2+}/$calmodulin. *J Biol Chem* **274**: 13748–13751.

Missiaen L, Taylor CW, Berridge MJ. 1992. Luminal Ca^{2+} promoting spontaneous Ca^{2+} release from inositol trisphosphate-sensitive stores of rat hepatocytes. *J Physiol* **455**: 623–640.

Miyakawa T, Maeda A, Yamazawa T, Hirose K, Kurosaki T, Iino M. 1999. Encoding of Ca^{2+} signals by differential expression of IP_3 receptor subtypes. *EMBO J* **18**: 1303–1308.

Miyakawa T, Mizushima A, Hirose K, Yamazawa T, Bezprozvanny I, Kurosaki T, Iino M. 2001. Ca^{2+}-sensor region of IP_3 receptor controls intracellular Ca^{2+} signaling. *EMBO J* **20**: 1674–1680.

Mohr FC, Hershey PEC, Zimányi I, Pessah IN. 1993. Regulation of inositol 1,4,5-trisphosphate receptors in rat basophilic leukemia cells. I. Multiple conformational states of the receptor in a microsomal preparation. *Biochim Biophys Acta* **1147**: 105–114.

Muallem S, Pandol SJ, Beeker TG. 1989. Hormone-evoked calcium release from intracellular stores is a quantal process. *J Biol Chem* **264**: 205–212.

Nadif Kasri N, Bultynck G, Sienaert I, Callewaert G, Erneux C, Missiaen L, Parys JB, De Smedt H. 2002. The role of calmodulin for inositol 1,4,5-trisphosphate receptor function. *Biochim Biophys Acta* **1600**: 19–31.

Nadif Kasri N, Holmes AM, Bultynck G, Parys JB, Bootman MD, Rietdorf K, Missiaen L, McDonald F, De Smedt H, Conway SJ, et al. 2004. Regulation of InsP$_3$ receptor activity by neuronal Ca^{2+}-binding proteins. *EMBO J* **23:** 312–321.

Nadif Kasri N, Torok K, Galione A, Garnham C, Callewaert G, Missiaen L, Parys JB, De Smedt H. 2006. Endogenously bound calmodulin is essential for the function of the inositol 1,4,5-trisphosphate receptor. *J Biol Chem* **281:** 8332–8338.

Nilius B, Owsianik G, Voets T, Peters JA. 2007. Transient receptor potential cation channels in disease. *Physiol Rev* **87:** 165–217.

Nosyreva E, Miyakawa T, Wang Z, Glouchankova L, Iino M, Bezprozvanny I. 2002. The high-affinity calcium-calmodulin-binding site does not play a role in the modulation of type 1 inositol 1,4,5-trisphosphate receptor function by calcium and calmodulin. *Biochem J* **365:** 659–667.

Nunn DL, Taylor CW. 1992. Luminal Ca^{2+} increases the sensitivity of Ca^{2+} stores to inositol 1,4,5-trisphosphate. *Mol Pharmacol* **41:** 115–119.

Oancea E, Meyer T. 1996. Reversible desensitization of inositol trisphosphate-induced calcium release provides a mechanism for repetitive calcium spikes. *J Biol Chem* **271:** 17253–17260.

Oda T, Yano M, Yamamoto T, Tokuhisa T, Okuda S, Doi M, Ohkusa T, Ikeda Y, Kobayashi S, Ikemoto N, et al. 2005. Defective regulation of interdomain interactions within the ryanodine receptor plays a key role in the pathogenesis of heart failure. *Circulation* **111:** 3400–3410.

Oldershaw KA, Nunn DL, Taylor CW. 1991. Quantal Ca^{2+} mobilization stimulated by inositol 1,4,5-trisphosphate in permeabilized hepatocytes. *Biochem J* **278:** 705–708.

Oldershaw KA, Taylor CW. 1993. Luminal Ca^{2+} increases the affinity of inositol 1,4,5-trisphosphate for its receptor. *Biochem J* **292:** 631–633.

Orlova EV, Serysheva II, van Heel M, Hamilton SL, Chiu W. 1996. Two structural configurations of the skeletal muscle calcium release channel. *Nat Struct Biol* **3:** 547–552.

Pantazaka E, Taylor CW. 2010. Targeting of inositol 1,4,5-trisphosphate receptor to the endoplasmic reticulum by its first transmembrane domain. *Biochem J* **425:** 61–69.

Park CY, Hoover PJ, Mullins FM, Bachhawat P, Covington ED, Raunser S, Walz T, Garcia KC, Dolmetsch RE, Lewis RS. 2009. STIM1 clusters and activates CRAC channels via direct binding of a cytosolic domain to Orai1. *Cell* **136:** 876–890

Parker AKT, Gergely FV, Taylor CW. 2004. Targeting of inositol 1,4,5-trisphosphate receptors to the endoplasmic reticulum by multiple signals within their transmembrane domains. *J Biol Chem* **279:** 23797–23805.

Parker I, Choi J, Yao Y. 1996. Elementary events of InsP$_3$-induced Ca^{2+} liberation in *Xenopus* oocytes: hot spots, puffs and blips. *Cell Calcium* **20:** 105–121.

Parker I, Ivorra I. 1990. Localized all-or-nothing calcium liberation by inositol trisphosphate. *Science* **250:** 977–979.

Parker I, Ivorra I. 1991. Caffeine inhibits inositol trisphosphate-mediated liberation of intracellular calcium in *Xenopus* oocytes. *J Physiol* **433:** 229–240.

Parys JB, Missiaen L, De Smedt H, Casteels R. 1993. Loading dependence of inositol 1,4,5-trisphosphate-induced Ca^{2+} release in the clonal cell line A7r5. *J Biol Chem* **268:** 25206–25212.

Parys JB, Sernett SW, DeLisle S, Snyder PM, Welsh MJ, Campbell KP. 1992. Isolation, characterization, and localization of the inositol 1,4,5-trisphosphate receptor protein in *Xenopus laevis* oocytes. *J Biol Chem* **267:** 18776–18782.

Patel S, Marchant JS, Brailoiu E. 2010. Two-pore channels: regulation by NAADP and customized roles in triggering calcium signals. *Cell Calcium* **47:** 480–490.

Patel S, Morris SA, Adkins CE, O'Beirne G, Taylor CW. 1997. Ca^{2+}-independent inhibition of inositol trisphosphate receptors by calmodulin: Redistribution of calmodulin as a possible means of regulating Ca^{2+} mobilization. *Proc Natl Acad Sci USA* **94:** 11627–11632.

Patterson RL, Boehning D, Snyder SH. 2004. Inositol 1,4,5-trisphosphate receptors as signal integrators. *Annu Rev Biochem* **73:** 437–465.

Petersen OH, Burdakov D, Tepikin AV. 1999. Polarity in intracellular calcium signalling. *Bioessays* **21:** 851–860.

Pietri F, Hilly M, Mauger J-P. 1990. Calcium mediates the interconversion between two states of the liver inositol 1,4,5-trisphosphate receptor. *J Biol Chem* **265:** 17478–17485.

Pinton P, Pozzan T, Rizzuto R. 1998. The Golgi apparatus is an inositol 1,4,5-trisphosphate-sensitive Ca^{2+} store, with functional properties distinct from those of the endoplasmic reticulum. *EMBO J* **17:** 5298–5308.

Putney JW Jr. 1997. *Capacitative calcium entry*. RGLandes Company, Austin, Texas, USA.

Putney JW, Poggioli J, Weiss SJ. 1981. Receptor regulation of calcium release and calcium permeability in parotid gland cells. *Philos Trans R Soc London [Biol]* **296:** 37–45.

Rahman T-U, Skupin A, Falcke M, Taylor CW. 2009. Clustering of IP$_3$ receptors by IP$_3$ retunes their regulation by IP$_3$ and Ca^{2+}. *Nature* **458:** 655–659.

Rahman T, Taylor CW. 2009. Dynamic regulation of IP$_3$ receptor clustering and activity by IP$_3$. *Channels* **3:** 336–332.

Ramos-Franco J, Bare D, Caenepeel S, Nani A, Fill M, Mignery G. 2000. Single-channel function of recombinant type 2 inositol 1,4,5-trisphosphate receptor. *Biophys J* **79:** 1388–1399.

Ramos-Franco J, Caenepeel S, Fill M, Mignery G. 1998a. Single channel function of recombinant type-1 inositol 1,4,5-trisphosphate receptor ligand binding domain splice variants. *Biophys J* **75:** 2783–2793.

Ramos-Franco J, Fill M, Mignery GA. 1998b. Isoform-specific function of single inositol 1,4,5-trisphosphate receptor channels. *Biophys J* **75:** 834–839.

Ramos-Franco J, Galvan D, Mignery GA, Fill M. 1999. Location of the permeation pathway in the recombinant type-1 inositol 1,4,5-trisphosphate receptor. *J Gen Physiol* **114:** 243–250.

Rasmussen H. 1970. Cell communcation, calcium ion, and cyclic adenosine monophosphate. *Science* **170:** 404–412.

Richardson A, Taylor CW. 1993. Effects of Ca^{2+} chelators on purified inositol 1,4,5-trisphosphate (InsP$_3$) receptors and InsP$_3$-stimulated Ca^{2+} mobilization. *J Biol Chem* **268:** 11528–11533.

Ridgway EB, Gilkey JC, Jaffe LF. 1977. Free calcium increases explosively in activating medaka eggs. *Proc Natl Acad Sci USA* **74:** 623–627.

Ringer S. 1883. A further contribution regarding the influence of the different constituents of the blood on the contraction of the heart. *J Physiol* **4:** 29–42.

Rizzuto R, Brini M, Murgia M, Pozzan T. 1993. Microdomains with high Ca^{2+} close to IP$_3$-sensitive channels that are sensed by neighbouring mitochondria. *Science* **262:** 744–747.

Rodney GG, Moore CP, Williams BY, Zhang J-Z, Krol J, Pedersen SE, Hamilton SL. 2001. Calcium binding to calmodulin leads to an N-terminal shift in its binding site on the ryanodine receptor. *J Biol Chem* **276:** 2069–2074.

Rossi AM, Riley AM, Potter BVL, Taylor CW. 2010. Adenophostins: high-affinity agonists of IP$_3$ receptors. *Curr Topics Membr* **66:** 209–233.

Rossi AM, Riley AM, Tovey SC, Rahman T, Dellis O, Taylor EJA, Veresov VG, Potter BVL, Taylor CW. 2009. Synthetic partial agonists reveal key steps in IP$_3$ receptor activation. *Nat Chem Biol* **5:** 631–639.

Samso M, Feng W, Pessah IN, Allen PD. 2009. Coordinated movement of cytoplasmic and transmembrane domains of RyR1 upon gating. *PLoS Biol* **7:** e85

Samso M, Wagenknecht T, Allen PD. 2005. Internal structure and visualization of transmembrane domains of the RyR1calcium release channel by cryo-EM. *Nat Struct Mol Biol* **6:** 539–544.

Sato C, Hamada K, Ogura T, Miyazawa A, Iwasaki K, Hiroaki Y, Tani K, Terauchi A, Fujiyoshi Y, Mikoshiba K. 2004. Inositol 1,4,5-trisphosphate receptor contains multiple cavities and L-shaped ligand-binding domains. *J Mol Biol* **336:** 155–164.

Schug ZT, da Fonseca PC, Bhanumathy CD, Wagner L 2nd, Zhang X, Bailey B, Morris EP, Yule DI, Joseph SK. 2008. Molecular characterization of the inositol 1,4,5-trisphosphate receptor pore-forming segment. *J Biol Chem* **283:** 2939–2948.

Schug ZT, Joseph SK. 2006. The role of the S4-S5 linker and C-terminal tail in inositol 1,4,5-trisphosphate receptor function. *J Biol Chem* **281:** 24431–24440.

Serysheva II, Bare DJ, Ludtke SJ, Kettlun CS, Chiu W, Mignery GA. 2003. Structure of the type 1 inositol 1,4,5-trisphosphate receptor revealed by cryomicroscopy. *J Biol Chem* **278:** 21319–21322.

Serysheva II, Hamilton SL, Chiu W, Ludtke SJ. 2005. Structure of a Ca^{2+} release channel at 14Å resolution. *J Struct Biol* **345:** 427–431.

Shuttleworth TJ. 1992. Ca^{2+} release from inositol trisphosphate-sensitive stores is not modulated by intraluminal [Ca^{2+}]. *J Biol Chem* **267:** 3573–3576.

Sienaert I, De Smedt H, Parys JB, Missiaen L, Vanlingen S, Sipma H, Casteels R. 1996. Characterization of a cytosolic and a luminal Ca^{2+} binding site in the type I inositol 1,4,5-trisphosphate receptor. *J Biol Chem* **271:** 27005–27012.

Sienaert I, Kasri NN, Vanlingen S, Parys J, Callewaert G, Missiaen L, De Smedt H. 2002. Localization and function of a calmodulin/apocalmodulin binding domain in the N-terminal part of the type 1 inositol 1,4,5-trisphosphate receptor. *Biochem J* **365:** 269–277.

Sienaert I, Missiaen L, De Smedt H, Parys JB, Sipma H, Casteels R. 1997. Molecular and functional evidence for multiple Ca^{2+}-binding domains on the type 1 inositol 1,4,5-trisphosphate receptor *J Biol Chem* **272:** 25899–25906.

Smith JS, Coronado R, Meissner G. 1985. Sarcoplasmic reticulum contains adenine nucleotide-activated calcium channels. *Nature* **316:** 446–449.

Solovyova N, Fernyhough P, Glazner G, Verkhratsky A. 2002. Xestospongin C empties the ER calcium store but does not inhibit InsP$_3$-induced Ca^{2+} release in cultured dorsal root ganglia neurones. *Cell Calcium* **32:** 49–52.

Spät A, Bradford PG, McKinney JS, Rubin RP, Putney JW Jr. 1986. A saturable receptor for ^{32}P-inositol-1,4,5-trisphosphate in hepatocytes and neutrophils. *Nature* **319:** 514–516.

Stehno-Bittel L, Lückhoff A, Clapham DE. 1995. Calcium release from the nucleus by InsP$_3$ receptor channels. *Neuron* **14:** 163–167.

Streb H, Irvine RF, Berridge MJ, Schulz I. 1983. Release of Ca^{2+} from a nonmitochondrial store of pancreatic acinar cells by inositol-1,4,5-trisphosphate. *Nature* **306:** 67–69.

Striggow F, Ehrlich BE. 1996. The inositol 1,4,5-trisphosphate receptor of cerebellum. Mn^{2+} permeability and regulation by cytosolic Mn^{2+}. *J Gen Physiol* **108:** 115–124.

Suematsu E, Hirata M, Hashimoto T, Kuriyama H. 1984. Inositol 1,4,5-trisphosphate releases Ca^{2+} from intracellular store sites in skinned single cells of porcine coronary artery. *Biochem Biophys Res Commun* **120:** 481–485.

Sun Y, Taylor CW. 2008. A calmodulin antagonist reveals a calmodulin-independent interdomain interaction essential for activation of inositol 1,4,5-trisphosphate receptors. *Biochem J* **416:** 243–253.

Supattapone S, Worley PF, Baraban JM, Snyder SH. 1988. Solubilization, purification, and characterization of an inositol trisphosphate receptor. *J Biol Chem* **263:** 1530–1534.

Sureshan KM, Riley AM, Rossi AM, Tovey SC, Dedos SG, Taylor CW, Potter BVL. 2009. Activation of IP$_3$ receptors by synthetic bisphosphate ligands. *Chem Commun* **14:** 1204–1206.

Sutko JL, Ito K, Kenyon JL. 1985. Ryanodine: a modifier of sarcoplasmic reticulum calcium release in striated muscle. *FASEB J* **44:** 2984–2988.

Swatton JE, Morris SA, Cardy TJA, Taylor CW. 1999. Type 3 inositol trisphosphate receptors in RINm5F cells are biphasically regulated by cytosolic Ca^{2+} and mediate quantal Ca^{2+} mobilization. *Biochem J* **344:** 55–60.

Swatton JE, Taylor CW. 2002. Fast biphasic regulation of type 3 inositol trisphosphate receptors by cytosolic calcium. *J Biol Chem* **277:** 17571–17579.

Szlufcik K, Bultynck G, Callewaert G, Missiaen L, Parys JB, De Smedt H. 2006. The suppressor domain of inositol 1,4,5-trisphosphate receptor plays an essential role in the protection against apoptosis. *Cell Calcium* **39:** 325–336.

Takeshima H, Nishimura S, Matsumoto T, Ishida H, Kangawa K, Minamino N, Matsuo H, Ueda M, Hanaoka

M, Hirose T, et al. 1989. Primary structure and expression from complementary DNA of skeletal muscle ryanodine receptor. *Nature* **339:** 439–445.

Tanimura A, Turner RJ. 1996. Calcium release in HSY cells conforms to a steady-state mechanism involving regulation of the inositol 1,4,5-trisphosphate receptor Ca^{2+} channel by luminal $[Ca^{2+}]$. *J Cell Biol* **132:** 607–616.

Tateishi H, Yano M, Mochizuki M, Suetomi T, Ono M, Xu X, Uchinoumi H, Okuda S, Oda T, Kobayashi S, et al. 2009. Defective domain-domain interactions within the ryanodine receptor as a critical cause of diastolic Ca^{2+} leak in failing hearts. *Cardiovasc Res* **81:** 536–545.

Taylor CW, da Fonseca PCA, Morris EP. 2004. IP_3 receptors: the search for structure. *Trends Biochem Sci* **29:** 210–219.

Taylor CW, Genazzani AA, Morris SA. 1999. Expression of inositol trisphosphate receptors. *Cell Calcium* **26:** 237–251.

Taylor CW, Laude AJ. 2002. IP_3 receptors and their regulation by calmodulin and cytosolic Ca^{2+}. *Cell Calcium* **32:** 321–334.

Taylor CW, Potter BVL. 1990. The size of inositol 1,4,5-trisphosphate-sensitive Ca^{2+} stores depends on inositol 1,4,5-trisphosphate concentration. *Biochem J* **266:** 189–194.

Taylor CW, Prole DL, Rahman T. 2009a. Ca^{2+} channels on the move. *Biochemistry* **48:** 12062–12080.

Taylor CW, Ur-Rahman T, Pantazaka E. 2009b. Targeting and clustering of IP_3 receptors: key determinants of spatially organized Ca^{2+} signals. *Chaos* **19:** 037102-037101–037102-037110.

Terentyev D, Nori A, Santoro M, Viatchenko-Karpinski S, Kubalova Z, Gyorke I, Terentyeva R, Vedamoorthyrao S, Blom NA, Valle G, et al. 2006. Abnormal interactions of calsequestrin with the ryanodine receptor calcium release channel complex linked to exercise-induced sudden cardiac death. *Circ Res* **98:** 1151–1158.

Thrower EC, Lea EJA, Dawson AP. 1998. The effects of free $[Ca^{2+}]$ on the cytosolic face of inositol (1,4,5)-trisphosphate receptor at the single channel level. *Biochem J* **330:** 559–564.

Tovey SC, Dedos SG, Rahman T, Taylor EJA, Pantazaka E, Taylor CW. 2010. Regulation of inositol 1,4,5-trisphosphate receptors by cAMP independent of cAMP-dependent protein kinase. *J Biol Chem* **285:** 12979–12989.

Tripathy A, Meissner G. 1996. Sarcoplasmic reticulum lumenal Ca^{2+} has access to cytosolic activation and inactivation sites of skeletal muscle Ca^{2+} release channel. *Biophys J* **70:** 2600–2615.

Tsien RY. 1980. New calcium indicators and buffers with high selectivity against magnesium and protons: design, synthesis, and properties of prototype structures. *Biochemistry* **19:** 2396–2404.

Tsien RY. 1981. A non-disruptive technique for loading calcium buffers and indicators into cells. *Nature* **290:** 527–528.

Tu H, Miyakawa T, Wang Z, Glouchankova L, Iino M, Bezprozvanny I. 2002. Functional characterization of the type 1 inositol 1,4,5-trisphosphate receptor coupling doamain SII($\pm$) splice variants and the *Opisthotonos* mutant form. *Biophys J* **82:** 1995–2004.

Tu H, Nosyreva E, Miyakawa T, Wang Z, Mizushima A, Iino M, Bezprozvanny I. 2003. Functional and biochemical analysis of the type 1 inositol (1,4,5)-trisphosphate receptor calcium sensor. *Biophys J* **85:** 290–299.

Tu H, Wang Z, Bezprozvanny I. 2005a. Modulation of mammalian inositol 1,4,5-trisphosphate receptor isoforms by calcium: a role of calcium sensor region. *Biophys J* **88:** 1056–1069.

Tu H, Wang Z, Nosyreva E, De Smedt H, Bezprozvanny I. 2005b. Functional characterization of mammalian inositol 1,4,5-trisphosphate receptor isoforms. *Biophys J* **88:** 1046–1055.

Uchida K, Miyauchi H, Furuichi T, Michikawa T, Mikoshiba K. 2003. Critical regions for activation gating of the inositol 1,4,5-trisphosphate receptor. *J Biol Chem* **278:** 16551–16560.

van de Put FHMM, De Pont JJHHM, Willems PHGM. 1994. Heterogeneity between intracellular Ca^{2+} stores as the underlying principle of quantal Ca^{2+} release by inositol 1,4,5-trisphosphate in permeabilized pancreatic acinar cells. *J Biol Chem* **269:** 12438–12443.

Varney MA, Rivera J, Lopez Bernal A, Watson SP. 1990. Are there subtypes of the inositol 1,4,5-trisphosphate receptor? *Biochem J* **269:** 211–216.

Wang R, Chen W, Cai S, Zhang J, Bolstad J, Wagenknecht T, Liu Z, Chen SR. 2007. Localization of an NH_2-terminal disease-causing mutation hot spot to the "clamp" region in the three-dimensional structure of the cardiac ryanodine receptor. *J Biol Chem* **282:** 17785–17793.

Wang R, Zhang L, Bolstad J, Diao N, Brown C, Ruest L, Welch W, Williams AJ, Chen SR. 2003. Residue Gln4863 within a predicted transmembrane sequence of the Ca^{2+} release channel (ryanodine receptor) is critical for ryanodine interaction. *J Biol Chem* **278:** 51557–51565.

Wang Y, Xu L, Pasek DA, Gillespie D, Meissner G. 2005. Probing the role of negatively charged amino acid residues in ion permeation of skeletal muscle ryanodine receptor. *Biophys J* **89:** 256–265.

Watras J, Bezprozvanny I, Ehrlich BE. 1991. Inositol 1,4,5-trisphosphate-gated channels in cerebellum: presence of multiple conductance states. *J Neurosci* **11:** 3239–3245.

Williams AJ, West DJ, Sitsapesan R. 2001. Light at the end of the Ca^{2+}-release channel tunnel: structures and mechanisms involved in ion translocation in ryanodine receptor channels. *Quart Rev Biophys* **34:** 61–104.

Williams JA. 1980. Regulation of pancreatic acinar cell function by intracellular calcium. *Am J Physiol* **238:** G269–G279.

Willoughby D, Cooper DM. 2007. Organization and Ca^{2+} regulation of adenylyl cyclases in cAMP microdomains. *Physiol Rev* **87:** 965–1010.

Wolfram F, Morris E, Taylor CW. 2010. Three-dimensional structure of recombinant type 1 inositol 1,4,5-trisphosphate receptor. *Biochem J* **428:** 483–489.

Woods NM, Cuthbertson KSR, Cobbold PH. 1986. Repetitive transient rises in cytoplasmic free calcium in hormone-stimulated hepatocytes. *Nature* **319:** 600–602.

Worley PF, Baraban JM, Supattapone S, Wilson VS, Snyder SH. 1987. Characterization of inositol trisphosphate receptor binding in brain. Regulation by pH and calcium. *J Biol Chem* **262:** 12132–12136.

Yamada M, Miyawaki A, Saito K, Yamamoto-Hino M, Ryo Y, Furuichi T, Mikoshiba K. 1995. The calmodulin-binding domain in the mouse type 1 inositol 1,4,5-trisphosphate receptor. *Biochem J* **308:** 83–88.

Yang J, McBride S, Mak D-OD, Vardi N, Palczewski K, Haeseleer F, Foskett JK. 2002. Identification of a family of calcium sensors as protein ligands of the inositol trisphosphate receptor Ca^{2+} release channels. *Proc Natl Acad Sci USA* **99:** 7711–7716.

Yoneshima H, Miyawaki A, Michikawa T, Furuichi T, Mikoshiba K. 1997. Ca^{2+} differentially regulates ligand-affinity states of type 1 and 3 inositol 1,4,5-trisphosphate receptors. *Biochem J* **322:** 591–596.

Zeng W, Mak DD, Li Q, Shin DM, Foskett JK, Muallem S. 2003. A new mode of Ca^{2+} signaling by G protein-coupled receptors: gating of IP_3 receptor Ca^{2+} release channels by G$\beta\gamma$. *Curr Biol* **13:** 872–876.

Zhang X, Joseph SK. 2001. Effect of mutation of a calmodulin-binding sites on Ca^{2+} regulation of inositol trisphosphate receptors. *Biochem J* **360:** 395–400.

Zhu MX, Ma J, Parrington J, Calcraft PJ, Galione A, Evans AM. 2010. Calcium signaling via two-pore channels: local or global, that is the question. *Am J Physiol* **298:** C430–C441.

NAADP Receptors

Antony Galione

Department of Pharmacology, University of Oxford, Oxford OX1 3QT, United Kingdom

Correspondence: antony.galione@pharm.ox.ac.uk

Of the established Ca^{2+} mobilizing messengers, NAADP is arguably the most tantalizing. It is the most potent, often efficacious at low nanomolar concentrations. Recent studies have identified a new class of calcium release channel, the two-pore channels (TPCs), as the likely targets for NAADP. These channels are endolysosomal in localization where they mediate local Ca^{2+} release, and have highlighted a new role of acidic organelles as targets for messenger-evoked Ca^{2+} mobilization. Three distinct roles of TPCs have been identified. The first is to effect local Ca^{2+} release that may play a role in endolysosomal function including vesicular fusion and trafficking. The second is to trigger global calcium release by recruiting Ca^{2+}-induced Ca^{2+} release (CICR) channels at lysosomal-ER junctions. The third is to regulate plasma membrane excitability by the targeting of Ca^{2+} release from appropriately positioned subplasma membrane stores to regulate plasma membrane Ca^{2+}-activated channels. In this review, I discuss the role of NAADP-mediated Ca^{2+} release from endolysosomal stores as a widespread trigger for intracellular calcium signaling mechanisms, and how studies of TPCs are beginning to enhance our understanding of the central role of lysosomes in Ca^{2+} signaling.

Calcium is the most evolutionarily ubiquitous of intracellular signals and controls cellular mechanisms as diverse as cellular motility, membrane fusion, ion channel function, enzyme activity, and gene expression (Berridge et al. 2003). Free cytoplasmic calcium levels are kept under tight control by pumps, exchangers, and buffering mechanisms including storage by organelles (Pozzan et al. 1994). Ca^{2+} signals may be elicited when these mechanisms are transiently overwhelmed by the opening of calcium permeable channels at the plasma membrane or in membranes of calcium-storing organelles. Chronic activation of such channels may lead to cell death, for example, through the activation of apoptotic signaling cascades (Berridge et al. 1998). Many cell surface receptors are linked to signaling pathways that lead to the mobilization of calcium from intracellular storage organelles through the activation of specific Ca^{2+} release channels (Clapham 1995). Three major intracellular messengers have been established to link cell stimulation with organellar Ca^{2+} release: inositol trisphosphate (IP_3), cyclic adenosine diphosphate ribose (cADPR), and nicotinic acid adenine nucleotide diphosphate (NAADP) (Bootman et al. 2002).

DISCOVERY OF NAADP AS A Ca^{2+} MOBILIZING MOLECULE

NAADP was discovered as a contaminant of commercial batches of β-NADP$^+$ by Lee and colleagues while they were investigating the effects of various pyridine nucleotides on calcium release from sea urchin egg homogenates (Clapper et al. 1987). The rationale for this was that at fertilization in sea urchin eggs, dramatic changes in pyridine nucleotide levels occur at a similar time to the generation of the calcium wave. Egg homogenates can be simply prepared from eggs and are remarkably stable, even after freezing, and sequester, and robustly release calcium when challenged with messengers and drugs (Morgan and Galione 2008). Three distinct calcium release mechanisms were shown. These were the early days of IP$_3$, and IP$_3$ was found to release calcium from microsomal stores. In addition, two metabolites of pyridine nucleotides, an enzyme-activated metabolite related to NAD$^+$, subsequently identified as cADPR (Lee et al. 1989), and alkaline-treated NADP, later shown to be NAADP (Lee and Aarhus 1995), were found to release Ca^{2+} from different subcellular nonmitochondrial fractions from egg homogenate (Fig. 1). A key feature of each mechanism is their display of homologous desensitization underscoring the independence of each of the three mechanisms.

NAADP AS A Ca^{2+} MOBILIZING MESSENGER

NAADP is the most potent of Ca^{2+} mobilizing messengers described, typically efficacious at pM or low nM concentrations. A growing number of cellular stimuli and activation of cell surface receptors have been found to be coupled to increases in NAADP levels, confirming its role as an intracellular messenger (Churchill et al. 2003; Masgrau et al. 2003; Rutter 2003; Yamasaki et al. 2005; Galione 2006; Gasser et al. 2006; Kim et al. 2008). Mediation of calcium signaling by NAADP has been implicated by two approaches: inhibition of agonist-evoked calcium signals by prior self-inactivation of the

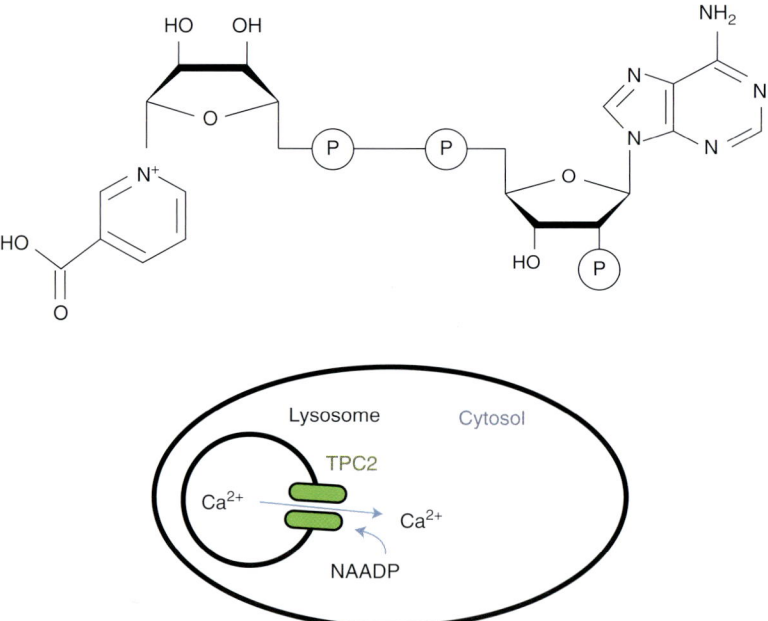

Figure 1. Structure and function of NAADP. NAADP differs from β-NADP in that the base nicotinic acid is substituted for nicotinamide (upper panel). NAADP, unlike NADP, is a potent Ca^{2+} mobilizing agent and interacts with two-pore channels in the membranes of lysosomes (lower panel).

NAADP receptor or NAADP receptor pharmacological blocker (Naylor et al. 2009) and measurements of cellular NAADP levels. Measurements of NAADP have been performed using either a radioreceptor assay, based on the high affinity NAADP binding protein of sea urchin eggs (Lewis et al. 2007), or by using a cycling assay of coupled enzyme reactions resulting in fluorescent resorufin production (Graeff and Lee 2002). Although some receptors appear to couple selectively to NAADP production, increasingly it is becoming apparent that several receptors couple to multiple Ca^{2+} mobilizing messengers (Aley et al. 2010), and this may be the norm.

Ca^{2+} STORES TARGETED BY NAADP

Accumulating evidence suggests that the primary Ca^{2+} stores targeted by NAADP are separate from the endoplasmic reticulum and are members of a group known as acidic organelles. The initial evidence for this came from the study of sea urchin eggs and was subsequently extended to mammalian cells.

Sea Urchin Eggs

The initial report of NAADP-evoked Ca^{2+} release using alkaline-activated NADP showed that the subcellular fraction reactive to NAADP in egg homogenates was largely separate from the microsomal/ER fraction sensitive to IP_3 and cADPR (Clapper et al. 1987). Abrogation of Ca^{2+} storage by the ER SERCA inhibitor, thapsigargin, while inhibiting Ca^{2+} release by either IP_3 or cADPR, only partially reduced Ca^{2+} release evoked by NAADP in both sea urchin egg homogenates (Genazzani and Galione 1996) and intact eggs (Churchill and Galione 2001a). Visualization of two separate Ca^{2+} stores was observed in elegant sea urchin egg stratification studies (Lee and Aarhus 2000). Stratification by centrifugation of intact eggs results in eggs forming elongated structures with different organelles separating to different "poles." Uniform photolysis of caged derivatives of Ca^{2+} mobilizing messengers resulted in IP_3 and cADPR releasing Ca^{2+} from the nuclear pole where ER accumulated, whereas NAADP released Ca^{2+} from the opposite end of the structure. These experiments are consistent with the primary Ca^{2+} store targeted by NAADP as being distinct from the ER.

In a series of important experiments using pharmacological analyses and subcellular fractionation, lysosomal-related organelles were implicated as the primary target organelle for NAADP-evoked Ca^{2+} release in sea urchin eggs (Churchill et al. 2002). Acidic stores, such as lysosomes, have been shown to sequester Ca^{2+} by mechanisms dependent on their low luminal pH (Patel and Docampo 2010). Inhibition of the vacuolar H^+-ATPase by bafilomycin decreases proton uptake into acidic stores; if their membranes are sufficiently leaky to protons, this leads to the alkalinization of their lumen. Uptake of Ca^{2+} into these stores appears to be dependent on the maintenance of the proton gradient because bafilomycin and protonophores inhibit Ca^{2+} storage by these organelles, although the detailed mechanisms are not well understood. A dense membrane fraction from sea urchin egg homogenates was isolated from a percoll gradient and consisted of "reserve granules." This fraction was enriched with lysosomal markers and supported ATP-dependent Ca^{2+} sequestration, which was inhibited by preincubation with bafilomycin or the protonophore, nigericin, but not thapsigargin. This fraction was found to contain [^{32}P]NAADP binding sites, and displayed NAADP but not IP_3/cADPR-evoked Ca^{2+} release. Reserve granules from sea urchin eggs are lysosome-related organelles. In intact sea urchin eggs, treatment with the lysosomotropic agent, glycyl-phenylalanine 2-naphthylamide (GPN), caused the reversible lysis of lysotracker-stained vesicles, resulting in a series of small-amplitude cytoplasmic Ca^{2+} signals, consistent with their role as Ca^{2+} stores. Importantly, GPN treatment in either intact eggs or egg homogenates selectively abolished NAADP-evoked Ca^{2+} release with little effect on Ca^{2+} release by either IP_3 or cADPR. From these data it was proposed that in sea urchin the primary target of NAADP are acidic stores rather than the endoplasmic reticulum. Consistent with this, experiments in sea urchin egg homogenates employing luminal pH indicators such as acridine orange or lysosensor also have shown that NAADP uniquely among Ca^{2+} mobilizing messengers also causes the alkalinization of store lumena, representing another possible signaling mechanism for this molecule (Morgan and Galione 2007b).

Mammalian Cells

Following these studies in sea urchin eggs, it was shown that NAADP also targeted acidic stores in a wide range of mammalian cells and in response to a variety of cellular stimuli (Mitchell et al. 2003; Kinnear et al. 2004; Yamasaki et al. 2004; Galione 2006; Gerasimenko et al. 2006; Menteyne et al. 2006; Zhang et al. 2006; Macgregor et al. 2007; Gambara et al. 2008; Jardin et al. 2008; Kim et al. 2008; Lloyd-Evans et al. 2008; Brailoiu et al. 2009b; Pandey et al. 2009; Thai et al. 2009; Dickinson et al. 2010; Zhang et al. 2010).

DESENSITIZATION OF NAADP-EVOKED Ca^{2+} RELEASE

The Ca^{2+} release mechanism activated by NAADP shows unusual and profound inactivation properties. One major area of confusion in this field is that the inactivation properties of Ca^{2+} release varies markedly between sea urchin egg and mammalian systems, which we have termed type I and type II, respectively (Morgan and Galione 2008) (Fig. 2).

Sea urchin eggs: The initial report demonstrating the efficacy of NAADP as a Ca^{2+} mobilizing molecule showed that NAADP released Ca^{2+} by a mechanism independent of IP_3 or ryanodine receptors (RyRs), based on each of these mechanisms showing homologous desensitization (Lee and Aarhus 1995). After NAADP stimulated Ca^{2+} release in egg homogenates, they became refractory to subsequent challenge with NAADP, but still responded to either IP_3 or cADPR. This was the first piece of evidence that NAADP activated a novel Ca^{2+} release channel, distinct from the principal Ca^{2+} release channels on the endoplasmic reticulum.

Further analysis of the phenomenon of self-inactivation of NAADP-evoked Ca^{2+} release in sea urchin eggs and homogenates revealed several profound and unusual features. A surprising finding was that pM concentrations of NAADP, although subthreshold for triggering Ca^{2+} release in egg homogenates, were able to inactivate completely the NAADP Ca^{2+} release mechanism to a subsequent challenge by nM concentration of NAADP that would normally evoke a maximal Ca^{2+} release (Aarhus et al. 1996; Genazzani et al. 1996). The extent of inactivation was dependent on both the concentration and duration of incubation (Genazzani et al. 1996; Genazzani et al. 1997b). Mechanisms of inactivation of the NAADP receptor are not understood, but may be related to the apparent irreversible binding of [^{32}P]NAADP. The radioligand appears to become occluded on binding in a time-dependent manner (Aarhus et al. 1996). Studies with the selective NAADP receptor antagonist, Ned-19, (Naylor et al. 2009) and its analogs (Rosen et al. 2009) have led to the proposal that there are two distinct binding sites for NAADP. The first is high affinity, whose occupancy leads to slow inactivation of the receptor, and a second lower affinity site that leads to rapid channel opening. Ned-20 blocks inactivation, but not activation of Ca^{2+} release by NAADP (Rosen et al. 2009).

Mammalian cells: There are key differences between desensitization of the NAADP receptor between sea urchin eggs, in which subthreshold concentrations of NAADP can fully inactivate the NAADP-sensitive Ca^{2+} release mechanism; whereas in a mammalian cell, high concentrations of NAADP are needed for full inactivation, which can occur in the apparent absence of receptor activation. The first report of NAADP action as a Ca^{2+}-mobilizing agent in a mammalian cell was in the pancreatic acinar cell (Cancela et al. 1999), which was also the system in which IP_3 was first shown to mobilize Ca^{2+} from nonmitochondrial stores (Streb et al. 1983). Using whole cell patch and measuring Ca^{2+}-activated currents, we found that a pipet concentration of 10 μM NAADP failed to elicit any responses. However, we noticed that after intracellular application of this concentration of NAADP, cholecystokinin (CCK), which usually increases cytosolic Ca^{2+} at pM

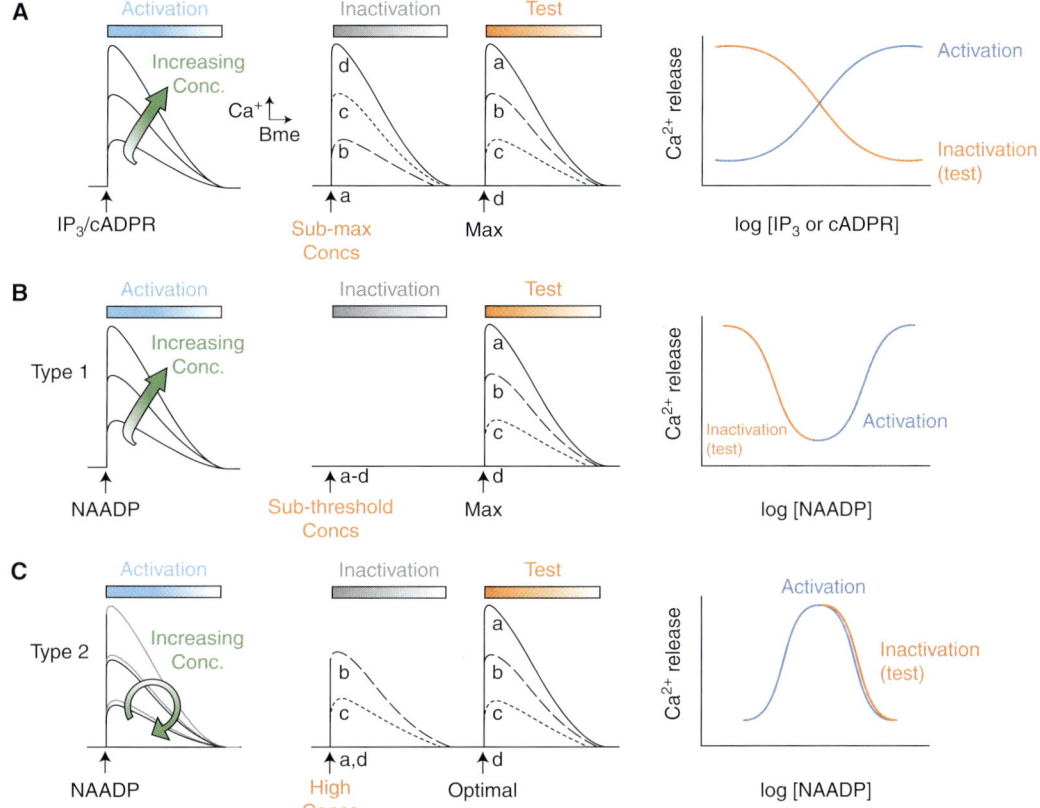

Figure 2. Differences between desensitization of mammalian and sea urchin NAADP receptors. Left-hand traces depict stylized Ca^{2+} dye fluorescence traces in response to messengers. Right-hand panels represent concentration-response graphs for activation or inactivation of intracellular Ca^{2+} release channels. (*A*) IP3 and cADPR-mediated Ca^{2+} release. Left-hand traces depict stylized Ca^{2+} dye fluorescence traces in response to various conditions. Right-hand panels represent concentration-response graphs for activation or inactivation. IP3 and cADPR demonstrate simple, monophasic concentration-response curves for activation (blue). Inactivation by ever-increasing submaximal concentrations (grey period) is revealed by the reciprocal diminution of a maximal "test" challenge (orange). (*B*) Desensitization of sea urchin NAADP receptors (Type I desensitization). The blue left-hand traces show stylized Ca^{2+} dye fluorescence traces in response to increasing concentrations of NAADP, which increases Ca^{2+} release, represented by a classical sigmoid log concentration-response curve (blue line, right-hand panel). However, preincubation with subthreshold concentrations of NAADP that do not evoke Ca^{2+} release desensitizes Ca^{2+} release in a time and concentration manner by subsequent challenge by a normally maximal NAADP (test) concentration (central panel, and orange curve, right panel). (*C*) Desensitization of mammalian NAADP receptors (Type 2 desensitization). Increasing concentrations of NAADP enhances Ca^{2+} release to a maximum (left and center panels). Thereafter, increasing concentrations of NAADP evoke progressively smaller Ca^{2+} release to a point when no Ca^{2+} release is evoked at high NAADP concentrations. This "bell-shaped" or hormetic log concentration-response curve is shown in the right-hand panel (blue curve). Modified from Morgan and Galione (2008).

concentrations by mobilizing Ca^{2+} from intracellular stores, now failed to evoke any response. We speculated that we may have inactivated the NAADP-evoked Ca^{2+} release mechanism that could be a key component of the CCK signal transduction mechanism. We therefore tried a range of NAADP concentrations and found that concentrations of NAADP as low as 50 nM in the pipet, elicited robust oscillatory responses, similar to those evoked by CCK in untreated cells. The concentration-response relationship for NAADP was found to be "bell-shaped," with maximal responses occurring at around 100 nM NAADP, whereas with concentrations in excess of 1 µM no effects were seen. Using caged NAADP, we showed that photolysis of this compound also evokes a series of spikes in Ca^{2+}-activated currents, which were suppressed in the presence of supramicromolar concentrations of free NAADP in the patch pipet. Bell-shaped concentration-response curves seem to be a major hallmark of mammalian NAADP-induced Ca^{2+} release. A subsequent study in a Jurkat T-cell line showed maximal Ca^{2+} release upon microinjection of around 100 nM NAADP. However, concentrations of >1 µM failed to elicit any response per se whilst inhibiting T-cell receptor activation (Berg et al. 2000). A number of further studies in different cell types used this phenomenon to implicate NAADP in the Ca^{2+} signal transduction pathways activated by various stimuli in the absence of selective NAADP antagonists at that time. These include glucose-evoked Ca^{2+} spiking in MIN6 cells (Masgrau et al. 2003), ET1-evoked Ca^{2+} release in pulmonary vascular smooth myocytes (Kinnear et al. 2004), and β_1 adrenoreceptor enhancement of Ca^{2+} signaling and contractility in ventricular cardiac myocytes (Macgregor et al. 2007).

PHARMACOLOGICAL PROPERTIES OF NAADP RECEPTORS

The pharmacology of NAADP-evoked Ca^{2+} release, initially investigated in sea urchin egg systems, showed major differences with the known Ca^{2+} release mechanisms in the ER. In egg homogenates, NAADP-evoked Ca^{2+} release was unaffected by the competitive IP_3R inhibitor, heparin, or by ryanodine or eight-substituted cADPR analogs that antagonize RyR-mediated Ca^{2+} release. An initial report that thio-NADP was a selective antagonist of NAADP (Chini et al. 1995) was subsequently explained by inactivation of the NAADP-sensitive Ca^{2+} release mechanism by traces of contaminating NAADP (Dickey et al. 1998).

A number of channel blockers were found to inhibit NAADP-evoked Ca^{2+} release selectively in sea urchin egg homogenates with little effect on either IP_3 or cADPR-mediated Ca^{2+} release (Genazzani et al. 1997a). These included voltage-gated Ca^{2+} channel (VGCC) blockers such as diltiazem, nifedipine, and D600 (although greater concentrations were required to block NAADP-evoked Ca^{2+} release than VGCCs), and purinoceptor antagonists such as aspyridoxal-phosphate-6-azophenyl-2',4'-disulfonate (PPADS) also display a degree of NAADP antagonism (Billington and Genazzani 2007). Because the NAADP receptor effectively discriminates between NAADP and NADP, which differs only by the substitution of a nicotinic acid moiety instead of nicotinamide, nicotinic acid analogs were developed that antagonize NAADP-induced Ca^{2+} release. These include CMA008 (Dowden et al. 2006) and BZ194 (Dammermann et al. 2009), which also have the advantage of being membrane permeant. Recently, a series of novel compounds have been identified by in silico screening strategies based on the three-dimensional shape and electrostatic properties of NAADP that are the most potent of NAADP antagonists developed so far (Naylor et al. 2009; Rosen et al. 2009). Ned-19, the founding member of these analogs, is becoming the most widely used antagonist on account of its potency, membrane permeability, and selectivity (Naylor et al. 2009; Rosen et al. 2009; Thai et al. 2009; Aley et al. 2010).

Interestingly, Ned-19 analogs have been used to dissect the activation and inactivation effects of NAADP at the sea urchin egg NAADP receptor (Rosen et al. 2009). Ned-20, which differs only from Ned-19 by the para rather than ortho position of a fluorine, blocks the inactivation of NAADP-sensitive Ca^{2+} release mechanism by subthreshold NAADP concentrations for Ca^{2+} release, without affecting NAADP-evoked Ca^{2+} release by higher NAADP concentrations and inhibits high affinity [^{32}P]NAADP binding to egg membranes (Rosen et al. 2009). These findings are consistent with multiple binding sites for the sea urchin egg NAADP receptor, with high affinity sites leading to inactivation and lower affinity sites leading to activation.

TWO-PORE CHANNELS

A family of novel intracellular channels termed two-pore channels (TPCs) have emerged as the leading candidates for NAADP-gated Ca^{2+} release channels. The founding member of this family, TPC1, was cloned in 2000 from a rat kidney cDNA library in a search for novel members of voltage-gated cation channels (Ishibashi et al. 2000). The putative channel had only a 20% homology with the transmembrane domains of the α subunit of voltage-gated Na^+ and Ca^{2+} channels, but the highest homology with a deposited sequence of a putative Ca^{2+} channel from the plant *Arabidopsis thaliana*. Subsequent analysis of the plant clone, AtTPC1, implicated a role for this protein in Ca^{2+} transport and signaling when expressed in yeast and *Arabidopsis* (Furuichi et al. 2001), and a role in germination and stomatal physiology as a component of the slow vacuolar channel (Peiter et al. 2005). The putative channel, rather than having four repeats of six transmembrane segments, as for voltage-gated Na^+ and Ca^{2+} channels, only has two. Thus in effect, the protein is the equivalent of half a Na^+ or Ca^{2+} channel, and may represent an ancestral form that has been duplicated later in evolution to give rise to the four domain channels (Fig. 3).

IDENTIFICATION OF TWO-PORE CHANNELS AS NAADP RECEPTORS

Two clues as to the candidature of TPCs as NAADP receptors emerged in the last few years. Michael Zhu, searching for novel TRP family membranes in around 2000, had cloned a second member of the TPC family, termed TPC2, and found that when heterologously expressed in HEK293 cells, it localized with the lysosomal marker, LAMP1. The second was the further analysis of AtTPC1 function by Sanders and colleagues, showing that AtTPC1 localized to plant vacuoles, the major plant acidic organelle and the functional equivalent of lysosomes in plants (Peiter et al. 2005). The localization of TPCs to acidic stores, and the partial pharmacological overlap of NAADP receptors with voltage-gated Ca^{2+} channels and TRP proteins, which show homologies with TPCs, made these proteins credible candidates as the elusive

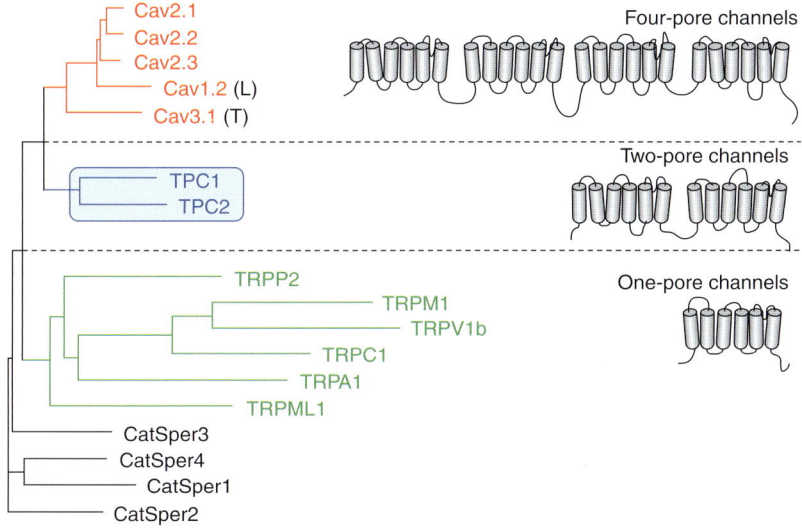

Figure 3. Phylogenetic tree for human two-pore channels and their relationship with voltage-gated Ca^{2+} channels and TRP members. It is likely that voltage-gated Ca^{2+} channels have arisen from two rounds of tandem duplication in evolution. Thus, TPCs having 12 transmembrane domains (12TM) may be considered ancient intermediate proteins between TRP channels (6TM), such as CatSpers in sperm or mucolipins or polycystins, and voltage-gated Ca^{2+} channels (24TM).

NAADP receptor. Over four years or so from 2005, we worked extensively with Zhu and collaborators, to test exhaustively the hypothesis that TPCs represented a family of NAADP-gated intracellular channels from several key standpoints. First, we examined the subcellular localization of the human TPC1 and TPC2 isoforms in HEK293 cells. In addition, because the genomes of many species, but not human or rodent, also express a third isoform, TPC3 (Cai and Patel 2010; Zhu et al. 2010), we also expressed the chicken TPC3 to examine its subcellular distribution in HEK293 cells (Calcraft et al. 2009). All three TPCs localize to the endolysosomal system with no apparent expression in Golgi, mitochondria, or ER. Only TPC2 consistently colocalized with the lysosomal marker, LAMP2, but not early or late endosomes. In contrast, TPC1 and TPC3 predominantly were expressed in endosomal and other unidentified compartments, but with only sparse colocalization with lysosomal markers. In HEK293 cells, TPCs are endogenously expressed at low levels, and endogenous TPC2 was also immunolocalized to lysosomes. Overexpression of the human HsTPC2 was associated with increased specific [^{32}P]NAADP binding to HEK293 cell membranes and immunoprecipitated TPC2 proteins. Both high and low affinity binding sites were manifest in membranes from TPC2-overexpressing cells, with K_d of 5 nM and 7 μM, remarkably similar to endogenous binding in membranes from mouse liver, a tissue which has a particularly high expression of TPCs. Photolysis of caged NAADP in patched wild-type HEK293 cells elicited a small Ca^{2+} response, whereas in cells stably overexpressing TPC2, a large biphasic Ca^{2+} response was evoked on NAADP uncaging or dialysis: an initial pacemaker-like ramp of Ca^{2+} was followed by a larger and faster transient Ca^{2+} release. Bafilomycin treatment abolished both phases of the Ca^{2+} response, whereas the IP$_3$R antagonist heparin blocked the second phase alone. This finding is consistent with the "trigger" hypothesis for a mode of NAADP action, whereby NAADP evokes a localized Ca^{2+} signal by mobilizing bafilomycin-sensitive acidic stores, which is then globalized by recruiting Ca^{2+}-induced Ca^{2+} release (CICR) from nearby ER, in this case by activating IP$_3$Rs. The concentration-response relationship between NAADP and Ca^{2+} release was of the characteristic bell-shape for NAADP in mammalian cells, with maximal Ca^{2+} release occurring at between 10 nM and 1 μM, whereas 1 mM was without effect. Importantly, shRNA against TPC2 completely abolished the response to NAADP. We also created $Tpc2^{-/-}$ mice, and found that NAADP-evoked activation of oscillatory Ca^{2+}-dependent cation currents in pancreatic β cells seen in wild-type cells, were abolished in cells from the knockout mice.

In contrast to TPC2, we found that HEK cells stably expressed with HsTPC1 evoked only a localized Ca^{2+} release in response to NAADP, which failed to globalize, as was the case for TPC2. One possibility is that the endosomal localization of TPC1 means that there is little close apposition with ER so that coupling with CICR channels is weaker. Two subsequent publications broadly confirmed our findings (Brailoiu et al. 2009a; Zong et al. 2009).

PROPERTIES OF ENDOGENOUS TPCs FROM SEA URCHIN EGGS

The properties of heterologously expressed mammalian TPCs made them strong candidates as NAADP receptors. However, most of the studies of NAADP-mediated Ca^{2+} release and [^{32}P]NAADP binding sites have been performed in sea urchin egg preparations, where the Ca^{2+} mobilizing effects of NAADP were first discovered. It was important to ascertain whether sea urchin eggs express TPCs and whether they functioned as NAADP receptors. Screening of the genome of the sea urchin *Strongylocentrotus purpuratus* revealed three TPC isoforms that were cloned from ovaries that displayed around 30% sequence homology between the isoforms (Brailoiu et al. 2010; Ruas et al. 2010). Importantly, immunoprecipitation of TPCs from solubilized egg membranes with polyclonal antibodies raised against each of the three TPC isoforms of TPCs produced immunocomplexes that specifically bound [^{32}P]NAADP with K_ds of around 1 nM. Binding of [^{32}P]NAADP to the immunocomplexes mirrored all the key features of binding to intact egg membranes, including K^+-dependent irreversibility and a similar binding selectivity for NAADP over NADP. These data provided compelling evidence that TPCs form complexes that can explain all the properties of [^{32}P]NAADP binding sites previously characterized from sea urchin egg preparations. As with their mammalian homologs, heterologous expression of the sea urchin TPC1 and TPC2 isoforms in HEK293 cells enhanced NAADP-evoked Ca^{2+} release from acidic Ca^{2+} stores, which was amplified by recruitment of IP$_3$Rs, although coupling between TPC1 and IP$_3$Rs appeared looser. In contrast, TPC3 actually suppressed the small NAADP-evoked response observed in control cells and also abolished the enhancement in cells stably transfected with TPC2 (Fig. 4). This effect of TPC3 is puzzling for several reasons. The effect of TPC3 cannot be accounted by a general dysregulation of acidic Ca^{2+} stores since measurement of both Ca^{2+} storage and luminal pH do not appear to be altered in cells overexpressing TPC3 expressing cells. Another possibility is that TPC3 has a dominant negative effect, perhaps by forming heterodimers, because it is likely, given the proposed structure of TPCs, that functional channels would form dimers. Indeed, homodimerization of human TPC2 has been reported (Zong et al. 2009), but given the differing subcellular localizations of each of the TPCs, at least when heterologously expressed, it is unclear whether heterodimerization can explain TPC3 suppression of NAADP-evoked Ca^{2+} release.

SINGLE-CHANNEL PROPERTIES OF HUMAN TPCs

Although TPCs are emerging as promising candidates as NAADP-gated Ca^{2+} release channels in the endolysosomal system, it is important to characterize their biophysical channel properties to show that they do indeed function in this way. However, their localization in organelles presents several problems because they are not readily amenable for electrophysiological analysis as for channels resident at the plasma membrane, and there is no evidence at present that they cycle to the plasma membrane as for other Ca^{2+} release channels (Taylor et al. 2009). The traditional way of studying organellar channels is their reconstitution into artificial bilayers for single channel analysis, as exemplified for IP$_3$R (Ehrlich and Watras 1988) and RyR (Lai et al. 1988) single-channel studies; although for ER channels, nuclear envelope patching has gained increasing popularity (Mak and Foskett 1997). In a preliminary report, immunopurified human TPC2 was reconstituted into lipid bilayers and shown to form

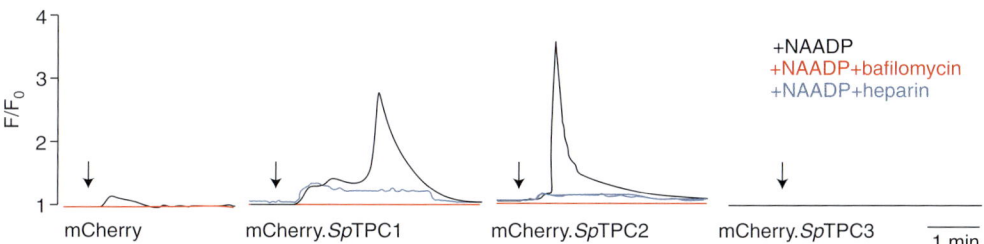

Figure 4. NAADP-mediated Ca^{2+} release in HEK293 cells expressing each of the three sea urchin TPC isoforms. Representative Ca^{2+} traces of cells dialyzed with NAADP (100 nM) and fura-2 via patch pipette in whole-cell configuration, in absence or presence of bafilomycin A1 (1 μM) or the IP$_3$R antagonist, heparin (200 μg/ml). Arrows indicate break-in. In wild-type cells, only a small endogenous response to NAADP was seen. In SpTPC1 and SpTPC2 cells, NAADP-evoked biphasic responses. The first component was from acidic stores (bafilomycin-sensitive), whereas the second phase, which requires the first to trigger it, is caused by recruitment of IP$_3$Rs (heparin-sensitive). TPC3 expression suppresses the endogenous response. Models for NAADP-triggered Ca^{2+} responses, based on interaction between different organelles (circle, lysosome and network, ER) are also shown above each series of traces. Modified from Ruas et al. (2010).

NAADP-gated cation conductances (Pitt et al. 2010). Channels were generally silent until application of NAADP to the *cis* or cytoplasmic face of the bilayer, and the channels showed a selectivity for cations with conductances of around 300 pS and 15 pS for K^+ and Ca^{2+} ions as the conducting species. Interestingly, NAADP sensitivity may be regulated by store filling with Ca^{2+}, because NAADP sensitivity was markedly dependent on *trans* or luminal Ca^{2+}, with the EC_{50} for NAADP-evoked enhancement of open probability decreasing from 500 nM to 5 nM as luminal Ca^{2+} increased to 200 μM. This is in the range of reported luminal free Ca^{2+} levels in lysosomes (Christensen et al. 2002; Lloyd-Evans et al. 2008). Thus, fluctuations in luminal Ca^{2+} because of cycles of release and uptake of Ca^{2+} could be important determinants of the effects of NAADP on Ca^{2+} release, and offers one explanation for how constant NAADP levels may elicit trains of Ca^{2+} spikes, as widely observed in various cell types (Cancela et al. 1999). Another variable is luminal pH of acidic stores, since NAADP has also been found to alkalinize acidic stores in sea urchin eggs and homogenates (Morgan and Galione 2007a; Morgan and Galione 2007b), and it is possible that luminal pH has significant effects on TPC2 channel properties. Importantly, the NAADP antagonist was also found to block single channel TPC2 currents (Pitt et al. 2010). However, it should be stressed here that although the immunopurified TPC complexes both form NAADP-gated Ca^{2+} channels (Pitt et al. 2010) and bind $[^{32}P]$NAADP (Calcraft et al. 2009; Ruas et al. 2010), the possibility remains that NAADP could interact with an accessory protein associated with TPCs instead of a direct interaction with TPC proteins themselves (Galione et al. 2009).

A single-channel analysis of NAADP-gated channels has also been performed from lysosomal enriched fractions derived from liver (Zhang and Li 2007) and bovine coronary vascular smooth muscle (Zhang et al. 2009). These channels conducted Cs^+ and were sensitive to NAADP, with open probabilities displaying a bell-shaped concentration dependence, and with maximum P_o occurring at 1 μM NAADP in both preparations. The pharmacology was consistent with previous studies of NAADP-evoked Ca^{2+} release, with block by VGCC antagonists, PPADS, and also amiloride. Interestingly, P_o was increased at acidic pH. In contrast to the situation in most mammalian cells examined so far, pretreatment with concentrations of NAADP as low as 0.5 nM blocked subsequent channel openings by higher NAADP concentrations, as seen for sea urchin egg receptors and in liver (Mandi et al. 2006). The identity of these channels were ascribed to mucolipin-1 (TRPML-1), a lysosomal TRP channel linked to the lysosomal storage disease, mucolipidosis IV, on the basis of a blocking effect of an anti-TPRML1 antibody and reduction of channel activity from cells treated with an siRNA TPRML1 construct. However, the identity of TRPML1 as an NAADP receptor candidate remains controversial (Pryor et al. 2006). In addition, a recent report suggests that NAADP may increase levels of a short variant of a TRPML2 transcript in lymphoid cells (Samie et al. 2009), underscoring the likely complex interactions between lysosomal channels.

INTERACTIONS OF NAADP AND OTHER Ca^{2+} SIGNALING PATHWAYS

NAADP-evoked Ca^{2+} release from lysosomes appears to be small and highly localized. Given the dynamic properties of these organelles, they are ideally suited to be targeted to the vicinity of Ca^{2+}-regulated effectors. Three modes of NAADP-mediated Ca^{2+} signaling mechanisms have been highlighted (Fig. 5).

NAADP and Lysosomal-ER Interactions

Organelle interactions in Ca^{2+}-signaling is not a new concept. For example, Ca^{2+}-microdomains may arise around sites of ER Ca^{2+} release, and neighboring organelles may be profoundly affected physiologically. Indeed ER-mitochondrial interactions have been well studied in the context of IP_3R and RyR-mediated Ca^{2+}-release (Rizzuto et al. 1998; Csordas et al. 2001), which impacts on mitochondrial metabolism and apoptotic pathways.

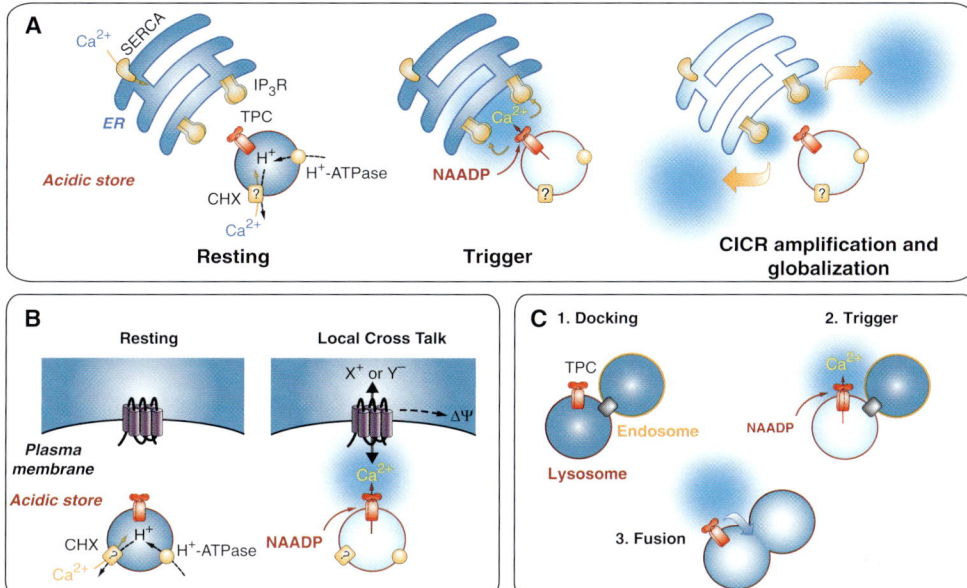

Figure 5. Three modes of NAADP-mediated Ca^{2+} signaling. (*A*) NAADP is a local trigger mechanism for detonating global CICR responses from the ER. (*B*) Local Ca^{2+} release by NAADP from acidic stores positioned under the plasma membrane may regulate membrane excitability (excitable cells) or ion fluxes (nonexcitable cells) by modulating Ca^{2+}-activated plasma membrane channels. (*C*) NAADP regulates local cytoplasmic Ca^{2+}/pH and luminal Ca^{2+}/pH in endolysosomal compartments that may regulate vesicular fusion of late endosomes/lysosomes. Modified from Galione et al. (2009) and Ruas et al. (2010).

The protein mitofusin 2 has been proposed to tether ER and mitochondrial membranes (de Brito and Scorrano 2008).

NAADP-evoked Ca^{2+} release and its effects on Ca^{2+}-release channels on the ER/SR was first noted in pancreatic acinar cells (Cancela et al. 1999). This phenomenon, whereby a localized microdomain of Ca^{2+} release from acidic stores triggers a larger release from the ER, is widely observed in both the sea urchin egg and in many types of mammalian cell, and is one of the fundamental principles of NAADP-mediated Ca^{2+} signaling. The trigger hypothesis was formulated by the finding that NAADP-evoked responses in pancreatic acinar cells could be blocked by either heparin or ryanodine, as well as self-inactivation of the NAADP receptor with NAADP itself (Cancela et al. 1999). This was visualized in the larger sea urchin egg by detailed imaging studies (Churchill and Galione 2000; Churchill and Galione 2001b; Churchill and Galione 2001a). NAADP was found to act as a local messenger to form Ca^{2+} gradients across the cell based on NAADP diffusion. These gradients could be amplified and globalized by CICR through the recruitment of IP_3R and RyR-dependent mechanisms on the ER. Because of the distinct self-inactivation properties of NAADP receptors, subsequent NAADP-evoked Ca^{2+}-signaling patterns only occur in regions of the cell where NAADP had not previously evoked a response (Churchill and Galione 2001b). This effect lasts for many minutes, representing a basic type of spatiotemporal memory in terms of the generation of Ca^{2+}-signal patterning. As well as spatial complexities, NAADP could produce temporal patterns in Ca^{2+} signals by the uptake of Ca^{2+} released from NAADP-sensitive stores into the ER to produce a series of Ca^{2+} spikes dependent on IP_3R and RyRs (Churchill and Galione 2001a).

In pulmonary vascular smooth muscle cells, NAADP and the vasoactive hormone, endothelin-1, evoke a localized Ca^{2+} release from lysosomes at lysosomal-SR junctions, which is then amplified and globalized by a mechanism dependent on recruitment of RyRs on the SR (Kinnear et al. 2004; Kinnear et al. 2008). Similar results have been reported in coronary smooth muscle myocytes (Zhang et al. 2006), and also implicated for early Fas signaling processes, which eventually lead to apoptosis (Zhang et al. 2010).

In Jurkat T cells, NAADP triggers Ca^{2+} release, which can be amplified by RyRs and IP$_3$Rs, but it has been proposed in this system that RyR1 may be the primary target of NAADP on the ER (Dammermann and Guse 2005; Dammermann et al. 2009). A role for RyR as the direct target for NAADP has also been proposed in pancreatic acinar cell ER/nuclear membranes, although other evidence points to direct activation of acidic stores (Yamasaki et al. 2004; Menteyne et al. 2006) followed by amplification by CICR. Such discrepancies are not surprising given the small release of Ca^{2+} released by lysosomes that TPC studies have revealed (Calcraft et al. 2009; Ruas et al. 2010), with amplification by ER mechanisms providing much larger Ca^{2+} signals. Thus in small cells, dissection of contributory Ca^{2+} release mechanisms can prove difficult (Galione and Petersen 2005), but employment of emerging molecular insights and tools may prove insightful (Galione et al. 2009).

Modulation of Plasma Membrane Excitability

As well as their involvement in organelle communication, NAADP and TPCs appear to play an important role in regulating ion fluxes across the plasma membrane and hence also excitability of excitable cells. NAADP has been shown to stimulate Ca^{2+} influx across the plasma membrane of several cell types including starfish oocytes (Moccia et al. 2003; Moccia et al. 2006), sea urchin eggs (Churchill et al. 2003), where it uniquely among Ca^{2+} mobilizing messengers mediates the polyspermic blocking "cortical flash," and Jurkat T cells (Langhorst et al. 2004). What is not clear is whether NAADP directly activates plasma membrane channels or whether NAADP first releases Ca^{2+} from intracellular stores, which then leads to activation of plasma membrane conductances. Indeed, at present, there is no evidence for TPC localization at the plasma membrane.

However, local NAADP-evoked Ca^{2+} release from acidic stores in the vicinity of the plasma membrane has been shown in several cell types to open Ca^{2+}-activated ion channels. This was first shown in non-excitable pancreatic acinar cells, where activation of such channels is likely to contribute to fluid secretion (Cancela et al. 1999). However, this may be a major mechanism in excitable cells. In pancreatic β cells, NAADP also evokes Ca^{2+}-dependent currents, which may contribute to glucose-mediated depolarization of the cells during stimulus-secretion coupling (Naylor et al. 2009), and which are absent in cells derived from $Tpc2^{-/-}$ mice (Calcraft et al. 2009). In neurons from the rat medulla oblongata (Brailoiu et al. 2009b), NAADP also depolarizes cells through a mechanism dependent on Ca^{2+} release from acidic stores.

NAADP and its Receptors in Endolysosomal Physiology

NAADP may be unique among Ca^{2+} mobilizing messengers in that in contrast to IP$_3$ or cADPR, it may in most cases directly evoke Ca^{2+} release from the endolysosomal system. NAADP-regulated TPCs are a new group of channels that are targeted to the endolysosomal system, along with mucolipins (Dong et al. 2010), P2X$_4$ receptors (Qureshi et al. 2007), and TRPM2 (Lange et al. 2009), all of which are likely to influence the ionic environment in acidic organelles. Interestingly, TRPM2 channels have also been proposed to be NAADP receptors (Beck et al. 2006); however, they have much lower affinities for NAADP, in the high μM range. TRPM2 channels could provide local Ca^{2+} signals that may directly impinge on the pleiotropic roles of the endolysosomal system including lysosomal biogenesis, vesicular trafficking and transport, and autophagy. Both local and luminal Ca^{2+} is important for many of these processes including homotypic fusion processes of endosomes and heterotypic fusions of late endosomes with lysosomes, as well as condensation of luminal contents (Piper and Luzio 2004; Luzio et al. 2007); release of Ca^{2+} from endolysosomal stores is thought to be a crucial regulatory mechanism. Overexpression of TPCs in HEK293 causes profound changes in trafficking, lysosomal size, and distribution as observed in certain lysosomal storage diseases (Ruas et al. 2010). These effects can be ameliorated by treatment with the NAADP antagonist, Ned-19. These data are suggestive of a major role for NAADP and TPC proteins in the regulation of luminal Ca^{2+}, Ca^{2+} release, and local Ca^{2+} signaling in endolysosomal physiology, and are thus likely to be key regulators of trafficking, autophagy, and other functions of these organelles.

Conclusions: Why Have Multiple Messengers for Ca^{2+} Release?

Over the last decade or so, NAADP has joined IP_3 and cADPR as a major Ca^{2+} mobilizing messenger. A major question in Ca^{2+} signaling research is how ubiquitous Ca^{2+} signals can encode specificity. A general view is that the complex spatial and temporal patterns of Ca^{2+} signals widely observed in cells are key to understanding this problem. The coordination of Ca^{2+} signals by multiple messengers acting at differentially distributed target Ca^{2+} release channels with different properties offers one possible solution. For example, NAADP-evoked Ca^{2+} release leads to neuronal cell differentiation (Brailoiu et al. 2006), whereas cADPR-mediated Ca^{2+} release leads to cell proliferation, but delays differentiation (Yue et al. 2009). On the other hand, activation of certain cell surface receptors may produce different combinations of messengers that are required to mimic the specific Ca^{2+} signaling patterns evoked by the particular receptor agonist (Cancela et al. 2002; Yamasaki et al. 2005), thus increasing the repertoire of cellular responses mediated by Ca^{2+}.

The emerging view that NAADP directly targets acidic stores rather than the ER is an important new principle in Ca^{2+} signaling and cellular homeostasis, and allows NAADP to evoke distinct Ca^{2+} signals from those directly mobilizing Ca^{2+} from the ER. This was initially proposed on the basis of pharmacological studies, but the identification of endolysosomal TPC proteins as major targets for NAADP has begun to cement this hypothesis in molecular terms. Three major consequences of NAADP-evoked Ca^{2+} release have been identified. The unifying principle is that NAADP, by mobilizing acidic stores, leads to localized Ca^{2+} signals that may trigger key cellular responses. Depending on the subcellular localization of these stores, there are fundamentally different consequences of NAADP-mediated Ca^{2+} release. For stores proximal to the plasma membrane, Ca^{2+}-activated plasma channels may be activated. Such ion fluxes produced in nonexcitable cells may, for example, be important in fluid secretion. In excitable cells, depolarization and changes in membrane excitability may result. For stores apposed to the ER, NAADP-evoked Ca^{2+} release from acidic stores may trigger globalized Ca^{2+} responses by activating IP_3Rs or RyRs by CICR. The third major aspect is the regulation of luminal Ca^{2+} and pH, as well as local Ca^{2+} signals in the endolysosomal system that may have a major impact on the many roles of these organelles in key cellular processes that they control, including vesicular trafficking, autophagy, apoptosis, and autolysis, as well as their role in fighting infection. Cellular stimuli may be selectively coupled to NAADP signaling pathways, or as is commonly observed, to multiple messenger pathways, either providing distinct patterns of Ca^{2+} signals leading to specific responses.

The establishment of a role of the endolysosomal system in Ca^{2+} signaling, the identification of specific Ca^{2+} release channels of acidic organelles as the targets for NAADP, open up new possibilities for a better understanding of the mechanisms of cellular Ca^{2+} signaling and how this goes awry in disease, and its control and pharmacological manipulation.

ACKNOWLEDGMENTS

This work in AG's laboratory is funded by the Wellcome Trust, and AG is a Principal Investigator of the British Heart Foundation Centre of Research Excellence at the University of Oxford.

I would like to thank Dr. Anthony Morgan for his helpful discussion and help with preparing the figures.

REFERENCES

Aarhus R, Dickey DM, Graeff RM, Gee KR, Walseth TF, Lee HC. 1996. Activation and inactivation of Ca^{2+} release by $NAADP^+$. *J Biol Chem* **271:** 8513–8516.

Aley PK, Noh HJ, Gao X, Tica AA, Brailoiu E, Churchill GC. 2010. A functional role for nicotinic acid adenine dinucleotide phosphate (NAADP) in oxytocin-mediated contraction of uterine smooth muscle from rat. *J Pharmacol Exp Ther* **333:** 726–735.

Beck A, Kolisek M, Bagley LA, Fleig A, Penner R. 2006. Nicotinic acid adenine dinucleotide phosphate and cyclic ADP-ribose regulate TRPM2 channels in T lymphocytes. *FASEB J* **20:** 962–964.

Berg I, Potter BV, Mayr GW, Guse AH. 2000. Nicotinic acid adenine dinucleotide phosphate ($NAADP^+$) is an essential regulator of T-lymphocyte Ca^{2+}-signaling. *J Cell Biol* **150:** 581–588.

Berridge MJ, Bootman MD, Lipp P. 1998. Calcium–a life and death signal. *Nature* **395:** 645–648.

Berridge MJ, Bootman MD, Roderick HL. 2003. Calcium: Calcium signalling: dynamics, homeostasis and remodelling. *Nat Rev Mol Cell Biol* **4:** 517–529.

Billington RA, Genazzani AA. 2007. PPADS is a reversible competitive antagonist of the NAADP receptor. *Cell Calcium* **41:** 505–511.

Bootman MD, Berridge MJ, Roderick HL. 2002. Calcium signalling: More messengers, more channels, more complexity. *Curr Biol* **12:** R563–565.

Brailoiu GC, Brailoiu E, Parkesh R, Galione A, Churchill GC, Patel S, Dun NJ. 2009b. NAADP-mediated channel 'chatter' in neurons of the rat medulla oblongata. *Biochem J* **419:** 91–97.

Brailoiu E, Churamani D, Cai X, Schrlau MG, Brailoiu GC, Gao X, Hooper R, Boulware MJ, Dun NJ, Marchant JS, et al. 2009a. Essential requirement for two-pore channel 1 in NAADP-mediated calcium signaling. *J Cell Biol* **186:** 201–209.

Brailoiu E, Churamani D, Pandey V, Brailoiu GC, Tuluc F, Patel S, Dun NJ. 2006. Messenger-specific role for nicotinic acid adenine dinucleotide phosphate in neuronal differentiation. *J Biol Chem* **281:** 15923–15928.

Brailoiu E, Hooper R, Cai X, Brailoiu GC, Keebler MV, Dun NJ, Marchant JS, Patel S. 2010. An ancestral deuterostome family of two-pore channels mediates nicotinic acid adenine dinucleotide phosphate-dependent calcium release from acidic organelles. *J Biol Chem* **285:** 2897–2901.

Cai X, Patel S. 2010. Degeneration of an intracellular ion channel in the primate lineage by relaxation of selective constraints. *Mol Biol Evol* (in press).

Calcraft PJ, Ruas M, Pan Z, Cheng X, Arredouani A, Hao X, Tang J, Rietdorf K, Teboul L, Chuang KT, et al. 2009. NAADP mobilizes calcium from acidic organelles through two-pore channels. *Nature* **459:** 596–600.

Cancela JM, Churchill GC, Galione A. 1999. Coordination of agonist-induced Ca^{2+}-signalling patterns by NAADP in pancreatic acinar cells. *Nature* **398:** 74–76.

Cancela JM, Van Coppenolle F, Galione A, Tepikin AV, Petersen OH. 2002. Transformation of local Ca^{2+} spikes to global Ca^{2+} transients: The combinatorial roles of multiple Ca^{2+} releasing messengers. *EMBO J* **21:** 909–919.

Chini EN, Beers KW, Dousa TP. 1995. Nicotinate adenine dinucleotide phosphate (NAADP) triggers a specific calcium release system in sea urchin eggs. *J Biol Chem* **270:** 3216–3223.

Christensen KA, Myers JT, Swanson JA. 2002. pH-dependent regulation of lysosomal calcium in macrophages. *J Cell Sci* **115:** 599–607.

Churchill GC, Galione A. 2000. Spatial control of Ca^{2+} signaling by nicotinic acid adenine dinucleotide phosphate diffusion and gradients. *J Biol Chem* **275:** 38687–38692.

Churchill GC, Galione A. 2001a. NAADP induces Ca^{2+} oscillations via a two-pool mechanism by priming IP3- and cADPR-sensitive Ca^{2+} stores. *EMBO J* **20:** 2666–2671.

Churchill GC, Galione A. 2001b. Prolonged inactivation of nicotinic acid adenine dinucleotide phosphate-induced Ca^{2+} release mediates a spatiotemporal Ca^{2+} memory. *J Biol Chem* **276:** 11223–11225.

Churchill GC, Okada Y, Thomas JM, Genazzani AA, Patel S, Galione A. 2002. NAADP mobilizes Ca^{2+} from reserve granules, lysosome-related organelles, in sea urchin eggs. *Cell* **111:** 703–708.

Churchill GC, O'Neill JS, Masgrau R, Patel S, Thomas JM, Genazzani AA, Galione A. 2003. Sperm deliver a new second messenger: NAADP. *Curr Biol* **13:** 125–128.

Clapham DE. 1995. Calcium signaling. *Cell* **80:** 259–268.

Clapper DL, Walseth TF, Dargie PJ, Lee HC. 1987. Pyridine nucleotide metabolites stimulate calcium release from sea urchin egg microsomes desensitized to inositol trisphosphate. *J Biol Chem* **262:** 9561–9568.

Csordas G, Thomas AP, Hajnoczky G. 2001. Calcium signal transmission between ryanodine receptors and mitochondria in cardiac muscle. *Trends Cardiovasc Med* **11:** 269–275.

Dammermann W, Guse AH. 2005. Functional ryanodine receptor expression is required for NAADP-mediated local Ca^{2+} signaling in T-lymphocytes. *J Biol Chem* **280:** 21394–21399.

Dammermann W, Zhang B, Nebel M, Cordiglieri C, Odoardi F, Kirchberger T, Kawakami N, Dowden J, Schmid F, Dornmair K, et al. 2009. NAADP-mediated Ca^{2+} signaling via type 1 ryanodine receptor in T cells revealed by a synthetic NAADP antagonist. *Proc Natl Acad Sci* **106:** 10678–10683.

de Brito OM, Scorrano L. 2008. Mitofusin 2 tethers endoplasmic reticulum to mitochondria. *Nature* **456:** 605–610.

Dickey DM, Aarhus R, Walseth TF, Lee HC. 1998. Thio-NADP is not an antagonist of NAADP. *Cell Biochem Biophys* **28:** 63–73.

Dickinson GD, Churchill GC, Brailoiu E, Patel S. 2010. Deviant NAADP-mediated Ca^{2+}-signalling upon lysosome proliferation. *J Biol Chem* **285:** 13321–13325.

Dong XP, Wang X, Xu H. 2010. TRP channels of intracellular membranes. *J Neurochem* **113**: 313–328.

Dowden J, Berridge G, Moreau C, Yamasaki M, Churchill GC, Potter BV, Galione A. 2006. Cell-permeant small-molecule modulators of NAADP-mediated Ca^{2+} release. *Chem Biol* **13**: 659–665.

Ehrlich BE, Watras J. 1988. Inositol 1,4,5-trisphosphate activates a channel from smooth muscle sarcoplasmic reticulum. *Nature* **336**: 583–586.

Furuichi T, Cunningham KW, Muto S. 2001. A putative two pore channel AtTPC1 mediates Ca^{2+} flux in Arabidopsis leaf cells. *Plant Cell Physiol* **42**: 900–905.

Galione A. 2006. NAADP, a new intracellular messenger that mobilizes Ca^{2+} from acidic stores. *Biochem Soc Trans* **34**: 922–926.

Galione A. 2008. Investigating cADPR and NAADP in intact and broken cell preparations. *Methods* **46**: 194–203.

Galione A, Petersen OH. 2005. The NAADP receptor: New receptors or new regulation? *Mol Interv* **5**: 73–79.

Galione A, Evans AM, Ma J, Parrington J, Arredouani A, Cheng X, Zhu MX. 2009. The acid test: The discovery of two-pore channels (TPCs) as NAADP-gated endolysosomal Ca^{2+} release channels. *Pflugers Arch* **458**: 869–876.

Gambara G, Billington RA, Debidda M, D'Alessio A, Palombi F, Ziparo E, Genazzani AA, Filippini A. 2008. NAADP-induced Ca^{2+} signaling in response to endothelin is via the receptor subtype B and requires the integrity of lipid rafts/caveolae. *J Cell Physiol* **216**: 396–404.

Gasser A, Bruhn S, Guse AH. 2006. Second messenger function of nicotinic acid adenine dinucleotide phosphate revealed by an improved enzymatic cycling assay. *J Biol Chem* **281**: 16906–16913.

Genazzani AA, Galione A. 1996. Nicotinic acid-adenine dinucleotide phosphate mobilizes Ca^{2+} from a thapsigargin-insensitive pool. *Biochem J* **315**: 721–725.

Genazzani AA, Empson RM, Galione A. 1996. Unique inactivation properties of NAADP-sensitive Ca^{2+} release. *J Biol Chem* **271**: 11599–11602.

Genazzani AA, Mezna M, Dickey DM, Michelangeli F, Walseth TF, Galione A. 1997a. Pharmacological properties of the Ca^{2+}-release mechanism sensitive to NAADP in the sea urchin egg. *Br J Pharmacol* **121**: 1489–1495.

Genazzani AA, Mezna M, Summerhill RJ, Galione A, Michelangeli F. 1997b. Kinetic properties of nicotinic acid adenine dinucleotide phosphate-induced Ca^{2+} release. *J Biol Chem* **272**: 7669–7675.

Gerasimenko JV, Flowerdew SE, Voronina SG, Sukhomlin TK, Tepikin AV, Petersen OH, Gerasimenko OV. 2006. Bile acids induce Ca^{2+} release from both the endoplasmic reticulum and acidic intracellular calcium stores through activation of inositol trisphosphate receptors and ryanodine receptors. *J Biol Chem* **281**: 40154–40163.

Graeff R, Lee HC. 2002. A novel cycling assay for nicotinic acid-adenine dinucleotide phosphate with nanomolar sensitivity. *Biochem J* **367**: 163–168.

Ishibashi K, Suzuki M, Imai M. 2000. Molecular cloning of a novel form (two-repeat) protein related to voltage-gated sodium and calcium channels. *Biochem Biophys Res Commun* **270**: 370–376.

Jardin I, Lopez JJ, Pariente JA, Salido GM, Rosado JA. 2008. Intracellular calcium release from human platelets: Different messengers for multiple stores. *Trends Cardiovasc Med* **18**: 57–61.

Kim BJ, Park KH, Yim CY, Takasawa S, Okamoto H, Im MJ, Kim UH. 2008. Generation of nicotinic acid adenine dinucleotide phosphate and cyclic ADP-ribose by glucagon-like peptide-1 evokes Ca^{2+} signal that is essential for insulin secretion in mouse pancreatic islets. *Diabetes* **57**: 868–878.

Kinnear NP, Boittin FX, Thomas JM, Galione A, Evans AM. 2004. Lysosome-sarcoplasmic reticulum junctions. A trigger zone for calcium signaling by nicotinic acid adenine dinucleotide phosphate and endothelin-1. *J Biol Chem* **279**: 54319–54326.

Kinnear NP, Wyatt CN, Clark JH, Calcraft PJ, Fleischer S, Jeyakumar LH, Nixon GF, Evans AM. 2008. Lysosomes co-localize with ryanodine receptor subtype 3 to form a trigger zone for calcium signalling by NAADP in rat pulmonary arterial smooth muscle. *Cell Calcium* **44**: 190–201.

Lai FA, Erickson HP, Rousseau E, Liu QY, Meissner G. 1988. Purification and reconstitution of the calcium release channel from skeletal muscle. *Nature* **331**: 315–319.

Lange I, Yamamoto S, Partida-Sanchez S, Mori Y, Fleig A, Penner R. 2009. TRPM2 functions as a lysosomal Ca^{2+}-release channel in beta cells. *Sci Signal* **2**: pra23.

Langhorst MF, Schwarzmann N, Guse AH. 2004. Ca2+ release via ryanodine receptors and Ca^{2+} entry: Major mechanisms in NAADP-mediated Ca^{2+} signaling in T-lymphocytes. *Cell Signal* **16**: 1283–1289.

Lee HC, Aarhus R. 1995. A derivative of NADP mobilizes calcium stores insensitive to inositol trisphosphate and cyclic ADP-ribose. *J Biol Chem* **270**: 2152–2157.

Lee HC, Aarhus R. 2000. Functional visualization of the separate but interacting calcium stores sensitive to NAADP and cyclic ADP-ribose. *J Cell Sci* **113**: 4413–4420.

Lee HC, Walseth TF, Bratt GT, Hayes RN, Clapper DL. 1989. Structural determination of a cyclic metabolite of NAD with intracellular calcium-mobilizing activity. *J Biol Chem* **264**: 1608–1615.

Lewis AM, Masgrau R, Vasudevan SR, Yamasaki M, O'Neill JS, Garnham C, James K, Macdonald A, Ziegler M, Galione A, et al. 2007. Refinement of a radioreceptor binding assay for nicotinic acid adenine dinucleotide phosphate. *Anal Biochem* **371**: 26–36.

Lloyd-Evans E, Morgan AJ, He X, Smith DA, Elliot-Smith E, Sillence DJ, Churchill GC, Schuchman EH, Galione A, Platt FM. 2008. Niemann-Pick disease type C1 is a sphingosine storage disease that causes deregulation of lysosomal calcium. *Nat Med* **14**: 1247–1255.

Luzio JP, Bright NA, Pryor PR. 2007. The role of calcium and other ions in sorting and delivery in the late endocytic pathway. *Biochem Soc Trans* **35**: 1088–1091.

Macgregor A, Yamasaki M, Rakovic S, Sanders L, Parkesh R, Churchill GC, Galione A, Terrar DA. 2007. NAADP controls cross-talk between distinct Ca^{2+} stores in the heart. *J Biol Chem* **282**: 15302–15311.

Mak DO, Foskett JK. 1997. Single-channel kinetics, inactivation, and spatial distribution of inositol trisphosphate (IP3) receptors in Xenopus oocyte nucleus. *J Gen Physiol* **109**: 571–587.

Mandi M, Toth B, Timar G, Bak J. 2006. Ca^{2+} release triggered by NAADP in hepatocyte microsomes. *Biochem J* **395:** 233–238.

Masgrau R, Churchill GC, Morgan AJ, Ashcroft SJ, Galione A. 2003. NAADP: A new second messenger for glucose-induced Ca^{2+} responses in clonal pancreatic beta cells. *Curr Biol* **13:** 247–251.

Menteyne A, Burdakov D, Charpentier G, Petersen OH, Cancela JM. 2006. Generation of specific Ca^{2+} signals from Ca^{2+} stores and endocytosis by differential coupling to messengers. *Curr Biol* **16:** 1931–1937.

Mitchell KJ, Lai FA, Rutter GA. 2003. Ryanodine receptor type I and nicotinic acid adenine dinucleotide phosphate receptors mediate Ca^{2+} release from insulin-containing vesicles in living pancreatic beta-cells (MIN6). *J Biol Chem* **278:** 11057–11064.

Moccia F, Billington RA, Santella L. 2006. Pharmacological characterization of NAADP-induced Ca^{2+} signals in starfish oocytes. *Biochem Biophys Res Commun* **348:** 329–336.

Moccia F, Lim D, Nusco GA, Ercolano E, Santella L. 2003. NAADP activates a Ca^{2+} current that is dependent on F-actin cytoskeleton. *FASEB J* **17:** 1907–1909.

Morgan AJ, Galione A. 2007a. Fertilization and nicotinic acid adenine dinucleotide phosphate induce pH changes in acidic Ca^{2+} stores in sea urchin eggs. *J Biol Chem* **282:** 37730–37737.

Morgan AJ, Galione A. 2007b. NAADP induces pH changes in the lumen of acidic Ca^{2+} stores. *Biochem J* **402:** 301–310.

Naylor E, Arredouani A, Vasudevan SR, Lewis AM, Parkesh R, Mizote A, Rosen D, Thomas JM, Izumi M, Ganesan A, et al. 2009. Identification of a chemical probe for NAADP by virtual screening. *Nat Chem Biol* **5:** 220–226.

Pandey V, Chuang CC, Lewis AM, Aley PK, Brailoiu E, Dun NJ, Churchill GC, Patel S. 2009. Recruitment of NAADP-sensitive acidic Ca^{2+} stores by glutamate. *Biochem J* **422:** 503–512.

Patel S, Docampo R. 2010. Acidic calcium stores open for business: Expanding the potential for intracellular Ca^{2+} signaling. *Trends Cell Biol* **20:** 277–286.

Peiter E, Maathuis FJ, Mills LN, Knight H, Pelloux J, Hetherington AM, Sanders D. 2005. The vacuolar Ca^{2+}-activated channel TPC1 regulates germination and stomatal movement. *Nature* **434:** 404–408.

Piper RC, Luzio JP. 2004. CUPpling calcium to lysosomal biogenesis. *Trends Cell Biol* **14:** 471–473.

Pitt SJ, Funnell T, Sitsapesan M, Venturi E, Rietdorf K, Ruas M, Ganesan A, Gosain R, Churchill GC, Zhu MX, et al. 2010. TPC2 is a novel NAADP-sensitive Ca^{2+}-release channel, operating as a dual sensor of luminal pH and Ca^{2+}. *J Biol Chem* M110.156927[pii]10.1074/jbc.M110.156927 (in press).

Pozzan T, Rizzuto R, Volpe P, Meldolesi J. 1994. Molecular and cellular physiology of intracellular calcium stores. *Physiological Reviews* **74:** 595–636.

Pryor PR, Reimann F, Gribble FM, Luzio JP. 2006. Mucolipin-1 is a lysosomal membrane protein required for intracellular lactosylceramide traffic. *Traffic* **7:** 1388–1398.

Qureshi OS, Paramasivam A, Yu JC, Murrell-Lagnado RD. 2007. Regulation of P2X4 receptors by lysosomal targeting, glycan protection and exocytosis. *J Cell Sci* **120:** 3838–3849.

Rizzuto R, Pinton P, Carrington W, Fay FS, Fogarty KE, Lifshitz LM, Tuft RA, Pozzan T. 1998. Close contacts with the endoplasmic reticulum as determinants of mitochondrial Ca^{2+} responses. *Science* **280:** 1763–1766.

Rosen D, Lewis AM, Mizote A, Thomas JM, Aley PK, Vasudevan SR, Parkesh R, Galione A, Izumi M, Ganesan A, et al. 2009. Analogues of the nicotinic acid adenine dinucleotide phosphate (NAADP) antagonist Ned-19 indicate two binding sites on the NAADP receptor. *J Biol Chem* **284:** 34930–34934.

Ruas M, Rietdorf K, Arredouani A, Davis LC, Lloyd-Evans E, Koegel H, Funnell TM, Morgan AJ, Ward JA, Watanabe K, et al. 2010. Purified TPC isoforms form NAADP receptors with distinct roles for Ca^{2+} signaling and endolysosomal trafficking. *Curr Biol* **20:** 703–709.

Rutter GA. 2003. Calcium signalling: NAADP comes out of the shadows. *Biochem J* **373:** e3–4.

Samie MA, Grimm C, Evans JA, Curcio-Morelli C, Heller S, Slaugenhaupt SA, Cuajungco MP. 2009. The tissue-specific expression of TRPML2 (MCOLN-2) gene is influenced by the presence of TRPML1. *Pflugers Arch* **459:** 79–91.

Streb H, Irvine RF, Berridge MJ, Schulz I. 1983. Release of Ca^{2+} from a nonmitochondrial intracellular store in pancreatic acinar cells by inositol-1,4,5-triphosphate. *Nature* **306:** 67–69.

Taylor CW, Prole DL, Rahman T. 2009. Ca^{2+} channels on the move. *Biochemistry* **48:** 12062–12080.

Thai TL, Churchill GC, Arendshorst WJ. 2009. NAADP receptors mediate calcium signaling stimulated by endothelin-1 and norepinephrine in renal afferent arterioles. *Am J Physiol Renal Physiol* **297:** F510–516.

Yamasaki M, Masgrau R, Morgan AJ, Churchill GC, Patel S, Ashcroft SJ, Galione A. 2004. Organelle selection determines agonist-specific Ca^{2+} signals in pancreatic acinar and beta cells. *J Biol Chem* **279:** 7234–7240.

Yamasaki M, Thomas JM, Churchill GC, Garnham C, Lewis AM, Cancela JM, Patel S, Galione A. 2005. Role of NAADP and cADPR in the induction and maintenance of agonist-evoked Ca^{2+} spiking in mouse pancreatic acinar cells. *Curr Biol* **15:** 874–878.

Yue J, Wei W, Lam CM, Zhao YJ, Dong M, Zhang LR, Zhang LH, Lee HC. 2009. CD38/cADPR/Ca^{2+} pathway promotes cell proliferation and delays nerve growth factor-induced differentiation in PC12 cells. *J Biol Chem* **284:** 29335–29342.

Zhang F, Li PL. 2007. Reconstitution and characterization of a nicotinic acid adenine dinucleotide phosphate (NAADP)-sensitive Ca^{2+} release channel from liver lysosomes of rats. *J Biol Chem* **282:** 25259–25269.

Zhang F, Xia M, Li PL. 2010. Lysosome-dependent Ca^{2+} release respsone to fas activation in coronary arterial myocytes through NAADP: evidence from cd38 gene knockouts. *Am J Physiol Cell Physiol* **298:** C1209–C1216.

Zhang F, Jin S, Yi F, Li PL. 2009. TRP-ML1 functions as a lysosomal NAADP-sensitive Ca^{2+} release channel in coronary arterial myocytes. *J Cell Mol Med* **13:** 3174–3185.

Zhang F, Zhang G, Zhang AY, Koeberl MJ, Wallander E, Li PL. 2006. Production of NAADP and its role in Ca^{2+} mobilization associated with lysosomes in coronary arterial myocytes. *Am J Physiol Heart Circ Physiol* **291:** H274–282.

Zhu MX, Ma J, Parrington J, Galione A, Mark Evans A. 2010. TPCs: Endolysosomal channels for Ca^{2+} mobilization from acidic organelles triggered by NAADP. *FEBS Lett* **584:** 1966–1974.

Zong X, Schieder M, Cuny H, Fenske S, Gruner C, Rotzer K, Griesbeck O, Harz H, Biel M, Wahl-Schott C. 2009. The two-pore channel TPCN2 mediates NAADP-dependent Ca^{2+}-release from lysosomal stores. *Pflugers Arch* **458:** 891–899.

Calcium Oscillations

Geneviève Dupont[1], Laurent Combettes[2], Gary S. Bird[3], and James W. Putney[3]

[1]Unité de Chronobiologie Théorique, Université Libre de Bruxelles, Faculté des Sciences CP231, B-1050 Brussels, Belgium

[2]Institut National de la Santé et de la Recherche Médicale Unité U442, Université de Paris-Sud, 91405 Orsay, France

[3]National Institute of Environmental Health Sciences–NIH, Department of Health and Human Services, Research Triangle Park, North Carolina 27709

Correspondence: susanh@bcm.edu

Calcium signaling results from a complex interplay between activation and inactivation of intracellular and extracellular calcium permeable channels. This complexity is obvious from the pattern of calcium signals observed with modest, physiological concentrations of calcium-mobilizing agonists, which typically present as sequential regenerative discharges of stored calcium, a process referred to as calcium oscillations. In this review, we discuss recent advances in understanding the underlying mechanism of calcium oscillations through the power of mathematical modeling. We also summarize recent findings on the role of calcium entry through store-operated channels in sustaining calcium oscillations and in the mechanism by which calcium oscillations couple to downstream effectors.

Calcium ions participate in a multiplicity of physiological and pathological functions. Among the most intensely studied, and the major focus of this article, is the role of Ca^{2+} as a cellular signal. Elevations in cytoplasmic Ca^{2+} mediate a plethora of cellular responses, ranging from extremely rapid events (muscle contraction, neurosecretion), to slower more subtle responses (cell division, differentiation, apoptosis). In contrast to most cellular signals, it is a relatively simple matter to observe changes in cytoplasmic Ca^{2+} in real time in living cells. As a result, the truly complex nature of Ca^{2+} signaling pathways has been revealed. The challenge is to understand what regulates these signals and what the biological significance of their complexity is.

In the majority of laboratory experiments examining effects of various stimulants on Ca^{2+} signaling, supramaximal concentrations of activating agonists are employed resulting in rapid, robust, and often sustained increases in cytoplasmic Ca^{2+}. It has long been appreciated that these signals result from a coordinated release of intracellular stores and increased Ca^{2+} influx across the plasma membrane (Bohr, 1973; Putney et al. 1981). The intracellular release of Ca^{2+} most commonly results from the Ca^{2+} releasing action of the phospholipase C-derived second messenger, inositol 1,4,5-trisphosphate

(InsP$_3$) (Streb et al. 1983), whereas the entry of Ca^{2+} is because of the activation of store-operated channels in the plasma membrane (Putney 1986). However, it is becoming increasingly clear that these large sustained elevations seldom occur with physiological levels of stimulants. Rather the more common pattern of Ca^{2+} signaling, in both excitable and nonexcitable cells is a pattern of periodic discharges and/or entry of Ca^{2+}. In excitable cells, such as the heart for example, these may be comprised of, or initiated by regenerative all-or-none plasma membrane channel activation, the Ca^{2+} action potential (Tsien et al. 1986) with amplification by intracellular Ca^{2+} release (Fabiato 1983). In nonexcitable cells, these spikes of cytoplasmic Ca^{2+} arise from regenerative discharge of stored Ca^{2+}, a process generally termed Ca^{2+} oscillations (Prince and Berridge 1973; Woods et al. 1986). Like Ca^{2+} action potentials, these all-or-none discharges of Ca^{2+} represent a form of excitable behavior of the intracellular Ca^{2+} release signaling mechanism. However, because it is not possible to easily monitor and control the transmembrane chemical and biophysical parameters, as is the case for excitable plasma membrane behavior, it has been more difficult to fully understand the basic mechanisms by which these Ca^{2+} oscillations arise. Thus, although the question has been exhaustively studied for well over twenty years, there is still uncertainty and controversy over the underlying processes that give rise to Ca^{2+} oscillations. A number of reviews have discussed these issues at some length (Berridge and Galione 1988; Rink and Jacob 1989; Berridge 1990; Petersen and Wakui 1990; Berridge 1991; Cuthbertson and Cobbold 1991; Meyer and Stryer 1991; Hellman et al. 1992; Tepikin and Petersen 1992; Thomas et al. 1992; Dupont and Goldbeter 1993; Keizer 1993; Sneyd et al. 1994; Li et al. 1995; Thomas et al. 1996; Shuttleworth 1999; Lewis 2003; Dupont et al. 2007). In the current treatment, we have chosen to focus on two important aspects of Ca^{2+} oscillations. First, we review the available evidence for various computational models of Ca^{2+} oscillations that employ a quantitative approach to validate or repudiate specific mechanisms. Second, we consider the interrelationship between Ca^{2+} oscillations and plasma membrane Ca^{2+} influx mechanisms, with the view that we may learn more of the physiological function that these intracellular discharges of Ca^{2+} provide.

COMPUTATIONAL MODELS FOR Ca^{2+} OSCILLATIONS

Since the first observations of Ca^{2+} oscillations, the experimental investigation of their molecular mechanism has been accompanied by numerous modeling approaches. One of the reasons for this is that rhythmic phenomena are known to rely on specific, nonlinear feedback processes that cannot readily be fully approached by intuitive and qualitative reasoning. Likewise, cAMP oscillations, circadian rhythms, cell cycle-related variations of the activity of cyclin associated kinases or the tumor-associated p53/mdm2 loop are other oscillatory phenomena in biology whose investigation largely benefits from a modeling approach (Goldbeter 2008). In the field of Ca^{2+} dynamics, modeling was also promoted by the fact that cytosolic Ca^{2+} was initially the only measurable variable of the system. This also prompted investigators to use modeling to identify the main messenger responsible for intra- or intercellular wave propagation when those spatially organized phenomena were reported in a variety of cell types (Thomas et al. 1996; Dupont et al. 2007). More recently, sub-cellular Ca^{2+} increases because of the opening of a small number of Ca^{2+} channels have also been simulated computationally. In this case, models offer the possibility to make the link between properties of the channels measured with electrophysiology and their behavior in the cytoplasm. In addition, simulations are required to understand how the regularity that is observed at the whole cell level (oscillations and waves) can emerge from the random behavior inherent to any ion channel.

One of the most attractive features about models is their ability to make experimentally testable predictions. That Ca^{2+} oscillations can occur in the presence of a constant level of InsP$_3$ and that the self-amplification of Ca^{2+} release from the ER into the cytoplasm lies at the basis of Ca^{2+} oscillations was for example first predicted theoretically (Goldbeter et al. 1990). This regulation is known as Ca^{2+}-induced Ca^{2+} release (CICR). However, this early model assumed that Ca^{2+} oscillations require the existence of 2 types of pools, some sensitive to $InsP_3$ and others possessing RyR and thereby sensitive to Ca^{2+}. This turned out not to be necessary as the $InsP_3R$ is itself sensitive to both Ca^{2+} and $InsP_3$.

In the past 20 years, many computational models have been developed. They differ by the precise oscillations that they aim to describe, as determined both by the cell type and by the agonist. Models also vary according to the level of description, from microscopic (in which case stochastic modeling must be used) to macroscopic (corresponding to a deterministic description). Another, somewhat intermediate level of description allowing one to take easily spatial and stochastic aspects into account is known as "threshold models" (see below). In each case, several models differing by the specific underlying assumptions have been proposed. Such a classification is presented in Table 1, together with some typical models in each category. Although in this review we focus on $InsP_3$- regulated Ca^{2+} oscillations, we have also indicated in Table 1 models for Ca^{2+} oscillations because of plasma-membrane voltage-gated Ca^{2+} channels or ryanodine receptors. Such models should be kept in mind when investigating $InsP_3$-induced Ca^{2+} increases, especially in view of the fact that they could provide a secondary oscillatory mechanism in some cell types. For example, in pancreatic acinar cells (Ventura and Sneyd 2006) or airway smooth muscle cells (Wang et al. 2008); the interplay between $InsP_3R$ and RyR plays a major physiological role.

Models for Ca^{2+} Oscillations With or Without $InsP_3$ Oscillations

As the $InsP_3R$ is biphasically regulated by Ca^{2+} (Bezprozvanny et al. 1991; Finch et al. 1991), models that describe the changes of states of this receptor can on their own account for Ca^{2+} oscillations. Such $InsP_3R$-based models are numerous and mainly differ by their level of details, the most detailed being useful to dissect the quantitative effect of possible changes in kinetic constants and affinities. The simplest ones allow for a better understanding of the essence of the observed phenomena, as exemplified by the early CICR model mentioned above (Goldbeter et al. 1990). As another example, simplified models point to the fact that for oscillations to occur, activation by Ca^{2+} has to be much faster than inhibition. Some of these $InsP_3R$-based models are indicated in Table 1 (class: $InsP_3R$, constant $InsP_3$) and readers should turn to (Sneyd and Falcke 2005) for a more comprehensive review of this type of model. All of these models face a common problem: the period of Ca^{2+} oscillations is imposed by the time taken by the receptor to recover from Ca^{2+}-induced inhibition. However, this rate constant has been estimated experimentally to be of the order of approximately 10 seconds in vitro (Finch et al. 1991; Combettes et al. 1994) and of a few seconds in vivo (Fraiman et al. 2006). These time delays are significantly shorter than the observed periods of Ca^{2+} oscillations that frequently exceed 1 minute. Thus, modeling here brings to light a clearly important limitation in our understanding of the mechanism of Ca^{2+} oscillations.

Such a discrepancy could either be explained by the existence of an additional control of the $InsP_3R$ activity, such as agonist-induced PKA-dependent phosphorylation (LeBeau et al. 1999) or $InsP_3$- induced inactivation (Hajnóczky and Thomas 1994), or suggest that although CICR at the level of the $InsP_3R$ has the potential to generate oscillations, it is not the main oscillatory mechanism in vivo. Periodic variations in the concentration of $InsP_3$ could indeed also drive Ca^{2+} oscillations, in which case the period would be imposed by the rates of $InsP_3$ metabolism (right column in Table 1). This assumption was in fact at the basis of some early models for agonist-induced Ca^{2+} oscillations (Meyer and Stryer 1988; Cuthbertson and Chay 1991). In both models, cytosolic Ca^{2+} and $InsP_3$ act as cross-catalytic messengers because $InsP_3$ triggers Ca^{2+} release from the ER, which in turn activates $InsP_3$ synthesis because

Table 1. Schematic classification of the main types of computational models for intracellular Ca^{2+} oscillations.

	Channels-based models				$InsP_3$ metabolism-based models
			$InsP_3R$		
	Voltage-gated	RyR	Constant $InsP_3$	Passive $InsP_3$ oscillations	Active $InsP_3$ oscillations
Deterministic	Fioretti et al. 2005; Zeng et al. 2009	Keizer and Levine 1996; Tang and Othmer 1994	De Young and Keizer 1992; Li and Rinzel 1994; Atri et al. 1993; Dupont and Swillens 1996; Tang et al. 1996; Bezprozvanny 1994	Dupont and Erneux 1997	Meyer and Stryer 1988; Cuthbertson and Chay 1991
				De Pitta et al. 2009; Kummer et al. 2000; Höfer et al. 2002	
Threshold			Dawson et al. 1999; Coombes et al. 2004; Thul et al. 2008		
Stochastic		Zahradnikova and Zahradnik 1996	Falcke 2004; Shuai et al. 2009; Williams et al. 2008		Kummer et al. 2005

In each category, only representative examples are indicated. Models indicated in the frame overlapping two columns are based on two distinct oscillatory mechanisms, one based on the $InsP_3R$ and one based on $InsP_3$ metabolism.

PLC is assumed to be stimulated by physiological levels of Ca^{2+}. This regulation of PLC activity by Ca^{2+} in the 0.1–1 μM range has been observed for the δ (Allen et al. 1997) and ζ isoforms (Kouchi et al. 2004) but not for the β (Renard et al. 1987), which is the isoform coupled to the G-protein pathway that is most frequently associated with hormone stimulation. However, these models differ in their assumptions about the nature of the negative feedback process required to switch off the Ca^{2+} rise. Meyer and Stryer (Meyer and Stryer 1988) assumed that Ca^{2+} started to decrease in the cytoplasm because of rapid pumping into the mitochondria, whereas Cuthbertson and Chay (Cuthbertson and Chay 1991) presumed that Ca^{2+}-activated PKC would down-regulate the G-proteins transducing receptor stimulation to PLC activation. Models for Ca^{2+} oscillations induced by $InsP_3$ oscillations fell somewhat into disfavor when the biphasic Ca^{2+} sensitivity of the $InsP_3R$ was discovered. Still, some models drew attention to the fact that oscillations in $InsP_3$ could occur because of the activation of $InsP_3$ catabolism by Ca^{2+}. As one of the $InsP_3$- metabolizing enzymes, the inositol 1,4,5-trisphosphate 3-kinase, is stimulated by the Ca^{2+}/calmodulin complex, each peak in Ca^{2+} induces a decrease in $InsP_3$ (Dupont and Erneux 1997). Models taking this regulation into account predict the concomitant occurrence of $InsP_3$ and Ca^{2+} oscillations. Interestingly, they also predict that these passive $InsP_3$ oscillations do not significantly affect the timing of the Ca^{2+} spikes (Dupont and Erneux 1997; Dupont et al. 2003; Tanimura et al. 2009).

$InsP_3$ oscillations were reported in some cell types by monitoring the translocation of green fluorescent protein (GFP) tagged to the pleckstrin homology (PH) domain of PLC, thus reawakening interest for models based on oscillatory production of InsP$_3$ (Taylor and Thorn 2001). Interestingly, concomitant oscillations of both InsP$_3$ and Ca^{2+} have been mostly observed in cell lines expressing the mGluR5 receptor. These glutamate-induced Ca^{2+} oscillations have unusual characteristics: most importantly, oscillations occur over a wide range of agonist concentration, their frequency is practically insensitive to the level of stimulation, and they are inhibited by PKC inhibitors. It is thus plausible that depending on the receptor type, different oscillatory mechanisms would prevail. In recent models for such types of oscillations, both oscillatory mechanisms (i.e., InsP$_3$R-based and InsP$_3$ metabolism-based) are considered at the same time (Kummer et al. 2000; Hofer et al. 2002; De Pitta M. et al. 2009). This allows the period of InsP$_3$R-based Ca^{2+} oscillations to be controlled by the rate of InsP$_3$ synthesis, through the regulation of either PLC or PKC by Ca^{2+}.

Modeling has also been used to define experimental tests that could discriminate between an InsP$_3$R-based and an InsP$_3$ metabolism-based mechanism. Sneyd et al. (2006) proposed to perturb agonist-induced Ca^{2+} oscillations by the direct, artificial release of InsP$_3$ in the cytoplasm (flash photolysis of caged InsP$_3$). As shown in Figure 1, the pattern of the Ca^{2+} rise after such a spike very much depends on the underlying oscillatory mechanism. In a model where InsP$_3$ metabolism is at the basis of Ca^{2+} oscillations, this sudden increase in InsP$_3$ will provoke a delay in the occurrence of the next Ca^{2+} spike, which corresponds to the time required for the level of InsP$_3$ to go back to its normal range of concentrations during oscillatory cycles. Once this is done, the situation is similar to the prepulse one and no change in frequency is observed as clearly shown in the left panel of Figure 1. The situation is drastically different for Ca^{2+} oscillations occurring with a constant level of InsP$_3$ because of the sequential activation and inhibition of the InsP$_3$R. In the framework of such a mechanism, the frequency of Ca^{2+} oscillations directly depends on the (constant) level of InsP$_3$. Thus, a sudden increase in InsP$_3$ during agonist-induced Ca^{2+} oscillations provokes a transient rise in frequency (Fig. 1, right panel). The interspike interval then progressively decreases to the period of the unperturbed

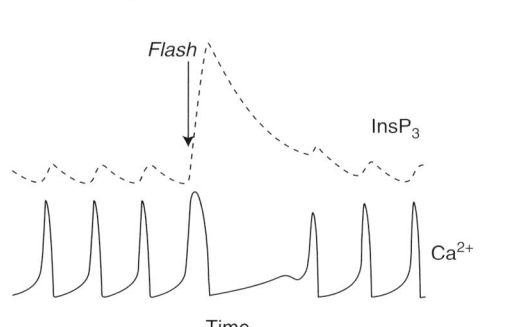
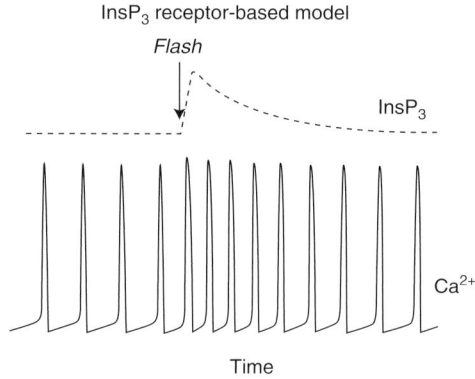

Figure 1. Schematic representation of the protocol proposed by Sneyd et al. (2006) to discriminate between an InsP$_3$R-based or an InsP$_3$ metabolism-based mechanism for Ca^{2+} oscillations. If phospholipase C (PLC) activity is stimulated by Ca^{2+} (*left* panel), oscillations in InsP$_3$ must accompany Ca^{2+} oscillations. If some InsP$_3$ is exogenously added during Ca^{2+} oscillations, it will delay the next Ca^{2+} spikes, without significant change in the frequency of Ca^{2+} oscillations. In contrast (*right* panel), if Ca^{2+} oscillations rely on successive cycles of activation/inhibition of the InsP$_3$ receptor (InsP$_3$R), Ca^{2+} oscillations can occur with a constant level of InsP$_3$. In this case, the addition of InsP$_3$ during oscillations will provoke a transient rise in the frequency of Ca^{2+} oscillations, with a progressive return to the original frequency. See text and Sneyd et al. (2006) for details.

system. The number of spikes necessary to the resettlement of the prepulse periodicity increases with the amount of InsP$_3$ released into the cell (Sneyd et al. [2006]; see also Chatton et al. [1998]).

To test this hypothesis, InsP$_3$ was released by flash photolysis in methacholine-stimulated pancreatic acinar cells and in carbachol-stimulated smooth muscle cells. In pancreatic acinar cells, liberation of InsP$_3$ during agonist-induced Ca^{2+} oscillations provoked a delay similar to the one shown in the left panel of Figure 1, suggesting that PLC activation by Ca^{2+} plays a predominant role in the oscillatory mechanism in this cell type. In contrast, the release of InsP$_3$ during methacholine-induced Ca^{2+} oscillations in airway smooth muscle cells provokes a transient acceleration of Ca^{2+} oscillations, as that seen in the right panel of Figure 1. It is thus concluded that in this cell type, Ca^{2+} oscillations rely on the InsP$_3$R dynamics. More recently, Swann and Yu (Swann and Yu 2008) have applied the same testing protocol to fertilization-induced Ca^{2+} oscillations in mouse eggs. As in panel A of Figure 1, the InsP$_3$ pulse induced an immediate Ca^{2+} spike followed by a delay longer than the period of oscillations before the next one. The return to the prepulse periodicity was straightforward. This is compatible with a mechanism whereby InsP$_3$ metabolism drives Ca^{2+} oscillation, in agreement with the fact that PLCζ, the PLC isoform triggering Ca^{2+} oscillations in mammalian eggs, is activated by physiological Ca^{2+} levels (Saunders et al. 2002; Dupont and Dumollard 2004).

A second indirect method to assess the involvement of InsP$_3$ dynamics in the core oscillatory mechanism has been tested in CHO cells (Politi et al. 2006). This method is based on the slowing down of InsP$_3$ dynamics with an InsP$_3$- binding protein that acts as an "InsP$_3$ buffer." As buffers change the kinetics but not the steady states, this compound would only affect Ca^{2+} oscillations relying on InsP$_3$ metabolism (as InsP$_3$R-based oscillations occur with a constant level of InsP$_3$). Moreover, the slowing-down of InsP$_3$ variations is assumed to affect intracellular dynamics only if InsP$_3$ oscillations rely on activation of InsP$_3$ synthesis by Ca^{2+} (PLC) and not stimulation of InsP$_3$ catabolism by Ca^{2+} (3-kinase). This protocol was applied in ATP-stimulated CHO cells transfected with an InsP$_3$- binding protein. These cells showed a dose-dependent quenching of Ca^{2+} oscillations, suggesting a PLC-based oscillatory mechanism in this cell type.

Models for Ca^{2+} Oscillations Taking Stochastic Aspects Into Account

All of the models discussed in the previous section are deterministic. This means that the effects of fluctuations (because of internal noise and microscopic inhomogeneities) are neglected. This is the common approach in modeling biochemical and chemical systems when the numbers of molecules involved in the process under interest are sufficiently large. In this case, stochastic fluctuations do not affect the average behavior of the ensemble, mainly because they statistically cancel each other out. For example, in electrophysiology, it is well known that the behavior of a few channels is random, but that the dynamics of neurons are very well described by deterministic equations of the Hodgkin-Huxley type.

An impressive number of studies have been devoted to the imaging of the Ca^{2+} releasing activity of a small number of InsP$_3$ receptors in vivo. These can occur either spontaneously or at submaximal InsP$_3$ concentrations. As expected, these events appear to be inherently stochastic. Their amplitude and the interval among them vary significantly under the same conditions, allowing only for a statistical description. The smallest observed events, called blips, involve a rise in cytosolic Ca^{2+} of about 40 nM and last in average 70 ms. These blips are believed to correspond to successive openings of a single InsP$_3$ receptor in the cytoplasm. This is possible because of the poor diffusing properties of Ca^{2+} in the cytoplasm allowing for rapid rebinding of Ca^{2+} on the channel activating sites, as indicated by quantitative models (Swillens et al. 1998). Alternatively, openings of one InsP$_3$R may trigger other InsP$_3$Rs within a cluster to generate slightly larger Ca^{2+} increases known as Ca^{2+} puffs. The rise in Ca^{2+} is then about 200 nM and lasts approximately 300 ms. Blips and puffs have been extensively studied in HeLa cells (Thomas et al. 2000; Bootman et al. 2002) and *Xenopus* oocytes (Marchant et al. 1999; Smith and Parker 2009). These elementary events have been modeled using a stochastic description of the dynamics of the InsP$_3$R (Swillens et al. 1999; Williams et al. 2008; Smith et al. 2009). These have for example allowed prediction of the approximate number of InsP$_3$Rs present in a cluster site (Swillens et al. 1999). As we will see below, this notion of clustering of InsP$_3$ receptors, necessary to explain the existence of puffs, has important implications in our understanding of global Ca^{2+} signals at the cell level.

In experiments, a rise in the level of InsP$_3$ transforms stochastic, elementary Ca^{2+} increases into regular, periodic Ca^{2+} increases propagating as waves in the cytoplasm. This transition would correspond to the fact that the rise in InsP$_3$ leads to an increase in the number of channels participating in the Ca^{2+} dynamics. Thus, the effect of fluctuations would become rather small, allowing the transition into a deterministic regime (i.e., a regime in which the behavior of the system can be predicted, as opposed to random processes). To test this hypothesis, we have performed a statistical analysis of the regularity of Ca^{2+} oscillations in noradrenaline-stimulated hepatocytes and found that the coefficient of variation of the period lies between 10% and 15%. Stochastic simulations taking into account realistic numbers of InsP$_3$Rs (about 6000 in a typical hepatocyte) accounted for such variability, if the receptors are assumed to be grouped in clusters of a few tens of channels (Dupont et al. 2008; Dupont and Combettes 2009). This supports the idea that repetitive Ca^{2+} spiking can be described as a deterministic oscillator. However, as the number of clusters is rather low, this oscillator is perturbed by noise leading to the 10-15% variation in the period. In agreement with this view, this coefficient of variation decreases with increasing levels of InsP$_3$, as the number of active channels increases. That this is the case both in the model and in hepatocytes is shown in Figure 2.

It has also been proposed that even at the cellular level, Ca^{2+} dynamics are intrinsically stochastic. The reason for that would lie in the spatial arrangement of the InsP$_3$Rs in clusters spaced from each other by a few microns. As Ca^{2+} is a poorly diffusible messenger in the cytoplasm, the Ca^{2+} rise occurring at one puff site would be unable to activate release from adjacent sites. This absence of communication would prevent global signaling. Thus, Ca^{2+} waves could be initiated only if, by chance, a sufficient number of clusters of InsP$_3$Rs become active at the same time. This process, called nucleation, would lead to a Ca^{2+} increase that is large enough to activate all the InsP$_3$- bound InsP$_3$Rs and generate a Ca^{2+} spike. In this type of modeling, the variation of the period is of the order of the period itself (Falcke 2004; Skupin et al. 2008). Interestingly, in this framework, spike initiation that corresponds to the time required for the concomitant opening of a few adjacent cluster sites can take very long as it is a random event. This would provide a possible explanation to the long periodicity of Ca^{2+} oscillations as compared with the characteristic kinetic parameters of InsP$_3$R dynamics (see above).

Stochastic simulations taking spatial aspects into account are computationally extremely expensive, and, in fact, can only be performed when assuming drastic simplifications. In this framework, "threshold" models can easily be used to simulate spatial propagation of Ca^{2+} waves while taking stochastic effects into account. Threshold models, sometimes referred to "fire-diffuse-fire" models (Dawson et al. 1999; Thul et al. 2008) are based on the concept of excitability. The idea is that the cell consists on a set of Ca^{2+} releasing sites spaced by region of the cytoplasm where Ca^{2+} can only be diffused or be pumped back in the ER. At each releasing site, release will occur when Ca^{2+} exceeds a threshold. These simulations lead to saltatory waves that much resemble experimental observations. Interestingly, the value of the threshold may be chosen to fluctuate to approximate the stochastic gating of receptors (Coombes et al. 2004). These studies allowed investigation of how noise can shape the dynamics of intracellular Ca^{2+} waves.

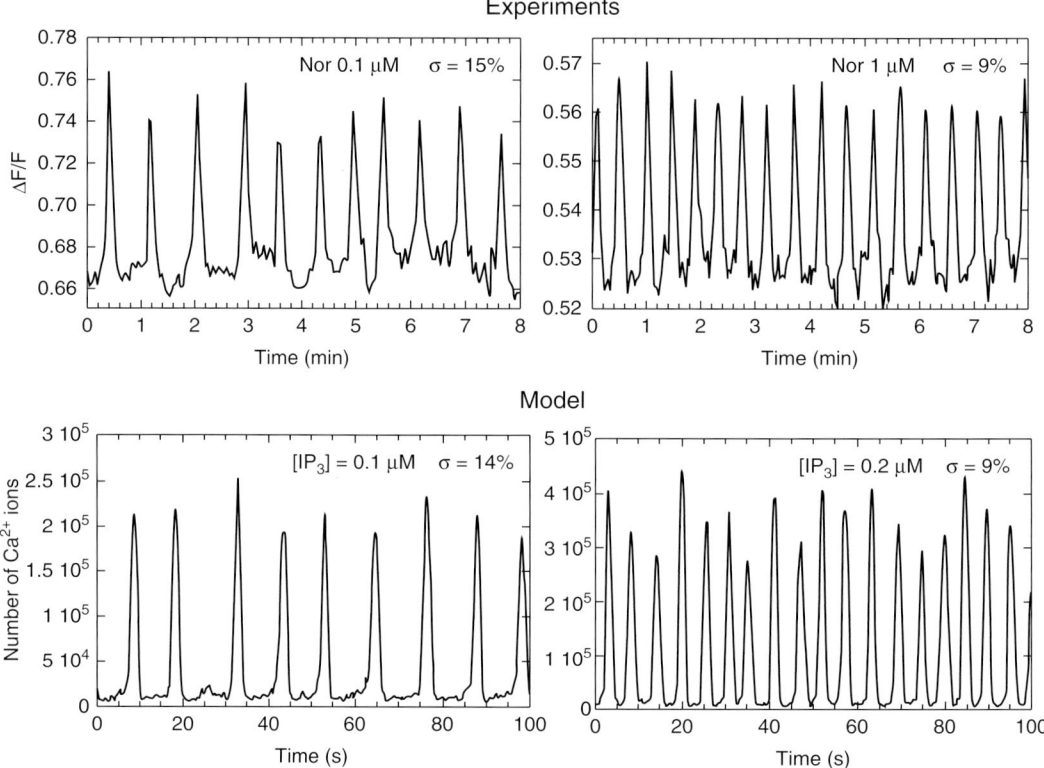

Figure 2. Comparison among Ca^{2+} oscillations observed in hepatocytes stimulated by noradrenaline (*upper* panels) and simulated by a stochastic Gillespie's algorithm taking into account realistic numbers of clusters of InsP$_3$ receptors (*lower* panels). The model is based on the assumption that Ca^{2+} dynamics occur in a deterministic regime, in which fluctuations are visible because of the rather low number of clusters. Both in the model and in experiments, the variability decreases when the frequency increases. See Dupont et al. (2008) for details.

Perspectives in Modeling Ca^{2+} Oscillations

The detailed characteristics of Ca^{2+} oscillations vary considerably from one cell type to another. In view of the large number of physiological responses mediated by Ca^{2+}, these changes may have significant implications. Modeling can be used to capture some detailed and quantitative understanding of this cell-to-cell variability. As an important factor, cells differ in their levels of expression of the three isoforms of the InsP$_3$R. This can be considered in models by simulating three distinct populations of channels, differing in their regulatory properties by Ca^{2+} and InsP$_3$. Simulations then point to the fact that modest changes in the regulatory properties of the InsP$_3$R can lead to significantly different oscillatory patterns, even if the bell-shaped dependence of the open probability on the level of Ca^{2+} are only slightly altered (Dupont and Combettes 2006). In particular, the model shows that the robustness of Ca^{2+} oscillations is clearly isoform-dependent, with type 2 being the most robust. This agrees with experiments wherein the levels of expression of the various InsP$_3$R subtypes have been genetically modified in DT40, HeLa, and COS-7 cells (Miyakawa et al. 1999; Hattori et al. 2004).

Surprisingly, many factors that are known to alter Ca^{2+} oscillations have not yet been extensively considered in models. For example, despite the large number of experimental studies devoted to the mechanism of Ca^{2+} entry (Putney and Bird 2008), most models for

Ca^{2+} oscillations in nonexcitable cells consider a closed system where Ca^{2+} exchanges are limited to fluxes between the cytoplasm and the ER. Mitochondrial Ca^{2+} handling is also known to alter cellular Ca^{2+} signals in the cytoplasm (Halestrap 2009). Although some models have been developed to explain the pumping and releasing properties of suspensions of mitochondria (Selivanov et al. 1998), their implication in intact cells have rarely been investigated theoretically (Marhl et al. 1998; Fall and Keizer 2001). As a last example, much remains to be performed in the field of intercellular Ca^{2+} wave propagation in which the oscillatory signal is coordinated at the organ level. Although quite well understood at the level of the communication among a few cells (Sneyd et al. 1995; Dupont et al. 2000; Hofer et al. 2002; Gracheva and Gunton 2003), signal transmission on large populations of cells remains puzzling. Such communication has vital implications as in the case of liver regeneration (Nicou et al. 2007) or in the brain (Haas et al. 2006).

INTERPLAY BETWEEN Ca^{2+} ENTRY AND Ca^{2+} RELEASE DURING Ca^{2+} OSCILLATIONS

The release of Ca^{2+} from intracellular stores, whether by maximal or submaximal concentrations of agonists, is generally accompanied by an increased influx of Ca^{2+} across the plasma membrane (Putney et al. 1981). Ca^{2+} oscillations run down in the absence of extracellular Ca^{2+}, suggesting a requirement for Ca^{2+} influx for their maintenance. However, at least in some nonexcitable cells, Ca^{2+} influx does not appear to be required to drive the oscillations. This can be shown by use of a technique that we have termed "lanthanide insulation" (Bird and Putney 2005). Relatively high concentrations (mM) of lanthanides (Gd^{3+}, La^{3+}) effectively inhibit both Ca^{2+} influx and Ca^{2+} extrusion at the plasma membrane (Van Breemen et al. 1972). Thus, in the presence of these high lanthanide concentrations, [Ca^{2+}]i signals are sustained, even in the absence of extracellular Ca^{2+} (Kwan et al. 1990; Bird and Putney 2005). This is also true for Ca^{2+} oscillations; lanthanide insulation permits sustained oscillations in the absence of extracellular Ca^{2+} (Sneyd et al. 2004; Bird and Putney 2005; Di Capite et al. 2009) (Fig. 3). This will be important in considering arguments about the physiological function of Ca^{2+} oscillations in a subsequent section.

In most cell types, especially in nonexcitable cells, release of store Ca^{2+} activates influx through store-operated Ca^{2+} (SOC) channels (Putney 1986; Parekh and Putney 2005). The most extensively studied and characterized

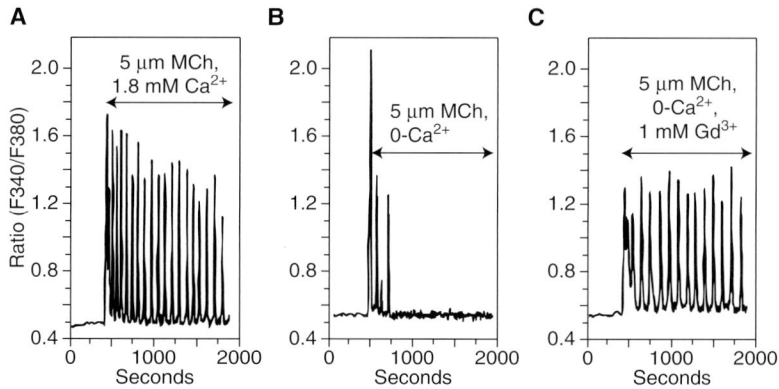

Figure 3. Lanthanide insulation renders Ca^{2+} oscillations independent of extracellular Ca^{2+}. (A) 5 μM methacholine (MCh) induces sustained oscillations in a HEK293 cell. (B) In the absence of extracellular Ca^{2+}, oscillations are not sustained. (C) In the presence of 1 mM Gd^{3+}, oscillations are sustained, even in the absence of extracellular Ca^{2+}. These panels illustrate responses of single cells. The statistical evaluation of multiple cells analyzed with this procedure is summarized in Bird and Putney (2005).

SOC current is the calcium-release-activated-calcium current (I_{crac}) (Hoth and Penner 1992; Parekh and Putney 2005). The channels underlying I_{crac} have thus been referred to as CRAC channels, which may represent a specific type of SOC channel. The properties of CRAC channels include high Ca^{2+} selectivity and very low single channel conductance (Parekh and Putney 2005). However, both store-operated Ca^{2+} fluxes (generally measured by use of fluorescent Ca^{2+} indicators) as well as I_{crac} have been most commonly investigated utilizing strategies that produce extensive, nearly complete depletion of endoplasmic Ca^{2+} stores, rather than under conditions of modest depletion as expected during Ca^{2+} oscillations. Indeed, the nature of the Ca^{2+} influx mechanism that supports Ca^{2+} oscillations has been the subject of some debate. Shuttleworth suggested that, rather than store operated channels, channels activated by arachidonic acid is necessary to maintain oscillations (Shuttleworth 1999). Such channels clearly exist and the Ca^{2+}-selective current underlying arachidonic acid-activated entry, termed I_{arc}, has been well characterized (Shuttleworth et al. 2004). The physiological function of these channels is less clear, however. The idea that ARC channels are involved in Ca^{2+} entry was based largely on pharmacological evidence utilizing a phospholipase A_2 inhibitor of questionable specificity (Bird and Putney 2005). On the other hand, use of Gd^{3+} (at low μM concentrations) and 2-aminoethyldiphenyl borate (2APB), two known inhibitors of SOC channels, caused rundown of muscarinic receptor-induced Ca^{2+} oscillations in HEK293 cells in a manner indistinguishable from that seen by simple omission of extracellular Ca^{2+} (Bird and Putney 2005).

Store-Operated Ca^{2+} Channels

The major molecular components of the SOC entry pathway are STIM (STIM1 and 2) and Orai (Orai1, 2, and 3) (Frischauf et al. 2008). STIM1 and 2 reside in the endoplasmic reticulum where they function as sensors of endoplasmic reticulum Ca^{2+} content. Both proteins have a Ca^{2+}-binding EF-hand motif in the N-terminus, directed toward the lumen of the endoplasmic reticulum (Dziadek and Johnstone 2007). A drop in endoplasmic reticulum Ca^{2+} content causes Ca^{2+} to dissociate from STIM1. This results in a conformational change (Zheng et al. 2008) that permits self-association of STIM molecules (Liou et al. 2007). In this poorly defined aggregated state, STIM localizes to near plasma membrane junctions (Orci et al. 2009) where it appears to be capable of directly interacting with Orai subunits of the store-operated channels (Park et al. 2009). There are also a number of publications implicating members of the TRPC (canonical transient receptor potential) cation channel family as candidate SOC channels (Birnbaumer et al. 2000; Rosado and Sage 2000; Abeele et al. 2003; Albert and Large 2003; Beech 2005; Ambudkar 2006; Huang et al. 2006). However, some of the published findings implicating TRPCs as SOC channels could not be reproduced (DeHaven et al. 2009). Nonetheless, it is clear that activation of phospholipase C activates Ca^{2+}-permeable TRPC channels (Vazquez et al. 2004) and it is thus possible that they will play a role in Ca^{2+} oscillations. In the HEK293 cell model, and consistent with the pharmacological data, knockdown by RNAi of either STIM1 or Orai1 in HEK293 essentially completely abrogated Ca^{2+} oscillations (Wedel et al. 2007). However, these proteins are also thought to play a role in the arachidonic acid pathway (Shuttleworth et al. 2007; Mignen et al. 2008). The role of STIM1 in ARC channel activation is quite different from that for SOC channels. ARC channels require STIM1 in the plasma membrane (Shuttleworth et al. 2004), whereas for the SOC channels, STIM1 is only necessary in the endoplasmic reticulum (Mercer et al. 2006). Thus, it is significant that loss of sustained oscillations following RNAi knockdown of STIM1 could be reversed by expression of a STIM1 construct modified to preclude its translocation to the plasma membrane (Wedel et al. 2007). In aggregate, the evidence convincingly supports the view that in the HEK293 cell model, the Ca^{2+} entry underlying Ca^{2+} oscillations is store-operated Ca^{2+} entry (Putney and Bird 2008).

During the process of Ca^{2+} oscillations, the quantity of Ca^{2+} discharged with each spike can be quite small (Bird and Putney 2005). Thus, it is presumed that when even a small amount of Ca^{2+} is lost from the endoplasmic reticulum, STIM is activated and causes Orai channels to open sufficiently to maintain intracellular Ca^{2+} stores. In this context, it is important to consider the sensitivities of the two STIM proteins, STIM1 and STIM2, to Ca^{2+} store depletion. When stores are gradually depleted of Ca^{2+}, STIM2 begins to move to the plasma membrane before STIM1 (Brandman et al. 2007; Bird et al. 2009). In addition, overexpression of STIM2, but not STIM1, results in constitutive Ca^{2+} entry (Soboloff et al. 2006b; Brandman et al. 2007; Parvez et al. 2008; Bird et al. 2009). These findings suggest that STIM2 is responsive to a very small degree of Ca^{2+} store depletion and is apparently partially active under resting conditions. STIM1 on the otherhand, appears to require a substantial degree of Ca^{2+} store depletion before it can be activated. It was thus somewhat surprising that when the relative roles of STIM1 and STIM2 in Ca^{2+} oscillations were assessed, it was STIM1 rather than STIM2 that was required (Bird et al. 2009).

In the HEK293 cell model used for these studies, Western analysis clearly showed the presence of STIM2 in quantities similar to those for STIM1 (Bird et al. 2009).Why then did the STIM2 play no significant role in supporting the Ca^{2+} oscillations? The answer is probably that STIM2 is a very inefficient activator of Orai channels. When both STIM2 and STIM1 are overexpressed in cells, STIM2 appears to act as an inhibitor of STIM1 (Soboloff et al. 2006a), as would be expected if its interaction with Orai channels is relatively ineffective. This should not be surprising if in fact STIM2 is already partially active under basal conditions. Nonetheless, when STIM2 is overexpressed in cells, along with overexpressed Orai1, large STIM2-dependent Ca^{2+} influx signals can be observed (Parvez et al. 2008). However, the stoichiometry of STIM2 and Orai1 is unknown in this condition, such that the relative efficiency of STIM2 as an activator of Orai1 cannot be evaluated. To compare the efficiency of STIM1 and STIM2 under conditions of similar stoichiometry with Orai1, the EF-hand mutants of STIM1 and STIM2 were transiently transfected into HEK293 cells, but without transfecting the cells with additional Orai1 (Bird et al. 2009). In this condition, STIM1 or STIM2 is presumed to be in considerable excess relative to the native Orai1. The EF-hand mutants of STIM1 and 2 are constitutively active, precluding the need for store depletion and thereby eliminating any contribution from endogenous STIM proteins. The result of this experiment was a robust constitutive entry of Ca^{2+} in the cells expressing the EF-hand mutant of STIM1, but no constitutive entry in cells expressing the EF-hand mutant of STIM2, consistent with the idea that STIM2 is a much poorer activator of Orai1 channels (Bird et al. 2009).

Thus, at least in HEK293 cells, the Ca^{2+} entry that supports Ca^{2+} oscillations is triggered by STIM1 interacting with and activating plasma membrane Orai1 channels. This suggests that during the oscillations, the endoplasmic reticulum Ca^{2+} concentration falls into the range in which STIM1 is activated. This presumably occurs in a small domain of the endoplasmic reticulum situated in close proximity to the plasma membrane, and this is consistent with earlier suggestions that there is a subcompartment of the endoplasmic reticulum that is specifically involved in regulating plasma membrane SOC channels (Ribeiro and Putney 1996; Parekh et al. 1997; Orci et al. 2009).

Physiological Relevance of SOC Channels for Cellular Signaling

The aforementioned scenario is relevant to a long standing question in the Ca^{2+} signaling field: what is the functional significance of Ca^{2+} oscillations? The generally accepted answer to this question is that Ca^{2+} oscillations provide a digital signal to downstream effectors and this is advantageous at low levels of signaling because of the high signal-to-noise discrimination of digital information encoding (Meyer et al. 1992; Thomas et al. 1996; Berridge 1997; Dolmetsch et al. 1998). Implicit in this conclusion is that the effectors downstream of the Ca^{2+} oscillations must have at least moderately elevated thresholds for responding, providing high signal-to-noise and preventing any ambiguity in distinguishing a true signal from random fluctuations. One well established target of Ca^{2+} released from the endoplasmic reticulum through $InsP_3$ receptors is the mitochondria (Rizzuto et al. 1992; Hajnóczky et al. 1995; Rizzuto et al. 1998; Rizzuto and Pozzan 2006). The mitochondrial uniporter, responsible for accumulation of Ca^{2+}, has a sensitivity to Ca^{2+} well above the physiological range of global Ca^{2+} changes (Blaustein et al. 1977; Burgess et al. 1983). Yet, the positioning of uptake sites close to sites of release, provides a kind of intracellular synapse assuring that bursts of local Ca^{2+} release are read by mitochondria, resulting in the activation of important Ca^{2+}-regulated mitochondrial enzymes (Csordas et al. 2006).

But in addition to mitochondrial function, Ca^{2+} signaling, most commonly through some kind of oscillatory mechanism, regulates a large variety of downstream effectors. There is evidence that some of these are tightly coupled to Ca^{2+} entering through SOC channels. One of the most extensively documented and investigated examples is the requirement for Ca^{2+} influx through CRAC channels for activation of calcineurin/NFAT signaling in T-lymphocytes (Oh-Hora 2009). Because activation of this pathway is known to require Ca^{2+} elevation for hours, it is implicitly understood that Ca^{2+} influx is necessary to maintain signaling over such a prolonged period. And it is clear that signaling fails in the absence of Ca^{2+} influx through CRAC channels (Feske et al. 2006). However, it is not clear whether Ca^{2+} entry provides the Ca^{2+} directly responsible for activation of calcineurin/NFAT signaling, or rather serves to maintain intracellular stores such that continued release through $InsP_3$ receptors can activate the pathway. There are a few studies that have experimentally shown the specific requirement of Ca^{2+} entering through SOC channels. Perhaps the first such Ca^{2+}-sensitive effectors found to be closely coupled to SOC entry were some of the Ca^{2+}-regulated adenylyl cyclases (Cooper et al. 1994). Another example elegantly showed in a study from Parekh's laboratory is cytoplasmic phospholipase A2 (cPLA2) (Chang et al. 2006), an important Ca^{2+}-regulated effector in many cells in the immune system and elsewhere. In the mast cell line RBL-1, Ca^{2+} store depletion activates plasma membrane CRAC channels, which leads to cPLA2 activation. The activation of cPLA2 was mediated by extracellular signal regulated kinases (ERKs). The ERKs are apparently activated by protein kinase C α, which is the likely target of Ca^{2+} entering through the CRAC channels (Chang et al. 2006). In this same study, it was shown that activation of the early gene, c-*fos*, was also closely coupled to Ca^{2+} entering through CRAC channels but was not dependent on ERK activation. The dependence of these responses specifically on Ca^{2+} entry was shown by use of the "lanthanide insulation" technique described above (Bird and Putney 2006). Thus, in the absence of extracellular Ca^{2+} but in the presence of high concentrations of lanthanides, depletion of Ca^{2+} stores with thapsigargin induced large sustained elevations of cytoplasmic Ca^{2+}, yet cPLA2 was not induced (Chang et al. 2008). This result indicated that the global rise in intracellular Ca^{2+} was not important for activation of gene expression, rather it was the Ca^{2+} specifically entering through the plasma membrane CRAC channels.

The above-discussed examples all used supramaximal stimuli to produce large, sustained increases in Ca^{2+} influx. In one study, the role of Ca^{2+} influx in activating gene expression during Ca^{2+} oscillations was investigated. Di Capite et al. (2009), following on the earlier demonstration of c-*fos* induction by SOC entry, examined the activation of c-*fos* expression in response to Ca^{2+} oscillations induced by modest concentrations of leukotriene C4 applied to RBL-1 cells. By use of the "lanthanide insulation" technique, oscillations were obtained in the absence of extracellular Ca^{2+}, but in the presence of high lanthanum and these oscillations were sustained and indistinguishable from those in the presence of extracellular Ca^{2+}. However, c-*fos* was only induced in the presence of extracellular Ca^{2+} when Ca^{2+} influx could occur through the CRAC channels. Thus, it may be concluded that in this instance the global rise in Ca^{2+} associated with Ca^{2+} oscillations is irrelevant for activation of the pathway that leads to c-*fos* activation. Rather, Ca^{2+} entering through the plasma membrane CRAC channels couples specifically to the initial steps in this signaling pathway (Di Capite et al. 2009).

The generality of this scenario remains to be shown. Indeed, gene expression can clearly be turned on in some experimental situations by elevations in global Ca^{2+} (Dolmetsch et al. 1998). Also, it is clear that Ca^{2+} released by IP_3Rs can also be coupled to specific signaling pathways. The clearest such example is the regulation of carbohydrate and energy metabolism through close IP_3R-mitochondrial coupling (Rizzuto and Pozzan 2006). Nonetheless, a specialized signaling function of SOC channels fits nicely with the previously related story of Ca^{2+} entry signaling by STIM1 during oscillations. If in fact Ca^{2+} entering through SOC channels can provide the key signal for downstream pathways, then it is not so surprising that STIM1 and not STIM2, provides the link between the oscillations and the SOC channels. It has for some time been supposed that the digital nature of cytoplasmic Ca^{2+} oscillations provides a high signal-to-noise signal input to downstream Ca^{2+}-regulated effectors, obviating the possibility of small unintentional signaling fluctuations. Because STIM1 requires a threshold of Ca^{2+} store depletion for activation, this then means that a Ca^{2+} oscillation, producing a transient but apparently substantial drop in ER luminal Ca^{2+}, will produce a transient activation of STIM1 and an incremental activation of the SOC channel. Small fluctuations in ER Ca^{2+} will not affect STIM1, preventing unintentional activation of SOC channels. However, small fluctuations in ER Ca^{2+} can activate STIM2 which will produce small activation of SOC channels sufficient to keep stores filled, but insufficient to activate downstream effectors. With this reasoning, the rise in cytoplasmic Ca^{2+} that is experimentally observed during Ca^{2+} oscillations is not directly relevant to the signaling pathway, except insofar as it reflects the digital drop in ER Ca^{2+} and subsequent activation of SOC channels. By similar reasoning, one would speculate that in some cell types, or in response to specific stimuli, SOC channels do not function to maintain intracellular stores so that $InsP_3$-induced release can activate downstream signaling; rather one might conclude that during Ca^{2+} oscillations, $InsP_3$-induced release functions to drive ER Ca^{2+} into the STIM1 sensing range, resulting in SOC channel opening and activation of downstream signaling. This would require not only spatial organization of the immediate downstream Ca^{2+} sensor, but also some degree of localized endoplasmic reticulum depletion because during a single Ca^{2+} oscillation, the extent of global endoplasmic reticulum depletion is quite small.

CONCLUSIONS

In this review, we have analyzed two major aspects of Ca^{2+} oscillations. First, we show how the use of computational models of Ca^{2+} oscillations can lead to a number of important conclusions about mechanisms of generation of oscillations. Models indeed allow us to conceptualize and quantify intuitive reasoning, which is particularly useful for oscillatory phenomena. We also emphasize the need for specific modeling approaches depending on the cell type, the stimulus and the specific aspect of Ca^{2+} signaling of interest. Second, we review recent findings indicating an important role for store-operated Ca^{2+} entry in Ca^{2+} oscillations and particularly in the physiological mechanism by which the digital information contained in the oscillations is linked to downstream effectors. Hopefully, continued research in these two areas will further increase our understanding of the mechanisms and meaning of the fascinating phenomena, Ca^{2+} oscillations.

ACKNOWLEDGMENTS

Work discussed in this review originating in the laboratory of JP and GB was supported by the Intramural Program, National Institutes of Health. GD is Maître de Recherche at the Belgian FNRS. GD acknowledges support from the Fonds de la Recherche Scientifique Médicale (grant No. 3.4636.04), the European Union through the Network of Excellence BioSim (Contract No. LSHB-CT-2004-005137), and the Belgian Program on Interuniversity Attraction Poles, initiated by the Belgian Federal Science Policy Office, project No.P6/25 (BIOMAGNET). LC acknowledges support from ANR (RPV07094LSA) and PNR in Hepatogastroenterology. LC is supported by an interface contract between Inserm and AP-HP (LE Kremlin-Bicetre Hospital). This work was supported by a PHC Tournesol 2009 program.

REFERENCES

Abeele FV, Shuba Y, Roudbaraki M, Lemonnier L, Vanoverberghe K, Mariot P, Skryma R, Prevarskaya N. 2003. Store-operated Ca^{2+} channels in prostate cancer epithelial cells: function, regulation, and role in carcinogenesis. *Cell Calcium* **33:** 357–373.

Albert AP, Large WA. 2003. Store-operated Ca^{2+}-permeable non-selective cation channels in smooth muscle cells. *Cell Calcium* **33:** 345–356.

Allen V, Swigart P, Cheung R, Cockcroft S, Katan M. 1997. Regulation of inositol lipid-specific phospholipase cδ by changes in Ca^{2+} ion concentrations. *Biochem J* **327:** 545–552.

Ambudkar IS. 2006. Ca^{2+} signaling microdomains: Platforms for the assembly and regulation of TRPC channels. *Trends Pharmacol Sci* **27:** 25–32.

Atri A, Amundson J, Clapham D, Sneyd J. 1993. A single-pool model for intracellular calcium oscillations and waves in the Xenopus laevis oocyte. *Biophys J* **65:** 1727–1739.

Beech DJ. 2005. Emerging functions of 10 types of TRP cationic channel in vascular smooth muscle. *Clin Exp Pharmacol Physiol* **32:** 597–603.

Berridge MJ. 1990. Calcium oscillations. *J Biol Chem* **265:** 9583–9586.

Berridge MJ. 1991. Cytoplasmic calcium oscillations: A two pool model. *Cell Calcium* **12:** 63–72.

Berridge MJ. 1997. Elementary and global aspects of calcium signalling. *J Physiol (Lond)* **499:** 291–306.

Berridge MJ, Galione A. 1988. Cytosolic calcium oscillators. *FASEB J* **2:** 3074–3082.

Bezprozvanny I. 1994. Theoretical analysis of calcium wave propagation based on inositol (1,4,5)-trisphosphate ($InsP_3$) receptor functional properties. *Cell Calcium* **16:** 151–166.

Bezprozvanny I, Watras J, Ehrlich BE. 1991. Bell-shaped calcium-response curves of $Ins(1,4,5)P_3$ and calcium-gated channels from endoplasmic reticulum of cerebellum. *Nature* **351:** 751–754.

Bird GS, Putney JW. 2005. Capacitative calcium entry supports calcium oscillations in human embryonic kidney cells. *J Physiol* **562:** 697–706.

Bird GS, Putney JW. 2006. Fluorescent indicators—facts and artifacts. In *Calcium Signaling* (ed. JW Putney), pp. 51–84. CRC Press, Boca Raton.

Bird GS, Hwang SY, Smyth JT, Fukushima M, Boyles RR, Putney JW Jr. 2009. STIM1 is a calcium sensor specialized for digital signaling. *Curr Biol* **19:** 1724–1729.

Birnbaumer L, Boulay G, Brown D, Jiang M, Dietrich A, Mikoshiba K, Zhu X, Qin N. 2000. Mechanism of capacitative Ca^{2+} entry (CCE): interaction between IP3 receptor and TRP links the internal calcium storage compartment to plasma membrane CCE channels. *Recent Prog Horm Res* **55:** 127–161.

Blaustein MP, Kendrick NC, Fried RC, Ratzlaff RW. 1977. Calcium metabolism at the mammalian presynaptic nerve terminal: Lessons from the synaptsome. In *Society for Neuroscience Symposia, Vol. II. Approaches to the Cell Biology of Neurons* (ed. MW Cowan, JA Ferrendelli), pp. 172–194. Society for Neuroscience, Bethesda, MD.

Bohr DF. 1973. Vascular smooth muscle updated. *Circ Res* **32:** 665–672.

Bootman MD, Berridge MJ, Roderick HL. 2002. Calcium signalling: more messengers, more channels, more complexity. *Curr Biol* **12:** R563–R565.

Brandman O, Liou J, Park WS, Meyer T. 2007. STIM2 is a feedback regulator that stabilizes basal cytosolic and endoplasmic reticulum $Ca^{(2+)}$ levels. *Cell* **131:** 1327–1339.

Burgess GM, McKinney JS, Fabiato A, Leslie BA, Putney JW. 1983. Calcium pools in saponin-permeabilized guinea-pig hepatocytes. *J Biol Chem* **258:** 15336–15345.

Chang WC, Capite JDi, Singaravelu K, Nelson C, Halse V, Parekh AB. 2008. Local Ca^{2+} Influx through Ca^{2+} Release-activated Ca^{2+} (CRAC) Channels Stimulates Production of an Intracellular Messenger and an Intercellular Pro-inflammatory Signal. *J Biol Chem* **283:** 4622–4631.

Chang WC, Nelson C, Parekh AB. 2006. Ca^{2+} influx through CRAC channels activates cytosolic phospholipase A2, leukotriene C4 secretion, and expression of c-fos through ERK-dependent and -independent pathways in mast cells. *FASEB J* **20:** 2381–2383.

Chatton JY, Cao Y, Stucki JW. 1998. Perturbation of myo-inositol-1,4,5-trisphosphate levels during agonist-induced Ca^{2+} oscillations. *Biophys J* **74:** 523–531.

Combettes LHannaert-Merah Z, Coquil J-F, Rousseau C, Claret M, Swillens S, Champeil P. 1994. Rapid filtration studies of the effect of cytosolic Ca^{2+} on inositol 1,4,5-trisphosphate-induced $^{45}Ca^{2+}$ release from cerebellar microsomes. *J Biol Chem* **269:** 17561–17571.

Coombes S, Hinch R, Timofeeva Y. 2004. Receptors, sparks, and waves in a fire-diffuse-fire framework for calcium release. *Prog Biophys Mol Biol* **85:** 197–216.

Cooper DMF, Yoshimura M, Zhang Y, Chiono M, Mahey R. 1994. Capacitative Ca^{2+} entry regulates Ca^{2+}-sensitive adenylyl cyclases. *Biochem J* **297:** 437–440.

Csordas G, Renken C, Varnai P, Walter L, Weaver D, Buttle KF, Balla T, Mannella CA, Hajnoczky G. 2006. Structural and functional features and significance of the physical linkage between ER and mitochondria. *J Cell Biol* **174:** 915–921.

Cuthbertson KS, Chay TR. 1991. Modeling receptor-controlled intracellular calcium oscillators. *Cell Calcium* **12:** 97–109.

Cuthbertson KSR, Cobbold PH. 1991. Oscillations in cell calcium (Collected papers and reviews). *Cell Calcium* **12:** 61–268.

Dawson SP, Keizer J, Pearson JE. 1999. Fire-diffuse-fire model of dynamics of intracellular calcium waves. *Proc Natl Acad Sci* **96:** 6060–6063.

DeHaven WI, Jones BF, Petranka JG, Smyth JT, Tomita T, Bird GS, Putney JW. 2009. TRPC channels function independently of STIM1 and Orai1. *J Physiol* **587:** 2275–2298.

De Pitta M, Goldberg M, Volman V, Berry H, Ben-Jacob E. 2009. Glutamate regulation of calcium and IP(3) oscillating and pulsating dynamics in astrocytes. *J Biol Phys* **35:** 383–411.

De Young GW, Keizer J. 1992. A single-pool inositol 1,4,5-trisphosphate-receptor-based model for agonist-stimulated oscillations in Ca^{2+} concentration. *Proc Nat Acad Sci* **89:** 9895–9899.

DiCapite J, Ng SW, Parekh AB. 2009. Decoding of cytoplasmic $Ca^{(2+)}$ oscillations through the spatial signature drives gene expression. *Curr Biol* **19:** 853–858.

Dolmetsch RE, Xu KL, Lewis RS. 1998. Calcium oscillations increase the efficiency and specificity of gene expression. *Nature* **392:** 933–936.

Dupont G, Goldbeter A. 1993. One-pool model for Ca^{2+} oscillations involving Ca^{2+} and inositol 1,4,5-trisphosphate as co-agonists for Ca^{2+} release. *Cell Calcium* **14:** 311–322.

Dupont G, Swillens S. 1996. Quantal release, incremental detection, and long-period Ca^{2+} oscillations in a model based on regulatory Ca^{2+}-binding sites along the permeation pathway. *Biophys J* **71:** 1714–1722.

Dupont G, Erneux C. 1997. Simulations of the effects of inositol 1,4,5-trisphosphate 3-kinase and 5-phosphatase activities on Ca^{2+} oscillations. *Cell Calcium* **22:** 321–331.

Dupont G, Dumollard R. 2004. Simulation of calcium waves in ascidian eggs: insights into the origin of the pacemaker sites and the possible nature of the sperm factor. *J Cell Sci* **117:** 4313–4323.

Dupont G, Combettes L. 2006. Modeling the effect of specific inositol 1,4,5-trisphosphate receptor isoforms on cellular Ca^{2+} signals. *Biol Cell* **98:** 171–182.

Dupont G, Combettes L. 2009. What can we learn from the irregularity of Ca^{2+} oscillations?. *Chaos* **19:** 037112.

Dupont G, Combettes L, Leybaert L. 2007. Calcium dynamics: spatio-temporal organization from the subcellular to the organ level. *Int Rev Cytol* **261:** 193–245.

Dupont G, Bou-Lovergne A, Combettes L. 2008. Stochastic Aspects of Oscillatory Ca^{2+} Dynamics in Hepatocytes. *Biophys J* **95:** 2193–2202.

Dupont G, Koukoui O, Clair C, Erneux C, Swillens S, Combettes L. 2003. Ca^{2+} oscillations in hepatocytes do not require the modulation of InsP$_3$ 3-kinase activity by Ca^{2+}. *FEBS Lett* **534:** 101–105.

Dupont G, Tordjmann T, Clair C, Swillens S, Claret M, Combettes L. 2000. Mechanism of receptor-oriented intercellular calcium wave propagation in hepatocytes. *FASEB J* **14:** 279–289.

Dziadek MA, Johnstone LS. 2007. Biochemical properties and cellular localisation of STIM proteins. *Cell Calcium* **42:** 123–132.

Fabiato A. 1983. Calcium-induced release of calcium from the cardiac sarcoplasmic reticulum. *Am J Physiol* **245:** C1–C4.

Falcke M. 2004. Reading the patterns in living cells—the physics of Ca^{2+} signaling. *Adv Physics* **53:** 255–440.

Fall CP, Keizer JE. 2001. Mitochondrial modulation of intracellular Ca^{2+} signaling. *J Theor Biol* **210:** 151–165.

Feske S, Gwack Y, Prakriya M, Srikanth S, Puppel SH, Tanasa B, Hogan PG, Lewis RS, Daly M, Rao A. 2006. A mutation in Orai1 causes immune deficiency by abrogating CRAC channel function. *Nature* **441:** 179–185.

Finch EA, Turner TJ, Goldin SM. 1991. Calcium as a co-agonist of inositol 1,4,5-trisphosphate-induced calcium release. *Science* **252:** 443–446.

Fioretti B, Franciolini F, Catacuzzeno L. 2005. A model of intracellular Ca^{2+} oscillations based on the activity of the intermediate-conductance Ca^{2+}-activated K^+ channels. *Biophys Chem* **113:** 17–23.

Fraiman D, Pando B, Dargan S, Parker I, Dawson SP. 2006. Analysis of puff dynamics in oocytes: interdependence of puff amplitude and interpuff interval. *Biophys J* **90:** 3897–3907.

Frischauf I, Schindl R, Derler I, Bergsmann J, Fahrner M, Romanin C. 2008. The STIM/Orai coupling machinery. *Channels (Austin)* **2:** 261–268.

Goldbeter A. 2008. Biological rhythms: clocks for all times. *Curr Biol* **18:** R751–R753.

Goldbeter A, Dupont G, Berridge MJ. 1990. Minimal model for signal-induced Ca^{2+} oscillations and for their frequency encoding through protein phosphorylation. *Proc Nat Acad Sci* **87:** 1461–1465.

Gracheva ME, Gunton JD. 2003. Intercellular communication via intracellular calcium oscillations. *J Theor Biol* **221:** 513–518.

Haas B, Schipke CG, Peters O, Sohl G, Willecke K, Kettenmann H. 2006. Activity-dependent ATP-waves in the mouse neocortex are independent from astrocytic calcium waves. *Cereb Cortex* **16:** 237–246.

Hajnóczky G, Thomas AP. 1994. The inositol trisphosphate calcium channel is inactivated by inositol trisphosphate. *Nature* **370:** 474–477.

Hajnóczky G, Robb-Gaspers LD, Seitz MB, Thomas AP. 1995. Decoding of cytosolic calcium oscillations in the mitochondria. *Cell* **82:** 415–424.

Halestrap AP. 2009. Mitochondrial calcium in health and disease. *Biochim Biophys Acta* **1787:** 1289–1290.

Hattori M, Suzuki AZ, Higo T, Miyauchi H, Michikawa T, Nakamura T, Inoue T, Mikoshiba K. 2004. Distinct roles of inositol 1,4,5-trisphosphate receptor types 1 and 3 in Ca^{2+} signaling. *J Biol Chem* **279:** 11967–11975.

Hellman B, Gylfe E, Grapengiesser E, Lund P-E, Berts A. 1992. Cytoplasmic Ca^{2+} oscillations in pancreatic b-cells. *Biochim Biophys Acta* **1113:** 295–305.

Hofer T, Venance L, Giaume C. 2002. Control and plasticity of intercellular calcium waves in astrocytes: A modeling approach. *J Neurosci* **22:** 4850–4859.

Hoth M, Penner R. 1992. Depletion of intracellular calcium stores activates a calcium current in mast cells. *Nature* **355:** 353–355.

Huang GN, Zeng W, Kim JY, Yuan JP, Han L, Muallem S, Worley PF. 2006. STIM1 carboxyl-terminus activates native SOC, Icrac and TRPC1 channels. *Nat Cell Biol* **8:** 1003–1010.

Keizer J, Levine L. 1996. Ryanodine receptor adaptation and Ca^{2+}(-)induced Ca^{2+} release-dependent Ca^{2+} oscillations. *Biophys J* **71:** 3477–3487.

Keizer J. 1993. Calcium oscillations and waves: Is the IP_3R Ca^{2+} channel the culprit? *Biophys J* **65:** 1359–1361.

Kouchi Z, Fukami K, Shikano T, Oda S, Nakamura Y, Takenawa T, Miyazaki S. 2004. Recombinant phospholipase Czeta has high Ca^{2+} sensitivity and induces Ca^{2+} oscillations in mouse eggs. *J Biol Chem* **279:** 10408–10412.

Kummer U, Olsen LF, Dixon CJ, Green AK, Bornberg-Bauer E, Baier G. 2000. Switching from simple to complex oscillations in calcium signaling. *Biophys J* **79:** 1188–1195.

Kummer U, Krajnc B, Pahle J, Green AK, Dixon CJ, Marhl M. 2005. Transition from stochastic to deterministic behavior in calcium oscillations. *Biophys J* **89:** 1603–1611.

Kwan CY, Takemura H, Obie JF, Thastrup O, Putney JW. 1990. Effects of methacholine, thapsigargin and La^{3+} on plasmalemmal and intracellular Ca^{2+} transport in lacrimal acinar cells. *Am J Physiol* **258:** C1006–C1015.

LeBeau AP, Yule DI, Groblewski GE, Sneyd J. 1999. Agonist-dependent phosphorylation of the inositol 1,4,5-trisphosphate receptor. A possible mechanism for agonist-specific calcium oscillations in pancreatic acinar cells. *J Gen Physiol* **113:** 851–871.

Lewis RS. 2003. Calcium oscillations in T-cells: Mechanisms and consequences for gene expression. *Biochem Soc Trans* **31:** 925–929.

Li YX, Rinzel J. 1994. Equations for $InsP_3$ receptor-mediated $[Ca^{2+}]i$ oscillations derived from a detailed kinetic model: a Hodgkin-Huxley like formalism. *J Theor Biol* **166:** 461–473.

Li Y-X, Keizer J, Stojilkovic SS, Rinzel J. 1995. Ca^{2+} excitability of the ER membrane: an explanation for IP_3-induced Ca^{2+} oscillations. *Am J Physiol* **269:** C1079–C1092.

Liou J, Fivaz M, Inoue T, Meyer T. 2007. Live-cell imaging reveals sequential oligomerization and local plasma membrane targeting of stromal interaction molecule 1 after Ca^{2+} store depletion. *Proc Natl Acad Sci* **104:** 9301–9306.

Marchant J, Callamaras N, Parker I. 1999. Initiation of IP_3-mediated Ca^{2+} waves in *Xenopus* oocytes. *EMBO J* **18:** 5285–5299.

Marhl M, Schuster S, Brumen M. 1998. Mitochondria as an important factor in the maintenance of constant amplitudes of cytosolic calcium oscillations. *Biophys Chem* **71:** 125–132.

Mercer JC, DeHaven WI, Smyth JT, Wedel B, Boyles RR, Bird GS, Putney JW. 2006. Large store-operated calcium-selected currents due to co-expression of orai1 or orai2 with the intracellular calcium sensor, stim1. *J Biol Chem* **281:** 24979–24990.

Meyer T, Stryer L. 1988. Molecular model for receptor-stimulated calcium spiking. *Proc Nat Acad Sci* **85:** 5051–5055.

Meyer T, Stryer L. 1991. Calcium spiking. *Annu Rev Biophys Biophys Chem* **20:** 153–174.

Meyer T, Hanson PI, Stryer L, Schulman H. 1992. Calmodulin trapping by calcium-calmodulin-dependent protein kinase. *Science* **256:** 1199–1202.

Mignen O, Thompson JL, Shuttleworth TJ. 2008. Both Orai1 and Orai3 are essential components of the arachidonate-regulated Ca^{2+}-selective (ARC) channels. *J Physiol (Lond)* **586:** 185–195.

Miyakawa T, Maeda A, Yamazawa T, Hirose K, Kurosaki T, Iino M. 1999. Encoding of Ca^{2+} signals by differential expression of IP_3 receptor subtypes. *EMBO J* **18:** 1303–1308.

Nicou A, Serriere V, Hilly M, Prigent S, Combettes L, Guillon G, Tordjmann T. 2007. Remodeling of calcium signaling during liver regeneration in the rat. *J Hepatol* **46:** 247–256.

Oh-Hora M. 2009. Calcium signaling in the development and function of T-lineage cells. *Immunol Rev* **231:** 210–224.

Orci L, Ravazzola M, Coadic MLe, Shen Ww, Demaurex N, Cosson P. 2009. STIM1-induced precortical and cortical subdomains of the endoplasmic reticulum. *Proc Natl Acad Sci* **106:** 19358–19362.

Parekh AB, Putney JW. 2005. Store-operated calcium channels. *Physiol Rev* **85:** 757–810.

Parekh AB, Fleig A, Penner R. 1997. The store-operated calcium current I_{CRAC}: Nonlinear activation by $InsP_3$ and dissociation from calcium release. *Cell* **89:** 973–980.

Park CY, Hoover PJ, Mullins FM, Bachhawat P, Covington ED, Raunser S, Walz T, Garcia KC, Dolmetsch RE, Lewis RS. 2009. STIM1 clusters and activates CRAC channels via direct binding of a cytosolic domain to Orai1. *Cell* **136:** 876–890.

Parvez S, Beck A, Peinelt C, Soboloff J, Lis A, Monteilh-Zoller M, Gill DL, Fleig A, Penner R. 2008. STIM2 protein mediates distinct store-dependent and store-independent modes of CRAC channel activation. *FASEB J* **22:** 752–761.

Petersen OH, Wakui M. 1990. Oscillating intracellular Ca^{2+} signals evoked by activation of receptors linked to inositol lipid hydrolysis: Mechanism of generation. *J Membrane Biol* **118:** 93–105.

Politi A, Gaspers LD, Thomas AP, Hofer T. 2006. Models of IP_3 and Ca^{2+} oscillations: Frequency encoding and identification of underlying feedbacks. *Biophys J* **90:** 3120–3133.

Prince WT, Berridge MJ. 1973. The role of calcium in the action of 5-hydroxytryptamine and cyclic AMP on salivary glands. *J Exp Biol* **56:** 367–384.

Putney JW. 1986. A model for receptor-regulated calcium entry. *Cell Calcium* **7:** 1–12.

Putney JW, Bird GS. 2008. Cytoplasmic calcium oscillations and store-operated calcium influx. *J Physiol* **586:** 3055–3059.

Putney JW, Poggioli J, Weiss SJ. 1981. Receptor regulation of calcium release and calcium permeability in parotid gland cells. *Phil Trans R Soc Lond B* **296:** 37–45.

Renard D, Poggioli J, Berthon B, Claret M. 1987. How far does phospholipase C activity depend on the cell calcium concentration? A study in intact cells. *Biochem J* **243:** 391–398.

Ribeiro CMP, Putney JW. 1996. Differential effects of protein kinase C activation on calcium storage and capacitative calcium entry in NIH 3T3 cells. *J Biol Chem* **271:** 21522–21528.

Rink TJ, Jacob R. 1989. Calcium oscillations in non-excitable cells. *Trends Neurosci* **12:** 43–46.

Rizzuto R, Pozzan T. 2006. Microdomains of intracellular Ca^{2+}: Molecular determinants and functional consequences. *Physiol Rev* **86:** 369–408.

Rizzuto R, Pinton P, Carrington W, Fay FS, Fogarty KE, Lifshitz LM, Tuft RA, Pozzan T. 1998. Close contacts with the endoplasmic reticulum as determinants of mitochondrial Ca^{2+} responses. *Science* **280:** 1763–1766.

Rizzuto R, Simpson AWM, Brini M, Pozzan T. 1992. Rapid changes of mitochondrial Ca^{2+} revealed by specifically targeted recombinant aequorin. *Nature* **358:** 325–327.

Rosado JA, Sage SO. 2000. A role for the actin cytoskeleton in the initiation and maintenance of store-mediated calcium entry in human platelets. *Trends Cardiovasc Med* **10:** 327–332.

Saunders CM, Larman MG, Parrington J, Cox LJ, Royse J, Blayney LM, Swann K, Lai FA. 2002. PLC zeta: A sperm-specific trigger of $Ca^{(2+)}$ oscillations in eggs and embryo development. *Development* **129:** 3533–3544.

Selivanov VA, Ichas F, Holmuhamedov EL, Jouaville LS, Evtodienko YV, Mazat JP. 1998. A model of mitochondrial $Ca^{(2+)}$-induced Ca^{2+} release simulating the Ca^{2+} oscillations and spikes generated by mitochondria. *Biophys Chem* **72:** 111–121.

Shuai JW, Yang DP, Pearson JE, Rudiger S. 2009. An investigation of models of the IP_3R channel in Xenopus oocyte. *Chaos* **19:** 037105.

Shuttleworth TJ. 1999. What drives calcium entry during $[Ca^{2+}]_i$ oscillations? - challenging the capacitative model. *Cell Calcium* **25:** 237–246.

Shuttleworth TJ, Thompson JL, Mignen O. 2004. ARC channels: A novel pathway for receptor-activated calcium entry. *Physiology (Bethesda)* **19:** 355–361.

Shuttleworth TJ, Thompson JL, Mignen O. 2007. STIM1 and the noncapacitative ARC channels. *Cell Calcium* **42:** 183–191.

Skupin A, Kettenmann H, Winkler U, Wartenberg M, Sauer H, Tovey SC, Taylor CW, Falcke M. 2008. How does intracellular Ca^{2+} oscillate: By chance or by the clock? *Biophys J* **94:** 2404–2411.

Smith IF, Parker I. 2009. Imaging the quantal substructure of single IP_3R channel activity during Ca^{2+} puffs in intact mammalian cells. *Proc Natl Acad Sci* **106:** 6404–6409.

Smith IF, Wiltgen SM, Shuai J, Parker I. 2009. $Ca^{(2+)}$ puffs originate from preestablished stable clusters of inositol trisphosphate receptors. *Sci Signal* **2:** ra77.

Sneyd J, Falcke M. 2005. Models of the inositol trisphosphate receptor. *Prog Biophys Mol Biol* **89:** 207–245.

Sneyd J, Charles AC, Sanderson MJ. 1994. A model for the propagation of intercellular calcium waves. *Am J Physiol* **266:** C293–C302.

Sneyd J, Tsaneva-Atanasova K, Reznikov V, Bai Y, Sanderson MJ, Yule DI. 2006. A method for determining the dependence of calcium oscillations on inositol trisphosphate oscillations. *Proc Natl Acad Sci* **103:** 1675–1680.

Sneyd J, Tsaneva-Atanasova K, Yule DI, Thompson JL, Shuttleworth TJ. 2004. Control of calcium oscillations by membrane fluxes. *Proc Natl Acad Sci* **101:** 1392–1396.

Sneyd J, Wetton BT, Charles AC, Sanderson MJ. 1995. Intercellular calcium waves mediated by diffusion of inositol trisphosphate: A two-dimensional model. *Am J Physiol* **268:** C1537–C1545.

Soboloff J, Spassova MA, Hewavitharana T, He LP, Xu W, Johnstone LS, Dziadek MA, Gill DL. 2006a. STIM2 is an inhibitor of STIM1-mediated store-operated Ca^{2+} entry. *Curr Biol* **16:** 1465–1470.

Soboloff J, Spassova MA, Tang XD, Hewavitharana T, Xu W, Gill DL. 2006b. Orai1 and STIM reconstitute store-operated calcium channel function. *J Biol Chem* **281:** 20661–20665.

Streb H, Irvine RF, Berridge MJ, Schulz I. 1983. Release of Ca^{2+} from a nonmitochondrial store in pancreatic cells by inositol-1,4,5-trisphosphate. *Nature* **306:** 67–68.

Swann K, Yu Y. 2008. The dynamics of calcium oscillations that activate mammalian eggs. *Int J Dev Biol* **52:** 585–594.

Swillens S, Champeil P, Combettes L, Dupont G. 1998. Stochastic simulation of a single inositol 1,4,5-trisphosphate-sensitive Ca^{2+} channel reveals repetitive openings during 'blip-like' Ca^{2+} transients. *Cell Calcium* **23:** 291–302.

Swillens S, Dupont G, Combettes L, Champeil P. 1999. From calcium blips to calcium puffs: theoretical analysis of the requirements for interchannel communication. *Proc Natl Acad Sci* **96:** 13750–13755.

Tang Y, Othmer HG. 1994. A model of calcium dynamics in cardiac myocytes based on the kinetics of ryanodine-sensitive calcium channels. *Biophys J* **67:** 2223–2235.

Tang Y, Stephenson JL, Othmer HG. 1996. Simplification and analysis of models of calcium dynamics based on IP3-sensitive calcium channel kinetics. *Biophys J* **70:** 246–263.

Tanimura A, Morita T, Nezu A, Shitara A, Hashimoto N, Tojyo Y. 2009. Use of fluorescence resonance energy transfer-based biosensors for the quantitative analysis of inositol 1,4,5-trisphosphate dynamics in calcium oscillations. *J Biol Chem* **284:** 8910–8917.

Taylor CW, Thorn P. 2001. Calcium signalling: IP_3 rises again . . . and again. *Current Biol* **11:** R352–R355.

Tepikin AV, Petersen OH. 1992. Mechanisms of cellular calcium oscillations in secretory cells. *Biochim Biophys Acta* **1137:** 197–207.

Thomas AP, Renard DC, Rooney TA. 1992. Spatial organization of Ca^{2+} signaling and $Ins(1,4,5)P_3$ action. *Adv Second Messenger Phosphoprotein Res* **26:** 225–263.

Thomas AP, Bird GStJ, Hajnóczky G, Robb-Gaspers LD, Putney JW. 1996. Spatial and temporal aspects of cellular calcium signalling. *FASEB J* **10:** 1505–1517.

Thomas D, Lipp P, Tovey SC, Berridge MJ, Li W, Tsien RY, Bootman MD. 2000. Microscopic properties of elementary Ca^{2+} release sites in non-excitable cells. *Current Biol* **10:** 8–15.

Thul R, Smith GD, Coombes S. 2008. A bidomain threshold model of propagating calcium waves. *J Math Biol* **56:** 435–463.

Tsien RW, Bean BP, Hess P, Lansman JB, Nilius B, Nowycky MC. 1986. Mechanisms of calcium channel modulation by β-adrenergic agents and dihydropyridine calcium agonists. *J Mol Cell Cardiol* **18:** 691–710.

Van Breemen C, Farinas B, Gerba P, McNaughton ED. 1972. Excitation-contraction coupling in rabbit aorta studied by the lanthanum method for measuring cellular calcium influx. *Circ Res* **30:** 44–54.

Vazquez G, Wedel BJ, Aziz O, Trebak M, Putney JW. 2004. The mammalian TRPC cation channels. *Biochim Biophys Acta* **1742:** 21–36.

Ventura AC, Sneyd J. 2006. Calcium oscillations and waves generated by multiple release mechanisms in pancreatic acinar cells. *Bull Math Biol* **68:** 2205–2231.

Wang I, Politi AZ, Tania N, Bai Y, Sanderson MJ, Sneyd J. 2008. A mathematical model of airway and pulmonary arteriole smooth muscle. *Biophys J* **94:** 2053–2064.

Wedel B, Boyles RR, Putney JW, Bird GS. 2007. Role of the store-operated calcium entry proteins, Stim1 and Orai1, in muscarinic-cholinergic receptor stimulated calcium oscillations in human embryonic kidney cells. *J Physiol* **579:** 679–689.

Williams GS, Molinelli EJ, Smith GD. 2008. Modeling local and global intracellular calcium responses mediated by diffusely distributed inositol 1,4,5-trisphosphate receptors. *J Theor Biol* **253:** 170–188.

Woods NM, Cuthbertson KS, Cobbold PH. 1986. Repetitive transient rises in cytoplasmic free calcium in hormone-stimulated hepatocytes. *Nature* **319:** 600–602.

Zahradnikova A, Zahradnik I. 1996. A minimal gating model for the cardiac calcium release channel. *Biophys J* **71:** 2996–3012.

Zen S, Li B, Zeng S, Chen S. 2009. Simulation of spontaneous Ca^{2+} oscillations in astrocytes mediated by voltage-gated calcium channels. *Biophys J* **97:** 2429–2437.

Zheng L, Stathopulos PB, Li GY, Ikura M. 2008. Biophysical characterization of the EF-hand and SAM domain containing Ca^{2+} sensory region of STIM1 and STIM2. *Biochem Biophys Res Commun* **369:** 240–246.

Cytosolic Ca^{2+} Buffers

Beat Schwaller

Unit of Anatomy, Department of Medicine, University of Fribourg, CH-1700 Fribourg, Switzerland

Correspondence: beat.schwaller@unifr.ch

"Ca^{2+} buffers," a class of cytosolic Ca^{2+}-binding proteins, act as modulators of short-lived intracellular Ca^{2+} signals; they affect both the temporal and spatial aspects of these transient increases in [Ca^{2+}]$_i$. Examples of Ca^{2+} buffers include parvalbumins (α and β isoforms), calbindin-D9k, calbindin-D28k, and calretinin. Besides their proven Ca^{2+} buffer function, some might additionally have Ca^{2+} sensor functions. Ca^{2+} buffers have to be viewed as one of the components implicated in the precise regulation of Ca^{2+} signaling and Ca^{2+} homeostasis. Each cell is equipped with proteins, including Ca^{2+} channels, transporters, and pumps that, together with the Ca^{2+} buffers, shape the intracellular Ca^{2+} signals. All of these molecules are not only functionally coupled, but their expression is likely to be regulated in a Ca^{2+}-dependent manner to maintain normal Ca^{2+} signaling, even in the absence or malfunctioning of one of the components.

DEFINITION OF A CYTOSOLIC Ca^{2+} BUFFER

Molecules serving as chelators for Ca^{2+} ions must contain negatively charged groups arranged in such a way as to fulfill the necessary geometrical constraints for chemical coordination. Proteins with appropriately spaced acidic side-chain residues (e.g., glutamate, aspartate) and/or backbone carbonyl groups provide the "cage" in which a Ca^{2+} ion may fit in. Evolutionarily well-conserved protein families differing in the way the Ca^{2+} ions are bound include the annexins, the C2-domain proteins, the EF-hand proteins, the pentraxins, the vitamin-K-dependent proteins, and the intraorganellar low-affinity, high-capacity Ca^{2+}-binding proteins; for more details on the structural diversity of EF-hand motifs, see Gifford et al. (2007), and for other Ca^{2+}-binding sites, see Bindreither and Lackner (2009). However, the term Ca^{2+} buffer is applied only to a small subset of cytosolic proteins of the EF-hand family, including parvalbumins (PV; alpha and beta isoforms), calbindin-D9k (CB-D9k), calbindin-D28k (CB-D28k), and calretinin (CR). The majority of EF-hand proteins belong to the group of "Ca^{2+} sensors"; that is, binding of Ca^{2+} ions induces a conformational change, which permits them to interact with specific targets in a Ca^{2+}-regulated manner. The prototypical examples of Ca^{2+} sensors are calmodulin (Chin and Means 2000) and proteins of the S100 family. However, if present at sufficiently high concentrations, Ca^{2+} sensors may also function as Ca^{2+} buffers. Importantly, cytosolic Ca^{2+} buffers do not act as buffers in analogy to chemical buffers such as pH buffers. The latter serve to clamp the pH to a predetermined value in such a way that the addition of an acid or a base elicits only a minor change in the pH of the solution. Therefore, the buffering capacity is highest when the pH is close to the pK value of the corresponding acid/base pair. The situation is fundamentally different for the cytosolic Ca^{2+} buffers. Under basal conditions, $[Ca^{2+}]_i$ is in the order of 20–100 nM. Yet, the dissociation constants for Ca^{2+} ($K_{D,Ca}$) of most Ca^{2+} buffers are almost one order of magnitude larger, ≈200 nM–1.5 μM (Table 1). Thus, in a resting cell, Ca^{2+} buffers are at large in their Ca^{2+}-free state, ready to bind Ca^{2+} ions whenever $[Ca^{2+}]_i$ increases. As a result of their specific properties, Ca^{2+}

Table 1. Properties of selected Ca^{2+}-binding proteins (adapted from Schwaller 2009).

	α PV	β PV (OM)	CB-D9k	CB-D28k	CR
Ca^{2+}-binding sites (functional)	3 (2)	3 (2)	2 (2)	6 (4)	6 (5)
Ca^{2+}-specific/ mixed Ca^{2+}/ Mg^{2+} sites	0/2	1/1	2/0	4/0[a]	5/0
$K_{D,Ca}$ (nM)	4–9[b]	Mixed: 42–45[d] Ca^{2+}-specific: 590–780[d]	K_{D1} ≈ 200–500[e] K_{D2} ≈ 60–300	high aff. (h)[f] K_{D1} ≈ 180–240 medium aff. (m) K_{D2} ≈ 410–510	$K_{D(T)}$ 28 μM[g] $K_{D(R)}$ 68 $K_{D(app)}$ 1.4 μM EF5: 36 μM
$K_{D,Mg}$	≈ 30 μM[b]	160–250 μM[d]		714 μM[a]	4.5 mM[h]
$K_{D,Ca(app)}$ (nM) at $[Mg^{2+}]$ of 0.5–1 mM	150–250[c]	230–310[c]			
$k_{on,Ca}$ (μM^{-1}s^{-1})	6[b]		up to 1000[i]	h sites ≈ 12[f] m sites ≈ 82	T sites: 1.8[g] R sites: 310 site EF5: 7.3
$k_{on,Mg}$ (μM^{-1}s^{-1})	0.1–1[c]				
cooperativity	no[k]	? yes[l]	yes[m]	yes n_H ≈ 1.1–1.2[a]	yes n_H ≈ 1.3–1.9[g]
$D_{Cabuffer}$ (μm^{2}s^{-1})	37–43[n] ~12			>100[o] ≈ 25	≈ 25[p]

[a]Although considered as Ca^{2+}-specific, at physiological $[Mg^{2+}]_i$, the apparent Ca^{2+} affinity is approximately two-fold lower (Berggard et al. 2002a).
[b](Eberhard and Erne 1994).
[c]$K_{D,Ca}$ and k_{on} for PV are $[Mg^{2+}]$ dependent; calculated values are estimates at $[Mg^{2+}]_i$ 0.6–0.9 mM.
[d](Cox et al. 1990).
[e](Linse et al. 1991).
[f]CB-D28k has high-affinity (h) and medium-affinity (m) sites, the stochiometry h/m is either 2/2 or 3/1 (Nagerl et al. 2000).
[g]For details, see text and Faas et al. (2007).
[h](Stevens and Rogers 1997).
[i]The value represents the diffusion limit, assuming a maximal Ca^{2+} diffusion rate of ≈ 200 μm^{2}s^{-1} (Martin et al. 1990).
[k]PV has 2 essentially identical Ca^{2+}-binding sites with n_H close to 1.
[l](Yin et al. 2006).
[m](Akke et al. 1991).
[n]The diffusion coefficients $D_{Cabuffer}$ for PV in muscle myoplasm (37 μm^{2}s^{-1}) (Maughan and Godt 1999) and Purkinje cell dendrites of 43 μm^{2}s^{-1}; (Schmidt et al. 2003a); smaller values are measured in PC soma and axons (≈12 μm^{2}s^{-1}); (Schmidt et al. 2007a).
[o]CB-D28k's mobility in PC dendrites is 26 μm^{2}s^{-1} (Schmidt et al. 2005), clearly slower than in water (Gabso et al. 1997), and also slower than PV in PC dendrites.
[p]Estimation based on the similar size of CB-D28k and CR.

buffers act in different ways to modulate the spatiotemporal aspects of cytosolic Ca^{2+} signals. How a given Ca^{2+} buffer affects intracellular Ca^{2+} signals depends on several parameters, including the intracellular concentration (Intracellular Concentration), the affinity for Ca^{2+} and other metal ions (Metal-binding Affinities), the kinetics of Ca^{2+} binding and release (Metal-binding Kinetics), and the intracellular mobility (Mobility and Interaction with Ligands). When measuring the total Ca^{2+}-buffering capacity (κ_S) of a cell, often the distinction is made between mobile and immobile buffers (Zhou and Neher 1993). The latter ones are defined as molecules capable of binding cytosolic Ca^{2+} that are not washed out when, for example, the plasma membrane is patched with a pipette. Very little is known about the molecular identity of immobile Ca^{2+} buffers, except their relatively low Ca^{2+} affinity; presumably, they are made up of cytosol-exposed stretches of membrane proteins and/or membrane-associated proteins with a rather low affinity for Ca^{2+} ions. Additionally, negatively charged phospholipid headgroups of the inner leaflet of the plasma membrane serve as "weak" Ca^{2+} chelators (McLaughlin et al. 1981). Thus, the slowly mobile or immobile buffers, together with mobile Ca^{2+} buffers, are responsible for the rather slow diffusion of Ca^{2+} ions inside a cell (Mobility and Interaction with Ligands).

IMPORTANT PARAMETERS TO CHARACTERIZE Ca^{2+} BUFFERS

Intracellular Concentration

The difficulty in obtaining reliable values for Ca^{2+}-buffer concentrations is linked to the fact that those cells that strongly express Ca^{2+} buffers are frequently only a subset of cells (often with complex morphologies) within composite tissues; for example, in a subset of neurons in the brain, in specific segments of the kidney nephron, etc. The concentration of PV is as high as 1.5 mM in the superfast swimbladder muscle of toadfish (Tikunov and Rome 2009) and approximately 1 mM in mouse fast-twitch muscle, while it is lower in other muscles and highly correlated with the speed of muscle relaxation (Heizmann et al. 1982). Within different neuron subpopulations, PV is on average one order of magnitude lower (50–150 μM): 80 μM in mouse (Schmidt et al. 2003b) and 120 μM in rat (Hackney et al. 2005) Purkinje cells; 150 μM in mouse cerebellar interneurons (Collin et al. 2005), and 100–300 μM in inner and outer hair cells from inner ear (Hackney et al. 2005). The concentration of mammalian β PV called oncomodulin (OM) is particularly high in rat cochlear outer hair cells (2–3 mM) (Hackney et al. 2005). High concentrations of CR (1.2 mM) are also present in tall saccular hair cells of the frog (Edmonds et al. 2000). In other cells, calretinin concentration is much lower: approximately 20 μM and 35 μM in rat inner and outer hair cells, respectively (Hackney et al. 2005), and 30–40 μM in mouse cerebellar granule cells (Gall et al. 2003). The concentration of CB-D28k is 150–360 μM in Purkinje cells (for a review, see Schwaller et al. 2002), and 40–50 μM in mature hippocampal dentate gyrus granule cells, CA3 interneurons, and in CA1 pyramidal cells (Muller et al. 2005). In cells expressing various mobile and immobile Ca^{2+} buffers of unknown identities and Ca^{2+}-binding properties, the concept of the Ca^{2+}-binding ratio of endogenous buffers has proven to be useful. The ratio of buffer-bound Ca^{2+} changes over free Ca^{2+} changes ($\kappa_S \sim [Ca^{2+} \text{ buffer}]/K_{D,Ca}$) serves as a measure to compare the Ca^{2+}-buffering capacity of different cell types. According to the single compartment and linear approximation model (Neher 1998), motor neurons with low cytosolic Ca^{2+} buffer expression also have a very low Ca^{2+}-buffering capacity ($\kappa_S < 50$) (Lips and Keller 1998). Hippocampal principal neurons and PV-expressing interneurons have κ_S values in the order of 60 and 150, respectively (Lee et al. 2000b). The highest Ca^{2+}-buffering capacity (κ_S of 900–2000) is seen in cerebellar Purkinje cells expressing high levels of CB-D28k and PV (Fierro and Llano 1996).

Metal-binding Affinities

Structurally, different types of EF-hand domains exist: the canonical EF-hand domain, comprising a Ca^{2+}-binding loop of 12 amino acids, in which Ca^{2+} ions are mostly coordinated via oxygen atoms of carboxyl side chain groups, and several non-canonical ones (Gifford et al. 2007). Typical for proteins of the S100 family, including CB-D9k (Marenholz et al. 2004), is the presence of a noncanonical loop, termed pseudo (Ψ) EF-hand, consisting of a loop of 14 amino acids with an inside-out conformation compared to the canonical loop; that is, Ca^{2+}-coordination is preferentially provided by backbone carbonyl groups (Nelson et al. 2002). Functionally, two types of EF-hand Ca^{2+}-binding sites are discernable due to their different selectivity and affinity for Ca^{2+} and Mg^{2+} ions (Celio et al. 1996). The Ca^{2+}-specific sites display affinities for Ca^{2+} (K_{Ca}) in the order of 10^{-3}–10^{-7} M and significantly lower ones for Mg^{2+} ($K_{Mg} = 10^{-1}$–10^{-2} M). The proteins CB-D28k and CR have 4 and 5 functional Ca^{2+}-binding sites of this type, respectively. The mixed Ca^{2+}/Mg^{2+} sites bind Ca^{2+} with high and Mg^{2+} with moderate affinity in a competitive manner (dissociation constants: $K_{Ca} = 10^{-7}$–10^{-9} M; $K_{Mg} = 10^{-3}$–10^{-5} M) (Table 1). PV is the prototypical example of a protein with two mixed sites. Based on PV's affinities for Ca^{2+} and Mg^{2+} and with $[Ca^{2+}]_i$ (of approximately 50 nM) and $[Mg^{2+}]_i$ (0.5–1 mM) inside a cell under basal conditions, PV's two Ca^{2+}/Mg^{2+} sites are, to a large degree, occupied by Mg^{2+}. In most proteins, EF-hand domains are paired; two helix-Ca^{2+}-binding loop-helix regions are connected by a short stretch of 5–10 amino acid residues and form a functional unit. Aside from providing structural stability of EF-hand domains, due to the close apposition also of the Ca^{2+}-binding loops in such a tandem domain (Fig. 1), binding of a Ca^{2+} ion to one loop may allosterically affect the other, both with respect to affinity and binding kinetics (Faas et al. 2007; Nelson et al. 2002). As a result of this building principle, the majority of EF-hand proteins have an even number of Ca^{2+}-binding domains (2 in CB-D9k, 6 in CB-D28k and CR); the uneven number (3) for PV (alpha and beta) is, together with the group of penta-EF-hand Ca^{2+}-binding proteins (Maki et al. 2002), rather the exception.

Metal-binding Kinetics

The majority of Ca^{2+} buffers have dissociation constants ($K_{D,Ca}$) in the low micromolar range (Table 1). Thus, in a resting cell (e.g., muscle fibers, neurons), Ca^{2+} buffers are mostly in the Ca^{2+}-free form. It is the kinetics of various Ca^{2+} buffers that strongly affect the spatiotemporal aspects of Ca^{2+} signals, in particular, in excitable cells (Schmidt et al. 2007b; Schmidt and Eilers 2009; Schwaller 2009). $[Ca^{2+}]_i$ transients last for tens of milliseconds to several hundred milliseconds and, thus, are differently modulated by kinetically distinct Ca^{2+} buffers. Their on-rates for Ca^{2+} binding (k_{on}) vary by more than two orders of magnitude. Typical fast buffers with Ca^{2+}-specific sites, including CB-D9k and troponin C, have k_{on} rates $> 10^8$ $M^{-1}s^{-1}$ comparable to the on-rate of the synthetic buffer BAPTA. At the other end of the kinetic scale, on-rates for the slow-onset buffer PV are $\approx 3 \times 10^6$ $M^{-1}s^{-1}$ under physiological conditions with $[Mg^{2+}]_i$ of ≈ 0.5–1.0 mM. PV's slow on-rate is the result of PV's Mg^{2+}/Ca^{2+} binding sites, where the rate for Ca^{2+}-binding is determined by the slow Mg^{2+} off-rate (Lee et al. 2000c) (Table 1). Of importance, in a nonphysiological setting, that is, in the absence of Mg^{2+}, the on-rate of Ca^{2+} binding to PV is very rapid (1.08×10^8 $M^{-1}s^{-1}$), almost as fast as the "fast" buffers (Lee et al. 2000c).

Since all endogenous Ca^{2+} buffers have more than one Ca^{2+}-binding site, a given protein may have sites with different affinities and kinetics; results for CB-D28k and CR are summarized here. CB-D28k contains two types of binding sites differing in $K_{D,Ca}$ and k_{on}: One type is a high-affinity site ($K_{D,Ca} \approx 200$ nM) that binds Ca^{2+} with a k_{on} comparable to that of EGTA ($\approx 1 \times 10^7$ $M^{-1}s^{-1}$), and the second type binds Ca^{2+} with intermediate affinity ($K_{D,Ca} \approx 400$–500 nM), but with an approximately 8-fold faster k_{on} (Table 1). The ratio between high:intermediate affinity sites is either 3:1 or 2:2. The experimental data could be equally well modeled with either stochiometry for the two types of binding sites (Nagerl et al. 2000). For determining CR's kinetic Ca^{2+}-binding properties in vitro, cooperativity was included

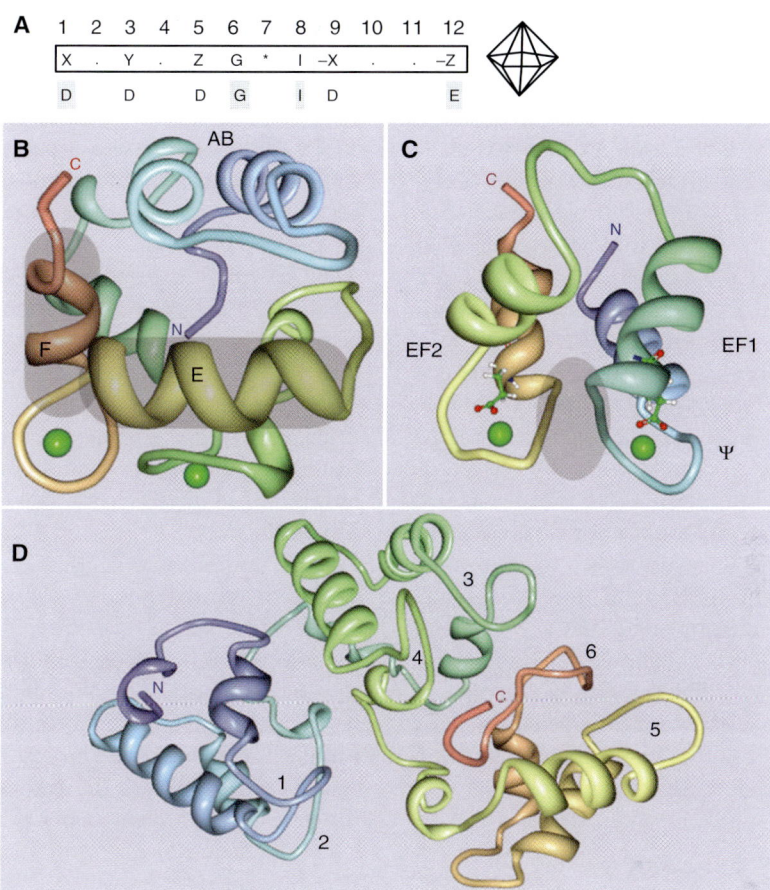

Figure 1. 3D-structures of selected EF-hand Ca^{2+} buffers. (*A*) Consensus sequence of the canonical EF-hand Ca^{2+}-binding loop of 12 amino acids. Amino acids X, Y, Z, and −Z provide side-chain oxygen ligands, * provides the backbone carbonyl oxygen, and at −X, a water molecule is hydrogen-bonded to a loop residue. Amino acids most often present at a given position are shown below, and shaded residues are the most conserved ones (Marsden et al. 1990). At positions X and −Z, Asp (D) and Glu (E) are generally present, respectively. The seven oxygen ligands coordinating the Ca^{2+} ion are located at the seven corners of a pentagonal bipyramid, and the Ca^{2+} ion (not shown) is in the center (right). (*B*) Solution structure of Ca^{2+}-bound human α PV; PDB: 1RJV. Both the CD domain (green) and EF domain (yellow/red) bind one Ca^{2+} ion each (green spheres) in canonical Ca^{2+}-binding loops of 12 amino acids. The orthogonally oriented helices E and F (gray-shaded) are connected by the Ca^{2+}-binding loop. Both Ca^{2+}-binding sites in PV are of the Ca^{2+}/Mg^{2+} mixed type. The N-terminal AB domain (blue) is necessary for protein stability. (*C*) NMR solution structure of bovine CB-D9k; PDB: 1B1G. The shown structure takes into account the Ca^{2+} ions and explicit solvent molecules. The N-terminal domain EF1 is a pseudo (Ψ) EF-hand with a larger loop of 14 amino acids, while the second domain (EF2) has a canonical Ca^{2+}-binding loop of 12 amino acids. In both loops, the Glu residue at the position −Z with the 2 carboxyl oxygen atoms (red) serves as a bidentate ligand representing two corners of the pentagonal bipyramid. This residue, most often Glu (rarely Asp), is a critical determinant for the Ca^{2+} affinity of the entire loop; Ca^{2+} ions are shown as green spheres. The two Ca^{2+}-binding loops are in close proximity and stabilized via short β-type interactions (gray-shaded area). (*D*) 3D NMR structure of CB-D28k; PDB: 2G9B. CB-D28k has a relatively compact structure comprising three Ca^{2+}-binding units, each unit consisting of a pair of EF-hands. Ca^{2+}-binding is restricted to the Ca^{2+}-binding loop 1 in the N-terminal unit (blue), to both loops 3 and 4 in the middle unit (green), and to loop 5 in the C-terminal pair (yellow/red). EF-hands 2 and 6 are nonfunctional, with respect to Ca^{2+}-binding. The Ca^{2+}-binding loops flanked by two almost perpendicular alpha-helical regions are numbered from 1 to 6. Images *B*–*D* were generated with PDB ProteinWorkshop 1.50 (Moreland et al. 2005).

in the mathematical model (Table 1). CR contains four 4 high-affinity, cooperative binding sites (Schwaller et al. 1997; Stevens and Rogers 1997) organized into two indistinguishable pairs, probably 1 & 2 and 3 & 4 (Faas et al. 2007) and one independent low-affinity Ca^{2+}-binding site (EF5) with an intrinsic dissociation constant (K'_D) of ≈ 0.5 mM (Schwaller et al. 1997). Within cooperative pairs, the two binding sites influence each other in an allosteric manner. In a pair, initially both binding sites are in a T- (tense) state, with a low affinity and slow binding rate. When the first Ca^{2+} ion is bound to a pair, the unoccupied site changes to an R- (relaxed) state, characterized by a high affinity and a fast binding rate. This leads to a gradual increase in the Ca^{2+} association rate (k_{on}) as $[Ca^{2+}]_i$ increases from a concentration in a resting neuron (≈ 50 nM) to approximately 1–10 μM after opening of Ca^{2+} channels (Schwaller 2009). Therefore, the kinetics of Ca^{2+} buffering of CR powerfully depends on the prevailing Ca^{2+} concentration prior to a perturbation, resulting in non-linear Ca^{2+} buffering by CR (for more details, see Fig. 3 in Schwaller 2009 and Faas et al. 2007).

Mobility and Interaction with Ligands

In an aqueous solution, the mobility defined as the diffusion coefficient (D) of a molecule is approximately proportional to the hydrodynamic radius (i.e., approximately proportional to the relative molecular mass; M_r). For relatively large molecules, such as dextrans and globular proteins, D should be proportional to the inverse cubic root of M_r. In Purkinje cells, PV is freely mobile, but D can vary considerably in different compartments: ≈ 12 $\mu m^2/s$ in axons, somata, and nuclei (Schmidt et al. 2007a) versus ≈ 43 $\mu m^2/s$ in dendrites (Schmidt et al. 2003a), most likely as the result of different cytoplasmic properties (e.g., tortuosity; i.e., diffusion in a porous medium). The latter value is very similar to PV's mobility in frog myoplasm, 43 and 32 $\mu m^2/s$ for transverse and longitudinal diffusion, respectively (Maughan and Godt 1999), but clearly smaller than in an aqueous solution: 140 $\mu m^2/s$ (Feher 1984). As expected from the larger M_r of CB-D28k (≈ 29 kDa) compared to PV (≈ 12 kDa), D in Purkinje cell dendrites is smaller: 26 $\mu m^2/s$ (Schmidt et al. 2005). How the presence of a Ca^{2+} buffer affects the spatiotemporal aspects of a cytosolic Ca^{2+} transient not only depends on the mobility of the Ca^{2+} buffer, but also on the mobility of the free Ca^{2+} ions. In the cytosol isolated from Xenopus laevis oocytes, the diffusion coefficient D of inositol 1,4,5-trisphophate ($InsP_3$) is much bigger than that of Ca^{2+} ions in a solution with $[Ca^{2+}]_i$ of 90 nM: 283 $\mu m^2/s$ versus 13 $\mu m^2/s$, respectively (Allbritton et al. 1992). Even when $[Ca^{2+}]_i$ is increased to 1 μM, presumably saturating slowly mobile and immobile buffers, D_{Ca} only reaches a value of 65 $\mu m^2/s$. This indicates that the slow diffusion of Ca^{2+} ions in a resting cell ($[Ca^{2+}]_i \approx 50$ nM) is caused by slowly mobile or immobile buffers "acting like velcro" for Ca^{2+} ions, limiting the effective range of an unbuffered free Ca^{2+} ion to ≈ 0.1 μm. Thus, the range of Ca^{2+} can be increased by buffered diffusion, that is, mobile Ca^{2+} buffers acting as shuttles transporting Ca^{2+} through the "mesh of immobile buffers" (Schmidt et al. 2007b).

Parvalbumins

Structural Aspects of Parvalbumins

The first prototypical structure of an EF-hand domain was determined in PV (Kretsinger and Nockolds 1973), an atypical EF-hand protein; the protein ($M_r \approx 12$ kDa; human gene symbol: *PVALB*) has an uneven number (3) of EF-hand domains, and the Ca^{2+}-binding sites are Ca^{2+}/Mg^{2+} mixed sites (Table 1). While the two C-terminal domains CD and EF are functional metal-binding sites, the N-terminal AB site is necessary for the PV's stability (Fig. 1) (Cox et al. (1999). The CD and EF domains form a pair consisting of two helix-loop-helix regions linked by a short stretch of 5–10 amino acid residues. In both sites, the Ca^{2+} ion in the center of the loop is coordinated by seven ligands sitting in the corners of a pentagonal bipyramid (Swain et al. 1989). Results based on solution structures of α PV and β PV (Babini et al. 2004) in the Ca^{2+}-loaded and the apo (metal-free) form are summarized: (I) PV's Ca^{2+}-loaded EF-hand domains and the linker region connecting the CD and EF domains are rather rigid structures; also the N- and C-termini of PV have a low intrinsic mobility (Baldellon et al. 1998); (II) Differences in the structure of apo- and Ca^{2+}-loaded forms of rat PV are small, mostly confined to the loop region. Thus, Ca^{2+} binding does not require major structural rearrangements (Henzl and Tanner 2008); (III) The first two points also hold true for the Ca^{2+}/Mg^{2+} (EF) site in rat β PV, while the noncanonical CD site undergoes significant structural alterations, when Ca^{2+} is removed from β PV (Henzl and Tanner 2007). Thus, the global rigidity of α PV favors this molecule to serve as a "simple" Ca^{2+} buffer, while the Ca^{2+}-dependent conformational changes in β PV may provide β PV also with a Ca^{2+} sensor function.

Functional Aspects of Parvalbumin and Oncomodulin

Cells with high PV expression levels include a subset of mostly GABA-ergic neurons, fast-twitch muscle fibers, and epithelial cells in the early distal convoluted tubule (DCT1) in nephrons of the kidney. PV's slow-onset Ca^{2+}-binding properties affect Ca^{2+} transients in a particular way: The rate in rise in $[Ca^{2+}]_i$ is hardly affected, but the initial rate of $[Ca^{2+}]_i$ decay is increased. In the later phase of the decay, the unbinding of Ca^{2+} ions from PV prolongs the late phase of the $[Ca^{2+}]_i$ decay. Thus, PV's hallmark is the conversion of a monoexponential $[Ca^{2+}]_i$ decay into a biexponential one (Collin et al. 2005; Lee et al. 2000c). In fast-twitch muscles, this increases muscle relaxation of an electrically induced muscle twitch, while barely affecting the contraction phase. In PV knockout mice ($PV^{-/-}$), twitch half-relaxation rates in fast muscles are slower than in the PV-expressing wild-type (WT) muscles (Schwaller et al. 1999). Conversely, the overexpression of PV by injection of *Pvalb* cDNA into the rat slow-twitch muscle, soleus, significantly increases the speed of relaxation, without affecting the contraction (Muntener et al. 1995). *Pvalb* gene delivery in rat heart in vivo increases the rate of heart relaxation in normal hearts and in an animal model of slowed cardiac muscle relaxation (Szatkowski et al. 2001). Thus, PV or genetically "tuned" PV variants are discussed as potential tools to enhance cardiac diastolic function (Rodenbaugh et al. 2007; Wang and Metzger 2008). An often-neglected aspect is the role of Ca^{2+} buffers in acting as transient Ca^{2+} sources, prolonging the $[Ca^{2+}]_i$ decay. In fast-twitch muscles subjected to long tetanic contractions, PV saturates with Ca^{2+} and consequently slows down relaxation (Raymackers et al. 2000). The slow decay component mediated by Ca^{2+}-bound PV also leads to a robust, PV-dependent, delayed transmitter release at cerebellar interneuron–interneuron synapses subsequent to presynaptic bursts of action potentials (Collin et al. 2005).

The PV-mediated, biexponential $[Ca^{2+}]_i$ decay is observed in: (I) PV-injected chromaffin cells (Lee et al. 2000c); (II) PV-containing hippocampal interneurons (Lee et al. 2000b); (III) PV-expressing molecular layer interneurons (MLI) in the cerebellum (Collin et al. 2005); (IV) presynaptic terminals of the calyx of Held (Muller et al. 2007); and (V) in Purkinje cell dendrites (Schmidt et al. 2007a). PV's acceleration of the early phase of $[Ca^{2+}]_i$ decay limits or slows down the buildup of residual $[Ca^{2+}]_i$ in presynaptic terminals, thus affecting short-term plasticity. The effect is most pronounced at timepoints, when $[Ca^{2+}]_i$ decay curves in the presence or absence of PV show the largest differences. PV's effect on decreasing/preventing paired-pulse facilitation at synapses between MLI and Purkinje cells is most pronounced at ≈33 Hz (Caillard et al. 2000). Also in the presynaptic terminals of the calyx of Held, PV accelerates the decay of spatially averaged $[Ca^{2+}]_i$ and paired-pulse facilitation (Muller et al. 2007). In hippocampal PV-interneurons, differences in paired-pulse modulation between WT and $PV^{-/-}$ mice are apparent only when trains at 33, 50, and 100 Hz are delivered (Vreugdenhil et al. 2003), likely due to differences in components of the Ca^{2+} signaling toolkit and/or more efficient presynaptic Ca^{2+} extrusion mechanisms. The largest relative effect of PV in preventing facilitation is seen at approximately

33 Hz, within the range of gamma frequency (30–80 Hz) oscillations. As a result, the power of kainite-induced gamma oscillations in area CA3 in vitro is approximately 3-fold higher in $PV^{-/-}$ versus WT tissue. This can be explained by an increased facilitation of GABA release at persistent high frequencies. In accordance with the hypothesis that changes in the inhibitory activity of PV neurons in the neocortex—often critically involved in strong perisomatic inhibition—may be a major mechanism underlying epileptic seizures (Mihaly et al. 1997). $PV^{-/-}$ mice have a lower threshold for pentylenetetrazole (PTZ)-induced seizures (Schwaller et al. 2004). The subpopulation of PV-immunoreactive (PV-ir) neurons is critically involved in controlling the output of principal neurons (Freund 2003); moreover, PV is not only a marker for these GABA-ergic interneurons, but contributes to controlling the network activity. The absence of PV in the cerebellar circuitry leads to the emergence of 160-Hz oscillations in vivo sustained by synchronous, rhythmic-firing Purkinje cells aligned along the parallel fiber axis (Servais et al. 2005). Also, $PV^{-/-}$ Purkinje cell-firing properties are different from WT ones: The complex spike duration and the spike pause are decreased, and the simple spike-firing rate is increased. These differences in firing properties, together with the oscillations, are the likely cause for the mild locomotor phenotype, that is, a slight impairment of motor coordination/motor learning (Farre-Castany et al. 2007).

Much less is known about OM's specific function. OM is present in the organ of Corti (Thalmann et al. 1995), more precisely in cochlear outer hair cells (Sakaguchi et al. 1998) in gerbil, rat, and mouse. An extracellular role for OM in retinal ganglion cell regeneration (Yin et al. 2006) was reported, but see also Hauk et al. (2008) and Schwaller (2009). An up-regulation of OM occurs in $PV^{-/-}$ mice in a sparse subpopulation of neurons in the thalamus and in the dentate gyrus, as well as in partly varicose axons in the diencephalon (Csillik et al. 2010). The functional significance of ectopic OM expression and the exact identity of neurons expressing OM in $PV^{-/-}$ mice remain to be shown.

Structural and Functional Aspects of Calbindin-D9k

CB-D9k (human gene symbol: *S100G*) is the smallest protein with four alpha-helical regions forming an EF-hand pair consisting of a canonical (EF2) and a noncanonical/pseudo (EF1) EF-hand domain, joined by a linker region of 10 amino acids (Fig. 1). The tandem domain is stabilized by a short beta-type interaction between the two Ca^{2+}-binding loops (Kordel et al. 1993). The Ca^{2+}-binding affinities of individual subdomains are several orders of magnitude lower than for the corresponding sites within the intact protein. Thus, EF-hands organized in tandem domains are the physiological relevant structures (Finn et al. 1992; Nelson et al. 2002). $K_{D,Ca}$ values are almost identical for both sites (Table 1), and the two Ca^{2+} ions bind with positive cooperativity (Linse et al. 1991). CB-D9k undergoes Ca^{2+}-induced conformational changes; they are, however, less pronounced than in the prototypical sensor calmodulin. This, together with no identified binding partners, indicates that CB-D9k most likely functions as a Ca^{2+} buffer, rather than a Ca^{2+} sensor (Skelton et al. 1994).

In a rat kidney, CB-D9k is expressed in the loop of Henle, the distal convoluted tubule, and in intercalated cells of the collecting duct (Bindels et al. 1991a). In a mouse, CB-D9k expression is present in late distal convoluted tubules (Lee et al. 2006) and the connecting tubules. Strong expression of CB-D9k is restricted to the first 2 cm of the duodenum (Huybers et al. 2007). CB-D9k is assumed to be a freely mobile molecule in the cytoplasm of specific epithelial cells of the kidney and duodenum. Its Ca^{2+}-binding properties, the regulation by $1,25(OH)_2$ vitamin D3 in the intestine, and CB-D9k's relative electrophoretic mobility led to the name calbindin-D9k (Kallfelz et al. 1967). In other tissues, CB-D9k expression is regulated also in other ways (Choi and Jeung 2008), for example, by estrogen in the uterus (Darwish et al. 1991) or by PTH in mouse primary renal tubular cells (Cao et al. 2002). Suggested functions of CB-D9k include a role in the regulation of Ca^{2+} transport processes across epithelial cells (Bindels et al. 1991b), but also as a Ca^{2+} buffer/shuttle optimally tuned for transcellular Ca^{2+} transport (Choi and Jeung 2008).

Calbindin-D28k

Structural and General Aspects of CB-D28k

CB-D28k ($M_r \approx 29$ kDa; human gene symbol: *CALB1*) has six EF-hand domains, four of which bind Ca^{2+} with medium/high affinity (Cheung et al. 1993). EF-hand 2 is nonfunctional and under physiological conditions EF6 most likely is as well. The four medium/high affinity sites (Nagerl et al. 2000) are considered Ca^{2+}-specific, albeit low affinity Mg^{2+} binding ($K_{D,Mg} \approx 700$ μM) to the same sites at physiological $[Mg^{2+}]_i$ decreases the apparent Ca^{2+} affinity approximately two-fold; additionally, Mg^{2+} binding increases the cooperativity of Ca^{2+} binding (Berggard et al. 2002a). The NMR solution structure of Ca^{2+}-bound rat CB-D28k reveals that it consists of a single, almost globular (ellipsoid) fold with six distinguishable EF-hand domains (Kojetin et al. 2006) (Fig. 1). The on-rates of CB-D28k's fast binding sites ($k_{on} \approx 8. \times 10^7 M^{-1}s^{-1}$) are fast enough to affect the early rising phase of Ca^{2+} transients; the peak amplitude is significantly decreased in WT Purkinje cells, when compared to Purkinje cells from CB-D28k$^{-/-}$ mice (Airaksinen et al. 1997). The fact that the time to peak is not significantly different in CB-D28k$^{-/-}$ and WT Purkinje cells indicates that the initial rise in $[Ca^{2+}]_i$ is principally governed by the properties (density, kinetics) of the Ca^{2+} channels. In Purkinje cell dendrites, CB-D28k acts as a fast Ca^{2+} buffer for the first approximately 100 ms, reducing the peak $[Ca^{2+}]_i$ amplitude to about one half, while later on prolonging the decay by acting as a Ca^{2+} source (Schmidt et al. 2003b).

Although principally considered as a freely mobile protein, CB-D28k binds to several identified target proteins including Ran-binding protein (RanBP) M (Lutz et al. 2003); caspase-3 (Bellido et al. 2000); 3′,5′-cyclic nucleotide phosphodiesterase (Reisner et al. 1992); plasma membrane ATPase (Morgan et al. 1986); L-type Ca^{2+} channel α subunit (*CANAC1C*) (Christakos et al. 2007); *myo*-inositol monophosphatase (IMPase) (Berggard et al. 2002b); and in the kidney to TRPV5 (Lambers et al. 2006). In most cases, binding studies were performed in vitro; the physiological implications of these interactions are not clear yet. In dendrites and spines of Purkinje cells, approximately 20% of CB-D28k molecules are temporarily, that is, for several seconds, immobile by their binding to IMPase, a key enzyme of the $InsP_3$-signaling cascade, and the fraction of immobilized CB-D28k increases by climbing fiber stimulation (Schmidt et al. 2005). In summary, the above findings, together with CB-D28k's Ca^{2+}-dependent conformational changes, indicate additional Ca^{2+} sensor functions (Berggard et al. 2002a).

Functional Aspects of CB-D28k

Reported functions for CB-D28k include a role in Ca^{2+} resorption in the kidney (Boros et al. 2009) and modulation of insulin production and secretion in pancreatic beta cells (Reddy et al. 1997; Sooy et al. 1999). Data on a putative neuroprotective role against excitotoxicity have not yet resulted in a consistent picture (for a review, see Schwaller et al. 2002 and Schwaller 2009). Results obtained in CB-D28k$^{-/-}$ mice are summarized. At first glance, these mice show no phenotype related to development, the general morphology of the nervous system, the visual (Wassle et al. 1998) and auditory (Airaksinen et al. 2000) systems, or behavior under standard housing conditions (Airaksinen et al. 1997). CB-D28k$^{-/-}$ mice show a mild—however more severe than PV$^{-/-}$ mice—impairment in motor coordination/motor learning (Airaksinen et al. 1997; Farre-Castany et al. 2007), likely resulting from the 160-Hz oscillations in the cerebellum (Cheron et al. 2004; Servais et al. 2005). The motor coordination phenotype is due to CB-D28k's absence in Purkinje cells, since this phenotype and alterations in Purkinje cell physiology also occur in mice with Purkinje cell-specific *Calb1* ablation (Barski et al. 2003). At the cellular level, short-term plasticity between cortical multipolar bursting cells and pyramidal cells, or at the mossy fiber-CA3 pyramidal cell synapse in the hippocampus, is affected by CB-D28k (Blatow et al. 2003). The rapid saturation of presynaptic CB-D28k transiently decreases the Ca^{2+} buffering capacity, leading to enhanced facilitation (Blatow et al. 2003) by a mechanism called "facilitation by buffer saturation" (Maeda et al. 1999; Neher 1998). Ca^{2+} buffers, such as CB-D28k, also affect Ca^{2+}-dependent inactivation (CDI) of voltage-dependent Ca^{2+} currents (I_{Ca}). In dentate gyrus granule cells with low or absent CB-D28k expression resulting from Ammon's horn sclerosis in humans (AHS) (Nagerl and Mody 1998) or in mice with *Calb1* gene ablation (Klapstein et al. 1998), CDI is increased, compared to CB-D28k-expressing granule cells, thereby decreasing the total Ca^{2+} load. Increased CDI in the absence of CB-D28k may be viewed as a protective/homeostatic mechanism to limit Ca^{2+} influx in order to augment the resistance against excitotoxicity and to protect the surviving neurons.

Structural and Functional Aspects of Calretinin

Human calretinin ($M_r \approx 31$ kDa; gene symbol: *CALB2*) consists of 271 amino acids and has 6 EF-hand domains, five of which are able to bind Ca^{2+} ions (Schwaller et al. 1997; Stevens and Rogers 1997). Structural data (NMR) is available only for the N-terminal 100 amino acids of rat CR comprising EF-hand domains 1 and 2 (Palczewska et al. 2001). As in CB-D28k, the two domains form a relatively tight structure. CR's Ca^{2+}-dependent conformational changes, together with results from other in vitro studies, suggest that CR also may have Ca^{2+}-sensor functions (Billing-Marczak and Kuznicki 1999).

A role for CR in neuroprotection against glutamate toxicity was postulated, but evidence is most often indirect or obtained in model systems in vitro (D'Orlando et al. 2001; Lukas and Jones 1994; Pike and Cotman 1995). For more details, see Schwaller 2009. $CR^{-/-}$ mice show impaired long-term potentiation (LTP) in the hippocampus (Gurden et al. 1998; Schurmans et al. 1997). While the effect on hippocampal LTP is indirect, the uniform expression of CR in cerebellar granule cells, together with the stereotypic cerebellar organization, has allowed for a detailed investigation of CR in vivo. In $CR^{-/-}$ granule cells, the excitability is increased, they show faster action potentials, and, under conditions generating repetitive spike discharges, show enhanced increases in frequency with injected currents (Gall et al. 2003). This leads to altered Ca^{2+} homeostasis in Purkinje cells. The firing properties of Purkinje cells are altered in alert $CR^{-/-}$ mice: The simple spike-firing rate increased the complex spike duration and the spike pause is shorter (Schiffmann et al. 1999). As in $PV^{-/-}$ and CB-D28k$^{-/-}$ mice, alert $CR^{-/-}$ mice show 160 Hz oscillations (Cheron et al. 2004) that appear phenotypically as an impairment of motor coordination. In alert "rescue" mice, where CR in $CR^{-/-}$ mice is selectively re-expressed in granule cells, granule cell excitability, as well as Purkinje cell firing, resembles that in WT mice. As a consequence, neither 160-Hz oscillations, nor motor coordination impairment, are detected in the rescue mice (Bearzatto et al. 2006).

The similarity of the oscillations and motor coordination deficits in mice deficient for either one of the three CaBPs points toward an effect at the cerebellar network level. In all three knockout strains, the oscillations are temporarily reduced by blocking of: (I) gap junctions between interneurons; (II) N-methyl-D-aspartate receptors; or (III) GABA$_A$ receptors. This indicates that oscillations emerge via a mechanism that synchronizes assemblies of Purkinje cells (mediated by parallel fiber excitation) and the network of chemically-coupled MLI. In addition, recurrent Purkinje cell collaterals (Orduz and Llano 2007) may be implicated in these oscillations (de Solages et al. 2008).

COMPARISON OF THE PHYSIOLOGICAL EFFECTS BROUGHT ABOUT BY CYTOSOLIC Ca^{2+} BUFFERS AND IMMOBILE "Ca^{2+} BUFFERS/STORES," IN PARTICULAR BY MITOCHONDRIA

Cytosolic Ca^{2+} buffers may be viewed as transitory Ca^{2+} sinks/stores, and together with the plasmalemmal extrusion systems, Ca^{2+} uptake into ER compartments and mitochondria serve as a cell's Ca^{2+} "off mechanisms"; Ca^{2+}-loaded buffers, together with release from Ca^{2+}-filled organelles (ER, mitochondria), subserve as the intracellular "on mechanisms" (Berridge et al. 2003). Here, I briefly put side-by-side the role of mitochondria in Ca^{2+} buffering/sequestration with the role of mobile cytosolic Ca^{2+} buffers. The comparison is primarily focused on excitable cells, mostly neurons (for a more detailed role on mitochondria and Ca^{2+} signaling, see Rimessi et al. 2008 and Szabadkai and Duchen 2008). In the large glutamatergic presynaptic terminals of the calyx of Held, mitochondria contribute to increase the rate in $[Ca^{2+}]_i$ decay, when peak $[Ca^{2+}]_i$ is >2.5 μM (Kim et al. 2005). This mitochondria-mediated increase in $[Ca^{2+}]_i$ decay closely resembles the action of the "slow buffer" PV in the same terminals (Muller et al. 2007). At the physiological level, this delayed buffering by PV and mitochondria affects plasticity of synaptic transmission. More importantly, it has an effect on both short-term facilitation and short-term depression. Blocking mitochondrial Ca^{2+} uptake in the calyx of Held slows down the recovery from synaptic depression (Billups and Forsythe 2002), an effect that can be reverted by the addition of 1 mM EGTA. Slow release of mitochondrial Ca^{2+}, but not from ER stores, leads to the post-tetanic potentiation (PTP) in motor axons contacting the opener muscle of the crayfish *Procambarus clarkii* leg (Tang and Zucker 1997). In analogy, at synapses between molecular layer interneurons (MLI), release of Ca^{2+}, likely from Ca^{2+}-bound PV, increases delayed transmitter release after an AP train (Collin et al. 2005). Both PV and mitochondria hardly affect basal synaptic transmission and show similar effects with respect to short-term modulation. The increased removal of intracellular Ca^{2+} by PV prevents the buildup of residual $[Ca^{2+}]$ and thus reduces paired-pulse facilitation at the calyx of Held and in MLI axon terminals; these findings were deduced by comparing $PV^{-/-}$ and WT mice (Collin et al. 2005; Muller et al. 2007). The reduced density of mitochondria in presynaptic axon terminals of synaptophilin knockout mice ($snph^{-/-}$) affects short-term facilitation in cultured $snph^{-/-}$ hippocampal neurons. While basal synaptic transmission evidenced by single EPSCs and miniature AMPA currents is unaltered, facilitation is increased (Kang et al. 2008). This effect closely resembles the one observed at hippocampal interneuron/CA1 pyramidal neuron synapses in $PV^{-/-}$ mice, where facilitation of IPSCs is augmented at stimulation frequencies >33 Hz (Vreugdenhil et al. 2003).

Both mitochondria and cytosolic Ca^{2+} buffers have an effect on the spreading of intracellular Ca^{2+} waves. While in *Xenopus laevis* oocytes, energized mitochondria promote the propagation of Ca^{2+} waves (Boitier et al. 1999), mitochondria in astrocytes limit the rate and extent of Ca^{2+} wave propagation (Jouaville et al. 1995). Also, Ca^{2+} buffers affect Ca^{2+} waves. The fast buffer CR promotes the spreading of $InsP_3$-evoked Ca^{2+} signals in oocytes, while the slow buffer PV shortens the duration of these Ca^{2+} signals and restricts the global responses to discrete localized events (puffs) (Dargan et al. 2004). In summary: (I) Both mitochondria and cytosolic Ca^{2+} buffers participate in the shaping of Ca^{2+} signals in presynaptic terminals and consequently have an effect on short-term modulation of synaptic plasticity, that is, facilitation, potentiation, and depression; (II) They also affect the spreading of Ca^{2+} waves; the effect depends on the cell type, on the kinetic properties of the cytosolic Ca^{2+} buffers, and also on ER luminal regulatory mechanisms involving the luminal Ca^{2+}-binding protein calreticulin (Camacho and Lechleiter 1995); (III) In the systems investigated so far, PV and mitochondria mostly behave as slow-onset buffers, rarely affecting the maximal amplitude of Ca^{2+} signals, but increasing the rate of $[Ca^{2+}]_i$ decay. In presynaptic terminals, the time window most strongly affected by the Ca^{2+} buffering action of PV and mitochondria is $\approx 10-200$ ms after peak $[Ca^{2+}]_i$; (IV) The physiological effect of Ca^{2+} buffering/sequestering by mitochondria and Ca^{2+} buffers is dependent on the cell type, morphology of involved compartments (e.g., presynaptic terminal, soma) and, importantly, the contribution of all other components from the Ca^{2+}

signaling toolkit; (V) Evidently, PV and mitochondria cannot completely replace one another with respect to Ca^{2+} buffering. They are still different with respect to several parameters: mobility, Ca^{2+} storing capacity, effects of Ca^{2+} binding/uptake on metabolism, Mg^{2+} effects, kinetics of Ca^{2+} binding/release, kinetics of synthesis/degradation, etc. The finding that mitochondria volume and PV expression levels are inversely correlated in several systems is discussed later in this article.

THE Ca^{2+} HOMEOSTASOME

How can a simple change in $[Ca^{2+}]_i$ observed in, for example, muscle contraction, neurotransmission, or cell cycle regulation be used by cells to elicit the correct downstream events, as diverse as membrane fusion of neurosecretory vesicles with the plasma membrane or activation/repression of genes? The obvious parameters are the amplitude of the Ca^{2+} signal and the duration or the frequency at which these signals are generated. Subtler regulations comprise the cell morphology, where Ca^{2+} signals are restricted to certain regions: dendrites, soma, or axon terminals of nerve cells. Finally, molecules implicated in Ca^{2+} signaling may be spatially restricted; for example, Ca^{2+} channel subunits in active zones (Bucurenciu et al. 2008). To achieve the necessary precision of Ca^{2+} signals, cells require an accurately tunable system for regulating $[Ca^{2+}]_i$. Opening of plasma membrane Ca^{2+} channels or Ca^{2+} release from internal stores results in an initial increase in $[Ca^{2+}]_i$. The shape of the Ca^{2+} signal in the cytosol, both with respect to space and time, is then modulated by immobile and, if present, by mobile Ca^{2+} buffers. Finally, extrusion systems such as plasma membrane Ca^{2+} pumps (PMCA), the Na^+/Ca^{2+} exchanger (NCX), and organellar uptake by the ER and/or mitochondria restore the initial situation with respect to $[Ca^{2+}]_i$. All of the above components are part of a cell's "Ca^{2+}-signaling toolkit" that is able to regulate the expression of its own components necessary for accurate and cell-specific Ca^{2+} signaling (Berridge et al. 2003; Schwaller 2009). One of the gene regulators is Ca^{2+} itself, and effects are mediated by Ca^{2+}/calmodulin-dependent kinases (CaMK) and Ca^{2+}-regulated phosphatases (e.g., calcineurin). As an example, long-term survival of cultured cerebellar granule cells is dependent on accurate Ca^{2+} signals necessitating temporal changes in the transcription of Ca^{2+}-signaling toolkit components: IP_3R and PMCAs 2 and 3 are up-regulated, while a PMCA4 splice variant and plasma membrane NCX2 are down-regulated in a calcineurin-dependent manner (Carafoli et al. 1999). Such adaptative/homeostatic mechanisms are also induced if Ca^{2+}-signaling toolkit components are functionally compromised (e.g., in genetic diseases) or purposely eliminated in knockout mice. The network of molecules implicated in Ca^{2+} signaling, homeostasis, and its own regulation is termed the Ca^{2+} homeostasome (Schwaller 2007, 2009).

Results on the modulation of the Ca^{2+} homeostasome brought about by altered expression of CaBPs (e.g., in transgenic mice) are summarized. The most surprising finding is that in essentially all CaBP knockout mice and in a given cell type, the deleted CaBP is not compensated by up-regulating expression of one of the more than 240 other EF-hand family members. That is, in the subset of identified "PV-ergic" neurons (e.g., Purkinje cells, stellate, and basket cells in the cerebellum), none of the other Ca^{2+} buffers (CB-D28k, CB-D9k, or CR) are expressed in the above-mentioned neuron subtypes of $PV^{-/-}$ mice (Schwaller et al. 2004). The same holds true for CR-immunoreactive or CB-D28k-immunoreactive neurons in the respective knockout strains, CB-$D28k^{-/-}$ and $CR^{-/-}$ (Airaksinen et al. 1997; Schiffmann et al. 1999). The most notable exception is the up-regulation of CB-D9k in epithelial kidney cells in CB-$D28k^{-/-}$ mice (Zheng et al. 2004). Two plausible explanations for the absence of compensation/homeostatic mechanisms at the level of other Ca^{2+} buffers are presented: (I) Neurons once committed to express a certain Ca^{2+} buffer have permanently inactivated/repressed the promoter for other Ca^{2+} buffers; (II) The specific properties (affinities, kinetics, cooperativity, mobility) of any other Ca^{2+} buffer would not be adequate to restore "normal" Ca^{2+} signaling (Schwaller 2009). Thus, if not at the level of other Ca^{2+} buffers, how do cells cope with the absence of a particular Ca^{2+} buffer? Also, does the absence or impairment of other components of the Ca^{2+}-signaling toolkit affect the expression of Ca^{2+} buffers?

Purkinje cells are characterized by extensive Ca^{2+} signaling and high expression levels of PV and CB-D28k and, thus, are well-suited to address these questions. The two Ca^{2+} buffers are present in the soma, axon, dendrites, and spines, indicating that they are principally mobile proteins. While PV is freely mobile in all compartments (Schmidt et al. 2007a), a fraction of CB-D28k molecules is immobilized in dendrites and spines by its binding to IMPase (Schmidt et al. 2005). The most striking alterations in the absence of PV are the morphological changes observed in the soma. The volume of mitochondria, Ca^{2+} sequestrating organelles that also serve as transient Ca^{2+} stores (Billups and Forsythe 2002; Murchison and Griffith 2000), is increased by about 40% selectively in a narrow compartment underneath the plasma membrane (Chen et al. 2006). Concomitantly, the subplasmalemmal smooth ER compartment is decreased (Fig. 2). These changes in the soma don't occur in CB-D28k$^{-/-}$ Purkinje cells. In the latter, subtle changes in the spine morphology are evident: Spines are longer and spine head volume is increased (Vecellio et al. 2000). In spiny pyramidal cells, spine heads are considered as separate biochemical compartments with negligible Ca^{2+} diffusion via the spine neck (Sabatini et al. 2002). However, modeling studies in Purkinje cell spines have revealed that Ca^{2+} buffers are not only involved in modulating the shape of Ca^{2+} transients within the spines, but together with the spine neck geometry also define the amount of Ca^{2+} ions that may reach the parental dendrite and lead to activation of Ca^{2+}-/CaM-dependent signaling cascades (Schmidt and Eilers 2009). Of the two principal Purkinje cell Ca^{2+} buffers, mostly CB-D28k is involved in spino-dendritic coupling by buffered Ca^{2+} diffusion, while the contribution of PV is minute. This is a likely explanation for the unaltered PV$^{-/-}$ Purkinje cell spine morphology. The absence of CB-D28k and PV not only affects Purkinje cell morphology, but also components directly involved in Ca^{2+} signaling. $Ca_v2.1$ (P/Q type) channels are the major voltage-operated Ca^{2+} channels of mature Purkinje cells ($>$90% of the whole-cell voltage-gated Ca^{2+} current) and regulate Ca^{2+} signaling and excitability of these cells. These channels are regulated by Ca^{2+}-dependent feedback mechanisms consisting of both Ca^{2+}-dependent facilitation (CDF) and inactivation (CDI). While the former process is essentially mediated by the Ca^{2+} sensor calmodulin (CaM) (Lee et al. 2000a), CDI is modulated by synthetic Ca^{2+} buffers (EGTA, BAPTA) and by the Ca^{2+} buffers PV and CB-D28k in vitro (Kreiner and Lee 2005). Of note, PV and CB-D28k affect $Ca_v2.1$ channel function differently than the synthetic buffers EGTA and BAPTA, often presumed to serve as close substitutes for endogenous Ca^{2+} buffers. CDI of $Ca_v2.1$ channel is assumed to depend on intracellular Ca^{2+} microdomains around Ca^{2+} channels. These microdomains are expected to be differently affected by various Ca^{2+} buffers, which in turn specifically influence the inactivation properties of the Ca^{2+} channels. Contrary to the expectation based on in vitro experiments, CDI in Purkinje cells of double knockout (CB-D28k$^{-/-}$PV$^{-/-}$) mice is not increased. However, P-type currents recorded in these cells exhibit increased voltage-dependent inactivation as the result of a decreased expression of the auxiliary $Ca_v\beta2a$ subunit compared to WT neurons (Kreiner et al. 2010) (Fig. 2). This, together with the observation that spontaneous action potentials are not different in CB-D28k$^{-/-}$PV$^{-/-}$ and WT Purkinje cells, indicates that increased inactivation due to molecular switching of $Ca_v2.1$ beta subunits may preserve normal activity-dependent Ca^{2+} signals in the absence of PV and CB-D28k.

A crosstalk between the regulation of a Ca^{2+} channel and Ca^{2+} buffer expression in Purkinje cells is also observed in leaner mice that have a mutation in the pore-forming alpha subunit of $Ca_v2.1$. The strongly-reduced $Ca_v2.1$ Ca^{2+} channel function leads to adaptive changes consisting of a diminished rapid Ca^{2+} buffering/

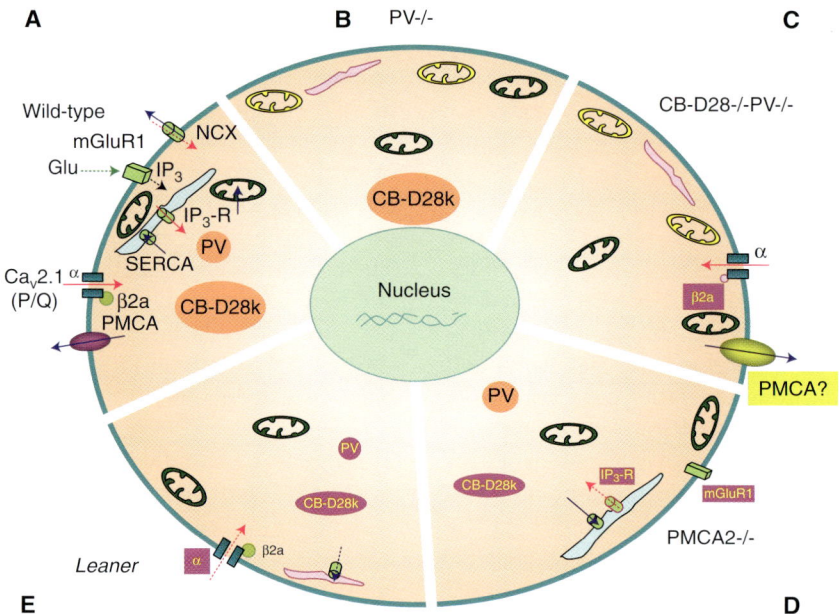

Figure 2. Homeostatic/adaptive changes in the soma of Purkinje cells (PC) caused by malfunctioning or elimination of Ca^{2+}-signaling toolkit component(s); regulation by the Ca^{2+} homeostasome. (*A*) A detailed situation is depicted for wild-type mice. Increases in $[Ca^{2+}]_i$ (red arrows) result from influx via $Ca_V2.1$ (P/Q) channels or release from internal stores (light blue) via the IP_3 receptor. IP_3 is generated by the activation of metabotropic glutamate receptors (mGluR). Ca^{2+} removal systems (blue arrows) include PMCA and NCX in the plasma membrane, SERCA pumps, and mitochondria (green). Identified Ca^{2+}-signaling toolkit components including organelles, which are up- or down-regulated, are marked in yellow and magenta, respectively. (*B*) In $PV^{-/-}$, PC subplasmalemmal mitochondria are increased, while ER volume directly underneath the plasma membrane is decreased. (*C*) In addition to the changes observed in $PV^{-/-}$, in PC lacking both, PV and CB-D28k the auxiliary $Ca_V\beta2a$ subunit of $Ca_V2.1$ (P/Q), is decreased, leading to increased voltage-dependent inactivation of P-type currents. Model studies indicate an up-regulation of Ca^{2+} extrusion systems, possibly PMCA. (*D*) In $PMCA2^{-/-}$ PC, expression of mGluR1 and of IP_3 receptor type 1 (IP3R1), responsible for the Ca^{2+} release from ER stores, is decreased. Also, the cytosolic Ca^{2+} buffering capacity mediated by CB-D28k is decreased. (*E*) In leaner mice PC that are characterized by strongly attenuated $Ca_V2.1$ Ca^{2+} channel function, the rapid Ca^{2+} buffering/sequestering capacity is reduced: PV and CB-D28k are down-regulated and (subplasmalemmal) ER is decreased/impaired, leading to reduced Ca^{2+} uptake.

sequestering capacity of Purkinje cells (Dove et al. 2000). The Ca^{2+} buffering capacity is less than 50% compared to Purkinje cells from WT mice, due to reduced PV and CB-D28k expression levels (Fig. 2). In addition, reduced Ca^{2+} uptake by the (likely subplasmalemmal) ER further contributes to the reduced buffering ability of leaner mice Purkinje cells (Murchison et al. 2002). Also, impairment of proteins responsible for Ca^{2+} extrusion in Purkinje cells activates the Ca^{2+} homeostasome. In Purkinje cells of mice with a mutation in PMCA2 characterized by a reduced Ca^{2+} extrusion, the rise in $[Ca^{2+}]_i$ during high K^+-induced depolarization is decreased. This is indicative of a Ca_V channel down-regulation likely to regulate $[Ca^{2+}]_i$ toward normal homeostasis (Ueno et al. 2002). $PMCA2^{-/-}$ mice have decreased expression levels of CB-D28k (Hu et al. 2006), metabotropic glutamate receptor 1 (mGluR1), and of $InsP_3$ receptor type 1 (IP3R1), responsible for the Ca^{2+} release from ER stores (Kurnellas et al. 2007) (Fig. 2). Again, this reduction in mGluR1-mediated $[Ca^{2+}]_i$ elevation may be viewed as an adaptive mechanism to cope with the reduced Ca^{2+} extrusion. The authors suggest that "the decrease in the expression of mGluR1 and its downstream effectors and perturbations in the mGluR1-signaling complex in the absence of PMCA2 may cumulatively result in aberrant mGluR signaling in Purkinje neurons, leading to cerebellar deficits in the PMCA2-null mouse." However, the severely distorted PMCA2$^{-/-}$ Purkinje cell morphology (smaller cell body, distorted dendritic tree) may also be a likely cause for the ataxic phenotype (Empson et al. 2007). In addition to the identified changes occurring in Purkinje cells of CB-D28k$^{-/-}$PV$^{-/-}$ mice, an up-regulation of Ca^{2+} extrusion/uptake mechanisms was hypothesized, since the decay of dendritic Ca^{2+} signals could be accurately fitted only when applying a two-fold higher Ca^{2+} extrusion rate, compared to the rates sufficient to model the Ca^{2+} transients in PV$^{-/-}$ and WT Purkinje cells (Schmidt et al. 2003b). Currently, no experimental data is available to account for the increased dendritic [Ca^{2+}]$_i$ decay in CB-D28k$^{-/-}$PV$^{-/-}$ Purkinje cells; putative candidates are PMCA isoforms, NCX isoforms, or increased uptake into stores.

What are the evidences that the changes discussed above are not "simple" compensation mechanisms, but may be considered as "truly homeostatic" mechanisms? One argument is the generality of the mechanism and a second one the reciprocality as exemplified for the relationship between PV content and mitochondrial volume. In the absence of PV in PV$^{-/-}$ mice, an up-regulation of mitochondria occurs not only in PV-ergic Purkinje cells (Chen et al. 2006) or cerebellar stellate and basket cells (B Schwaller, unpubl.), but is also seen in PV$^{-/-}$ fast-twitch muscle fibers (Chen et al. 2001). The latter are characterized by high PV expression levels in WT mice. Vice versa, in transgenic mice ectopically expressing PV in striatal neurons (Van Den Bosch et al. 2002), a neuron subpopulation normally not expressing PV, the mitochondrial volume is decreased by almost 50% (Maetzler et al. 2004). This reduction accounts for the heightened excitotoxic injury provoked by a local injection of ibotenic acid. A last example of reciprocality: elimination of PV and CB-D28k from Purkinje cells alters Ca$_v$2.1 channel function (Kreiner et al. 2010), while a reduced Ca^{2+} influx due to a mutation in the Ca$_v$2.1 channel down-regulates PV and CB-D28k (Dove et al. 2000). The elucidation of the pathways and molecular mechanisms responsible for the regulation of the Ca^{2+} homeostasome remains an exciting topic for future research.

ACKNOWLEDGMENTS

I would like to thank Thomas Henzi and Walter-Vincent Blum, University of Fribourg, for helpful comments.

REFERENCES

Airaksinen L, Virkkala J, Aarnisalo A, Meyer M, Ylikoski J, Airaksinen MS. 2000. Lack of calbindin-D28k does not affect hearing level or survival of hair cells in acoustic trauma. *ORL J Otorhinolaryngol Relat Spec* **62:** 9–12.

Airaksinen MS, Eilers J, Garaschuk O, Thoenen H, Konnerth A, Meyer M. 1997. Ataxia and altered dendritic calcium signaling in mice carrying a targeted null mutation of the calbindin D28k gene. *Proc Natl Acad Sci U S A* **94:** 1488–1493.

Akke M, Forsen S, Chazin WJ. 1991. Molecular basis for co-operativity in Ca^{2+} binding to calbindin D9k. 1H nuclear magnetic resonance studies of (Cd^{2+})1-bovine calbindin D9k. *J Mol Biol* **220:** 173–189.

Allbritton NL, Meyer T, Stryer L. 1992. Range of messenger action of calcium ion and inositol 1,4,5,-trisphophate. *Science* **258:** 1812–1815.

Babini E, Bertini I, Capozzi F, Del Bianco C, Hollender D, Kiss T, Luchinat C, Quattrone A. 2004. Solution structure of human beta-parvalbumin and structural comparison with its paralog alpha-parvalbumin and with their rat orthologs. *Biochemistry* **43:** 16076–16085.

Baldellon C, Alattia JR, Strub MP, Pauls T, Berchtold MW, Cave A, Padilla A. 1998. N-15 NMR relaxation studies of calcium-loaded parvalbumin show tight dynamics compared to those of other EF-hand proteins. *Biochemistry* **37:** 9964–9975.

Barski JJ, Hartmann J, Rose CR, Hoebeek F, Morl K, Noll-Hussong M, De Zeeuw CI, Konnerth A, Meyer M. 2003. Calbindin in cerebellar Purkinje cells is a critical determinant of the precision of motor coordination. *J Neurosci* **23:** 3469–3477.

Bearzatto B, Servais L, Roussel C, Gall D, Baba-Aissa F, Schurmans S, de Kerchove d'Exaerde A, Cheron G, Schiffmann SN. 2006. Targeted calretinin expression in granule cells of calretinin-null mice restores normal cerebellar functions. *FASEB J* **20:** 380–382.

Bellido T, Huening M, Raval-Pandya M, Manolagas SC, Christakos S. 2000. Calbindin-D28k is expressed in osteoblastic cells and suppresses their apoptosis by inhibiting caspase-3 activity. *J Biol Chem* **275:** 26328–26332.

Berggard T, Miron S, Onnerfjord P, Thulin E, Akerfeldt KS, Enghild JJ, Akke M, Linse S. 2002a. Calbindin D28k exhibits properties characteristic of a Ca^{2+} sensor. *J Biol Chem* **277:** 16662–16672.

Berggard T, Szczepankiewicz O, Thulin E, Linse S. 2002b. Myo-inositol monophosphatase is an activated target of calbindin D28k. *J Biol Chem* **277:** 41954–41959.

Berridge MJ, Bootman MD, Roderick HL. 2003. Calcium signalling: dynamics, homeostasis and remodelling. *Nat Rev Mol Cell Biol* **4:** 517–529.

Billing-Marczak K, Kuznicki J. 1999. Calretinin–sensor or buffer–function still unclear. *Pol J Pharmacol* **51:** 173–178.

Billups B, Forsythe ID. 2002. Presynaptic mitochondrial calcium sequestration influences transmission at mammalian central synapses. *J Neurosci* **22:** 5840–5847.

Boitier E, Rea R, Duchen MR. 1999. Mitochondria exert a negative feedback on the propagation of intracellular Ca^{2+} waves in rat cortical astrocytes. *J Cell Biol* **145:** 795–808.

Bindels RJ, Hartog A, Timmermans JA, van Os CH. 1991a. Immunocytochemical localization of calbindin-D28k, calbindin-D9k and parvalbumin in rat kidney. *Contrib Nephrol* **91:** 7–13.

Bindels RJ, Timmermans JA, Hartog A, Coers W, van Os CH. 1991b. Calbindin-D9k and parvalbumin are exclusively located along basolateral membranes in rat distal nephron. *J Am Soc Nephrol* **2:** 1122–1129.

Bindreither D, Lackner P. 2009. Structural diversity of calcium binding sites. *Gen Physiol Biophys* **28:** F82–F88.

Blatow M, Caputi A, Burnashev N, Monyer H, Rozov A. 2003. Ca^{2+} buffer saturation underlies paired pulse facilitation in calbindin-D28k-containing terminals. *Neuron* **38:** 79–88.

Boros S, Bindels RJ, Hoenderop JG. 2009. Active Ca^{2+} reabsorption in the connecting tubule. *Pflugers Arch* **458:** 99–109.

Bucurenciu I, Kulik A, Schwaller B, Frotscher M, Jonas P. 2008. Nanodomain Coupling between Ca^{2+} Channels and Ca^{2+} Sensors Promotes Fast and Efficient Transmitter Release at a Cortical GABAergic Synapse. *Neuron* **57:** 536–545.

Caillard O, Moreno H, Schwaller B, Llano I, Celio MR, Marty A. 2000. Role of the calcium-binding protein parvalbumin in short-term synaptic plasticity. *Proc Natl Acad Sci U S A* **97:** 13372–13377.

Camacho P, Lechleiter JD. 1995. Calreticulin inhibits repetitive intracellular Ca^{2+} waves. *Cell* **82:** 765–771.

Cao LP, Bolt MJ, Wei M, Sitrin MD, Chun Li Y. 2002. Regulation of calbindin-D9k expression by 1,25-dihydroxyvitamin D(3) and parathyroid hormone in mouse primary renal tubular cells. *Arch Biochem Biophys* **400:** 118–124.

Carafoli E, Genazzani A, Guerini D. 1999. Calcium controls the transcription of its own transporters and channels in developing neurons. *Biochem Biophys Res Commun* **266:** 624–632.

Celio M, Pauls T, Schwaller B (ed.) 1996. *Guidebook to the Calcium-Binding Proteins*. Oxford University Press, Oxford.

Chen G, Carroll S, Racay P, Dick J, Pette D, Traub I, Vrbova G, Eggli P, Celio M, Schwaller B. 2001. Deficiency in parvalbumin increases fatigue resistance in fast-twitch muscle and upregulates mitochondria. *Am J Physiol (Cell Physiol)* **281:** C114–C122.

Chen G, Racay P, Bichet S, Celio MR, Eggli P, Schwaller B. 2006. Deficiency in parvalbumin, but not in calbindin D-28k upregulates mitochondrial volume and decreases smooth endoplasmic reticulum surface selectively in a peripheral, subplasmalemmal region in the soma of Purkinje cells. *Neuroscience* **142:** 97–105.

Cheron G, Gall D, Servais L, Dan B, Maex R, Schiffmann SN. 2004. Inactivation of calcium-binding protein genes induces 160 Hz oscillations in the cerebellar cortex of alert mice. *J Neurosci* **24:** 434–441.

Cheung WT, Richards DE, Rogers JH. 1993. Calcium binding by chick calretinin and rat calbindin D28k synthesised in bacteria. *Eur J Biochem* **215:** 401–410.

Chin D, Means AR. 2000. Calmodulin: a prototypical calcium sensor. *Trends Cell Biol* **10:** 322–328.

Choi KC, Jeung EB. 2008. Molecular mechanism of regulation of the calcium-binding protein calbindin-D(9k), and its physiological role(s) in mammals: a review of current research. *J Cell Mol Med* **12:** 409–420.

Christakos S, Dhawan P, Peng X, Obukhov AG, Nowycky MC, Benn BS, Zhong Y, Liu Y, Shen Q. 2007. New insights into the function and regulation of vitamin D target proteins. *J Steroid Biochem Mol Biol* **103:** 405–410.

Collin T, Chat M, Lucas MG, Moreno H, Racay P, Schwaller B, Marty A, Llano I. 2005. Developmental changes in parvalbumin regulate presynaptic Ca^{2+} signaling. *J Neurosci* **25:** 96–107.

Cox JA, Milos M, MacManus JP. 1990. Calcium- and magnesium-binding properties of oncomodulin. Direct binding studies and microcalorimetry. *J Biol Chem* **265:** 6633–6637.

Cox JA, Durussel I, Scott DJ, Berchtold MW. 1999. Remodeling of the AB site of rat parvalbumin and oncomodulin into a canonical EF-hand. *Eur J Biochem* **264:** 790–799.

Csillik B, Schwaller B, Mihaly A, Henzi T, Losonczic E, Knyihar-Csillik E. 2010. Upregulated expression of oncomodulin, the beta isoform of parvalbumin, in perikarya and axons in the diencephalon of parvalbumin knockout mice. *Neuroscience* **165:** 749–757.

D'Orlando C, Fellay B, Schwaller B, Salicio V, Bloc A, Gotzos V, Celio MR. 2001. Calretinin and calbindin D-28k delay the onset of cell death after excitotoxic stimulation in transfected P19 cells. *Brain Res* **909:** 145–158.

Dargan SL, Schwaller B, Parker I. 2004. Spatiotemporal patterning of IP3-mediated Ca^{2+} signals in Xenopus oocytes by Ca^{2+}-binding proteins. *J Physiol* **556:** 447–461.

Darwish H, Krisinger J, Furlow JD, Smith C, Murdoch FE, DeLuca HF. 1991. An estrogen-responsive element mediates the transcriptional regulation of calbindin D-9K gene in rat uterus. *J Biol Chem* **266:** 551–558.

de Solages C, Szapiro G, Brunel N, Hakim V, Isope P, Buisseret P, Rousseau C, Barbour B, Lena C. 2008. High-frequency organization and synchrony of activity in the purkinje cell layer of the cerebellum. *Neuron* **58:** 775–788.

Dove LS, Nahm SS, Murchison D, Abbott LC, Griffith WH. 2000. Altered calcium homeostasis in cerebellar Purkinje cells of leaner mutant mice. *J Neurophysiol* **84:** 513–524.

Eberhard M, Erne P. 1994. Calcium and magnesium binding to rat parvalbumin. *Eur J Biochem* **222:** 21–26.

Edmonds B, Reyes R, Schwaller B, Roberts WM. 2000. Calretinin modifies presynaptic calcium signaling in frog saccular hair cells. *Nat Neurosci* **3:** 786–790.

Empson RM, Garside ML, Knopfel T. 2007. Plasma membrane Ca^{2+} ATPase 2 contributes to short-term synapse plasticity at the parallel fiber to Purkinje neuron synapse. *J Neurosci* **27:** 3753–3758.

Faas GC, Schwaller B, Vergara JL, Mody I. 2007. Resolving the fast kinetics of cooperative binding: Ca^{2+} buffering by calretinin. *PLoS Biol* **5:** e311. doi: 310.1371/journal.pbio.0050311.

Farre-Castany MA, Schwaller B, Gregory P, Barski J, Mariethoz C, Eriksson JL, Tetko IV, Wolfer D, Celio MR, Schmutz I, Albrecht U, Villa AE. 2007. Differences in locomotor behavior revealed in mice deficient for the calcium-binding proteins parvalbumin, calbindin D-28k or both. *Behav Brain Res* **178:** 250–261.

Feher JJ. 1984. Measurement of facilitated calcium diffusion by a soluble calcium-binding protein. *Biochim Biophys Acta* **773:** 91–98.

Fierro L, Llano I. 1996. High endogenous calcium buffering in Purkinje cells from rat cerebellar slices. *J Physiol* **496:** 617–625.

Finn BE, Kordel J, Thulin E, Sellers P, Forsen S. 1992. Dissection of calbindin D9k into two Ca^{2+}-binding subdomains by a combination of mutagenesis and chemical cleavage. *FEBS Lett* **298:** 211–214.

Freund TF. 2003. Interneuron Diversity series: Rhythm and mood in perisomatic inhibition. *Trends Neurosci* **26:** 489–495.

Gabso M, Neher E, Spira ME. 1997. Low mobility of the Ca^{2+} buffers in axons of cultured Aplysia neurons. *Neuron* **18:** 473–481.

Gall D, Roussel C, Susa I, D'Angelo E, Rossi P, Bearzatto B, Galas MC, Blum D, Schurmans S, Schiffmann SN. 2003. Altered neuronal excitability in cerebellar granule cells of mice lacking calretinin. *J Neurosci* **23:** 9320–9327.

Gifford JL, Walsh MP, Vogel HJ. 2007. Structures and metal-ion-binding properties of the Ca^{2+}-binding helix-loop-helix EF-hand motifs. *Biochem J* **405:** 199–221.

Gurden H, Schiffmann SN, Lemaire M, Bohme GA, Parmentier M, Schurmans S. 1998. Calretinin expression as a critical component in the control of dentate gyrus long-term potentiation induction in mice. *Eur J Neurosci* **10:** 3029–3033.

Hackney CM, Mahendrasingam S, Penn A, Fettiplace R. 2005. The concentrations of calcium buffering proteins in mammalian cochlear hair cells. *J Neurosci* **25:** 7867–7875.

Hauk TG, Muller A, Lee J, Schwendener R, Fischer D. 2008. Neuroprotective and axon growth promoting effects of intraocular inflammation do not depend on oncomodulin or the presence of large numbers of activated macrophages. *Exp Neurol* **209:** 469–482.

Heizmann CW, Berchtold MW, Rowlerson AM. 1982. Correlation of parvalbumin concentration with relaxation speed in mammalian muscles. *Proc Natl Acad Sci U S A* **79:** 7243–7247.

Henzl MT, Tanner JJ. 2007. Solution structure of Ca^{2+}-free rat beta-parvalbumin (oncomodulin). *Protein Sci* **16:** 1914–1926.

Henzl MT, Tanner JJ. 2008. Solution structure of Ca^{2+}-free rat alpha-parvalbumin. *Protein Sci* **17:** 431–438.

Hu J, Qian J, Borisov O, Pan S, Li Y, Liu T, Deng L, Wannemacher K, Kurnellas M, Patterson C, Elkabes S, Li H. 2006. Optimized proteomic analysis of a mouse model of cerebellar dysfunction using amine-specific isobaric tags. *Proteomics* **6:** 4321–4334.

Huybers S, Naber TH, Bindels RJ, Hoenderop JG. 2007. Prednisolone-induced Ca^{2+} malabsorption is caused by diminished expression of the epithelial Ca^{2+} channel TRPV6. *Am J Physiol Gastrointest Liver Physiol* **292:** G92–G97.

Jouaville LS, Ichas F, Holmuhamedov EL, Camacho P, Lechleiter JD. 1995. Synchronization of calcium waves by mitochondrial substrates in Xenopus laevis oocytes. *Nature* **377:** 438–441.

Kallfelz FA, Taylor AN, Wasserman RH. 1967. Vitamin D-induced calcium binding factor in rat intestinal mucosa. *Proc Soc Exp Biol Med* **125:** 54–58.

Kang JS, Tian JH, Pan PY, Zald P, Li C, Deng C, Sheng ZH. 2008. Docking of axonal mitochondria by syntaphilin controls their mobility and affects short-term facilitation. *Cell* **132:** 137–148.

Kim MH, Korogod N, Schneggenburger R, Ho WK, Lee SH. 2005. Interplay between Na^+/Ca^{2+} exchangers and mitochondria in Ca^{2+} clearance at the calyx of Held. *J Neurosci* **25:** 6057–6065.

Klapstein GJ, Vietla S, Lieberman DN, Gray PA, Airaksinen MS, Thoenen H, Meyer M, Mody I. 1998. Calbindin-D28k fails to protect hippocampal neurons against ischemia in spite of its cytoplasmic calcium buffering properties: evidence from calbindin-D28k knockout mice. *Neuroscience* **85:** 361–373.

Kojetin DJ, Venters RA, Kordys DR, Thompson RJ, Kumar R, Cavanagh J. 2006. Structure, binding interface and hydrophobic transitions of Ca^{2+}-loaded calbindin-D(28K). *Nat Struct Mol Biol* **13:** 641–647.

Kordel J, Skelton NJ, Akke M, Chazin WJ. 1993. High-resolution structure of calcium-loaded calbindin D9k. *J Mol Biol* **231:** 711–734.

Kreiner L, Lee A. 2005. Endogenous and exogenous Ca^{2+} buffers differentially modulate Ca^{2+}-dependent inactivation of $Ca_V 2.1$ Ca^{2+} channels. *J Biol Chem* **281:** 4691–4698.

Kreiner L, Christel CJ, Benveniste M, Schwaller B, Lee A. 2010. Compensatory regulation of Cav2.1 Ca^{2+} channels in cerebellar Purkinje neurons lacking parvalbumin and calbindin D-28k. *J Neurophysiol* **103:** 371–381.

Kretsinger RH, Nockolds CE. 1973. Carp muscle calcium-binding protein. II. Structure determination and general description. *J Biol Chem* **248:** 3313–3326.

Kurnellas MP, Lee AK, Li H, Deng L, Ehrlich DJ, Elkabes S. 2007. Molecular alterations in the cerebellum of the plasma membrane calcium ATPase 2 (PMCA2)-null mouse indicate abnormalities in Purkinje neurons. *Mol Cell Neurosci* **34:** 178–188.

Lambers TT, Mahieu F, Oancea E, Hoofd L, de Lange F, Mensenkamp AR, Voets T, Nilius B, Clapham DE,

Hoenderop JG, Bindels RJ. 2006. Calbindin-D28K dynamically controls TRPV5-mediated Ca^{2+} transport. *Embo J* **25**: 2978–2988.

Lee A, Scheuer T, Catterall WA. 2000a. Ca^{2+}/calmodulin-dependent facilitation and inactivation of P/Q-type Ca^{2+} channels. *J Neurosci* **20**: 6830–6838.

Lee GS, Choi KC, Jeung EB. 2006. Glucocorticoids differentially regulate expression of duodenal and renal calbindin-D9k through glucocorticoid receptor-mediated pathway in mouse model. *Am J Physiol Endocrinol Metab* **290**: E299–E307.

Lee SH, Rosenmund C, Schwaller B, Neher E. 2000b. Differences in Ca^{2+} buffering properties between excitatory and inhibitory hippocampal neurons from the rat. *J Physiol* **525**: 405–418.

Lee SH, Schwaller B, Neher E. 2000c. Kinetics of Ca^{2+} binding to parvalbumin in bovine chromaffin cells: implications for $[Ca^{2+}]$ transients of neuronal dendrites. *J Physiol* **525**: 419–432.

Linse S, Johansson C, Brodin P, Grundstrom T, Drakenberg T, Forsen S. 1991. Electrostatic contributions to the binding of Ca^{2+} in calbindin D9k. *Biochemistry* **30**: 154–162.

Lips MB, Keller BU. 1998. Endogenous calcium buffering in motoneurones of the nucleus hypoglossus from mouse. *J Physiol (Lond)* **511**: 105–117.

Lukas W, Jones KA. 1994. Cortical neurons containing calretinin are selectively resistant to calcium overload and excitotoxicity. *Neuroscience* **61**: 307–316.

Lutz W, Frank EM, Craig TA, Thompson R, Venters RA, Kojetin D, Cavanagh J, Kumar R. 2003. Calbindin D28K interacts with Ran-binding protein M: identification of interacting domains by NMR spectroscopy. *Biochem Biophys Res Commun* **303**: 1186–1192.

Maeda H, Ellis-Davies GC, Ito K, Miyashita Y, Kasai H. 1999. Supralinear Ca^{2+} signaling by cooperative and mobile Ca^{2+} buffering in Purkinje neurons. *Neuron* **24**: 989–1002.

Maetzler W, Nitsch C, Bendfeldt K, Racay P, Vollenweider F, Schwaller B. 2004. Ectopic parvalbumin expression in mouse forebrain neurons increases excitotoxic injury provoked by ibotenic acid injection into the striatum. *Exp Neurol* **186**: 78–88.

Maki M, Kitaura Y, Satoh H, Ohkouchi S, Shibata H. 2002. Structures, functions and molecular evolution of the penta-EF-hand Ca^{2+}-binding proteins. *Biochim Biophys Acta* **1600**: 51–60.

Marenholz I, Heizmann CW, Fritz G. 2004. S100 proteins in mouse and man: from evolution to function and pathology (including an update of the nomenclature). *Biochem Biophys Res Commun* **322**: 1111–1122.

Marsden BJ, Shaw GS, Sykes BD. 1990. Calcium binding proteins. Elucidating the contributions to calcium affinity from an analysis of species variants and peptide fragments. *Biochem Cell Biol* **68**: 587–601.

Martin SR, Linse S, Johansson C, Bayley PM, Forsen S. 1990. Protein surface charges and Ca^{2+} binding to individual sites in calbindin D9k: stopped-flow studies. *Biochemistry* **29**: 4188–4193.

Maughan DW, Godt RE. 1999. Parvalbumin concentration and diffusion coefficient in frog myoplasm. *J Muscle Res Cell Motil* **20**: 199–209.

McLaughlin S, Mulrine N, Gresalfi T, Vaio G, McLaughlin A. 1981. Adsorption of divalent cations to bilayer membranes containing phosphatidylserine. *J Gen Physiol* **77**: 445–473.

Mihaly A, Szente M, Dubravcsik Z, Boda B, Kiraly E, Nagy T, Domonkos A. 1997. Parvalbumin- and calbindin-containing neurons express c-fos protein in primary and secondary (mirror) epileptic foci of the rat neocortex. *Brain Res* **761**: 135–145.

Moreland JL, Gramada A, Buzko OV, Zhang Q, Bourne PE. 2005. The Molecular Biology Toolkit (MBT): A Modular Platform for Developing Molecular Visualization Applications. *BMC Bioinformatics* **6**:21.

Morgan DW, Welton AF, Heick AE, Christakos S. 1986. Specific in vitro activation of Ca,Mg-ATPase by vitamin D-dependent rat renal calcium binding protein (calbindin D28K). *Biochem Biophys Res Commun* **138**: 547–553.

Muller A, Kukley M, Stausberg P, Beck H, Muller W, Dietrich D. 2005. Endogenous Ca^{2+} buffer concentration and Ca^{2+} microdomains in hippocampal neurons. *J Neurosci* **25**: 558–565.

Muller M, Felmy F, Schwaller B, Schneggenburger R. 2007. Parvalbumin is a mobile presynaptic Ca^{2+} buffer in the calyx of held that accelerates the decay of Ca^{2+} and short-term facilitation. *J Neurosci* **27**: 2261–2271.

Muntener M, Kaser L, Weber J, Berchtold MW. 1995. Increase of skeletal muscle relaxation speed by direct injection of parvalbumin cDNA. *Proc Natl Acad Sci U S A* **92**: 6504–6508.

Murchison D, Griffith WH. 2000. Mitochondria buffer non-toxic calcium loads and release calcium through the mitochondrial permeability transition pore and sodium/calcium exchanger in rat basal forebrain neurons. *Brain Res* **854**: 139–151.

Murchison D, Dove LS, Abbott LC, Griffith WH. 2002. Homeostatic compensation maintains Ca^{2+} signaling functions in Purkinje neurons in the leaner mutant mouse. *Cerebellum* **1**: 119–127.

Nagerl UV, Mody I. 1998. Calcium-dependent inactivation of high-threshold calcium currents in human dentate gyrus granule cells. *J Physiol* **509**: 39–45.

Nagerl UV, Novo D, Mody I, Vergara JL. 2000. Binding kinetics of calbindin-D(28k) determined by flash photolysis of caged Ca^{2+}. *Biophys J* **79**: 3009–3018.

Neher E. 1998. Usefulness and limitations of linear approximations to the understanding of Ca^{2+} signals. *Cell Calcium* **24**: 345–357.

Nelson MR, Thulin E, Fagan PA, Forsen S, Chazin WJ. 2002. The EF-hand domain: a globally cooperative structural unit. *Protein Sci* **11**: 198–205.

Orduz D, Llano I. 2007. Recurrent axon collaterals underlie facilitating synapses between cerebellar Purkinje cells. *Proc Natl Acad Sci U S A* **104**: 17831–17836.

Palczewska M, Groves P, Ambrus A, Kaleta A, Kover KE, Batta G, Kuznicki J. 2001. Structural and biochemical characterization of neuronal calretinin domain I-II (residues 1–100). Comparison to homologous calbindin D28k domain I-II (residues 1–93). *Eur J Biochem* **268**: 6229–6237.

Pike CJ, Cotman CW. 1995. Calretinin-immunoreactivity neurons are resistant to B-amyloid toxicity in vitro. *Brain Res* **671**: 293–298.

Raymackers JM, Gailly P, Schoor MC, Pette D, Schwaller B, Hunziker W, Celio MR, Gillis JM. 2000. Tetanus relaxation of fast skeletal muscles of the mouse made parvalbumin deficient by gene inactivation. *J Physiol* **527**: 355–364.

Reddy D, Pollock AS, Clark SA, Sooy K, Vasavada RC, Stewart AF, Honeyman T, Christakos S. 1997. Transfection and overexpression of the calcium binding protein calbindin-D28k results in a stimulatory effect on insulin synthesis in a rat beta cell line (RIN 1046–38). *Proc Natl Acad Sci U S A* **94**: 1961–1966.

Reisner PD, Christakos S, Vanaman TC. 1992. In vitro enzyme activation with calbindin-D28k, the vitamin D-dependent 28 kDa calcium binding protein. *FEBS Lett* **297**: 127–131.

Rimessi A, Giorgi C, Pinton P, Rizzuto R. 2008. The versatility of mitochondrial calcium signals: from stimulation of cell metabolism to induction of cell death. *Biochim Biophys Acta* **1777**: 808–816.

Rodenbaugh DW, Wang W, Davis J, Edwards T, Potter JD, Metzger JM. 2007. Parvalbumin isoforms differentially accelerate cardiac myocyte relaxation kinetics in an animal model of diastolic dysfunction. *Am J Physiol Heart Circ Physiol* **293**: H1705–H1713.

Sabatini BL, Oertner TG, Svoboda K. 2002. The life cycle of Ca^{2+} ions in dendritic spines. *Neuron* **33**: 439–452.

Sakaguchi N, Henzl MT, Thalmann I, Thalmann R, Schulte BA. 1998. Oncomodulin is expressed exclusively by outer hair cells in the organ of Corti. *J Histochem Cytochem* **46**: 29–40.

Schiffmann SN, Cheron G, Lohof A, d'Alcantara P, Meyer M, Parmentier M, Schurmans S. 1999. Impaired motor coordination and Purkinje cell excitability in mice lacking calretinin. *Proc Natl Acad Sci U S A* **96**: 5257–5262.

Schmidt H, Brown EB, Schwaller B, Eilers J. 2003a. Diffusional mobility of parvalbumin in spiny dendrites of cerebellar purkinje neurons quantified by fluorescence recovery after photobleaching. *Biophys J* **84**: 2599–2608.

Schmidt H, Stiefel KM, Racay P, Schwaller B, Eilers J. 2003b. Mutational analysis of dendritic Ca^{2+} kinetics in rodent Purkinje cells: role of parvalbumin and calbindin D28K. *J Physiol* **551**: 13–32.

Schmidt H, Schwaller B, Eilers J. 2005. Calbindin D28k targets myo-inositol monophosphatase in spines and dendrites of cerebellar Purkinje neurons. *Proc Natl Acad Sci U S A* **102**: 5850–5855.

Schmidt H, Arendt O, Brown EB, Schwaller B, Eilers J. 2007a. Parvalbumin is freely mobile in axons, somata and nuclei of cerebellar Purkinje neurones. *J Neurochem* **100**: 727–735.

Schmidt H, Kunerth S, Wilms C, Strotmann R, Eilers J. 2007b. Spino-dendritic cross-talk in rodent Purkinje neurons mediated by endogenous Ca^{2+}-binding proteins. *J Physiol* **581**: 619–629.

Schmidt H, Eilers J. 2009. Spine neck geometry determines spino-dendritic cross-talk in the presence of mobile endogenous calcium binding proteins. *J Comput Neurosci* **27**: 229–243.

Schurmans S, Schiffmann SN, Gurden H, Lemaire M, Lipp H-P, Schwam V, Pochet R, Imperato A, Böhme GA, Parmentier M. 1997. Impaired LTP induction in the dentate gyrus of calretinin-deficient mice. *Proc. Natl. Acad. Sci.* **94**: 10415–10420.

Schwaller B, Durussel I, Jermann D, Herrmann B, Cox JA. 1997. Comparison of the Ca^{2+}-binding properties of human recombinant calretinin-22k and calretinin. *J Biol Chem* **272**: 29663–29671.

Schwaller B, Dick J, Dhoot G, Carroll S, Vrbova G, Nicotera P, Pette D, Wyss A, Bluethmann H, Hunziker W, Celio MR. 1999. Prolonged contraction-relaxation cycle of fast-twitch muscles in parvalbumin knockout mice. *Am J Physiol (Cell Physiol)* **276**: C395–403.

Schwaller B, Meyer M, Schiffmann S. 2002. 'New' functions for 'old' proteins: the role of the calcium-binding proteins calbindin D-28k, calretinin and parvalbumin, in cerebellar physiology. Studies with knockout mice. *Cerebellum* **1**: 241–258.

Schwaller B, Tetko IV, Tandon P, Silveira DC, Vreugdenhil M, Henzi T, Potier MC, Celio MR, Villa AE. 2004. Parvalbumin deficiency affects network properties resulting in increased susceptibility to epileptic seizures. *Mol Cell Neurosci* **25**: 650–663.

Schwaller B. 2007. Emerging Functions of the "Ca2+ Buffers" Parvalbumin, Calbindin D-28k and Calretinin in the Brain. In *Handbook of Neurochemistry and Molecular Neurobiology Neural Protein Metabolism and function* (ed. A Lajtha, N Banik), pp. 197–222. Springer, New York.

Schwaller B. 2009. The continuing disappearance of "pure" Ca^{2+} buffers. *Cell Mol Life Sci* **66**: 275–300.

Servais L, Bearzatto B, Schwaller B, Dumont M, De Saedeleer C, Dan B, Barski JJ, Schiffmann SN, Cheron G. 2005. Mono- and dual-frequency fast cerebellar oscillation in mice lacking parvalbumin and/or calbindin D-28k. *Eur J Neurosci* **22**: 861–870.

Skelton NJ, Kördel J, Akke M, Forsén S, Chazin WJ. 1994. Signal transduction versus buffering activity in Ca++-binding proteins. *Nat Struct Biol* **1**: 239–245.

Sooy K, Schermerhorn T, Noda M, Surana M, Rhoten WB, Meyer M, Fleischer N, Sharp GW, Christakos S. 1999. Calbindin-D(28k) controls $[Ca^{2+}]_i$ and insulin release. Evidence obtained from calbindin-d(28k) knockout mice and beta cell lines. *J Biol Chem* **274**: 34343–34349.

Stevens J, Rogers JH. 1997. Chick calretinin: purification, composition, and metal binding activity of native and recombinant forms. *Protein Expr Purif* **9**: 171–181.

Swain AL, Kretsinger RH, Amma EL. 1989. Restrained least squares refinement of native (calcium) and cadmium-substituted carp parvalbumin using X-ray crystallographic data at 1.6-A resolution. *J Biol Chem* **264**: 16620–16628.

Szabadkai G, Duchen MR. 2008. Mitochondria: the hub of cellular Ca^{2+} signaling. *Physiology* **23**: 84–94.

Szatkowski ML, Westfall MV, Gomez CA, Wahr PA, Michele DE, DelloRusso C, Turner II, Hong KE, Albayya FP, Metzger JM. 2001. In vivo acceleration of heart relaxation performance by parvalbumin gene delivery. *J Clin Invest* **107**: 191–198.

Tang Y, Zucker RS. 1997. Mitochondrial involvement in post-tetanic potentiation of synaptic transmission. *Neuron* **18**: 483–491.

Thalmann I, Shibasaki O, Comegys TH, Henzl MT, Senarita M, Thalmann R. 1995. Detection of a beta-parvalbumin isoform in the mammalian inner ear. *Biochem Biophys Res Commun* **215**: 142–147.

Tikunov BA, Rome LC. 2009. Is high concentration of parvalbumin a requirement for superfast relaxation? *J Muscle Res Cell Motil* **30**: 57–65.

Ueno T, Kameyama K, Hirata M, Ogawa M, Hatsuse H, Takagaki Y, Ohmura M, Osawa N, Kudo Y. 2002. A mouse with a point mutation in plasma membrane Ca^{2+}-ATPase isoform 2 gene showed the reduced Ca^{2+} influx in cerebellar neurons. *Neurosci Res* **42**: 287–297.

Van Den Bosch L, Schwaller B, Vleminckx V, Meijers B, Stork S, Ruehlicke T, Van Houtte E, Klaassen H, Celio MR, Missiaen L, Robberecht W, Berchtold MW. 2002. Protective effect of parvalbumin on excitotoxic motor neuron death. *Exp Neurol* **174**: 150–161.

Vecellio M, Schwaller B, Meyer M, Hunziker W, Celio MR. 2000. Alterations in Purkinje cell spines of calbindin D-28 k and parvalbumin knock-out mice. *Eur J Neurosci* **12**: 945–954.

Vreugdenhil M, Jefferys JG, Celio MR, Schwaller B. 2003. Parvalbumin-deficiency facilitates repetitive IPSCs and gamma oscillations in the hippocampus. *J Neurophysiol* **89**: 1414–1422.

Wang W, Metzger JM. 2008. Parvalbumin isoforms for enhancing cardiac diastolic function. *Cell Biochem Biophys* **51**: 1–8.

Wassle H, Peichl L, Airaksinen MS, Meyer M. 1998. Calcium-binding proteins in the retina of a calbindin-null mutant mouse. *Cell Tissue Res* **292**: 211–218.

Yin Y, Henzl MT, Lorber B, Nakazawa T, Thomas TT, Jiang F, Langer R, Benowitz LI. 2006. Oncomodulin is a macrophage-derived signal for axon regeneration in retinal ganglion cells. *Nat Neurosci* **9**: 843–852.

Zheng W, Xie Y, Li G, Kong J, Feng JQ, Li YC. 2004. Critical role of calbindin-D28k in calcium homeostasis revealed by mice lacking both vitamin D receptor and calbindin-D28k. *J Biol Chem* **279**: 52406–52413.

Zhou Z, Neher E. 1993. Mobile and immobile calcium buffers in bovine adrenal chromaffin cells. *J Physiol* **469**: 245–273.

Organellar Calcium Buffers

Daniel Prins and Marek Michalak

Department of Biochemistry, School of Molecular and Systems Medicine, University of Alberta, Edmonton, Alberta T6G 2H7, Canada

Correspondence: Marek.Michalak@ualberta.ca

Ca^{2+} is an important intracellular messenger affecting many diverse processes. In eukaryotic cells, Ca^{2+} storage is achieved within specific intracellular organelles, especially the endoplasmic/sarcoplasmic reticulum, in which Ca^{2+} is buffered by specific proteins known as Ca^{2+} buffers. Ca^{2+} buffers are a diverse group of proteins, varying in their affinities and capacities for Ca^{2+}, but they typically also carry out other functions within the cell. The wide range of organelles containing Ca^{2+} and the evidence supporting cross-talk between these organelles suggest the existence of a dynamic network of organellar Ca^{2+} signaling, mediated by a variety of organellar Ca^{2+} buffers.

INTRACELLULAR Ca^{2+} DYNAMICS

Ca^{2+} serves as an intracellular messenger in various cellular processes, including muscle contraction, gene expression, and fertilization (Berridge et al. 2003). To use Ca^{2+}, the cell requires a readily mobilizable source of Ca^{2+}, the majority of which is found within the lumen of the endoplasmic reticulum (ER) and/or sarcoplasmic reticulum (SR), but is also located in the Golgi apparatus, peroxisomes, mitochondria, and endolysosomal compartments. It is not surprising that Ca^{2+} levels are of central importance in dictating the function of proteins that reside in intracellular organelles. Although some Ca^{2+} exists as free ions within these compartments, much of it is buffered by specific proteins, simply known as Ca^{2+} buffers. However, these proteins are diverse in terms of structure, oligomerization, affinity and capacity for Ca^{2+}, and physical basis for binding Ca^{2+} ions, one commonality across Ca^{2+} buffers is that they also serve other roles within the cell. These roles include catalyzing the correct folding of other cellular proteins, regulating Ca^{2+} release and retention, and communicating information about Ca^{2+} levels within organelles to other proteins.

ENDOPLASMIC RETICULUM

The ER is a multifunctional organelle within the eukaryotic cell that serves as the single largest Ca^{2+} store inside nonstriated muscle cells. The ER is also responsible for functions as diverse as protein synthesis and posttranslational modification, lipid and steroid metabolism, and drug detoxification (Michalak and Opas 2009). Within the ER lumen, the total concentration of Ca^{2+} is approximately 1 mM, with free Ca^{2+} in the range of approximately 200 μM and the remainder buffered via ER resident proteins (Michalak and Opas 2009). Most ER Ca^{2+}

buffering proteins also serve as ER chaperones or folding enzymes, responsible for correctly protein folding that are transiting through the ER.

Calreticulin

Calreticulin, a 46-kDa ER luminal resident protein, is responsible for buffering up to 50% of ER Ca^{2+} in nonmuscle cells (Nakamura et al. 2001a; Nakamura et al. 2001b). Structurally, calreticulin consists of three distinct domains: N, which is the amino-terminal and implicated (together with the P-domain) in chaperone function; P, which is central, proline-rich, and a structural backbone; and C, which is the carboxy-terminal and critical for Ca^{2+} buffering. The N-domain of calnexin, which is homologous to calreticulin, is primarily β-sheet and globular, with high homology to the structure of plant lectins, suggesting a role in the binding of monoglucosylated substrates to calreticulin as part of its chaperone role (Schrag et al. 2001). Recent work using small angle X-ray scattering (SAXS) showed that the N-domain of calreticulin itself is indeed globular and fits well onto modeled calnexin (Norgaard Toft et al. 2008). The N-domain conformation is dynamic and is stabilized by oligosaccharide binding (Saito et al. 1999; Conte et al. 2007) and the binding of Ca^{2+} at a high-affinity site (Corbett et al. 2000; Conte et al. 2007), though this binding does not affect its affinity for oligosaccharides (Conte et al. 2007).

The P-domain of calreticulin is proline-rich, which suggests that it may show conformational flexibility. Its sequence contains two pairs of repeated amino acid sequences, 1 and 2, in the order 111222 (Fliegel et al. 1989). The P-domain adopts an extended conformation with antiparallel β-sheets between the repeated amino acid sequences; the domain as a whole protrudes out from the N- and C-domains (Ellgaard et al. 2001a; Ellgaard et al. 2001b). The extended protrusion requires a β-hairpin turn at amino acid residues 238 to 241; small angle X-ray scattering (SAXS) analyses indicate that this is in a spiral-like conformation (Norgaard Toft et al. 2008). TROSY-NMR experiments showed that the tip of the P-domain protrusion accounts for the binding site of ERp57, an oxidoreductase-folding enzyme (Frickel et al. 2002).

The C-domain of calreticulin is enriched in negatively charged amino acid residues responsible for its Ca^{2+}-buffering capabilities (Nakamura et al. 2001b). It binds Ca^{2+} with high capacity (25 mol of Ca^{2+} per mol of protein) and low affinity ($K_d = 2$ mM) (Nakamura et al. 2001b). The conformation of this region is highly dependent on variations in Ca^{2+} concentrations within a physiological range (Corbett et al. 2000). SAXS studies indicate that the C-domain of calreticulin may be globular (Norgaard Toft et al. 2008). Ca^{2+} binding stabilizes the C-domain into a more compact, α-helical conformation; the Ca^{2+} concentration required to induce this change, 400 μM, is well within the range of concentrations to which calreticulin would be exposed physiologically (Villamil Giraldo et al. 2009).

Calreticulin Gain-of-Function and Loss-of-Function

Understanding the roles calreticulin and Ca^{2+} play at cellular and organismal levels requires accounting for both its role as a chaperone and as an ER luminal Ca^{2+} buffer. Cell culture and animal models of calreticulin deficiency (loss-of-function) and overexpression (gain-of-function) have shown how tight control of protein folding and Ca^{2+} homeostasis, as exerted by calreticulin, are necessary for proper function and development.

In mice, calreticulin deficiency (loss-of-function) is lethal at embryonic day 14.5 because of impaired cardiogenesis, manifested as abnormally thin ventricular walls and improper myofibrillogenesis (Mesaeli et al. 1999). The insufficiency in this pathway can be traced to a lack of nuclear translocation of NF-AT3 (nuclear factor of activated T-cells), which is activated by calcineurin, a Ca^{2+}-dependent phosphatase. $crt^{-/-}$ cells showed no cytoplasmic Ca^{2+} spike in response to the agonist bradykinin, indicating that the IP_3 pathway to stimulate release of Ca^{2+} from the ER was affected

(Mesaeli et al. 1999). Embryonic lethality could be rescued via heterologous expression of a constitutively active mutant of calcineurin in the heart, demonstrating that calreticulin's role in cardiogenesis depends on its regulation of intracellular Ca^{2+} dynamics from the ER lumen (Guo et al. 2002). Calreticulin may regulate the IP_3R in a Ca^{2+}-dependent manner (Camacho and Lechleiter 1995; Naaby-Hansen et al. 2001). Furthermore, in a cell culture model, cardiomyocytes derived from $crt^{-/-}$ embryonic stem cells showed lower rates of myofibrillar development (Li et al. 2002), thought to involve transcriptional pathways regulated by Ca^{2+} (Lynch et al. 2006). The critical role of Ca^{2+} signaling is underscored by the observation that myofibrillar development could be rescued in $crt^{-/-}$ cells by transient ionomycin treatment (Li et al. 2002). Taken together, these investigations show that the absence of calreticulin severely affects Ca^{2+}-regulated pathways, via both calreticulin's Ca^{2+} buffering and its regulation of protein folding.

Calreticulin overexpression (gain-of-function) is also detrimental to the maintenance of Ca^{2+} homeostasis and the molecular pathways it regulates. Overexpressing calreticulin within the lumen of the ER increases the amount of Ca^{2+} that can be released after inhibition of SERCA via thapsigargin, showing that calreticulin does in fact buffer readily mobilizable Ca^{2+} within the ER (Mery et al. 1996). Furthermore, overexpression of calreticulin delays the process of store-operated Ca^{2+} entry if ER luminal depletion is incomplete but not when depletion is complete, again suggesting that calreticulin buffers a significant amount of Ca^{2+} within the lumen (Fasolato et al. 1998). At a cellular level, augmented calreticulin levels increase the susceptibility of SERCA2a to oxidative stress (Ihara et al. 2005). In the murine heart, targeted overexpression of calreticulin and concomitant increased ER Ca^{2+} capacity results in impaired gap junctions, aberrant Ca^{2+} handling, and arrhythmia, culminating in heart block and death (Nakamura et al. 2001a; Hattori et al. 2007). Interestingly, calreticulin overexpression also results in impaired synthesis of MEF2C (myocyte enhancer factor 2C), a transcription factor implicated in cardiac development, suggesting that calreticulin may play multiple roles in controlling downstream gene transcription (Hattori et al. 2007).

Immunoglobulin Binding Protein BiP/GRP78

BiP (immunoglobulin binding protein), also known as GRP78, similarly to calreticulin, is an ER luminal resident protein known to play an important role in binding to unfolded proteins and assisting in the attainment of the correct conformations. Its most prominent roles are as a regulator of the unfolded protein response, a player in ER stress, and a crucial component of the protein translocation machinery (Dudek et al. 2009). BiP/GRP78 deficiency (loss-of-function) is extremely detrimental and is lethal at embryonic day 3.5 in mice (Luo et al. 2006). In addition to its protein binding functions, BiP/GRP78 serves as an important luminal Ca^{2+} buffer, likely responsible for buffering approximately 25% of the total ER Ca^{2+} load (Lievremont et al. 1997). As BiP/GRP78 is expressed at a higher level than is calreticulin, its Ca^{2+} binding should be considered low capacity, approximately 1–2 mol of Ca^{2+} per mol of protein, and low affinity (Lievremont et al. 1997; Lamb et al. 2006). BiP/GRP78 contains an ATPase domain through which it harnesses energy to fold its client proteins (Dudek et al. 2009). Intriguingly, the affinity of BiP/GRP78 for Ca^{2+} is altered by its binding to ATP or ADP, suggesting interplay between ER Ca^{2+} filling and the chaperone activity of BiP/GRP78 (Lamb et al. 2006). Indeed, prolonged diminishment of ER Ca^{2+} stores can abrogate interactions between BiP/GRP78 and its client proteins (Suzuki et al. 1991) and its ATPase activity is increased when Ca^{2+} levels are low (Kassenbrock and Kelly 1989). BiP/GRP78 also plays a role in the protein translocation machinery, both in binding to unfolded proteins to maintain and prevent their misfolding (Dudek et al. 2009) and, importantly, in closing the Sec61 channel to maintain the ER Ca^{2+} permeability barrier both before and after translocation (Haigh and Johnson 2002; Alder et al.

2005). BiP/GRP78 is also a critical component of the unfolded protein response (Rutkowski and Kaufman 2004) and regulates ER associated degradation (ERAD) (Hebert et al. 2009). In summary, BiP/GRP78 is a crucial ER Ca^{2+} buffer, accounting for about one quarter of the ER's buffering capacity, and also shows Ca^{2+}-dependent regulation of chaperone activity.

GRP94

GRP94 (glucose regulated protein of 94-kDa, endoplasmin, CaBP4) is an ER resident protein with a Ca^{2+} buffering role (Macer and Koch 1988); (Van et al. 1989). It binds Ca^{2+} with high capacity (15 to 28 mol of Ca^{2+} per mol of protein) (Macer and Koch 1988; Van et al. 1989). Ca^{2+} binding can be further divided into four sites of higher affinity ($K_d = 2.0$ μM) and eleven sites of lower affinity ($K_d = 600$ μM) (Van et al. 1989; Ying and Flatmark 2006). Structurally, GRP94 undergoes a conformational change to be less α-helical in the presence of 100 μM Ca^{2+} (Van et al. 1989; Ying and Flatmark 2006). GRP94 binds to ER luminal peptides and this binding is increased in the absence of Ca^{2+}, consistent with its role as a response to ER stress (Ying and Flatmark 2006; Biswas et al. 2007). Moreover, GRP94 binding to ATP induces a conformational change and subsequent dimerization. This enables recognition of an immature client protein, whereas attainment of the correct folding by the client protein leads to yet another shape change in GRP94 causing release of the substrate (Immormino et al. 2004; Rosser et al. 2004). Intriguingly, GRP94 is protective in cardiac pathologies, reducing cardiomyocyte necrosis in response to artificially induced Ca^{2+} overload or simulated ischemic insult (Vitadello et al. 2003). Furthermore, GRP94 expression is increased in hearts undergoing prolonged atrial fibrillation, hypothesized to have a protective role against injury (Vitadello et al. 2001). GRP94 also plays a role in apoptosis, particularly when related to perturbations in Ca^{2+} homeostasis. The hepatitis C virus E2 protein blocks apoptosis by inducing overexpression of GRP94: artificially increasing GRP94 levels blocks apoptosis, whereas siRNA knockdown of GRP94 eliminates the antiapoptotic effect of HCV E2 (Lee et al. 2008). In pancreatic cancer patients, heightened expression of GRP94 correlated with a worsened prognosis because of GRP94's antiapoptotic effect (Pan et al. 2009). Hypoxic conditions are known to upregulate GRP94, underscoring its importance during tumor development (Paris et al. 2005). GRP94 has been shown to regulate apoptosis via stabilization of Ca^{2+} homeostasis (Bando et al. 2004), suggesting that its antiapoptotic effects are a consequence of its Ca^{2+} buffering rather than its peptide-binding activity. GRP94 also protects against cell death induced by ischemia/reperfusion injuries (Bando et al. 2003).

Protein Disulfide Isomerase

Protein disulfide isomerase (PDI) is an ER luminal protein that is capable of isomerizing disulfide bonds on proteins transiting through the ER. It has long been known to bind Ca^{2+} (Macer and Koch 1988) with high capacity (19 mol of Ca^{2+} per mol of protein) (Lebeche et al. 1994); (Lucero et al. 1994). Heterologous expression of PDI in Chinese hamster ovary cells increased the Ca^{2+} stored within ER microsomes (Lucero et al. 1998). Structurally, the carboxy-terminal region of PDI is enriched in paired acidic residues; removal of these residues significantly reduces the amount of Ca^{2+} that PDI can bind (Lucero and Kaminer 1999). In general, PDI was enzymatically more active in the presence of increased Ca^{2+} (Lucero and Kaminer 1999), again showing modulation of enzyme activity by ER Ca^{2+} levels.

ERp72, a PDI-Like Protein

ERp72, an ER luminal protein with thioredoxin-like motifs, is homologous to the rat protein CaBP2 (calcium-binding protein 2) (Van et al. 1993). Its primary function appears to be as a molecular chaperone (Nigam et al. 1994) where it isomerizes disulfide bonds (Rupp et al. 1994). Though it does bind Ca^{2+}, the chaperone activity of ERp72 is unaffected by Ca^{2+} concentrations (Rupp et al. 1994) and overexpression of ERp72 does not increase ER Ca^{2+} stores, suggesting its protein folding activity is more important than its Ca^{2+} binding (Lievremont et al. 1997).

The ER, as the principal intracellular Ca^{2+} store in nonstriated muscle cells, has evolved numerous proteins with the capability to buffer Ca^{2+} with variable capacities and affinities. Most Ca^{2+}-buffering proteins also serve as key modulators of protein folding, both upstream, as in calreticulin, which acts on unfolded proteins to ensure their correct folding, and downstream, as in BiP/GRP78, a key player in the unfolded protein response. In addition, most Ca^{2+} buffers are capable of dynamic responses to variations in ER Ca^{2+} levels, particularly with respect to conformation. It is therefore clear that ER Ca^{2+} buffering capacity is inextricably linked to other cellular processes that are the responsibility of the ER.

SARCOPLASMIC RETICULUM Ca^{2+} STORES

The sarcoplasmic reticulum (SR) is an organelle, closely related to the ER, that is found in cardiac, skeletal, and smooth muscle (Michalak and Opas 2009). Its major function is in regulation of Ca^{2+} fluxes responsible for muscular contraction by serving as the principal intracellular Ca^{2+} store within muscle cells; total SR Ca^{2+} levels, similarly to the ER, are in the millimolar range (Michalak and Opas 2009).

The most abundant Ca^{2+}-binding protein within the SR is calsequestrin, a high capacity Ca^{2+} binding protein. There are two isoforms of calsequestrin, skeletal muscle (calsequestrin-1, or Casq1) and cardiac muscle (calsequestrin-2, or Casq2) (Murphy et al. 2009). Interestingly, although fast-twitch muscle contains almost exclusively calsequestrin-1 isoform, slow-twitch muscle fibers also contain some calsequestrin-2 (Murphy et al. 2009). Calsequestrins across numerous species show high ($>75\%$) homology (Beard et al. 2004); moreover, the two isoforms show high homology to each other (Wei et al. 2009b). Binding of Ca^{2+} to calsequestrin is based on extensive stretches of acidic amino acids within the carboxy-terminal region and is high capacity, with calculated values of mol of Ca^{2+} bound per mol of cardiac calsequestrin ranging from 18 (Slupsky et al. 1987) to 60 (Park et al. 2004). Skeletal muscle calsequestrin binds more Ca^{2+} than its cardiac muscle counterpart: up to 80 mol of Ca^{2+} per mol of protein for skeletal muscle calsequestrin compared to 60 mol per mol of protein for cardiac muscle calsequestrin at saturating concentrations of Ca^{2+} (Park et al. 2004); (Wei et al. 2009b). Ca^{2+} binding to both isoforms of calsequestrin significantly affect the conformation of the protein, making it much more compact and playing a role in oligomerization (Ikemoto et al. 1972; Mitchell et al. 1988; He et al. 1993).

The crystal structure of calsequestrin revealed the presence of three almost identical domains, each of which shows high homology to *Escherichia coli* thioredoxin motifs, suggesting that calsequestrin may be yet another Ca^{2+}-binding protein with the ability to fold other proteins (Wang et al. 1998). In vivo, skeletal muscle calsequestrin consists of long, ribbon-like polymers that were described via electron microscopy (though not yet identified as calsequestrin) as early as 1970 (Franzini-Armstrong 1970). In the presence of Ca^{2+}, calsequestrin forms two different types of dimers, front-to-front and back-to-back. Both types are stabilized by salt bridges and Ca^{2+} ions binding into the negatively charged pocket between two copies of the protein (Wang et al. 1998). Back-to-back dimerization of skeletal calsequestrin occurs first; at a lower Ca^{2+} level (10 μM) whereas at a higher Ca^{2+} concentration (1 mM), calsequestrin forms front-to-front dimers followed by formation of linear polymers (Wang et al. 1998). Electron tomography shows that calsequestrin polymers are physically connected to the junctional region of the SR membrane in muscle cells (Franzini-Armstrong 1970; Wagenknecht et al. 2002). Polymerization behavior differs between the two calsequestrin isoforms: in vitro in the presence of 1 mM Ca^{2+}, calsequestrin-1 is mostly polymerized whereas calsequestrin-2 is primarily monomeric or dimeric (Park et al. 2003; Wei et al. 2009b). Calsequestrin-2's polymerization may be inhibited by its longer carboxy-terminal tail interfering with the Ca^{2+}-binding pocket (Park et al. 2004; Wei et al. 2009b). These differences in polymerization are thought to be related to the varying Ca^{2+} capacities of the two isoforms, though it is unclear if polymerization affects Ca^{2+} capacity or vice versa (Wei et al. 2009b).

Calsequestrin is also known to interact with the RyR. These interactions are thought to be mediated by the transmembrane proteins junctin and triadin, though recent results suggest that only junctin may be crucial for the regulation of RyR proteins by calsequestrin in skeletal muscle (Wei et al. 2009a). Interactions between calsequestrin and the RyR depend on the luminal Ca^{2+} concentration (Gyorke et al. 2004). In cardiac systems, calsequestrin-2 interacts with triadin to inhibit RyR2 channel opening, possibly regulated by SR luminal Ca^{2+} (Terentyev et al. 2007). Interactions between calsequestrin and RyR provide a range of sensitivity over the range of physiological Ca^{2+} concentration (Qin et al. 2008). The skeletal muscle isoform of calsequestrin also interacts with RyR proteins, though in a different and more complex fashion than does the cardiac isoform. Calsequestrin-1 is known to undergo reversible phosphorylation at low Ca^{2+} concentrations corresponding to depleted SR Ca^{2+} stores. Phosphorylated calsequestrin strongly interacts with junctin and hence inhibits the RyR1 channel (Beard et al. 2008). Junctin, but not triadin, is required for skeletal calsequestrin to exert regulatory control over RyR1 channels (Wei et al. 2009a). At resting SR Ca^{2+} levels, skeletal calsequestrin inhibits RyR1; by contrast, cardiac calsequestrin activates both RyR1 and RyR2 (Wei et al. 2009b).

Mice deficient in the *Casq2* gene have only 11% lower SR Ca^{2+} storage in cardiac muscle cells (Knollmann et al. 2006). The hearts of these mice display increased SR Ca^{2+} leak, especially after catecholaminergic stimulation, causing ventricular arrhythmias (Knollmann et al. 2006). Similarly, mutated forms of human cardiac calsequestrin cause certain forms of CPVT (catecholamine-induced polymorphic ventricular tachycardia), a disease associated with polymorphic ventricular tachycardias in response to adrenergic stimulation and exercise, often culminating in sudden death (Terentyev et al. 2006). The exact mechanisms linking mutated cardiac muscle calsequestrin to impaired Ca^{2+} handling phenotypes vary from mutation to mutation. One human mutant of calsequestrin, $G^{112}+5X$, is a frame-shift mutation leading to premature termination; this protein is incapable of binding Ca^{2+} (di Barletta et al. 2006) or polymerizing (Terentyev et al. 2008). A different mutant of cardiac calsequestrin associated with CPVT, $R^{33}Q$, shows normal Ca^{2+} binding, but has lost the capability to inhibit RyR on emptying of SR Ca^{2+} stores, both at rest and after stimulation (Terentyev et al. 2006; Terentyev et al. 2008). The $R^{33}Q$ mutation also impairs front-to-front dimerization, but not to such an extent as to abrogate polymerization (Valle et al. 2008); by contrast, the $L^{167}H$ mutation completely eliminates polymer formation (Valle et al. 2008). The $D^{307}H$ mutation almost entirely eliminates the Ca^{2+} sensitivity of calsequestrin's conformation and also impairs its binding to junctin, preventing interactions with the RyR channel (Houle et al. 2004; Viatchenko-Karpinski et al. 2004; Kalyanasundaram et al. 2009). From these results, it is clear that calsequestrin's importance is not limited to its buffering of Ca^{2+} but also its interaction with, and regulation of, other constituents of the Ca^{2+} release pathway. This conclusion is underscored by the fact that a 25% reduction in calsequestrin levels does not affect SR Ca^{2+} levels, but does increase Ca^{2+} leak via RyR channels, pointing to a regulatory function of calsequestrin on RyR (Chopra et al. 2007).

A mouse model deficient in skeletal muscle calsequestrin (*Casq1*) has also been generated (Paolini et al. 2007). The absence of calsequestrin-1 is not lethal and surprisingly has only a slight effect on the contraction of skeletal muscles (Paolini et al. 2007). Absence of calsequestrin correlates to reduced Ca^{2+} release from the SR (because of impaired Ca^{2+} accumulation near RyRs; partially compensated for by increased expression of release channels) and reduced Ca^{2+} reuptake by the SR (because of impaired Ca^{2+} buffering) (Paolini et al. 2007). *Casq1* null mice have only 20% of the SR Ca^{2+} stores of wild-type mice, suggesting that the role of skeletal calsequestrin in fast-twitch muscles may be primarily its Ca^{2+} buffering, required to keep free luminal Ca^{2+} concentrations low, thus reducing SR Ca^{2+} leaking (Murphy et al. 2009). The increased susceptibility of *Casq1* null mice to heat stroke (Protasi et al. 2009) is likely a consequence of increased SR Ca^{2+} leakage in fast-twitch muscle fibers (Murphy et al. 2009). In mouse models of Duchenne muscular dystrophy, the most affected muscle fibers are associated with decreased levels of calsequestrin and, consequently, impaired Ca^{2+} handling (Pertille et al. 2009).

Experiments with calsequestrin overexpression (gain-of-function) have solidified its role as the key Ca^{2+} buffer within the lumen of the SR. Mouse models specifically overexpressing cardiac muscle calsequestrin in the heart showed cardiac hypertrophy and increased quantities of Ca^{2+} stored within the SR, both supportive of a role for calsequestrin as a Ca^{2+} buffer (Jones et al. 1998; Sato et al. 1998; Schmidt et al. 2000). Overexpression of calsequestrin increases SR Ca^{2+} stores and impairs restoration of free Ca^{2+} levels within the SR lumen (Terentyev et al. 2003; Terentyev et al. 2008), indicating that calsequestrin does buffer Ca^{2+}. Interestingly, the effects of overexpressing calsequestrin were similar to those of introducing a chemical Ca^{2+} buffer into the SR lumen (Terentyev et al. 2002; Terentyev et al. 2003), indicating that calsequestrin's functions in vivo include both Ca^{2+} buffering and regulation of the RyR proteins.

Calsequestrin, the primary Ca^{2+} buffer within the SR, is a fascinating protein to consider in terms of Ca^{2+} binding. Its conformation, oligomerization state, and interactions with other proteins all vary with physiological concentrations of Ca^{2+}. Although it is unquestionably a Ca^{2+} buffer, equally important is its regulation of SR Ca^{2+} release via communications with the RyR. As calsequestrin overexpression can be mimicked via introduction of a nonprotein Ca^{2+} buffer, it seems evident that although some copies of calsequestrin are involved in interactions with the RyR, the vast majority of calsequestrin is simply concerned with Ca^{2+} buffering in vivo.

Minor SR Ca^{2+} Buffering Proteins

The minor SR Ca^{2+} buffering protein HRC (histidine-rich Ca^{2+} binding protein) is a 165-kDa protein first identified and characterized in 1989 (Hofmann et al. 1989; Hofmann et al. 1991). HRC binds Ca^{2+} with high capacity and low affinity and, like calsequestrin, exists as a multimer within the SR lumen (Suk et al. 1999). However, in contrast to calsequestrin, which exists as higher order oligomers in the presence of increasing levels of Ca^{2+}, HRC dissociates from pentamers to dimers and trimers when Ca^{2+} levels are elevated (Suk et al. 1999). Moreover, unlike calsequestrin, HRC is less tightly folded and more sensitive to trypsin digestion in the presence of Ca^{2+} (Suk et al. 1999). HRC is known to be directly involved in Ca^{2+} binding, as shown through studies that correlated overexpression of HRC with impaired SR Ca^{2+} uptake (Gregory et al. 2006) and increased total SR Ca^{2+} stores (Kim et al. 2003). Early evidence showed that HRC, via its glutamate-enriched carboxy-terminal region, interacts with triadin in a Ca^{2+}-dependent manner to anchor HRC to the junctional membrane (Sacchetto et al. 1999; Lee et al. 2001; Sacchetto et al. 2001); this led to the hypothesis that HRC may also be responsible for regulating RyR activity through its interactions with triadin. HRC also interacts with SERCA in cardiac muscle (Arvanitis et al. 2007), leading to the intriguing hypothesis that HRC may link Ca^{2+} release (via triadin) and Ca^{2+} uptake (via SERCA) in the SR lumen (Pritchard and Kranias 2009). HRC is implicated in cardiovascular disease: overexpression (gain-of-function) of HRC provides protection against heart damage induced by ischemia/reperfusion (Zhou et al. 2007) whereas a $S^{96}A$ mutation in HRC has been identified in human patients to correlate with decreased survival in idiopathic dilated cardiomyopathy (Arvanitis et al. 2008). Mice deficient (loss-of-function) in HRC are more liable to develop cardiac hypertrophy when treated with isoproterenol (Jaehnig et al. 2006).

Junctate, a 33-kDa protein localized to the SR membranes, is an alternative splice product of the same gene that produces junctin (Treves

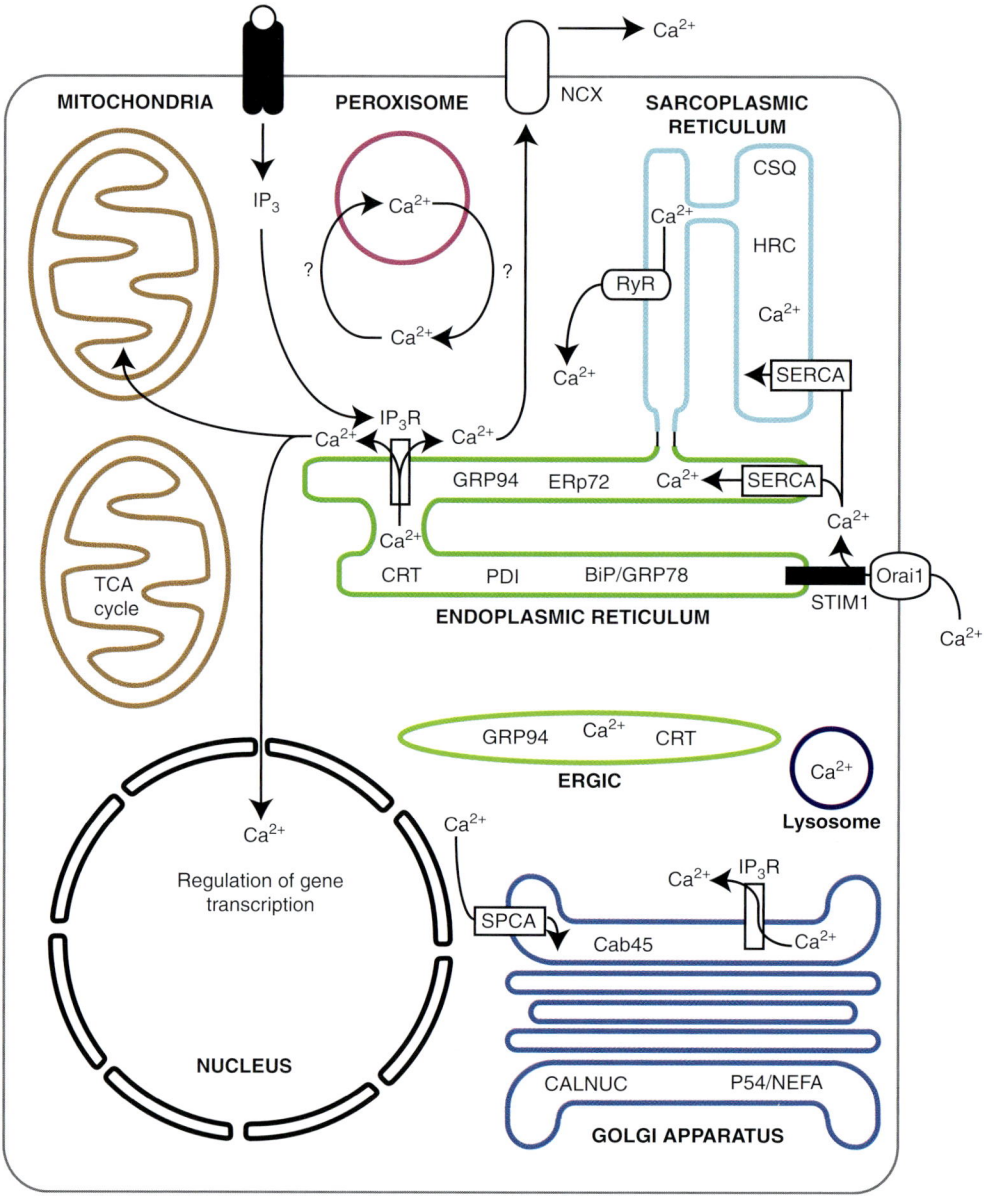

Figure 1. Organellar Ca^{2+} buffering and intracellular Ca^{2+} dynamics. Ca^{2+} is stored within several different organelles including the endoplasmic reticulum (ER), ERGIC, the Golgi apparatus, mitochondria, and peroxisomes. A typical Ca^{2+} release pathway is shown at *top*: an extracellular ligand binds to its receptor, leading to production of IP_3, which binds to the IP_3R on the ER membrane, stimulating release of Ca^{2+} from the ER lumen. Note that the Golgi apparatus also contains IP_3R molecules and thus may contribute to Ca^{2+} release from stores; Golgi Ca^{2+} uptake occurs via SPCA pumps. As shown at *left*, Ca^{2+} released from the ER has several different fates, including regulation of gene transcription within the nucleus; uptake by mitochondria, which are typically closely apposed to the ER network and where Ca^{2+} can affect metabolism; or extrusion from the cell via Na^+/Ca^{2+} exchanger plasma membrane proteins. Peroxisomes are known to maintain Ca^{2+} at higher levels than those of the cytoplasm, but how this gradient is developed and maintained is not yet known. Ca^{2+} levels are elevated in the ER, in the SR, in the ERGIC, and in the Golgi apparatus, with most Ca^{2+} bound to buffering proteins, as shown within these organelles. (*See facing page for legend.*)

et al. 2000). It is a high capacity, moderate affinity Ca^{2+}-binding protein (Treves et al. 2000); consequently, overexpressing junctate in skeletal muscle increases total SR Ca^{2+} stores (Divet et al. 2007). In addition to its luminal domain binding Ca^{2+} directly, in the ER, junctate interacts with IP_3R proteins and TRPC3 (transient receptor potential channel) channels to regulate Ca^{2+} homeostasis, both in response to agonists and to store depletion (Treves et al. 2004). In the SR of mouse cardiomyocytes, junctate interacts with SERCA2a, further linking it with Ca^{2+} handling (Kwon and Kim do 2009). Overexpression of junctate in the mouse heart leads to impaired Ca^{2+} transients, culminating in cardiac hypertrophy and arrhythmias (Hong et al. 2008).

GOLGI Ca^{2+} STORES

Although the ER and SR are accepted as the major organellar Ca^{2+} buffers, several other subcellular organelles, including the Golgi apparatus, mitochondria, peroxisomes, and endosomes/lysosomes, are known to maintain Ca^{2+} at significantly higher levels than those of the cytoplasm.

The Golgi apparatus was identified as containing Ca^{2+} in the millimolar range during the 1990s (Chandra et al. 1994; Pezzati et al. 1997). Golgi Ca^{2+} stores can be released in response to $InsP_3$-triggered pathways and were shown to be developed by proteins residing within the Golgi, instead of existing solely because of vesicular trafficking from the Ca^{2+}-rich ER (Pinton et al. 1998). Ca^{2+} is bound within the Golgi by three proteins, Cab45, P54/NEFA, and CALNUC (nucleobindin), of which the Ca^{2+} buffering capabilities of CALNUC are the most important and have been the most extensively investigated. Cab45 is a 45-kDa soluble protein with six Ca^{2+}-binding EF-hands; interestingly, it was the first soluble protein shown to be retained in the Golgi lumen (Scherer et al. 1996), though later results show Cab45 or its variants are also found in the cytoplasm (Lam et al. 2007) and secreted by pancreatic cancer cells (Gronborg et al. 2006). P54/NEFA (DNA binding/EF hand/acidic amino acid rich region protein) contains two EF-hands, but is thought to function in a regulatory role rather than as a Ca^{2+} buffer (Morel-Huaux et al. 2002). Another Golgi Ca^{2+}-binding protein, CALNUC, shows some homology to calreticulin and is tightly associated with the inner membrane of the Golgi apparatus (Lin et al. 1998). CALNUC contains two EF-hand motifs, one of which has high affinity for Ca^{2+}, making the protein a low capacity, high-affinity Ca^{2+} buffer (Lin et al. 1999). CALNUC is extremely abundant in the Golgi lumen (approximately 0.4% of total Golgi protein), suggesting that it may be the principal Golgi Ca^{2+} store (Lin et al. 1999). The apparent affinity of CALNUC for Ca^{2+} has been reported as 6–7 μM (Lin et al. 1999; Kanuru et al. 2009). CALNUC is tightly folded in the presence of Ca^{2+} and that loss of Ca^{2+} exposes a large hydrophobic surface (de Alba and Tjandra 2004). They further postulate that this surface serves to modulate interactions with other proteins and that it may serve as both a Ca^{2+} buffer and a Ca^{2+} sensor (de Alba and Tjandra 2004). CALNUC has also been shown to be secreted (Lavoie et al. 2002) and may play a role in bone development (Petersson et al. 2004).

Figure 1. (*Continued*) The store-operated Ca^{2+} entry (SOC) is also represented. In response to depleted ER Ca^{2+} stores, an ER luminal Ca^{2+} sensor, STIM1, oligomerizes and migrates to subplasmalemmal punctae where it communicates with Orai1. Orai1 functions as a plasma membrane Ca^{2+} channel that allows for Ca^{2+} entry from the extracellular milieu into the cytoplasm, where it is taken up by SERCA into the ER lumen. BiP/GRP78 binding protein/glucose-regulated protein of 78-kDa; CRT, calreticulin; ERGIC, endoplasmic reticulum/Golgi intermediate complex; GRP94, glucose-regulated protein of 94-kDa; IP_3, inositol trisphosphate; IP_3R, inositol trisphosphate receptor; NCX, Na^+/Ca^{2+} exchanger; P54/NEFA, DNA binding/EF hand/acidic amino acid rich region protein; PDI, protein disulfide isomerase; SERCA, sarcoplasmic/endoplasmic reticulum Ca^{2+}-ATPase; SPCA, secretory pathway Ca^{2+}-transport ATPase; STIM1, stromal interacting molecule 1.

The importance of Golgi Ca^{2+} signaling is shown by the effects of its dysfunction. Hailey-Hailey disease, characterized by skin blistering and lesions, is caused by mutations in the gene encoding SPCA1, the secretory pathway Ca^{2+} ATPase responsible for maintenance of Golgi Ca^{2+} gradients (Hu et al. 2000; Sudbrak et al. 2000). Furthermore, reduced SPCA1 expression in a mouse model caused impaired neural polarity (Sepulveda et al. 2009).

ENDOPLASMIC RETICULUM GOLGI INTERMEDIATE COMPLEX

There is growing evidence that the ERGIC (endoplasmic reticulum Golgi intermediate complex) plays a role in Ca^{2+} storage, as would be expected for a compartment located between two Ca^{2+}-enriched organelles. The ERGIC contains SERCA and significant quantities of GRP94, usually thought of as an ER resident protein, allowing ERGIC to take up Ca^{2+} and buffer it (Ying et al. 2002). Calreticulin has also been detected within the ERGIC and may contribute to its Ca^{2+} buffering (Zuber et al. 2000).

ERGIC and the Golgi apparatus, as the two organelles following the ER in the secretory pathway, also contain Ca^{2+} at significantly higher levels than those found in the cytoplasm. It is therefore intriguing that both compartments have been shown to take up and buffer Ca^{2+} independently of trafficking from the ER. The proteins responsible for Ca^{2+} buffering within ERGIC and the Golgi apparatus are, as with many of the previously discussed Ca^{2+} buffers, multifunctional and responsible for communicating information about Ca^{2+} levels to other proteins.

MITOCHONDRIA

Mitochondria, the cellular organelles responsible primarily for energy generation, are also intracellular Ca^{2+} stores. Within the matrix of mitochondria, Ca^{2+} is stored not bound to Ca^{2+} buffering proteins but instead precipitated out as an insoluble salt, $CaPO_4$. The handling of Ca^{2+} by mitochondria is too complex to be covered in great detail here, but one aspect worth highlighting is how mitochondria communicate with the ER with respect to Ca^{2+} fluxes. Release of Ca^{2+} from the ER results in elevated cytoplasmic Ca^{2+} levels, after which mitochondria take up Ca^{2+} resulting in elevated matrix Ca^{2+} levels (Rizzuto et al. 1992). ER and mitochondria are often tightly apposed in regions known as mitochondria-associated membranes (MAM); the interactions between these two organelles are regulated by cytoplasmic Ca^{2+} levels (Wang et al. 2000). Interestingly, resting cytoplasmic Ca^{2+} levels promote dissociation of the two organelles whereas higher Ca^{2+} levels enhance their association, suggesting that mitochondria may act as buffers to soak up Ca^{2+} released from the ER lumen (Wang et al. 2000). The Ca^{2+} cross-talk between the ER and mitochondria is known to be implicated in cell death and, consequently, may be a target in cancer therapy (Rizzuto et al. 2009).

PEROXISOMES

Recent results indicate that peroxisomes are capable of taking up and storing Ca^{2+} within their lumens at concentrations much higher than those found in the cytoplasm (Raychaudhury et al. 2006; Lasorsa et al. 2008). However, it is not yet known how Ca^{2+} is stored within these organelles and whether there exist specific peroxisomal Ca^{2+} buffering proteins.

ENDOLYSOSOMAL COMPARTMENT

Recent results show that nicotinic acid adenine dinucleotide phosphate (NAADP) acts as a second messenger to mobilize intracellular Ca^{2+} stores through actions on two-pore channels (TPCs) in endosomal membranes (Calcraft et al. 2009). Endosomes and lysosomes may thus be considered to be Ca^{2+} storage organelles; the importance of lysosomal Ca^{2+} is shown by Niemann-Pick type C1 disease, a neurodegenerative lysosomal storage disease caused by perturbations in lysosomal Ca^{2+} handling (Lloyd-Evans et al. 2008). However, the mechanism of lysosomal Ca^{2+} buffering has not been described.

CONCLUSION

Eukaryotic cells use Ca^{2+} as an intracellular signal to regulate a wealth of processes; it is thus unsurprising to see that these cells have evolved equally varied proteins to buffer Ca^{2+} within the cell. The importance of these proteins is shown by consequences of their loss or mutation, which frequently result in cardiac and/or muscular phenotypes, emphasizing the importance of Ca^{2+} in muscular contraction. One characteristic uniting organellar Ca^{2+} buffers, which display such variety in their Ca^{2+} binding and their responses to Ca^{2+}, is their multifunctionality. Instead of being limited to serving as a passive sponge for Ca^{2+} within intracellular organelles, Ca^{2+}-buffering proteins are responsible for a variety of processes, including protein folding, regulation of apoptosis, and regulating Ca^{2+} release pathways. It thus may be limiting to name these proteins simply Ca^{2+} buffers, as this reduces their functions to just one, whereas they typically have other roles that may trump Ca^{2+} buffering in importance. Furthermore, the versatility of Ca^{2+} as a messenger suggests that the presence of proteins that respond to Ca^{2+} within so many organelles correlates to a network of Ca^{2+} signaling linking subcellular organelles, as shown in Figure 1.

ACKNOWLEDGMENTS

Work in our laboratory is supported by grants from the Canadian Institutes of Health Research Heart (MOP-53050. MOP-15415, MOP-15291) and Stroke Foundation of Alberta, Alberta Innovates–Heath Sciences. D. Prins is supported by the Canadian Institutes of Health Research Frederick Banting and Charles Best Canada Graduate Scholarship–Master's Award and a Studentship Award from the Alberta Innovates–Health Sciences.

ABBREVIATIONS USED

ER, endoplasmic reticulum; IP_3, inositol-1,4,5-trisphosphate; IP_3R, inositol-1,4,5-trisphosphate receptor; PDI, protein disulphide isomerase; RyR, ryanodine receptor; SERCA, sarcoplasmic/endoplasmic reticulum Ca^{2+} ATPase; SOCE, store-operated Ca^{2+} entry; SR, sarcoplasmic reticulum; STIM1, stromal-interacting molecule-1.

REFERENCES

Alder NN, Shen Y, Brodsky JL, Hendershot LM, Johnson AE. 2005. The molecular mechanisms underlying BiP-mediated gating of the Sec61 translocon of the endoplasmic reticulum. *J Cell Biol* **168:** 389–399.

Arvanitis DA, Sanoudou D, Kolokathis F, Vafiadaki E, Papalouka V, Kontrogianni-Konstantopoulos A, Theodorakis GN, Paraskevaidis IA, Adamopoulos S, Dorn GWII, et al. 2008. The Ser96Ala variant in histidine-rich calcium-binding protein is associated with life-threatening ventricular arrhythmias in idiopathic dilated cardiomyopathy. *Eur Heart J* **29:** 2514–2525.

Arvanitis DA, Vafiadaki E, Fan GC, Mitton BA, Gregory KN, Del Monte F, Kontrogianni-Konstantopoulos A, Sanoudou D, Kranias EG. 2007. Histidine-rich Ca-binding protein interacts with sarcoplasmic reticulum Ca-ATPase. *Am J Physiol Heart Circ Physiol* **293:** H1581–H1589.

Bando Y, Katayama T, Aleshin AN, Manabe T, Tohyama M. 2004. GRP94 reduces cell death in SH-SY5Y cells perturbated calcium homeostasis. *Apoptosis* **9:** 501–508.

Bando Y, Katayama T, Kasai K, Taniguchi M, Tamatani M, Tohyama M. 2003. GRP94 (94-kDa glucose-regulated protein) suppresses ischemic neuronal cell death against ischemia/reperfusion injury. *Eur J Neurosci* **18:** 829–840.

Beard NA, Laver DR, Dulhunty AF. 2004. Calsequestrin and the calcium release channel of skeletal and cardiac muscle. *Prog Biophys Mol Biol* **85:** 33–69.

Beard NA, Wei L, Cheung SN, Kimura T, Varsanyi M, Dulhunty AF. 2008. Phosphorylation of skeletal muscle calsequestrin enhances its Ca^{2+} binding capacity and promotes its association with junctin. *Cell Calcium* **44:** 363–373.

Berridge MJ, Bootman MD, Roderick HL. 2003. Calcium signalling: dynamics, homeostasis and remodelling. *Nat Rev Mol Cell Biol* **4:** 517–529.

Biswas C, Ostrovsky O, Makarewich CA, Wanderling S, Gidalevitz T, Argon Y. 2007. The peptide-binding activity of GRP94 is regulated by calcium. *Biochem J* **405:** 233–241.

Calcraft PJ, Ruas M, Pan Z, Cheng X, Arredouani A, Hao X, Tang J, Rietdorf K, Teboul L, Chuang KT, et al. 2009. NAADP mobilizes calcium from acidic organelles through two-pore channels. *Nature* **459:** 596–600.

Camacho P, Lechleiter JD. 1995. Calreticulin inhibits repetitive intracellular Ca^{2+} waves. *Cell* **82:** 765–771.

Chandra S, Fewtrell C, Millard PJ, Sandison DR, Webb WW, Morrison GH. 1994. Imaging of total intracellular calcium and calcium influx and efflux in individual resting and stimulated tumor mast cells using ion microscopy. *J Biol Chem* **269:** 15186–15194.

Chopra N, Kannankeril PJ, Yang T, Hlaing T, Holinstat I, Ettensohn K, Pfeifer K, Akin B, Jones LR, Franzini-Armstrong C, et al. 2007. Modest reductions of cardiac calsequestrin increase sarcoplasmic reticulum Ca^{2+} leak independent of luminal Ca^{2+} and trigger ventricular arrhythmias in mice. *Circ Res* **101:** 617–626.

Conte IL, Keith N, Gutierrez-Gonzalez C, Parodi AJ, Caramelo JJ. 2007. The interplay between calcium and the in vitro lectin and chaperone activities of calreticulin. *Biochemistry* **46:** 4671–4680.

Corbett EF, Michalak KM, Oikawa K, Johnson S, Campbell ID, Eggleton P, Kay C, Michalak M. 2000. The conformation of calreticulin is influenced by the endoplasmic reticulum luminal environment. *J Biol Chem* **275:** 27177–27185.

de Alba E, Tjandra N. 2004. Structural studies on the Ca^{2+}-binding domain of human nucleobindin (calnuc). *Biochemistry* **43:** 10039–10049.

di Barletta MR, Viatchenko-Karpinski S, Nori A, Memmi M, Terentyev D, Turcato F, Valle G, Rizzi N, Napolitano C, Gyorke S, et al. 2006. Clinical phenotype and functional characterization of CASQ2 mutations associated with catecholaminergic polymorphic ventricular tachycardia. *Circulation* **114:** 1012–1019.

Divet A, Paesante S, Grasso C, Cavagna D, Tiveron C, Paolini C, Protasi F, Huchet-Cadiou C, Treves S, Zorzato F. 2007. Increased Ca^{2+} storage capacity of the skeletal muscle sarcoplasmic reticulum of transgenic mice over-expressing membrane bound calcium binding protein junctate. *J Cell Physiol* **213:** 464–474.

Dudek J, Benedix J, Cappel S, Greiner M, Jalal C, Muller L, Zimmermann R. 2009. Functions and pathologies of BiP and its interaction partners. *Cell Mol Life Sci* **66:** 1556–1569.

Ellgaard L, Riek R, Braun D, Herrmann T, Helenius A, Wuthrich K. 2001a. Three-dimensional structure topology of the calreticulin P-domain based on NMR assignment. *FEBS Lett* **488:** 69–73.

Ellgaard L, Riek R, Herrmann T, Guntert P, Braun D, Helenius A, Wuthrich K. 2001b. NMR structure of the calreticulin P-domain. *Proc Natl Acad Sci* **98:** 3133–3138.

Fasolato C, Pizzo P, Pozzan T. 1998. Delayed activation of the store-operated calcium current induced by calreticulin overexpression in RBL-1 cells. *Mol Biol Cell* **9:** 1513–1522.

Fliegel L, Burns K, MacLennan DH, Reithmeier RA, Michalak M. 1989. Molecular cloning of the high affinity calcium-binding protein (calreticulin) of skeletal muscle sarcoplasmic reticulum. *J Biol Chem* **264:** 21522–21528.

Franzini-Armstrong C. 1970. Studies of the triad : I. Structure of the Junction in Frog Twitch Fibers. *J Cell Biol* **47:** 488–499.

Frickel EM, Riek R, Jelesarov I, Helenius A, Wuthrich K, Ellgaard L. 2002. TROSY-NMR reveals interaction between ERp57 and the tip of the calreticulin P-domain. *Proc Natl Acad Sci* **99:** 1954–1959.

Gregory KN, Ginsburg KS, Bodi I, Hahn H, Marreez YM, Song Q, Padmanabhan PA, Mitton BA, Waggoner JR, Del Monte F, et al. 2006. Histidine-rich Ca binding protein: a regulator of sarcoplasmic reticulum calcium sequestration and cardiac function. *J Mol Cell Cardiol* **40:** 653–665.

Gronborg M, Kristiansen TZ, Iwahori A, Chang R, Reddy R, Sato N, Molina H, Jensen ON, Hruban RH, Goggins MG, et al. 2006. Biomarker discovery from pancreatic cancer secretome using a differential proteomic approach. *Mol Cell Proteomics* **5:** 157–171.

Guo L, Nakamura K, Lynch J, Opas M, Olson EN, Agellon LB, Michalak M. 2002. Cardiac-specific expression of calcineurin reverses embryonic lethality in calreticulin-deficient mouse. *J Biol Chem* **277:** 50776–50779.

Gyorke I, Hester N, Jones LR, Gyorke S. 2004. The role of calsequestrin, triadin, and junctin in conferring cardiac ryanodine receptor responsiveness to luminal calcium. *Biophys J* **86:** 2121–2128.

Haigh NG, Johnson AE. 2002. A new role for BiP: closing the aqueous translocon pore during protein integration into the ER membrane. *J Cell Biol* **156:** 261–270.

Hattori K, Nakamura K, Hisatomi Y, Matsumoto S, Suzuki M, Harvey RP, Kurihara H, Hattori S, Yamamoto T, Michalak M, et al. 2007. Arrhythmia induced by spatiotemporal overexpression of calreticulin in the heart. *Mol Genet Metab* **91:** 285–293.

He Z, Dunker AK, Wesson CR, Trumble WR. 1993. Ca^{2+}-induced folding and aggregation of skeletal muscle sarcoplasmic reticulum calsequestrin. The involvement of the trifluoperazine-binding site. *J Biol Chem* **268:** 24635–24641.

Hebert DN, Bernasconi R, Molinari M. 2009. ERAD substrates: Which way out? *Semin Cell Dev Biol* **21:** 526–532.

Hofmann SL, Goldstein JL, Orth K, Moomaw CR, Slaughter CA, Brown MS. 1989. Molecular cloning of a histidine-rich Ca^{2+}-binding protein of sarcoplasmic reticulum that contains highly conserved repeated elements. *J Biol Chem* **264:** 18083–18090.

Hofmann SL, Topham M, Hsieh CL, Francke U. 1991. cDNA and genomic cloning of HRC, a human sarcoplasmic reticulum protein, and localization of the gene to human chromosome 19 and mouse chromosome 7. *Genomics* **9:** 656–669.

Hong CS, Kwon SJ, Cho MC, Kwak YG, Ha KC, Hong B, Li H, Chae SW, Chai OH, Song CH, et al. 2008. Overexpression of junctate induces cardiac hypertrophy and arrhythmia via altered calcium handling. *J Mol Cell Cardiol* **44:** 672–682.

Houle TD, Ram ML, Cala SE. 2004. Calsequestrin mutant D307H exhibits depressed binding to its protein targets and a depressed response to calcium. *Cardiovasc Res* **64:** 227–233.

Hu Z, Bonifas JM, Beech J, Bench G, Shigihara T, Ogawa H, Ikeda S, Mauro T, Epstein EH Jr. 2000. Mutations in ATP2C1, encoding a calcium pump, cause Hailey-Hailey disease. *Nat Genet* **24:** 61–65.

Ihara Y, Kageyama K, Kondo T. 2005. Overexpression of calreticulin sensitizes SERCA2a to oxidative stress. *Biochem Biophys Res Commun* **329:** 1343–1349.

Ikemoto N, Bhatnagar GM, Nagy B, Gergely J. 1972. Interaction of divalent cations with the 55,000-dalton protein component of the sarcoplasmic reticulum. Studies of fluorescence and circular dichroism. *J Biol Chem* **247:** 7835–7837.

Immormino RM, Dollins DE, Shaffer PL, Soldano KL, Walker MA, Gewirth DT. 2004. Ligand-induced conformational shift in the N-terminal domain of GRP94, an Hsp90 chaperone. *J Biol Chem* **279:** 46162–46171.

Jaehnig EJ, Heidt AB, Greene SB, Cornelissen I, Black BL. 2006. Increased susceptibility to isoproterenol-induced cardiac hypertrophy and impaired weight gain in mice lacking the histidine-rich calcium-binding protein. *Mol Cell Biol* **26:** 9315–9326.

Jones LR, Suzuki YJ, Wang W, Kobayashi YM, Ramesh V, Franzini-Armstrong C, Cleemann L, Morad M. 1998. Regulation of Ca^{2+} signaling in transgenic mouse cardiac myocytes overexpressing calsequestrin. *J Clin Invest* **101:** 1385–1393.

Kalyanasundaram A, Bal NC, Franzini Armstrong C, Knollmann BC, Periasamy M. 2009. The calsequestrin mutation CASQ2D307H does not affect protein stability and targeting to the JSR but compromises its dynamic regulation of calcium buffering. *J Biol Chem* **285:** 3076–3083.

Kanuru M, Samuel JJ, Balivada LM, Aradhyam GK. 2009. Ion-binding properties of Calnuc, Ca^{2+} versus Mg^{2+}-Calnuc adopts additional and unusual Ca^{2+}-binding sites upon interaction with G-protein. *FEBS J* **276:** 2529–2546.

Kassenbrock CK, Kelly RB. 1989. Interaction of heavy chain binding protein (BiP/GRP78) with adenine nucleotides. *EMBO J* **8:** 1461–1467.

Kim E, Shin DW, Hong CS, Jeong D, Kim DH, Park WJ. 2003. Increased Ca^{2+} storage capacity in the sarcoplasmic reticulum by overexpression of HRC (histidine-rich Ca^{2+} binding protein). *Biochem Biophys Res Commun* **300:** 192–196.

Knollmann BC, Chopra N, Hlaing T, Akin B, Yang T, Ettensohn K, Knollmann BE, Horton KD, Weissman NJ, Holinstat I, et al. 2006. Casq2 deletion causes sarcoplasmic reticulum volume increase, premature Ca^{2+} release, and catecholaminergic polymorphic ventricular tachycardia. *J Clin Invest* **116:** 2510–2520.

Kwon SJ, Kim do H. 2009. Characterization of junctate-SERCA2a interaction in murine cardiomyocyte. *Biochem Biophys Res Commun* **390:** 1389–1394.

Lam PP, Hyvarinen K, Kauppi M, Cosen-Binker L, Laitinen S, Keranen S, Gaisano HY, Olkkonen VM. 2007. A cytosolic splice variant of Cab45 interacts with Munc18b and impacts on amylase secretion by pancreatic acini. *Mol Biol Cell* **18:** 2473–2480.

Lamb HK, Mee C, Xu W, Liu L, Blond S, Cooper A, Charles IG, Hawkins AR. 2006. The affinity of a major Ca^{2+} binding site on GRP78 is differentially enhanced by ADP and ATP. *J Biol Chem* **281:** 8796–8805.

Lasorsa FM, Pinton P, Palmieri L, Scarcia P, Rottensteiner H, Rizzuto R, Palmieri F. 2008. Peroxisomes as novel players in cell calcium homeostasis. *J Biol Chem* **283:** 15300–15308.

Lavoie C, Meerloo T, Lin P, Farquhar MG. 2002. Calnuc, an EF-hand Ca^{2+}-binding protein, is stored and processed in the Golgi and secreted by the constitutive-like pathway in AtT20 cells. *Mol Endocrinol* **16:** 2462–2474.

Lebeche D, Lucero HA, Kaminer B. 1994. Calcium binding properties of rabbit liver protein disulfide isomerase. *Biochem Biophys Res Commun* **202:** 556–561.

Lee HG, Kang H, Kim DH, Park WJ. 2001. Interaction of HRC (histidine-rich Ca^{2+}-binding protein) and triadin in the lumen of sarcoplasmic reticulum. *J Biol Chem* **276:** 39533–39538.

Lee SH, Song R, Lee MN, Kim CS, Lee H, Kong YY, Kim H, Jang SK. 2008. A molecular chaperone glucose-regulated protein 94 blocks apoptosis induced by virus infection. *Hepatology* **47:** 854–866.

Li J, Puceat M, Perez-Terzic C, Mery A, Nakamura K, Michalak M, Krause KH, Jaconi ME. 2002. Calreticulin reveals a critical Ca^{2+} checkpoint in cardiac myofibrillogenesis. *J Cell Biol* **158:** 103–113.

Lievremont JP, Rizzuto R, Hendershot L, Meldolesi J. 1997. BiP, a major chaperone protein of the endoplasmic reticulum lumen, plays a direct and important role in the storage of the rapidly exchanging pool of Ca^{2+}. *J Biol Chem* **272:** 30873–30879.

Lin P, Le-Niculescu H, Hofmeister R, McCaffery JM, Jin M, Hennemann H, McQuistan T, De Vries L, Farquhar MG. 1998. The mammalian calcium-binding protein, nucleobindin (CALNUC), is a Golgi resident protein. *J Cell Biol* **141:** 1515–1527.

Lin P, Yao Y, Hofmeister R, Tsien RY, Farquhar MG. 1999. Overexpression of CALNUC (nucleobindin) increases agonist and thapsigargin releasable Ca^{2+} storage in the Golgi. *J Cell Biol* **145:** 279–289.

Lloyd-Evans E, Morgan AJ, He X, Smith DA, Elliot-Smith E, Sillence DJ, Churchill GC, Schuchman EH, Galione A, Platt FM. 2008. Niemann-Pick disease type C1 is a sphingosine storage disease that causes deregulation of lysosomal calcium. *Nat Med* **14:** 1247–1255.

Lucero HA, Kaminer B. 1999. The role of calcium on the activity of ERcalcistorin/protein-disulfide isomerase and the significance of the C-terminal and its calcium binding. A comparison with mammalian protein-disulfide isomerase. *J Biol Chem* **274:** 3243–3251.

Lucero HA, Lebeche D, Kaminer B. 1994. ERcalcistorin/protein disulfide isomerase (PDI). Sequence determination and expression of a cDNA clone encoding a calcium storage protein with PDI activity from endoplasmic reticulum of the sea urchin egg. *J Biol Chem* **269:** 23112–23119.

Lucero HA, Lebeche D, Kaminer B. 1998. ERcalcistorin/protein-disulfide isomerase acts as a calcium storage protein in the endoplasmic reticulum of a living cell. Comparison with calreticulin and calsequestrin. *J Biol Chem* **273:** 9857–9863.

Luo S, Mao C, Lee B, Lee AS. 2006. GRP78/BiP is required for cell proliferation and protecting the inner cell mass from apoptosis during early mouse embryonic development. *Mol Cell Biol* **26:** 5688–5697.

Lynch JM, Chilibeck K, Qui Y, Michalak M. 2006. Assembling pieces of the cardiac puzzle; calreticulin and calcium-dependent pathways in cardiac development, health, and disease. *Trends Cardiovasc Med* **16:** 65–69.

Macer DR, Koch GL. 1988. Identification of a set of calcium-binding proteins in reticuloplasm, the luminal content of the endoplasmic reticulum. *J Cell Sci* **91:** 61–70.

Mery L, Mesaeli N, Michalak M, Opas M, Lew DP, Krause KH. 1996. Overexpression of calreticulin increases intracellular Ca^{2+} storage and decreases store-operated Ca^{2+} influx. *J Biol Chem* **271:** 9332–9339.

Mesaeli N, Nakamura K, Zvaritch E, Dickie P, Dziak E, Krause KH, Opas M, MacLennan DH, Michalak M. 1999. Calreticulin is essential for cardiac development. *J Cell Biol* **144:** 857–868.

Michalak M, Opas M. 2009. Endoplasmic and sarcoplasmic reticulum in the heart. *Trends Cell Biol* **19**: 253–259.

Mitchell RD, Simmerman HK, Jones LR. 1988. Ca^{2+} binding effects on protein conformation and protein interactions of canine cardiac calsequestrin. *J Biol Chem* **263**: 1376–1381.

Morel-Huaux VM, Pypaert M, Wouters S, Tartakoff AM, Jurgan U, Gevaert K, Courtoy PJ. 2002. The calcium-binding protein p54/NEFA is a novel luminal resident of medial Golgi cisternae that traffics independently of mannosidase II. *Eur J Cell Biol* **81**: 87–100.

Murphy RM, Larkins NT, Mollica JP, Beard NA, Lamb GD. 2009. Calsequestrin content and SERCA determine normal and maximal Ca^{2+} storage levels in sarcoplasmic reticulum of fast- and slow-twitch fibres of rat. *J Physiol* **587**: 443–460.

Naaby-Hansen S, Wolkowicz MJ, Klotz K, Bush LA, Westbrook VA, Shibahara H, Shetty J, Coonrod SA, Reddi PP, Shannon J, et al. 2001. Co-localization of the inositol 1,4,5-trisphosphate receptor and calreticulin in the equatorial segment and in membrane bounded vesicles in the cytoplasmic droplet of human spermatozoa. *Mol Hum Reprod* **7**: 923–933.

Nakamura K, Robertson M, Liu G, Dickie P, Guo JQ, Duff HJ, Opas M, Kavanagh K, Michalak M. 2001a. Complete heart block and sudden death in mice overexpressing calreticulin. *J Clin Invest* **107**: 1245–1253.

Nakamura K, Zuppini A, Arnaudeau S, Lynch J, Ahsan I, Krause R, Papp S, De Smedt H, Parys JB, Muller-Esterl W, et al. 2001b. Functional specialization of calreticulin domains. *J Cell Biol* **154**: 961–972.

Nigam SK, Goldberg AL, Ho S, Rohde MF, Bush KT, Sherman M. 1994. A set of endoplasmic reticulum proteins possessing properties of molecular chaperones includes Ca^{2+}-binding proteins and members of the thioredoxin superfamily. *J Biol Chem* **269**: 1744–1749.

Norgaard Toft K, Larsen N, Steen Jorgensen F, Hojrup P, Houen G, Vestergaard B. 2008. Small angle X-ray scattering study of calreticulin reveals conformational plasticity. *Biochim Biophys Acta* **1784**: 1265–1270.

Pan Z, Erkan M, Streit S, Friess H, Kleeff J. 2009. Silencing of GRP94 expression promotes apoptosis in pancreatic cancer cells. *Int J Oncol* **35**: 823–828.

Paolini C, Quarta M, Nori A, Boncompagni S, Canato M, Volpe P, Allen PD, Reggiani C, Protasi F. 2007. Reorganized stores and impaired calcium handling in skeletal muscle of mice lacking calsequestrin-1. *J Physiol* **583**: 767–784.

Paris S, Denis H, Delaive E, Dieu M, Dumont V, Ninane N, Raes M, Michiels C. 2005. Up-regulation of 94-kDa glucose-regulated protein by hypoxia-inducible factor-1 in human endothelial cells in response to hypoxia. *FEBS Lett* **579**: 105–114.

Park H, Park IY, Kim E, Youn B, Fields K, Dunker AK, Kang C. 2004. Comparing skeletal and cardiac calsequestrin structures and their calcium binding: a proposed mechanism for coupled calcium binding and protein polymerization. *J Biol Chem* **279**: 18026–18033.

Park H, Wu S, Dunker AK, Kang C. 2003. Polymerization of calsequestrin. Implications for Ca^{2+} regulation. *J Biol Chem* **278**: 16176–16182.

Pertille A, de Carvalho CL, Matsumura CY, Neto HS, Marques MJ. 2009. Calcium-binding proteins in skeletal muscles of the mdx mice: potential role in the pathogenesis of Duchenne muscular dystrophy. *Int J Exp Pathol* **91**: 63–71.

Petersson U, Somogyi E, Reinholt FP, Karlsson T, Sugars RV, Wendel M. 2004. Nucleobindin is produced by bone cells and secreted into the osteoid, with a potential role as a modulator of matrix maturation. *Bone* **34**: 949–960.

Pezzati R, Bossi M, Podini P, Meldolesi J, Grohovaz F. 1997. High-resolution calcium mapping of the endoplasmic reticulum-Golgi-exocytic membrane system. Electron energy loss imaging analysis of quick frozen-freeze dried PC12 cells. *Mol Biol Cell* **8**: 1501–1512.

Pinton P, Pozzan T, Rizzuto R. 1998. The Golgi apparatus is an inositol 1,4,5-trisphosphate-sensitive Ca^{2+} store, with functional properties distinct from those of the endoplasmic reticulum. *EMBO J* **17**: 5298–5308.

Pritchard TJ, Kranias EG. 2009. Junctin and the histidine-rich Ca^{2+} binding protein: potential roles in heart failure and arrhythmogenesis. *J Physiol* **587**: 3125–3133.

Protasi F, Paolini C, Dainese M. 2009. Calsequestrin-1: a new candidate gene for malignant hyperthermia and exertional/environmental heat stroke. *J Physiol* **587**: 3095–3100.

Qin J, Valle G, Nani A, Nori A, Rizzi N, Priori SG, Volpe P, Fill M. 2008. Luminal Ca^{2+} regulation of single cardiac ryanodine receptors: insights provided by calsequestrin and its mutants. *J Gen Physiol* **131**: 325–334.

Raychaudhury B, Gupta S, Banerjee S, Datta SC. 2006. Peroxisome is a reservoir of intracellular calcium. *Biochim Biophys Acta* **1760**: 989–992.

Rizzuto R, Marchi S, Bonora M, Aguiari P, Bononi A, De Stefani D, Giorgi C, Leo S, Rimessi A, Siviero R, et al. 2009. Ca^{2+} transfer from the ER to mitochondria: when, how and why. *Biochim Biophys Acta* **1787**: 1342–1351.

Rizzuto R, Simpson AW, Brini M, Pozzan T. 1992. Rapid changes of mitochondrial Ca^{2+} revealed by specifically targeted recombinant aequorin. *Nature* **358**: 325–327.

Rosser MF, Trotta BM, Marshall MR, Berwin B, Nicchitta CV. 2004. Adenosine nucleotides and the regulation of GRP94-client protein interactions. *Biochemistry* **43**: 8835–8845.

Rupp K, Birnbach U, Lundstrom J, Van PN, Soling HD. 1994. Effects of CaBP2, the rat analog of ERp72, and of CaBP1 on the refolding of denatured reduced proteins. Comparison with protein disulfide isomerase. *J Biol Chem* **269**: 2501–2507.

Rutkowski DT, Kaufman RJ. 2004. A trip to the ER: coping with stress. *Trends Cell Biol* **14**: 20–28.

Sacchetto R, Damiani E, Turcato F, Nori A, Margreth A. 2001. Ca^{2+}-dependent interaction of triadin with histidine-rich Ca^{2+}-binding protein carboxyl-terminal region. *Biochem Biophys Res Commun* **289**: 1125–1134.

Sacchetto R, Turcato F, Damiani E, Margreth A. 1999. Interaction of triadin with histidine-rich Ca^{2+}-binding protein at the triadic junction in skeletal muscle fibers. *J Muscle Res Cell Motil* **20**: 403–415.

Saito Y, Ihara Y, Leach MR, Cohen-Doyle MF, Williams DB. 1999. Calreticulin functions in vitro as a molecular

Sato Y, Ferguson DG, Sako H, Dorn GWII, Kadambi VJ, Yatani A, Hoit BD, Walsh RA, Kranias EG. 1998. Cardiac-specific overexpression of mouse cardiac calsequestrin is associated with depressed cardiovascular function and hypertrophy in transgenic mice. *J Biol Chem* **273**: 28470–28477.

Scherer PE, Lederkremer GZ, Williams S, Fogliano M, Baldini G, Lodish HF. 1996. Cab45, a novel Ca^{2+}-binding protein localized to the Golgi lumen. *J Cell Biol* **133**: 257–268.

Schmidt AG, Kadambi VJ, Ball N, Sato Y, Walsh RA, Kranias EG, Hoit BD. 2000. Cardiac-specific overexpression of calsequestrin results in left ventricular hypertrophy, depressed force-frequency relation and pulsus alternans in vivo. *J Mol Cell Cardiol* **32**: 1735–1744.

Schrag JD, Bergeron JJ, Li Y, Borisova S, Hahn M, Thomas DY, Cygler M. 2001. The Structure of calnexin, an ER chaperone involved in quality control of protein folding. *Mol Cell* **8**: 633–644.

Sepulveda MR, Vanoevelen J, Raeymaekers L, Mata AM, Wuytack F. 2009. Silencing the SPCA1 (secretory pathway Ca^{2+}-ATPase isoform 1) impairs Ca^{2+} homeostasis in the Golgi and disturbs neural polarity. *J Neurosci* **29**: 12174–12182.

Slupsky JR, Ohnishi M, Carpenter MR, Reithmeier RA. 1987. Characterization of cardiac calsequestrin. *Biochemistry* **26**: 6539–6544.

Sudbrak R, Brown J, Dobson-Stone C, Carter S, Ramser J, White J, Healy E, Dissanayake M, Larregue M, Perrussel M, et al. 2000. Hailey-Hailey disease is caused by mutations in ATP2C1 encoding a novel Ca^{2+} pump. *Hum Mol Genet* **9**: 1131–1140.

Suk JY, Kim YS, Park WJ. 1999. HRC (histidine-rich Ca^{2+} binding protein) resides in the lumen of sarcoplasmic reticulum as a multimer. *Biochem Biophys Res Commun* **263**: 667–671.

Suzuki CK, Bonifacino JS, Lin AY, Davis MM, Klausner RD. 1991. Regulating the retention of T-cell receptor alpha chain variants within the endoplasmic reticulum: Ca^{2+}-dependent association with BiP. *J Cell Biol* **114**: 189–205.

Terentyev D, Kubalova Z, Valle G, Nori A, Vedamoorthyrao S, Terentyeva R, Viatchenko-Karpinski S, Bers DM, Williams SC, Volpe P, et al. 2008. Modulation of SR Ca release by luminal Ca and calsequestrin in cardiac myocytes: effects of CASQ2 mutations linked to sudden cardiac death. *Biophys J* **95**: 2037–2048.

Terentyev D, Nori A, Santoro M, Viatchenko-Karpinski S, Kubalova Z, Gyorke I, Terentyeva R, Vedamoorthyrao S, Blom NA, Valle G, et al. 2006. Abnormal interactions of calsequestrin with the ryanodine receptor calcium release channel complex linked to exercise-induced sudden cardiac death. *Circ Res* **98**: 1151–1158.

Terentyev D, Viatchenko-Karpinski S, Gyorke I, Volpe P, Williams SC, Gyorke S. 2003. Calsequestrin determines the functional size and stability of cardiac intracellular calcium stores: Mechanism for hereditary arrhythmia. *Proc Natl Acad Sci* **100**: 11759–11764.

Terentyev D, Viatchenko-Karpinski S, Valdivia HH, Escobar AL, Gyorke S. 2002. Luminal Ca^{2+} controls termination and refractory behavior of Ca^{2+}-induced Ca^{2+} release in cardiac myocytes. *Circ Res* **91**: 414–420.

Terentyev D, Viatchenko-Karpinski S, Vedamoorthyrao S, Oduru S, Gyorke I, Williams SC, Gyorke S. 2007. Protein protein interactions between triadin and calsequestrin are involved in modulation of sarcoplasmic reticulum calcium release in cardiac myocytes. *J Physiol* **583**: 71–80.

Treves S, Feriotto G, Moccagatta L, Gambari R, Zorzato F. 2000. Molecular cloning, expression, functional characterization, chromosomal localization, and gene structure of junctate, a novel integral calcium binding protein of sarco(endo)plasmic reticulum membrane. *J Biol Chem* **275**: 39555–39568.

Treves S, Franzini-Armstrong C, Moccagatta L, Arnoult C, Grasso C, Schrum A, Ducreux S, Zhu MX, Mikoshiba K, Girard T, et al. 2004. Junctate is a key element in calcium entry induced by activation of InsP3 receptors and/or calcium store depletion. *J Cell Biol* **166**: 537–548.

Valle G, Galla D, Nori A, Priori SG, Gyorke S, de Filippis V, Volpe P. 2008. Catecholaminergic polymorphic ventricular tachycardia-related mutations R33Q and L167H alter calcium sensitivity of human cardiac calsequestrin. *Biochem J* **413**: 291–303.

Van PN, Peter F, Soling HD. 1989. Four intracisternal calcium-binding glycoproteins from rat liver microsomes with high affinity for calcium. No indication for calsequestrin-like proteins in inositol 1,4,5-trisphosphate-sensitive calcium sequestering rat liver vesicles. *J Biol Chem* **264**: 17494–17501.

Van PN, Rupp K, Lampen A, Soling HD. 1993. CaBP2 is a rat homolog of ERp72 with proteindisulfide isomerase activity. *Eur J Biochem* **213**: 789–795.

Viatchenko-Karpinski S, Terentyev D, Gyorke I, Terentyeva R, Volpe P, Priori SG, Napolitano C, Nori A, Williams SC, Gyorke S. 2004. Abnormal calcium signaling and sudden cardiac death associated with mutation of calsequestrin. *Circ Res* **94**: 471–477.

Villamil Giraldo AM, Lopez Medus M, Gonzalez Lebrero M, Pagano RS, Labriola CA, Landolfo L, Delfino JM, Parodi AJ, Caramelo JJ. 2009. The structure of calreticulin C-terminal domain is modulated by physiological variations of calcium concentration. *J Biol Chem* **285**: 4544–4553.

Vitadello M, Ausma J, Borgers M, Gambino A, Casarotto DC, Gorza L. 2001. Increased myocardial GRP94 amounts during sustained atrial fibrillation: a protective response? *Circulation* **103**: 2201–2206.

Vitadello M, Penzo D, Petronilli V, Michieli G, Gomirato S, Menabo R, Di Lisa F, Gorza L. 2003. Overexpression of the stress protein Grp94 reduces cardiomyocyte necrosis due to calcium overload and simulated ischemia. *FASEB J* **17**: 923–925.

Wagenknecht T, Hsieh CE, Rath BK, Fleischer S, Marko M. 2002. Electron tomography of frozen-hydrated isolated triad junctions. *Biophys J* **83**: 2491–2501.

Wang HJ, Guay G, Pogan L, Sauve R, Nabi IR. 2000. Calcium regulates the association between mitochondria and a smooth subdomain of the endoplasmic reticulum. *J Cell Biol* **150**: 1489–1498.

Wang S, Trumble WR, Liao H, Wesson CR, Dunker AK, Kang CH. 1998. Crystal structure of calsequestrin from rabbit skeletal muscle sarcoplasmic reticulum. *Nat Struct Biol* **5**: 476–483.

Wei L, Gallant EM, Dulhunty AF, Beard NA. 2009a. Junctin and triadin each activate skeletal ryanodine receptors but junctin alone mediates functional interactions with calsequestrin. *Int J Biochem Cell Biol* **41**: 2214–2224.

Wei L, Hanna AD, Beard NA, Dulhunty AF. 2009b. Unique isoform-specific properties of calsequestrin in the heart and skeletal muscle. *Cell Calcium* **45**: 474–484.

Ying M, Flatmark T. 2006. Binding of the viral immunogenic octapeptide VSV8 to native glucose-regulated protein Grp94 (gp96) and its inhibition by the physiological ligands ATP and Ca^{2+}. *FEBS J* **273**: 513–522.

Ying M, Sannerud R, Flatmark T, Saraste J. 2002. Colocalization of Ca^{2+}-ATPase and GRP94 with p58 and the effects of thapsigargin on protein recycling suggest the participation of the pre-Golgi intermediate compartment in intracellular Ca^{2+} storage. *Eur J Cell Biol* **81**: 469–483.

Zhou X, Fan GC, Ren X, Waggoner JR, Gregory KN, Chen G, Jones WK, Kranias EG. 2007. Overexpression of histidine-rich Ca-binding protein protects against ischemia/reperfusion-induced cardiac injury. *Cardiovasc Res* **75**: 487–497.

Zuber C, Spiro MJ, Guhl B, Spiro RG, Roth J. 2000. Golgi apparatus immunolocalization of endomannosidase suggests post-endoplasmic reticulum glucose trimming: implications for quality control. *Mol Biol Cell* **11**: 4227–4240.

The Plasma Membrane Ca^{2+} ATPase and the Plasma Membrane Sodium Calcium Exchanger Cooperate in the Regulation of Cell Calcium

Marisa Brini[1] and Ernesto Carafoli[2]

[1]Department of Biological Chemistry, University of Padova, 35131 Padova, Italy
[2]Venetian Institute of Molecular Medicine, 35129 Padova, Italy
Correspondence: marisa.brini@unipd.it

Calcium is an ambivalent signal: It is essential for the correct functioning of cell life, but may also become dangerous to it. The plasma membrane Ca^{2+} ATPase (PMCA) and the plasma membrane Na^+/Ca^{2+} exchanger (NCX) are the two mechanisms responsible for Ca^{2+} extrusion. The NCX has low Ca^{2+} affinity but high capacity for Ca^{2+} transport, whereas the PMCA has a high Ca^{2+} affinity but low transport capacity for it. Thus, traditionally, the PMCA pump has been attributed a housekeeping role in maintaining cytosolic Ca^{2+}, and the NCX the dynamic role of counteracting large cytosolic Ca^{2+} variations (especially in excitable cells). This view of the roles of the two Ca^{2+} extrusion systems has been recently revised, as the specific functional properties of the numerous PMCA isoforms and splicing variants suggests that they may have evolved to cover both the basal Ca^{2+} regulation (in the 100 nM range) and the Ca^{2+} transients generated by cell stimulation (in the µM range).

Ca^{2+} controls critical cellular responses in all eukaryotic organisms. It controls both short-term biological processes that occur in milliseconds, such as muscle contraction, as well as long-term processes that require longer times, such as cell proliferation and organ development. The specificity of cellular Ca^{2+} signals is controlled by a sophisticated "toolkit" comprising numerous ion channels, pumps, and exchangers that drive the fluxes of Ca^{2+} ions across the plasma membrane and across the membranes of intracellular organelles (Berridge et al. 2003).

The plasma membrane contains several types of channels that mediate Ca^{2+} entry from the extracellular ambient, and two systems for Ca^{2+} extrusion: a low affinity, high capacity Na^+/Ca^{2+} exchanger (NCX), and a high-affinity, low-capacity Ca^{2+}-ATPase (the plasma membrane Ca^{2+} pump (PMCA)) (Fig. 1). The type of channels and the relative proportions of NCX and PMCA vary with the cell type, the NCX being particularly abundant in excitable tissues, e.g., heart and brain. The regulated opening of the Ca^{2+} channels by either voltage gating, interaction with ligands or the emptying of intracellular stores, allows a limited amount of Ca^{2+} to enter the cell to transmit signals to its designated targets. Thereafter, the Ca^{2+} transients must be dissipated: its extrusion from the

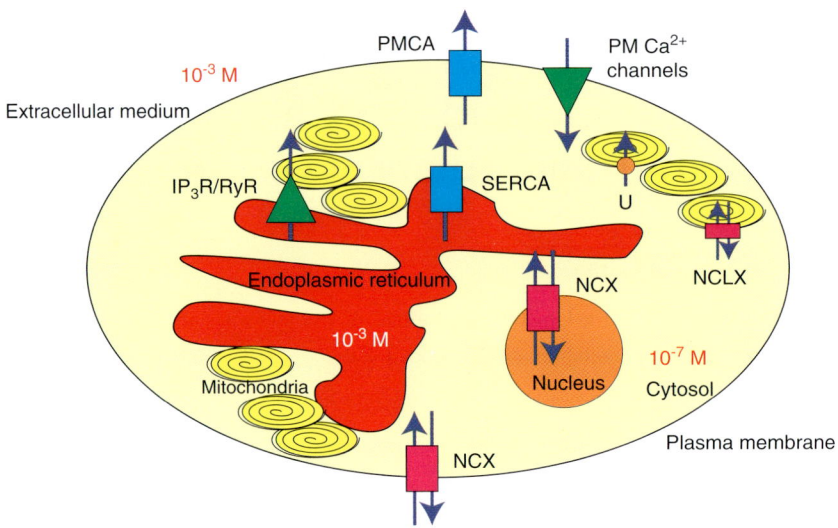

Figure 1. A schematic representation of the structures involved in cellular Ca^{2+} homeostasis. The model shows a cell with its Ca^{2+}-transporting systems: Ca^{2+}-ATPases (plasma membrane and sarco/endoplasmic reticulum, PMCA and SERCA), plasma membrane (PM) Ca^{2+} channels, Na^+/Ca^{2+} exchangers (NCX and NCLX), 1,4,5-triphosphate receptor (IP_3R) and ryanodine receptor (RyR), the electrophoretic mitochondrial uptake uniporter (U). Mitochondria are drawn as yellow ellipses, nucleus as orange circle and endoplasmic reticulum is colored in red. The different Ca^{2+}-transporting systems cooperate to maintain the Ca^{2+} concentration gradient between the extracellular and the intracellular ambient.

cell is mediated by the NCX and the PMCA pump, but Ca^{2+} is also restored to basal levels by sequestration in the endo/sarcoplasmic reticulum via the SERCA pump and in the mitochondria by the electrophoretic uniporter. The NCX has also been found at the inner membrane of the nuclear envelope (NE) and has been proposed to mediate Ca^{2+} flux between the nucleoplasm and the NE (Xie et al. 2002), and then to the ER (Wu et al. 2009) in neuronal and certain other cell types. Ca^{2+} binding proteins also contributed to Ca^{2+} buffering: In this review, we will not cover them, as we will only discuss the systems that extrude Ca^{2+} out of the cell.

The PMCA pump is a minor component of the total protein of the plasma membrane (less than 0.1% of it). Quantitatively, it is overshadowed by the more powerful NCX in excitable tissue like heart; however, even cells in which the NCX predominates, the PMCA pump is likely to be the fine tuner of cytosolic Ca^{2+}, as it can operate in a concentration range in which the low affinity NCX is relatively very inefficient.

The PMCA was discovered in erythrocytes (Schatzmann 1966), and was then described and characterized in numerous other cell types. It was purified in 1979 using a calmodulin affinity column (Niggli et al. 1979), and cloned about 10 years later (Shull and Greeb 1988; Verma et al. 1988). It shows the same essential membrane topology properties of the SERCA pump. Molecular modeling work using the structure of the SERCA pump as a template (Toyoshima et al. 2000) predicts the same general features of the latter, with 10 transmembrane domains and the large cytosolic headpiece divided into the three main cytosolic A, N, and P domains. The Na^+/Ca^{2+} cotransport process was discovered at about the same time as PMCA by two independent groups working on heart (Reuter and Seitz 1968) and on the squid giant axon (Baker et al. 1969). The exchanger was cloned in 1990 (Nicoll et al. 1990). The sequence was initially predicted to correspond to a protein with 11 transmembrane domains and one large cytosolic loop linking transmembrane domain five and six but a revised model predicting only nine transmembrane domains is now generally accepted.

PLASMA MEMBRANE CALCIUM PUMP

Structural and Regulatory Characteristics

The PMCA pump belongs to the family of P-type ATPases, which are characterized by the temporary conservation of ATP energy in the form of a phosphorylated enzyme intermediate (hence P-type) formed between the γ − phosphate of hydrolyzed ATP and an invariant D-residue in a highly conserved sequence of the pump molecules: SDKTGT (L/I/V/M) (T/I/S). The pump operates with a 1:1 Ca^{2+}/ATP stoichiometry as a Ca^{2+}: H^+ exchanger: the matter of Ca^{2+}/H^+ stoichiometry is still controversial (Niggli et al. 1982; Hao et al. 1994). The pump has high Ca^{2+} affinity and a multitude of agents that modulate it. The K_d of the pump for Ca^{2+}, which is 10 to 20 μM in the resting state, decreases to less than 1 μM following calmodulin interaction (Brini and Carafoli 2009). Acidic phospholipids also increase the Ca^{2+} affinity of the pump. Even if the molecular mechanism of the activation by phospholipids is obscure, it could be physiologically meaningful: It has been calculated that in the membrane environment the pump is probably permanently activated by acidic phospholipids to about 50% of its maximal activity (Niggli et al. 1981).

Structurally, the pump is predicted to consist of 10 transmembrane domains, two large intracellular loops, and of amino- and carboxy-terminal cytoplasmic tails (Fig. 2). The 90-residue amino-terminal tail contains a consensus-binding site for the 14-3-3 protein, which plays an inhibitory role on pump activity (Rimessi et al. 2005). The first cytosolic loop between transmembrane domains two and three is considered the "transducer domain." It contains one of the two sites that mediate the activation by acididic phospholipids (Zvaritch et al. 1990) and one of the two sites for the autoinhibitory interaction with the calmodulin binding domain. This loop also contains site A, which is one of the two main sites of the alternative

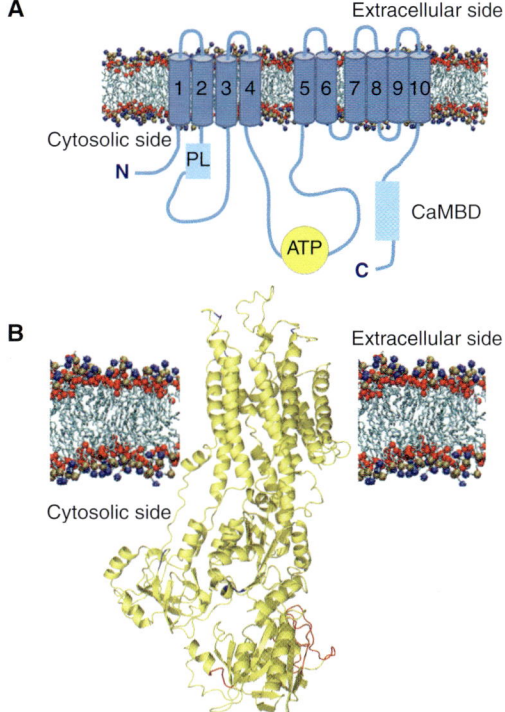

Figure 2. (A) Topology model of PMCA. The pump is organized in the membrane with ten transmembrane domains connected on the external side by short loops. The cytosolic portion of the pump contains the catalytic center and other functionally important domains. The ATP binding site is indicated with a yellow circle. Acidic phospholipid binding domain (PL) and Calmodulin binding domain (CaMBD) are represented with pale blue boxes. (B) Deduced 3D structure of the PMCA pump. The three-dimensional structure of the PMCA pump has been obtained modeling it on that of the SERCA pump (Toyoshima et al. 2000). The image is a kind gift of Dr. M. Hilge (Nijmegen, Holland).

splicing that generates pump variants. Interestingly, the A-splice-site-insert also plays a key role in the targeting of PMCA to specialized portion of the plasma membrane of polarized cells (Chicka and Strehler 2003).

The cytosolic loop that connects transmembrane domains four and five contains the catalytic center that includes the ATP binding site and the aspartate residue that forms the acyl phosphate intermediate. It also contains the second binding site for the carboxy-terminal calmodulin-binding domain. The carboxy-terminal tail contains the main regulatory sites for the activity of the pump: the calmodulin binding domain (which also binds acidic phospholipids), consensus sites for protein kinases A (PKA), and C (PKC) and high affinity allosteric Ca^{2+} binding sites. The calmodulin-binding domain of the PMCA pump acts as an autoinhibitory domain, binding to "receptor" sites in the first and second cytoplasmic loops of the pump. Calmodulin would remove the domain from the binding sites, freeing the pump from autoinhibition.

The carboxy-terminal region of the pump is also the target of isoform-specific phosphorylations by PKA and PKC. Phosphorylation could influence the regulation of the pumps either directly (e.g., by de-inhibiting it) or indirectly (e.g., via interference with calmodulin regulation). Recent studies have shown that PKC affects the various PMCA isoforms and splicing variants in different ways according to variations in the carboxy-terminal sequence: the phosphorylation may activate the pump, inhibit it, or have no effects (Strehler and Zacharias 2001). As for PKA, its consensus site has only been described in isoform one of the pump. The PMCA pump is also a substrate of intracellular proteases, as its carboxy-terminal tail contains target sequences for the Ca^{2+} dependent protease calpain (Guerini et al. 2003) and for caspase 1 and 3. Calpain cleaves the pump in the calmodulin binding sequence, leading to permanent activation of the pump. Both activation and inactivation have been described for caspases 1 and 3 action on PMCA2 and 4 (Schwab et al. 2002; Paszty et al. 2005).

The carboxy-terminal domain of the pump also contains an important site of alternative splicing, site C. C-site splicing occurs in all pump isoforms (albeit with varying degree of complexity) and causes the inclusion of one (or two) additional exons, or of portion of exons. The insertion has structural and functional consequences: its leads to the truncation of the pump because of a frame shift in the coding region, and alters, for instance, the pH sensitivity of the interaction of calmodulin with the calmodulin binding domain. In general, the truncated isoforms have much decreased affinity for calmodulin (Enyedi et al. 1994). Surprisingly, however, the full length and the truncated variants expressed in living cells appear to have the same ability to restore the cytosolic Ca^{2+} transient generated by cell stimulation: the finding suggests that in vivo, calmodulin may not be the only factor regulating PMCA activity (Brini et al. 2003).

The carboxy-terminal domain also mediates the interaction with PDZ domains of different proteins, and mediates PMCA dimerization (which also activates the pump (Kosk-Kosicka and Bzdega 1988; Vorherr et al. 1991)). The interaction with PDZ domains is apparently limited to the full-length splice isoforms, as the carboxy-terminally truncated variants lack the consensus site for its binding. PMCA has been reported to interact with PDZ domain-containing proteins involved in the recruitment and maintenance of membrane proteins in specific membrane domains, such as members of membrane associate guanylate kinase (MAGUK) family, i.e., SAP, CASK, the Na/H exchanger regulatory factor 2, PISP and the Ania3/Homer protein (reviewed in (Brini and Carafoli 2009)). These proteins could contribute to targeting and retention of PMCA in specialized domains. The increased local pump concentration could regulate Ca^{2+} signaling. The PDZ binding domain also directs the interaction of one of the PMCA-isoforms (4b) with nitric oxide synthase 1 (NOS-1) in a ternary complex with α-syntropin. The latter would bind to a domain further upstream in the pump sequence (Williams et al. 2006). The double interaction would thus down-regulate the production of NO by decreasing Ca^{2+} in the immediate vicinity of the synthase: i.e., the pump could function as a modulator of signal transduction, in addition to regulating cell Ca^{2+} (Cartwright et al. 2009).

Isoforms and Tissue Distribution

The four basic PMCA isoforms (the PMCA is the product of four separate genes) have tissue-specific expression (Brini and Carafoli 2009). Although PMCA1 and PMCA4 are expressed in most tissues, PMCA2 and 3 are found in a restricted number of tissues, among them in brain, striated muscle and the mammary gland. In particular, PMCA2 is abundant in specialized cells such as cerebellar Purkinje neurons and cochlear hair cells, and it is also present in uterus and in lactating mammary glands. It is also expressed in significant amounts in liver and kidney. PMCA3 distribution is more restricted, the choroid plexus expressing significant amounts of it.

The transcript of each of the four genes encoding PMCA pumps is subject to alternative splicing: The sites in which it occurs are named A, B, and C. Of the large number of splice variants theoretically possible about 30 have being detected at the RNA or protein levels (Brini and Carafoli 2009). Alternative splicing at site A and C occurs for all four isoforms (however, PMCA1 is never spliced at site A). Alternative splicing at site B has been described only for human isoform one and four and it could be artifactual. Splice site A is located upstream of one of the regulatory binding sites for acidic phospholipids and involves up to three exons that can be optionally inserted or excluded. In PMCA3 and 4 a single exon may be included or excluded in the mature mRNA, thus generating the x and z splice variants, respectively. The situation in PMCA2 is more complex: in the human pump, three exons of 33, 60, and 42 nt, alternatively used, could in principle originate eight different variants. However, only four of them have so far been detected as mRNAs in a variety of tissues. Variant w includes all three exons, variant z excludes all three exons, variant x includes the 42 nt exon, and variant y includes the 33 and 60 nt exons. This alternative splicing was shown to be essential in the apical targeting of PMCA, and thus could also be involved in the regulation of the pump distribution in specialized portions of the plasma membrane (directly or through the interaction with other molecular determinants). Considering that splice site A is contiguous to one of the two phospholipid-binding domains, it could in principle effect the phospholipid regulation of PMCA isoforms by altering the accessibility of the binding domain to acidic phospholipids. As mentioned, the splicing at site B may be an artifact, as its occurrence in vivo has been a point of controversy. It has been proposed to remove a carboxy-terminal 108 nt segment, leading to the loss of the 10th transmembrane domain and causing a reorganization of the pump topology (Preiano et al. 1996). Splice site C is located in the carboxy-terminal of PMCA protein, within the calmodulin binding domain: the inclusion of a large exon (154–175 nt) alters the carboxy-terminal half of the domain in all isoforms, and changes the reading frame of the remaining carboxy-terminal tail. This splicing is thus responsible for the generation of variant *a*, truncated after the first 12 residues of the calmodulin binding domain. The variant has reduced responsiveness to calmodulin with respect to the full-length variant (termed variant *b*). Lower Ca-calmodulin affinity and a significantly elevated basal pumping activity are general characteristics of the *a* splice variants in comparison with the *b* variants (Preiano et al. 1996; Elwess et al. 1997; Caride et al. 2007).

Interestingly, the expression of the PMCA pumps is transcriptionally regulated by Ca^{2+} itself (Zacharias and Strehler 1996; Carafoli et al. 1999). The phenomenon has been studied on cultured neurons and is correlated to their maturation. In cerebellar granule cells, as Ca^{2+} concentration in the cytoplasm increases, the pattern of expression of the PMCA pumps undergoes a dramatic rearrangement: isoform two and three become strongly up-regulated, whereas isoform one undergoes a splicing switch that generates high levels of the truncated *a* variant. In contrast, PMCA4 disappears in a process that is calcineurin dependent (Guerini et al. 2000). The reprogramming of the transcription of PMCAs is essential to the survival of the developing neurons: the regulation of PMCA expression may thus be critical to the survival of cells exposed to pathological increases in intracellular Ca^{2+}. Parallel amplification of Ca^{2+} influx and efflux pathways may enable differentiated neurons to precisely localize Ca^{2+} signals in time and space (Usachev et al. 2001). Similarly, in rat hippocampal neurons, transcripts for all isoforms of the PMCA pump have been found to be strongly up-regulated during the second week in culture, the overall increase in Ca^{2+} extrusion systems being accompanied by changes in the expression and cellular localization of different isoforms. The accumulation of PMCAs in dendrites and dendritic spines coincided with the functional maturation of these neurons, underlining the importance of the proper spatial organization of the Ca^{2+} extrusion systems in synaptic function and development (Kip et al. 2006).

Role in Physiology and Pathology

PMCA pumps are expressed ubiquitously, thus, they are unlikely candidates as molecules that could mediate cell specific processes. However, the existence of so many PMCA variants (basic isoforms and alternative spliced forms), the finding on knockout mice for the specific PMCA isoforms, and studies on the mechanisms modulating their cell/tissue specific distribution indicate that PMCA pumps may play important roles as signaling molecules, in addition to having a constitutive role as Ca^{2+} housekeeping enzymes.

The ablation of isoforms two and four in mice has indeed revealed that they are important in specialized biological processes. PMCA2 plays an essential role in the hearing process, because it dissipates with peculiar kinetics the Ca^{2+} transients generated by the opening of the mechanoelectrical transduction (MET) channels in the stereocilia of cochlear hair cells. Mutations in the PMCA2 pumps in hereditary deafness impair the longer-term export of Ca^{2+} from hair cells, but not their ability to respond to a sudden increase of Ca^{2+}. This suggests that the mutated pumps reduce the extracellular Ca^{2+} gradient in the endolymph surrounding the hair cells, compromising the adaptation process responsible for the control of the opening of MET channels (Ficarella et al. 2007; Spiden et al. 2008). Interestingly, the PMCA2 isoform has another peculiar function in mammary gland: its activity is responsible for the high content of Ca^{2+} in the milk. PMCA2 is strongly up-regulated in lactating mammary gland, and PMCA2 null mice have 40% less Ca^{2+} in the milk (Reinhardt et al. 2004).

PMCA4 is crucial to male fertility: in mice, its ablation reduces sperm motility, probably resulting from Ca^{2+} overload and mitochondrial damage (Schuh et al. 2004). Interestingly, PMCA4 is also essential for the modulation of the activity of the calcium/calmodulin dependent neuronal nitric oxide synthase (nNOS) (Schuh et al. 2001) and thus, in turn, controls nitric oxide production, which is important in the regulation of excitation-contraction coupling of the heart. It has been shown that PMCA4b regulates cardiac contractility in vivo through its interaction with nNOS (Oceandy et al. 2007).

PMCA defects have been described in a large number of disease conditions, among them those linked to the oxidative stress in several tissues, particularly brain, in brain ischemia, diabetes, atherosclerosis, aging, neurodegenerative diseases, and various disturbances of Ca^{2+} metabolism. PMCA defects have also been reported in various cancer types (reviewed in Monteith et al. 2007), in line with accumulating evidence on the involvement of Ca^{2+} homeostasis dysfunction in the process of malignant transformation. The genetic diseases involving the PMCA pump are best characterized: a number of dysfunctions linked to the ablation, or to partial disruptions, including point mutations, of its genes, have been described in mice. So far, the only identified spontaneous human disease related to a PMCA pump (isoform 2) defect is the form of hereditary deafness described above (Schultz et al. 2005; Ficarella et al. 2007).

Genetic Manipulations

Knockout mice have been developed and the phenotypes studied for PMCA1, PMCA2, and PMCA4, but not yet for PMCA3. As this isoform is widely expressed in developing embryos, it may be essential for normal gestation.

The ablation of the PMCA1 gene has resulted in early embryonic lethality in homozygotes, underlying the essential role of PMCA1 in the early stages of development and in organogenesis, in line with its suggested housekeeping function (Okunade et al. 2004). Heterozygotes, by contrast, appeared healthy, however, loss of one copy of the PMCA1 gene in mice with a PMCA4 null background led to increased propensity to apoptosis in the smooth muscles of blood vessels, possibly because of Ca^{2+} overload (Okunade et al. 2004). The ablation of the PMCA2 gene revealed its involvement in the hearing process: the phenotype of the PMCA2 knockout mice (Kozel et al. 1998) was characterized by balance impairment and hearing loss. The study of the vestibular inner ear revealed the absence of the otoconia and the progressive degeneration of the hair cells beginning 10 days after birth. The PMCA2 pump is probably also important to general neuronal Ca^{2+} homeostasis because PMCA2 null mice showed a significant reduction in the number of spinal cord motor neurons (Kurnellas MP 2005) and abnormalities in Purkinje neurons (Kurnellas MP 2007). However, the balance defect was apparently not due, as could in principle have been expected, or was only partially due, to a cerebellar dysfunction. It was apparently related to the absence of the Ca-carbonate crystals of the otoconia embedded in the otolithic membrane overlying the sensory epithelium that sense gravity and linear acceleration in the inner ear. In keeping with the high level of expression of PMCA2 in the mammary gland, the ablation of its gene, in addition to generating the hearing loss phenotype, also strongly reduced the concentration of Ca^{2+} in the milk (Reinhardt et al. 2004).

The ablation of the PMCA4 gene failed to cause a very evident general pathological phenotype (Schuh et al. 2004), but local defects were present. This is interesting, as PMCA4 is also widely expressed, and has also been proposed to play a housekeeping role. However, it has now become evident that PMCA4 plays more specialized roles, and is not only essential for the general function of controlling Ca^{2+} homeostasis in all cells. One prominent defect caused by PMCA4 dysfunction was male infertility, reflecting the dominance of the PMCA4 pump in the testis, where it represents more than 90% of the total PMCA protein (Prasad et al. 2004). The ablation of the PMCA4 gene produced other localized dysfunctions, for instance in the modulation of the Ca^{2+} signals in B-lymphocytes (Chen et al. 2004), and in the contractility of vascular or bladder smooth muscles (Okunade et al. 2004).

Studies on mice overexpressing PMCA4 isoform had revealed other important functions for this isoform. Early studies, in which the PMCA4 pump had been expressed specifically in the myocardium of the rat, and in vascular smooth muscle cells in mice, had indicated a role in the modulation of myocardial growth and hypertrophy (Hammes et al. 1998). They have also indicated a role in the regulation of the peripheral vascular tone, as the mice displayed increased peripheral blood pressure (Gros et al. 2003; Schuh et al. 2003).

SODIUM/CALCIUM EXCHANGER

Structural and Regulatory Characteristics

NCX accomplishes Ca^{2+} extrusion by using the electrochemical gradient of Na^+: during each cycle three Na^+ ions enter the cell and one Ca^{2+} ion is extruded against its gradient (for comprehensive reviews see Blaustein and Lederer [1999]; Lytton [2007]). In addition to being transported, cytoplasmic Na^+ and Ca^{2+} ions also regulate exchanger activity. Binding of Ca^{2+} ions to sites located in the cytosolic loop generally activate the exchanger, whereas binding of Na^+ ions has been shown to deactivate it: the physiological importance and molecular mechanisms underlying the regulation remain unclear. The operation of NCX is fully reversible, and the direction of the movement of the transported ions depends entirely on the electrochemical gradients of Na^+ and Ca^{2+} and on the number of ions that bind to the molecule and are transported. Under resting cellular ionic condition and membrane potential, the NCX acts to extrude Ca^{2+} from the cytoplasm. However, in the heart, when the plasma membrane becomes depolarized during systole, and the Na^+ levels rise because of the opening of the plasma membrane voltage–operated Na^+ channels, the exchanger reverses its operation and mediates Ca^{2+} entry. The influx of Ca^{2+} may play an important regulatory role in the excitation-contraction process by influencing the gating of voltage-operated Ca^{2+} channels and by altering the SR Ca^{2+} load.

A membrane topology model based on isoform one of the NCX now predicts nine transmembrane α-helices that divide the molecule in a amino-terminal portion composed of the first five transmembrane domains, and a carboxy-terminal portion composed of the last four transmembrane domains (Fig. 3A). These two portions are separated by a cytosolic loop of about 500 residues that contains the site for NCX1 regulation, and is thus the target for the development of inhibitory compounds. NCX1 also contains two regions of internal repeats: the α repeats (α1 and α2 in Fig. 3A) are involved in ion binding and transport; the β repeats in the large intracellular loop are involved in binding of regulatory Ca^{2+} (CBD1 and CBD2 in Fig. 3). The β repeats have been structural defined by crystallization (Nicoll et al. 2006) and NMR studies (Hilge et al. 2006).

The first portion of the large cytosolic loop (close to transmembrane domain five) is an amphipathic sequence, called XIP (exchanger inhibitor peptide) because the addition of a peptide corresponding to this sequence inhibits the exchanger. This first portion of the region is responsible for the regulation by Na^+ and the acidic phospholipids. The second portion of the loop contains the two Ca^{2+} binding domains (CBD1 and CBD2) that are arranged in an antiparallel fashion, and connected through a third domain designated as CLD (α-catenin like domain) to the membrane portion of the NCX (Fig. 3B). The CBD1 is the primary Ca^{2+} sensor that detects small cytosolic Ca^{2+} increases and undergoes large structural changes that activate the exchanger. The CBD2 undergoes instead modest structural alterations and binds Ca^{2+} only at elevated concentrations (Hilge et al. 2006; Nicoll et al. 2006). The carboxy-terminal end of the large cytosolic loop contains a hydrophobic and proline-rich sequence that had been originally modeled as a transmembrane domain. Recent evidence has positioned it on the cytosolic side of the membrane.

In addition to the two transported species, other regulatory agents of the exchanger include the intracellular pH, metabolic components (e.g., ATP, phosphatydyl inositol 4,5 bisphosphate), protein kinases PKA and PKC, redox agents, hydroxyl radicals, H_2O_2, dithiothreitol, O^{2-}, Fe^{3+}, Fe^{2+}, Cu^{2+}, and OH^- (Doering and Lederer 1993; Matsuoka et al. 1995; Matsuoka et al. 1997; Iwamoto et al. 1998; He et al. 2000).

As is the case for the PMCA, a number of cytoskeletal proteins have been shown to interact

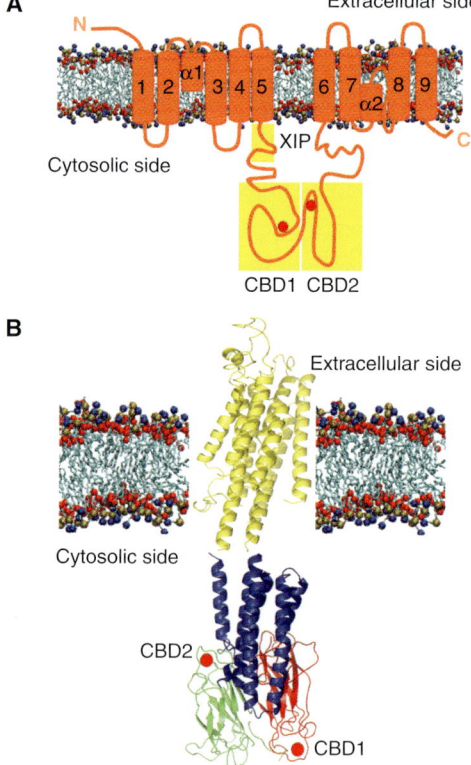

Figure 3. (A) Topology model of NCX. The nine transmembrane domains comprise the two putative transport repeat regions: α-1 and α-2. XIP region, CBD1 and CBD2 are indicated by yellow boxes. The red spheres indicate the Ca^{2+} binding sites position (B) Hypothetical structural model of the intact NCX. The nine transmembrane domains are shown as yellow α-helices, the CLD as blue α-helices and the CBD1 and CBD2 β repeats as red and green β strands, respectively. The red spheres indicate the Ca^{2+} binding sites position. The image is a kind gift of Dr. M. Hilge (Nijmegen, Holland).

with the NCX. The large cytosolic loop interacts with several proteins: the 14-3-3 protein, phosphorylated PLM (phospholemman, a member of a family of transport regulators, best known as modulators of Na, K-ATPase activity) and calcineurin: they all have inhibitory function on NCX activity (Katanosaka et al. 2005; Pulina et al. 2006; Wang et al. 2006; Zhang et al. 2006). Other proteins have also been shown to interact with the NCX: ankyrin (Li et al. 1993), the filamentous actin network (Condrescu and Reeves 2006), caveolin 3 (in the heart) (Bossuyt et al. 2002; Camors et al. 2006), and caveolin 1 and 2 (in neuronal cells) (Cha et al. 2004). These interactions have been proposed to regulate NCX activity by modulating the membrane NCX localization and by recruiting it to specialized portions of the plasma membrane as a part of macromolecular complexes (Schulze et al. 2003; Lencesova et al. 2004), similarly to PMCA, which can be recruited through the interaction with nNOS to form a specialized Ca^{2+} concentration microdomain.

The phosphorylation of NCX1 remains a topic of much controversy: several studies have claimed that PKA and/or PKC, operating through adrenergic stimulation, could modulate NCX function. The regulation could also be indirect and mediated by interactions with other proteins in macromolecular complexes.

Isoforms and Tissue Distribution

The NCX family belongs to a larger superfamily of related Ca^{2+}/cation antiporter genes that has five major branches. Three of them relate to the Na^+/Ca^{2+} transport in mammals: NCX (SLC8 family), NCKX (Na^+/Ca^{2+}-K^+ exchangers, SLC24 family) and the CCX (Ca^{2+}/anion exchangers), which contains the Na^+/Ca^{2+}-Li^+ exchanger (NCLX). NCKX proteins play a crucial role in regulating Ca^{2+} fluxes during photoreceptors adaptation, synaptic plasticity, and skin pigmentation (for a comprehensive review see Lytton [2007]). NCLX is the single mammalian member of the phylogenetically ancestral CCX family. It has been initially characterized as a novel plasma membrane Na^+/Ca^{2+} exchanger that catalyzes Na^+ or Li^+-dependent Ca^{2+} transport, but very recent evidence indicates that it is the long elusive mitochondrial Na^+/Ca^{2+} exchanger (Palty 2009).

The SLC8 family includes three basic proteins, NCX1, NCX2, and NCX3, encoded by three separate genes. A fourth member, described in teleost fish, is not present in the mammalian genome (Marshall et al. 2005). NCX1 was originally characterized and cloned from heart, but its expression is almost ubiquitous, with high levels in brain and kidney. NCX2 and NCX3 are selectively expressed in the brain and skeletal muscle, and in some neuronal populations, respectively. NCX1 is the best-characterized member of the family. Its expression is controlled by alternative promoters in a tissue-specific and transcription factor-specific manner: one of them is specific to heart, one to kidney, and one to the other tissues. The transcripts of NCX1 gene are alternatively spliced at two sites: the first is located in the 5′-untranslated region and does not change the structure of the mature protein. The second is located within the coding region and leads to the introduction of two mutually exclusive exons (exon A in excitable tissues such as brain and muscle, and exon B in nonexcitable tissues), and of four other exons, thus potentially encoding a large number of NCX1 isoforms with differences in the cytosolic loop. Interestingly, giant excised patch clamp experiments have shown that exon A containing isoforms, in contrast to exon B containing variants, may not have intracellular Na^+-dependent inactivation (Dyck et al. 1999). NCX3, but probably not NCX2, is alternatively spliced to generate variability in the cytoplasmic region corresponding to that of NCX1. No distinctive properties have been described for the different splicing isoforms, suggesting that their specific function is probably related to different spatial targeting.

Importantly, as is the case for the PMCA pump, the transcription of NCX genes is also differentially regulated by Ca^{2+} during development, and also in adult neurons (Carafoli et al. 1999). NCX2 and NCX3 transcripts are downregulated and up-regulated in cerebellar granule neurons, respectively, under the conditions of intracellular Ca^{2+}-increase that reprogram the pattern of transcription of the PMCA pumps. The down-regulation of NCX2 transcription in cerebellar granules is calcineurin dependent (Li et al. 2000), whereas that of NCX3 in model neuroblastoma neurons is regulated by Ca^{2+} through the transcriptional repressor DREAM (Gomez-Villafuertes et al. 2005).

Role in Physiology and Pathology

As the NCX is not an enzyme, the qualitative measurement of its activity has always been difficult and the role of NCX in physiological conditions has been mostly inferred from theoretical considerations and from observations obtained on isolated systems in artificial conditions, or with invasive approaches such as giant patch techniques. More recently, the availability of selective inhibitors (Iwamoto 2007) and genetic manipulation approaches have permitted the dissection of many physiological aspects of NCX activity. The role of NCX has so far been best characterized in the heart: contraction is initiated by a small influx of Ca^{2+} from the extracellular ambient that induces a larger release of Ca^{2+} from the SR; relaxation requires Ca^{2+} removal, which is achieved mainly by the extrusion activity of NCX1 and the pumping back of Ca^{2+} in the SR lumen by the SERCA pump. Ca^{2+} extrusion through NCX1 also generates an inward depolarizing current (because of Na^+ entry and to the uneven charge translocation) which may contribute to the shaping of the action potential, and/or counteracts the Ca^{2+} entry pathway by reducing its driving force. The Ca^{2+} entry during the reverse mode operation of NCX1 (induced by signals that generated Na^+ entry, i.e., through ionotropic glutamate receptors) has been described to contribute to Ca^{2+} influx. These actions probably do not modify the single action potential, but they have been suggested to play a possible role in the refilling of the stores with Ca^{2+}, thanks to the formation of protein complexes and functional coupling between plasma membrane microdomains and the ER lumen mediated by local Ca^{2+} gradients (Blaustein and Golovina 2001). In addition to the heart, where its function has received great attention, numerous studies have underlined the role of the NCX in other tissues as well. The NCX1 has been shown to play a special role in the kidney, where the Ca^{2+} filtered at the glomerulus is reabsorbed passively along the proximal nephron. However, at the distal nephron level, vitamin D_3 and parathyroid hormone regulate Ca^{2+} transport. NCX1 is highly expressed in the basolateral membrane of the distal nephron, and helps regulating the active transcellular Ca^{2+} reabsorption, thus contributing to the regulation of systemic Ca^{2+} levels rather than to the intracellular levels. The role of the exchanger in other tissues has not been analyzed in the same detail as it has in the heart (or kidney); however, in the last few years its role in the brain has received increasing attention, particularly as a means to protect neurons from ischemic damage. The specific contribution of each of the three basic NCX isoforms cannot be established with certainly, because they are coexpressed in brain neurons. However, the widespread and abundant expression of NCX2 has indicated a key role for it in neuronal Ca^{2+} homeostasis. NCX3 also appears to be important to neurons: its specific cleavage during brain ischemia and in neurons undergoing excitotoxicity has shown that it plays a critical role in protecting them from ischemic insults (Bano et al. 2005; Annunziato et al. 2007). However, the issue still has controversial aspects: evidence has been provided that during brain ischemia changes in membrane potential and unregulated Ca^{2+} entry could activate the NCX Ca^{2+} entry-mode, and thus contribute to Ca^{2+} overload and eventual neurodegeneration (Kintner et al. 2007). Other work has suggested instead that during excitotoxic insults, the activation of calpain leads to the cleavage of NCX and thus to the impairment of the Ca^{2+} extrusion process, culminating in excitotoxic death. Thus, the inhibition (or the down-regulation) of the exchanger could be neuroprotective (Luo et al. 2007), but could also transform a Ca^{2+} transient elicited by nonexcitotoxic stimulation into a lethal Ca^{2+} overload (Bano et al. 2005; Annunziato et al. 2007).

Considering the essential role of NCX in regulating Ca^{2+} homeostasis in a variety of tissues, the importance of understanding how changes in its activity could contribute to pathological situations is obvious. No genetic disease has so far been associated to NCX mutations, but a number of pathologies have been related to its malfunctioning. Again, the best documented studies are related to heart dysfunctions: NCX deficiencies have been claimed to play a role both in cardiac arrhythmias and in the ischemia/reperfusion injury. They can contribute to the former in two ways: through changes in the balance between the forward-mode of Ca^{2+} efflux and the reverse-mode of Ca^{2+} influx consequent on the deregulation of intracellular Na^+, or through conditions that could lead to Ca^{2+} overload in the SR, and thus to spontaneous Ca^{2+} release from it that could generate extrasystolic events: the extra Ca^{2+} release could enhance NCX activity, thus generating a depolarizing current that can contribute to the elongation of the action potential. Up-regulation of NCX expression at the transcriptional level has been described in cardiac hypertrophy, ischemia, and failure (Kent et al. 1993). However, it is difficult to establish whether the up-regulation is the cause of the dysfunction or rather an adaptive mechanism that, in the end, leads to cardiac hypertrophy and heart failure.

In ischemia/reperfusion injury the action of NCX is still linked to its reverse mode of operation: although other factors could contribute, the removal of H^+ ions that accumulate during the ischemic period by the Na^+/H^+ exchanger could enhance the intracellular Na^+, thus activating Ca^{2+} entry, and eventually causing SR Ca^{2+} overload.

Genetic Manipulations

In the last few years, various NCX knockout mice have been generated. Homozygous NCX1 KO mice were not vital and died during embryonic development, possibly because of heart failure (Wakimoto et al. 2000). However, the cardiac-specific knockout of NCX1 was not lethal (Henderson et al. 2004) suggesting that the cause of lethality in NCX1 KO may have been extra-cardiac, and that NCX2 and 3 could not compensate for the absence of NCX1. The cardiac-specific NCX1 KO mice presented only a mild deficit in cardiac function that culminated in hypertrophy and heart failure only as the mice aged. Ca^{2+} measurements on myocytes revealed that the reduced Ca^{2+} clearance in the absence of NCX1 led to enhanced Ca^{2+} dependent inactivation of L-type Ca^{2+} channels. These myocytes also displayed a shorter action potential because of hyperpolarization caused probably by the increase in the expression of K^+ channel subunits. The outcome of these modifications was a reduction of about 80% of Ca^{2+} fluxes through the plasma membrane: however, this reduction did not impair the Ca^{2+} release from the SR and contractility because of compensation mechanisms involving the gain of excitation-contraction coupling.

NCX2 KO mice displayed impairment in several hippocampal-dependent learning and memory tasks (Jeon et al. 2003). Larger presynaptic Ca^{2+} transients evoked an increase in the neurotransmitter release, and the increased postsynaptic Ca^{2+} transient enhanced long-term potentiation: the findings are consistent with the predominant role of NCX2 in pre- and postsynaptic Ca^{2+} clearance.

Mice lacking the NCX3 gene, which is highly expressed in cerebellum and in the peripheral nervous system, showed reduced motor activity and weakness of forelimb muscles (Sokolow et al. 2004). In addition, skeletal muscle defects were also observed, in line with the finding of abundant NCX3 expression in skeletal muscle.

CONCLUDING REMARKS

To respond dynamically to the changing needs of Ca^{2+} signaling, cells must be able to precisely control the type, amount, localization, and activation of Ca^{2+} transporters. This control is performed in two ways: through the variable expression of Ca^{2+} transporters with different biochemical characteristics, and through the fine modulation of their function by expressing differently active isoforms in response to the local or temporal cell demands. The cell also modulates the activity of its transporters by choosing specific molecular partners. This increases the complexity of the controlling machinery, but protects the cells from "the Ca^{2+} damage" that would inevitably follow the general failure of the Ca^{2+} controlling operation. The absence, or malfunction, of specific Ca^{2+} transport proteins would induce a confined defect: although generating functional discomfort, it would frequently still be compatible with cell life.

ACKNOWLEDGMENTS

The original work described in this review by the authors has been supported over the years by grants from the Italian Ministry of University and Research (FIRB2001 to E.C., PRIN 2003 and 2005 to M.B), the Telethon Foundation (Project GGP04169 to M.B.), the FP6 program of the European Union (FP6 Network of Excellence NeuroNe, LSH-2003-2.1.3-3 to E.C. and Integrated Project Eurohear LSHG-CT-20054-512063, to E.C.), the Human Frontier Science Program Organization to E.C., the Italian National Research Council (CNR) to M.B.

REFERENCES

Annunziato L, Pignataro G, Boscia F, Sirabella R, Formisano L, Saggese M, Cuomo O, Gala R, Secondo A, Viggiano D, et al. 2007. ncx1, ncx2, and ncx3 gene product expression and function in neuronal anoxia and brain ischemia. *Ann N Y Acad Sci* **1099:** 413–426.

Baker PF, Blaustein MP, Hodgkin AL, Steinhardt RA. 1969. The influence of calcium on sodium efflux in squid axons. *J Physiol* **200:** 431–458.

Bano D, Young KW, Guerin CJ, Lefeuvre R, Rothwell NJ, Naldini L, Rizzuto R, Carafoli E, Nicotera P. 2005. Cleavage of the plasma membrane Na^+/Ca^{2+} exchanger in excitotoxicity. *Cell* **120:** 275–285.

Berridge MJ, Bootman MD, Roderick HL. 2003. Calcium signalling: dynamics, homeostasis and remodelling. *Nat Rev Mol Cell Biol* **4:** 517–529.

Blaustein MP, Golovina VA. 2001. Structural complexity and functional diversity of endoplasmic reticulum Ca(2+) stores. *Trends Neurosci* **24:** 602–608.

Blaustein MP, Lederer WJ. 1999. Sodium/calcium exchange: its physiological implications. *Physiol Rev* **79:** 763–854.

Bossuyt J, Taylor BE, James-Kracke M, Hale CC. 2002. Evidence for cardiac sodium-calcium exchanger association with caveolin-3. *FEBS Lett* **511:** 113–117.

Brini M, Carafoli E. 2009. Calcium pumps in health and disease. *Physiol Rev* **89:** 1341–1378.

Brini M, Coletto L, Pierobon N, Kraev N, Guerini D, Carafoli E. 2003. A comparative functional analysis of plasma membrane Ca2+ pump isoforms in intact cells. *J Biol Chem* **278:** 24500–24508.

Camors E, Charue D, Trouve P, Monceau V, Loyer X, Russo-Marie F, Charlemagne D. 2006. Association of annexin A5 with Na^+/Ca^{2+} exchanger and caveolin-3 in non-failing and failing human heart. *J Mol Cell Cardiol* **40:** 47–55.

Carafoli E, Genazzani A, Guerini D. 1999. Calcium controls the transcription of its own transporters and channels in developing neurons. *Biochem Biophys Res Commun* **266:** 624–632.

Caride AJ, Filoteo AG, Penniston JT, Strehler EE. 2007. The plasma membrane Ca^{2+} pump isoform 4a differs from isoform 4b in the mechanism of calmodulin binding and activation kinetics: implications for Ca^{2+} signaling. *J Biol Chem* **282:** 25640–25648.

Cartwright EJ, Oceandy D, Neyses L. 2009. Physiological implications of the interaction between the plasma membrane calcium pump and nNOS. *Pflugers Arch* **457:** 665–671.

Cha SH, Shin SY, Jung SY, Kim YT, Park YJ, Kwak JO, Kim HW, Suh CK. 2004. Evidence for Na^+/Ca^{2+} exchanger 1 association with caveolin-1 and -2 in C6 glioma cells. *IUBMB Life* **56:** 621–627.

Chen J, McLean PA, Neel BG, Okunade G, Shull GE, Wortis HH. 2004. CD22 attenuates calcium signaling by potentiating plasma membrane calcium-ATPase activity. *Nat Immunol* **5:** 651–657.

Chicka MC, Strehler EE. 2003. Alternative splicing of the first intracellular loop of plasma membrane Ca^{2+}-ATPase isoform 2 alters its membrane targeting. *J Biol Chem* **278:** 18464–18470.

Condrescu M, Reeves JP. 2006. Actin-dependent regulation of the cardiac $Na^{(+)}/Ca^{(2+)}$ exchanger. *Am J Physiol Cell Physiol* **290:** C691–C701.

Doering AE, Lederer WJ. 1993. The mechanism by which cytoplasmic protons inhibit the sodium-calcium exchanger in guinea-pig heart cells. *J Physiol* **466:** 481–499.

Dyck C, Omelchenko A, Elias CL, Quednau BD, Philipson KD, Hnatowich M, Hryshko LV. 1999. Ionic regulatory properties of brain and kidney splice variants of the NCX1 $Na^{(+)}$-$Ca^{(2+)}$ exchanger. *J Gen Physiol* **114:** 701–711.

Elwess NL, Filoteo AG, Enyedi A, Penniston JT. 1997. Plasma membrane Ca^{2+} pump isoforms 2a and 2b are unusually responsive to calmodulin and Ca^{2+}. *J Biol Chem* **272:** 17981–17986.

Enyedi A, Verma AK, Heim R, Adamo HP, Filoteo AG, Strehler EE, Penniston JT. 1994. The Ca^{2+} affinity of the plasma membrane Ca^{2+} pump is controlled by alternative splicing. *J Biol Chem* **269:** 41–43.

Ficarella R, Di Leva F, Bortolozzi M, Ortolano S, Donaudy F, Petrillo M, Melchionda S, Lelli A, Domi T, Fedrizzi L, et al. 2007. A functional study of plasma-membrane calcium-pump isoform 2 mutants causing digenic deafness. *Proc Natl Acad Sci* **104:** 1516–1521.

Gomez-Villafuertes R, Torres B, Barrio J, Savignac M, Gabellini N, Rizzato F, Pintado B, Gutierrez-Adan A, Mellstrom B, Carafoli E, et al. 2005. Downstream regulatory element antagonist modulator regulates Ca^{2+} homeostasis and viability in cerebellar neurons. *J Neurosci* **25:** 10822–10830.

Gros R, Afroze T, You XM, Kabir G, Van Wert R, Kalair W, Hoque AE, Mungrue IN, Husain M. 2003. Plasma membrane calcium ATPase overexpression in arterial smooth muscle increases vasomotor responsiveness and blood pressure. *Circ Res* **93:** 614–621.

Guerini D, Pan B, Carafoli E. 2003. Expression, purification, and characterization of isoform 1 of the plasma membrane Ca^{2+} pump: focus on calpain sensitivity. *J Biol Chem* **278:** 38141–38148.

Guerini D, Wang X, Li L, Genazzani A, Carafoli E. 2000. Calcineurin controls the expression of isoform 4CII of the plasma membrane $Ca^{(2+)}$ pump in neurons. *J Biol Chem* **275:** 3706–3712.

Hammes A, Oberdorf-Maass S, Rother T, Nething K, Gollnick F, Linz KW, Meyer R, Hu K, Han H, Gaudron P, et al. 1998. Overexpression of the sarcolemmal calcium pump in the myocardium of transgenic rats. *Circ Res* **83:** 877–888.

Hao L, Rigaud JL, Inesi G. 1994. Ca^{2+}/H^+ countertransport and electrogenicity in proteoliposomes containing erythrocyte plasma membrane Ca-ATPase and exogenous lipids. *J Biol Chem* **269:** 14268–14275.

He Z, Feng S, Tong Q, Hilgemann DW, Philipson KD. 2000. Interaction of PIP(2) with the XIP region of the cardiac Na/Ca exchanger. *Am J Physiol Cell Physiol* **278:** C661–C666.

Henderson SA, Goldhaber JI, So JM, Han T, Motter C, Ngo A, Chantawansri C, Ritter MR, Friedlander M, Nicoll DA, et al. 2004. Functional adult myocardium in the absence of Na^+-Ca^{2+} exchange: cardiac-specific knockout of NCX1. *Circ Res* **95:** 604–611.

Hilge M, Aelen J, Vuister GW. 2006. Ca^{2+} regulation in the Na^+/Ca^{2+} exchanger involves two markedly different Ca^{2+} sensors. *Mol Cell* **22:** 15–25.

Iwamoto T. 2007. Na^+/Ca^{2+} exchange as a drug target—insights from molecular pharmacology and genetic engineering. *Ann N Y Acad Sci* **1099:** 516–528.

Iwamoto T, Pan Y, Nakamura TY, Wakabayashi S, Shigekawa M. 1998. Protein kinase C-dependent regulation of Na^+/Ca^{2+} exchanger isoforms NCX1 and NCX3 does not require their direct phosphorylation. *Biochemistry* **37:** 17230–17238.

Jeon D, Yang YM, Jeong MJ, Philipson KD, Rhim H, Shin HS. 2003. Enhanced learning and memory in mice lacking Na^+/Ca^{2+} exchanger 2. *Neuron* **38:** 965–976.

Katanosaka Y, Iwata Y, Kobayashi Y, Shibasaki F, Wakabayashi S, Shigekawa M. 2005. Calcineurin inhibits Na^+/Ca^{2+} exchange in phenylephrine-treated hypertrophic cardiomyocytes. *J Biol Chem* **280:** 5764–5772.

Kent RL, Rozich JD, McCollam PL, McDermott DE, Thacker UF, Menick DR, McDermott PJ, Cooper Gt. 1993. Rapid expression of the $Na^{(+)}$-Ca^{2+} exchanger in response to cardiac pressure overload. *Am J Physiol* **265:** H1024–H1029.

Kintner DB, Luo J, Gerdts J, Ballard AJ, Shull GE, Sun D. 2007. Role of Na^+-K^+-Cl^- cotransport and Na^+/Ca^{2+} exchange in mitochondrial dysfunction in astrocytes following in vitro ischemia. *Am J Physiol Cell Physiol* **292:** C1113–C1122.

Kip SN, Gray NW, Burette A, Canbay A, Weinberg RJ, Strehler EE. 2006. Changes in the expression of plasma membrane calcium extrusion systems during the maturation of hippocampal neurons. *Hippocampus* **16:** 20–34.

Kosk-Kosicka D, Bzdega T. 1988. Activation of the erythrocyte Ca^{2+}-ATPase by either self-association or interaction with calmodulin. *J Biol Chem* **263:** 18184–18189.

Kozel PJ, Friedman RA, Erway LC, Yamoah EN, Liu LH, Riddle T, Duffy JJ, Doetschman T, Miller ML, Cardell EL, et al. 1998. Balance and hearing deficits in mice with a null mutation in the gene encoding plasma membrane Ca^{2+}-ATPase isoform 2. *J Biol Chem* **273:** 18693–18696.

Kurnellas MP, Lee AK, Li H, Deng L, Ehrlich DJ, Elkabes S. 2007. Molecular alterations in the cerebellum of the plasma membrane calcium ATPase 2 (PMCA2)-null mouse indicate abnormalities in Purkinje neurons. *Mol Cell Neurosci* **34:** 178–188.

Kurnellas MP, Nicot A, Shull GE, Elkabes S. 2005. Plasma membrane calcium ATPase deficiency causes neuronal pathology in the spinal cord: a potential mechanism for neurodegeneration in multiple sclerosis and spinal cord injury. *FASEB J* **19:** 298–300.

Lencesova L, O'Neill A, Resneck WG, Bloch RJ, Blaustein MP. 2004. Plasma membrane-cytoskeleton-endoplasmic reticulum complexes in neurons and astrocytes. *J Biol Chem* **279:** 2885–2893.

Li L, Guerini D, Carafoli E. 2000. Calcineurin controls the transcription of Na^+/Ca^{2+} exchanger isoforms in developing cerebellar neurons. *J Biol Chem* **275:** 20903–20910.

Li ZP, Burke EP, Frank JS, Bennett V, Philipson KD. 1993. The cardiac Na^+-Ca^{2+} exchanger binds to the cytoskeletal protein ankyrin. *J Biol Chem* **268:** 11489–11491.

Luo J, Wang Y, Chen X, Chen H, Kintner DB, Shull GE, Philipson KD, Sun D. 2007. Increased tolerance to ischemic neuronal damage by knockdown of Na^+-Ca^{2+} exchanger isoform 1. *Ann N Y Acad Sci* **1099:** 292–305.

Lytton J. 2007. Na^+/Ca^{2+} exchangers: three mammalian gene families control Ca^{2+} transport. *Biochem J* **406:** 365–382.

Marshall CR, Pan TC, Le HD, Omelchenko A, Hwang PP, Hryshko LV, Tibbits GF. 2005. cDNA cloning and expression of the cardiac Na^+/Ca^{2+} exchanger from Mozambique tilapia (Oreochromis mossambicus) reveal a teleost membrane transporter with mammalian temperature dependence. *J Biol Chem* **280:** 28903–28911.

Matsuoka S, Nicoll DA, He Z, Philipson KD. 1997. Regulation of cardiac $Na^{(+)}$-Ca^{2+} exchanger by the endogenous XIP region. *J Gen Physiol* **109:** 273–286.

Matsuoka S, Nicoll DA, Hryshko LV, Levitsky DO, Weiss JN, Philipson KD. 1995. Regulation of the cardiac $Na^{(+)}$-Ca^{2+} exchanger by Ca^{2+}. Mutational analysis of the $Ca^{(2+)}$-binding domain. *J Gen Physiol* **105:** 403–420.

Monteith GR, McAndrew D, Faddy HM, Roberts-Thomson SJ. 2007. Calcium and cancer: targeting Ca^{2+} transport. *Nat Rev Cancer* **7:** 519–530.

Nicoll DA, Longoni S, Philipson KD. 1990. Molecular cloning and functional expression of the cardiac sarcolemmal $Na^{(+)}$-Ca^{2+} exchanger. *Science* **250:** 562–565.

Nicoll DA, Sawaya MR, Kwon S, Cascio D, Philipson KD, Abramson J. 2006. The crystal structure of the primary

Ca^{2+} sensor of the Na$^+$/Ca^{2+} exchanger reveals a novel Ca^{2+} binding motif. *J Biol Chem* **281:** 21577–21581.

Niggli V, Adunyah ES, Carafoli E. 1981. Acidic phospholipids, unsaturated fatty acids, and limited proteolysis mimic the effect of calmodulin on the purified erythrocyte Ca^{2+}-ATPase. *J Biol Chem* **256:** 8588–8592.

Niggli V, Penniston JT, Carafoli E. 1979. Purification of the (Ca^{2+}-Mg^{2+})-ATPase from human erythrocyte membranes using a calmodulin affinity column. *J Biol Chem* **254:** 9955–9958.

Niggli V, Sigel E, Carafoli E. 1982. The purified Ca^{2+} pump of human erythrocyte membranes catalyzes an electroneutral Ca^{2+}-H+ exchange in reconstituted liposomal systems. *J Biol Chem* **257:** 2350–2356.

Oceandy D, Cartwright EJ, Emerson M, Prehar S, Baudoin FM, Zi M, Alatwi N, Venetucci L, Schuh K, Williams JC, et al. 2007. Neuronal nitric oxide synthase signaling in the heart is regulated by the sarcolemmal calcium pump 4b. *Circulation* **115:** 483–492.

Okunade GW, Miller ML, Pyne GJ, Sutliff RL, O'Connor KT, Neumann JC, Andringa A, Miller DA, Prasad V, Doetschman T, et al. 2004. Targeted ablation of plasma membrane Ca^{2+}-ATPase (PMCA) 1 and 4 indicates a major housekeeping function for PMCA1 and a critical role in hyperactivated sperm motility and male fertility for PMCA4. *J Biol Chem* **279:** 33742–33750.

Palty R, Silverman WF, Hershfinkel M, Caporale T, Sensi SL, Parnis J, Nolte C, Fishman D, Shoshan-Barmatz V, Herrmann S, et al. 2009. NCLX is an essential component of mitochondrial Na$^+$/Ca^{2+} exchange. *Proc Natl Acad Sci* **107:** 436–441.

Paszty K, Antalffy G, Penheiter AR, Homolya L, Padanyi R, Ilias A, Filoteo AG, Penniston JT, Enyedi A. 2005. The caspase-3 cleavage product of the plasma membrane Ca^{2+}-ATPase 4b is activated and appropriately targeted. *Biochem J* **391:** 687–692.

Prasad V, Okunade GW, Miller ML, Shull GE. 2004. Phenotypes of SERCA and PMCA knockout mice. *Biochem Biophys Res Commun* **322:** 1192–1203.

Preiano BS, Guerini D, Carafoli E. 1996. Expression and functional characterization of isoforms 4 of the plasma membrane calcium pump. *Biochemistry* **35:** 7946–7953.

Pulina MV, Rizzuto R, Brini M, Carafoli E. 2006. Inhibitory interaction of the plasma membrane Na$^+$/Ca^{2+} exchangers with the 14-3-3 proteins. *J Biol Chem* **281:** 19645–19654.

Reinhardt TA, Lippolis JD, Shull GE, Horst RL. 2004. Null mutation in the gene encoding plasma membrane Ca^{2+}-ATPase isoform 2 impairs calcium transport into milk. *J Biol Chem* **279:** 42369–42373.

Reuter H, Seitz N. 1968. The dependence of calcium efflux from cardiac muscle on temperature and external ion composition. *J Physiol* **195:** 451–470.

Rimessi A, Coletto L, Pinton P, Rizzuto R, Brini M, Carafoli E. 2005. Inhibitory interaction of the 14-3-3{epsilon} protein with isoform 4 of the plasma membrane Ca$^{(2+)}$-ATPase pump. *J Biol Chem* **280:** 37195–37203.

Schatzmann HJ. 1966. ATP-dependent Ca^{++}-extrusion from human red cells. *Experientia* **22:** 364–365.

Schuh K, Cartwright EJ, Jankevics E, Bundschu K, Liebermann J, Williams JC, Armesilla AL, Emerson M, Oceandy D, Knobeloch KP, et al. 2004. Plasma membrane Ca^{2+} ATPase 4 is required for sperm motility and male fertility. *J Biol Chem* **279:** 28220–28226.

Schuh K, Quaschning T, Knauer S, Hu K, Kocak S, Roethlein N, Neyses L. 2003. Regulation of vascular tone in animals overexpressing the sarcolemmal calcium pump. *J Biol Chem* **278:** 41246–41252.

Schuh K, Uldrijan S, Telkamp M, Rothlein N, Neyses L. 2001. The plasmamembrane calmodulin-dependent calcium pump: a major regulator of nitric oxide synthase I. *J Cell Biol* **155:** 201–205.

Schultz JM, Yang Y, Caride AJ, Filoteo AG, Penheiter AR, Lagziel A, Morell RJ, Mohiddin SA, Fananapazir L, Madeo AC, et al. 2005. Modification of human hearing loss by plasma-membrane calcium pump PMCA2. *N Engl J Med* **352:** 1557–1564.

Schulze DH, Muqhal M, Lederer WJ, Ruknudin AM. 2003. Sodium/calcium exchanger (NCX1) macromolecular complex. *J Biol Chem* **278:** 28849–28855.

Schwab BL, Guerini D, Didszun C, Bano D, Ferrando-May E, Fava E, Tam J, Xu D, Xanthoudakis S, Nicholson DW, et al. 2002. Cleavage of plasma membrane calcium pumps by caspases: a link between apoptosis and necrosis. *Cell Death Differ* **9:** 818–831.

Shull GE, Greeb J. 1988. Molecular cloning of two isoforms of the plasma membrane Ca^{2+}-transporting ATPase from rat brain. Structural and functional domains exhibit similarity to Na$^+$,K$^+$- and other cation transport ATPases. *J Biol Chem* **263:** 8646–8657.

Sokolow S, Manto M, Gailly P, Molgo J, Vandebrouck C, Vanderwinden JM, Herchuelz A, Schurmans S. 2004. Impaired neuromuscular transmission and skeletal muscle fiber necrosis in mice lacking Na/Ca exchanger 3. *J Clin Invest* **113:** 265–273.

Spiden SL, Bortolozzi M, Di Leva F, de Angelis MH, Fuchs H, Lim D, Ortolano S, Ingham NJ, Brini M, Carafoli E, et al. 2008. The novel mouse mutation Oblivion inactivates the PMCA2 pump and causes progressive hearing loss. *PLoS Genet* **4:** e1000238

Strehler EE, Zacharias DA. 2001. Role of alternative splicing in generating isoform diversity among plasma membrane calcium pumps. *Physiol Rev* **81:** 21–50.

Toyoshima C, Nakasako M, Nomura H, Ogawa H. 2000. Crystal structure of the calcium pump of sarcoplasmic reticulum at 2.6 A resolution. *Nature* **405:** 647–655.

Usachev YM, Toutenhoofd SL, Goellner GM, Strehler EE, Thayer SA. 2001. Differentiation induces up-regulation of plasma membrane Ca$^{(2+)}$-ATPase and concomitant increase in Ca$^{(2+)}$ efflux in human neuroblastoma cell line IMR-32. *J Neurochem* **76:** 1756–1765.

Verma AK, Filoteo AG, Stanford DR, Wieben ED, Penniston JT, Strehler EE, Fischer R, Heim R, Vogel G, Mathews S, et al. 1988. Complete primary structure of a human plasma membrane Ca^{2+} pump. *J Biol Chem* **263:** 14152–14159.

Vorherr T, Kessler T, Hofmann F, Carafoli E. 1991. The calmodulin-binding domain mediates the self-association of the plasma membrane Ca^{2+} pump. *J Biol Chem* **266:** 22–27.

Wakimoto K, Kobayashi K, Kuro OM, Yao A, Iwamoto T, Yanaka N, Kita S, Nishida A, Azuma S, Toyoda Y, et al. 2000. Targeted disruption of Na$^+$/Ca^{2+} exchanger gene leads to cardiomyocyte apoptosis and defects in heartbeat. *J Biol Chem* **275:** 36991–36998.

Wang J, Zhang XQ, Ahlers BA, Carl LL, Song J, Rothblum LI, Stahl RC, Carey DJ, Cheung JY. 2006. Cytoplasmic tail of phospholemman interacts with the intracellular loop of the cardiac Na^+/Ca^{2+} exchanger. *J Biol Chem* **281:** 32004–32014.

Williams JC, Armesilla AL, Mohamed TM, Hagarty CL, McIntyre FH, Schomburg S, Zaki AO, Oceandy D, Cartwright EJ, Buch MH, et al. 2006. The sarcolemmal calcium pump, α-1 syntrophin, and neuronal nitric-oxide synthase are parts of a macromolecular protein complex. *J Biol Chem* **281:** 23341–23348.

Wu G, Xie X, Lu ZH, Ledeen RW. 2009. Sodium-calcium exchanger complexed with GM1 ganglioside in nuclear membrane transfers calcium from nucleoplasm to endoplasmic reticulum. *Proc Natl Acad Sci* **106:** 10829–10834.

Xie X, Wu G, Lu ZH, Ledeen RW. 2002. Potentiation of a sodium-calcium exchanger in the nuclear envelope by nuclear GM1 ganglioside. *J Neurochem* **81:** 1185–1195.

Zacharias DA, Strehler EE. 1996. Change in plasma membrane $Ca^{2(+)}$-ATPase splice-variant expression in response to a rise in intracellular Ca^{2+}. *Curr Biol* **6:** 1642–1652.

Zhang XQ, Ahlers BA, Tucker AL, Song J, Wang J, Moorman JR, Mounsey JP, Carl LL, Rothblum LI, Cheung JY. 2006. Phospholemman inhibition of the cardiac Na^+/Ca^{2+} exchanger. Role of phosphorylation. *J Biol Chem* **281:** 7784–7792.

Zvaritch E, James P, Vorherr T, Falchetto R, Modyanov N, Carafoli E. 1990. Mapping of functional domains in the plasma membrane Ca^{2+} pump using trypsin proteolysis. *Biochemistry* **29:** 8070–8076.

The Ca^{2+} Pumps of the Endoplasmic Reticulum and Golgi Apparatus

Ilse Vandecaetsbeek, Peter Vangheluwe, Luc Raeymaekers, Frank Wuytack, and Jo Vanoevelen

Laboratory of Ca^{2+}-Transport ATPases, Department of Molecular Cell Biology, K.U. Leuven, B-3000 Leuven, Belgium

Correspondence: Frank.Wuytack@med.kuleuven.be

The various splice variants of the three SERCA- and the two SPCA-pump genes in higher vertebrates encode P-type ATPases of the P_{2A} group found respectively in the membranes of the endoplasmic reticulum and the secretory pathway. Of these, SERCA2b and SPCA1a represent the housekeeping isoforms. The SERCA2b form is characterized by a luminal carboxy terminus imposing a higher affinity for cytosolic Ca^{2+} compared to the other SERCAs. This is mediated by intramembrane and luminal interactions of this extension with the pump. Other known affinity modulators like phospholamban and sarcolipin decrease the affinity for Ca^{2+}. The number of proteins reported to interact with SERCA is rapidly growing. Here, we limit the discussion to those for which the interaction site with the ATPase is specified: HAX-1, calumenin, histidine-rich Ca^{2+}-binding protein, and indirectly calreticulin, calnexin, and ERp57. The role of the phylogenetically older and structurally simpler SPCAs as transporters of Ca^{2+}, but also of Mn^{2+}, is also addressed.

All cells invest a considerable part of their total energy budget in active transport to keep up transmembrane (TM) ion gradients (Rolfe and Brown 1997). Prokaryotes already evolved P-type ion-transport ATPases/ion pumps to that aim (Axelsen and Palmgren 1998). The name P-type refers to the transient transfer of the γ-phosphate group of ATP to a highly conserved aspartate group in the enzyme forming a phospho-intermediate. This autophosphorylation is an important step in the pump's catalytic cycle (Kuhlbrandt 2004). Based on amino-acid sequence comparisons and on the exon/intron layout of the corresponding genes, three types of P-type Ca^{2+} pumps can be discerned in Eumetazoa: the SERCA-, the SPCA-, and the PMCA-type. Whereas ancestral representatives of each type are recognized in some Eubacteria and Archaea, it is also remarkable that some Eukaryotes have apparently lost either SERCA or SPCA pumps. Yeast for instance lacks SERCA pumps whereas plants thrive well without SPCAs (Mills et al. 2008). The SERCA pumps, which are found in the endoplasmic reticulum (ER) or in the sarcoplasmic reticulum (SR) of eukaryotic cells and the evolutionary older secretory pathway ATPases (SPCA) found in the Golgi apparatus, are closely related to each other and together belong to the P_{2A} subfamily. They form the topic of this review. The plasma-membrane Ca^{2+}-pumps (PMCA), on the other hand, appear to be phylogenetically the oldest of the three and form the P_{2B}-subfamily branch. PMCAs are addressed in an article by Brini and Carafoli (2009). Some further information on the evolution of the three types of ATPases was recently reviewed by Palmgren and Axelsen (1998) and Vangheluwe et al. (2009). Of the three families, only SERCA pumps translocate two Ca^{2+} ions and hydrolyze one ATP for each catalytic turnover. They possess two Ca^{2+}-transport sites: site I and site II; the numbers specify the sequence of filling of the respective sites. The single Ca^{2+}-binding site of the SPCA and PMCA pumps structurally corresponds to site II of SERCA (Toyoshima 2009).

THE UBIQUITOUS SERCA2 Ca^{2+} PUMP

SERCA2 Splicing Variants

Vertebrates generate multiple SERCA isoforms as a result of alternative processing of the transcripts of three paralogous SERCA genes *(ATP2A1-3)* (Brini and Carafoli 2009). Invertebrates typically have only a single SERCA gene that is orthologous to the vertebrate housekeeping SERCA2. The two major vertebrate SERCA2 protein isoforms are the housekeeping SERCA2b and the more specialized SERCA2a isoform. The latter is found in slow skeletal muscle and cardiac muscle, but is also expressed in lower amounts in smooth muscle and in neuronal cells (Vandecaetsbeek et al. 2009a). Recently novel SERCA2c (Dally et al. 2006) and SERCA2d (Kimura et al. 2005) isoforms were discovered in the heart, but are expressed at low levels and their physiological meaning remains to be further explored.

Physiological Role of SERCA2

The housekeeping SERCA2b Ca^{2+} pump serves a dual role. By translocating Ca^{2+} from the cytosol into the lumen of the ER, it restores the cytosolic Ca^{2+} concentration to its low resting level (circa 100 nM). At the same time, SERCA2b maintains a sufficiently high (circa 500 μM) luminal ER Ca^{2+} concentration. The ER not only serves as a useful Ca^{2+} store for the release of Ca^{2+} that activates an impressive number of cellular activities (e.g., contraction, fertilization, insulin release, etc.) but it also creates the luminal environment necessary for almost all local enzyme activities (such as protein folding and synthesis of lipids and steroids) and that controls cell fate (proliferation, apoptosis, growth, or differentiation) (Wuytack et al. 2002).

The muscle variant SERCA2a removes the Ca^{2+} stimulus for contraction by pumping myoplasmic Ca^{2+} into the SR and thereby determines the Ca^{2+} load of the SR, which in turn determines the amount of Ca^{2+} that can be released for the next contraction. Together, SERCA2a is a major determinant of the speed and force of cardiac contraction and relaxation (Periasamy and Huke 2001). SERCA2 expression is reduced in end-stage heart failure, contributing to an impaired contractility of the heart (Hasenfuss et al. 1994).

Ablation in mice of the two *Atp2a2* alleles is incompatible with life (Periasamy et al. 1999). But in light of the central role SERCA2a exerts in the heart, it is quite surprising that in an inducible cardiac-specific knock-out mouse model at 4 weeks following *Atp2a2* gene deletion, cardiac function remained near normal in spite of the drop of the myocardial SERCA2 levels below 5% of controls (Andersson et al. 2009). However, end-stage heart failure developed at 7 weeks. These results show the remarkable power of a compensatory (albeit ultimately failing) response to such a major acute reduction in SERCA2 function (Andersson et al. 2009). The effect of heterozygous knock-out of *Atp2a2* in mice is also paralleled by compensatory responses, with only slight impact on cardiac contractility and relaxation without eliciting cardiac disease (Periasamy et al. 1999; Ji et al. 2000). With age, these heterozygotes are prone to develop squamous cell tumors, which supports the notion that altered Ca^{2+} homeostasis plays a significant role in cancer (Liu et al. 2001; Prasad et al. 2005). Likewise, humans lacking one functional *ATP2A2* allele do not develop cardiomyopathy (Tavadia et al.

2001), but the effect of reduced Ca^{2+} uptake activity is manifested in keratinocytes, where it triggers the onset of the skin disorder of Darier (Sakuntabhai et al. 1999).

Whereas previous studies suggest that changes in SERCA2 expression levels are reasonably well tolerated in the heart (Ji et al. 2000; Tavadia et al. 2001), other studies point to a more critical regulation of the apparent affinity of the Ca^{2+} pump for cytosolic Ca^{2+} ions (MacLennan and Kranias 2003; Vandecaetsbeek et al. 2009a; Sipido and Vangheluwe 2010). For normal cardiac function, the affinity of SERCA2a in the cardiac SR needs to be controlled within a tight window (Vangheluwe et al. 2005a; Vandecaetsbeek et al. 2009a). Genetic manipulations in mouse that lead to the expression of the high Ca^{2+}-affinity variant SERCA2b in the cardiomyocyte instead of the normal SERCA2a, triggers cardiac hypertrophy and heart failure (Ver Heyen et al. 2001; Vangheluwe et al. 2006b). Likewise, in humans (Haghighi et al. 2003), but not in mice (Luo et al. 1994), the increased Ca^{2+} affinity resulting from the absence of phospholamban (PLN, i.e., an affinity modulator of the pump, discussed below) triggers heart failure (Haghighi et al. 2003). On the contrary, a chronic reduction in the Ca^{2+} affinity triggered by a higher activity of PLN is also associated with dilated cardiomyopathy in humans (Haghighi et al. 2001; Schmitt et al. 2003; Haghighi et al. 2006).

The Ca^{2+}-Pumping Mechanism

Ten years ago, the first high-resolution crystal structure of the fast-twitch skeletal-muscle isoform SERCA1a was published (Toyoshima et al. 2000). Since then, we have been spoilt by high-resolution crystal structures of SERCA1a in nine different conformations, yielding detailed molecular insights of the Ca^{2+}-pumping process (reviewed in Moller et al. 2005; Toyoshima 2008; Toyoshima 2009). In addition, structures of other archetypical P-type ATPases (Na^+/K^+-ATPase [Morth et al. 2007; Shinoda et al. 2009] and H^+-ATPase [Pedersen et al. 2007]) were reported. The basic structure of these P-type ATPases is very well conserved, even if the overall sequence similarity is low (Fig. 1). Three cytosolic domains can be recognized in the P-type ATPases: a nucleotide-binding (N), phosphorylation (P), and actuator (A) domain (Fig. 1). ATP binds on the N-domain, whereas the P-domain drives ATP hydrolysis leading to phosphorylation of a highly conserved aspartate in the P-domain. The A-domain then contains a conserved glutamate that catalyzes the dephosphorylation of the P-domain (Kuhlbrandt 2004; Vangheluwe et al. 2009). The large headpiece is intimately connected with and partially embedded in the TM region that contains the ion-binding sites. This connection assures tight coupling between ATP hydrolysis in the cytosolic domains and ion transport across the membrane. Surprisingly, the overall structure of the TM region is also highly conserved with only subtle differences accounting for ion specificity (Gadsby 2007). The accessibility of the TM Ca^{2+}-binding sites in SERCA1a is controlled by both a cytosolic and a luminal gate, which are under control of the phosphorylation and dephosphorylation events, respectively, in the headpiece (Moller et al. 2005; Toyoshima 2008; Toyoshima 2009). Moreover, a feedback mechanism associated with ion binding guarantees that ATP hydrolysis can only occur when ions are bound. This tight coupling assures that first the cytosolic gate closes and Ca^{2+} ions are occluded before ATP hydrolysis and opening of the luminal gate can occur (Moller et al. 2005; Toyoshima 2008; Toyoshima 2009). This allows Ca^{2+} ions to be pumped against an almost 10000-fold gradient across the ER/SR membrane (Toyoshima 2009).

Structure of the Ubiquitous SERCA2b Pump

Although the ubiquitous SERCA2b pump shares an overall 85% sequence identity with SERCA1a, which points to a common Ca^{2+}-pumping mechanism (Toyoshima 2009), three related properties discriminate the SERCA2b isoform from SERCA1a or SERCA2a: the characteristic two-fold higher affinity for cytosolic Ca^{2+} ions, the lower maximal turnover rate

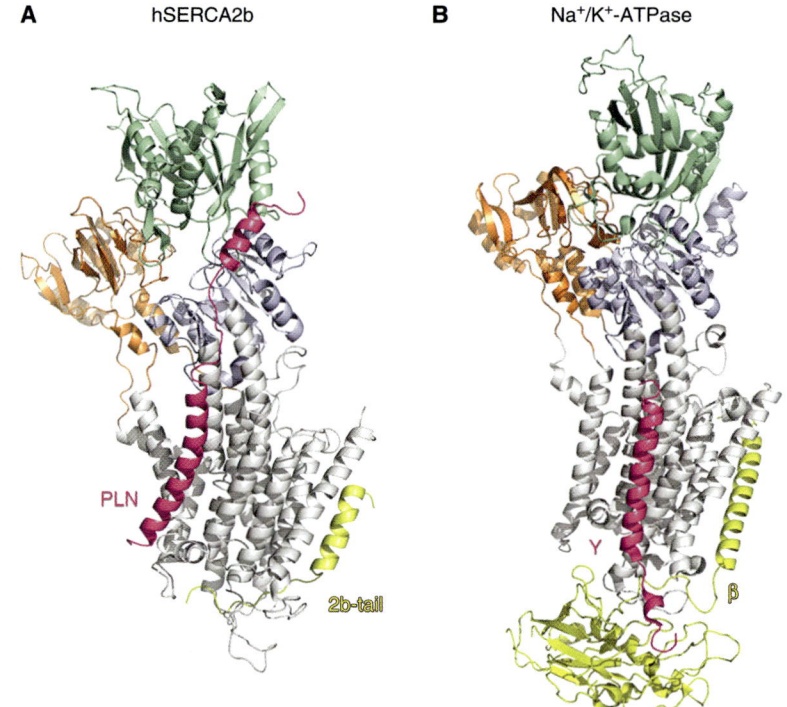

Figure 1. Interesting structural similarities between SERCA2b and Na$^+$/K$^+$-ATPase. (*A*) The PLN NMR structure (Seidel et al. 2008) and the carboxyl terminus of SERCA2b (Vandecaetsbeek et al. 2009b) modeled on the E2 crystal structure of rabbit SERCA1a (2AGV) (Obara et al. 2005). (*B*) Crystal structure of the pig renal Na$^+$/K$^+$-ATPase α-subunit (2ZXE) (Shinoda et al. 2009) in the E2 conformation, together with its regulatory β- and γ-subunits. Interesting similarities exist between the binding sites of the regulatory β- and γ-subunits on the Na$^+$/K$^+$-ATPase and, respectively, the 2b-tail and PLN on the SERCA1a pump. Orange: A-domain; Blue: P-domain; Green: N-domain; Gray: TM-domain. PLN, phospholamban; SLN, sarcolipin; 2b-tail, SERCA2b carboxyl terminus.

and the presence of a unique carboxy-terminal extension (2b-tail) comprising an additional TM segment (TM11) and a luminal extension (LE) (Lytton et al. 1992; Verboomen et al. 1994; Dode et al. 2003; Vandecaetsbeek et al. 2009b). Functional measurements on SERCA2b mutants and SERCA1a-2b chimeras revealed that both of these regions contribute to the functional effect of the 2b-tail (Verboomen et al. 1994; Vandecaetsbeek et al. 2009b). Based on the known SERCA1a crystal structures and the solved NMR structure of TM11, a structural model for SERCA2b was proposed that is backed up by extensive mutagenesis results (see Fig. 1A in Vandecaetsbeek et al. 2009b). According to that model, TM11 is interacting with TM7 and TM10 of the Ca^{2+} ATPase, a relatively immobile part of the pump. A groove between luminal loops L5-6 and L7-8 is opened at the luminal side of TM11, for the descent of LE. This displacement allows that the peptide consisting of the last four, crucial amino-acids at the pump's carboxyl terminus (1039-MFWS) reaches a luminal binding pocket that is formed by the five luminal loops of the pump (Vandecaetsbeek et al. 2009b). This intramolecular interaction stabilizes the pump in the Ca^{2+}-bound E1 conformation with high-affinity binding sites facing the cytosol. Mathematical modeling confirmed that this could explain the increased apparent affinity for Ca^{2+} (Vandecaetsbeek et al. 2009b). Moreover, the experimentally observed slower E1-P to E2-P and E2-P to E2 transitions (Dode et al. 2003) are tightly coupled to extensive rearrangements of the proposed luminal docking site of the 2b-tail (Vandecaetsbeek et al. 2009b). How the short TM11 α-helix alters the enzymatic properties at the distant and relatively immobile TM helices TM7 and TM10, remains to be clarified.

Regulators of the ER Ca^{2+} Pump

Given its central position in cellular Ca^{2+} homeostasis, the activity of SERCA2 is prone to tight regulation. At least a dozen of different proteins were suggested to regulate SERCA2 activity (previously reviewed in Vangheluwe et al. 2005a; Vandecaetsbeek et al. 2009a). This suggests that as for the intracellular Ca^{2+} channels inositol-1,4,5-trisphosphate receptor (IP3R) or the ryanodine receptor (RyR) (Foskett et al. 2007), SERCA2 might form a multiprotein complex varying in composition in different cell types. However, because of its smaller size and the requirement to undergo major conformational changes during its enzymatic cycle, formation of a macromolecular SERCA complex is probably more restricted.

It is of some concern that studies on the effect of putative SERCA modulators often rely on overexpression, which on itself can lead to ER stress via the unfolded protein response (UPR) that includes up-regulation of SERCA2b expression and activity (Caspersen et al. 2000). In addition, the effect of these modulators is almost never confined to SERCA because they are nearly always part of the Ca^{2+} signalome also affecting Ca^{2+} release channels. Finally, direct interaction of these regulators with the pump is often documented by immunoprecipitation, which for TM proteins is technically very challenging. The thriving literature of putative SERCA regulators should therefore be viewed with caution as long as the interaction site is not properly identified. Here, we will only focus on those regulators for which the binding site on the Ca^{2+} pump is defined and well-documented (Fig. 2).

Phospholamban and Sarcolipin

The related small TM proteins PLN and sarcolipin (SLN) are the best-studied regulators of the SERCA pump (reviewed in MacLennan and Kranias 2003; Vangheluwe et al. 2006a; Bhupathy et al. 2007; Periasamy et al. 2008). In contrast to the 2b-tail, these proteins interact with the pump to reduce the apparent affinity for cytosolic Ca^{2+} ions, which inhibits overall Ca^{2+} transport (Lee 2003; MacLennan and Kranias 2003). In vivo, PLN is mainly coexpressed with SERCA2a in the heart, smooth muscle, and slow-twitch skeletal-muscle fibers. During the β-adrenergic response in cardiac muscle, phosphorylation of PLN by protein kinase A and/or Ca^{2+}-calmodulin kinase II (CaMKII) promotes dissociation of the complex, which reverses the inhibition of SERCA2a (reviewed in MacLennan and Kranias 2003). Dissociated PLN also exists in a stable but inactive, pentameric state, which is promoted by phosphorylation (Kimura et al. 1997). PLN-SERCA2a dissociation causes a dramatic increase in SR Ca^{2+} transport leading to improved cardiac contraction and relaxation (Luo et al. 1994). Studies in numerous PLN animal models further showed its central role in cardiac contractility (reviewed in MacLennan and Kranias 2003). Moreover, human PLN mutations leading to either a chronic increase like L39stop (Haghighi et al. 2003) or decrease like R14del (Haghighi et al. 2006) or R9C (Schmitt et al. 2003) of the apparent Ca^{2+} affinity of the pump trigger the onset of dilated cardiomyopathy and heart failure at a young age. In line with the effect of the SERCA2a→b isoform switch (Vangheluwe et al. 2006b), these studies further indicate that regulating the Ca^{2+} affinity of the pump is of vital importance to maintain normal cardiac function and development (Vangheluwe et al. 2006a). This appears to be more important in humans than in mice (Haghighi et al. 2003; MacLennan et al. 2003; Zhao et al. 2006). More recent studies suggest that the regulation of the pump by PLN phosphorylation is crucial for maintaining some cardiac reserve to prevent heart failure (Schmitt et al. 2009). This is in line with an increased morbidity and mortality in

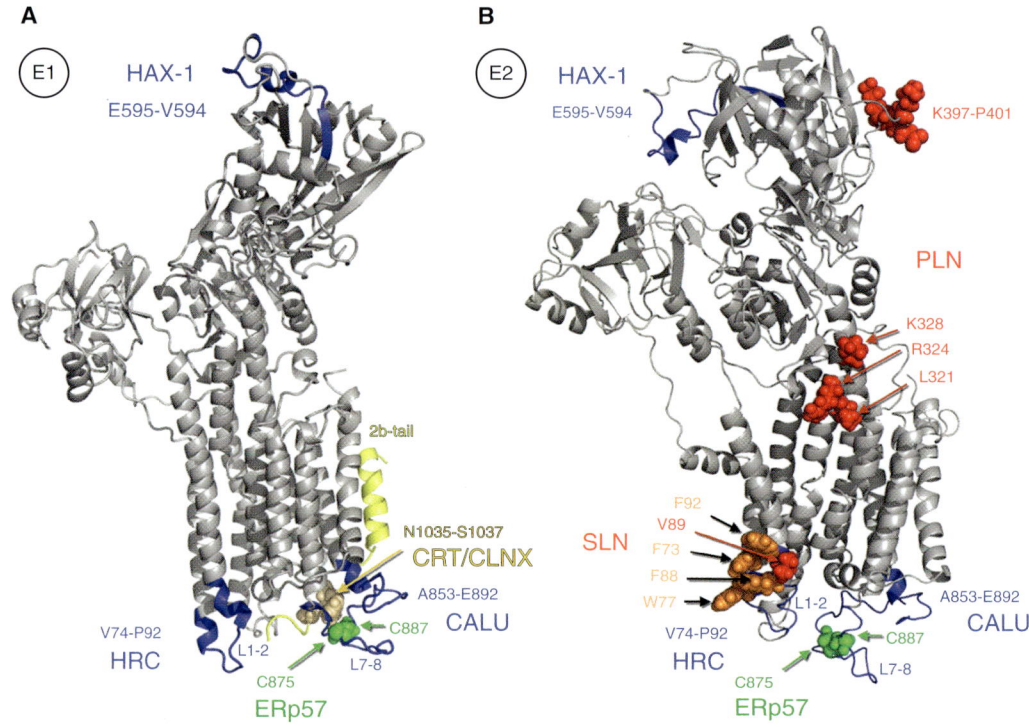

Figure 2. Interaction sites of different SERCA regulators. Different interaction sites are depicted on the crystal structure of rSERCA1a in the E1 conformation (1SU4) (Toyoshima et al. 2000) (*A*) and in E2 (2AGV) (Obara et al. 2005) (*B*). Note that PLN and SLN only interact in E2, and the 2b-tail predominantly in E1, and therefore are only depicted in the respective conformations. CALU, Calumenin; PLN, phospholamban; SLN, sarcolipin; HAX-1, HS1-associated protein; CRT, calreticulin; CLNX, calnexin; ERp57, endoplasmic reticulum thiol-disulfide oxidoreductase; HRC, histidine-rich Ca^{2+}-binding protein; 2b-tail, SERCA2b carboxyl terminus.

heart failure patients with a lower response to β-agonists (Wu et al. 2004; Kobayashi et al. 2008).

PLN inhibits the SERCA2a and SERCA2b isoforms to the same extent (Verboomen et al. 1992), thus occupying a different affinity-regulating site on the Ca^{2+} pump than the 2b-tail (see Fig. 1A in Vandecaetsbeek et al. 2009b). In fact, extensive crosslinking, site-directed mutagenesis and structural modeling studies have shown that residues in both the cytoplasmic and the TM portions of PLN are involved in direct interaction with SERCA2a (Fig. 2B) (James et al. 1989; Kimura et al. 1996; Asahi et al. 1999; Asahi et al. 2001; Toyoshima et al. 2003). First proof of the direct interaction between SERCA and PLN came from a homobifunctional crosslink between a lysine in the N-domain of SERCA2a (in the region 397-401) and a lysine in the cytosolic region of PLN (K3) (James et al. 1989). To date, evidence for at least three sites of close association between SERCA1a and PLN was provided by robust homobifunctional crosslinking: between V89C positioned on TM2 of SERCA1a with V49C (Toyoshima et al. 2003), between L321C at the cytosol-membrane boundary of SERCA1a TM4 with N27C (Toyoshima et al. 2003), and between K328C in the cytosolic domain with Q23C (Morita et al. 2008). Additional heterobifunctional crosslinks were observed between the SERCA2a isoform and PLN, but unexpectedly and in apparent contrast with earlier studies, no such crosslinks were observed involving K3 of PLN (Chen et al. 2003). Phosphorylation of PLN or high Ca^{2+}

concentrations lead to the (partial) dissociation of the PLN-SERCA2a complex preventing crosslinking (Chen et al. 2007; Morita et al. 2008). Together, these studies showed that the interaction of the PLN TM region occurs in a hydrophobic cleft only present in the Ca^{2+}-free E2 conformation that is formed by TM2,4,6,9 (Toyoshima et al. 2003). This interaction occurs at the border between the highly mobile (TM1-6) and the immobile (TM7-10) parts of the pump and inhibits the closing of the cleft during the transition from E2 to E1. The profound effect of PLN phosphorylation on the functional and physical interaction with the Ca^{2+} pump already indicates that the cytosolic interaction with the N-domain could be equally important. This is further corroborated by the functional effect of mutating the cytosolic domain of PLN (reviewed in MacLennan et al. 2003). Phosphorylation would partially unwind the cytosolic region, indicating an order-to-disorder transition (Metcalfe et al. 2005; Karim et al. 2006), which would prevent more distant interactions such as a crucial H-bridge between R324 and Q26 (Traaseth et al. 2006; Traaseth et al. 2008). Together this would loosen the interaction or even cause a complete dissociation of the PLN-SERCA2a complex.

Although several lines of evidence indicate that monomeric PLN is the active species (Kimura et al. 1997), recent structural observations indicate that PLN pentamers might also interact with the pump, although at a different site (close to TM3) than the monomer. It remains unknown whether this serves a physiological function (Stokes et al. 2006).

SLN appeared to act as the functional counterpart of PLN in fast-twitch skeletal-muscle. But SLN is also found together with PLN in the atria of the heart (Minamisawa et al. 2003; Vangheluwe et al. 2005b; Babu et al. 2007a), where it modulates the activity of the SERCA2a pump and is under control of β-adrenergic stimulation (Babu et al. 2007b), presumably via CaMKII-dependent phosphorylation of T5 (Bhupathy et al. 2009). By analogy, the conservation in sequence, structure and dynamics between SLN and PLN suggest that SLN would fit into the same hydrophobic groove as PLN having similar regulatory properties (Traaseth et al. 2008). The aromatic residues of the highly conserved luminal extension RSYQY of SLN are functionally relevant (Odermatt et al. 1998) and would interact with aromatic residues on the face of luminal loop L1-2 of SERCA (possibly with the side chains of F73, W77, F88 or F92), opposite to that which constitutes the luminal interaction site of the 2b-tail (Fig. 2) (Hughes et al. 2007). TM1 undergoes a strong upward movement during the enzymatic cycle, which might be affected by this interaction. Notably, this SLN luminal tail is also crucial for proper integration of SLN in the membrane (Gramolini et al. 2004).

PLN and SLN would fit together side-by-side into the same TM cleft TM2,4,6,9, leading to a tighter functional interaction (Fig. 2B) (Asahi et al. 2003). This would explain the super-inhibitory properties of the PLN-SLN heterodimers observed in vitro (Asahi et al. 2002). Given the functional importance of the cytosolic domain of PLN and luminal extension of SLN, an additional stabilization of the complex might arise from their combined interaction with the pump (Hughes et al. 2007). So far, there is no clear evidence for this super-inhibition under physiological circumstances in the atria of the heart where both SERCA regulators are found (Bhupathy et al. 2007; Periasamy et al. 2008; Vandecaetsbeek et al. 2009a).

Surprisingly, the proposed positions of the 2b-tail and PLN/SLN on the Ca^{2+} pump strikingly mirrors the observed interaction site of the Na^+/K^+-ATPase β- and γ-subunits (Fig. 1) (Toyoshima et al. 2003; Morth et al. 2007; Vandecaetsbeek et al. 2009b). Although these modulators evolved independently from each other, they seem to occupy similar binding sites on the corresponding pump sharing similar molecular mechanisms. In all cases, the functional effect is related to a combined interaction of a TM region and luminal or cytosolic extensions with the pump, which might stabilize one of the conformational intermediates of the enzyme (Vandecaetsbeek et al. 2009b). Notably, the site of interaction of the γ-subunit was determined from the E2 Na^+/K^+-ATPase crystal structure (Morth et al. 2007; Shinoda et al.

2009), but in contrast to earlier modeling of PLN on SERCA1a (Toyoshima et al. 2003) and the γ-subunit on Na^+/K^+-ATPase (Li et al. 2004), the binding occurs on TM9, at the outside of the proposed cleft.

Antiapoptotic Proteins HAX-1 and Bcl-2

The HS1-associated protein HAX-1 (35 kDa) is an integral membrane protein normally residing in the outer mitochondrial membrane (Suzuki et al. 1997; Vafiadaki et al. 2009b). It interacts with a multitude of proteins. It was proposed that depending on the available interaction partners, the subcellular localization and functional properties of HAX-1 might vary among different tissues (Vafiadaki et al. 2009b). Recently, PLN was identified via yeast two-hybrid screen and GST-pull-down experiments as a novel interaction partner of HAX-1 (Vafiadaki et al. 2007). The site of the HAX-1 and PLN interaction is well documented and is confined to the regions 203-245 of HAX-1 and 16-22 of PLN, overlapping with the PLN phosphorylation sites (Vafiadaki et al. 2007). The direct association between HAX-1 and PLN was further established in vivo (Zhao et al. 2009). HAX-1 serves an inhibitory role on basal contractility of the heart by stabilizing the PLN monomers and lowering the apparent Ca^{2+} affinity of SERCA2a. Notably, this effect is reversed during β-adrenergic stimulation (Zhao et al. 2009).

The HAX-1 GST-pull-down experiments also detected SERCA2a, which implies that PLN can interact simultaneously with HAX-1 and SERCA2a, notably with similar binding affinities (K_D of 0.70 μM and 1 μM, respectively (Kimura and Inui 2002; Vafiadaki et al. 2007). The HAX-1 interaction is confined to residues 575-594 in the SERCA2 N-domain, enclosing an accessible and highly conserved loop, on the opposite site of the proposed cytosolic PLN interaction region 397-401 (Fig. 2) (Vafiadaki et al. 2009a). Whether this interaction also occurs in the physiological setting of the heart remains to be investigated.

The preferential mitochondrial localization of HAX-1 in HEK-293 cells can be changed to an ER distribution on cotransfection with PLN (Vafiadaki et al. 2007), but not on cotransfection with SERCA1a or SERCA2 (Vafiadaki et al. 2009a). Interaction of HAX-1 in the outer membrane of the mitochondria and with the ER-based SERCA could be possible at the ER-mitochondrial nexus sites, which are considered crucial for eliciting apoptosis. HAX-1 overexpression in HEK-293 cells results in a posttranscriptional downregulation of SERCA2 protein levels. The resulting lower ER Ca^{2+} content could explain the antiapoptotic role of HAX-1 (Vafiadaki et al. 2009a). In addition, because of its association with PLN and SERCA2 on one hand, and interaction with caspase-9 on the other hand, HAX-1 might link two Ca^{2+}-regulated processes in the heart: contractility and cell survival (Han et al. 2006).

These observations on HAX-1 are remarkably parallel to the effects of Bcl-2, another antiapoptotic protein (reviewed in Vafiadaki et al. 2009b; Vandecaetsbeek et al. 2009a). Bcl-2 is also located in the mitochondria and can be found in the ER, where it is able to interact with SERCA2. However, the putative interaction site of Bcl-2 on the pump remains to be defined, and how Bcl-2 affects ER Ca^{2+} reuptake remains somewhat controversial. Experimental evidence supports different alternatives: a) the interaction between SERCA and Bcl-2 inactivates the pump, presumably by destabilizing the protein (Dremina et al. 2004), b) Bcl-2 would regulate the SERCA expression levels (Kuo et al. 1998; Vanden Abeele et al. 2002), and c) Bcl-2 could inactivate SERCA by extraction of the ATPase from caveolae-related domains in the SR (Dremina et al. 2006).

SERCA Complexes Involving Luminal Proteins Calreticulin, Calnexin, and ERp57

Two of the earliest proposed SERCA2b interactors are the lectin molecular chaperones: the 46-kDa ER luminal Ca^{2+}-binding calreticulin (CRT) and its 90-kDa homolog the type-I ER integral protein calnexin (CLNX). Both proteins contain a globular N-domain involved in glucose or oligo-saccharide binding, an extended P-domain mediating ERp57 binding and an acidic Ca^{2+}-binding C-domain (Michalak et al. 2009). The C-domain of CRT can bind 25 mol of Ca^{2+} with low (2 mM) affinity (Baksh and Michalak 1991) and thus CRT complexes over half of all ER luminal Ca^{2+}. Luminal Ca^{2+} buffering by CLNX is much less pronounced because it contains much less Ca^{2+}-binding sites and its acidic carboxyl terminus protrudes into the cytosol. The direct interaction between these luminal ER Ca^{2+} buffers and the Ca^{2+} pump and release channels might represent an elegant feed-back system that controls ER Ca^{2+} filling (John et al. 1998; Roderick et al. 2000).

According to some early reports Ca^{2+}-loaded CRT or CLNX would interact with the N-linked carbohydrates inserted on residues 1035-NFS in the isoform-specific luminal extension of SERCA2b (Fig. 2A) (John et al. 1998). Although this is a consensus N-glycosylation site (N1035), glycosylation was never experimentally observed (John et al. 1998; Roderick et al. 2000; Vandecaetsbeek et al. 2009b). The lack of glycosylation does however not a priori exclude CLNX or CRT binding to SERCA because these ER chaperones can occasionally also bind nonglycosylated targets (Roderick et al. 2000; Ireland et al. 2008). The interaction with CRT or CLNX would exert an inhibitory effect on the Ca^{2+}-wave propagation in *Xenopus oocytes* (John et al. 1998; Roderick et al. 2000). However, mutants in this site retain normal Ca^{2+}-dependent ATPase-activity when overexpressed in COS cells (Vandecaetsbeek et al. 2009b). According to the SERCA2b molecular model, the 2b-tail is buried in luminal loops of the pump making its interaction with other proteins less likely (Vandecaetsbeek et al. 2009b).

ERp57, a member of the PDI family with thio-oxidoreductase activity catalyzing disulfide-bond formation of glycoproteins (Ni and Lee 2007) is recruited into the SERCA2b-chaperone complex and establishes a disulfide bridge between C875 and C887 in L7-8 of SERCA2 (Fig. 2) (Li and Camacho 2004). According to the proposed model, SERCA2 with an oxidized loop (S-S bridge is present) would be inhibited and remain so as long as ERp57 is bound (Li and Camacho 2004). The conclusion that reduced C875 and C887 in L7-8 are required for full SERCA2 activity is difficult to reconcile with the observation that mutations of either or both of the cysteine residues resulted in a loss of transport without loss of Ca^{2+}-dependent ATPase activity in SERCA1 (Daiho et al. 2001). Note that these cysteine residues are conserved in SERCA1-3, and that the C875G mutation is a known Darier mutant (Ruiz-Perez et al. 1999). Finally, we want to remark that ERp57 does not require interactions with CLNX and CRT to recognize its substrate (Zhang et al. 2009) and that CRT binds to SERCA2a oxidatively damaged by H_2O_2 treatment, which leads to SERCA degradation via a proteasome-dependent pathway (Ihara et al. 2005).

SERCA-Calumenin Interaction

Calumenin (CALU; 50 kDa) is a ubiquitously expressed protein, conserved from invertebrates to vertebrates, which is found in the lumen of the ER and SR (Sahoo et al. 2009). Because of its nonconsensus ER-retention signal, the protein can escape from the ER and even be secreted (Vorum et al. 1999). CALU belongs to the CREC family, which members share multiple EF-hand Ca^{2+}-binding motifs (Honore 2009). CALU binds in its Ca^{2+}-loaded form to the luminal domain of SERCA2 and presumably also the other SERCAs. GST-pull-down experiments with the different luminal loops of the pump showed that CALU interacts with L7-8 of the ATPase (presumably region 853-892, Fig. 2), i.e., close to or overlapping with the ERp57 interaction area, but apparently on the other side of the 2b-tail interaction site. CALU prefers the Ca^{2+}-bound E1 conformation of SERCA, and when bound decreases the apparent Ca^{2+} affinity of the ATPase (Sahoo et al. 2009). Overexpression of CALU in rat neonatal cardiomyocytes reduced SR Ca^{2+} uptake and decreased fractional release. Thus, interaction with the ryanodine receptor RyR2 is also suggested from these experiments. CALU would be essential during the early stages of development, similar to other Ca^{2+}-binding ER chaperone proteins like CRT, and ERp57. Much lower levels of CALU than calsequestrin are present in the adult heart.

Of note, the longest luminal loop L7-8 of SERCA2 apparently is the interaction site of several regulators (Fig. 2): the 2b-tail (Vandecaetsbeek et al. 2009b), ERp57 (Li and Camacho 2004) and CALU (Sahoo et al. 2009). Also, the extracellular loop L7-8 of the Na^+/K^+-ATPase α-subunit is functionally interacting with the extracellular region of the β-subunit (Morth et al. 2007). The long L7-8 would predominantly serve a regulatory function, because Ca^{2+} transport is supported with a much shorter L7-8, as in the closely related SPCA Ca^{2+} pump (Vangheluwe et al. 2009). This loop may regulate the Ca^{2+}-binding affinity of SERCA2 through modulation of the Ca^{2+}-binding pocket in TM8 (a true Ca^{2+}-affinity effect) or via stabilization of an intermediate of the pump exerting a kinetic effect on the apparent Ca^{2+} affinity (Vandecaetsbeek et al. 2009b).

Histidine-Rich Ca^{2+}-Binding Protein

Another luminal Ca^{2+}-binding protein that interacts with SERCA2 is the histidine-rich Ca^{2+}-binding protein (HRC; 170 kDa), which shows an inhibitory interaction with the luminal domain of SERCA2 where it binds to L1-2 (region 74-90, Fig. 2) (Arvanitis et al. 2007). Note that this site potentially overlaps with the binding site of SLN or the luminal extension of the 2b-tail (Hughes et al. 2007). HRC binds Ca^{2+} with high capacity, but low affinity (Hofmann et al. 1989; Picello et al. 1992). HRC shares similarities with calsequestrin, the major SR Ca^{2+} buffer protein, but is much less abundant (1% of skeletal muscle SR) (Damiani et al. 1997; Pritchard and Kranias 2009). Using different regions HRC binds in a Ca^{2+}-dependent manner with the SERCA pump and with triadin, which is part of the RyR Ca^{2+}-release complex (Pritchard and Kranias 2009). If the Ca^{2+} load in the SR is low, HRC would interact with SERCA. If HRC becomes saturated with Ca^{2+}, it dissociates from SERCA and interacts with triadin to modulate Ca^{2+} release (Arvanitis et al. 2007). This dual interaction would ensure a cross-talk between SR Ca^{2+} uptake and release in the heart (Pritchard and Kranias 2009). However, the functional effect of HRC on SERCA2a is less clear. Overexpression of HRC in mouse results in depressed cardiomyocyte Ca^{2+} uptake (Gregory et al. 2006), indicating that HRC would inhibit SERCA2 activity. The fact that such inhibition would occur at low SR Ca^{2+}, when high activity should be more appropriate to refill the SR, is somewhat counter-intuitive. Direct measurements of SERCA activity and cardiomyocyte SR Ca^{2+} handling in the presence and absence of HRC are needed to clarify this further.

Other SERCA Isoforms

SERCA1

SERCA1 represents a highly specialized pump isoform which, with the notable exception of brown adipose tissue (de Meis 2003), that is, a cell type embryologically closely related to muscle (Enerback 2009), appears to be almost exclusively expressed in fast skeletal muscle fibers of all vertebrates from fish to mammals. Expression of SERCA1 is spatially controlled by the type of innervation the muscle fiber receives (Hamalainen and Pette 1997). Humans and some large animals tolerate the absence of SERCA1 reasonably well as is seen in some forms of human Brody myopathy (Odermatt et al. 1996) and in congenital pseudomyotonia in Chianina cattle (Drogemuller et al. 2008), but the lack of SERCA1 is lethal in mice (Pan et al. 2003) and zebra fish (Hirata et al. 2004).

The transcript of the *ATP2A1* gene can be processed into two different SERCA1 mRNAs coding for an adult SERCA1a and for SERCA1b, a form found only in neonatal or regenerating muscle (Zador et al. 2007). In SERCA1b, a highly-conserved octapeptide (-DPEDERRK) replaces the carboxy-terminal Gly residue of SERCA1a. The physiological and functional relevance of this extension remains unknown (Maruyama and MacLennan 1988; Zador et al. 2007). Insertion of the aberrant isoform into the ER reduces the ER Ca^{2+} concentration and induces apoptosis (Chami et al. 2000; Chami et al. 2001).

SERCA3

SERCA3 represents the last described and most enigmatic member of the SERCA family. It shows a limited cell-specific and differentiation-stage dependent expression pattern and a bewildering number of splice variants. At least six different variants in human (SERCA3a-f) are known, three in mice (SERCA3a-c) and two in rats (SERCA3a,b/c) (Dally et al. 2009). High expression of SERCA3 is found in various types of blood cells including lymphocytes, platelets, and mast cells, in endothelial cells, in epithelia of the intestinal or respiratory tract and in cerebellar Purkinje neurons (Wuytack et al. 1994; Baba-Aissa et al. 1996a). It should be mentioned, however, that in these cells SERCA3 is always coexpressed with the housekeeping SERCA2b isoform (Papp et al. 1991; Wootton and Michelangeli 2006).

All six SERCA3 splice variants present a 5- to 10-fold lower apparent affinity for cytosolic Ca^{2+} than SERCA2b (Chandrasekera et al. 2009). The obvious question that then arises is what the meaning is of the coexpression in a cell of the high-affinity SERCA2b with a low-affinity SERCA3. Especially, SERCA3 knockout mice do not display any overt phenotype, further questioning the physiological importance of SERCA3.

Cells belonging to the hematopoietic lineage and epithelial or endocrine secretory cells are endowed with a complex Ca^{2+}-signaling network (Guse et al. 1993). SERCA3 would here help to shape spatiotemporal cytosolic Ca^{2+} oscillation patterns (Arredouani et al. 2002). A differential subcellular localization of SERCA3 versus SERCA2, whereby SERCA3 would then most likely face an environment with locally higher Ca^{2+} concentration would also help in this respect. In epithelial cells, SERCA3 resides in a distinct subcellular localization positioned more at the basal region of the cell (Lee et al. 1997; Petersen 2003). A complex subcellular distribution of various SERCA3 splice variants was also described in human cardiomyocytes, although the expression levels of the various splice variants must be rather low (Dally et al. 2009). Of these, SERCA3f was found close to the plasma membrane and to be up-regulated in human failing heart (Dally et al. 2009).

In human platelets, SERCA3 is thought to reside in membranes of an acidic lysosome-related Ca^{2+} store, from which it can possibly be released via NAADP-gated two-pore channels (Calcraft et al. 2009; Brailoiu et al. 2010) whereas SERCA2b is confined to the so-called dense tubular system. The latter store is derived from the ER and its Ca^{2+} can be discharged by IP3R-mediated Ca^{2+}-release (Juska et al. 2008). On Ca^{2+} depletion, each of both types of stores activates its own store-operated Ca^{2+}-entry mechanism (SOCE) (Redondo et al. 2008b), although in the case of the acidic store SOCE appears to be more pronounced (Rosado et al. 2004). SOCE thereby relies on the formation of macromolecular complexes involving the respective SERCA isoforms. Complexes of SERCA3 and IP3R-2 in the acidic store and of a transient receptor potential channel TRPC1–TRPC6 heterodimer in the adjacent plasma membrane have been shown (Redondo et al. 2008a). On depletion of the acidic Ca^{2+} stores in platelets with thrombin or with a combination of thapsigargin and ionomycin, SERCA3 also forms complexes with STIM1 and Orai1 (Lopez et al. 2008).

Yet another indication for a specific role of SERCA3 in cellular Ca^{2+} signaling is found in its specific up-regulation during cell differentiation. Differentiation of vascular endothelium (Mountian et al. 1999), myeloid cells (Launay et al. 1999) or colon epithelial cells (Gelebart et al. 2002) is accompanied by an up-regulation of SERCA3 rather than of SERCA2b. Conversely, on malignant transformation colon cells loose their SERCA3 expression (Brouland et al. 2005) and both Epstein-Barr virus-mediated immortalization of B-lymphocytes with its accompanying lymphomagenesis and normal B-lymphocyte activation in lymph nodes are also paralleled by SERCA3 down-regulation (Dellis et al. 2009).

A number of reported germ-line mutations in the *ATP2A3* gene may predispose to cancer development (Korosec et al. 2008; Korosec et al. 2009). Presumably, haploinsufficiency of this gene underlies this predisposition. Remarkably, one of these mutants is also more frequently found in type II diabetic patients (Varadi et al. 1999).

Normal SERCA3 activity in vascular endothelium (Liu et al. 1997) and in respiratory epithelium (Kao et al. 1999) is important for relaxation of the adjacent smooth muscle as shown by defects in the relaxation in SERCA3 KO mice. The reported higher resistance of SERCA3 versus SERCA2b to oxidative damage might be considered as a meaningful adaptation in these local environments (Grover et al. 2003).

SPCAs

The SPCAs, together with the SERCAs, are responsible for loading the Golgi complex and the secretory compartment with Ca^{2+}. In contrast to SERCAs, SPCAs are also equipped to transport Mn^{2+} and thus supply this essential trace metal to the Golgi lumen. A number of comprehensive reviews have been published recently by our group (Vanoevelen et al. 2007; Vangheluwe et al. 2009) and by others (Dhitavat et al. 2004; Foggia and Hovnanian 2004; Brini and Carafoli 2009).

Short History

The archetypal member of the SPCA family was independently discovered in yeast (*Saccharomyces cerevisiae*) by two laboratories and named Plasma membrane ATPase-related, or Pmr1. Smith et al. (Smith et al. 1985) cloned *PMR1* by complementation of "super-secreting" yeast mutants (*ssc*) while Serrano et al. (Serrano et al. 1986) identified the same gene by hybridization with a *PMA1* (plasma-membrane H^+-ATPase) probe. Later on, homologs were studied in many animal species and in other fungi because of its value for biotechnology (efficient secretion of heterologously expressed proteins).

In humans, the *ATP2C1* gene-encoding SPCA1 was mapped to chromosome 3 and gained interest when it proved to be the gene that causes Hailey-Hailey disease (OMIM 169600), an acantholytic skin disease (Hu et al. 2000; Sudbrak et al. 2000).

A novel paralogue, *ATP2C2*, was found in the genome of higher vertebrates. Its protein product SPCA2 was characterized independently by two groups (Vanoevelen et al. 2005; Xiang et al. 2005). Its expression pattern suggests a more specific cellular role.

Structural Aspects of SPCAs

SPCAs differ from SERCAs mainly by the presence of only one ion-binding site (corresponding to site II in SERCA1). The structure of this site and its access pathway is probably affected by more distant residues and the packing of the TM helices allowing also for the transport of Mn^{2+} with high affinity (Wei et al. 1999; Wei et al. 2000; Van Baelen et al. 2001; Vangheluwe et al. 2009). In SPCAs, the E1 conformation is stabilized with respect to E2, explaining the observation that SPCAs have much higher apparent affinity for the transported ions than SERCAs (Dode et al. 2006). Compared to SERCA1a, structures of the two SPCA isoforms are more compact as shown by the shorter luminal and cytosolic loops (Fig. 3) (Vangheluwe et al. 2009). As indicated above, at least some of these longer loops of the SERCA pump represent specific binding sites for regulatory proteins. The homology models of SPCA1 and SPCA2 look almost identical (Fig. 3). Only minor differences are apparent, especially in the amino terminus and carboxyl terminus. In Pmr1, the amino terminus contains an EF-hand-like motif that binds Ca^{2+} and is crucial for Ca^{2+} transport (Wei et al. 1999). Although the EF-hand like motif in hSPCA1 is even more degenerate compared to PMR1, $^{45}Ca^{2+}$-overlay experiments on the GST-purified amino terminus of hSPCA1 also indicated the binding of Ca^{2+} (Vanoevelen, unpublished).

Expression Pattern

SPCA1

SPCA1 is the housekeeping Ca^{2+} and Mn^{2+} pump of the secretory pathway because it is expressed in all cell types studied. However,

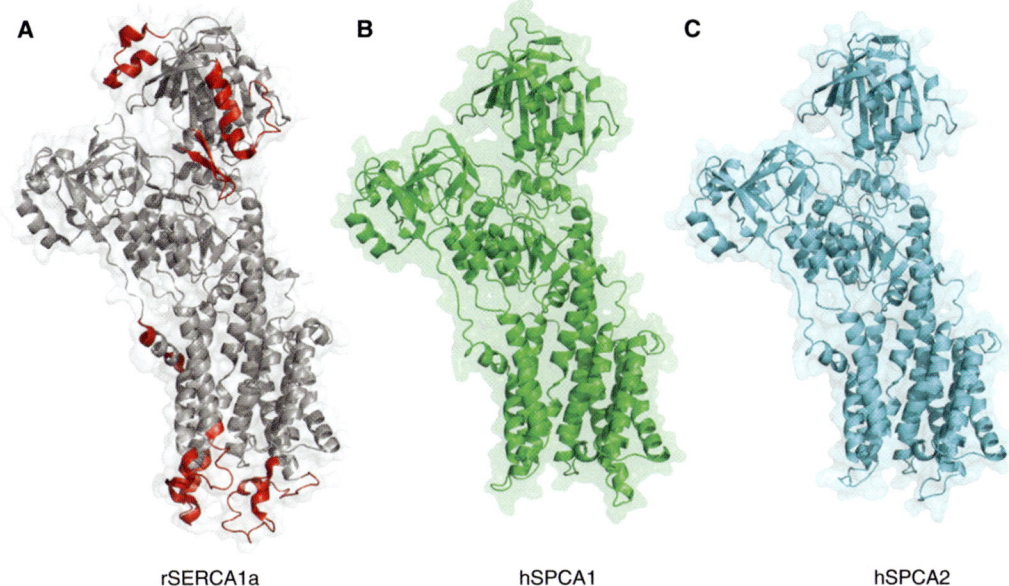

Figure 3. Comparison between the rSERCA1a, hSPCA1, and hSPCA2 structures. Homology models of hSPCA1 (*B*) and hSPCA2 (*C*) based on the E2 rSERCA1a structure (*A*) (1WPG) (Toyoshima et al. 2004). Homology models were obtained from the SWISS-MODEL repository (Kiefer et al. 2009). SPCA1 and SPCA2 are very similar, but in general more compact than SERCA1a. The longer loops in SERCA are indicated in red and are predominantly found in the N-domain and in the luminal loops.

different laboratories described different relative expression levels in various tissues. Wootton et al. (Wootton et al. 2004) observed much higher mRNA and protein expression in rat brain and testis than in other tissues whereas this difference was not observed in the corresponding tissues of humans (Hu et al. 2000; Vanoevelen et al. 2005).

The human *ATP2C1* gene transcript is alternatively spliced, giving rise to different protein products. Although there has been some confusion about the various splice variants, Fairclough et al. presented a unifying study describing four isoforms (Fairclough et al. 2003). The corresponding proteins are termed SPCA1a-d and only differ in their carboxyl termini. Three splice variants SPCA1a, b, and d are functional whereas SPCA1c, which is truncated within the last TM segment, is nonfunctional and rapidly degraded (Dode et al. 2006). Exploring the *ATP2C1* gene structure in the database points to the interesting peculiarity that the terminal exon of SPCA1b overlaps with the coding region of the neighboring gene *Asteroid 1* whose open reading frame is oriented in the opposite direction with respect to that of *ATP2C1*.

The yeast Pmr1 is localized in the Golgi apparatus possibly restricted to some of its subcompartments (Antebi and Fink 1992). SPCA from *Caenorhabditis elegans* heterologously expressed in COS-1 cells (Van Baelen et al. 2001) and the human SPCA1 expressed in CHO cells (Ton et al. 2002) showed a localization largely coinciding with Golgi markers. It is now well established that both overexpressed SPCA and the endogenous SPCAs in a whole range of cell types are present in the Golgi compartment (reviewed in Missiaen et al. 2007).

In human spermatozoa, SPCA1 displays an unusual subcellular distribution: it is found in the area behind the nucleus extending into the midpiece. SPCA1 is believed to be the only intracellular Ca^{2+} pump in these cells because both functional and immunocytochemical tests failed to show the presence of SERCAs (Harper et al. 2005). A similar picture arises from sea-urchin sperm cells, which lack SERCAs. Their SPCAs are located in the zone occupied by the single giant mitochondrion where also the main ATPases involved in Ca^{2+}-store filling are situated (Gunaratne and Vacquier 2006, 2007).

In the fly (*Drosophila melanogaster*), three SPCA splice-variants (SPoCk-A; SPoCk-B; SPoCk-C) are expressed. Of these isoforms, only SPoCk-A is targeted to the Golgi apparatus. The subcellular localization of SPoCk-B and SPoCk-C is less clear and unexpected targeting to, respectively, the ER and the peroxisomes was reported (Southall et al. 2006). Furthermore, expression of the SPoCk-C variant was shown to be sexually dimorphic (Southall et al. 2006).

Expression analysis in developing mouse brain showed that SPCA1 expression is prominent and at constant levels during the entire development of brain cortex, hippocampus, and cerebellum. In spite of the apparently unchanged expression levels, SPCA-associated Ca^{2+}-ATPase activity increased with the stage of development (Sepulveda et al. 2008). SPCA1 was localized in Golgi stacks of the soma and the initial part of the primary dendritic trunk in main cortical, hippocampal and cerebellar neurons, and is present from the earliest postnatal stages onward. Although SPCA1 expression has been reported in different glial cultures (Murin et al. 2006), other efforts to show SPCA- or SERCA-pump expression in glial cells in nervous tissue were unsuccessful (Baba-Aissa et al. 1996b; Sepulveda et al. 2007; Sepulveda et al. 2008). Because glial cells express high levels of the Mn^{2+}-dependent glutamine synthetase (Wedler and Denman 1984), the low levels of SPCAs argues against a role of SPCAs in Mn^{2+} uptake. However, in rat brain SPCA1 is upregulated following Mn^{2+} exposure (Zhang et al. 2005), which would be compatible with a role in Mn^{2+} detoxification, as also observed in yeast (Lapinskas et al. 1995).

SPCA2

Screening of the genome databases shows that besides the ancestral housekeeping *ATP2C1* gene, a second paralogue, *ATP2C2*, emerged in the genomes of vertebrates higher than fish. The corresponding gene is also lacking in invertebrates.

In human tissues, SPCA2 expression is more restricted than that of SPCA1, suggesting a more specialized physiological function of the former. Its mRNA is most abundant throughout the gastrointestinal tract, in trachea, thyroid, salivary gland, mammary gland and in prostate (Vanoevelen et al. 2005). It is striking that SPCA2 is most abundantly expressed in cells possessing a highly active secretion system like the mammary gland cells during lactation (Faddy et al. 2008) and the mucin-secreting goblet cells in human colon (Dmitriev et al. 2005; Vanoevelen et al. 2005). This indicates an important role for SPCA2 in protein secretion. However, reported SPCA2 expression in keratinocytes and hippocampal neurons does not fit this picture. These data on mRNA expression should however be confirmed at the protein level. So far, the presently available antibodies could only show SPCA2 expression in cultured hippocampal neurons (Mattiazzi et al. 2005), in the colon (Vanoevelen et al. 2005), in the secretory acini of the mouse mammary gland (Faddy et al. 2008) and in neutrophil granulocytes (Baron et al. 2009).

The precise subcellular localization of SPCA2 is not completely unambiguous. In human goblet cells, both SPCA2 and SPCA1 colocalized with Golgi markers in a compact structure near the apical pole of the nucleus (Vanoevelen et al. 2005). In addition, on heterologous expression in COS-1 cells, SPCA2 appeared predominantly in the Golgi area (Missiaen et al. 2007). In cultured mouse hippocampal neurons, however, SPCA2 staining showed a punctate distribution in the cell body and in the dendrites (Xiang et al. 2005). Although in neurons the Golgi apparatus does in general appear as a more fragmented structure, SPCA2 only partially colocalized with the *trans*-Golgi marker TGN38. It was therefore argued that in hippocampal neurons SPCA2 is, at least partially localized in downstream, post-Golgi segments of the secretory pathway (Xiang et al. 2005). Taken together, the available data indicate that SPCA2 can be found in the Golgi complex and in more downstream compartments of the secretory pathway.

Role of SPCAs in Cellular Physiology

Insights from PMR1 Mutants in Yeast

Although homozygous null mutations in the *ATP2C1* gene encoding SPCA1 seem to be lethal in mammals (Okunade et al. 2007), they are tolerated in lower eukaryotes, including fungi and *C. elegans* (Rudolph et al. 1989; Cho et al. 2005), where compensatory mechanisms presumably suffice to allow viability. An attractive model for understanding such mechanisms is the yeast orthologue Pmr1. *PMR1* mutants in yeast display pleiotropic changes in Ca^{2+}-dependent growth (Antebi and Fink 1992), secretion of unprocessed proteins (Antebi and Fink 1992), outer-chain glycosylation (Rudolph et al. 1989), Mn^{2+} tolerance (Lapinskas et al. 1995), salt tolerance (Park et al. 2001), cell shape (Cortes et al. 2004), virulence (Bates et al. 2005) and viability (Agaphonov et al. 2007). The characterization of the diverse *PMR1*-mutant phenotypes in yeast has been invaluable in providing the basis for studies on the role of metazoan SPCA orthologues. Some of these studies will be discussed in the following parts.

Studies in Cell Systems

Van Baelen et al. used RNA interference to understand the role of SPCA1 in HeLa cells (Van Baelen et al. 2003). Luminal $[Ca^{2+}]$ measurements using Golgi-targeted aequorin showed that endogenous SPCA1 was responsible for Ca^{2+} uptake in a subcompartment of the Golgi. On knock-down, the frequency of histamine- induced baseline Ca^{2+}-oscillations was reduced, indicating that in these cells a SPCA1-related Ca^{2+}-store may affect cytosolic Ca^{2+} signals.

SPCA1 also seems to be an important component of Ca^{2+} signaling in insulin-secreting cells (Mitchell et al. 2004). Knock-down of SPCA1 diminished Ca^{2+} uptake into the ER and in dense-core secretory vesicles, increased Ca^{2+} influx through L-type Ca^{2+} channels and increased the response to glucose. The time course of glucose-induced Ca^{2+} oscillations was also modified (Mitchell et al. 2004).

The same approach in cell lines expressing misfolded proteins revealed defects in protein processing and degradation (Ramos-Castaneda et al. 2005). Furthermore, SPCA1 deficiency rendered cells hypersensitive to ER stress.

Down-regulating SPCA1 in neurons compromises differentiation. The affected neurons displayed increased numbers of neurites of reduced length as compared to control cells. Additionally, Golgi Ca^{2+}-signalling was disturbed and trafficking of proteins through the Golgi was also hampered (Sepulveda et al. 2009). It is also known that both expression and activity of SPCA1 changes on ischemic events in the brain (Pavlikova et al. 2009).

Studies in Other Model Organisms

Knockdown of SPCA1 in *C. elegans* rendered the worms highly sensitive to Ca^{2+}-deficient and Mn^{2+}-enriched conditions and made them more resistant to oxidative stress (Cho et al. 2005). These defects are reminiscent of the mutant phenotype observed in yeast, as discussed earlier.

Using a genetically transmissible RNA-interference strategy in *Drosophila*, Southall et al. also showed aberrant Ca^{2+} signaling combined with defective neuropeptide-stimulated diuresis in the Malpighian tubes of transgenic flies (Southall et al. 2006).

Expression levels and activity of SPCAs change in response to altered physiological needs. In response to changes in glucose concentration, SPCA1 expression levels significantly increased in smooth muscle cells cultured in high-glucose medium versus normal medium. Functional consequences consisted of increased ATPase activity and altered thapsigargin-insensitive AVP (arginine-vasopressin)-induced cytosolic Ca^{2+} transients. These results indicate that SPCA can play a role in Ca^{2+} uptake within smooth muscle cells (Lai and Michelangeli 2009).

Expression of SPCA1 and especially SPCA2 rapidly adapts to lactation. SPCA2 is up-regulated 35-fold whereas SPCA1 expression only rises two-fold. These results clearly suggest an important role for SPCA2 specifically (in addition to PMCA2) in the transport of high amounts of proteins and Ca^{2+} into milk (Faddy et al. 2008). Conversely, on mammary gland involution the expression level of both pumps is reduced 80–95% in an early phase and subsequently up-regulated again to meet normal physiological needs (Reinhardt and Lippolis 2008).

The description of the phenotype of $SPCA1^{-/-}$ mice has shown the important housekeeping function of SPCA (Okunade et al. 2007). Homozygous mutant mice died in utero before gestation day 10.5. The animals showed growth retardation and had an open rostral neural tube. At the subcellular level, the Golgi membranes were dilated, expanded in amount and with fewer stacked leaflets. In addition, the number of Golgi-associated vesicles was increased although processing and trafficking of proteins in the secretory pathway was apparently normal. Apoptosis was increased and a large increase of cytoplasmic lipids was observed, consistent with impaired handling of lipids by the Golgi complex. The authors introduced the concept of Golgi stress to summarize these defects (Okunade et al. 2007). Adult SPCA1 heterozygous mice were found to have an increased incidence of squamous cell tumors of epithelial cells in the skin and esophagus (Okunade et al. 2007). In addition, SERCA2 heterozygous mice developed such tumors (Graef et al. 2001). The development of squamous cell tumors in aged $ATP2A2^{+/-}$ and $ATP2C1^{+/-}$ mice indicates that SERCA2 and SPCA1 haploinsufficiency predisposes murine keratinocytes to neoplasia. The possible links between Ca^{2+}-transporting proteins and cancer have been reviewed in detail by Monteith et al. (Monteith et al. 2007).

SPCAs and Human Disease

Hailey-Hailey disease (OMIM 169600) is an autosomal-dominant skin disease caused by the loss of one functional copy of the *ATP2C1* gene encoding SPCA1 (Hu et al. 2000; Sudbrak et al. 2000). It is characterized by an increased propensity for the formation of erosive and oozing skin lesions in the flexural areas (Hailey and Hailey 1939) from the second decade of life on. In recent years, a large number of causative mutations have been described (Cialfi et al. 2009). One cannot miss the remarkable parallels between the inactivation of one allele of the *ATP2A2* or *ATP2C1* genes causing very similar dermatological problems, respectively, Darier and Hailey-Hailey disease (Dhitavat et al. 2004). However, in contrast to keratinocytes of Darier patients, keratinocytes of Hailey-Hailey patients show an abnormal response to extracellular Ca^{2+}. Apparently, Darier keratinocytes behave normally in this respect because SPCA1 is up-regulated and can compensate for the partial loss of SERCA2 function (Foggia et al. 2006).

Very recently, *ATP2C2* in addition to the *CMIP* (c-maf inducing protein) gene has genetically been linked to both a human developmental disorder termed specific language impairment (SLI) and to phonological short-term memory. Detailed analysis indicates that both genes are independently involved. This study provides molecular evidence for a role of phonological short-term memory in language acquisition (Newbury et al. 2009).

CONCLUSIONS

SERCA and SPCA pumps help to establish and maintain low cytosolic and high luminal free Ca^{2+} concentration in respectively the ER and the organelles of the secretory pathway. Failure to keep this vital Ca^{2+} gradient results in ER stress, Golgi stress and cell death. It is thus physiologically important to maintain the activity of the pump, which is mainly accomplished by meticulously controlling the affinity of the pump for Ca^{2+}. To that extent, the cell has at its disposal several SERCA isoforms displaying differences in Ca^{2+} affinity and of affinity modulators of the pump, such as phospholamban and sarcolipin. Furthermore, multitudes of additional SERCA2 modulators were recently identified, although more work is needed to clarify their functional and physiological roles.

The role of the SPCA pumps in the secretory pathway is less well understood, but a remarkable property of SPCA is its ability to transport Mn^{2+}. Transport of Mn^{2+} from the cytosol to the lumen of the secretory pathway organelles provides these with a necessary cofactor for several of the resident enzymes and may be important for Mn^{2+} detoxification of the cells.

ACKNOWLEDGMENTS

P.V. and J.V. are Postdoctoral Fellows of the Fonds voor Wetenschappelijk Onderzoek (F.W.O.)—Vlaanderen (Research Foundation—Flanders). This work was also supported by the Interuniversity Attraction Poles Program—Belgian Science Policy IUAP P6/28, and by the F.W.O.—Vlaanderen G.0646.08 (to F.W.).

REFERENCES

Agaphonov MO, Plotnikova TA, Fokina AV, Romanova NV, Packeiser AN, Kang HA, Ter-Avanesyan MD. 2007. Inactivation of the Hansenula polymorpha PMR1 gene affects cell viability and functioning of the secretory pathway. *FEMS Yeast Res* **7:** 1145–1152.

Andersson KB, Birkeland JA, Finsen AV, Louch WE, Sjaastad I, Wang Y, Chen J, Molkentin JD, Chien KR, Sejersted OM, et al. 2009. Moderate heart dysfunction in mice with inducible cardiomyocyte-specific excision of the Serca2 gene. *J Mol Cell Cardiol* **47:** 180–187.

Antebi A, Fink GR. 1992. The yeast Ca^{2+}-ATPase homologue, PMR1, is required for normal Golgi function and localizes in a novel Golgi-like distribution. *Mol Biol Cell* **3:** 633–654.

Arredouani A, Guiot Y, Jonas JC, Liu LH, Nenquin M, Pertusa JA, Rahier J, Rolland JF, Shull GE, Stevens M, et al. 2002. SERCA3 ablation does not impair insulin secretion but suggests distinct roles of different sarcoendoplasmic reticulum Ca^{2+} pumps for Ca^{2+} homeostasis in pancreatic β-cells. *Diabetes* **51:** 3245–3253.

Arvanitis DA, Vafiadaki E, Fan GC, Mitton BA, Gregory KN, Del Monte F, Kontrogianni-Konstantopoulos A, Sanoudou D, Kranias EG. 2007. Histidine-rich Ca^{2+}-binding protein interacts with sarcoplasmic reticulum Ca^{2+}-ATPase. *Am J Physiol Heart Circ Physiol* **293:** H1581–H1589.

Asahi M, Green NM, Kurzydlowski K, Tada M, MacLennan DH. 2001. Phospholamban domain IB forms an interaction site with the loop between transmembrane helices M6 and M7 of sarco(endo)plasmic reticulum Ca^{2+} ATPases. *Proc Natl Acad Sci* **98:** 10061–10066.

Asahi M, Kimura Y, Kurzydlowski K, Tada M, MacLennan DH. 1999. Transmembrane helix M6 in sarco(endo)plasmic reticulum Ca^{2+}-ATPase forms a functional interaction site with phospholamban. Evidence for physical interactions at other sites. *J Biol Chem* **274:** 32855–32862.

Asahi M, Kurzydlowski K, Tada M, MacLennan DH. 2002. Sarcolipin inhibits polymerization of phospholamban to induce superinhibition of sarco(endo)plasmic reticulum Ca^{2+}-ATPases (SERCAs). *J Biol Chem* **277:** 26725–26728.

Asahi M, Sugita Y, Kurzydlowski K, De Leon S, Tada M, Toyoshima C, MacLennan DH. 2003. Sarcolipin regulates sarco(endo)plasmic reticulum Ca^{2+}-ATPase (SERCA) by binding to transmembrane helices alone or in association with phospholamban. *Proc Natl Acad Sci* **100:** 5040–5045.

Axelsen KB, Palmgren MG. 1998. Evolution of substrate specificities in the P-type ATPase superfamily. *J Mol Evol* **46:** 84–101.

Baba-Aissa F, Raeymaekers L, Wuytack F, Callewaert G, Dode L, Missiaen L, Casteels R. 1996a. Purkinje neurons express the SERCA3 isoform of the organellar type Ca^{2+}-transport ATPase. *Brain Res Mol Brain Res* **41:** 169–174.

Baba-Aissa F, Raeymaekers L, Wuytack F, De Greef C, Missiaen L, Casteels R. 1996b. Distribution of the organellar Ca^{2+} transport ATPase SERCA2 isoforms in the cat brain. *Brain Res* **743:** 141–153.

Babu GJ, Bhupathy P, Carnes CA, Billman GE, Periasamy M. 2007a. Differential expression of sarcolipin protein during muscle development and cardiac pathophysiology. *J Mol Cell Cardiol* **43:** 215–222.

Babu GJ, Bhupathy P, Timofeyev V, Petrashevskaya NN, Reiser PJ, Chiamvimonvat N, Periasamy M. 2007b. Ablation of sarcolipin enhances sarcoplasmic reticulum calcium transport and atrial contractility. *Proc Natl Acad Sci* **104:** 17867–17872.

Baksh S, Michalak M. 1991. Expression of calreticulin in Escherichia coli and identification of its Ca^{2+} binding domains. *J Biol Chem* **266:** 21458–21465.

Baron S, Struyf S, Wuytack F, Van Damme J, Missiaen L, Raeymaekers L, Vanoevelen J. 2009. Contribution of intracellular Ca^{2+} stores to Ca^{2+} signaling during chemokinesis of human neutrophil granulocytes. *Biochim Biophys Acta* **1793:** 1041–1049.

Bates S, MacCallum DM, Bertram G, Munro CA, Hughes HB, Buurman ET, Brown AJ, Odds FC, Gow NA. 2005. Candida albicans Pmr1p, a secretory pathway P-type Ca^{2+}/Mn^{2+}-ATPase, is required for glycosylation and virulence. *J Biol Chem* **280:** 23408–23415.

Bhupathy P, Babu GJ, Periasamy M. 2007. Sarcolipin and phospholamban as regulators of cardiac sarcoplasmic reticulum Ca^{2+} ATPase. *J Mol Cell Cardiol* **42:** 903–911.

Bhupathy P, Babu GJ, Ito M, Periasamy M. 2009. Threonine-5 at the N-terminus can modulate sarcolipin function in cardiac myocytes. *J Mol Cell Cardiol* **47:** 723–729.

Brailoiu E, Hooper R, Cai X, Brailoiu GC, Keebler MV, Dun NJ, Marchant JS, Patel S. 2010. An ancestral deuterostome family of two-pore channels mediates nicotinic acid adenine dinucleotide phosphate-dependent calcium release from acidic organelles. *J Biol Chem* **285:** 2897–2901.

Brini M, Carafoli E. 2009. Calcium pumps in health and disease. *Physiol Rev* **89:** 1341–1378.

Brouland JP, Gelebart P, Kovacs T, Enouf J, Grossmann J, Papp B. 2005. The loss of sarco/endoplasmic reticulum calcium transport ATPase 3 expression is an early event during the multistep process of colon carcinogenesis. *Am J Pathol* **167:** 233–242.

Calcraft PJ, Ruas M, Pan Z, Cheng X, Arredouani A, Hao X, Tang J, Rietdorf K, Teboul L, Chuang KT, et al. 2009. NAADP mobilizes calcium from acidic organelles through two-pore channels. *Nature* **459:** 596–600.

Caspersen C, Pedersen PS, Treiman M. 2000. The sarco/endoplasmic reticulum calcium-ATPase 2b is an endoplasmic reticulum stress-inducible protein. *J Biol Chem* **275:** 22363–22372.

Chami M, Gozuacik D, Lagorce D, Brini M, Falson P, Peaucellier G, Pinton P, Lecoeur H, Gougeon ML, le Maire M, et al. 2001. SERCA1 truncated proteins unable to pump calcium reduce the endoplasmic reticulum calcium concentration and induce apoptosis. *J Cell Biol* **153:** 1301–1314.

Chami M, Gozuacik D, Saigo K, Capiod T, Falson P, Lecoeur H, Urashima T, Beckmann J, Gougeon ML, Claret M, et al. 2000. Hepatitis B virus-related insertional mutagenesis implicates SERCA1 gene in the control of apoptosis. *Oncogene* **19:** 2877–2886.

Chandrasekera PC, Kargacin ME, Deans JP, Lytton J. 2009. Determination of apparent calcium affinity for endogenously expressed human sarco(endo)plasmic reticulum calcium-ATPase isoform SERCA3. *Am J Physiol Cell Physiol* **296:** C1105–C1114.

Chen Z, Akin BL, Jones LR. 2007. Mechanism of reversal of phospholamban inhibition of the cardiac Ca^{2+}-ATPase by protein kinase A and by anti-phospholamban monoclonal antibody 2D12. *J Biol Chem* **282:** 20968–20976.

Chen Z, Stokes DL, Rice WJ, Jones LR. 2003. Spatial and dynamic interactions between phospholamban and the canine cardiac Ca^{2+} pump revealed with use of heterobifunctional cross-linking agents. *J Biol Chem* **278:** 48348–48356.

Cho JH, Ko KM, Singaravelu G, Ahnn J. 2005. Caenorhabditis elegans PMR1, a P-type calcium ATPase, is important for calcium/manganese homeostasis and oxidative stress response. *FEBS Lett* **579:** 778–782.

Cialfi S, Oliviero C, Ceccarelli S, Marchese C, Barbieri L, Biolcati G, Uccelletti D, Palleschi C, Barboni L, De Bernardo C, et al. 2009. Complex multipathways alterations and oxidative stress are associated with Hailey-Hailey disease. *Br J Dermatol* **162:** 518–526.

Cortes JC, Katoh-Fukui R, Moto K, Ribas JC, Ishiguro J. 2004. Schizosaccharomyces pombe Pmr1p is essential for cell wall integrity and is required for polarized cell growth and cytokinesis. *Eukaryot Cell* **3:** 1124–1135.

Daiho T, Yamasaki K, Saino T, Kamidochi M, Satoh K, Iizuka H, Suzuki H. 2001. Mutations of either or both Cys876 and Cys888 residues of sarcoplasmic reticulum Ca^{2+}-ATPase result in a complete loss of Ca^{2+} transport activity without a loss of Ca^{2+}-dependent ATPase activity. *J Biol Chem* **276:** 32771–32778.

Dally S, Bredoux R, Corvazier E, Andersen JP, Clausen JD, Dode L, Fanchaouy M, Gelebart P, Monceau V, Del Monte F, et al. 2006. Ca^{2+}-ATPases in non-failing and failing heart: evidence for a novel cardiac sarco/endoplasmic reticulum Ca^{2+}-ATPase 2 isoform (SERCA2c). *Biochem J* **395:** 249–258.

Dally S, Monceau V, Corvazier E, Bredoux R, Raies A, Bobe R, del Monte F, Enouf J. 2009. Compartmentalized expression of three novel sarco/endoplasmic reticulum Ca^{2+}-ATPase 3 isoforms including the switch to ER stress, SERCA3f, in non-failing and failing human heart. *Cell Calcium* **45:** 144–154.

Damiani E, Tobaldin G, Bortoloso E, Margreth A. 1997. Functional behaviour of the ryanodine receptor/Ca^{2+}-release channel in vesiculated derivatives of the junctional membrane of terminal cisternae of rabbit fast muscle sarcoplasmic reticulum. *Cell Calcium* **22:** 129–150.

de Meis L. 2003. Brown adipose tissue Ca^{2+}-ATPase: uncoupled ATP hydrolysis and thermogenic activity. *J Biol Chem* **278:** 41856–41861.

Dellis O, Arbabian A, Brouland JP, Kovacs T, Rowe M, Chomienne C, Joab I, Papp B. 2009. Modulation of B-cell endoplasmic reticulum calcium homeostasis by Epstein-Barr virus latent membrane protein-1. *Mol Cancer* **8:** 59.

Dhitavat J, Fairclough RJ, Hovnanian A, Burge SM. 2004. Calcium pumps and keratinocytes: lessons from Darier's disease and Hailey-Hailey disease. *Br J Dermatol* **150:** 821–828.

Dmitriev RI, Pestov NB, Korneenko TV, Kostina MB, Shakhparonov MI. 2005. Characterization of Second Isoform of Secretory Pathway Ca^{2+}/Mn^{2+}-ATPase. *J Gen Physiol* **126:** 71a–72a.

Dode L, Andersen JP, Leslie N, Dhitavat J, Vilsen B, Hovnanian A. 2003. Dissection of the functional differences between sarco(endo)plasmic reticulum Ca^{2+}-ATPase (SERCA) 1 and 2 isoforms and characterization of Darier disease (SERCA2) mutants by steady-state and transient kinetic analyses. *J Biol Chem* **278:** 47877–47889.

Dode L, Andersen JP, Vanoevelen J, Raeymaekers L, Missiaen L, Vilsen B, Wuytack F. 2006. Dissection of the functional differences between human secretory pathway Ca^{2+}/Mn^{2+}-ATPase (SPCA) 1 and 2 isoenzymes by steady-state and transient kinetic analyses. *J Biol Chem* **281:** 3182–3189.

Dremina ES, Sharov VS, Schoneich C. 2006. Displacement of SERCA from SR lipid caveolae-related domains by Bcl-2: a possible mechanism for SERCA inactivation. *Biochemistry* **45:** 175–184.

Dremina ES, Sharov VS, Kumar K, Zaidi A, Michaelis EK, Schoneich C. 2004. Anti-apoptotic protein Bcl-2 interacts with and destabilizes the sarcoplasmic/endoplasmic reticulum Ca^{2+}-ATPase (SERCA). *Biochem J* **383:** 361–370.

Drogemuller C, Drogemuller M, Leeb T, Mascarello F, Testoni S, Rossi M, Gentile A, Damiani E, Sacchetto R. 2008. Identification of a missense mutation in the bovine ATP2A1 gene in congenital pseudomyotonia of Chianina cattle: An animal model of human Brody disease. *Genomics* **92:** 474–477.

Enerback S. 2009. The origins of brown adipose tissue. *N Engl J Med* **360:** 2021–2023.

Faddy HM, Smart CE, Xu R, Lee GY, Kenny PA, Feng M, Rao R, Brown MA, Bissell MJ, Roberts-Thomson SJ, et al. 2008. Localization of plasma membrane and secretory calcium pumps in the mammary gland. *Biochem Biophys Res Commun* **369:** 977–981.

Fairclough RJ, Dode L, Vanoevelen J, Andersen JP, Missiaen L, Raeymaekers L, Wuytack F, Hovnanian A. 2003. Effect of Hailey-Hailey Disease mutations on the function of a new variant of human secretory pathway Ca^{2+}/Mn^{2+}-ATPase (hSPCA1). *J Biol Chem* **278:** 24721–24730.

Foggia L, Hovnanian A. 2004. Calcium pump disorders of the skin. *Am J Med Genet C Semin Med Genet* **131C:** 20–31.

Foggia L, Aronchik I, Aberg K, Brown B, Hovnanian A, Mauro TM. 2006. Activity of the hSPCA1 Golgi Ca^{2+} pump is essential for Ca^{2+}-mediated Ca^{2+} response and cell viability in Darier disease. *J Cell Sci* **119:** 671–679.

Foskett JK, White C, Cheung KH, Mak DO. 2007. Inositol trisphosphate receptor Ca^{2+} release channels. *Physiol Rev* **87:** 593–658.

Gadsby DC. 2007. Structural biology: ion pumps made crystal clear. *Nature* **450:** 957–959.

Gelebart P, Kovacs T, Brouland JP, van Gorp R, Grossmann J, Rivard N, Panis Y, Martin V, Bredoux R, Enouf J, et al. 2002. Expression of endomembrane calcium pumps in colon and gastric cancer cells. Induction of SERCA3 expression during differentiation. *J Biol Chem* **277:** 26310–26320.

Graef IA, Chen F, Chen L, Kuo A, Crabtree GR. 2001. Signals transduced by Ca^{2+}/calcineurin and NFATc3/c4 pattern the developing vasculature. *Cell* **105:** 863–875.

Gramolini AO, Kislinger T, Asahi M, Li W, Emili A, MacLennan DH. 2004. Sarcolipin retention in the endoplasmic reticulum depends on its C-terminal RSYQY sequence and its interaction with sarco(endo)plasmic Ca^{2+}-ATPases. *Proc Natl Acad Sci* **101:** 16807–16812.

Gregory KN, Ginsburg KS, Bodi I, Hahn H, Marreez YM, Song Q, Padmanabhan PA, Mitton BA, Waggoner JR, Del Monte F, et al. 2006. Histidine-rich Ca^{2+} binding protein: a regulator of sarcoplasmic reticulum calcium sequestration and cardiac function. *J Mol Cell Cardiol* **40:** 653–665.

Grover AK, Kwan CY, Samson SE. 2003. Effects of peroxynitrite on sarco/endoplasmic reticulum Ca^{2+} pump isoforms SERCA2b and SERCA3a. *Am J Physiol Cell Physiol* **285:** C1537–C1543.

Gunaratne JH, Vacquier VD. 2006. Evidence for a secretory pathway Ca^{2+}-ATPase in sea urchin spermatozoa. *FEBS Lett* **580:** 3900–3904.

Gunaratne JH, Vacquier VD. 2007. Sequence, annotation and developmental expression of the sea urchin Ca^{2+} ATPase family. *Gene* **397:** 67–75.

Guse AH, Roth E, Emmrich F. 1993. Intracellular Ca^{2+} pools in Jurkat T-lymphocytes. *Biochem J* **291:** 447–451.

Haghighi K, Kolokathis F, Gramolini AO, Waggoner JR, Pater L, Lynch RA, Fan GC, Tsiapras D, Parekh RR, Dorn GW, et al. 2006. A mutation in the human phospholamban gene, deleting arginine 14, results in lethal, hereditary cardiomyopathy. *Proc Natl Acad Sci* **103:** 1388–1393.

Haghighi K, Kolokathis F, Pater L, Lynch RA, Asahi M, Gramolini AO, Fan GC, Tsiapras D, Hahn HS, Adamopoulos S, et al. 2003. Human phospholamban null results in lethal dilated cardiomyopathy revealing a critical difference between mouse and human. *J Clin Invest* **111:** 869–876.

Haghighi K, Schmidt AG, Hoit BD, Brittsan AG, Yatani A, Lester JW, Zhai J, Kimura Y, Dorn GW, MacLennan DH, et al. 2001. Superinhibition of sarcoplasmic reticulum function by phospholamban induces cardiac contractile failure. *J Biol Chem* **276:** 24145–24152.

Hailey HW, Hailey HE. 1939. Familial benign chronic pemphigus. *Arch Dermatol Syphilol* **39:** 679–685.

Hamalainen N, Pette D. 1997. Coordinated fast-to-slow transitions of myosin and SERCA isoforms in chronically stimulated muscles of euthyroid and hyperthyroid rabbits. *J Muscle Res Cell Motil* **18:** 545–554.

Han Y, Chen YS, Liu Z, Bodyak N, Rigor D, Bisping E, Pu WT, Kang PM. 2006. Overexpression of HAX-1 protects cardiac myocytes from apoptosis through caspase-9 inhibition. *Circ Res* **99:** 415–423.

Harper C, Wootton L, Michelangeli F, Lefievre L, Barratt C, Publicover S. 2005. Secretory pathway Ca^{2+}-ATPase (SPCA1) Ca^{2+} pumps, not SERCAs, regulate complex $[Ca^{(2+)}]_i$ signals in human spermatozoa. *J Cell Sci* **118:** 1673–1685.

Hasenfuss G, Reinecke H, Studer R, Meyer M, Pieske B, Holtz J, Holubarsch C, Posival H, Just H, Drexler H. 1994. Relation between myocardial function and expression of sarcoplasmic reticulum Ca^{2+}-ATPase in failing and nonfailing human myocardium. *Circ Res* **75:** 434–442.

Hirata H, Saint-Amant L, Waterbury J, Cui W, Zhou W, Li Q, Goldman D, Granato M, Kuwada JY. 2004. accordion, a zebrafish behavioral mutant, has a muscle relaxation defect due to a mutation in the ATPase Ca^{2+} pump SERCA1. *Development* **131:** 5457–5468.

Hofmann SL, Goldstein JL, Orth K, Moomaw CR, Slaughter CA, Brown MS. 1989. Molecular cloning of a histidine-rich Ca^{2+}-binding protein of sarcoplasmic reticulum that contains highly conserved repeated elements. *J Biol Chem* **264:** 18083–18090.

Honore B. 2009. The rapidly expanding CREC protein family: members, localization, function, and role in disease. *Bioessays* **31:** 262–277.

Hu Z, Bonifas JM, Beech J, Bench G, Shigihara T, Ogawa H, Ikeda S, Mauro T, Epstein EHJr. 2000. Mutations in ATP2C1, encoding a calcium pump, cause Hailey-Hailey disease. *Nat Genet* **24:** 61–65.

Hughes E, Clayton JC, Kitmitto A, Esmann M, Middleton DA. 2007. Solid-state NMR and functional measurements indicate that the conserved tyrosine residues of sarcolipin are involved directly in the inhibition of SERCA1. *J Biol Chem* **282:** 26603–26613.

Ihara Y, Kageyama K, Kondo T. 2005. Overexpression of calreticulin sensitizes SERCA2a to oxidative stress. *Biochem Biophys Res Commun* **329:** 1343–1349.

Ireland BS, Brockmeier U, Howe CM, Elliott T, Williams DB. 2008. Lectin-deficient calreticulin retains full functionality as a chaperone for class I histocompatibility molecules. *Mol Biol Cell* **19:** 2413–2423.

James P, Inui M, Tada M, Chiesi M, Carafoli E. 1989. Nature and site of phospholamban regulation of the Ca^{2+} pump of sarcoplasmic reticulum. *Nature* **342:** 90–92.

Ji Y, Lalli MJ, Babu GJ, Xu Y, Kirkpatrick DL, Liu LH, Chiamvimonvat N, Walsh RA, Shull GE, Periasamy M. 2000. Disruption of a single copy of the SERCA2 gene results in altered Ca^{2+} homeostasis and cardiomyocyte function. *J Biol Chem* **275:** 38073–38080.

John LM, Lechleiter JD, Camacho P. 1998. Differential modulation of SERCA2 isoforms by calreticulin. *J Cell Biol* **142:** 963–973.

Juska A, Jardin I, Rosado JA. 2008. Physical properties of two types of calcium stores and SERCAs in human platelets. *Mol Cell Biochem* **311:** 9–18.

Kao J, Fortner CN, Liu LH, Shull GE, Paul RJ. 1999. Ablation of the SERCA3 gene alters epithelium-dependent relaxation in mouse tracheal smooth muscle. *Am J Physiol* **277:** L264–L270.

Karim CB, Zhang Z, Howard EC, Torgersen KD, Thomas DD. 2006. Phosphorylation-dependent conformational switch in spin-labeled phospholamban bound to SERCA. *J Mol Biol* **358:** 1032–1040.

Kiefer F, Arnold K, Kunzli M, Bordoli L, Schwede T. 2009. The SWISS-MODEL Repository and associated resources. *Nucleic Acids Res* **37:** D387–D392.

Kimura T, Nakamori M, Lueck JD, Pouliquin P, Aoike F, Fujimura H, Dirksen RT, Takahashi MP, Dulhunty AF, Sakoda S. 2005. Altered mRNA splicing of the skeletal muscle ryanodine receptor and sarcoplasmic/endoplasmic reticulum Ca^{2+}-ATPase in myotonic dystrophy type 1. *Hum Mol Genet* **14:** 2189–2200.

Kimura Y, Inui M. 2002. Reconstitution of the cytoplasmic interaction between phospholamban and Ca^{2+}-ATPase of cardiac sarcoplasmic reticulum. *Mol Pharmacol* **61:** 667–673.

Kimura Y, Kurzydlowski K, Tada M, MacLennan DH. 1996. Phospholamban regulates the Ca^{2+}-ATPase through intramembrane interactions. *J Biol Chem* **271:** 21726–21731.

Kimura Y, Kurzydlowski K, Tada M, MacLennan DH. 1997. Phospholamban inhibitory function is activated by depolymerization. *J Biol Chem* **272:** 15061–15064.

Kobayashi M, Izawa H, Cheng XW, Asano H, Hirashiki A, Unno K, Ohshima S, Yamada T, Murase Y, Kato TS, et al. 2008. Dobutamine stress testing as a diagnostic tool for evaluation of myocardial contractile reserve in asymptomatic or mildly symptomatic patients with dilated cardiomyopathy. *J Am Coll Cardiol Img* **1:** 718–726.

Korosec B, Glavac D, Volavsek M, Ravnik-Glavac M. 2008. Alterations in genes encoding sarcoplasmic-endoplasmic reticulum Ca^{2+} pumps in association with head and neck squamous cell carcinoma. *Cancer Genet Cytogenet* **181:** 112–118.

Korosec B, Glavac D, Volavsek M, Ravnik-Glavac M. 2009. ATP2A3 gene is involved in cancer susceptibility. *Cancer Genet Cytogenet* **188:** 88–94.

Kuhlbrandt W. 2004. Biology, structure and mechanism of P-type ATPases. *Nat Rev Mol Cell Biol* **5:** 282–295.

Kuo TH, Kim HR, Zhu L, Yu Y, Lin HM, Tsang W. 1998. Modulation of endoplasmic reticulum calcium pump by Bcl-2. *Oncogene* **17:** 1903–1910.

Lai P, Michelangeli F. 2009. Changes in expression and activity of the secretory pathway Ca^{2+} ATPase 1 (SPCA1) in A7r5 vascular smooth muscle cells cultured at different glucose concentrations. *Biosci Rep* **29:** 397–404.

Lapinskas PJ, Cunningham KW, Liu XF, Fink GR, Culotta VC. 1995. Mutations in PMR1 suppress oxidative damage in yeast cells lacking superoxide dismutase. *Mol Cell Biol* **15:** 1382–1388.

Launay S, Gianni M, Kovacs T, Bredoux R, Bruel A, Gelebart P, Zassadowski F, Chomienne C, Enouf J, Papp B. 1999. Lineage-specific modulation of calcium pump expression during myeloid differentiation. *Blood* **93:** 4395–4405.

Lee AG. 2003. How phospholamban could affect the apparent affinity of Ca^{2+}-ATPase for Ca^{2+} in kinetic experiments. *FEBS Lett* **551:** 37–41.

Lee MG, Xu X, Zeng W, Diaz J, Kuo TH, Wuytack F, Racymaekers L, Muallem S. 1997. Polarized expression of Ca^{2+} pumps in pancreatic and salivary gland cells. Role in initiation and propagation of [Ca^{2+}]i waves. *J Biol Chem* **272:** 15771–15776.

Li Y, Camacho P. 2004. Ca^{2+}-dependent redox modulation of SERCA 2b by ERp57. *J Cell Biol* **164:** 35–46.

Li C, Grosdidier A, Crambert G, Horisberger JD, Michielin O, Geering K. 2004. Structural and functional interaction sites between Na$^+$,K$^+$-ATPase and FXYD proteins. *J Biol Chem* **279:** 38895–38902.

Liu LH, Boivin GP, Prasad V, Periasamy M, Shull GE. 2001. Squamous cell tumors in mice heterozygous for a null allele of Atp2a2, encoding the sarco(endo)plasmic reticulum Ca^{2+}-ATPase isoform 2 Ca^{2+} pump. *J Biol Chem* **276:** 26737–26740.

Liu LH, Paul RJ, Sutliff RL, Miller ML, Lorenz JN, Pun RY, Duffy JJ, Doetschman T, Kimura Y, MacLennan DH, et al. 1997. Defective endothelium-dependent relaxation of vascular smooth muscle and endothelial cell Ca^{2+} signaling in mice lacking sarco(endo)plasmic reticulum Ca^{2+}-ATPase isoform 3. *J Biol Chem* **272:** 30538–30545.

Lopez JJ, Jardin I, Bobe R, Pariente JA, Enouf J, Salido GM, Rosado JA. 2008. STIM1 regulates acidic Ca^{2+} store refilling by interaction with SERCA3 in human platelets. *Biochem Pharmacol* **75:** 2157–2164.

Luo W, Grupp IL, Harrer J, Ponniah S, Grupp G, Duffy JJ, Doetschman T, Kranias EG. 1994. Targeted ablation of the phospholamban gene is associated with markedly enhanced myocardial contractility and loss of β-agonist stimulation. *Circ Res* **75:** 401–409.

Lytton J, Westlin M, Burk SE, Shull GE, MacLennan DH. 1992. Functional comparisons between isoforms of the sarcoplasmic or endoplasmic reticulum family of calcium pumps. *J Biol Chem* **267:** 14483–14489.

MacLennan DH, Kranias EG. 2003. Phospholamban: a crucial regulator of cardiac contractility. *Nat Rev Mol Cell Biol* **4:** 566–577.

MacLennan DH, Asahi M, Tupling AR. 2003. The regulation of SERCA-type pumps by phospholamban and sarcolipin. *Ann N Y Acad Sci* **986:** 472–480.

Maruyama K, MacLennan DH. 1988. Mutation of aspartic acid-351, lysine-352, and lysine-515 alters the Ca^{2+} transport activity of the Ca^{2+}-ATPase expressed in COS-1 cells. *Proc Natl Acad Sci* **85:** 3314–3318.

Mattiazzi A, Mundina-Weilenmann C, Guoxiang C, Vittone L, Kranias E. 2005. Role of phospholamban phosphorylation on Thr17 in cardiac physiological and pathological conditions. *Cardiovasc Res* **68:** 366–375.

Metcalfe EE, Traaseth NJ, Veglia G. 2005. Serine 16 phosphorylation induces an order-to-disorder transition in monomeric phospholamban. *Biochemistry* **44:** 4386–4396.

Michalak M, Groenendyk J, Szabo E, Gold LI, Opas M. 2009. Calreticulin, a multi-process calcium-buffering chaperone of the endoplasmic reticulum. *Biochem J* **417:** 651–666.

Mills RF, Doherty ML, Lopez-Marques RL, Weimar T, Dupree P, Palmgren MG, Pittman JK, Williams LE. 2008. ECA3, a Golgi-localized P2A-type ATPase, plays a crucial role in manganese nutrition in Arabidopsis. *Plant Physiol* **146:** 116–128.

Minamisawa S, Wang Y, Chen J, Ishikawa Y, Chien KR, Matsuoka R. 2003. Atrial chamber-specific expression of sarcolipin is regulated during development and hypertrophic remodeling. *J Biol Chem* **278:** 9570–9575.

Missiaen L, Dode L, Vanoevelen J, Raeymaekers L, Wuytack F. 2007. Calcium in the Golgi apparatus. *Cell Calcium* **41:** 405–416.

Mitchell KJ, Tsuboi T, Rutter GA. 2004. Role for Plasma Membrane-Related Ca^{2+}-ATPase-1 (ATP2C1) in Pancreatic β-Cell Ca^{2+} Homeostasis Revealed by RNA Silencing. *Diabetes* **53:** 393–400.

Moller JV, Nissen P, Sorensen TL, le Maire M. 2005. Transport mechanism of the sarcoplasmic reticulum Ca^{2+}-ATPase pump. *Curr Opin Struct Biol* **15:** 387–393.

Monteith GR, McAndrew D, Faddy HM, Roberts-Thomson SJ. 2007. Calcium and cancer: targeting Ca^{2+} transport. *Nat Rev Cancer* **7:** 519–530.

Morita T, Hussain D, Asahi M, Tsuda T, Kurzydlowski K, Toyoshima C, MacLennan DH. 2008. Interaction sites among phospholamban, sarcolipin, and the sarco(endo)plasmic reticulum Ca^{2+}-ATPase. *Biochem Biophys Res Commun* **369:** 188–194.

Morth JP, Pedersen BP, Toustrup-Jensen MS, Sorensen TL, Petersen J, Andersen JP, Vilsen B, Nissen P. 2007. Crystal structure of the sodium-potassium pump. *Nature* **450:** 1043–1049.

Mountian I, Manolopoulos VG, De Smedt H, Parys JB, Missiaen L, Wuytack F. 1999. Expression patterns of sarco/endoplasmic reticulum Ca^{2+}-ATPase and inositol 1,4,5-trisphosphate receptor isoforms in vascular endothelial cells. *Cell Calcium* **25:** 371–380.

Murin R, Verleysdonk S, Raeymaekers L, Kaplan P, Lehotsky J. 2006. Distribution of secretory pathway Ca^{2+} ATPase (SPCA1) in neuronal and glial cell cultures. *Cell Mol Neurobiol* **26:** 1355–1365.

Newbury DF, Winchester L, Addis L, Paracchini S, Buckingham LL, Clark A, Cohen W, Cowie H, Dworzynski K, Everitt A, et al. 2009. CMIP and ATP2C2 modulate phonological short-term memory in language impairment. *Am J Hum Genet* **85:** 264–272.

Ni M, Lee AS. 2007. ER chaperones in mammalian development and human diseases. *FEBS Lett* **581:** 3641–3651.

Obara K, Miyashita N, Xu C, Toyoshima I, Sugita Y, Inesi G, Toyoshima C. 2005. Structural role of countertransport revealed in Ca^{2+} pump crystal structure in the absence of Ca^{2+}. *Proc Natl Acad Sci* **102:** 14489–14496.

Odermatt A, Becker S, Khanna VK, Kurzydlowski K, Leisner E, Pette D, MacLennan DH. 1998. Sarcolipin regulates the activity of SERCA1, the fast-twitch skeletal muscle sarcoplasmic reticulum Ca^{2+}-ATPase. *J Biol Chem* **273:** 12360–12369.

Odermatt A, Taschner PE, Khanna VK, Busch HF, Karpati G, Jablecki CK, Breuning MH, MacLennan DH. 1996. Mutations in the gene-encoding SERCA1, the fast-twitch skeletal muscle sarcoplasmic reticulum Ca^{2+} ATPase, are associated with Brody disease. *Nat Genet* **14:** 191–194.

Okunade GW, Miller M, Azhar M, Andringa A, Sanford LP, Doetschman T, Prasad V, Shull GE. 2007. Loss of the Atp2c1 secretory pathway Ca^{2+}-ATPase (SPCA1) in mice causes Golgi stress, apoptosis, and midgestational death in homozygous embryos and squamous cell tumors in adult heterozygotes. *J Biol Chem* **282:** 26517–26527.

Palmgren MG, Axelsen KB. 1998. Evolution of P-type ATPases. *Biochim Biophys Acta* **1365:** 37–45.

Pan Y, Zvaritch E, Tupling AR, Rice WJ, de Leon S, Rudnicki M, McKerlie C, Banwell BL, MacLennan DH. 2003. Targeted disruption of the ATP2A1 gene encoding the sarco(endo)plasmic reticulum Ca^{2+} ATPase isoform 1 (SERCA1) impairs diaphragm function and is lethal in neonatal mice. *J Biol Chem* **278:** 13367–13375.

Papp B, Enyedi A, Kovacs T, Sarkadi B, Wuytack F, Thastrup O, Gardos G, Bredoux R, Levy-Toledano S, Enouf J. 1991. Demonstration of two forms of calcium pumps by thapsigargin inhibition and radioimmunoblotting in platelet membrane vesicles. *J Biol Chem* **266:** 14593–14596.

Park SY, Seo SB, Lee SJ, Na JG, Kim YJ. 2001. Mutation in PMR1, a Ca^{2+}-ATPase in Golgi, confers salt tolerance in Saccharomyces cerevisiae by inducing expression of PMR2, a Na^{+}-ATPase in plasma membrane. *J Biol Chem* **276:** 28694–28699.

Pavlikova M, Tatarkova Z, Sivonova M, Kaplan P, Krizanova O, Lehotsky J. 2009. Alterations induced by ischemic preconditioning on secretory pathways Ca^{2+}-ATPase (SPCA) gene expression and oxidative damage after global cerebral ischemia/reperfusion in rats. *Cell Mol Neurobiol* **29:** 909–916.

Pedersen BP, Buch-Pedersen MJ, Morth JP, Palmgren MG, Nissen P. 2007. Crystal structure of the plasma membrane proton pump. *Nature* **450:** 1111–1114.

Periasamy M, Bhupathy P, Babu GJ. 2008. Regulation of sarcoplasmic reticulum Ca^{2+} ATPase pump expression and its relevance to cardiac muscle physiology and pathology. *Cardiovasc Res* **77:** 265–273.

Periasamy M, Huke S. 2001. SERCA pump level is a critical determinant of Ca^{2+} homeostasis and cardiac contractility. *J Mol Cell Cardiol* **33:** 1053–1063.

Periasamy M, Reed TD, Liu LH, Ji Y, Loukianov E, Paul RJ, Nieman ML, Riddle T, Duffy JJ, Doetschman T, et al. 1999. Impaired cardiac performance in heterozygous mice with a null mutation in the sarco(endo)plasmic reticulum Ca^{2+}-ATPase isoform 2 (SERCA2) gene. *J Biol Chem* **274:** 2556–2562.

Petersen OH. 2003. Localization and regulation of Ca^{2+} entry and exit pathways in exocrine gland cells. *Cell Calcium* **33:** 337–344.

Picello E, Damiani E, Margreth A. 1992. Low-affinity Ca^{2+}-binding sites versus Zn^{2+}-binding sites in histidine-rich Ca^{2+}-binding protein of skeletal muscle sarcoplasmic reticulum. *Biochem Biophys Res Commun* **186:** 659–667.

Prasad V, Boivin GP, Miller ML, Liu LH, Erwin CR, Warner BW, Shull GE. 2005. Haploinsufficiency of Atp2a2, encoding the sarco(endo)plasmic reticulum Ca^{2+}-ATPase isoform 2 Ca^{2+} pump, predisposes mice to squamous cell tumors via a novel mode of cancer susceptibility. *Cancer Res* **65:** 8655–8661.

Pritchard TJ, Kranias EG. 2009. Junctin and the histidine-rich Ca^{2+} binding protein: potential roles in heart failure and arrhythmogenesis. *J Physiol* **587:** 3125–3133.

Ramos-Castaneda J, Park YN, Liu M, Hauser K, Rudolph H, Shull GE, Jonkman MF, Mori K, Ikeda S, Ogawa H, et al. 2005. Deficiency of ATP2C1, a Golgi ion pump, induces secretory pathway defects in endoplasmic reticulum (ER)-associated degradation and sensitivity to ER stress. *J Biol Chem* **280:** 9467–9473.

Redondo PC, Jardin I, Lopez JJ, Salido GM, Rosado JA. 2008a. Intracellular Ca^{2+} store depletion induces the formation of macromolecular complexes involving hTRPC1, hTRPC6, the type II IP3 receptor and SERCA3 in human platelets. *Biochim Biophys Acta* **1783:** 1163–1176.

Redondo PC, Salido GM, Pariente JA, Sage SO, Rosado JA. 2008b. SERCA2b and 3 play a regulatory role in store-operated calcium entry in human platelets. *Cell Signal* **20:** 337–346.

Reinhardt TA, Lippolis JD. 2008. Mammary gland involution is associated with rapid down regulation of major mammary Ca^{2+}-ATPases. *Biochem Biophys Res Commun* **378:** 99–102.

Roderick HL, Lechleiter JD, Camacho P. 2000. Cytosolic phosphorylation of calnexin controls intracellular Ca^{2+} oscillations via an interaction with SERCA2b. *J Cell Biol* **149:** 1235–1248.

Rolfe DF, Brown GC. 1997. Cellular energy utilization and molecular origin of standard metabolic rate in mammals. *Physiol Rev* **77:** 731–758.

Rosado JA, Lopez JJ, Harper AG, Harper MT, Redondo PC, Pariente JA, Sage SO, Salido GM. 2004. Two pathways for store-mediated calcium entry differentially dependent on the actin cytoskeleton in human platelets. *J Biol Chem* **279:** 29231–29235.

Rudolph HK, Antebi A, Fink GR, Buckley CM, Dorman TE, LeVitre J, Davidow LS, Mao JI, Moir DT. 1989. The yeast secretory pathway is perturbed by mutations in PMR1, a member of a Ca^{2+} ATPase family. *Cell* **58:** 133–145.

Ruiz-Perez VL, Carter SA, Healy E, Todd C, Rees JL, Steijlen PM, Carmichael AJ, Lewis HM, Hohl D, Itin P, et al. 1999. ATP2A2 mutations in Darier's disease: variant cutaneous phenotypes are associated with missense mutations, but neuropsychiatric features are independent of mutation class. *Hum Mol Genet* **8:** 1621–1630.

Sahoo SK, Kim T, Kang GB, Lee JG, Eom SH, Kim do H. 2009. Characterization of calumenin-SERCA2 interaction in mouse cardiac sarcoplasmic reticulum. *J Biol Chem* **284:** 31109–31121.

Sakuntabhai A, Ruiz-Perez V, Carter S, Jacobsen N, Burge S, Monk S, Smith M, Munro CS, O'Donovan M, Craddock N, et al. 1999. Mutations in ATP2A2, encoding a Ca^{2+} pump, cause Darier disease. *Nat Genet* **21:** 271–277.

Schmitt JP, Ahmad F, Lorenz K, Hein L, Schulz S, Asahi M, Maclennan DH, Seidman CE, Seidman JG, Lohse MJ. 2009. Alterations of phospholamban function can exhibit cardiotoxic effects independent of excessive sarcoplasmic reticulum Ca^{2+}-ATPase inhibition. *Circulation* **119:** 436–444.

Schmitt JP, Kamisago M, Asahi M, Li GH, Ahmad F, Mende U, Kranias EG, MacLennan DH, Seidman JG, Seidman CE. 2003. Dilated cardiomyopathy and heart failure caused by a mutation in phospholamban. *Science* **299:** 1410–1413.

Seidel K, Andronesi OC, Krebs J, Griesinger C, Young HS, Becker S, Baldus M. 2008. Structural characterization of Ca^{2+}-ATPase-bound phospholamban in lipid bilayers by solid-state nuclear magnetic resonance (NMR) spectroscopy. *Biochemistry* **47:** 4369–4376.

Sepulveda MR, Berrocal M, Marcos D, Wuytack F, Mata AM. 2007. Functional and immunocytochemical evidence for the expression and localization of the secretory pathway Ca^{2+}-ATPase isoform 1 (SPCA1) in cerebellum relative to other Ca^{2+} pumps. *J Neurochem* **103:** 1009–1018.

Sepulveda MR, Marcos D, Berrocal M, Raeymaekers L, Mata AM, Wuytack F. 2008. Activity and localization of the Secretory Pathway Ca^{2+}-ATPase isoform 1 (SPCA1) in different areas of the mouse brain during postnatal development. *Mol Cell Neurosci* **38:** 461–473.

Sepulveda MR, Vanoevelen J, Raeymaekers L, Mata AM, Wuytack F. 2009. Silencing the SPCA1 (secretory pathway Ca^{2+}-ATPase isoform 1) impairs Ca^{2+} homeostasis in the Golgi and disturbs neural polarity. *J Neurosci* **29:** 12174–12182.

Serrano R, Kielland-Brandt MC, Fink GR. 1986. Yeast plasma membrane ATPase is essential for growth and has homology with (Na$^+$+K$^+$), K$^+$- and Ca^{2+}-ATPases. *Nature* **319:** 689–693.

Shinoda T, Ogawa H, Cornelius F, Toyoshima C. 2009. Crystal structure of the sodium-potassium pump at 2.4 A resolution. *Nature* **459:** 446–450.

Sipido KR, Vangheluwe P. 2010. Targeting sarcoplasmic reticulum Ca^{2+} uptake to improve heart failure: hit or miss. *Circ Res* **106:** 230–233.

Smith RA, Duncan MJ, Moir DT. 1985. Heterologous protein secretion from yeast. *Science* **229:** 1219–1224.

Southall TD, Terhzaz S, Cabrero P, Chintapalli VR, Evans JM, Dow JAT, Davies S-A. 2006. Novel subcellular locations and functions for secretory pathway Ca^{2+}/Mn^{2+}-ATPases. *Physiol Genomics* **26:** 35–45.

Stokes DL, Pomfret AJ, Rice WJ, Glaves JP, Young HS. 2006. Interactions between Ca^{2+}-ATPase and the pentameric form of phospholamban in two-dimensional co-crystals. *Biophys J* **90:** 4213–4223.

Sudbrak R, Brown J, Dobson-Stone C, Carter S, Ramser J, White J, Healy E, Dissanayake M, Larregue M, Perrussel M, et al. 2000. Hailey-Hailey disease is caused by mutations in ATP2C1 encoding a novel Ca^{2+} pump. *Hum Mol Genet* **9:** 1131–1140.

Suzuki Y, Demoliere C, Kitamura D, Takeshita H, Deuschle U, Watanabe T. 1997. HAX-1, a novel intracellular protein, localized on mitochondria, directly associates with HS1, a substrate of Src family tyrosine kinases. *J Immunol* **158:** 2736–2744.

Tavadia S, Tait RC, McDonagh TA, Munro CS. 2001. Platelet and cardiac function in Darier's disease. *Clin Exp Dermatol* **26:** 696–699.

Ton VK, Mandal D, Vahadji C, Rao R. 2002. Functional expression in yeast of the human secretory pathway Ca^{2+}, Mn^{2+}-ATPase defective in Hailey-Hailey disease. *J Biol Chem* **277:** 6422–6427.

Toyoshima C. 2008. Structural aspects of ion pumping by Ca^{2+}-ATPase of sarcoplasmic reticulum. *Arch Biochem Biophys* **476:** 3–11.

Toyoshima C. 2009. How Ca^{2+}-ATPase pumps ions across the sarcoplasmic reticulum membrane. *Biochim Biophys Acta* **1793:** 941–946.

Toyoshima C, Nomura H, Tsuda T. 2004. Lumenal gating mechanism revealed in calcium pump crystal structures with phosphate analogues. *Nature* **432:** 361–368.

Toyoshima C, Asahi M, Sugita Y, Khanna R, Tsuda T, MacLennan DH. 2003. Modeling of the inhibitory interaction of phospholamban with the Ca^{2+} ATPase. *Proc Natl Acad Sci* **100:** 467–472.

Toyoshima C, Nakasako M, Nomura H, Ogawa H. 2000. Crystal structure of the calcium pump of sarcoplasmic reticulum at 2.6 A resolution. *Nature* **405:** 647–655.

Traaseth NJ, Thomas DD, Veglia G. 2006. Effects of Ser16 phosphorylation on the allosteric transitions of phospholamban/Ca^{2+}-ATPase complex. *J Mol Biol* **358:** 1041–1050.

Traaseth NJ, Ha KN, Verardi R, Shi L, Buffy JJ, Masterson LR, Veglia G. 2008. Structural and dynamic basis of phospholamban and sarcolipin inhibition of Ca^{2+}-ATPase. *Biochemistry* **47:** 3–13.

Vafiadaki E, Arvanitis DA, Pagakis SN, Papalouka V, Sanoudou D, Kontrogianni-Konstantopoulos A, Kranias EG. 2009a. The Anti-apoptotic Protein HAX-1 Interacts with SERCA2 and Regulates Its Protein Levels to Promote Cell Survival. *Mol Biol Cell* **20:** 306–318.

Vafiadaki E, Papalouka V, Arvanitis DA, Kranias EG, Sanoudou D. 2009b. The role of SERCA2a/PLN complex, Ca^{2+} homeostasis, and anti-apoptotic proteins in determining cell fate. *Pflugers Arch* **457:** 687–700.

Vafiadaki E, Sanoudou D, Arvanitis DA, Catino DH, Kranias EG, Kontrogianni-Konstantopoulos A. 2007. Phospholamban interacts with HAX-1, a mitochondrial protein with anti-apoptotic function. *J Mol Biol* **367:** 65–79.

Van Baelen K, Vanoevelen J, Callewaert G, Parys JB, De Smedt H, Raeymaekers L, Rizzuto R, Missiaen L, Wuytack F. 2003. The contribution of the SPCA1 Ca^{2+} pump to the Ca^{2+} accumulation in the Golgi apparatus of HeLa cells assessed via RNA-mediated interference. *Biochem Biophys Res Commun* **306:** 430–436.

Van Baelen K, Vanoevelen J, Missiaen L, Raeymaekers L, Wuytack F. 2001. The Golgi PMR1 P-type ATPase of *Caenorhabditis elegans*. Identification of the gene and demonstration of calcium and manganese transport. *J Biol Chem* **276:** 10683–10691.

Vandecaetsbeek I, Raeymaekers L, Wuytack F, Vangheluwe P. 2009a. Factors controlling the activity of the SERCA2a pump in the normal and failing heart. *Biofactors* **35:** 484–499.

Vandecaetsbeek I, Trekels M, De Maeyer M, Ceulemans H, Lescrinier E, Raeymaekers L, Wuytack F, Vangheluwe P. 2009b. Structural basis for the high Ca^{2+} affinity of the ubiquitous SERCA2b Ca^{2+} pump. *Proc Natl Acad Sci* **106:** 18533–18538.

Vanden Abeele F, Skryma R, Shuba Y, Van Coppenolle F, Slomianny C, Roudbaraki M, Mauroy B, Wuytack F, Prevarskaya N. 2002. Bcl-2-dependent modulation of Ca^{2+} homeostasis and store-operated channels in prostate cancer cells. *Cancer Cell* **1:** 169–179.

Vangheluwe P, Raeymaekers L, Dode L, Wuytack F. 2005a. Modulating sarco(endo)plasmic reticulum Ca^{2+} ATPase 2 (SERCA2) activity: cell biological implications. *Cell Calcium* **38:** 291–302.

Vangheluwe P, Schuermans M, Zador E, Waelkens E, Raeymaekers L, Wuytack F. 2005b. Sarcolipin and phospholamban mRNA and protein expression in cardiac and skeletal muscle of different species. *Biochem J* **389:** 151–159.

Vangheluwe P, Sepulveda MR, Missiaen L, Raeymaekers L, Wuytack F, Vanoevelen J. 2009. Intracellular Ca^{2+}- and Mn^{2+}-transport ATPases. *Chem Rev* **109:** 4733–4759.

Vangheluwe P, Sipido KR, Raeymaekers L, Wuytack F. 2006a. New perspectives on the role of SERCA2's Ca^{2+} affinity in cardiac function. *Biochim Biophys Acta* **1763:** 1216–1228.

Vangheluwe P, Tjwa M, Van Den Bergh A, Louch WE, Beullens M, Dode L, Carmeliet P, Kranias E, Herijgers P, Sipido KR, et al. 2006b. A SERCA2 pump with an increased Ca^{2+} affinity can lead to severe cardiac hypertrophy, stress intolerance and reduced life span. *J Mol Cell Cardiol* **41:** 308–317.

Vanoevelen J, Dode L, Raeymaekers L, Wuytack F, Missiaen L. 2007. Diseases involving the Golgi calcium pump. *Subcell Biochem* **45:** 385–404.

Vanoevelen J, Dode L, Van Baelen K, Fairclough RJ, Missiaen L, Raeymaekers L, Wuytack F. 2005. The secretory pathway Ca^{2+}/Mn^{2+}-ATPase 2 is a Golgi-localized pump with high affinity for Ca^{2+} ions. *J Biol Chem* **280:** 22800–22808.

Varadi A, Lebel L, Hashim Y, Mehta Z, Ashcroft SJ, Turner R. 1999. Sequence variants of the sarco(endo)plasmic reticulum Ca^{2+}-transport ATPase 3 gene (SERCA3) in Caucasian type II diabetic patients (UK Prospective Diabetes Study 48). *Diabetologia* **42:** 1240–1243.

Ver Heyen M, Heymans S, Antoons G, Reed T, Periasamy M, Awede B, Lebacq J, Vangheluwe P, Dewerchin M, Collen D, et al. 2001. Replacement of the muscle-specific sarcoplasmic reticulum Ca^{2+}-ATPase isoform SERCA2a by the nonmuscle SERCA2b homologue causes mild concentric hypertrophy and impairs contraction-relaxation of the heart. *Circ Res* **89:** 838–846.

Verboomen H, Wuytack F, De Smedt H, Himpens B, Casteels R. 1992. Functional difference between SERCA2a and SERCA2b Ca^{2+} pumps and their modulation by phospholamban. *Biochem J* **286:** 591–595.

Verboomen H, Wuytack F, Van den Bosch L, Mertens L, Casteels R. 1994. The functional importance of the extreme C-terminal tail in the gene 2 organellar Ca^{2+}-transport ATPase (SERCA2a/b). *Biochem J* **303:** 979–984.

Vorum H, Hager H, Christensen BM, Nielsen S, Honore B. 1999. Human calumenin localizes to the secretory pathway and is secreted to the medium. *Exp Cell Res* **248:** 473–481.

Wedler FC, Denman RB. 1984. Glutamine synthetase: the major Mn(II) enzyme in mammalian brain. *Curr Top Cell Regul* **24:** 153–169.

Wei Y, Chen J, Rosas G, Tompkins DA, Holt PA, Rao R. 2000. Phenotypic screening of mutations in Pmr1, the yeast secretory pathway Ca^{2+}/Mn^{2+}-ATPase, reveals residues critical for ion selectivity and transport. *J Biol Chem* **275:** 23927–23932.

Wei Y, Marchi V, Wang R, Rao R. 1999. An N-terminal EF hand-like motif modulates ion transport by Pmr1, the yeast Golgi Ca^{2+}/Mn^{2+}-ATPase. *Biochemistry* **38:** 14534–14541.

Wootton LL, Argent CC, Wheatley M, Michelangeli F. 2004. The expression, activity and localisation of the secretory pathway Ca^{2+}-ATPase (SPCA1) in different mammalian tissues. *Biochim Biophys Acta* **1664:** 189–197.

Wootton LL, Michelangeli F. 2006. The effects of the phenylalanine 256 to valine mutation on the sensitivity of sarcoplasmic/endoplasmic reticulum Ca^{2+} ATPase (SERCA) Ca^{2+} pump isoforms 1, 2, and 3 to thapsigargin and other inhibitors. *J Biol Chem* **281:** 6970–6976.

Wu WC, Bhavsar JH, Aziz GF, Sadaniantz A. 2004. An overview of stress echocardiography in the study of patients with dilated or hypertrophic cardiomyopathy. *Echocardiography* **21:** 467–475.

Wuytack F, Papp B, Verboomen H, Raeymaekers L, Dode L, Bobe R, Enouf J, Bokkala S, Authi KS, Casteels R. 1994. A sarco/endoplasmic reticulum Ca^{2+}-ATPase 3-type Ca^{2+} pump is expressed in platelets, in lymphoid cells, and in mast cells. *J Biol Chem* **269:** 1410–1416.

Wuytack F, Raeymaekers L, Missiaen L. 2002. Molecular physiology of the SERCA and SPCA pumps. *Cell Calcium* **32:** 279–305.

Xiang M, Mohamalawari D, Rao R. 2005. A novel isoform of the secretory pathway Ca^{2+},Mn^{2+}-ATPase, hSPCA2, has unusual properties and is expressed in the brain. *J Biol Chem* **280:** 11608–11614.

Zador E, Vangheluwe P, Wuytack F. 2007. The expression of the neonatal sarcoplasmic reticulum Ca^{2+} pump (SERCA1b) hints to a role in muscle growth and development. *Cell Calcium* **41:** 379–388.

Zhang S, Fu J, Zhou Z. 2005. Changes in the brain mitochondrial proteome of male Sprague-Dawley rats treated with manganese chloride. *Toxicol Appl Pharmacol* **202:** 13–17.

Zhang Y, Kozlov G, Pocanschi CL, Brockmeier U, Ireland BS, Maattanen P, Howe C, Elliott T, Gehring K, Williams DB. 2009. ERp57 does not require interactions with calnexin and calreticulin to promote assembly of class I histocompatibility molecules, and it enhances peptide loading independently of its redox activity. *J Biol Chem* **284:** 10160–10173.

Zhao W, Waggoner JR, Zhang ZG, Lam CK, Han P, Qian J, Schroder PM, Mitton B, Kontrogianni-Konstantopoulos A, Robia SL, et al. 2009. The anti-apoptotic protein HAX-1 is a regulator of cardiac function. *Proc Natl Acad Sci* **106:** 20776–20781.

Zhao W, Yuan Q, Qian J, Waggoner JR, Pathak A, Chu G, Mitton B, Sun X, Jin J, Braz JC, et al. 2006. The presence of Lys27 instead of Asn27 in human phospholamban promotes sarcoplasmic reticulum Ca^{2+}-ATPase superinhibition and cardiac remodeling. *Circulation* **113:** 995–1004.

The Diversity of Calcium Sensor Proteins in the Regulation of Neuronal Function

Hannah V. McCue, Lee P. Haynes, and Robert D. Burgoyne

The Physiological Laboratory, School of Biomedical Sciences, University of Liverpool, Liverpool L69 3BX, United Kingdom

Correspondence: burgoyne@liv.ac.uk

Calcium signaling in neurons as in other cell types mediates changes in gene expression, cell growth, development, survival, and cell death. However, neuronal Ca^{2+} signaling processes have become adapted to modulate the function of other important pathways including axon outgrowth and changes in synaptic strength. Ca^{2+} plays a key role as the trigger for fast neurotransmitter release. The ubiquitous Ca^{2+} sensor calmodulin is involved in various aspects of neuronal regulation. The mechanisms by which changes in intracellular Ca^{2+} concentration in neurons can bring about such diverse responses has, however, become a topic of widespread interest that has recently focused on the roles of specialized neuronal Ca^{2+} sensors. In this article, we summarize synaptotagmins in neurotransmitter release, the neuronal roles of calmodulin, and the functional significance of the NCS and the CaBP/calneuron protein families of neuronal Ca^{2+} sensors.

Calcium signaling in many cell types can mediate changes in gene expression, cell growth, development, survival, and cell death. However, neuronal calcium signaling processes have become adapted to modulate the function of important pathways in the brain, including neuronal survival, axon outgrowth, and changes in synaptic strength. Changes in the concentration of intracellular free Ca^{2+} ($[Ca^{2+}]i$) are essential for the transmission of information through the nervous system as the trigger for neurotransmitter release at synapses. In addition, alterations in $[Ca^{2+}]i$ can lead to a wide range of different physiological changes that can modify neuronal functions over time scales of milliseconds through tens of minutes to days or longer (Berridge 1998). Many of these processes have been shown to be dependent upon the particular route of Ca^{2+} entry into the cell. It has long been known that the physiological outcome from a change in $[Ca^{2+}]i$ depends on its location, amplitude, and duration. The importance of location becomes even more pronounced in neurons because of their complex and extended morphologies. $[Ca^{2+}]i$ also regulates neuronal development and neuronal survival (Spitzer 2006). In addition, modifications to Ca^{2+} signaling pathways have been suggested to underlie various neuropathological disorders (Braunewell 2005; Berridge 2010).

Highly localized Ca^{2+} elevations (Augustine et al. 2003) formed following Ca^{2+} entry though voltage-gated Ca^{2+} channels (VGCCs) lead to synaptic vesicle fusion with the presynaptic membrane and thereby allow neurotransmitter release within less than a millisecond. Differently localized and timed Ca^{2+} signals can, for example, result in changes to the properties of the VGCCs (Catterall and Few 2008) or lead to changes in gene expression (Bito et al. 1997). Postsynaptic Ca^{2+} signals arising from activation of NMDA receptors give rise to two important processes in synaptic plasticity, long term potentiation (LTP) and long term depression (LTD). LTP and LTD are examples of the way synaptic transmission can change synaptic efficacy and are thought to be important in modulating learning and memory. Importantly, the Ca^{2+} signals that bring about either LTP or LTD differ only in their timing and duration. LTP is triggered by Ca^{2+} signals on the micromolar scale for shorter durations, whereas LTD is triggered by changes in $[Ca^{2+}]i$ on the nanomolar scale for longer durations (Yang et al. 1999). Specific Ca^{2+} signals are likely to be decoded by different Ca^{2+} sensor proteins. These are proteins that undergo a conformational change on Ca^{2+} binding and then interact with and regulate various target proteins. Among those Ca^{2+} sensors that are important for neuronal function are the synaptotagmins that control neurotransmitter release (Chapman 2008), the ubiquitous EF-hand containing sensor calmodulin that has many neuronal roles, and the more recently discovered neuronal EF-hand containing proteins, including the neuronal calcium sensor (NCS) protein (Burgoyne 2007) and the calcium-binding protein (CaBP)/calneuron (Haeseleer et al. 2002) families. We will briefly review synaptotagmins and the neuronal functions of calmodulin but concentrate on the NCS and CaBP families of Ca^{2+} sensors.

SYNAPTOTAGMINS AND NEUROTRANSMITTER RELEASE

Synaptotagmins are transmembrane proteins mostly found associated with synaptic and secretory vesicles. There are multiple known isoforms of synaptotagmin (Craxton 2004) of which synaptotagmin I is the best studied. The role of synaptotagmins in neurotransmitter release has been the subject of intense investigations, which have been extensively reviewed (Chapman 2008; Rizo and Rosenmund 2008; Sudhof and Rothman 2009) and so only a brief outline is given here. Synaptotagmins bind Ca^{2+} with relatively low affinity (Kd > 10 μM) through their two C2 domains (C2A and C2B) (Shao et al. 1998; Fernandez et al. 2001), which are functional in many but not all synaptotagmin isoforms. Ca^{2+} binding by C2 domains requires coordination of Ca^{2+} by both the protein and membrane lipids and this lipid interaction is a key aspect for its function. In synaptotagmin I, the C2A and C2B domains (Fig. 1) bind three and two Ca^{2+} ions, respectively (Shao et al. 1998; Fernandez et al. 2001). It is now well established that synaptotagmin I is a key sensor for evoked, synchronous neurotransmitter release in many classes of neurons (Fernandez-Chacon et al. 2001). Structure–function studies based on expression of specific mutants have been carried out in mice, worms, and flies. For example, disruption of Ca^{2+} binding to the C2B domain of synaptotagmin I has

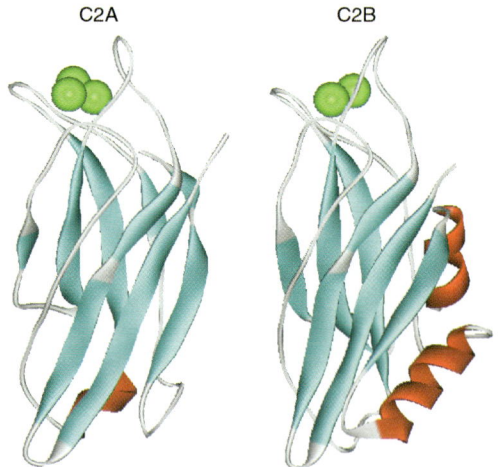

Figure 1. Structures of the C2A and C2B domains of synaptotagmin I. The structures show the isolated C2 domains in their Ca^{2+}-loaded state with the bound Ca^{2+} ions shown in green. The coordinates for the structures for the C2A and C2B domains come from the PDB files 1BYN and 1K5W, respectively.

been shown to have a more deleterious effect than disruption of Ca^{2+} binding to its C2A domain (Mackler et al. 2002; Robinson et al. 2002). The details of exactly how it triggers exocytosis and the function of other syntaptotagmin isoforms remain controversial. Membrane fusion requires the pairing and interaction of so-called SNARE proteins on vesicle and target membranes (Sollner et al. 1993). These can assemble into a SNARE complex that may form the minimal fusion machinery. For synaptic vesicle and neuroendocrine exocytosis, the SNARE proteins are SNAP-25, syntaxin 1, and synaptobrevin. In the case of regulated exocytosis, such as in neurotransmitter release, vesicle fusion is tightly regulated and requires a Ca^{2+} signal for activation. Ca^{2+} entry through VGCCs leading to Ca^{2+} elevation in local microdomains close to the mouth of the Ca^{2+} channel is able to trigger very rapid (within less than 1 ms) fusion of synaptic vesicles. Synaptotagmin can bind to both syntaxin and SNAP-25, and fast neurotransmitter release requires synaptotagmin (Geppert et al. 1994) probably prebound to assembled or partially assembled SNARE complexes (Schiavo et al. 1997; Rickman et al. 2006) so that Ca^{2+}-induced interaction with phospholipids can occur rapidly (Xue et al. 2008). It is still under debate how important synaptotagmin is in vesicle docking (de Wit et al. 2009) and how it acts at the plasma membrane in fusion itself (Tang et al. 2006; Hui et al. 2009). Synaptotagmin could act as a brake on fusion that is relieved on Ca^{2+} binding or have a positive role in membrane fusion (Chicka et al. 2008). A recent focus has been on the combined role of synaptotagmin and another SNARE interacting protein complexin in timing synaptic vesicle fusion (Sudhof and Rothman 2009), but much still remains to be learnt about the molecular basis of its function. While synaptotagmin is a key sensor for evoked neurotransmitter release, an alternative C2-domain containing protein, Doc2b, has been identified as a Ca^{2+} sensor for spontaneous neurotransmitter release (Groffen et al. 2010). Like synaptotagmin, Doc2b appears to function via interaction with SNARE complexes.

NEURONAL FUNCTIONS OF CALMODULIN

Calmodulin is a ubiquitously expressed Ca^{2+}-binding protein that can bind four Ca^{2+} ions through its four EF-hand domains (Chattopadhyaya et al. 1992). This protein has been highly conserved throughout evolution, is found in all eukaryotes, and is 100% identical across all vertebrates at the amino acid level. It is involved in the regulation of many essential physiological processes including cell motility, exocytosis, cytoskeletal assembly, and modulation of intracellular Ca^{2+} concentrations. The first two EF-hands of calmodulin form an amino-terminal globular domain that is joined by a flexible linker to a highly homologous carboxy-terminal region encompassing the third and fourth EF-hands. The carboxy-terminal pair of EF-hands has a much higher affinity for Ca^{2+} than the amino-terminal pair, which allows the two domains to behave independently at varying Ca^{2+} concentrations (Tadross et al. 2008). The highly flexible linker between the two domains can be bent dramatically upon binding to target proteins (Fig. 2) and is an essential property of calmodulin, which permits this protein to interact with a large and diverse array of interacting partners. The significant conformational changes on binding to its targets (Fallon et al. 2005) can increase its affinity for Ca^{2+}.

Calmodulin is present in brain at high concentrations (up to ~100 μM). In addition to its more general functions, calmodulin also has a series of specific roles in transducing Ca^{2+} signals in neurons, including, for example, in the regulation of glutamate receptors (O'Connor 1999), ion channels (Saimi and Kung 2002), and proteins in signaling pathways such as neuronal nitric oxide synthase, and it can affect synaptic plasticity (Lisman et al. 2002; Xia and Storm 2005). One key direct function of calmodulin is in regulating the activity of VGCCs by binding to channel subunits (Catterall and Few 2008). Ca^{2+}-binding to VGCC-associated calmodulin can have a range of effects on channel function, including mediating Ca^{2+}-dependent facilitation or Ca^{2+}-inactivation (Lee et al. 2000; DeMaria et al. 2001; Catterall and Few 2008;

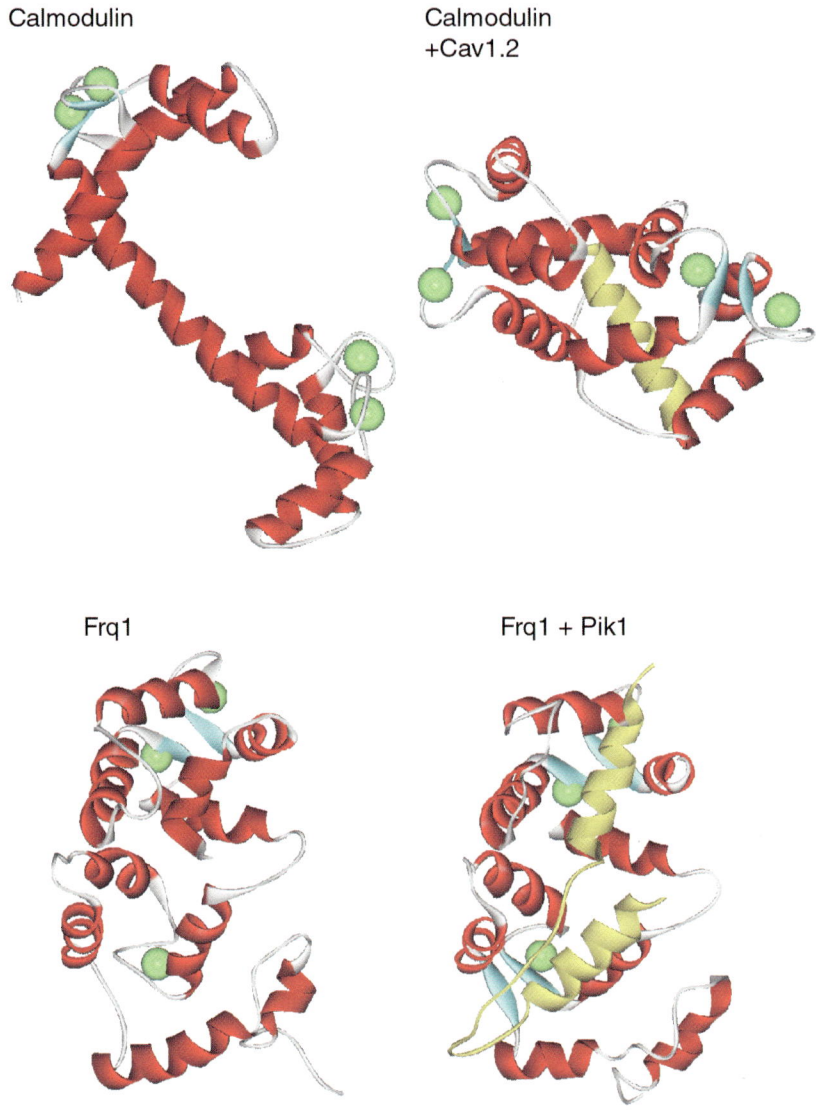

Figure 2. Comparison of the structures of Ca^{2+}-loaded calmodulin and yeast frequenin with and without bound target peptides. The structures at the top are of Ca^{2+}-bound calmodulin alone (PDB 1CLL) or in a complex with the IQ-like domain of the $Ca_v1.2$ Ca^{2+}-channel α-subunit (PDB 2F3Z). The structures at the bottom are of the Ca^{2+}-bound yeast frequenin (Frq1) alone (PDB 1FPW) or in a complex with the binding domain from Pik1 (PDB 2JU0). In each of the complexes, the target peptide is shown in yellow.

Liu et al. 2010). Calmodulin is also constitutively associated with and regulates opening of Ca^{2+}-activated potassium channels (Xia 1998; Schumacher et al. 2001) and other types of potassium channels (Wen and Levitan 2002). Two other major modes of action of calmodulin are exerted indirectly through its target proteins Ca^{2+}/calmodulin-dependent kinases (CaMKs) and calcineurin. CaMKs contribute to a number of regulatory pathways involving, for example, phosphorylation of AMPA receptors (Barria et al. 1997) and the nuclear transcription factor CREB (Deisseroth et al. 1998). Calmodulin also positively regulates presynaptic release probability and this is mediated via activation of CaMKII (Pang et al. 2010). The Ca^{2+}-activated phosphatase calcineurin can dephosphorylate a wide range of neuronal proteins, leading to direct effects and effects through changes in gene transcription following activation of the transcription factor NFAT and its translocation into the nucleus. Calcineurin has also been implicated, for example, in synaptic plasticity (Malleret et al. 2001; Xia and Storm 2005). Although many aspects of neuronal function are known to be regulated by calmodulin, proteins related to calmodulin have been discovered in recent years, which are enriched or expressed exclusively in neurons. Duplication and diversification of the calmodulin gene may have given rise to these neuronal calcium sensing proteins so that they can carry out specific neuronal functions in higher organisms.

NCS PROTEIN FAMILY

Whereas calmodulin is ubiquitously expressed, the expression of other calcium sensing proteins can be restricted to particular tissues and cell types. A good example of this is the neuronal calcium sensor (NCS) family of proteins, which are primarily expressed in neurons or retinal photoreceptors. The NCS family of proteins is related to calmodulin but have distinct properties that allow them to carry out nonredundant roles that do not overlap with the functions of calmodulin. Members of the NCS protein family have been implicated in the regulation of neurotransmitter release, regulation of cell-surface receptors and ion channels, control of gene transcription, cell growth, and survival (Burgoyne 2007). The NCS proteins are encoded by a family of 14 genes in mammals with greater diversity stemming from alternative splicing of transcripts from a number of the genes. All NCS gene products harbor four EF-hand motifs and display limited similarity ($<20\%$) to calmodulin (Burgoyne 2004).

NCS-1 is the most widely expressed of the NCS proteins and is thought to be the primordial NCS protein. The protein was first discovered (as frequenin) in *Drosophila melanogaster* (Pongs et al. 1993), where there are two very closely related genes (Sanchez-Gracia et al. 2010). Although initially thought to be neuronal specific (Nef et al. 1995), NCS-1 has also been identified in *Saccharomyces cerevisiae* (Hendricks et al. 1999). After this first appearance of NCS-1 in yeast, there has been a steady increase in the diversity of this family throughout evolution, which roughly correlates with increasing organism complexity. Five classes of NCS proteins have now been identified in higher organisms termed classes A-E (Burgoyne 2007). Class A contains NCS-1, which is present in yeast and all higher organisms. Class B consists of the visinin-like proteins (VSNLs), which appear first in *Caenorhabditis elegans*. Classes C and D evolved with the appearance of fish and comprise recoverin and the guanylyl-cyclase-activating proteins (GCAPs), respectively. Finally, class E contains the K^+ channel-interacting proteins (KChIPs), which are found in insects and evolutionary subsequent species (Burgoyne 2004). Mammals have a single NCS-1, five VSNL proteins (hippocalcin, neurocalcin δ, and VILIPs1-3), a single recoverin, three GCAPs, and four KChIPs. Expression of the recoverins and GCAPs is restricted to the retina, whereas the rest of the NCS family is found in varied neuronal populations (Burgoyne 2007). Although localization and expression studies have proven difficult because of cross-reactivity of antibodies, it has been established that certain neurons express several or all of the NCS proteins, but in general, the expression profile for each of the NCS proteins is unique (Paterlini et al. 2000; Rhodes et al. 2004). This suggests that despite the high sequence homology between the proteins, each is likely to perform distinct functions in specific cell types (Burgoyne and Weiss 2001).

Unlike calmodulin, not all EF-hands are functional in the NCS proteins and the most amino-terminal EF-hand is unable to bind Ca^{2+} in all family members. In the case of recoverin and KChIP1, only two of its four EF-hand motifs are functional in Ca^{2+} binding (Burgoyne et al. 2004; Burgoyne 2007). Unlike the dumbbell structure of calmodulin, the NCS proteins are compact and globular when Ca^{2+}-bound and they undergo limited conformational change on binding to their target proteins (Ames et al. 2006; Pioletti et al. 2006; Strahl et al. 2007; Wang et al. 2007) (Fig. 2). NCS proteins also differ from calmodulin in that many have motifs that allow membrane association. KChIP1 and all the members of classes A–D are N-myristoylated, whereas certain KChIP2, KChIP3, and KChIP4 isoforms harbor palmitoylation motifs. In some cases, the membrane association conferred by these moieties is dynamically regulated by Ca^{2+} binding when a sequestered mysristoyl chain becomes solvent-exposed following a Ca^{2+}-driven shift in conformation as originally described for recoverin (Ames et al. 1997). VSNL proteins are also cytosolic at resting $[Ca^{2+}]_i$ but localize to the plasma membrane or Golgi complex upon Ca^{2+} elevation (O'Callaghan et al. 2002; Spilker et al. 2002; O'Callaghan et al. 2003b). Each of the NCS proteins displays distinct subcellular localizations, which are in part determined by additional interactions with specific phosphoinositides mediated by basic amino-terminal residues immediately proximal to the site of acylation (O'Callaghan et al. 2003a; O'Callaghan et al. 2005).

NCS proteins are multifunctional regulators of various proteins involved in processes ranging from trafficking and ion channel modulation to gene transcription (Burgoyne 2004), and the function of NCS-1 in particular has been intensively studied. NCS-1, the primordial NCS protein, is highly evolutionarily conserved, retaining 59% identity with its yeast ortholog, frequenin. It displays a high Ca^{2+}-binding affinity and is able to respond to small fluctuations in $[Ca^{2+}]_i$. NCS-1 is amino-terminally myristoylated and is constitutively associated with membranes including plasma and Golgi membranes (O'Callaghan et al. 2002), although in some cell lines, NCS-1 has been found to be partially cytosolic (de Barry et al. 2006) and it is able to rapidly exchange between membrane and cytosolic pools (Handley et al. 2010). In contrast to all other NCS family members, NCS-1 is not neuron specific and is expressed in neuroendocrine cells (McFerran et al. 1998) and at low levels in several nonneuronal cell types. NCS-1 has three functional EF-hand motifs, which have differing cation specificities. It has been suggested that under resting conditions when $[Ca^{2+}]_i$ is low (≤ 0.1 μM), EF2 and EF3 are Mg^{++} bound, whereas EF4 is a Ca^{2+} specific binding site and remains vacant. In the Mg^{++} bound state, NCS-1 adopts a conformation, which reduces exposure of hydrophobic regions. This may be important in the prevention of nonspecific interactions in the absence of a specific Ca^{2+}-signal. In the presence of elevated $[Ca^{2+}]_i$, EF2 and EF3 become Ca^{2+}-occupied, simultaneously followed by Ca^{2+} binding to EF4 (Aravind et al. 2008). The Mg^{++} bound form of NCS-1 has a fivefold lower affinity for Ca^{2+} than the Mg^{++}-free/Ca^{2+}-free apo-form. This implies that Mg^{++} binding permits significant modulation of NCS-1 and is important in fine tuning its Ca^{2+}-sensing properties (Aravind et al. 2008; Mikhaylova et al. 2009).

Much current understanding concerning the function of NCS-1 derives from overexpression or knockout studies. Overexpression in *Drosophila* caused a frequency-dependent facilitation of neurotransmitter release (Pongs et al. 1993) and its importance for neurotransmissions has been confirmed by knockout of the two *Drosophila* frequenin genes (Dason et al. 2009). In *Xenopus*, overexpression caused enhanced spontaneous and evoked transmission at neuromuscular junctions (Olafsson et al. 1995) and over-expression was also found to increase Ca^{2+}-dependent exocytosis of dense core granules in PC12 cells (McFerran et al. 1998) and to enhance associative learning and memory in *Caenorhabditis elegans* (Gomez et al. 2001; Hilfiker 2003).

Knockout of NCS-1 (Frq1) in the yeast *Saccharomyces cerevisiae* is lethal because of its requirement for the activation of Pik1, one of the two yeast phosphatidylinositol-4 kinases (PI4Ks) (Hendricks et al. 1999). NCS-1 can also interact with the mammalian Golgi enzyme PI4KIIIβ and enhances its activity three- to 10-fold (Taverna et al. 2002; Haynes et al. 2005; de Barry et al. 2006). The interaction with Golgi-associated PI4KIIIβ suggests that it may regulate secretion through the modulation of phosphatidylinositol-dependent trafficking steps (Hendricks et al. 1999; Zhao et al. 2001; Haynes et al. 2005). In support of this, NCS-1 has also been demonstrated to associate with another PI4KIIIβ regulator ARF1, a small GTPase critical to multiple trafficking steps in mammalian cells (Haynes et al. 2005; Haynes et al. 2007).

Knockout of NCS-1 in other organisms is not lethal but does generate specific phenotypes. In *Dictyostelium discoideum*, loss of NCS-1 function alters developmental rate (Coukell et al. 2004) and in *C. elegans* results in impaired learning and memory (Gomez et al. 2001). Knockdown of one of the two NCS-1 genes in zebrafish, ncs-1, prevents formation of the semicircular canals of the inner ear (Blasiole et al. 2005). The signaling pathway involving NCS-1, ARF1, and PI4KIIIβ (Haynes et al. 2005) modulates the secretion of components important for the development of the vestibular apparatus of the inner ear (Petko et al. 2009). Knockdown of NCS-1 or expression of a dominant–negative inhibitor based on an EF-hand mutation (Weiss et al. 2000) disrupted the induction of long-term depression in rat cortical neurons (Jo et al. 2008).

Many different binding partners have been identified for NCS-1 (Haynes et al. 2006; Haynes et al. 2007) (Fig. 3) and in some cases these interactions overlap with those of calmodulin (Schaad et al. 1996). This overlap using in vitro binding assays may not be physiologically meaningful because of substantial lower affinity of NCS-1 for known calmodulin targets (Schaad et al. 1996; Fitzgerald et al. 2008). NCS-1 has a higher affinity for calcium than calmodulin and therefore may preferentially interact with certain binding partners when the amplitude of a Ca^{2+}-signal falls below the threshold for activation of calmodulin. For example, both calmodulin and NCS-1 have been shown to interact with and desensitize dopamine D2 receptors but are likely to mediate their effects at different $[Ca^{2+}]_i$ (Kabbani et al. 2002; Woods et al. 2008). Functional analyses have confirmed that NCS-1 is a regulator of D2 receptors and that this function modulates learning in mice (Saab et al. 2009). Other NCS-1 target proteins (Fig. 3) appear to be specific for NCS-1 (Haynes et al. 2006). Various studies have implicated NCS-1 in the regulation of VGCCs (Weiss et al. 2000; Tsujimoto et al. 2002; Dason et al. 2009) but there is as yet no evidence for a direct interaction. Interestingly, in *Drosophila*, the effects of NCS-1 on both neurotransmission and nerve-terminal growth can be explained by a functional interaction with the VGCC *cacophony*, which is related to the mammalian P/Q-type VGCCs (Dason et al. 2009). In contrast, in this study, there was no evidence for an essential functional interaction with the fly PI4KIIIβ ortholog *four wheel drive*.

The example of NCS-1 illustrates how evolutionary pressures have fine-tuned Ca^{2+} sensors to carry out specialized neuronal functions. The individual properties of NCS-1 allow this protein to localize to discrete domains within the cell and interact with distinct target proteins under conditions of Ca^{2+} stimulation, which would not activate the archetypal Ca^{2+} sensor, calmodulin. Although NCS-1 has been found not to be neuronal specific and may carry out more generalized functions conserved through evolution in organisms from yeast onwards, further adaptive mutations have given rise to many more members of the NCS protein family, each tasked with dedicated functions.

Much less is known about the VSNL or class B proteins, although these appear to modulate various signal transduction pathways such as cyclic nucleotide and MAPK signaling (Braunewell and Klein-Szanto 2009). VILIP-1 has been found to regulate a class of purinergic receptors (Chaumont et al. 2008). They have been shown to have effects on gene expression and are also involved in traffic of proteins to the plasma membrane (Lin et al. 2002; Brackmann et al. 2005). Hippocalcin has been suggested to be involved as a Ca^{2+} sensor in long term depression in hippocampal neurons (Palmer et al. 2005) and shows a Ca^{2+}/myristoyl switch for translocation within such neurons (Markova et al. 2008). It has also been implicated in protection from neuronal apoptosis (Mercer et al. 2000; Korhonen et al. 2004).

Recoverin is expressed exclusively in the retina and is believed to have a role in light

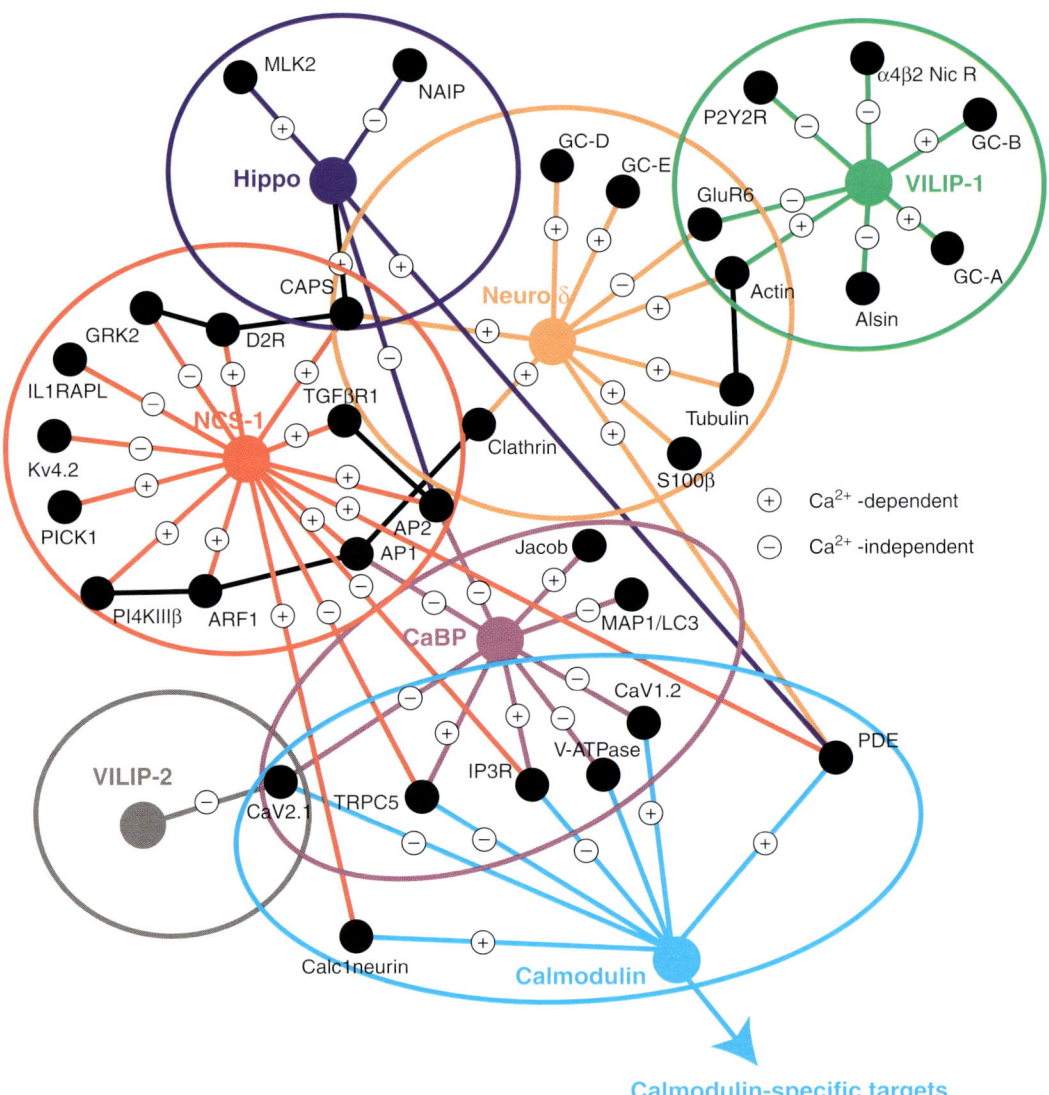

Figure 3. An interaction map showing protein–protein interactions made by some NCS proteins and CaBP1 compared to calmodulin. Known protein interactions for CABP1, hippocalcin, NCS-1, neurocalcin δ, VILIP1, and VILIP2. Links indicate where these target proteins have also been found to interact with calmodulin. It is also indicated whether these interactions require the Ca^{2+}-bound form of the protein or not.

adaptation and can enhance visual sensitivity (Polans et al. 1996; Sampath et al. 2005). Recoverin is found primarily in rod and cone cells of the retina (Yamagata et al. 1990; Dizhoor et al. 1991). Recoverin was predicted to prolong the lifetime of photolyzed rhodopsin by inhibiting its phosphorylation by rhodopsin kinase to extend the light response (Chen et al. 1995; Klenshin et al. 1995). The function of recoverin has been controversial and this hypothesis may be oversimplified. Discrepancies have been noted regarding the $[Ca^{2+}]_i$ required for rhodopsin kinase interaction, which may lie outside normal physiological limits but analysis of recoverin knockout mice have shown changes in photoresponses consistent with a physiological role in inhibition of rhodopsin kinase (Makino et al. 2004).

The structure of recoverin has been extensively studied by X-ray crystallography and NMR studies to interrogate its structure in its Ca^{2+}-bound and -free forms (Flaherty et al. 1993; Ames et al. 1995; Tanaka et al. 1995; Ames et al. 1997; Ames et al. 2002; Weiergraber et al. 2003). Recoverin is composed of two distinct domains connected through a bent linker and forms a compact structure in the absence of Ca^{2+}. Unlike other NCS proteins, recoverin has only two functional EF-hand motifs. Upon binding of Ca^{2+}, the amino-terminal domain comprising EF-1 and EF-2 rotates through 45° relative to the carboxy-terminal domain driving extrusion of its buried myristoyl group to permit association with membranes and revealing a hydrophobic surface, which can mediate interaction with the target protein rhodopsin kinase (Ames et al. 2006). The residues involved in the interaction of the myristoyl group with the hydrophobic pocket are also conserved in the other members of the NCS family, however not all of the other family members display this Ca^{2+}/myristoyl switch (O'Callaghan et al. 2002; Stephen et al. 2007). NCS-1 and KChIP1 expose a similar hydrophobic surface upon Ca^{2+}-binding, which could be similarly important for target interactions (Bourne et al. 2001; Scannevin et al. 2004; Zhou et al. 2004b; Pioletti et al. 2006). In contrast, other NCS proteins are able to interact with certain binding proteins in the absence of Ca^{2+} and therefore Ca^{2+}-driven exposure of a hydrophobic surface cannot be the sole mechanism by which these proteins bind to effectors. Although extensive structural characterization of recoverin may go some way to inform an understanding of the general conserved structures of members of the NCS family, subtle differences in "active" surface residues of the individual proteins gives rise to their ability to interact specifically with a wide range of binding partners.

GCAPs are the only known activators of retinal guanylyl cyclases (GCs) (Palczewski et al. 2004) and are known to be physiological regulators of light adaptation (Mendez et al. 2001; Burns et al. 2002; Howes et al. 2002; Pennesi et al. 2003). They are unusual in that they activate GCs when in their Ca^{2+}-free form but become inhibitors of GCs at higher Ca^{2+} concentrations (Dizhoor and Hurley 1996). GCAP3 is expressed in cone cells, whereas GCAP1 and GCAP2 are expressed in rod cells, and despite GCAP1 and GCAP2 having the same function in the same cell type, the two proteins have different Ca^{2+} binding affinities for GC activation. This means that both proteins are required for GC activation over the full physiological Ca^{2+} concentration range, thus maximizing the dynamic range of GC activity (Koch 2006). The GCAPs are an example of how calcium sensors have become adapted to increase the dynamic Ca^{2+} sensitivity of important regulatory mechanisms in specialized cell types (Palczewski et al. 2004).

KChIPs have been found to associate with transient voltage-gated potassium channels of the Kv4 family (An et al. 2000) and can stimulate their traffic to the plasma membrane (Hasdemir et al. 2005). Four KChIP genes and a large number of expressed splice variants are present in mammals (Pruunsild and Timmusk 2005). Knockout of KChIP1 has revealed a potential role in the GABAergic inhibitory system (Xiong et al. 2009). The KChIPs are expressed predominantly in the brain but KChIP2 is also expressed in the heart, and knockout of KChIP2 causes a complete loss of calcium-dependent transient outward potassium currents and susceptibility to ventricular tachycardia (Kuo et al. 2001). KChIP3 is also known as DREAM or calsenilin, and has documented roles in transcriptional regulation (Carrion et al. 1999; Mellstrom and Naranjo 2001) and in the processing of presenilins and amyloid precursor protein, which are important in the pathogenesis of Alzheimer's disease (Buxbaum et al. 1998; Jo et al. 2004). Despite KChIP3 being implicating in three quite different functions, it is likely that these are all physiologically relevant. KChIP3 knockout mice show reduced responses in acute pain models because of changes in prodynorphin synthesis (Cheng et al. 2002), decreased β-amyloid production, and physiological defects consistent with changes to the Kv4 channels

(Lilliehook et al. 2003). Although many of the KChIPs and their isoforms may have overlapping functions, some differences between them are beginning to emerge (Holmqvist et al. 2002; Venn et al. 2008).

In support of key roles for the NCS family in higher organisms, a number of recent studies have implicated these proteins in the pathological progression of human neurological diseases in addition to the potential link with Alzheimer's disease via the interaction of KChIP3 with presenilins. VILIP1 may have a role in Alzheimer's disease because of an association with amyloid plaques in diseased brains (Schnurra et al. 2001). NCS-1 has been found to be up-regulated in patients with schizophrenia and bipolar disorders (Koh et al. 2003) and also interacts with interleukin-1 receptor accessory protein-like (IL1RAPL), a protein, which when mutated results in X-linked mental retardation (Bahi et al. 2003). The effects conferred by NCS proteins in all of these diseases would appear to be dependent on the up- or down-regulation of their expression. As yet few genetic links have been established between NCS proteins and the aforementioned diseases, suggesting epigenetic effects may be responsible. One idea is that epigenetic mediated alterations in NCS protein function may contribute to cognitive impairments observed in neurodegenerative states. Targeting of NCS protein function through this novel route may offer a future therapeutic approach for such diseases (Braunewell 2005).

The NCS protein family has evolved to carry out specialized neuronal functions that are separate to those of calmodulin. When attempting to decipher precisely why these proteins are particularly well adapted for carrying out functions in neurons, it is therefore relevant to compare their properties to those of calmodulin. Of note is their approximately 10-fold higher affinity for Ca^{2+} when compared to calmodulin. This higher affinity would allow the NCS proteins to be activated at much lower Ca^{2+} concentrations and, in combination with calmodulin, extends the dynamic range over which Ca^{2+} can regulate neuronal processes. In this way, responses to very slight or more dramatic changes in $[Ca^{2+}]_i$ would depend on which populations of calcium binding proteins are activated under particular conditions (Burgoyne and Weiss 2001). As mentioned previously, the individual expression patterns and subcellular localization of each of the NCS proteins is also likely to represent a key factor in their specific roles in neuronal cell signaling. The characteristic amino-terminal myristoylation or palmitoylation modifications, which allow these proteins to associate with membranes, may spatially partition them to relevant subcellular sites within the cell, leading to a faster and more efficient response to particular Ca^{2+} signals. Specific physiological outcomes will be determined by their distinct target proteins. The various members of the NCS family arose at points in evolution corresponding to increasing neuronal sophistication in higher animals. As such, these proteins represent an example of how the properties of calcium binding proteins have been fine-tuned to act in specific neuronal signaling pathways.

CaBP Family

The CaBPs are a relatively recently discovered family of EF-hand containing Ca^{2+}-binding proteins, which are only found in vertebrates (Haeseleer et al. 2000). They represent another example of a diverse family of Ca^{2+}-sensors capable of regulating discrete processes in the nervous systems of higher organisms. The CaBPs share sequence homology with calmodulin and also display a similar structural arrangement of EF-hand motifs. Each of the CaBPs has four EF hands, although, like the NCS proteins, they display different patterns of EF-hand inactivation (Fig. 4). In CaBPs 1–5, the second EF-hand motif is inactive with the exception of CaBP3, which also has an inactive EF-1 motif. CaBP3 is believed to represent a pseudogene as only the mRNA has been detected in cells and as yet no protein product has been found (Haeseleer et al. 2000). Two proteins were named CaBP7 and CaBP8 (Haeseleer et al. 2002), but bioinformatic analysis is more consistent with them being a conserved and distinct subfamily of CaBPs (McCue et al. 2010).

Figure 4. Schematic diagram showing the domain structure of calmodulin and members of the CaBP/calneuron protein family. Active EF-hand motifs are shown in red and inactive EF-hand motifs are shown in pink. Compared to calmodulin, the CaBPs have an extended linker region between their first EF-hand pair and their second EF-hand pair (shown in black). CaBP1 and CaBP2 have an N-myristoylation site (shown in blue). CaBP1 and CaBP2 have alternative splice sites at their N-terminus, which give rise to long and short isoforms (shown in orange). Calneurons 1 and 2 possess a 38 amino acid extension at their C-terminus (shown in purple).

We will therefore refer to them by their alternative names, calneuron 2 and calneuron 1, respectively (Wu et al. 2001; Mikhaylova et al. 2006). The calneurons, by contrast to the CaBPs, have a different pattern of EF-hand inactivation with active EF-hands 1 and 2 and inactive EF-hands 3 and 4 (Mikhaylova et al. 2006). The CaBPs also differ from calmodulin in that their central α helical linker domain connecting the carboxy- and amino-terminal EF-hand pairs is extended by four amino acid residues. This has been suggested to allow these proteins to interact with unique targets (Haeseleer et al. 2000).

A major difference compared with calmodulin is the ability of CaBP 1 and 2 and calneurons 1 and 2 to target to specific cellular membranes (McCue et al. 2009). CaBP 1 and 2 are amino-terminally myristoylated, which permits localization to the plasma membrane and Golgi apparatus (Haeseleer et al. 2000; Haynes et al. 2004). The precise amino-terminal sequence to which the myristoyl group is attached is also important in the targeting of these two proteins, as exemplified by the long and short splice isoforms generated from their genes, which show subtle differences in their localization. CaBP1-Long localizes predominantly to the Golgi and also displays some cytosolic localization, whereas CaBP1-Short localizes most prominently to the plasma membrane and to Golgi structures (Haeseleer et al. 2000; McCue et al. 2009). Alternative splicing of the CaBP1 gene generates a third protein product, caldendrin (Seidenbecher et al. 1998). This splice isoform is significantly larger than either CaBP1-Long or CaBP1-Short because of an amino-terminal extension, but caldendrin mRNA lacks the exon required for N-myristoylation and as a result the protein displays a markedly different subcellular localization to its shorter relatives.

N-terminal acylation is important in the localization of some CaBPs, but the calneurons appear to be targeted via a different mechanism. Like CaBP1 and CaBP2, calneurons 1 and 2 localize to internal membranes that colabel with Golgi-specific markers and also to vesicular structures (McCue et al. 2009; Mikhaylova et al. 2009). Calneurons 1 and 2 do not possess an amino-terminal myristoylation motif and differ from the rest of the CaBP family because of a 38-amino acid extension at their carboxyl terminus. Analysis of this sequence revealed a predicted C-terminal transmembrane domain with a cytosolic amino terminus. The carboxy-terminal domain resembles tail-anchor motifs and directly localizes calneurons 1 and 2 to membranes particularly of the trans-Golgi network (McCue et al. 2009).

To date, only the structure of CaBP1-Short has been solved (Wingard et al. 2005; Li et al. 2009). This structural information may provide insight into the structures of the rest of the CaBPs. Analogy to calmodulin would suggest that the CaBPs should adopt a dumbbell like tertiary conformation consisting of an amino-terminal domain containing EF-1 and EF-2 and a carboxy-terminal domain containing EF-3 and EF-4, connected by a central linker. NMR analysis revealed that CaBP1 does indeed have two independent, noninteracting domains, joined by a flexible linker (Wingard et al. 2005). Investigation into the effects of Mg^{++} and Ca^{2+} binding has shown that as predicted the second EF hand of CaBP1 is incapable of binding divalent cations. EF-3 and EF-4 bind to both Mg^{++} and to Ca^{2+}, whereas EF-1 is thought to be constitutively Mg^{++} bound. Binding to either Mg^{++} or Ca^{2+} induces distinct conformations of this protein. Mg^{++} binding results in a global conformational change, whereas Ca^{2+} binding results in only a localized change in EF-3 and EF-4. These two conformational states may allow CaBP1 to interact with different target molecules driven by the ratio of Mg^{++} to Ca^{2+} (Wingard et al. 2005; Li et al. 2009). The Mg^{++} bound form of CaBP1 is similar to that of apo-calmodulin, but the Ca^{2+} bound form appears markedly different. This is perhaps unsurprising as neither of the amino-terminal EF-hands of CaBP1 bind to Ca^{2+} under saturating conditions and only EF-1 binds to Mg^{++}. This results in a constitutively closed conformation of the amino-terminal domain, whereas the carboxy-terminal domain can switch to an open conformation upon Ca^{2+} binding to EF-3 and EF-4. Comparison of the carboxy-terminal domain with that of calmodulin, however, still reveals differences in exposed hydrophobic residues thought to mediate target interactions (Wingard et al. 2005).

The structural differences between calmodulin and CaBP1 may go some way to explaining how they impose differing effects on the same target molecules. For instance, both CaBP1 and calmodulin bind to L-type Ca^{2+} channels with calmodulin causing Ca^{2+}-induced channel closure but CaBP1 promoting channel opening (Zhou et al. 2004a; Zhou et al. 2005). Both calmodulin and CaBP1 also regulate inositol 1,4,5-trisphosphate receptors (IP_3Rs) (Yang et al. 2002; Haynes et al. 2004; Kasri et al. 2004) with CaBP1 binding the type I IP_3R with 100-fold higher affinity than calmodulin. This high affinity binding may result from the exposure of a distinct hydrophobic patch revealed in the carboxyl terminus of CaBP1 upon Ca^{2+}-binding (Haynes et al. 2004; Li et al. 2009). This unique surface hydrophobicity profile is likely to be important for the specialization of CaBP1 function in the brain and retina, and the existence of splice isoforms is also likely to further fine-tune the actions of this Ca^{2+} sensor. The differing expression patterns, subcellular targeting mechanisms, and Ca^{2+} binding properties of the various members of the CaBP protein family would allow them to carry out highly specialized regulatory roles modulating important Ca^{2+}-channels in the central nervous system.

The majority of studies to date on CaBP1 have examined the functions of the longest splice isoform caldendrin and it is not yet clear whether the other splice isoforms of CaBP1 can carry out the same functions. CaBP1-Long and -Short have been found to have roles in the regulation of various Ca^{2+}-channels including P/Q-type ($Ca_V2.1$) channels (Lee et al. 2002), L-type ($Ca_V1.2$) channels (Zhou et al. 2005; Cui et al. 2007), TRPC5 channels (Kinoshita-Kawada et al. 2005), and IP_3Rs (Yang et al. 2002), which they apparently inhibit (Haynes et al. 2004; Kasri et al. 2004). The interaction of $Ca_V2.1$ with CaBP1 appears to rely acutely upon amino-terminal myristoylation. Wild type, myristoylated, CaBP1-Long enhances channel inactivation and shifts the activation range to more depolarizing voltages (Lee et al. 2002). An N-myristoylation mutant, however, was unable to mediate these effects and instead modulated channels in a similar fashion to calmodulin (Few et al. 2005). Differential modulation of L-type channels depending on the splice isoform of CaBP1 has also been observed. CaBP1-Short has been shown to completely inhibit inactivation of $Ca_V1.2$ channels (Zhou et al. 2005), but caldendrin causes a more modest suppression and signals through a different set of molecular determinants (Tippens and Lee 2007). This suggests that the subcellular localization of CaBP1 splice variants is important for their function and there are likely to be individual roles for each protein. Interactions of caldendrin with other types of proteins have also been reported, such as its interaction with light chain 3 of MAP1A/B, a microtubule cytoskeletal protein (Seidenbecher et al. 2004), and an interaction with myo1c, a member of the myosin-1 family of motor proteins (Tang et al. 2007). Finally, a role for caldendrin in NMDA receptor (NMDAR) signaling has been reported via an interaction with a novel neuronal protein, Jacob. Upon extrasynaptic NMDAR activation, Jacob translocates to the nucleus to influence CREB activity, resulting in the stripping of synaptic contacts and an associated simplification of dendritic architecture. Synaptic NMDAR mediated synpatodendritic $[Ca^{2+}]_i$ elevation induces caldendrin binding to Jacob, inhibiting nuclear trafficking and maintaining dendritic organization. This interaction represents a novel mechanism of synapse to nucleus communication and highlights the important roles of CaBP family members in the mammalian central nervous system (Dieterich et al. 2008).

Little information is available concerning the function of CaBP2 apart from in vitro studies suggesting that it might stimulate CaMK activity (Cui et al. 2007). Initially, CaBP2 was detected exclusively in the retina, although it has also been identified in auditory inner hair cells (Cui et al. 2007). CaBP5 was also detected in inner hair cells as well as in the retina, but in contrast to CaBP2, was found to have a modest inhibitory effect on the inactivation of $Ca_V1.3$ channels in transfected cells (Cui et al. 2007). Little is known about the functions of CaBP5, but knockout mice displayed reduced sensitivity of retinal ganglion cells to light responses, implicating CaBP5 in phototransduction pathways. CaBP5 was also found to interact with and suppress calcium-dependent inactivation of $Ca_V1.2$ channels (Rieke et al. 2008).

CaBP4 is the most extensively characterized of the CaBP family. It is expressed in the retina, where it localizes to synaptic terminals and has also been detected in auditory inner hair cells. CaBP4 modulates voltage gated Ca^{2+}-channels and directly associates with the carboxyl terminus of the $Ca_V1.4$ α1 pore-forming subunit, shifting the activation range of the channel to more hyperpolarized voltages in transfected cells (Haeseleer et al. 2004). CaBP4 has also been shown to eliminate Ca^{2+}-dependent inactivation of $Ca_V1.3$ channels, which is likely to be important in the modulation of these channels in inner hair cells, where Ca^{2+}-dependent inactivation is weak or absent, probably allowing the audition of sustained sounds (Yang et al. 2006). A stronger inhibitory effect has been noted for CaBP1, however, suggesting that CaBP4 may not be the key Ca^{2+} sensor involved in this process (Cui et al. 2007). The function of CaBP4 is modulated by protein kinase C ζ in the retina, with increased CaBP4 phosphorylation in light-adapted tissue. Phosphorylation prolongs Ca^{2+} currents through $Ca_V1.3$ channels and suggests that light-stimulated phosphorylation of CaBP4 might help to regulate presynaptic Ca^{2+} signals in photoreceptors (Lee et al. 2007). CaBP4 has also been implicated in neurotransmitter release at synaptic terminals because of its interaction with unc119, a synaptic photoreceptor protein important for neurotransmitter release and maintenance of the nervous system (Haeseleer 2008). Knockout of CaBP4 results in mice with abnormalities in retinal function, where rod bipolar responses are approximately 100 times lower than those observed in wild-type animals (Haeseleer et al. 2004).

The functions of calneurons 1 and 2 have only recently begun to be investigated in detail. Both have been found to inhibit the activity of PI4KIIIβ at low or resting $[Ca^{2+}]_i$. Overexpression of the proteins was also found to inhibit Golgi-to-plasma membrane trafficking, caused enlargement of the trans-Golgi network (TGN), and reduced the number of Piccolo-Bassoon positive transport vesicles. A molecular switch for the production of phosphoinositides at the TGN is thought to be created by the opposing roles of NCS-1 and calneurons 1 or 2. At elevated Ca^{2+} levels, NCS-1 preferentially binds to PI4KIIIβ over the calneurons, activating the enzyme to drive enhanced TGN-to-plasma membrane trafficking (Mikhaylova et al. 2009). Patch clamping experiments have shown that over-expressed calneuron 1 can inhibit N-type Ca^{2+}-channel currents in 293T cells and this inhibition was not observed with a truncated calneuron 1 lacking its hydrophobic C-terminus, suggesting normal localisation is important in carrying out this function (Shih et al. 2009).

CaBPs have been directly or indirectly implicated in multiple neuronal diseases. Post-mortem brains of chronic schizophrenics have lower numbers of caldendrin-immunoreactive neurons, which express the protein at a much higher level. This loss of caldendrin in some neurons and up-regulation in others is likely to profoundly change synapto-dendritic signalling in schizophrenic patients (Bernstein et al. 2007). Changes in the distribution of caldendrin have also been observed in kainate-induced epileptic seizures in rats. Caldendrin translocates to the postsynaptic density only in rats that suffered epileptic seizures and may implicate the protein in the pathophysiology of the disease (Smalla et al. 2003). CaBP4 function has been convincingly linked to disease and mutations in this gene generate defects in retinal function. Knockout of CaBP4 was shown to cause a phenotype similar to that of incomplete congenital stationary night blindness patients (Haeseleer et al. 2004) and mutations in CaBP4 can cause autosomal recessive night blindness (Zeitz et al. 2006). Patients with mutations in the CaBP4 gene have been identified, which display congenital stationary night blindness. However, some patients with mutations display different phenotypes (Zeitz et al. 2006). In particular, a novel homozygous nonsense mutation has been reported in two siblings that resulted in severely reduced cone function but only negligible effects on rod function (Littink et al. 2009). It appears, therefore, that genetic mutations in CaBP4 underlie cone-rod synaptic disorders (Littink et al. 2009).

CONCLUDING REMARKS

It has become increasingly clear that a full understanding of how specific aspects of neuronal function are regulated in response to spatially and temporally distinct Ca^{2+} signals will require a detailed knowledge of the Ca^{2+} sensors involved. Some of these, like synaptotagmin, are specialized for particular neuronal functions, whereas others such as calmodulin may be involved in multiple cellular processes. In recent years, much has been learnt about the functions of the NCS family of Ca^{2+} sensors, although the functions of some of the family members are still unknown. Nor is it clear what the significance is of the multiple genes and splice variants of these proteins. The CaBPs have so far been much less studied and much remains to be learnt about the functions of each of these sensors. Further advances will require new insights into the molecular targets of each of the Ca^{2+} sensors, the molecular basis for their regulation of these targets, and more detailed dissection of the functional roles of each protein in identified neurons.

ACKNOWLEDGMENTS

HVM was supported by a Wellcome Trust Prize PhD Studentship.

REFERENCES

Ames JB, Hamashima N, Molchanova T. 2002. Structure and calcium-binding studies of a recoverin mutant (E85Q) in an allosteric intermediate state. *Biochemistry* **41:** 5776–5787.

Ames JB, Ishima R, Tanaka T, Gordon JI, Stryer L, Ikura M. 1997. Molecular mechanics of calcium-myristoyl switches. *Nature* **389:** 198–202.

Ames JB, Levay K, Wingard JN, Lusin JD, Slepak VZ. 2006. Structural basis for calcium-induced inhibition of rhodopsin kinase by recoverin. *J Biol Chem* **281:** 37237–37245.

Ames JB, Tanaka T, Ikura M, Stryer L. 1995. Nuclear magnetic resonance evidence for Ca^{2+}-induced extrusion of the myristoyl group of recoverin. *J Biol Chem* **270:** 30909–30913.

An WF, Bowlby MR, Bett M, Cao J, Ling HP, Mendoza G, Hinson JW, Mattsson KI, Strassle BW, Trimmer JS, et al. 2000. Modulation of A-type potassium channels by a family of calcium sensors. *Nature* **403:** 553–556.

Aravind P, Chandra K, Reddy PP, Jeromin A, Chary KV, Sharma Y. 2008. Regulatory and structural EF-hand motifs of neuronal calcium sensor-1: Mg^{2+} modulates Ca^{2+} binding, Ca^{2+}-induced conformational changes, and equilibrium unfolding transitions. *J Mol Biol* **376:** 1100–1115.

Augustine GJ, Santamaria F, Tanaka K. 2003. Local calcium signaling in neurons. *Neuron* **40:** 331–346.

Bahi N, Friocourt G, Carrié A, Graham ME, Weiss JL, Chafey P, Fauchereau F, Burgoyne RD, Chelly J. 2003. IL1 receptor accessory protein like, a protein involved in X-linked mental retardation, interacts with Neuronal Calcium Sensor-1 and regulates exocytosis. *Human Mol Genetics* **12:** 1415–1425.

Barria A, Muller D, Derkach V, Griffith LC, Soderling TR. 1997. Regulatory phosphorylation of AMPA-type glutamate receptors by CaM-KII during long-term potentiation. *Science* **276:** 2042–2045.

Bernstein HG, Sahin J, Smalla KH, Gundelfinger ED, Bogerts B, Kreutz MR. 2007. A reduced number of cortical neurons show increased Caldendrin protein levels in chronic schizophrenia. *Schizophr Res* **96:** 246–256.

Berridge MJ. 1998. Neuronal calcium signalling. *Neuron* **21:** 13–26.

Berridge MJ. 2010. Calcium hypothesis of Alzheimer's disease. *Pflugers Arch* **459:** 441–449.

Bito H, Deisseroth K, Tsien RW. 1997. Ca^{2+}-dependent regulation in neuronal gene expression. *Current Opinion in Neurobiol* **7:** 419–429.

Blasiole B, Kabbani N, Boehmler W, Thisse B, Thisse C, Canfield V, Levenson R. 2005. Neuronal calcium sensor-1 gene ncs-1 is essential for semicircular canal formation in zebrafish inner ear. *J Neurobiol* **64:** 285–297.

Bourne Y, Dannenberg J, Pollmann V, Marchot P, Pongs O. 2001. Immunocytochemical localisation and crystal structure of human frequenin (neuronal calcium sensor 1). *J Biol Chem* **276:** 11949–11955.

Brackmann M, Schuchmann S, Anand R, Braunewell KH. 2005. Neuronal Ca^{2+} sensor protein VILIP-1 affects cGMP signalling of guanylyl cyclase B by regulating clathrin-dependent receptor recycling in hippocampal neurons. *J Cell Sci* **118:** 2495–2505.

Braunewell K-H. 2005. The darker side of Ca^{2+} signaling by neuronal Ca^{2+}-sensor proteins: From Alzheimer's disease to cancer. *Trends Pharmacol Sci* **26:** 345–351.

Braunewell KH, Klein-Szanto AJ. 2009. Visinin-like proteins (VSNLs): Interaction partners and emerging functions in signal transduction of a subfamily of neuronal Ca^{2+}-sensor proteins. *Cell Tissue Res* **335:** 301–316.

Burgoyne RD. 2004. The neuronal calcium-sensor proteins. *Biochim Biophys Acta* **1742:** 59–68.

Burgoyne RD. 2007. Neuronal calcium sensor proteins: Generating diversity in neuronal Ca^{2+} signalling. *Nat Rev Neurosci* **8:** 182–193.

Burgoyne RD, Weiss JL. 2001. The neuronal calcium sensor family of Ca^{2+}-binding proteins. *Biochem J* **353:** 1–12.

Burgoyne RD, O'Callaghan DW, Hasdemir B, Haynes LP, Tepikin AV. 2004. Neuronal calcium sensor proteins: Multitalented regulators of neuronal function. *Trends Neurosci* **27:** 203–209.

Burns ME, Mendez A, Chen J, Baylor DA. 2002. Dynamics of cyclic AMP synthesis in retinal rods. *Neuron* **36:** 81–91.

Buxbaum JD, Choi EK, Luo YX, Lilliehook C, Crowley AC, Merriam DE, Wasco W. 1998. Calsenilin: A calcium-binding protein that interacts with the presenilins and regulates the levels of a presenilin fragment. *Nature Medicine* **4:** 1177–1181.

Carrion AM, Link WA, Ledo F, Mellstrom B, Naranjo JR. 1999. DREAM is a Ca^{2+}-regulated transcriptional repressor. *Nature* **398:** 80–84.

Catterall WA, Few AP. 2008. Calcium channel regulation and presynaptic plasticity. *Neuron* **59:** 882–901.

Chapman ER. 2008. How does synaptotagmin trigger neurotransmitter release? *Annu Rev Biochem* **77:** 615–641.

Chattopadhyaya R, Meador WE, Means AR, Quiocho FA. 1992. Calmodulin structure refined at 1.7 A resolution. *J Mol Biol* **228:** 1177–1192.

Chaumont S, Compan V, Toulme E, Richler E, Housley GD, Rassendren F, Khakh BS. 2008. Regulation of P2X2 receptors by the neuronal calcium sensor VILIP1. *Sci Signal* **1:** ra8.

Chen CK, Inglese J, Lefkowitz RJ, Hurley JB. 1995. Ca^{2+}-dependent interaction of recoverin with rhodopsin kinase. *J Biol Chem* **270:** 18060–18066.

Cheng H-YM, Pitcher GM, Laviolette SR, Whishaw IQ, Tong KI, Kockeritz LK, Wada T, Joza NA, Crackower M, Goncalves J, et al. 2002. DREAM is a critical transcriptional repressor for pain modulation. *Cell* **108:** 31–43.

Chicka MC, Hui E, Liu H, Chapman ER. 2008. Synaptotagmin arrests the SNARE complex before triggering fast, efficient membrane fusion in response to Ca2+. *Nat Struct Mol Biol* **15:** 827–835.

Coukell B, Cameron A, Perusini S, Shim K. 2004. Disruption of the NCS-1/frequenin-related ncsA gene in *Dictyostelium discoideum* accelerates development. *Develop Growth Differ* **46:** 449–458.

Craxton M. 2004. Synaptotagmin gene content of the sequenced genomes. *BMC Genomics* **5:** 43.

Cui G, Meyer AC, Calin-Jageman I, Neef J, Haeseleer F, Moser T, Lee A. 2007. Ca2+-binding proteins tune Ca^{2+}-feedback to Cav1.3 channels in mouse auditory hair cells. *J Physiol* **585:** 791–803.

Dason JS, Romero-Pozuelo J, Marin L, Iyengar BG, Klose MK, Ferrus A, Atwood HL. 2009. Frequenin/NCS-1 and the Ca^{2+}-channel {α} 1-subunit co-regulate synaptic transmission and nerve-terminal growth. *J Cell Sci* **122:** 4109–4121.

de Barry J, Janoshazi A, Dupont JL, Procksch O, Chasserot-Golaz S, Jeromin A, Vitale N. 2006. Functional implication of neuronal calcium sensor-1 and PI4 kinase-β interaction in regulated exocytosis of PC12 cells. *J Biol Chem* **281:** 18098–18111.

de Wit H, Walter AM, Milosevic I, Gulyas-Kovacs A, Riedel D, Sorensen JB, Verhage M. 2009. Synaptotagmin-1 docks secretory vesicles to syntaxin-1/SNAP-25 acceptor complexes. *Cell* **138:** 935–946.

Deisseroth K, Heist EK, Tsien RW. 1998. Translocation of calmodulin to the nucleus supports CREB phosphorylation in hippocampal neurons. *Nature* **392:** 198–202.

DeMaria CD, Soong TW, Alselkhan BA, Alvania RS, Yue DT. 2001. Calmodulin bifurcates the local Ca^{2+} signal that modulates P/Q-type Ca^{2+} channels. *Nature* **411:** 484–489.

Dieterich DC, Karpova A, Mikhaylova M, Zdobnova I, Konig I, Landwehr M, Kreutz M, Smalla KH, Richter K, Landgraf P, et al. 2008. Caldendrin-Jacob: A protein liaison that couples NMDA receptor signalling to the nucleus. *PLoS Biol* **6:** e34.

Dizhoor AM, Hurley JB. 1996. Inactivation of EF-hands makes GCAP-2 (p24) a constitutive activator of photoreceptor guanylyl cyclase by preventing a Ca^{2+}-induced "activator-to-inhibitor" transition. *J Biol Chem* **271:** 19346–19350.

Dizhoor AM, Ray S, Kumar S, Niemi G, Spencer M, Brolley D, Walsh KA, Philipov PP, Hurley JB, Stryer L. 1991. Recoverin: A calcium sensitive activator of retinal rod guanylate cyclase. *Science* **251:** 915–918.

Fallon JL, Halling DB, Hamilton SL, Quiocho FA. 2005. Structure of calmodulin bound to the hydrophobic IQ domain of the cardiac Ca(v)1.2 calcium channel. *Structure* **13:** 1881–1886.

Fernandez-Chacon R, Konigstorfer A, Gerber SH, Garcia J, Matos MF, Stevens CF, Brose N, Rizo J, Rosenmund C, Sudhof TC. 2001. Synaptotagmin I functions as a calcium regulator of release probability. *Nature* **410:** 41–49.

Fernandez I, Arac D, Ubach J, Gerber SH, Shin O-h, Gao Y, Anderson RGW, Sudhof TC, Rizo J. 2001. Three-dimensional structure of the synaptotagmin 1 C2B-domain: Synaptotagmin 1 as a phospholipid binding machine. *Neuron* **32:** 1057–1069.

Few AP, Lautermilch NJ, Westenbroek RE, Scheuer T, Catterall WA. 2005. Differential regulation of $Ca_v2.1$ channels by calcium binding protein 1 and visinin-like protein-2 requires N-terminal myristoylation. *J Neurosci* **25:** 7071–7080.

Fitzgerald DJ, Burgoyne RD, Haynes LP. 2008. Neuronal calcium sensor proteins are unable to modulate NFAT activation in mammalian cells. *Biochim Biophys Acta* **1780:** 240–248.

Flaherty KM, Zoulya S, Stryer L, McKay DB. 1993. 3-Dimensional structure of recoverin, a calcium sensor in vision. *Cell* **75:** 709–716.

Geppert M, Goda Y, Hammer RE, Li C, Rosahl TW, Stevens CF, Sudhof TC. 1994. Synaptotagmin I: A major Ca^{2+} sensor for transmitter release at a central synapse. *Cell* **79:** 717–727.

Gomez M, De Castro E, Guarin E, Sasakura H, Kuhara A, Mori I, Bartfai T, Bargmann CI, Nef P. 2001. Ca^{2+} signalling via the neuronal calcium sensor-1 regulates associative learning and memory in *C.elegans*. *Neuron* **30:** 241–248.

Groffen AJ, Martens S, Arazola RD, Cornelisse LN, Lozovaya N, de Jong AP, Goriounova NA, Habets RL, Takai Y, Borst JG, et al. 2010. Doc2b is a high-affinity Ca2+ sensor for spontaneous neurotransmitter release. *Science* **327:** 1614–1618.

Haeseleer F. 2008. Interaction and colocalization of CaBP4 and Unc119 (MRG4) in photoreceptors. *Investigative Ophthalmol Visual Sci* **49:** 2366–2375.

Haeseleer F, Imanishi Y, Maeda T, Possin DE, Maeda A, Lee A, Rieke F, Palczewski K. 2004. Essential role of Ca^{2+}-binding protein 4, a $Ca_v1.4$ channel regulator in photoreceptor synaptic function. *Nature Neurosci* **7:** 1079–1087.

Haeseleer F, Imanishi Y, Sokal I, Filipek S, Palczewski K. 2002. Calcium-binding proteins: Intracellular sensors from the calmodulin superfamily. *Biochem Biophys Res Comm* **290:** 615–623.

Haeseleer F, Sokal I, Verlinde CLMJ, Erdjument-Bromage H, Tempst P, Pronin AN, Benovic JL, Fariss RN, Palczewski K. 2000. Five members of a novel Ca^{2+} binding protein (CABP) subfamily with similarity to calmodulin. *J Biol Chem* **275:** 1247–1260.

Handley MT, Lian LY, Haynes LP, Burgoyne RD. 2010. Structural and functional deficits in a Neuronal Calcium Sensor-1 mutant identified in a case of Autistic Spectrum Disorder. *PLoS ONE* **5:** e10534.

Hasdemir B, Fitzgerald DJ, Prior IA, Tepikin AV, Burgoyne RD. 2005. Traffic of Kv4 K+ channels mediated by KChIP1 is via a novel post-ER vesicular pathway. *J Cell Biol* **171:** 459–469.

Haynes LP, Tepikin AV, Burgoyne RD. 2004. Calcium Binding Protein 1 is an inhibitor of agonist-evoked, inositol 1,4,5-trisphophate-mediated calcium signalling. *J Biol Chem* **279:** 547–555.

Haynes LP, Thomas GMH, Burgoyne RD. 2005. Interaction of neuronal calcium sensor-1 and ARF1 allows bidirectional control of PI(4) kinase and TGN-plasma membrane traffic. *J Biol Chem* **280:** 6047–6054.

Haynes LP, Fitzgerald DJ, Wareing B, O'Callaghan DW, Morgan A, Burgoyne RD. 2006. Analysis of the interacting partners of the neuronal calcium-binding proteins L-CaBP1, hippocalcin, NCS-1 and neurocalcin. *Proteomics* **6:** 1822–1832.

Haynes LP, Sherwood MW, Dolman NJ, Burgoyne RD. 2007. Specificity, promiscuity and localization of ARF protein interactions with NCS-1 and phosphatidylinositol-4 kinase-IIIβ. *Traffic* **8:** 1080–1092.

Hendricks KB, Wang BQ, Schnieders EA, Thorner J. 1999. Yeast homologue of neuronal frequenin is a regulator of phosphatidylinositol-4-OH kinase. *Nature Cell Biology* **1:** 234–241.

Hilfiker S. 2003. Neuronal calcium sensor-1: A multifunctional regulator of secretion. *Biochem Soc Trans* **31:** 828–832.

Holmqvist MH, Cao J, Hernandez-Pineda R, Jacobson MD, Carroll KI, Sung MA, Betty M, Ge P, Gilbride KJ, Brown ME, et al. 2002. Elimination of fast inactivation in Kv4 A-type potassium channels by an auxiliary subunit domain. *Proc Natl Acad Sci USA* **99:** 1035–1040.

Howes KA, Pennesi ME, Sokal I, Church-Kopish J, Schmidt B, Margolis D, Frederick JM, Rieke F, Palczewski K, Wu SM, et al. 2002. GCAP1 rescues rod photoreceptor response in GCAP1/GCAP2 knockout mice. *EMBO Journal* **21:** 1545–1554.

Hui E, Johnson CP, Yao J, Dunning FM, Chapman ER. 2009. Synaptotagmin-mediated bending of the target membrane is a critical step in Ca(2+)-regulated fusion. *Cell* **138:** 709–721.

Jo J, Heon S, Kim MJ, Son GH, Park Y, Henley JM, Weiss JL, Sheng M, Collingridge GL, Cho K. 2008. Metabotropic glutamate receptor-mediated LTD involves two interacting Ca^{2+} sensors, NCS-1 and PICK1. *Neuron* **60:** 1095–1111.

Jo D-G, Jang J, Kim B-J, Lundkvist J, Jung Y-K. 2004. Overexpression of calsenilin enhances g-secretase activity. *Neuroscience letters* **378:** 59–64.

Kabbani N, Negyessy L, Lin R, Goldman-Rakic P, Levenson R. 2002. Interaction with the neuronal calcium sensor NCS-1 mediates desensitization of the D2 dopamine receptor. *J Neurosci* **22:** 8476–8486.

Kasri NN, Holmes AM, Bultynck G, Parys JB, Bootman MD, Rietdorf K, Missiaen L, McDonald F, Smedt HD, Conway SJ, et al. 2004. Regulation of $InsP_3$ receptor activity by neuronal Ca^{2+}-binding proteins. *EMBO J* **23:** 1–10.

Kinoshita-Kawada M, Tang J, Xiao R, Kaneko S, Foskett JK, Zhu MX. 2005. Inhibition of TRPC5 channels by Ca^{2+} binding protein 1 in *Xenopus* oocytes. *Pflugers Arch* **450:** 345–354.

Klenshin VA, Calvert PD, Bownds MD. 1995. Inhibition of rhodopsin kinase by recoverin. *J Biol Chem* **270:** 16147–16152.

Koch K-W. 2006. GCAPs, the classical neuronal calcium sensors in the retina. A Ca^{2+}-relay model of guanylate cyclase activation. *Calcium Binding Proteins* **1:** 3–6.

Koh PO, Undie AS, Kabbani N, Levenson R, Goldman-Rakic PS, Lidow MS. 2003. Up-regulation of neuronal calcium sensor-1 (NCS-1) in the prefrontal cortex of schizophrenic and bipolar patients. *Proc Natl Acad Sci* **100:** 313–317.

Korhonen L, Hansson I, Kukkonen JP, Brannvall K, Kobayashi M, Takamatsu K, Lindholm D. 2004. Hippocalcin protects against caspase-12-induced and age-dependent neuronal degeneration. *Mol Cell Neurosci* **28:** 85–95.

Kuo H-C, Cheng C-F, Clark RB, Lin JJ-C, Lin JL-C, Hoshijima M, Nguyen-Tran VTB, Gu Y, Ikeda Y, Chu P-H, et al. 2001. A defect in the Kv channel-interacting protein 2 (KChIP2) gene leads to a complete loss of I_{to} and confers susceptibility to ventricular tachycardia. *Cell* **107:** 801–813.

Lee A, Scheuer T, Catterall WA. 2000. Ca^{2+}/calmodulin-dependent facilitation and inactivation of P/Q-type Ca^{2+} channels. *J Neurosci* **20:** 6830–6838.

Lee A, Jimenez A, Cui G, Haeseleer F. 2007. Phosphorylation of the Ca^{2+}-binding protein CaBP4 by protein kinase C zeta in photoreceptors. *J Neurosci* **27:** 12743–12754.

Lee A, Westenbroek RE, Haeseleer F, Palczewski K, Scheuer T, Catterall WA. 2002. Differential modulation of Cav2.1 channels by calmodulin and Ca^{2+}-binding protein 1. *Nature Neurosci* **5:** 210–217.

Li C, Chan J, Haeseleer F, Mikoshiba K, Palczewski K, Ikura M, Ames JB. 2009. Structural insights into Ca2+-dependent regulation of inositol 1,4,5-trisphosphate receptors by CaBP1. *J Biol Chem* **284:** 2472–2481.

Lilliehook C, Bozdagi O, Yao J, Gomez-Ramirez M, Zaidi NF, Wasco W, Gandy S, Santucci AC, Haroutunian V, Huntley GW, et al. 2003. Altered Ab formation and long-term potentiation in a calsenilin knock-out. *J Neurosci* **23:** 9097–9106.

Lin L, Jeanclos EM, Treuil M, Braunewell K-H, Gundelfinger ED, Anand R. 2002. The calcium sensor protein visinin-like protein-1 modulates the surface expression and agonist sensitivity of the a4b2 nicotinic acetylcholine receptor. *J Biol Chem* **277:** 41872–41878.

Lisman J, Schulman H, Cline H. 2002. The molecular basis of CaMKII function in synaptic and behavioural memory. *Nat Rev Neurosci* **3:** 175–190.

Littink KW, van Genderen MM, Collin RW, Roosing S, de Brouwer AP, Riemslag FC, Venselaar H, Thiadens AA, Hoyng CB, Rohrschneider K, et al. 2009. A novel homozygous nonsense mutation in CABP4 causes congenital cone-rod synaptic disorder. *Investigative Ophthalmol Visual Sci* **50:** 2344–2350.

Liu X, Yang PS, Yang W, Yue DT. 2010. Enzyme-inhibitor-like tuning of Ca^{2+} channel connectivity with calmodulin. *Nature* **463:** 968–972.

Mackler JM, Drummond JA, Loewen CA, Robinson IM, Reist NE. 2002. The C2B Ca^{2+}-binding motif of synaptotagmin is required for synaptic transmission in vivo. *Nature* **418:** 340–344.

Makino CL, Dodd RL, Chen J, Burns ME, Roca A, Simon MI, Baylor DA. 2004. Recoverin regulates light-dependent phosphodiesterase activity in retinal rods. *J Gen Physiol* **123:** 729–741.

Malleret G, Haditsch U, Genoux D, Jones MW, Bliss TV, Vanhoose AM, Weitlauf C, Kandel ER, Winder DG, Mansuy IM. 2001. Inducible and reversible enhancement of learning, memory, and long-term potentiation by genetic inhibition of calcineurin. *Cell* **104:** 675–686.

Markova O, Fitzgerald D, Stepanyuk A, Dovgan A, Cherkas V, Tepikin A, Burgoyne RD, Belan P. 2008. Hippocalcin signaling via site-specific translocation in hippocampal neurons. *Neuroscience letters* **442:** 152–157.

McCue HV, Burgoyne RD, Haynes LP. 2009. Membrane targeting of the EF-hand containing calcium-sensing proteins CaBP7 and CaBP8. *Biochem Biophys Res Commun* **380:** 825–831.

McCue HV, Haynes LP, Burgoyne RD. 2010. Bioinformatic analysis of CaBP/calneuron proteins reveals a family of highly conserved vertebrate Ca^{2+}-binding proteins. *BMC Res Notes* **3:** 118.

McFerran BW, Graham ME, Burgoyne RD. 1998. NCS-1, the mammalian homologue of frequenin is expressed in chromaffin and PC12 cells and regulates neurosecretion from dense-core granules. *J Biol Chem* **273:** 22768–22772.

Mellstrom B, Naranjo JR. 2001. Ca^{2+}-dependent transcriptional repression and derepression: DREAM, a direct effector. *Seminars Cell Develop Biol* **12:** 59–63.

Mendez A, Burns ME, Sokal I, Dizhoor AM, Baehr W, Palczewski K, Baylor DA, Chen J. 2001. Role of guanylate cyclase-activating proteins (GCAPs) in setting the flash sensitivity of rod photoreceptors. *Proc Natl Acad Sci* **98:** 9948–9953.

Mercer WA, Korhonen L, Skoglosa Y, Olssen P-A, Kukknen JP, Lindholm D. 2000. NAIP interacts with hippocalcin and protects neurons against calcium-induced cell death through caspase-3-dependent and -independent pathways. *EMBO J* **19:** 3597–3607.

Mikhaylova M, Reddy PP, Munsch T, Landgraf P, Suman SK, Smalla KH, Gundelfinger ED, Sharma Y, Kreutz MR.

2009. Calneurons provide a calcium threshold for trans-Golgi network to plasma membrane trafficking. *Proc Natl Acad Sci* **106:** 9093–9098.

Mikhaylova M, Sharma Y, Reissner C, Nagel F, Aravind P, Rajini B, Smalla KH, Gundelfinger ED, Kreutz MR. 2006. Neuronal Ca^{2+} signaling via caldendrin and calneurons. *Biochim Biophys Acta* **1763:** 1229–1237.

Nef S, Fiumelli H, de Castro E, Raes M-B, Nef P. 1995. Identification of a neuronal calcium sensor (NCS-1) possibly involved in the regulation of receptor phosphorylation. *J Receptor Signal Trans* **15:** 365–378.

O'Callaghan DW, Tepikin AV, Burgoyne RD. 2003b. Dynamics and calcium-sensitivity of the Ca^{2+}-myristoyl switch protein hippocalcin in living cells. *J Cell Biol* **163:** 715–721.

O'Callaghan DW, Haynes LP, Burgoyne RD. 2005. High-affinity interaction of the N-terminal myristoylation motif of the neuronal calcium sensor protein hippocalcin with phosphatidylinositol 4,5-bisphosphate. *Biochem J* **391:** 231–238.

O'Callaghan DW, Hasdemir B, Leighton M, Burgoyne RD. 2003a. Residues within the myristoylation motif determine intracellular targeting of the neuronal Ca^{2+} sensor protein KChIP1 to post-ER transport vesicles and traffic of Kv4 K+ channels. *J Cell Sci* **116:** 4833–4845.

O'Callaghan DW, Ivings L, Weiss JL, Ashby MC, Tepikin AV, Burgoyne RD. 2002. Differential use of myristoyl groups on neuronal calcium sensor proteins as a determinant of spatio-temporal aspects of Ca^{2+}-signal transduction. *J BiolChem* **277:** 14227–14237.

O'Connor V, El Far O, Bofill-Cardona E, Nanoff C, Freissmuth M, Karschin A, Airas JM, Betz H, Boehm S. 1999. Calmodulin dependence of presynaptic metabotropic glutamate receptor signalling. *Science* **286:** 1180–1184.

Olafsson P, Wang T, Lu B. 1995. Molecular cloning and functional characterisation of the *Xenopus* Ca^{2+} binding protein frequenin. *Proc Natl Acad Sci* **92:** 8001–8005.

Palczewski K, Sokal I, Baehr W. 2004. Guanylate cyclase-activating proteins: Structure, function and diversity. *Biochem Biophys Res Comm* **322:** 1123–1130.

Palmer CL, Lim W, Hastie PG, Toward M, Korolchuk VI, Burbidge SA, Banting G, Collingridge GL, Isaac JT, Henley JM. 2005. Hippocalcin functions as a calcium sensor in hippocampal LTD. *Neuron* **47:** 487–494.

Pang ZP, Cao P, Xu W, Sudhof TC. 2010. Calmodulin controls synaptic strength via presynaptic activation of calmodulin kinase II. *J Neurosci* **30:** 4132–4142.

Paterlini M, Revilla V, Grant AL, Wisden W. 2000. Expression of the neuronal calcium sensor protein family in the rat brain. *Neurosci* **99:** 205–216.

Pennesi ME, Howes KA, Baehr W, Wu SM. 2003. Guanylate cyclase-activating protein (GCAP) 1 rescues cone recovery kinetics in GCAP1/GCAP2 knockout mice. *Proc Natl Acad Sci* **100:** 6783–6788.

Petko JA, Kabbani N, Frey C, Woll M, Hickey K, Craig M, Canfield VA, Levenson R. 2009. Proteomic and functional analysis of NCS-1 binding proteins reveals novel signaling pathways required for inner ear development in zebrafish. *BMC Neurosci* **10:** 27.

Pioletti M, Findeisen F, Hura GL, Minor DL. 2006. Three-dimensional structure of the KChIP1–Kv4.3 T1 complex reveals a cross-shaped octamer. *Nature Struct Mol Biol* **13:** 987–995.

Polans A, Baehr W, Palczewski K. 1996. Turned on by Ca2+! The physiology and pathology of Ca2+-binding proteins in the retina. *Trends Neurosci* **19:** 547–554.

Pongs O, Lindemeier J, Zhu XR, Theil T, Endelkamp D, Krah-Jentgens I, Lambrecht H-G, Koch KW, Schwemer J, Rivosecchi R, et al. 1993. Frequenin - A novel calcium-binding protein that modulates synaptic efficacy in the *Drosophila* nervous system. *Neuron* **11:** 15–28.

Pruunsild P, Timmusk T. 2005. Structure, alternative splicing, and expression of the human and mouse KCNIP gene family. *Genomics* **86:** 581–593.

Rhodes KJ, Carroll KI, Sung MA, Doliveira LC, Monaghan MM, Burke SL, Strassle BW, Buchwalder L, Menegola M, Cao J, et al. 2004. KChIPs and Kv4asubunitsSA as integral components of A-type potassium channels in mammalian brain. *J Neurosci* **24:** 7903–7915.

Rickman C, Jiménez JL, Graham ME, Archer DA, Soloviev M, Burgoyne RD, Davletov B. 2006. Conserved prefusion protein assembly in regulated exocytosis. *Mol Biol Cell* **17:** 283–294.

Rieke F, Lee A, Haeseleer F. 2008. Characterization of Ca^{2+}-binding protein 5 knockout mouse retina. *Investigative Ophthalmol Visual Sci* **49:** 5126–5135.

Rizo J, Rosenmund C. 2008. Synaptic vesicle fusion. *Nat Struct Mol Biol* **15:** 665–674.

Robinson IM, Ranjan R, Schwarz TL. 2002. Synaptotagmins I and IV promote transmitter release independently of Ca^{2+} binding in the C2A domain. *Nature* **418:** 336–340.

Saab BJ, Georgiou J, Nath A, Lee FJ, Wang M, Michalon A, Liu F, Mansuy IM, Roder JC. 2009. NCS-1 in the dentate gyrus promotes exploration, synaptic plasticity, and rapid acquisition of spatial memory. *Neuron* **63:** 643–656.

Saimi Y, Kung C. 2002. Calmodulin as an ion channel subunit. *Ann Rev Physiol* **64:** 289–311.

Sampath A, Strissel KJ, Elias R, Arshavsky VY, McGinnis JF, Rieke F, Hurley JB. 2005. Recoverin improves rod-mediated vision by enhancing signal transmission in the mouse retina. *Neuron* **46:** 413–420.

Sanchez-Gracia A, Romero-Pozuelo J, Ferrus A. 2010. Two frequenins in *Drosophila*: unveiling the evolutionary history of an unusual neuronal calcium sensor (NCS) duplication. *BMC Evolutionary Biol* **10:** 54.

Scannevin RH, Wang K-W, Jow F, Megules J, Kospco DC, Edris W, Carroll KC, Lu Q, Xu W, Xu Z, et al. 2004. Two N-terminal domains of Kv4 K+ channels regulate binding to and modulation by KChIP1. *Neuron* **41:** 587–598.

Schaad NC, De Castro E, Nef S, Hegi S, Hinrichsen R, Martone ME, Ellisman MH, Sikkink R, Sygush J, Nef P. 1996. Direct modulation of calmodulin targets by the neuronal calcium sensor NCS-1. *Proc Natl Acad Sci* **93:** 9253–9258.

Schiavo G, Stenbeck G, Rothman JE, Sollner TH. 1997. Binding of the synaptic vesicle v-SNARE, synaptotagmin, to the plasma membrane t-SNARE, SNAP-25 can explain docked vesicles at neurotoxin-treated synapses. *Proc Natl Acad Sci* **94:** 997–1001.

Schnurra I, Bernstein H-G, Riederer P, Braunewell K-H. 2001. The neuronal calcium sensor protein VILIP-1 is associated with amyloid plaques and promotes cell death and tau phosphorylation in vitro: A link between calcium sensors and alzheimers disease? *Neurobiology Disease* **8:** 900–909.

Schumacher MA, Rivard AF, Bachinger HP, Adelman JP. 2001. Structure of the gating domain of a Ca^{2+}-activated K^+ channel complexed with Ca^{2+}/calmodulin. *Nature* **410:** 1120–1124.

Seidenbecher CI, Landwehr M, Smalla K-H, Kreutz M, Dieterich DC, Zuschratter W, Reissner C, Hammarback JA, Bockers TM, Gundelfinger ED, et al. 2004. Caldendrin but not calmodulin binds to light chain 3 of MAP1A/B: An association with the microtubule cytoskeleton highlighting exclusive binding partners for neuronal Ca^{2+}-sensor proteins. *J Mol Biol* **336:** 957–970.

Seidenbecher CI, Langnaese K, Sanmarti-Vila L, Boeckers TM, Smalla K-H, Sabel BA, Garner CC, Gundelfinger ED, Kreutz MR. 1998. Caldendrin, a novel neuronal calcium-binding protein confined to the somatodendritic compartment. *J Biol Chem* **273:** 21324–21331.

Shao X, Fernandez I, Sudhof TC, Rizo J. 1998. Solution structures of the Ca^{2+}-free and Ca^{2+}-bound C2A domain of synaptotagmin I: Does Ca^{2+} induce a conformational change? *Biochem* **37:** 16106–16115.

Shih PY, Lin CL, Cheng PW, Liao JH, Pan CY. 2009. Calneuron I inhibits Ca^{2+} channel activity in bovine chromaffin cells. *Biochem Biophys Res Commun* **388:** 549–553.

Smalla K-H, Seidenbecher CI, Tischmeyer W, Schicknick H, Wyneken U, Bockers TM, Gundelfinger ED, Kreutz MR. 2003. Kainate-induced epileptic seizures induce a recruitment of caldendrin to the postsynaptic density in rat brain. *Mol Brain Res* **116:** 159–162.

Sollner T, Whiteheart SW, Brunner M, Erdjument-Bromage H, Geromanos S, Tempst P, Rothman JE. 1993. SNAP receptors implicated in vesicle targeting and fusion. *Nature* **362:** 318–324.

Spilker C, Dresbach T, Braunewell K-H. 2002. Reversible translocation and activity-dependent localisation of the calcium-myristoyl switch protein VILIP-1 to different membrane compartments in living hippocampal neurons. *J Neurosci* **22:** 7331–7339.

Spitzer NC. 2006. Electrical activity in early neuronal development. *Nature* **444:** 707–712.

Stephen R, Bereta G, Golczak M, Palczewski K, Sousa MC. 2007. Stabilizing function for myristoyl group revealed by the crystal structure of a neuronal calcium sensor, guanylate cyclase-activating protein 1. *Structure* **15:** 1392–1402.

Strahl T, Huttner IG, Lusin JD, Osawa M, King D, Thorner J, Ames JB. 2007. Structural insights into activation of phosphatidylinositol 4-kinase (pik1) by yeast frequenin (Frq1). *J Biol Chem* **282:** 30949–30959.

Sudhof TC, Rothman JE. 2009. Membrane fusion: Grappling with SNARE and SM proteins. *Science* **323:** 474–477.

Tadross MR, Dick IE, Yue DT. 2008. Mechanism of local and global Ca^{2+} sensing by calmodulin in complex with a Ca^{2+} channel. *Cell* **133:** 1228–1240.

Tanaka T, Ames JB, Harvey TS, Stryer L, Ikura M. 1995. Sequestration of the membrane targeting myristoyl group of recoverin in the calcium-free state. *Nature* **376:** 444–447.

Tang N, Lin T, Yang J, Foskett JK, Ostap EM. 2007. CIB1 and CaBP1 bind to the myo1c regulatory domain. *J Muscle Res Cell Motility* **28:** 285–291.

Tang J, Maximov A, Shin OH, Dai H, Rizo J, Sudhof TC. 2006. A complexin/synaptotagmin 1 switch controls fast synaptic vesicle exocytosis. *Cell* **126:** 1175–1187.

Taverna E, Francolini M, Jeromin A, Hilfiker S, Roder J, Rosa P. 2002. Neuronal calcium sensor 1 and phosphatidylinositol 4-OH kinase β interact in neuronal cells and are translocated to membranes during nucleotide-evoked exocytosis. *J Cell Sci* **115:** 3909–3922.

Tippens AL, Lee A. 2007. Caldendrin, a neuron-specific modulator of Cav/1.2 (L-type) Ca2+ channels. *J Biol Chem* **282:** 8464–8473.

Tsujimoto T, Jeromin A, Satoh N, Roder JC, Takahashi T. 2002. Neuronal calcium sensor 1 and activity-dependent facilitation of P/Q-type calcium channel currents at presynaptic nerve terminals. *Science* **295:** 2276–2279.

Venn N, Haynes LP, Burgoyne RD. 2008. Specific effects of KChIP3/calsenilin/DREAM but not KChIPs1, 2 and 4 on calcium signalling and regulated secretion in PC12 cells. *Biochem J* **413:** 71–80.

Wang H, Yan Y, Liu Q, Huang Y, Shen Y, Chen L, Chen Y, Yang Q, Hao Q, Wang K, et al. 2007. Structural basis for modulation of Kv4 K^+ channels by auxiliary KChIP subunits. *Nat Neurosci* **10:** 32–39.

Weiergraber OH, Senin II, Philippov PP, Granzin J, Koch K-W. 2003. Impact of N-terminal myristoylation on the Ca^{2+}-dependent conformational transition in recoverin. *J Biol Chem* **278:** 22972–22979.

Weiss JL, Archer DA, Burgoyne RD. 2000. NCS-1/frequenin functions in an autocrine pathway regulating Ca^{2+} channels in bovine adrenal chromaffin cells. *J Biol Chem* **275:** 40082–40087.

Wen H, Levitan IB. 2002. Calmodulin is an auxiliary subunit of KCNQ2/3 potassium channels. *J Neurosci* **22:** 7991–8001.

Wingard JN, Chan J, Bosanac I, Haeseleer F, Palczewski K, Ikura M, Ames JB. 2005. Structural analysis of Mg^{2+} and Ca^{2+} binding to CaBP1, a neuron-specific regulator of calcium channels. *J Biol Chem* **280:** 37461–37470.

Woods AS, Marcellino D, Jackson SN, Franco R, Ferre S, Agnati LF, Fuxe K. 2008. How calmodulin interacts with the adenosine A(2A) and the dopamine D(2) receptors. *J Proteome Res* **7:** 3428–3434.

Wu YQ, Lin X, Liu CM, Jamrich M, Shaffer LG. 2001. Identification of a human brain-specific gene, calneuron 1, a new member of the calmodulin superfamily. *Mol Genet Metab* **72:** 343–350.

Xia Z, Storm DR. 2005. The role of calmodulin as a signal integrator for synaptic plasticity. *Nat Rev Neurosci* **6:** 267–276.

Xia X-M, Fakler B, Rivard A, Wayman G, Johnson-Pais T, Keen JE, Ishii T, Hirschberg B, Bond CT, Lutsenko S, Maylie J, Adelman JP. 1998. Mechanisms of calcium gating in small conductance calcium activated potassium channels. *Nature* **395:** 503–507.

Xiong H, Xia K, Li B, Zhao G, Zhang Z. 2009. KChIP1: A potential modulator to GABAergic system. *Acta Biochim Biophys Sin (Shanghai)* **41:** 295–300.

Xue M, Ma C, Craig TK, Rosenmund C, Rizo J. 2008. The Janus-faced nature of the C(2)B domain is fundamental for synaptotagmin-1 function. *Nat Struct Mol Biol* **15:** 1160–1168.

Yamagata K, Goto K, Kuo CH, Kondo H, Miki N. 1990. Visinin: A novel calcium binding protein expressed in retianl cone cells. *Neuron* **4:** 469–476.

Yang S-N, Tang Y-G, Zucker RS. 1999. Selective induction of LTP and LTD by postsynaptic $[Ca^{2+}]_i$ elevation. *J Neurophysiol* **81:** 781–787.

Yang PS, Alseikhan BA, Hiel H, Grant L, Mori MX, Yang W, Fuchs PA, Yue DT. 2006. Switching of Ca^{2+}-dependent inactivation of Ca(V)1.3 channels by calcium binding proteins of auditory hair cells. *J Neurosci* **26:** 10677–10689.

Yang J, McBride S, Mak D-OD, Vardi N, Palczewski K, Haeseleer F, Foskett JK. 2002. Identification of a family of calcium sensors as protein ligands of inositol trisphosphate receptor Ca^{2+} release channels. *Proc Natl Acad Sci* **99:** 7711–7716.

Zeitz C, Kloeckener-Gruissem B, Forster U, Kohl S, Magyar I, Wissinger B, Matyas G, Borruat FX, Schorderet DF, Zrenner E, et al. 2006. Mutations in CABP4, the gene encoding the Ca^{2+}-binding protein 4, cause autosomal recessive night blindness. *Am J Hum Genet* **79:** 657–667.

Zhao X, Varnai P, Tuymetovna G, Balla A, Toth ZE, Oker-Blom C, Roder J, Jeromin A, Balla T. 2001. Interaction of neuronal calcium sensor-1 (NCS-1) with phosphatidylinositol 4-kinase β stimulates lipid kinase activity and affects membrane trafficking in COS-7 cells. *J Biol Chem* **276:** 40183–40189.

Zhou H, Kim S-A, Kirk EA, Tippens AL, Sun H, Haeseleer F, Lee A. 2004a. Ca^{2+}-binding protein-1 facilitates and forms a postsynaptic complex with $Ca_v1.2$ (L-type) Ca^{2+} channels. *J Neurosci* **24:** 4698–4708.

Zhou H, Yu K, McCoy KL, Lee A. 2005. Molecular mechanism for divergent regulation of $Ca_v1.2$ Ca^{2+} channels by calmodulin and Ca^{2+}-binding protein-1. *J BiolChem* **280:** 29612–29619.

Zhou W, Qian Y, Kunjilwar K, Pfaffinger PJ, Choe S. 2004b. Structural insights into the functional interaction of KChIP1 with shal-type K^+ channels. *Neuron* **41:** 573–586.

Regulation by Ca^{2+}-Signaling Pathways of Adenylyl Cyclases

Michelle L. Halls and Dermot M.F. Cooper

Department of Pharmacology, University of Cambridge, Cambridge CB2 1PD, United Kingdom

Correspondence: dmfc2@cam.ac.uk

Interplay between the signaling pathways of the intracellular second messengers, cAMP and Ca^{2+}, has vital consequences for numerous essential physiological processes. Although cAMP can impact on Ca^{2+}-homeostasis at many levels, Ca^{2+} either directly, or indirectly (via calmodulin [CaM], CaM-binding proteins, protein kinase C [PKC] or $G\beta\gamma$ subunits) may also regulate cAMP synthesis. Here, we have evaluated the evidence for regulation of adenylyl cyclases (ACs) by Ca^{2+}-signaling pathways, with an emphasis on verification of this regulation in a physiological context. The effects of compartmentalization and protein signaling complexes on the regulation of AC activity by Ca^{2+}-signaling pathways are also addressed. Major gaps are apparent in the interactions that have been assumed, revealing a need to comprehensively clarify the effects of Ca^{2+} signaling on individual ACs, so that the important ramifications of this critical interplay between Ca^{2+} and cAMP are fully appreciated.

OVERVIEW

Cyclic AMP (cAMP) and calcium (Ca^{2+}) are arguably the prototypical second messengers that control cellular homeostasis. Whereas, for instance, nitric oxide and cyclic GMP may serve essential functions in a number of physiological situations, cAMP and Ca^{2+} are the only truly ubiquitous second messengers. Significantly, it also happens that each of the mammalian adenylyl cyclases (ACs), which are the synthetic sources of cAMP, are potentially regulated by some aspect of the Ca^{2+}-signaling pathway—either directly by Ca^{2+} and/or calmodulin (CaM) or indirectly by CaM kinase (CaMK), protein kinase C (PKC), or calcineurin (CaN), all of which are potentially activated either when $[Ca^{2+}]i$ is increased, or as a result of stimulation of the phospholipase C (PLC) pathway (reviewed in Sunahara et al. 1996; Willoughby and Cooper 2007; Sadana and Dessauer 2009). In addition $\beta\gamma$ subunits of G-proteins liberated in response to $G\alpha_q$-coupled receptors can potentially regulate six of the nine membrane-bound AC species (reviewed in Sunahara et al. 1996; Willoughby and Cooper 2007; Sadana and Dessauer 2009). This susceptibility of cAMP production to regulation by the Ca^{2+}-signaling pathway may reflect a remnant control by Ca^{2+} over the presumed newer second messenger cAMP, a developmental sophistication or convergent evolution.[1] Whatever the origin of this interaction it is important to consider that Ca^{2+} is never elevated without a possible consequence—either positive or negative—for cAMP levels emanating from any of the ACs. Conversely, it is also noteworthy that cAMP itself impacts on Ca^{2+}-elevation at numerous levels—ranging from direct effects of cAMP on hyperpolarization-activated cyclic nucleotide-gated (HCN) channels and cyclic nucleotide-gated (CNG) channels, to effects of protein kinase A (PKA) and exchange protein directly activated by cAMP (EPAC) on numerous aspects of Ca^{2+}-signaling, including inositol trisphosphate ($InsP_3$) receptors (reviewed in Straub et al. 2004), voltage-gated Ca^{2+} channels (VGCCs) (reviewed in Dai et al. 2009), etc., so that nonlinearity and great complexity is to be the expected norm for the concentration profile of both messengers. An extension of this interaction may be that targets of these second messengers respond to not readily discernible integrals of their respective concentrations and certainly not to gross elevations or declines in the levels of the messengers at some cumulative time-point (which tends to be the experimentalists' approach). This notion elaborates on the proposal made almost 30 years ago, by Howard Rasmussen, of Ca^{2+} and cAMP as "synarchic" messengers (Rasmussen 1981). He, along with Michael Berridge, pointed out that the two systems were rarely independent but were often antagonistic, sometimes synergistic or occasionally redundant (Berridge 1975). Obviously, at the time that Rasmussen and Berridge were discussing synarchic messengers, there was no appreciation of the molecular identities or the multiplicities of any of the components and interactions between the two pathways at numerous early steps. Furthermore, the spatial and temporal complexity of which we are now aware was unknown, and so resolving the problem (or indeed understanding the potential) arising from the integration of these two ubiquitous second messengers is now infinitely more complex and challenging.

Consequently, if we are to seek to understand the role played by cAMP (or Ca^{2+}) it becomes essential to be able to think about both messengers in equivalent temporal and spatial dimensions. Given the additional developing recognition that cAMP signaling and ACs are highly organized within cells, it seems important to acknowledge that we really know very little about the detailed control by Ca^{2+} or cAMP of cellular processes. Until recently the types of evidence to be gathered to implicate cAMP in a process—actually first promulgated by Sutherland and Rall (Robison et al 1971) were (1) the hormone should stimulate AC in membranes, (2) the hormone should affect cAMP levels in intact cells, (3) inhibition of phosphodiesterase (PDE) should mimic the effect of putative cAMP-linked hormones, and (4) exogenous cAMP should mimic the effect of putative cAMP-linked hormones.[2] In the light of current knowledge we must now recognize that these are naïve, and in some cases, impossible conditions to fulfill, for reasons that will be expanded on in this article.

Against this backdrop, the major purpose of this review is to address the impact of Ca^{2+}-signaling on each of the mammalian ACs (1) this requires a serious assessment of the evidence for how all of the various potentially Ca^{2+}-regulated ACs are actually regulated as a consequence of activation of Ca^{2+}-signaling pathways, (2) an assessment of their physiological role in terms of their susceptibility to Ca^{2+}-regulation, (3) to summarize what is known about the spatial constraints that may be in place to ensure or preclude this regulation, which will include summarizing what is known of cAMP compartments (these should be viewed to be both dynamic and regulatable, in the manner

---

[1]Not only the synthesis, but also the degradation, of cAMP can be affected by $[Ca^{2+}]i$ as a consequence of the activation of one of the members of the phosphodiesterase family, PDE1. This enzyme is not widespread but it can influence cAMP levels where it is expressed.

[2]Application of exogenous cAMP to cells is not dissimilar to applying ionophore or sustained high K^+ depolarization to mimic the effect of physiological elevation of Ca^{2+}; all of the potential spatial and temporal information is cast aside. Such an experiment will only apparently "work" if the readout is so gross that the limitations are not apparent.

of, for instance, focal adhesion complexes), and (4) to argue for an assessment of changes in these second messengers in real time and comparable spaces. The major emphasis will be on the first point because this has not been seriously evaluated in recent years; the other points have received much more recent attention (reviewed in Hanoune and Defer 2001; Willoughby and Cooper 2007; Sadana and Dessauer 2009).

REGULATORY SUSCEPTIBILITIES OF ACs

Nine mammalian membrane-localized ACs have been identified. They share a common, complex 12-membrane-spanning architecture interspersed with two "Walker A" ATP binding motifs that associate to form the catalytic domain (Fig. 1).

Their similarity to the ATP binding cassette superfamily of transporters has been noted and discussed in the still unresolved context of whether they might function as transporters (Krupinski et al. 1989; Willoughby and Cooper 2007). Their largely conserved catalytic domains (C1a and C2a) has led to the conclusion, based on the crystal structure analysis of an AC5C1a/AC2C2a couple, that activation is achieved by promoting an open conformation of this domain (Tesmer et al. 1997; Tesmer et al. 1999; Mou et al. 2009). There is also a modicum of evidence that ACs may at least dimerize, if not assemble in higher order structures, and that the complex transmembrane architecture may permit or indeed assist such oligomerization, which has been speculated to provide an additional ability of the ACs to form associations with other transmembrane proteins (reviewed in Cooper and Crossthwaite 2006).

Based largely on sequence analysis and relatedness, dendrogram presentation of the nine membrane-bound ACs has facilitated their division into four families: 1, 3, and 8; 2, 4, and 7; 5 and 6; and 9 (Fig. 2).

Studies expressing the cloned lead family members (but not all species, see later) in vitro suggested a pattern of regulatory susceptibilities that concurred with this familial organization as displayed in Table 1. This represents a consensus between earlier and recent reviews that reflects the widely accepted regulation of these enzymes.

A cursory examination of Table 1 reveals that all of the nine membrane-bound ACs are potentially regulated as a consequence of the activation of Ca^{2+}-signaling pathways—either directly by Ca^{2+}, or Ca^{2+}-bound CaM, slightly less directly by CaMK or CaN, or indirectly by PKC, or by $\beta\gamma$ subunits of G-proteins released by $G\alpha_q$-linked GPCRs (Fig. 3). We have noted this potential interaction previously (Cooper et al. 1995; Willoughby and Cooper 2007) and proposed that it formed a major opportunity

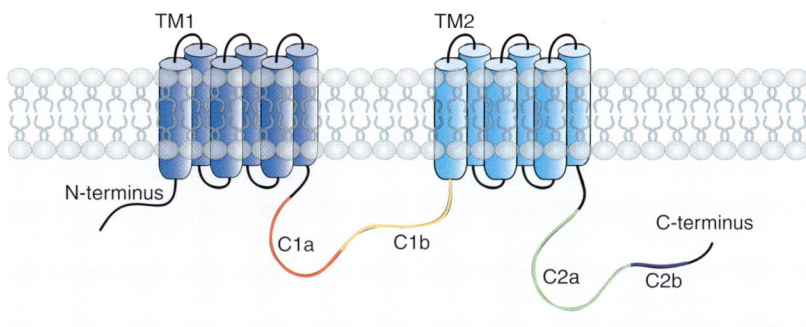

Figure 1. General structural domains of the nine membrane-bound ACs. Each of the nine membrane-bound ACs consist of two transmembrane clusters (TM1 and TM2) each consisting of six membrane-spanning domains. TM1 and TM2 are joined by an intracellular loop containing the C1a and C1b regions. Following TM2 is a long intracellular tail, containing the C2a and C2b regions before the carboxyl terminus.

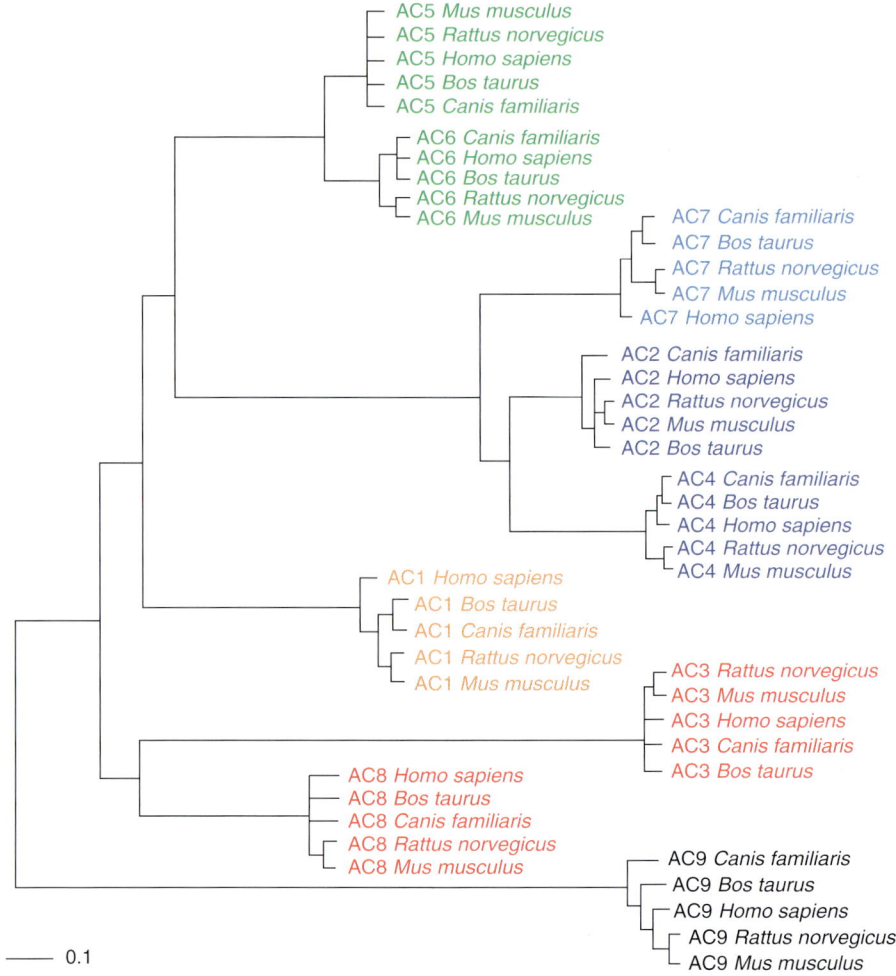

Figure 2. Phylogenetic tree of the nine membrane-bound ACs. The sequences of the nine membrane-bound ACs, from five species (human, rat, mouse, dog, and cow) were analyzed for relatedness and a phylogenetic tree was constructed using the Phylogeny.fr server (Castresana 2000; Guindon and Gascuel 2003; Edgar 2004; Anisimova and Gascuel 2006; Chevenet et al. 2006; Dereeper et al. 2008). The branch length is proportional to the number of substitutions per site.

for harmonizing the activities of the two systems in a realization and extension of the earlier discussions of Rasmussen and Berridge.

Because of the potential importance of these interactions, a major purpose of this review is to assess how convincing it is that any of these modes of regulation are actually used—particularly in any physiological context. We therefore performed a detailed review of the literature, and then assessed the evidence in a rather objective but unweighted manner, for the forms of regulation outlined in Table 1. An obvious caveat for this style of analysis is that on a rigorous critical review, the findings from one original research paper may outweigh the conclusions presented in multiple, less robust experimental studies. Nevertheless, here, we are merely looking for a consistency and reproducibility in the existing evidence for the different forms of regulation of ACs by Ca^{2+}-signaling pathways. The literature was categorized based on the experimental design of the study, and thus designated

Table 1. The "consensus" regulation of ACs by the Ca^{2+}-signaling pathway.[a]

Isoform	Type of regulation			
	Ca^{2+}	CaMK/CaN	PKC	Gβγ
AC1	Activation (CaM)	Inhibition (CaMKIV)	No Effect/Activation	Inhibition
AC2	No Effect		Activation	Activation
AC3	Activation (CaM)	Inhibition (CaMKII)	No Effect/Activation	No Effect/Inhibition
AC4	No Effect		Activation	Activation
AC5	Inhibition		Activation	No Effect/Activation
AC6	Inhibition		Activation/Inhibition	No Effect/Activation
AC7	No Effect		Activation	No Effect/Activation
AC8	Activation (CaM)		No Effect	Inhibition
AC9		Inhibition (CaN)	Inhibition	No Effect

Some ACs have two effects listed for a particular form of Ca^{2+} regulation—this is because of differing conclusions drawn from the two reviews from which the table was compiled.

[a]The table lists the regulation of each of the nine membrane-bound ACs by Ca^{2+}, either directly or indirectly via CaMK, CaN, PKC, or Gβγ. This consensus view of AC regulation was compiled from two recent and comprehensive reviews: Willoughby and Cooper (2007) and Sadana and Dessauer (2009), and the conclusions do not differ significantly from earlier summaries.

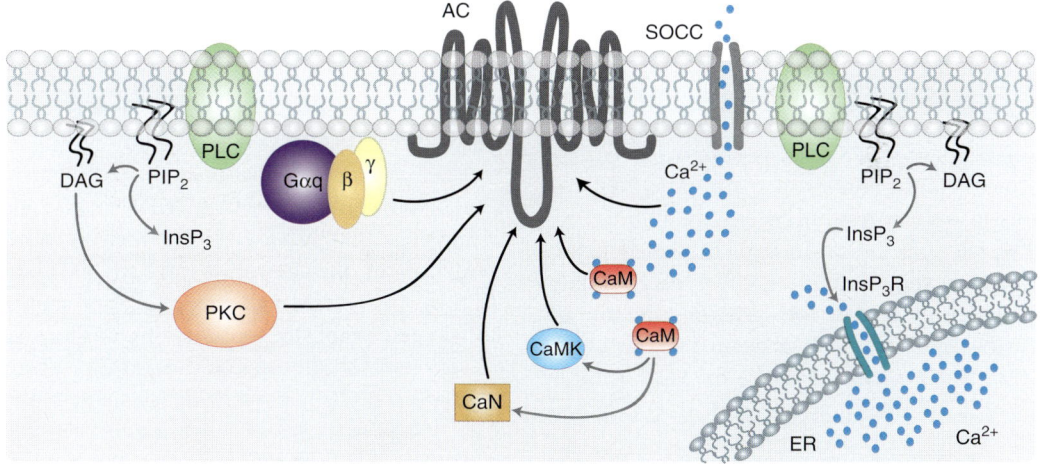

Figure 3. Regulation of ACs by Ca^{2+}. Ca^{2+} can directly, and indirectly regulate the nine membrane-bound ACs. Submicromolar [Ca^{2+}] can directly regulate AC, and some ACs specifically respond to Ca^{2+} derived from capacitative Ca^{2+} entry (CCE) from store-operated Ca^{2+} channels (SOCCs). Ca^{2+} can also regulate AC by binding calmodulin (CaM), and the Ca^{2+}/CaM complex can then affect AC activity. Ca^{2+}-bound CaM can also activate Ca^{2+}/calmodulin-activated kinase (CaMK) and calcineurin (CaN), both of which may regulate AC. More indirectly, Gβγ subunits from Gα$_q$ linked receptors can also regulate AC activity. In addition, Gα$_q$ can activate phospholipase C (PLC), which converts phosphatidylinositol 4,5-bisphosphate (PIP$_2$) to diacylglycerol (DAG) and inositol trisphosphate (IP$_3$). DAG activates protein kinase C (PKC), which can also modulate the activity of AC; InsP$_3$ binds to and activates its receptors (InsP$_3$R) on the endoplasmic reticulum (ER), thereby releasing Ca^{2+} from the ER stores into the cytoplasm. This emptying of the ER Ca^{2+} stores triggers extracellular Ca^{2+} entry by SOCCs.

as conducted in either overexpression or endogenous systems. Following this primary classification, the data was further defined as either binding- or phosphorylation-based, biochemical or pharmacological evidence.

In this analysis we draw attention to the level of proof that has been achieved and highlight the need for clarification of these issues to begin to understand this most critical aspect of cellular signaling.

REGULATION OF ACs BY Ca^{2+}-SIGNALING PATHWAYS

Direct Regulation by Ca^{2+}

The nine AC isoforms are classically grouped according to the ability of Ca^{2+} to regulate their activity. It has been accepted that submicromolar Ca^{2+} via CaM activates AC1, AC3, and AC8, whereas submicromolar Ca^{2+} alone inhibits AC5 and AC6, but has no effect on AC2, AC4, AC7, or AC9. The forms of evidence for Ca^{2+} regulation of each AC are detailed in Table 2.

Inhibition of ACs by Ca^{2+}: AC5 and AC6

The paradigm for Ca^{2+}-mediated inhibition of AC5 and AC6 in the submicromolar range[3] (independently of CaM) is supported largely by biochemical studies of membrane preparations in overexpression systems (Guillou et al. 1999; Hu et al. 2002). Indeed the crystal structure of a high affinity Ca^{2+}-pyrophosphate (PPi) complex with the catalytic domain of AC5 has recently been presented (Mou et al. 2009). Much of the evidence for Ca^{2+} inhibition of AC6 comes from experiments conducted with whole cell overexpression systems, although there are also a good number of papers reporting modulation of AC6 activity by Ca^{2+} in an endogenous setting (Boyajian et al. 1989; Garritsen et al. 1992; Yu et al. 1993; Grunberger et al. 2006; Tang et al. 2008). Although the substantiation of Ca^{2+} inhibition of the highly similar AC5 is less extensive (six studies in overexpression systems, and three in endogenous systems), it is uniform in its contention. Thus, there is good evidence for the inhibition of both AC5 and AC6 by Ca^{2+}, and the conclusion can be considered uncontentious both in vitro and in vivo.

Direct Activation of ACs by Ca^{2+}/CaM: AC1, AC3, and AC8

The most evidence for Ca^{2+} regulation of AC activity exists for the stimulation of AC1 and AC8 via CaM. The majority of this evidence comes from overexpression systems, using purified proteins, membrane preparations, or whole cells. Despite this, the direct binding of CaM to AC1 or AC8 has only been shown a few times for each enzyme (Vorherr et al. 1993; Gu and Cooper 1999; Masada et al. 2009). More importantly, and almost uniquely, there exists for these well-studied ACs, evidence of Ca^{2+}/CaM regulation in endogenous systems; much of this work was facilitated by the generation of AC1 and AC8 knockout mice, and the abundant expression of these isoforms in brain. However, a majority of this evidence still relies on experiments conducted in isolated membranes, with only two (AC1) or three (AC8) studies using intact primary cells (Villacres et al. 1998; Watson et al. 2000; Cioffi et al. 2002; Trubey et al. 2006; Wang et al. 2007). Nevertheless, there is robust evidence for the activation of both AC1 and AC8 by Ca^{2+}/CaM and this regulation can be considered established both in vitro and in vivo.

Far fewer studies have addressed Ca^{2+}/CaM regulation of AC3; some evidence suggests that Ca^{2+}/CaM activates AC3, although the effect is nothing like as pronounced as for AC1 and AC8, and is also conditional on AC3 activation. Although the effect is supported in endogenous systems (Mamluk et al. 1999; Hoffert et al. 2005), the available evidence is not as uniform as it is for Ca^{2+}/CaM activation of AC1 and AC8. In fact one study reports only minimal AC3 activation with high concentrations of Ca^{2+} (above the submicromolar concentrations that affect AC1, AC5, AC6, and AC8), and AC3 activation does not occur in the intact cell in

---

[3]All ACs are inhibited by Ca^{2+} in the submillimolar concentration range; this is believed to be because of a simple competition between Ca^{2+} and Mg^{++} at the catalytic site (Cooper 1991).

Table 2. Evidence for direct regulation of ACs by Ca^{2+}.[a]

Isoform	Action	Binding	Overexpression systems		Endogenous systems		Conclusions
			Biochemical evidence	Pharmacological evidence	Biochemical evidence	Pharmacological evidence	
AC1	Binding	+ CaM					Activation
	Activation		+++	+++	+++	+	
AC2	No Effect		++			+	No effect
AC3	No Effect		+	+			?
	Activation		+	+	+	+	
AC4	No Effect		+				No effect
AC5	Inhibition		++	+	+	+	Inhibition
AC6	Activation			+			Inhibition
	Inhibition		++	+++	+	+	
AC7	No Effect		+				No effect
AC8	Binding	+ CaM					Activation
	Activation		+++	+++	+++	++	
AC9	No Effect		+	+			No Effect

The accepted paradigm for Ca^{2+} regulation of each AC is listed in the first column, followed by the type of evidence provided in the literature: + 1–2 papers, ++ 3–5 papers, and +++ >6 papers. Overexpression systems refer to generic cell lines overexpressing recombinant proteins, and endogenous systems refer to primary cell lines or tissues endogenously expressing the AC of interest. Binding-based studies include coimmunoprecipitation or pull-down experiments; biochemical evidence refers to experiments conducted with membrane preparations, and pharmacological evidence refers to experiments conducted using intact cells. The conclusions that can be reasonably drawn following this analysis are listed in the final column: larger type indicates a uniform conclusion from >2 papers, smaller type indicates either a uniform conclusion from ≤2 papers or contradictory evidence substantially outweighed by the conclusion, ? indicates contradictory evidence that prevents a conclusion. (In this and other tables, "papers" refers to original experimental research papers that reported an effect, and not to the number of subsequent citations.)

[a]The table details the volume and type of evidence in the literature for direct regulation of ACs by Ca^{2+}, or by Ca^{2+}/CaM.

response to capacitative Ca^{2+} entry (CCE) unlike the case for AC1 and AC8 (Fagan et al. 1996).

AC2, AC4, AC7, and AC9 Are Not Directly Regulated by Ca^{2+}

The extent of verification of the Ca^{2+}-insensitivity of the remaining ACs is fairly even. A few highly cited papers, and one conducted in an endogenous setting (Feinstein et al. 1991; Lustig et al. 1993; Guillou et al. 1999; Hu et al. 2002) report no effect of Ca^{2+} on membrane preparations overexpressing AC2. Similarly, only one (though highly cited) study (Gao and Gilman 1991) details no effect of Ca^{2+} on membrane preparations overexpressing AC4. The lack of effect of Ca^{2+} on AC7 is again reported by only one paper (Crossthwaite et al. 2005), with the study conducted in membrane preparations overexpressing AC7. For AC9, the evidence is derived from two studies in overexpression systems, one of which is also highly cited (Premont et al. 1996). Thus although the evidence presented in these studies is uniform, it is sparse and predominantly derived from membrane preparations of overexpression systems. Nevertheless it agrees with the "inferred" view.

Indirect Ca^{2+} Regulation via CaMKII, CaMKIV, and CaN

Ca^{2+} can also indirectly, via CaM, regulate ACs via the protein kinases CaMKII, CaMKIV, or the protein phosphatase, CaN (Table 3). Of the nine membrane-bound AC isoforms, only AC1, AC3, and AC9 have been studied in any detail in terms of their regulation by these Ca^{2+}/CaM dependent proteins.

Table 3. Evidence for CaMKII, CaMKIV, or CaN regulation of ACs.[a]

Isoform	Action	Overexpression systems			Endogenous systems			Conclusions
		Phos.	Biochemical evidence	Pharmacological evidence	Phos.	Biochemical evidence	Pharmacological evidence	
AC1	Phos. (CaMKIV)	+						
	No Effect (CaMKII)							No Effect (CaMKII)
	Inhibition (CaMKIV)		+	+				Inhibition (CaMKIV)
AC2								
AC3	Phos.	+			+			
	Inhibition (CaMKII)			++				Inhibition (CaMKII)
AC8	No Effect (CaMKII and CaMKIV)			+			++	No Effect (CaMKII and CaMKIV)
AC9	Activation (CaMKII)			+				Activation (CaMKII)
	Inhibition (CaN)			++			+	Inhibition (CaN)

The accepted paradigm for regulation of each AC is listed in the first column, followed by the type of evidence provided in the literature: + 1–2 papers, ++ 3–5 papers, and +++ >6 papers. AC4, AC5, AC6, and AC7 do not appear on this table, as no evidence can be found for their regulation by CaN, or CaMK. Overexpression systems refer to generic cell lines overexpressing recombinant proteins, and endogenous systems refer to primary cell lines or tissues endogenously expressing the AC of interest. Phosphorylation-based studies include ^{32}P-incorporation experiments, biochemical evidence refers to experiments conducted with membrane preparations, and pharmacological evidence refers to experiments conducted using intact cells. The conclusions that can be reasonably drawn following this analysis are listed in the final column: large type indicates a uniform conclusion from >2 papers, smaller type indicates either a uniform conclusion from ≤2 papers or contradictory evidence substantially outweighed by the conclusion. Phos., indicates phosphorylation.
[a]The table details the volume and type of evidence in the literature for regulation of ACs by CaMKII, CaMKIV or CaN.

Table 4. Evidence for PKC regulation of ACs.[a]

Isoform	Action	Overexpression systems			Endogenous systems			Conclusions
		Binding or Phos.	Biochemical evidence	Pharmacological evidence	Binding or Phos.	Biochemical evidence	Pharmacological evidence	
AC1	No Effect			+				?
	Inhibition			+				
AC2	Phos.	++ P	+++	+++	+ P		+	Activation
AC3	Activation			+++			+	Activation
	Activation			+				
AC4	Binding		+	+	+ B			?
	No Effect			+				
	Activation		+					
	Inhibition							
AC5	Phos.	+ P		+				Activation
	No Effect		+	++			+	
	Activation							
AC6	Binding or Phos.	+ B,P	++	+				Inhibition
	No Effect			+				
	Inhibition							
AC7	Phos.	+ P	+	++			+	Activation
	Activation							
AC8								
AC9	Inhibition			+				Inhibition

The accepted paradigm for PKC regulation of each AC is listed in the first column, followed by the type of evidence provided in the literature: + 1–2 papers, ++ 3–5 papers, and +++ >6 papers. Overexpression systems refer to generic cell lines overexpressing recombinant proteins, and endogenous systems refer to primary cell lines or tissues endogenously expressing the AC. Binding- or phosphorylation-based studies include coimmunoprecipitation, pull-down or ^{32}P-incorporation experiments; biochemical evidence refers to experiments conducted with membrane preparations, and pharmacological evidence refers to experiments conducted using intact cells. The conclusions that can be reasonably drawn following this analysis are listed in the final column: large type indicates a uniform conclusion from >2 papers, smaller type indicates either a uniform conclusion from ≤2 papers or contradictory evidence substantially outweighed by the conclusion, ? indicates contradictory evidence. B, indicates binding evidence, Phos. or P, indicates evidence of phosphorylation.

[a]The table details the volume and type of evidence in the literature for regulation of ACs by PKC.

Inhibition of AC1 and AC3 by CaMKs

Most evidence exists for the inhibition of AC3 by CaMKII; in both overexpression (Wei et al. 1996), and endogenous systems (Wei et al. 1998), CaMKII has been shown to phosphorylate AC3 at Ser^{1076} thus inhibiting cAMP synthesis. However, further support for this assertion solely depends on the use of CaMKII inhibitors and concurrent demonstration of AC3 expression. Nevertheless, the uniformity of reports does suggest that AC3 is inhibited by CaMKII. Much less evidence exists for the inhibition of AC1 by CaMKIV; only one paper (in an overexpression system) shows phosphorylation of AC1 by CaMKIV, and subsequent inhibition of Ca^{2+}/CaM-stimulated AC1 following CaMKIV overexpression (Wayman et al. 1996). The same study also shows a lack of effect of CaMKII on AC1, and no effect of CaMKII or CaMKIV on AC8-stimulated cAMP.

AC9: Activation by CaMKII and Inhibition by CaN

Five studies in whole cells overexpressing AC9, and one in primary rat anterior pituitary corticotrophs (Antoni et al. 2003), report inhibition of this AC by CaN. All of these studies depend on the use of CaN inhibitors, combined with AC9 expression. Only one study has provided evidence for CaMKII-mediated potentiation of cAMP generated following stimulation of AC9, which again relied on the use of CaM and CaMKII inhibitors (Cumbay and Watts 2005). Thus although there is conformity within these studies, suggesting activation of AC9 by CaMKII but inhibition via CaN, the extent of this support is quite sparse.

Indirect Regulation by PKC

The elevation of PKC activity is one consequence of stimulation of PLC by GPCRs, in addition to the production of InsP$_3$ and Ca^{2+}-mobilization (Fig. 3). Much evidence exists for PKC activation of AC2, although the majority of this evidence comes from either membrane preparations or whole cells overexpressing the enzyme, with activation of PKC induced by phorbol esters and then often precluded using PKC inhibitors. Importantly there are also four studies, one from tissue endogenously expressing AC2 (Chakrabarti et al. 1998), that show increased phosphorylation of the AC in response to PKC activation (Jacobowitz and Iyengar 1994; Zimmermann and Taussig 1996; Böl et al. 1997a; Böl et al. 1997b). There is less, and conflicting, evidence for PKC activation of AC4; all conducted in overexpression systems, two studies show no effect (Jacobowitz et al. 1993; Zimmermann and Taussig 1996), one shows inhibition (Zimmermann and Taussig 1996) and another stimulation (Marjamaki et al. 1997), depending on the coactivator used. However one study shows binding of PKC to AC4 in an endogenous setting (Rhim et al. 2006). For PKC activation of AC7, there is only a small amount of evidence, although the reports are all consistent, and there is good evidence for both phosphorylation of AC7 in an overexpression system (Nelson et al. 2003), and activation of endogenously expressed enzyme (Haslauer et al. 1998; Antoni et al. 2003; Lariviere et al. 2007). Thus although a detailed review of the literature clearly confirms the prevailing view of activation of both AC2 and AC7 by PKC, the evidence regarding PKC-activation of AC4 is contradictory and does not permit a concrete conclusion.

The Remaining AC Isoforms

The Ca^{2+}-inhibitable AC5 and AC6 have also been studied in terms of their PKC susceptibility. The evidence is conflicting for the regulation of both ACs by PKC, despite data showing binding and/or phosphorylation of the enzymes (Kawabe et al. 1994; Iwami et al. 1995; Lai et al. 1997; Lin et al. 2002; Rhim et al. 2006). Consequently, a detailed review of the literature does not allow a simple conclusion regarding the regulation of AC5 or AC6 by PKC.

Even less evidence addresses the PKC regulation of AC1, AC3, AC8, and AC9. Two different studies report conflicting effects of PKC on AC1 (Jacobowitz et al. 1993; Yoshimura and Cooper 1993). A similar paucity of evidence exists for activation of AC3 by PKC, despite consistency of these reports, and additional evidence for PKC activation of AC3 in an endogenous setting (Mamluk et al. 1999). There are no reports of an effect of PKC on AC8, and only two studies report PKC-mediated inhibition of AC9 (Cumbay and Watts 2004; Cumbay and Watts 2005). Although the volume of reports is sparse, it appears likely that PKC can activate AC3, but inhibits AC9. The conflicting nature of the evidence regarding PKC regulation of AC1, and the lack of reports regarding potential regulation of AC8, prevent any conclusions regarding the PKC regulation of these two ACs.

Indirect Ca^{2+} Regulation via G$\beta\gamma$

The ACs classically designated as PKC-activated, AC2, AC4, and AC7, are also defined by their conditional activation by G$\beta\gamma$ subunits. Similarly, the group of ACs classically activated by Ca^{2+}/CaM (AC1, AC3, and AC8), are concurrently defined by G$\beta\gamma$-mediated inhibition. A detailed review of the literature regarding G$\beta\gamma$ regulation of ACs is summarized in Table 5.

ACs Conditionally Activated by G$\beta\gamma$: AC2, AC4, and AC7

The most evidence for G$\beta\gamma$ regulation of AC, in terms of both volume of studies and citation record, refers to the activation of AC2 by G$\beta\gamma$ subunits. There are 23 studies in isolated membranes or whole cells overexpressing AC2 that document this activation, and some of these are very frequently cited (Tang and Gilman 1991; Federman et al.1992). An additional four studies show direct G$\beta\gamma$ binding to AC2, and this has also been reported in endogenous expression systems (Wang et al. 2005; Wang and Burns 2006). In stark contrast, there is much less evidence for the activation of AC4 by G$\beta\gamma$ subunits, with only two studies conducted in overexpression systems (Gao and Gilman 1991; Marjamaki et al. 1997). However, evidence for both binding and activation in an endogenous setting also exists (Belevych et al. 2001; Wang et al. 2005; Wang and Burns 2006). The scant evidence for AC7 activation by G$\beta\gamma$ subunits consists of one well-cited study in whole cells overexpressing the enzyme (Yoshimura et al. 1996). Thus there is a substantial volume and consistency of studies reporting activation of AC2 by G$\beta\gamma$ subunits, and this conclusion appears to hold for AC4 and AC7, despite a much narrower evidence base.

Inhibition of AC1, AC3, and AC8 by G$\beta\gamma$

The second-most verified G$\beta\gamma$-mediated regulation of AC, is the inhibition of AC1 by G$\beta\gamma$ subunits. Eleven studies document this inhibition in overexpression systems, in addition to one that reports the direct binding of G$\beta\gamma$ to the enzyme (Weitmann et al. 2001). Furthermore there is a highly cited paper that shows inhibition of ACs by G$\beta\gamma$ in membranes from brain (Chen et al. 1995), although this does not distinguish between AC1 and AC8. Apart from this study, only one other documents AC8 inhibition by G$\beta\gamma$ subunits, which was conducted in whole cells overexpressing the enzyme (Steiner et al. 2006). The evidence for AC3 regulation by G$\beta\gamma$ is conflicting and sparse; two studies in overexpression systems have reported no effect (Tang and Gilman 1991; Marjamaki et al. 1997), whereas another reports inhibition (Diel et al. 2006). There is no evidence from endogenous systems. Thus although the inhibition of AC1 by G$\beta\gamma$ represents a robust paradigm, G$\beta\gamma$ mediated inhibition of AC8 is less substantiated, and no firm conclusions can be made regarding the regulation of AC3 by G$\beta\gamma$ subunits.

Effects of G$\beta\gamma$ on AC5, AC6, and AC9

There is conflicting evidence for an effect of G$\beta\gamma$ subunits on both AC5 and AC6, with no effect, activation and inhibition all reported. Binding of G$\beta\gamma$ subunits to AC5 and AC6 has been reported in both overexpression (Sadana et al. 2009) and endogenous systems (Gao et al. 2007), although the latter involved a yeast-two hybrid screen of a mouse brain cDNA library. The most highly cited papers report the activation of AC5 by G$\beta\gamma$ subunits (Avidor-Reiss et al. 1996), and no effect of G$\beta\gamma$ on AC6 (Premont et al. 1992); however, these reports are

Table 5. Evidence for Gβγ regulation of ACs.[a]

Isoform	Action	Overexpression systems			Endogenous systems			Conclusions
		Binding	Biochemical evidence	Pharmacological evidence	Binding	Biochemical evidence	Pharmacological evidence	
AC1	Binding	+						
	Inhibition		+++					Inhibition
AC2	Binding	++						
	Activation		+++	+++			+	Activation
AC3	No Effect		+	+				
	Inhibition		+					?
AC4	Binding							
	Activation		+	+		+		Activation
AC5	Binding	+			+			
	No Effect		+					
	Activation		+	+				
	Inhibition			+				?
AC6	Binding	+			+			
	No Effect		+					
	Activation		+					
	Inhibition			+				?
AC7	Activation			+		+		Activation
AC8	Inhibition			+				Inhibition
AC9	No Effect		+					No Effect

The accepted paradigm for Gβγ regulation of each AC is listed in the first column, followed by the type of evidence provided in the literature: + 1–2 papers, ++ 3–5 papers, and +++ >6 papers. Overexpression systems refer to generic cell lines overexpressing recombinant proteins, and endogenous systems refers to primary cell lines or tissues endogenously expressing the AC of interest. Binding-based studies include coimmunoprecipitation or pull-down experiments; biochemical evidence refers to experiments conducted with membrane preparations, and pharmacological evidence refers to experiments conducted using intact cells. The conclusions that can be reasonably drawn following this analysis are listed in the final column: large type indicates a uniform conclusion from >2 papers, smaller type indicates either a uniform conclusion from ≤2 papers or contradictory evidence substantially outweighed by the conclusion, ? indicates contradictory evidence.

[a]The table details the volume and type of evidence in the literature for regulation of ACs by Gβγ.

contradicted by an additional five studies. Although the reports for no effect of Gβγ subunits on AC9 activity are consistent, only two studies have shown this, both in overexpression systems (Premont et al. 1996; Cumbay and Watts 2005). Hence no concrete statements can be made regarding the effects of Gβγ on regulation of AC5 or AC6. In contrast, there appears to be no effect of Gβγ on AC9.

OVERALL SUMMARY OF EVIDENCE FOR THE REGULATION OF Ca^{2+}-SIGNALING PATHWAYS

Table 6 summarizes the evidence for direct and indirect regulation of ACs by Ca^{2+} emanating from our assessment of the literature.[4] It seems clear that AC5 and AC6 are inhibited by Ca^{2+}, AC1 and AC8 are activated by Ca^{2+}/CaM, AC2 is activated by PKC and Gβγ, and AC1 is inhibited by Gβγ. Further, it is likely that AC3 is conditionally activated by Ca^{2+}/CaM and activated by PKC but inhibited by CaMKII, AC5 is activated by PKC, AC6 is inhibited by PKC, AC7 is activated by PKC, and AC9 is inhibited by CaN. The regulation of the remaining ACs by Ca^{2+}, PKC and Gβγ remains uncertain.[5]

The initial classification of the nine membrane-bound ACs into subgroups based on sequence similarities, while initially useful and constructive on a gross scale, is perhaps too vague for the amount and type of evidence present in the literature, at least in terms of their susceptibility to regulation by Ca^{2+}.[6] Thus although AC1, AC3, and AC8 are activated to varying extents by Ca^{2+}/CaM, AC1 is also inhibited by Gβγ subunits, and AC3 is inhibited by CaMKII and activated by PKC. Similarly,

although both AC5 and AC6 are inhibited by Ca^{2+}, they have opposing regulation by PKC (AC5 is activated, whereas AC6 is inhibited). Even more surprisingly, although AC2 is activated by PKC and Gβγ, only AC7 (but not AC4) is also activated by PKC, and neither AC4 nor AC7 appears to be activated by Gβγ.

An additional confounding factor in this analysis is the likely sensitivity of the ACs to particular isoforms of the regulating proteins, specifically in the case of regulation by PKC and Gβγ subunits. It would not be surprising for an AC to be differentially regulated by distinct isoforms of PKC, or combinations of Gβγ subunits. Thus some of the more conflicting evidence compiled here may be resolved by clarifying the precise PKC and Gβγ entities that were used, because studies using unidentified PKC or Gβγ preparations can obviously be ambiguous. Additionally, the background cell type will be a critical determinant in susceptibility of the AC isoforms to Ca^{2+}-signaling pathways. This will not only influence the available complement of regulatory proteins, including varied PKC isoforms or Gβγ subunit combinations, but may also dictate the AC response to a particular stimulus because of differential compartmentalization of Ca^{2+}-signaling pathway components. Again, resolution of the effects of cell type on the regulation of AC activity may further illuminate the analysis conducted here.

Nevertheless, although the current state-of-the-art classification of the nine membrane-bound ACs, based on amino acid sequences, does generally apply on a gross scale, it becomes significantly less concrete, and quite sparse, when examined in detail. Consequently there is a strong case, given the potential importance of the regulation, for a systematic and controlled analysis of each AC.

PHYSIOLOGICAL ROLES FOR THE Ca^{2+}-DEPENDENCY OF ACs

Of course, even when we are convinced that these enzymes may be regulated as outlined above in vivo is there any evidence that this regulation is used or is an important part of the physiological role of these ACs?

---

[4]We intended to be relatively comprehensive in this review, although we have missed some studies, but hopefully not with any unrepresentative consequences.

[5]In some cases the ACs have not been studied in sufficient detail, whereas in other cases the situation is indeed far from certain based on conflicting studies, as outlined in the literature review.

[6]It should be recognized that the classification of the ACs into families based on sequence similarity leaves ample room for significant differences between members of the same family.

Table 6. Evidence-based summary of the regulation of ACs by the Ca^{2+}-signaling pathway.[a]

Isoform	Type of regulation			
	Ca^{2+}	CaMK/CaN	PKC	Gβγ
AC1	**Activation (CaM)**	–	?	Inhibition
AC2	No Effect		**Activation**	**Activation**
AC3	?	Inhibition (CaMKII)	Activation	?
AC4	–		?	–
AC5	**Inhibition (Direct)**		Activation	?
AC6	**Inhibition (Direct)**		Inhibition	?
AC7	–		Activation	–
AC8	**Activation (CaM)**	–		–
AC9	–	Inhibition (CaN)	–	–

Large text indicates strong support for the regulation, small text indicates consistent but weaker support, – indicates that although uniform, the limited evidence available prevents concrete conclusions, and ? indicates that the available evidence is conflicting and prevents any robust conclusions.

[a]The table shows the effects of Ca^{2+}, CaM, CaMK, CaN, PKC, and Gβγ on the nine membrane-bound ACs that can be robustly supported by the available literature.

AC1 is exclusively neuronal in its disposition and so whether it is regulated in an endogenous setting is not simple to resolve. Knock-out studies suggest a clear role in various models of learning and memory, addiction, stress-responses, and pain. In particular, AC1 knockout mice show decreased Ca^{2+}/CaM stimulated AC activity in the hippocampus and cerebellum, which correlates with reduced long-term potentiation (Wu et al. 1995; Storm et al. 1998). These studies clearly indicate an essential physiological role for AC1, but its Ca^{2+}-sensitivity is best attested to by the exuberant axon model, in which appropriate pathfinding does not occur in the absence of neuronal excitability, Ca^{2+}-entry or in AC1 knockout mice (Nicol et al. 2007). The latter study strongly suggests that stimulation of AC1 by Ca^{2+} does occur and that it is essential for the effects of cAMP in that context. A knock-in AC1 that was not regulated by Ca^{2+}/CaM would cement these conclusions, if it was expressed appropriately.

AC8 knockout mice have a clear array of defects, with a predominant impairment in learning and memory; this is also associated with decreased Ca^{2+}/CaM stimulated AC activity in the hippocampus, hypothalamus, thalamus, and brainstem (Schaefer et al. 2000). So, clearly these enzymes are important; however, whether their sensitivity to Ca^{2+} is important in a physiological context remains to be absolutely shown.

Other studies examining the effect of AC5 and AC6 knockout have shown definite physiological consequences, but none of these can be directly attributed to the Ca^{2+}-inhibitability of these enzymes (reviewed in Sadana and Dessauer 2009). AC5 knockout mice have altered pain responses, attenuated motor function and altered cardiac function. It had been speculated that the predominance of AC5 and AC6 in cardiac tissue could contribute significantly to the rhythmicity of sympathetic control of inotropy. Specifically, the AC5 knockout mice have decreased left ventricular ejection fraction, and attenuated baroreflexes and this is associated with a loss of acetylcholine mediated $Gα_i$ inhibition of AC activity, and a reduced Ca^{2+}-mediated inhibition of cAMP (Okumura et al. 2003). AC6 knockout mice show decreased left ventricular function, and this is also associated with a decreased Ca^{2+}-mediated inhibition of cAMP (Tang et al. 2008).

There are also physiological effects of AC3 knockout; the mice do not show intermale aggressiveness or male sexual behaviors, and this is associated with decreased cAMP in response to pheromones (Wang et al. 2006). Thus, although there are definitive consequences of knockout of AC1, AC3, AC5, AC6, and AC8, apart from AC1, whether the physiological effects shown are caused by the Ca^{2+}-dependency of the ACs remains to be determined.

FUNCTIONAL COMPARTMENTALIZATION OF Ca^{2+} AND cAMP SIGNALING

From the foregoing it is reasonably clear that AC1 and AC8, and AC5 and AC6 are stimulated and inhibited respectively by Ca^{2+}-rises in vitro and in the intact cell. However, and of even more regulatory significance, they are also highly discerning for the form of Ca^{2+}-rise to which they will respond.

AC Regulation by Ca^{2+} Entry

Elsewhere, and in detail, we have summarized how CCE triggered by either GPCR-linked agonists or by passive store depletion regulates each of the directly Ca^{2+}-regulatable ACs, either when expressed heterologously or when endogenously expressed (Cooper 2003; Willoughby and Cooper 2007). In our lab we have tended to use CCE that is triggered by passive store depletion rather than by agonists, but in various studies carbachol (CCh), bradykinin, or thyrotropin-releasing hormone (TRH) have also been used (Boyajian et al. 1991; Wachten et al. 2010; Willoughby et al. 2010). However because the consequences associated with agonist-triggered store depletion via PLC stimulation, such as PKC activation or the liberation of $\beta\gamma$ subunits of G-proteins, can have confounding additional effects on various AC species (as outlined above) the results of passive store depletion seem more straightforward.

Despite regulation by CCE, these enzymes are extremely unresponsive to ionophore mediated Ca^{2+}-entry. Early reports of responses to high concentrations of ionophore and external $CaCl_2$ were likely due directly to the triggering of CCE that is caused by store permeabilization and a subsequent CCE component of the Ca^{2+}-rise. A large body of data not reviewed here has established the dependence on CCE (Willoughby and Cooper 2007). Such is the dependence of the Ca^{2+}-sensitive ACs on CCE that we and others have used CCE-induced changes in cAMP as a measure for separating CCE from other Ca^{2+}-entry processes (Shuttleworth and Thompson 1999; Martin and Cooper 2006) and further, to study potential candidates that participate in CCE, such as Orai and STIM1 (the heterologous expression of which leads to increased regulation of these ACs by CCE; Martin et al. 2009).

This ability of the ACs to be regulated by specific forms of physiological Ca^{2+}-rise strongly suggests a selective affinity of the enzymes for the immediate environments of Ca^{2+} channels if not an actual affinity for Ca^{2+} channel subunits. In fact in addition to CCE, these ACs can also be regulated by voltage-gated Ca^{2+}-entry, though this topic has been far less well studied (Chetkovich et al. 1991; Yu et al. 1993; Fagan et al. 2000), and overexpression of CNG channels can promote the inhibition of AC6 by Ca^{2+}-entry (Fagan et al. 1999). The properties of Ca^{2+} channel subunits to form heterogeneous (and poorly understood) associations, e.g., various TRPC1/TRPM1/Orai combinations, or the various combinations of α, β, γ, and ε subunits possible in VGCCs, may also allow such subunits to associate with other membrane-inserted proteins (such as ACs), which are themselves capable of multivalency (Cooper and Crossthwaite 2006). Nevertheless, no credible associations between ACs and putative channel components have yet been revealed.

One puzzle implicit in the reliance of Ca^{2+}-regulatable ACs on CCE is the nature of the Ca^{2+} signal to which these ACs respond. In vitro (steady-state) studies suggest a susceptibility to concentrations of Ca^{2+} in the just submicromolar range; however significantly higher concentrations would be anticipated in the immediate vicinity of Ca^{2+} channels. We have speculated elsewhere on this apparent anachronism (Willoughby and Cooper 2007). The simplest reconciliation may be that ACs in intact cells specifically respond to the kinetic upstrokes in Ca^{2+} concentrations that occur at near-by channels—not to the steady state levels that are subsequently achieved.

AC Regulation by Agonist-Triggered Ca^{2+} Release

This mode of Ca^{2+} rise causes a small stimulation of both AC1 and AC8 when they are heterologously expressed. This is a fraction of the effect of CCE in the case of AC8 but almost equal to the effect of CCE in the case of AC1 (Masada et al. 2009). There is also evidence from endogenous systems that AC8 is stimulated by Ca^{2+} release mediated by TRH in GH3B6 cells (Wachten et al. 2010) but the effect of entry was not explored in that study.

From the foregoing it is clear that both the Ca^{2+}-sensitive ACs, and the machinery for Ca^{2+}-entry must be organized in such a way as to ensure the specific regulation of the ACs. Whether there is any compartmentalization or organization with regard to regulation of ACs by PKC, CaMK, CaN, or Gβγ subunits has never been addressed, but there is no reason to imagine that such organization may not occur. Below we outline some of our speculations on the molecular mechanisms or organizational constraints necessitated by the precise Ca^{2+}-regulation of ACs.

NATURE OF THE COMPARTMENTS FOR Ca^{2+} AND cAMP

Implicit in the regulation of cAMP by Ca^{2+}-signaling pathways is a degree of organizational constraint; we are now beginning to learn something of the elements of these constraints. Although it is well known that compartmentalization is intrinsic to Ca^{2+}, the rapid diffusion of Ca^{2+} makes these compartments extremely transient. Nevertheless at sites of Ca^{2+}-release or Ca^{2+}-entry, steep declines in concentration from extremely high to ambient can be both predicted and observed. Imaging studies have revealed sparks, sparklets, scintillas, puffs, or blinks (reviewed in Berridge 2006; Rizzuto and Pozzan 2006). Although Ca^{2+}-buffering can also play a significant theoretical role in the diffusion of Ca^{2+}, few practical consequences of buffering by proteins such as calbindin have been shown. There is of course gross compartmentalization of Ca^{2+} within ER stores and the mitochondria. In addition, assembly of Ca^{2+}-entry complexes in response to store depletion, such as the puncta formed by Orai, and STIM1 (along with other elements, e.g., TRPC1) show a degree of subcellular compartmentalization. Direct compartmentalized signaling by Ca^{2+} is obviously a major regulatory device in situations such as vesicle release sites at synapses, etc. Based on the foregoing discussion of the selective regulation by Ca^{2+}-entry of Ca^{2+}-sensitive ACs, it can be anticipated that ACs cluster at such locations.

Compartmentalized signaling occurs at a different level for cAMP. As with channels, the source of cAMP—in this case the ACs—is the first opportunity for compartmentalization of the second messenger. PDEs are a further early opportunity to impact on these compartments, because the presence or absence of a PDE will impact heavily on the diffusion range of cAMP (reviewed in Fischmeister et al. 2006; Zaccolo 2006; Houslay 2009). Ca^{2+}-stimulated PDE1 is potentially particularly relevant, because in this case Ca^{2+}-rises will also impact on the PDE-limited diffusion of cAMP. Curiously, very few studies have considered the role of PDE1 in controlling cAMP (Evans et al. 1984) but it does appear that for continual activation of PDE1, sustained CCE is required (Goraya et al. 2004).

At a very gross level some ACs are localized in lipid raft domains of the plasma membrane whereas others are excluded (Ostrom et al. 2002; Crossthwaite et al. 2005). However, the issue of lipid rafts had been controversial based on oversimple assumptions and variability between systems. In situations such as the immunologic synapse or postsynaptic densities, clearly stable domains are encountered that differ in their lipid composition from the rest of the plasma membrane. On the other hand dynamically changing lipid inhomogeneities also occur, which can have variably observed half-lives and are variably populated by signaling molecules. Encounters between such rafts may promote or facilitate productive regulatory interactions between resident proteins. These transient associations could potentially promote cAMP hotspots by concentrating, e.g., GPCR and G-proteins or other regulatory factors such as the elements of the CCE apparatus, with ACs and facilitating their interaction. In the case of the nine membrane-bound ACs, the acutely Ca^{2+}-regulated species AC1, AC8, AC5, and AC6 are all localized in rafts, as assessed by a variety of criteria and preparation methods (Fagan et al. 2000; Ostrom et al. 2002; Pagano et al. 2009; reviewed in Patel et al. 2008). In addition, the elements of the CCE apparatus, Orai and STIM1, can also be found in lipid rafts (Pani et al. 2008). However, the conclusion that STIM1 associates with lipid rafts was recently contested in elegant functional studies of CCE in live cells (DeHaven et al. 2009). Furthermore, the studies that originally suggested an essential functional association of Ca^{2+}-regulated ACs with lipid rafts, showed only a loss of the regulation of the ACs following cholesterol depletion, with no associated change in CCE (Fagan et al 2000; Smith et al. 2002). These conflicting studies of the residence or otherwise of the CCE apparatus in lipid rafts underline the difficulties in comparing biochemical and cell biological assessments of raft components.

Nevertheless, there are clear indications that ACs can act as organizers of their own domain. Most of the ACs (i.e., AC1, AC2, AC3, AC5, AC6, AC8, and AC9) bind A-kinase anchoring proteins (AKAPs), and AKAPs can also bind PDEs as well as PKC, CaN, and various channels (reviewed in Dessauer 2009). In the brain, AC5 associates with a complex including AKAP79, PKA, and AMPA receptors to facilitate glutamate-stimulated AC5 activity (Efendiev et al. 2010), and AC2 associates with Yotiao (AKAP9) which inhibits the AC (Piggot et al. 2008). In the heart AC5 associates with mAKAPβ and PKA, which attenuates AC5 activity (Kapiloff et al. 2009). In pancreatic and neuronal systems, AC8 associates with AKAP79, which limits the sensitivity of the AC to stimulation by CCE (Willoughby et al. 2010). AC8 also associates with protein phosphatase PP2A, which is itself a scaffolding protein (Cooper and Crossthwaite 2006). Signaling hubs that are built around the ACs ensure fidelity in cAMP signaling, and, in the case of the neuronal system, ensure that Ca^{2+} very efficiently regulates the AC that organizes the scaffold. Other data also suggest that ACs can interact with the cytoskeleton and thereby potentially regulate cellular dynamics (reviewed in Patel et al. 2008). In this context it is interesting that emptying of Ca^{2+} stores by treatment with 1 mM TPEN (an intra-ER Zn^{2+}/Ca^{2+}-chelator) or 5 μM ionomycin results in increased cAMP accumulation in a colonic cell line. Sequestration of free Ca^{2+} within the ER, or extrusion of Ca^{2+} from the ER in these cells (which do not express a Ca^{2+}-stimulable AC1 or AC8) resulted in a slow increase in cAMP accumulation, independently of CCE (i.e., the effect occurred in the absence of extracellular Ca^{2+}) and cytosolic Ca^{2+} (thereby excluding direct effects of Ca^{2+} on AC activity), but in a manner and time scale that appeared to coincide with STIM1 translocation (Lefkimmiatis et al. 2009; Roy et al. 2010). The time course of this response does not match the rapid effects of extracellular Ca^{2+} on agonist- or thapsigargin-store depleted cells expressing either Ca^{2+}-stimulable or Ca^{2+}-inhibitable ACs; nor were such Ca^{2+}-independent effects on AC activity seen during store-depletion before the addition of extracellular Ca^{2+} in such systems (Fagan et al. 1998; Willoughby and Cooper 2006). An intriguing mechanism is thus suggested whereby the gross cellular architectural changes that occur following treatments with high concentrations of TPEN and ionomycin, which include STIM1 translocation to the plasma membrane, leads to increased cAMP accumulation. Conceivably the increased cAMP accumulation results from a relief from an undefined inhibitory constraint, e.g., between AC and the cytoskeleton or plasma membrane, that accompanies the major cellular transitions that are associated with store depletion and STIM1 translocation. A similar slow effect, presumed to reflect disinhibition of ACs, is seen when cells are deprived of cholesterol (Fagan et al. 2000; Pagano et al. 2009), thereby disrupting lipid rafts and their associated protein complexes.

EXPLORING AC ACTIVITY AT THE SINGLE CELL LEVEL

Given the susceptibility of most of the nine membrane-bound ACs to regulation by various aspects of Ca^{2+}-signaling as discussed up to now, along with the essential compartmentalization associated with this regulation, it seems essential to adopt very discerning methods in order to establish whether or how these enzymes are regulated. In addition, given that Ca^{2+}-levels can oscillate in a nonhomogenous manner from cell to cell, it is essential to adopt single cell methods to determine whether the various Ca^{2+}-sensitive ACs can actually respond in an oscillatory manner to such Ca^{2+}-signals.

Elsewhere (reviewed in Nikolaev and Lohse 2006; Willoughby and Cooper 2008), methods have been described that allow the measurement of cAMP in single cells; these are all based on the natural targets of cAMP, i.e., PKA, EPAC, and CNG and HCN channels, and they all have specific advantages and applications. In addition modification of these sensors to allow targeting to specific AC microdomains has facilitated the beginning of an explicit exploration of these microenvironments (Wachten et al. 2010). Such methods are also potentially particularly powerful in dynamic situations, e.g., in moving cells.

cAMP oscillations had been predicted based on the exposure of Ca^{2+}-sensitive ACs to oscillations in Ca^{2+} concentrations (e.g., in cardiac tissue; reviewed in Cooper et al. 1995). Examples of cAMP oscillations have now been described in a number of systems by multiple single-cell cAMP imaging methods (Gorbunova et al. 2002; Landa et al. 2005; Dyachok et al. 2006; Willoughby and Cooper 2006), and a variety of mechanisms for the oscillations in cAMP have been advanced. It is still not clear whether cAMP oscillations generate functional consequences in a physiological setting, although one study clearly indicates that cAMP oscillations derived from Ca^{2+}-stimulation of AC1 are required for exuberant axonal pathfinding (Nicol et al. 2007).

FUTURE DIRECTIONS

This review of the regulation of ACs by activation of Ca^{2+}-signaling pathways has been surprising from a few viewpoints. Whereas it seems clear that AC1, AC8 and AC5, and AC6 are directly regulated by Ca^{2+}/CaM and Ca^{2+}, respectively, both in vitro and as a result of meaningful elevations in intracellular Ca^{2+}, the regulatory influence of the Ca^{2+}-signaling pathway on the other ACs, particularly in terms of the role of PKC, Gβγ subunits, CaN, CaMKII, and CaMKIV, is far from established. This is not to suggest that the evidence is unconvincing, just that the subject has received so little attention. Given the potential importance of these interactions and the undoubted potential magnitude of the effects, this is a striking oversight in signaling awareness. It would seem extremely worthwhile to attempt to fill these gaps. Obviously, if we do not comprehensively appreciate how these enzymes are regulated, we cannot begin to contemplate their potential physiological significance. The problem is compounded by the growing awareness of the complexity of compartmentalized signaling and the extreme paucity of information on the organization and cellular distribution of, for instance, PKC. Nevertheless, the relevant tools are available in terms of both single cell techniques and intracellular targeting of probes; the philosophical will may understandably be lacking given the magnitude of the information that is likely to be uncovered. The comfortable naïveté that accompanies two-dimensional portrayals of signaling cascades is of course troubled by confronting the likely complexity of real cell signaling. Against this background, it seems essential to engage with these issues and to develop further conceptual and experimental frameworks for addressing these interactions. Whereas a "systems" approach conjures up images of the mechanistically unimaginative, the true and considerable dimensions of the problem may demand at least a flirtation with such approaches (Xu et al. 2010). In any event there is little doubt that these two crucial systems are critically intertwined, that there are highly significant consequences for this interaction, and we do need to address and respond to this complexity.

ACKNOWLEDGMENTS

MLH is a National Health and Medical Research Council of Australia Overseas Biomedical Fellow (519581), DMFC is a Royal Society Wolfson

Research Fellow. The work in the authors' laboratory is supported by the Wellcome Trust (RG31760). We thank Dr Debbie Willoughby and Dr. Nana Masada for careful revision of the manuscript.

REFERENCES

Anisimova M, Gascuel O. 2006. Approximate likelihood-ratio test for branches: A fast, accurate, and powerful alternative. *Syst Biol* **55:** 539–552.

Antoni FA, Sosunov AA, Haunso A, Paterson JM, Simpson J. 2003. Short-term plasticity of cyclic adenosine 3′,5′-monophosphate signaling in anterior pituitary corticotrope cells: the role of adenylyl cyclase isotypes. *Mol Endocrinol* **17:** 692–703.

Avidor-Reiss T, Nevo I, Levy R, Pfeuffer T, Vogel Z. 1996. Chronic opioid treatment induces adenylyl cyclase V superactivation. Involvement of Gβγ. *J Biol Chem* **271:** 21309–21315.

Böl G-F, Hülster A, Pfeuffer T. 1997b. Adenylyl cyclase type II is stimulated by PKC via C-terminal phosphorylation. *Biochim Biophys Acta* **1358:** 307–313.

Böl G-F, Gros C, Hülster A, Bösel A, Pfeuffer T. 1997a. Phorbol ester-induced sensitisation of adenylyl cyclase type II is related to phosphorylation of threonine 1057. *Biochem Biophys Res Comm* **237:** 251–256.

Belevych AE, Sims C, Harvey RD. 2001. ACh-induced rebound stimulation of L-type Ca(2+) current in guinea-pig ventricular myocytes, mediated by Gβγ-dependent activation of adenylyl cyclase. *J Physiol* **536:** 677–692.

Berridge MJ. 2006. Calcium microdomains: Organization and function. *Cell Calcium* **40:** 405–412.

Berridge MJ. 1975. The interaction of cyclic nucleotides and calcium in the control of cellular activity. *Adv Cyclic Nucleotide Res* **6:** 1–98.

Boyajian CL, Garritsen A, Cooper DMF. 1991. Bradykinin stimulates Ca^{2+} mobilization in NCB-20 cells leading to direct inhibition of adenylyl cyclase. A novel mechanism for inhibition of cAMP production. *J Biol Chem* **266:** 4995–5003.

Boyajian CL, Bickford-Wimer P, Kim MB, Freedman R, Cooper DMF. 1989. Pertussis toxin lesioning of the nucleus caudate-putamen attenuates adenylate cyclase inhibition and alters neuronal electrophysiological activity. *Brain Res* **495:** 66–74.

Castresana J. 2000. Selection of conserved blocks from multiple alignments for their use in phylogenetic analysis. *Mol Biol Evol* **17:** 540–552.

Chakrabarti S, Wang L, Tang WJ, Gintzler AR. 1998. Chronic morphine augments adenylyl cyclase phosphorylation: Relevance to altered signaling during tolerance/dependence. *Mol Pharmacol* **54:** 949–953.

Chen J, DeVivo M, Dingus J, Harry A, Li J, Sui J, Carty DJ, Blank JL, Exton JH, Stoffel RH, et al. 1995. A region of adenylyl cyclase 2 critical for regulation by G protein βγ subunits. *Science* **268:** 1166–1169.

Chetkovich DM, Gray R, Johnston D, Sweatt JD. 1991. N-methyl-D-aspartate receptor activation increases cAMP levels and voltage-gated Ca^{2+} channel activity in area CA1 of hippocampus. *Proc Natl Acad Sci* **88:** 6467–6471.

Chevenet F, Brun C, Banuls A-L, Jacq B, Christen R. 2006. TreeDyn: Towards dynamic graphics and annotations for analyses of trees. *BMC Bioinformatics* **7:** 439–447.

Cioffi DL, Moore TM, Schaack J, Creighton JR, Cooper DMF, Stevens T. 2002. Dominant regulation of interendothelial cell gap formation by calcium-inhibited type 6 adenylyl cyclase. *J Cell Biol* **157:** 1267–1278.

Cooper DMF. 1991. Inhibition of adenylate cyclase by Ca^{2+} – a counterpart to stimulation by Ca^{2+}/calmodulin. *Biochem J* **278:** 903–904.

Cooper DMF. 2003. Regulation and organization of adenylyl cyclases and cAMP. *Biochem J* **375:** 517–529.

Cooper DMF, Crossthwaite AJ. 2006. Higher-order organization and regulation of adenylyl cyclases. *Trends Pharmacol Sci* **27:** 426–431.

Cooper DMF, Mons N, Karpen JW. 1995. Adenylyl cyclases and the interaction between calcium and cAMP signalling. *Nature* **374:** 421–424.

Crossthwaite AJ, Seebacher T, Masada N, Ciruela A, Dufraux K, Schultz JE, Cooper DMF. 2005. The cytosolic domains of Ca^{2+}-sensitive adenylyl cyclases dictate their targeting to plasma membrane lipid rafts. *J Biol Chem* **280:** 6380–6391.

Cumbay MG, Watts VJ. 2004. Novel regulatory properties of human type 9 adenylate cyclase. *J Pharmacol Exp Ther* **310:** 108–115.

Cumbay MG, Watts VJ. 2005. Gαq potentiation of adenylate cyclase type 9 activity through a Ca^{2+}/calmodulin-dependent pathway. *Biochem Pharmacol* **69:** 1247–1256.

Dai S, Hall DD, Hell JW. 2009. Supramolecular assemblies and localized regulation of voltage-gated ion channels. *Physiol Rev* **89:** 411–452.

DeHaven WI, Jones BF, Petranka JG, Smyth JT, Tomita T, Bird GS, Putney JW. 2009. TRPC channels function independently of STIM1 and Orai1. *J Physiol* **587:** 2275–2298.

Dereeper A, Guignon V, Blanc G, Audic S, Buffet S, Chevenet F, Dufayard J-F, Guindon S, Lefort V, Lescot M, et al. 2008. Phylogeny.fr: Robust phylogenetic analysis for the non-specialist. *Nucl Acids Res* **36:** W465–469.

Dessauer CW. 2009. Adenylyl cyclase-A-kinase anchoring protein complexes: the next dimensions in cAMP signaling. *Mol Pharmacol* **76:** 1256–1264.

Diel S, Klass K, Wittig B, Kleuss C. 2006. Gβγ activation site in adenylyl cyclase type II. Adenylyl cyclase type III is inhibited by Gβγ. *J Biol Chem* **281:** 288–294.

Dyachok O, Isakov Y, Sågetorp J, Tengholm A. 2006. Oscillations of cyclic AMP in hormone-stimulates insulin-secreting β-cells. *Nature* **439:** 349–352.

Edgar RC. 2004. MUSCLE: multiple sequence alignment with high accuracy and high throughput. *Nucl Acids Res* **32:** 1792–1797.

Efendiev R, Samelson BK, Nguyen BT, Phatarpekar PV, Baameur F, Scott JD, Dessauer CW. 2010. AKAP79 interacts with multiple adenylyl cyclase (AC) isoforms and scaffolds AC 5 and 6 to AMPA receptors. *J Biol Chem* doi 101074/jbcM110109769.

Evans T, Smith MM, Tanner LI, Harden TK. 1984. Muscarinic cholinergic receptors of two cell lines that regulate cyclic AMP metabolism by different molecular mechanisms. *Mol Pharmacol* **26**: 395–404.

Fagan KA, Mahey R, Cooper DMF. 1996. Functional co-localization of transfected Ca^{2+}-stimulable adenylyl cyclases with capacitative Ca^{2+} entry sites. *J Biol Chem* **271**: 12438–12444.

Fagan KA, Mons N, Cooper DMF. 1998. Dependence of the Ca^{2+}-inhibitable adenylyl cyclase of C6-2B glioma cells on capacitative Ca^{2+} entry. *J Biol Chem* **273**: 9297–9305.

Fagan KA, Smith KE, Cooper DMF. 2000. Regulation of the Ca^{2+}-inhibitable adenylyl cyclase type VI by capacitative Ca^{2+} entry requires localization in cholesterol-rich domains. *J Biol Chem* **275**: 26530–26537.

Fagan KA, Graf RA, Tolman S, Schaack J, Cooper DMF. 2000. Regulation of a Ca^{2+}-sensitive adenylyl cyclase in an excitable cell. Role of voltage-gated versus capacitative Ca^{2+} entry. *J Biol Chem* **275**: 40187–40194.

Fagan KA, Rich TC, Tolman S, Schaack J, Karpen JW, Cooper DMF. 1999. Adenovirus-mediated expression of an olfactory cyclic nucleotide-gated channel regulates the endogenous Ca^{2+}-inhibitable adenylyl cyclase in C6-2B glioma cells. *J Biol Chem* **274**: 12445–12453.

Federman AD, Conklin BR, Schrader KA, Reed RR, Bourne HR. 1992. Hormonal stimulation of adenylyl cyclase through Gi-protein βγ subunits. *Nature* **356**: 159–161.

Feinstein PG, Schrader KA, Bakalyar HA, Tang WJ, Krupinski J, Gilman AG, Reed RR. 1991. Molecular cloning and characterization of a Ca^{2+}/calmodulin-insensitive adenylyl cyclase from rat brain. *Proc Natl Acad Sci* **88**: 10173–10177.

Fischmeister R, Castro LRV, Abi-Gerges A, Rochais F, Jurevicius J, Leroy J, Vandecasteele G. 2006. Compartmentation of cyclic nucleotide signaling in the heart: The role of cyclic nucleotide phosphodiesterases. *Circ Res* **99**: 816–828.

Gao BN, Gilman AG. 1991. Cloning and expression of a widely distributed (type IV) adenylyl cyclase. *Proc Natl Acad Sci* **88**: 10178–10182.

Gao X, Sadana R, Dessauer CW, Patel TB. 2007. Conditional stimulation of type V and VI adenylyl cyclases by G protein βγ subunits. *J Biol Chem* **282**: 294–302.

Garritsen A, Zhang Y, Firestone JA, Browning MD, Cooper DMF. 1992. Inhibition of cyclic AMP accumulation in intact NCB-20 cells as a direct result of elevation of cytosolic Ca^{2+}. *J Neurochem* **59**: 1630–1639.

Goraya TA, Masada N, Ciruela A, Cooper DMF. 2004. Sustained entry of Ca^{2+} is required to activate Ca^{2+}-calmodulin-dependent phosphodiesterase 1A. *J Biol Chem* **279**: 40494–40504.

Gorbunova YV, Spitzer NC. 2002. Dynamic interactions of cyclic AMP transients and spontaneous Ca^{2+} spikes. *Nature* **418**: 93–96.

Grunberger C, Obermayer B, Klar J, Kurtz A, Schweda F. 2006. The Calcium paradoxon of renin release: Calcium suppresses renin exocytosis by inhibition of calcium-dependent adenylate cyclases AC5 and AC6. *Circ Res* **99**: 1197–1206.

Gu C, Cooper DMF. 1999. Calmodulin-binding sites on adenylyl cyclase type VIII. *J Biol Chem* **274**: 8012–8021.

Guillou JL, Nakata H, Cooper DMF. 1999. Inhibition by calcium of mammalian adenylyl cyclases. *J Biol Chem* **274**: 35539–35545.

Guindon S, Gascuel O. 2003. A simple, fast, and accurate algorithm to estimate large phylogenies by maximum likelihood. *Syst Biol* **52**: 696–704.

Hanoune J, Defer N. 2001. Regulation and role of adenylyl cyclase isoforms. *Annu Rev Pharmacol Toxicol* **41**: 145–174.

Haslauer M, Baltensperger K, Porzig H. 1998. Thrombin and phorbol esters potentiate Gs-mediated cAMP formation in intact human erythroid progenitors via two synergistic signaling pathways converging on adenylyl cyclase type VII. *Mol Pharmacol* **53**: 837–845.

Hoffert JD, Chou CL, Fenton RA, Knepper MA. 2005. Calmodulin is required for vasopressin-stimulated increase in cyclic AMP production in inner medullary collecting duct. *J Biol Chem* **280**: 13624–13630.

Houslay MD. 2009. Underpinning compartmentalised cAMP signalling through targeted cAMP breakdown. *Trends Biochem Sci* **35**: 91–100.

Hu B, Nakata H, Gu C, De Beer T, Cooper DMF. 2002. A critical interplay between Ca^{2+} inhibition and activation by Mg^{2+} of AC5 revealed by mutants and chimeric constructs. *J Biol Chem* **277**: 33139–33147.

Iwami G, Kawabe J, Ebina T, Cannon PJ, Homcy CJ, Ishikawa Y. 1995. Regulation of adenylyl cyclase by protein kinase A. *J Biol Chem* **270**: 12481–12484.

Jacobowitz O, Iyengar R. 1994. Phorbol ester-induced stimulation and phosphorylation of adenylyl cyclase 2. *Proc Natl Acad Sci* **91**: 10630–10634.

Jacobowitz O, Chen J, Premont RT, Iyengar R. 1993. Stimulation of specific types of Gs-stimulated adenylyl cyclases by phorbol ester treatment. *J Biol Chem* **268**: 3829–3832.

Kapiloff MS, Piggott LA, Sadana R, Li J, Heredia LA, Henson E, Efendiev R, Dessauer CW. 2009. An adenylyl cyclase-mAKAPβ signaling complex regulates cAMP levels in cardiac myocytes. *J Biol Chem* **284**: 23540–23546.

Kawabe J, Iwami G, Ebina T, Ohno S, Katada T, Ueda Y, Homcy CJ, Ishikawa Y. 1994. Differential activation of adenylyl cyclase by protein kinase C isoenzymes. *J Biol Chem* **269**: 16554–16558.

Krupinski J, Coussen F, Bakalyar HA, Tang WJ, Feinstein PG, Orth K, Slaughter C, Reed RR, Gilman AG. 1989. Adenylyl cyclase amino acid sequence: Possible channel- or transporter-like structure. *Science* **244**: 1558–1564.

Lai HL, Yang TH, Messing RO, Ching YH, Lin SC, Chern Y. 1997. Protein kinase C inhibits adenylyl cyclase type VI activity during desensitization of the A2a-adenosine receptor-mediated cAMP response. *J Biol Chem* **272**: 4970–4977.

Landa LR, Harbeck M, Kailhara K, Chepurny O, Kitiphongspattana K, Graf O, Nikolaev VO, Lohse MJ, Holz GG, Roe MW. 2005. Interplay of Ca^{2+} and cAMP signaling in the insulin-secreting MIN6 β-cell line. *J Biol Chem* **280**: 31294–31302.

Lariviere S, Garrel G, Simon V, Soh J-W, Laverriere J-N, Counis R, Cohen-Tannoudji J. 2007. Gonadotropin-releasing hormone couples to 3',5'-cyclic adenosine-5'-monophosphate pathway through novel protein kinase

C-δ and -ε in LβT2 gonadotrope cells. *Endocrinology* **148:** 1099–1107.

Lefkimmiatis K, Srikanthan M, Maiellaro I, Moyer MP, Curci S, Hofer AM. 2009. Store-operated cyclic AMP signalling mediated by STIM1. *Nat Cell Biol* **11:** 433–442.

Lin TH, Lai HL, Kao YY, Sun CN, Hwang MJ, Chern Y. 2002. Protein kinase C inhibits type VI adenylyl cyclase by phosphorylating the regulatory N domain and two catalytic C1 and C2 domains. *J Biol Chem* **277:** 15721–15728.

Lustig KD, Conklin BR, Herzmark P, Taussig R, Bourne HR. 1993. Type II adenylylcyclase integrates coincident signals from Gs, Gi, and Gq. *J Biol Chem* **268:** 13900–13905.

Ma H, Green RD. 1992. Modulation of cardiac cyclic AMP metabolism by adenosine receptor agonists and antagonists. *Mol Pharmacol* **42:** 831–837.

Mamluk R, Defer N, Hanoune J, Meidan R. 1999. Molecular identification of adenylyl cyclase 3 in bovine corpus luteum and its regulation by prostaglandin F2α-induced signaling pathways. *Endocrinology* **140:** 4601–4608.

Marjamaki A, Sato M, Bouet-Alard R, Yang Q, Limon-Boulez I, Legrand C, Lanier SM. 1997. Factors determining the specificity of signal transduction by guanine nucleotide-binding protein-coupled receptors. Integration of stimulatory and inhibitory input to the effector adenylyl cyclase. *J Biol Chem* **272:** 16466–16473.

Martin AC, Cooper DMF. 2006. Capacitative and 1-oleyl-2-acetyl-sn-glycerol-activated Ca^{2+} entry distinguished using adenylyl cyclase type 8. *Mol Pharmacol* **70:** 769–777.

Martin AC, Willoughby D, Ciruela A, Ayling LJ, Pagano M, Wachten S, Tengholm A, Cooper DMF. 2009. Capacitative Ca^{2+} entry via Orai1 and stromal interacting molecule 1 (STIM1) regulates adenylyl cyclase type 8. *Mol Pharmacol* **75:** 830–842.

Masada N, Ciruela A, Macdougall DA, Cooper DMF. 2009. Distinct mechanisms of regulation by Ca^{2+}/calmodulin of type 1 and 8 adenylyl cyclases support their different physiological roles. *J Biol Chem* **284:** 4451–4463.

Mou T-C, Masada N, Cooper DMF, Sprang SR. 2009. Structural basis for inhibition of mammalian adenylyl cyclase by calcium. *Biochemistry* **48:** 3387–3397.

Nelson EJ, Hellevuo K, Yoshimura M, Tabakoff B. 2003. Ethanol-induced phosphorylation and potentiation of the activity of type 7 adenylyl cyclase. Involvement of protein kinase C δ. *J Biol Chem* **278:** 4552–4560.

Nicol X, Voyatzis S, Muzerelle A, Narboux-Neme N, Sudhof TC, Miles R, Gaspar P. 2007. cAMP oscillations and retinal activity are permissive for ephrin signaling during the establishment of the retinotopic map. *Nat Neurosci* **10:** 340–347.

Nikolaev VO, Lohse MJ. 2006. Monitoring of cAMP synthesis and degradation in living cells. *Physiology* **21:** 86–92.

Okumura S, Kawabe J-I, Yatani A, Takagi G, Lee M-C, Hong C, Liu J, Takagi I, Sadoshima J, Vatner DE, et al. 2003. Type 5 adenylyl cyclase disruption alters not only sympathetic but also parasympathetic and calcium-mediated cardiac regulation. *Circ Res* **93:** 364–371.

Ostrom RS, Liu X, Head BP, Gregorian C, Seashotlz TM, Insel PA. 2002. Localization of adenylyl cyclase isoforms and G protein-coupled receptors in vascular smooth muscle cells: expression in caveolin-rich and noncaveolin domains. *Mol Pharmacol* **62:** 983–992.

Pagano M, Clynes MA, Masada N, Ciruela A, Ayling L-J, Wachten S, Cooper DMF. 2009. Insights into the residence in lipid rafts of adenylyl cyclase AC8 and its regulation by capacitative calcium entry. *Am J Physiol Cell Physiol* **296:** C607–619.

Pani B, Ong HL, Liu X, Rauser K, Ambudkar IS, Singh BB. 2008. Lipid rafts determine clustering of STIM1 in endoplasmic reticulum-plasma membrane junctions and regulation of store-operated Ca^{2+} entry (SOCE). *J Biol Chem* **283:** 17333–17340.

Patel HH, Murray F, Insel PA. 2008. Caveolae as organizers of pharmacologically relevant signal transduction molecules. *Annu Rev Pharmacol Toxicol* **48:** 359–391.

Piggott LA, Bauman AL, Scott JD, Dessauer CW. 2008. The A-kinase anchoring protein Yotiao binds and regulates adenylyl cyclase in brain. *Proc Natl Acad Sci* **105:** 13835–13840.

Premont RT, Chen J, Ma HW, Ponnapalli M, Iyengar R. 1992. Two members of a widely expressed subfamily of hormone-stimulated adenylyl cyclases. *Proc Natl Acad Sci* **89:** 9809–9813.

Premont RT, Matsuoka I, Mattei MG, Pouille Y, Defer N, Hanoune J. 1996. Identification and characterization of a widely expressed form of adenylyl cyclase. *J Biol Chem* **271:** 13900–13907.

Rasmussen H. 1981. In *Calcium and cAMP as synarchic messengers.* John Wiley and Sons, Inc., New Jersey.

Rhim J-H, Jang I-S, Yeo E-J, Song K-Y, Park SC. 2006. Role of protein kinase C-dependent A-kinase anchoring proteins in lysophosphatidic acid-induced cAMP signaling in human diploid fibroblasts. *Aging Cell* **5:** 451–461.

Rizzuto R, Pozzan T. 2006. Microdomains of intracellular Ca^{2+}: Molecular determinants and functional consequences. *Physiol Rev* **86:** 369–408.

Robison GA, Butcher RW, Sutherland EW. 1971. In *Cyclic AMP* Academic Press, New York.

Roy J, Lefkimmiatis K, Moyer MP, Curci S, Hofer AM. 2010. The ω-3 fatty acid eicosapentaenoic acid elicits cAMP generation in colonic epithelial cells via a 'store-operated' mechanism. *Am J Physiol Gastrointest Liver Physiol* **299:** G715–G722.

Sadana R, Dessauer CW. 2009. Physiological roles for G protein-regulated adenylyl cyclase isoforms: Insights from knockout and overexpression studies. *Neurosignals* **17:** 5–22.

Sadana R, Dascal N, Dessauer CW. 2009. N terminus of type 5 adenylyl cyclase scaffolds Gs heterotrimer. *Mol Pharmacol* **76:** 1256–1264.

Schaefer ML, Wong ST, Wozniak DF, Muglia LM, Liauw JA, Zhuo M, Nardi A, Hartman RE, Vogt SK, Luedke CE, et al. 2000. Altered stress-induced anxiety in adenylyl cyclase type VIII-deficient mice. *J Neurosci* **20:** 4809–4820.

Shuttleworth TJ, Thompson JL. 1999. Discriminating between capacitative and arachidonate-activated Ca^{2+} entry pathways in HEK293 cells. *J Biol Chem* **274:** 31174–31178.

Smith KE, Chen G, Fagan KA, Hu B, Cooper DMF. 2002. Residence of adenylyl cyclase type 8 in caveolae is necessary but not sufficient for regulation by capacitative Ca^{2+} entry. *J Biol Chem* **277:** 6025–6031.

Steiner D, Saya D, Schallmach E, Simonds WF, Vogel Z. 2006. Adenylyl cyclase type-VIII activity is regulated by Gβγ subunits. *Cell Signal* **18:** 62–68.

Storm DR, Hansel C, Hacker B, Parent A, Linden DJ. 1998. Impaired cerebellar long-term potentiation in type I adenylyl cyclase mutant mice. *Neuron* **20:** 1199–1210.

Straub SV, Wagner LE, Bruce JI, Yule DI. 2004. Modulation of cytosolic calcium signaling by protein kinase A-mediated phosphorylation of inositol 1,4,5-trisphosphate receptors. *Biol Res* **37:** 593–602.

Sunahara RK, Dessauer CW, Gilman AG. 1996. Complexity and diversity of mammalian adenylyl cyclases. *Annu Rev Pharmacol Toxicol* **36:** 461–480.

Tang WJ, Gilman AG. 1991. Type-specific regulation of adenylyl cyclase by G protein βγ subunits. *Science* **254:** 1500–1503.

Tang T, Gao MH, Lai NC, Firth AL, Takahashi T, Guo T, Yuan JX-J, Roth DM, Hammond HK. 2008. Adenylyl cyclase type 6 deletion decreases left ventricular function via impaired calcium handling. *Circulation* **117:** 61–69.

Tesmer JJ, Sunahara RK, Gilman AG, Sprang SR. 1997. Crystal structure of the catalytic domains of adenylyl cyclase in a complex with Gsα.GTPγS. *Science* **278:** 1907–1916.

Tesmer JJ, Sunahara RK, Johnson RA, Gosselin G, Gilman AG, Sprang SR. 1999. Two-metal-ion catalysis in adenylyl cyclase. *Science* **285:** 756–760.

Trubey KR, Culpepper S, Maruyama Y, Kinnamon SC, Chaudhari N. 2006. Tastants evoke cAMP signal in taste buds that is independent of calcium signaling. *Am J Physiol Cell Physiol* **291:** C237–244.

Villacres EC, Wong ST, Chavkin C, Storm DR. 1998. Type I adenylyl cyclase mutant mice have impaired mossy fiber long-term potentiation. *J Neurosci* **18:** 3186–3194.

Vorherr T, Knopfel L, Hofmann F, Mollner S, Pfeuffer T, Carafoli E. 1993. The calmodulin binding domain of nitric oxide synthase and adenylyl cyclase. *Biochemistry* **32:** 6081–6088.

Wachten S, Masada N, Ayling L-J, Ciruela A, Nikolaev VO, Lohse MJ, Cooper DMF. 2010. Distinct pools of cAMP centre on different isoforms of adenylyl cyclase in pituitary-derived GH3B6 cells. *J Cell Sci* **123:** 95–106.

Wang H-Y, Burns LH. 2006. Gβγ that interacts with adenylyl cyclase in opioid tolerance originates from a Gs protein. *J Neurobiol* **66:** 1302–1310.

Wang H, Gong B, Vadakkan KI, Toyoda H, Kaang BK, Zhuo M. 2007. Genetic evidence for adenylyl cyclase 1 as a target for preventing neuronal excitotoxicity mediated by N-methyl-D-aspartate receptors. *J Biol Chem* **282:** 1507–1517.

Wang HY, Friedman E, Olmstead MC, Burns LH. 2005. Ultra-low-dose naloxone suppresses opioid tolerance, dependence and associated changes in μ opioid receptor-G protein coupling and Gβγ signaling. *Neuroscience* **135:** 247–261.

Watson EL, Jacobson KL, Singh JC, Idzerda R, Ott SM, DiJulio DH, Wong ST, Storm DR. 2000. The type 8 adenylyl cyclase is critical for Ca^{2+} stimulation of cAMP accumulation in mouse parotid acini. *J Biol Chem* **275:** 14691–14699.

Wayman G, Wei J, Wong S, Storm D. 1996. Regulation of type I adenylyl cyclase by calmodulin kinase IV in vivo. *Mol Cell Biol* **16:** 6075–6082.

Wei J, Wayman G, Storm DR. 1996. Phosphorylation and inhibition of type III adenylyl cyclase by calmodulin-dependent protein kinase II in vivo. *J Biol Chem* **271:** 24231–24235.

Wei J, Zhao AZ, Chan GCK, Baker LP, Impey S, Beavo JA, Storm DR. 1998. Phosphorylation and inhibition of olfactory adenylyl cyclase by CaM kinase II in neurons: A mechanism for attenuation of olfactory signals. *Neuron* **21:** 495–504.

Weitmann S, Schultz G, Kleuss C. 2001. Adenylyl cyclase type II domains involved in Gβγ stimulation. *Biochemistry* **40:** 10853–10858.

Willoughby D, Cooper DMF. 2006. Ca^{2+} stimulation of adenylyl cyclase generates dynamic oscillations in cyclic AMP. *J Cell Sci* **119:** 828–836.

Willoughby D, Cooper DMF. 2007. Organization and Ca^{2+} regulation of adenylyl cyclases in cAMP microdomains. *Physiol Rev* **87:** 965–1010.

Willoughby D, Cooper DMF. 2008. Live-cell imaging of cAMP dynamics. *Nat Methods* **5:** 29–36.

Willoughby D, Masada N, Wachten S, Pagano M, Halls ML, Everett KL, Ciruela A, Cooper DMF. 2010. A-kinase anchoring protein 79/150 interacts with adenylyl cyclase type 8 and regulates Ca^{2+}-dependent cAMP synthesis in pancreatic and neuronal systems. *J Biol Chem* doi 101074/jbcM110120725.

Wu ZL, Thomas SA, Villacres EC, Xia Z, Simmons ML, Chavkin C, Palmiter RD, Storm DR. 1995. Altered behavior and long-term potentiation in type I adenylyl cyclase mutant mice. *Proc Natl Acad Sci* **92:** 220–224.

Xu TR, Vyshemirsky V, Gormand A, von Kriegsheim A, Girolami M, Baillie GS, Ketley D, Dunlop AJ, Milligan G, Houslay MD, et al. 2010. Inferring signaling pathway topologies from multiple perturbation measurements of specific biochemical species. *Sci Signal* **3:** ra20.

Yoshimura M, Cooper DMF. 1993. Type-specific stimulation of adenylyl cyclase by protein kinase C. *J Biol Chem* **268:** 4604–4607.

Yoshimura M, Ikeda H, Tabakoff B. 1996. μ-Opioid receptors inhibit dopamine-stimulated activity of type V adenylyl cyclase but enhance dopamine-stimulated activity of type VII adenylyl cyclase. *Mol Pharmacol* **50:** 43–51.

Yu HJ, Ma H, Green RD. 1993. Calcium entry via L-type calcium channels acts as a negative regulator of adenylyl cyclase activity and cyclic AMP in cardiac myocytes. *Mol Pharmacol* **44:** 689–693.

Zaccolo M. 2006. Phosphodiesterases and compartmentalized cAMP signalling in the heart. *Eur J Cell Biol* **85:** 693–697.

Zimmermann G, Taussig R. 1996. Protein kinase C alters the responsiveness of adenylyl cyclases to G protein α and βγ subunits. *J Biol Chem* **271:** 27161–27166.

Protein Kinase C: The "Masters" of Calcium and Lipid

Peter Lipp[1] and Gregor Reither[2]

[1]Institute for Molecular Cell Biology, Medical Faculty, Saarland University, Homburg/Saar, Germany
[2]Cell Biology and Biophysics Unit, EMBL, Heidelberg, Germany

Correspondence: Peter.Lipp@uniklinikum-saarland.de

The coordinated and physiological behavior of living cells in an organism critically depends on their ability to interact with surrounding cells and with the extracellular space. For this, cells have to interpret incoming stimuli, correctly process the signals, and produce meaningful responses. A major part of such signaling mechanisms is the translation of incoming stimuli into intracellularly understandable signals, usually represented by second messengers or second-messenger systems. Two key second messengers, namely the calcium ion and signaling lipids, albeit extremely different in nature, play an important and often synergistic role in such signaling cascades. In this report, we will shed some light on an entire family of protein kinases, the protein kinases C, that are perfectly designed to exactly decode these two second messengers in all of their properties and convey the signaling content to downstream processes within the cell.

Once generated, second messengers relay their information content in a plethora of properties, including time, quantity (i.e., concentration), space (i.e., subcellular distribution), and interestingly into any combination of these three characteristics. Nevertheless, such information is meaningless for the cell unless it has a toolkit of read-out systems that can actually interpret such second-messenger properties and relate them further downstream into complex signaling networks, or directly to effector systems. An important system is the family of protein kinase Cs (PKCs) that can read-out lipid signals alone, or combine the ability to read-out simultaneous lipid and Ca^{2+} signals. A common denominator of all PKCs is the property to convey signals downstream by phosphorylation of additional signaling partners or effector proteins. We will briefly introduce the PKC subfamilies with particular emphasis on their signaling ability, discuss the important sensing domains, and their properties, before concentrating on sensing details of the subfamily of conventional PKCs and their role in signal integration in greater depth.

THE PKC FAMILY

In a mammalian cell, members of the PKC family represent $\sim 2\%$ of the entire kinome (Mellor and Parker 1998; Manning et al. 2002) and display an almost ubiquitous expression throughout the human body. The PKC family comprises 10 family members that are grouped into three subfamilies based on their particular domain composition and arrangement (see Fig. 1). Despite their widespread expression, the assignment of particular isoforms to specific functions has been difficult because PKCs display a signaling network of their own. The activity of novel PKCs (nPKCs) is, for example, modulated by the activity of atypical PKCs (aPKCs) (Parekh et al. 2000). In recent years, the application of knockout (KO) mouse models has proven seminal in unraveling some of the functions that can now be assigned to a particular PKC isoform, for example (Leitges 2007). However, results from KO approaches have proven difficult to interpret. For example, although the nPKC isoform, PKCδ, is known to be expressed ubiquitously, the KO mouse, deficient of PKCδ in all cells, showed no obvious phenotype (Leitges et al. 2001; Miyamoto et al. 2002). Only extensive analysis revealed mild phenotypes in specialized cell types, such as cells from the bone marrow (Leitges et al. 2002). The extensive networking among PKC members might thus render the interpretation of such mouse models rather difficult.

Very often enzymes gain substrate specificity via interaction of their activity domain (or nearby domains) and particular substrates. Unfortunately, for PKCs the situation appears to be more complex because their kinase domain, despite smaller differences, appears to be rather unspecific toward substrates (Kennelly and Krebs 1991; Pearce et al. 2010). It is generally believed that the PKC consensus phosphorylation sequence varies little between PKC isoforms. Large overlapping phosphorylation can be found for the same target between different families of AGC kinases (Zhang et al. 2006). From these findings, the following question is rather obvious: How do PKCs gain specificity if the kinase domains' contribution to specific phosphorylation appears rather low? This question becomes even more central when one considers that the 10 PKC family members share overlap in terms of substrate specificity, but phosphorylate specific targets. Although unique functions have been assigned to particular PKCs, this information does not illuminate the mechanisms underlying their specificity.

An answer to the question of specificity can partly be found in the regulatory domain(s) of PKCs. Although, depending on the particular isoform, the regulatory domain(s) bind lipids and in some cases also Ca^{2+} ions, how exactly PKCs achieve specificity is still rather unclear. Among other domains, PKCs contain C1 domain(s) (C1a and C1b) that might contribute specificity, but they bind lipids rather than substrates. It should be noted here though, that some evidence for putative direct C1–protein interaction was reported recently (Colon-Gonzalez and Kazanietz 2006). In addition, conventional PKCs (cPKCs) contain a Ca^{2+}-binding C2 domain, but Ca^{2+} ions of course do not serve as substrates. Thus, we suggest that the lack of substrate specificity attributed to the different PKC kinase domains is actually an important feature for PKCs, because they ought not to express specificity through classical enzyme–substrate interactions but rather through the mechanism of targeting. This means that PKCs follow the idea of local signaling domains in an overwhelmingly dynamic way. Typically, PKCs are only activated within a local signaling domain where specific lipids (such as DAGs) are produced to allow binding to C1 domain(s), or where Ca^{2+} signaling occurs to trigger a complex choreography of binding steps (see below, and for a recent review see Rosse et al. 2010).

ACTIVATION OF PKCs

To understand substrate-specific interactions, we need to have a close look into the mechanisms that lead to activation of PKCs. The steps leading to activation can be divided into two major episodes that are shared by almost all PKC members: (1) maturation and priming and (2) acute activation of the enzymatic activity. In this review, we will not focus on the first episode(s), because a recent review by Alexandra Newton has explained such steps in detail (Newton 2010).

In brief, once the translation of the PKC protein has been accomplished, highly ordered steps of phosphorylations on the protein itself are needed to gain activation competence. The

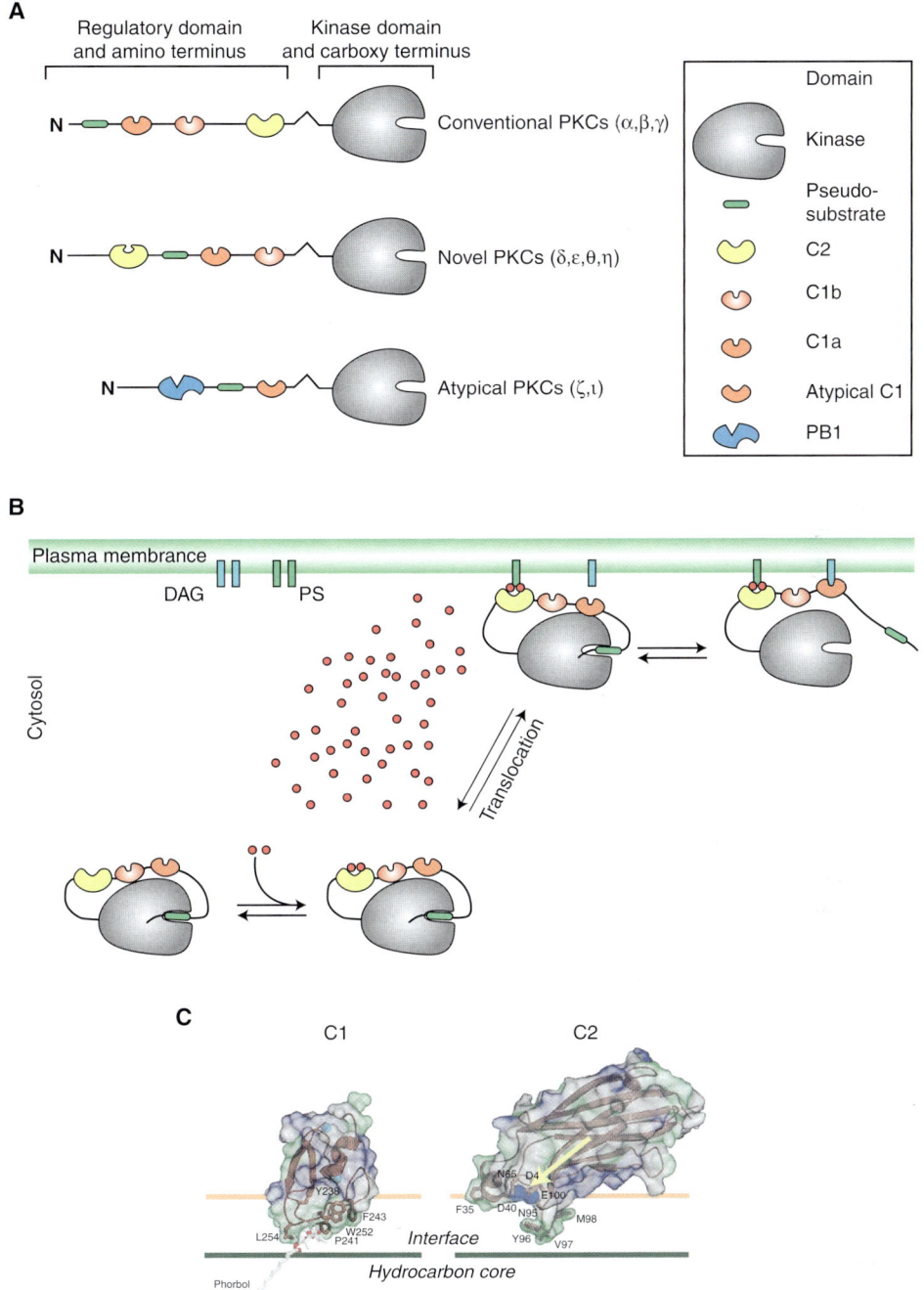

Figure 1. Basic properties of protein kinases C. (*A*) The general domain structure of the three subfamilies of the PKCs. The *inset* explains the symbols used. (*B*) After maturation and priming steps, activation of the kinase activity involves sequential binding of Ca^{2+} to the C2 domain, translocation to the plasma membrane, and binding of the C1a domain to DAG. (*C*) Interaction of the C1 domain with lipids of the plasma membrane (*left*) is much more intimate than for the Ca^{2+} (blue spheres marked with an arrow) loaded C2 domain (*right*). (*C*, Adapted from Hurley and Misra 2000; reprinted with permission from Annual Reviews, Inc. © 2000.)

current understanding is that both maturation and priming might be regulated by an entire group of cofactors, including, but not restricted to, partners stabilizing certain spatial arrangements of the immature PKC protein, such as chaperones (Gould et al. 2009). These stable intermediates are essential for proper phosphorylation to occur. Additional necessary phosphorylation steps involve the activity of the kinase PDK-1 that specifically modifies residues in the activation loop. These phosphorylation events serve a twofold function: (1) stabilizing the structure of the kinase domain and (2) specific priming of the enzymatic activity of the PKC molecule. Such PDK-1-mediated modifications are characteristic for all PKC family members. The particular importance of PDK-1 for PKC was shown by Balendran and coworkers, who reported that PDK-1 knockdown down-regulates PKC levels most likely by increased proteolysis of immature PKCs (Balendran et al. 2000). Nevertheless, all these maturation and priming steps only represent foreplay leading to the real action—the induction of the kinase activity itself.

For all PKC family members, activation of the kinase domain involves dissociation of an autoinhibitory domain from the kinase activity center. These domains contain amino acid sequences mimicking substrates for the kinase domain, but lack the ability to be phosphorylated. Although for cPKCs this pseudosubstrate is almost amino-terminal, for nPKCs and aPKCs, it is buried deeper in the regulatory part of the PKC molecule (see Fig. 1).

The members of the nPKC subfamily contain C2 domains, but for these C2 domains the required acidic Ca^{2+}-binding pocket is absent, thus their Ca^{2+}-binding affinity is too low to be relevant physiologically (reviewed in Hurley and Misra 2000). Therefore, nPKCs solely rely on lipid binding of their C1 domains. Both C1 domains, C1a and C1b, display different binding affinities for DAGs and/or phorbol esters. The C1a domain is usually associated with DAG binding, whereas the C1b domain conveys phorbol ester binding (Colon-Gonzalez and Kazanietz 2006). Interestingly, even among nPKCs, differences between this distribution of properties has been reported. Although the C1 domains of PKCε share an apparent equal binding to DAG and phorbol esters, the C1a domain of PKCδ apparently prefers DAG to phorbol esters (Stahelin et al. 2005a). In all instances, binding of the C1 domain(s) to their appropriate lipid partners releases the pseudosubstrate from the kinase-binding pocket leading to kinase activation.

The aPKCs represent a borderline PKC subfamily when considering the activation mechanism. Activation of aPKCs is not mediated by binding to DAG, phosphatidylserines, or phorbol esters because their C1 domain is altered (atypical C1 domain). Binding to these classical PKC activators in this atypical C1 domain is suppressed threefold: (1) The C1 domains structure is de-formed into a "flat surface" that is not able to penetrate into the lipid layer necessary for hydrophobic interactions, (2) the C1 domain cannot unzip to create the necessary lipid-binding pocket, and (3) hydrophilic amino acids disrupt the hydrophobicity of the binding pocket (Mott et al. 1996; Zhou et al. 2002). For a comparison between the varieties of C1 domains, please refer to a recent review (Colon-Gonzalez and Kazanietz 2006). Nevertheless, recently, binding of the atypical C1 domain to other lipid partners such as phosphatidic acid, ceramide, and PIP_3 has been implicated (Hirai and Chida 2003). Such interaction with the lipid releases autoinhibition, but still requires activation steps in the form of phosphorylation events at the PKC molecule itself, where PDK-1 again appears to prime the kinase and autophosphorylation finally results in increased kinase activity (Hirai and Chida 2003). In addition to such lipid-dependent phosphorylation of the atypical PKCs, direct protein–protein interactions through their PB1 domain can modulate their activity (see Fig. 1) (Puls et al. 1997; Qiu et al. 2000).

The Dynasty of Conventional PKCs: Decoding Ca^{2+} and Lipids

The cPKCs display the most complex and dynamic activation patterns among all PKC family members. Although maturation and priming of cPKCs follows a very similar scheme as described above and reviewed in Newton (2010), cPKCs reside in the cytosol during resting periods waiting to be triggered by rises in intracellular Ca^{2+} concentration. In this section, we will have a closer look at various aspects of how cPKCs decode Ca^{2+} and lipid signals, whereas the following section highlights feedback and cPKCs' influence on Ca^{2+} signaling.

More than anywhere else in the web of PKCs, the activation of cPKCs depends on its distribution; "location, location, location" is critical for specificity of action. As visualized in a seminal report by Oancea and Meyer (1998), cPKCs are positioned at the crossing of two important cellular signaling routes: receptor-mediated activation of PLCs (and consequently production of DAG) and liberation of Ca^{2+} from internal stores or influx from the extracellular space.

To perform these tasks, cPKCs contain a toolset of two domain structures (namely C1 and C2) that enable the molecules to sense DAG/phorbol ester production and Ca^{2+} increases, respectively (Newton and Johnson 1998). In this section, we summarize our current knowledge about the complex interaction between cPKCs and Ca^{2+} with particular respect to how the Ca^{2+} signaling toolkit (Berridge et al. 2000) can be read-out by cPKCs and referred downstream in the signaling cascade.

As already mentioned for the other PKC subfamilies, we have to consider the particular way cPKCs are activated to understand how cPKCs are indeed able to interpret both Ca^{2+} and DAG signals in living cells. Figure 1 gives a summary of our current notion on the final activation steps after maturation and priming of the molecule itself. Under resting conditions (i.e., basal Ca^{2+} concentration of $\sim$100 nM), cPKCs distribute in the cytosol with basically no membrane binding. It should be noted that in the absence of Ca^{2+} membrane binding of cPKCs does not occur, even when DAG is present in the plasma membrane. There might be two reasons for this: (1) Negative charges in the Ca^{2+}-binding pocket might actually repel the molecule from the plasma membrane; in particular, when considering the large amount of negatively charged head groups of phospholipids in the inner leaflet of the plasma membrane and (2) in the Ca^{2+}-free form, the conformation of the C1/C2 tandem domain might prevent the C1 domain(s) from encountering DAG in the plasma membrane (Stahelin et al. 2005b).

Interaction Between C2 Domains and Phospholipids

Although binding between DAG and the C1 domain(s) appears reasonably straightforward, there are at least three different mechanisms for the Ca^{2+}-dependent C2 membrane interaction (see also Hurley and Misra 2000).

Conformational Changes Induced by Ligation of Ca^{2+}: Binding of Ca^{2+} ions at the three Ca^{2+}-binding regions (CBR1-3) induces a conformational change in the C2 domains in such a way that phospholipid binding will become possible. There are at least two lines of evidence that support that notion. Growing crystals of C2 domains is only possible in the state where C2 is occupied by a single Ca^{2+}; binding of the second one renders the crystal unstable (Sutton and Sprang 1998). Although the C2 domains of PKCα and PKCγ only bind 2 Ca^{2+}, PKCβ is able to bind 3 Ca^{2+} in a cooperative manner, also suggesting major conformational changes in the C2 domains during successive Ca^{2+} binding (Kohout et al. 2002). In some studies, such changes extend well beyond the C2 domain, giving rise to further conformational changes in the C1 and pseudosubstrate domains (Bolsover et al. 2003).

Bridging between C2 Domain and Phospholipid in Plasma Membrane: In comparison to the electrostatic role of Ca^{2+} detailed below, the notion of Ca^{2+} bridging assumes that both partners, the CBRs of the C2 and the head groups of the phospholipids, coordinate the arrangement of the bound Ca^{2+} ions (Dessen et al. 1999; Verdaguer et al. 1999). Similar Ca^{2+} bridging had been described for another Ca^{2+}-sensing and membrane-binding protein family, the annexins (Swairjo et al. 1995).

Electrostatic Interaction between Ca^{2+}-C2 and Plasma Membrane: As described above, the binding pocket for Ca^{2+}, flanked by the CBRs, presents the plasma membrane with an array of negative charges that not only discourage membrane phospholipid interactions, but almost results in repelling actions between the C2 domain and phospholipids (Rizo and Südhof 1998; Ubach et al. 1998). Chelating Ca^{2+} ions by the CBRs switches this behavior entirely, because in the Ca^{2+}-bound form, the surface charge has now changed from negative to positive and the Ca^{2+}-C2 domain now becomes very much attracted to phospholipids. A seminal study by Kohout and coworkers has shown the enormous rapidity with which cPKCs associate with membrane after a stepwise Ca^{2+} increase (Kohout et al. 2002; Reither et al. 2006).

Seeing Is Believing: Imaging cPKC Dynamics in Living Cells

The advent of sophisticated imaging techniques together with the approach of fusing cPKCs with fluorescent proteins (FPs) allowed the direct visualization and analysis of cPKC dynamics in living cells (Meyer and Oancea 2000; see also Fig. 2A). Here, the group around Tobias Meyer has performed groundbreaking work in not only demonstrating rapid cPKC membrane translocation, but also characterizing such behavior in detail using mutational approaches (Oancea and Meyer 1998). Applying line-scan analysis of cPKC translocation together with line-scan FRAP approaches, Michael Schaefer and coworkers showed that rapid translocation of cPKCs to the plasma membrane does not necessarily require specific transport processes or specific receptors at the plasma membrane, free diffusion was sufficient to explain the observed translocation kinetics (Schaefer et al. 2001). We recently demonstrated that Ca^{2+}-dependent membrane association also occurs in mutated PKCα with a dysfunctional DAG-binding domain (Reither et al. 2006). By rendering the C1 domain DAG insensitive, we were able to show that Ca^{2+} association to the C2 domain is both sufficient and necessary for rapid membrane association (Reither et al. 2006).

Most of the studies reporting cPKC dynamics in living cells concentrate on analyzing global changes of the cPKC distribution (e.g., Oancea and Meyer 1998; for more details see Violin and Newton 2003). But if cPKCs are indeed versatile decoding machines for Ca^{2+} then the question arises as to whether they are able to sense the entire Ca^{2+} signaling toolkit, from the global, homogeneous Ca^{2+} signals, to the rapidly propagating Ca^{2+} wave, and finally spatially restricted Ca^{2+} signals, including elementary Ca^{2+} transients, such as Ca^{2+} puffs (Bootman et al. 1997a,b; Lipp et al. 1997; Koizumi et al. 1999).

First indications for cPKCs' ability to translocate to the plasma membrane in a spatially restricted manner was provided by Tobias Meyers group (Codazzi et al. 2001) in a study using PKCγ fusion proteins with FP tags. They analyzed translocation behavior of PKCγ in astrocytes using total internal reflection microscopy (TIRF) to restrict fluorescence detection to the basal plasma membrane (Fig. 2A). In the same report, the authors showed that similarly restricted translocations could also be found when expressing C2-FP constructs alone, suggesting that they were looking at the effects of local Ca^{2+} signals (Fig. 2Ab) (Codazzi et al. 2001). Nevertheless, their very low temporal resolution (around 6 s per image) together with the lack of a simultaneous Ca^{2+} recording prevented a direct link between these two events.

Recently, we were able for the first time to provide compelling evidence that PKCα was able to decode Ca^{2+} signals generated by various stimulation strengths. Threshold stimulation of nonexcitable cells with Ca^{2+}-mobilizing agonists leads to a breakdown of the global signal into spatially restricted Ca^{2+} transients (Bootman et al. 1997b; Berridge et al. 2000). Figure 2B depicts such data, indicating that threshold stimulation (5 μM ATP) resulted in a spatially restricted Ca^{2+} signal (Fig. 2Bc, lower row, left) whereas supra-threshold stimulation caused global Ca^{2+} increases (Fig. 2Bc, lower row, right). Importantly, translocation of the PKCα fusion protein very closely resembled the spatiotemporal properties of the underlying Ca^{2+} transients (Fig. 2Bc, upper row). Such data suggested that activation (i.e., membrane

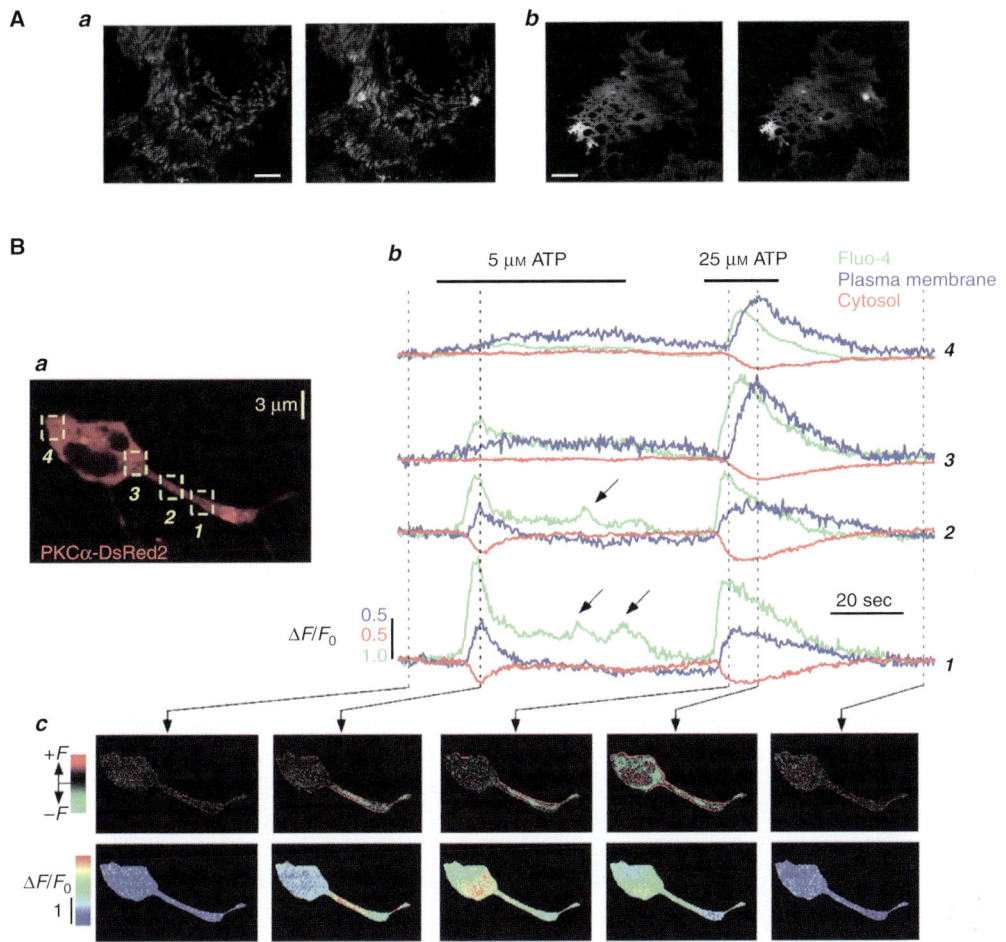

Figure 2. Spatially restricted cPKC translocations are driven by localized Ca^{2+} signals. (A) Astrocytes were imaged by TIRF microscopy and stimulated by 100 μM glutamate. Spatially restricted plasma membrane accumulation of GFP-PKCγ (a) and GFP-C2 (b) can be observed. Scale bar, 5 μm. (Adapted from Codazzi et al. 2001; reprinted with permission from Elsevier © 2001.) (B) HEK cells expressing PKCα-DsRed2 were loaded with fluo4 and stimulated with threshold (5 μM ATP) and suprathreshold (25 μM ATP) agonist concentrations. Spatially restricted Ca^{2+} signals (Bc, lower row, 5 μM) resulted in localized PKC translocation to the plasma membrane (Bc, upper row, 5 μM) while global Ca^{2+} (Bc, lower row, 25 μM) induced global translocation (Bc, upper row, 25 μM). Numbers in Bb correspond to regions of interest in Ba. (Adapted from Reither et al. 2006; reprinted with permission from The Rockefeller University Press © 2006.)

association) and deactivation (i.e., membrane dissociation) might actually be extremely dynamic and short lived, a notion we will discuss below in greater detail.

We and others have reported on the breakdown of global Ca^{2+} transients into elementary Ca^{2+} signals, referred to as Ca^{2+} sparks and Ca^{2+} puffs for ryanodine receptor and $InsP_3$ receptor signals, respectively (reviewed in Lipp and Niggli 1996; Berridge et al. 1999). It is well acknowledged that such elementary Ca^{2+} transients serve as building blocks for more complex Ca^{2+} transients, including Ca^{2+} waves but whether these spatially and temporally restricted signals may have a local signaling role on their own is still unclear. For a universally applied Ca^{2+} read-out system, cPKCs should thus also be able to sense these Ca^{2+}

transients and translate them into highly localized translocation events. By using rapid 2D confocal imaging of PKCα-FP-expressing cells and threshold stimulations, we were able to show elementary Ca^{2+} release-mediated local translocation events (LTEs) (see Fig. 3) (Reither et al. 2006). Thus, a possible signaling role can be directly assigned to elementary Ca^{2+} release sites close to the plasma membrane (≤ 1.5 μm): they might serve as a tool for cells to induce spatially restricted signaling cascades by local recruitment of cPKCs and ensuing downstream events. Such versatility in reading-out of local Ca^{2+} signals has as of yet only directly been shown for cPKCs, although similar abilities have been suggested for another very important Ca^{2+} read-out sensor: calmodulin (Deisseroth et al. 1998; Wheeler et al. 2008).

Terminating cPKC Signaling

Equally important to initiating cPKC signaling by rises in intracellular Ca^{2+} is the termination of such potent signaling molecules. For this, it would be beneficial to be able to follow cPKC phosphorylation in living cells.

Two such approaches have been established so far, referred to as CKAR (Violin et al. 2003; Gallegos et al. 2006) and KCP (Schleifenbaum et al. 2004; Brumbaugh et al. 2006). Both have used the fluorescence resonance energy transfer (FRET) approach with different phosphorylation sensors. Roger Tsien and Alexandra Newton have incorporated a construct comprised of a consensus sequence and a specific binding domain recognizing the phosphorylated PKC consensus sequence (Violin et al. 2003). Alternatively, Carsten Schultz's group have used conformational bending based on multiple phosphorylations in a kind of "hinge" region of the sensor (Schleifenbaum et al. 2004). In both cases, PKC-dependent phosphorylation of the sensor results in a conformational rearrangement of CFP and YFP at both ends of the sensors and thus changing the FRET ratio that can be measured.

In the original report using the CKAR probe, the investigators were able to show that

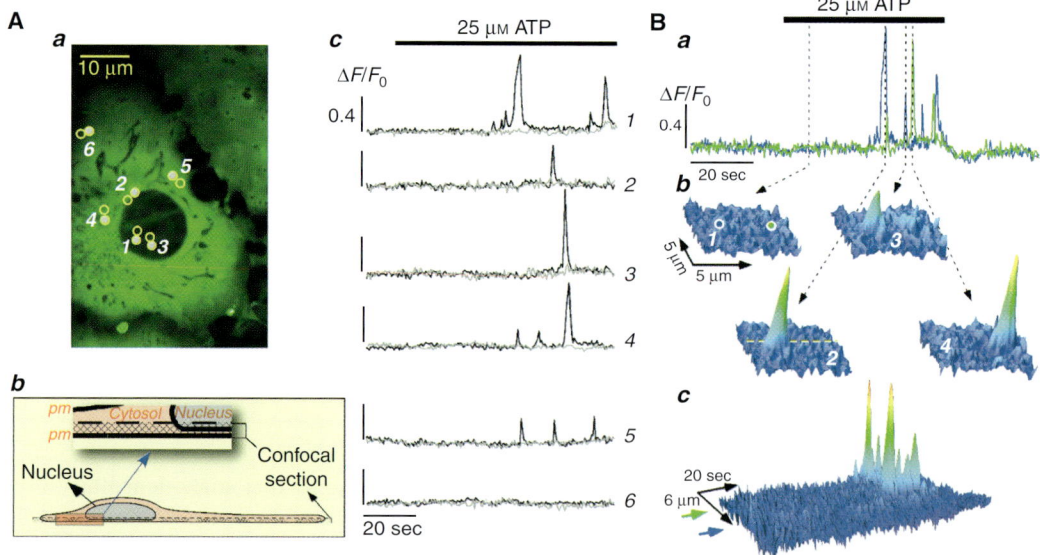

Figure 3. Local translocation events (LTEs) in PKCα-expressing cells. (*A*) LTEs were highly spatially restricted and occurred randomly in the cell. (*Aa,c*) Open circles represent LTE locations whereas gray circles display neighboring locations. (*Ab*) Arrangement of confocal recording. (*B*) Spatiotemporal properties of LTEs. The pseudo-line-scan images in *Bc* were constructed along the dashed line indicated in *Bb2*. (*B*, Adapted from Reither et al. 2006; reprinted with permission from The Rockefeller University Press © 2006.)

oscillations in Ca^{2+} are directly translated into oscillations of substrate phosphorylation (Violin et al. 2003); a novel and important finding. When analyzing their results in greater detail, especially the temporal relationship between Ca^{2+} increases and phosphorylation to switch on and off, they found a time delay between these two events of only 10 s, which was not far off the temporal resolution of their recording. From these findings, one has to conclude that the switching of the cPKCs from off to on, and vice versa, is almost instantaneous, even considering that the phosphorylation probe only senses the steady-state between phosphorylation events (via PKCs) and dephosphorylation by phosphatases. But how are cPKC switched off?

In biochemical studies using cPKC-C2 domains and analyzing their membrane association and dissociation kinetics, Newton's and Kohout's groups have found very fast kinetics for both processs (Nalefski and Newton 2001; Kohout et al. 2002). Although the k_{on} rate for membrane association was $\sim 10^{10}$ $M^{-1}sec^{-1}$, membrane dissociation k_{off} was still very fast (~ 150 sec^{-1}) (Nalefski and Newton 2001). Because the Ca^{2+} dissociation was in the same order of magnitude (Kohout et al. 2002), Ca^{2+} unbinding and membrane dissociation are quasi-simultaneous. But, cPKCs also interact with the plasma membrane through C1 domain(s) that are believed to display much slower kinetics for membrane interactions. A recent biochemical study has indeed shown that the isolated C1 domain of PKCβ exhibits much slower kinetics. Here k_{on} and k_{off} were measured to be 10^9 $M^{-1}sec^{-1}$ and 0.1 sec^{-1}, respectively (Dries and Newton 2008). It has to be mentioned here that Dries and Newton found two distinct, kinetically different membrane-binding modes of C1 domains: a weak but fast binding/unbinding to phosphatidylserine and a high affinity but slow interaction with DAG/PMA (Dries and Newton 2008).

Thus, up to now we can draw the following picture (cf. Fig. 1): Membrane association in cPKCs appears to be solely driven by C2-mediated initial membrane contacts with anionic phospholipids, whereas retention of the cPKC at the membrane is dominated by DAG–C1 interactions that apparently not only provide the energy for pseudosubstrate dissociation and kinase activation but also massively increase the likelihood of kinase–substrate interactions by restricting the "search dimensions" from the 3D cytosol to the flat inner surface of the plasma membrane (i.e., 2D) (Kholodenko et al. 2000).

Although all those findings describe basic biochemical properties of cPKCs, the question arises whether any of such properties are reflected in the behavior of cPKCs in the living cell? For this, again, imaging techniques have provided compelling evidence for the notion put forward above. Flash photolysis of caged-Ca^{2+} compounds in living cells while simultaneously monitoring Ca^{2+} and PKCα translocation has supported the extremely fast membrane association of the kinase (Reither et al. 2006). Substantial membrane accumulation could be monitored less than 250 msec after photolytic Ca^{2+} increase.

When considering the kinetics of DAG production measured during agonist stimulation, which has been reported to show no, or only mild, oscillations (Codazzi et al. 2001; Bartlett et al. 2005), DAG production is rather unlikely to cause phosphorylation cycles of cPKCs as reported previously (Violin et al. 2003). As suggested earlier, such oscillations coincide with Ca^{2+} oscillations, which in turn should also be responsible for switching off kinase activity. But what exactly is the rate-limiting step for returning kinase activity to basal levels? Ca^{2+} levels alone cannot be the sole driving force behind that, because during stimulation with a Ca^{2+}-mobilizing agonist, DAG is also being produced. It appears feasible to assume that despite rapid decreases in Ca^{2+} at the plasma membrane, it will be DAG-C1 dissociation that will dominate membrane dissociation and eventually switching off the kinase activity. By analyzing membrane-dissociation kinetics in living cells, we have provided recent evidence that the C1–membrane interaction (presumably via DAG) is a major determinant of this process (Reither et al. 2006). Similar to stop-flow analysis in a spectrometer we have used photolysis of a caged-Ca^{2+} buffer (diazo-2) to decrease Ca^{2+} in a quasi-stepwise manner. In living cells, we were able to identify that membrane dissociation of PKCα was indeed slow and characterized by a bi-exponential kinetic. When we rendered the C1a domain DAG insensitive, the same experimental approach revealed fast membrane dissociation described by a single, rapid exponential decay of membrane fluorescence (Reither et al. 2006). The resulting speed of membrane dissociation was increased by more than an order of magnitude, providing compelling evidence for our notion put forward above. The fact that, in the wild-type PKCα, membrane dissociation was bi-exponential also strongly suggests that even during normal cellular stimulation there are two pools of membrane-bound cPKCs: (1) a loosely bound population, where binding is only mediated by weaker PKC–anionic phospholipid interaction mediated by Ca^{2+}-C2 and C1a domains, and (2) a tightly bound cPKC population, which interact via C1a-DAG binding and most likely represent the activated kinase population.

It is therefore feasible to assume that activation and inactivation of cPKCs are primarily driven by the DAG-C1a-binding kinetics, which appear rate limiting in both processes (Nalefski and Newton 2001; Dries and Newton 2008), but that the C2–membrane interaction is the major determinant for the localization of cPKCs to the plasma membrane. Thus, the spatiotemporal properties of the underlying Ca^{2+} signals determine the strength, timing, and localization of membrane association of cPKCs, but the strength of the agonist-mediated DAG production appears to be the determinant for the strength of cPKC activation. In this respect, the answer to the question of whether cPKC translocation driven by elementary Ca^{2+} signals (LTEs) does indeed signal downstream is still unknown and will only be answered when we have phosphorylation sensors available with a signal-to-noise ratio that allows sensing their FRET changes in cellular microdomains.

Up to now we have only discussed one part of the Ca^{2+}–cPKC interaction, namely how Ca^{2+} influences cPKC activity, but we have to consider a dialog between Ca^{2+} and cPKCs because in return cPKCs' activity can also modulate Ca^{2+} signaling, which will be the topic of the following section.

CONVENTIONAL PKCs AND BEYOND: cPKCs AND THEIR REGULATORY INFLUENCE ON Ca^{2+} HANDLING

In this section, we focus on the immediate functional feedback with respect to the modulation of Ca^{2+} regulators. The long-term effects of cPKC-dependent regulation of gene expression will not be considered here. In a Ca^{2+} centric view of cellular signaling, we would like to distinguish direct regulators of Ca^{2+} such as Ca^{2+} channels as positive modules together with the counteracting pumps from second level regulators (regulator of regulators). In the following, the effect of cPKCs on the first- and second-level regulators will be discussed.

First Level of Regulation

At the first level of regulation, we consider RyRs, $InsP_3Rs$, and other Ca^{2+} channels as main positive regulators, whereas SERCA, PMCA, and NCX are main effectors of Ca^{2+} clearance (Uhlen and Fritz 2010). Consistent with the fact that cPKCs main "operation" field is the plasma membrane; most of the known effectors are membrane associated or in the vicinity of it.

cPKC–PMCA

The plasma membrane ATPase (PMCA) has a PKCγ phosphorylation site at its carboxyl terminus that interferes with the inhibitory binding of calmodulin (Wang et al. 1991). In addition, PKCα is relevant for the up-regulation of PMCA activity (Penniston and Enyedi 1998).

cPKC–TRPs

As already highlighted in a previous contribution of this series (Gees et al. 2010), Ca^{2+} entry via TRPs is tightly regulated by cPKCs. A comprehensive overview of further regulatory action was provided recently (Venkatachalam and Montell 2007). Don Gill's group showed a down-regulation of the TRPC family by PKC-dependent phosphorylation of the channel protein itself (Venkatachalam et al. 2003). Several members of the TRPVs are up-regulated by cPKCs. Nevertheless, one has to consider that the situation is more complex because the activity of TRPCs and TRPVs is also modulated by PLC-dependent hydrolysis of PIP_2. Because PLCs are regulated by PKCs (see below), this relationship displays yet another level of PKC-dependent regulation of TRPs. The other subfamilies of TRP channels seem to only weakly depend on PKC activity, concluding that within the superfamily of TRPs the main effect of cPKC activity might be transduced by TRPC and TRPV.

cPKC–NCX

Na^+/Ca^{2+} exchangers together with SERCA pumps have the main negative effect on the cytosolic Ca^{2+} concentration (Berridge 2009). There is strong experimental evidence for NCX's activity, up-regulation by cPKC-dependent phosphorylation (Schulze et al. 2003). cPKCs seem to be integrated in a regulatory complex of various enzymes, such as additional kinases and phosphatases (Ruknudin et al. 2007).

Second Level of Regulation

Beside these direct regulators, there are several protein families, which are regulated by cPKCs and have a mediate effect on Ca^{2+} handling. This second level of regulation also integrates more PKC-independent actions. Therefore, the direct effect of cPKCs is often difficult to separate, especially because the various protein families are in a continuous dynamic state of intimate interaction. This renders predictions rather difficult. Nevertheless, cPKCs can be recognized as a signaling hub translating the increase in cytosolic Ca^{2+} in activity changes of a plethora of substrate proteins. Here we want to highlight the ubiquitously involved proteins that feed back Ca^{2+}-dependent PKC activity onto Ca^{2+} handling (see Fig. 4).

cPKC–PLC

PLCs, especially the members of the β subfamily, are known to be phosphorylated by PKCα (Strassheim and Williams 2000; Xu et al. 2004). The phosphorylation of PLCβ1 at S887 has an inhibitory effect. Regulatory effects of

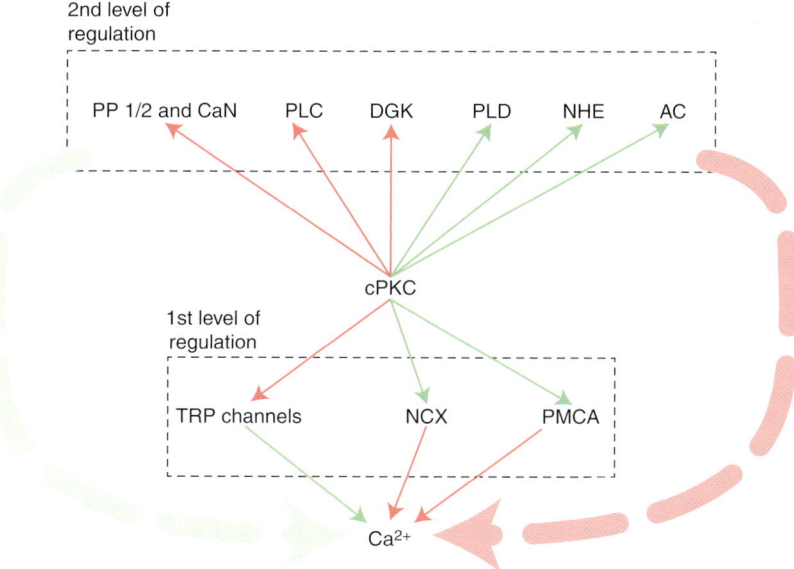

Figure 4. The activity of cPKCs is not only modulated by Ca^{2+} but in turn affects the activity of a plethora of control mechanisms that influence intracellular Ca^{2+} handling. Green arrows indicate positive effects, whereas the red color is indicative of negative effects.

PKCα on other isoforms such as PLCδ1, even though it might not be a direct effect, are reported as negative (Fujii et al. 2009). The regulation of PLCs interferes with signaling events dependent on PIP_2, DAG, and IP_3.

cPKC–PLD

The regulation of phospholipases D (PLD) by cPKCs opens another branch to the signaling effects of phosphatidic acid (PA). Several lines of evidence for a stimulatory effect of PKCα on PLDs have been reported (Hornia et al. 1999; Han et al. 2002; Chen and Exton 2004; Lee et al. 2009).

cPKC–DGK

Interestingly, there is a second route to influence PA signaling via diacylglycerol kinases (DGK). For DGKδ and DGKζ down-regulation by PKCα was reported (Luo et al. 2003; Imai et al. 2004).

cPKC–AC

The family of adenylate cyclases (AC) consists of nine membrane located isoforms. The effect of PKCs on ACs has already been well described by Halls and Cooper within this series of reviews (Halls and Cooper 2011). Despite the fact that not all of the ACs are up-regulated by cPKC phosphorylation, we would still consider a positive effect on the generation of cAMP as the main outcome. Experimental evidence for direct activation by PKCα was found for AC2 and AC5 (Kawabe et al. 1994; Zimmermann and Taussig 1996). The cPKC-driven increase of cAMP levels brings in a whole subnetwork of Ca^{2+} regulating events downstream from protein kinases A (Reiken et al. 2003; Schulze et al. 2003) and small GTPases via guanine exchange factor EPAC (Schmidt et al. 2001; Pereira et al. 2007).

cPKC–NHE

Another global effect by changing the intracellular pH on signaling could be mediated by the cPKC-dependent up-regulation of the sodium/proton exchangers (NHE). Even though there is only sparse evidence for a direct phosphorylation, supportive evidence has been presented for PKCβ-dependent up-regulation via another interactor (Takahashi et al. 1999; Itoh et al. 2005). The positive effect is also supported by experiments using cPKC-specific inhibitors to down-regulate NHEs (Pederson et al. 2002).

cPKC–Protein Phosphatases

Kinase activity is counteracted by phosphatases (see above). Interestingly, kinases and phosphatases inhibit each other resulting in a stabilizing competition (Srivastava et al. 2002; duBell and Rogers 2004). The effect of cPKCs is mediated by phosphorylation and activation of the inhibitory subunits of CPI-17 (Kolosova et al. 2004; Zemlickova et al. 2004). In addition, there is a report suggesting a competing effect of inhibitory PKA phosphorylation I-1 (Braz et al. 2004). Furthermore, evidence has been presented for down-regulation of calcineurin (Tung 1986; Hashimoto and Soderling 1989). We consider a transient down-regulation by cPKC activity because protein phosphatases remove their own inhibitory phosphates resulting in a positive feedback loop for global phosphatase activity. This will influence the whole kinase-phosphatase balance even though we assume that there might be different outcomes at confined local complex assemblies.

Taking all these feedback loops between cPKCs and global Ca^{2+} handling together appears to suggest, that cPKCs do impose mainly negative effect via different immediate and mediate regulatory branches. This is already supported by data from different groups (Young et al. 2002; Venkatachalam et al. 2003; Fontainhas et al. 2005; Sakwe et al. 2005).

INTEGRATION OF IMMEDIATE AND INTERMEDIATE cPKC-Ca^{2+} HANDLING REGULATORY LOOPS

As suggested already in an earlier contribution to this series (Dupont et al. 2010), theoretical models help to provide insights into the regulatory mechanisms of signaling networks. To test the effect of cPKCs on Ca^{2+} handling in detail and under the condition of a larger set of Ca^{2+} regulating entities we used the commercially available model DynaCellNet1.0 (www.deplecto.de). This model comprises 45 nodes downstream from G protein coupled receptors that are interconnected by 305 functional relations. The advantage of this modeling approach is that—despite the fact that models reduce complexity—it provides a system still with the characteristics of a complex system. The modeled Ca^{2+} transients are calculated in dependence of the changing states of the 44 other nodes. DynaCellNet1.0 provides semiquantitative predictions and is mainly a tool to test and generate hypotheses. To investigate the full range of cPKC-dependent regulation we arbitrarily fixed the cPKC activity at different levels, from very low (0) to massive (140). This setup will not directly be transferable into experiments, because fixing levels of enzyme activity is not possible, but it provides an overview of different states of cPKC activity beyond knockdown and overexpression.

The calculated Ca^{2+} responses (Fig. 5) result from a stimulation of ATP sensitive purinergic receptors (Wettschureck and Offermanns 2005). Beside others, all of the above discussed protein families are integrated. The response under "free-running, physiological" conditions, with no restrictions on cPKCs activity, is shown as control (Fig. 5A–D, black graphs). Despite a slight increase in peak amplitude the main effect of a reduction of cPKC activity

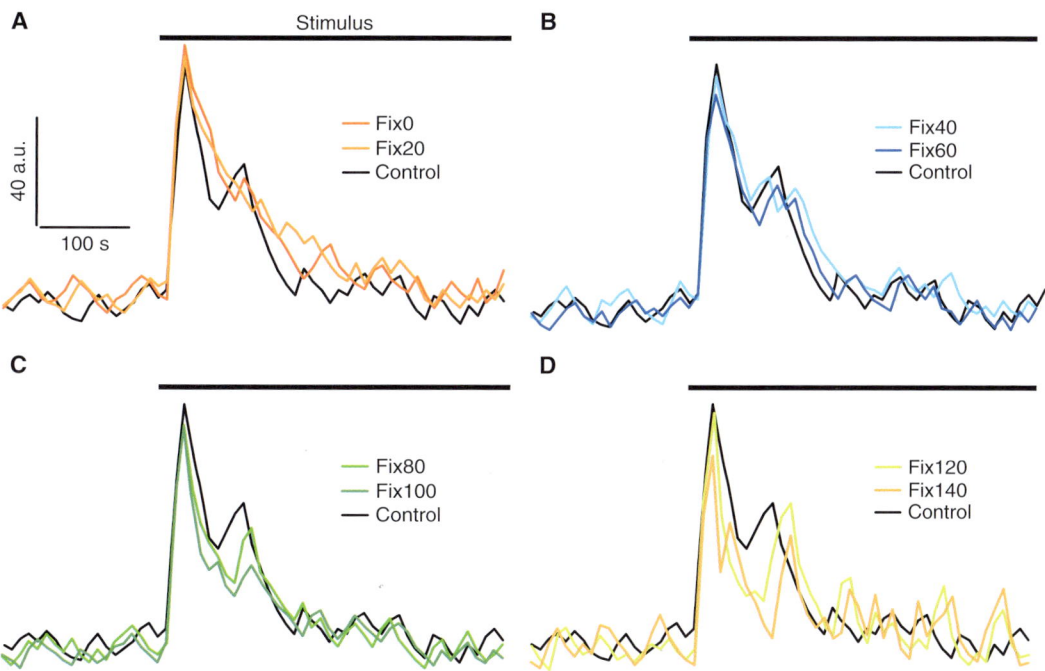

Figure 5. Modeled Ca^{2+} transients at different cPKC activity levels. The Ca^{2+} transient under control condition (A–D; black graph) was calculated with no restrictions on cPKC activity as a result of stimulation of purinergic receptors after 200 sec with ATP. (A) At low levels (fix0 and fix20) of cPKC activity the primary peak is increased in amplitude and Ca^{2+} removal is slowed down and less noisy. (B) Ca^{2+} transients modeled with medium cPKC activity levels (fix40 and fix60) are not distinguishable from control conditions. (C) At high levels of cPKC activity (fix80 and fix100) Ca^{2+} removal is accelerated resulting in a reduced duration and amplitude of the initial transient (fix100). (D) Extremely high levels (fix120 and fix140) of cPKC activity result in a decreased primary peak and an increase of noise in the tail of the transient most likely corresponding to additional Ca^{2+} oscillations.

levels is an attenuated clearance of cytosolic Ca^{2+} (Fig. 5A). A similar effect was observed by Ulianich and coworkers (2006). As expected further activation of cPKCs result into an accelerated clearance and a decrease in peak amplitude. Compared to published data, the effect is not very pronounced (Young et al. 2002; Venkatachalam et al. 2003; Sakwe et al. 2005), but the results shown in these reports are based on phorbol ester treatments, which might not exclusively be a result of PKC activity and/or will result in "unphysiological" activity levels and phosphorylation events (Kazanietz 2002). When we model a phorbol ester treatment the model shows a much more pronounced decrease in the amplitude of the Ca^{2+} transient. An interesting aspect of cPKC regulation is highlighted by the model at extremely high levels of cPKC activity (Fig. 5D). Apart from a massive down-regulation by the first level of regulation, the positive intermediate effects from the second level of regulation become more prominent. The increase of oscillatory behavior results from an "echo" of high cPKC activity in the system. The straightforward notion would be to assume a stimulating effect via cAMP-PKA-EPAC together with a reincrease in protein phosphatase activity via PLD and PA.

CONCLUSIONS

In the context of cellular Ca^{2+} signaling, cPKCs have the role of an important modulator, acting on the fine-tuning of Ca^{2+} responses. The question whether the "echo" plays a role might become relevant under patho-physiological conditions were either cPKC expression is up-regulated or one of the alternative routes (cAMP-PKA-EPAC or PLD-PA) has a pronounced effect because of changes in the composition of the different regulatory feedback loops.

REFERENCES

Balendran A, Hare GR, Kieloch A, Williams MR, Alessi DR. 2000. Further evidence that 3-phosphoinositide-dependent protein kinase-1 (PDK1) is required for the stability and phosphorylation of protein kinase C (PKC) isoforms. *FEBS Lett* **484:** 217–223.

Bartlett PJ, Young KW, Nahorski SR, Challiss RA. 2005. Single cell analysis and temporal profiling of agonist-mediated inositol 1,4,5-trisphosphate, Ca^{2+}, diacylglycerol, and protein kinase C signaling using fluorescent biosensors. *J Biol Chem* **280:** 21837–21846.

Berridge MJ. 2009. Module 5: Off mechanisms. *Cell signalling biology* Portland Press, Ltd, London.

Berridge M, Lipp P, Bootman M. 1999. Calcium signalling. *Curr Biol* **9:** R157–R159.

Berridge MJ, Lipp P, Bootman MD. 2000. The versatility and universality of calcium signalling. *Nat Rev Mol Cell Biol* **1:** 11–21.

Bolsover SR, Gomez-Fernandez JC, Corbalan-Garcia S. 2003. Role of the Ca^{2+}/phosphatidylserine binding region of the C2 domain in the translocation of protein kinase Cα to the plasma membrane. *J Biol Chem* **278:** 10282–10290.

Bootman M, Niggli E, Berridge M, Lipp P. 1997a. Imaging the hierarchical Ca^{2+} signalling system in HeLa cells. *J Physiol* **499:** 307–314.

Bootman MD, Berridge MJ, Lipp P. 1997b. Cooking with calcium: The recipes for composing global signals from elementary events. *Cell* **91:** 367–373.

Braz JC, Gregory K, Pathak A, Zhao W, Sahin B, Klevitsky R, Kimball TF, Lorenz JN, Nairn AC, Liggett SB, et al. 2004. PKC-α regulates cardiac contractility and propensity toward heart failure. *Nat Med* **10:** 248–254.

Brumbaugh J, Schleifenbaum A, Gasch A, Sattler M, Schultz C. 2006. A dual parameter FRET probe for measuring PKC and PKA activity in living cells. *J Am Chem Soc* **128:** 24–25.

Chen JS, Exton JH. 2004. Regulation of phospholipase D2 activity by protein kinase Cα. *J Biol Chem* **279:** 22076–22083.

Codazzi F, Teruel MN, Meyer T. 2001. Control of astrocyte Ca^{2+} oscillations and waves by oscillating translocation and activation of protein kinase C. *Curr Biol* **11:** 1089–1097.

Colon-Gonzalez F, Kazanietz MG. 2006. C1 domains exposed: From diacylglycerol binding to protein–protein interactions. *Biochim Biophys Acta* **1761:** 827–837.

Deisseroth K, Heist-Tsien RW. 1998. Translocation of calmodulin to the nucleus supports CREB phosphorylation in hippocampal neurons. *Nature* **392:** 198–202.

Dessen A, Tang J, Schmidt H, Stahl M, Clark JD, Seehra J, Somers WS. 1999. Crystal structure of human cytosolic phospholipase A2 reveals a novel topology and catalytic mechanism. *Cell* **97:** 349–360.

Dries DR, Newton AC. 2008. Kinetic analysis of the interaction of the C1 domain of protein kinase C with lipid membranes by stopped-flow spectroscopy. *J Biol Chem* **283:** 7885–7893.

duBell WH, Rogers TB. 2004. Protein phosphatase 1 and an opposing protein kinase regulate steady-state L-type Ca^{2+} current in mouse cardiac myocytes. *J Physiol* **556:** 79–93.

Dupont G, Combettes L, Birds GS, Putney JW. 2011. Calcium oscillations. *Cold Spring Harb Perspect Biol* **3:** a004226.

Fontainhas AM, Obukhov AG, Nowycky MC. 2005. Protein kinase Cα modulates depolarizaton-evoked changes of intracellular Ca^{2+} concentration in a rat pheochromocytoma cell line. *Neuroscience* **133**: 393–403.

Fujii M, Yi KS, Kim MJ, Ha SH, Ryu SH, Suh PG, Yagisawa H. 2009. Phosphorylation of phospholipase C-δ 1 regulates its enzymatic activity. *J Cell Biochem* **108**: 638–650.

Gallegos LL, Kunkel MT, Newton AC. 2006. Targeting protein kinase C activity reporter to discrete intracellular regions reveals spatiotemporal differences in agonist-dependent signaling. *J Biol Chem* **281**: 30947–30956.

Gees M, Colsoul B, Nilius B. 2010. The role of transient receptor potential cation channels in Ca^{2+} signaling. *Cold Spring Harb Perspect Biol* **2**: a003962.

Gould CM, Kannan N, Taylor SS, Newton AC. 2009. The chaperones Hsp90 and Cdc37 mediate the maturation and stabilization of protein kinase C through a conserved PXXP motif in the C-terminal tail. *J Biol Chem* **284**: 4921–4935.

Halls ML, Cooper DM. 2011. Regulation by Ca^{2+}-signaling pathways of adenylyl cyclases. *Cold Spring Harb Perspect Biol* **3**: a004143.

Han JM, Kim Y, Lee JS, Lee CS, Lee BD, Ohba M, Kuroki T, Suh PG, Ryu SH. 2002. Localization of phospholipase D1 to caveolin-enriched membrane via palmitoylation: Implications for epidermal growth factor signaling. *Mol Biol Cell* **13**: 3976–3988.

Hashimoto Y, Soderling TR. 1989. Regulation of calcineurin by phosphorylation. Identification of the regulatory site phosphorylated by Ca^{2+}/calmodulin-dependent protein kinase II and protein kinase C. *J Biol Chem* **264**: 16524–16529.

Hirai T, Chida K. 2003. Protein kinase Cζ (PKCζ): Activation mechanisms and cellular functions. *J Biochem* **133**: 1–7.

Hornia A, Lu Z, Sukezane T, Zhong M, Joseph T, Frankel P, Foster DA. 1999. Antagonistic effects of protein kinase C α and δ on both transformation and phospholipase D activity mediated by the epidermal growth factor receptor. *Mol Cell Biol* **19**: 7672–7680.

Hurley JH, Misra S. 2000. Signaling and subcellular targeting by membrane-binding domains. *Annu Rev Biophys Biomol Struct* **29**: 49–79.

Imai S, Kai M, Yamada K, Kanoh H, Sakane F. 2004. The plasma membrane translocation of diacylglycerol kinase δ1 is negatively regulated by conventional protein kinase C–dependent phosphorylation at Ser-22 and Ser-26 within the pleckstrin homology domain. *Biochem J* **382**: 957–966.

Itoh S, Ding B, Bains CP, Wang N, Takeishi Y, Jalili T, King GL, Walsh RA, Yan C, Abe J. 2005. Role of p90 ribosomal S6 kinase (p90RSK) in reactive oxygen species and protein kinase C β (PKC-β)–mediated cardiac troponin I phosphorylation. *J Biol Chem* **280**: 24135–24142.

Kawabe J, Iwami G, Ebina T, Ohno S, Katada T, Ueda Y, Homcy CJ, Ishikawa Y. 1994. Differential activation of adenylyl cyclase by protein kinase C isoenzymes. *J Biol Chem* **269**: 16554–16558.

Kazanietz MG. 2002. Novel "nonkinase" phorbol ester receptors: The C1 domain connection. *Mol Pharmacol* **61**: 759–767.

Kennelly PJ, Krebs EG. 1991. Consensus sequences as substrate specificity determinants for protein kinases and protein phosphatases. *J Biol Chem* **266**: 15555–15558.

Kholodenko BN, Hoek JB, Westerhoff HV. 2000. Why cytoplasmic signalling proteins should be recruited to cell membranes. *Trends Cell Biol* **10**: 173–178.

Kohout SC, Corbalan-Garcia S, Torrecillas A, Gomez-Fernandez JC, Falke JJ. 2002. C2 domains of protein kinase C isoforms α, β, and γ: Activation parameters and calcium stoichiometries of the membrane-bound state. *Biochemistry* **41**: 11411–11424.

Koizumi S, Bootman MD, Bobanovic LK, Schell MJ, Berridge MJ, Lipp P. 1999. Characterization of elementary Ca^{2+} release signals in NGF-differentiated PC12 cells and hippocampal neurons. *Neuron* **22**: 125–137.

Kolosova IA, Ma SF, Adyshev DM, Wang P, Ohba M, Natarajan V, Garcia JG, Verin AD. 2004. Role of CPI-17 in the regulation of endothelial cytoskeleton. *Am J Physiol Lung Cell Mol Physiol* **287**: L970–980.

Lee CS, Kim KL, Jang JH, Choi YS, Suh PG, Ryu SH. 2009. The roles of phospholipase D in EGFR signaling. *Biochim Biophys Acta* **1791**: 862–868.

Leitges M. 2007. Functional PKC in vivo analysis using deficient mouse models. *Biochem Soc Trans* **35**: 1018–1020.

Leitges M, Sanz L, Martin P, Duran A, Braun U, Garcia JF, Camacho F, Diaz-Meco MT, Rennert PD, Moscat J. 2001. Targeted disruption of the ζPKC gene results in the impairment of the NF-κB pathway. *Mol Cell* **8**: 771–780.

Leitges M, Gimborn K, Elis W, Kalesnikoff J, Hughes MR, Krystal G, Huber M. 2002. Protein kinase C-δ is a negative regulator of antigen-induced mast cell degranulation. *Mol Cell Biol* **22**: 3970–3980.

Lipp P, Niggli E. 1996. A hierarchical concept of cellular and subcellular Ca^{2+}-signalling. *Prog Biophys Mol Biol* **65**: 265–296.

Lipp P, Thomas D, Berridge MJ, Bootman MD. 1997. Nuclear calcium signalling by individual cytoplasmic calcium puffs. *EMBO J* **16**: 7166–7173.

Luo B, Prescott SM, Topham MK. 2003. Protein kinase C α phosphorylates and negatively regulates diacylglycerol kinase ζ. *J Biol Chem* **278**: 39542–39547.

Manning G, Whyte DB, Martinez R, Hunter T, Sudarsanam S. 2002. The protein kinase complement of the human genome. *Science* **298**: 1912–1934.

Mellor H, Parker PJ. 1998. The extended protein kinase C superfamily. *Biochem J* **332**: 281–292.

Meyer T, Oancea E. 2000. Studies of signal transduction events using chimeras to green fluorescent protein. *Methods Enzymol* **327**: 500–513.

Miyamoto A, Nakayama K, Imaki H, Hirose S, Jiang Y, Abe M, Tsukiyama T, Nagahama H, Ohno S, Hatakeyama S, et al. 2002. Increased proliferation of B cells and autoimmunity in mice lacking protein kinase Cδ. *Nature* **416**: 865–869.

Mott HR, Carpenter JW, Zhong S, Ghosh S, Bell RM, Campbell SL. 1996. The solution structure of the Raf-1 cysteine-rich domain: A novel ras and phospholipid binding site. *Proc Natl Acad Sci* **93**: 8312–8317.

Nalefski EA, Newton AC. 2001. Membrane binding kinetics of protein kinase C βII mediated by the C2 domain. *Biochemistry* **40:** 13216–13229.

Newton AC. 2010. Protein kinase C: Poised to signal. *Am J Physiol Endocrinol Metab* **298:** E395–E402.

Newton AC, Johnson JE. 1998. Protein kinase C: A paradigm for regulation of protein function by two membrane-targeting modules. *Biochim Biophys Acta* **1376:** 155–172.

Oancea E, Meyer T. 1998. Protein kinase C as a molecular machine for decoding calcium and diacylglycerol signals. *Cell* **95:** 307–318.

Parekh DB, Ziegler W, Parker PJ. 2000. Multiple pathways control protein kinase C phosphorylation. *Embo J* **19:** 496–503.

Pearce LR, Komander D, Alessi DR. 2010. The nuts and bolts of AGC protein kinases. *Nat Rev Mol Cell Biol* **11:** 9–22.

Pederson SF, Varming C, Christensen ST, Hoffmann EK. 2002. Mechanisms of activation of NHE by cell shrinkage and by calyculin A in Ehrlich ascites tumor cells. *J Membr Biol* **189:** 67–81.

Penniston JT, Enyedi A. 1998. Modulation of the plasma membrane Ca^{2+} pump. *J Membr Biol* **165:** 101–109.

Pereira L, Metrich M, Fernandez-Velasco M, Lucas A, Leroy J, Perrier R, Morel E, Fischmeister R, Richard S, Benitah JP, et al. 2007. The cAMP binding protein Epac modulates Ca^{2+} sparks by a Ca^{2+}/calmodulin kinase signalling pathway in rat cardiac myocytes. *J Physiol* **583:** 685–694.

Puls A, Schmidt S, Grawe F, Stabel S. 1997. Interaction of protein kinase C ζ with ZIP, a novel protein kinase C-binding protein. *Proc Natl Acad Sci* **94:** 6191–6196.

Qiu RG, Abo A, Steven Martin G. 2000. A human homolog of the *C. elegans* polarity determinant Par-6 links Rac and Cdc42 to PKCζ signaling and cell transformation. *Curr Biol* **10:** 697–707.

Reiken S, Gaburjakova M, Guatimosim S, Gomez AM, D'Armiento J, Burkhoff D, Wang J, Vassort G, Lederer WJ, Marks AR. 2003. Protein kinase A phosphorylation of the cardiac calcium release channel (ryanodine receptor) in normal and failing hearts. Role of phosphatases and response to isoproterenol. *J Biol Chem* **278:** 444–453.

Reither G, Schaefer M, Lipp P. 2006. PKCα: A versatile key for decoding the cellular calcium toolkit. *J Cell Biol* **174:** 521–533.

Rizo J, Sudhof TC. 1998. C2-domains, structure and function of a universal Ca^{2+}-binding domain. *J Biol Chem* **273:** 15879–15882.

Rosse C, Linch M, Kermorgant S, Cameron AJ, Boeckeler K, Parker PJ. 2010. PKC and the control of localized signal dynamics. *Nat Rev Mol Cell Biol* **11:** 103–112.

Ruknudin AM, Wei SK, Haigney MC, Lederer WJ, Schulze DH. 2007. Phosphorylation and other conundrums of Na/Ca exchanger, NCX1. *Ann NY Acad Sci* **1099:** 103–118.

Sakwe AM, Rask L, Gylfe E. 2005. Protein kinase C modulates agonist-sensitive release of Ca^{2+} from internal stores in HEK293 cells overexpressing the calcium sensing receptor. *J Biol Chem* **280:** 4436–4441.

Schaefer M, Albrecht N, Hofmann T, Gudermann T, Schultz G. 2001. Diffusion-limited translocation mechanism of protein kinase C isotypes. *FASEB J* **15:** 1634–1636.

Schleifenbaum A, Stier G, Gasch A, Sattler M, Schultz C. 2004. Genetically encoded FRET probe for PKC activity based on pleckstrin. *J Am Chem Soc* **126:** 11786–11787.

Schmidt M, Evellin S, Weernink PA, von Dorp F, Rehmann H, Lomasney JW, Jakobs KH. 2001. A new phospholipase-C-calcium signalling pathway mediated by cyclic AMP and a Rap GTPase. *Nat Cell Biol* **3:** 1020–1024.

Schulze DH, Muqhal M, Lederer WJ, Ruknudin AM. 2003. Sodium/calcium exchanger (NCX1) macromolecular complex. *J Biol Chem* **278:** 28849–28855.

Srivastava J, Goris J, Dilworth SM, Parker PJ. 2002. Dephosphorylation of PKCδ by protein phosphatase 2Ac and its inhibition by nucleotides. *FEBS Lett* **516:** 265–269.

Stahelin RV, Digman MA, Medkova M, Ananthanarayanan B, Melowic HR, Rafter JD, Cho W. 2005a. Diacylglycerol-induced membrane targeting and activation of protein kinase Cε: Mechanistic differences between protein kinases Cδ and Cε. *J Biol Chem* **280:** 19784–19793.

Stahelin RV, Wang J, Blatner NR, Rafter JD, Murray D, Cho W. 2005b. The origin of C1A–C2 interdomain interactions in protein kinase Cα. *J Biol Chem* **280:** 36452–36463.

Strassheim D, Williams CL. 2000. P2Y2 purinergic and M3 muscarinic acetylcholine receptors activate different phospholipase C-β isoforms that are uniquely susceptible to protein kinase C–dependent phosphorylation and inactivation. *J Biol Chem* **275:** 39767–39772.

Sutton RB, Sprang SR. 1998. Structure of the protein kinase Cβ phospholipid-binding C2 domain complexed with Ca^{2+}. *Structure* **6:** 1395–1405.

Swairjo MA, Concha NO, Kaetzel MA, Dedman JR, Seaton BA. 1995. Ca^{2+}-bridging mechanism and phospholipid head group recognition in the membrane-binding protein annexin V. *Nat Struct Biol* **2:** 968–974.

Takahashi E, Abe J, Gallis B, Aebersold R, Spring DJ, Krebs EG, Berk BC. 1999. p90(RSK) is a serum-stimulated Na^+/H^+ exchanger isoform-1 kinase. Regulatory phosphorylation of serine 703 of Na^+/H^+ exchanger isoform-1. *J Biol Chem* **274:** 20206–20214.

Tung HY. 1986. Phosphorylation of the calmodulin-dependent protein phosphatase by protein kinase C. *Biochem Biophys Res Commun* **138:** 783–788.

Uhlen P, Fritz N. 2010. Biochemistry of calcium oscillations. *Biochem Biophys Res Commun* **396:** 28–32.

Ubach J, Zhang X, Shao X, Sudhof TC, Rizo J. 1998. Ca^{2+} binding to synaptotagmin: How many Ca^{2+} ions bind to the tip of a C2-domain? *EMBO J* **17:** 3921–3930.

Ulianich L, Elia MG, Treglia AS, Muscella A, Di Jeso B, Storelli C, Marsigliante S. 2006. The sarcoplasmic-endoplasmic reticulum Ca^{2+} ATPase 2b regulates the Ca^{2+} transients elicited by P2Y2 activation in PC Cl3 thyroid cells. *J Endocrinol* **190:** 641–649.

Venkatachalam K, Montell C. 2007. TRP channels. *Annu Rev Biochem* **76:** 387–417.

Venkatachalam K, Zheng F, Gill DL. 2003. Regulation of canonical transient receptor potential (TRPC) channel function by diacylglycerol and protein kinase C. *J Biol Chem* **278:** 29031–29040.

Verdaguer N, Corbalan-Garcia S, Ochoa WF, Fita I, Gomez-Fernandez JC. 1999. Ca^{2+} bridges the C2 membrane-binding domain of protein kinase Cα directly to phosphatidylserine. *EMBO J* **18:** 6329–6338.

Violin JD, Newton AC. 2003. Pathway illuminated: Visualizing protein kinase C signaling. *IUBMB Life* **55:** 653–660.

Violin JD, Zhang J, Tsien RY, Newton AC. 2003. A genetically encoded fluorescent reporter reveals oscillatory phosphorylation by protein kinase C. *J Cell Biol* **161:** 899–909.

Wang KK, Wright LC, Machan CL, Allen BG, Conigrave AD, Roufogalis BD. 1991. Protein kinase C phosphorylates the carboxyl terminus of the plasma membrane Ca^{2+}-ATPase from human erythrocytes. *J Biol Chem* **266:** 9078–9085.

Wettschureck N, Offermanns S. 2005. Mammalian G proteins and their cell type specific functions. *Physiol Rev* **85:** 1159–1204.

Wheeler DG, Barrett CF, Groth RD, Safa P, Tsien RW. 2008. CaMKII locally encodes L-type channel activity to signal to nuclear CREB in excitation-transcription coupling. *J Cell Biol* **183:** 849–863.

Xu P, Wang J, Kodavatiganti R, Zeng Y, Kass IS. 2004. Activation of protein kinase C contributes to the isoflurane-induced improvement of functional and metabolic recovery in isolated ischemic rat hearts. *Anesth Analg* **99:** 993–1000.

Young SH, Wu SV, Rozengurt E. 2002. Ca^{2+}-stimulated Ca^{2+} oscillations produced by the Ca^{2+}-sensing receptor require negative feedback by protein kinase C. *J Biol Chem* **277:** 46871–46876.

Zemlickova E, Johannes FJ, Aitken A, Dubois T. 2004. Association of CPI-17 with protein kinase C and casein kinase I. *Biochem Biophys Res Commun* **316:** 39–47.

Zhang HH, Lipovsky AI, Dibble CC, Sahin M, Manning BD. 2006. S6K1 regulates GSK3 under conditions of mTOR-dependent feedback inhibition of Akt. *Mol Cell* **24:** 185–197.

Zhou M, Horita DA, Waugh DS, Byrd RA, Morrison DK. 2002. Solution structure and functional analysis of the cysteine-rich C1 domain of kinase suppressor of Ras (KSR). *J Mol Biol* **315:** 435–446.

Zimmermann G, Taussig R. 1996. Protein kinase C alters the responsiveness of adenylyl cyclases to G protein α and $\beta\gamma$ subunits. *J Biol Chem* **271:** 27161–27166.

Ca^{2+} Signaling During Mammalian Fertilization: Requirements, Players, and Adaptations

Takuya Wakai[1], Veerle Vanderheyden[2], and Rafael A. Fissore[1]

[1]Department of Veterinary and Animal Sciences, University of Massachusetts, Amherst, Massachusetts 01003
[2]Laboratory of Molecular and Cellular Signaling, Department of Molecular Cell Biology, K.U. Leuven, Campus Gasthuisberg, B-3000 Leuven, Belgium

Correspondence: rfissore@vasci.umass.edu

Changes in the intracellular concentration of calcium ($[Ca^{2+}]_i$) represent a vital signaling mechanism enabling communication among cells and between cells and the environment. The initiation of embryo development depends on a $[Ca^{2+}]_i$ increase(s) in the egg, which is generally induced during fertilization. The $[Ca^{2+}]_i$ increase signals egg activation, which is the first stage in embryo development, and that consist of biochemical and structural changes that transform eggs into zygotes. The spatiotemporal patterns of $[Ca^{2+}]_i$ at fertilization show variability, most likely reflecting adaptations to fertilizing conditions and to the duration of embryonic cell cycles. In mammals, the focus of this review, the fertilization $[Ca^{2+}]_i$ signal displays unique properties in that it is initiated after gamete fusion by release of a sperm-derived factor and by periodic and extended $[Ca^{2+}]_i$ responses. Here, we will discuss the events of egg activation regulated by increases in $[Ca^{2+}]_i$, the possible downstream targets that effect these egg activation events, and the property and identity of molecules both in sperm and eggs that underpin the initiation and persistence of the $[Ca^{2+}]_i$ responses in these species.

An increase in the intracellular concentration of calcium ($[Ca^{2+}]_i$) underlies the initiation, progression and/or completion of a wide variety of cellular processes, including fertilization, muscle contraction, secretion, cell division, and apoptosis (Berridge et al. 2000). To survive and proliferate, cells and organisms must communicate, and changes in $[Ca^{2+}]_i$ allow them to quickly respond to environmental, nutritional, or ligand challenges with responses that regulate cell fate and function. Cells devote significant amounts of their energy reserves to create and maintain ionic gradients between extracellular and intracellular milieus and also within the latter, thereby allowing brief alterations in these gradients to have profound signaling effects. In the case of Ca^{2+}, myriad proteins have acquired the ability to bind Ca^{2+}, which allows them to interpret and transform these elevations into cellular functions. This review will examine the cellular modifications induced by $[Ca^{2+}]_i$ changes during fertilization in mature mammalian oocytes, henceforth referred to as eggs.

Oocytes during maturation ready themselves for fertilization and the initiation of embryogenesis. During this transition, oocytes undergo changes that include the resumption and progression of meiosis, the development of polyspermy-preventing mechanisms, the reorganization of the cytoskeleton with spindle formation and displacement to the cortex, and the translation, accumulation, and degradation of specific mRNAs and proteins involved in development (Horner and Wolfner 2008b). In most species, and in all mammals, a $[Ca^{2+}]_i$ signal is responsible for breaking the meiosis-imposed developmental pause, causing egg activation, which is the first stage of embryo development (Whitaker 2006; Horner and Wolfner 2008b). The egg activating $[Ca^{2+}]_i$ signal is generally associated with sperm-egg fusion, which occurs at different stages of meiosis depending on the species (Stricker 1999), although in insects, where fertilization is dissociated from activation and where embryos can develop parthenogenetically, the presumed $[Ca^{2+}]_i$ increase is thought to be induced by mechanical stimulation during ovulation/oviductal transport (Page and Orr-Weaver 1997; Horner and Wolfner 2008a).

The $[Ca^{2+}]_i$ responses that underlie egg activation offer a great deal of diversity regarding their spatiotemporal configuration, reflecting both the plasticity of the Ca^{2+} signaling machinery as well as the dissimilar Ca^{2+} requirements for egg activation among species. Generally speaking, species can be categorized either as displaying a single $[Ca^{2+}]_i$ increase, which is the case of sea urchins, starfish, frogs, and fish, or showing multiple $[Ca^{2+}]_i$ changes, also known as oscillations, which is the case of nemertian worms, ascidians, and mammals (Stricker 1999; Miyazaki and Ito 2006). Elucidation of the signaling cascades and identification of the molecules/receptor(s) that initiate the Ca^{2+} signal at fertilization has proven elusive, and this review will not dwell on that literature; readers are referred to excellent recent reviews on the subject (Whitaker 2006; Parrington et al. 2007). Nonetheless, research has found that Src-family kinases (SFKs) and phospholipase Cγ (PLCγ) are involved in the activation of the phosphoinositide pathway and production of inositol 1,4,5-trisphosphate (IP_3) during fertilization in sea urchins, starfish, and frogs, which reflects the contribution of a plasma membrane receptor/signaling complex (Giusti et al. 1999; Sato et al. 2000). Remarkably, a receptor responsible for recruiting and activating SFKs during fertilization remains undiscovered (Mahbub Hasan et al. 2005). Similarly, it has proved difficult to uncover how the sperm initiates oscillations. Research now suggests that this may be accomplished by a novel mechanism whereby the signaling molecule/cargo, known as the sperm factor (SF), is released by the sperm into the ooplasm after fusion of the gametes. Importantly, the SF is not IP_3 or Ca^{2+} but rather it contains a protein moiety (Swann 1990; Wu et al. 1997; Kyozuka et al. 1998; Harada et al. 2007). To date, only the mammalian SF's molecular identity has been resolved, and found to be another member of the PLC family, a novel sperm-specific isoform named PLCζ (Saunders et al. 2002). This review will examine the literature on mammalian PLCζs and will focus as well on the egg molecules that are required to initiate and sustain $[Ca^{2+}]_i$ oscillations in these species.

EGG ACTIVATION

Following the resumption of meiosis during maturation, vertebrate eggs arrest at the metaphase stage of the second meiosis (MII). Sperm entry induces the resumption and completion of meiosis, release of cortical granules (CG), progression into interphase and pronuclear (PN) formation (Fig. 1A); these phenomena, which make possible the transition from egg to embryo, are collectively known as "egg activation" (Schultz and Kopf 1995; Stricker 1999; Ducibella et al. 2002). As stated earlier, an increase in $[Ca^{2+}]_i$ is the universal trigger of egg activation in all species studied to date (Stricker 1999), and in mammals this signal adopts a pattern of brief but periodical increases in $[Ca^{2+}]_i$ that last for several hours after sperm entry (Miyazaki et al. 1986). The spatiotemporal pattern of these $[Ca^{2+}]_i$ responses is decoded by downstream effectors, underpinning the distinct cellular events. We briefly review the events of egg activation that are controlled by

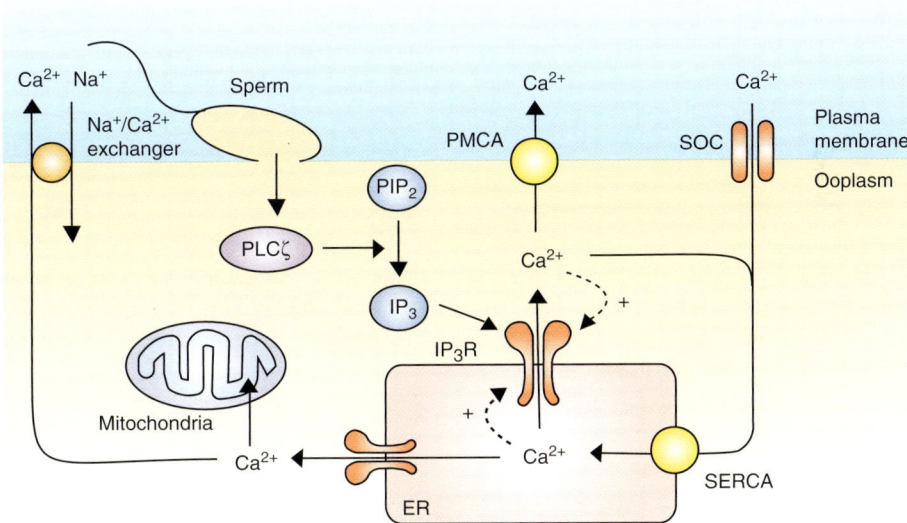

Figure 1. Temporal course of activation events in mouse eggs with a characteristic $[Ca^{2+}]_i$ response and candidate molecules involved in $[Ca^{2+}]_i$ oscillations and Ca^{2+} homeostasis. (A) Main cellular events of egg activation and approximate time in hours (hr) required for their completion after sperm entry. (B) A typical pattern of $[Ca^{2+}]_i$ oscillations associated with fertilization or with injection of PLCζ cRNA. Note that recordings were terminated prematurely. (C) On fusion, the sperm delivers phospholipase C (PLC)ζ, which hydrolyzes phosphatidylinositol 4,5-bisphosphate (PIP_2) into inositol 1,4,5-trisphospahte (IP_3) and dyacyl glycerol (DAG). IP_3 binds its receptor, IP_3R1, causing Ca^{2+} release out of the endoplasmic reticulum (ER). Following Ca^{2+} release, basal $[Ca^{2+}]_i$ levels are regulated by the combined action of the sarcoendoplasmic reticulum Ca^{2+} ATPase (SERCA), plasma membrane Ca^{2+} pump (PMCA), Na/Ca^{2+} exchanger and mitochondria. Store operated Ca^{2+} channels (SOC) are proposed to mediate Ca^{2+} influx required to fill the ER and maintain oscillations. Broken lines suggest feedback action of Ca^{2+} on IP_3R1.

$[Ca^{2+}]_i$ increases and the underlying molecular effectors.

Events of Egg Activation Require $[Ca^{2+}]_i$ Increases

Release from the MII arrest is an early and necessary event of egg activation, as it allows completion of meiosis, establishment of euploidy and progression into interphase with DNA synthesis. The MII block is imposed at the conclusion of maturation prior to ovulation by the action of the cytostatic factor (CSF) (Masui and Markert 1971). The CSF constrains the activity of the anaphase promoting factor (APC), an E3 ubiquitin ligase (Tunquist and Maller 2003), which is responsible for the ubiquitination and degradation of cyclin B (CycB). CycB and the cyclin-dependent kinase 1 (Cdk1, also known as cdc2) are the molecular components of the maturation promoting factor (MPF) (Swenson et al. 1986; Draetta et al. 1989), and inhibition of CycB degradation by the CSF stabilizes MPF, which results in MII arrest, as MPF activity correlates with metaphase (Masui and Markert 1971). Research has shown that endogenous meiotic inhibitor 2 (Emi2) is an inhibitory component of APC that underlies the MII arrest (Schmidt et al. 2005; Tung et al. 2005; Shoji et al. 2006). In MII oocytes, inhibition of APC also ensures that persistent levels of the separase inhibitor securin prevents sister chromatid separation (Madgwick et al. 2004; Nabti et al. 2008).

The association between intracellular ionic increases and release from meiotic arrest was first proposed at the end of the nineteenth century by Loeb and colleagues who observed that initiation of development in sea urchin eggs was possible simply by varying the concentration and composition of the fertilizing medium (Loeb 1907). Subsequently, it was learned that $[Ca^{2+}]_i$ levels dramatically change after fertilization (Mazia 1935), which focused attention on the role of this ion. Steinhardt and colleagues showed the dominant role of Ca^{2+}, as they were able to promote parthenogenetic development in a variety of species by exposing eggs to Ca^{2+} ionophores (Steinhardt et al. 1974).

How $[Ca^{2+}]_i$ increases induce release from the MII arrest remained unknown for decades. Research showed that $[Ca^{2+}]_i$ increases per se were unable to induce CycB degradation and that instead it required binding to a calmodulin-sensitive enzyme (Lorca et al. 1993), which was later shown to be Ca^{2+}-calmodulin-dependent protein kinase II (CaMKII). Nonetheless, how CaMKII relieved the CSF-imposed MII arrest remained unknown. Adding to the confusion was the finding that despite the need for a $[Ca^{2+}]_i$ increase, Xenopus egg extracts depleted of Polo-like kinase 1 (Plx1), a kinase known until then more for its role on spindle organization, were unable to overcome the MII arrest (Descombes and Nigg 1998). This quandary was solved when it was discovered that Emi2 is a key component of CSF and that Emi2 phosphorylation by Plx1, which causes its degradation, is required for CycB proteasomal degradation (Schmidt et al. 2005). Subsequent studies found that binding of Plx1 to Emi2 requires a preceding phosphorylation by CaMKII, thereby molecularly linking the need for $[Ca^{2+}]_i$ and Plx1 to exit MII.

Although the aforementioned pathway was elucidated in Xenopus eggs and extracts, mouse eggs rely on similar mechanisms to enter and exit the MII arrest. Most prominently, it is well documented that CycB undergoes degradation during fertilization (Nixon et al. 2002) and that each of the sperm-induced $[Ca^{2+}]_i$ increases is accompanied by a parallel increase in CaMKII activity (Markoulaki et al. 2003). The role of CaMKII on mammalian egg activation was convincingly shown by studies in the mouse, in which expression of constitutive active forms of CaMKII into eggs initiated all events of egg activation, except CG exocytosis, and promoted development to the blastocyst stage (Madgwick et al. 2005; Knott et al. 2006). Conversely, depletion of the CaMKIIγ isoform abrogated the ability of these eggs to exit MII in response to $[Ca^{2+}]_i$ stimulation (Backs et al. 2010; Chang et al. 2009), causing infertility. Research also implicated Emi2 in MII arrest in the mouse, as inhibition of Emi2 synthesis prevents cyclin B1 accumulation during maturation (Madgwick et al. 2006), which causes spontaneous activation (Shoji et al. 2006). The role of Plk1 in mouse MII arrest remains unexplored, although our preliminary data show that treatment of eggs with BI2536, a new and selective Plk1 inhibitor, prevents CycB degradation and MII exit in eggs treated with $SrCl_2$ (data not shown). In Xenopus, Plx1 phosphorylates xEmi2 within a phosphodegron motif, after which xEmi2 is rapidly targeted for degradation, but this motif, or its canonical replacements, is absent in the mouse homolog (Perry and Verlhac 2008). Although the role of Ca^{2+} and the molecular pathways required for MII exit and embryo development are conserved in vertebrate eggs, it is presently unknown how Plk1 regulates Emi2 function in mammals.

CG exocytosis is another event of egg activation that depends on Ca^{2+} release (Kline and Kline 1992a). CG release underlies, at least in part, the cortical remodeling that occurs after fertilization (Sardet et al. 2002), and modifies the components of zona pellucida to prevent polyspermy, thereby ensuring the formation of a diploid zygote. Although it was believed that the effects of Ca^{2+} on CGs were transduced by activation of protein kinase C (PKC), as PKC agonists promoted CG release and other activation events, later studies using PKC inhibitors failed to prevent fertilization-associated CG release (Ducibella and LeFevre 1997). Importantly, the widespread expression of PKC isoforms in oocytes (Gallicano et al. 1997; Eliyahu et al. 2001; Page Baluch et al. 2004), along with their distinct cellular distribution (Viveiros et al. 2001; Page Baluch et al. 2004), and the implications of their impact on Ca^{2+} influx (Halet 2004), suggest important roles for these enzymes in setting off embryo development. CaMKII was also expected to participate in CG exocytosis, although the aforementioned studies using constitutively active forms of the protein (Knott et al. 2006) or eggs devoid of CaMKII have ruled out this possibility (Backs et al. 2010; Chang et al. 2009). Recent studies have implicated myosin light chain kinase (MLCK), another Ca^{2+}-dependent kinase, as being involved in CG exocytosis in mouse fertilization, as pharmacological inhibitors greatly diminished their release in response to Ca^{2+} stimulation (Matson et al. 2006). The role of MLCK on CG exocytosis is not unexpected, as myosin II, a direct target of MLKC, and actin microfilaments are involved in cortical reorganization in the mouse (Simerly et al. 1998; Deng et al. 2005) and zebrafish eggs (Becker and Hart 1999). Importantly, the molecular regulation of MLCK needs to be determined, as besides its requirement for Ca^{2+}, it is highly sensitive to phosphorylation, and kinases such as ERK and Rho that are active during meiosis might have regulatory roles (Deng et al. 2005). In summary, the molecular effectors for several events of egg activation downstream of Ca^{2+} have been uncovered over the last decade. Although important gaps remain, the requirement for $[Ca^{2+}]_i$ oscillations for initiation of mammalian development is unambiguous.

Single versus Multiple $[Ca^{2+}]_i$ Increases

The early ionophore studies hinted to a pivotal role for Ca^{2+} in the initiation of development, especially with regard to the increases in $[Ca^{2+}]_i$, but not in K^+ or pH, induced all early and late events of egg activation (Steinhardt and Epel 1974). Nevertheless, whether or not such changes happened during normal fertilization and how their inhibition affected development was unknown. Evidence soon accumulated, first using the luminescent protein "aequorin" synthesized by Shimomura and colleagues (Shimomura and Johnson 1970) that explosive $[Ca^{2+}]_i$ increases accompanied fertilization in medaka fish eggs (Ridgway et al. 1977) and in sea urchin eggs (Steinhardt et al. 1977). Unlike the single $[Ca^{2+}]_i$ increases detected in these early recordings, measurements of $[Ca^{2+}]_i$ changes in mammals revealed that their eggs displayed $[Ca^{2+}]_i$ oscillations (Cuthbertson et al. 1981; Miyazaki and Igusa 1981). Although oscillatory $[Ca^{2+}]_i$ responses were subsequently reported in non-mammalian species, mammalian eggs are the only ones whose oscillations extend for over several hours (Stricker 1999). Further, research soon followed demonstrating that abrogation of fertilization-associated $[Ca^{2+}]_i$ increases, which was accomplished with the Ca^{2+} chelator BAPTA, prevented all events of egg activation and prevented the initiation of development (Kline and Kline 1992a). Together, these results confirmed the widespread role of Ca^{2+} as the activation signal for development.

Although the elevation of $[Ca^{2+}]_i$ is ubiquitous in fertilization, the presence of long-lasting oscillations is a hallmark of mammalian fertilization (Fig. 1B). Remarkably, the developmental advantages and underlying molecular changes associated with these oscillations remain unclear. Research by Ducibella et al. underscored the varying sensitivities of egg activation events to $[Ca^{2+}]_i$ increases. For instance, most events of egg activation, such as CG exocytosis, meiotic resumption and recruitment of maternal mRNAs require fewer $[Ca^{2+}]_i$

increases for initiation than for completion, and early events such as CG exocytosis and release from MII arrest require fewer $[Ca^{2+}]_i$ responses for completion than later events, such as PN formation and recruitment of maternal mRNAs (Ducibella et al. 2002). In this context, oscillations make sense, especially to promote CycB degradation and inactivation of MPF, as CycB synthesis is continuous (Nixon et al. 2002; Marangos and Carroll 2004) and a single $[Ca^{2+}]_i$ increase would be unable to promote its complete degradation (Nixon et al. 2002). Nevertheless, if exit of MII arrest is overcome either by a single, overwhelming $[Ca^{2+}]_i$ increase induced by an electrical pulse (Ozil et al. 2005) or by expression of a constitutively active form of CaMKII, development to the blastocyst stage is only mildly impaired (Madgwick et al. 2005; Ozil et al. 2005; Knott et al. 2006). This apparent lack of impact of $[Ca^{2+}]_i$ oscillations on preimplantation development is in contrast to another report showing that both premature termination or excessive Ca^{2+} stimulation negatively impacts preimplantation and postimplantation development, and alters embryonic gene expression (Ozil et al. 2006). Similar research documented that parthenogenotes generated without a $[Ca^{2+}]_i$ increase by exposing eggs to cycloheximide, a protein synthesis inhibitor, showed altered gene expression and poor development to the blastocyst stage (Rogers et al. 2006). Nonetheless, development of these embryos was rescued to the same extent by exposure to a single or multiple $[Ca^{2+}]_i$ increases, casting doubts on the beneficial effects of $[Ca^{2+}]_i$ oscillations on development.

One way that multiple $[Ca^{2+}]_i$ elevations may pose a developmental advantage is by specifically stimulating embryonic gene expression. The recruitment of maternal mRNAs, which mediates new protein synthesis after fertilization, takes place during the period of oscillations and is susceptible to the magnitude of the $[Ca^{2+}]_i$ stimulation; more pulses more protein synthesis (Ducibella et al. 2002). To this end, one of the two transcripts identified after fertilization is cyclin A (Oh et al. 1997; Fuchimoto et al. 2001), which participates in the activation of the embryonic genome (Fuchimoto et al. 2001). Thus, based on research in hippocampal neurons, the suggestion was made that pulsatile activation of CaMKII may underlie the enhanced gene expression observed after repeated $[Ca^{2+}]_i$ pulses (Ducibella et al. 2006). Subsequent research, however, showed that recruitment of mRNAs could occur independently of this kinase (Backs et al. 2010). Furthermore, it might not be under the exclusive control of Ca^{2+}, as in the absence of cell cycle progression, fertilization-initiated oscillations failed to induce recruitment of mRNAs (Backs et al. 2010). Therefore, it might that the total magnitude of the $[Ca^{2+}]_i$ increase, as proposed by Ozil and colleagues (Ducibella et al. 2006; Ozil et al. 2006) rather than the temporal pattern of $[Ca^{2+}]_i$ increases is the determinant factor of egg activation in mammals. Nonetheless, oscillations may be necessary, as besides signaling the stepwise completion of all events of egg activation, it might be the only manner whereby mammalian eggs can attain a Ca^{2+} signal of sufficient magnitude to ensure CycB degradation without undermining other cellular functions.

MOLECULAR PLAYERS RESPONSIBLE FOR $[Ca^{2+}]_i$ OSCILLATIONS DURING FERTILIZATION

The $[Ca^{2+}]_i$ oscillations that underlie egg activation in mammals rely on molecular players widely characterized in other cellular systems in which they mediate $[Ca^{2+}]_i$ responses induced by a variety of agonists such as hormones, growth factors and antigen-presenting mechanisms (Berridge et al. 2000; Clapham 2007). In gametes however, the function and regulation of some of these molecules has been adapted to respond to the unique requirements of fertilization. For instance, oocytes require weeks or months of preparation before being ready for fertilization, because interruptions in the cell cycle are imposed during meiosis to synchronize oocyte and follicular growth before reinitiating meiosis and ovulation. It is believed that during this growth phase, oocytes do not require $[Ca^{2+}]_i$ elevations, and Ca^{2+} release mechanisms are quiescent (Carroll et al. 1994). Importantly, these mechanisms are quickly reactivated in fully-grown oocytes after receiving an LH surge, which is the endocrinological signal that induces oocyte maturation from the germinal vesicle (GV) stage. During this process, which may last from 12 to 48 hr according to the species, the oocytes' Ca^{2+} release mechanisms undergo reprogramming and optimization so that fertilization can initiate $[Ca^{2+}]_i$ oscillations. The sperm also undergoes a protracted preparation, undergoing changes during transport through the male reproductive tract and more closely as it approaches the site of fertilization in the female tract (Suarez 2008b). Remarkably, some of these changes also involve $[Ca^{2+}]_i$ increases (Suarez 2008a), although they occur while preserving the sperms' Ca^{2+} activating signal.

Two molecules stand out in mammalian fertilization as central to the initiation and maintenance of $[Ca^{2+}]_i$ oscillations; namely, the IP_3R1 receptor in eggs and PLCζ in the sperm. Here we will describe the evidence supporting their role in mammalian fertilization, focusing on regulatory mechanisms and highlighting some of the unanswered questions regarding their regulation. We will also review other molecular mechanisms required to maintain oscillations, especially those affecting Ca^{2+} influx whose function in eggs has not been widely investigated (Fig. 1C).

IP_3R1

IP_3R1 in MII Eggs

The IP_3R is the main intracellular Ca^{2+}-release channel of almost all mammalian cell types and is located in the endoplasmic reticulum (ER), the cells' main Ca^{2+} reservoir (reviewed in Berridge et al. [2000]; Bootman et al. [2001]). The IP_3R is a large protein (>250 kDa) and functions as a tetramer (>1000 kDa). Each monomer consists of more than 2600 amino acids and can be broadly divided into three regions, a cytosolic amino-terminal domain that binds IP_3, a regulatory domain that contains multiple regulatory sites for Ca^{2+}, ATP, and other modulatory molecules/proteins (MacKrill 1999; Patterson et al. 2004) and a carboxy-terminal channel domain that contains six transmembrane domains and a short cytosolic tail. As described by Taylor and Tovey (2010), the activation and opening of the IP_3R1 requires binding by both Ca^{2+} and IP_3, and the regulation of IP_3-induced Ca^{2+} release (IICR) by Ca^{2+} adopts a bell-shape form, as IICR is stimulated at low $[Ca^{2+}]_i$ and inhibited at high $[Ca^{2+}]_i$ (Taylor and Tovey 2010; Iino 1990; Finch et al. 1991). This dual regulation of IP_3R1 by Ca^{2+} and IP_3 makes it especially suited to support long lasting oscillations.

There exists three IP_3R isoforms (reviewed in (Berridge et al. [2000]), and mammalian oocytes and eggs and their surrounding cells express all isoforms (Fissore et al. 1999a; Fissore et al. 1999b; Diaz-Munoz et al. 2008). Importantly, oocytes and eggs overwhelmingly express the type I IP_3R isoform (Kume et al. 1997; Fissore et al. 1999a; Jellerette et al. 2000; Iwasaki et al. 2002). The initial suggestion that IP_3R may play a role during fertilization arose from studies in sea urchin eggs in which an increase in phosphoinositide metabolism accompanied fertilization (Turner et al. 1984), an observation that was soon followed by the demonstration that injection of IP_3 triggered Ca^{2+} release (Clapper and Lee 1985) and cortical granule exocytosis (Turner et al. 1986). Studies followed in hamster oocytes in which injection of IP_3 and guanine nucleotides initiated repeated Ca^{2+} release from intracellular stores (Miyazaki 1988). Purification and identification of the IP_3R protein from the cerebellum occurred in the late 1980s (Maeda et al. 1988; Furuichi et al. 1989), and confirmation of its significance in mammalian fertilization took place soon after when both the initiation of $[Ca^{2+}]_i$ oscillations (Miyazaki et al. 1992) and egg activation (Xu et al. 1994) were prevented by injection of a functional blocking antibody raised against the Carboxy-terminal end of mouse IP_3R1. Subsequent studies confirmed the role of IP_3R1 in fertilization in other species (Parys et al. 1994; Thomas et al. 1998; Yoshida et al. 1998; Runft et al. 1999; Goud et al. 2002; Iwasaki et al. 2002).

Fertilization-associated $[Ca^{2+}]_i$ oscillations in mouse zygotes undergo changes during the transition from the MII stage into interphase, becoming initially less frequent before ceasing altogether at the time of PN formation (Jones et al. 1995; Kono et al. 1996; Deguchi et al. 2000). During this transition, the IP$_3$R1 undergo several modifications and it is possible that either singly or collectively these influence the pattern of oscillations. For example, mammalian eggs richly express IP$_3$R1, as only 20 mouse eggs are required for its detection by Western blotting (Parrington et al. 1998; Jellerette et al. 2000) and within 4 h after sperm entry the IP$_3$R1 mass is reduced approximately to a half (Parrington et al. 1998; Deguchi et al. 2000; Kurokawa and Fissore 2003). Moreover, recent research shows that IP$_3$R1 degradation alone can explain the widening of the $[Ca^{2+}]_i$ intervals, although not the termination of oscillations (Lee et al.). Changes in IP$_3$R1 localization may also affect the pattern of oscillations. In eggs, the IP$_3$R1 and the ER are organized in clusters near the cortex, a location that might facilitate the initiation of $[Ca^{2+}]_i$ oscillations, as the PLCζ concentration may be higher in this area after sperm-egg fusion. The accumulation of ER clusters in the cortex may also enhance IP$_3$R1 sensitivity, as $[Ca^{2+}]_i$ oscillations originate from the hemisphere opposite to the MII spindle where ER/IP$_3$R1 clusters are particularly dense (Kline et al. 1999; Dumollard et al. 2004). Interestingly, in *Xenopus*, IP$_3$R1s that are more sensitive move to the cortex from the subcortex as oocytes progress to the MII stage (Boulware and Marchant 2005). Importantly, in the mouse, ER (FitzHarris et al. 2003) and possibly IP$_3$R1 (our unpublished data) cortical clusters disappear ahead of the termination of the oscillations, suggesting that they are not required for the persistence of oscillations. Nevertheless, the precise distribution of IP$_3$R1 in eggs suggest an important role during fertilization, which may correspond to the need for localized high amplitude $[Ca^{2+}]_i$ increases to facilitate CG release to prevent polyspermy (McAvey et al. 2002).

IP$_3$R1 function may also be regulated by phosphorylation. Not surprisingly, the first report describing IP$_3$R1 phosphorylation in eggs suggested an association with cell cycle kinases (Jellerette et al. 2004), which play a prominent role in the MII arrest. IP$_3$R1 phosphorylation in mouse eggs was first characterized using an antibody that identifies proteins phosphorylated at the MPM-2 epitope, which consists of phosphorylated serines(S)/threonines(T) next to prolines(P) surrounded by hydrophobic amino acids (Westendorf et al. 1994). IP$_3$R1 becomes phosphorylated at a MPM-2-detectable epitope during maturation (Lee et al. 2006) reaching maximal reactivity at the MII stage (Fig. 2). Following egg activation, it becomes gradually dephosphorylated and phosphorylation is not regained at first mitosis (Lee et al. 2006). The responsible kinases for IP$_3$R1 MPM-2 phosphorylation remain to be determined, although several M-phase kinases, such as polo-like kinase 1 (Plk1), mitogen-activated protein kinase (MAPK), and Cdk1 can all

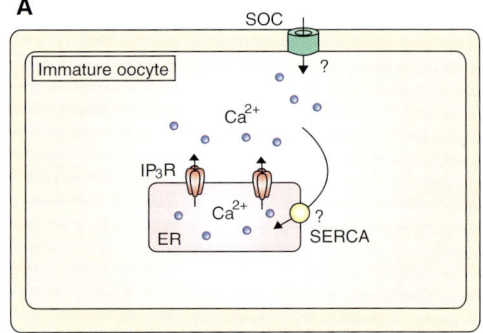

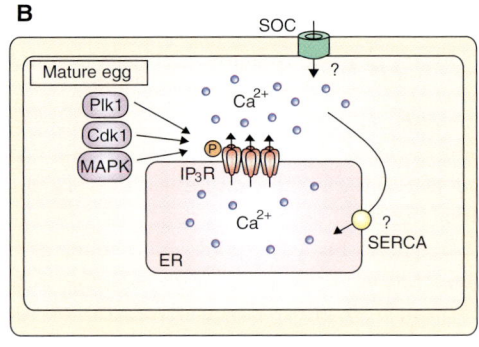

Figure 2. IP$_3$R1-mediated Ca^{2+} release increases during mouse oocyte maturation. Several factors may contribute to this, including the increased Ca^{2+} content of the stores, IP$_3$R1 organization into clusters and IP$_3$R1 phosphorylation. Question marks suggest mechanisms that are suspected to contribute to Ca^{2+} influx and increased Ca^{2+} store content. Phosphorylating kinases are Polo-like kinase-1, Cyclin-dependent kinase-1, and Mitogen-associated protein kinase.

phosphorylate this epitope (Joughin et al. 2009). Our studies in mouse oocytes using pharmacological inhibitors revealed that Plk1 might be involved in MPM-2 phosphorylation early during maturation (Ito et al. 2008; Vanderheyden et al. 2009) and MAPK during the MI to MII transition, although whether they directly phosphorylate the receptor remains to be shown. Research in DT40 B-cell lymphocytes showed IP$_3$R1 phosphorylation by MAPK at S^{436}, a residue that lies in the receptor's ligand binding domain within a consensus site for the kinase; using back cross phosphorylation studies, this group also showed IP$_3$R1 MAPK phosphorylation of mouse MII eggs (Bai et al. 2006). This study did not examine the role of this phosphorylation in eggs, although in microsome preparations MAPK IP$_3$R1 phosphorylation decreased IP$_3$ binding and Ca^{2+} release (Bai et al. 2006). In contrast to this, phosphorylation of IP$_3$R1 in somatic cells by Cdk1, which was observed to occur in several conserved Cdk1 motifs under in vitro and in vivo conditions, enhanced IP$_3$ binding and Ca^{2+} release (Malathi et al. 2003; Malathi et al. 2005). IP$_3$R1 phosphorylation within Cdk1 and MAPK consensus sites was reported in *Xenopus* oocytes and independent activation of these kinases increased IP$_3$R-mediated Ca^{2+} release (Sun et al. 2009). Notably, an earlier study in mouse zygotes had dismissed the role of MPF and MAPK on [Ca^{2+}]$_i$ oscillations, as the continuation was unaltered by the decline in MPF activity, which occurs at the time of second PB extrusion, or after inhibition of the MAPK pathway with U0126 (Marangos et al. 2003). Importantly, the phosphorylation status of IP$_3$R1 was not examined in that study, and subsequent results found that IP$_3$R1 MPM-2 phosphorylation outlasts MPF activity (Lee et al. 2006) and that 4 hours U0126 exposure does not eliminate MPM-2 IP$_3$R1 reactivity (our unpublished observations). A possible interpretation for these results is that IP$_3$R1 phosphorylation, and the phosphorylation of other M-phase substrates in eggs is safeguarded by suppression of phosphatase(s) activity, a function that has been attributed to the Greatwall kinase in mitotic cells and *Xenopus* egg extracts (Castilho et al. 2009). Therefore, accumulating evidence suggests a regulatory role for phosphorylation on IP$_3$R1 function during fertilization, although the responsible kinases, phosphorylation sites and their impact on IP$_3$R1 function remain to be clarified.

IP$_3$R1 in Maturing Oocytes

The precise spatio-temporal pattern of sperm-associated [Ca^{2+}]$_i$ responses in vertebrate eggs is established during oocyte maturation. For example, in vitro fertilized mouse GV oocytes show fewer [Ca^{2+}]$_i$ oscillations and each [Ca^{2+}]$_i$ increase shows reduced duration and amplitude than those observed in fertilized MII eggs (Jones et al. 1995a; Mehlmann et al. 1996). The molecular events underlying these changes are not understood, although changes in IP$_3$R1 sensitivity, i.e., the receptor's ability to conduct Ca^{2+} in response to IP$_3$, are thought to be involved. Importantly, studies to elucidate these mechanisms are needed, but given the recalcitrant nature of some of these changes in MII eggs, it is suggested that they should be performed during maturation.

As discussed above, IP$_3$R1 phosphorylation during maturation by M-phase kinases is thought to enhance Ca^{2+} release in eggs. MPM-2 reactivity, which is used as a marker of their activity, is first evidenced in IP$_3$R1 at the time of GV breakdown (GVBD) and persists until the MII stage, a period that closely coincides with the increased function of IP$_3$R1 during maturation (Mehlmann and Kline 1994). Inhibition of Plk1, a kinase involved in the activation of Cdk1, reduced and delayed MPM-2 IP$_3$R1 reactivity and decreased [Ca^{2+}]$_i$ release through IP$_3$R1 at the GVBD stage (Ito et al. 2008a; Vanderheyden et al. 2009). Nonetheless, the persistent presence of BI2556, a nonreversible and specific Plk1 inhibitor, did not eliminate MPM-2 IP$_3$R1 reactivity, which even experienced a partial recovery, suggesting that other kinases, possibly Cdk1, might phosphorylate IP$_3$R1 (our unpublished observation). Investigation of the role of Cdk1 on IP$_3$R1 phosphorylation in mouse oocytes/eggs is hindered by the findings that roscovitine, a specific Cdk1 inhibitor, indirectly inhibits Plk1 activity (our unpublished results) and greatly reduces Ca^{2+} store content ($[Ca^{2+}]_{ER}$) (Deng and Shen 2000), thereby compromising the interpretation of the results. Inhibition of the MAPK pathway, which does not affect the aforementioned kinases and only mildly affects IP$_3$R1 MPM-2 phosphorylation after the MI stage, greatly reduces $[Ca^{2+}]_i$ responses (Lee et al. 2006). It is therefore possible that IP$_3$R1 phosphorylation by MAPK is not recognized by the MPM-2 antibody or, alternatively, that the MAPK pathway affects other aspects of Ca^{2+} homeostasis. In this regard, one study found an altered Ca^{2+} content in U1026-treated oocytes (Matson and Ducibella 2007) whereas the other did not (Lee et al. 2006). Therefore, additional mutational studies are needed to clarify the impact of M-phase kinases on IP$_3$R1-mediated Ca^{2+} release in mammalian oocytes and eggs.

Besides M-phase kinases, numerous studies in somatic cells have shown that IP$_3$R isoforms can be phosphorylated by various, more wide-ranging kinases that generally increase IP$_3$-R-mediated Ca^{2+} release (Bezprozvanny 2005; Vanderheyden et al. 2009), although phosphorylation by PKB reportedly reduces IP$_3$R-mediated Ca^{2+} release (Szado et al. 2008). The most commonly implicated kinases include PKA, PKC, and CaMKII, all of which have important physiological functions in oocytes and eggs (Ducibella and Fissore 2008). Extensive phosphopeptide mapping combined with substrate specific antibodies in *Xenopus* oocytes found that IP$_3$R1 is uniformly phosphorylated throughout maturation in both PKA consensus motifs, whereas PKC sites seemed unperturbed (Sun et al. 2009). Whether or not PKA IP$_3$R1 phosphorylation has functional consequences in oocytes is unknown, although in somatic cells this phosphorylation has been associated with increased IP$_3$R1 activity (DeSouza et al. 2002) and reduced $[Ca^{2+}]_{ER}$, which may reportedly underlie the antiapoptotic effects of some members of the Bcl-2 family of proteins (Oakes et al. 2005). Moreover, it has been suggested that these proteins modify the PKA-associated IP$_3$R1 phosphorylation status (Oakes et al. 2005), although research from other laboratories has not confirmed this mechanism of action of Bcl-2 family protein on IP$_3$R1 function (Rong et al. 2008; Rong et al. 2009). It is worth noting that $[Ca^{2+}]_{ER}$ is low in GV oocytes in spite of persistent Ca^{2+} influx, as evident by the continuous spontaneous oscillations at this stage (Carroll and Swann 1992). On GVBD, however, $[Ca^{2+}]_{ER}$ undergoes a marked increase (Jones et al. 1995a), which occurs concurrently with the termination of the oscillations that implies suppression/reduction of Ca^{2+} influx. Given that cAMP levels decrease at GVBD (Norris et al. 2009), it is therefore possible that a Ca^{2+} leak mechanism regulated by PKA IP$_3$R1 phosphorylation may be implicated in Ca^{2+} homeostasis during oocyte maturation. Future studies should examine whether PKA-mediated IP$_3$R1 phosphorylation changes during mouse oocyte maturation.

The differential redistribution of ER/IP$_3$R1 may also enhance IP$_3$R1 function during oocyte maturation. Before the initiation of maturation, the ER in mouse oocytes shows a homogeneous distribution with slight accumulation around the GV, although by the MII stage, the ER displays a fine tubular network appearance with dense accumulation in the cortex (Mehlmann et al. 1995), which is thought to facilitate the initiation of sperm-induced $[Ca^{2+}]_i$ oscillations (Kline et al. 1999). The dramatic reorganization ensues at about the time of GVBD and is underpinned by distinct components of the cytoskeleton (FitzHarris et al. 2007), as the migration of the ER toward the condensing chromosomes is dependent on microtubules, whereas its dispersal from the MI spindle to the egg's cortex relies on actin microfilaments (FitzHarris et al. 2003). The IP$_3$R1 are also organized in cortical clusters at the MII stage (Mehlmann et al. 1996; Fissore et al. 1999a; Ito et al. 2008a), although it remains to be established whether the same cytoskeletal mechanisms that control ER organization control IP$_3$R1 distribution. Curiously, in spite of the large remodeling that the oocyte's ER undertakes, it is unknown whether this reorganization affects IP$_3$R1's sensitivity and the ability to initiate and support oscillations in mammalian eggs. Thus, preventing ER and possibly IP$_3$R1 cortical cluster organization with appropriate cytoskeleton inhibitors would help elucidate the influence of their reorganization on IP_3R1 function in oocytes.

Lastly, changes in other cytoplasmic parameters are likely to contribute to increase IP_3R1 sensitivity in oocytes. For example, the increase in $[Ca^{2+}]_{ER}$ during maturation (Kline and Kline 1992a; Jones et al. 1995a) may not only increase the amount of available Ca^{2+} for release, but may also increase the receptor's sensitivity (Missiaen et al. 1991). The increase in $[Ca^{2+}]_{ER}$ is likely the result of careful regulation of the pathways that control Ca^{2+} influx and efflux, which in *Xenopus* oocytes are known to be actively regulated during maturation (El-Jouni et al. 2005; Yu et al. 2009). Remarkably, the molecular identity and significance of these mechanisms in mammalian oocytes remain largely unknown and will be discussed below.

Ca^{2+} Homeostasis in Oocytes and Eggs

$[Ca^{2+}]_i$ Clearing Mechanisms

$[Ca^{2+}]_i$ oscillations in mammals continue for long periods that can exceed 20 h (Fissore et al. 1992; Sun et al. 1992). For $[Ca^{2+}]_i$ responses to continue without attenuation following a $[Ca^{2+}]_i$ increase, $[Ca^{2+}]_i$ levels need to be returned to baseline and stores refilled in anticipation of the next $[Ca^{2+}]_i$ response. To bring $[Ca^{2+}]_i$ to baseline, cells either return free cytosolic Ca^{2+} into the ER by the action of the sarco-endoplasmic reticulum Ca^{2+} ATPases (SERCAs), and/or extrude it by the action of plasma membrane (PM) Ca^{2+} ATPases (PMCAs) and Na^+/Ca^{2+} exchangers (Berridge et al. 2000; Bootman et al. 2001). Few studies have addressed the function of these molecules in mammalian oocytes/eggs, although the presence of SERCA2b can be surmised by the alteration of $[Ca^{2+}]_i$ levels caused by exposure to thapsigargin, an inhibitor of SERCA (Kline and Kline 1992b; Lawrence and Cuthbertson 1995; Machaty et al. 2002). Exposure of MII eggs to thapsigargin causes a slow and steady increase in $[Ca^{2+}]_i$ followed by a protracted decline, whereas in fertilized eggs it prevents the continuation of oscillations (Kline and Kline 1992b). Importantly, the molecular presence and cellular distribution of SERCA2b has not yet been examined in mammalian oocytes, although transcripts have been found in GV and MII stage oocytes (Su et al. 2007). In *Xenopus* oocytes, expression of the SERCA2 protein was documented by immunofluorescence and it was shown to undergo reorganization similar to that described for IP_3R (El-Jouni et al. 2005). Given that the levels of $[Ca^{2+}]_{ER}$ change dramatically during maturation, it is possible that SERCA activity may be actively regulated during this process. In somatic cells, SERCA activity can be regulated by different mechanisms, including binding to regulatory proteins such phospholamban, sarcolipin, and by several posttranslational modifications (reviewed in Brini and Carafoli [2009]). An earlier report in *Xenopus* oocytes showed that SERCA2b activity could be regulated by association with the ER chaperone protein, calnexin, which inhibited the pump's activity; phosphorylation of calnexin relieved both the association with SERCA and its inhibition (Roderick et al. 2000). Although additional studies are needed to understand the conservation of this mechanism, it is worth noting that addition of roscovitine prematurely, albeit reversibly, terminates $[Ca^{2+}]_i$ oscillations during fertilization (Deng and Shen 2000). Although the inhibitor's target was not elucidated, $[Ca^{2+}]_{ER}$ levels were severely depleted, suggesting an effect either on SERCA activity or on the Ca^{2+} influx mechanism(s). Future studies should explore the pathways involved in regulation of SERCA activity in mammalian eggs, as its function in sustaining long-term $[Ca^{2+}]_i$ oscillations after fertilization.

Besides the ER sequestration of Ca^{2+}, cytosolic $[Ca^{2+}]_i$ can also be returned to baseline by the action of PMCA and the Na^+/Ca^{2+} exchanger, which release Ca^{2+} into the external media. The functional activity of Na/Ca^{2+} exchanger was shown in mouse eggs by two different reports (Pepperell et al. 1999; Carroll 2000). It was shown that elimination of Na^+ from the external media caused $[Ca^{2+}]_i$ responses, or accelerated existing ones, and these responses were ascribed to reverse mode Na^+-Ca^{2+} exchange. In spite of the initial changes, even in the absence of external Na^+, $[Ca^{2+}]_i$ levels returned to baseline levels, implying that the action of PMCA may be more physiologically relevant (Carroll 2000). The molecular presence of PMCA has not been documented in mammalian oocytes/eggs, although in *Xenopus* oocytes PMCA1 seems to contribute to the shaping of $[Ca^{2+}]_i$ responses (El-Jouni et al. 2005; El-Jouni et al. 2008). For instance, in GV oocytes the presence of 1 mM La^{3+}, which is known to inhibit PMCA function, delay the half-time decay of a $[Ca^{2+}]_i$ increase induced by iononomycin, but the same treatment was without effect in MII eggs, suggesting that the pump's activity is down regulated during maturation (El-Jouni et al. 2005). Research by these investigators also showed that most of the PMCA in the plasma membrane becomes internalized during maturation explaining, at least in part, the lower $[Ca^{2+}]_i$ clearing capacity of eggs versus oocytes (El-Jouni et al. 2005). Nevertheless, it remains to be shown how PMCA plays a role in mammalian fertilization and what are the putative regulatory mechanisms, as complete internalization seems unlikely in this system, given that $[Ca^{2+}]_i$ increases occur uninterruptedly for hours in these eggs.

The mitochondria may also contribute to regulate baseline $[Ca^{2+}]_i$ in the presence of oscillations (Duchen 2000; Rizzuto et al. 2000), as they can sequester Ca^{2+} into the matrix thereby decreasing the overall cytosolic Ca^{2+} load (Rizzuto et al. 1998). Despite early evidence to the contrary (Liu et al. 2001), this does not seem to be the main function in eggs, as inhibition of Ca^{2+} mitochondrial uptake does not immediately terminate sperm-initiated oscillations (Dumollard et al. 2004). Instead, and possibly because of its vicinity to the IP$_3$R1/ER, the Ca^{2+}-driven ATP output may be the mitochondria's most critical contribution to Ca^{2+} homeostasis in MII eggs, as it maintains SERCA activity, which is required to sustain sperm-triggered Ca^{2+} oscillations (Dumollard et al. 2004).

Ca^{2+} Influx Mechanisms

Given that a fraction of Ca^{2+} from each $[Ca^{2+}]_i$ increase is secreted out of the egg by the action of PMCA or the Na^+/Ca^{2+} exchanger, external Ca^{2+} must be taken in to maintain $[Ca^{2+}]_{ER}$. Ca^{2+} influx plays a pivotal role in fertilization, as sperm-initiated $[Ca^{2+}]_i$ oscillations cease prematurely in the absence of external Ca^{2+} (Igusa and Miyazaki 1983; Igusa et al. 1983; Winston et al. 1995). Nevertheless, the molecules that mediate Ca^{2+} influx and their regulation remain poorly characterized. Oocytes/eggs and somatic cells use several Ca^{2+} influx mechanisms, including receptor-operated channels (ROCs) and voltage-operated Ca^{2+} channels (VOCs) (Berridge et al. 2000; Smyth et al. 2006), the last of which is active in mammalian oocytes and eggs (Tosti and Boni 2004). Notably, although changes in membrane potential accompany fertilization in mammals, several findings suggest that they might not be causally linked to the replenishment of the stores, as $[Ca^{2+}]_i$ increases precede changes in membrane potential (Igusa et al. 1983), Ca^{2+} influx continues between $[Ca^{2+}]_i$ increases (McGuinness et al. 1996) and, in the mouse, the changes in membrane potential are almost imperceptible (Igusa et al. 1983). These findings raise the prospect that Ca^{2+} influx in oocytes may be attained, at least in part, by a different mechanism(s). Store-operated Ca^{2+} entry (SOCE), which is associated with $[Ca^{2+}]_{ER}$ levels (Putney 1986), may fulfill this role in oocytes/eggs. The presence of store operated Ca^{2+} channels (SOC) to mediate SOCE and their electrophysiological properties were surmised in mast cells and in T-cells more than 10 years (Lewis and Cahalan 1989; Zweifach and Lewis 1993), although their molecular identity remained elusive until recently. Using a small RNA interference (RNAi) screen, two groups found that STIM1 was required for SOCE, as its reduction decreased Ca^{2+} influx in response to thapsigargin (Liou et al. 2005; Roos et al. 2005). Given that STIM1 lacks an obvious channel, the search was on to find the required channel partner protein, one of which was quickly identified as Orai1 (Feske et al. 2006; Vig et al. 2006). Besides the recent demonstration of molecular coupling at the cellular level between STIM1 and Orai1 (Park et al. 2009), spontaneous mutations of these proteins in humans have been linked to related immune response-related diseases, conclusively implicating their function in the same

Ca^{2+} influx pathway (Feske et al. 2005; Picard et al. 2009). For general information about Stim and Orai family of proteins, see Lewis 2011.

Evidence for SOCE in mammalian eggs was first observed after the application of thapsigargin, which caused a large Ca^{2+} influx after adding Ca^{2+} back to the media, (Kline and Kline 1992b; Machaty et al. 2002). Subsequent studies implicated SOCE in fertilization, as using the manganese-quenching technique it was found that in mouse eggs the initiation of each [Ca^{2+}]$_i$ increase coincided with divalent cation influx (McGuinness et al. 1996). Although SOCE was also described in human eggs (Martin-Romero et al. 2008), the understanding of the molecular underpinning of this influx in mammalian oocytes remains poor.

Transient receptor potential (TRP) ion channels (Venkatachalam and Montell 2007), which show widespread cellular distribution and display numerous regulatory mechanisms, were considered as possible mediators of Ca^{2+} influx in eggs. Expression at the transcript level was noted in porcine and mouse oocytes for several of the TRP family members (Machaty et al. 2002; Su et al. 2007), although evidence for their involvement in fertilization has yet to materialize. In contrast, two recent manuscripts examined the expression and function of STIM1 in oocytes. In porcine oocytes, STIM1 expression was detected at the mRNA level, and over-expression or knock down of STIM1 enhanced/reduced, respectively, thapsigargin-promoted Ca^{2+} influx. Expression of YFP-tagged STIM1 suggested ER localization and "puncta" reorganization in these oocytes, although more conclusive studies are needed (Koh et al. 2009). In mouse eggs, STIM1 was detected by western blotting, although the apparent molecular weight (Gomez-Fernandez et al. 2009) seems lower compared to published data in mouse somatic cells (Manji et al. 2000). Further, the detection of endogenous STIM1 by immunofluorescence revealed large patches (Gomez-Fernandez et al. 2009), which seem disproportionate to the reportedly low abundance of this protein in most cell types (Park et al. 2009). Lastly, whereas transcripts of Orai1 and two have been detected in mouse oocytes and eggs (Su et al. 2007), their protein expression has not been confirmed, and therefore their involvement in Ca^{2+} influx during mammalian fertilization remains to be shown.

A better understanding of SOCE's molecular effectors and regulatory mechanisms already exists in *Xenopus* oocytes and eggs (Machaca and Haun 2002; Yu et al. 2009). Initial research showed inactivation of SOCE, which is manifested by the uncoupling of Ca^{2+} store depletion and Ca^{2+} influx, around the time of GVBD (Machaca and Haun 2002). Although the inactivating mechanism was not known, it was determined to be associated with the activities of the M-phase kinases that regulate GVBD (Machaca and Haun, 2002). A follow up study found that during GVBD Orai1 is internalized from the plasma membrane, and STIM1's ability to form clusters and puncta is obliterated, which together disable SOCE (Yu et al. 2009). Earlier, an uncoupling between Ca^{2+} influx and Ca^{2+} store content was reported in somatic cells during mitosis (Preston et al. 1991). Those findings were recently extended, and SOCE inactivation during mitosis was associated with Stim1 phosphorylation by Cdk1, which prevents its rearrangement and precludes coupling and activation of Orai1 (Smyth et al. 2009). STIM1 phosphorylation by Cdk1 was also noted in the foregoing *Xenopus* study, although it was deemed to have minor impact on STIM1 function (Yu et al. 2009). It is noteworthy that a conserved Cdk1 phosphorylation site present in mammalian STIM1 is absent from *Xenopus* STIM1. Importantly, unlike the previous examples, SOCE is operational during the MII stage of mammalian fertilization. Therefore, future studies should examine the regulatory mechanism(s) that control SOCE during maturation and fertilization in mammals.

PLCζ

There has been much debate and speculation as to the mechanism(s) that triggers [Ca^{2+}]$_i$ oscillations during mammalian fertilization. Several excellent recent reviews have addressed this topic in depth (Swann et al. 2006; Parrington et al. 2007; Horner and Wolfner 2008b) and therefore only the most salient and outstanding aspects of PLCζ will be discussed here. As noted earlier, research in a variety of species including mammals showed that fertilization-associated $[Ca^{2+}]_i$ responses require the same agonists and signaling cascades that cause Ca^{2+} release in somatic cells (Miyazaki et al. 1993; Miyazaki and Ito 2006). Nevertheless, although stimulation of these pathways induced $[Ca^{2+}]_i$ responses, they failed to reproduce the pattern of $[Ca^{2+}]_i$ oscillations associated with mammalian fertilization, leaving open the possibility that a different mechanism may underpin oscillations in these species (Swann et al. 1989). Observations first in sea urchin eggs and then in ascidian eggs noted that injection of sperm extracts caused PM currents similar to that observed in fertilization (Dale et al. 1985; Dale 1988). Subsequently, studies in mammals showed that injections of sperm extracts initiated fertilization-like oscillations and egg activation (Stice and Robl 1990; Swann 1990). Based on these results and in light of the protracted nature of the $[Ca^{2+}]_i$ oscillations, which can vastly exceed the interaction time of gametes at the PM, a hypothesis was proposed whereby a SF acts as the trigger of oscillations after fusion of the gametes (Swann and Lai 1997). Although this hypothesis was received with skepticism, support for it grew steadily, as injection of sperm extracts initiated oscillations in several mammalian and nonmammalian species (Stricker 1997; Wu et al. 1997). Furthermore, physiological support for this concept was provided both when intracytoplasmic sperm injection into eggs (ICSI) resulted in the birth of young (Palermo et al. 1992), and the subsequent demonstration that ICSI initiated fertilization-like oscillations in several mammalian species (Tesarik and Testart 1994; Nakano et al. 1997; Kurokawa and Fissore 2003; Malcuit et al. 2006). Together, these studies consolidated the concept of the SF as the initiator of oscillations in mammalian eggs, although identification of the active principle would have to wait for another decade.

Identification of PLCζ

The search for the SF's active component(s) was the subject of intense interest and it was not without some false starts. A turning point came when studies using sea urchin egg extracts and in vitro PLC assays revealed that cytosolic preparations from mammalian sperm possessed high PLC activity, which was nearly twice as high as the activity present in other tissues (Parrington et al. 1999; Jones et al. 2000; Rice et al. 2000). In addition, it was discovered that the sperm's PLC activity displayed high sensitivity to Ca^{2+}, meaning that it shows near maximal activity in the presence of basal $[Ca^{2+}]$ concentrations (Rice et al. 2000), which in most cells are of ~ 100 nM (Clapham 2007). This feature made the putative SF a credible candidate to trigger oscillations, because to attain high specific activity most PLCs require $[Ca^{2+}]$ concentrations in excess of $1 \mu M$ (Rebecchi and Pentyala 2000; Nomikos et al. 2005), concentrations that are not compatible with MII arrest. It was therefore not surprising that injection of recombinant proteins representing most of the known isoforms expressed in sperm (Choi et al. 2001; Fukami et al. 2001) failed to initiate oscillations in mouse eggs (Parrington et al. 2002), or if they did, they did so at nonphysiological concentrations (Mehlmann et al. 2001). Hence, it became evident that if a PLC were to be the SF, it had to be a novel PLC. Toward this end, the novel sperm-specific PLCζ (Saunders et al. 2002) was identified in a PLC homology screen of mouse testis expressed sequence tags. Data in the latter study and in follow up reports provided strong evidence to support the concept that PLCζ is the pivotal, and possibly exclusive, initiator of $[Ca^{2+}]_i$ oscillations in mammals. Specifically, injection of recombinant PLCζ (Fujimoto et al. 2004; Kouchi et al. 2004) or PLCζ cRNA evoked sperm-like oscillations in mouse (Saunders et al. 2002), rat (Ito et al. 2008b), human (Rogers et al. 2004), bovine (Malcuit et al. 2005; Ross et al. 2008), porcine (Yoneda et al. 2006), and equine (Bedford-Guaus et al. 2008) eggs. In vitro PLC assays, using recombinant PLCζ confirmed the enzyme's high sensitivity to Ca^{2+}, which render it nearly fully active at basal $[Ca^{2+}]_i$ concentrations (Kouchi et al. 2004). Immunolocalization studies localized PLCζ to the postacrosomal region of mouse sperm (Fujimoto et al. 2004) and to the equatorial area of bull and human sperm (Yoon and Fissore 2007; Grasa et al. 2008; Yoon et al. 2008), regions that first come in contact with the ooplasm following gamete fusion, respectively (Sutovsky et al. 2003).

Recent evidence linking PLCζ expression and fertility further strengthened the role of PLCζ as the initiator of $[Ca^{2+}]_i$ oscillations in mammals. One study examined the ability of sperm from patients with repeated ICSI failure to initiate $[Ca^{2+}]_i$ oscillations in mouse eggs. The sperm from a few of these patients were incapable of initiating $[Ca^{2+}]_i$ responses, and examination of PLCζ expression by immunofluorescence and by Western blotting found reduced/absent levels of the enzyme in these sperm (Yoon et al. 2008). The results suggest that the inability of these sperm to activate eggs might be the main cause of their infertility. Consistent with this notion, studies have shown that the infertility of patients with globozoospermia, an affliction where even after ICSI most patients remain sterile, can be overcome by ICSI followed by Ca^{2+} ionophore-aided egg activation (Taylor et al. 2010; Heindryckx et al. 2005). A second study in patients with ICSI failure found that, in addition to reduced expression of PLCζ, a point mutation was identified that compromises PLCζ's ability to initiate $[Ca^{2+}]_i$ oscillations (Heytens et al. 2009). Collectively, the evidence supporting PLCζ as the mammalian SF is compelling. Nevertheless, questions remain regarding its expression during spermatogenesis and storage in sperm, its mechanism of release into the ooplasm, and mechanism(s) of activation once in the egg.

Despite evidence that PLCζ serves as the principal trigger of oscillations in mammals, research has unearthed species-specific differences that might prove useful in elucidating how PLCζ is regulated during fertilization. For example, although mouse PLCζ, which is the most studied, accumulates into the nucleus following PN formation (Saunders et al. 2002; Yoda et al. 2004), none of the other PLCζ isoforms tested display this localization despite sharing a nuclear localization signal (Cooney et al.; Ito et al. 2008b). There seems also to be significant differences in specific activity. For instance, based on the concentrations of cRNAs required to initiate oscillations, human PLCζ seems ∼40-fold more active than mouse PLCζ, which itself is significantly more active than the rat enzyme (Cox et al. 2002; Rogers et al. 2004; Ito et al. 2008b). Although the role of these species-specific variations has not been explored carefully, it is tempting to speculate that they are the result of adaptations to promote the optimal activation signal. To this end, it is revealing that the species with the weakest PLCζ, the rat, has the easiest oocytes to activate (Zernicka-Goetz 1991; Ito et al. 2007). Future studies should examine whether an inverse association exists between expression levels/activity of PLCζ in sperm and IP_3R1 sensitivity/strength of the CSF-arresting machinery in eggs. Similarly, future studies should elucidate the molecular changes that underlie the differences in PLCζ activity among species. For example, despite missing the pleckstrin homology (PH) domain, PLCζ shows the modular organization characteristic of other PLCs, which consists of 4 EF hand Ca^{2+}-binding domains, X and Y catalytic domains, and the Ca^{2+}-dependent phospholipid-binding C2 domain (Rebecchi and Pentyala 2000). The EF-hand domains, and especially the EF3-hand domain, have been suggested to confer the high Ca^{2+} sensitivity of PLCζ through in vitro studies (Kouchi et al. 2005; Nomikos et al. 2005); whether sequence differences in this or other EF-hand domains underlie PLCζ species-specific differences should be examined.

CONCLUSIONS

The study of the Ca^{2+} mechanisms that underlie fertilization in mammals has resulted in important contributions to the Ca^{2+} signaling field in general and to the field of fertilization in particular. For example, the indispensable role of IP_3R1-mediated Ca^{2+} release in regulating cellular functions was unequivocally shown in mouse fertilization (Miyazaki et al. 1992). Likewise, the discovery of the SF's active component, PLCζ (Saunders et al. 2002), not only provided evidence for a novel way of activating Ca^{2+} signaling in a host cell, but also added a new member with unique properties to the all important family of PLC enzymes. Importantly, and despite progress in the role of these two molecules in fertilization, we are still unaware of their fine regulatory mechanisms. For example, IP_3R1 function is greatly optimized during oocyte maturation, but the precise underlying molecular mechanisms responsible for these changes remain undetermined. Similarly, how the seemingly constitutive activity of PLCζ is provisionally restrained in the sperm and how its expression is regulated during spermatogenesis are questions that need addressing. Lastly, although $[Ca^{2+}]_i$ oscillations trigger mammalian development, we remain uninformed of the regulation of SERCA, which recycles Ca^{2+} into the ER, and of the molecules that underpin Ca^{2+} influx, which sustain the oscillations. Identification and elucidation of these regulatory mechanisms in oocytes will deepen our understating of fertilization, information that could be then used in the clinic for the diagnosis of infertility, and to enhance developmental competence of embryos generated by a variety of Assisted Reproductive Technology procedures.

ACKNOWLEDGMENTS

This work was supported by grants from the USDA and the NIH/N.I.C.H.D-HD051872 to R.A.F. The authors wish to thank all members of the Fissore lab for their generous contributions and suggestions. We apologize to those whose work was not cited because of space limitations.

REFERENCES

Backs J, Stein P, Backs T, Duncan FE, Grueter CE, McAnally J, Qi X, Schultz RM, Olson EN. 2010. The gamma isoform of CaM kinase II controls mouse egg activation by regulating cell cycle resumption. *Proc Natl Acad Sci* **107:** 81–86.

Bai GR, Yang LH, Huang XY, Sun FZ. 2006. Inositol 1,4,5-trisphosphate receptor type 1 phosphorylation and regulation by extracellular signal-regulated kinase. *Biochem Biophys Res Commun* **348:** 1319–1327.

Becker KA, Hart NH. 1999. Reorganization of filamentous actin and myosin-II in zebrafish eggs correlates temporally and spatially with cortical granule exocytosis. *J Cell Sci* **112 (Pt 1):** 97–110.

Bedford-Guaus SJ, Yoon SY, Fissore RA, Choi YH, Hinrichs K. 2008. Microinjection of mouse phospholipase C zeta complementary RNA into mare oocytes induces long-lasting intracellular calcium oscillations and embryonic development. *Reprod Fertil Dev* **20:** 875–883.

Berridge MJ, Lipp P, Bootman MD. 2000. The versatility and universality of calcium signalling. *Nat Rev Mol Cell Biol* **1:** 11–21.

Bezprozvanny I. 2005. The inositol 1,4,5-trisphosphate receptors. *Cell Calcium* **38:** 261–272.

Bootman M, Collins T, Peppiatt C, Prothero L, MacKenzie L, De Smet P, Travers M, Tovey S, Seo J, Berridge M, et al. 2001. Calcium signalling—an overview. *Semin Cell Dev Biol* **12:** 3–10.

Boulware MJ, Marchant JS. 2005. IP3 receptor activity is differentially regulated in endoplasmic reticulum subdomains during oocyte maturation. *Curr Biol* **15:** 765–770.

Brini M, Carafoli E. 2009. Calcium pumps in health and disease. *Physiol Rev* **89:** 1341–1378.

Carroll J. 2000. Na+-Ca2+ exchange in mouse oocytes: modifications in the regulation of intracellular free Ca2+ during oocyte maturation. *J Reprod Fertil* **118:** 337–342.

Carroll J, Swann K. 1992. Spontaneous cytosolic calcium oscillations driven by inositol trisphosphate occur during in vitro maturation of mouse oocytes. *J Biol Chem* **267:** 11196–11201.

Carroll J, Swann K, Whittingham D, Whitaker M. 1994. Spatiotemporal dynamics of intracellular [Ca2+]i oscillations during the growth and meiotic maturation of mouse oocytes. *Development* **120:** 3507–3517.

Castilho PV, Williams BC, Mochida S, Zhao Y, Goldberg ML. 2009. The M phase kinase Greatwall (Gwl) promotes inactivation of PP2A/B55delta, a phosphatase directed against CDK phosphosites. *Mol Biol Cell* **20:** 4777–4789.

Chang HY, Minahan K, Merriman JA, Jones KT. 2009. Calmodulin-dependent protein kinase gamma 3 (CamKIIgamma3) mediates the cell cycle resumption of metaphase II eggs in mouse. *Development* **136:** 4077–4081.

Choi D, Lee E, Hwang S, Jun K, Kim D, Yoon BK, Shin HS, Lee JH. 2001. The biological significance of phospholipase C beta 1 gene mutation in mouse sperm in the acrosome reaction, fertilization, and embryo development. *J Assist Reprod Genet* **18:** 305–310.

Clapham DE. 2007. Calcium signaling. *Cell* **131:** 1047–1058.

Clapper DL, Lee HC. 1985. Inositol trisphosphate induces calcium release from nonmitochondrial stores in sea urchin egg homogenates. *J Biol Chem* **260:** 13947–13954.

Cooney MA, Malcuit C, Cheon B, Holland MK, Fissore RA, D'Cruz NT. 2010. Species-specific differences in the activity and nuclear localization of murine and bovine phospholipase C, Zeta 1. *Biol Reprod* **83:** 92–101.

Cox LJ, Larman MG, Saunders CM, Hashimoto K, Swann K, Lai FA. 2002. Sperm phospholipase Czeta from humans and cynomolgus monkeys triggers Ca^{2+} oscillations, activation and development of mouse oocytes. *Reproduction* **124:** 611–623.

Cuthbertson KS, Whittingham DG, Cobbold PH. 1981. Free Ca^{2+} increases in exponential phases during mouse oocyte activation. *Nature* **294:** 754–757.

Dale B. 1988. Primary and secondary messengers in the activation of ascidian eggs. *Exp Cell Res* **177:** 205–211.

Dale B, DeFelice LJ, Ehrenstein G. 1985. Injection of a soluble sperm fraction into sea-urchin eggs triggers the cortical reaction. *Experientia* **41:** 1068–1070.

Deguchi R, Shirakawa H, Oda S, Mohri T, Miyazaki S. 2000. Spatiotemporal analysis of $Ca^{(2+)}$ waves in relation to the sperm entry site and animal-vegetal axis during $Ca^{(2+)}$ oscillations in fertilized mouse eggs. *Dev Biol* **218:** 299–313.

Deng MQ, Shen SS. 2000. A specific inhibitor of p34(cdc2)/cyclin B suppresses fertilization-induced calcium oscillations in mouse eggs. *Biol Reprod* **62:** 873–878.

Deng M, Williams CJ, Schultz RM. 2005. Role of MAP kinase and myosin light chain kinase in chromosome-induced development of mouse egg polarity. *Dev Biol* **278:** 358–366.

Descombes P, Nigg EA. 1998. The polo-like kinase Plx1 is required for M phase exit and destruction of mitotic regulators in *Xenopus* egg extracts. *Embo J* **17:** 1328–1335.

DeSouza N, Reiken S, Ondrias K, Yang YM, Matkovich S, Marks AR. 2002. Protein kinase A and two phosphatases are components of the inositol 1,4,5-trisphosphate receptor macromolecular signaling complex. *J Biol Chem* **277:** 39397–39400.

Diaz-Munoz M, de la Rosa Santander P, Juarez-Espinosa AB, Arellano RO, Morales-Tlalpan V. 2008. Granulosa cells express three inositol 1,4,5-trisphosphate receptor isoforms: cytoplasmic and nuclear Ca2+ mobilization. *Reprod Biol Endocrinol* **6:** 60.

Draetta G, Luca F, Westendorf J, Brizuela L, Ruderman J, Beach D. 1989. Cdc2 protein kinase is complexed with both cyclin A and B: evidence for proteolytic inactivation of MPF. *Cell* **56:** 829–838.

Duchen MR. 2000. Mitochondria and calcium: from cell signalling to cell death. *J Physiol* **529 Pt 1:** 57–68.

Ducibella T, Fissore R. 2008. The roles of Ca^{2+}, downstream protein kinases, and oscillatory signaling in regulating fertilization and the activation of development. *Dev Biol* **315:** 257–279.

Ducibella T, LeFevre L. 1997. Study of protein kinase C antagonists on cortical granule exocytosis and cell-cycle resumption in fertilized mouse eggs. *Mol Reprod Dev* **46:** 216–226.

Ducibella T, Schultz RM, Ozil JP. 2006. Role of calcium signals in early development. *Semin Cell Dev Biol* **17:** 324–332.

Ducibella T, Huneau D, Angelichio E, Xu Z, Schultz RM, Kopf GS, Fissore R, Madoux S, Ozil JP. 2002. Egg-to-embryo transition is driven by differential responses to $Ca^{(2+)}$ oscillation number. *Dev Biol* **250:** 280–291.

Dumollard R, Marangos P, Fitzharris G, Swann K, Duchen M, Carroll J. 2004. Sperm-triggered $[Ca^{2+}]$ oscillations and Ca^{2+} homeostasis in the mouse egg have an absolute requirement for mitochondrial ATP production. *Development* **131:** 3057–3067.

El-Jouni W, Haun S, Machaca K. 2008. Internalization of plasma membrane Ca^{2+}-ATPase during *Xenopus* oocyte maturation. *Dev Biol* **324:** 99–107.

El-Jouni W, Jang B, Haun S, Machaca K. 2005. Calcium signaling differentiation during *Xenopus* oocyte maturation. *Dev Biol* **288:** 514–525.

Eliyahu E, Kaplan-Kraicer R, Shalgi R. 2001. PKC in eggs and embryos. *Front Biosci* **6:** D785–D791.

Feske S, Gwack Y, Prakriya M, Srikanth S, Puppel SH, Tanasa B, Hogan PG, Lewis RS, Daly M, Rao A. 2006. A mutation in Orai1 causes immune deficiency by abrogating CRAC channel function. *Nature* **441:** 179–185.

Feske S, Prakriya M, Rao A, Lewis RS. 2005. A severe defect in CRAC Ca^{2+} channel activation and altered K^+ channel gating in T cells from immunodeficient patients. *J Exp Med* **202:** 651–662.

Finch EA, Turner TJ, Goldin SM. 1991. Calcium as a co-agonist of inositol 1,4,5-trisphosphate-induced calcium release. *Science* **252:** 443–446.

Fissore RA, Dobrinsky JR, Balise JJ, Duby RT, Robl JM. 1992. Patterns of intracellular Ca^{2+} concentrations in fertilized bovine eggs. *Biol Reprod* **47:** 960–969.

Fissore RA, Reis MM, Palermo GD. 1999b. Isolation of the Ca2+ releasing component(s) of mammalian sperm extracts: the search continues. *Mol Hum Reprod* **5:** 189–192.

Fissore RA, Longo FJ, Anderson E, Parys JB, Ducibella T. 1999a. Differential distribution of inositol trisphosphate receptor isoforms in mouse oocytes. *Biol Reprod* **60:** 49–57.

FitzHarris G, Marangos P, Carroll J. 2003. Cell cycle-dependent regulation of structure of endoplasmic reticulum and inositol 1,4,5-trisphosphate-induced Ca^{2+} release in mouse oocytes and embryos. *Mol Biol Cell* **14:** 288–301.

FitzHarris G, Marangos P, Carroll J. 2007. Changes in endoplasmic reticulum structure during mouse oocyte maturation are controlled by the cytoskeleton and cytoplasmic dynein. *Dev Biol* **305:** 133–144.

Fuchimoto D, Mizukoshi A, Schultz RM, Sakai S, Aoki F. 2001. Posttranscriptional regulation of cyclin A1 and cyclin A2 during mouse oocyte meiotic maturation and preimplantation development. *Biol Reprod* **65:** 986–993.

Fujimoto S, Yoshida N, Fukui T, Amanai M, Isobe T, Itagaki C, Izumi T, Perry AC. 2004. Mammalian phospholipase Czeta induces oocyte activation from the sperm perinuclear matrix. *Dev Biol* **274:** 370–383.

Fukami K, Nakao K, Inoue T, Kataoka Y, Kurokawa M, Fissore R, Nakamura K, Katsuki M, Mikoshiba K, Yoshida N, et al. 2001. Requirement of phospholipase Cdelta4 for the zona pellucida-induced acrosome reaction. *Science* **292:** 920–923.

Furuichi T, Yoshikawa S, Miyawaki A, Wada K, Maeda N, Mikoshiba K. 1989. Primary structure and functional expression of the inositol 1,4,5-trisphosphate-binding protein P400. *Nature* **342:** 32–38.

Gallicano GI, Yousef MC, Capco DG. 1997. PKC—a pivotal regulator of early development. *Bioessays* **19:** 29–36.

Giusti AF, Carroll DJ, Abassi YA, Terasaki M, Foltz KR, Jaffe LA. 1999. Requirement of a Src family kinase for initiating calcium release at fertilization in starfish eggs. *J Biol Chem* **274:** 29318–29322.

Gomez-Fernandez C, Pozo-Guisado E, Ganan-Parra M, Perianes MJ, Alvarez IS, Martin-Romero FJ. 2009. Relocalization of STIM1 in mouse oocytes at fertilization: early involvement of store-operated calcium entry. *Reproduction* **138:** 211–221.

Goud PT, Goud AP, Leybaert L, Van Oostveldt P, Mikoshiba K, Diamond MP, Dhont M. 2002. Inositol 1,4,5-trisphosphate receptor function in human oocytes: calcium responses and oocyte activation-related phenomena induced by photolytic release of InsP(3) are blocked by a specific antibody to the type I receptor. *Mol Hum Reprod* **8:** 912–918.

Grasa P, Coward K, Young C, Parrington J. 2008. The pattern of localization of the putative oocyte activation factor, phospholipase Czeta, in uncapacitated, capacitated, and ionophore-treated human spermatozoa. *Hum Reprod* **23:** 2513–2522.

Halet G. 2004. PKC signaling at fertilization in mammalian eggs. *Biochim Biophys Acta* **1742:** 185–189.

Harada Y, Matsumoto T, Hirahara S, Nakashima A, Ueno S, Oda S, Miyazaki S, Iwao Y. 2007. Characterization of a sperm factor for egg activation at fertilization of the newt Cynops pyrrhogaster. *Dev Biol* **306:** 797–808.

Heindryckx B, Van der Elst J, De Sutter P, Dhont M. 2005. Treatment option for sperm- or oocyte-related fertilization failure: assisted oocyte activation following diagnostic heterologous ICSI. *Hum Reprod* **20:** 2237–2241.

Heytens E, Parrington J, Coward K, Young C, Lambrecht S, Yoon S, Fissore R, Hamer R, Deane C, Ruas M, et al. 2009. Reduced amounts and abnormal forms of phospholipase C zeta (PLCzeta) in spermatozoa from infertile men. *Hum Reprod* **24:** 2417–2428.

Horner VL, Wolfner MF. 2008a. Mechanical stimulation by osmotic and hydrostatic pressure activates Drosophila oocytes in vitro in a calcium-dependent manner. *Dev Biol* **316:** 100–109.

Horner VL, Wolfner MF. 2008b. Transitioning from egg to embryo: triggers and mechanisms of egg activation. *Dev Dyn* **237:** 527–544.

Igusa Y, Miyazaki S. 1983. Effects of altered extracellular and intracellular calcium concentration on hyperpolarizing responses of the hamster egg. *J Physiol* **340:** 611–632.

Igusa Y, Miyazaki S, Yamashita N. 1983. Periodic hyperpolarizing responses in hamster and mouse eggs fertilized with mouse sperm. *J Physiol* **340:** 633–647.

Iino M. 1990. Biphasic Ca2+ dependence of inositol 1,4,5-trisphosphate-induced Ca release in smooth muscle cells of the guinea pig taenia caeci. *J Gen Physiol* **95:** 1103–1122.

Ito J, Shimada M, Hochi S, Hirabayashi M. 2007. Involvement of Ca2+-dependent proteasome in the degradation of both cyclin B1 and Mos during spontaneous activation of matured rat oocytes. *Theriogenology* **67:** 475–485.

Ito M, Shikano T, Oda S, Horiguchi T, Tanimoto S, Awaji T, Mitani H, Miyazaki S. 2008b. Difference in Ca2+ oscillation-inducing activity and nuclear translocation ability of PLCZ1, an egg-activating sperm factor candidate, between mouse, rat, human, and medaka fish. *Biol Reprod* **78:** 1081–1090.

Ito J, Yoon SY, Lee B, Vanderheyden V, Vermassen E, Wojcikiewicz R, Alfandari D, De Smedt H, Parys JB, Fissore RA. 2008a. Inositol 1,4,5-trisphosphate receptor 1, a widespread Ca^{2+} channel, is a novel substrate of polo-like kinase 1 in eggs. *Dev Biol* **320:** 402–413.

Iwasaki H, Chiba K, Uchiyama T, Yoshikawa F, Suzuki F, Ikeda M, Furuichi T, Mikoshiba K. 2002. Molecular characterization of the starfish inositol 1,4,5-trisphosphate receptor and its role during oocyte maturation and fertilization. *J Biol Chem* **277:** 2763–2772.

Jellerette T, He CL, Wu H, Parys JB, Fissore RA. 2000. Down-regulation of the inositol 1,4,5-trisphosphate receptor in mouse eggs following fertilization or parthenogenetic activation. *Dev Biol* **223:** 238–250.

Jellerette T, Kurokawa M, Lee B, Malcuit C, Yoon SY, Smyth J, Vermassen E, De Smedt H, Parys JB, Fissore RA. 2004. Cell cycle-coupled $[Ca^{2+}](i)$ oscillations in mouse zygotes and function of the inositol 1,4,5-trisphosphate receptor-1. *Dev Biol* **274:** 94–109.

Jones KT, Carroll J, Whittingham DG. 1995a. Ionomycin, thapsigargin, ryanodine, and sperm induced Ca^{2+} release increase during meiotic maturation of mouse oocytes. *J Biol Chem* **270:** 6671–6677.

Jones KT, Carroll J, Merriman JA, Whittingham DG, Kono T. 1995. Repetitive sperm-induced Ca^{2+} transients in mouse oocytes are cell cycle dependent. *Development* **121:** 3259–3266.

Jones KT, Matsuda M, Parrington J, Katan M, Swann K. 2000. Different Ca^{2+}-releasing abilities of sperm extracts compared with tissue extracts and phospholipase C isoforms in sea urchin egg homogenate and mouse eggs. *Biochem J* **346 (Pt 3):** 743–749.

Joughin BA, Naegle KM, Huang PH, Yaffe MB, Lauffenburger DA, White FM. 2009. An integrated comparative phosphoproteomic and bioinformatic approach reveals a novel class of MPM-2 motifs upregulated in EGFRvIII-expressing glioblastoma cells. *Mol Biosyst* **5:** 59–67.

Kline D, Kline JT. 1992a. Repetitive calcium transients and the role of calcium in exocytosis and cell cycle activation in the mouse egg. *Dev Biol* **149:** 80–89.

Kline D, Kline JT. 1992b. Thapsigargin activates a calcium influx pathway in the unfertilized mouse egg and suppresses repetitive calcium transients in the fertilized egg. *J Biol Chem* **267:** 17624–17630.

Kline D, Mehlmann L, Fox C, Terasaki M. 1999. The cortical endoplasmic reticulum (ER) of the mouse egg: localization of ER clusters in relation to the generation of repetitive calcium waves. *Dev Biol* **215:** 431–442.

Knott JG, Gardner AJ, Madgwick S, Jones KT, Williams CJ, Schultz RM. 2006. Calmodulin-dependent protein kinase II triggers mouse egg activation and embryo development in the absence of Ca^{2+} oscillations. *Dev Biol* **296:** 388–395.

Koh S, Lee K, Wang C, Cabot RA, Machaty Z. 2009. STIM1 regulates store-operated $Ca^{(2+)}$ entry in oocytes. *Dev Biol* **330:** 368–376.

Kono T, Jones KT, Bos-Mikich A, Whittingham DG, Carroll J. 1996. A cell cycle-associated change in Ca^{2+} releasing activity leads to the generation of Ca^{2+} transients in mouse embryos during the first mitotic division. *J Cell Biol* **132**: 915–923.

Kouchi Z, Fukami K, Shikano T, Oda S, Nakamura Y, Takenawa T, Miyazaki S. 2004. Recombinant phospholipase Czeta has high Ca^{2+} sensitivity and induces Ca^{2+} oscillations in mouse eggs. *J Biol Chem* **279**: 10408–10412.

Kouchi Z, Shikano T, Nakamura Y, Shirakawa H, Fukami K, Miyazaki S. 2005. The role of EF-hand domains and C2 domain in regulation of enzymatic activity of phospholipase Czeta. *J Biol Chem* **280**: 21015–21021.

Kume S, Muto A, Okano H, Mikoshiba K. 1997. Developmental expression of the inositol 1,4,5-trisphosphate receptor and localization of inositol 1,4,5-trisphosphate during early embryogenesis in *Xenopus laevis*. *Mech Dev* **66**: 157–168.

Kurokawa M, Fissore RA. 2003. ICSI-generated mouse zygotes exhibit altered calcium oscillations, inositol 1,4,5-trisphosphate receptor-1 down-regulation, and embryo development. *Mol Hum Reprod* **9**: 523–533.

Kyozuka K, Deguchi R, Mohri T, Miyazaki S. 1998. Injection of sperm extract mimics spatiotemporal dynamics of Ca^{2+} responses and progression of meiosis at fertilization of ascidian oocytes. *Development* **125**: 4099–4105.

Lawrence YM, Cuthbertson KS. 1995. Thapsigargin induces cytoplasmic free Ca^{2+} oscillations in mouse oocytes. *Cell Calcium* **17**: 154–164.

Lee B, Vermassen E, Yoon SY, Vanderheyden V, Ito J, Alfandari D, De Smedt H, Parys JB, Fissore RA. 2006. Phosphorylation of IP3R1 and the regulation of $[Ca^{2+}]i$ responses at fertilization: a role for the MAP kinase pathway. *Development* **133**: 4355–4365.

Lee B, Yoon SY, Malcuit C, Parys JB, Fissore RA. Inositol 1,4,5-trisphosphate receptor 1 degradation in mouse eggs and impact on $[Ca^{2+}]i$ oscillations. *J Cell Physiol* **222**: 238–247.

Lewis RS, Cahalan MD. 1989. Mitogen-induced oscillations of cytosolic Ca^{2+} and transmembrane Ca^{2+} current in human leukemic T cells. *Cell Regul* **1**: 99–112.

Lewis RS. 2011. Store-operated calcium channels: New perspectives on mechanism and function. *Cold Spring Harb Perspect Biol* doi:10.1101/cshperspect.a003970.

Liou J, Kim ML, Heo WD, Jones JT, Myers JW, Ferrell JE Jr, Meyer T. 2005. STIM is a Ca^{2+} sensor essential for Ca^{2+} store-depletion-triggered Ca^{2+} influx. *Curr Biol* **15**: 1235–1241.

Liu L, Hammar K, Smith PJ, Inoue S, Keefe DL. 2001. Mitochondrial modulation of calcium signaling at the initiation of development. *Cell Calcium* **30**: 423–433.

Loeb J. 1907. On the chemical character of the process of fertilization and its bearing upon the theory of life phenomena. *Science* **26**: 425–437.

Lorca T, Cruzalegui FH, Fesquet D, Cavadore JC, Mery J, Means A, Doree M. 1993. Calmodulin-dependent protein kinase II mediates inactivation of MPF and CSF upon fertilization of *Xenopus* eggs. *Nature* **366**: 270–273.

Machaca K, Haun S. 2002. Induction of maturation-promoting factor during *Xenopus* oocyte maturation uncouples $Ca(^{2+})$ store depletion from store-operated $Ca(^{2+})$ entry. *J Cell Biol* **156**: 75–85.

Machaty Z, Ramsoondar JJ, Bonk AJ, Bondioli KR, Prather RS. 2002. Capacitative calcium entry mechanism in porcine oocytes. *Biol Reprod* **66**: 667–674.

MacKrill JJ. 1999. Protein-protein interactions in intracellular Ca^{2+}-release channel function. *Biochem J* **337** (**Pt 3**): 345–361.

Madgwick S, Levasseur M, Jones KT. 2005. Calmodulin-dependent protein kinase II, and not protein kinase C, is sufficient for triggering cell-cycle resumption in mammalian eggs. *J Cell Sci* **118**: 3849–3859.

Madgwick S, Hansen DV, Levasseur M, Jackson PK, Jones KT. 2006. Mouse Emi2 is required to enter meiosis II by reestablishing cyclin B1 during interkinesis. *J Cell Biol* **174**: 791–801.

Madgwick S, Nixon VL, Chang HY, Herbert M, Levasseur M, Jones KT. 2004. Maintenance of sister chromatid attachment in mouse eggs through maturation-promoting factor activity. *Dev Biol* **275**: 68–81.

Maeda N, Niinobe M, Nakahira K, Mikoshiba K. 1988. Purification and characterization of P400 protein, a glycoprotein characteristic of Purkinje cell, from mouse cerebellum. *J Neurochem* **51**: 1724–1730.

Mahbub Hasan AK, Sato K, Sakakibara K, Ou Z, Iwasaki T, Ueda Y, Fukami Y. 2005. Uroplakin III, a novel Src substrate in *Xenopus* egg rafts, is a target for sperm protease essential for fertilization. *Dev Biol* **286**: 483–492.

Malathi K, Kohyama S, Ho M, Soghoian D, Li X, Silane M, Berenstein A, Jayaraman T. 2003. Inositol 1,4,5-trisphosphate receptor (type 1) phosphorylation and modulation by Cdc2. *J Cell Biochem* **90**: 1186–1196.

Malathi K, Li X, Krizanova O, Ondrias K, Sperber K, Ablamunits V, Jayaraman T. 2005. Cdc2/cyclin B1 interacts with and modulates inositol 1,4,5-trisphosphate receptor (type 1) functions. *J Immunol* **175**: 6205–6210.

Malcuit C, Knott JG, He C, Wainwright T, Parys JB, Robl JM, Fissore RA. 2005. Fertilization and inositol 1,4,5-trisphosphate (IP3)-induced calcium release in type-1 inositol 1,4,5-trisphosphate receptor down-regulated bovine eggs. *Biol Reprod* **73**: 2–13.

Malcuit C, Maserati M, Takahashi Y, Page R, Fissore RA. 2006. Intracytoplasmic sperm injection in the bovine induces abnormal $[Ca^{2+}]i$ responses and oocyte activation. *Reprod Fertil Dev* **18**: 39–51.

Manji SS, Parker NJ, Williams RT, van Stekelenburg L, Pearson RB, Dziadek M, Smith PJ. 2000. STIM1: a novel phosphoprotein located at the cell surface. *Biochim Biophys Acta* **1481**: 147–155.

Marangos P, Carroll J. 2004. Fertilization and InsP$_3$-induced Ca^{2+} release stimulate a persistent increase in the rate of degradation of cyclin B1 specifically in mature mouse oocytes. *Dev Biol* **272**: 26–38.

Marangos P, FitzHarris G, Carroll J. 2003. Ca^{2+} oscillations at fertilization in mammals are regulated by the formation of pronuclei. *Development* **130**: 1461–1472.

Markoulaki S, Matson S, Abbott AL, Ducibella T. 2003. Oscillatory CaMKII activity in mouse egg activation. *Dev Biol* **258**: 464–474.

Martin-Romero FJ, Ortiz-de-Galisteo JR, Lara-Laranjeira J, Dominguez-Arroyo JA, Gonzalez-Carrera E, Alvarez IS. 2008. Store-operated calcium entry in human oocytes and sensitivity to oxidative stress. *Biol Reprod* **78**: 307–315.

Masui Y, Markert CL. 1971. Cytoplasmic control of nuclear behavior during meiotic maturation of frog oocytes. *J Exp Zool* **177**: 129–145.

Matson S, Ducibella T. 2007. The MEK inhibitor, U0126, alters fertilization-induced [Ca^{2+}]i oscillation parameters and secretion: differential effects associated with in vivo and in vitro meiotic maturation. *Dev Biol* **306**: 538–548.

Matson S, Markoulaki S, Ducibella T. 2006. Antagonists of myosin light chain kinase and of myosin II inhibit specific events of egg activation in fertilized mouse eggs. *Biol Reprod* **74**: 169–176.

Mazia D. 1937. The release of calcium in Arbacia eggs on fertilization. *J Cell Comp Physiol* **10**: 291–304.

McAvey BA, Wortzman GB, Williams CJ, Evans JP. 2002. Involvement of calcium signaling and the actin cytoskeleton in the membrane block to polyspermy in mouse eggs. *Biol Reprod* **67**: 1342–1352.

McGuinness OM, Moreton RB, Johnson MH, Berridge MJ. 1996. A direct measurement of increased divalent cation influx in fertilised mouse oocytes. *Development* **122**: 2199–2206.

Mehlmann LM, Kline D. 1994. Regulation of intracellular calcium in the mouse egg: calcium release in response to sperm or inositol trisphosphate is enhanced after meiotic maturation. *Biology of Reproduction* **51**: 1088–1098.

Mehlmann LM, Mikoshiba K, Kline D. 1996. Redistribution and increase in cortical inositol 1,4,5-trisphosphate receptors after meiotic maturation of the mouse oocyte. *Dev Biol* **180**: 489–498.

Mehlmann LM, Chattopadhyay A, Carpenter G, Jaffe LA. 2001. Evidence that phospholipase C from the sperm is not responsible for initiating Ca$^{(2+)}$ release at fertilization in mouse eggs. *Dev Biol* **236**: 492–501.

Mehlmann LM, Terasaki M, Jaffe LA, Kline D. 1995. Reorganization of the endoplasmic reticulum during meiotic maturation of the mouse oocyte. *Dev Biol* **170**: 607–615.

Missiaen L, Taylor CW, Berridge MJ. 1991. Spontaneous calcium release from inositol trisphosphate-sensitive calcium stores. *Nature* **352**: 241–244.

Miyazaki S. 1988. Inositol 1,4,5-trisphosphate-induced calcium release and guanine nucleotide-binding protein-mediated periodic calcium increases in golden hamster eggs. *J Cell Biol* **106**: 345–353.

Miyazaki S, Igusa Y. 1981. Fertilization potential in golden hamster eggs consists of recurring hyperpolarizations. *Nature* **290**: 702–704.

Miyazaki S, Ito M. 2006. Calcium signals for egg activation in mammals. *J Pharmacol Sci* **100**: 545–552.

Miyazaki S, Hashimoto N, Yoshimoto Y, Kishimoto T, Igusa Y, Hiramoto Y. 1986. Temporal and spatial dynamics of the periodic increase in intracellular free calcium at fertilization of golden hamster eggs. *Dev Biol* **118**: 259–267.

Miyazaki S, Shirakawa H, Nakada K, Honda Y. 1993. Essential role of the inositol 1,4,5-trisphosphate receptor/Ca^{2+} release channel in Ca^{2+} waves and Ca^{2+} oscillations at fertilization of mammalian eggs. *Dev Biol* **158**: 62–78.

Miyazaki S, Yuzaki M, Nakada K, Shirakawa H, Nakanishi S, Nakade S, Mikoshiba K. 1992. Block of Ca^{2+} wave and Ca^{2+} oscillation by antibody to the inositol 1,4,5-trisphosphate receptor in fertilized hamster eggs. *Science* **257**: 251–255.

Nabti I, Reis A, Levasseur M, Stemmann O, Jones KT. 2008. Securin and not CDK1/cyclin B1 regulates sister chromatid disjunction during meiosis II in mouse eggs. *Dev Biol* **321**: 379–386.

Nakano Y, Shirakawa H, Mitsuhashi N, Kuwabara Y, Miyazaki S. 1997. Spatiotemporal dynamics of intracellular calcium in the mouse egg injected with a spermatozoon. *Mol Hum Reprod* **3**: 1087–1093.

Nixon VL, Levasseur M, McDougall A, Jones KT. 2002. Ca$^{(2+)}$ oscillations promote APC/C-dependent cyclin B1 degradation during metaphase arrest and completion of meiosis in fertilizing mouse eggs. *Curr Biol* **12**: 746–750.

Nomikos M, Blayney LM, Larman MG, Campbell K, Rossbach A, Saunders CM, Swann K, Lai FA. 2005. Role of phospholipase C-zeta domains in Ca^{2+}-dependent phosphatidylinositol 4,5-bisphosphate hydrolysis and cytoplasmic Ca^{2+} oscillations. *J Biol Chem* **280**: 31011–31018.

Norris RP, Ratzan WJ, Freudzon M, Mehlmann LM, Krall J, Movsesian MA, Wang H, Ke H, Nikolaev VO, Jaffe LA. 2009. Cyclic GMP from the surrounding somatic cells regulates cyclic AMP and meiosis in the mouse oocyte. *Development* **136**: 1869–1878.

Oakes SA, Scorrano L, Opferman JT, Bassik MC, Nishino M, Pozzan T, Korsmeyer SJ. 2005. Proapoptotic BAX and BAK regulate the type 1 inositol trisphosphate receptor and calcium leak from the endoplasmic reticulum. *Proc Natl Acad Sci* **102**: 105–110.

Oh B, Hwang SY, Solter D, Knowles BB. 1997. Spindlin, a major maternal transcript expressed in the mouse during the transition from oocyte to embryo. *Development* **124**: 493–503.

Ozil JP, Banrezes B, Toth S, Pan H, Schultz RM. 2006. Ca^{2+} oscillatory pattern in fertilized mouse eggs affects gene expression and development to term. *Dev Biol* **300**: 534–544.

Ozil JP, Markoulaki S, Toth S, Matson S, Banrezes B, Knott JG, Schultz RM, Huneau D, Ducibella T. 2005. Egg activation events are regulated by the duration of a sustained [Ca^{2+}]cyt signal in the mouse. *Dev Biol* **282**: 39–54.

Page AW, Orr-Weaver TL. 1997. Activation of the meiotic divisions in *Drosophila* oocytes. *Dev Biol* **183**: 195–207.

Page Baluch D, Koeneman BA, Hatch KR, McGaughey RW, Capco DG. 2004. PKC isotypes in post-activated and fertilized mouse eggs: association with the meiotic spindle. *Dev Biol* **274**: 45–55.

Palermo G, Joris H, Devroey P, Van Steirteghem AC. 1992. Pregnancies after intracytoplasmic injection of single spermatozoon into an oocyte. *Lancet* **340**: 17–18.

Park CY, Hoover PJ, Mullins FM, Bachhawat P, Covington ED, Raunser S, Walz T, Garcia KC, Dolmetsch RE, Lewis RS. 2009. STIM1 clusters and activates CRAC channels via direct binding of a cytosolic domain to Orai1. *Cell* **136**: 876–890.

Parrington J, Brind S, De Smedt H, Gangeswaran R, Lai FA, Wojcikiewicz R, Carroll J. 1998. Expression of inositol

1,4,5-trisphosphate receptors in mouse oocytes and early embryos: the type I isoform is upregulated in oocytes and downregulated after fertilization. *Dev Biol* **203:** 451–461.

Parrington J, Davis LC, Galione A, Wessel G. 2007. Flipping the switch: how a sperm activates the egg at fertilization. *Dev Dyn* **236:** 2027–2038.

Parrington J, Jones KT, Lai A, Swann K. 1999. The soluble sperm factor that causes Ca^{2+} release from sea-urchin (Lytechinus pictus) egg homogenates also triggers Ca^{2+} oscillations after injection into mouse eggs. *Biochem J* **341 (Pt 1):** 1–4.

Parrington J, Jones ML, Tunwell R, Devader C, Katan M, Swann K. 2002. Phospholipase C isoforms in mammalian spermatozoa: potential components of the sperm factor that causes Ca^{2+} release in eggs. *Reproduction* **123:** 31–39.

Parys JB, McPherson SM, Mathews L, Campbell KP, Longo FJ. 1994. Presence of inositol 1,4,5-trisphosphate receptor, calreticulin, and calsequestrin in eggs of sea urchins and *Xenopus laevis*. *Dev Biol* **161:** 466–476.

Patterson RL, Boehning D, Snyder SH. 2004. Inositol 1,4,5-trisphosphate receptors as signal integrators. *Annu Rev Biochem* **73:** 437–465.

Pepperell JR, Kommineni K, Buradagunta S, Smith PJ, Keefe DL. 1999. Transmembrane regulation of intracellular calcium by a plasma membrane sodium/calcium exchanger in mouse ova. *Biol Reprod* **60:** 1137–1143.

Perry AC, Verlhac MH. 2008. Second meiotic arrest and exit in frogs and mice. *EMBO Rep* **9:** 246–251.

Picard C, McCarl C, Papolos A, Khalil S, Luthy K, Hivroz C, LeDeist F, Rieux-Laucat F, Rechavi G, Rao A, et al. 2009. STIM1 mutation associated with a syndrome of immunodeficiency and autoimmunity. *N Engl J Med* **360:** 1971–1980.

Preston SF, Sha'afi RI, Berlin RD. 1991. Regulation of Ca^{2+} influx during mitosis: Ca^{2+} influx and depletion of intracellular Ca^{2+} stores are coupled in interphase but not mitosis. *Cell Regul* **2:** 915–925.

Putney JW Jr. 1986. A model for receptor-regulated calcium entry. *Cell Calcium* **7:** 1–12.

Rebecchi MJ, Pentyala SN. 2000. Structure, function, and control of phosphoinositide-specific phospholipase C. *Physiol Rev* **80:** 1291–1335.

Rice A, Parrington J, Jones KT, Swann K. 2000. Mammalian sperm contain a $Ca^{(2+)}$-sensitive phospholipase C activity that can generate InsP(3) from PIP(2) associated with intracellular organelles. *Dev Biol* **228:** 125–135.

Ridgway EB, Gilkey JC, Jaffe LF. 1977. Free calcium increases explosively in activating medaka eggs. *Proc Natl Acad Sci* **74:** 623–627.

Rizzuto R, Bernardi P, Pozzan T. 2000. Mitochondria as all-round players of the calcium game. *J Physiol* **529 Pt 1:** 37–47.

Rizzuto R, Pinton P, Carrington W, Fay FS, Fogarty KE, Lifshitz LM, Tuft RA, Pozzan T. 1998. Close contacts with the endoplasmic reticulum as determinants of mitochondrial Ca^{2+} responses. *Science* **280:** 1763–1766.

Roderick HL, Lechleiter JD, Camacho P. 2000. Cytosolic phosphorylation of calnexin controls intracellular $Ca^{(2+)}$ oscillations via an interaction with SERCA2b. *J Cell Biol* **149:** 1235–1248.

Rogers NT, Halet G, Piao Y, Carroll J, Ko MS, Swann K. 2006. The absence of a $Ca^{(2+)}$ signal during mouse egg activation can affect parthenogenetic preimplantation development, gene expression patterns, and blastocyst quality. *Reproduction* **132:** 45–57.

Rogers NT, Hobson E, Pickering S, Lai FA, Braude P, Swann K. 2004. Phospholipase Czeta causes Ca^{2+} oscillations and parthenogenetic activation of human oocytes. *Reproduction* **128:** 697–702.

Rong Y, Aromolaran A, Bultynck G, Zhong F, Li X, McColl K, Matsuyama S, Herlitze S, Roderick H, Bootman MW, et al. 2008. Targeting Bcl-2-IP_3 receptor interaction to reverse Bcl-2's inhibition of apoptotic calcium signals. *Mol Cell* **31:** 255–265.

Rong YP, Bultynck G, Aromolaran AS, Zhong F, Parys JB, De Smedt H, Mignery GA, Roderick HL, Bootman MD, Distelhorst CW. 2009. The BH4 domain of Bcl-2 inhibits ER calcium release and apoptosis by binding the regulatory and coupling domain of the IP_3 receptor. *Proc Natl Acad Sci* **106:** 14397–14402.

Roos J, DiGregorio P, Yeromin A, Ohlsen K, Lioudyno M, Zhang S, Safrina O, Kozak J, Wagner S, Cahalan MA, et al. 2005. STIM1, an essential and conserved component of store-operated Ca^{2+} channel function. *J Cell Biol* **169:** 435–445.

Ross PJ, Beyhan Z, Iager AE, Yoon SY, Malcuit C, Schellander K, Fissore RA, Cibelli JB. 2008. Parthenogenetic activation of bovine oocytes using bovine and murine phospholipase C zeta. *BMC Dev Biol* **8:** 16.

Runft LL, Watras J, Jaffe LA. 1999. Calcium release at fertilization of *Xenopus* eggs requires type I IP(3) receptors, but not SH2 domain-mediated activation of PLCgamma or G(q)-mediated activation of PLCbeta. *Dev Biol* **214:** 399–411.

Sardet C, Prodon F, Dumollard R, Chang P, Chenevert J. 2002. Structure and function of the egg cortex from oogenesis through fertilization. *Dev Biol* **241:** 1–23.

Sato K, Tokmakov AA, Iwasaki T, Fukami Y. 2000. Tyrosine kinase-dependent activation of phospholipase Cgamma is required for calcium transient in *Xenopus* egg fertilization. *Dev Biol* **224:** 453–469.

Saunders CM, Larman MG, Parrington J, Cox LJ, Royse J, Blayney LM, Swann K, Lai FA. 2002. PLC zeta: a sperm-specific trigger of $Ca^{(2+)}$ oscillations in eggs and embryo development. *Development* **129:** 3533–3544.

Schmidt A, Duncan PI, Rauh NR, Sauer G, Fry AM, Nigg EA, Mayer TU. 2005. *Xenopus* polo-like kinase Plx1 regulates XErp1, a novel inhibitor of APC/C activity. *Genes Dev* **19:** 502–513.

Schultz RM, Kopf GS. 1995. Molecular basis of mammalian egg activation. *Curr Top Dev Biol* **30:** 21–62.

Shimomura O, Johnson FH. 1970. Calcium binding, quantum yield, and emitting molecule in aequorin bioluminescence. *Nature* **227:** 1356–1357.

Shoji S, Yoshida N, Amanai M, Ohgishi M, Fukui T, Fujimoto S, Nakano Y, Kajikawa E, Perry AC. 2006. Mammalian Emi2 mediates cytostatic arrest and transduces the signal for meiotic exit via Cdc20. *EMBO J* **25:** 834–845.

Simerly C, Nowak G, de Lanerolle P, Schatten G. 1998. Differential expression and functions of cortical myosin IIA and IIB isotypes during meiotic maturation, fertilization, and mitosis in mouse oocytes and embryos. *Mol Biol Cell* **9:** 2509–2525.

Smyth JT, Dehaven WI, Jones BF, Mercer JC, Trebak M, Vazquez G, Putney JW Jr. 2006. Emerging perspectives in store-operated Ca^{2+} entry: roles of Orai, Stim and TRP. *Biochim Biophys Acta* **1763:** 1147–1160.

Smyth JT, Petranka JG, Boyles RR, DeHaven WI, Fukushima M, Johnson KL, Williams JG, Putney JW Jr. 2009. Phosphorylation of STIM1 underlies suppression of store-operated calcium entry during mitosis. *Nat Cell Biol* **11:** 1465–1472.

Steinhardt RA, Epel D. 1974. Activation of sea-urchin eggs by a calcium ionophore. *Proc Natl Acad Sci* **71:** 1915–1919.

Steinhardt R, Zucker R, Schatten G. 1977. Intracellular calcium release at fertilization in the sea urchin egg. *Dev Biol* **58:** 185–196.

Steinhardt RA, Epel D, Carroll EJ Jr, Yanagimachi R. 1974. Is calcium ionophore a universal activator for unfertilised eggs? *Nature* **252:** 41–43.

Stice SL, Robl JM. 1990. Activation of mammalian oocytes by a factor obtained from rabbit sperm. *Mol Reprod Dev* **25:** 272–280.

Stricker SA. 1997. Intracellular injections of a soluble sperm factor trigger calcium oscillations and meiotic maturation in unfertilized oocytes of a marine worm. *Dev Biol* **186:** 185–201.

Stricker SA. 1999. Comparative biology of calcium signaling during fertilization and egg activation in animals. *Dev Biol* **211:** 157–176.

Su YQ, Sugiura K, Woo Y, Wigglesworth K, Kamdar S, Affourtit J, Eppig JJ. 2007. Selective degradation of transcripts during meiotic maturation of mouse oocytes. *Dev Biol* **302:** 104–117.

Suarez SS. 2008a. Control of hyperactivation in sperm. *Hum Reprod Update* **14:** 647–657.

Suarez SS. 2008b. Regulation of sperm storage and movement in the mammalian oviduct. *Int J Dev Biol* **52:** 455–462.

Sun L, Haun S, Jones RC, Edmondson RD, Machaca K. 2009. Kinase-dependent regulation of IP3-dependent Ca^{2+} release during oocyte maturation. *J Biol Chem* **284:** 20184–20196.

Sun FZ, Hoyland J, Huang X, Mason W, Moor RM. 1992. A comparison of intracellular changes in porcine eggs after fertilization and electroactivation. *Development* **115:** 947–956.

Sutovsky P, Manandhar G, Wu A, Oko R. 2003. Interactions of sperm perinuclear theca with the oocyte: implications for oocyte activation, anti-polyspermy defense, and assisted reproduction. *Microsc Res Tech* **61:** 362–378.

Swann K. 1990. A cytosolic sperm factor stimulates repetitive calcium increases and mimics fertilization in hamster eggs. *Development* **110:** 1295–1302.

Swann K, Lai FA. 1997. A novel signalling mechanism for generating Ca^{2+} oscillations at fertilization in mammals. *Bioessays* **19:** 371–378.

Swann K, Igusa Y, Miyazaki S. 1989. Evidence for an inhibitory effect of protein kinase C on G-protein-mediated repetitive calcium transients in hamster eggs. *EMBO J* **8:** 3711–3718.

Swann K, Saunders CM, Rogers NT, Lai FA. 2006. PLCzeta (zeta): a sperm protein that triggers Ca2+ oscillations and egg activation in mammals. *Semin Cell Dev Biol* **17:** 264–273.

Swenson KI, Farrell KM, Ruderman JV. 1986. The clam embryo protein cyclin A induces entry into M phase and the resumption of meiosis in *Xenopus* oocytes. *Cell* **47:** 861–870.

Szado T, Vanderheyden V, Parys J, De Smedt H, Rietdorf K, Kotelevets L, Chastre E, Khan F, Landegren U, Soderberg OL, et al. 2008. Phosphorylation of inositol 1,4,5-trisphosphate receptors by protein kinase B/Akt inhibits Ca^{2+} release and apoptosis. *Proc Natl Acad Sci* **105:** 2427–2432.

Taylor SL, Yoon SY, Morshedi MS, Lacey DR, Jellerette T, Fissore RA, Oehninger S. 2010. Complete globozoospermia associated with PLCzeta deficiency treated with calcium ionophore and ICSI results in pregnancy. *Reprod Biomed Online* **20:** 559–564.

Taylor CW, Tovey SC. 2010. IP_3 receptors: toward understanding their activation. *Cold Spring Harb Perspect* doi:10.1101/cshperspect.a004010.

Tesarik J, Testart J. 1994. Treatment of sperm-injected human oocytes with Ca^{2+} ionophore supports the development of Ca^{2+} oscillations. *Biol Reprod* **51:** 385–391.

Thomas TW, Eckberg WR, Dube F, Galione A. 1998. Mechanisms of calcium release and sequestration in eggs of Chaetopterus pergamentaceus. *Cell Calcium* **24:** 285–292.

Tosti E, Boni R. 2004. Electrical events during gamete maturation and fertilization in animals and humans. *Hum Reprod Update* **10:** 53–65.

Tung JJ, Hansen DV, Ban KH, Loktev AV, Summers MK, Adler JR 3rd, Jackson PK. 2005. A role for the anaphase-promoting complex inhibitor Emi2/XErp1, a homolog of early mitotic inhibitor 1, in cytostatic factor arrest of *Xenopus* eggs. *Proc Natl Acad Sci* **102:** 4318–4323.

Tunquist BJ, Maller JL. 2003. Under arrest: cytostatic factor (CSF)-mediated metaphase arrest in vertebrate eggs. *Genes Dev* **17:** 683–710.

Turner PR, Jaffe LA, Fein A. 1986. Regulation of cortical vesicle exocytosis in sea urchin eggs by inositol 1,4,5-trisphosphate and GTP-binding protein. *J Cell Biol* **102:** 70–76.

Turner PR, Sheetz MP, Jaffe LA. 1984. Fertilization increases the polyphosphoinositide content of sea urchin eggs. *Nature* **310:** 414–415.

Vanderheyden V, Wakai T, Bultynck G, De Smedt H, Parys JB, Fissore RA. 2009. Regulation of inositol 1,4,5-trisphosphate receptor type 1 function during oocyte maturation by MPM-2 phosphorylation. *Cell Calcium* **46:** 56–64.

Venkatachalam K, Montell C. 2007. TRP channels. *Annu Rev Biochem* **76:** 387–417.

Vig M, Peinelt C, Beck A, Koomoa D, Rabah D, Koblan-Huberson M, Kraft S, Turner H, Fleig A, Penner RP, et al. 2006. CRACM1 is a plasma membrane protein essential for store-operated Ca^{2+} entry. *Science* **312:** 1220–1223.

Viveiros MM, Hirao Y, Eppig JJ. 2001. Evidence that protein kinase C (PKC) participates in the meiosis I to meiosis II transition in mouse oocytes. *Dev Biol* **235:** 330–342.

Westendorf JM, Rao PN, Gerace L. 1994. Cloning of cDNAs for M-phase phosphoproteins recognized by the MPM2 monoclonal antibody and determination of the phosphorylated epitope. *Proc Natl Acad Sci* **91:** 714–718.

Whitaker M. 2006. Calcium at fertilization and in early development. *Physiol Rev* **86:** 25–88.

Winston NJ, McGuinness O, Johnson MH, Maro B. 1995. The exit of mouse oocytes from meiotic M-phase requires an intact spindle during intracellular calcium release. *J Cell Sci* **108 (Pt 1):** 143–151.

Wu H, He CL, Fissore RA. 1997. Injection of a porcine sperm factor triggers calcium oscillations in mouse oocytes and bovine eggs. *Mol Reprod Dev* **46:** 176–189.

Xu Z, Kopf GS, Schultz RM. 1994. Involvement of inositol 1,4,5-trisphosphate-mediated Ca^{2+} release in early and late events of mouse egg activation. *Development* **120:** 1851–1859.

Yoda A, Oda S, Shikano T, Kouchi Z, Awaji T, Shirakawa H, Kinoshita K, Miyazaki S. 2004. Ca^{2+} oscillation-inducing phospholipase C zeta expressed in mouse eggs is accumulated to the pronucleus during egg activation. *Dev Biol* **268:** 245–257.

Yoneda A, Kashima M, Yoshida S, Terada K, Nakagawa S, Sakamoto A, Hayakawa K, Suzuki K, Ueda J, Watanabe T. 2006. Molecular cloning, testicular postnatal expression, and oocyte-activating potential of porcine phospholipase Czeta. *Reproduction* **132:** 393–401.

Yoon SY, Fissore RA. 2007. Release of phospholipase C zeta and $[Ca^{2+}]i$ oscillation-inducing activity during mammalian fertilization. *Reproduction* **134:** 695–704.

Yoon S, Jellerette T, Salicioni A, Lee H, Yoo M, Coward K, Parrington J, Grow D, Cibelli J, Visconti PA, et al. 2008. Human sperm devoid of PLC, zeta 1 fail to induce $Ca(^{2+})$ release and are unable to initiate the first step of embryo development. *J Clin Invest* **118:** 3671–3681.

Yoshida M, Sensui N, Inoue T, Morisawa M, Mikoshiba K. 1998. Role of two series of Ca^{2+} oscillations in activation of ascidian eggs. *Dev Biol* **203:** 122–133.

Yu F, Sun L, Machaca K. 2009. Orai1 internalization and STIM1 clustering inhibition modulate SOCE inactivation during meiosis. *Proc Natl Acad Sci* **106:** 17401–17406.

Zernicka-Goetz M. 1991. Spontaneous and induced activation of rat oocytes. *Mol Reprod Dev* **28:** 169–176.

Zweifach A, Lewis RS. 1993. Mitogen-regulated Ca^{2+} current of T lymphocytes is activated by depletion of intracellular Ca^{2+} stores. *Proc Natl Acad Sci* **90:** 6295–6299.

Calcium Signaling in Neuronal Development

Sheila S. Rosenberg and Nicholas C. Spitzer

Neurobiology Section, Division of Biological Sciences, Kavli Institute for Brain and Mind, University of California at San Diego, La Jolla, California 92093

Correspondence: nspitzer@ucsd.edu

The development of the nervous system involves the generation of a stunningly diverse array of neuronal subtypes that enable complex information processing and behavioral outputs. Deciphering how the nervous system acquires and interprets information and orchestrates behaviors will be greatly enhanced by the identification of distinct neuronal circuits and by an understanding of how these circuits are formed, changed, and/or maintained over time. Addressing these challenging questions depends in part on the ability to accurately identify and characterize the unique neuronal subtypes that comprise individual circuits. Distinguishing characteristics of neuronal subgroups include but are not limited to neurotransmitter phenotype, dendritic morphology, the identity of synaptic partners, and the expression of constellations of subgroup-specific proteins, including ion channel subtypes.

How is the diversity of neuronal subtypes generated during development? Intrinsic genetic programs, extracellular signals, and both spontaneous and experience-dependent activity all contribute to the development of a functional nervous system. Here we focus on the role of activity, and specifically calcium-mediated signaling, in the generation of neuronal diversity. Brief elevations of intracellular calcium levels, also referred to as calcium transients, have been implicated in the regulation of various stages of neuronal development, including proliferation, migration, differentiation, and survival. Calcium-mediated regulation of neuronal proliferation and migration has been recently reviewed (Spitzer 2006; Platel et al. 2008). Here we address the role of calcium transients in neuronal differentiation. Calcium-mediated signaling contributes to the specification of neuronal subtype through the regulation of axonal pathfinding, dendritic growth and arborization, and specification of neurotransmitter subtype (Fig. 1).

How is it that a single cation can regulate so many different aspects of differentiation? In addition, how does calcium-mediated regulation of each distinguishing neuronal feature lead to specific phenotypic characteristics in different cell types? Although the answers to these questions are not yet fully resolved, innovative work from many groups has helped establish that the spatial and temporal characteristics of activity are important for conferring specificity in calcium-mediated signaling. Therefore we begin this review by describing characteristics of calcium transients in the embryonic nervous system. Next, we review evidence demonstrating that different patterns of calcium transients

Copyright © 2012 Cold Spring Harbor Laboratory Press; all rights reserved
Cite this article as *Cold Spring Harb Perspect Biol* doi: 10.1101/cshperspect.a004259

S.S. Rosenberg and N.C. Spitzer

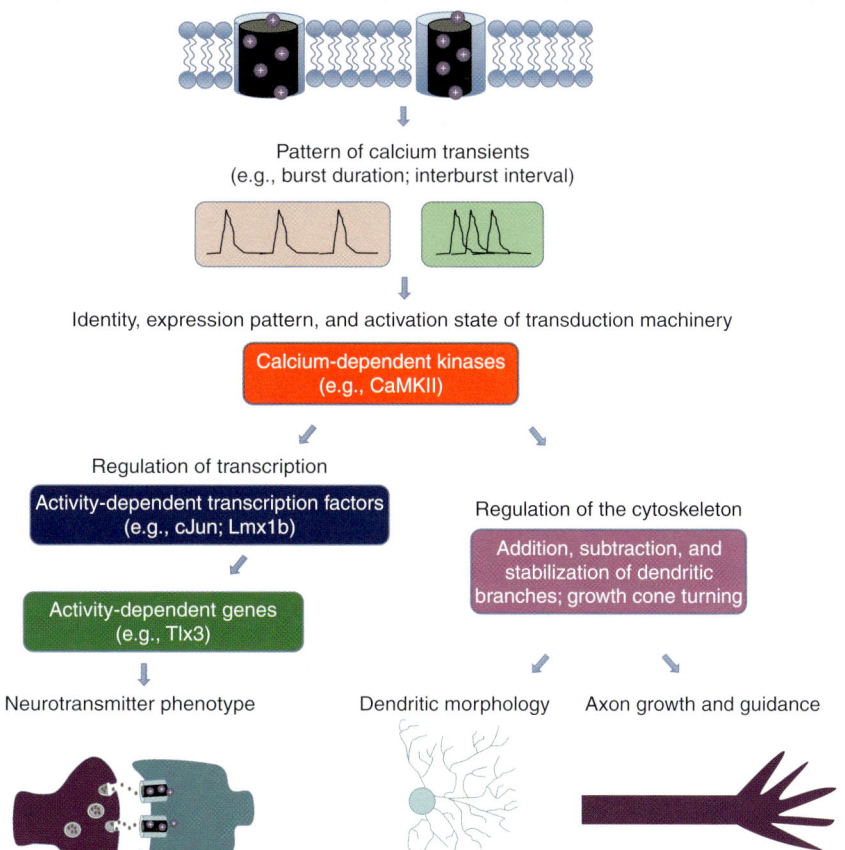

Figure 1. Providing specificity for calcium signaling in neuronal differentiation. Calcium transients direct neuronal differentiation by regulating neurotransmitter phenotype, dendritic morphology, and axonal growth and guidance. Factors dictating intracellular calcium dynamics include the subcellular location of ion channels within the neuron and the neuron-specific constellation of ion channels and receptors expressed by individual cells. The location and identity of these channels and receptors influence the timing and frequency of calcium transients and determine whether the changes in calcium concentration occur in a global or localized fashion. Spatiotemporal patterns of calcium transients select the downstream mechanisms involved in neuronal differentiation. Calcium transients activate enzymes that transduce ionic signals into biochemical ones. These enzymes impact differentiation either through transcriptional mechanisms or by the regulation of cytoskeletal dynamics. Activation or repression of transcription factors controls neurotransmitter expression whereas cytoskeletal remodeling regulates axon and dendrite morphogenesis.

regulate distinct aspects of neuronal differentiation. Additionally, we discuss some downstream mechanisms involved in calcium-mediated effects on neuronal differentiation, including the activation of specific transcription factors and cytoskeletal rearrangements. We also review some of the identified calcium-binding partners and specific downstream targets that help to convey specificity in calcium signaling. The studies discussed here show that the temporal dynamics of calcium transients and the mode of calcium entry into the neuron influence the specific effects of calcium signaling in neuronal development. Based on these findings, it is likely that the mechanisms regulating calcium channel expression and localization play a fundamental role in regulating calcium-driven differentiation and the determination of neuronal subtype. Therefore, we conclude with findings that provide insight into calcium-dependent mechanisms regulating the subcellular targeting of calcium-permeable channels.

SPATIOTEMPORAL CHARACTERISTICS OF CALCIUM TRANSIENTS DURING EARLY NEURONAL MATURATION

Both spontaneous and experience-driven patterns of activity influence nervous system development. In experience-mediated activity, the number and frequency of calcium transients are responsible for conveying information about the stimulus intensity. Different patterns of calcium spike activity, and the resulting variations in ionic flux, lead to specific outcomes for neuronal development and plasticity. Similarly, the timing of expression of spontaneous activity is important for multiple aspects of neuronal development. Spontaneous activity, occurring in the absence of sensory input, takes place both before and after the formation of synaptic connections. The mechanisms involved in spontaneous synaptically mediated activity have been recently reviewed (Blankenship and Feller 2010).

We focus here on spontaneous activity present in the nervous system prior to the formation of synaptic connections. Remarkably, these early forms of activity have significant roles in directing neuronal differentiation. In an effort to understand the functional importance of early forms of spontaneous activity, calcium transients have been characterized in embryonic *Xenopus* spinal neurons beginning at the time of neural tube closure. These neurons show two distinct types of calcium transients: calcium spikes and calcium waves (Gu et al. 1994; Spitzer et al. 1994; Gu and Spitzer 1995, 1997). Calcium spikes involve a rapid increase in calcium concentration generated by calcium-dependent action potentials and calcium-induced calcium release, quickly followed by a stereotyped double exponential decay with time constants of $\sim$10 sec and $\sim$3 min. As measured by calcium imaging with Fura-2, calcium spikes raise intracellular calcium from 10^{-7} to 10^{-6} M, in $<$5 sec (Gu et al. 1994).

In contrast to the stereotyped pattern of calcium spikes, calcium waves show variability in duration, with rise and decay times lasting on the order of half a minute to several minutes in duration. Calcium waves also show a more moderate increase in fluorescence intensity when assessed with calcium imaging. In addition to these differences in amplitude and duration, spikes and waves also differ in their spatial localization within neurons. Consistent with their initiation by calcium-dependent action potentials, calcium spikes propagate rapidly throughout the cell. In contrast, calcium waves occur in both the soma and the growth cone, but frequently remain sequestered in the compartment in which they are generated. These two types of calcium transients also exert different effects on neuronal differentiation. Calcium spikes regulate neurotransmitter specification, whereas calcium waves regulate the rate of axon extension, indicating that the spatiotemporal features of spontaneous activity may provide functional specificity.

Spontaneous activity prior to the formation of synaptic networks has been observed in multiple areas of the nervous system and in multiple species. In the developing mammalian brain, spontaneous transients are present in slices of the visual, somatosensory, and frontal cortex at postnatal day 0 (Yuste et al. 1992; O'Donovan 1999). In the developing chick retina, calcium transients appear as early as embryonic day eight (E8) (Catsicas et al. 1998). These signals are also present in cultured chick dorsal root ganglion (DRG) neurons isolated at E8–E11 (Gomez et al. 1995). These neurons generate calcium transients in their growth cones that regulate the rate of growth cone migration, consistent with the idea that the subcellular location of activity patterns influences the effect of calcium signaling on neuronal differentiation. In the following sections, we discuss studies that provide additional insight into the spatiotemporal characteristics of calcium transients and the types of downstream mechanisms that confer specificity in the regulation of neuronal differentiation.

CALCIUM-MEDIATED SPECIFICATION OF NEUROTRANSMITTER PHENOTYPE

Early studies of cultured embryonic *Xenopus* spinal neurons showed a role for calcium spikes in the regulation of GABA immunoreactivity. Elimination of calcium spikes by the removal of extracellular calcium, blockade of calcium channels, or by the addition of the calcium chelator BAPTA, significantly reduced the number of GABAergic neurons (Holliday et al. 1991; Gu and Spitzer 1995). In the absence of extracellular calcium, artificial calcium spiking generated by pulses of KCl and $CaCl_2$ was sufficient to rescue the GABAergic phenotype (Gu and Spitzer 1995). These findings show that spontaneous calcium spikes are necessary and sufficient for the specification of neurotransmitter expression in developing *Xenopus* spinal neurons. What is the mechanism by which calcium signaling regulates GABA expression? In spinal neurons cultured at early stages of development, the release of calcium from intracellular stores is necessary for the generation of calcium spikes. Experimental depletion of intracellular calcium stores leads to a reduction in calcium spiking, and a decrease in the number of GABAergic cells (Holliday et al. 1991). These studies indicate that in addition to the influx of extracellular calcium, release of intracellular calcium stores is also necessary for neuronal differentiation. Together, these findings lead to a model in which the influx of extracellular calcium induces the release of calcium from intracellular stores, resulting in an increase in the number of GABAergic cells. Calcium-mediated regulation of GABA expression appears to occur via a frequency-dependent mechanism, as shown by removing extracellular calcium and stimulating cultured neurons with different frequencies of calcium spikes (Gu and Spitzer 1995; Watt et al. 2000). Spike stimulation mimicking the frequency of spontaneous transients was shown to be most effective in replicating the effects of spontaneous transients on neurotransmitter expression.

The regulation of GABAergic phenotype by calcium spike frequency prompts a number of interesting questions. How does the frequency of calcium spikes modulate neurotransmitter specification? Is spike frequency responsible for regulating gene expression? Are diverse frequencies of calcium transients involved in generating the diversity of neuronal subtypes present in the nervous system? Do specific frequencies elicit the expression of specific neurotransmitters in different neuronal subtypes? Several of these questions were addressed by imaging the calcium transients of specific classes of neurons in the developing *Xenopus* spinal cord (Borodinsky et al. 2004). Four different neuronal subtypes—Rohon–Beard sensory neurons, ventral interneurons, ventral motor neurons, and dorsolateral interneurons—show unique patterns of spontaneous calcium spiking and disparate neurotransmitter phenotypes. For example, in the developmental period prior to synapse formation, Rohon–Beard neurons show a relatively low spike frequency and express glutamate, whereas motor neurons show a gradual increase in spike frequency and express acetylcholine. These results indicate a potential relationship between spiking activity and the expression of various neurotransmitters, which is consistent with activity-dependent regulation of GABA expression. The role of spontaneous activity in regulating neurotransmitter phenotype was investigated using ion channel misexpression to manipulate activity and the resulting calcium transients (Borodinsky et al. 2004). Injection of voltage-gated sodium channel transcripts into *Xenopus* embryos at the two-cell stage results in a global increase in spiking activity and is accompanied by a decrease in the number of neurons expressing the excitatory transmitters glutamate and acetylcholine, as well as a corresponding increase in the number of neurons expressing the inhibitory transmitters glycine and GABA. In contrast, injection of inwardly rectifying potassium channel transcripts results in an overall decrease in the extent of calcium spiking, accompanied by a decrease in the numbers of neurons expressing inhibitory transmitters and an increase in the expression of excitatory transmitters by neurons other than sensory and motor neurons. These results indicate the existence of a homeostatic relationship between calcium signaling and neurotransmitter expression (Fig. 2A).

Changes in neurotransmitter specification that serve to readjust the relative degree of excitation and inhibition represent an intriguing strategy by which the nervous system can preserve a homeostatic balance in overall neuronal activity. Additionally, the activity-dependent acquisition of a novel neurotransmitter profile reveals a previously unexpected degree of plasticity in neuronal phenotype. This plasticity is also reflected in the matching of appropriate postsynaptic receptors to neurons expressing novel neurotransmitters (Borodinsky and Spitzer 2007). Studies of the developing *Xenopus* neuromuscular junction (NMJ) reveal that multiple neurotransmitter receptors are initially expressed in skeletal muscle. As development proceeds, muscles are innervated by neurons expressing acetylcholine, and receptors not responding to acetylcholine are normally down-regulated. Calcium-spike mediated expression of novel neurotransmitters results in the retention of alternate receptor subtypes at the NMJ. These findings support the likelihood of a functionally relevant role for neurons that undergo activity-induced acquisition of novel neurotransmitters.

Activity-dependent regulation of neurotransmitter phenotype raises new questions about the mechanisms involved in calcium-mediated neuronal specification. What are the triggers for spontaneous activity? Paracrine action of GABA and glutamate, secreted prior to synapse formation, drives spontaneous activity in the *Xenopus* neural tube (Root et al. 2008), and expression of the β-subunit of the sodium pump appears to be required for the generation of spontaneous calcium spikes (Chang and Spitzer 2009). What are the downstream mechanisms responsible for activity-dependent neurotransmitter respecification? Recent studies show that the activity-dependent regulation of GABAergic and glutamatergic expression in *Xenopus tropicalis* is mediated through the expression of the *tlx3* homeobox gene (Marek et al. 2010). In mice, *tlx3* has been shown to act as a transcriptional selector gene, responsible for both the inhibition of GABAergic development and the up-regulation of the glutamatergic phenotype in the dorsal horn of the embryonic spinal cord (Cheng et al. 2004, 2005). In *Xenopus*, morpholino knockdown of *tlx3* expression was sufficient to block the decrease in GABAergic neurons and corresponding increase in glutamatergic neurons that results from inhibition of activity. Additionally, overexpression of *tlx3* blocked the increase in GABA-expressing cells and decrease in glutamate-expressing cells that results from a global increase in activity. These results indicate that *tlx3* expression mediates the activity-dependent regulation of glutamatergic and GABAergic cell fate. The use of luciferase transcription assays showed that the effect of calcium activity on *tlx3* expression is mediated through the activity-dependent phosphorylation of the cJun transcription factor. An increase in activity leads to an increase in cJun phosphorylation and activation of cJun, which suppresses *tlx3* expression by binding to a noncanonical cAMP responsive element (CRE) within the *tlx3* promoter. This results in an increase in the number of GABAergic cells and a decrease in the number of glutamatergic neurons. In contrast, a decrease in activity causes a decrease in the number of neurons showing phosphorylated cJun, and a decrease in the number of GABAergic cells and an increase in the number of glutamatergic cells (Fig. 2B). Together, these findings show that spontaneous activity and

A Activity-dependent changes in calcium signaling regulate neurotransmitter specification

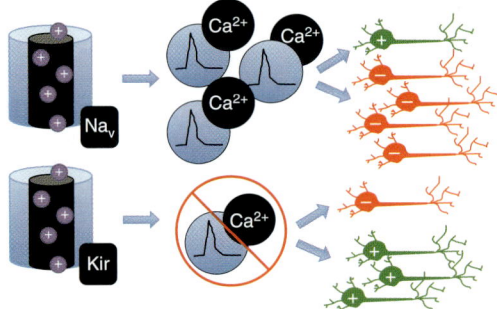

B Calcium signaling modulates transcriptional regulation of neurotransmitter specification

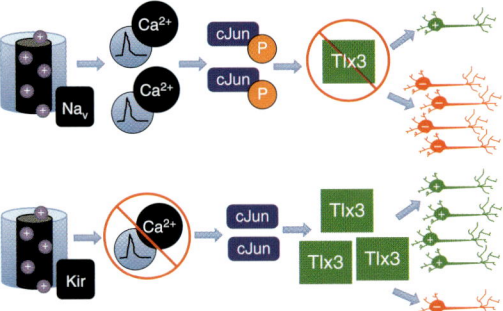

C Experience-dependent calcium signaling regulates neurotransmitter respecification

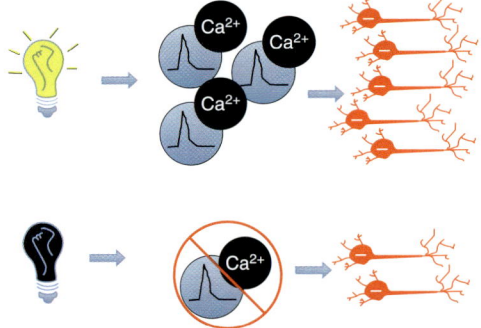

Figure 2. Activity-dependent neurotransmitter specification at early stages of neuronal development. (*A*) Prior to synapse formation in the embryonic spinal cord, overexpression of voltage-gated sodium channels (Na$_v$) increases the incidence and frequency of calcium spikes that lead to a decrease in the number of neurons expressing excitatory transmitters (in green) and an increase in the number of neurons expressing inhibitory transmitters (in red). Overexpression of inward rectifier potassium channels (Kir) decreases calcium spiking and produces the opposite effect on transmitter specification. (*B*) Increasing the incidence and frequency of calcium spikes in the embryonic spinal cord leads to phosphorylation of the cJun transcription factor that binds to the promoter of the *tlx3* selector gene and suppresses its expression; this results in a decrease in the number of glutamatergic neurons (in green) and an increase in the number of GABAergic neurons (in red). Suppressing spike production leads to dephosphorylation of cJun that no longer represses *tlx3* expression; this results in an increase in the number of glutamatergic neurons and a decrease in the number of GABAergic neurons. (*C*) After synapse formation in the postembryonic brain, bright light illumination increases calcium spike activity that leads to an increase in the number of hypothalamic neurons that express dopamine (in red), which acts as an inhibitory transmitter. Dark exposure decreases calcium spike activity that leads to a decrease in the number of dopaminergic neurons.

the resulting calcium transients mediate the activation of a transcription factor that regulates a downstream selector gene involved in the determination of neurotransmitter expression. These results illustrate the importance of activity-dependent transcriptional regulation in controlling neuronal differentiation.

Activity-dependent transcriptional regulation of neurotransmitter expression is not restricted to spinal neurons. Additional studies show that activity-dependent regulation of the transcription factor Lmx1b controls the extent of serotonergic specification in neurons present in the raphe nucleus of the *Xenopus* hindbrain (Demarque and Spitzer 2010). Increased calcium spiking results in a decrease in the number of neurons expressing Lmx1b and a corresponding decrease in the number of serotonergic neurons. In contrast, a decrease in calcium spiking activity leads to an increase in Lmx1b expression and an increase in the number of serotonergic neurons in the hindbrain. Importantly, this study shows that the activity-dependent induction of additional serotonergic neurons leads to a change in behavioral output—specifically in the swimming behavior of the *Xenopus* larvae. Serotonin has a key role in limiting the duration of swimming episodes (Sillar et al. 1998), suggesting that an increase in the number of serotonergic cells will lead to a decrease in the duration in swimming episodes. Consistent with this hypothesis, the presence of additional serotonergic neurons generated either by experimental reduction of calcium spiking or overexpression of Lmx1b is sufficient to reduce both the duration of fictive swimming firing patterns recorded from ventral nerve roots and the duration of swimming episodes performed by free swimming larvae (Demarque and Spitzer 2010). In contrast, either an increase in activity or a reduction in Lmx1b expression is sufficient to extend the duration of both fictive and free swimming episodes. These findings show that experimentally induced alterations in calcium spiking can modulate behavior by redirecting the differentiated neurotransmitter fate of neuronal subtypes.

The experimental manipulation of calcium spiking in developing *Xenopus* embryos raises the important question of whether the effects of activity on neurotransmitter phenotype are restricted to early embryonic spontaneous activity or whether these changes can also be mediated by experience-driven changes in activity occurring after synapse formation. Studies of the circuitry underlying the camouflage behavior of *Xenopus* larvae tested whether experience-mediated changes in calcium spiking can influence neurotransmitter expression. Exposure of larvae either to light or to a white background activates dopaminergic neurons in the ventral suprachiasmatic nucleus (VSC), which regulate the downstream pathways responsible for changes in skin pigmentation. Exposing larvae to light for a period of several hours leads to an increase in the number of dopaminergic neurons, which is blocked by pharmacological inhibition of calcium spiking in the VSC (Fig. 2C) (Dulcis and Spitzer 2008). This increase in the dopaminergic population occurs through the acquisition of dopamine expression by existing neurons and does not involve the generation of new neurons. These findings indicate that experience-driven calcium spiking is sufficient to alter the number of dopaminergic neurons, demonstrating that experience-mediated activity can also influence neurotransmitter phenotype. Importantly, these experience-generated dopaminergic neurons are functionally sufficient to mediate camouflage behavior after the selective elimination of the original dopaminergic population.

In addition to experience-mediated activity, the binding of extracellular ligands can also modulate intracellular calcium levels. Furthermore, extracellular retrograde signals from axonal targets have been shown to influence neurotransmitter specification (Schotzinger and Landis 1988, 1990). Is it possible that these signals mediate neurotransmitter specification via regulation of calcium signaling? In developing rodents, sympathetic neurons innervating the sweat glands and the periosteum are initially adrenergic. Following innervation, these neurons down-regulate noradrenaline and instead adopt a cholinergic phenotype. Soluble factors derived from the innervated target tissues contribute to this switch in neuronal identity

(Habecker et al. 1997; Asmus et al. 2001). Although the exact identity of these soluble factors remains elusive, these signaling molecules appear to belong to the IL-6 family of neuropoietic cytokines (Rao and Landis 1990). The selective in vivo deletion of the neuropoietic cytokine receptor subunit gp130 in noradrenergic neurons results in a decrease in the number of cholinergic neurons innervating the sweat glands, although the total number of innervating neurons is unchanged (Stanke et al. 2006). Additional studies of hippocampal neurons show that elevated levels of IL-6 both in vivo and in vitro modulate the expression of L-type calcium channels and alter the dynamics of calcium signaling and network activity (Nelson et al. 2004; Vereyken et al. 2007). These results indicate that signaling through the IL-6 receptor gp130 may regulate neurotransmitter phenotype by modulating calcium signaling.

THE ROLE OF ACTIVITY AND CALCIUM SIGNALING IN DENDRITIC ARBORIZATION

Establishment and plasticity of dendritic morphology represent important aspects of neuronal differentiation. The relative growth and branching of dendritic trees influences the number of synaptic connections that a given neuron can form. Subgroups of neurons are characterized by distinct dendritic morphologies, which include variations in both the length and number of dendritic branches (Parrish et al. 2007; Spruston 2008). A role for activity-dependent regulation of dendritic arborization has been described for numerous neuronal subtypes (Wong and Ghosh 2002), and activity-mediated calcium signaling has been implicated in regulating the plasticity of dendritic branching and growth. Increases in intracellular calcium levels, resulting from the activation of NMDA receptors, promote an increase in dendritic branching in both rodent and amphibian neurons (Rajan and Cline 1998; Chevaleyre et al. 2002).

In cultures of developing *Xenopus* tectal neurons, application of APV selectively disrupts NMDA-mediated transmission and decreases the rate of dendritic growth, limiting both the extension of existing branches and the addition of new branches (Rajan and Cline 1998). In more mature cultured tectal neurons, the activation of both NMDA and AMPA receptors is necessary for dendritic growth and the maintenance of existing branches. Furthermore, the stability of dendritic arbors in *Xenopus* tectal neurons in vivo is regulated by the activity of the calcium-activated kinase, calcium-calmodulin-dependent kinase II, CaMKII (Wu and Cline 1998). As tectal neurons mature and migrate away from their site of generation in the ventricular layer of the tectum, the rate of dendritic growth declines, coincident with an increase in the expression of CaMKII. Overexpression of constitutively active CaMKII in developing neurons leads to a decrease in the rate of dendritic growth. Additionally, more mature tectal neurons show enhanced dendritic growth in the presence of the CaMKII inhibitor, KN-93. CaMKII regulates the rate of dendritic growth by inhibiting the addition and turnover of new dendritic branches. These results prompt further inquiry into the mechanisms regulating CaMKII expression and activity, as well as the CaMKII substrates and downstream pathways controlling dendritic growth and plasticity.

CaMKII is expressed in a graded fashion in the *Xenopus* tectum, with high expression restricted to the rostral-lateral region and little to no expression in the caudal-medial region where tectal neurons are initially formed. The CaMKII expression gradient may be influenced by inhibitory factors expressed in the ventricular layers of the tectum that facilitate increased dendritic growth by restricting CaMKII expression in newly generated neurons. However, both monocular deprivation and the induction of neuronal hyperexcitability in rodents result in activity-dependent regulation of CaMKII expression (Hanson and Schulman 1992; Hudmon and Schulman 2002). The maturation and migration of *Xenopus* tectal neurons is accompanied by an increase in the density of AMPA receptors at developing synapses and a corresponding decline in the number of silent synapses containing predominantly NDMA receptors (Wu and Cline 1998). This developmental shift in neuronal signaling capacity enables alterations in the extent of calcium influx, which could directly influence the activity-dependent regulation of CaMKII expression. It seems plausible that the establishment of the tectal CaMKII expression gradient results from a combinatorial effect of activity-dependent and extrinsic molecular mechanisms, which can both exert a degree of transcriptional control.

Dendritic branching is also regulated by both depolarization and the application of the peptidergic signaling molecules oxytocin and vasopressin in rat supraoptic nucleus (SON) slice cultures (Chevaleyre et al. 2002). Increased dendritic branching in SON neurons can be induced by the application of either NMDA or oxytocin and vasopressin, as well as by depolarization with high levels of extracellular potassium. Additionally, the injection of newborn rat pups with antagonists to either NMDA receptors or oxytocin/vasopressin receptors leads to a decrease in dendritic branching in the SON. In slice cultures, the increase in dendritic branching mediated by NMDA, oxytocin/vasopressin, and depolarization, requires both the activity of voltage-gated calcium channels and the release of calcium from intracellular stores. However, depolarization in the presence of either NMDA or oxytocin/vasopressin antagonists is not sufficient to increase dendritic branching. This result indicates that the regulation of dendritic branching requires multiple intracellular events downstream from NMDA and peptide-mediated signaling. These findings underscore the likelihood of a combinatorial role for both activity-dependent and activity-independent mechanisms in the calcium-mediated regulation of dendritic growth and plasticity.

CALCIUM-MEDIATED REGULATION OF AXON OUTGROWTH AND PATHFINDING

Neurotransmitter expression and dendritic morphology represent distinguishing features of different neuronal subtypes. The location and identity of a neuron's downstream synaptic partners provide an additional basis for neuronal characterization. Calcium signaling contributes to the regulation of both the rate of axon outgrowth and the directional navigation of the growth cone en route to the axon's target.

Live imaging studies performed in the developing *Xenopus* spinal cord indicate that the rate of axon outgrowth is regulated by the frequency of spontaneous calcium transients in the growth cone (Gomez and Spitzer 1999). The neuronal subtypes evaluated, including sensory Rohon–Beard neurons, ventral motor neurons, and two different classes of interneurons, show distinct frequencies of calcium transients, which are correlated with different rates of axon outgrowth. Neurons showing a high frequency of calcium transients experience a slow rate of axon outgrowth, whereas neurons with low frequency calcium activity show rapid rates of extension. Blocking or increasing the frequency of these calcium waves shows that they are both necessary and sufficient to regulate axon outgrowth. The frequency of the calcium transients within a given neuronal subtype is not fixed, and instead is subject to change depending on the location of the growth cone. This observation is consistent with studies demonstrating that gradients of extracellular guidance cues influence the calcium activity within neuronal growth cones (Hong et al. 2000; Henley et al. 2004). Additionally, the *Xenopus* live imaging findings are supported by studies of cultured *Xenopus* spinal neurons, which show that an optimal frequency of calcium transients regulates neurite extension (Gu and Spitzer 1995).

Recent work reveals that calcium-mediated regulation of axon outgrowth can also be controlled by kainate receptor-activated signaling. In cultured hippocampal neurons, electrical activity acts downstream from kainate receptor activation to decrease the rate of growth cone extension and axon outgrowth (Ibarretxe et al. 2007). The removal of extracellular calcium and the depletion of neuronal intracellular calcium stores prevent the inhibition of axon outgrowth that occurs on kainate application. These results indicate that an increase in intracellular calcium is necessary for kainate receptor-mediated inhibition of axon outgrowth.

Although kainate application and direct electrical stimulation inhibit growth cone motility in a significant percentage of cultured neurons, the growth rate of some cells remains unimpaired. This finding indicates that divergent neuronal subtypes within hippocampal cultures may respond to extracellular and activity-dependent cues in a heterogeneous fashion. Differential responses to activity-dependent signaling could allow neuronal subtypes to extend their axons to targets in disparate locations.

The selection of a neuron's target—whether it is a neuron or another cell type—is determined to a large extent by the direction of axon outgrowth, which is governed by the response of the axonal growth cone to extracellular axon guidance cues. Guidance molecules include both attractive and repulsive cues that interact with receptors on the growth cone to stimulate turning, which ultimately dictates the direction of growth. The selection of an axonal target is therefore heavily influenced by the specific repertoire of guidance receptors expressed by developing axons. How does the interaction between guidance cues and their corresponding receptors mediate directional changes in the trajectory of the growth cone? Extracellular gradients of both attractive and repulsive guidance cues have been shown to affect growth cone calcium and cyclic nucleotide levels differentially depending on whether the guidance cue is attractive or repulsive (Hong et al. 2000; Henley et al. 2004). Calcium transients in growth cone filopodia regulate turning onto preferred biological substrates (Gomez et al. 2001). Some of the downstream signaling mechanisms responsible for growth cone turning have recently been reviewed (Gomez and Zheng 2006; Zheng and Poo 2007; O'Donnell et al. 2009; Kolodkin and Tessier-Lavigne 2010). Neurotransmitters may also act as guidance molecules. Early studies show that growth cones can turn toward gradients of acetylcholine or glutamate in a calcium-dependent manner in vitro (Zheng et al. 1994, 1996). In vivo, endocannabinoids regulate connectivity in the rodent cortex (Berghuis et al. 2007) and serotonin modulation of responsiveness to Netrin-1 may account for its disruption of sensory maps of thalamocortical axons (Bonnin et al. 2007). However, the role of calcium signaling in these cases is unclear.

THE RELATIONSHIP BETWEEN ACTIVITY AND ION CHANNEL EXPRESSION

The studies described above indicate that activity-dependent calcium signaling plays a significant role in regulating multiple aspects of neuronal differentiation. Many of these studies suggest that the influx of calcium into the neuron plays a critical role in inducing the release of intracellular calcium stores, allowing the progression of subsequent downstream events that are required for differentiation. A common theme in the work described here is the importance of both the spatial and temporal pattern of calcium transients in providing specificity for calcium-mediated regulation of distinct elements of neuronal development. The expression patterns and subcellular locations of calcium-permeable channels are instrumental in determining the spatiotemporal dynamics of calcium influx. Calcium enters the cell primarily through voltage-sensitive calcium channels, ionotropic glutamate receptors, and transient receptor potential (TRP) channels. Understanding the mechanisms that regulate the expression and localization of these channels can provide additional insight into the process of neuronal differentiation and the generation of specific neuronal subtypes.

Many of the studies described in this review suggest that neuronal phenotype is less hard-wired than previously thought and that activity-dependent events can modulate neuronal phenotype throughout an extended period of development. Developmental changes in the expression profile of calcium-permeable channels provide a potential mechanism by which calcium signaling could play different roles at different times in the development of a single cell or cell type. For example, developmental studies of cultured mouse hippocampal neurons show that $Ca_v1.2$, the most abundant neuronal L-type calcium channel, is initially expressed throughout the entire neuron (Obermair et al. 2004). As the neuron develops, the expression of $Ca_v1.2$ is down-regulated in the axon, particularly in the more distal parts. In the mature neuron, the expression of $Ca_v1.2$ is restricted to the somatic and dendritic compartments of the cell. These observations are intriguing because they present a potential scenario in which the early expression of $Ca_v1.2$ in the axon, particularly in the growth cone, could play a role in the regulation of calcium-mediated axonal guidance and outgrowth, in combination with TRP channels (Wang and Poo 2005). Temporal regulation of the expression of calcium-permeable channels could provide a useful strategy for selectively altering the nature of calcium signaling for different developmental events. Further studies are required to determine whether developmental changes in calcium channel expression patterns provide specificity for calcium-mediated effects on neuronal differentiation. It is also necessary to elucidate whether temporal changes in the expression of calcium channels occur in multiple neuronal subtypes, in addition to mouse hippocampal neurons.

Observation of transient expression of $Ca_v1.2$ in the axons of hippocampal neurons prompts further inquiry into the mechanisms responsible for the compartment specific targeting and subsequent compartmental restriction of $Ca_v1.2$. The targeting of $Ca_v1.2$ in cultured rat hippocampal neurons is regulated by the interaction of the Ca_v α1c subunit with the calcium-binding protein, calmodulin (Wang et al. 2007). Intriguingly, the calmodulin-dependent trafficking of $Ca_v1.2$ to distal dendrites is enhanced following up-regulation of intracellular calcium. Mechanistically, this provides a means by which calcium signaling in the neuron can regulate future calcium signaling events. For example, influx of calcium leading to a greater level of $Ca_v1.2$ insertion in the membrane and trafficking to more distal dendrites, could act as a positive feedback mechanism to promote additional subtype-specific calcium-mediated events influencing dendritic development. In this way, calcium signaling could act as a positive feedback mechanism to regulate differentiation in a subset of developing cells, aiding the specification of different neuronal subtypes. Additionally, calcium-mediated changes in calcium channel trafficking could underlie activity-dependent changes in neurotransmitter phenotype. For instance, up-regulation of calcium signaling promoted by the introduction of voltage-gated sodium channel constructs could lead to an increase in the membrane targeting of $Ca_v1.2$ channels. Increased calcium influx as a result of additional $Ca_v1.2$ channels could evoke changes in excitation-transcriptional signaling that may be responsible for the phosphorylation of transcription factors like cJun and the activity-dependent expression of genes such as Tlx3. Future studies will delineate further roles of calcium signaling in neuronal development and shed light on the myriad ways in which the influx of calcium at the membrane can mediate changes in the expression of genes, the structure of the cytoskeleton, and the subcellular targeting of proteins such as ion-permeable channels.

FREQUENCY CODING BY CALCIUM TRANSIENTS

The distinct developmental roles of the calcium transients described above indicate that the temporal dynamics of calcium fluctuations are reliably decoded and translated into specific effects on neuronal differentiation. In the mature nervous system, neuronal action potentials, resulting from both evoked and spontaneous activity, are all-or-none events that are characterized by a stereotyped change in membrane polarization and ion flux. Because the magnitude of depolarization does not provide information about the nature of the stimulus, the nervous system often relies on frequency coding to translate the temporal patterns of spiking activity into useful information. Similarly, in the developing nervous system, the rate of neurite extension is regulated by the frequency of spontaneous calcium transients (Gu and Spitzer 1995). In addition, the frequency of calcium transients regulates the expression levels of GAD67 (an enzyme that synthesizes GABA), and the number of GABAergic neurons.

How is information encoded in the temporal pattern of action potential firing and calcium transients? Temporal variability is present in the duration of bursts of single spikes and in the length of time between bursts—the interburst interval. How do neurons decode these temporal features of spiking into effects on neuronal development and plasticity? Future studies deciphering neuronal decoding strategies will provide insight into the regulation of neuronal development and specification, as well the formation and plasticity of neuronal circuits.

CONCLUDING THOUGHTS

Elucidating the role of calcium signaling in neuronal development provides a better understanding of the formation and function of healthy neuronal circuits, and may also provide key insight into the pathogenesis of neurodegenerative and neurodevelopmental disorders. Parkinson's disease, Huntington's disease, and amyotrophic lateral sclerosis are characterized by the death or dysfunction of specific neuronal subtypes. Similarly, subtype-specific deficits have also been hypothesized to play a role in neurodevelopmental disorders such as autism and schizophrenia. Further exploration of the calcium-mediated mechanisms responsible for neuronal subtype specification may identify new treatments for neurological disorders.

REFERENCES

Asmus SE, Tian H, Landis SC. 2001. Induction of cholinergic function in cultured sympathetic neurons by periosteal cells: Cellular mechanisms. *Dev Biol* **235:** 1–11.

Berghuis P, Rajnicek AM, Morozov YM, Ross RA, Mulder J, Urban GM, Monory K, Marsicano G, Matteoli M, Canty A, et al. 2007. Hardwiring the brain: Endocannabinoids shape neuronal connectivity. *Science* **316:** 1212–1216.

Blankenship AG, Feller MB. 2010. Mechanisms underlying spontaneous patterned activity in developing neural circuits. *Nat Rev Neurosci* **11:** 18–29.

Bonnin A, Torii M, Wang L, Rakic P, Levitt P. 2007. Serotonin modulates the response of embryonic thalamocortical axons to netrin-1. *Nat Neurosci* **10:** 588–597.

Borodinsky LN, Spitzer NC. 2007. Activity-dependent neurotransmitter-receptor matching at the neuromuscular junction. *Proc Natl Acad Sci* **104:** 335–340.

Borodinsky LN, Root CM, Cronin JA, Sann SB, Gu X, Spitzer NC. 2004. Activity-dependent homeostatic specification of transmitter expression in embryonic neurons. *Nature* **429:** 523–530.

Catsicas M, Bonness V, Becker D, Mobbs P. 1998. Spontaneous Ca^{2+} transients and their transmission in the developing chick retina. *Curr Biol* **8:** 283–286.

Chang LW, Spitzer NC. 2009. Spontaneous calcium spike activity in embryonic spinal neurons is regulated by developmental expression of the Na^+, K^+-ATPase β3 subunit. *J Neurosci* **29:** 7877–7885.

Cheng L, Arata A, Mizuguchi R, Qian Y, Karunaratne A, Gray PA, Arata S, Shirasawa S, Bouchard M, Luo P, et al. 2004. Tlx3 and Tlx1 are post-mitotic selector genes determining glutamatergic over GABAergic cell fates. *Nat Neurosci* **7:** 510–517.

Cheng L, Samad OA, Xu Y, Mizuguchi R, Luo P, Shirasawa S, Goulding M, Ma Q. 2005. Lbx1 and Tlx3 are opposing switches in determining GABAergic versus glutamatergic transmitter phenotypes. *Nat Neurosci* **8:** 1510–1515.

Chevaleyre V, Moos FC, Desarmenien MG. 2002. Interplay between presynaptic and postsynaptic activities is required for dendritic plasticity and synaptogenesis in the supraoptic nucleus. *J Neurosci* **22:** 265–273.

Demarque M, Spitzer NC. 2010. Activity-dependent expression of Lmx1b regulates specification of serotonergic neurons modulating swimming behavior. *Neuron* **67:** 321–334.

Dulcis D, Spitzer NC. 2008. Illumination controls differentiation of dopamine neurons regulating behaviour. *Nature* **456:** 195–201.

Gomez TM, Spitzer NC. 1999. In vivo regulation of axon extension and pathfinding by growth-cone calcium transients. *Nature* **397:** 350–355.

Gomez TM, Zheng JQ. 2006. The molecular basis for calcium-dependent axon pathfinding. *Nat Rev Neurosci* **7:** 115–125.

Gomez TM, Snow DM, Letourneau PC. 1995. Characterization of spontaneous calcium transients in nerve growth cones and their effect on growth cone migration. *Neuron* **14:** 1233–1246.

Gomez TM, Robles E, Poo M, Spitzer NC. 2001. Filopodial calcium transients promote substrate-dependent growth cone turning. *Science* **291:** 1983–1987.

Gu X, Spitzer NC. 1995. Distinct aspects of neuronal differentiation encoded by frequency of spontaneous Ca^{2+} transients. *Nature* **375:** 784–787.

Gu X, Spitzer NC. 1997. Breaking the code: Regulation of neuronal differentiation by spontaneous calcium transients. *Dev Neurosci* **19:** 33–41.

Gu X, Olson EC, Spitzer NC. 1994. Spontaneous neuronal calcium spikes and waves during early differentiation. *J Neurosci* **14:** 6325–6335.

Habecker BA, Symes AJ, Stahl N, Francis NJ, Economides A, Fink JS, Yancopoulos GD, Landis SC. 1997. A sweat gland-derived differentiation activity acts through known cytokine signaling pathways. *J Biol Chem* **272:** 30421–30428.

Hanson PI, Schulman H. 1992. Neuronal Ca^{2+}/calmodulin-dependent protein kinases. *Annu Rev Biochem* **61:** 559–601.

Henley JR, Huang KH, Wang D, Poo MM. 2004. Calcium mediates bidirectional growth cone turning induced by myelin-associated glycoprotein. *Neuron* **44:** 909–916.

Holliday J, Adams RJ, Sejnowski TJ, Spitzer NC. 1991. Calcium-induced release of calcium regulates differentiation of cultured spinal neurons. *Neuron* **7:** 787–796.

Hong K, Nishiyama M, Henley J, Tessier-Lavigne M, Poo M. 2000. Calcium signalling in the guidance of nerve growth by netrin-1. *Nature* **403:** 93–98.

Hudmon A, Schulman H. 2002. Neuronal Ca^{2+}/calmodulin-dependent protein kinase II: The role of structure and autoregulation in cellular function. *Annu Rev Biochem* **71:** 473–510.

Ibarretxe G, Perrais D, Jaskolski F, Vimeney A, Mulle C. 2007. Fast regulation of axonal growth cone motility by electrical activity. *J Neurosci* **27:** 7684–7695.

Kolodkin AL, Tessier-Lavigne M. 2010. Mechanisms and molecules of neuronal wiring: A primer. *Cold Spring Harb Perspect Biol* doi: 101101/cshperspecta001727.

Marek KW, Kurtz LM, Spitzer NC. 2010. cJun integrates calcium activity and tlx3 expression to regulate neurotransmitter specification. *Nat Neurosci* **13:** 944–950.

Nelson TE, Netzeband JG, Gruol DL. 2004. Chronic interleukin-6 exposure alters metabotropic glutamate receptor-activated calcium signalling in cerebellar Purkinje neurons. *Eur J Neurosci* **20:** 2387–2400.

Obermair GJ, Szabo Z, Bourinet E, Flucher BE. 2004. Differential targeting of the L-type Ca^{2+} channel $\alpha 1C$ ($Ca_V 1.2$) to synaptic and extrasynaptic compartments in hippocampal neurons. *Eur J Neurosci* **19:** 2109–2122.

O'Donnell M, Chance RK, Bashaw GJ. 2009. Axon growth and guidance: Receptor regulation and signal transduction. *Annu Rev Neurosci* **32:** 383–412.

O'Donovan MJ. 1999. The origin of spontaneous activity in developing networks of the vertebrate nervous system. *Curr Opin Neurobiol* **9:** 94–104.

Parrish JZ, Emoto K, Kim MD, Jan YN. 2007. Mechanisms that regulate establishment, maintenance, and remodeling of dendritic fields. *Annu Rev Neurosci* **30:** 399–423.

Platel JC, Dave KA, Bordey A. 2008. Control of neuroblast production and migration by converging GABA and glutamate signals in the postnatal forebrain. *J Physiol* **586:** 3739–3743.

Rajan I, Cline HT. 1998. Glutamate receptor activity is required for normal development of tectal cell dendrites in vivo. *J Neurosci* **18:** 7836–7846.

Rao MS, Landis SC. 1990. Characterization of a target-derived neuronal cholinergic differentiation factor. *Neuron* **5:** 899–910.

Root CM, Velazquez-Ulloa NA, Monsalve GC, Minakova E, Spitzer NC. 2008. Embryonically expressed GABA and glutamate drive electrical activity regulating neurotransmitter specification. *J Neurosci* **28:** 4777–4784.

Schotzinger RJ, Landis SC. 1988. Cholinergic phenotype developed by noradrenergic sympathetic neurons after innervation of a novel cholinergic target in vivo. *Nature* **335:** 637–639.

Schotzinger RJ, Landis SC. 1990. Acquisition of cholinergic and peptidergic properties by sympathetic innervation of rat sweat glands requires interaction with normal target. *Neuron* **5:** 91–100.

Sillar KT, Reith CA, McDearmid JR. 1998. Development and aminergic neuromodulation of a spinal locomotor network controlling swimming in *Xenopus* larvae. *Ann NY Acad Sci* **860:** 318–332.

Spitzer NC. 2006. Electrical activity in early neuronal development. *Nature* **444:** 707–712.

Spitzer NC, Gu X, Olson E. 1994. Action potentials, calcium transients and the control of differentiation of excitable cells. *Curr Opin Neurobiol* **4:** 70–77.

Spruston N. 2008. Pyramidal neurons: Dendritic structure and synaptic integration. *Nat Rev Neurosci* **9:** 206–221.

Stanke M, Duong CV, Pape M, Geissen M, Burbach G, Deller T, Gascan H, Otto C, Parlato R, Schutz G, et al. 2006. Target-dependent specification of the neurotransmitter phenotype: Cholinergic differentiation of sympathetic neurons is mediated in vivo by gp 130 signaling. *Development* **133:** 141–150.

Vereyken EJ, Bajova H, Chow S, de Graan PN, Gruol DL. 2007. Chronic interleukin-6 alters the level of synaptic proteins in hippocampus in culture and in vivo. *Eur J Neurosci* **25:** 3605–3616.

Wang GX, Poo MM. 2005. Requirement of TRPC channels in netrin-1-induced chemotropic turning of nerve growth cones. *Nature* **434:** 898–904.

Wang HG, George MS, Kim J, Wang C, Pitt GS. 2007. Ca^{2+}/calmodulin regulates trafficking of $Ca_V 1.2$ Ca^{2+} channels in cultured hippocampal neurons. *J Neurosci* **27:** 9086–9093.

Watt SD, Gu X, Smith RD, Spitzer NC. 2000. Specific frequencies of spontaneous Ca^{2+} transients upregulate GAD 67 transcripts in embryonic spinal neurons. *Mol Cell Neurosci* **16:** 376–387.

Wong RO, Ghosh A. 2002. Activity-dependent regulation of dendritic growth and patterning. *Nat Rev Neurosci* **3:** 803–812.

Wu GY, Cline HT. 1998. Stabilization of dendritic arbor structure in vivo by CaMKII. *Science* **279:** 222–226.

Yuste R, Peinado A, Katz LC. 1992. Neuronal domains in developing neocortex. *Science* **257:** 665–669.

Zheng JQ, Poo MM. 2007. Calcium signaling in neuronal motility. *Annu Rev Cell Dev Biol* **23:** 375–404.

Zheng JQ, Felder M, Connor JA, Poo MM. 1994. Turning of nerve growth cones induced by neurotransmitters. *Nature* **368:** 140–144.

Zheng JQ, Wan JJ, Poo MM. 1996. Essential role of filopodia in chemotropic turning of nerve growth cone induced by a glutamate gradient. *J Neurosci* **16:** 1140–1149.

Visualization of Ca^{2+} Signaling During Embryonic Skeletal Muscle Formation in Vertebrates

Sarah E. Webb and Andrew L. Miller

Section of Biochemistry and Cell Biology, and State Key Laboratory of Molecular Neuroscience, Division of Life Science, The Hong Kong University of Science and Technology, Clear Water Bay, Kowloon, Hong Kong SAR, PRC

Correspondence: almiller@ust.hk

Dynamic changes in cytosolic and nuclear Ca^{2+} concentration are reported to play a critical regulatory role in different aspects of skeletal muscle development and differentiation. Here we review our current knowledge of the spatial dynamics of Ca^{2+} signals generated during muscle development in mouse, rat, and *Xenopus* myocytes in culture, in the exposed myotome of dissected *Xenopus* embryos, and in intact normally developing zebrafish. It is becoming clear that subcellular domains, either membrane-bound or otherwise, may have their own Ca^{2+} signaling signatures. Thus, to understand the roles played by myogenic Ca^{2+} signaling, we must consider (1) the triggers and targets within these signaling domains; (2) interdomain signaling, and (3) how these Ca^{2+} signals integrate with other signaling networks involved in myogenesis. Imaging techniques that are currently available to provide direct visualization of these Ca^{2+} signals are also described.

The recognition of Ca^{2+} as a key regulator of muscle contraction dates back to Sydney Ringer's seminal observations in the latter part of the 19th Century (Ringer 1883; Ringer 1886; Ringer and Buxton 1887; see reviews by Martonosi 2000; Szent-Györgyi 2004). More recently, evidence is steadily accumulating to support the proposition that Ca^{2+} also plays a necessary and essential role in regulating embryonic muscle development and differentiation (Flucher and Andrews 1993; Ferrari et al. 1996; Lorenzon et al. 1997; Ferrari and Spitzer 1998, 1999; Wu et al. 2000; Powell et al. 2001; Jaimovich and Carrasco 2002; Li et al. 2004; Brennan et al. 2005; Harris et al. 2005; Campbell et al. 2006; Terry et al. 2006; Fujita et al. 2007; and see reviews by Berchtold et al. 2000; Ferrari et al. 2006; Al-Shanti and Stewart 2009). What is currently lacking, however, is extensive direct visualization of the spatial dynamics of the Ca^{2+} signals generated by developing and differentiating muscle cells. This is especially so concerning in situ studies. The object of this article, therefore, is to review and report the current state of our understanding concerning the spatial nature of Ca^{2+} signaling during embryonic muscle development, especially from an in vivo perspective, and to suggest possible directions for future research. The focus of our article is embryonic skeletal muscle development because of this being an area of significant current interest. Several of the basic observations reported, however, may also be common to cardiac muscle development and in some cases to smooth muscle development. What the recent development of reliable imaging techniques has most certainly done, is to add an extra dimension of complexity to understanding the roles played by Ca^{2+} signaling in skeletal muscle development. For example, it is clear that membrane-bound subcellular compartments, such as the nucleus (Jaimovich and Carrasco 2002), may have endogenous Ca^{2+} signaling activities, as do specific cytoplasmic domains, such as the subsarcolemmal space (Campbell et al. 2006). How these Ca^{2+} signals interact with specific down-stream targets within their particular domain, and how they might serve to communicate information among domains, will most certainly be one of the future challenges in elucidating the Ca^{2+}-mediated regulation of muscle development.

Any methodology used to study the properties of biological molecules and how they interact during development should ideally provide spatial information, because researchers increasingly need to integrate data about the interactions that underlie a biological process (such as differentiation) with information regarding the precise location within cells or an embryo where these interactions take place. Current Ca^{2+} imaging techniques are beginning to provide us with this spatial information, and are thus opening up exciting new avenues of investigation in our quest to understand the signaling pathways that regulate muscle development (Table 1).

Ca^{2+} IMAGING METHODOLOGIES USED IN BOTH IN VIVO AND IN VITRO STUDIES

Ca^{2+} signals have been visualized during muscle formation using both fluorescent and luminescent Ca^{2+}-sensitive reporters. Fluorescent reporters can be incorporated into cells via microinjection or infusion from whole-cell patch pipettes. Alternatively, many fluorescent Ca^{2+} probes are also available as acetoxymethyl (AM) esters. In this form, the probes are membrane permeable and insensitive to Ca^{2+}. Thus, cells and tissues can simply be incubated with the probe, which is dissolved in normal bathing medium. Once inside the cell, intracellular esterases then hydrolyse the AM moiety, and thus reveal the active Ca^{2+}-sensitive reporter (Tsien 1981). Unfortunately, using these methods, it is not possible to target the reporters to particular organelles, cells, or tissues in vivo. However, a few years ago a number of genetically encoded fluorescent Ca^{2+} reporters that are based on GFP or its spectral variants, were designed. The first family of genetic Ca^{2+} indicators, the cameleons, were based on the Ca^{2+}-sensitive protein, calmodulin, and relied on a fluorescence resonance energy transfer to occur between cyan fluorescent protein (CFP) and yellow fluorescent protein (YFP) in the presence of Ca^{2+} (Miyawaki et al. 1997; Persechini et al. 1997). More recently, several other genetically encoded fluorescent Ca^{2+} reporters, such as camgaroo-2, inverse pericam and DRIP (DsRed2-referenced Inverse Pericam), have been engineered (Baird et al. 1999; Hasan et al. 2004; Shimozono et al. 2004). Most of the groups who have imaged Ca^{2+} signals during skeletal muscle formation in cells, or in dissected or intact animals, used single-wavelength fluorescent Ca^{2+} reporters that are excited by visible light and either incubated the cells/tissues in the AM ester form, e.g. fluo-3 AM (Ferrari and Spitzer 1999; Jaimovich et al. 2000; Carrasco et al. 2003) or else microinjected individual cells, e.g., with oregon green 488 BAPTA-1 dextran (10 kDa), fluo-4 dextran (10 kDa), or calcium green-1 dextran (Brennan et al. 2005; Cheung et al. 2010). Although a transgenic line of *Caenorhabditis elegans* that expresses cameleon specifically in the pharyngeal muscle has been reported (Kerr et al. 2000), Ca^{2+} transients were only imaged in the adult worms during feeding behavior and not during the formation of the muscle.

The majority of reports that describe fluorescence Ca^{2+} imaging of developing muscle cells, exposed myotome and intact animals use confocal laser scanning microscopy to visualize

Table 1. Examples of developing muscle systems in which Ca^{2+} signals have been directly visualized and spatial information is provided.

Animal	Intact animals/Cells in culture	Ca^{2+} reporter	Reporter Loading Protocol	Reference
Rat	1° cultures prepared from hind limb muscle of neonatal rat pups	Fluo 3-AM	Cells incubated in 5.4 μM reporter for 30 min at 25°C.	Jaimovich et al. 2000
Mouse	Myotubes grown from C2C12 subclone of the C2 mouse muscle cell line	Fluo 3-AM	Incubated in 5 μM reporter plus 0.1% pluronic F-127 for 1 h at r.t.	Flucher and Andrews 1993
	Myotubes isolated from the intercostal muscles of E18 wild-type and RyR type 3-null mice.	Fluo 3-AM	Cells incubated with 4 μM for 30 min at r.t.	Conklin et al. 1999b
	Myotubes in culture prepared from newborn mice.	Fluo 3-AM	Cells incubated in 10 μM for 20 min.	Shirokova et al. 1999
	1° cultures prepared from hind limb muscle from newborn mice.	Fluo 3-AM	Cells incubated in 5.4 μM reporter for 30 min at 25°C.	Powell et al. 2001
	Embryonic day 18 (E18) isolated diaphragm muscle fibers	Fluo 4-AM	Incubated in 10 μM reporter for 30 min.	Chun et al. 2003
Chick	Myotubes prepared from leg or breast of 11-day chick embryos	Fluo 3-AM	Incubated in 5 μM reporter plus 0.1% pluronic F-127 for 1 h at r.t.	Flucher and Andrews 1993
	Myoblasts isolated from thigh muscle of E12 embryos.	Fluo 3-AM	1 mM stock was diluted 1:200 with 0.2% pluronic F-127. Cells were incubated for 60 min at r.t. in the dark.	Tabata et al. 2006
Xenopus	Exposed myotome in dissected embryo	Fluo-3 AM	Incubated dissected tissue in 10 μM reporter for 30–60 min.	Ferrari and Spitzer 1999
	1° myocyte cultures prepared from stage 15 *Xenopus* embryos.	Fluo-4 AM	Cells incubated in 2 μM reporter plus 0.01% pluronic F-127 for 60 min.	Campbell et al. 2006
Zebrafish	Intact animals	Calcium green-1 dextran (10S)	Reporter at 20 mM was injected into a single blastomere between the 32- and 128-cell stage.	Zimprich et al. 1998
	Intact animals	Oregon Green 488 BAPTA dextran	Single blastomeres from 32-cell stage embryos injected with reporter (i.c. 100 μM) and tetramethylrhodamine dextran (i.c. 40 μM).	Ashworth et al. 2001
	Intact animals	Oregon Green 488 BAPTA dextran	Microinjected with rhodamine dextran to give an intracellular concentration of ∼40 μM.	Ashworth 2004
	Intact animals	Aequorin	[a]Embryos injected with 700 pg aeq-mRNA at the 1-cell stage and then incubated with 50 μM f-coelenterazine from the 64-cell stage.	Cheung et al. 2006
	Intact animals	Aequorin	Transgenic fish that express apoaequorin in the skeletal muscles were incubated with 50 μM f-coelenterazine from the 8-cell stage.	Cheung et al. 2010

[a]Expression of aequorin was ubiquitous but it was suggested that the Ca^{2+} signals visualized in the trunk at the approximately 8–20-somite stage and at ∼47 hpf might play a role in muscle development.

the Ca^{2+} signals (Shirokova et al. 1999; Conklin et al. 1999a, 1999b; Ferrari and Spitzer 1999; Jaimovich et al. 2000; Ashworth et al. 2001; Powell et al. 2001; Chun et al. 2003). In the past, this imaging method tended to be somewhat incompatible with long-duration Ca^{2+} imaging because of photobleaching of reporters and phototoxicity caused by the excitation light, which led to a disruption of development. The development of Ca^{2+} reporters that can be excited with less damaging, longer wavelengths of light, as well as the application of experimental protocols that minimize the exposure of the reporter to excitation, have greatly helped to offset these constraints. However, for short duration imaging experiments, fluorescent confocal imaging has excellent spatial resolution and so can be used to accurately determine the organelles, cells or tissues responsible for generating particular Ca^{2+} transients.

Although fluorescent Ca^{2+} reporters require light excitation, luminescent Ca^{2+} reporters do not. The most common luminescent Ca^{2+} reporter used is aequorin, which consists of a 21-kDa protein, apoaequorin, in combination with a hydrophobic prosthetic group, coelenterazine, and oxygen. Although so-called "native" aequorin was originally extracted and purified from *Aequoria* spp. jellyfish (Shimomura et al. 1962), the successful production of the apoaequorin protein component by molecular cloning techniques (Inouye et al. 1985; Prasher et al. 1987), in addition to the development of synthetic analogues of coelenterazine (Shimomura et al. 1988, 1989, 1990), meant that a variety of recombinant semi-synthetic aequorins, with improved sensitivity to Ca^{2+} ions, are now available. Some of the first applications of native aequorin were in the study of Ca^{2+} transients in isolated crab (*Maia squinado*) muscle fibers (Ashley 1969) and the classical study of isolated barnacle (*Balanus nubilus*) muscle fibers (Ashley and Ridgeway 1970). The main advantages of aequorins are that they are nontoxic; do not buffer Ca^{2+} to any great extent; are available at a variety of different sensitivities; involve ultra-low background signal (i.e., there is no autoluminescence, analogous to autofluorescence); show inherent contrast enhancement (i.e., the light increases as the approximate second power of the Ca^{2+} concentration); have very long life times in the cytosol; and have a very wide dynamic range (from 0.01–10 μM in vivo; Miller et al. 1994). Thus, aequorin imaging is an excellent technique to use when imaging Ca^{2+} signals over periods of hours, rather than minutes. This is a particular advantage when studying developmental Ca^{2+} signaling over a period of days. The advantages of aequorin, however, are offset by that fact that the images obtained have a low spatial resolution, especially in the z-axis. In addition, the Ca^{2+} signals generated in embryos at the later stages of development (for example, when the muscles are forming) are difficult to image because recombinant aequorin injected into embryos at the one-cell stage is almost used up by approximately 24 hours postfertilization (hpf; Webb and Miller 2000). It is also clearly impractical to microinject the aequorin into thousands of individual cells in a later-stage embryo. One way to overcome this restriction is to express the protein component of the aequorin complex (i.e., apoaequorin), via genetic means, and then to reconstitute the active aequorin protein complex in vivo by loading the aequorin complex cofactor, coelenterazine, into the embryos separately.

In 2006, we developed a technique in which we transiently and ubiquitously expressed apoaequorin in zebrafish embryos by microinjecting an apoaequorin-mRNA (*aeq*-mRNA) into 1-cell stage embryos (Cheung et al. 2006). Although this transient aequorin expression approach successfully extended the aequorin-based Ca^{2+} imaging window by an additional 24 hours (to ~48 hpf), the *aeq*-mRNA was gradually degraded in the injected embryos resulting in a steady decline in the production of the apoaequorin (Cheung et al. 2006). Furthermore, as the expression of aequorin was ubiquitous, it was impossible to identify the precise groups of cells, tissues or organ anlage that were generating a particular signal in more complex, later-stage embryos (Cheung et al. 2006). Thus, most recently, we have successfully generated a line of transgenic zebrafish that stably expresses apoaequorin specifically in the musculature using a muscle-specific α-actin promoter (Cheung et al. 2010). In this transgenic line, an EGFP marker was also coexpressed with the apoaequorin from a bicistronic mRNA using the internal ribosome entry site (IRES) sequence. Thus, the expression level and distribution of EGFP reflected the expression level and distribution of apoaequorin in the embryos. Using this transgenic α- actin aeq line of fish, it has been possible to continuously image Ca^{2+} signals specifically from the developing musculature in zebrafish embryos up to ∼52 hpf (Cheung et al. 2010).

Currently, several different types of equipment are commercially available that can be used to detect or visualize aequorin-generated luminescence. Several companies supply relatively inexpensive equipment that provides simple temporal Ca^{2+} information via a photomultiplier tube (PMT). However, only a small number of very specialist companies build the sort of custom-designed imaging systems that provide both temporal and spatial luminescent information, as well as bright-field and fluorescence images, thus enabling the correlation of Ca^{2+} signaling events with morphological features and other cellular changes. All of these imaging systems are built around some form of position-sensitive photon detector. We have two such systems in our laboratory, which were both designed by Mr. Eric Karplus of Science Wares, Inc. (Falmouth, Massachusetts); the first we have had for approximately 15 years and is based around a resistive anode imaging photon detector (RA-IPD; Photek Inc.; Miller et al. 1994; Webb et al. 1997), whereas the other, which has just been commissioned, is based around an electron multiplying charge coupled device (EMCCD; Andor Inc.; Webb et al. 2010). Figure 1 illustrates a block diagram of the RA-IPD-based imaging system, which was used to acquire Ca^{2+} signaling information from the developing skeletal musculature of the intact zebrafish embryo shown in Figure 2Bi.

Most recently, we have developed a complementary luminescent and fluorescent Ca^{2+} imaging approach, in which we use aequorin imaging initially, to provide a two-dimensional, low resolution imaging map of Ca^{2+} signaling over extended time-periods. From the information acquired from aequorin-based imaging, we then use a fluorescent Ca^{2+} reporter in conjunction with confocal imaging, to explore the specific characteristics of the Ca^{2+} transients at much higher spatial resolution. Aequorin-based imaging tells us "where" and "when" to look, and then we investigate any interesting Ca^{2+} signals observed in further detail with fluorescent confocal imaging. This approach is especially useful when imaging Ca^{2+} signaling events in the relatively large, three-dimensional zebrafish embryos.

VISUALIZATION OF Ca^{2+} SIGNALS IN DIFFERENT ANIMAL SYSTEMS

The regulation of $[Ca^{2+}]_i$ has been reported to play a critical role in many aspects of muscle development (myogenesis) and function. Although the majority of Ca^{2+}-muscle studies has been conducted in vitro, with cells in culture, a number of more recent studies have been conducted in vivo, i.e., using intact, normally developing animals. In the following sections we will describe some of seminal Ca^{2+} imaging studies performed using mouse and rat myotubes in culture, in intact normally developing zebrafish embryos, and in the developing myotome of *Xenopus*.

Ca^{2+} Transients Visualized in Mouse and Rat Myotubes

Flucher and Andrews (1993) were one of the first to image Ca^{2+} transients in developing myotubes. They used the C2C12 mouse muscle cell-line as well as preparing primary cultures of both chicken embryo and rat fetal thigh muscle, thus giving a variety of myotubes that differentiated at different rates. Using the fluorescent Ca^{2+} reporter, fluo-3, they showed that in well-differentiated myotubes, spontaneous and electrically-induced contractions were accompanied by transient increases in $[Ca^{2+}]_i$ that were generated synchronously throughout the myotube (Flucher and Andrews 1993). In young myotubes, however, a different type of spontaneous Ca^{2+} release was observed. These

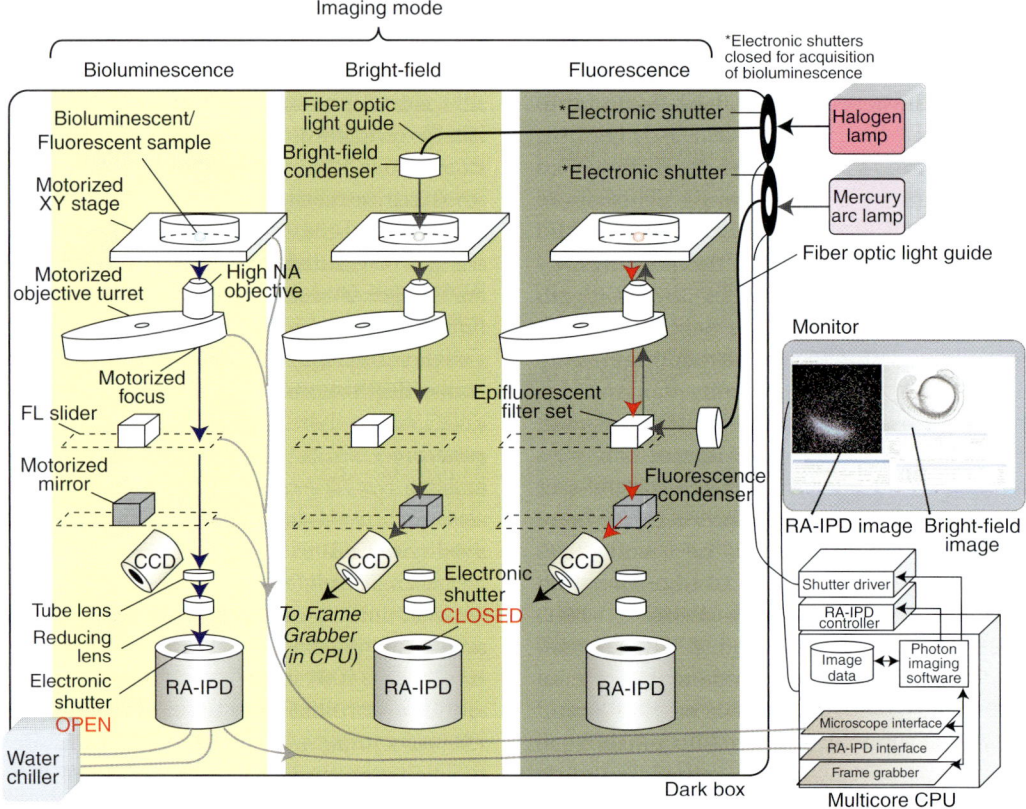

Figure 1. Schematic representation of our luminescence imaging system. This system can be used to acquire bright-field and fluorescence imaging information (via a CCD camera) as well as collect bioluminescence data (via a resistive-anode Imaging Photon Detector; RA-IPD). A high level of automation for the microscope makes it possible to rapidly switch among the various imaging modes, and makes it possible to have the computer run automated acquisition sequences over extended periods, for example overnight. The motorized focus allows the computer to acquire image stacks in any imaging mode for three-dimensional reconstructions. This system was designed and built by Science Wares Inc., Falmouth, Massachusetts.

were brief (i.e., ~50–200 ms) Ca^{2+} transients that were localized to small regions (i.e., ~20–50 μm) within the myotubes. These so-called "fast-localized Ca^{2+} transients" were not associated with contractions, were insensitive to electrical stimulation, and occurred in consistent locations and in the absence of extracellular Ca^{2+}. Flucher and Andrews (1993) also showed that in noncontracting myotubes, a low concentration of caffeine stimulated the generation of Ca^{2+} waves, and these were accompanied by a localized contraction. These authors suggested that the different types of Ca^{2+} transients might reflect the stage of differentiation of the myotubes, with the fast-localized Ca^{2+} transients being a temporary event occurring in the younger cells before the functional differentiation of excitation-contraction (E-C) coupling is accomplished.

Lorenzon et al. (1997) also imaged Ca^{2+} transients in differentiating C2C12 mouse cells. Using fura-2 AM as the Ca^{2+} reporter, they showed that distinct patterns of Ca^{2+} transients could again be visualized at different times during myogenesis, and that they could be induced following a change from growth medium to fusion medium. They described these transients as spontaneous $[Ca^{2+}]_i$ oscillations, which developed in microdomains of 5–10 μm in length along the myotube and then propagated within the myotubes as waves with a velocity ~20 μm/s. They also described localized $[Ca^{2+}]_i$ spikes, some of which were restricted to a length of just 2–3 μm, whereas others expanded to lengths of up to 20 μm along the myotube; and global $[Ca^{2+}]_i$ spikes, which were generated in the entire myotube and occurred mainly in mature myotubes. The smaller Ca^{2+} spikes (i.e., those restricted to lengths of ~2–3 μm) were suggested to resemble the Ca^{2+} sparks previously described in heart muscle (Cheng et al. 1993). Lorenzon et al. (1997) also showed that at the start of differentiation, i.e., before the voltage-operated Ca^{2+} channels (VOCCs) and ryanodine receptors (RyRs) were coupled, the transients were caused by the release of Ca^{2+} from an intracellular source, but when the VOCCs and RyRs were coupled later in myogenesis there was a switch to a membrane oscillatory mechanism, with spontaneous depolarizations of the membrane causing the observed global Ca^{2+} transients. Conklin et al. (1999a) and Shirokova et al. (1999) also described Ca^{2+} sparks in myotube cultures prepared from wild-type and ryanodine receptor type 3 (RyR3) knockout embryonic and newborn mice, respectively that had been loaded with fluo-3 AM.

It has been shown that in embryonic mammalian skeletal muscle, RyR3s are coexpressed with RyR1s at high levels; however, during the postnatal period the expression of RyR3 declines to very low levels (Flucher et al. 1999). In addition, although RyR1s are coupled to dihydropyridine receptors (DHPRs) to provide a voltage-dependent SR Ca^{2+} release mechanism, which is independent of Ca^{2+} flux across the sarcolemma, RyR3s are activated by voltage-independent mechanisms, for example via Ca^{2+} induced Ca^{2+} release, rather than being coupled to DHPRs (Fessenden et al. 2000; Protasi et al. 2000). Conklin et al. (1999a) showed that when the RyR3 channels are absent then RyR1 channels play a role in generating the Ca^{2+} sparks, however in wild-type embryonic muscle both RyR1s and RyR3s contribute to the Ca^{2+} release for a particular Ca^{2+} spark.

Conklin et al. (1999b) also investigated the kinetics of the Ca^{2+} sparks generated in myotubes prepared from mouse embryos lacking either the β_{1a} subunit or the α_{1S} subunit of the DHPR. They showed that E-C coupling is not essential for the generation of Ca^{2+} sparks in these cells, as sparks were still observed when either the β_{1a} or α_{1S} subunits of the DHPR were absent. As the properties of the Ca^{2+} sparks generated in the DHPR β_{1a} or α_{1S} deficient cells were very different, however, Conklin et al. (1999b) suggested that the subunits might regulate the RyRs independently when the DHPR voltage sensor is nonfunctional. Furthermore, Pisaniello et al. (2003) showed that the expression of both RyR1s and RyR3s corresponds to that of the muscle-specific differentiation marker, sarcomeric myosin. They also showed that Ca^{2+} release via RyRs seems to be required for the differentiation of fetal myoblasts as treatment with ryanodine inhibited the formation of myotubes and the expression of myosin in the fetal muscle cells.

Jaimovich et al. (2000) and Powell et al. (2001) analyzed the Ca^{2+} waves propagating through primary cultures of newborn rat and mouse myotubes loaded with fluo 3-AM. They showed that there are both fast and slow propagating Ca^{2+} waves both of which are triggered by high K^+ and are independent of external Ca^{2+}. They also showed that the former is associated with RyRs and E-C coupling, whereas the latter resulted in an increase in nucleoplasmic Ca^{2+} but not in contraction. An example of a slow Ca^{2+} wave propagating through a rat myotube from nucleus to nucleus is shown in Figure 2A. They showed that the treatment of myotubes with nifedipine or 2-APB completely blocked the slow nucleoplasmic Ca^{2+} wave, which suggests that DHPRs are the main membrane voltage sensors for this Ca^{2+} signal, and that the signal is generated mainly by Ca^{2+} release from internal stores via IP$_3$Rs (Powell et al. 2001; Jaimovich and Carrasco 2002; Araya et al. 2003). Furthermore, this slow IP$_3$R-mediated Ca^{2+} transient was reported to be linked to developmental gene expression, in which K^+-induced depolarization of mouse myotubes resulted in the rapid (i.e., within 30 s – 10 min)

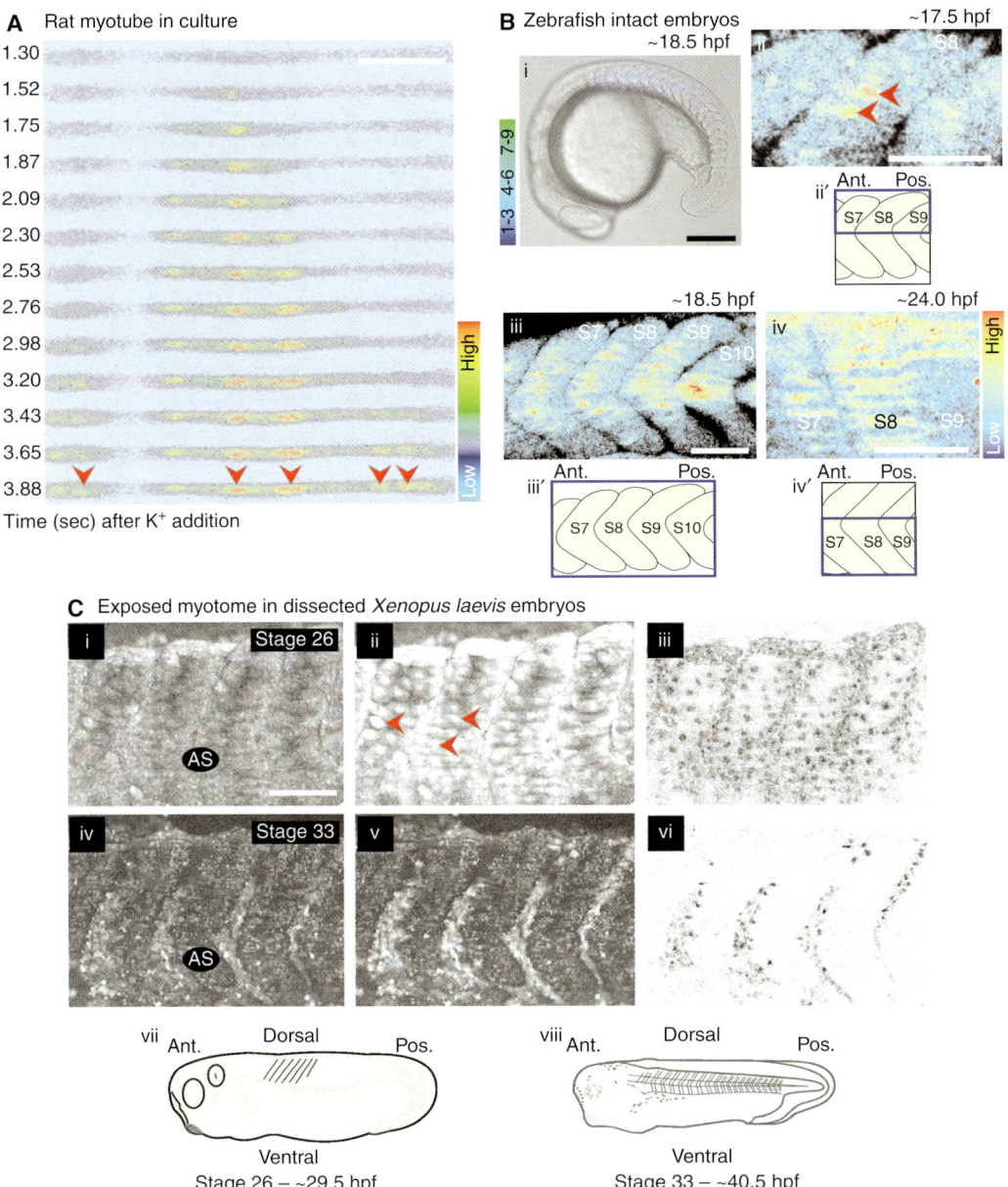

Figure 2. Examples of the spatial patterns of Ca^{2+} signals observed in rat myotubes, in intact normally developing zebrafish embryos and following photoactivation of caged IP_3 in exposed myotome in dissected *Xenopus* embryos. (*A*) A slow Ca^{2+} wave propagating in a rat myotube that had been preloaded with fluo 3-AM. Series of fluorescence images taken at the times indicated after the addition of 47 mM K^+ to the incubation medium. High-K^+ medium was added after two previous waves were elicited, and the corresponding high-K^+ solutions were replaced by normal medium. The relatively low level of fluorescence elicited by this slow wave shows that the Ca^{2+} signal is located mainly in the cell nuclei and that it is propagated in sequence, from one nucleus to another. Red arrowheads indicate the elevated Ca^{2+} in the nuclei. Scale bar is 50 μm. (*B*) The Ca^{2+} transients observed in the developing trunk muscles of intact zebrafish embryos, visualized by (*Bi*) luminescence and (*Bii–Biv*) fluorescence imaging methodologies (modified from Cheung et al. 2010). (*See facing page for legend.*)

phosphorylation of the mitogen activated kinases, ERK 1/2 and the transcription factor, CREB, as well as the expression (within 5–15 min) of various early genes (Powell et al. 2001; Jaimovich and Carrasco 2002; Carrasco et al. 2003; Cárdenas et al. 2005).

Ca^{2+} Signaling During the Formation of Muscle in Zebrafish Embryos

Zebrafish embryos provide an excellent model system for studying embryogenesis in vivo. A combination of their optical transparency, large size, and relatively rapid development makes the embryos especially useful for Ca^{2+} imaging.

In zebrafish, as with other vertebrates, the skeletal muscles contain two main types of muscle fibers. Slow muscle fibers generate low-force but long duration contractions and they are localized largely in the lateral periphery of the myotome in a wedge, which is centered on the horizontal myoseptum. On the other hand, fast muscle fibers generate high-force, short duration contractions and are located in the deep part of the myotome. Located between the slow and fast muscle fibers there are intermediate fibers which produce both high-force and long duration contractions (van Raamsdonk et al. 1978; Devoto et al. 1996). A timeline of the key events that occur during the development and innervation of the skeletal muscle in zebrafish is shown in Figure 3, starting with the localization of the slow and fast muscle cell precursors in the marginal zone of the blastoderm during the gastrula period at ∼6 hpf and ending with the onset of swimming at 26–27 hpf.

The slow muscle cells develop from adaxial cells, a population of muscle precursor cells that lie adjacent to the notochord before segmentation begins (Devoto et al. 1996; Barresi et al. 2001). The adaxial cells are formed from the paraxial mesoderm when the notochord precursors secrete signaling molecules such as Hedgehog proteins (Currie and Ingham 1996; Blagden et al. 1997; Stickney et al. 2000). Shortly after somite formation the adaxial cells, which are initially cuboidal in shape, elongate to extend across the anterior-posterior axis of the somite and then the majority migrate medially to laterally through the somite to form a superficial monolayer of approximately 20 cells at the

Figure 2. (Continued) (Bi) A representative example of the spatial pattern of Ca^{2+} transients generated in the trunk of an apoaequorin expressing transgenic embryo at ∼18.5 hpf. This image represents 10 seconds of accumulated light superimposed on to the appropriate bright-field image. Color scale indicates luminescent flux in photons/pixel. Scale bar is 200 μm. (Bii–Biv) Representative examples of Ca^{2+} signals generated in the trunk of embryos at (Bii) ∼17.5 hpf (∼17-somite stage), (Biii) ∼18.5 hpf (∼18-somite stage), and (Biv) ∼24 hpf (∼ prim-5 stage) as visualized by confocal microscopy using calcium green-1 dextran. Red arrowheads in panel Bii indicate elevated Ca^{2+} in the nuclei. The color scale represents the level of $[Ca^{2+}]_i$, where red indicates a high level and blue indicates a low level. Ant. and Pos. are anterior and posterior, respectively. S7, S8, S9, and S10 are somites 7–10. Scale bars are 50 μm. Panels Bii'–Biv' are schematics to show the location in the trunk where the single confocal sections were acquired. (C) Developmental expression of functional IP_3Rs in the anterior somites (AS) from Xenopus embryos at (Ci–Ciii) stage 26 and (Civ–Cvi) stage 33. Panels Ci and Civ show fluo-3 fluorescence prior to IP_3 uncaging; panels Cii and Cv show the Ca^{2+} elevations induced after illumination with UV light for 5 seconds to uncage the IP_3; and panels Ciii and Cvi show the relative difference in fluorescence pre- and post-IP_3 photolysis in inverted format. Red arrowheads in panel Cii indicate elevated Ca^{2+} in the nuclei. Scale bar is 250 μm. Panels Cvii and Cviii show schematics of intact Xenopus embryos at stage 26 and 33, respectively. Ant. and Pos. are anterior and posterior, respectively. (A, Reprinted, with permission, from Jaimovich et al. [2000], American Physiological Society; C, Reprinted, with permission, from Ferrari and Spitzer 1999, Elsevier.)

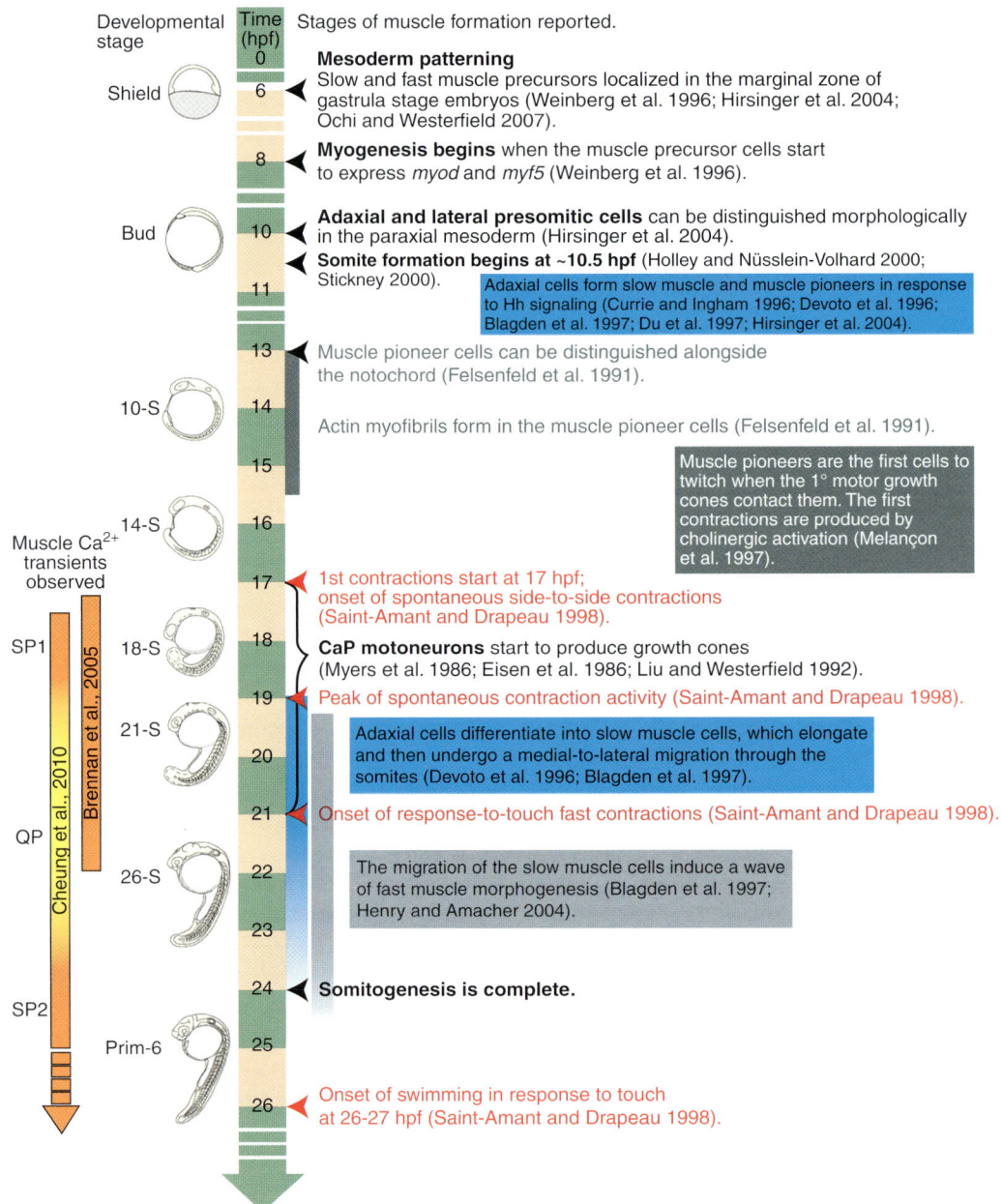

Figure 3. A timeline of muscle development in zebrafish. This illustrates the main events that have been reported to occur during mesodermal patterning, somitogenesis, myogenesis and the onset of muscle function from ~5 hpf to ~27 hpf. Included with this developmental timeline are the patterns of Ca^{2+} signals observed in the forming trunk muscles. Ca^{2+} signaling periods 1 and 2 (SP1 and SP2) and the Ca^{2+} signaling quiet period (QP) observed in slow muscle cells during development (Cheung et al. 2010) are labeled.

periphery of the myotome (Devoto et al. 1996; Du et al. 1997). Devoto et al. (1996) showed that in the rostral trunk this process takes approximately 5 hours (i.e., from ∼17 hpf to ∼23 hpf or the 16– to 28–somite stages), during which time the adaxial cells differentiate into slow muscle cells. A subgroup of adaxial cells do not migrate but remain in a medial location within the somite. These are called muscle pioneer cells and they form the horizontal myoseptum, which separates the dorsal and ventral regions of the myotome (Halpern et al. 1993). The other, nonadaxial, cells in the myotome (called the lateral presomitic cells) also do not migrate and these differentiate into the fast muscle cells via a process induced by the medial-to-lateral migration of the slow muscle cells through the somite (Devoto et al. 1996; Wolff et al. 2003; Henry and Amacher 2004).

The first spontaneous contractions occur in the trunk muscle at ∼17 hpf (i.e., 16-somite stage) (Liu and Westerfield 1992; Saint-Amant and Drapeau 1998; Drapeau et al. 2002), when the individual myotome blocks are innervated by axons of the first of the developing primary motoneurons, the caudal primary motoneurons (Eisen et al. 1986; Myers et al. 1986; Westerfield et al. 1986; Brennan et al. 2005). These initial spontaneous muscle contractions are then followed sequentially by touch-evoked rapid coils (starting at ∼21 hpf) and then organized swimming behavior (starting at ∼30 hpf) (Saint-Amant and Drapeau 1998).

Use of Aequorin to Image Ca^{2+} Transients in the Developing Zebrafish Myotome

Créton et al. (1998) was one of the first groups to report the visualization of Ca^{2+} signals in the developing trunk of zebrafish embryos. They loaded embryos with aequorin at the single-cell stage so that it was partitioned into all the subsequent progeny, and in this way, they were able to image Ca^{2+} signals continuously during early development. At the 18–22 somite stages (i.e., ∼18–20 hpf), Créton et al. (1998) visualized regions of high Ca^{2+} in the trunk and suggested that these signals might arise from the early muscle movements of the embryos. Webb and Miller (2003) also used aequorin imaging and described Ca^{2+} signals in the developing trunk of embryos at the 18-somite stage (∼18 hpf). In this report, by using a relatively low luminescence accumulation time (i.e., just 60 s), the dynamic nature of the trunk Ca^{2+} signals was clearly shown.

As aequorin is largely used up by approximately 24 hours when it is injected into embryos at the single-cell stage, the level of luminescence generated during the latter stages of muscle differentiation tends to be relatively low, and continues to fall to barely detectable levels. However, by injecting apoaequorin mRNA into embryos at the single-cell stage to express the protein component of aequorin transiently in embryos, and then reconstituting the active aequorin in vivo by incubating embryos in the coelenterazine cofactor, Cheung et al. (2006) were able to extend the aequorin-based Ca^{2+} imaging window by approximately 24 hours and they described several different types of Ca^{2+} signals in the developing trunk, for example during the Segmentation Period at ∼12 hpf (i.e., the 6-somite stage) and at ∼18–19 hpf (i.e., the 18–20-somite stage), as well as during the Pharyngula Period at ∼27–36 hpf. Cheung et al. (2006) described the first signal as being a highly localized Ca^{2+} transient with a duration of approximately 60–120 seconds, and it was suggested that it might correlate with somite formation. They described the second signal as being a relatively widespread and long lasting (i.e., ∼10 min) Ca^{2+} transient generated by the approximately 12–14 anterior-most somites but not by the approximately 6 posterior somites. They suggested that this transient might play a role in some aspect of muscle development but with the long duration, it was unlikely to be involved in E-C coupling. They described the third type of signal as being a series of rapid Ca^{2+} transients, each lasting in the order of just a few seconds and suggested that they may result from the more rapid contractions that have been reported to occur in the older embryos (Saint-Amant and Drapeau 1998).

Although the transient aequorin expression approach successfully extends the aequorin-based Ca^{2+} imaging window to ∼48 hpf, the aeq-mRNA is gradually degraded in the injected embryos, which results in a steady decline in the production of apoaequorin (Cheung et al. 2006). Furthermore, as the expression of aequorin is ubiquitous in mRNA-injected embryos, it is difficult to identify the groups of cells, tissues or organ anlage that are generating a particular signal in the more complex, later stage embryos (Cheung et al. 2006). Thus, most recently, we developed a transgenic line of zebrafish, in which a muscle-specific α-actin promoter was used to target the apoaequorin specifically to the musculature (Cheung et al. 2010). Imaging intact embryos revealed two distinct periods of long-range Ca^{2+} signaling in the developing trunk musculature: between ∼17.5 and 19.5 hpf and after ∼23 hpf. Figure 2B shows a representative example of a Ca^{2+} signal generated at ∼18.5 hpf, i.e., in the first Ca^{2+} signaling period. These signals were only generated between somites 2 and 12, i.e., they were never observed to arise in the posterior 6 somites. This is somewhat similar to the Ca^{2+} signals generated at 18–19 hpf that were reported for embryos in which apoaequorin was expressed transiently (Cheung et al. 2006).

Ashworth and Brennan (2005) presented preliminary data from intact zebrafish embryos describing the injection at the 16-cell stage of an expression plasmid encoding GFP-aequorin under the control of the immediate early cytomegalovirus (CMV) promoter. This was followed by incubation with coelenterazine for an hour at the end of the Gastrula Period to reconstitute active aequorin. At the 28-somite stage (∼24 hpf), using a PMT they were then able to record some Ca^{2+} activity and identify the expression of GFP-aequorin in some muscle cells via subsequent fluorescent microscopy.

The Use of Fluorescent Ca^{2+} Reporters to Image Ca^{2+} Transients in the Developing Myotome of Zebrafish

Fluorescent Ca^{2+} reporters have also been used to visualize Ca^{2+} transients in the developing musculature of zebrafish embryos. Ashworth et al. (2001) visualized Ca^{2+} transients in spontaneously contracting muscle cells within intact embryos at 22 hpf using the fluorescent Ca^{2+} reporter, Oregon green 488 BAPTA dextran. They also showed that in the presence of the Ca^{2+} buffer, BAPTA dextran, the amplitude of the Ca^{2+} transients was significantly reduced but not completely inhibited during muscle contraction. Ashworth (2004) also showed the changes in intracellular Ca^{2+} during the contraction of a single muscle fiber within an intact zebrafish embryo, this time at 24 hpf. In this report, cells were loaded with both Oregon green 488 BAPTA dextran and rhodamine dextran, with the latter reporter being insensitive to changes in Ca^{2+} but revealing changes in fluorescence that occur because of other factors, for example, because of cell movement during muscle contraction. This methodology resulted in a more accurate measure of the Ca^{2+} changes that take place during muscle cell contraction because of the elimination of fluorescence artifacts (Ashworth 2004).

Most recently, Brennan et al. (2005) injected embryos with either Oregon green 488 BAPTA dextran, which is considered to be a high-affinity Ca^{2+} indicator, or Fluo-4 dextran, a low affinity Ca^{2+} reporter, in conjunction with tetramethylrhodamine dextran, a Ca^{2+} insensitive fluorescent reporter similar to rhodamine dextran. They then recorded Ca^{2+} signals in the developing slow muscle cells at a time when the first neuromuscular contacts are known to form (i.e., from 16–22 hpf). Brennan et al. (2005) reported that the first Ca^{2+} signals were detected in the Oregon green 488 BAPTA dextran-loaded embryos at ∼17.25 hpf and these corresponded to the first contractions of these muscle cells in the embryo. Ca^{2+} transients were also detected in the Fluo-4 dextran-labeled muscle cells but these transients were not observed until 18–19 hpf. Brennan et al. (2005) also characterized the developmental distribution of RyRs in the somites and myotome using in situ hybridization and immunohistochemistry. They showed that at the 10-somite stage, the RyR gene is expressed in the muscle pioneer cells and adaxial cells of the newly formed somites. By 20 hpf to 24 hpf, bands of RyR protein were clearly observed in the developing slow muscle cells. Brennan et al. (2005)

treated embryos at ~20–20.5 hpf with the L-type Ca^{2+} channel blocker, nifedipine, and with the RyR blockers, dantrolene and ryanodine, and showed that all three drugs resulted in a decrease in the number of contracting embryos. They also showed that these spontaneous contractions were inhibited, and Ca^{2+} transients were blocked, by treatment with the acetylcholine receptor inhibitor, α-bungarotoxin. In addition, ryanodine and α-bungarotoxin treatment also disrupted the organization of the myofibrils in the slow muscle cells. Brennan et al. (2005) thus proposed a critical role for nerve-mediated Ca^{2+} signals in the formation of physiologically functional slow muscle during zebrafish development.

Representative examples of the Ca^{2+} signals visualized in the developing slow muscle cells of intact zebrafish embryos are shown in Figure 2Bii–2Biv (modified from Cheung et al. 2010). These embryos were loaded at the one-cell stage with calcium green-1 dextran and then visualized by confocal microscopy at ~17.5 hpf, ~18.5 hpf, and ~24 hpf. At ~17.5 hpf and ~18.5 hpf the Ca^{2+} signals appeared to be generated predominantly in the nucleus and peri-nuclear region of each individual muscle cell, with less signal being generated in the cytoplasm (Fig. 2Bii, 2Biii). At ~24 hpf, however, the Ca^{2+} signals appear to be largely cytosolic (Fig. 2Biv).

Ca^{2+} TRANSIENTS VISUALIZED IN CULTURED *XENOPUS* MYOCYTES AND IN EXPOSED MYOTOME

Ferrari et al. (1996) reported that spontaneous Ca^{2+} signals are generated during the first 15 hours in culture in primary cultures of *Xenopus* embryonic myocytes. They loaded young cultures (i.e., within 4–6 h of culture) with the AM ester forms of fura-2 and fluo-3 to measure resting Ca^{2+} and Ca^{2+} transients, respectively. In mixed myocyte/neuron cultures, young myocytes (i.e., at 0–4 h in culture) were shown to generate regular Ca^{2+} transients at a frequency of 7–10 per hour for approximately the first 2 h, after which transients were produced less regularly at a rate of approximately 6 per hour.

In addition, myocytes only generated spontaneous transient elevations of Ca^{2+} for the first 15 hours of culture. Furthermore, cultures containing both myocytes and neurons had more transient-generating cells than myocyte cultures cultured without neurons. The Ca^{2+} signals were shown to be generated by Ca^{2+} release from intracellular stores via RyRs as treatment with caffeine, a RyR agonist, was shown to stimulate the generation of Ca^{2+} transients. In addition, transients were abolished following treatment with an inhibitory concentration (100 μM) of ryanodine. Furthermore, application of the Ca^{2+} chelator, BAPTA-AM significantly disrupted myofibril organization and sarcomere assembly. Ferrari et al. (1998) subsequently confirmed that the release of Ca^{2+} from RyR stores is necessary for myofibillogenesis as treatment with 100 μM ryanodine also disrupted the formation of the myofibrils and reduced the number of sarcomeres by disrupting the assembly of the myosin thick filament. Ferrari et al. (1998) showed that the inhibition of Ca^{2+}/calmodulin-dependent myosin light chain kinase (MLCK) resulted in a disruption of sarcomeric organization, and they suggested that RyRs and MLCK might play key roles in the formation of the contractile apparatus during myocyte development. It was subsequently shown that the spontaneous Ca^{2+} signals generated in cultured *Xenopus* myocytes also play a role in the organization of titin and capZ during the assembly of the actin thin filaments at early stages of sarcomere assembly (Li et al. 2004; Harris et al. 2005) but are not involved in the association between titin and myosin (Harris et al. 2005). Li et al. (2004) also proposed a model highlighting just some of the steps in sarcomere assembly that might be regulated by Ca^{2+}. For thin filament assembly, these included Ca^{2+} activation of titin kinase in the formation of the Z-discs, of troponin/tropomyosin in thin filament capping, and of MLCK for thick filament assembly.

In addition to their visualization of Ca^{2+} transients in differentiating myocytes in culture, Ferrari and Spitzer (1999) also showed that similar Ca^{2+} dynamics occur in embryos in which the ectoderm and dermatome had been

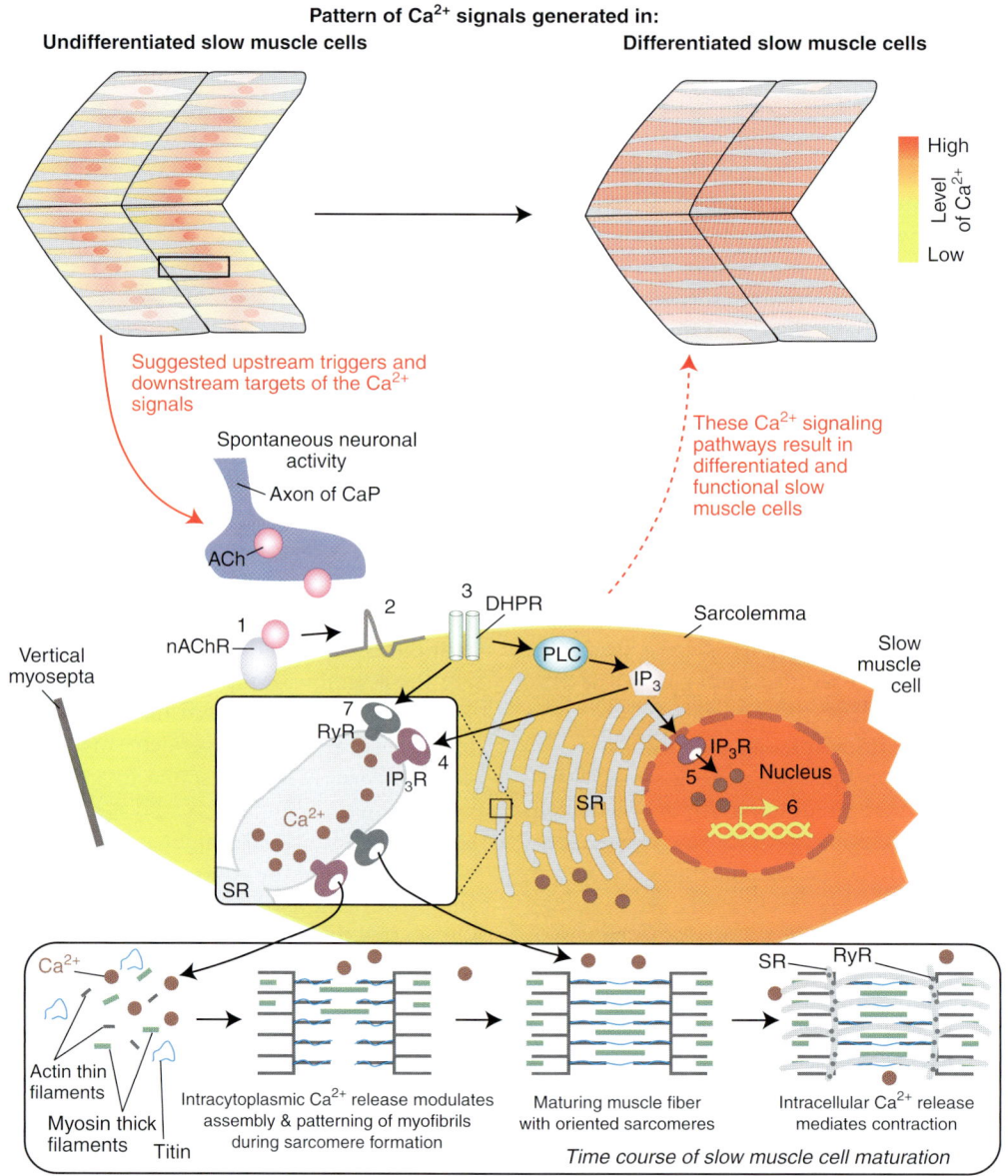

Figure 4. Schematic diagram of the proposed signaling cascade for the generation of the 17–18 hpf Ca^{2+} signals in slow muscle cells during myofibrillogenesis in zebrafish embryos. The time when the myotome Ca^{2+} transients are first observed (i.e., at ∼17.5 hpf) corresponds to the time when the myotomes are first innervated by the caudal primary (CaP) axons (Eisen et al. 1986; Myers et al. 1986; Liu and Westerfield 1992). (*1*) Acetylcholine (ACh), released from the axons, binds to nicotinic acetylcholine receptors (nAChRs) on the surface of the slow muscle cells (SMCs), which (*2*) induces depolarization of the sarcolemma (Behra et al. 2002; Brennan et al. 2005). This depolarization activates dihydropyridine receptors (DHPRs) (*3*), that in turn stimulate the release of Ca^{2+} (see red circles) from intracellular stores. The Ca^{2+} signals are generated by Ca^{2+} release via both IP$_3$Rs and RyRs (Ferrari and Spitzer 1999; Jaimovich et al. 2000). It has been proposed that DHPRs can activate the production of IP$_3$ from phospholipase C (PLC) and then induce Ca^{2+} release via IP$_3$Rs from the sarcoplasmic reticulum (*4*) and the perinuclear cisternae (*5*) (Jaimovich et al. 2000; Powell et al. 2001; Jaimovich and Carrasco 2002; Araya et al. 2003). (*See facing page for legend.*)

removed on one side to expose the somites. The dissected embryos, which were at stage 23/24 (i.e., with 12–15 formed somites) were loaded with fluo-3 AM prior to time-lapse confocal imaging. Three main types of spontaneous Ca^{2+} signals were observed: those arising in single cells, those arising in groups of cells and those propagating as waves in the ventrolateral region of the myotome. They reported that these correlated with both somite formation and with myocyte maturation (Ferrari and Spitzer 1999). They also showed the functional distribution of both RyR and IP_3R stores in intact myotome by inducing increases in Ca^{2+} in response to incubation with caffeine and photoactivation of caged IP_3, respectively. An example of the Ca^{2+} signals elicited in the myotome of stage 26 and stage 33 embryos following photoactivation of caged IP_3 is shown in Figure 2C. At stage 26, functional IP_3 receptors were expressed in the muscle cells of anterior somites (AS), with the most prominent expression occurring in the nuclei. By stage 33, however, these more mature muscle cells no longer responded to the generation of IP_3 although Ca^{2+} elevations were observed in nonmuscle cells located along the somite borders. It was suggested that although immature myocytes have both RyR- and IP_3R-activated Ca^{2+} stores, there is a down-regulation of the IP_3Rs as these cells differentiate. This down-regulation of IP_3R activity and up-regulation of RyR activity as skeletal muscle cells mature has also been reported from intact zebrafish embryos (Cheung et al. 2010). A somewhat similar situation has been reported during the development of cardiac myocytes, where it has been shown that IP_3R expression occurs before the appearance of RyRs in early embryos (Kockskämper et al. 2008). In mature cardiac myocytes on the other hand, IP_3Rs are expressed at approximately 100-fold lower levels than RyRs. This suggests a similar switch in importance of the Ca^{2+} release mechanism as the cardiac myocytes mature (Llewelyn Roderick and Bootman 2007). In addition, it has been suggested that Ca^{2+} release via IP_3Rs may initiate Ca^{2+}-induced Ca^{2+} release from the RyR-activated store, as the generation of propagating waves of Ca^{2+} in the *Xenopus* myotome ceases coincidently with the down-regulation of IP_3R expression (Ferrari and Spitzer 1999).

Most recently, a new class of spontaneous Ca^{2+} transients has been described in cultured *Xenopus* myocytes in addition to those previously described (Ferrari et al. 1996; 1998; Ferrari and Spitzer 1999). Using high-speed confocal imaging in line scanning mode, Campbell et al. (2006) described a series of short duration transients, or SDTs (i.e., with a mean duration of ~ 2 s), which propagate in the subsarcolemmal space and are generated from RyR stores located in this restricted cytoplasmic domain. Campbell et al. (2006) thus called the previously described Ca^{2+} transients, which had a duration of approximately 80-second, long duration transients, or LDTs. Although other Ca^{2+} transients with short durations have been previously reported, i.e., Ca^{2+} sparks and fast localized transients (FLTs) generated in embryonic mouse skeletal muscle (Conklin et al. 1999a; Chun et al. 2003) and frog skeletal muscle fibers (Hollingworth et al. 2001; Lacampagne et al. 2003), it was proposed that SDTs represent a separate, new class of Ca^{2+} transients (Campbell et al. 2006). In addition, although SDTs arise before myofibrillogenesis

Figure 4. (*Continued*) Moreover, it has been shown that elevations of nuclear Ca^{2+} can activate gene expression in skeletal muscle cells (6) (Carrasco et al. 2003; Cárdenas et al. 2005). In addition, it has been proposed that direct coupling of DHPRs with juxtaposed RyRs, which are localized along the interface of the sarcoplasmic reticulum (SR) and T-tubule, at the terminal cisternae of the SR (Ward and Lederer 2005), can trigger Ca^{2+} release through RyRs (7) (Schneider and Chandler 1973; Flucher et al. 1999). Ca^{2+} released via RyRs is reported to play a role in various aspects of sarcomere assembly as well as regulating muscle contraction (Li et al. 2004; Brennan et al. 2005; Harris et al. 2005).

and persist during the time of sarcomere assembly (i.e., similar to the Ca^{2+} sparks and FLTs), they do not appear to be necessary for the assembly process. On the other hand, LDTs are generated throughout the whole cytosol of myocytes rather than being restricted to the subsarcolemmal space. Like SDTs, LDTs arise prior to myofibrillogenesis but they stop before the onset of sarcomere assembly. Furthermore, Campbell et al. (2006) showed that sarcomere assembly could be blocked by extending the period of LDT generation via treatment with caffeine and they suggested that LDTs might play a role in activating sarcoplasmic regulatory pathways and the regulation of gene expression.

CONCLUSIONS

From processes as diverse as developmental gene activation (Wu et al. 2000; McKinsey et al. 2000; Berridge et al. 2003; Carrasco et al. 2003; Kubis et al. 2003; and reviewed by Berchtold et al. 2000; Al-Shanti and Stewart 2009) to sarcomeric protein assembly (Ferrari and Spitzer 1998, 1999; Li et al. 2004; Brennan et al. 2005; Harris et al. 2005; Campbell et al. 2006; Fujita et al. 2007; and reviewed by Ferrari et al. 2006), Ca^{2+} transients have been correlated with specific events and processes during muscle development and thus assigned numerous possible regulatory roles (Fig. 4). It is encouraging that many aspects reported from imaging isolated muscle cells have also been observed in intact, or near-intact, systems. This feature has been nicely illustrated by data reported from *Xenopus*, in which Ca^{2+} signals from near-intact embryos (where the ectoderm and dermatome were removed to expose the underlying somites; Ferrari and Spitzer, 1999) closely resembled those from cultured embryonic myocytes (Ferrari et al. 1996). Furthermore, the generation of distinct subcellular Ca^{2+} domains in cultured skeletal muscle cells that are either membrane-bound or otherwise (Jaimovich et al. 2000; Cárdenas et al. 2005) has also been shown to occur in slow muscle cells of intact zebrafish embryos (Cheung et al. 2010). We suggest, therefore, that the use of both in vitro and in vivo models will continue to serve as complementary systems; in which the former is a relatively simple experimental set-up that is easy to image, whereas the latter can be used to confirm in vitro results as well as to extend the investigation of Ca^{2+} signaling in muscle cell development into the normal in situ environment. For example, a recent study using intact zebrafish embryos has shown that the attachment of slow muscle cells to the vertical myosepta is dependent on RyR-generated Ca^{2+} transients (Cheung et al. 2010). Furthermore, maintaining the in situ environment will also allow the exploration of the relationship between Ca^{2+} dynamics, endogenous biophysical forces, and the differentiation of skeletal muscle cells. One of the key challenges during the next decade will be to identify the precise upstream triggers and down-stream targets of the multitude of Ca^{2+} signaling events that occur during the differentiation of a variety of different types of muscle cells. A second challenge will be to understand how the Ca^{2+}-mediated muscle-assembly instructions are integrated into multiple signaling networks during muscle development (Ochi and Westerfield 2007): For example with IP_3 signaling pathways (Kockskämper et al. 2008; Berridge 2009); with Wnt signaling pathways (Cossu and Borello 1999; Ling et al. 2009); and with MAP kinase signaling pathways (Gotoh and Nishida 1995; Lluis et al. 2006), to name but a few. Furthermore, we suggest that temporal data, although being helpful and supportive, will not, however, tell the whole story. Four-dimensional Ca^{2+} imaging information, i.e., the three spatial dimensions plus time, will be an essential component in extending our understanding of muscle development and differentiation.

ACKNOWLEDGMENTS

We would like to thank Enrique Jaimovich, Michael B. Ferrari, and Nicholas C. Spitzer who kindly gave us permission to use their previously published work. Special thanks also to Osamu Shimomura for his generous support of aequorin-based imaging over the years. We acknowledge financial support from Hong Kong Research

Grants Council GRF awards: HKUST-6416/06M, HKUST-661707 and HKUST-662109, and from Hong Kong University Grants Committee award: SEG_HKUST01.

REFERENCES

Al-Shanti N, Stewart CE. 2009. Ca^{2+}/calmodulin-dependent transcriptional pathways: potential mediators of skeletal muscle growth and development. *Biol Rev* **84:** 637–652.

Araya R, Liberona JL, Cárdenas JC, Riveros N, Estrada M, Powell JA, Carrasco MA, Jaimovich E. 2003. Dihydropyridine receptors as voltage sensors for a depolarization-evoked IP_3R-mediated, slow calcium signal in skeletal muscle cells. *J Gen Physiol* **121:** 3–16.

Ashley CC. 1969. Aequorin-monitored calcium transients in single *Maia* muscle fibres. *J Physiol* **203:** 32–33.

Ashley CC, Ridgeway EB. 1970. On the relationships between membrane potential, calcium transient and tension in single barnacle muscle fibers. *J Physiol* **209:** 105–130.

Ashworth R. 2004. Approaches to measuring calcium in zebrafish: Focus on neuronal development. *Cell Calcium* **35:** 393–402.

Ashworth R, Brennan C. 2005. Use of transgenic zebrafish reporter lines to study calcium signaling in development. *Brief Funct Gen Proteo* **4:** 186–194.

Ashworth R, Zimprich F, Bolsover SR. 2001. Buffering intracellular calcium disrupts motoneuron development in intact zebrafish embryos. *Dev Brain Res* **129:** 169–179.

Baird GS, Zacharias DA, Tsien RY. 1999. Circular permutation and receptor insertion within green fluorescent proteins. *Proc Natl Acad Sci USA* **96:** 11241–11246.

Barresi MJ, D'Angelo JA, Hernández LP, Devoto SH. 2001. Distinct mechanisms regulate slow-muscle development. *Curr Biol* **11:** 1432–1438.

Behra M, Cousin X, Bertrand C, Vonesch JL, Biellmann D, Chatonnet A, Strähle U. 2002. Acetylcholinesterase is required for neuronal and muscular development in the zebrafish embryo. *Nature Neurosci* **5:** 111–118.

Berchtold MW, Brinkmeier H, Müntener M. 2000. Calcium ion in skeletal muscle: Its crucial role for muscle function, plasticity, and disease. *Physiol Rev* **80:** 1215–1265.

Berridge MJ. 2009. Inositol trisphosphate and calcium signaling mechanisms. *Biochim Biophy Acta* **1793:** 933–940.

Berridge MJ, Bootman MD, Llewelyn Roderick H. 2003. Calcium signaling: dynamics, homeostasis, and remodeling. *Nature Rev Mol Cell Biol* **4:** 517–529.

Blagden CS, Currie PD, Ingham PW, Hughes SM. 1997. Notochord induction of zebrafish slow muscle mediated by Sonic hedgehog. *Genes Dev* **11:** 2163–2175.

Brennan C, Mangoli M, Dyer CEF, Ashworth R. 2005. Acetylcholine and calcium signaling regulates muscle fibre formation in the zebrafish embryo. *J Cell Sci* **118:** 5181–5190.

Campbell NR, Podugu SP, Ferrari MB. 2006. Spatiotemporal characterization of short versus long duration calcium transients in embryonic muscle and their role in myofibrillogenesis. *Dev Biol* **292:** 253–264.

Cárdenas C, Liberona JL, Molgó J, Colasante C, Mignery GA, Jaimovich E. 2005. Nuclear inositol 1,4,5-trisphosphate receptors regulate local Ca^{2+} transients and modulate cAMP response element binding protein phosphorylation. *J Cell Sci* **118:** 3131–3140.

Carrasco MA, Riveros N, Ríos J, Müller M, Torres F, Pineda J, Lantadilla S, Jaimovich E. 2003. Depolarization-induced slow calcium transients activate early genes in skeletal muscle cells. *Am J Physiol* **284:** C1438–C1447.

Cheng H, Lederer WJ, Cannell MB. 1993. Calcium sparks: elementary events underlying excitation-contraction coupling in heart muscle. *Science* **262:** 740–744.

Cheung CY, Webb SE, Love DR, Miller AL. 2010. Visualization, characterization, and modulation of Ca^{2+} signaling during the development of slow muscle cells in intact zebrafish embryos. *Int J Dev Biol* doi: 10.1387/ijdb.103160cc.

Cheung CY, Webb SE, Meng A, Miller AL. 2006. Transient expression of apoaequorin in zebrafish embryos: extending the ability to image calcium transients during later stages of development. *Int J Dev Biol* **50:** 561–569.

Chun LG, Ward CW, Schneider MF. 2003. Ca^{2+} sparks are initiated by Ca^{2+} entry in embryonic mouse skeletal muscle and decrease in frequency postnatally. *Am J Physiol* **285:** C686–C697.

Conklin MW, Barone V, Sorrentino V, Coronado R. 1999a. Contribution of ryanodine receptor type 3 to Ca^{2+} sparks in embryonic mouse skeletal muscle. *Biophys J* **77:** 1394–1403.

Conklin MW, Powers P, Gregg RG, Coronado R. 1999b. Ca^{2+} sparks in embryonic mouse skeletal muscle selectively deficient in dihydropyridine receptor α_{1s} or β_{1a} subunits. *Biophys J* **76:** 657–669.

Cossu G, Borello U. 1999. Wnt signaling and the activation of myogenesis in mammals. *EMBO J* **18:** 6867–6872.

Créton R, Speksnijder JE, Jaffe LF. 1998. Patterns of free calcium in zebrafish embryos. *J Cell Sci* **111:** 1613–1622.

Currie PD, Ingham PW. 1996. Induction of a specific muscle cell type by a hedgehog-like protein in zebrafish. *Nature* **382:** 452–455.

Devoto SH, Melancon E, Eisen JS, Westerfield M. 1996. Identification of separate slow and fast muscle precursor cells in vivo, prior to somite formation. *Development* **122:** 3371–3780.

Drapeau P, Saint-Amant L, Buss RR, Chong M, McDearmid JR, Brustein E. 2002. Development of the locomotor network in zebrafish. *Prog Neurobiol* **68:** 85–111.

Du SJ, Devoto SH, Westerfield M, Moon RT. 1997. *hedgehog* and *TGF-β* gene families. *J Cell Biol* **139:** 145–156.

Eisen JS, Myers PZ, Westerfield M. 1986. Pathway selection by growth cones of identified motoneurones in live zebrafish embryos. *Nature* **320:** 269–271.

Felsenfeld AL, Curry M, Kimmel CB. 1991. The *fub-1* mutation blocks initial myofibril formation in zebrafish muscle pioneer cells. *Dev Biol* **148:** 23–30.

Ferrari MB, Spitzer NC. 1999. Calcium signaling in the developing *Xenopus* myotome. *Dev Biol* **213:** 269–282.

Ferrari MB, Podugu S, Eskew JD. 2006. Assembling the myofibril: coordinating contractile cable construction with calcium. *Cell Biochem Biophys* **45:** 317–337.

Ferrari MB, Rohrbough J, Spitzer NC. 1996. Spontaneous calcium transients regulate myofibrillogenesis in embryonic *Xenopus* myocytes. *Dev Biol* **178:** 484–497.

Ferrari MB, Ribbeck K, Hagler DJ, Spitzer NC. 1998. A calcium signaling cascade essential for myosin thick filament assembly in *Xenopus* myocytes. *J Cell Biol* **141:** 1349–1356.

Fessenden JD, Wang Y, Moore RA, Chen SRW, Allen PD, Pessah IN. 2000. Divergent functional properties of ryanodine receptor types 1 and 3 expressed in a myogenic cell line. *Biophys J* **79:** 2509–2525.

Flucher BE, Andrews SB. 1993. Characterization of spontaneous and action potential-induced calcium transients in developing myotubes in vitro. *Cell Motil Cytoskel* **25:** 143–157.

Flucher BE, Conti A, Takeshima H, Sorrentino V. 1999. Type 3 and type 1 ryanodine receptors are localized in triads of the same mammalian skeletal muscle fibers. *J Cell Biol* **146:** 621–629.

Fujita H. Nedachi T, Kanzaki M. 2007. Accelerated de novo sarcomere assembly by electric pulse stimulation in C2C12 myotubes. *Exp Cell Res* **313:** 1853–1865.

Gotoh Y, Nishida E. 1995. Activation mechanism and function of the MAP kinase cascade. *Mol Reprod Dev* **42:** 486–493.

Halpern ME, Ho RK, Walker C, Kimmel CB. 1993. Induction of muscle pioneers and floor plate is distinguished by the zebrafish *no tail* mutation. *Cell* **75:** 99–111.

Harris BN, Li H, Terry M, Ferrari MB. 2005. Calcium transients regulate titin organization during myofibrillogenesis. *Cell Motil Cytoskel* **60:** 129–139.

Hasan MT, Friedrich RW, Euler T, Larkum ME, Giese G, Both M, Duebel J, Waters J, Bujard H, Griesbeck O, et al. 2004. Functional fluorescent Ca^{2+} indicator proteins in transgenic mice under TET control. *PLoS Biol* **2:** 763–775.

Henry CA, Amacher SL. 2004. Zebrafish slow muscle cell migration induces a wave of fast muscle morphogenesis. *Dev Cell* **7:** 917–923.

Hirsinger E, Stellabotte F, Devoto SH, Westerfield M. 2004. Hedgehog signaling is required for commitment but not initial induction of slow muscle precursors. *Dev Biol* **275:** 143–157.

Holley SA, Nüsslein-Volhard C. 2000. Somitogenesis in zebrafish. *Curr Top Dev Biol* **47:** 247–277.

Hollingworth S, Peet J, Chandler WK, Baylor SM. 2001. Calcium sparks in intact skeletal muscle fibers of the frog. *J Gen Physiol* **118:** 653–678.

Inouye S, Noguchi M, Sakaki Y, Takagi Y, Miyata T, Iwanaga S, Miyata T, Tsuji FI. 1985. Cloning and sequence analysis of cDNA for the luminescent protein aequorin. *Proc Natl Acad Sci USA* **82:** 3154–3158.

Jaimovich E, Carrasco MA. 2002. IP3 dependent Ca^{2+} signals in muscle cells are involved in regulation of gene expression. *Biol Res* **35:** 195–202.

Jaimovich E, Reyes R, Liberona JL, Powell JA. 2000. IP3 receptors, IP3 transients, and nucleus-associated Ca^{2+} signals in cultured skeletal muscle. *Am J Physiol Cell Physiol* **278:** C998–C1010.

Kerr R, Lev-Ram V, Baird G, Vincent P, Tsien R, Schafer WR. 2000. Optical imaging of calcium transients in neurons and pharyngeal muscle of *C. elegans*. *Neuron* **26:** 583–594.

Kockskämper J, Zima AV, Llewelyn Roderick H, Pieske B, Blatter LA, Bootman MD. 2008. Emerging roles of inositol 1,4,5-trisphosphate signaling in cardiac myocytes. *J Mol Cell Cardiol* **45:** 128–147.

Kubis HP, Hanke N, Scheibe RJ, Meissner JD, Gros G. 2003. Ca^{2+} transients activate calcinurin/NFATc1 and initiate fast-to-slow transformation in a primary skeletal muscle culture. *Am J Physiol* **285:** C56–C63.

Lacampagne A, Ward CW, Klein MG, Schneider MF. 2003. Time course of individual Ca^{2+} sparks in frog skeletal muscle recorded at high time resolution. *J Gen Physiol* **118:** 653–678.

Li H, Cook JD, Terry M, Spitzer NG, Ferrari MB. 2004. Calcium transients regulate patterned actin assembly during myofibrillogenesis. *Dev Dyn* **229:** 231–242.

Ling L, Nurcombe V, Cool SM. 2009. Wnt signaling controls the fate of mesenchymal stem cells. *Gene* **433:** 1–7.

Liu DW, Westerfield M. 1992. Clustering of muscle acetylcholine receptors requires motoneurons in live embryos, but not in cell culture. *J Neurosci* **12:** 1859–1866.

Llewelyn Roderick H, Bootman MD. 2007. Pacemaking, arrhythmias, inotropy and hypertrophy: the many possible facets of IP3 signalling in cardiac myocytes. *J Physiol* **581:** 883–884.

Lluis F, Perdiguero E, Nebreda AR, Muñoz-Cánoves P. 2006. Regulation of skeletal muscle gene expression by p38 MAP kinases. *Trends Cell Biol* **16:** 36–44.

Lorenzon P, Giovannelli A, Ragozzino D, Eusebi F, Ruzzier F. 1997. Spontaneous and repetitive calcium transients in C2C12 mouse myotubes during in vivo myogenesis. *Eur J Neurosci* **9:** 800–808.

Martonosi AN. 2000. Animal electricity, Ca^{2+} and muscle contraction. A brief history of muscle research. *Acta Biochim Polon* **47:** 493–516.

McKinsey TA, Zhang CL, Olson EN. 2000. Activation of myocyte enhancer factor-2 transcription factor by calcium/calmodulin-dependent protein kinase-stimulated binding of 14-3-3 to histone deacetylase 5. *Proc Natl Acad Sci USA* **19:** 14400–14405.

Melançon E, Liu DW, Westerfield M, Eisen JS. 1997. Pathfinding by identified zebrafish motoneurons in the absence of muscle pioneers. *J Neurosci* **17:** 7796–7804.

Miller AL, Karplus E, Jaffe LF. 1994. Imaging $[Ca^{2+}]$ with aequorin using a photon imaging detector. *Meth Cell Biol* **40:** 305–338.

Miyawaki A, Llopis J, Heim R, McCaffery JM, Adams JA, Ikura M, Tsien RY. 1997. Fluorescent indicators for Ca^{2+} based on green fluorescent proteins and calmodulin. *Nature* **388:** 882–887.

Myers PZ, Eisen JS, Westerfield M. 1986. Development and axonal outgrowth of identified motoneurons in the zebrafish. *J Neurosci* **6:** 2278–2289.

Ochi H, Westerfield M. 2007. Signaling networks that regulate muscle development: Lessons from zebrafish. *Develop Growth Differ* **49:** 1–11.

Persechini A, Lynch JA, Romoser VA. 1997. Novel fluorescent indicator proteins for monitoring free intracellular Ca^{2+}. *Cell Calcium* **22:** 209–216.

Pisaniello A, Serra C, Rossi D, Vivarelli E, Sorrentino V, Molinaro M, Bouche M. 2003. The block of ryanodine receptors selectively inhibits fetal myoblast differentiation. *J Cell Sci* **116:** 1589–1597.

Powell JA, Carrasco MA, Adams DS, Drouet B, Rios J, Müller M, Estrada M, Jaimovich E. 2001. IP$_3$ receptor function and localization in myotubes: An unexplored Ca^{2+} signaling pathway in skeletal muscle. *J Cell Sci* **114:** 3673–3683.

Prasher DC, McCann RO, Longiaru M, Cormier MJ. 1987. Sequence comparisons of complementary DNAs encoding aequorin isotypes. *Biochem* **26:** 1326–1332.

Protasi F, Takekura H, Wang Y, Chen SR, Meissner G, Allen PD, Franzini-Armstrong C. 2000. RYR1and RYR3 have different roles in the assembly of calcium release units of skeletal muscle. *Biophys J* **79:** 2494–2508.

Ringer S. 1883. A further contribution regarding the influence of the different constituents of the blood on the contraction of the heart. *J Physiol (London)* **4:** 29–42.

Ringer S. 1886. Further experiments regarding the influence of small quantities of lime, potassium and other salts on muscular tissues. *J Physiol (London)* **7:** 291–308.

Ringer S, Buxton LW. 1887. Concerning the action of calcium, potassium, and sodium salts upon the eel's heart and upon skeletal muscle of the frog. *J Physiol (London)* **8:** 15–19.

Saint-Amant L, Drapeau P. 1998. Time course of the development of motor behaviors in the zebrafish embryo. *J Neurobiol* **37:** 622–632.

Schneider MF, Chandler WK. 1973. Voltage dependent charge movement of skeletal muscle: a possible step in excitation-contraction coupling. *Nature* **242:** 244–246.

Shimomura O, Johnson FH, Saiga Y. 1962. Extraction, purification and properties of aequorin, a bioluminescence protein from the luminous hydromedusan, *Aequorea J Cell Comp Physiol* **59:** 223–240.

Shimomura O, Musicki B, Kishi Y. 1988. Semi-synthetic aequorin: An improved tool for the measurement of calcium ion concentration. *Biochem J* **251:** 405–410.

Shimomura O, Musicki B, Kishi Y. 1989. Semi-synthetic aequorins with improved sensitivity to Ca^{2+} ions. *Biochem J* **261:** 913–920.

Shimomura O, Inouye S, Musicki B, Kishi Y. 1990. Recombinant aequorin and recombinant semi-synthetic aequorins: Cellular Ca^{2+} indicators. *Biochem J* **270:** 309–312.

Shimozono S, Fukano T, Kimura KD, Mori I, Kirini Y, Miyawaki A. 2004. Slow Ca^{2+} dynamics in pharyngeal muscles in *Caenorhabditis elegans* during fast pumping. *EMBO Rep* **5:** 521–526.

Shirokova N, Shirokov R, Rossi D, González A, Kirsch WG, García J, Sorrentino V, Ríos E. 1999. Spatially segregated control of Ca^{2+} release in developing skeletal muscle of mice. *J Physiol* **521:** 483–495.

Stickney HL, Barresi MJ, Devoto SH. 2000. Somite development in zebrafish. *Dev Dyn* **219:** 287–303.

Szent-Györgyi AG. 2004. Milestones in Physiology. The early history of the biochemistry of muscle contraction. *J Gen Physiol* **123:** 631–641.

Tabata S, Takemura Y, Kobayashi M, Ikeda M, Nishimura S, Sato Y, Tatsumi R, Ikeuchi Y, Iwamoto H. 2006. Mechanism of Ca^{2+} release in myoblasts derived from chicken embryos. *J Electron Micros* **55:** 265–271.

Terry M, Walker DD, Ferrari MB. 2006. Protein phosphatase activity is necessary for myofibrillogenesis. *Cell Biochem Biophys* **45:** 265–278.

Tsien RY. 1981. A non-disruptive technique for loading calcium buffers and indicators into cells. *Nature* **290:** 527–528.

Van Raamsdonk W, Pool CW, de Kronnie G. 1978. Differentiation of muscle fiber types in the teleost *Brachydanio rerio*. *Anat Embryol* **153:** 137–155.

Ward CW, Lederer WJ. 2005. Ghost sparks. *Nature Cell Biol* **7:** 457–459.

Webb SE, Miller AL. 2000. Calcium signaling during zebrafish embryonic development. *BioEssays* **22:** 113–123.

Webb SE, Miller AL. 2003. Calcium signaling during embryonic development. *Nature Rev Mol Cell Biol* **4:** 539–551.

Webb SE, Lee KW, Karplus E, Miller AL. 1997. Localized calcium transients accompany furrow positioning, propagation, and deepening during the early cleavage period of zebrafish embryos. *Dev Biol* **192:** 78–92.

Webb SE, Rogers KL, Karplus E, Miller AL. 2010. The use of aequorins to record and visualize Ca^{2+} dynamics from subcellular microdomains to whole organisms. *Meth Cell Biol* **99:** 263–300.

Weinberg ES, Allende ML, Kelly CS, Abdelhamid A, Murakami T, Andermann P, Doerre OG, Grunwald DJ, Riggleman B. 1996. Developmental regulation of zebrafish *MyoD* in wild-type, *no tail* and *spadetail* embryos. *Development* **122:** 271–280.

Westerfield M, McMurray JV, Eisen JS. 1986. Identified motoneurons and their innervation of axial muscles in the zebrafish. *J Neurosci* **6:** 2267–2277.

Wolff C, Roy S, Ingham PW. 2003. Multiple muscle cell identities induced by distinct levels and timing of hedgehog activity in the zebrafish embryo. *Curr Biol* **13:** 1169–1181.

Wu H, Naya FJ, McKinsey TA, Mercer B, Shelton JM, Chin ER, Simard AR, Michel RN, Bassel-Duby R, Olson EN, et al. 2000. MEF2 responds to multiple calcium-regulated signals in the control of skeletal muscle fiber type. *EMBO J* **19:** 1963–1973.

Zimprich F, Ashworth R, Bolsover S. 1998. Real-time measurements of calcium dynamics in neurons developing in situ within zebrafish embryos. *Pfugers Arch Eur J Physiol* **436:** 489–493.

Calcium Signaling in Synapse-to-Nucleus Communication

Anna M. Hagenston and Hilmar Bading

CellNetworks—Cluster of Excellence, Department of Neurobiology, and the Interdisciplinary Center for Neurosciences (IZN), University of Heidelberg, 69120 Heidelberg, Germany

Correspondence: Hilmar.Bading@uni-hd.de

Changes in the intracellular concentration of calcium ions in neurons are involved in neurite growth, development, and remodeling, regulation of neuronal excitability, increases and decreases in the strength of synaptic connections, and the activation of survival and programmed cell death pathways. An important aspect of the signals that trigger these processes is that they are frequently initiated in the form of glutamatergic neurotransmission within dendritic trees, while their completion involves specific changes in the patterns of genes expressed within neuronal nuclei. Accordingly, two prominent aims of research concerned with calcium signaling in neurons are determination of the mechanisms governing information conveyance between synapse and nucleus, and discovery of the rules dictating translation of specific patterns of inputs into appropriate and specific transcriptional responses. In this article, we present an overview of the avenues by which glutamatergic excitation of dendrites may be communicated to the neuronal nucleus and the primary calcium-dependent signaling pathways by which synaptic activity can invoke changes in neuronal gene expression programs.

The significance of intracellular calcium (Ca^{2+}) increases for the regulation of gene expression in neurons is well established (Bading et al. 1997; Chawla 2002; Greer and Greenberg 2008; Redmond 2008; Hardingham and Bading 2010). Activity-dependent changes in gene expression in neurons participate in a broad range of processes and behaviors, including synaptic activity-induced acquired neuroprotection (Hardingham et al. 2002; Papadia et al. 2005; Zhang et al. 2007b, 2009; Lau and Bading 2009), activity-dependent regulation of synapse number, size, and function (Flavell et al. 2006; Shalizi et al. 2006; Saneyoshi et al. 2010; Mauceri et al. 2011), modulation of dendritic complexity (Redmond et al. 2002; Mauceri et al. 2011), growth factor signaling (Gall and Lauterborn 1992; Castren et al. 1998; Greenberg et al. 2009), regulation of long-term changes in synaptic strength (Korzus et al. 2004; Limback-Stokin et al. 2004; Raymond and Redman 2006), and memory consolidation (Bailey et al. 1996; Pittenger et al. 2002; Limback-Stokin et al. 2004; Wood et al. 2005; Etkin et al. 2006; Mauceri et al. 2011). Increases in the intracellular concentration of Ca^{2+} ($[Ca^{2+}]_i$) that might trigger gene transcription as a consequence of neuronal activity are mediated by

Ca^{2+} entry through L-type voltage-dependent Ca^{2+} channels (VDCCs) and ligand-gated channels such as N-methyl-D-aspartate receptors (NMDARs) (Bading et al. 1993, 1995) and α-amino-3-hydroxyl-5-methyl-4-isoxazolepropionate receptors (AMPARs) (Jonas and Burnashev 1995; Cohen and Greenberg 2008), and by release of Ca^{2+} from the endoplasmic reticulum (ER) (Fig. 1) (Hardingham and Bading 1998; Power and Sah 2002, 2007; Stutzmann et al. 2003; Watanabe et al. 2006; Hagenston et al. 2008). Ca^{2+} transients that invade the cell nucleus appear to play a particularly important role in activity-dependent transcription (Hardingham et al. 1997, 2001b; Zhang et al. 2007b, 2009) and are critical for several long-lasting adaptive responses, including acquired neuroprotection and memory formation (Bading 2000; Limback-Stokin et al. 2004; Papadia et al. 2005; Zhang et al. 2009, 2011; Mauceri et al. 2011).

Molecular studies of Ca^{2+}-regulated genomic responses have largely focused on the interactions between synaptic excitation, the particular sources of cytoplasmic $[Ca^{2+}]_i$ increases, the stimulation of downstream identified signaling cascades, and the activation or repression of specific transcription factor targets. Prominent among the Ca^{2+} sources, Ca^{2+}-dependent second messenger cascades, and transcription factors are NMDARs and L-type VDCCs, Ca^{2+}/calmodulin-dependent protein kinase (CaMK) and Ras/mitogen activated protein kinase (Ras/MAPK) signaling cascades, and the transcription factor cyclic AMP response element binding protein (CREB) as well as the ternary complex factor (TCF)/serum response factor (SRF) transcription factor complex (Fig. 2) (Bading et al. 1993; Xia et al. 1996; Cruzalegui and Bading 2000; Greer and Greenberg 2008). Briefly summarized, synaptically released glutamate binds to and activates NMDARs and AMPARs, leading to a depolarization of the postsynaptic membrane and the activation of L-type VDCCs. Ca^{2+} influx through synapse-associated dendritic NMDARs and L-type VDCCs is sensed by calmodulin (CaM), which initiates a chain of signaling events including stimulation of CaMK and Ras/MAPK cascades. The culmination of these and other kinase cascades is the phosphorylation and activation of a variety of transcription factors, most notably CREB, but also TCF, nuclear factor κB (NF-κB), myocyte enhancer factor 2 (MEF2), and others (Cruzalegui and Bading 2000; Greer and Greenberg 2008). Several important genomic responses following neuronal activity are controlled by Ca^{2+} signals in the nucleus (Bading 2000; Zhang et al. 2009). For example, increases in nuclear $[Ca^{2+}]_i$ stimulate gene transcription indirectly through the nuclear resident Ca^{2+}/calmodulin-dependent protein kinase IV (CaMKIV) (Hardingham et al. 2001b) or directly by interaction with regulators of gene transcription such as transcription factor downstream regulatory element antagonistic modulator (DREAM) (Carrion et al. 1999; Ledo et al. 2002).

Cumulating evidence suggests that, in addition to regulating transcription via its influence on specific transcription factors, Ca^{2+}-dependent modulation of genomic responses also operates at a global level by influencing chromatin structure. Nuclear Ca^{2+} signals acting via CaMKIV, for instance, stimulate the histone acetyltransferase CREB binding protein (CBP). Activated CBP, which interacts physically with numerous transcription factors (e.g., CREB) to promote transcription, influences gene transcription by catalyzing histone acetylation and subsequent chromatin decondensation (Chawla et al. 1998; Cruzalegui and Bading 2000; Alarcon et al. 2004; Korzus et al. 2004; Bedford et al. 2010). In a similar vein, nuclear Ca^{2+} signals induce the subcellular redistribution of class II histone deacetylases (HDACs), enzymes whose activity leads to chromatin compression and diminished accessibility by several transcription factors to their target binding sequences (Chawla et al. 2003; Tian et al. 2009, 2010b). Recent results indicate that also de novo DNA methylation, a mechanism stimulating global chromatin remodeling and gene transcription, is controlled by a neuronal activity and Ca^{2+}-dependent mechanism (Chen et al. 2003; Martinowich et al. 2003; Zhou et al. 2006; Skene et al. 2010). Ca^{2+}-dependent synapse-to-nucleus communication

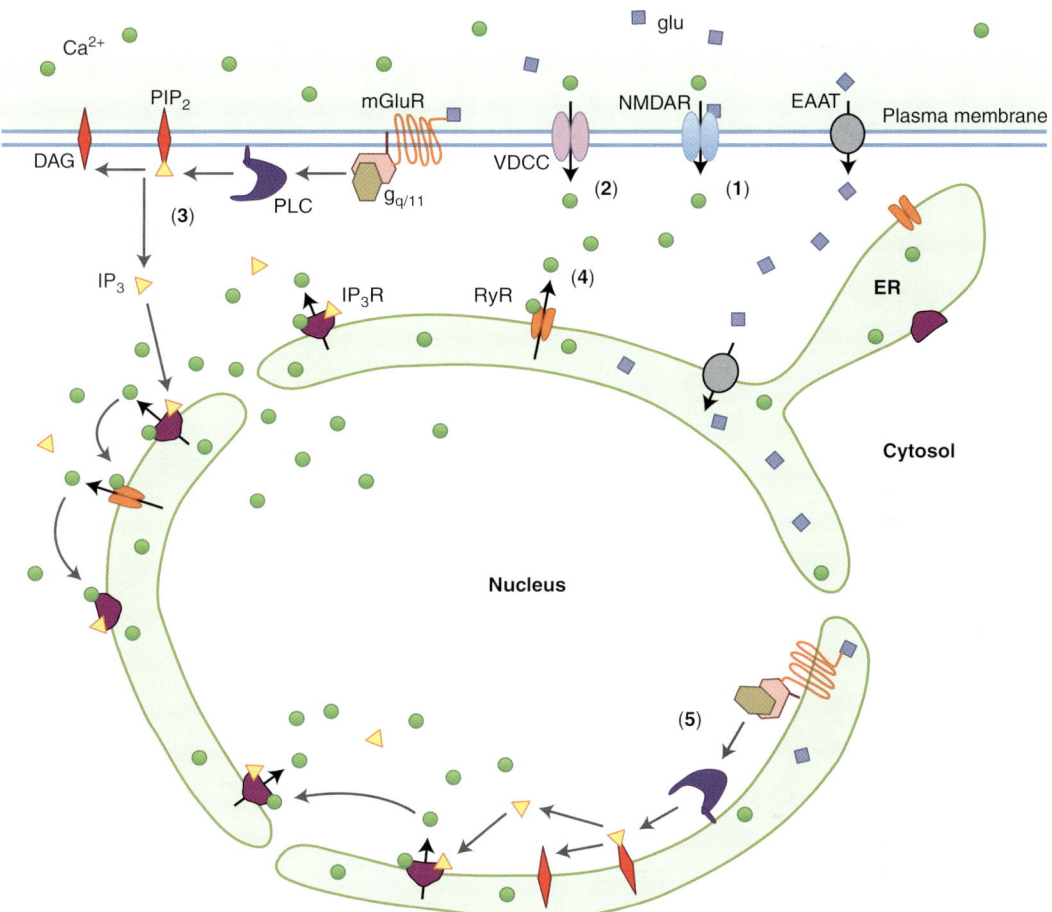

Figure 1. Sources of synaptic activity induced calcium signals. Here we consider five distinct routes by which Ca^{2+} may enter the neuronal cytoplasm. (1) Ca^{2+} may enter the cell from the extracellular space via ionotropic glutamate receptors, particularly NMDARs. (2) Ca^{2+} may also pass from the extracellular space into the cytoplasm by way of VDCCs, most notably the dihydropine-sensitive class of high voltage-activated, or L-type, VDCCs. (3) Stimulation of mGluRs can trigger release of Ca^{2+} into the cytoplasm from intracellular Ca^{2+} stores like ER: synaptically released glutamate activates mGluRs, which are coupled via $G_{q/11}$ GTP-binding proteins to PLC. Activated PLC cleaves membrane-bound PIP_2 to yield DAG and soluble IP_3, which may then diffuse to and activate IP_3Rs on the ER membrane. (4) Cytosolic Ca^{2+} signals originating from any of the ligand-gated glutamate receptors, VDCCs, or from IP_3Rs can be amplified via the Ca^{2+}-dependent activation of RyRs and subsequent internal release of Ca^{2+} from intracellular stores. (5) Synaptically released glutamate may activate mGluRs on the inner nuclear envelope subsequent to being taken up by EAATs on the plasma membrane first into the cytosol, and then by EAATs on the nuclear envelope into the nuclear lumen. Stimulation of intranuclear mGluRs may consequently lead to the release of Ca^{2+} directly into the nucleus from IP_3Rs localized on the inner nuclear envelope.

therefore involves changes both in the activity of specific transcription factor targets as well as chromatin structure-dependent changes in the accessibility of transcription factors to their respective genomic binding sites.

One of the basic functions subserved by synaptic activity and Ca^{2+}-dependent transcriptional regulation is the experience-dependent induction of persistent changes in synaptic efficacy. Long-term potentiation (LTP), an activity

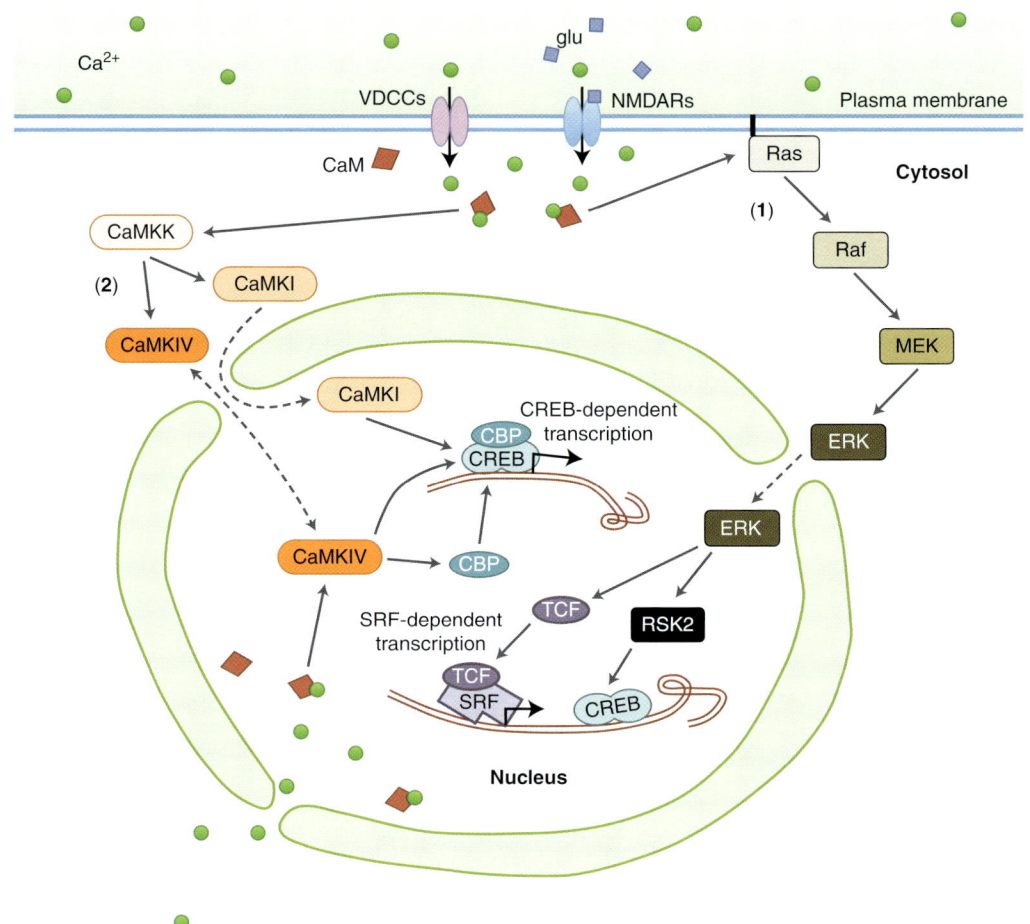

Figure 2. Activity-dependent gene expression in neurons is regulated by distinct calcium signaling pathways. Distinct activity-dependent genomic responses may be explained by the differential activation of regulatory signaling cascades and transcription factors. (1) Ca^{2+} influx through L-type VDCCs and synaptically activated NMDARs binds to and activates CaM, leading to activation of the GTP-binding protein Ras and subsequent induction of the Ras/MAPK signaling pathway. Ras binds and activates the mitogen activated protein kinase kinase kinase Raf. Raf in turn phosphoyrlates and activates MEK, which phosphoyrlates and activates ERK 1 and 2. Following its translocation into the nucleus, ERK can phosphorylate intermediate kinases like RSK2, which in turn phosphorylate CREB. ERK may also phosphorylate and activate the TCF-family transcription factors, which upon forming a complex with SRF may initiate gene transcription downstream from SRF binding sites. (2) CaM activation by Ca^{2+} entering through NMDARs and VDCCs, but especially through L-type VDCCs, also stimulates the CaMK signal transduction cascade. Ca^{2+}/CaM binds to and activates CaMKK. Next, CaMKK phosphorylates and activates cytoplasmic CaMKI, which can in turn translocate into the nucleus (Sakagami et al. 2005; Wayman et al. 2008). CaMKK also phosphorylates Ca^{2+}/CaM-bound CaMKIV. CaMKIV and perhaps also CaMKI can phosphorylate the transcription factor CREB (Wayman et al. 2008). Stimulation of CREB-mediated transcription, however, requires the additional activation by CaMKIV of the CREB coactivator CBP. It follows that VDCC and NMDAR-mediated $[Ca^{2+}]_i$ increases and subsequent activation of Ras/MAPK pathways are sufficient to induce TCF/SRF-dependent transcription. However, activation of CREB-dependent transcription necessitates the additional Ca^{2+}-dependent stimulation of CaMKIV and CBP by nuclear Ca^{2+}.

induced persistent increase in synaptic efficacy, is thought to underlie certain forms of learning and memory in the mammalian brain. A critical event leading to the induction of LTP is the postsynaptic influx of Ca^{2+} (Raymond 2007). The persistence of LTP and that of other simple forms of synaptic plasticity is dependent on gene transcription taking place within approximately two hours of the instigating event (Nguyen et al. 1994; Arnold et al. 2005). Although it is unclear exactly which genes are critical for LTP, it is well established that genomic responses following synaptic activity and Ca^{2+} entry vary according to the selection of transcription factors these signals target (Greer and Greenberg 2008). Which transcription factors are activated depends, in turn, on the source and characteristics of stimulatory $[Ca^{2+}]_i$ increases and on the selection of intracellular signals they induce (Bading et al. 1993, 1997; Dudek and Fields 2001; Hardingham et al. 2002; Greer and Greenberg 2008). Ca^{2+} influx originating at L-type VDCCs and synaptic NMDARs, for example, couples to both CaMK and Ras/MAPK signaling pathways and can thus lead to the phosphorylation of both CREB and the SRF-associated factor, TCF (Cruzalegui and Bading 2000). However, while MAPK signaling is sufficient to stimulate TCF/SRF-dependent transcription, CREB-dependent transcription requires the additional nuclear Ca^{2+}-dependent activation of CaMKIV (Fig. 2) (Hardingham et al. 1997, 2001a,b; Chawla et al. 1998).

A second role for synaptic activity-dependent transcriptional regulation is in activating survival-promoting genes that enable neurons to build up a neuroprotective shield. Activity-dependent neuronal survival is induced by Ca^{2+} entry through synaptic NMDARs and requires—for neuroprotection to be long-lasting—that Ca^{2+} transients invade the cell nucleus, suggesting a mechanism involving CREB-mediated transcription (Wang et al. 1995; Hardingham et al. 2001b; Arnold et al. 2005; Zhang et al. 2007b; Hardingham 2009; Dick and Bading 2010). Procedures that interfere with electrical activity and compromise NMDAR function or nuclear Ca^{2+} signaling can have deleterious effects on neuron health. For example, the selective blockade of nuclear Ca^{2+} signaling prevents cultured hippocampal neurons from building up antiapoptotic activity upon synaptic NMDAR stimulation (Hardingham et al. 2002; Arnold et al. 2005; Papadia et al. 2005; Hardingham 2009; Lau and Bading 2009; Dick and Bading 2010). Conversely, enhancing neuronal firing and synaptic NMDAR activity is neuroprotective: networks of cultured hippocampal neurons that have experienced periods of action potential bursting causing Ca^{2+} entry through synaptic NMDARs show enhanced resistance to cell death-inducing stimuli (Hardingham et al. 2001b; Hardingham 2009; Lau and Bading 2009; Dick and Bading 2010). On the other hand, stimulation of extrasynaptic NMDARs (ES-NMDARs) counters these effects and activates proapoptotic signaling cascades (Hardingham et al. 2002, 2009; Vanhoutte and Bading 2003; Papadia et al. 2005; Zhang et al. 2007b, 2009, 2011; Leveille et al. 2008; Stanika et al. 2009; Wahl et al. 2009). In vitro studies using cultured neurons have revealed that two mechanistically distinct processes mediate the activity-dependent survival afforded by Ca^{2+} entry through synaptic NMDARs. One involves the PI3K-AKT pathway (reviewed in Hardingham 2009). The second requires Ca^{2+} signaling to the nucleus and the activation or inhibition of a variety of different target genes (Hardingham et al. 2001b, 2009; Francis et al. 2004; Lau and Bading 2009; Dick and Bading 2010). Within the pool of nuclear Ca^{2+}-stimulated survival genes are a set of ten induced genes shown to provide neurons with a neuroprotective shield both in cell culture and in animal models of neurodegeneration (Zhang et al. 2009). Some of these genes, which were termed "Activity-Regulated Inhibitors of Death," or AID genes, may be induced by activation of the transcription factor CREB, and appear to protect neurons via a common process that yields mitochondria more resistant to cellular stressors and toxins (Zhang et al. 2007b, 2009, 2011; Lau and Bading 2009; Leveille et al. 2010). Another synaptic NMDAR-, nuclear Ca^{2+}-, and CREB-dependent prosurvival gene is brain-derived neurotrophic factor

(BDNF), a synaptic plasticity associated neurotrophic factor whose expression can both protect neurons from future insult and rescue neurons from cell death induced by synaptic activity blockade (Favaron et al. 1993; Hardingham et al. 2002; Hansen et al. 2004; Jiang et al. 2005; Zhang et al. 2009).

LOCAL Ca^{2+} SIGNALING HAS GLOBAL FUNCTIONAL CONSEQUENCES

The predominance of excitatory synaptic transmission in the central nervous system occurs at dendritic protuberances called synaptic spines. These spines are distributed throughout the dendritic arbors of hippocampal and cortical pyramidal neurons. However, the soma, the first ~100 μm of the primary apical dendrite, and the first 30–50 μm of the basal dendrites of these neurons are nearly devoid of spines (Spruston and Mcbain 2007). Indeed, the mean distance between the soma and the first synaptic spine is estimated to be 40 μm (Bannister and Larkman 1995). Moreover, the most distal spines are situated on terminal branches many 100s of μm away from the soma (Spruston and Mcbain 2007). In view of the tight control placed on cytoplasmic Ca^{2+} transients by the numerous Ca^{2+} buffers, pumps, exchangers, and other Ca^{2+}-binding proteins in neurons, these morphological features pose an interesting problem for Ca^{2+}-dependent synapse-to-nucleus communication. At least three Ca^{2+}-dependent routes have been proposed by which neurons may surmount this problem and successfully communicate synaptic excitation from spine to nucleus. Briefly, these are the diffusion and/or active transport and subsequent translocation of Ca^{2+}-regulated proteins from their synaptic sites of activation into the nucleus (Fig. 3A) (Deisseroth et al. 1998; Mermelstein et al. 2001; Otis et al. 2006; Wiegert et al. 2007; Lai et al. 2008; Jordan and Kreutz 2009; Zehorai et al. 2010); the influx of Ca^{2+} through somatic and perisomatic VDCCs during excitatory postsynaptic potential (EPSP)- and action potential-associated depolarizations (Fig. 3B) (Westenbroek et al. 1990; Murphy et al. 1991;

Bading et al. 1993, 1995; Dudek and Fields 2002; Saha and Dudek 2008); and propagating inositol trisphosphate (IP_3) receptor (IP_3R)-dependent waves of internal Ca^{2+} release that invade the somatonuclear compartment (Fig. 3C) (Hardingham et al. 2001b; Power and Sah 2002, 2007; Watanabe et al. 2006; Hagenston et al. 2008). An additional means by which synaptic activity could be conveyed to the nucleus involves the passive propagation of electrical potentials along ER membranes to the nuclear envelope (Fig. 3D) (Shemer et al. 2008).

The first step in excitatory neurotransmission is the release of glutamate into the cleft between a presynaptic terminal and the electron-dense region of its apposing synaptic spine known as the postsynaptic density. This rich structure is composed of scaffolding molecules, cell adhesion proteins, glutamate receptors, VDCCs, and a wide array of signal transducers involved in the regulation of synaptic function, gene transcription, and memory (Okabe 2007). Therefore, in addition to generating EPSPs, the postsynaptic reception of glutamate by AMPARs and NMDARs may set into motion a wealth of intracellular signaling cascades through local interactions with second messengers. A central player in this relay is Ca^{2+}, and its most prominent source in the postsynaptic density is the NMDAR (Cole et al. 1989; Xia et al. 1996; Pinato et al. 2009). NMDAR-mediated $[Ca^{2+}]_i$ increases in dendritic spines may be amplified by ryanodine receptor (RyR)-dependent Ca^{2+}-induced Ca^{2+} release from spine ER (Emptage et al. 1999; Raymond and Redman 2006). Imaging studies suggest nonetheless that synaptic NMDAR-dependent Ca^{2+} transients are essentially confined within spine heads, with little to no invasion of dendritic shafts (Yuste et al. 2000; Nakamura et al. 2002; Sabatini et al. 2002; Noguchi et al. 2005). If not via the generation of global Ca^{2+} signals, synaptic NMDARs must communicate with the nucleus via locally activated Ca^{2+}-regulated proteins and processes in the postsynaptic density. This idea is supported by studies employing exogenously applied EGTA. EGTA is a Ca^{2+} buffer that allows

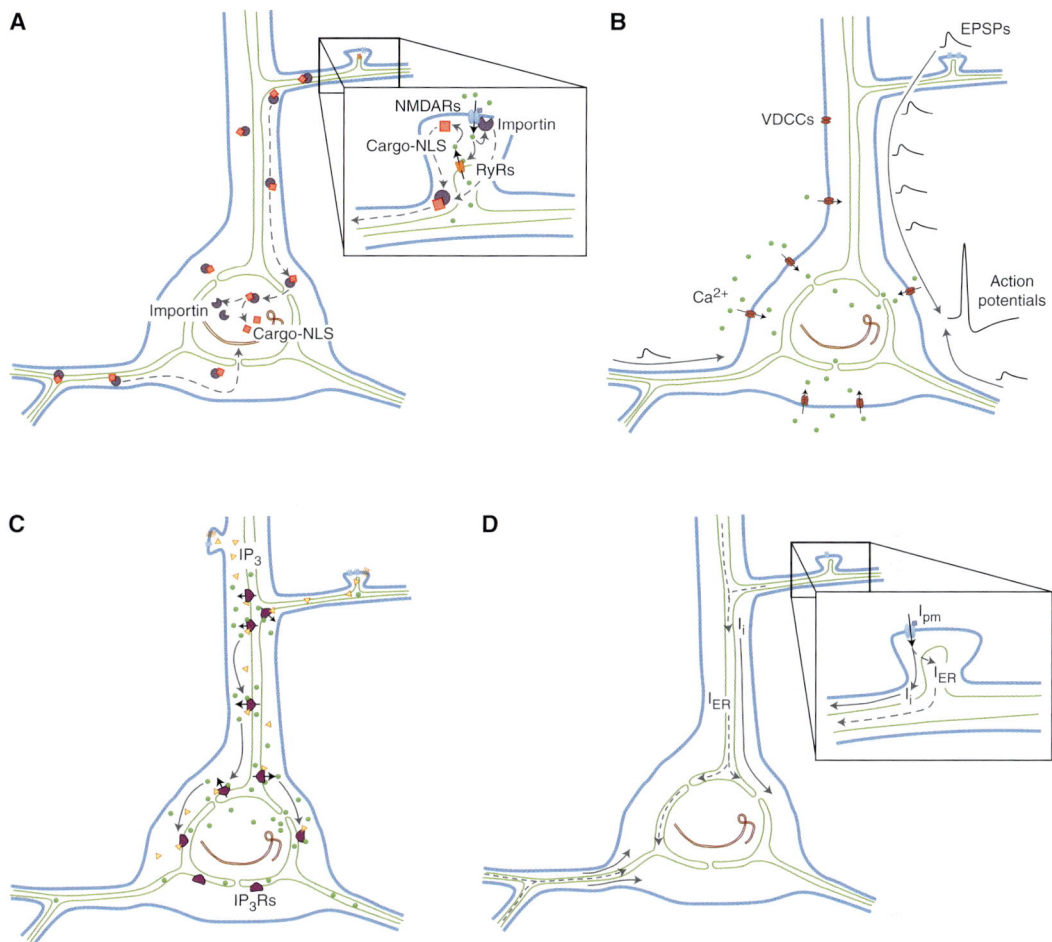

Figure 3. Possible routes of communication between synapse and nucleus. Schematic illustration of four pathways by which information impinging onto synapses may be conveyed to the nucleus. (*A*) Glutamatergic activation of synaptic AMPARs and NMDARs trigger increases in Ca^{2+} within synaptic spines. These Ca^{2+} increases may be amplified locally by Ca^{2+}-induced Ca^{2+} release via RyRs on spinous ER. $[Ca^{2+}]_i$ increases within spines trigger the activation of numerous NLS-harboring second messengers (cargo-NLS) and the mobilization of importin. Importin/cargo complexes then diffuse or are actively transported through the dendritic cytosol to invade the soma and nucleus. (*B*) Synaptic activity evokes synaptic EPSPs, which travel along the plasma membrane to the soma. When the summation of inputs exceeds firing threshold, an action potential is triggered. Membrane depolarizations accompanying EPSPs and action potentials activate somatic and perisomatic L-type VDCCs, which allow Ca^{2+} to flow down its concentration gradient from the extracellular space into the soma, from where it crosses the nuclear envelope to invade the nucleus. (*C*) Robust stimulation of glutamatergic afferents can result in the activation of perisynaptic $G_{q/11}$-protein-coupled glutamate receptors, or mGluRs, which in turn trigger a signaling cascade that results in the production of the second messenger IP_3. IP_3 subsequently binds to and activates IP_3Rs on the ER membrane, causing their Ca^{2+}-permeable channels to open and allowing Ca^{2+} to flow down its concentration gradient from the ER lumen into the cytosol. Ca^{2+}, an IP_3R coagonist, may then stimulate neighboring IP_3Rs, thus triggering a propagating wave of regenerative internal Ca^{2+} release. Robust Ca^{2+} waves that invade the soma are able to freely cross the nuclear envelope and generate robust nuclear Ca^{2+} transients. (*D*) Stimulation of ligand-operated and voltage-dependent ion channels in the plasma membrane of spines evokes local currents across the plasma membrane, I_{pm}. (*Legend continued on next page.*)

the generation of submembranous $[Ca^{2+}]_i$ increases near the mouth of Ca^{2+} influx channels, but prevents the spread of free Ca^{2+} from synaptic spines into the dendritic shaft and soma (Stern 1992; Deisseroth et al. 1996; Hardingham et al. 2001a). Introduction of EGTA into cultured hippocampal neurons thus effectively blocks global Ca^{2+} signaling, but permits the generation of cytoplasmic Ca^{2+} transients in the immediate vicinity of NMDARs. Under these conditions, synaptic activity has been observed to both induce Ras/MAPK signaling and trigger the phosphorylation of CREB (Deisseroth et al. 1996; Hardingham et al. 2001a).

MOBILE SECOND MESSENGERS IN SYNAPSE-TO-NUCLEUS COMMUNICATION

The diffusion and/or active transport of calcium-regulated proteins between synaptic NMDAR-associated Ca^{2+} signaling microdomains in the postsynaptic density and their targets in the nucleus represent attractive mechanisms for synapse-to-nucleus communication (Fig. 3A) (Deisseroth et al. 1998; Mermelstein et al. 2001; Thompson et al. 2004; Otis et al. 2006; Dieterich et al. 2008; Lai et al. 2008; Jordan and Kreutz 2009). The hypothesis that synaptic activity is relayed to the nucleus via mobile second messengers is supported by studies showing that NMDARs interact directly or indirectly with several different signaling molecules implicated in neuronal activity-dependent gene transcription, including CaM, CaMKs, protein phosphatase 1 (PP1), calcineurin, and members of the Ras/MAPK signaling cascade (Seidenbecher et al. 1998; Graef et al. 1999; Westphal et al. 1999; Zuhlke et al. 1999; Husi et al. 2000; Dolmetsch et al. 2001; Tolias et al. 2005; Wheeler et al. 2008), as well as by studies showing that synaptic activation of NMDARs and VDCCs induces the active transport and nuclear import of a broad range of synaptically localized proteins involved in transcriptional regulation, including the transcription factors NF-κB, nuclear factor of activated T-cells (NFAT), and CREB2 (Graef et al. 1999; Wellmann et al. 2001; Meffert et al. 2003; Tomida et al. 2003; Fagerlund et al. 2008; Lai et al. 2008; Jordan and Kreutz 2009). In this regard, it is particularly interesting that NMDARs have been observed to interact also in an activity-dependent fashion with importin α (Thompson et al. 2004; Jeffrey et al. 2009), a soluble transport receptor that mediates the transport and/or nuclear import of a variety of nuclear localization signal (NLS)-containing cargoes from distal dendritic locations to the nucleus (Otis et al. 2006; Jordan et al. 2007; Dieterich et al. 2008; Lai et al. 2008). More specifically, the NR1-1a NMDAR subunit and importin α were recently reported to colocalize within synaptic spines and to coimmunoprecipitate in extracts derived from hippocampal neuronal cultures and acute slices (Jeffrey et al. 2009). These interactions were found to depend on the binding of importin α to an NLS sequence in the cytoplasmic tail of NR1-1a, and to be negatively regulated by phosphorylation of nearby residues via the Ca^{2+}-dependent kinase protein kinase C. Accordingly, treatments that stimulate synaptic NMDARs and up-regulate protein kinase C activity—glutamate stimulation of neuronal cultures or high-frequency, LTP-inducing stimulation of presynaptic afferents in acute brain slices—were observed to disrupt the coimmunoprecipitation

---

Figure 3. (*Continued*) Either as a result of capacitative conduction of plasma membrane potential fluctuations combined with Ca^{2+} uptake into the ER via sarcoendoplasmic reticulum Ca^{2+} ATPases, or as a consequence of the release of Ca^{2+} via RyRs and IP$_3$Rs localized to spines and dendrites, synaptic currents are predicted to evoke an electrotonic signal across and along the ER membrane, such that synaptic currents flow simultaneously through the cytosol (I_i) and within the ER (I_{ER}) toward the soma. Summation of many such excitatory electrotonic signals can give rise to an EPSP-like depolarization across the nuclear envelope. Importantly, action potentials initiating at the soma would also result in propagating electrotonic signals along the ER. In this case, however, the resultant changes in potential across the nuclear envelope would be hyperpolarizing, thereby distinguishing these signals from others that originate distally in synaptic spines.

and colocalization of NR1-1a with importin α (Jeffrey et al. 2009). These findings suggest a mechanism by which importin α may be made available in a Ca^{2+}- and activity-dependent fashion to facilitate the transport and translocation of synaptically localized NLS-harboring proteins such as NFAT to the nucleus.

Might such a mechanism underlie synaptic activity regulated genomic responses through the Ras/MAPK signaling cascade? Until recently, there were no indications either that the MAPKs harbor an NLS enabling binding to importins or that they might interact with another NLS-containing protein (Lange et al. 2007; Zehorai et al. 2010). These findings made it difficult to clarify evidence suggesting that the nuclear translocation of extracellular signal-regulated kinase (ERK), a key player in Ras/MAPK signaling within the nucleus, was mediated by both facilitated diffusion and active transport mechanisms (Adachi et al. 1999; Xu and Massague 2004; Zehorai et al. 2010). A recent study introduced the possibility, however, that nuclear import of ERK may be controlled by its interaction with a member of the importin β family of nuclear import proteins (Chuderland et al. 2008). In this study, the authors identified a novel nuclear transport sequence in ERK's kinase insert domain that enabled ERK to bind importin β and diffuse across the nuclear envelope. ERK's interaction with importin β necessitated its prior phosphorylation by MAPK/ERK kinase (MEK) and consequent release from anchoring proteins into the cytoplasm (Chuderland et al. 2008). These findings suggest that importin-dependent transport is likely to contribute to the somatonuclear translocation of active, diffusible ERK. They do not, however, resolve the question of whether ERK translocation from distal dendritic processes into the soma is likewise facilitated by an active transport mechanism involving importins (Thompson et al. 2004; Otis et al. 2006; Lai et al. 2008). We would suggest that this is an unlikely scenario. NMDAR-dependent $[Ca^{2+}]_i$ increases activate the Ras/MAPK pathway through a Ca^{2+} pool in their immediate vicinity (Hardingham et al. 2001a), and synaptically activated Ras spreads along dendrites by diffusion, reaching distances of only ∼ 10 μm from its spine of origin before inactivating (Harvey et al. 2008). In an investigation employing exogenously expressed ERK fused to a photoactivable and photobleachable probe, it was found that synaptic activity similarly mobilizes the Ras effectors ERK 1 and 2, and that their trafficking in dendrites likewise proceeds by passive diffusion (Wiegert et al. 2007). Significantly, ERK was observed in this study to have a very limited reach within dendrites, and showed a length constant of only ∼30 μm (Wiegert et al. 2007). These findings indicate that successful signal propagation to the nucleus by the Ras/MAPK pathway most probably depends on the distance between the nucleus and the site of activated synapses. Thus, while perisomatic synaptic NMDAR activation and subsequent ERK mobilization may suffice to trigger new ERK-dependent nuclear responses, distal synaptic NMDARs are unlikely to stimulate transcription through the Ras/MAPK signaling cascade (Wiegert et al. 2007).

Phosphorylation and activation of CREB following synaptic NMDAR activity is achieved by the convergence of at least two Ca^{2+}-dependent kinase pathways: CaMKIV in the nucleus mediates rapid CREB phosphorylation whereas the Ras/MAPK pathway acting through ERK promotes CREB phosphorylation in a slower, but more long-lasting manner (Hardingham et al. 2001a; Wu et al. 2001). For its part, CaMKIV activation requires that it bind Ca^{2+}-activated calmodulin (Ca^{2+}/CaM) and be phosphorylated by cytosolic Ca^{2+}/calmodulin-dependent protein kinase kinase (CaMKK) to achieve an enzymatically active state (Anderson et al. 1998; Means 2000; Sakagami et al. 2000). The reported constitutive localization of CaMKIV within the nucleus (Jensen et al. 1991; Nakamura et al. 1995; Bito et al. 1996) combined with its requisite involvement in the activation of CREB-mediated transcription (Chawla et al. 1998) raised the possibility that CREB induction via CaMKIV involves the translocation of a synaptically activated Ca^{2+} effector such as CaM or CaMKK. Stimulus-triggered $[Ca^{2+}]_i$ increases have been reported to induce the rapid translocation of both endogenous Ca^{2+}/CaM and fluorescently labeled Ca^{2+}/CaM from the cytoplasm to the nucleus of cultured hippocampal neurons, possibly in a complex with CaMKK (Deisseroth et al. 1998; Mermelstein et al. 2001). However, these findings are inconsistent both with the observed prominent nuclear localization of CaM in cortical and hippocampal neurons in vivo and in vitro (Caceres et al. 1983; Hoskins et al. 1986; Hardingham et al. 2001b), and with the lack of evidence supporting CaMKK nuclear localization or translocation (Deisseroth et al. 1998; Sakagami et al. 2000). Moreover, in a similar series of experiments, no evidence was found for signal-regulated translocation either of endogenous CaM or of microinjected fluorescently labeled CaM (Hardingham et al. 2001b). Instead, it was observed that rapid CaMK-dependent CREB phosphorylation can be evoked even when nuclear transport is blocked, and that an increase in nuclear $[Ca^{2+}]_i$ suffices to induce this response (Hardingham et al. 2001b).

Following these studies, a model has emerged for the activation of CaMKIV, which involves not the nuclear translocation of CaM, but rather the generation of nuclear calcium transients. Accordingly, the initial stimulation of CaMKIV begins with an increase in nuclear $[Ca^{2+}]_i$ and a physical association between CaMKIV and nuclearly localized Ca^{2+}/CaM (Hardingham et al. 2001b; Anderson et al. 2004). Binding of Ca^{2+}/CaM by CaMKIV is accompanied by its disengagement from the negative regulator protein phosphatase 2a (Park and Soderling 1995; Westphal et al. 1998; Kasahara et al. 1999; Anderson et al. 2004) and is followed by its phosphorylation and activation through CaMKK (Anderson et al. 1998, 2004; Chow et al. 2005). Interestingly, although CaMKK is triggered by $[Ca^{2+}]_i$ increases, it also shows a substantial level of activity at basal $[Ca^{2+}]_i$ (Anderson et al. 1998). It follows that synaptic activity-associated nuclear $[Ca^{2+}]_i$ increases may be sufficient for stimulating CaMKIV-dependent CREB phosphorylation (Chawla et al. 1998; Hardingham et al. 2001b). Yet where does the phosphorylation of CaMKIV occur? Although CaMKIV is localized to the nucleus, it has been observed to dynamically redistribute between cytoplasmic and nuclear compartments (Matthews et al. 1994; Lalonde et al. 2004; Lemrow et al. 2004; Kotera et al. 2005). CaMKK, on the other hand, shows a strictly cytoplasmic distribution (Anderson et al. 1998; Sakagami et al. 2000). Thus, CaMKIV appears to be phosphorylated by CaMKK in the cytoplasm (Lemrow et al. 2004), after which it is transported into the nucleus, perhaps via a mechanism involving importin α-mediated facilitated diffusion (Matthews et al. 1994; Lemrow et al. 2004; Kotera et al. 2005).

VDCCs are important for the neuronal activity-dependent regulation of gene expression. Neurons express a variety of VDCCs having varying pharmacological sensitivity, activation thresholds, conductances, and localizations (Catterall et al. 2005; Vacher et al. 2008). Of these, the L-type family of high voltage-activated VDCCs plays the most prominent role in postsynaptic signal transduction to the nucleus. These Ca^{2+} channels are expressed in the soma and proximal dendrites at both synaptic and extrasynaptic locations (Westenbroek et al. 1990; Hell et al. 1993; Obermair et al. 2004; Leitch et al. 2009), and are strongly implicated in the somatodendritic $[Ca^{2+}]_i$ increases that are generated in response to action potentials (Westenbroek et al. 1990; Liu et al. 2003; Vacher et al. 2008). In view of their somatodendritic localization, it may be tempting to conclude that the involvement of L-type channels in transcriptional regulation can be accounted for solely through the global Ca^{2+} transients associated with their activation. Numerous studies have shown, however, as is the case for NMDARs, that Ca^{2+} can act locally near the mouth of the channel to trigger transcription-relevant events (Deisseroth et al. 1996; Dolmetsch et al. 2001; Hardingham et al. 2001a; Weick et al. 2003; Wheeler et al. 2008). This reliance of VDCC-mediated genomic responses on local Ca^{2+} signaling suggests that L-type channels reside within Ca^{2+} signaling microdomains in close physical proximity to Ca^{2+}-dependent molecular machinery. Indeed, not only are L-type VDCCs localized to postsynaptic densities in synaptic spines, where they interact with CaM, CaMKs, the Ca^{2+}-dependent protein phosphatase calcineurin, and other Ca^{2+}-dependent second messengers (Graef et al. 1999; Weick et al. 2003; Greer and Greenberg 2008; Vacher et al. 2008; Wheeler et al. 2008; Dai et al. 2009), the L-type channel also harbors a CaM-binding motif on its carboxy-terminus enabling the tethering of CaM within several nanometers of its Ca^{2+}-conductive pore (Zuhlke and Reuter 1998; Dolmetsch et al. 2001). Consistent with these findings, L-type VDCC-dependent phosphorylation of transcriptional activators has been reported to depend—at least in part—on the activation of synaptically localized Ca^{2+}/CaM-dependent kinases (Deisseroth et al. 1998; Dolmetsch et al. 2001; Weick et al. 2003; Wheeler et al. 2008). An additional means by which synaptic L-type channels can control genomic responses is via local interactions with the protein kinase A anchoring protein (AKAP79/150) and the protein phosphatase calcineurin (Gray et al. 1998; Oliveria et al. 2007). In particular, anchoring of calcineurin to L-type VDCCs by AKAP79/150 is thought to facilitate its activation by L-type VDCC-mediated Ca^{2+} influx. Ca^{2+}-activated calcineurin, in turn, can dephosphorylate the transcription factor NFAT, unmasking its NLS and leading to nuclear import (Beals et al. 1997; Graef et al. 1999; Greer and Greenberg 2008; Vashishta et al. 2009).

EXTRASYNAPTIC NMDA RECEPTORS ANTAGONIZE SYNAPSE-TO-NUCLEUS COMMUNICATION

Up to this point, we have concentrated on delineating the routes by which glutamatergic synaptic input may be conveyed to the nucleus. Just as significant, however, are the extrasynaptic routes by which glutamatergic excitation-transcription coupling is controlled. In the brain, signal reception by extrasynaptic glutamate receptors, particularly ES-NMDARs, has been suggested to play a pivotal role in the neurodegenerative processes associated with cerebral ischemia, seizure, and traumatic brain injury. ES-NMDARs are also implicated in the etiology of several neurodegenerative disorders, including Huntington's disease, Alzheimer's disease, and amyotrophic lateral sclerosis (Hardingham and Bading 2003, 2010; Vanhoutte and Bading 2003; Arundine and Tymianski 2004; Bossy-Wetzel et al. 2004; Bezprozvanny 2007; Kalia et al. 2008; Hardingham 2009; Milnerwood et al. 2010). The functional importance of ES-NMDARs in these disorders has been proposed to follow from their influence on cell survival and cell death pathways.

Although Ca^{2+} entry through synaptic NMDARs promotes neuronal survival via the activation of a neuroprotective gene program and suppression of apoptotic cascades, Ca^{2+} influx through ES-NMDARs both opposes cell survival pathways and triggers cell death signals leading to neuron loss (Hardingham et al. 2002, 2009; Vanhoutte and Bading 2003; Papadia et al. 2005; Zhang et al. 2007b, 2009; Leveille et al. 2008; Stanika et al. 2009; Wahl et al. 2009). For example, synaptic NMDAR activation both inhibits transcription of the proapoptotic gene *Puma* and triggers the nuclear Ca^{2+}- and CREB-dependent expression of AID genes and the prosurvival immediate early gene, *BDNF* (Favaron et al. 1993; Hardingham et al. 2002; Hansen et al. 2004; Jiang et al. 2005; Zhang et al. 2007b, 2009; Lau and Bading 2009; Leveille et al. 2010). Conversely, stimulation of extrasynaptic receptors is linked to up-regulated expression of the proapoptotic putative Ca^{2+}-activated chloride channel Clca1 (Zhang et al. 2007b; Wahl et al. 2009). Moreover, ES-NMDARs activate a dephosphorylating CREB shut-off pathway and antagonize synaptic activity induced BDNF expression and cell survival (Hardingham and Bading 2002; Hardingham et al. 2002; Vanhoutte and Bading 2003; Papadia et al. 2005; Ivanov et al. 2006; Zhang et al. 2007b).

One important messenger in synaptic NMDAR- and Ca^{2+}-dependent coupling to CREB is ERK. Although Ca^{2+} influx through synaptic NMDARs induces the phosphorylation and mobilization of ERK, Ca^{2+} entry via ES-NMDARs instead leads to ERK inactivation and retention at the plasma membrane, thus preventing its interaction with and activation of nuclear transcription factor targets (Cruzalegui and Bading 2000; Krapivinsky et al. 2003;

Ivanov et al. 2006; Leveille et al. 2008; Zehorai et al. 2010). Consequently, stimulation of the total (synaptic and extrasynaptic) NMDAR pool triggers a reduced level of ERK activity and associated transcription factor activation (Bading and Greenberg 1991; Chandler et al. 2001; Hardingham et al. 2001a, 2002; Krapivinsky et al. 2003; Ivanov et al. 2006; Leveille et al. 2008; Gao et al. 2010). Such antagonism by ES-NMDARs of synaptic NMDAR-dependent signaling to the nucleus extends to a number of additional signaling pathways. For example, ES-NMDAR signaling has been observed to disrupt the nucleocytoplasmic shuttling of class II HDACs, thereby preventing disinhibition of prosurvival transcription factors MEF2 and CREB in response to synaptic NMDAR stimulation (Chawla et al. 2003; Vanhoutte and Bading 2003; Bolger and Yao 2005). Furthermore, synaptic NMDARs signals promote the phosphorylation and subsequent nuclear export of the forkhead box O (FOXO) family of proapoptotic transcription factors, and also inhibit subsequent FOXO nuclear import. Stimulation of ES-NMDARs, on the other hand, induces rapid translocation of FOXO from the cytoplasm into the nucleus (Brunet et al. 1999; Gilley et al. 2003; Al-Mubarak et al. 2009; Hardingham 2009; Dick and Bading 2010). A picture is thus emerging in which neurons' genomic responses to Ca^{2+} entry through NMDARs depends on the subcellular localization of these receptors (Fig. 4). The particular means by which ES-NMDARs antagonize intracellular signals induced by synaptic NMDARs remains to be clearly delineated. One intriguing possibility is that the two receptor populations experience distinct biochemical environments harboring different sets of Ca^{2+}-dependent effector molecules (Hardingham and Bading 2010).

NUCLEAR Ca^{2+} SIGNALING IN SYNAPSE-TO-NUCLEUS COMMUNICATION

An increasing number of studies suggest that $[Ca^{2+}]_i$ increases within the nucleus represent an essential component of synapse-to-nucleus communication: nuclear Ca^{2+} signals have been implicated in the regulation of neuronal gene expression programs (Chawla et al. 1998; Carrion et al. 1999; Ledo et al. 2000; Hardingham et al. 2001b; Ledo et al. 2002; Zhang et al. 2009) and are a critical component in the molecular machinery that translates different sets of stimuli into distinct transcriptional responses (Hardingham et al. 1997; Mellstrom et al. 2004). Nuclear Ca^{2+} signaling is linked also to a number of neuronal functions that have been shown to depend on regulated gene expression (Limback-Stokin et al. 2004; Papadia et al. 2005; Raymond and Redman 2006; Schneider et al. 2007; Alexander et al. 2009; Fontan-Lozano et al. 2009). For instance, 100 Hz and θ-burst synaptic stimulation of afferent fibers—both of which are known to induce transcription-dependent LTP—evoke robust nuclear Ca^{2+} transients (Raymond and Redman 2002, 2006; Johenning and Holthoff 2007; Power and Sah 2007). Conversely, blockade of somatonuclear $[Ca^{2+}]_i$ increases during stimulation prevents the development of transcription-dependent, but not transcription-independent LTP (Raymond and Redman 2002, 2006), whereas antagonism of nuclear Ca^{2+}/CaM signaling inhibits both neuronal activity-induced gene transcription and transcription-dependent LTP, and selectively prevents the formation of transcription-dependent long-term memory (Limback-Stokin et al. 2004). The means by which nuclear Ca^{2+} signaling controls memory processes include the transcriptional regulation of vascular endothelial growth factor D (VEGFD), a mitogen for endothelial cells whose expression modulates dendritic length and complexity (Mauceri et al. 2011). Nuclear Ca^{2+}/CaM signaling is also critically important for the activation of a neuroprotective gene program, and its inhibition has been linked to an increased susceptibility to cellular stressors (Papadia et al. 2005; Zhang et al. 2007b, 2009; Lau and Bading 2009; Dick and Bading 2010).

The first indication that nuclear Ca^{2+} might play a part distinct from cytoplasmic Ca^{2+} in the neuronal activity-dependent regulation of gene expression was gained by an investigation employing a dextran-conjugated Ca^{2+} chelator (Hardingham et al. 1997). AtT20 cells, into whose nuclei this chelator had been injected,

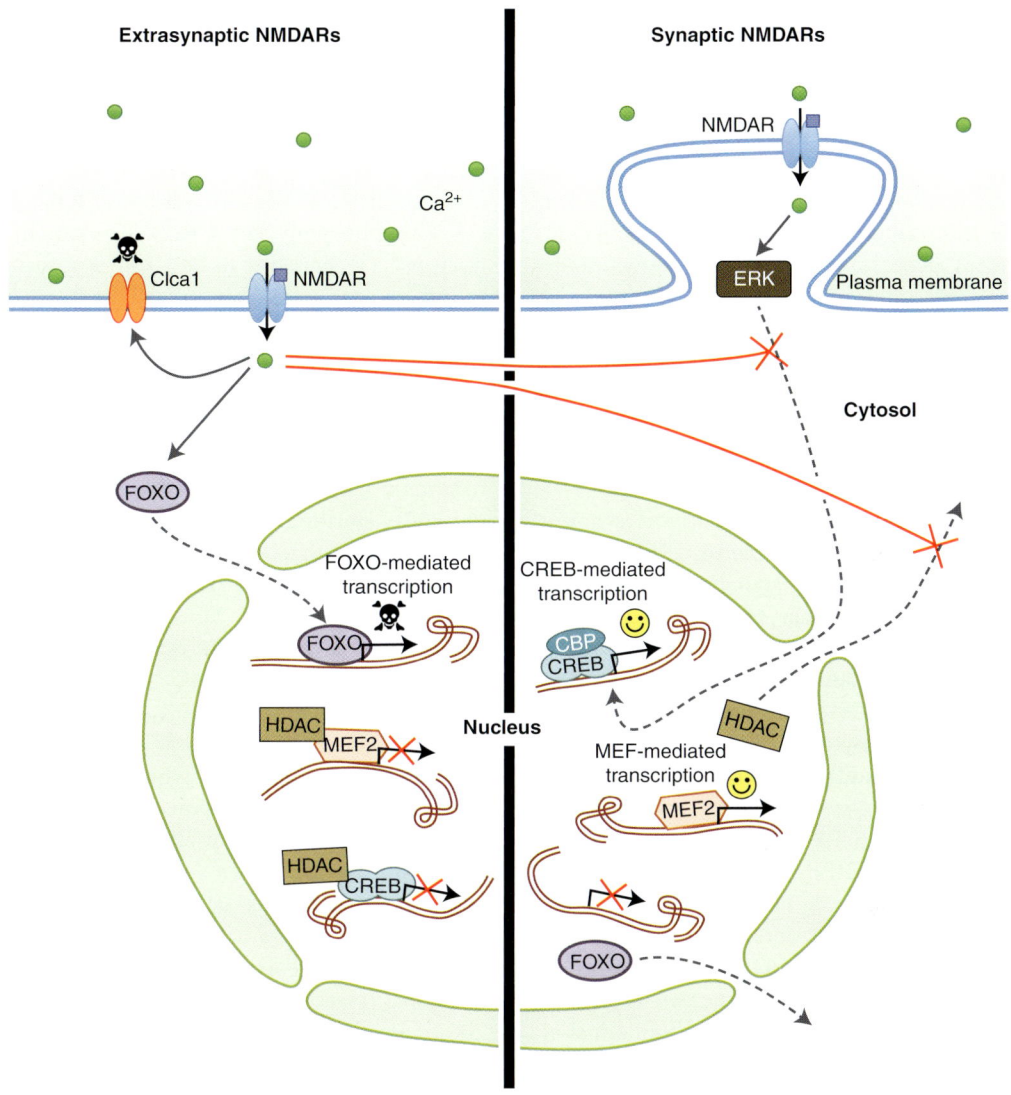

Figure 4. Extrasynaptic NMDARs antagonize synaptic NMDAR-dependent synapse-to-nucleus signaling. NMDAR-dependent Ca^{2+} signaling promotes neuronal survival, but can also cause cell degeneration and apoptotic death. The location of the activated NMDAR pool specifies the transcriptional response. In particular, Ca^{2+} influx through synaptic NMDARs leads to the activation and nuclear translocation of ERK followed by phosphorylation of the neuroprotection-associated transcription factor, CREB, and transcription of prosurvival gene products including the prosurvival neurotrophic factor BDNF. Synaptic NMDAR activity also induces the nuclear export of HDACs and of the proapoptotic transcription factor FOXO. Export of HDACs allows for the acetylation and disinhibition of transcription mediated by prosurvival transcription factors such as MEF2 and CREB. Conversely, ES-NMDAR-mediated Ca^{2+} entry triggers transcription of the proapoptotic Ca^{2+}-activated chloride channel Clca1, triggers the rapid dephosphorylation and inactivation of ERK, thus preventing its nuclear translocation, interferes with the nuclear export of HDACs, and triggers the rapid nuclear import of FOXO proteins. Thus, whereas activation of synaptic NMDARs promotes the transcription of neuroprotective target genes via CREB and MEF2, ES-NMDAR activation actively represses CREB- and MEF-dependent gene transcription, and leads to the transcription of genes associated with neuronal degeneration and apoptotic cell death.

showed intact cytoplasmic Ca^{2+} transients, but significantly reduced nuclear Ca^{2+} signals in response to stimulation of plasma membrane Ca^{2+} channels. Analysis of transcriptional responses subsequently revealed that the reduction in nuclear Ca^{2+} signaling could be specifically linked to an inhibition of CREB-dependent gene transcription. In contrast, TCF/SRF-dependent gene transcription was shown to be contingent only on the generation of $[Ca^{2+}]_i$ increases in the cytoplasm (Hardingham et al. 1997). A follow-up study subsequently unveiled a critical role for nuclear Ca^{2+} in CREB-dependent gene transcription stimulated by the activity of synaptic NMDARs in neurons (Hardingham et al. 2001b).

How does nuclear Ca^{2+} support synaptic NMDAR-mediated signaling to CREB? Synaptic activity-dependent signaling to CREB is achieved primarily via two different Ca^{2+}-dependent kinase cascades: the CaMK and the Ras/MAPK pathways (Chawla et al. 1998; Hardingham et al. 1999). Although both pathways cause phosphorylation of CREB on its activator site serine 133, additional stimulation of CBP, a transcriptional coactivator that interacts with phosphorylated CREB, is also crucial for the induction of CREB-mediated transcription (Chawla et al. 1998; Hu et al. 1999; Impey et al. 2002). The induction of CBP, in turn, critically depends on CaMKIV, the activation of which can be achieved by an increase in nuclear $[Ca^{2+}]_i$ (Chawla et al. 1998; Hardingham et al. 1999; Impey et al. 2002). CREB-mediated transcription triggered by synaptic NMDARs thus necessitates convergence of the Ras/MAPK signaling cascade with a nuclear Ca^{2+}- and CaMKIV-dependent signal to CBP. It follows that the presence or absence of nuclear $[Ca^{2+}]_i$ increases during up-regulated NMDAR-dependent Ras/MAPK signaling has the capacity to determine whether these may induce CREB-mediated gene transcription. Nuclear Ca^{2+} signaling is thus a critical determinant of whether the activation of synaptic NMDARs leads to solely TCF/SRF-, or both TCF/SRF- and CREB-dependent transcription (see Fig. 2) (Hardingham et al. 1997, 2001a,b; Chawla et al. 1998).

It has been proposed that $[Ca^{2+}]_i$ increases must originate at synapses to trigger neuronal activity-dependent CREB-mediated transcriptional responses (Murphy et al. 1991; Deisseroth et al. 1996). An important point therefore is that CaMKIV, which resides in the nucleus, not only is necessary for the activation of CBP, but can also phosphorylate CREB. Consequently, nuclear Ca^{2+} signaling to CaMKIV is sufficient to induce gene transcription via CREB (Hardingham et al. 2001b). The duration of nuclear $[Ca^{2+}]_i$ increases, a feature which correlates with the duration of CBP activation and CREB phosphorylation, is an important factor for determining the magnitude of genomic responses (Bito et al. 1996; Chawla and Bading 2001; Hardingham et al. 2001b). It would therefore seem that—whether arising from synaptic NMDARs, L-type VDCCs, or intracellular Ca^{2+} stores—any Ca^{2+} signal that invades the nucleus and is of sufficient duration and magnitude to stimulate CaMKIV, may likewise be sufficient to trigger CREB-mediated gene transcription.

An additional means by which nuclear Ca^{2+} can regulate genomic responses is via a direct interaction with the transcriptional repressor DREAM. DREAM is a Ca^{2+}- and DNA-binding protein that inhibits gene transcription by restricting the access of stimulatory transcription factors like CREB to their DNA-binding domains (Carrion et al. 1999; Ledo et al. 2002). Binding of Ca^{2+} to DREAM allows it to dissociate from its DNA binding site (Carrion et al. 1999; Ledo et al. 2000, 2002). Thus, nuclear Ca^{2+} acting through DREAM can function as a signal for transcriptional disinhibition. New research implicates DREAM also in synaptic plasticity and memory consolidation. For example, DREAM knockout mice show enhanced CREB-dependent gene transcription, enhanced LTP, and improved performance in a variety of behavioral assays for transcription-dependent long-term memory (Lilliehook et al. 2003; Alexander et al. 2009; Fontan-Lozano et al. 2009). Notably, these mice continued to show improved memory formation during aging, and were found to manifest a marked decrease in aging-associated pathological changes in the hippocampus (Fontan-Lozano et al. 2009). Moreover, using electrophoretic mobility shift assays with hippocampal nuclear extracts, it was observed that DREAM binding to DNA decreased following training on a hippocampus-dependent behavioral task (Fontan-Lozano et al. 2009). These findings underline the critical involvement of nuclear Ca^{2+}-dependent control of genomic responses through direct interactions with transcriptional regulators like DREAM in the coupling of excitatory stimuli to memory formation and cognitive functioning.

EPIGENETIC MECHANISMS IN SYNAPTIC ACTIVITY AND Ca^{2+}-DEPENDENT TRANSCRIPTIONAL REGULATION

Chromatin remodeling is now emerging as a powerful means by which synaptic activity and intracellular Ca^{2+} signaling can induce genomic responses (Fig. 5). Dynamic changes in chromatin structure are brought about by an array of histone posttranslational modifications including phosphorylation, methylation, and acetylation. These changes are implicated in gene transcription, synaptic plasticity, memory consolidation, and neuronal survival (Alarcon et al. 2004; Korzus et al. 2004; Jiang et al. 2006; Oliveira et al. 2007; Lubin et al. 2008; Koshibu et al. 2009), and their dysregulation is causatively linked to both developmental and neurodegenerative disorders (Petrij et al. 1995; Jorgensen and Bird 2002; Alarcon et al. 2004; Jiang et al. 2006; Zhou et al. 2006; Fischer et al. 2007; Duclot et al. 2010).

Histone acetylation, which results in chromatin expansion and increased access to transcription factor binding sites, is controlled by histone acetyltransferases (HATs) and HDACs. One of the means by which neuronal activity-dependent Ca^{2+} signaling can influence transcription factor activity is through the dynamic regulation of these enzymes. For example, synaptic activity-dependent $[Ca^{2+}]_i$ increases act through CaMKIV to stimulate CBP, a CREB coactivator and HAT that interacts with numerous transcription factors (Chawla et al. 1998; Sterner and Berger 2000; Impey et al. 2002;

Bedford et al. 2010). In a similar vein, NMDAR-mediated Ca^{2+} signaling through the Ras/MAPK pathway may stimulate gene transcription via p300, a CBP analog HAT (Sterner and Berger 2000; Legube and Trouche 2003; Li et al. 2003; Oliveira et al. 2006, 2007; Bedford et al. 2010). Synaptic activation of NMDARs and L-type VDCCs shapes transcriptional responses also by triggering the CaMK-dependent nuclear export of the class II HDAC and MEF2 corepressor, HDAC5, an effect that is blocked either by costimulation of ES-NMDARs or by pharmacological induction of cyclic AMP signaling (McKinsey et al. 2000; Chawla et al. 2003; Belfield et al. 2006). These and other findings suggest that transcription factor phosphorylation and histone acetylation may represent parallel targets of synapse-to-nucleus Ca^{2+}-dependent signaling through the CaMK and Ras/MAPK pathways. Indeed, a number of recent studies have established CBP-, p300-, CaMK-, and/or NMDAR/Ras/MAPK-dependent histone modifications as critical regulatory mechanisms controlling the synaptic plasticity gene BDNF, the induction of LTP, and the consolidation of memory (Hardingham et al. 1999; Impey et al. 2002; Chawla et al. 2003; Chwang et al. 2006, 2007; Oliveira et al. 2006, 2007; Wood et al. 2006; Chandramohan et al. 2008; Brami-Cherrier et al. 2009; Tian et al. 2009, 2010a,b; Duclot et al. 2010).

Synaptic activity and Ca^{2+}-dependent regulation of gene expression may further be achieved via the modulation of cofactors that facilitate the association of HATs and HDACs with DNA. An example is methyl-CpG binding protein 2 (MeCP2), a transcriptional regulator that has been shown to suppress gene expression by recruiting HDAC-containing protein complexes to the genome (Jones et al. 1998; Nan et al. 1998; Jorgensen and Bird 2002; Chen et al. 2003; Martinowich et al. 2003; Zhou et al. 2006; Tao et al. 2009). Synaptic activity and Ca^{2+}-dependent regulation of MeCP2 can be accomplished via its phosphorylation by CaMKs (Zhou et al. 2006; Tao et al. 2009). In particular, the Ca^{2+}-dependent phosphorylation of MeCP2 has been shown both to trigger transcriptional derepression of known

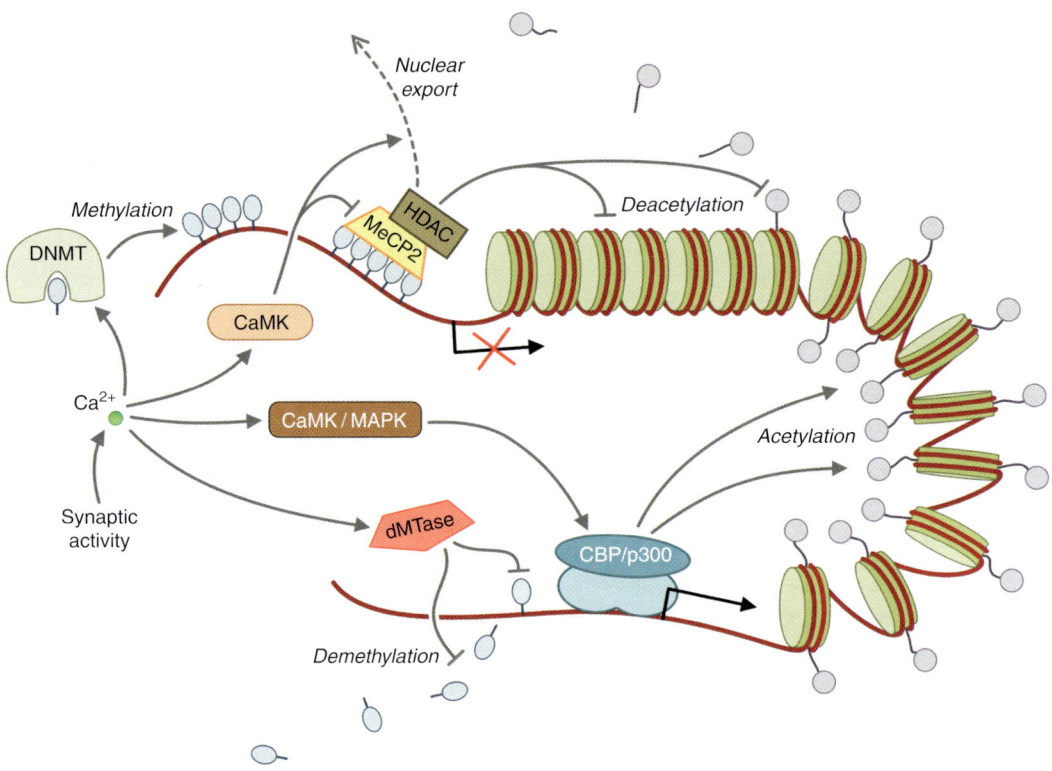

Figure 5. Epigenetic mechanisms in synaptic activity and Ca^{2+}-dependent transcriptional regulation. Synaptic activity-dependent $[Ca^{2+}]_i$ increases may influence chromatin structure—and therewith access of transcription factors and regulatory enzymes to their specific interaction domains—through influences on histone acetylation and DNA methylation. Synaptic activity-dependent Ca^{2+} signals acting through CaM and CAMK and Ras/MAPK signaling cascades can, for instance, activate HATs such as CBP and p300. These transcriptional coactivators catalyze the posttranslational acetylation of histones, leading to chromatin relaxation, increased access of transcription factors to their respective DNA binding domains, and up-regulation of gene transcription in affected genomic regions. Alternately, nuclear Ca^{2+} increases associated with synaptic activity may trigger the CAMK-dependent phosphorylation and dissociation of methyl CpG- binding proteins like MeCP2 from methylated sequences of DNA as well as the nuclear export of the HDACs with which they interact. HDAC activity results in the deacetylation of histones and the subsequent compression of chromatin, leading to a suppression of transcriptional activity. MeCP2 dissociation and HDAC export thus induce a derepression of gene transcription. Ca^{2+} influx associated with synaptic activity is linked also to de novo DNA methylation via DNMTs such as DNMT3a, and to active DNA demethylation via dMTases such as GADD45a. Note that, although not shown here, histone phosphorylation, ubiquitination, and methylation also contribute to the epigenetic regulation of gene expression (Levenson and Sweatt 2006), and may likewise be involved in Ca^{2+}-dependent synapse-to-nucleus communication.

plasticity-associated genes via a mechanism that involves its dissociation from the genome followed by an up-regulation of histone acetylation (Chen et al. 2003; Martinowich et al. 2003; Zhou et al. 2006; Tao et al. 2009) and to activate gene transcription by recruiting transcription factors such as CREB1 to the genome (Chahrour et al. 2008).

An additional mechanism by which synaptic activity and associated $[Ca^{2+}]_i$ increases might stimulate the genomic responses required for neuronal activity-induced acquired neuroprotection, synaptic plasticity, and long-term memory is via the dynamic regulation of DNA methylation (Levenson et al. 2006; Barreto et al. 2007; Miller and Sweatt 2007; Zhang et al. 2007b, 2009; Lubin et al. 2008; Miller et al. 2008, 2010; Feng et al. 2010). De novo DNA methylation mediated by DNA methyltransferases (DNMTs) such as DNMT3a has, for instance, been shown to have a critical function in the molecular signaling cascades that underlie synaptic plasticity and memory formation (Levenson et al. 2006; Feng et al. 2010). Moreover, it has been reported that LTP-inducing stimulation in vitro and fear conditioning training in vivo induce rapid methylation changes in the promoter regions of specific plasticity regulated genes such as protein phosphatase 1, brain-derived neurotrophic factor, and reelin (Martinowich et al. 2003; Miller and Sweatt 2007; Lubin et al. 2008). Importantly, pharmacological disruption of Ca^{2+} signaling via NMDARs or of DNMTs was observed not only to compromise the induction of LTP and the consolidation of long-term memory but also to block stimulus-induced methylation changes (Levenson et al. 2006; Miller and Sweatt 2007; Lubin et al. 2008; Miller et al. 2008, 2010). In combination with the recent identification of growth arrest and DNA damage inducible protein 45 α (GADD45α), a putative stress-induced nuclear DNA demethylase (DMTase) reported to relieve gene silencing by erasing methylation marks (Barreto et al. 2007; Rai et al. 2008; Ma et al. 2009), as a target of synaptic activity-dependent nuclear Ca^{2+}/CaM signaling (Zhang et al. 2007a, 2009), these findings suggest that Ca^{2+} may indeed participate in the regulation of DNA methylation. Clearly, much work remains in identifying the particular signaling pathways linking neuronal activity-dependent $[Ca^{2+}]_i$ increases to dynamic DNA methylation and chromatin remodeling. We nonetheless anticipate that this developing field will yield many exciting and important insights into the mechanisms underlying synapse-to-nucleus communication.

Ca^{2+} REGULATION OF NUCLEAR PORE PERMEABILITY

Next to its function as a second messenger in the governance of neuronal activity-dependent genomic responses, Ca^{2+} entering the nucleus may have the added function of regulating nuclear permeability to second messengers. Nucleocytoplasmic transport of soluble molecules takes place at the nuclear pore, a large macromolecular complex consisting of eight ion channels and a large central passage (Akey 1989; Otis et al. 2006). Passage of molecules through this complex depends both on size and on the presence or absence of appropriate localization sequences that enable interaction with nuclear importing proteins. Accordingly, molecules larger in size than 40–60 kD need a signaling sequence to cross the nuclear envelope, and molecules smaller than 4–10 kD diffuse freely between cytosol and nucleus. Intermediate-sized molecules (10–40 kD) do not require a targeting signal, although their passage may be regulated at the level of the nuclear pore (reviewed in O'Brien et al. 2007). Two Ca^{2+}-dependent mechanisms have been proposed for such regulation. On one hand, the filling state of the nuclear lumen, which is contiguous with the ER lumen, may influence the size of the nuclear pore. Accordingly, depletion of nuclear Ca^{2+} stores in a variety of cell types has been shown to induce a conformational change of the nuclear pore complex such that it excludes molecules greater in size than approximately 10 kD (Greber and Gerace 1995; Stehno-Bittel et al. 1995; Perez-Terzic et al. 1999; Wang and Clapham 1999; Erickson et al. 2006; Stoffler et al. 2006). Alternately, a recent study suggests that Ca^{2+} may alter nuclear permeability cytoplasmically: stimulation of $G_{q/11}$-coupled receptors and subsequent internal Ca^{2+} release were shown to rapidly and briefly increase nuclear permeability in a liver cell line (O'Brien et al. 2007). Interestingly, this increase in permeability—for which an increase in $[Ca^{2+}]_i$ near the nuclear envelope was both necessary and sufficient—was unidirectional from cytosol to nucleus. These findings raise the very intriguing possibility that, in addition to regulating genomic responsiveness via interactions with nuclearly localized effectors, somatonuclear Ca^{2+} signals may simultaneously promote the facilitated diffusion of second messengers into the nucleus.

GENERATING NUCLEAR Ca^{2+} SIGNALS

With respect to the generation of nuclear Ca^{2+} signals, it is important to note that the nuclear envelope does not present a spatial barrier for the transmission of Ca^{2+} transients between somatic and nuclear compartments (Danker et al. 2001; Shahin et al. 2001; Eder and Bading 2007). Cytoplasmic $[Ca^{2+}]_i$ increases that reach the nuclear envelope are thus poised to participate in nuclear Ca^{2+}-dependent transcriptional regulation. Likely the most obvious candidates for generating robust nuclear Ca^{2+} signals are L-type VDCCs (Fig. 3B). In fact, somatic action potentials, which generate robust L-type VDCC-mediated Ca^{2+} transients in the cytoplasm and nucleus (Johenning and Holthoff 2007), have been reported to induce CREB phosphorylation, CREB-dependent up-regulated gene expression, and transcription-dependent synaptic plasticity (Dudek and Fields 2002). Investigations aimed at identifying specific Ca^{2+} sources involved in the induction of synaptic plasticity also point to L-type VDCCs and somatic Ca^{2+} increases specifically as critical mediators of transcription-dependent plasticity (Raymond and Redman 2002, 2006). By way of their subcellular localization on neuronal somata and proximal dendrites, these channels are appropriately positioned to generate robust somatonuclear $[Ca^{2+}]_i$ increases in response to neuronal activity-dependent depolarizations of the plasma membrane (Westenbroek et al. 1990; Obermair et al. 2004; Leitch et al. 2009). Interestingly, the activation kinetics of L-type channels has been suggested to enable their preferred responsiveness to trains of EPSPs over action potentials (Mermelstein et al. 2000; Liu et al. 2003). This proposed ability of L-type VDCCs to discriminate between different forms of depolarizing stimuli could explain why synaptic activity has been observed to be more effective in causing Ca^{2+}-dependent CREB phosphorylation than action potentials alone (Deisseroth et al. 1996; Mermelstein et al. 2000), and suggests that L-type channels on the somatic membrane might contribute to synapse-to-nucleus communication in the form of a synaptic activity-dependent nuclear Ca^{2+} signal.

Another particularly interesting candidate for the Ca^{2+}-dependent conveyance of information between synapse and nucleus is mGluR-mediated, IP_3R-dependent Ca^{2+} release from intracellular stores (Fig. 3C). IP_3R-dependent internal Ca^{2+} release has been implicated in CREB-mediated gene transcription (Hardingham et al. 2001b; Hu et al. 2002; Carrasco et al. 2004; Mao et al. 2008) as well as in the long-term potentiation and depression of synaptic strength, the stimulation of neurite outgrowth, and the activation of trophic and apoptotic signaling cascades (Berridge 1998; Verkhratsky and Petersen 1998; Korkotian and Segal 1999; Yeckel et al. 1999; Berridge et al. 2000; Rose and Konnerth 2001; Vanderklish and Edelman 2002; Raymond and Redman 2006). Neuronal $[Ca^{2+}]_i$ increases caused by the release of Ca^{2+} from IP_3-sensitive intracellular stores are unique when compared to $[Ca^{2+}]_i$ increases generated by VDCC- or NMDAR-dependent Ca^{2+} influx in a number of important ways. Perhaps the most notable of these is the ability of internal Ca^{2+} release to propagate as a wave from its dendritic site of initiation to distant regions of the neuron including the soma and nucleus (Jaffe and Brown 1994; Pozzo-Miller et al. 1996; Power and Sah 2002, 2007; Larkum et al. 2003; Watanabe et al. 2006; Hagenston et al. 2008). These waves of Ca^{2+} release are predominantly observed in the primary apical dendrites of pyramidal neurons, in which they regularly cover distances upward of 50 μm (Nakamura et al. 1999, 2002; Power and Sah 2007; Hagenston et al. 2008). In contrast, $[Ca^{2+}]_i$ increases resulting from NMDAR activation are predominantly observed in highly spinous apical oblique and basal dendrites, and are almost negligible in the primary apical dendrite or soma (Schiller et al. 2000; Nakamura et al. 2002; Raymond and Redman 2006). Moreover, Ca^{2+} waves can regularly attain amplitudes of up to 10 μM, approximately 5- to 15-fold greater than $[Ca^{2+}]_i$ increases evoked by trains of action potentials (Nakamura et al. 1999; Power and Sah 2002, 2007; Larkum et al. 2003). Additionally, whereas VGCC- and NMDAR-associated $[Ca^{2+}]_i$ increases turn on at the start of a stimulus or spike train and typically begin to decay at the end of that train (Pozzo-Miller et al. 1996), the rising phase of synaptically activated Ca^{2+} release generally initiates with a delay of tens to hundreds of milliseconds, can easily outlast the duration of synaptic stimulation, and may even be as great as hundreds of milliseconds in any one location (Pozzo-Miller et al. 1996; Kapur et al. 2001; Larkum et al. 2003; Power and Sah 2007). The decay of release-associated $[Ca^{2+}]_i$ transients is similarly prolonged, presumably because of the slow inactivation kinetics of the IP_3 receptor (Li et al. 1995; Taylor 1998; Foskett et al. 2007). Thus, the total duration of internal Ca^{2+} release-associated $[Ca^{2+}]_i$ increases is typically on the order of hundreds of milliseconds to longer than one second (Nakamura et al. 1999; Larkum et al. 2003). In sum, mGluR-mediated IP_3R-dependent Ca^{2+} waves have the combined characteristics of large amplitude, long duration, and ability to propagate long distances along the primary apical dendrite into the soma and nucleus. These features make Ca^{2+} waves ideally suited to evoke large, long-lasting nuclear Ca^{2+} signals and robust transcriptional responses in response to synaptic activity (Chawla and Bading 2001; Power and Sah 2002, 2007; Hagenston et al. 2008).

Under what conditions are Ca^{2+} waves likely to play a role in nuclear Ca^{2+} signaling? Investigations of IP_3R function in reduced cellular and bilayer preparations, modeling studies, and examinations of Ca^{2+} waves in nonneuronal cells describe a number of possible influences on the extent of Ca^{2+} wave propagation (e.g., Parker and Ivorra 1990; Bezprozvanny et al. 1991; Bootman et al. 1997; Callamaras et al. 1998; Verkhratsky and Petersen 1998; Solovyova and Verkhratsky 2002; Friel 2004; Foskett et al. 2007). These include $[Ca^{2+}]_i$ at the time of IP_3R activation by IP_3, the concentration and distribution of mobilized IP_3 in the cytosol, the size of the readily releasable ER Ca^{2+} pool, and the biophysical state of IP_3Rs. These potential influences translate into a number of different factors likely to regulate the propagation of mGluR-mediated, IP_3R-dependent Ca^{2+} waves toward and into the somata of neurons, many of which have been validated by Ca^{2+} imaging data in neurons. These factors include the generation of action potentials and VDCC-dependent $[Ca^{2+}]_i$ increases coincident with IP_3-mobilizing stimuli (Nakamura et al. 1999; Yamamoto et al. 2000; Larkum et al. 2003; Hagenston et al. 2008), the number and distribution of activated synapses and the activity of neuromodulatory G protein-coupled receptors linked to IP_3 generation (Lezcano and Bergson 2002; Nakamura et al. 2002; Tang et al. 2003; Watanabe et al. 2006; Dai et al. 2008; Hagenston et al. 2008; Fitzpatrick et al. 2009), the activity history of stimulated neurons (Jaffe and Brown 1994; Pozzo-Miller et al. 1996; Morikawa et al. 2000; Rae et al. 2000; Larkum et al. 2003; Stutzmann et al. 2003; Gulledge and Stuart 2005; Power and Sah 2005; Hong and Ross 2007; Hagenston et al. 2008), and the balance of activated kinases and phosphatases that target the IP_3R. Protein kinase A in particular is promising in this respect, as its activity is linked not only to IP_3R phosphorylation, but also to increased IP_3R sensitivity to IP_3 and to augmented internal Ca^{2+} release (Wojcikiewicz and Luo 1998; Bugrim 1999; Tang et al. 2003; Patterson et al. 2004; Straub et al. 2004; Bezprozvanny 2005; Foskett et al. 2007; AM Hagenston and M Yeckel, unpubl.). Following from these studies, we predict that IP_3R-dependent Ca^{2+} waves are most likely to contribute to nuclear Ca^{2+} signaling when elicited by the robust and simultaneous activation of synapses on many proximal apical oblique branches during and/or following a train of action potentials, and particularly when protein kinase A signaling to IP_3Rs has been engaged.

An additional mechanism has recently emerged by which synaptic activity may give rise to nuclear $[Ca^{2+}]_i$ increases. This means of synapse-to-nucleus communication involves the import of glutamate through excitatory amino acid transporters (EAATs) and the subsequent activation of mGluRs located on the inner nuclear membrane (Fig. 1) (O'Malley et al. 2003; Jong et al. 2005, 2007; Sheldon and Robinson 2007; Kumar et al. 2008). More specifically, synaptically released glutamate is proposed to be transported by extra- and perisynaptic EAATs into the cytosol and diffuse to the cell soma, in which it is transported by nuclear EAATs into the nuclear lumen. There it binds to and activates nuclear resident mGluRs, leading to the liberation of IP_3 into the nucleus where it can bind IP_3Rs on the inner nuclear envelope. Thus, glutamate uptake at the synapse is proposed to give rise to Ca^{2+} release from the nuclear lumen directly into the nucleus (O'Malley et al. 2003; Jong et al. 2005, 2007; Sheldon and Robinson 2007; Kumar et al. 2008). This model is bolstered by data documenting the presence and functional responsiveness of EAATs, mGluRs, and IP_3Rs on the inner nuclear membranes of a variety of cell types including dissociated neurons (Humbert et al. 1996; Echevarria et al. 2003; Gerasimenko et al. 2003; O'Malley et al. 2003; Jong et al. 2005, 2007; Marchenko et al. 2005; Quesada and Verdugo 2005; Marchenko and Thomas 2006; Sheldon and Robinson 2007; Kumar et al. 2008; Bootman et al. 2009; Rodrigues et al. 2009). It remains to be seen, however, whether and under which conditions internal Ca^{2+} release mediated by intranuclear mGluRs and IP_3Rs can be evoked in intact brain tissue. One possibility is that particularly robust stimulus trains such as those known to induce LTP lead to spillover of glutamate, which is then taken up by perisynaptic and extrasynaptic EAATs and shuttled to the nucleus. Subsequent intranuclear Ca^{2+} release might act to ensure the generation of a Ca^{2+} signal concomitant with the import of synaptically activated second messengers. As such, this pathway could support NMDAR- and Ras/MAPK-mediated stimulation of CREB-dependent transcription, synaptic plasticity, and neuronal activity-dependent neuroprotection. Alternatively, this proposed signaling pathway to the nucleus might be most robustly induced when extrasynaptic glutamatergic signaling is elevated. As such, it could contribute to the genomic responses and prodeath activity associated with ES-NMDAR activation, ischemia, and seizure (Hardingham et al. 2002; Liu et al. 2007; Sierra-Paredes and Sierra-Marcuno 2007; Zhang et al. 2007b; Leveille et al. 2008; Scimemi et al. 2009; Wahl et al. 2009).

THE ENDOPLASMIC RETICULUM AS A CONDUIT FROM SYNAPSE TO NUCLEUS

ER in neurons, the "neuron within a neuron," forms a continuous network that extends into all parts of the cell from the nuclear envelope to the most distal dendritic and axonal processes (Berridge 1998). In addition to enabling the propagation of Ca^{2+} waves over long distances in the dendrites and soma, this ubiquitous distribution of ER may allow for postsynaptic signal propagation either via the diffusion of second messengers embedded in or attached to its membranes (Jordan and Kreutz 2009) or in the form of rapid intraluminal Ca^{2+} redistributions within and between dendritic and somatonuclear compartments (Park et al. 2000, 2008; Choi et al. 2006). This latter transit system, which is called Ca^{2+} tunneling, has been credited with the ability of dendrites and spines sparse in ER to persistently respond to repetitive Ca^{2+} mobilizing inputs (Park et al. 2000; Choi et al. 2006). Conversely, ER Ca^{2+} tunneling is suggested to result in depletion of the somatonuclear internal store following strong and long-lasting stimulation of Ca^{2+} release in multiple dendritic compartments (Choi et al. 2006; Park et al. 2008). In view of the influence exerted by luminal Ca^{2+} on nuclear pore permeability (Greber and Gerace 1995; Stehno-Bittel et al. 1995; Lee et al. 1998; Danker et al. 2001), IP_3R activity (Missiaen et al. 1992; Bezprozvanny and Ehrlich 1994; Thrower et al. 2000), and the driving force of Ca^{2+} ions leaving the ER (e.g., Solovyova et al. 2002), and of the potential for somatonuclear internal Ca^{2+} release to regulate CaMKIV activity and nuclear pore permeability, such a signal could have important consequences for the regulation of gene expression programs.

A recent study introduces another particularly intriguing ER-dependent mechanism by which signals may be transmitted from synapse to nucleus: as passively propagating electrotonic potentials along ER membranes (Fig. 3D) (Shemer et al. 2008). This proposal is grounded on the basis that, like the plasma membrane, the ER membrane is an effective charge separator and may thus support the generation and propagation of trans-ER depolarizations and hyperpolarizations. Such potential fluctuations are likely to occur at ER in synaptic spines and dendrites either as a result of capacitative conduction of plasma membrane potential fluctuations combined with Ca^{2+} uptake into the ER via sarcoendoplasmic reticulum Ca^{2+} ATPases, or as a consequence of the release of Ca^{2+} via RyRs and IP_3Rs localized to spines and dendrites. Alternately, ER potentials may be evoked in the soma during action potentials and in response to internal Ca^{2+} release. Using a computational "cable in cable" model, it was shown that electrotonic potentials across the ER—and associated current flow within the ER—can be rapidly and effectively transmitted both from synapse to soma, and from the soma into the dendritic arbor (Shemer et al. 2008). Significantly, the investigators observe that these two modes of signal propagation have opposite effects on the transmembrane potential across the nuclear envelope and suggest on this basis that electrotonic signal propagation along ER membranes may account for the ability of the nucleus to differentiate between synaptic signals originating at dendritic spines and backpropagating action potentials generated in the soma (Shemer et al. 2008). That said, it remains to be determined what the specific consequences of nuclear envelope depolarizations or hyperpolarizations might be. One possible scenario involves the activation of putative intracellular voltage-dependent cation channels (Schmid et al. 1990; Martin and Ashley 1993; Mazzanti et al. 2001; Bkaily et al. 2009; Matzke et al. 2010). Alternatively, nuclear envelope depolarizations may regulate signaling proteins that have been proposed to lead to intranuclear IP_3 mobilization (Billups et al. 2006; Ryglewski et al. 2007; Bootman et al. 2009; Liu et al. 2009; see above). Among other possibilities, changes in nuclear potential might also modulate nuclear pore permeability (Matzke and Matzke 1991; Mazzanti et al. 2001; Matzke et al. 2010). Although these possible consequences of a changing nuclear potential remain to be tested, their potential to influence nuclear Ca^{2+} signaling and neuronal activity-dependent gene transcription nonetheless makes the conduction and summation of electrotonic potentials along ER and nuclear membranes a compelling prospective route for synapse-to-nucleus communication.

CONCLUDING REMARKS

We have attempted here to provide a brief overview of the means by which synaptic excitation may be conveyed to the nucleus to influence neuronal activity-dependent gene transcription, particularly as it relates to synaptic plasticity, learning and memory, and neuronal survival and death processes. Just as the findings we describe inform our understanding of basic neuronal functions and functional responses, so too do they serve the greater aims of neuroscience and cellular signaling research to unravel the causes of neurocognitive disorders and to provide direction for the eventual development of possible therapeutic interventions. Indeed, dysregulations of the many of the proteins and processes involved in the generation of neuronal activity-dependent genomic responses we have described here have already been implicated in a wide range of psychiatric, developmental, and neurodegenerative disorders, including schizophrenia, depression, drug addiction, Rett syndrome, fragile X mental retardation, Alzheimer's disease, Huntington's disease, and ischemic stroke (Hong et al. 2005; Tsankova et al. 2007; Abel and Zukin 2008; Cohen and Greenberg 2008; Graff and Mansuy 2009; Kramer and van Bokhoven 2009; Reichenberg et al. 2009). Given the importance of nuclear calcium for the health of neurons and their ability to undergo activity-induced adaptations, a dysfunction in the generation of calcium transients and/or propagation to the nucleus (termed "nuclear calciopathy"; Zhang et al. 2011) may be a common denominator of both cognitive impairments and neurodegenerative processes. Thus, just as we are motivated to continue developing a more complete understanding of intracellular Ca^{2+} signaling and its role in the dialog between synapse and nucleus, so too can we be driven by the knowledge that our findings may lead the way to an improved understanding of those changes responsible for the altered emotional and conceptual responses that define mental illness, the impaired learning and memory that accompanies developmental cognitive disorders, and the cell death associated with neurodegenerative diseases.

ACKNOWLEDGMENTS

We wish to thank A.M. Oliveira for helpful discussions during the preparation of this manuscript. A.M.H. was supported by a postdoctoral fellowship from CellNetworks—Cluster of Excellence at the University of Heidelberg (EXC81). H.B. is a member of CellNetworks—Cluster of Excellence (EXC81), and was supported by the Alexander von Humboldt Foundation, the ERC Advanced Grant, the Deutsche Forschungsgemeinschaft, the EU Network of Excellence NeuroNE, and the EU Project GRIPANNT.

REFERENCES

Abel T, Zukin RS. 2008. Epigenetic targets of HDAC inhibition in neurodegenerative and psychiatric disorders. *Curr Opin Pharmacol* **8:** 57–64.

Adachi M, Fukuda M, Nishida E. 1999. Two co-existing mechanisms for nuclear import of MAP kinase: Passive diffusion of a monomer and active transport of a dimer. *EMBO J* **18:** 5347–5358.

Akey CW. 1989. Interactions and structure of the nuclear pore complex revealed by cryo-electron microscopy. *J Cell Biol* **109:** 955–970.

Alarcon JM, Malleret G, Touzani K, Vronskaya S, Ishii S, Kandel ER, Barco A. 2004. Chromatin acetylation, memory, and LTP are impaired in CBP$^{+/-}$ mice: A model for the cognitive deficit in Rubinstein-Taybi syndrome and its amelioration. *Neuron* **42:** 947–959.

Alexander JC, McDermott CM, Tunur T, Rands V, Stelly C, Karhson D, Bowlby MR, An WF, Sweatt JD, Schrader LA. 2009. The role of calsenilin/DREAM/KChIP3 in contextual fear conditioning. *Learn Mem* **16:** 167–177.

Al-Mubarak B, Soriano FX, Hardingham GE. 2009. Synaptic NMDAR activity suppresses FOXO1 expression via a cis-acting FOXO binding site: FOXO1 is a FOXO target gene. *Channels* **3:** 233–238.

Anderson KA, Means RL, Huang QH, Kemp BE, Goldstein EG, Selbert MA, Edelman AM, Fremeau RT, Means AR. 1998. Components of a calmodulin-dependent protein kinase cascade. Molecular cloning, functional characterization and cellular localization of $Ca^{2+}/$calmodulin-dependent protein kinase kinase β. *J Biol Chem* **273:** 31880–31889.

Anderson KA, Noeldner PK, Reece K, Wadzinski BE, Means AR. 2004. Regulation and function of the calcium/calmodulin-dependent protein kinase IV/protein serine/threonine phosphatase 2A signaling complex. *J Biol Chem* **279:** 31708–31716.

Arnold FJ, Hofmann F, Bengtson CP, Wittmann M, Vanhoutte P, Bading H. 2005. Microelectrode array recordings of cultured hippocampal networks reveal a simple model for transcription and protein synthesis-dependent plasticity. *J Physiol* **564:** 3–19.

Arundine M, Tymianski M. 2004. Molecular mechanisms of glutamate-dependent neurodegeneration in ischemia and traumatic brain injury. *Cell Mol Life Sci* **61:** 657–668.

Bading H. 2000. Transcription-dependent neuronal plasticity the nuclear calcium hypothesis. *Eur J Biochem* **267:** 5280–5283.

Bading H, Greenberg ME. 1991. Stimulation of protein tyrosine phosphorylation by NMDA receptor activation. *Science* **253:** 912–914.

Bading H, Ginty DD, Greenberg ME. 1993. Regulation of gene expression in hippocampal neurons by distinct calcium signaling pathways. *Science* **260:** 181–186.

Bading H, Segal MM, Sucher NJ, Dudek H, Lipton SA, Greenberg ME. 1995. N-methyl-D-aspartate receptors are critical for mediating the effects of glutamate on intracellular calcium concentration and immediate early gene expression in cultured hippocampal neurons. *Neuroscience* **64:** 653–664.

Bading H, Hardingham GE, Johnson CM, Chawla S. 1997. Gene regulation by nuclear and cytoplasmic calcium signals. *Biochem Biophys Res Commun* **236:** 541–543.

Bailey CH, Bartsch D, Kandel ER. 1996. Toward a molecular definition of long-term memory storage. *Proc Natl Acad Sci* **93:** 13445–13452.

Bannister NJ, Larkman AU. 1995. Dendritic morphology of CA1 pyramidal neurones from the rat hippocampus: II. Spine distributions. *J Comp Neurol* **360:** 161–171.

Barreto G, Schafer A, Marhold J, Stach D, Swaminathan SK, Handa V, Doderlein G, Maltry N, Wu W, Lyko F, et al. 2007. Gadd45a promotes epigenetic gene activation by repair-mediated DNA demethylation. *Nature* **445:** 671–675.

Beals CR, Clipstone NA, Ho SN, Crabtree GR. 1997. Nuclear localization of NF-ATc by a calcineurin-dependent, cyclosporin-sensitive intramolecular interaction. *Genes Dev* **11:** 824–834.

Bedford DC, Kasper LH, Fukuyama T, Brindle PK. 2010. Target gene context influences the transcriptional requirement for the KAT3 family of CBP and p300 histone acetyltransferases. *Epigenetics* **5:** 9–15.

Belfield JL, Whittaker C, Cader MZ, Chawla S. 2006. Differential effects of Ca^{2+} and cAMP on transcription mediated by MEF2D and cAMP-response element-binding protein in hippocampal neurons. *J Biol Chem* **281:** 27724–27732.

Berridge MJ. 1998. Neuronal calcium signaling. *Neuron* **21:** 13–26.

Berridge MJ, Lipp P, Bootman MD. 2000. The versatility and universality of calcium signalling. *Nat Rev Mol Cell Biol* **1:** 11–21.

Bezprozvanny I. 2005. The inositol 1,4,5-trisphosphate receptors. *Cell Calcium* **38:** 261–272.

Bezprozvanny I. 2007. Inositol 1,4,5-tripshosphate receptor, calcium signalling and Huntington's disease. *Subcell Biochem* **45:** 323–335.

Bezprozvanny I, Ehrlich BE. 1994. Inositol (1,4,5)-trisphosphate (InsP3)-gated Ca channels from cerebellum: Conduction properties for divalent cations and regulation by intraluminal calcium. *J Gen Physiol* **104:** 821–856.

Bezprozvanny I, Watras J, Ehrlich BE. 1991. Bell-shaped calcium-response curves of Ins(1,4,5)P3- and calcium-gated channels from endoplasmic reticulum of cerebellum. *Nature* **351:** 751–754.

Billups D, Billups B, Challiss RA, Nahorski SR. 2006. Modulation of Gq-protein-coupled inositol trisphosphate and Ca^{2+} signaling by the membrane potential. *J Neurosci* **26:** 9983–9995.

Bito H, Deisseroth K, Tsien RW. 1996. CREB phosphorylation and dephosphorylation: A Ca^{2+}- and stimulus duration-dependent switch for hippocampal gene expression. *Cell* **87:** 1203–1214.

Bkaily G, Avedanian L, Jacques D. 2009. Nuclear membrane receptors and channels as targets for drug development in cardiovascular diseases. *Can J Physiol Pharmacol* **87:** 108–119.

Bolger TA, Yao TP. 2005. Intracellular trafficking of histone deacetylase 4 regulates neuronal cell death. *J Neurosci* **25:** 9544–9553.

Bootman M, Niggli E, Berridge M, Lipp P. 1997. Imaging the hierarchical Ca^{2+} signalling system in HeLa cells. *J Physiol* **4992:** 307–314.

Bootman MD, Fearnley C, Smyrnias I, Macdonald F, Roderick HL. 2009. An update on nuclear calcium signalling. *J Cell Sci* **122:** 2337–2350.

Bossy-Wetzel E, Schwarzenbacher R, Lipton SA. 2004. Molecular pathways to neurodegeneration. *Nat Med* **10** (Suppl): S2–S9.

Brami-Cherrier K, Roze E, Girault JA, Betuing S, Caboche J. 2009. Role of the ERK/MSK1 signalling pathway in chromatin remodelling and brain responses to drugs of abuse. *J Neurochem* **108:** 1323–1335.

Brunet A, Bonni A, Zigmond MJ, Lin MZ, Juo P, Hu LS, Anderson MJ, Arden KC, Blenis J, Greenberg ME. 1999. Akt promotes cell survival by phosphorylating and inhibiting a Forkhead transcription factor. *Cell* **96:** 857–868.

Bugrim AE. 1999. Regulation of Ca^{2+} release by cAMP-dependent protein kinase. A mechanism for agonist-specific calcium signaling? *Cell Calcium* **25:** 219–226.

Caceres A, Bender P, Snavely L, Rebhun LI, Steward O. 1983. Distribution and subcellular localization of calmodulin in adult and developing brain tissue. *Neuroscience* **10:** 449–461.

Callamaras N, Marchant JS, Sun XP, Parker I. 1998. Activation and co-ordination of InsP3-mediated elementary Ca^{2+} events during global Ca^{2+} signals in *Xenopus* oocytes. *J Physiol* **5091:** 81–91.

Carrasco MA, Jaimovich E, Kemmerling U, Hidalgo C. 2004. Signal transduction and gene expression regulated by calcium release from internal stores in excitable cells. *Biol Res* **37:** 701–712.

Carrion AM, Link WA, Ledo F, Mellstrom B, Naranjo JR. 1999. DREAM is a Ca^{2+}-regulated transcriptional repressor. *Nature* **398:** 80–84.

Castren E, Berninger B, Leingartner A, Lindholm D. 1998. Regulation of brain-derived neurotrophic factor mRNA levels in hippocampus by neuronal activity. *Prog Brain Res* **117:** 57–64.

Catterall WA, Perez-Reyes E, Snutch TP, Striessnig J. 2005. International Union of Pharmacology. XLVIII. Nomenclature and structure–function relationships of voltage-gated calcium channels. *Pharmacol Rev* **57:** 411–425.

Chahrour M, Jung SY, Shaw C, Zhou X, Wong ST, Qin J, Zoghbi HY. 2008. MeCP2, a key contributor to neurological disease, activates and represses transcription. *Science* **320:** 1224–1229.

Chandler LJ, Sutton G, Dorairaj NR, Norwood D. 2001. N-methyl D-aspartate receptor-mediated bidirectional control of extracellular signal-regulated kinase activity in cortical neuronal cultures. *J Biol Chem* **276:** 2627–2636.

Chandramohan Y, Droste SK, Arthur JS, Reul JM. 2008. The forced swimming-induced behavioural immobility response involves histone H3 phospho-acetylation and c-Fos induction in dentate gyrus granule neurons via activation of the N-methyl-D-aspartate/extracellular signal-regulated kinase/mitogen- and stress-activated kinase signalling pathway. *Eur J Neurosci* **27:** 2701–2713.

Chawla S. 2002. Regulation of gene expression by Ca^{2+} signals in neuronal cells. *Eur J Pharmacol* **447:** 131–140.

Chawla S, Bading H. 2001. CREB/CBP and SRE-interacting transcriptional regulators are fast on-off switches: Duration of calcium transients specifies the magnitude of transcriptional responses. *J Neurochem* **79:** 849–858.

Chawla S, Hardingham GE, Quinn DR, Bading H. 1998. CBP: A signal-regulated transcriptional coactivator controlled by nuclear calcium and CaM kinase IV. *Science* **281:** 1505–1509.

Chawla S, Vanhoutte P, Arnold FJ, Huang CL, Bading H. 2003. Neuronal activity-dependent nucleocytoplasmic shuttling of HDAC4 and HDAC5. *J Neurochem* **85:** 151–159.

Chen WG, Chang Q, Lin Y, Meissner A, West AE, Griffith EC, Jaenisch R, Greenberg ME. 2003. Derepression of BDNF transcription involves calcium-dependent phosphorylation of MeCP2. *Science* **302:** 885–889.

Choi YM, Kim SH, Chung S, Uhm DY, Park MK. 2006. Regional interaction of endoplasmic reticulum Ca^{2+} signals between soma and dendrites through rapid luminal Ca^{2+} diffusion. *J Neurosci* **26:** 12127–12136.

Chow FA, Anderson KA, Noeldner PK, Means AR. 2005. The autonomous activity of calcium/calmodulin-dependent protein kinase IV is required for its role in transcription. *J Biol Chem* **280:** 20530–20538.

Chuderland D, Konson A, Seger R. 2008. Identification and characterization of a general nuclear translocation signal in signaling proteins. *Mol Cell* **31:** 850–861.

Chwang WB, O'Riordan KJ, Levenson JM, Sweatt JD. 2006. ERK/MAPK regulates hippocampal histone phosphorylation following contextual fear conditioning. *Learn Mem* **13**: 322–328.

Chwang WB, Arthur JS, Schumacher A, Sweatt JD. 2007. The nuclear kinase mitogen- and stress-activated protein kinase 1 regulates hippocampal chromatin remodeling in memory formation. *J Neurosci* **27**: 12732–12742.

Cohen S, Greenberg ME. 2008. Communication between the synapse and the nucleus in neuronal development, plasticity, and disease. *Annu Rev Cell Dev Biol* **24**: 183–209.

Cole AJ, Saffen DW, Baraban JM, Worley PF. 1989. Rapid increase of an immediate early gene messenger RNA in hippocampal neurons by synaptic NMDA receptor activation. *Nature* **340**: 474–476.

Cruzalegui FH, Bading H. 2000. Calcium-regulated protein kinase cascades and their transcription factor targets. *Cell Mol Life Sci* **57**: 402–410.

Dai R, Ali MK, Lezcano N, Bergson C. 2008. A crucial role for cAMP and protein kinase a in d1 dopamine receptor regulated intracellular calcium transients. *Neurosignals* **16**: 112–123.

Dai S, Hall DD, Hell JW. 2009. Supramolecular assemblies and localized regulation of voltage-gated ion channels. *Physiol Rev* **89**: 411–452.

Danker T, Shahin V, Schlune A, Schafer C, Oberleithner H. 2001. Electrophoretic plugging of nuclear pores by using the nuclear hourglass technique. *J Membr Biol* **184**: 91–99.

Deisseroth K, Bito H, Tsien RW. 1996. Signaling from synapse to nucleus: Postsynaptic CREB phosphorylation during multiple forms of hippocampal synaptic plasticity. *Neuron* **16**: 89–101.

Deisseroth K, Heist EK, Tsien RW. 1998. Translocation of calmodulin to the nucleus supports CREB phosphorylation in hippocampal neurons. *Nature* **392**: 198–202.

Dick O, Bading H. 2010. Synaptic activity and nuclear calcium signaling protect hippocampal neurons from death signal-associated nuclear translocation of FoxO3a induced by extrasynaptic N-methyl-D-aspartate receptors. *J Biol Chem* **285**: 19354–19361.

Dieterich DC, Karpova A, Mikhaylova M, Zdobnova I, Konig I, Landwehr M, Kreutz M, Smalla KH, Richter K, Landgraf P, et al. 2008. Caldendrin-Jacob: A protein liaison that couples NMDA receptor signalling to the nucleus. *PLoS Biol* **6**: e34.

Dolmetsch RE, Pajvani U, Fife K, Spotts JM, Greenberg ME. 2001. Signaling to the nucleus by an L-type calcium channel-calmodulin complex through the MAP kinase pathway. *Science* **294**: 333–339.

Duclot F, Meffre J, Jacquet C, Gongora C, Maurice T. 2010. Mice knock out for the histone acetyltransferase p300/CREB binding protein-associated factor develop a resistance to amyloid toxicity. *Neuroscience* **167**: 850–863.

Dudek SM, Fields RD. 2001. Mitogen-activated protein kinase/extracellular signal-regulated kinase activation in somatodendritic compartments: Roles of action potentials, frequency, and mode of calcium entry. *J Neurosci* **21**: RC122.

Dudek SM, Fields RD. 2002. Somatic action potentials are sufficient for late-phase LTP-related cell signaling. *Proc Natl Acad Sci* **99**: 3962–3967.

Echevarria W, Leite MF, Guerra MT, Zipfel WR, Nathanson MH. 2003. Regulation of calcium signals in the nucleus by a nucleoplasmic reticulum. *Nat Cell Biol* **5**: 440–446.

Eder A, Bading H. 2007. Calcium signals can freely cross the nuclear envelope in hippocampal neurons: Somatic calcium increases generate nuclear calcium transients. *BMC Neurosci* **8**: 57.

Emptage N, Bliss TV, Fine A. 1999. Single synaptic events evoke NMDA receptor-mediated release of calcium from internal stores in hippocampal dendritic spines. *Neuron* **22**: 115–124.

Erickson ES, Mooren OL, Moore D, Krogmeier JR, Dunn RC. 2006. The role of nuclear envelope calcium in modifying nuclear pore complex structure. *Can J Physiol Pharmacol* **84**: 309–318.

Etkin A, Alarcon JM, Weisberg SP, Touzani K, Huang YY, Nordheim A, Kandel ER. 2006. A role in learning for SRF: Deletion in the adult forebrain disrupts LTD and the formation of an immediate memory of a novel context. *Neuron* **50**: 127–143.

Fagerlund R, Melen K, Cao X, Julkunen I. 2008. NF-κB p52, RelB and c-Rel are transported into the nucleus via a subset of importin α molecules. *Cell Signal* **20**: 1442–1451.

Favaron M, Manev RM, Rimland JM, Candeo P, Beccaro M, Manev H. 1993. NMDA-stimulated expression of BDNF mRNA in cultured cerebellar granule neurones. *Neuroreport* **4**: 1171–1174.

Feng J, Zhou Y, Campbell SL, Le T, Li E, Sweatt JD, Silva AJ, Fan G. 2010. Dnmt1 and Dnmt3a maintain DNA methylation and regulate synaptic function in adult forebrain neurons. *Nat Neurosci* **13**: 423–430.

Fischer A, Sananbenesi F, Wang X, Dobbin M, Tsai LH. 2007. Recovery of learning and memory is associated with chromatin remodelling. *Nature* **447**: 178–182.

Fitzpatrick JS, Hagenston AM, Hertle DN, Gipson KE, Bertetto-D'Angelo L, Yeckel MF. 2009. Inositol-1,4,5-trisphosphate receptor-mediated Ca^{2+} waves in pyramidal neuron dendrites propagate through hot spots and cold spots. *J Physiol* **587**: 1439–1459.

Flavell SW, Cowan CW, Kim TK, Greer PL, Lin Y, Paradis S, Griffith EC, Hu LS, Chen C, Greenberg ME. 2006. Activity-dependent regulation of MEF2 transcription factors suppresses excitatory synapse number. *Science* **311**: 1008–1012.

Fontan-Lozano A, Romero-Granados R, del-Pozo-Martin Y, Suarez-Pereira I, Delgado-Garcia JM, Penninger JM, Carrion AM. 2009. Lack of DREAM protein enhances learning and memory and slows brain aging. *Curr Biol* **19**: 54–60.

Foskett JK, White C, Cheung KH, Mak DO. 2007. Inositol trisphosphate receptor Ca^{2+} release channels. *Physiol Rev* **87**: 593–658.

Francis JS, Dragunow M, During MJ. 2004. Over expression of ATF-3 protects rat hippocampal neurons from in vivo injection of kainic acid. *Brain Res Mol Brain Res* **124**: 199–203.

Friel D. 2004. Interplay between ER Ca^{2+} uptake and release fluxes in neurons and its impact on $[Ca^{2+}]$ dynamics. *Biol Res* **37**: 665–674.

Gall C, Lauterborn J. 1992. The dentate gyrus: A model system for studies of neurotrophin regulation. *Epilepsy Res Suppl* **7**: 171–185.

Gao C, Gill MB, Tronson NC, Guedea AL, Guzman YF, Huh KH, Corcoran KA, Swanson GT, Radulovic J. 2010. Hippocampal NMDA receptor subunits differentially regulate fear memory formation and neuronal signal propagation. *Hippocampus* **20**: 1072–1082.

Gerasimenko JV, Maruyama Y, Yano K, Dolman NJ, Tepikin AV, Petersen OH, Gerasimenko OV. 2003. NAADP mobilizes Ca^{2+} from a thapsigargin-sensitive store in the nuclear envelope by activating ryanodine receptors. *J Cell Biol* **163**: 271–282.

Gilley J, Coffer PJ, Ham J. 2003. FOXO transcription factors directly activate bim gene expression and promote apoptosis in sympathetic neurons. *J Cell Biol* **162**: 613–622.

Graef IA, Mermelstein PG, Stankunas K, Neilson JR, Deisseroth K, Tsien RW, Crabtree GR. 1999. L-type calcium channels and GSK-3 regulate the activity of NF-ATc4 in hippocampal neurons. *Nature* **401**: 703–708.

Graff J, Mansuy IM. 2009. Epigenetic dysregulation in cognitive disorders. *Eur J Neurosci* **30**: 1–8.

Gray PC, Johnson BD, Westenbroek RE, Hays LG, Yates JR III, Scheuer T, Catterall WA, Murphy BJ. 1998. Primary structure and function of an A kinase anchoring protein associated with calcium channels. *Neuron* **20**: 1017–1026.

Greber UF, Gerace L. 1995. Depletion of calcium from the lumen of endoplasmic reticulum reversibly inhibits passive diffusion and signal-mediated transport into the nucleus. *J Cell Biol* **128**: 5–14.

Greenberg ME, Xu B, Lu B, Hempstead BL. 2009. New insights in the biology of BDNF synthesis and release: Implications in CNS function. *J Neurosci* **29**: 12764–12767.

Greer PL, Greenberg ME. 2008. From synapse to nucleus: Calcium-dependent gene transcription in the control of synapse development and function. *Neuron* **59**: 846–860.

Gulledge AT, Stuart GJ. 2005. Cholinergic inhibition of neocortical pyramidal neurons. *J Neurosci* **25**: 10308–10320.

Hagenston AM, Fitzpatrick JS, Yeckel MF. 2008. MGluR-mediated calcium waves that invade the soma regulate firing in layer V medial prefrontal cortical pyramidal neurons. *Cereb Cortex* **18**: 407–423.

Hansen HH, Briem T, Dzietko M, Sifringer M, Voss A, Rzeski W, Zdzisinska B, Thor F, Heumann R, Stepulak A, et al. 2004. Mechanisms leading to disseminated apoptosis following NMDA receptor blockade in the developing rat brain. *Neurobiol Dis* **16**: 440–453.

Hardingham GE. 2009. Coupling of the NMDA receptor to neuroprotective and neurodestructive events. *Biochem Soc Trans* **37**: 1147–1160.

Hardingham GE, Bading H. 1998. Nuclear calcium: a key regulator of gene expression. *Biometals* **11**: 345–358.

Hardingham GE, Bading H. 2002. Coupling of extrasynaptic NMDA receptors to a CREB shut-off pathway is developmentally regulated. *Biochim Biophys Acta* **1600**: 148–153.

Hardingham GE, Bading H. 2003. The Yin and Yang of NMDA receptor signalling. *Trends Neurosci* **26**: 81–89.

Hardingham GE, Bading H. 2010. Synaptic versus extrasynaptic NMDA receptor signalling: implications for neurodegenerative disorders. *Nat Rev Neurosci* **11**: 682–696.

Hardingham GE, Chawla S, Johnson CM, Bading H. 1997. Distinct functions of nuclear and cytoplasmic calcium in the control of gene expression. *Nature* **385**: 260–265.

Hardingham GE, Chawla S, Cruzalegui FH, Bading H. 1999. Control of recruitment and transcription-activating function of CBP determines gene regulation by NMDA receptors and L-type calcium channels. *Neuron* **22**: 789–798.

Hardingham GE, Arnold FJ, Bading H. 2001a. A calcium microdomain near NMDA receptors: on switch for ERK-dependent synapse-to-nucleus communication. *Nat Neurosci* **4**: 565–566.

Hardingham GE, Arnold FJ, Bading H. 2001b. Nuclear calcium signaling controls CREB-mediated gene expression triggered by synaptic activity. *Nat Neurosci* **4**: 261–267.

Hardingham GE, Fukunaga Y, Bading H. 2002. Extrasynaptic NMDARs oppose synaptic NMDARs by triggering CREB shut-off and cell death pathways. *Nat Neurosci* **5**: 405–414.

Harvey CD, Yasuda R, Zhong H, Svoboda K. 2008. The spread of Ras activity triggered by activation of a single dendritic spine. *Science* **321**: 136–140.

Hell JW, Westenbroek RE, Warner C, Ahlijanian MK, Prystay W, Gilbert MM, Snutch TP, Catterall WA. 1993. Identification and differential subcellular localization of the neuronal class C and class D L-type calcium channel α 1 subunits. *J Cell Biol* **123**: 949–962.

Hong M, Ross WN. 2007. Priming of intracellular calcium stores in rat CA1 pyramidal neurons. *J Physiol* **584**: 75–87.

Hong EJ, West AE, Greenberg ME. 2005. Transcriptional control of cognitive development. *Curr Opin Neurobiol* **15**: 21–28.

Hoskins B, Burton CK, Liu DD, Porter AB, Ho IK. 1986. Regional and subcellular calmodulin content of rat brain. *J Neurochem* **46**: 303–304.

Hu SC, Chrivia J, Ghosh A. 1999. Regulation of CBP-mediated transcription by neuronal calcium signaling. *Neuron* **22**: 799–808.

Hu M, Liu QS, Chang KT, Berg DK. 2002. Nicotinic regulation of CREB activation in hippocampal neurons by glutamatergic and nonglutamatergic pathways. *Mol Cell Neurosci* **21**: 616–625.

Humbert JP, Matter N, Artault JC, Koppler P, Malviya AN. 1996. Inositol 1,4,5-trisphosphate receptor is located to the inner nuclear membrane vindicating regulation of nuclear calcium signaling by inositol 1,4,5-trisphosphate. Discrete distribution of inositol phosphate receptors to inner and outer nuclear membranes. *J Biol Chem* **271**: 478–485.

Husi H, Ward MA, Choudhary JS, Blackstock WP, Grant SG. 2000. Proteomic analysis of NMDA receptor-adhesion protein signaling complexes. *Nat Neurosci* **3**: 661–669.

Impey S, Fong AL, Wang Y, Cardinaux JR, Fass DM, Obrietan K, Wayman GA, Storm DR, Soderling TR, Goodman RH. 2002. Phosphorylation of CBP mediates transcriptional activation by neural activity and CaM kinase IV. *Neuron* **34**: 235–244.

Ivanov A, Pellegrino C, Rama S, Dumalska I, Salyha Y, Ben-Ari Y, Medina I. 2006. Opposing role of synaptic and extrasynaptic NMDA receptors in regulation of the extracellular signal-regulated kinases (ERK) activity in cultured rat hippocampal neurons. *J Physiol* **572:** 789–798.

Jaffe DB, Brown TH. 1994. Metabotropic glutamate receptor activation induces calcium waves within hippocampal dendrites. *J Neurophysiol* **72:** 471–474.

Jeffrey RA, Ch'ng TH, O'Dell TJ, Martin KC. 2009. Activity-dependent anchoring of importin α at the synapse involves regulated binding to the cytoplasmic tail of the NR1-1a subunit of the NMDA receptor. *J Neurosci* **29:** 15613–15620.

Jensen KF, Ohmstede CA, Fisher RS, Sahyoun N. 1991. Nuclear and axonal localization of Ca^{2+}/calmodulin-dependent protein kinase type Gr in rat cerebellar cortex. *Proc Natl Acad Sci* **88:** 2850–2853.

Jiang X, Tian F, Mearow K, Okagaki P, Lipsky RH, Marini AM. 2005. The excitoprotective effect of N-methyl-D-aspartate receptors is mediated by a brain-derived neurotrophic factor autocrine loop in cultured hippocampal neurons. *J Neurochem* **94:** 713–722.

Jiang H, Poirier MA, Liang Y, Pei Z, Weiskittel CE, Smith WW, DeFranco DB, Ross CA. 2006. Depletion of CBP is directly linked with cellular toxicity caused by mutant huntingtin. *Neurobiol Dis* **23:** 543–551.

Johenning FW, Holthoff K. 2007. Nuclear calcium signals during L-LTP induction do not predict the degree of synaptic potentiation. *Cell Calcium* **41:** 271–283.

Jonas P, Burnashev N. 1995. Molecular mechanisms controlling calcium entry through AMPA-type glutamate receptor channels. *Neuron* **15:** 987–990.

Jones PL, Veenstra GJ, Wade PA, Vermaak D, Kass SU, Landsberger N, Strouboulis J, Wolffe AP. 1998. Methylated DNA and MeCP2 recruit histone deacetylase to repress transcription. *Nat Genet* **19:** 187–191.

Jong YJ, Kumar V, Kingston AE, Romano C, O'Malley KL. 2005. Functional metabotropic glutamate receptors on nuclei from brain and primary cultured striatal neurons. Role of transporters in delivering ligand. *J Biol Chem* **280:** 30469–30480.

Jong YJ, Schwetye KE, O'Malley KL. 2007. Nuclear localization of functional metabotropic glutamate receptor mGlu1 in HEK293 cells and cortical neurons: role in nuclear calcium mobilization and development. *J Neurochem* **101:** 458–469.

Jordan BA, Kreutz MR. 2009. Nucleocytoplasmic protein shuttling: the direct route in synapse-to-nucleus signaling. *Trends Neurosci* **32:** 392–401.

Jordan BA, Fernholz BD, Khatri L, Ziff EB. 2007. Activity-dependent AIDA-1 nuclear signaling regulates nucleolar numbers and protein synthesis in neurons. *Nat Neurosci* **10:** 427–435.

Jorgensen HF, Bird A. 2002. MeCP2 and other methyl-CpG binding proteins. *Ment Retard Dev Disabil Res Rev* **8:** 87–93.

Kalia LV, Kalia SK, Salter MW. 2008. NMDA receptors in clinical neurology: Excitatory times ahead. *Lancet Neurol* **7:** 742–755.

Kapur A, Yeckel M, Johnston D. 2001. Hippocampal mossy fiber activity evokes Ca2+ release in CA3 pyramidal neurons via a metabotropic glutamate receptor pathway. *Neuroscience* **107:** 59–69.

Kasahara J, Fukunaga K, Miyamoto E. 1999. Differential effects of a calcineurin inhibitor on glutamate-induced phosphorylation of Ca^{2+}/calmodulin-dependent protein kinases in cultured rat hippocampal neurons. *J Biol Chem* **274:** 9061–9067.

Korkotian E, Segal M. 1999. Release of calcium from stores alters the morphology of dendritic spines in cultured hippocampal neurons. *Proc Natl Acad Sci* **96:** 12068–12072.

Korzus E, Rosenfeld MG, Mayford M. 2004. CBP histone acetyltransferase activity is a critical component of memory consolidation. *Neuron* **42:** 961–972.

Koshibu K, Graff J, Beullens M, Heitz FD, Berchtold D, Russig H, Farinelli M, Bollen M, Mansuy IM. 2009. Protein phosphatase 1 regulates the histone code for long-term memory. *J Neurosci* **29:** 13079–13089.

Kotera I, Sekimoto T, Miyamoto Y, Saiwaki T, Nagoshi E, Sakagami H, Kondo H, Yoneda Y. 2005. Importin α transports CaMKIV to the nucleus without utilizing importin β. *EMBO J* **24:** 942–951.

Kramer JM, van Bokhoven H. 2009. Genetic and epigenetic defects in mental retardation. *Int J Biochem Cell Biol* **41:** 96–107.

Krapivinsky G, Krapivinsky L, Manasian Y, Ivanov A, Tyzio R, Pellegrino C, Ben-Ari Y, Clapham DE, Medina I. 2003. The NMDA receptor is coupled to the ERK pathway by a direct interaction between NR2B and RasGRF1. *Neuron* **40:** 775–784.

Kumar V, Jong YJ, O'Malley KL. 2008. Activated nuclear metabotropic glutamate receptor mGlu5 couples to nuclear Gq/11 proteins to generate inositol 1,4,5-trisphosphate-mediated nuclear Ca^{2+} release. *J Biol Chem* **283:** 14072–14083.

Lai KO, Zhao Y, Ch'ng TH, Martin KC. 2008. Importin-mediated retrograde transport of CREB2 from distal processes to the nucleus in neurons. *Proc Natl Acad Sci* **105:** 17175–17180.

Lalonde J, Lachance PE, Chaudhuri A. 2004. Monocular enucleation induces nuclear localization of calcium/calmodulin-dependent protein kinase IV in cortical interneurons of adult monkey area V1. *J Neurosci* **24:** 554–564.

Lange A, Mills RE, Lange CJ, Stewart M, Devine SE, Corbett AH. 2007. Classical nuclear localization signals: definition, function, and interaction with importin α. *J Biol Chem* **282:** 5101–5105.

Larkum ME, Watanabe S, Nakamura T, Lasser-Ross N, Ross WN. 2003. Synaptically activated Ca^{2+} waves in layer 2/3 and layer 5 rat neocortical pyramidal neurons. *J Physiol* **549:** 471–488.

Lau D, Bading H. 2009. Synaptic activity-mediated suppression of p53 and induction of nuclear calcium-regulated neuroprotective genes promote survival through inhibition of mitochondrial permeability transition. *J Neurosci* **29:** 4420–4429.

Ledo F, Carrion AM, Link WA, Mellstrom B, Naranjo JR. 2000. DREAM-αCREM interaction via leucine-charged domains derepresses downstream regulatory element-dependent transcription. *Mol Cell Biol* **20:** 9120–9126.

Ledo F, Kremer L, Mellstrom B, Naranjo JR. 2002. Ca^{2+}-dependent block of CREB-CBP transcription by repressor DREAM. *EMBO J* **21**: 4583–4592.

Lee MA, Dunn RC, Clapham DE, Stehno-Bittel L. 1998. Calcium regulation of nuclear pore permeability. *Cell Calcium* **23**: 91–101.

Legube G, Trouche D. 2003. Regulating histone acetyltransferases and deacetylases. *EMBO Rep* **4**: 944–947.

Leitch B, Szostek A, Lin R, Shevtsova O. 2009. Subcellular distribution of L-type calcium channel subtypes in rat hippocampal neurons. *Neuroscience* **164**: 641–657.

Lemrow SM, Anderson KA, Joseph JD, Ribar TJ, Noeldner PK, Means AR. 2004. Catalytic activity is required for calcium/calmodulin-dependent protein kinase IV to enter the nucleus. *J Biol Chem* **279**: 11664–11671.

Leveille F, El Gaamouch F, Gouix E, Lecocq M, Lobner D, Nicole O, Buisson A. 2008. Neuronal viability is controlled by a functional relation between synaptic and extrasynaptic NMDA receptors. *FASEB J* **22**: 4258–4271.

Leveille F, Papadia S, Fricker M, Bell KF, Soriano FX, Martel MA, Puddifoot C, Habel M, Wyllie DJ, Ikonomidou C, et al. 2010. Suppression of the intrinsic apoptosis pathway by synaptic activity. *J Neurosci* **30**: 2623–2635.

Levenson JM, Sweatt JD. 2006. Epigenetic mechanisms: a common theme in vertebrate and invertebrate memory formation. *Cell Mol Life Sci* **63**: 1009–1016.

Levenson JM, Roth TL, Lubin FD, Miller CA, Huang IC, Desai P, Malone LM, Sweatt JD. 2006. Evidence that DNA (cytosine-5) methyltransferase regulates synaptic plasticity in the hippocampus. *J Biol Chem* **281**: 15763–15773.

Lezcano N, Bergson C. 2002. D1/D5 dopamine receptors stimulate intracellular calcium release in primary cultures of neocortical and hippocampal neurons. *J Neurophysiol* **87**: 2167–2175.

Li YX, Keizer J, Stojilkovic SS, Rinzel J. 1995. Ca^{2+} excitability of the ER membrane: an explanation for IP3-induced Ca^{2+} oscillations. *Am J Physiol* **269**: C1079–1092.

Li QJ, Yang SH, Maeda Y, Sladek FM, Sharrocks AD, Martins-Green M. 2003. MAP kinase phosphorylation-dependent activation of Elk-1 leads to activation of the co-activator 300. *EMBO J* **22**: 281–291.

Lilliehook C, Bozdagi O, Yao J, Gomez-Ramirez M, Zaidi NF, Wasco W, Gandy S, Santucci AC, Haroutunian V, Huntley GW, et al. 2003. Altered Aβ formation and long-term potentiation in a calsenilin knock-out. *J Neurosci* **23**: 9097–9106.

Limback-Stokin K, Korzus E, Nagaoka-Yasuda R, Mayford M. 2004. Nuclear calcium/calmodulin regulates memory consolidation. *J Neurosci* **24**: 10858–10867.

Liu Z, Ren J, Murphy TH. 2003. Decoding of synaptic voltage waveforms by specific classes of recombinant high-threshold Ca^{2+} channels. *J Physiol* **553**: 473–488.

Liu Y, Wong TP, Aarts M, Rooyakkers A, Liu L, Lai TW, Wu DC, Lu J, Tymianski M, Craig AM, et al. 2007. NMDA receptor subunits have differential roles in mediating excitotoxic neuronal death both in vitro and in vivo. *J Neurosci* **27**: 2846–2857.

Liu QH, Zheng YM, Korde AS, Yadav VR, Rathore R, Wess J, Wang YX. 2009. Membrane depolarization causes a direct activation of G protein-coupled receptors leading to local Ca^{2+} release in smooth muscle. *Proc Natl Acad Sci U S A* **106**: 11418–11423.

Lubin FD, Roth TL, Sweatt JD. 2008. Epigenetic regulation of BDNF gene transcription in the consolidation of fear memory. *J Neurosci* **28**: 10576–10586.

Ma DK, Guo JU, Ming GL, Song H. 2009. DNA excision repair proteins and Gadd45 as molecular players for active DNA demethylation. *Cell Cycle* **8**: 1526–1531.

Mao LM, Zhang GC, Liu XY, Fibuch EE, Wang JQ. 2008. Group I metabotropic glutamate receptor-mediated gene expression in striatal neurons. *Neurochem Res* **33**: 1920–1924.

Marchenko SM, Thomas RC. 2006. Nuclear Ca^{2+} signalling in cerebellar Purkinje neurons. *Cerebellum* **5**: 36–42.

Marchenko SM, Yarotskyy VV, Kovalenko TN, Kostyuk PG, Thomas RC. 2005. Spontaneously active and InsP3-activated ion channels in cell nuclei from rat cerebellar Purkinje and granule neurones. *J Physiol* **565**: 897–910.

Martin C, Ashley RH. 1993. Reconstitution of a voltage-activated calcium conducting cation channel from brain microsomes. *Cell Calcium* **14**: 427–438.

Martinowich K, Hattori D, Wu H, Fouse S, He F, Hu Y, Fan G, Sun YE. 2003. DNA methylation-related chromatin remodeling in activity-dependent BDNF gene regulation. *Science* **302**: 890–893.

Matthews RP, Guthrie CR, Wailes LM, Zhao X, Means AR, McKnight GS. 1994. Calcium/calmodulin-dependent protein kinase types II and IV differentially regulate CREB-dependent gene expression. *Mol Cell Biol* **14**: 6107–6116.

Matzke AJM, Matzke MA. 1991. The electrical properties of the nuclear envelope, and their possible role in the regulation of eukaryotic gene expression. *Bioelectrochem Bioenerg* **25**: 357–370.

Matzke AJ, Weiger TM, Matzke M. 2010. Ion channels at the nucleus: Electrophysiology meets the genome. *Mol Plant* **3**: 642–652.

Mauceri D, Freitag HE, Oliveira AM, Bengtson CP, Bading H. 2011. Nuclear calcium-VEGFD signaling controls maintenance of dendrite arborization necessary for memory formation. *Neuron* **71**: 117–130.

Mazzanti M, Bustamante JO, Oberleithner H. 2001. Electrical dimension of the nuclear envelope. *Physiol Rev* **81**: 1–19.

McKinsey TA, Zhang CL, Olson EN. 2000. Activation of the myocyte enhancer factor-2 transcription factor by calcium/calmodulin-dependent protein kinase-stimulated binding of 14-3-3 to histone deacetylase 5. *Proc Natl Acad Sci* **97**: 14400–14405.

Means AR. 2000. Regulatory cascades involving calmodulin-dependent protein kinases. *Mol Endocrinol* **14**: 4–13.

Meffert MK, Chang JM, Wiltgen BJ, Fanselow MS, Baltimore D. 2003. NF-κB functions in synaptic signaling and behavior. *Nat Neurosci* **6**: 1072–1078.

Mellstrom B, Torres B, Link WA, Naranjo JR. 2004. The BDNF gene: Exemplifying complexity in Ca^{2+}-dependent gene expression. *Crit Rev Neurobiol* **16**: 43–49.

Mermelstein PG, Bito H, Deisseroth K, Tsien RW. 2000. Critical dependence of cAMP response element-binding protein phosphorylation on L-type calcium channels supports a selective response to EPSPs in preference to action potentials. *J Neurosci* **20:** 266–273.

Mermelstein PG, Deisseroth K, Dasgupta N, Isaksen AL, Tsien RW. 2001. Calmodulin priming: nuclear translocation of a calmodulin complex and the memory of prior neuronal activity. *Proc Natl Acad Sci* **98:** 15342–15347.

Miller CA, Sweatt JD. 2007. Covalent modification of DNA regulates memory formation. *Neuron* **53:** 857–869.

Miller CA, Campbell SL, Sweatt JD. 2008. DNA methylation and histone acetylation work in concert to regulate memory formation and synaptic plasticity. *Neurobiol Learn Mem* **89:** 599–603.

Miller CA, Gavin CF, White JA, Parrish RR, Honasoge A, Yancey CR, Rivera IM, Rubio MD, Rumbaugh G, Sweatt JD. 2010. Cortical DNA methylation maintains remote memory. *Nat Neurosci* **13:** 664–666.

Milnerwood AJ, Gladding CM, Pouladi MA, Kaufman AM, Hines RM, Boyd JD, Ko RW, Vasuta OC, Graham RK, Hayden MR, et al. 2010. Early increase in extrasynaptic NMDA receptor signaling and expression contributes to phenotype onset in Huntington's disease mice. *Neuron* **65:** 178–190.

Missiaen L, De Smedt H, Droogmans G, Casteels R. 1992. Ca^{2+} release induced by inositol 1,4,5-trisphosphate is a steady-state phenomenon controlled by luminal Ca^{2+} in permeabilized cells. *Nature* **357:** 599–602.

Morikawa H, Imani F, Khodakhah K, Williams JT. 2000. Inositol 1,4,5-triphosphate-evoked responses in midbrain dopamine neurons. *J Neurosci* **20:** RC103.

Murphy TH, Worley PF, Baraban JM. 1991. L-type voltage-sensitive calcium channels mediate synaptic activation of immediate early genes. *Neuron* **7:** 625–635.

Nakamura Y, Okuno S, Sato F, Fujisawa H. 1995. An immunohistochemical study of Ca^{2+}/calmodulin-dependent protein kinase IV in the rat central nervous system: Light and electron microscopic observations. *Neuroscience* **68:** 181–194.

Nakamura T, Barbara JG, Nakamura K, Ross WN. 1999. Synergistic release of Ca^{2+} from IP3-sensitive stores evoked by synaptic activation of mGluRs paired with backpropagating action potentials. *Neuron* **24:** 727–737.

Nakamura T, Lasser-Ross N, Nakamura K, Ross WN. 2002. Spatial segregation and interaction of calcium signalling mechanisms in rat hippocampal CA1 pyramidal neurons. *J Physiol* **543:** 465–480.

Nan X, Ng HH, Johnson CA, Laherty CD, Turner BM, Eisenman RN, Bird A. 1998. Transcriptional repression by the methyl-CpG-binding protein MeCP2 involves a histone deacetylase complex. *Nature* **393:** 386–389.

Nguyen PV, Abel T, Kandel ER. 1994. Requirement of a critical period of transcription for induction of a late phase of LTP. *Science* **265:** 1104–1107.

Noguchi J, Matsuzaki M, Ellis-Davies GC, Kasai H. 2005. Spine-neck geometry determines NMDA receptor-dependent Ca^{2+} signaling in dendrites. *Neuron* **46:** 609–622.

Obermair GJ, Szabo Z, Bourinet E, Flucher BE. 2004. Differential targeting of the L-type Ca^{2+} channel α 1C (CaV1.2) to synaptic and extrasynaptic compartments in hippocampal neurons. *Eur J Neurosci* **19:** 2109–2122.

O'Brien EM, Gomes DA, Sehgal S, Nathanson MH. 2007. Hormonal regulation of nuclear permeability. *J Biol Chem* **282:** 4210–4217.

Okabe S. 2007. Molecular anatomy of the postsynaptic density. *Mol Cell Neurosci* **34:** 503–518.

Oliveira AM, Abel T, Brindle PK, Wood MA. 2006. Differential role for CBP and p300 CREB-binding domain in motor skill learning. *Behav Neurosci* **120:** 724–729.

Oliveira AM, Wood MA, McDonough CB, Abel T. 2007. Transgenic mice expressing an inhibitory truncated form of p300 exhibit long-term memory deficits. *Learn Mem* **14:** 564–572.

Oliveria SF, Dell'Acqua ML, Sather WA. 2007. AKAP79/150 anchoring of calcineurin controls neuronal L-type Ca^{2+} channel activity and nuclear signaling. *Neuron* **55:** 261–275.

O'Malley KL, Jong YJ, Gonchar Y, Burkhalter A, Romano C. 2003. Activation of metabotropic glutamate receptor mGlu5 on nuclear membranes mediates intranuclear Ca^{2+} changes in heterologous cell types and neurons. *J Biol Chem* **278:** 28210–28219.

Otis KO, Thompson KR, Martin KC. 2006. Importin-mediated nuclear transport in neurons. *Curr Opin Neurobiol* **16:** 329–335.

Papadia S, Stevenson P, Hardingham NR, Bading H, Hardingham GE. 2005. Nuclear Ca^{2+} and the cAMP response element-binding protein family mediate a late phase of activity-dependent neuroprotection. *J Neurosci* **25:** 4279–4287.

Park IK, Soderling TR. 1995. Activation of Ca^{2+}/calmodulin-dependent protein kinase (CaM-kinase) IV by CaM-kinase kinase in Jurkat T lymphocytes. *J Biol Chem* **270:** 30464–30469.

Park MK, Petersen OH, Tepikin AV. 2000. The endoplasmic reticulum as one continuous Ca^{2+} pool: visualization of rapid Ca^{2+} movements and equilibration. *EMBO J* **19:** 5729–5739.

Park MK, Choi YM, Kang YK, Petersen OH. 2008. The endoplasmic reticulum as an integrator of multiple dendritic events. *Neuroscientist* **14:** 68–77.

Parker I, Ivorra I. 1990. Localized all-or-none calcium liberation by inositol trisphosphate. *Science* **250:** 977–979.

Patterson RL, Boehning D, Snyder SH. 2004. Inositol 1,4,5-trisphosphate receptors as signal integrators. *Annu Rev Biochem* **73:** 437–465.

Perez-Terzic C, Gacy AM, Bortolon R, Dzeja PP, Puceat M, Jaconi M, Prendergast FG, Terzic A. 1999. Structural plasticity of the cardiac nuclear pore complex in response to regulators of nuclear import. *Circ Res* **84:** 1292–1301.

Petrij F, Giles RH, Dauwerse HG, Saris JJ, Hennekam RC, Masuno M, Tommerup N, van Ommen GJ, Goodman RH, Peters DJ, et al. 1995. Rubinstein-Taybi syndrome caused by mutations in the transcriptional co-activator CBP. *Nature* **376:** 348–351.

Pinato G, Pegoraro S, Iacono G, Ruaro ME, Torre V. 2009. Calcium control of gene regulation in rat hippocampal neuronal cultures. *J Cell Physiol* **220:** 727–747.

Pittenger C, Huang YY, Paletzki RF, Bourtchouladze R, Scanlin H, Vronskaya S, Kandel ER. 2002. Reversible inhibition of CREB/ATF transcription factors in region

CA1 of the dorsal hippocampus disrupts hippocampus-dependent spatial memory. *Neuron* **34**: 447–462.

Power JM, Sah P. 2002. Nuclear calcium signaling evoked by cholinergic stimulation in hippocampal CA1 pyramidal neurons. *J Neurosci* **22**: 3454–3462.

Power JM, Sah P. 2005. Intracellular calcium store filling by an L-type calcium current in the basolateral amygdala at subthreshold membrane potentials. *J Physiol* **562**: 439–453.

Power JM, Sah P. 2007. Distribution of IP3-mediated calcium responses and their role in nuclear signalling in rat basolateral amygdala neurons. *J Physiol* **580**: 835–857.

Pozzo-Miller LD, Petrozzino JJ, Golarai G, Connor JA. 1996. Ca^{2+} release from intracellular stores induced by afferent stimulation of CA3 pyramidal neurons in hippocampal slices. *J Neurophysiol* **76**: 554–562.

Quesada I, Verdugo P. 2005. InsP3 signaling induces pulse-modulated Ca^{2+} signals in the nucleus of airway epithelial ciliated cells. *Biophys J* **88**: 3946–3953.

Rae MG, Martin DJ, Collingridge GL, Irving AJ. 2000. Role of Ca^{2+} stores in metabotropic L-glutamate receptor-mediated supralinear Ca^{2+} signaling in rat hippocampal neurons. *J Neurosci* **20**: 8628–8636.

Rai K, Huggins IJ, James SR, Karpf AR, Jones DA, Cairns BR. 2008. DNA demethylation in zebrafish involves the coupling of a deaminase, a glycosylase, and gadd45. *Cell* **135**: 1201–1212.

Raymond CR. 2007. LTP forms 1, 2 and 3: Different mechanisms for the "long" in long-term potentiation. *Trends Neurosci* **30**: 167–175.

Raymond CR, Redman SJ. 2002. Different calcium sources are narrowly tuned to the induction of different forms of LTP. *J Neurophysiol* **88**: 249–255.

Raymond CR, Redman SJ. 2006. Spatial segregation of neuronal calcium signals encodes different forms of LTP in rat hippocampus. *J Physiol* **570**: 97–111.

Redmond L. 2008. Translating neuronal activity into dendrite elaboration: Signaling to the nucleus. *Neurosignals* **16**: 194–208.

Redmond L, Kashani AH, Ghosh A. 2002. Calcium regulation of dendritic growth via CaM kinase IV and CREB-mediated transcription. *Neuron* **34**: 999–1010.

Reichenberg A, Mill J, MacCabe JH. 2009. Epigenetics, genomic mutations and cognitive function. *Cogn Neuropsychiatry* **14**: 377–390.

Rodrigues MA, Gomes DA, Nathanson MH, Leite MF. 2009. Nuclear calcium signaling: a cell within a cell. *Braz J Med Biol Res* **42**: 17–20.

Rose CR, Konnerth A. 2001. Stores not just for storage. intracellular calcium release and synaptic plasticity. *Neuron* **31**: 519–522.

Ryglewski S, Pflueger HJ, Duch C. 2007. Expanding the neuron's calcium signaling repertoire: Intracellular calcium release via voltage-induced PLC and IP3R activation. *PLoS Biol* **5**: e66.

Sabatini BL, Oertner TG, Svoboda K. 2002. The life cycle of Ca^{2+} ions in dendritic spines. *Neuron* **33**: 439–452.

Saha RN, Dudek SM. 2008. Action potentials: to the nucleus and beyond. *Exp Biol Med (Maywood)* **233**: 385–393.

Sakagami H, Umemiya M, Saito S, Kondo H. 2000. Distinct immunohistochemical localization of two isoforms of Ca^{2+}/calmodulin-dependent protein kinase kinases in the adult rat brain. *Eur J Neurosci* **12**: 89–99.

Sakagami H, Kamata A, Nishimura H, Kasahara J, Owada Y, Takeuchi Y, Watanabe M, Fukunaga K, Kondo H. 2005. Prominent expression and activity-dependent nuclear translocation of Ca^{2+}/calmodulin-dependent protein kinase Iδ in hippocampal neurons. *Eur J Neurosci* **22**: 2697–2707.

Saneyoshi T, Fortin DA, Soderling TR. 2010. Regulation of spine and synapse formation by activity-dependent intracellular signaling pathways. *Curr Opin Neurobiol* **20**: 108–115.

Schiller J, Major G, Koester HJ, Schiller Y. 2000. NMDA spikes in basal dendrites of cortical pyramidal neurons. *Nature* **404**: 285–289.

Schmid A, Dehlinger-Kremer M, Schulz I, Gogelein H. 1990. Voltage-dependent InsP3-insensitive calcium channels in membranes of pancreatic endoplasmic reticulum vesicles. *Nature* **346**: 374–376.

Schneider M, Spanagel R, Zhang SJ, Bading H, Klugmann M. 2007. Adeno-associated virus (AAV)-mediated suppression of Ca^{2+}/calmodulin kinase IV activity in the nucleus accumbens modulates emotional behaviour in mice. *BMC Neurosci* **8**: 105.

Scimemi A, Tian H, Diamond JS. 2009. Neuronal transporters regulate glutamate clearance, NMDA receptor activation, and synaptic plasticity in the hippocampus. *J Neurosci* **29**: 14581–14595.

Seidenbecher CI, Langnaese K, Sanmarti-Vila L, Boeckers TM, Smalla KH, Sabel BA, Garner CC, Gundelfinger ED, Kreutz MR. 1998. Caldendrin, a novel neuronal calcium-binding protein confined to the somato-dendritic compartment. *J Biol Chem* **273**: 21324–21331.

Shahin V, Danker T, Enss K, Ossig R, Oberleithner H. 2001. Evidence for Ca^{2+}- and ATP-sensitive peripheral channels in nuclear pore complexes. *FASEB J* **15**: 1895–1901.

Shalizi A, Gaudilliere B, Yuan Z, Stegmuller J, Shirogane T, Ge Q, Tan Y, Schulman B, Harper JW, Bonni A. 2006. A calcium-regulated MEF2 sumoylation switch controls postsynaptic differentiation. *Science* **311**: 1012–1017.

Sheldon AL, Robinson MB. 2007. The role of glutamate transporters in neurodegenerative diseases and potential opportunities for intervention. *Neurochem Int* **51**: 333–355.

Shemer I, Brinne B, Tegner J, Grillner S. 2008. Electrotonic signals along intracellular membranes may interconnect dendritic spines and nucleus. *PLoS Comput Biol* **4**: e1000036.

Sierra-Paredes G, Sierra-Marcuno G. 2007. Extrasynaptic GABA and glutamate receptors in epilepsy. *CNS Neurol Disord Drug Targets* **6**: 288–300.

Skene PJ, Illingworth RS, Webb S, Kerr AR, James KD, Turner DJ, Andrews R, Bird AP. 2010. Neuronal MeCP2 is expressed at near histone-octamer levels and globally alters the chromatin state. *Mol Cell* **37**: 457–468.

Solovyova N, Verkhratsky A. 2002. Monitoring of free calcium in the neuronal endoplasmic reticulum: an overview of modern approaches. *J Neurosci Methods* **122**: 1–12.

Solovyova N, Veselovsky N, Toescu EC, Verkhratsky A. 2002. Ca^{2+} dynamics in the lumen of the endoplasmic reticulum in sensory neurons: direct visualization of Ca^{2+}-induced Ca^{2+} release triggered by physiological Ca^{2+} entry. *EMBO J* **21:** 622–630.

Spruston N, Mcbain C. 2007. Structural and functional properties of hippocampal neurons. *The Hippocampus book* (ed. P Andersen, et al.), pp. 133–201. Oxford University Press, New York.

Stanika RI, Pivovarova NB, Brantner CA, Watts CA, Winters CA, Andrews SB. 2009. Coupling diverse routes of calcium entry to mitochondrial dysfunction and glutamate excitotoxicity. *Proc Natl Acad Sci* **106:** 9854–9859.

Stehno-Bittel L, Perez-Terzic C, Clapham DE. 1995. Diffusion across the nuclear envelope inhibited by depletion of the nuclear Ca^{2+} store. *Science* **270:** 1835–1838.

Stern MD. 1992. Buffering of calcium in the vicinity of a channel pore. *Cell Calcium* **13:** 183–192.

Sterner DE, Berger SL. 2000. Acetylation of histones and transcription-related factors. *Microbiol Mol Biol Rev* **64:** 435–459.

Stoffler D, Schwarz-Herion K, Aebi U, Fahrenkrog B. 2006. Getting across the nuclear pore complex: New insights into nucleocytoplasmic transport. *Can J Physiol Pharmacol* **84:** 499–507.

Straub SV, Wagner LE II, Bruce JI, Yule DI. 2004. Modulation of cytosolic calcium signaling by protein kinase A-mediated phosphorylation of inositol 1,4,5-trisphosphate receptors. *Biol Res* **37:** 593–602.

Stutzmann GE, LaFerla FM, Parker I. 2003. Ca^{2+} signaling in mouse cortical neurons studied by two-photon imaging and photoreleased inositol triphosphate. *J Neurosci* **23:** 758–765.

Tang TS, Tu H, Wang Z, Bezprozvanny I. 2003. Modulation of type 1 inositol (1,4,5)-trisphosphate receptor function by protein kinase a and protein phosphatase 1α. *J Neurosci* **23:** 403–415.

Tao J, Hu K, Chang Q, Wu H, Sherman NE, Martinowich K, Klose RJ, Schanen C, Jaenisch R, Wang W, et al. 2009. Phosphorylation of MeCP2 at Serine 80 regulates its chromatin association and neurological function. *Proc Natl Acad Sci* **106:** 4882–4887.

Taylor CW. 1998. Inositol trisphosphate receptors: Ca^{2+}-modulated intracellular Ca^{2+} channels. *Biochim Biophys Acta* **1436:** 19–33.

Thompson KR, Otis KO, Chen DY, Zhao Y, O'Dell TJ, Martin KC. 2004. Synapse to nucleus signaling during long-term synaptic plasticity; a role for the classical active nuclear import pathway. *Neuron* **44:** 997–1009.

Thrower EC, Mobasheri H, Dargan S, Marius P, Lea EJ, Dawson AP. 2000. Interaction of luminal calcium and cytosolic ATP in the control of type 1 inositol (1,4,5)-trisphosphate receptor channels. *J Biol Chem* **275:** 36049–36055.

Tian F, Hu XZ, Wu X, Jiang H, Pan H, Marini AM, Lipsky RH. 2009. Dynamic chromatin remodeling events in hippocampal neurons are associated with NMDA receptor-mediated activation of Bdnf gene promoter 1. *J Neurochem* **109:** 1375–1388.

Tian F, Marini AM, Lipsky RH. 2010a. NMDA receptor activation induces differential epigenetic modification of Bdnf promoters in hippocampal neurons. *Amino Acids* **38:** 1067–1074.

Tian F, Marini AM, Lipsky RH. 2010b. Effects of histone deacetylase inhibitor Trichostatin A on epigenetic changes and transcriptional activation of Bdnf promoter 1 by rat hippocampal neurons. *Ann NY Acad Sci* **1199:** 186–193.

Tolias KF, Bikoff JB, Burette A, Paradis S, Harrar D, Tavazoie S, Weinberg RJ, Greenberg ME. 2005. The Rac1-GEF Tiam1 couples the NMDA receptor to the activity-dependent development of dendritic arbors and spines. *Neuron* **45:** 525–538.

Tomida T, Hirose K, Takizawa A, Shibasaki F, Iino M. 2003. NFAT functions as a working memory of Ca^{2+} signals in decoding Ca^{2+} oscillation. *EMBO J* **22:** 3825–3832.

Tsankova N, Renthal W, Kumar A, Nestler EJ. 2007. Epigenetic regulation in psychiatric disorders. *Nat Rev Neurosci* **8:** 355–367.

Vacher H, Mohapatra DP, Trimmer JS. 2008. Localization and targeting of voltage-dependent ion channels in mammalian central neurons. *Physiol Rev* **88:** 1407–1447.

Vanderklish PW, Edelman GM. 2002. Dendritic spines elongate after stimulation of group 1 metabotropic glutamate receptors in cultured hippocampal neurons. *Proc Natl Acad Sci* **99:** 1639–1644.

Vanhoutte P, Bading H. 2003. Opposing roles of synaptic and extrasynaptic NMDA receptors in neuronal calcium signalling and BDNF gene regulation. *Curr Opin Neurobiol* **13:** 366–371.

Vashishta A, Habas A, Pruunsild P, Zheng JJ, Timmusk T, Hetman M. 2009. Nuclear factor of activated T-cells isoform c4 (NFATc4/NFAT3) as a mediator of antiapoptotic transcription in NMDA receptor-stimulated cortical neurons. *J Neurosci* **29:** 15331–15340.

Verkhratsky AJ, Petersen OH. 1998. Neuronal calcium stores. *Cell Calcium* **24:** 333–343.

Wahl AS, Buchthal B, Rode F, Bomholt SF, Freitag HE, Hardingham GE, Ronn LC, Bading H. 2009. Hypoxic/ischemic conditions induce expression of the putative pro-death gene Clca1 via activation of extrasynaptic N-methyl-D-aspartate receptors. *Neuroscience* **158:** 344–352.

Wang H, Clapham DE. 1999. Conformational changes of the in situ nuclear pore complex. *Biophys J* **77:** 241–247.

Wang J, Campos B, Jamieson GA Jr, Kaetzel MA, Dedman JR. 1995. Functional elimination of calmodulin within the nucleus by targeted expression of an inhibitor peptide. *J Biol Chem* **270:** 30245–30248.

Watanabe S, Hong M, Lasser-Ross N, Ross WN. 2006. Modulation of calcium wave propagation in the dendrites and to the soma of rat hippocampal pyramidal neurons. *J Physiol* **575:** 455–468.

Wayman GA, Lee YS, Tokumitsu H, Silva AJ, Soderling TR. 2008. Calmodulin-kinases: modulators of neuronal development and plasticity. *Neuron* **59:** 914–931.

Weick JP, Groth RD, Isaksen AL, Mermelstein PG. 2003. Interactions with PDZ proteins are required for L-type calcium channels to activate cAMP response element-binding protein-dependent gene expression. *J Neurosci* **23:** 3446–3456.

Wellmann H, Kaltschmidt B, Kaltschmidt C. 2001. Retrograde transport of transcription factor NF-κB in living neurons. *J Biol Chem* **276:** 11821–11829.

Westenbroek RE, Ahlijanian MK, Catterall WA. 1990. Clustering of L-type Ca^{2+} channels at the base of major dendrites in hippocampal pyramidal neurons. *Nature* **347:** 281–284.

Westphal RS, Anderson KA, Means AR, Wadzinski BE. 1998. A signaling complex of Ca^{2+}-calmodulin-dependent protein kinase IV and protein phosphatase 2A. *Science* **280:** 1258–1261.

Westphal RS, Tavalin SJ, Lin JW, Alto NM, Fraser ID, Langeberg LK, Sheng M, Scott JD. 1999. Regulation of NMDA receptors by an associated phosphatase-kinase signaling complex. *Science* **285:** 93–96.

Wheeler DG, Barrett CF, Groth RD, Safa P, Tsien RW. 2008. CaMKII locally encodes L-type channel activity to signal to nuclear CREB in excitation-transcription coupling. *J Cell Biol* **183:** 849–863.

Wiegert JS, Bengtson CP, Bading H. 2007. Diffusion and not active transport underlies and limits ERK1/2 synapse-to-nucleus signaling in hippocampal neurons. *J Biol Chem* **282:** 29621–29633.

Wojcikiewicz RJ, Luo SG. 1998. Phosphorylation of inositol 1,4,5-trisphosphate receptors by cAMP-dependent protein kinase. Type I, II, and III receptors are differentially susceptible to phosphorylation and are phosphorylated in intact cells. *J Biol Chem* **273:** 5670–5677.

Wood MA, Kaplan MP, Park A, Blanchard EJ, Oliveira AM, Lombardi TL, Abel T. 2005. Transgenic mice expressing a truncated form of CREB-binding protein (CBP) exhibit deficits in hippocampal synaptic plasticity and memory storage. *Learn Mem* **12:** 111–119.

Wood MA, Attner MA, Oliveira AM, Brindle PK, Abel T. 2006. A transcription factor-binding domain of the coactivator CBP is essential for long-term memory and the expression of specific target genes. *Learn Mem* **13:** 609–617.

Wu GY, Deisseroth K, Tsien RW. 2001. Activity-dependent CREB phosphorylation: Convergence of a fast, sensitive calmodulin kinase pathway and a slow, less sensitive mitogen-activated protein kinase pathway. *Proc Natl Acad Sci* **98:** 2808–2813.

Xia Z, Dudek H, Miranti CK, Greenberg ME. 1996. Calcium influx via the NMDA receptor induces immediate early gene transcription by a MAP kinase/ERK-dependent mechanism. *J Neurosci* **16:** 5425–5436.

Xu L, Massague J. 2004. Nucleocytoplasmic shuttling of signal transducers. *Nat Rev Mol Cell Biol* **5:** 209–219.

Yamamoto K, Hashimoto K, Isomura Y, Shimohama S, Kato N. 2000. An IP3-assisted form of Ca^{2+}-induced Ca^{2+} release in neocortical neurons. *Neuroreport* **11:** 535–539.

Yeckel MF, Kapur A, Johnston D. 1999. Multiple forms of LTP in hippocampal CA3 neurons use a common postsynaptic mechanism. *Nat Neurosci* **2:** 625–633.

Yuste R, Majewska A, Holthoff K. 2000. From form to function: calcium compartmentalization in dendritic spines. *Nat Neurosci* **3:** 653–659.

Zehorai E, Yao Z, Plotnikov A, Seger R. 2010. The subcellular localization of MEK and ERK–a novel nuclear translocation signal (NTS) paves a way to the nucleus. *Mol Cell Endocrinol* **314:** 213–220.

Zhang F, Wang LP, Brauner M, Liewald JF, Kay K, Watzke N, Wood PG, Bamberg E, Nagel G, Gottschalk A, et al. 2007a. Multimodal fast optical interrogation of neural circuitry. *Nature* **446:** 633–639.

Zhang SJ, Steijaert MN, Lau D, Schutz G, Delucinge-Vivier C, Descombes P, Bading H. 2007b. Decoding NMDA receptor signaling: identification of genomic programs specifying neuronal survival and death. *Neuron* **53:** 549–562.

Zhang SJ, Zou M, Lu L, Lau D, Ditzel DA, Delucinge-Vivier C, Aso Y, Descombes P, Bading H. 2009. Nuclear calcium signaling controls expression of a large gene pool: Identification of a gene program for acquired neuroprotection induced by synaptic activity. *PLoS Genet* **5:** e1000604.

Zhang SJ, Buchthal B, Lau D, Hayer S, Dick O, Schwaninger M, Veltkamp R, Zou M, Weiss U, Bading H. 2011. A signaling cascade of nuclear calcium-CREB-ATF3 activated by synaptic NMDA receptors defines a gene repression module that protects against extrasynaptic NMDA receptor-induced neuronal cell death and ischemic brain damage. *J Neurosci* **31:** 4978–4990.

Zhou Z, Hong EJ, Cohen S, Zhao WN, Ho HY, Schmidt L, Chen WG, Lin Y, Savner E, Griffith EC, et al. 2006. Brain-specific phosphorylation of MeCP2 regulates activity-dependent Bdnf transcription, dendritic growth, and spine maturation. *Neuron* **52:** 255–269.

Zuhlke RD, Reuter H. 1998. Ca^{2+}-sensitive inactivation of L-type Ca^{2+} channels depends on multiple cytoplasmic amino acid sequences of the α1C subunit. *Proc Natl Acad Sci* **95:** 3287–3294.

Zuhlke RD, Pitt GS, Deisseroth K, Tsien RW, Reuter H. 1999. Calmodulin supports both inactivation and facilitation of L-type calcium channels. *Nature* **399:** 159–162.

Calcium Signaling in Cardiac Myocytes

Claire J. Fearnley[1,3], H. Llewelyn Roderick[1,2,3], and Martin D. Bootman[1,3]

[1]Laboratory of Signalling and Cell Fate, The Babraham Institute, Babraham, Cambridge CB22 3AT, United Kingdom
[2]Department of Pharmacology, University of Cambridge, Cambridge CB2 1PD, United Kingdom

Correspondence: Llewelyn.roderick@bbsrc.ac.uk; martin.bootman@bbsrc.ac.uk

Calcium (Ca^{2+}) is a critical regulator of cardiac myocyte function. Principally, Ca^{2+} is the link between the electrical signals that pervade the heart and contraction of the myocytes to propel blood. In addition, Ca^{2+} controls numerous other myocyte activities, including gene transcription. Cardiac Ca^{2+} signaling essentially relies on a few critical molecular players—ryanodine receptors, voltage-operated Ca^{2+} channels, and Ca^{2+} pumps/transporters. These moieties are responsible for generating Ca^{2+} signals upon cellular depolarization, recovery of Ca^{2+} signals following cellular contraction, and setting basal conditions. Whereas these are the central players underlying cardiac Ca^{2+} fluxes, networks of signaling mechanisms and accessory proteins impart complex regulation on cardiac Ca^{2+} signals. Subtle changes in components of the cardiac Ca^{2+} signaling machinery, albeit through mutation, disease, or chronic alteration of hemodynamic demand, can have profound consequences for the function and phenotype of myocytes. Here, we discuss mechanisms underlying Ca^{2+} signaling in ventricular and atrial myocytes. In particular, we describe the roles and regulation of key participants involved in Ca^{2+} signal generation and reversal.

OVERVIEW OF THE CARDIAC CYCLE

The mammalian heart is a complex organ consisting of four chambers—the left and right atria and the left and right ventricles. Through a highly coordinated series of events, the muscular heart pumps blood through the pulmonary and systemic vasculature (Fukuta and Little 2008). During diastole, all four chambers are relaxed. Systole is initiated by propagation of a depolarizing action potential from the sino-atrial node located in the apex of the right atrium, through the right and then the left atrium. This depolarization induces contraction of these chambers, forcing blood into the ventricles. On reaching the atrioventricular (AV) node, the depolarization pauses for a short time period (0.1 s in humans) to ensure completion of atrial systole. Importantly, the AV node acts as an electrical insulator between the atria and ventricles. The AV node prevents the transfer of aberrant contraction patterns to the ventricles, such as the spontaneous electrical activity occurring during atrial fibrillation. The lower portion of the AV node is designated the bundle of His, which then splits into the left and right branches, allowing activation of the left and right ventricles,

---

[3]All three authors contributed equally to this article.

respectively. These branches give rise to thin filaments called Purkinje fibers, composed of noncontractile cells that distribute the action potential to ventricular myocytes and enable the heart to contract in a coordinated fashion. Transduction of the depolarization signal through the His-Purkinje system causes ventricular systole. The contraction wave, traveling up from the ventricular base, expels blood into the pulmonary artery then on to the lungs, or through the aorta into the arterial system. Retrograde flow of blood is prevented by valves between the atria and ventricles.

As indicated above, the SA node situated at the apex of the right atrium is responsible for initiation of the cardiac action potential. At rest, the membrane potential starts around -70 mV (V_m) and slowly depolarizes until an action potential is triggered. Ca^{2+} signals may play a key role in action potential generation, although there is considerable debate regarding the major mechanisms controlling the rate of SA node depolarization (see Lakatta and DiFrancesco 2009). One primary component of SA node depolarization is known as I_f (f stands for funny) (Brown et al. 1979), an ion current mediated by hyperpolarizing-activated cyclic nucleotide-gated (HCN) channels (DiFrancesco 1993). Because this current is triggered by hyperpolarization, it is activated at the start of diastole and slowly declines throughout the pacemaker period. HCN channels are relatively nonselective, and they therefore generate an inward current depolarizing V_m toward the threshold for firing an action potential. Intracellular Ca^{2+} cycling has also been proposed to act as a primary regulator of SA node depolarization. Imaging SA node cells reveals spontaneous elementary Ca^{2+} signals known as Ca^{2+} sparks arising from the SR and preceding action potential generation (Huser et al. 2000). Ca^{2+} sparks reflect the concerted opening of a cluster of RyRs. The Ca^{2+} sparks activate sodium/calcium exchange (NCX), which promotes membrane depolarization because three Na^+ ions enter for each Ca^{2+} ion that leaves. T-type Ca^{2+} channels ("transient current;" Ca_v3) may also provide a source of Ca^{2+} for triggering Ca^{2+} sparks (Bogdanov et al. 2001; Berridge 2003). When V_m reaches a critical threshold (-40 to -50 mV), plasma membrane L-type Ca^{2+} channels are opened ($I_{Ca,L}$), allowing a large influx of Ca^{2+} into the cytosol and increasing the membrane potential to $\sim +10$ mV. It is this depolarization signal that is transmitted from the SA node through the cardiac conduction system, culminating in cardiac myocyte contraction. Within the SA node cells, $I_{Ca,L}$ activates an outward potassium current (I_K) which hyperpolarizes the membrane and curtails the action potential. The hyperpolarization leads to activation of I_f, T-type Ca^{2+} channels and Ca^{2+} sparks, to begin the next conduction cycle.

EXCITATION-CONTRACTION COUPLING (EC-COUPLING)

EC-coupling is the process pairing myocyte depolarization with mechanical contraction. Ca^{2+} is the critical intermediary (Bers 2008). Indeed, since Ringer's experiments more than a century ago, Ca^{2+} has been known to be an essential mediator of this process (Ringer 1883). As the action potential sweeps over the heart, the plasma membrane (sarcolemma) of each myocyte becomes depolarized (~ -90 mV to $\sim +20$ mV) thereby causing concerted opening of L-type VOCCs ("long-lasting current;" $Ca_v1.2$). Ca^{2+} flows via the VOCCs into a restricted space between the sarcolemma and the underlying sarcoplasmic reticulum (SR) known as the "junctional zone" or "dyadic cleft." The accumulation of Ca^{2+} ions during an action potential increases the Ca^{2+} concentration within this microdomain from ~ 100 nM to ~ 10 μM. This elementary Ca^{2+} influx signal, derived from the activation of VOCCs is known as a "Ca^{2+} sparklet" (Fig. 1) (Wang et al. 2001). The distribution of Ca^{2+} sparklet magnitudes suggests that one or several VOCCs can give rise to such signals within myocytes (Cheng and Wang 2002).

Ca^{2+} sparklets themselves are not adequate to cause substantial contraction. However, they are sufficient to induce opening of RyRs (type 2 RyRs) on the closely apposed SR, via a process known as "Ca^{2+}-induced Ca^{2+} release"

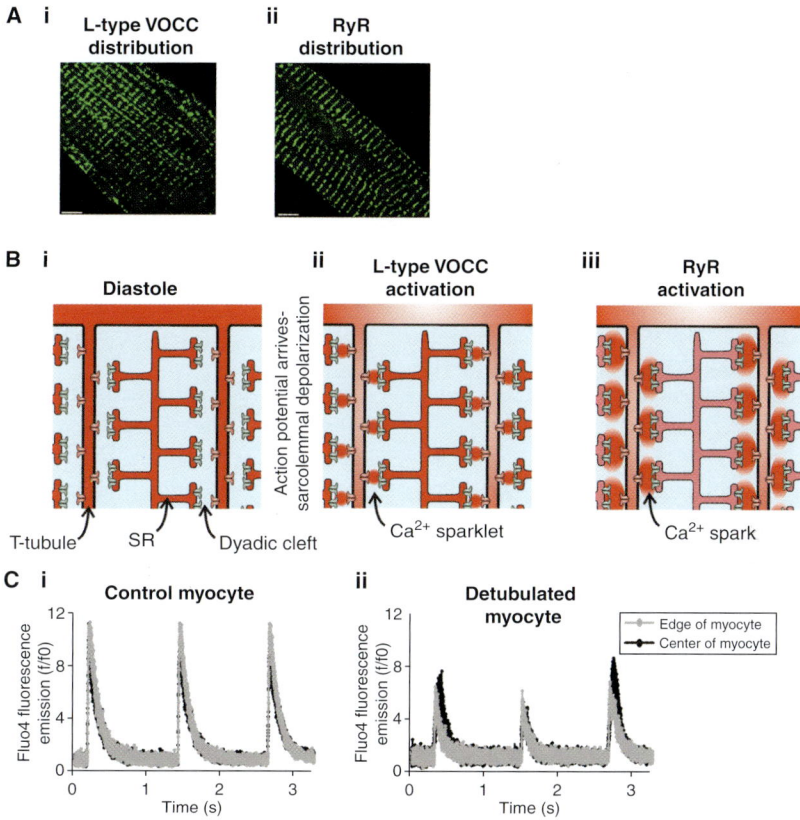

Figure 1. Excitation contraction coupling in ventricular myocytes. Panel *A* illustrates the distribution of L-type VOCCs (*Ai*) and type 2 RyRs (*Aii*) in a section of a ventricular myocyte. The distributions of these proteins are essentially overlapping at the level of the light microscope. Panel *B* is a cartoon sequence of events leading to the generation of a Ca^{2+} signal within a ventricular myocyte. A small section of a ventricular myocyte is depicted with two T-tubule projections (T-tubule spacing ~ 1.8 μm). During the diastolic phase (*Bi*), the L-type VOCCs (red channels on the T-tubule membranes) and RyRs (blue channels on SR membrane) are silent. Arrival of the action potential causes depolarization of the sarcolemma and activation of the L-type VOCCs thereby generating "Ca^{2+} sparklets" (*Bii*). The Ca^{2+} sparklets trigger activation of the RyRs thereby producing "Ca^{2+} sparks" (*Biii*). Panel *Ci* depicts the consistent, global Ca^{2+} responses observed in an electrically paced ventricular myocyte. The black and gray traces indicate the Ca^{2+} concentration (measured with fluo4) at the center and edge of the myocyte. The profile of the Ca^{2+} signal was essentially the same in both locations. Panel *Cii* illustrates what happens in a ventricular myocyte following detubulation (using formamide treatment). Detubulation decreases the amplitude of systolic Ca^{2+} transients, and provokes spatial heterogeneity of the resultant Ca^{2+} signals. The black and gray traces indicate the Ca^{2+} concentration at the center and edge of the myocyte. Whereas the Ca^{2+} responses in the edge of the myocyte were reasonably consistent, the signals in the center of the cell showed beat-to-beat variation in amplitude. Such Ca^{2+} signal alternans are a potential cause of cardiac arrhythmia.

(CICR). Seminal studies in the 1970s showed that RyR activity is dependent on cytosolic Ca^{2+} levels, with low concentrations (1–10 μM) being activatory and high concentrations (>10 μM) inhibiting the channel (Fabiato and Fabiato 1972; Fabiato 1983). The concentration of Ca^{2+} at the dyadic cleft following activation of L-type Ca^{2+} channels falls within the range of that required for channel activation, thus facilitating CICR.

The activation of a cluster of RyRs, and consequent mobilization of Ca^{2+} from the SR, produces an elementary Ca^{2+} release signal know as a "Ca^{2+} spark" (Fig. 1) (Cheng and

Lederer 2008). As with Ca^{2+} sparklets, Ca^{2+} spark magnitudes can vary, indicating that different numbers of RyRs participate in their generation. Release of Ca^{2+} through the RyRs increases the concentration of Ca^{2+} to >100 µM in the dyadic cleft. It is estimated that ~25 L-type Ca^{2+} channels and 100 RyRs are closely associated within the dyadic cleft to form a "couplon" (Bers and Guo 2005). Ca^{2+} ions diffuse out of the cleft to engage the contractile machinery, thereby promoting cell shortening to provide the force for pumping blood. During a single action potential, thousands of Ca^{2+} spark sites are simultaneously activated by their corresponding Ca^{2+} sparklet triggers (Cheng and Lederer 2008). Diffusion of Ca^{2+} ions, and their subsequent spatial and temporal summation, produces an average global Ca^{2+} increase of 500 nM to ~1 µM. Troponin C (TnC), the Ca^{2+}-binding component of the contractile filaments, is sensitive to Ca^{2+} concentrations over that range thereby allowing coupling of the AP-mediated Ca^{2+} transient and contraction. At the end of an action potential Ca^{2+} transients are rapidly terminated, and the cells return to resting diastolic levels in preparation for the next depolarization.

In addition to L-type VOCCs, contractile cardiac myocytes express a T-type current (Ca_v3 family), which is so named because of its transient nature. As described in the section on pacemaking above, this current is activated at a more negative membrane potential than the L-type (Nowycky et al. 1985). Although $I_{Ca,T}$ plays a role in depolarizing SA node cells, under normal physiological situations its role is negligible within ventricular and atrial cardiac myocytes as most Ca^{2+} enters through the L-type channels. However, T-type VOCC expression is up-regulated during cardiac hypertrophy, and they may provide a critical Ca^{2+} signal to drive hypertrophic remodeling. For example, $Ca_v3.2$ knockout mice did not display cardiac hypertrophy in response to pressure overload or angiotensin II (Chiang et al. 2009).

Skeletal muscle EC-coupling is also mediated by elevations in intracellular Ca^{2+}, and has some similarities to the cardiac scheme described above. The key difference in this tissue is Ca^{2+} sparklets do not trigger Ca^{2+} release from the SR. Rather, L-type VOCCs ($Ca_v1.1$) on the sarcolemma have a direct physical interaction with RyRs (type 1 RyRs) on underlying SR forming the triadic junction (Rios and Brum 1987; Block et al. 1988). Depolarization of the sarcolemma induces a conformational change in the VOCCs, which allosterically activates the RyRs.

REGULATION OF L-TYPE VOCCs

The channels and homeostatic processes underlying Ca^{2+} release and Ca^{2+} clearance are subject to regulation by multiple signaling pathways. These signaling pathways can rapidly alter the amplitude and/or spatial properties of myocyte Ca^{2+} signaling to acutely modulate cardiac output. For example, L-type VOCCs are subject to regulation by cAMP-dependent kinase (protein kinase A; PKA) downstream from β-adrenergic receptor stimulation, e.g., during the fight or flight response. PKA phosphorylation increases channel activity, thereby contributing to the increased cardiac contraction (positive inotropic response) evoked by β-adrenergic stimuli. PKA-mediated phosphorylation increases opening of the VOCC by a twofold mechanism—it increases both the number of channels in an activatable state, and their activation probability (Catterall 2000). PKA phosphorylates Ser-1928 in the amino terminal of the α_1 subunit (Perets et al. 1996), in addition to residues on the β_2 subunit (Curtis and Catterall 1985; Gerhardstein et al. 1999). Rapid dephosphorylation of the channel is provided by the Ser/Thr phosphatases 1 and 2A (PP1 and PP2A, respectively) (Kamp and Hell 2000).

L-type VOCCs are also phosphorylated by protein kinase C (PKC) in response to activation of G_q-coupled receptors, e.g., α_1-adrenergic, endothelin, and angiotensin II receptors. The targets for PKC phosphorylation are proposed to be two Thr residues in the amino terminal of the α_1 subunit (Shistik et al. 1998; McHugh et al. 2000). The effects of PKC phosphorylation are less clear than for PKA phosphorylation. It is proposed that the effect may be reliant on the particular PKC isoform activated, the expression of which varies in a complex developmental, species-dependent, and disease-regulated manner in the heart. Additional regulation of the L-type Ca^{2+} channel is provided by cGMP-dependent protein kinase G (PKG), which has been shown to have an inhibitory effect, thus opposing the effects of PKA (Hartzell and Fischmeister 1986; Abi-Gerges et al. 2001).

Inactivation of the L-type Ca^{2+} channel is mediated by both membrane repolarization and by Ca^{2+} itself (Ca^{2+}-dependent inactivation, CDI). The latter acts as a negative feedback loop, and is thought to be the more important of the two mechanisms (Lee et al. 1985). CDI is controlled by calmodulin (CaM), which is constitutively bound to the channel (Peterson et al. 1999; Qin et al. 1999). The CaM binding site is a canonical "IQ" CaM-binding motif within the carboxyl terminus of the α_1 subunit that binds the Ca^{2+}-free form of CaM (apo-CaM) (Rhoads and Friedberg 1997; Zuhlke and Reuter 1998). Ca^{2+} entering through L-type channels during EC-coupling binds to apo-CaM to form Ca-CaM, which inactivates the channel (Tang et al. 2003). Interestingly, Ca-CaM can also enhance Ca^{2+} entry through L-type Ca^{2+} channels by Ca^{2+}-dependent facilitation (CDF). CDF is also dependent on the IQ motif in the cytoplasmic tail of the α_{1C} subunit (Zuhlke et al. 1999). This capacity for dual regulation is caused by the presence of both high and low affinity Ca^{2+} binding sites (EF-hands) within CaM, in the carboxyl and amino terminal lobes, respectively. It is believed that CDI depends on Ca^{2+} bound to the amino terminal EF-hands, whereas CDF depends on the carboxyl terminal EF hands (DeMaria et al. 2001). Subsequently, it was revealed that the two lobes of CaM can detect Ca^{2+} arising from distinct sources—the high affinity carboxyl terminal site sensing local Ca^{2+} arising within the nanodomain of the channel mouth, whereas the low affinity amino terminal lobe detects global Ca^{2+} signals (Tadross et al. 2008). It therefore appears that Ca^{2+} binding to apo-CaM at the carboxyl terminus of L-type VOCCs potentiates Ca^{2+} entry once a Ca^{2+} sparklet is forming. However, once global Ca^{2+} is elevated, Ca^{2+} also binds to the amino terminal lobe of CaM leading to CDI and termination of Ca^{2+} entry.

Additional modification of $I_{Ca,L}$ is provided by Ca^{2+}/CaM-dependent kinase II (CaMKII), which potentiates the influx of Ca^{2+} (Anderson et al. 1994; Xiao et al. 1994; Yuan and Bers 1994; Wu et al. 2001b). CaMKII interacts with and phosphorylates the carboxyl terminal of the α_1 subunit (Hudmon et al. 2005). CaMKII remains tightly bound to the channel even in the absence of Ca^{2+}, although it is only active when it has Ca-CaM bound. Importantly, Ca-CaM can remain bound to CaMKII, and the kinase active, even after global Ca^{2+} has declined, thereby allowing CaMKII to act as a detector of Ca^{2+} spike frequency (Hudmon et al. 2005).

REGULATION OF RyRs AND CICR

There are three mammalian RyR isoforms (RyR1–3). RyR2 is predominant in cardiac myocytes, with significantly lesser amounts of RyR3. The RyR channel is a large homotetrameric assembly of ~ 2 megadaltons (each subunit has a molecular mass of ~ 560 kDa) (Lanner et al. 2010). Structural studies have revealed four-fold symmetry, with a four-leaf clover or mushroom morphology formed by the transmembrane domain and bulky cytoplasmic domain (Anderson et al. 1989; Serysheva 2004). Binding of Ca^{2+} is proposed to cause conformational changes that evoke channel gating in a mechanism similar to that of a camera iris, with twisting of the transmembrane regions opening the ion pore (Serysheva et al. 1999).

RyRs are bound by a multitude of accessory proteins, comprising a macromolecular signaling complex. These interactions determine the efficiency and specificity of signaling to and from RyRs, and between other signal transduction cascades. Moreover, scaffolding of proteins to RyRs recruits and concentrates important regulatory proteins in the junctional zone where they are ideally located to modulate and/or be regulated by EC-coupling. These interactions occur on both the lumenal and cytosolic face of the RyR (Lanner et al. 2010). In the lumen of the SR, RyRs interact with calsequestrin (CSQ), the major Ca^{2+} binding/storage protein of muscle. CSQ is a low-affinity Ca^{2+} storage protein that maintains the SR lumenal free Ca^{2+} concentration between 100–500 μM (Yano and Zarain-Herzberg 1994; Berridge 2002). The critical role of CSQ in Ca^{2+} storage was shown by the increase or decrease in SR Ca^{2+} load observed in experiments in which CSQ expression was enhanced or suppressed, respectively (Terentyev et al. 2003). In addition to acting as the major Ca^{2+} storage protein in cardiac muscle, CSQ also regulates RyR channel activity (Prins and Michalak 2011). CSQ reversibly changes between monomeric to oligomeric forms in response to changes in luminal Ca^{2+} concentration. It has been suggested that CSQ oligomers are present when the SR is replete, and that these mainly serve to buffer Ca^{2+}. However, when RyRs open and SR luminal Ca^{2+} declines, Ca^{2+} unbinds from calsequestrin and the oligomeric protein dissociates. The calsequestrin monomers bind to RyRs (via a protein intermediate called triadin) and inhibit channel activity. This is an important component of the mechanisms that terminate Ca^{2+} release during each heartbeat (Gyorke et al. 2009).

The key role of CSQ in regulating Ca^{2+} storage and RyR function is highlighted by a pathological condition known as catecholaminergic polymorphic ventricular tachycardia (CPVT) that is observed in patients with CSQ mutations. CPVT is a life-threatening form of cardiac dysrhythmia typically brought about by emotional or physical stress. Recessive CSQ mutations are found in ∼3% of CPVT patients (Katz et al. 2009). The effects of some CPVT-inducing CSQ mutations on calcium fluxes in cardiac myocytes are known. For example, mutation of the aspartate to histidine at residue 307 in CSQ (CSQ^{D307H}) causes decreased SR Ca^{2+} storage and release, and increases the frequency of delayed afterdepolarizations (DADs; spontaneous electrical depolarization of cardiac myocyte independent of the SA node-evoked AP) (Viatchenko-Karpinski et al. 2004). Another mutation, arginine to glutamic acid at residue 33 (CSQ^{R33Q}), decreases the interaction between the RyR and CSQ, resulting in abnormal regulation of the RyR by lumenal Ca^{2+} and increased Ca^{2+} release (Terentyev et al. 2006). A substantial proportion of CPVT patients (∼50%) express mutated RyRs that have altered association with accessory proteins.

CaM is an important regulator of the RyR in cardiac myocytes, both in its Ca^{2+} bound and Ca^{2+} free form (Ca-CaM and apo-CaM, respectively). The binding site for CaM on the RyR was mapped to the carboxyl terminal of the receptor by site-directed mutagenesis (Porter Moore et al. 1999a; Porter Moore et al. 1999b), in agreement with cryo-EM studies (Wagenknecht et al. 1997). Binding of CaM decreases Ca^{2+} efflux through these channels (Meissner and Henderson 1987). This is thought to be facilitated by a reduction in the opening probability and in the Ca^{2+}-dependent activation of the channel (Balshaw et al. 2001).

Further Ca^{2+}-dependent modification of RyRs is provided by CaMKII-dependent phosphorylation. Sequence analysis revealed six consensus phosphorylation sites in the RyR (Zucchi and Ronca-Testoni 1997). Both Ser-2809 (Witcher et al. 1991) and Ser-2815 (Wehrens et al. 2004) have been shown to be crucial for CaMKII-dependent phosphorylation. CaMKII phosphorylation increases RyR activity (Witcher et al. 1991; Wehrens et al. 2004), although some studies have contested this idea (Lokuta et al. 1995; Wu et al. 2001a). Phosphorylation by CaMKII is emerging as the dominant mode of regulation of RyR activity during adrenergic stimulation and in the greater contractility associated with increased frequency of myocyte contraction (Wu et al. 2009; Grimm and Brown 2010).

Ca^{2+} release through RyRs is regulated by interaction with FK binding proteins (FKBPs), named because of their binding of the immunosuppressant drug FK506. Cardiac myocytes express two isoforms of the 12 kDa FKBP, namely FKBP12 and FKBP12.6 (also known as calstabin1 and calstabin2, respectively). The cardiac RyR2 binds FKBP12.6 with a higher affinity (Timerman et al. 1996; Jeyakumar et al. 2001). FKBP12.6 binding stabilizes the coordinated gating of RyR subunits within a tetramer, thereby enabling channels to transition between the fully closed and the fully open state, while also shifting the Ca^{2+} dependence of channel opening to a higher Ca^{2+} concentration (Bers 2004). CPVT-inducing mutations within the type 2 RyR channel decrease the association between the receptor and FKBP12.6, although only under conditions of β-adrenergic stimulation. FKBP12.6 dissociation may lead to increased Ca^{2+} release from the SR (Wehrens et al. 2003).

RyRs can also be phosphorylated by PKA, which is tethered by an A kinase anchoring protein (mAKAP) (Marx et al. 2000). PKA-dependent phosphorylation of the receptor was reported to increase its responsiveness to Ca^{2+} (Valdivia et al. 1995), although the precise consequences of PKA-dependent phosphorylation have been controversial. A widely discussed model proposed that PKA phosphorylation of RyRs causes dissociation of FKBP12.6, thereby leading to increased probability of channel opening (Marx et al. 2000; Wehrens et al. 2003). This model of RyR regulation would not only be relevant physiologically, coupling β-adrenergic stimulation with enhanced Ca^{2+} release, but would also be important in disease conditions in which elevated PKA phosphorylation had been reported to decrease FKBP12.6-RyR associations and result in increased spontaneous diastolic RyR activity and DADs. However, the dependence of FKBP12.6 association with the RyR on PKA phosphorylation has been much disputed (Xiao et al. 2007). Indeed, a recent report provides substantial evidence that the association of FKBP12.6 interaction with the RyR is insensitive to the degree of PKA phosphorylation (Guo et al. 2010).

PKA may also regulate RyR activity by phosphorylation of Sorcin, another RyR-interacting protein. Sorcin is a ubiquitously expressed 22 kDa Ca^{2+} binding protein that inhibits Ca^{2+} release via the RyR. This inhibitory effect is lost following its phosphorylation by PKA (Lokuta et al. 1997).

A cAMP phosphodiesterase (PDE) has also been identified within the RyR macromolecular complex (specifically, the PDE4D3 isoform). The presence of this enzyme provides a mechanism to tightly regulate cAMP, and thus PKA activity, in the vicinity of the receptor. The levels of this PDE isoform have been reported to be reduced in failing hearts (Lehnart et al. 2005).

Type 2 RyRs are associated with phosphatases to mediate rapid receptor dephosphorylation and return it to basal levels of activity. For example, calcineurin is suggested to be an accessory protein of cardiac RyRs (Bandyopadhyay et al. 2000), and is proposed to decrease Ca^{2+} channel activity. In addition, the phosphatases PP1 and PP2A are associated with cardiac RyRs (Marx et al. 2001). PP1 associates with RyRs via an interaction with spinophilin, whereas PP2A binds to a targeting protein PR130, which then anchors it to RyRs (Marx et al. 2001).

CARDIAC MYOCYTE CONTRACTION

Contraction of cardiac myocytes is facilitated by myofilaments organized into sarcomeres, situated along the long axis of the cell. The sarcomere consists of myosin-containing thick filaments surrounded by a hexagonal array of thin filaments, which are made up of actin polymers and troponin/α-tropomyosin (Tn/Tm) regulatory units (Parmacek and Solaro 2004). Additionally, each thin filament is separated at the Z-line by an actin binding protein, α-actinin.

Every seventh actin monomer comprising the thin filament is bound to a Tn/Tm complex. Tn is composed of three subunits: TnC, Troponin I (TnI), and Troponin T (TnT) (Greaser and Gergely 1971). TnC has a similar structure to CaM, and contains Ca^{2+} binding EF-hands (Parmacek and Solaro 2004). TnI is an inhibitory subunit, and TnT constitutively interacts with Tm. Binding of Ca^{2+} to TnC causes a conformational change in associated TnI. This enables Tn/Tm to slide into the groove between actin monomers, allowing the myosin thick filament to bind actin, thus forming a crossbridge. By repetitive, transient actin-myosin interactions and utilizing energy from ATP hydrolysis, the two filaments slide relative to each other, thus shortening the cell. Coordinated shortening of the entire myocyte population by the spreading AP leads to cardiac contraction. As cytosolic Ca^{2+} levels decline, Ca^{2+} is released from TnC leading to crossbridge detachment, and the thick and thin filaments slide past each other back to their original positions. Cross-bridges cannot form between the filaments as they travel past in that direction.

Troponin proteins are regulated by phosphorylation, which alters their activity and thus affects cardiac myocyte contraction. PKA phosphorylation downstream from β-adrenergic stimulation is of particular importance, as this would facilitate altered cardiac contraction during, for example, physical exertion. Much work has been completed in elucidating the functional outcome of this modification, and the consensus of opinion is that PKA phosphorylation leads to increased cardiac contraction (for a review see Metzger and Westfall 2004). PKC phosphorylates regions on TnT and TnI. Early studies reported divergent results regarding the effects of PKC phosphorylation, although more recent studies point toward an inhibition of contractile function following PKC phosphorylation (Takeishi et al. 1998; Sumandea et al. 2004).

A large number of mutations have been identified in Ca^{2+}-dependent contractile proteins that are linked with specific cardiomyopathies (reviewed in Morimoto 2008). Clinically, cardiomyopathies can be divided into four main groups: hypertrophic cardiomyopathy (HCM), dilated cardiomyopathy (DCM), restrictive cardiomyopathy (RCM), and arrhythmogenic right ventricular dysplasia/cardiomyopathy (ARVD/C). HCM is characterized by thickening of the ventricular walls with accompanying decreases in ventricular chamber volume. In this condition, systolic function is preserved at the expense of diastolic function, which is responsible for symptoms of heart failure and death in these patients. Thickened ventricular walls are also observed in DCM, in addition to an increase in chamber volume. The clinical outcomes include systolic dysfunction, which also leads to heart failure and sudden cardiac death. RCM is linked to diastolic dysfunction, although there is little or no effect on systolic function or ventricular wall thickness. Mutations leading to all three of these myopathies have been identified in the contractile proteins of the sarcomere, although interestingly none have yet been found that are linked to ARVD/C (Morimoto 2008). Considering TnT, for example, 27 mutations have been identified that have been linked to HCM, which act by increasing the Ca^{2+} sensitivity of cardiac muscle contraction. Two TnT mutations have been linked to DCM, the functional consequence being decreased Ca^{2+} sensitivity and an increased affinity of TnT for tropomyosin. Thirty-three mutations in TnI have been found that are linked with HCM, DCM, and RCM, mainly causing increased Ca^{2+} sensitivity and impaired interaction of TnT with TnI. Regarding TnC, a HCM-causing mutation has been identified that abolishes its interaction with TnI, thus losing the altered Ca^{2+} sensitivity imparted by PKA-dependent phosphorylation. Thirteen tropomyosin mutations have been identified, leading to HCM and DCM in the manner described previously.

Ca^{2+} EFFLUX

EC-coupling events are short-lived—atrial and ventricular myocytes reach peak contraction within a few tens of milliseconds of action potential initiation at the SA node. After cytosolic Ca^{2+} has activated the contractile units, it is rapidly extruded from the cytosol in preparation for the following action potential (Shannon and Bers 2004). The main efflux mechanisms in cardiac myocytes are the plasma membrane NCX and the sarco/endoplasmic reticulum Ca^{2+} ATPase (SERCA) pump on the SR. The plasma membrane Ca^{2+} ATPase (PMCA) and the mitochondrial uniporter may also play a more minor role. The relative contribution of these mechanisms varies in a species-dependent manner, for example in rabbit ventricles ~70% of cytosolic Ca^{2+} is removed by SERCA2a, 28% by NCX, and the remaining 2% by the mitochondrial uniporter.

A similar pattern is apparent in ventricles from dog, cat, guinea pig, and human. Alternately, in mouse and rat ventricles, SERCA2a removes ~92% of the Ca^{2+}, leaving only ~7% for the NCX (Bassani et al. 1994; Bers 2001). The difference in the relative contributions of the Ca^{2+} clearance pathways is reflected by the observed kinetics of NCX activity between different species (Sham et al. 1995). NCX current density is lowest in myocytes from rat, and highest in those from guinea pig. The adaptability of this system was shown in a recent report showing minimal cardiac dysfunction in a cardiac-specific SERCA2 KO mouse (Andersson et al. 2009). Ca^{2+} fluxes were maintained by increased activity of the L-type Ca^{2+} channel and NCX, demonstrating that these proteins can compensate for a major reduction in SERCA levels (<5% SERCA2 protein remained in myocardial tissue 4 weeks after gene excision).

In its forward mode, NCX uses the electrochemical gradient across the sarcolemma to translocate three Na^+ ions into the cytosol and expel one Ca^{2+}. In this situation, NCX is a depolarizing current. If Na^+ levels are high, the exchanger may change to its reverse mode, and bring Ca^{2+} into the cell. The exchanger was discovered around 40 years ago in the squid giant axon (Baker et al. 1969) and in the mammalian heart (Reuter and Seitz 1968). Following cloning of the canine NCX (Nicoll et al. 1990), three isoforms were identified (NCX1-3). Of these, NCX1 (110 kDa) is predominant in cardiac myocytes. It is responsible for exporting an amount of Ca^{2+} approximately equivalent to the flux entering via L-type Ca^{2+} channels (Bridge et al. 1990). The adaptability of cardiac myocytes is again shown in the NCX KO mouse, in which EC-coupling is maintained by a compensatory reduction in Ca^{2+} influx (Pott et al. 2007).

Cardiac NCX is regulated by the concentration of cytoplasmic ions, specifically activation by Ca^{2+} (Hilgemann 1990) and inhibition by Na^+ (Hilgemann et al. 1992). Maintenance of the ionic gradient between the extracellular space and cytosol is critical for NCX activity. In particular, intracellular Na^+ must remain low for NCX activity. This is facilitated by the action of the sodium potassium ATPase, which extrudes Na^+ entering the cell during the action potential. Inhibition of this pump by cardiac glycosides such as digoxin and ouabain results in an accumulation of intracellular Na^+, leading to a suppression of NCX activity and thereby attenuating Ca^{2+} efflux. As a consequence, Ca^{2+} transient amplitude and myocyte contraction are increased. Because of these positive inotropic effects, digoxin has long been used as therapy for failing heart. However, this manipulation can also lead to arrhythmia and myocyte death.

NCX has been shown to be phosphorylated by PKC and PKA suggesting that its activity may be regulated by these posttranslational modifications (Iwamoto et al. 1996; Ruknudin et al. 2000). NCX interacts with a number of accessory proteins. In particular, it interacts with the transmembrane protein phospholemman, which exerts an inhibitory effect (Zhang et al. 2003; Ahlers et al. 2005; Cheung et al. 2007). Phosphorylation of phospholemman is reported to occur on residues within the cytoplasmic carboxyl terminal of the protein, specifically Ser-68, and acts to inhibit NCX (Song et al. 2005a; Zhang et al. 2006; Cheung et al. 2007). This appears to be mediated by PKC and not PKA (Zhang et al. 2006). NCX has been identified within a macromolecular complex containing PKA and its anchoring protein mAKAP, together with PKC and the phosphatases PP1 and PP2A (Schulze et al. 2003), providing a robust mechanism for the phosphorylation and regulation of NCX and/or phospholemman.

Ca^{2+} reuptake into the SR is mediated by the SERCA pump, which uses energy from ATP hydrolysis. Molecular cloning has identified three SERCA isoforms (SERCA1–3), all of which undergo alternative splicing. The alternately spliced isoforms of SERCA2 (SERCA2a and SERCA2b) possess distinct carboxyl terminal residues. SERCA isoforms differ in their relative affinity for Ca^{2+} and their transport rates (Lytton et al. 1992). SERCA2a, the main cardiac isoform, possesses a $K_{1/2}$ of ~0.4 μM; however, in vivo, this is ~0.9 μM because of its association with its regulatory transmembrane phosphoprotein, phospholamban (PLB). Cardiac SERCA pumps are usually situated on a region of the SR separate from the junctional zone where the RyRs are located. During recovery of a Ca^{2+} transient, Ca^{2+} is pumped into the SR at the location of the SERCA enzymes. The Ca^{2+} ions return to the dyadic SR lumen by tunneling through the SR network.

SERCA regulation in the heart is primarily governed by PLB. Inactive, phosphorylated PLB exists as a pentamer. It depolymerizes when not phosphorylated and can then interact with SERCA (Kimura et al. 1997). Unphosphorylated PLB monomers decrease the affinity of SERCA for Ca^{2+} (Tada et al. 1974). Subsequent phosphorylation of PLB prevents its interaction with SERCA, increasing the apparent activity of the pump and the rate of Ca^{2+} accumulation within the SR. PLB can be phosphorylated on three sites: Ser-16 by PKA, Thr-17 by CaMKII, and Ser-10 by PKC (Movsesian et al. 1984; Simmerman et al. 1986), facilitating regulation by a variety of signaling pathways. CaMKII is present in a multiprotein complex with SERCA2a, and can directly phosphorylate the Ca^{2+} pump (Toyofuku et al. 1994; Narayanan and Xu 1997) leading to enhanced activity (Xu and Narayanan 1999; Xu et al. 1999). Modulation of PLB-mediated SERCA inhibition is a major mechanism for acute enhancement of cardiac function following β-adrenergic receptor activation. As a result of reduced PLB interaction with the pump, the rates of clearance of the Ca^{2+} transient and relaxation is increased (positive lusitropic response), and SR store loading is enhanced. Because of greater Ca^{2+} within the store, the magnitude of Ca^{2+} fluxes is elevated thereby producing enhanced contraction.

SUBCELLULAR ORGANIZATION OF CARDIAC MYOCYTES

As described previously, the action potential generated at the SA node sweeps rapidly through the heart, coordinating cardiac contraction by activating atrial and then ventricular myocytes in synchrony. Within individual myocytes, the depolarization signal culminates in activation of the sarcomeric contractile units by elevating intracellular Ca^{2+}. The spatial properties of the Ca^{2+} increase depending on the structure of the different myocytes within the heart. For example, adult ventricular myocytes possess numerous invaginations of the sarcolemma, which form a regular array of inwardly directed membranous structures known as transverse tubules (T-tubule) (Fig. 1) (Song et al. 2005b). T-tubules are narrow (~200 nm diameter) and occur at regular intervals of ~2 μm (Brette and Orchard 2003). Additional branches project from the main T-tubules to give a complex network of sarcolemmal intrusions (Ayettey and Navaratnam 1978). T-tubules are a feature of mammalian ventricular myocytes, and are absent in the ventricles of birds (Bossen et al. 1978), reptiles, and amphibians (Bossen and Sommer 1984).

The presence of T-tubules facilitates homogenous Ca^{2+} transients during EC-coupling in ventricular myocytes. T-tubules serve to create dyadic junctions deep within the volume of a ventricular myocyte. In this way, Ca^{2+} sparks can be triggered simultaneously throughout a cell. The alternative to T-tubules is observed in neonatal myocytes and atrial myocytes, where dyadic junctions occur solely at the periphery of the cells (Fig. 2). T-tubules also act as important scaffolding regions for many of the proteins essential for Ca^{2+} signaling (Chase and Orchard 2011). For example, the sarcolemmal L-type Ca^{2+} channels and NCX are abundant on T-tubule membranes (Orchard and Brette 2008). It has been estimated that >75% of I_{Ca} flows into a myocyte through the T-tubules because of the high concentration of L-type Ca^{2+} channels in this region. As mentioned in the previous section, Ca^{2+}-dependent inactivation of I_{Ca} is one of the main mechanisms to curtail Ca^{2+} influx. This feedback mechanism appears to be more potent at T-tubules, meaning I_{Ca} at T-tubules is large but inactivates rapidly. In contrast, inactivation of I_{Ca} at the peripheral sarcolemma is slower overall, resulting in more Ca^{2+} entering the cell across the outer region of a cell than at T-tubule sites. This prolonged Ca^{2+} entry occurs during the

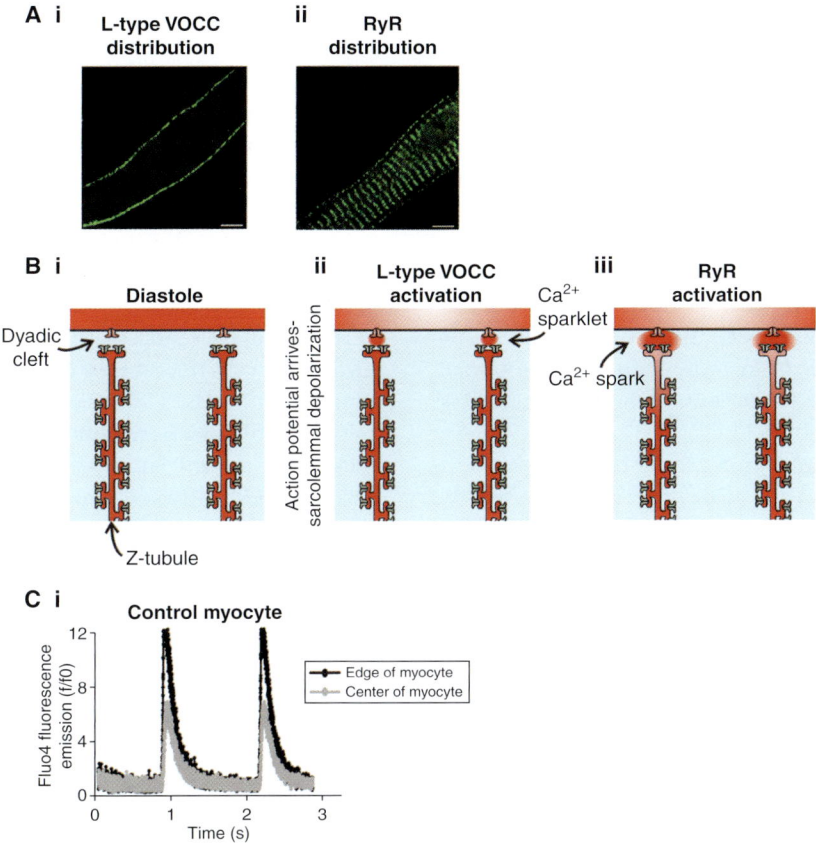

Figure 2. Excitation contraction coupling in atrial myocytes. Panel *A* illustrates the distribution of L-type VOCCs (*Ai*) and type 2 RyRs (*Aii*) in a section of an atrial myocyte. The pattern of L-type VOCC expression is clearly different from that in ventricular cells. The distribution of RyRs is similar to that in ventricular cells, except that there is an evident population of peripheral RyRs around the edge of the myocyte. Solely these peripheral RyRs align with the L-type VOCCs to produce functional dyads. Panel *B* is a cartoon sequence of events leading to the generation of a Ca^{2+} signal within an atrial myocyte. A small section of an atrial myocyte is depicted. There are no T-tubules, but instead two prominent SR tubules with a spacing of ∼1.8 μm. Such SR tubules have previously denoted as "Z-tubules," as they occupy the Z-line (just like T-tubules). During the diastolic phase (*Bi*), the L-type VOCCs (red channels on the T-tubule membranes) and RyRs (blue channels on SR membrane) are silent. Arrival of the action potential causes depolarization of the sarcolemma and activation of the L-type VOCCs thereby generating "Ca^{2+} sparklets" at the periphery of the cell (*Bii*). The Ca^{2+} sparklets trigger activation of nearby RyRs thereby producing "Ca^{2+} sparks" (*Biii*). Panel *Ci* depicts the gradient of Ca^{2+} typically observed during electrical pacing of atrial myocytes. The Ca^{2+} signal at the edge of the cell (black trace) is larger and more rapidly rising than the central response (gray trace). The extent to which the Ca^{2+} signal occurs in the center of the cell (and thereby causes contraction) is dependent on the inotropic status of the cell. Application of a β-adrenergic agonist can make atrial Ca^{2+} signals become homogenous.

latter stages of I_{Ca}, and it has been proposed to promote SR Ca^{2+} loading in preparation for the following next cycle of EC-coupling.

Cardiac myocytes display an important homeostatic principle known as "autoregulation" (Eisner et al. 1998). This is essentially a balance between I_{Ca}, Ca^{2+} release from the SR and Ca^{2+} efflux via NCX. Acute changes in any one of these fluxes will exert compensatory changes in the others that bring systolic Ca^{2+} transients back to normal levels. For example, under conditions of increased SR, Ca^{2+} release following RyR phosphorylation Ca^{2+} influx through I_{Ca} is reduced and Ca^{2+} efflux via NCX is increased, thus maintaining the steady-state systolic Ca^{2+} transient. Taking into consideration that Ca^{2+}-dependent inactivation following SR Ca^{2+} release occurs predominantly at the T-tubules, and NCX is also concentrated in this region, it is evident that autoregulation occurs mainly at the T-tubules. Because of the principle of autoregulation, prolonged changes in inotropic status require sustained changes in more than one component of the I_{Ca}/Ca^{2+} release/NCX triumvirate (Eisner et al. 2009).

The T-tubular network of atrial myocytes is generally not as well developed as that observed in ventricular myocytes, especially in the atrial myocytes of small mammals. The sarcolemmal L-type Ca^{2+} channels of atrial myocytes provide a triggering Ca^{2+} signal for a small population of "junctional" RyRs situated below the sarcolemma at the periphery of the cells (Bootman et al. 2006). The consequence of this arrangement is that the initial Ca^{2+} influx (I_{Ca}) activates Ca^{2+} sparks solely from the peripheral SR (Figs. 1 and 2). In the absence of a positive inotropic agonist, the Ca^{2+} signal remains confined to the cellular periphery and contraction is minimal (Bootman et al. 2011).

In addition to the junctional RyRs, atrial myocytes express a major population of nonjunctional RyRs that form a 3-dimensional lattice of Ca^{2+} release sites within the cells (the distribution of RyRs is actually very similar between ventricular and atrial myocytes—the location of L-type VOCCs is different) (Figs. 1 and 2) (Chen-Izu et al. 2006; Schulson et al. 2011). Because these nonjunctional RyRs are not located within a dyadic junction, they are not activated by I_{Ca} (Mackenzie et al. 2001). However, they can be activated by CICR if the peripheral Ca^{2+} signal is sufficient to act as a trigger for a Ca^{2+} wave. The depth that such centripetal Ca^{2+} waves propagate within an atrial myocyte determines the extent of contraction—the deeper a Ca^{2+} wave spreads the more contractile filaments will be engaged (Mackenzie et al. 2004). The extent of atrial myocyte contraction is regulated by controlling the spread of the centripetal Ca^{2+} waves. The atrial myocytes in some mammalian species display a relatively high degree of T-tubule membrane. Why some atrial cells should rely on T-tubules when others do not is unclear. However, it has been shown that the presence of T-tubules within individual atrial myocytes is correlated with cell diameter, suggesting that tubulation acts to coordinate Ca^{2+} signaling in larger cells (Smyrnias et al. 2010).

Neonatal rat ventricular myocytes are similar to atrial cells in that they lack a fully formed T-tubule network. The T-tubules appear progressively through development (Sedarat et al. 2000). Just as with atrial myocytes, the main region of VOCC-RyR interaction within a neonatal myocyte is at the periphery of the cell. Therefore, Ca^{2+} signals within a neonatal myocyte closely resemble those in an adult atrial cell, and change to become homogenous Ca^{2+} signals when T-tubules arise.

Ventricular myocytes from hearts that are progressing toward failure show decreased organization of their T-tubules, and loss of T-tubules (Fig. 3) (He et al. 2001). As a result, there is reduced coupling efficiency between the L-type VOCCs on the sarcolemma and RyRs on the underlying SR. The loss of T-tubule membranes means that dyadic junctions are lost and RyRs become "orphaned." This leads to abnormal EC-coupling characterized by an increased propensity for arrhythmia and decreased magnitude of the Ca^{2+} response and contraction (Louch et al. 2004). EC-coupling in detubulated ventricular myocytes resembles that observed in atrial myocytes, whereby Ca^{2+} signals are initiated at the plasma membrane and propagate through the myocyte by the relatively slow process of CICR (Smyrnias et al. 2010).

ADAPTATION TO DISEASE

In response to pressure or volume overload, damage, or genetic factors, the heart mounts an adaptive hypertrophic response. Because cardiac myocytes are terminally differentiated, enlargement of the heart results mainly from growth of existing myocytes and not as a result of proliferation. For many hypertrophic stimuli, the remodeling of the heart is initially beneficial by acting to increase cardiac output. However, under conditions in which stress persists, for example, because of hypertension, the remodeled heart undergoes a process known as "decompensation." Specifically, the thickness of the ventricle wall diminishes, and the ability of the heart to supply the cardiovascular requirements of the organism is lost. Such pathological cardiac hypertrophy has a poor prognosis and leads to cardiac failure and sudden death (Levy et al. 1990).

A widely supported hypothesis is that changes in Ca^{2+} cycling mediate the hypertrophic response (Molkentin 2006). Although initial modifications of Ca^{2+} handling are beneficial, as the heart progresses to failure they may contribute to pathology (Fig. 3) (Roderick et al. 2007). A general feature of calcium fluxes during adaptive hypertrophy is an increase in the amplitudes of the Ca^{2+} transient, and also of the Ca^{2+} sparks that underlie them. Contributing to this is an increase in SR store loading mediated by increased SERCA pump activity. SERCA activity is enhanced through several mechanisms: increased SERCA expression, decreased PLB expression, and increased PLB phosphorylation. A decrease in NCX current and an increase in RyR activity have also been detected during this stage of hypertrophy.

As the heart progresses to failure, further changes in Ca^{2+} regulation and flux are observed. Contributing to the decreased contractility of the failing heart is a general decrease in the amplitude of each action potential-evoked Ca^{2+} transient. Underlying this reduced Ca^{2+} signal are a number of factors, foremost of which is a decrease in SERCA activity (Fig. 3) (Hoshijima et al. 2006). SERCA activity may be modified through a reduction in its expression, increased PLB expression or decreased PLB phosphorylation. As a result of this decrease in SERCA function, diastolic Ca^{2+} is elevated (a common feature of the failing heart) while myocyte relaxation is prolonged and SR store content is diminished (Roderick et al. 2007). The key contribution of SERCA dysfunction to the decrease in cardiac function during failure has led to the development of viral-mediated SERCA overexpression for gene therapy (Lyon et al. 2011).

Failing hearts are also characterized by more arrhythmic Ca^{2+} signals. This is perhaps surprising given that SR store loading—a key determinant of RyR activity—is decreased. RyRs isolated from failing hearts and incorporated into lipid bilayers are more spontaneously active than RyR isolated from control hearts, perhaps explaining this conundrum (Kubalova et al. 2005). Indeed, myocytes from failing hearts showed decreased store loading, increased spontaneous RyR opening (observed as Ca^{2+} sparks) and decreased SR Ca^{2+} reuptake. Another mechanism for increased arrhythmic Ca^{2+} signaling during heart failure is the expression of inositol 1,4,5-trisphosphate ($InsP_3$) gated Ca^{2+} release channels ($InsP_3Rs$). $InsP_3Rs$ are typically >50-fold less abundant than RyR in healthy myocytes (Kockskamper et al. 2008). However, their expression increases significantly during hypertrophy and heart failure. In particular, their expression within dyadic junctions increases (Fig. 3) (Harzheim et al. 2009, 2010). The up-regulation of $InsP_3Rs$ provides a positive-feedback loop that can promote further hypertrophic remodeling (Nakayama et al. 2010). Although $InsP_3Rs$ are not capable of mounting significant Ca^{2+} signals by themselves, their location next to RyRs within dyadic junctions means that they can trigger CICR and thereby augment EC-coupling. The expression of $InsP_3Rs$ may be an initially beneficial adaptive aspect of myocyte remodeling in that they can help to increase systolic Ca^{2+} transients and evoke greater contraction. However, concomitant with this potentially beneficial aspect of enhanced $InsP_3R$ expression is an undesirable increase in the propensity of cells to show spontaneous Ca^{2+} release events that may cause arrhythmia. This is because of the fact that $InsP_3Rs$ are not solely tuned to the activation of I_{Ca}, but can open independently whenever cytosolic $InsP_3$ levels are sufficient (Harzheim et al. 2009).

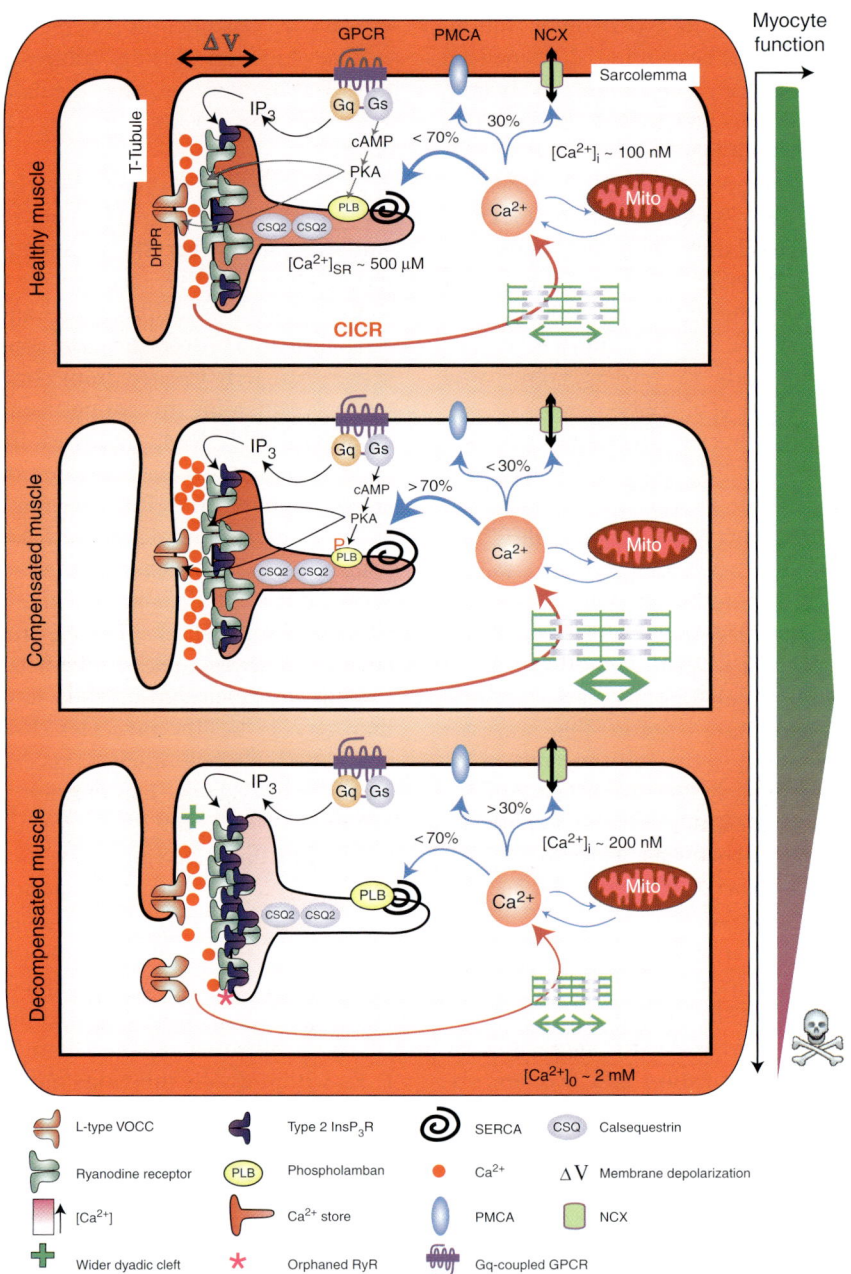

Figure 3. Excitation contraction coupling in healthy and decompensated hypertrophic cardiac muscle. The figure depicts the architecture and molecular composition of a cardiac dyad in a normal, healthy myocyte (*upper* panel), a compensated hypertrophic situation (*middle* panel), and a decompensated failing myocyte (*lower* panel). A key depicting the symbols used to represent the major players in EC-coupling is provided at the *bottom* of the figure. Depolarization of the plasma membrane results in Ca^{2+} influx through L-type voltage gated channels in the T-Tubule, which stimulates Ca^{2+} release via RyRs located on the juxtaposed SR. Following diffusion out of the dyadic cleft, Ca^{2+} encounters the contractile filaments causing myocyte contraction. Myocyte relaxation is then brought about by Ca^{2+} recycling back into the SR by the SERCA pump or extrusion across the plasma membrane via NCX. (*See facing page for legend.*)

The importance of cellular architecture to EC-coupling is highlighted by its contribution to Ca^{2+} dysregulation during cardiac failure. In myocytes from failing heart, the synchronicity of the initiation of the Ca^{2+} transient is reduced and the amplitude of the transient diminished (Gomez et al. 1997). Work from a number of laboratories using live and fixed cell staining approaches, have now established that T-tubules are lost or atrophied as hypertrophy develops, whereas the distribution of RyRs on the SR is unaffected (Gomez et al. 2001; Song et al. 2005b). As a consequence of this membrane remodeling, RyRs become progressively orphaned. Deficiencies in EC-coupling because of T-tubule remodeling have also been observed during the earlier stages of hypertrophy, albeit with less dramatic consequences (Xu et al. 2007). During this stage, the width of the dyadic cleft is marginally increased, altering the kinetic properties of coupling between membrane depolarization and Ca^{2+} release from the SR. Inotropic stimuli such as adrenaline, which serve to increase Ca^{2+} influx and sensitivity of RyRs to release, overcome this deficiency in EC-coupling thus indicating that the Ca^{2+} signaling machinery is still intact. Similarly, dyadic InsP$_3$Rs, which as indicated earlier are elevated during hypertrophy, may also contribute to overcoming the decreased efficiency of coupling.

CONCLUSION

The raison d'etre of cardiac myocytes is controlled contraction in response to repetitive electrical depolarization signals—a function that is fundamentally controlled by Ca^{2+}. Although there are a limited number of key players—I_{Ca}, NCX, SERCA, RyRs, TnC—involved in generating and reversing Ca^{2+} signals, they are subject to numerous levels of regulation and an array of interactions with other proteins. Furthermore, the cellular location of the Ca^{2+} signaling systems is critical in determining the spatial properties of the Ca^{2+} signals during EC-coupling. Cardiac Ca^{2+} signals can be acutely altered to provide rapid changes in cardiac output. The heart responds to long-term increased hemodynamic demand by remodeling. This remodeling encompasses both structural changes and altered gene expression. Depending on the stimulus, this can be an adaptive, reversible remodeling that promotes Ca^{2+} signaling and cardiac function. Alternatively, a heart can become committed to an

Figure 3. (*Continued*) Neurohormonal activation of Gα_q-coupled receptors leads to the formation of InsP$_3$, which stimulates Ca^{2+} release via InsP$_3$Rs located in the dyad. This InsP$_3$-stimulated Ca^{2+} release sensitizes neighboring RyRs, causing enhanced Ca^{2+} fluxes and increasing the frequency of arrhythmic events. Activation of β-adrenergic receptors increases intracellular cAMP, which activates PKA leading to phosphorylation of PLB. On phosphorylation, PLB dissociates from SERCA thereby enhancing Ca^{2+} transport activity. Compensated/adaptive hypertrophy is associated with enhanced EC-coupling. Significantly contributing to this phenotype is an increase in SR store loading. This is brought about by an up-regulation of SERCA activity mediated by either an increase in SERCA or decrease in PLB expression. Alternatively, as a result of PKA-dependent phosphorylation, PLB interaction with SERCA and suppression of Ca^{2+} transport may also be decreased. Increased SERCA activity also serves to increase the rate of relaxation thereby allowing more rapid cycles of myocyte contraction. The width of the dyadic cleft may also marginally increase at this stage of hypertrophic remodeling. However, IP$_3$Rs are up-regulated in the dyad during hypertrophy supplementing the Ca^{2+} signal arising via RyRs to possibly further support EC-coupling. During decompensated hypertrophy, myocyte architecture and protein expression are remodeled. Specifically, the width of the dyadic cleft is increased making it harder for Ca^{2+} arising via VOCCs to activate Ca^{2+} release from RyRs. T-Tubules also atrophy resulting in orphaned RyRs. The SERCA-PLB ratio is also modified to favor decreased SERCA activity. Notably, InsP$_3$R expression in the dyad is increased. As a result, more of the RyRs that are located in this region are close enough to InsP$_3$Rs to be affected by Ca^{2+} arising from them. Overactivation of these InsP$_3$Rs, for example, by the elevated levels of circulating ET-1 present during heart failure, promotes arrhythmias thereby contributing to the pathology associated with heart failure.

irreversible form of remodeling that causes progressively weaker and more arrhythmic Ca^{2+} signaling. The remodeling causes changes in Ca^{2+} signaling and vice versa. Therefore, within the context of the heart Ca^{2+} is very much a signal for both life and death.

REFERENCES

Abi-Gerges N, Fischmeister R, Mery PF. 2001. G protein-mediated inhibitory effect of a nitric oxide donor on the L-type Ca^{2+} current in rat ventricular myocytes. *J Physiol* **531:** 117–130.

Ahlers BA, Zhang XQ, Moorman JR, Rothblum LI, Carl LL, Song J, Wang J, Geddis LM, Tucker AL, Mounsey JP, et al. 2005. Identification of an endogenous inhibitor of the cardiac Na^+/Ca^{2+} exchanger, phospholemman. *J Biol Chem* **280:** 19875–19882.

Anderson K, Lai FA, Liu QY, Rousseau E, Erickson HP, Meissner G. 1989. Structural and functional characterization of the purified cardiac ryanodine receptor-Ca^{2+} release channel complex. *J Biol Chem* **264:** 1329–1335.

Andersson KB, Birkeland JA, Finsen AV, Louch WE, Sjaastad I, Wang Y, Chen J, Molkentin JD, Chien KR, Sejersted OM, et al. 2009. Moderate heart dysfunction in mice with inducible cardiomyocyte-specific excision of the Serca2 gene. *J Mol Cell Cardiol* **47:** 180–187.

Ayettey AS, Navaratnam V. 1978. The T-tubule system in the specialized and general myocardium of the rat. *J Anat* **127:** 125–140.

Baker PF, Blaustein MP, Hodgkin AL, Steinhardt RA. 1969. The influence of calcium on sodium efflux in squid axons. *J Physiol* **200:** 431–458.

Balshaw DM, Xu L, Yamaguchi N, Pasek DA, Meissner G. 2001. Calmodulin binding and inhibition of cardiac muscle calcium release channel (ryanodine receptor). *J Biol Chem* **276:** 20144–20153.

Bandyopadhyay A, Shin DW, Ahn JO, Kim DH. 2000. Calcineurin regulates ryanodine receptor/Ca^{2+}-release channels in rat heart. *Biochem J* **352** (Pt 1): 61–70.

Bassani JW, Bassani RA, Bers DM. 1994. Relaxation in rabbit and rat cardiac cells: Species-dependent differences in cellular mechanisms. *J Physiol* **476:** 279–293.

Berridge MJ. 2002. The endoplasmic reticulum: A multifunctional signaling organelle. *Cell Calcium* **32:** 235–249.

Berridge MJ. 2003. Cardiac calcium signalling. *Biochem Soc Trans* **31:** 930–933.

Bers DM. 2004. Macromolecular complexes regulating cardiac ryanodine receptor function. *J Mol Cell Cardiol* **37:** 417–429.

Bers DM. 2008. Calcium cycling and signaling in cardiac myocytes. *Annu Rev Physiol* **70:** 23–49.

Bers DM, Guo T. 2005. Calcium signaling in cardiac ventricular myocytes. *Ann NY Acad Sci* **1047:** 86–98.

Block BA, Imagawa T, Campbell KP, Franzini-Armstrong C. 1988. Structural evidence for direct interaction between the molecular components of the transverse tubule/sarcoplasmic reticulum junction in skeletal muscle. *J Cell Biol* **107:** 2587–2600.

Bogdanov KY, Vinogradova TM, Lakatta EG. 2001. Sinoatrial nodal cell ryanodine receptor and Na^+-Ca^{2+} exchanger: Molecular partners in pacemaker regulation. *Circ Res* **88:** 1254–1258.

Bootman MD, Higazi DR, Coombes S, Roderick HL. 2006. Calcium signalling during excitation-contraction coupling in mammalian atrial myocytes. *J Cell Sci* **119:** 3915–3925.

Bootman MD, Smyrnias I, Thul R, Coombes S, Roderick HL. 2011. Atrial cardiomyocyte calcium signalling. *Biochim Biophys Acta* **1813:** 922–934.

Bossen EH, Sommer JR. 1984. Comparative stereology of the lizard and frog myocardium. *Tissue Cell* **16:** 173–178.

Bossen EH, Sommer JR, Waugh RA. 1978. Comparative stereology of the mouse and finch left ventricle. *Tissue Cell* **10:** 773–784.

Brette F, Orchard C. 2003. T-tubule function in mammalian cardiac myocytes. *Circ Res* **92:** 1182–1192.

Bridge JH, Smolley JR, Spitzer KW. 1990. The relationship between charge movements associated with ICa and INa-Ca in cardiac myocytes. *Science* **248:** 376–378.

Brown HF, DiFrancesco D, Noble SJ. 1979. How does adrenaline accelerate the heart? *Nature* **280:** 235–236.

Catterall WA. 2000. Structure and regulation of voltage-gated Ca^{2+} channels. *Annu Rev Cell Dev Biol* **16:** 521–555.

Chase A, Orchard CH. 2011. Ca efflux via the sarcolemmal Ca ATPase occurs only in the t-tubules of rat ventricular myocytes. *J Mol Cell Cardiol* **50:** 187–193.

Chen-Izu Y, McCulle SL, Ward CW, Soeller C, Allen BM, Rabang C, Cannell MB, Balke CW, Izu LT. 2006. Three-dimensional distribution of ryanodine receptor clusters in cardiac myocytes. *Biophys J* **91:** 1–13.

Cheng H, Lederer WJ. 2008. Calcium sparks. *Physiol Rev* **88:** 1491–1545.

Cheng H, Wang SQ. 2002. Calcium signaling between sarcolemmal calcium channels and ryanodine receptors in heart cells. *Front Biosci* **7:** d1867–d1878.

Cheung JY, Rothblum LI, Moorman JR, Tucker AL, Song J, Ahlers BA, Carl LL, Wang J, Zhang XQ. 2007. Regulation of cardiac Na^+/Ca^{2+} exchanger by phospholemman. *Ann NY Acad Sci* **1099:** 119–134.

Chiang CS, Huang CH, Chieng H, Chang YT, Chang D, Chen JJ, Chen YC, Chen YH, Shin HS, Campbell KP, et al. 2009. The $Ca_v3.2$ T-type Ca^{2+} channel is required for pressure overload-induced cardiac hypertrophy in mice. *Circ Res* **104:** 522–530.

Curtis BM, Catterall WA. 1985. Phosphorylation of the calcium antagonist receptor of the voltage-sensitive calcium channel by cAMP-dependent protein kinase. *Proc Natl Acad Sci* **82:** 2528–2532.

DeMaria CD, Soong TW, Alseikhan BA, Alvania RS, Yue DT. 2001. Calmodulin bifurcates the local Ca^{2+} signal that modulates P/Q-type Ca^{2+} channels. *Nature* **411:** 484–489.

DiFrancesco D. 1993. Pacemaker mechanisms in cardiac tissue. *Annu Rev Physiol* **55:** 455–472.

Eisner DA, Trafford AW, Diaz ME, Overend CL, O'Neill SC. 1998. The control of Ca release from the cardiac sarcoplasmic reticulum: Regulation versus autoregulation. *Cardiovasc Res* **38**: 589–604.

Eisner DA, Kashimura T, O'Neill SC, Venetucci LA, Trafford AW. 2009. What role does modulation of the ryanodine receptor play in cardiac inotropy and arrhythmogenesis? *J Mol Cell Cardiol* **46**: 474–481.

Fabiato A. 1983. Calcium-induced release of calcium from the cardiac sarcoplasmic reticulum. *Am J Physiol* **245**: C1–C14.

Fabiato A, Fabiato F. 1972. Excitation-contraction coupling of isolated cardiac fibers with disrupted or closed sarcolemmas. Calcium-dependent cyclic and tonic contractions. *Circ Res* **31**: 293–307.

Fukuta H, Little WC. 2008. The cardiac cycle and the physiologic basis of left ventricular contraction, ejection, relaxation, and filling. *Heart Fail Clin* **4**: 1–11.

Gerhardstein BL, Puri TS, Chien AJ, Hosey MM. 1999. Identification of the sites phosphorylated by cyclic AMP-dependent protein kinase on the β_2 subunit of L-type voltage-dependent calcium channels. *Biochemistry* **38**: 10361–10370.

Gomez AM, Valdivia HH, Cheng H, Lederer MR, Santana LF, Cannell MB, McCune SA, Altschuld RA, Lederer WJ. 1997. Defective excitation-contraction coupling in experimental cardiac hypertrophy and heart failure. *Science* **276**: 800–806.

Gomez AM, Guatimosim S, Dilly KW, Vassort G, Lederer WJ. 2001. Heart failure after myocardial infarction: Altered excitation-contraction coupling. *Circulation* **104**: 688–693.

Greaser ML, Gergely J. 1971. Reconstitution of troponin activity from three protein components. *J Biol Chem* **246**: 4226–4233.

Grimm M, Brown JH. 2010. β-adrenergic receptor signaling in the heart: Role of CaMKII. *J Mol Cell Cardiol* **48**: 322–330.

Guo T, Cornea RL, Huke S, Camors E, Yang Y, Picht E, Fruen BR, Bers DM. 2010. Kinetics of FKBP12.6 binding to ryanodine receptors in permeabilized cardiac myocytes and effects on Ca sparks. *Circ Res* **106**: 1743–1752.

Gyorke S, Stevens SC, Terentyev D. 2009. Cardiac calsequestrin: Quest inside the SR. *J Physiol* **587**: 3091–3094.

Hartzell HC, Fischmeister R. 1986. Opposite effects of cyclic GMP and cyclic AMP on Ca^{2+} current in single heart cells. *Nature* **323**: 273–275.

Harzheim D, Movassagh M, Foo RS, Ritter O, Tashfeen A, Conway SJ, Bootman MD, Roderick HL. 2009. Increased InsP3Rs in the junctional sarcoplasmic reticulum augment Ca^{2+} transients and arrhythmias associated with cardiac hypertrophy. *Proc Natl Acad Sci* **106**: 11406–11411.

Harzheim D, Talasila A, Movassagh M, Foo RS, Figg N, Bootman MD, Roderick HL. 2010. Elevated InsP3R expression underlies enhanced calcium fluxes and spontaneous extra-systolic calcium release events in hypertrophic cardiac myocytes. *Channels* **4**: 67–71.

He J, Conklin MW, Foell JD, Wolff MR, Haworth RA, Coronado R, Kamp TJ. 2001. Reduction in density of transverse tubules and L-type Ca^{2+} channels in canine tachycardia-induced heart failure. *Cardiovasc Res* **49**: 298–307.

Hilgemann DW. 1990. Regulation and deregulation of cardiac Na^+-Ca^{2+} exchange in giant excised sarcolemmal membrane patches. *Nature* **344**: 242–245.

Hilgemann DW, Matsuoka S, Nagel GA, Collins A. 1992. Steady-state and dynamic properties of cardiac sodium-calcium exchange. Sodium-dependent inactivation. *J Gen Physiol* **100**: 905–932.

Hoshijima M, Knoll R, Pashmforoush M, Chien KR. 2006. Reversal of calcium cycling defects in advanced heart failure toward molecular therapy. *J Am Coll Cardiol* **48**: A15–A23.

Hudmon A, Schulman H, Kim J, Maltez JM, Tsien RW, Pitt GS. 2005. CaMKII tethers to L-type Ca^{2+} channels, establishing a local and dedicated integrator of Ca^{2+} signals for facilitation. *J Cell Biol* **171**: 537–547.

Huser J, Blatter LA, Lipsius SL. 2000. Intracellular Ca^{2+} release contributes to automaticity in cat atrial pacemaker cells. *J Physiol* **524** (Pt 2): 415–422.

Iwamoto T, Pan Y, Wakabayashi S, Imagawa T, Yamanaka HI, Shigekawa M. 1996. Phosphorylation-dependent regulation of cardiac Na^+/Ca^{2+} exchanger via protein kinase C. *J Biol Chem* **271**: 13609–13615.

Jeyakumar LH, Ballester L, Cheng DS, McIntyre JO, Chang P, Olivey HE, Rollins-Smith L, Barnett JV, Murray K, Xin HB, et al. 2001. FKBP binding characteristics of cardiac microsomes from diverse vertebrates. *Biochem Biophys Res Commun* **281**: 979–986.

Kamp TJ, Hell JW. 2000. Regulation of cardiac L-type calcium channels by protein kinase A and protein kinase C. *Circ Res* **87**: 1095–1102.

Katz G, Arad M, Eldar M. 2009. Catecholaminergic polymorphic ventricular tachycardia from bedside to bench and beyond. *Curr Prob Cardiol* **34**: 9–43.

Kimura Y, Kurzydlowski K, Tada M, MacLennan DH. 1997. Phospholamban inhibitory function is activated by depolymerization. *J Biol Chem* **272**: 15061–15064.

Kockskamper J, Zima AV, Roderick HL, Pieske B, Blatter LA, Bootman MD. 2008. Emerging roles of inositol 1,4,5-trisphosphate signaling in cardiac myocytes. *J Mol Cell Cardiol* **45**: 128–147.

Kubalova Z, Terentyev D, Viatchenko-Karpinski S, Nishijima Y, Gyorke I, Terentyeva R, da Cunha DN, Sridhar A, Feldman DS, Hamlin RL, et al. 2005. Abnormal intrastore calcium signaling in chronic heart failure. *Proc Natl Acad Sci* **102**: 14104–14109.

Lakatta EG, DiFrancesco D. 2009. What keeps us ticking: A funny current, a calcium clock, or both? *J Mol Cell Cardiol* **47**: 157–170.

Lanner JT, Georgiou DK, Joshi AD, Hamilton SL. 2010. Ryanodine receptors: Structure, expression, molecular details, and function in calcium release. *Cold Spring Harb Perspect Biol* **2**: a003996.

Lee KS, Marban E, Tsien RW. 1985. Inactivation of calcium channels in mammalian heart cells: Joint dependence on membrane potential and intracellular calcium. *J Physiol* **364**: 395–411.

Lehnart SE, Wehrens XH, Reiken S, Warrier S, Belevych AE, Harvey RD, Richter W, Jin SL, Conti M, Marks AR. 2005. Phosphodiesterase 4D deficiency in the ryanodine-receptor complex promotes heart failure and arrhythmias. *Cell* **123:** 25–35.

Levy D, Garrison RJ, Savage DD, Kannel WB, Castelli WP. 1990. Prognostic implications of echocardiographically determined left ventricular mass in the Framingham Heart Study. *N Engl J Med* **322:** 1561–1566.

Lokuta AJ, Rogers TB, Lederer WJ, Valdivia HH. 1995. Modulation of cardiac ryanodine receptors of swine and rabbit by a phosphorylation-dephosphorylation mechanism. *J Physiol* **487** (Pt 3)**:** 609–622.

Lokuta AJ, Meyers MB, Sander PR, Fishman GI, Valdivia HH. 1997. Modulation of cardiac ryanodine receptors by sorcin. *J Biol Chem* **272:** 25333–25338.

Louch WE, Bito V, Heinzel FR, Macianskiene R, Vanhaecke J, Flameng W, Mubagwa K, Sipido KR. 2004. Reduced synchrony of Ca^{2+} release with loss of T-tubules—a comparison to Ca^{2+} release in human failing cardiomyocytes. *Cardiovasc Res* **62:** 63–73.

Lyon AR, Bannister ML, Collins T, Pearce E, Sepehripour AH, Dubb SS, Garcia E, O'Gara P, Liang L, Kohlbrenner E, et al. 2011. SERCA2a gene transfer decreases SR calcium leak and reduces ventricular arrhythmias in a model of chronic heart failure. *Circ Arrhythm Electrophysiol* doi: 101161/CIRCEP110961615.

Lytton J, Westlin M, Burk SE, Shull GE, MacLennan DH. 1992. Functional comparisons between isoforms of the sarcoplasmic or endoplasmic reticulum family of calcium pumps. *J Biol Chem* **267:** 14483–14489.

Mackenzie L, Bootman MD, Berridge MJ, Lipp P. 2001. Predetermined recruitment of calcium release sites underlies excitation-contraction coupling in rat atrial myocytes. *J Physiol* **530:** 417–429.

Mackenzie L, Roderick HL, Berridge MJ, Conway SJ, Bootman MD. 2004. The spatial pattern of atrial cardiomyocyte calcium signalling modulates contraction. *J Cell Sci* **117:** 6327–6337.

Marx SO, Reiken S, Hisamatsu Y, Jayaraman T, Burkhoff D, Rosemblit N, Marks AR. 2000. PKA phosphorylation dissociates FKBP12.6 from the calcium release channel (ryanodine receptor): Defective regulation in failing hearts. *Cell* **101:** 365–376.

Marx SO, Reiken S, Hisamatsu Y, Gaburjakova M, Gaburjakova J, Yang YM, Rosemblit N, Marks AR. 2001. Phosphorylation-dependent regulation of ryanodine receptors: A novel role for leucine/isoleucine zippers. *J Cell Biol* **153:** 699–708.

McHugh D, Sharp EM, Scheuer T, Catterall WA. 2000. Inhibition of cardiac L-type calcium channels by protein kinase C phosphorylation of two sites in the N-terminal domain. *Proc Natl Acad Sci* **97:** 12334–12338.

Meissner G, Henderson JS. 1987. Rapid calcium release from cardiac sarcoplasmic reticulum vesicles is dependent on Ca^{2+} and is modulated by Mg^{2+}, adenine nucleotide, and calmodulin. *J Biol Chem* **262:** 3065–3073.

Metzger JM, Westfall MV. 2004. Covalent and noncovalent modification of thin filament action: The essential role of troponin in cardiac muscle regulation. *Circ Res* **94:** 146–158.

Molkentin JD. 2006. Dichotomy of Ca^{2+} in the heart: Contraction versus intracellular signaling. *J Clin Invest* **116:** 623–626.

Morimoto S. 2008. Sarcomeric proteins and inherited cardiomyopathies. *Cardiovasc Res* **77:** 659–666.

Movsesian MA, Nishikawa M, Adelstein RS. 1984. Phosphorylation of phospholamban by calcium-activated, phospholipid-dependent protein kinase. Stimulation of cardiac sarcoplasmic reticulum calcium uptake. *J Biol Chem* **259:** 8029–8032.

Nakayama H, Bodi I, Maillet M, DeSantiago J, Domeier TL, Mikoshiba K, Lorenz JN, Blatter LA, Bers DM, Molkentin JD. 2010. The IP3 receptor regulates cardiac hypertrophy in response to select stimuli. *Circ Res* **107:** 659–666.

Narayanan N, Xu A. 1997. Phosphorylation and regulation of the Ca^{2+}-pumping ATPase in cardiac sarcoplasmic reticulum by calcium/calmodulin-dependent protein kinase. *Basic Res Cardiol* **92** (Suppl 1)**:** 25–35.

Nicoll DA, Longoni S, Philipson KD. 1990. Molecular cloning and functional expression of the cardiac sarcolemmal Na^+-Ca^{2+} exchanger. *Science* **250:** 562–565.

Nowycky MC, Fox AP, Tsien RW. 1985. Three types of neuronal calcium channel with different calcium agonist sensitivity. *Nature* **316:** 440–443.

Orchard C, Brette F. 2008. T-tubules and sarcoplasmic reticulum function in cardiac ventricular myocytes. *Cardiovasc Res* **77:** 237–244.

Parmacek MS, Solaro RJ. 2004. Biology of the troponin complex in cardiac myocytes. *Prog Cardiovasc Dis* **47:** 159–176.

Perets T, Blumenstein Y, Shistik E, Lotan I, Dascal N. 1996. A potential site of functional modulation by protein kinase A in the cardiac Ca^{2+} channel α_{1C} subunit. *FEBS Lett* **384:** 189–192.

Peterson BZ, DeMaria CD, Adelman JP, Yue DT. 1999. Calmodulin is the Ca^{2+} sensor for Ca^{2+}-dependent inactivation of L-type calcium channels. *Neuron* **22:** 549–558.

Porter Moore C, Rodney G, Zhang JZ, Santacruz-Toloza L, Strasburg G, Hamilton SL. 1999a. Apocalmodulin and Ca^{2+} calmodulin bind to the same region on the skeletal muscle Ca^{2+} release channel. *Biochemistry* **38:** 8532–8537.

Porter Moore C, Zhang JZ, Hamilton SL. 1999b. A role for cysteine 3635 of RYR1 in redox modulation and calmodulin binding. *J Biol Chem* **274:** 36831–36834.

Pott C, Henderson SA, Goldhaber JI, Philipson KD. 2007. Na^+/Ca^{2+} exchanger knockout mice: Plasticity of cardiac excitation-contraction coupling. *Ann NY Acad Sci* **1099:** 270–275.

Prins D, Michalak M. 2011. Organellar calcium buffers. *Cold Spring Harb Perspect Biol* **3:** a004069.

Qin N, Olcese R, Bransby M, Lin T, Birnbaumer L. 1999. Ca^{2+}-induced inhibition of the cardiac Ca^{2+} channel depends on calmodulin. *Proc Natl Acad Sci* **96:** 2435–2438.

Reuter H, Seitz N. 1968. The dependence of calcium efflux from cardiac muscle on temperature and external ion composition. *J Physiol* **195:** 451–470.

Rhoads AR, Friedberg F. 1997. Sequence motifs for calmodulin recognition. *FASEB J* **11:** 331–340.

Ringer S. 1883. A further contribution regarding the influence of the different constituents of the blood on the contraction of the heart. *J Physiol* **4:** 29–42.3.

Rios E, Brum G. 1987. Involvement of dihydropyridine receptors in excitation-contraction coupling in skeletal muscle. *Nature* **325**: 717–720.

Roderick HL, Higazi DR, Smyrnias I, Fearnley C, Harzheim D, Bootman MD. 2007. Calcium in the heart: When it's good, it's very very good, but when it's bad, it's horrid. *Biochem Soc Trans* **35**: 957–961.

Ruknudin A, He S, Lederer WJ, Schulze DH. 2000. Functional differences between cardiac and renal isoforms of the rat Na^+-Ca^{2+} exchanger NCX1 expressed in Xenopus oocytes. *J Physiol* **529** (Pt 3): 599–610.

Schulson MN, Scriven DR, Fletcher P, Moore ED. 2011. Couplons in rat atria form distinct subgroups defined by their molecular partners. *J Cell Sci* **124**: 1167–1174.

Schulze DH, Muqhal M, Lederer WJ, Ruknudin AM. 2003. Sodium/calcium exchanger (NCX1) macromolecular complex. *J Biol Chem* **278**: 28849–28855.

Sedarat F, Xu L, Moore ED, Tibbits GF. 2000. Colocalization of dihydropyridine and ryanodine receptors in neonate rabbit heart using confocal microscopy. *Am J Physiol Heart Circ Physiol* **279**: H202–H209.

Serysheva II. 2004. Structural insights into excitation-contraction coupling by electron cryomicroscopy. *Biochemistry (Moscow)* **69**: 1226–1232.

Serysheva II, Schatz M, van Heel M, Chiu W, Hamilton SL. 1999. Structure of the skeletal muscle calcium release channel activated with Ca^{2+} and AMP-PCP. *Biophys J* **77**: 1936–1944.

Sham JS, Hatem SN, Morad M. 1995. Species differences in the activity of the Na^+-Ca^{2+} exchanger in mammalian cardiac myocytes. *J Physiol* **488** (Pt 3): 623–631.

Shannon TR, Bers DM. 2004. Integrated Ca^{2+} management in cardiac myocytes. *Ann NY Acad Sci* **1015**: 28–38.

Shistik E, Ivanina T, Blumenstein Y, Dascal N. 1998. Crucial role of N terminus in function of cardiac L-type Ca^{2+} channel and its modulation by protein kinase C. *J Biol Chem* **273**: 17901–17909.

Simmerman HK, Collins JH, Theibert JL, Wegener AD, Jones LR. 1986. Sequence analysis of phospholamban. Identification of phosphorylation sites and two major structural domains. *J Biol Chem* **261**: 13333–13341.

Smyrnias I, Mair W, Harzheim D, Walker SA, Roderick HL, Bootman MD. 2010. Comparison of the T-tubule system in adult rat ventricular and atrial myocytes, and its role in excitation-contraction coupling and inotropic stimulation. *Cell Calcium* **47**: 210–223.

Song J, Zhang XQ, Ahlers BA, Carl LL, Wang J, Rothblum LI, Stahl RC, Mounsey JP, Tucker AL, Moorman JR, et al. 2005a. Serine 68 of phospholemman is critical in modulation of contractility, $[Ca^{2+}]_i$ transients, and Na^+/Ca^{2+} exchange in adult rat cardiac myocytes. *Am J Physiol Heart Circ Physiol* **288**: H2342–H2354.

Song LS, Guatimosim S, Gomez-Viquez L, Sobie EA, Ziman A, Hartmann H, Lederer WJ. 2005b. Calcium biology of the transverse tubules in heart. *Ann NY Acad Sci* **1047**: 99–111.

Sumandea MP, Burkart EM, Kobayashi T, De Tombe PP, Solaro RJ. 2004. Molecular and integrated biology of thin filament protein phosphorylation in heart muscle. *Ann NY Acad Sci* **1015**: 39–52.

Tada M, Kirchberger MA, Repke DI, Katz AM. 1974. The stimulation of calcium transport in cardiac sarcoplasmic reticulum by adenosine $3':5'$-monophosphate-dependent protein kinase. *J Biol Chem* **249**: 6174–6180.

Tadross MR, Dick IE, Yue DT. 2008. Mechanism of local and global Ca^{2+} sensing by calmodulin in complex with a Ca^{2+} channel. *Cell* **133**: 1228–1240.

Takeishi Y, Chu G, Kirkpatrick DM, Li Z, Wakasaki H, Kranias EG, King GL, Walsh RA. 1998. In vivo phosphorylation of cardiac troponin I by protein kinase Cβ2 decreases cardiomyocyte calcium responsiveness and contractility in transgenic mouse hearts. *J Clin Invest* **102**: 72–78.

Tang W, Halling DB, Black DJ, Pate P, Zhang JZ, Pedersen S, Altschuld RA, Hamilton SL. 2003. Apocalmodulin and Ca^{2+} calmodulin-binding sites on the $Ca_V1.2$ channel. *Biophys J* **85**: 1538–1547.

Terentyev D, Viatchenko-Karpinski S, Gyorke I, Volpe P, Williams SC, Gyorke S. 2003. Calsequestrin determines the functional size and stability of cardiac intracellular calcium stores: Mechanism for hereditary arrhythmia. *Proc Natl Acad Sci* **100**: 11759–11764.

Terentyev D, Nori A, Santoro M, Viatchenko-Karpinski S, Kubalova Z, Gyorke I, Terentyeva R, Vedamoorthyrao S, Blom NA, Valle G, et al. 2006. Abnormal interactions of calsequestrin with the ryanodine receptor calcium release channel complex linked to exercise-induced sudden cardiac death. *Circ Res* **98**: 1151–1158.

Timerman AP, Onoue H, Xin HB, Barg S, Copello J, Wiederrecht G, Fleischer S. 1996. Selective binding of FKBP12.6 by the cardiac ryanodine receptor. *J Biol Chem* **271**: 20385–20391.

Toyofuku T, Curotto Kurzydlowski K, Narayanan N, MacLennan DH. 1994. Identification of Ser38 as the site in cardiac sarcoplasmic reticulum Ca^{2+}-ATPase that is phosphorylated by Ca^{2+}/calmodulin-dependent protein kinase. *J Biol Chem* **269**: 26492–26496.

Valdivia HH, Kaplan JH, Ellis-Davies GC, Lederer WJ. 1995. Rapid adaptation of cardiac ryanodine receptors: Modulation by Mg^{2+} and phosphorylation. *Science* **267**: 1997–2000.

Viatchenko-Karpinski S, Terentyev D, Gyorke I, Terentyeva R, Volpe P, Priori SG, Napolitano C, Nori A, Williams SC, Gyorke S. 2004. Abnormal calcium signaling and sudden cardiac death associated with mutation of calsequestrin. *Circ Res* **94**: 471–477.

Wagenknecht T, Radermacher M, Grassucci R, Berkowitz J, Xin HB, Fleischer S. 1997. Locations of calmodulin and FK506-binding protein on the three-dimensional architecture of the skeletal muscle ryanodine receptor. *J Biol Chem* **272**: 32463–32471.

Wang SQ, Song LS, Lakatta EG, Cheng H. 2001. Ca^{2+} signalling between single L-type Ca^{2+} channels and ryanodine receptors in heart cells. *Nature* **410**: 592–596.

Wehrens XH, Lehnart SE, Huang F, Vest JA, Reiken SR, Mohler PJ, Sun J, Guatimosim S, Song LS, Rosemblit N, et al. 2003. FKBP12.6 deficiency and defective calcium release channel (ryanodine receptor) function linked to exercise-induced sudden cardiac death. *Cell* **113**: 829–840.

Wehrens XH, Lehnart SE, Reiken SR, Marks AR. 2004. Ca^{2+}/calmodulin-dependent protein kinase II phosphorylation regulates the cardiac ryanodine receptor. *Circ Res* **94:** e61–e70.

Witcher DR, Kovacs RJ, Schulman H, Cefali DC, Jones LR. 1991. Unique phosphorylation site on the cardiac ryanodine receptor regulates calcium channel activity. *J Biol Chem* **266:** 11144–11152.

Wu Y, Colbran RJ, Anderson ME. 2001a. Calmodulin kinase is a molecular switch for cardiac excitation-contraction coupling. *Proc Natl Acad Sci* **98:** 2877–2881.

Wu Y, Dzhura I, Colbran RJ, Anderson ME. 2001b. Calmodulin kinase and a calmodulin-binding "IQ" domain facilitate L-type Ca^{2+} current in rabbit ventricular myocytes by a common mechanism. *J Physiol* **535:** 679–687.

Wu Y, Gao Z, Chen B, Koval OM, Singh MV, Guan X, Hund TJ, Kutschke W, Sarma S, Grumbach IM, et al. 2009. Calmodulin kinase II is required for fight or flight sinoatrial node physiology. *Proc Natl Acad Sci* **106:** 5972–5977.

Xiao RP, Cheng H, Lederer WJ, Suzuki T, Lakatta EG. 1994. Dual regulation of Ca^{2+}/calmodulin-dependent kinase II activity by membrane voltage and by calcium influx. *Proc Natl Acad Sci* **91:** 9659–9663.

Xiao J, Tian X, Jones PP, Bolstad J, Kong H, Wang R, Zhang L, Duff HJ, Gillis AM, Fleischer S, et al. 2007. Removal of FKBP12.6 does not alter the conductance and activation of the cardiac ryanodine receptor or the susceptibility to stress-induced ventricular arrhythmias. *J Biol Chem* **282:** 34828–34838.

Xu A, Narayanan N. 1999. Ca^{2+}/calmodulin-dependent phosphorylation of the Ca^{2+}-ATPase, uncoupled from phospholamban, stimulates Ca^{2+}-pumping in native cardiac sarcoplasmic reticulum. *Biochem Biophys Res Commun* **258:** 66–72.

Xu L, Tripathy A, Pasek DA, Meissner G. 1999. Ruthenium red modifies the cardiac and skeletal muscle Ca^{2+} release channels (ryanodine receptors) by multiple mechanisms. *J Biol Chem* **274:** 32680–32691.

Xu M, Zhou P, Xu SM, Liu Y, Feng X, Bai SH, Bai Y, Hao XM, Han Q, Zhang Y, et al. 2007. Intermolecular failure of L-type Ca^{2+} channel and ryanodine receptor signaling in hypertrophy. *PLoS Biol* **5:** e21.

Yano K, Zarain-Herzberg A. 1994. Sarcoplasmic reticulum calsequestrins: Structural and functional properties. *Mol Cell Biochem* **135:** 61–70.

Yuan W, Bers DM. 1994. Ca-dependent facilitation of cardiac Ca current is due to Ca-calmodulin-dependent protein kinase. *Am J Physiol* **267:** H982–H993.

Zhang XQ, Qureshi A, Song J, Carl LL, Tian Q, Stahl RC, Carey DJ, Rothblum LI, Cheung JY. 2003. Phospholemman modulates Na^+/Ca^{2+} exchange in adult rat cardiac myocytes. *Am J Physiol Heart Circ Physiol* **284:** H225–H233.

Zhang XQ, Ahlers BA, Tucker AL, Song J, Wang J, Moorman JR, Mounsey JP, Carl LL, Rothblum LI, Cheung JY. 2006. Phospholemman inhibition of the cardiac Na^+/Ca^{2+} exchanger. Role of phosphorylation. *J Biol Chem* **281:** 7784–7792.

Zucchi R, Ronca-Testoni S. 1997. The sarcoplasmic reticulum Ca^{2+} channel/ryanodine receptor: Modulation by endogenous effectors, drugs and disease states. *Pharmacol Rev* **49:** 1–51.

Zuhlke RD, Reuter H. 1998. Ca^{2+}-sensitive inactivation of L-type Ca^{2+} channels depends on multiple cytoplasmic amino acid sequences of the α_{1C} subunit. *Proc Natl Acad Sci* **95:** 3287–3294.

Zuhlke RD, Pitt GS, Deisseroth K, Tsien RW, Reuter H. 1999. Calmodulin supports both inactivation and facilitation of L-type calcium channels. *Nature* **399:** 159–162.

Calcium Signaling in Smooth Muscle

David C. Hill-Eubanks[1], Matthias E. Werner[2], Thomas J. Heppner[1], and Mark T. Nelson[1,2]

[1]Department of Pharmacology, College of Medicine, University of Vermont, Burlington, Vermont 05405
[2]Cardiovascular Medicine, Faculty of Medical and Human Sciences, University of Manchester, Manchester M13 9NT, United Kingdom

Correspondence: Mark.Nelson@uvm.edu

Changes in intracellular Ca^{2+} are central to the function of smooth muscle, which lines the walls of all hollow organs. These changes take a variety of forms, from sustained, cell-wide increases to temporally varying, localized changes. The nature of the Ca^{2+} signal is a reflection of the source of Ca^{2+} (extracellular or intracellular) and the molecular entity responsible for generating it. Depending on the specific channel involved and the detection technology employed, extracellular Ca^{2+} entry may be detected optically as graded elevations in intracellular Ca^{2+}, junctional Ca^{2+} transients, Ca^{2+} flashes, or Ca^{2+} sparklets, whereas release of Ca^{2+} from intracellular stores may manifest as Ca^{2+} sparks, Ca^{2+} puffs, or Ca^{2+} waves. These diverse Ca^{2+} signals collectively regulate a variety of functions. Some functions, such as contractility, are unique to smooth muscle; others are common to other excitable cells (e.g., modulation of membrane potential) and nonexcitable cells (e.g., regulation of gene expression).

SMOOTH MUSCLE

Smooth muscle cells form a continuous layer that lines the walls of the hollow organs of the body, such as blood vessels, intestines, urinary bladder, airways, lymphatics, penis, and uterus. A defining feature of smooth muscle cells is their ability to contract. This property reflects the excitable nature of these cells, which allows for membrane potential-dependent influx of calcium (Ca^{2+}) and the Ca^{2+}-dependent formation of cross-bridges between myosin and actin—the two major contractile proteins that drive contraction. The contractile property of smooth muscle plays an important functional role in these organs, notably by allowing dynamic changes in luminal volume. These changes may regulate the translational movement of the organ's contents, such as in the gastrointestinal tract, where the peristaltic action caused by sequential contraction of smooth muscle segments is responsible for the movement of food, and the urinary bladder, where smooth muscle in the wall relaxes during filling and contracts forcefully to expel urine during micturition. Uterine smooth muscle plays a similar role, relaxing during gestation to accommodate fetal growth and contracting vigorously during parturition. In the vasculature, the contractility of smooth muscle in the vessel wall is a primary determinant of blood pressure, which in turn controls blood flow and the distribution of nutrients and oxygen throughout the body.

Ca^{2+} SIGNALS IN SMOOTH MUSCLE

Smooth muscle contractility, and therefore hollow organ function, is regulated by changes in intracellular Ca^{2+} concentration ([Ca^{2+}]$_i$). These changes may take a variety of forms, the simplest of which is a "global" (cell-wide) increase of the type associated with Ca^{2+}-triggered actin-myosin cross-bridge formation and contraction. However, changes in intracellular Ca^{2+} may also be highly localized and often include a temporal component. Thus, Ca^{2+} changes may be sustained or transient, stationary or moving, or regularly repeating (Berridge 1997; Sanders 2001).

The nature of the Ca^{2+} signal depends to a large extent on the molecular mechanism responsible for generating it. Broadly speaking, there are two major mechanisms by which [Ca^{2+}]$_i$ is raised in smooth muscle: (1) entry of Ca^{2+} from the extracellular space, and (2) release of Ca^{2+} from intracellular stores. Influx of extracellular Ca^{2+} is mediated by ion channels in the plasmalemmal membrane, the most prominent of which is the voltage-dependent Ca^{2+} channel (VDCC). Nonselective cation channels, such as transient receptor potential (TRP) channels and ionotropic purinergic (P2X) receptors, are also potentially important extracellular Ca^{2+} entry pathways in smooth muscle cells. Although a number of intracellular organelles take up and release Ca^{2+}, the sarcoplasmic reticulum (SR) represents the largest pool of releasable Ca^{2+} in smooth muscle cells. In response to a variety of stimuli, Ca^{2+}-release channels in the SR, namely ryanodine receptors (RyR) and inositol trisphosphate receptors (IP$_3$Rs), mediate efflux of Ca^{2+} from the SR into the cytoplasm of the cell.

IMAGING INTRACELLULAR Ca^{2+}

Fluorescence, a property that allows certain molecules to absorb specific wavelengths of light and release energy in the form of light at a longer wavelength, has been exploited to produce numerous fluorescent indicators, including Ca^{2+}-sensitive fluorescent dyes. The fluorescence properties of Ca^{2+}-sensitive indicators change when Ca^{2+} is bound; thus, such dyes can be used to detect intracellular Ca^{2+} levels. Ion fluxes in smooth muscle can occur very rapidly—often in the millisecond range. This property has motivated the development of a number of very rapid fluorescent dyes that enable detection of such changes in ion concentrations with high temporal resolution.

Fluorescent dyes can be loaded into cells by microinjection, but are more commonly introduced by incubating isolated smooth muscle cells or intact tissue with the membrane-permeant acetoxymethyl (AM) ester of the dye. The AM form is readily taken up by cells, but is acted on by intracellular esterases that cleave the ester bond to release the free anion, which is not membrane permeant and is thus retained within the cell. Most fluorescent Ca^{2+} indicators are based on fluorophore-conjugated derivatives of the Ca^{2+} chelator, BAPTA (bis-[o-amino-phenoxy]-ethane-$N,N,N'N'$-tetraacetic acid), which is used for both ratiometric and nonratiometric Ca^{2+} applications (Tsien 1980). Ratiometric dyes are used to measure the intracellular concentration of Ca^{2+}. These dyes show a shift in the excitation (e.g., Fura-2) or emission (e.g., indole-1) spectrum according to the concentration of free or unbound Ca^{2+}. From the ratio of bound and unbound Ca^{2+}, the free intracellular Ca^{2+} concentration can be determined. Nonratiometric dyes, such as those in the Fluo family, show an increase in fluorescence quantum yield or intensity on binding Ca^{2+} and are useful for detecting qualitative changes in Ca^{2+} levels. Although nonratiometric dyes are not usually used for quantitative Ca^{2+} measurements, methods have been developed to calculate Ca^{2+} concentrations using single-wavelength fluorescence signals (Jaggar et al. 1998a; Maravall et al. 2000). One advantage of ratiometric dyes is that the ratio normalizes fluorescence variations caused by uneven cell thickness, dye distribution, dye leakage, or photobleaching—problems that are common to nonratiometric dyes. A disadvantage of ratiometric dyes is that they require excitation in the UV range. Some parameters to consider when selecting a fluorescent dye include (1) ion specificity,

(2) dissociation constant (K_d), (3) hardware suitability (excitation and emission spectra), (4) fluorescence intensity, (5) availability as an AM ester, and (6) sensitivity to photobleaching.

Ca^{2+} signals are typically imaged using laser-scanning confocal microsopes. In its simplest form, a confocal microscope system comprises three main components: (1) a light source; (2) optical and electronic components to manipulate, display, and analyze signals; and (3) a microscope. Lasers, coupled to either upright or inverted microscopes, are commonly used as the light source for confocal microscopes. Although gas lasers (e.g., Ar-ion, Kr-ion, HeNe) provide numerous lasing lines from UV to red, and are well suited for optimal excitation of fluorescent probes, solid-state lasers are increasingly being used because they offer the advantages of longer lifetime, lower power consumption, and compact size.

To record very rapid events or determine the kinetics of a Ca^{2+} event by confocal microscopy, researchers have typically measured Ca^{2+} fluxes using a line-scanning procedure. In line-scan mode, a single line is repeatedly scanned across the cell for a period of time. Each line is then aligned to form an image that is a plot of fluorescence along the scanned line versus time (Fig. 1). Although line scans are still used, the development of very sensitive CCD (charge-coupled device) cameras and rapid Ca^{2+}-sensitive dyes allows laser-scanning confocal systems to routinely achieve detailed spatial and temporal resolution, which is crucial for determining the origin of a Ca^{2+} event.

Confocal microscopy systems can also be used to measure changes in cellular Ca^{2+} caused by activation of photoprotected ("caged") compounds. In this technique, the concentration of ions, signaling molecules, or other biologically active compounds can be instantaneously changed by UV stimulation of cells loaded with their caged equivalent. The resulting changes in intracellular Ca^{2+} can be monitored using the same system. By employing caged Ca^{2+} or Ca^{2+}-releasing compounds (e.g., IP_3), this approach can also be used to directly elevate Ca^{2+} within a cell. This uncaging strategy facilitates the study of a particular signaling step independent of preceding steps in the signaling pathway.

Specialized techniques have also been developed for measuring Ca^{2+} signals in specific subcellular compartments. One such technique is total internal reflection fluorescence (TIRF) microscopy, which is used to selectively visualize the area immediately beneath the plasmalemma. In TIRF microscopy, an evanescent wave is created by the reflection of a laser beam at the interface between a glass coverslip and the cytoplasm of cells attached to the coverslip (Fig. 2). Because TIRF measurements

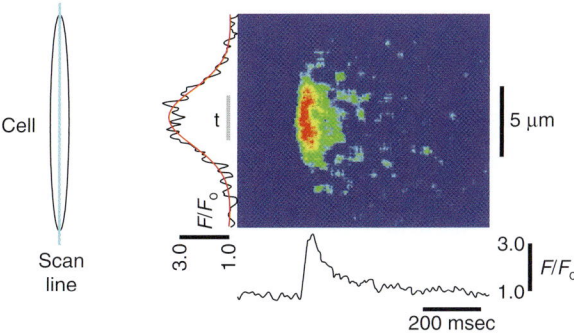

Figure 1. Line-scan imaging. The image shown demonstrates the time course of fractional fluorescence (F/F_0; *bottom*) and spatial distribution of the Ca^{2+} spark (*left*) fitted to a Gaussian distribution (red line). Gray bar labeled "t" indicates the region over which the fluorescence time course was averaged. Scan lines are displayed vertically in a continuous manner. (*Inset*) Orientation of scanning line. (Adapted from Bonev et al. 1997; reprinted with permission from The American Physiological Society © 1997.)

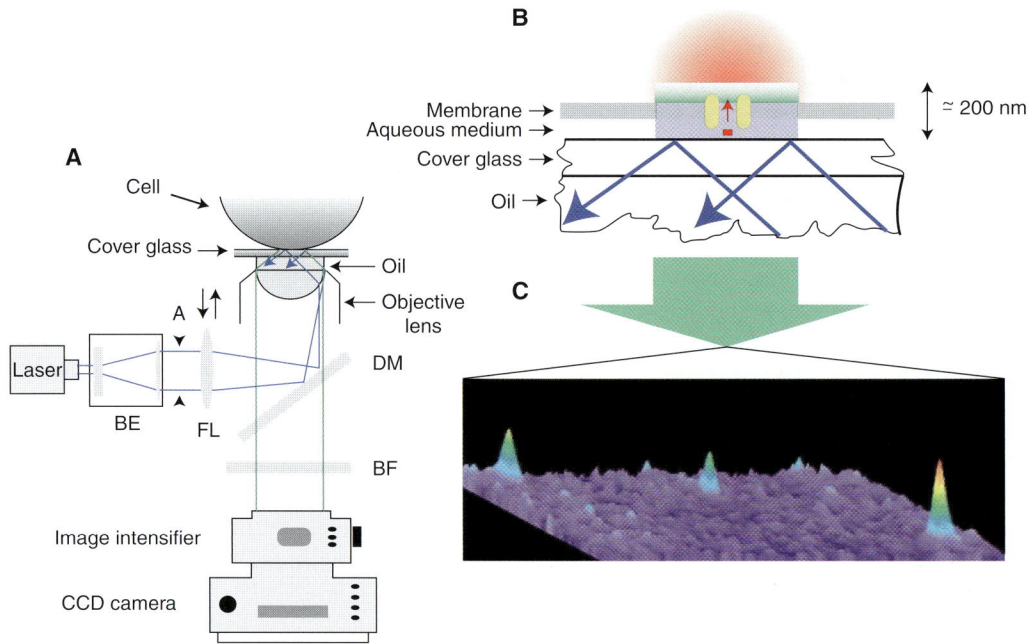

Figure 2. TIRF microscopy. (*A*) A schematic of the TIRF imaging system. A, adjustable rectangular knife-blade aperture; BE, beam expander; FL, focusing lens; BF, barrier filter; DM, dichroic mirror; CCD, charge-coupled device. (*B*) Imaging membrane-proximate fluorescent Ca^{2+} signals near an open channel by TIRF microscopy. (*C*) A single frame of a TIRF microscopy video image illustrating Ca^{2+} signals generated by three channels within an 80 × 80-μm patch of membrane. Increasing Ca^{2+} concentrations are indicated by both "warmer" colors and height. (Adapted from Demuro and Parker 2004; reprinted with permission from Elsevier © 2004.)

usually use nonratiometric dyes, they have the same limitations noted above for these fluorophores.

Ca^{2+} SIGNALS FROM OUTER SPACE: Ca^{2+} INFLUX

Signals Mediated by VDCCs

Membrane potential depolarization activates VDCCs—the major contributors to increases in $[Ca^{2+}]_i$. VDCCs are multisubunit complexes comprising a pore-forming α1 subunit and regulatory β, α2δ, and γ subunits (Curtis and Catterall 1984; Hosey et al. 1987; Leung et al. 1987; Vaghy et al. 1987). Most of the functional properties of the VDCC channel, including voltage sensitivity, Ca^{2+} permeability, Ca^{2+}-dependent inactivation, and sensitivity to pharmacological block by organic Ca^{2+} channel blockers, are attributable to the α1 subunit. The domain organization of the α1 subunit creates a pseudotetrameric structure in which the four repeat domains (I, II, III, IV), each composed of six transmembrane segments (S1–6) and intracellular amino- and carboxyl-termini (Catterall 2000; Jurkat-Rott and Lehmann-Horn 2004), are analogous to the individual subunits of structurally similar, tetrameric voltage-dependent potassium (K^+) channels. The S4 transmembrane segments of each domain serve as voltage sensors; in response to changes in membrane potential, they move outward and rotate, producing a conformational change that opens the pore (Catterall 2000).

The pore-forming α1 subunit is expressed as multiple splice variants with different regulatory and biophysical properties. Additional molecular diversity is provided by four different, variably spliced β subunits (Birnbaumer et al. 1998), which further modify VDCC biophysical properties and regulate surface expression of the α1 subunit; properties of the VDCC complex may be additionally modulated by splice variants of the α2δ regulatory subunit (Angelotti and Hofmann 1996).

VDCC-mediated currents are characterized by high voltage of activation, large single-channel conductance, and slow voltage-dependent inactivation. VDCC-mediated currents also display a characteristic sensitivity to dihydropyridines (Reuter 1983), a class of drugs used clinically in the treatment of hypertension (Nelson et al. 1990; Snutch et al. 2001). In vascular and visceral smooth muscle, these dihydropyridine-sensitive currents are attributable to the expression of the L-type VDCC pore-forming α1C subunit ($Ca_V1.2$) (Keef et al. 2001; Moosmang et al. 2003; Wegener et al. 2004); but in some smooth muscle types, the α1D ($Ca_V1.3$) subunit is expressed as well (Nikitina et al. 2007). Of the four β subunits, only two—β2 and β3—are clearly detected at the protein level in smooth muscle, although mRNA for all four isoforms is expressed (Hullin et al. 1992; Murakami et al. 2003).

There is also evidence for the expression of a dihydropyridine-insensitive Ca^{2+} current in some smooth muscle types. This T-type (transient) current is mediated by Ca_V3 pore-forming α subunits—primarily the $Ca_V3.1$ isoform in smooth muscle (Bielefeldt 1999; Perez-Reyes 2003). Expression of T-type channels varies between different smooth muscle types, but their presence often goes undetected because of their low levels of expression or because they are obscured by the specific recording conditions used. Moreover, the negative steady-state inactivation property of these channels results in their being half-inactivated at about −70 mV. Therefore, over the range of smooth muscle resting membrane potential (−50 to −30 mV), T-type channels may be completely inactivated. Because of these gating properties, the importance of T-type currents in the regulation of smooth muscle membrane potential is a matter of controversy. However, there is evidence from the rabbit urethra that, at least in this tissue, T-type channels regulate action potential frequency (Bradley et al. 2004).

It also has been suggested that voltage-dependent Ca^{2+} currents in smooth muscle with a T-type pharmacology may arise because of a $Ca_V3.1$ (and/or $Ca_V3.2$) splice variant with a more depolarized activation voltage (Kuo et al. 2010).

Three different Ca^{2+} signals mediated by VDCCs have been identified in smooth muscle: (1) global elevations in intracellular Ca^{2+}, (2) Ca^{2+} "flashes," and (3) Ca^{2+} "sparklets."

Global Ca^{2+} Signals

As the name suggests, global Ca^{2+} signals reflect changes in Ca^{2+} concentration that are essentially uniform throughout the cell. In smooth muscle, depolarization of the membrane by ∼15 mV from its resting potential (approximately −50 to −40 mV) elevates global Ca^{2+} to ∼300–400 nM, whereas hyperpolarization by ∼15 mV lowers Ca^{2+} to ∼100 nM. These signals are typically monitored using ratiometric dyes (e.g., Fura-2), which are ideally suited to measuring Ca^{2+} concentration over this range (K_d ∼300 nM). Thus, by modulating the steady-state open probability of VDCCs, slow changes in membrane potential can have profound, sustained effects on global intracellular Ca^{2+}. This is illustrated in Figure 3, which shows that the membrane potential depolarization that accompanies elevation of intraluminal pressure from 60 mmHg (Fig. 3A) to 100 mmHg (Fig. 3B) increases the global intracellular Ca^{2+} concentration in the vascular wall. Inhibition of VDCC channels with nisoldipine decreases global $[Ca^{2+}]_i$ (Fig. 3C). Changes in global Ca^{2+} are accompanied by changes in vascular diameter, underscoring the importance of global Ca^{2+} in regulating the contractile state of smooth muscle (see below).

Ca^{2+} Flashes

Some types of smooth muscle (e.g., urinary bladder, gallbladder, ureter) show action potentials. These rapid, transient changes in membrane potential, which show a characteristic temporal profile, are unique to excitable cells. However, unlike other excitable cells, such as

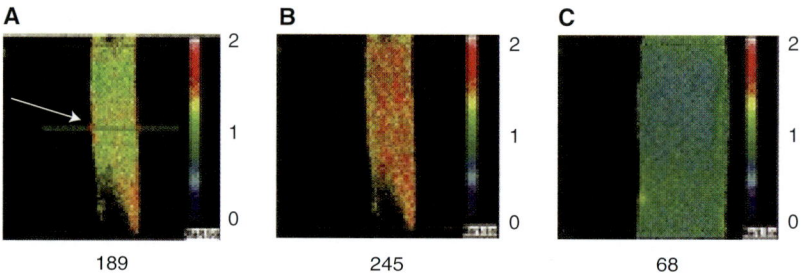

Figure 3. Global Ca^{2+}. Ca^{2+} images obtained from a rat basilar artery pressurized to (A) 60 mmHg, (B) 100 mmHg, and (C) 100 mmHg in the presence of nisoldipine. The numbers below each panel correspond to the Ca^{2+} concentration in the smooth muscle of the vascular wall, calculated from the ratio of Fura-2 fluorescence at 340 and 380 nm. Note contraction and dilation in B and C, respectively, relative to A. (Adapted from Knot and Nelson 1998; reprinted with permission from *The Journal of Physiology* © 1998.)

neurons and cardiac myocytes, where action potentials are initiated by activation of channels that predominantly mediate sodium (Na^+) influx, the upstroke of the action potential in smooth muscle reflects massive Ca^{2+} entry through VDCCs. The resulting rapid elevation of global intracellular Ca^{2+} can be detected optically as a Ca^{2+} flash that brightly and briefly lights up cells loaded with fluorescent Ca^{2+} indicators. A recording of a Ca^{2+} flash in urinary bladder smooth muscle is shown in Figure 4. Ca^{2+} flashes can be evoked by electrical field stimulation, but also occur spontaneously (as in the example shown), likely reflecting the spontaneous release of neurotransmitters from sympathetic (e.g., mesenteric arteries) or parasympathetic (e.g., urinary bladder) nerve terminals (Klockner and Isenberg 1985; Heppner et al. 2005). Note that in this example, Ca^{2+} was simultaneously elevated in two adjacent cells, indicating that these events may be coupled.

Ca^{2+} Sparklets

Cheng and colleagues first measured the local Ca^{2+} signal caused by the opening of a single L-type VDCC in cardiac muscle (Wang et al. 2001b), and referred to these events as Ca^{2+} sparklets. Ca^{2+} sparklets are also present in vascular smooth muscle (Fig. 5), where they are detected by TIRF microscopy as highly localized, dihydropyridine-sensitive, subplasmalemmal Ca^{2+}-release events reflecting the activity of an individual channel or cluster of channels (Navedo et al. 2005). The average area of a Ca^{2+} sparklet is $\sim 0.8~\mu m^2$ or $\sim 0.08\%$ of the surface membrane of a typical arterial smooth muscle cell (Santana et al. 2008). Ca^{2+} influx through Ca^{2+} sparklet sites is quantal, and the size of a given Ca^{2+} sparklet depends on the number of quanta activated. One quantal unit of Ca^{2+} release elevates $[Ca^{2+}]_i$ by about 35 nM. The specific locations of Ca^{2+} sparklets vary between cells, but within a cell, Ca^{2+} sparklets are predominantly stationary events that occur in specific regions of the sarcolemmal membrane (Fig. 5A–C). Both low- and high-activity Ca^{2+} sparklet sites have been described. The frequency of the latter type of event, termed a persistent sparklet, is increased by activation of protein kinase C (PKC), which recruits previously silent sites and increases the frequency of low-activity sites (Fig. 5D).

In heart muscle, where local coupling of Ca^{2+} influx through single VDCCs to RyRs is central to excitation-contraction coupling (Cannell et al. 1995; Lopez-Lopez et al. 1995), a Ca^{2+} sparklet can activate four to six nearby RyRs to cause a Ca^{2+} spark (see below). However, there is no evidence for this direct VDCC-to-RyR communication in smooth muscle. Instead, Ca^{2+} currents through VDCCs appear to activate RyRs indirectly through elevations in global Ca^{2+} and SR Ca^{2+} load (Collier et al. 2000; Herrera and Nelson 2002; Wellman and Nelson 2003; Essin and Gollasch 2009).

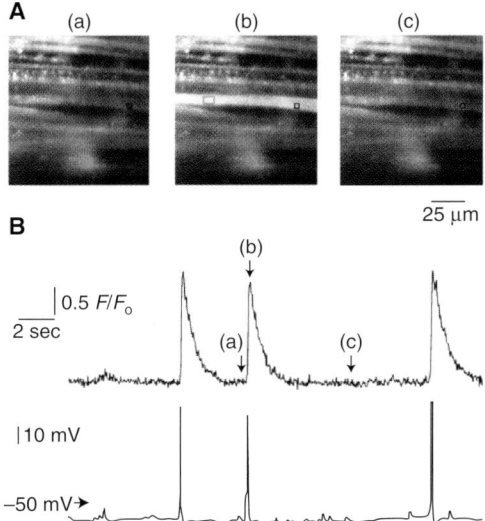

Figure 4. Ca^{2+} flashes. (A) Selected images recorded before (a), during (b), and after (c) a spontaneous action potential in a smooth muscle bundle loaded with Fluo-4 and impaled with a microelectrode (green rectangle). (B) Simultaneous recordings of changes in Ca^{2+}-activated fluorescence (*upper* trace) and voltage (*lower* trace) from a single bundle of urinary bladder smooth muscle. Changes in Ca^{2+}-activated fluorescence were measured from the red box in A; the letters a, b, and c denote the times at which the correspondingly labeled images in A were acquired. Note that each of the three action potentials induced a simultaneous increase in Ca^{2+}-activated fluorescence. (Adapted from Heppner et al. 2005; reprinted with permission from *The Journal of Physiology* © 2005.)

Signals Mediated by Store-Operated Ca^{2+} Channels

Extracellular Ca^{2+} influx in response to depletion of intracellular Ca^{2+} stores, a process termed store-operated Ca^{2+} entry (SOCE), is known to play an important role in a number of cell types, notably nonexcitable cells. However, the molecular mechanism underlying this coupling long remained elusive. Recent seminal work by a number of independent groups effectively resolved this question, clearly identifying ubiquitously expressed STIM proteins (Liou et al. 2005; Roos et al. 2005) as endoplasmic reticulum (ER) Ca^{2+} sensors, and members of the Orai family (Feske et al. 2006; Vig et al. 2006; Zhang et al. 2006) of transmembrane proteins as the entities responsible for mediating Ca^{2+} entry (reviewed in Varnai et al. 2009). These researchers showed that, in response to a decrease in ER Ca^{2+} concentrations, the low-affinity Ca^{2+}-binding STIM proteins aggregate to form discrete plasmalemmal-proximate clusters that tether Orai proteins. This physical coupling activates Orai, which is a highly selective Ca^{2+} channel, thereby promoting extracellular Ca^{2+} entry. The identification of the STIM-Orai mechanism has sparked renewed interest in investigating SOCE in smooth muscle (reviewed in Wang et al. 2008). These studies, most of which have been performed using cultured smooth muscle cells, have consistently shown that STIM and Orai family members are expressed in smooth muscle, and, under the conditions tested, are capable of functionally coupling store depletion to extracellular Ca^{2+} entry (Peel et al. 2006, 2008; Takahashi et al. 2007; Ng et al. 2010; Park Hopson et al. 2011). It has also been suggested that, in addition to promoting Orai activity, STIM1 negatively regulates VDCCs (Wang et al. 2010). Some recent studies have provided evidence for STIM-Orai coupling in native smooth muscle preparations, and have suggested a role for this mechanism in hypertension (Giachini et al. 2009, 2010). An optical signature of Orai-mediated Ca^{2+} influx has not been defined and additional research will be required to definitively establish the physiological relevance of this pathway in native smooth muscle tissues.

Signals Mediated by Nonselective Cation Channels

In contrast to VDCCs, which show a high selectivity for Ca^{2+} ions over monovalent cations, nonselective cation channels typically also allow influx of extracellular Na^+. Although channels of this type are permeable to Ca^{2+} and thus directly increase $[Ca^{2+}]_i$ to some degree, in many if not most cases, their major impact on $[Ca^{2+}]_i$ is indirect through Na^+-dependent membrane potential depolarization and activation of

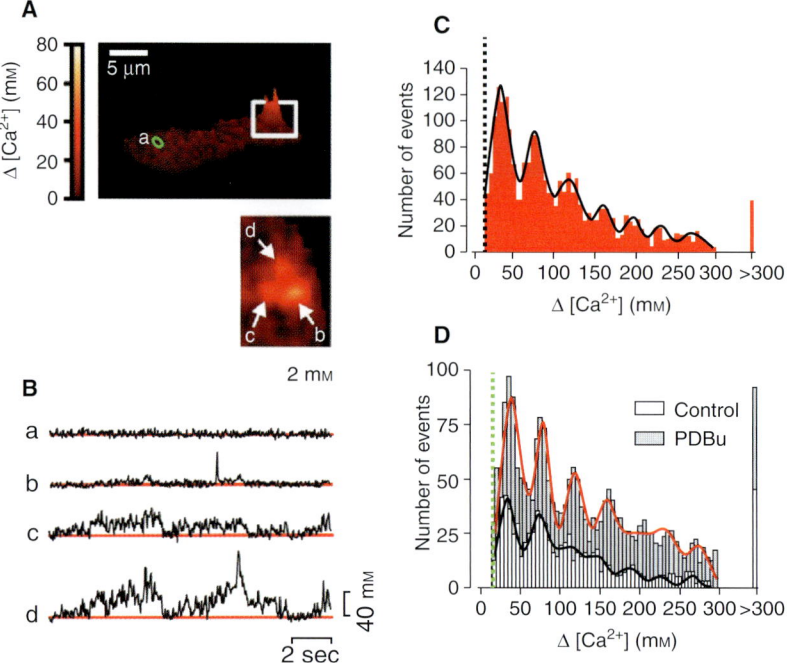

Figure 5. Ca^{2+} sparklets. (*A*) Surface plot of Ca^{2+} imaged in a freshly isolated arterial myocyte. (*Inset*) Higher magnification view of boxed area showing three active Ca^{2+} sparklet sites (2 mM extracellular Ca^{2+}). (*B*) Traces showing time course of changes in Ca^{2+} at sites a–d. (*C*) Amplitude histogram of Ca^{2+} sparklets (20 mM extracellular Ca^{2+}). (*D*) Amplitude histogram of Ca^{2+} sparklets (2 mM extracellular Ca^{2+}) in the absence and presence of the PKC activator phorbol 12,13-dibutyrate (PDBu). Solid lines in *C* and *D* are best fits to a Gaussian function. (Adapted from Navedo et al. 2005; reprinted with permission from The National Academy of Sciences © 2005.)

VDCCs. Several types of Ca^{2+}-permeable, nonselective cation channels are present in smooth muscle, including receptor-activated channels, mechanosensitive channels, tonically active channels, and channels activated by SR Ca^{2+} store depletion.

Receptor-Activated Cation Channels

Of the various nonselective cation channels expressed in smooth muscle, only the ATP-gated P2X receptor (P2XR) is clearly associated with an optically identifiable Ca^{2+} signal. The functional P2XR complex is thought to be a trimer (Aschrafi et al. 2004)—an unusual structural arrangement in ion-channel space where tetramers dominate. Each subunit contains intracellular amino- and carboxyl-termini and two membrane-spanning domains that are connected by a large extracellular domain (Khakh 2001; North 2002). Binding of ATP (three molecules per complex) to a site in the large extracellular domain induces a conformational change that results in rapid (milliseconds) opening of the pore.

These Ca^{2+}- and Na^{+}-permeable channels (Benham and Tsien 1987; Schneider et al. 1991) mediate a rapid local influx of Na^+ and Ca^{2+} at nerve–muscle junctions following activation by neurally released ATP (Lamont and Wier 2002; Lamont et al. 2006). The influx of Na^+ and Ca^{2+} creates an excitatory junction potential (EJP) that contributes directly to the increase in postjunctional excitability. Multiple lines of evidence, including studies using knockout mice, indicate that the $P2X_1$ receptor is the predominant P2X receptor isoform expressed in smooth muscle (Mulryan et al.

2000; Vial and Evans 2002; Lamont et al. 2006; Heppner et al. 2009).

Although most of the excitatory junction current (EJC) associated with P2X$_1$ receptor activation is carried by the more abundant Na$^+$ ions, Ca^{2+} influx is substantial. This influx can be detected optically in the form of local elementary purinergic-induced Ca^{2+} transients (Fig. 6). These events have been described in vas deferens (Brain et al. 2002), mesenteric arteries (Lamont and Wier 2002), and urinary bladder (Heppner et al. 2005), where they have been termed neuroeffector Ca^{2+} transients (NCTs), junctional Ca^{2+} transients (jCaTs), and nerve-evoked elementary purinergic Ca^{2+} transients, respectively. The kinetic properties of these purinergic Ca^{2+} signals are similar to one another and are clearly distinct from those of other local Ca^{2+} transients (Hill-Eubanks et al. 2010).

Other Nonselective Cation Channels

Smooth muscle cells express a number of nonselective cation channels of the TRP family. Although no signature signaling event associated with Ca^{2+} influx through TRP channels has been reported in the literature, given the relative selectivity of these channels for Ca^{2+} (and the high Ca^{2+} permeability and single-channel conductance of some TRP family members, notably TRPV), imaging methods that have been used to examine jCaT-like events (confocal microscopy) and/or VDCC-mediated sparklets (TIRF) may ultimately provide a means to optically detect Ca^{2+} influx through these channels.

Ca^{2+} SIGNALS FROM INNER SPACE: RELEASE OF Ca^{2+} FROM INTRACELLULAR STORES

The most important intracellular Ca^{2+} store in smooth muscle is the SR. Cytosolic Ca^{2+} is transported into the SR by the action of the SR/ER Ca^{2+} ATPase (SERCA). SERCA activity is negatively regulated by the protein phospholamban, a target of protein kinase A (PKA) and protein kinase G (PKG). On phosphorylation, the SERCA inhibitory activity of phospholamban is lost, increasing Ca^{2+} uptake and SR Ca^{2+} load. Free Ca^{2+} taken up by the SR is buffered by Ca^{2+}-binding proteins, which retain transported Ca^{2+} and reduce the free Ca^{2+}

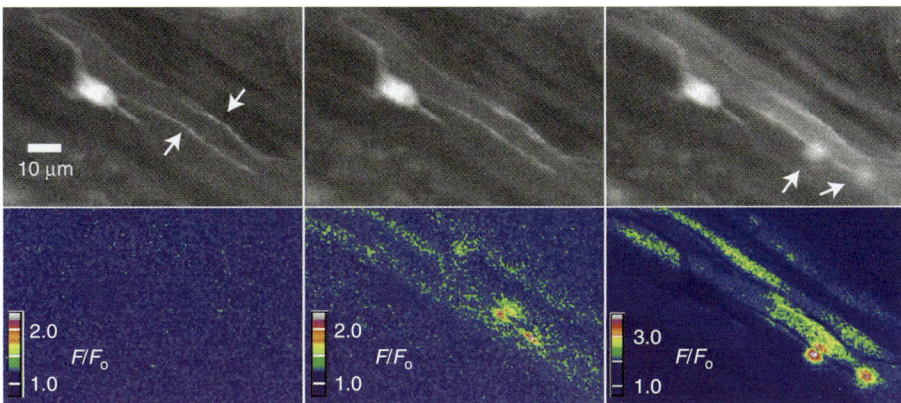

Figure 6. Elementary purinergic Ca^{2+} transients recorded from urinary bladder smooth muscle. Transient events were evoked by electrical field stimulation (2-sec train, 5 Hz; 37°C) in bladder strips loaded with Fluo-4 and scanned at a rate of 30 images/sec. The top three images illustrate nerve processes before (*top* panel, *left*, indicated by arrows) and after stimulation (*middle* and *right* panels). The three *bottom* panels illustrate color-coded ratios (F/F_0) of the images above. Intracellular Ca^{2+} increases first in the nerve fibers, and then two local Ca^{2+} transients are detected (*right* panel, *bottom* right quadrant, indicated by arrows). The activity of Ca^{2+} transients continued throughout the duration of field stimulation. (Unpublished data from Mark Nelson.)

concentration gradient, facilitating continued Ca^{2+} uptake from the cytoplasm (Pozzan et al. 1994).

Ca^{2+} sequestered in the SR may be delivered to the cytosol through RyRs and IP_3 receptors in the SR membrane. Although these two channel types are phenotypically similar on a superficial level (both function to release Ca^{2+} from SR stores) they are very different molecular entities with distinctive regulatory features and characteristic Ca^{2+}-release signatures.

Signals Mediated by RyRs: Ca^{2+} Sparks

There are three RyR subtypes (RyR1-3), each of which is expressed at varying levels in different smooth muscle tissues (Neylon et al. 1995; Yang et al. 2005; Prinz and Diener 2008). RyRs are large tetrameric complexes formed from ~560-kDa subunits. Each subunit contains four membrane-spanning domains, a large cytosol-facing amino-terminal region containing the Ca^{2+}-binding site as well as binding sites for numerous accessory proteins, and a short SR-luminal carboxy-terminal domain (reviewed in Zalk et al. 2007; Lanner et al. 2010).

Ca^{2+} flux through ryanodine receptors is detectable in the form of elementary release events termed Ca^{2+} sparks. First discovered in cardiac muscle (Cheng et al. 1993) and subsequently identified in skeletal (Klein et al. 1996) and smooth (Nelson et al. 1995) muscle, a Ca^{2+} spark represents the opening of a few (likely four to six) RyR channels in the SR membrane (Cheng and Lederer 2008). Ca^{2+} sparks have been detected in a wide variety of smooth muscle types, including those from arteries (Nelson et al. 1995), portal vein (Mironneau et al. 1996; Gordienko and Bolton 2002), urinary bladder (Herrera et al. 2001), ureter (Burdyga and Wray 2005), airway (Sieck et al. 1997), and the gastrointestinal tract (Gordienko et al. 1998). Ca^{2+} sparks in arterial smooth muscle can be detected in isolated myocytes as well as in intact pressurized arteries. Ca^{2+} sparks are rapid, transient, stationary events. The rise time of sparks in vascular smooth muscle and urinary bladder smooth muscle is ~20–40 msec (Nelson et al. 1995; Herrera et al. 2001); their spatial spread is ~12.6 µm, and this spread corresponds to ~1% of the surface membrane. The duration (half-time) of these events is ~50–60 msec; this contrasts with nerve-evoked purinergic Ca^{2+} transients, which have half-times of ~110–145 msec (Brain et al. 2002; Lamont and Wier 2002; Heppner et al. 2005). Current evidence indicates that smooth muscle Ca^{2+} sparks are attributable to activation of RyR2, although both RyR1 and RyR3 may influence spark activity (Vaithianathan et al. 2010).

In smooth muscle, localized increases in Ca^{2+} associated with Ca^{2+} sparks activate closely juxtaposed large-conductance, Ca^{2+}-activated K^+ (BK_{Ca}) channels in the plasma membrane. The K_d of BK_{Ca} channels for Ca^{2+} is ~20 µM at the physiological membrane potential of −40 mV. A single spark causes a local increase of $[Ca^{2+}]_i$ of 10–30 µM and activates about 30 nearby BK_{Ca} channels, increasing their open probability by approximately 100-fold (Jaggar et al. 2000; Perez et al. 2001). In smooth muscle from adult animals, there is a one-to-one relationship between sparks and BK_{Ca} channel–mediated transient outward currents, indicating that all spark sites are functionally coupled to BK_{Ca} channel clusters. In current-clamp mode, activation of BK_{Ca} channels by a single spark causes about a 20-mV hyperpolarization (Jaggar et al. 1998b).

Stimuli that increase SR Ca^{2+} load increase the frequency of Ca^{2+} sparks, but sparks also occur spontaneously. The outward currents associated with this latter activity are termed "spontaneous transient outward currents" or "STOCs" (Benham and Bolton 1986). An example depicting the time course and decay kinetics of a single Ca^{2+} spark event is presented in Figure 7A. Simultaneous electrophysiological recordings and traces showing analyzed Ca^{2+} signals (Fig. 7B) highlight the one-to-one relationship between sparks and STOCs.

In addition to activating BK_{Ca} channels to produce STOCs, Ca^{2+} sparks can also activate Ca^{2+}-sensitive chloride (Cl_{Ca}) channels to produce spontaneous transient inward currents (STICs) (Hogg et al. 1993). Where BK_{Ca} and Cl_{Ca} channels coexist, spontaneous transient outward/inward currents (STOICs) are

Calcium Signaling in Smooth Muscle

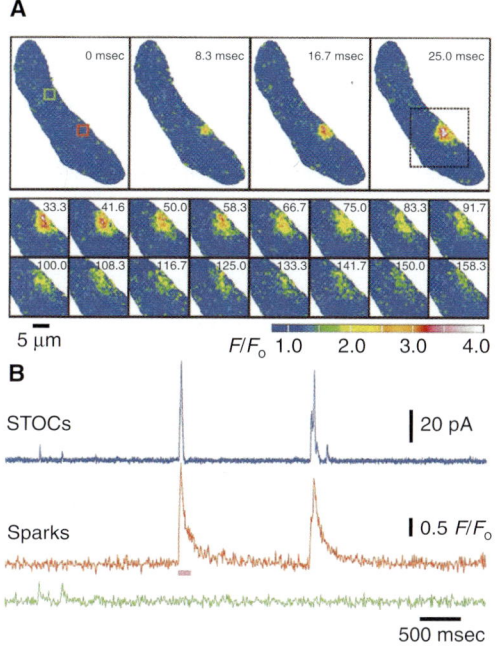

Figure 7. Ca^{2+} sparks in an isolated rat basilar artery myocyte. (A) (Top) Two-dimensional confocal images of an entire smooth muscle cell showing the time course of the fractional increase in Fluo-3 fluorescence (F/F_0) of a typical Ca^{2+} spark. (Bottom) Images obtained from the region of interest in the top-right panel (dotted box) depicting spark decay. Pseudocolor denotes relative Ca^{2+} levels as indicated by the bar. (B) Simultaneous measurements of STOCs and sparks at −40 mV highlighting the temporal association between the events. The pink bar below the sparks trace corresponds to the time period imaged in A. (Adapted from Perez et al. 1999; reprinted with permission from The Rockefeller University Press © 1999.)

produced (ZhuGe et al. 1998; Jaggar et al. 2000; Wellman and Nelson 2003). Both hyperpolarizing (STOCs) and depolarizing currents (STICs) modulate membrane potential and excitability, with STOCs being inhibitory and STICs being excitatory.

Signals Mediated by IP$_3$Rs: Ca^{2+} Waves

There are three IP$_3$ receptor subtypes (IP$_3$R1-3), each of which is expressed at varying levels in different smooth muscle tissues (Newton et al. 1994; Tasker et al. 1999; Boittin et al. 2000; Grayson et al. 2004). IP$_3$Rs are large homo- or heterotetrameric complexes formed from approximately 2700 to 2800 amino acid subunits. Each subunit contains six membrane-spanning domains, a large cytosol-facing amino-terminal region containing the IP$_3$ binding site, and a short cytosolic carboxy-terminal domain. The pore of the channel is formed by the coassociation of transmembrane domains 5 and 6 from each of the four subunits (Foskett et al. 2007).

Ca^{2+} release by IP$_3$Rs is regulated by two second messengers: IP$_3$ and Ca^{2+}. The prototypical signaling pathway that leads to elevation of IP$_3$ is activation of $G\alpha_{q/11}$-type G protein-coupled receptors. Among the most prominent agonists of this pathway in smooth muscle are neurohumoral vasoconstrictors. Activation of this pathway stimulates phospholipase C, resulting in hydrolysis of membrane-associated phosphoinositide 4,5-bisphosphate (PIP$_2$) into IP$_3$ and diacyclglycerol. The IP$_3$ generated by this pathway binds IP$_3$Rs and promotes channel gating, releasing Ca^{2+} into the cytosol. The Ca^{2+} released by IP$_3$Rs can reciprocally modulate IP$_3$R activity in two ways: as Ca^{2+} rises from low nanomolar basal levels to low micromolar levels in the vicinity of the channel, it activates IP$_3$Rs; at higher local levels, the channel becomes inactivated. These activation/inactivation properties together with regulation of channel function by multiple interacting factors, allow IP$_3$Rs to generate a large variety of temporally and spatially modulated Ca^{2+}-signaling patterns within the cell (Foskett et al. 2007).

In some smooth muscle types, IP$_3$R-mediated Ca^{2+} release is transient and localized. For example, smooth muscle cells from colon and portal vein show spontaneous Ca^{2+} spark-like events that are enhanced by IP$_3$ production and eliminated by IP$_3$R blockade (Bayguinov et al. 2000; Gordienko and Bolton 2002). These Ca^{2+} release events, termed Ca^{2+} "puffs," have a biophysical signature (e.g., kinetics, magnitude, spatial spread) that distinguishes them from the RyR-mediated Ca^{2+} sparks that are prominent in most other smooth muscle types (e.g., vascular smooth muscle, gallbladder, and urinary bladder) (Nelson et al. 1995;

Herrera et al. 2001; Pozo et al. 2002). Although Ca^{2+} puffs are unitary events, they can act as initiation sites for intracellular Ca^{2+} waves and thereby contribute to global Ca^{2+} signals (Bootman and Berridge 1996; Thomas et al. 1998).

A Ca^{2+} wave, defined as an increase in $[Ca^{2+}]_i$ that propagates across the entire smooth muscle cell from an initial site of release, is perhaps the most studied IP_3R-mediated Ca^{2+}-signaling event. First described by Iino in rat tail arteries (Fig. 8), Ca^{2+} waves are a common feature of vascular smooth muscle cells exposed to $G\alpha_{q/11}$-coupled vasoconstrictor agonists, such as UTP (Jaggar and Nelson 2000) and norepinephrine (Iino et al. 1994; Boittin et al. 1999; Miriel et al. 1999; Ruehlmann et al. 2000), or electrical field stimulation of perivascular nerves.

In the current view, Ca^{2+} waves reflect the activation/inactivation properties of IP_3Rs, and arise through a regenerative Ca^{2+}-induced Ca^{2+} release (CICR) mechanism. Ca^{2+} released from the SR acts on successive adjacent IP_3Rs or IP_3R clusters in a cascading fashion, creating a leading edge of Ca^{2+} elevation that traverses the length of the cell. When Ca^{2+} waves propagate toward each other and collide, they cancel each other out because of depletion of SR Ca^{2+} stores on either side of the collision site (Stevens et al. 1999). Therefore, a continuous cycling of SR Ca^{2+} release and reuptake maintains the Ca^{2+} waveform. Ca^{2+} released from SR stores is sufficient for further release and maintenance of Ca^{2+} waves, indicating that the mechanism of smooth muscle Ca^{2+} wave propagation is independent of extracellular Ca^{2+} entry (Boittin et al. 1999; Jaggar and Nelson 2000; Peng et al. 2001; Heppner et al. 2002). Repeating IP_3R-dependent Ca^{2+} signals may also take the form of whole-cell oscillations in $[Ca^{2+}]_i$.

The view of Ca^{2+} waves as strictly IP_3R-mediated events oversimplifies the true situation. In actuality, Ca^{2+} waves arise in response to activation of IP_3Rs and/or RyRs (Iino et al. 1994; Boittin et al. 1999; Hirose et al. 1999; Jaggar and Nelson 2000; Lee et al. 2002), and their properties as well as the relative contributions of IP_3Rs and RyRs may differ depending on the nature of the stimulus or tissue context. In some cases, these events appear to exclusively reflect the activity of RyRs. One such example is provided by rat cerebral arteries, where Nelson and colleagues (Heppner et al. 2002)

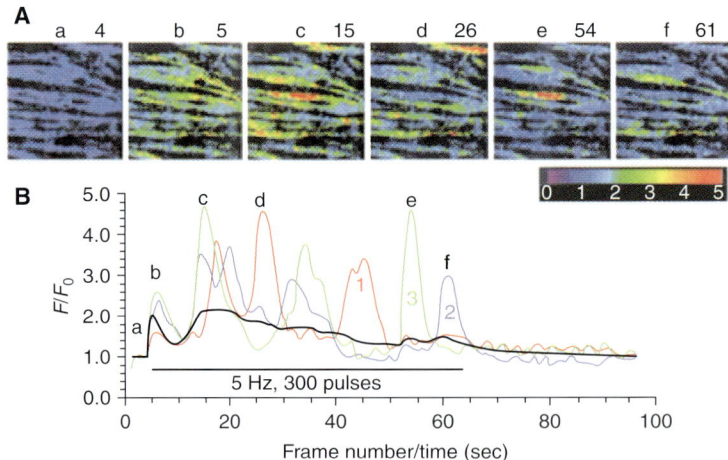

Figure 8. Ca^{2+} waves. (*A*) Two-dimensional pseudocolor confocal microscopic images of rat tail artery smooth muscle showing dynamic, recurrent changes in intracellular Ca^{2+} (measured by changes in the intensity of Fluo-3 fluorescence) following electrical stimulation of perivascular sympathetic nerves. Six of 96 consecutive frames collected at a rate of one frame per second are shown. (*B*) Red, blue, and green lines depict changes in fluorescence intensity as a function of time (*a–f* in *A*) in three selected regions of interest (white boxes in *a*). (Adapted from Iino et al. 1994; reprinted with permission from *The EMBO Journal* © 1994.)

have found that Ca^{2+} waves are induced by caffeine, which acts on RyRs but not IP_3Rs. Moreover, these events are insensitive to the nominally selective IP_3R blockers, xestospongin C and 2-aminoethoxydiphenyl borate (2-APB), but are completely eliminated by ryanodine. Interestingly, RyR-mediated Ca^{2+} signaling in this preparation is sharply dependent on pH. Increasing the pH of the bathing solution from 7.4 to 7.5 increased Ca^{2+} spark frequency by ~50%; above this pH, Ca^{2+} waves came to predominate, increasing approximately threefold between pH 7.5 and pH 7.6 and doubling again at pH 7.7–7.8.

FUNCTIONAL CORRELATES OF Ca^{2+} SIGNALS

Contractility

Smooth muscle contraction is driven by Ca^{2+}-calmodulin activation of myosin light chain kinase, which has a Ca^{2+} half-activation of ~400 nM (Stull et al. 1998). The gain of smooth muscle contraction to Ca^{2+} can be adjusted through regulation of myosin light chain phosphatase (Somlyo and Somlyo 2003; Mizuno et al. 2008).

Global Ca^{2+}

Membrane potential plays an important role in all excitable cells, including smooth muscle, where it regulates $[Ca^{2+}]_i$ and thereby smooth muscle contraction. The resting membrane potential of smooth muscle is approximately −50 to −40 mV, which is positive to the equilibrium potential for K^+ (E_K). In arterial smooth muscle, this membrane potential is sufficient to increase the steady-state open probability of VDCCs, elevate global intracellular Ca^{2+} from ~100 nM to ~200 nM, and cause a tonic constriction (Knot and Nelson 1998). As noted above, membrane potential hyperpolarization to −60 mV lowers Ca^{2+} to about 100 nM, and depolarization to about −30 mV elevates global Ca^{2+} to ~300–400 nM; these changes in global intracellular Ca^{2+} are sufficient to cause maximal dilation and constriction, respectively. The fundamental relationship between global Ca^{2+} and smooth muscle membrane potential is depicted in Figure 9, which

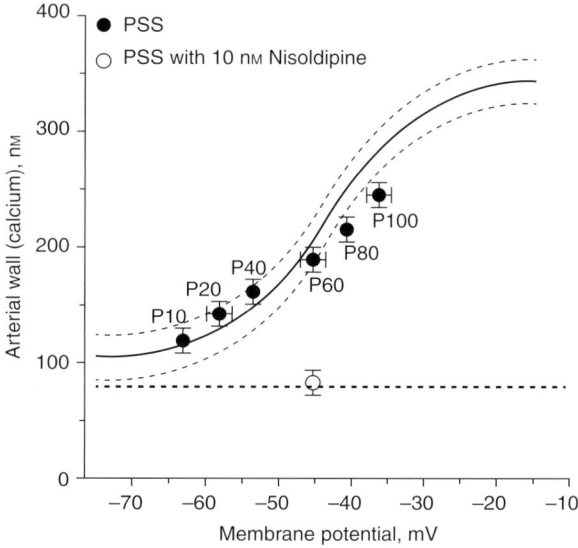

Figure 9. Intravascular pressure-membrane potential-$[Ca^{2+}]_i$ relationships. Fundamental relationships among intravascular pressure (P10–P100, in mm Hg), membrane potential, and arterial wall Ca^{2+}. (Adapted from Knot and Nelson 1998; reprinted with permission from *The Journal of Physiology* © 1998.)

also highlights the role of intravascular pressure as a physiological driver of changes in membrane potential. Oscillations in membrane potential caused by fluctuations in Ca^{2+} entry through VDCCs lead to vasomotion of the arterial wall.

Junctional Ca^{2+} Transients and Ca^{2+} Flashes

Nerve stimulation–evoked localized Ca^{2+} influx through $P2X_1R$ channels causes a depolarizing current carried by Na^+ and Ca^{2+} ions that activates VDCCs; in urinary bladder smooth muscle, this manifests as a Ca^{2+} flash. Coordinated flash activity among smooth muscle cells in a bundle leads to a transient contraction. Thus, junctional Ca^{2+} transients mediated by $P2X_1Rs$ may indirectly modulate contraction by triggering VDCC activity.

The bursts of local elevations in intracellular Ca^{2+} provided by junctional Ca^{2+} transients also have the potential to trigger activation of proximate RyRs in the SR through a CICR mechanism. The best evidence for the existence of such a mechanism comes from studies of the vas deferens by Cunnane and coworkers (Brain et al. 2003). In this preparation, the magnitude of neurally evoked, ATP-induced Ca^{2+} transients was reduced by ∼45% by inhibition of RyRs with ryanodine. Moreover, treatment with caffeine to increase RyR activity produced a 16-fold increase in the frequency of neurally evoked junctional Ca^{2+} transients. Collectively, these results argue that the neurally evoked Ca^{2+} signal associated with extracellular Ca^{2+} influx triggers—and merges with—a RyR-mediated Ca^{2+} signal, creating an optically detectable signal that reflects a summation of the two separate transient release events. The more modest inhibitory effect of ryanodine on jCaTs in mesenteric arteries (∼13%) (Lamont and Wier 2002) and the apparent absence of an effect of ryanodine on purinergic Ca^{2+} transients in rat urinary bladder (Heppner et al. 2005) suggest a degree of variability among tissues and species. How (or if) this communication from $P2X_1Rs$ to RyRs influences the contractile behavior of smooth muscle is not clear. The additional increment of Ca^{2+} may sum with $P2X_1R$ and VDCC-mediated Ca^{2+} to augment the transient contraction. Alternatively, CICR-activated RyRs could, in theory, couple to BK_{Ca} channels to oppose contraction. Depending on the relative speed of IP_3 production by concurrent activation of adrenergic or muscarinic receptors, it is also possible that Ca^{2+} influx through $P2X_1Rs$ could amplify local IP_3R activation by IP_3, a possibility that has not yet been explored experimentally.

Ca^{2+} Sparks

In cardiac and skeletal muscle, local Ca^{2+} entry through VDCCs activates proximate RyRs, producing Ca^{2+} sparks that summate to create a substantial increase in global Ca^{2+}; thus, Ca^{2+} sparks play a dominant role in contraction in these tissues. In smooth muscle, the molecular architecture is much different, resulting in unique linkages that create a phenotypically opposite functional outcome. In particular, the close physical coupling between VDCCs and RyRs that characterizes striated muscle cells is absent in smooth muscle cells. In its place is a close linkage between the RyR and plasma membrane BK_{Ca} channels, which are not expressed in cardiac or skeletal muscle cells. As a result of this unique architecture, Ca^{2+} released by RyRs in the SR in the form of sparks activates juxtaposed BK_{Ca} channels, promoting an outward K^+ current that hyperpolarizes the smooth muscle membrane and reduces VDCC activity. The resulting decrease in Ca^{2+} influx thus opposes VDCC-mediated smooth muscle contraction (Nelson et al. 1995; Perez et al. 1999; Jaggar et al. 2000). Under this scenario, Ca^{2+} influx through VDCCs initiates the BK_{Ca} channel-mediated feedback mechanism by enhancing RyR activity, by increasing global Ca^{2+} and SR Ca^{2+} stores (Collier et al. 2000; Herrera and Nelson 2002; Wellman and Nelson 2003; Essin and Gollasch 2009).

As is observed with $P2X_1$ agonists, simultaneous activation of RyRs by rapid addition of high levels of the RyR activator, caffeine, can cause global Ca^{2+} transients and a transient contraction (Wellman and Nelson 2003).

Ca^{2+} Waves

Ca^{2+} waves normally occur asynchronously in smooth muscle cells. However, in some vascular beds, Ca^{2+} waves may synchronize in neighboring arterial myocytes to initiate vasomotion (Peng et al. 2001), supporting the idea that Ca^{2+} waves can supply the Ca^{2+} needed for smooth muscle contraction (Kasai et al. 1997; Boittin et al. 1999; Mufti et al. 2010). Alternatively, Ca^{2+} waves can influence contractility indirectly through activation of Ca^{2+}-dependent ion channels located in the plasmalemmal membrane. For example, Ca^{2+} waves can activate Cl_{Ca} channels to promote membrane depolarization leading to enhanced Ca^{2+} entry through VDCCs (Mironneau et al. 1996). Ca^{2+} waves can also activate BK_{Ca} channels, thereby promoting membrane potential hyperpolarization, closure of VDCCs, and induction of smooth muscle relaxation (Young et al. 2001). Thus, while information on Ca^{2+} waves in smooth muscle continues to accumulate, the physiological function of these Ca^{2+} signals remains uncertain.

Ca^{2+}-DEPENDENT TRANSCRIPTION FACTOR ACTIVATION

Global Ca^{2+}: Excitation-Transcription Coupling

Excitation-contraction coupling, in which depolarization induces contraction through VDCC-mediated increases in intracellular Ca^{2+}, is paralleled by a conceptually similar mechanism that links depolarization-induced increases in $[Ca^{2+}]_i$ to activation of Ca^{2+}-sensitive transcription factors. This process, which has been termed excitation-transcription coupling, translates short-term Ca^{2+}-signaling and contractile events into long-term regulation of the smooth muscle cell transcriptome. Unlike cardiac and skeletal muscle cells, smooth muscle cells are highly plastic; their phenotype is maintained through dynamic regulation of gene expression in response to environmental cues (Owens 1995). Thus, excitation-transcription coupling serves to maintain the contractile phenotype by promoting the expression of smooth muscle-specific genes. It also provides a mechanism for phenotypic switching to a "synthetic" phenotype characterized by expression of genes that promote proliferation, matrix deposition, and other functions that come into play under pathological conditions and in the context of vessel repair and new vessel formation. Recent work from Owens and colleagues (Wamhoff et al. 2004) has shown that VDCC-mediated elevations in Ca^{2+} act through two distinct mechanisms to regulate the contractile and synthetic/proliferative phenotypes. In the first, depolarization-induced Ca^{2+} elevation induces SRF (serum response factor)-regulated smooth muscle–specific genes (e.g., myosin heavy chain, smooth muscle α-actin) through activation of Rho/Rho kinase and stimulation of myocardin, a potent coactivator of SRF first identified by Olson and colleagues (Wang et al. 2001a). In the second mechanism, elevated intracellular Ca^{2+} acts through calmodulin-dependent kinase (CaMK) to activate CREB (cAMP responsive element binding protein) and the immediately early gene, c-fos, which is involved in proliferative responses. A similar CaMK/CREB-dependent mechanism has been implicated in the VDCC-mediated induction of Egr-1 (Pulver-Kaste et al. 2006) and c-fos (Cartin et al. 2000) in native cerebral arteries, and TRP channels in gall bladder smooth muscle (Morales et al. 2007). In this latter study, a role for the phosphatase calcineurin was also suggested.

Spatially and Temporally Modulated Ca^{2+} Signals

Studies on the effects of Ca^{2+} signal modulation on transcription factor activation in smooth muscle cells are limited, but seminal work by Lewis and Tsien and colleagues (Dolmetsch et al. 1998; Li et al. 1998) in nonexcitable cells has shown a role for amplitude and frequency modulation of Ca^{2+} signals in differentially regulating the activity of Ca^{2+}-sensitive transcription factors. These researchers showed that large transient increases in $[Ca^{2+}]_i$ are sufficient to robustly activate NF-κB and c-Jun terminal kinase (JNK) but not NFAT (nuclear factor of activated T-cells), which is effectively activated by a sustained, graded increase in global intracellular Ca^{2+}. It has been further shown that activation of Ca^{2+}-sensitive transcription factors is modulated by oscillatory elevations in intracellular Ca^{2+}: for all transcription factors tested (NFAT, Oct/OAP, and NF-κB), high-frequency oscillations enhanced the efficacy of a given increase in Ca^{2+}, whereas low-frequency oscillations activated only NF-κB.

Ca^{2+}-Signaling Microdomains

Recent studies by Santana and coworkers suggest a model in which the scaffolding protein AKAP250 targets PKC and calcineurin to caveolin-containing membrane microdomains, where they associated with VDCCs to form a signaling unit capable of mediating persistent Ca^{2+} sparklets (Santana and Navedo 2009). These studies further indicate that persistent VDCC-mediated Ca^{2+} sparklets activate the Ca^{2+}-calmodulin-dependent transcription factor NFATc3 (Nieves-Cintron et al. 2008), which modulates expression of the Kv2.1 voltage-dependent K^+ channel and the β1 subunit of the BK_{Ca} channel in these cells (Amberg et al. 2004). By extension, similar complexes of $P2X_1Rs$ with kinases and phosphatases might form Ca^{2+}-signaling microdomains in postjunctional smooth muscle cell membranes, enabling nerve-evoked purinergic transients to regulate activation of NFAT or other Ca^{2+}-sensitive transcription factors.

CONCLUSIONS

In addition to global elevations in intracellular Ca^{2+} mediated by VDCCs, smooth muscle shows a variety of local Ca^{2+} signals, including Ca^{2+} sparks (RyRs), Ca^{2+} puffs (IP_3Rs), Ca^{2+} waves (IP_3Rs/RyRs), junctional Ca^{2+} transients ($P2X_1Rs$), Ca^{2+} flashes (VDCCs), and Ca^{2+} sparklets (VDCCs). Each signal represents the manifestation of different molecular circuits, which collectively serve to modulate membrane potential, contractility, and gene expression.

ACKNOWLEDGMENTS

The work was supported by NIH grants R37DK 053832, RO1 DK065947, RO1 HL44455, PO1 HL077378, P20 R016435, and RO1 HL098243; the Totman Trust for Medical Research; Research into Ageing (P332); The Royal Society (RG080197); and the British Heart Foundation (PG/07/115).

REFERENCES

Amberg GC, Rossow CF, Navedo MF, Santana LF. 2004. NFATc3 regulates Kv2.1 expression in arterial smooth muscle. *J Biol Chem* **279**: 47326–47334.

Angelotti T, Hofmann F. 1996. Tissue-specific expression of splice variants of the mouse voltage-gated calcium channel α2δ subunit. *FEBS Lett* **397**: 331–337.

Aschrafi A, Sadtler S, Niculescu C, Rettinger J, Schmalzing G. 2004. Trimeric architecture of homomeric P2X2 and heteromeric P2X1+2 receptor subtypes. *J Mol Biol* **342**: 333–343.

Bayguinov O, Hagen B, Bonev AD, Nelson MT, Sanders KM. 2000. Intracellular calcium events activated by ATP in murine colonic myocytes. *Am J Physiol Cell Physiol* **279**: C126–C135.

Benham CD, Bolton TB. 1986. Spontaneous transient outward currents in single visceral and vascular smooth muscle cells of the rabbit. *J Physiol* **381**: 385–406.

Benham CD, Tsien RW. 1987. A novel receptor-operated Ca^{2+}-permeable channel activated by ATP in smooth muscle. *Nature* **328**: 275–278.

Berridge MJ. 1997. Elementary and global aspects of calcium signalling. *J Physiol* **499**: 291–306.

Bielefeldt K. 1999. Molecular diversity of voltage-sensitive calcium channels in smooth muscle cells. *J Lab Clin Med* **133**: 469–477.

Birnbaumer L, Qin N, Olcese R, Tareilus E, Platano D, Costantin J, Stefani E. 1998. Structures and functions of calcium channel β subunits. *J Bioenerg Biomembr* **30**: 357–375.

Boittin FX, Macrez N, Halet G, Mironneau J. 1999. Norepinephrine-induced Ca^{2+} waves depend on InsP(3) and ryanodine receptor activation in vascular myocytes. *Am J Physiol* **277**: C139–C151.

Boittin FX, Coussin F, Morel JL, Halet G, Macrez N, Mironneau J. 2000. Ca^{2+} signals mediated by Ins(1,4,5)P(3)-gated channels in rat ureteric myocytes. *Biochem J* **349**: 323–332.

Bonev AD, Jaggar JH, Rubart M, Nelson MT. 1997. Activators of protein kinase C decrease Ca^{2+} spark frequency in smooth muscle cells from cerebral arteries. *Am J Physiol* **273**: C2090–C2095.

Bootman MD, Berridge MJ. 1996. Subcellular Ca^{2+} signals underlying waves and graded responses in HeLa cells. *Curr Biol* **6**: 855–865.

Bradley JE, Anderson UA, Woolsey SM, Thornbury KD, McHale NG, Hollywood MA. 2004. Characterization of T-type calcium current and its contribution to electrical activity in rabbit urethra. *Am J Physiol Cell Physiol* **286**: C1078–C1088.

Brain KL, Jackson VM, Trout SJ, Cunnane TC. 2002. Intermittent ATP release from nerve terminals elicits focal smooth muscle Ca^{2+} transients in mouse vas deferens. *J Physiol* **541**: 849–862.

Brain KL, Cuprian AM, Williams DJ, Cunnane TC. 2003. The sources and sequestration of Ca^{2+} contributing to neuroeffector Ca^{2+} transients in the mouse vas deferens. *J Physiol* **553:** 627–635.

Burdyga T, Wray S. 2005. Action potential refractory period in ureter smooth muscle is set by Ca sparks and BK channels. *Nature* **436:** 559–562.

Cannell MB, Cheng H, Lederer WJ. 1995. The control of calcium release in heart muscle. *Science* **268:** 1045–1049.

Cartin L, Lounsbury KM, Nelson MT. 2000. Coupling of Ca^{2+} to CREB activation and gene expression in intact cerebral arteries from mouse: Roles of ryanodine receptors and voltage-dependent Ca^{2+} channels. *Circ Res* **86:** 760–767.

Catterall WA. 2000. Structure and regulation of voltage-gated Ca^{2+} channels. *Annu Rev Cell Dev Biol* **16:** 521–555.

Cheng H, Lederer WJ. 2008. Calcium sparks. *Physiol Rev* **88:** 1491–1545.

Cheng H, Lederer WJ, Cannell MB. 1993. Calcium sparks: Elementary events underlying excitation-contraction coupling in heart muscle. *Science* **262:** 740–744.

Collier ML, Ji G, Wang Y, Kotlikoff MI. 2000. Calcium-induced calcium release in smooth muscle: Loose coupling between the action potential and calcium release. *J Gen Physiol* **115:** 653–662.

Curtis BM, Catterall WA. 1984. Purification of the calcium antagonist receptor of the voltage-sensitive calcium channel from skeletal muscle transverse tubules. *Biochemistry* **23:** 2113–2118.

Demuro A, Parker I. 2004. Imaging the activity and localization of single voltage-gated Ca^{2+} channels by total internal reflection fluorescence microscopy. *Biophys J* **86:** 3250–3259.

Dolmetsch RE, Xu K, Lewis RS. 1998. Calcium oscillations increase the efficiency and specificity of gene expression [see comments]. *Nature* **392:** 933–936.

Essin K, Gollasch M. 2009. Role of ryanodine receptor subtypes in initiation and formation of calcium sparks in arterial smooth muscle: Comparison with striated muscle. *J Biomed Biotechnol* **2009:** 135249.

Feske S, Gwack Y, Prakriya M, Srikanth S, Puppel SH, Tanasa B, Hogan PG, Lewis RS, Daly M, Rao A. 2006. A mutation in Orai1 causes immune deficiency by abrogating CRAC channel function. *Nature* **441:** 179–185.

Foskett JK, White C, Cheung KH, Mak DO. 2007. Inositol trisphosphate receptor Ca^{2+} release channels. *Physiol Rev* **87:** 593–658.

Giachini FR, Chiao CW, Carneiro FS, Lima VV, Carneiro ZN, Dorrance AM, Tostes RC, Webb RC. 2009. Increased activation of stromal interaction molecule-1/Orai-1 in aorta from hypertensive rats: A novel insight into vascular dysfunction. *Hypertension* **53:** 409–416.

Giachini FR, Webb RC, Tostes RC. 2010. STIM and Orai proteins: Players in sexual differences in hypertension-associated vascular dysfunction? *Clin Sci (Lond)* **118:** 391–396.

Gordienko DV, Bolton TB. 2002. Crosstalk between ryanodine receptors and IP(3) receptors as a factor shaping spontaneous Ca^{2+}-release events in rabbit portal vein myocytes. *J Physiol* **542:** 743–762.

Gordienko DV, Bolton TB, Cannell MB. 1998. Variability in spontaneous subcellular calcium release in guinea-pig ileum smooth muscle cells. *J Physiol* **507:** 707–720.

Grayson TH, Haddock RE, Murray TP, Wojcikiewicz RJ, Hill CE. 2004. Inositol 1,4,5-trisphosphate receptor subtypes are differentially distributed between smooth muscle and endothelial layers of rat arteries. *Cell Calcium* **36:** 447–458.

Heppner TJ, Bonev AD, Santana LF, Nelson MT. 2002. Alkaline pH shifts Ca^{2+} sparks to Ca^{2+} waves in smooth muscle cells of pressurized cerebral arteries. *Am J Physiol Heart Circ Physiol* **283:** H2169–H2176.

Heppner TJ, Bonev AD, Nelson MT. 2005. Elementary purinergic Ca^{2+} transients evoked by nerve stimulation in rat urinary bladder smooth muscle. *J Physiol* **564:** 201–212.

Heppner TJ, Werner ME, Nausch B, Vial C, Evans RJ, Nelson MT. 2009. Nerve-evoked purinergic signalling suppresses action potentials, Ca^{2+} flashes and contractility evoked by muscarinic receptor activation in mouse urinary bladder smooth muscle. *J Physiol* **587:** 5275–5288.

Herrera GM, Heppner TJ, Nelson MT. 2001. Voltage dependence of the coupling of Ca^{2+} sparks to BK(Ca) channels in urinary bladder smooth muscle. *Am J Physiol Cell Physiol* **280:** C481–C490.

Herrera GM, Nelson MT. 2002. Differential regulation of SK and BK channels by Ca^{2+} signals from Ca^{2+} channels and ryanodine receptors in guinea-pig urinary bladder myocytes. *J Physiol* **541:** 483–492.

Hill-Eubanks DC, Werner ME, Nelson MT. 2010. Local elementary purinergic-induced Ca^{2+} transients: From optical mapping of nerve activity to local Ca^{2+} signaling networks. *J Gen Physiol* **136:** 149–154.

Hirose K, Kadowaki S, Tanabe M, Takeshima H, Iino M. 1999. Spatiotemporal dynamics of inositol 1,4,5-trisphosphate that underlies complex Ca^{2+} mobilization patterns. *Science* **284:** 1527–1530.

Hogg RC, Wang Q, Helliwell RM, Large WA. 1993. Properties of spontaneous inward currents in rabbit pulmonary artery smooth muscle cells. *Pflugers Arch* **425:** 233–240.

Hosey MM, Barhanin J, Schmid A, Vandaele S, Ptasienski J, O'Callahan C, Cooper C, Lazdunski M. 1987. Photoaffinity labelling and phosphorylation of a 165 kilodalton peptide associated with dihydropyridine and phenylalkylamine-sensitive calcium channels. *Biochem Biophys Res Commun* **147:** 1137–1145.

Hullin R, Singer-Lahat D, Freichel M, Biel M, Dascal N, Hofmann F, Flockerzi V. 1992. Calcium channel β subunit heterogeneity: Functional expression of cloned cDNA from heart, aorta and brain. *EMBO J* **11:** 885–890.

Iino M, Kasai H, Yamazawa T. 1994. Visualization of neural control of intracellular Ca^{2+} concentration in single vascular smooth muscle cells in situ. *EMBO J* **13:** 5026–5031.

Jaggar JH, Nelson MT. 2000. Differential regulation of Ca^{2+} sparks and Ca^{2+} waves by UTP in rat cerebral artery smooth muscle cells. *Am J Physiol Cell Physiol* **279:** C1528–C1539.

Jaggar JH, Stevenson AS, Nelson MT. 1998a. Voltage dependence of Ca^{2+} sparks in intact cerebral arteries. *Am J Physiol* **274:** C1755–C1761.

Jaggar JH, Wellman GC, Heppner TJ, Porter VA, Perez GJ, Gollasch M, Kleppisch T, Rubart M, Stevenson AS, Lederer WJ, et al. 1998b. Ca^{2+} channels, ryanodine receptors and Ca^{2+}-activated K^+ channels: A functional unit for regulating arterial tone. *Acta Physiol Scand* **164:** 577–587.

Jaggar JH, Porter VA, Lederer WJ, Nelson MT. 2000. Calcium sparks in smooth muscle. *Am J Physiol Cell Physiol* **278:** C235–C256.

Jurkat-Rott K, Lehmann-Horn F. 2004. The impact of splice isoforms on voltage-gated calcium channel α1 subunits. *J Physiol* **554:** 609–619.

Kasai Y, Yamazawa T, Sakurai T, Taketani Y, Iino M. 1997. Endothelium-dependent frequency modulation of Ca^{2+} signalling in individual vascular smooth muscle cells of the rat. *J Physiol* **504:** 349–357.

Keef KD, Hume JR, Zhong J. 2001. Regulation of cardiac and smooth muscle Ca^{2+} channels ($Ca_V1.2a,b$) by protein kinases. *Am J Physiol Cell Physiol* **281:** C1743–C1756.

Khakh BS. 2001. Molecular physiology of P2X receptors and ATP signalling at synapses. *Nat Rev Neurosci* **2:** 165–174.

Klein MG, Cheng H, Santana LF, Jiang YH, Lederer WJ, Schneider MF. 1996. Two mechanisms of quantized calcium release in skeletal muscle. *Nature* **379:** 455–458.

Klockner U, Isenberg G. 1985. Action potentials and net membrane currents of isolated smooth muscle cells (urinary bladder of the guinea-pig). *Pflugers Arch* **405:** 329–339.

Knot HJ, Nelson MT. 1998. Regulation of arterial diameter and wall $[Ca^{2+}]$ in cerebral arteries of rat by membrane potential and intravascular pressure. *J Physiol* **508:** 199–209.

Kuo IY, Wolfle SE, Hill CE. 2010. T-type calcium channels and vascular function: The new kid on the block? *J Physiol* **589:** 783–795.

Lamont C, Wier WG. 2002. Evoked and spontaneous purinergic junctional Ca^{2+} transients (jCaTs) in rat small arteries. *Circ Res* **91:** 454–456.

Lamont C, Vial C, Evans RJ, Wier WG. 2006. P2X1 receptors mediate sympathetic postjunctional Ca^{2+} transients in mesenteric small arteries. *Am J Physiol Heart Circ Physiol* **291:** H3106–3113.

Lanner JT, Georgiou DK, Joshi AD, Hamilton SL. 2010. Ryanodine receptors: Structure, expression, molecular details, and function in calcium release. *Cold Spring Harb Perspect Biol* **2:** a003996.

Lee CH, Poburko D, Kuo KH, Seow CY, van Breemen C. 2002. Ca^{2+} oscillations, gradients, and homeostasis in vascular smooth muscle. *Am J Physiol Heart Circ Physiol* **282:** H1571–H1583.

Leung AT, Imagawa T, Campbell KP. 1987. Structural characterization of the 1,4-dihydropyridine receptor of the voltage-dependent Ca^{2+} channel from rabbit skeletal muscle. Evidence for two distinct high molecular weight subunits. *J Biol Chem* **262:** 7943–7946.

Li W, Llopis J, Whitney M, Zlokarnik G, Tsien RY. 1998. Cell-permeant caged InsP3 ester shows that Ca^{2+} spike frequency can optimize gene expression [see comments]. *Nature* **392:** 936–941.

Liou J, Kim ML, Heo WD, Jones JT, Myers JW, Ferrell JE Jr, Meyer T. 2005. STIM is a Ca^{2+} sensor essential for Ca^{2+}-store-depletion-triggered Ca^{2+} influx. *Curr Biol* **15:** 1235–1241.

Lopez-Lopez JR, Shacklock PS, Balke CW, Wier WG. 1995. Local calcium transients triggered by single L-type calcium channel currents in cardiac cells. *Science* **268:** 1042–1045.

Maravall M, Mainen ZF, Sabatini BL, Svoboda K. 2000. Estimating intracellular calcium concentrations and buffering without wavelength ratioing. *Biophys J* **78:** 2655–2667.

Miriel VA, Mauban JR, Blaustein MP, Wier WG. 1999. Local and cellular Ca^{2+} transients in smooth muscle of pressurized rat resistance arteries during myogenic and agonist stimulation. *J Physiol* **518 (Pt 3):** 815–824.

Mironneau J, Arnaudeau S, Macrez-Lepretre N, Boittin FX. 1996. Ca^{2+} sparks and Ca^{2+} waves activate different Ca^{2+}-dependent ion channels in single myocytes from rat portal vein. *Cell Calcium* **20:** 153–160.

Mizuno Y, Isotani E, Huang J, Ding H, Stull JT, Kamm KE. 2008. Myosin light chain kinase activation and calcium sensitization in smooth muscle in vivo. *Am J Physiol Cell Physiol* **295:** C358–C364.

Moosmang S, Schulla V, Welling A, Feil R, Feil S, Wegener JW, Hofmann F, Klugbauer N. 2003. Dominant role of smooth muscle L-type calcium channel $Ca_v1.2$ for blood pressure regulation. *EMBO J* **22:** 6027–6034.

Morales S, Diez A, Puyet A, Camello PJ, Camello-Almaraz C, Bautista JM, Pozo MJ. 2007. Calcium controls smooth muscle TRPC gene transcription via the CaMK/calcineurin-dependent pathways. *Am J Physiol Cell Physiol* **292:** C553–C563.

Mufti RE, Brett SE, Tran CH, Abd El-Rahman R, Anfinogenova Y, El-Yazbi A, Cole WC, Jones PP, Chen SR, Welsh DG. 2010. Intravascular pressure augments cerebral arterial constriction by inducing voltage-insensitive Ca^{2+} waves. *J Physiol* **588:** 3983–4005.

Mulryan K, Gitterman DP, Lewis CJ, Vial C, Leckie BJ, Cobb AL, Brown JE, Conley EC, Buell G, Pritchard CA, et al. 2000. Reduced vas deferens contraction and male infertility in mice lacking P2X1 receptors. *Nature* **403:** 86–89.

Murakami M, Yamamura H, Suzuki T, Kang MG, Ohya S, Murakami A, Miyoshi I, Sasano H, Muraki K, Hano T, et al. 2003. Modified cardiovascular L-type channels in mice lacking the voltage-dependent Ca^{2+} channel β3 subunit. *J Biol Chem* **278:** 43261–43267.

Navedo MF, Amberg GC, Votaw VS, Santana LF. 2005. Constitutively active L-type Ca^{2+} channels. *Proc Natl Acad Sci* **102:** 11112–11117.

Nelson MT, Patlak JB, Worley JF, Standen NB. 1990. Calcium channels, potassium channels, and voltage dependence of arterial smooth muscle tone. *Am J Physiol* **259:** C3–C18.

Nelson MT, Cheng H, Rubart M, Santana LF, Bonev AD, Knot HJ, Lederer WJ. 1995. Relaxation of arterial smooth muscle by calcium sparks [see comments]. *Science* **270:** 633–637.

Newton CL, Mignery GA, Sudhof TC. 1994. Co-expression in vertebrate tissues and cell lines of multiple inositol 1,4,5-trisphosphate (InsP3) receptors with distinct affinities for InsP3. *J Biol Chem* **269:** 28613–28619.

Neylon CB, Richards SM, Larsen MA, Agrotis A, Bobik A. 1995. Multiple types of ryanodine receptor/Ca^{2+} release channels are expressed in vascular smooth muscle. *Biochem Biophys Res Commun* **215**: 814–821.

Ng LC, Ramduny D, Airey JA, Singer CA, Keller PS, Shen XM, Tian H, Valencik M, Hume JR. 2010. Orai1 interacts with STIM1 and mediates capacitative Ca^{2+} entry in mouse pulmonary arterial smooth muscle cells. *Am J Physiol Cell Physiol* **299**: C1079–C1090.

Nieves-Cintron M, Amberg GC, Navedo MF, Molkentin JD, Santana LF. 2008. The control of Ca^{2+} influx and NFATc3 signaling in arterial smooth muscle during hypertension. *Proc Natl Acad Sci* **105**: 15623–15628.

Nikitina E, Zhang ZD, Kawashima A, Jahromi BS, Bouryi VA, Takahashi M, Xie A, Macdonald RL. 2007. Voltage-dependent calcium channels of dog basilar artery. *J Physiol* **580**: 523–541.

North RA. 2002. Molecular physiology of P2X receptors. *Physiol Rev* **82**: 1013–1067.

Owens GK. 1995. Regulation of differentiation of vascular smooth muscle cells. *Physiol Rev* **75**: 487–517.

Park Hopson K, Truelove J, Chun J, Wang Y, Waeber C. 2011. S1P activates Store-operated calcium entry via receptor and non receptor-mediated pathways in vascular smooth muscle cells. *Am J Physiol Cell Physiol* **300**: 919–926.

Peel SE, Liu B, Hall IP. 2006. A key role for STIM1 in store operated calcium channel activation in airway smooth muscle. *Respir Res* **7**: 119.

Peel SE, Liu B, Hall IP. 2008. ORAI and store-operated calcium influx in human airway smooth muscle cells. *Am J Respir Cell Mol Biol* **38**: 744–749.

Peng H, Matchkov V, Ivarsen A, Aalkjaer C, Nilsson H. 2001. Hypothesis for the initiation of vasomotion. *Circ Res* **88**: 810–815.

Perez GJ, Bonev AD, Patlak JB, Nelson MT. 1999. Functional coupling of ryanodine receptors to KCa channels in smooth muscle cells from rat cerebral arteries. *J Gen Physiol* **113**: 229–238.

Perez GJ, Bonev AD, Nelson MT. 2001. Micromolar Ca^{2+} from sparks activates Ca^{2+}-sensitive K^+ channels in rat cerebral artery smooth muscle. *Am J Physiol Cell Physiol* **281**: C1769–C1775.

Perez-Reyes E. 2003. Molecular physiology of low-voltage-activated t-type calcium channels. *Physiol Rev* **83**: 117–161.

Pozo MJ, Perez GJ, Nelson MT, Mawe GM. 2002. Ca^{2+} sparks and BK currents in gallbladder myocytes: Role in CCK-induced response. *Am J Physiol Gastrointest Liver Physiol* **282**: G165–G174.

Pozzan T, Rizzuto R, Volpe P, Meldolesi J. 1994. Molecular and cellular physiology of intracellular calcium stores. *Physiol Rev* **74**: 595–636.

Prinz G, Diener M. 2008. Characterization of ryanodine receptors in rat colonic epithelium. *Acta Physiol (Oxf)* **193**: 151–162.

Pulver-Kaste RA, Barlow CA, Bond J, Watson A, Penar PL, Tranmer B, Lounsbury KM. 2006. Ca^{2+} source-dependent transcription of CRE-containing genes in vascular smooth muscle. *Am J Physiol Heart Circ Physiol* **291**: 97–105.

Reuter H. 1983. Calcium channel modulation by neurotransmitters, enzymes and drugs. *Nature* **301**: 569–574.

Roos J, DiGregorio PJ, Yeromin AV, Ohlsen K, Lioudyno M, Zhang S, Safrina O, Kozak JA, Wagner SL, Cahalan MD, et al. 2005. STIM1, an essential and conserved component of store-operated Ca^{2+} channel function. *J Cell Biol* **169**: 435–445.

Ruehlmann DO, Lee CH, Poburko D, van Breemen C. 2000. Asynchronous Ca^{2+} waves in intact venous smooth muscle. *Circ Res* **86**: E72–E79.

Sanders KM. 2001. Invited review: Mechanisms of calcium handling in smooth muscles. *J Appl Physiol* **91**: 1438–1449.

Santana LF, Navedo MF. 2009. Molecular and biophysical mechanisms of Ca^{2+} sparklets in smooth muscle. *J Mol Cell Cardiol* **47**: 436–444.

Santana LF, Navedo MF, Amberg GC, Nieves-Cintron M, Votaw VS, Ufret-Vincenty CA. 2008. Calcium sparklets in arterial smooth muscle. *Clin Exp Pharmacol Physiol* **35**: 1121–1126.

Schneider P, Hopp HH, Isenberg G. 1991. Ca^{2+} influx through ATP-gated channels increments $[Ca^{2+}]_i$ and inactivates ICa in myocytes from guinea-pig urinary bladder. *J Physiol* **440**: 479–496.

Sieck GC, Kannan MS, Prakash YS. 1997. Heterogeneity in dynamic regulation of intracellular calcium in airway smooth muscle cells. *Can J Physiol Pharmacol* **75**: 878–888.

Snutch TP, Sutton KG, Zamponi GW. 2001. Voltage-dependent calcium channels—Beyond dihydropyridine antagonists. *Curr Opin Pharmacol* **1**: 11–16.

Somlyo AP, Somlyo AV. 2003. Ca^{2+} sensitivity of smooth muscle and nonmuscle myosin II: Modulated by G proteins, kinases, and myosin phosphatase. *Physiol Rev* **83**: 1325–1358.

Stevens RJ, Weinert JS, Publicover NG. 1999. Visualization of origins and propagation of excitation in canine gastric smooth muscle. *Am J Physiol* **277**: C448–C460.

Stull JT, Lin PJ, Krueger JK, Trewhella J, Zhi G. 1998. Myosin light chain kinase: Functional domains and structural motifs. *Acta Physiol Scand* **164**: 471–482.

Takahashi Y, Watanabe H, Murakami M, Ono K, Munehisa Y, Koyama T, Nobori K, Iijima T, Ito H. 2007. Functional role of stromal interaction molecule 1 (STIM1) in vascular smooth muscle cells. *Biochem Biophys Res Commun* **361**: 934–940.

Tasker PN, Michelangeli F, Nixon GF. 1999. Expression and distribution of the type 1 and type 3 inositol 1,4,5-trisphosphate receptor in developing vascular smooth muscle. *Circ Res* **84**: 536–542.

Thomas D, Lipp P, Berridge MJ, Bootman MD. 1998. Hormone-evoked elementary Ca^{2+} signals are not stereotypic, but reflect activation of different size channel clusters and variable recruitment of channels within a cluster. *J Biol Chem* **273**: 27130–27136.

Tsien RY. 1980. New calcium indicators and buffers with high selectivity against magnesium and protons: Design, synthesis, and properties of prototype structures. *Biochemistry* **19**: 2396–2404.

Vaghy PL, Williams JS, Schwartz A. 1987. Receptor pharmacology of calcium entry blocking agents. *Am J Cardiol* **59:** 9A–17A.

Vaithianathan T, Narayanan D, Asuncion-Chin MT, Jeyakumar LH, Liu J, Fleischer S, Jaggar JH, Dopico AM. 2010. Subtype identification and functional characterization of ryanodine receptors in rat cerebral artery myocytes. *Am J Physiol Cell Physiol* **299:** C264–C278.

Varnai P, Hunyady L, Balla T. 2009. STIM and Orai: The long-awaited constituents of store-operated calcium entry. *Trends Pharmacol Sci* **30:** 118–128.

Vial C, Evans RJ. 2002. P2X(1) receptor-deficient mice establish the native P2X receptor and a P2Y6-like receptor in arteries. *Mol Pharmacol* **62:** 1438–1445.

Vig M, Peinelt C, Beck A, Koomoa DL, Rabah D, Koblan-Huberson M, Kraft S, Turner H, Fleig A, Penner R, et al. 2006. CRACM1 is a plasma membrane protein essential for store-operated Ca^{2+} entry. *Science* **312:** 1220–1223.

Wamhoff BR, Bowles DK, McDonald OG, Sinha S, Somlyo AP, Somlyo AV, Owens GK. 2004. L-type voltage-gated Ca^{2+} channels modulate expression of smooth muscle differentiation marker genes via a rho kinase/myocardin/SRF-dependent mechanism. *Circ Res* **95:** 406–414.

Wang D, Chang PS, Wang Z, Sutherland L, Richardson JA, Small E, Krieg PA, Olson EN. 2001a. Activation of cardiac gene expression by myocardin, a transcriptional cofactor for serum response factor. *Cell* **105:** 851–862.

Wang SQ, Song LS, Lakatta EG, Cheng H. 2001b. Ca^{2+} signalling between single L-type Ca^{2+} channels and ryanodine receptors in heart cells. *Nature* **410:** 592–596.

Wang Y, Deng X, Hewavitharana T, Soboloff J, Gill DL. 2008. Stim, ORAI and TRPC channels in the control of calcium entry signals in smooth muscle. *Clin Exp Pharmacol Physiol* **35:** 1127–1133.

Wang Y, Deng X, Mancarella S, Hendron E, Eguchi S, Soboloff J, Tang XD, Gill DL. 2010. The calcium store sensor, STIM1, reciprocally controls Orai and $Ca_V1.2$ channels. *Science* **330:** 105–109.

Wegener JW, Schulla V, Lee TS, Koller A, Feil S, Feil R, Kleppisch T, Klugbauer N, Moosmang S, Welling A, et al. 2004. An essential role of $Ca_V1.2$ L-type calcium channel for urinary bladder function. *FASEB J* **18:** 1159–1161.

Wellman GC, Nelson MT. 2003. Signaling between SR and plasmalemma in smooth muscle: Sparks and the activation of Ca^{2+}-sensitive ion channels. *Cell Calcium* **34:** 211–229.

Yang XR, Lin MJ, Yip KP, Jeyakumar LH, Fleischer S, Leung GP, Sham JS. 2005. Multiple ryanodine receptor subtypes and heterogeneous ryanodine receptor-gated Ca^{2+} stores in pulmonary arterial smooth muscle cells. *Am J Physiol Lung Cell Mol Physiol* **289:** L338–L348.

Young RC, Schumann R, Zhang P. 2001. Intracellular calcium gradients in cultured human uterine smooth muscle: A functionally important subplasmalemmal space. *Cell Calcium* **29:** 183–189.

Zalk R, Lehnart SE, Marks AR. 2007. Modulation of the ryanodine receptor and intracellular calcium. *Annu Rev Biochem* **76:** 367–385.

Zhang SL, Yeromin AV, Zhang XH, Yu Y, Safrina O, Penna A, Roos J, Stauderman KA, Cahalan MD. 2006. Genome-wide RNAi screen of Ca^{2+} influx identifies genes that regulate Ca^{2+} release-activated Ca^{2+} channel activity. *Proc Natl Acad Sci* **103:** 9357–9362.

ZhuGe R, Sims SM, Tuft RA, Fogarty KE, Walsh JV Jr. 1998. Ca^{2+} sparks activate K^+ and Cl^- channels, resulting in spontaneous transient currents in guinea-pig tracheal myocytes. *J Physiol* **513:** 711–718.

Apoptosis and Autophagy: Decoding Calcium Signals that Mediate Life or Death

Michael W. Harr[1,4] and Clark W. Distelhorst[1,2,3]

[1]Division of Hematology and Oncology, Department of Medicine, Case Western Reserve University, Cleveland, Ohio 44106

[2]Department of Pharmacology, Case Western Reserve University, Cleveland, Ohio 44106

[3]Case Comprehensive Cancer Center, Case Western Reserve University, Cleveland, Ohio 44106

[4]Molecular Pharmacology and Chemistry Program, Sloan-Kettering Institute, Memorial Sloan-Kettering Cancer Center, New York, New York 10021

Correspondence: cwd@case.edu

Calcium is a versatile and dynamic 2nd messenger that is essential for the survival of all higher organisms. In cells that undergo activation or excitation, calcium is released from the endoplasmic/sarcoplasmic reticulum to activate calcium-dependent kinases and phosphatases, thereby regulating numerous cellular processes; for example, apoptosis and autophagy. In the case of apoptosis, endogenous ligands or pharmacological agents induce prolonged cytosolic calcium elevation, which in turn leads to cell death. In contrast, there is now evidence that calcium regulates autophagy by several mechanisms, and these may be important for maintaining cell survival. Here we summarize what is known about how calcium regulates these life and death decisions. We pay particular attention to pathways that have been described in lymphocytes and cardiomyocytes, as these systems provide optimal models for understanding calcium signaling in the context of normal cell physiology.

Apoptosis is a process of programmed cell death or suicide that occurs when cells have undergone irreversible stress or damage. It is required to maintain normal cell homeostasis or to eliminate a population of cells that may be harmful to the organism or unnecessary during organ development (Green 2003). For example, it is the primary mechanism by which potentially autoreactive T cells are eliminated from the immune system. There are two conventional apoptosis pathways: the extrinsic pathway, which is typically initiated by death receptors (e.g., Fas) on the plasma membrane and the intrinsic (mitochondrial) pathway, which involves permeabilization of the outer mitochondrial membrane followed by the release of cytochrome c. In this review, we primarily focus our attention on the intrinsic pathway due to the importance of intracellular calcium in the regulation of this process.

In brief, cytochrome c release stimulates apoptosis via its interaction with the protein Apaf-1, which in turn activates the initiator caspase-9 and the executioner caspase-3 (Green 2005). Caspases comprise a family of cysteine proteases that are essential for the classically observed cellular and biochemical characteristics of apoptosis, which include (but are not limited to) membrane blebbing, chromatin condensation, and DNA fragmentation. Another class of cysteine proteases, calpains, require calcium for their activation and are important mediators of apoptosis following ER stress. As discussed later in this review, calpains are reported to directly activate caspases, thus promoting apoptotic cell death independent of mitochondrial cytochrome c release. The following sections provide a more detailed explanation of the varied ways in which calcium signals induce cell death and are themselves regulated.

APOPTOSIS REGULATION BY ANTIGEN RECEPTORS: A MODEL FOR PROGRAMMED CELL DEATH

Much that is known about calcium signaling came from immunological studies using activated lymphocytes (Berridge 1997). Immature T cells are an ideal model for investigating apoptosis because they are programmed to die during development. This is evident by the fact that 95% of double positive (i.e., CD4/CD8) thymocytes undergo apoptosis as a consequence of negative selection (Starr et al. 2003). Apoptosis of thymocytes occurs when self antigen presented on thymic epithelial cells binds to T-cell receptors with strong avidity (Hogquist 2001). As depicted in Figure 1, ligation of the T-cell receptor activates a signaling pathway that results in autophosphorylation of Src family kinases Lck and Fyn, which are recruited to the plasma membrane to phosphorylate the zeta chain of the T-cell receptor (Latour and Veillette 2001; Mustelin and Tasken 2003; Palacios and Weiss 2004). Activation of these kinases facilitates the hydrolysis of phosphatidylinositol 4,5-bisphosphate by phospholipase Cγ, thereby generating diacylglycerol and inositol 1,4,5-trisphosphate (IP_3) (Lewis 2001). IP_3 mediates ER calcium release through the opening of IP_3 receptors (IP_3Rs), which in turn stimulates calcineurin-mediated activation of the nuclear factor for the activation of T cells (NFAT) (Gallo et al. 2006; Winslow and Crabtree 2005; Winslow et al. 2003). Cytosolic calcium release is also mediated by ryanodine receptors, which are calcium channels expressed in lymphocytes, cardiomyocytes, and neurons. It has been suggested that ryanodine receptors facilitate calcium flux in response to nicotinic acid adenine dinucleotide phosphate (NAADP) and cyclic adenosine dinucleotide phosphate ribose, both of which are produced during T-cell activation (Berg et al. 2000; Dammermann et al. 2009; Guse et al. 1999). However, unlike ADP ribose, the notion that NAADP is a direct activator of ryanodine receptors is not yet certain and may be context- or cell-type specific (Galione and Petersen 2005). Nevertheless, the generation of these 2nd messengers, along with IP_3, may be required for robust calcium elevation in response to antigen receptor stimulation.

In general, lymphocyte activation encodes distinct patterns of calcium signaling, which ultimately regulate cell proliferation, survival, and apoptosis (Berridge 1997; Lewis 2001) (Fig. 1). It has been proposed that strong agonist stimulation of T-cell receptors generates calcium transients that trigger apoptosis, whereas weak stimulation produces calcium oscillations that are needed for cell survival (Randriamampita and Trautmann 2003). The patterns of calcium elevation following strong and weak agonist stimulation are vastly different. For example, calcium transients can be detected in cells 1–2 minutes following activation of the T-cell receptor. They are generally synchronized and characterized by a broad peak that is high in amplitude. On the other hand, calcium oscillations consist of asynchronous and repetitious spikes that persist for as long as one hour following activation. Using fluorescence-activated cell sorting, we have shown that T cells with a high level of cytosolic calcium more readily undergo apoptosis compared to those that have lower levels (Zhong et al. 2006). Further, in those cells that undergo calcium oscillations in response to weak agonist stimulation, NFAT is rapidly de-phosphorylated, and this is associated with increased levels of IL-2 mRNA (Harr et al. 2009; Zhong et al. 2006). These observations are consistent with the strength of signal theory, which states that T cells undergo positive or negative selection

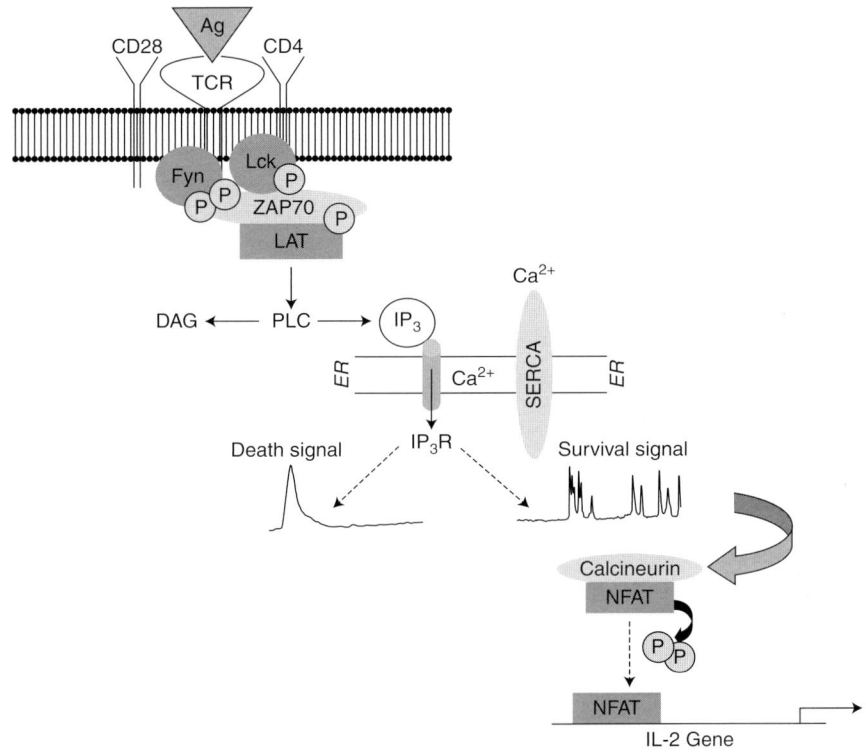

Figure 1. Calcium signaling mediated by T-cell activation. In a CD4 positive T cell, activation is induced by antigen binding to the T-cell receptor (TCR) and a co-stimulatory signal mediated by ligation of CD28. Src family kinases Fyn and Lck are activated by autophosphorylation and recruited to the plasma membrane to associate with CD3 (TCR) and CD4, respectively. This in turn leads to the phosphorylation of ZAP-70 and the adaptor protein LAT in order to activate phospholipase C and generate 2nd messengers IP$_3$ and diacylglycerol (DAG). IP$_3$ binds to the IP$_3$R, resulting in ER calcium release. Calcium is released into the cytosol by way of a single calcium transient (left), which is associated with cell death by apoptosis. On the other hand, calcium oscillations activate calcineurin, which dephosphorylates NFAT, thereby sending it to the nucleus to activate transcription of IL-2. The Sarcoplasmic/Endoplasmic Reticulum Calcium ATPase (SERCA) is responsible for maintaining the appropriate concentration of luminal calcium by actively transporting calcium across the ER membrane.

according to the avidity of T-cell receptor activation (Hogquist 2001; Mariathasan et al. 1998). This provides one example in which the amplitude and frequency of calcium signals encodes information that regulates apoptosis or cell survival.

APOPTOSIS REGULATION BY HORMONES AND OTHER SIGNALING MOLECULES

Corticosteroids

Knowledge of calcium-dependent apoptosis evolved from studies examining the effects of glucocorticoids on immature T cells. Glucocorticoids, such as cortisol, are physiological immunomodulatory hormones that regulate immune cell development. Pharmacologically, synthetic glucocorticoids, such as prednisone and dexamethasone, are widely used to treat autoimmune disease and cancer (e.g., leukemia and lymphoma) because of their ability to suppress the immune system and selectively kill immature lymphocytes, respectively. Glucocorticoid hormones are secreted from thymic epithelial cells to antagonize self-antigen recognition in immature thymocytes (Ashwell et al. 2000). Thus, glucocorticoids negatively regulate T-cell activation by attenuating T-cell receptor signaling (Baus et al. 1996; Lowenberg et al. 2005; Van Laethem et al. 2001). Conversely, T-cell receptor signaling can also inhibit glucocorticoid-induced apoptosis, a concept known as mutual antagonism (Ashwell et al. 2000; Jamieson and Yamamoto 2000; Tolosa and Ashwell 1999).

While studying these inhibitory effects on T-cell activation, our laboratory discovered that short-term treatment with glucocorticoids modulates T-cell receptor-mediated calcium elevation by converting calcium transients to oscillatory signals (Harr et al. 2009). However, prolonged glucocorticoid treatment with pharmacological concentrations of prednisone or dexamethasone results in thymocyte apoptosis by a mechanism that is dependent, in part, on de novo transcription (Herold et al. 2006). Importantly, glucocorticoid-induced apoptosis in thymocytes is consistently associated with a sustained rise in cytosolic calcium (Bian et al. 1997; Cohen and Duke 1984; Kaiser and Edelman 1977; Lam et al. 1993; McConkey et al. 1989; Orrenius et al. 1991). This increase in cytosolic calcium is associated with the classically observed biochemical characteristics of apoptosis, including DNA fragmentation and endonuclease activity. While there is evidence that cytosolic calcium elevation contributes to the induction of apoptosis, this mechanism has not been firmly established. Marks and colleagues found that anti-sense mediated knock-down of IP_3R1 protected cells from apoptosis induced by dexamethasone (Jayaraman and Marks 1997). However, we observed that dexamethasone-mediated up-regulation of IP_3Rs did not contribute to cytosolic calcium elevation or apoptosis following glucocorticoid treatment (Davis et al. 2008). An alternative theory is that dexamethasone down-regulates the sarcoplasmic endoplasmic reticulum ATPase (SERCA) that pumps calcium into the ER (Chai et al. 2009), thereby decreasing ER luminal calcium and inducing apoptosis.

Angiotensins

Angiotensins are a second class of hormone that induce calcium-dependent apoptosis, specifically Angiotensin II (Cigola et al. 1997; Kajstura et al. 1997; Palomeque et al. 2009; Yamada et al. 1996). Angiotensins are oligomeric peptides released in response to steroid hormones, such as glucocorticoids and estrogen. They are powerful vasoconstrictors, and consequently, angiotensin receptors are targets for antihypertensive medications (Gradman 2009). In a cardiomyocyte, muscle contraction is stimulated by the opening of an L-type calcium channel that enables calcium release via ryanodine receptors (Fig. 2). Angiotensins bind to their receptors (AT1 and AT2) resulting in the generation of IP_3 followed by transient calcium elevations (Mattiazzi 1997). While ryanodine receptors are more abundant than IP_3Rs in cardiomyocytes, both calcium and IP_3 are required for IP_3R channel opening. While still not universally accepted, it is likely that calcium release via neighboring ryanodine receptors facilitates IP_3R-opening, enabling both channels to function cooperatively in response to angiotensin ligands (Kockskamper et al. 2008).

Much like glucocorticoids, treatment with higher concentrations of angiotensin II results in apoptosis that can be blocked by receptor antagonists (Andreka et al. 2004). As depicted in Figure 2, stimulation of cardiomyocytes with angiotensin II causes an acute release of cytosolic calcium, and several reports suggest that calcium elevation contributes to apoptosis (Kajstura et al. 1997). For instance, verapamil, an L-type calcium channel blocker, inhibits angiotensin-induced apoptosis (Goldenberg et al. 2001). Further, ectopic expression of angiotensin receptors (AT1 and AT2) results in apoptosis by a calcium-dependent mechanism (Aranguiz-Urroz et al. 2009).

Testosterone

Testosterone is a steroid hormone that has rapid effects on cardiomyocytes. In one study, it was shown that testosterone increased cytoplasmic calcium concentrations in 1–7 minutes by an IP_3R-dependent fashion (Vicencio et al. 2006). A second study reported that testosterone signals activate the extracellular signal-regulated kinase (ERK), and this activation could be inhibited by 2-aminoethyldiphenyl borate and the phospholipase C inhibitor U-73122, suggesting

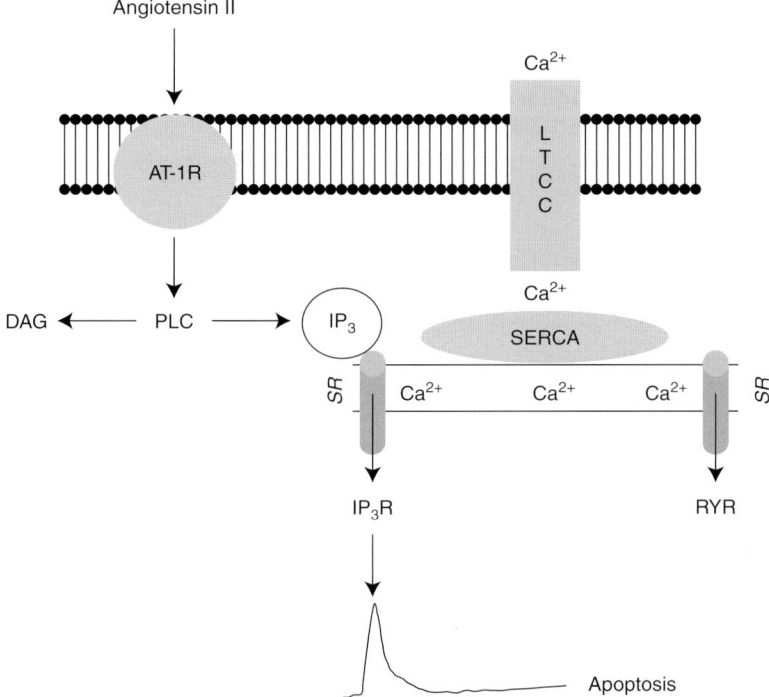

Figure 2. Calcium signaling mediated by angiotensin II hormone in a cardiomyocyte. In a cardiomyocyte, calcium signaling is mediated by the influx of calcium through L-type calcium channels (LTCC). The Sarcoplasmic/Endoplasmic Reticulum Calcium ATPase (SERCA) is responsible for maintaining the appropriate concentration of luminal calcium by actively transporting calcium ion across the SR membrane, while ryanodine receptors and IP$_3$Rs promote its release into the cytosol. Angiotensin, a peptide hormone, binds to the AT-1 receptor (AT-1R), a G-protein coupled receptor that activates phospholipase C (PLC) following GTP hydrolysis, thereby generating IP$_3$ and diacylglycerol (DAG). It should be noted that ryanodine receptors are 50- to 100-fold more abundant than IP$_3$Rs in cardiomyocytes. Therefore, calcium release via ryanodine receptors have shown to be much more robust compared to calcium responses that are mediated by IP$_3$Rs (Kockskamper et al. 2008). Nevertheless, there is unequivocal evidence for the contribution of IP$_3$Rs during SR-calcium release, which is likely due to synergy between the two calcium channels in mediating calcium-induced calcium release.

that testosterone generates IP$_3$ in cardiomyocytes (Altamirano et al. 2009). Similar rapid non-genomic effects of testosterone have been observed in T cells, where calcium influx was observed within seconds (Benten et al. 1997). However, it remains to be determined whether these calcium signals would eventually lead to apoptosis. Interestingly, in neuronal cells, low concentrations of testosterone result in calcium oscillations, whereas higher concentrations induce apoptosis, also by a mechanism that is IP$_3$R dependent (Estrada et al. 2006). Thus, steroid hormones can have direct apoptotic effects on multiple cells types, as is the case for glucocorticoids and androgens, and perhaps indirectly via glucocorticoid regulation of angiotensin.

Nitric Oxide

Nitric oxide is an endogenous signaling molecule that regulates muscle contraction, oxygen consumption, and mitochondrial metabolism in the heart (Massion et al. 2003). It is produced by a family of calcium-dependent enzymes called nitric oxide synthases (NOS). NOS enzymes have been found to localize to the sarcoplasmic reticulum and can inhibit L-type channel and ryanodine-receptor activity (Barouch et al. 2002; Sears et al. 2003; Xu et al. 1999). Further, there is evidence that NOS enzymes can inhibit calcium channel activity by S-nitrosylation of thiol residues (Davidson and Duchen 2006; Razavi et al. 2005), which in turn decreases mitochondrial calcium uptake, thereby preventing cytochrome c release and mitochondrial metabolism (Brooks et al. 2000; Dedkova and Blatter 2005; Khan and Hare 2003). In addition, nitric oxide can also attenuate apoptosis by nitrosylation of caspases on cysteine residues (Dimmeler et al. 1997). The ability of nitric oxide to prevent apoptosis in a calcium-dependent manner may be important for understanding certain pathophysiologies such as ischemia or reperfusion injury.

APOPTOSIS REGULATION BY PHARMACOLOGICAL AGENTS

Although calcium-mediated apoptosis can occur by physiological signals, there are multiple cytotoxic agents that function to disrupt calcium homeostasis leading to apoptotic cell death. Among these are thapsigargin, staurosporine, and cisplatin.

Thapsigargin

Thapsigargin decreases the ER calcium pool by inhibiting SERCA pumps, which results in ER stress and apoptosis (Lam et al. 1993). Apoptosis induced by thapsigargin occurs by a mechanism that is dependent on the activation of caspase-12, a mammalian protease that localizes to the ER and is important for mediating apoptosis in response to ER stress (Szegezdi et al. 2003). This is exemplified by experiments performed in mice in which caspase-12 had been deleted (Nakagawa et al. 2000). In vitro studies have demonstrated that a calpain activates caspase-12 leading to the subsequent activation of caspase-9 (Morishima et al. 2002; Nakagawa and Yuan 2000; Rao et al. 2002). These data suggest the possibility that apoptosis induced by thapsigargin can occur independently of cytochrome c release and is thus directly induced by calcium via calpain activation.

Staurosporine

Staurosporine is a natural apoptosis-inducing alkaloid originally isolated from Streptomyces *staurosporeus*. It directly provokes calcium leak from the ER by activating caspase-3 mediated cleavage of IP_3R1 (Hirota et al. 1999). Additionally, it was shown that cleavage of IP_3R1 contributed, in part, to the induction of apoptosis by accelerating calcium leak (Assefa et al. 2004; Verbert et al. 2008). In these experiments, transfection of a mutant IP_3R resistant to caspase-mediated cleavage partially inhibited apoptosis induction by staurosporine in B cells lacking wild type IP_3Rs. A recent study by Mikoshiba and colleagues further identified IP_3Rs as being important mediators of apoptosis induction by staurosporine. They determined that G protein-coupled receptor kinase interacting proteins (GITs) bind to IP_3Rs to inhibit their function and suppress apoptosis in the presence of staurosporine (Zhang et al. 2009b). Finally, staurosporine also promotes the activation of a mitochondrial calpain that positively regulates apoptosis (Norberg et al. 2008), thus illustrating similarities with ER stress-driven pathways.

Cisplatin

A third example is cisplatin, a platinum-based chemotherapeutic agent used to treat several types of cancer. Cisplatin also causes an IP_3R-dependent increase in cytosolic calcium and subsequent activation of a calpain prior to the induction of apoptosis (Mandic et al. 2003; Schrodl et al. 2009; Splettstoesser et al. 2007). Further, cisplatin treatment results in ER stress as suggested by increased expression of Grp78 and activation of caspase-12 (Mandic et al. 2003). Interestingly, IP_3R1 contributes to cisplatin sensitivity in bladder cancer, as knocking down its expression in cell lines mediates resistance to apoptosis (Tsunoda et al. 2005). These results collectively indicate that cisplatin induces apoptosis, in part, by disrupting calcium homeostasis in a variety of cell types.

APOPTOSIS REGULATION BY ANTI-APOPTOTIC Bcl-2 FAMILY PROTEINS

In each of the previous examples in which apoptosis is regulated by calcium, cell death can be readily inhibited by anti-apoptotic proteins such as B cell leukemia/lymphoma-2 (Bcl-2) and other Bcl-2 family members. As its name implies, Bcl-2 was first identified because it was overexpressed in B cell follicular lymphoma (Tsujimoto et al. 1985). Membership in the Bcl-2 family is defined by the presence of Bcl-2 homology domains (BH domains) (Chipuk et al. 2010). Bcl-2 has four BH domains. BH1, BH2, and BH3 are located within the C-terminal half, where they participate in forming a hydrophobic groove that binds and thereby inhibits proapoptotic family members. The BH4 domain is located near the N-terminus and connected to the C-terminal half of Bcl-2 by an unstructured loop, facilitating intra- and intermolecular interactions. It is now known that Bcl-2 is overexpressed in a number of cancers because of its ability to inhibit cell death and promote survival of malignant cells (Cory and Adams 2002). In fact, Bcl-2 localizes not only to the outer mitochondrial membrane but also to the ER, where it regulates IP$_3$-mediated calcium release. The observation that Bcl-2 regulates calcium release from the ER was initially made more than 15 years ago (Baffy et al. 1993; Lam et al. 1994) and the overall importance of Bcl-2 on the ER is exemplified in studies in which ER-targeted Bcl-2 inhibited apoptosis in response to agents that depolarize the mitochondrial membrane (Annis et al. 2001).

The Bcl-2-IP$_3$ Receptor Interaction

Our laboratory was the first to show that Bcl-2 directly interacts with IP$_3$Rs to inhibit IP$_3$-dependent calcium flux (Fig. 3) (Chen et al. 2004). This interaction, as well as an interaction of the Bcl-2 homologue Bcl-xL with the IP$_3$R, has subsequently been detected by a number of laboratories (Rong and Distelhorst 2007). Bcl-2 directly inhibits IP$_3$R channel opening in vitro in lipid bilayer experiments and also inhibits IP$_3$-induced calcium release in T cells (Chen et al. 2004; Zhong et al. 2006). We have

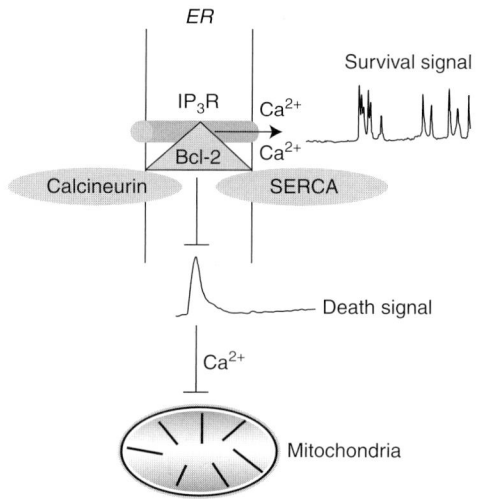

Figure 3. The Bcl-2-IP$_3$R interaction inhibits ER-calcium release. Bcl-2 localizes to the ER where it binds IP$_3$Rs to inhibit calcium transients. In T cells, calcium transients are activated in response to strong T-cell receptor ligation, which results in apoptosis that can be inhibited by Bcl-2. In contrast, calcium oscillations that are associated with cell survival are promoted by Bcl-2 and Bcl-xL. In addition, Bcl-2 regulates the level of ER luminal calcium by increasing membrane permeability or by interacting with the Sarcoplasmic/Endoplasmic Reticulum Calcium ATPase (SERCA). Bcl-2 also interacts with calcineurin, thereby forming a complex with both calcineurin and IP$_3$Rs on the ER membrane.

now further elucidated the mechanism of the Bcl-2-IP$_3$R interaction by demonstrating that the BH4 domain of Bcl-2 associates with the regulatory and coupling domain of IP$_3$R1, specifically an 80 amino acid sequence within domain 3 (Rong et al. 2008). Further analysis indicated that the BH4 domain was both necessary and sufficient to inhibit IP$_3$-mediated calcium signals and subsequently apoptosis in T cells (Rong et al. 2009). This is particularly interesting given that the BH4 domain of Bcl-2 and Bcl-xL seem to be involved in preventing apoptosis. Although our work has emphasized the interaction of Bcl-2 with the regulatory and coupling domain of the IP$_3$R, there is also evidence that Bcl-2 and/or Bcl-xL may interact with the C-terminus of the IP$_3$R, and through this interaction, enhance IP$_3$-mediated calcium oscillations (White et al. 2005). In light of these discoveries, there is now increased interest in using small molecules to inhibit Bcl-2 and enhance proapoptotic calcium transients. One example of this was the use of a Bcl-2 inhibitor HA14-1 that induced cytochrome c release and apoptosis by a calcium-dependent mechanism in myeloid leukemia (An et al. 2004).

Bcl-2 and IP$_3$R Phosphorylation

Other proteins within the Bcl-2-IP$_3$R complex may regulate ER calcium release by altering phosphorylation of IP$_3$Rs. As shown in Figure 3, Bcl-2 binds the phosphatase calcineurin (Erin et al. 2003; Shibasaki et al. 1997), and calcineurin also interacts with IP$_3$Rs (Cameron et al. 1995). Because IP$_3$R channel activity is positively regulated by phosphorylation (DeSouza et al. 2002), it is reasonable to speculate that Bcl-2 may facilitate the dephosphorylation of IP$_3$R by interacting with calcineurin. Although such a mechanism has not been definitively established, we have observed that IP$_3$R phosphorylation is decreased in Bcl-2 overexpressing T cells (Chen et al. 2004). Furthermore, Bcl-2 has been reported to regulate IP$_3$R phosphorylation in Bax/Bak double knockout cells (Oakes et al. 2005). Moreover, Xu et al. have implicated protein phosphatase 1 in the regulation of IP$_3$R phosphorylation by Bcl-2 (Xu et al. 2007). Thus, much remains to be determined regarding the specific kinases and phosphatases that regulate IP$_3$R channel opening, how these are regulated by Bcl-2-IP$_3$R interaction, and how they contribute overall to the regulation of calcium signals by Bcl-2 and its relatives.

Bcl-2 and Mitochondrial Cross Talk

Another important function of Bcl-2 and Bcl-xL is to inhibit calcium-mediated cross talk between ER and mitochondria (Kruman and Mattson 1999; Pinton et al. 2008). Because both organelles are in close proximity, calcium is rapidly taken up by mitochondria via the calcium uniporter on the outer mitochondrial membrane (Hajnoczky et al. 2006; Hanson et al. 2004; Rizzuto et al. 2003; Szalai et al. 1999; Szlufcik et al. 2006). Bcl-2 and Bcl-xL inhibit calcium redistribution to the mitochondria, thereby limiting calcium uptake (Hanson et al. 2008; Pinton et al. 2008) (Fig. 3). In the context of apoptosis, it was shown that Bcl-2 inhibited mitochondrial calcium uptake following IL-3 and serum withdrawal in hematopoietic cells and fibroblasts, respectively (Baffy et al. 1993; Magnelli et al. 1994). Interestingly, Bcl-2 inhibited apoptosis in cardiomyocytes exposed to ceramide and staurosporine, which caused depolarization of the mitochondrial membrane and cytochrome c release (Pacher and Hajnoczky 2001). A more recent study suggests that Bcl-2 inhibits calcium release through L-type channels, thereby preventing mitochondrial calcium uptake (Diaz-Prieto et al. 2008). Finally, Mcl-1, an anti-apoptotic protein overexpressed in myeloid and lymphoid leukemia, also blocks calcium redistribution following exposure to staurosporine (Minagawa et al. 2005).

Bcl-2 Regulation of Luminal Calcium

In contrast to experiments reporting that Bcl-2 inhibits IP$_3$R opening, there is also substantial evidence that Bcl-2 and Bcl-xL directly regulate the concentration of luminal calcium (Fig. 3). Initial studies showed that Bcl-2 increased membrane permeability, thereby resulting in calcium leak and decreased signaling in response to ATP (Foyouzi-Youssefi et al. 2000; Pinton et al. 2000). Another study documented that knocking down Bcl-2 prevented the loss of ER luminal calcium (Oakes et al. 2005). While these studies also cite the significance of the Bcl-2-IP$_3$R interaction in regulating the calcium pool, others have found that Bcl-2 depletes luminal calcium by interacting with SERCA (Dremina et al. 2004; Dremina et al. 2006; Vanden Abeele et al. 2002). In spite of these differences in experimental findings, it cannot be refuted that Bcl-2 localizes to the ER to inhibit calcium signaling. Thus, mechanistic differences may be attributed to cell type in which the relative expression and localization of Bcl-2 family members are considerably distinct. Accordingly, the effect of luminal calcium is dependent upon the predominant IP$_3$R isoform expressed, yet not necessary for Bcl-2 or Bcl-xL mediated effects on calcium signaling (Li et al. 2007).

Bcl-2 Regulation of Prosurvival Calcium Signals

Although Bcl-2 inhibits calcium transients that are associated with apoptosis, anti-apoptotic Bcl-2 family members can also enhance calcium oscillations that promote survival (Fig. 3). In T cells, Bcl-2 enhances calcium oscillations induced by weak T cell receptor stimulation and thus increases NFAT activation (Zhong et al. 2006). Similarly, Bcl-xL enhances prosurvival oscillations following weak ligand stimulation (Distelhorst and Zhong, unpubl). In DT40 B lymphocytes, Bcl-2, Bcl-xL, and Mcl-1 enhance calcium oscillations by sensitizing cells to lower concentrations of IP_3 (Eckenrode et al. 2010; Li et al. 2007; White et al. 2005). This in turn leads to accelerated mitochondrial metabolism and cell survival. Additional studies in other cell types support this hypothesis by demonstrating that overexpression of Bcl-2 enhances calcium oscillations in epithelial cells following stimulation with ATP and also in neuronal cells to facilitate survival (Jiao et al. 2005; Palmer et al. 2004).

APOPTOSIS REGULATION BY PROAPOPTOTIC Bcl-2 FAMILY PROTEINS

In contrast to Bcl-2, proapoptotic Bcl-2 family proteins are missing a classic BH4 domain, although recent findings suggest the presence of consensus BH4 sequence in proapoptotic family members (Chipuk et al. 2010). Multidomain BH proteins (i.e., BH1-3) include Bax and Bak, both of which are essential for apoptosis driven by the mitochondrial pathway. Multiple studies using Bax/Bak knockout models have demonstrated that the loss of these proteins confers resistance to numerous apoptotic stimuli (Wei et al. 2001).

Calcium Regulation by Bax and Bak

Like Bcl-2 and Bcl-xL, Bax and Bak also localize to the ER where they regulate calcium homeostasis. In Bax/Bak knockout cells, ER luminal calcium is decreased compared to wild type cells, which compromises ER calcium release as well as mitochondrial uptake (Oakes et al. 2005; Scorrano et al. 2003). Consistent with these observations, the Bax Inhibitor-1 protein facilitates ER calcium leak, depleting the available calcium pool (Chae et al. 2004; Kim et al. 2008). Although the mechanism by which Bax and Bak decrease luminal calcium has not been determined, it is generally inferred that Bax/Bak ordinarily prevents Bcl-2-mediated ER calcium leak, and thus their deficiency promotes depletion of ER luminal calcium by Bcl-2. Moreover, Bax alone is required for calcium elevation in response to cytotoxic agents, such as staurosporine (Nutt et al. 2002a). For example, reconstitution of Bax in a prostate cancer cell line augmented cytosolic calcium elevation and restored mitochondrial uptake. The same group also reported that Bax/Bak overexpression induced calcium elevation followed by cytochrome c release and apoptosis (Nutt et al. 2002b). Interestingly, when Bak is targeted to the ER, it facilitates cytosolic calcium elevation and activates caspase-12; yet this does not occur when Bak is specifically targeted to mitochondria (Zong et al. 2003). In addition, ER-targeted Bak requires calcium in conjunction with ER stress for apoptosis to occur (Klee et al. 2009). This suggests that the localization of Bax and/or Bak may determine which effector pathway is induced. It is possible that Bax/Bak localization to the ER is favored when cells undergo apoptosis induced by ER stress.

Calcium Regulation by BH3-only Proteins

Other proapoptotic Bcl-2 family members have only the BH3 domain and therefore are designated as "BH3-only" proteins. These include Bim, Bad, Bik, BNIP3, PUMA, and NOXA (Cory and Adams 2002). Like their multidomain counterparts, these proteins also localize to both ER and mitochondria. At the mitochondria, BH3-only proteins facilitate Bax/Bak oligomerization by two potential mechanisms. In brief, one model suggests they function to directly activate Bax and Bak, whereas another suggests they do so indirectly by sequestering anti-apoptotic Bcl-2 family members (Cheng et al. 2001; Willis et al. 2007). Although the mechanism remains controversial, it is now certain that BH3-only proteins are necessary for Bax/Bak activation and the induction of apoptosis.

While the role of these proteins at the mitochondria has been extensively studied, there are sufficient data to conclude that they also regulate apoptosis at the ER. For example, Bim translocates to the ER following ER stress and may be required for the activation of caspase-12 (Morishima et al. 2004). Thus, ER localization of Bim as well as Bak may be necessary for apoptosis in response to ER stress-inducing agents. Interestingly, dexamethasone induces Bim transcript and protein levels in T cells (Lu et al. 2006; Wang et al. 2003), a process that is associated with the elevation of cytosolic calcium and required for a robust apoptotic response. Engagement of T-cell receptors also stimulates de novo transcription of Bim by a calcium-dependent mechanism (Cante-Barrett et al. 2006). It has yet to be determined if this process contributes to apoptosis induced by T-cell receptor activation. Intriguingly, T cells deficient in Bim have impaired calcium release following their stimulation, and this is associated with increased binding of Bcl-2 to IP_3Rs (Ludwinski et al. 2009). Thus, another role of Bim may be to enhance calcium elevation by sequestering Bcl-2 away from IP_3Rs at the ER membrane.

Bik is a BH3-only protein that promotes ER calcium depletion in a Bax/Bak dependent manner (Mathai et al. 2005). Similarly, BNIP3 causes a leak of ER luminal calcium when selectively targeted to the ER (Zhang et al. 2009a). NOXA and PUMA are both p53 target genes that are up-regulated in response to genotoxic stress. A study by Shore and colleagues has shown that NOXA may function cooperatively with Bik to promote the activation of Bax and Bak (Germain et al. 2005). A recent study indicates that the mitochondrial targeting region of NOXA functions to increase mitochondrial permeability and release calcium. Interestingly, a peptide corresponding to this region was able to induce calcium-dependent cell death by necrosis (Seo et al. 2009). This demonstrates that BH3-only proteins are not only important for regulating calcium flux and homeostasis, but may also function to regulate other mechanisms of cell death. Collectively, the Bcl-2 family makes up a network of proteins and each contributes to the regulation of normal calcium homeostasis. It is clear that alterations in expression or localization of these proteins can have profound effects on cell viability by inducing apoptosis.

AUTOPHAGY

Autophagy is a process of self-eating whereby cellular organelles and proteins are phagocytosed in order to produce energy during metabolic stress (Levine and Klionsky 2004). It is an evolutionarily conserved physiological process that is thought to promote cell survival. Some cellular contexts in which autophagy may be induced include nutrient deprivation, hypoxia, ER stress, abnormal cell growth, and microbial infection (Mizushima et al. 2008). On the other hand, autophagy has also been shown to promote cell death under certain conditions and stimuli. In fact, autophagy is often referred to as type II programmed cell death (distinct from type I programmed cell death) because it does not require caspase activation or DNA fragmentation, which are classical characteristics of apoptosis (Levine and Yuan 2005). However, it is likely that both processes occur simultaneously, and thus, it is important to understand the signaling pathways that govern autophagy, especially when considering that many of the same mechanisms regulate apoptosis.

During autophagy, double membrane vesicles, or autophagosomes, fuse with the lysosome, leading to the degradation of cellular proteins (Mizushima 2007). An example of an autophagosome is illustrated by the electron micrograph in Figure 4; shown are autophagosomes from a WEHI 7.2 T cell stably expressing Bcl-2 and cultured in normal growth media containing dexamethasone. Glucocorticoids promote autophagy (Laane et al. 2009; Swerdlow et al. 2008), and this process is most evident when apoptosis is inhibited by Bcl-2 (Swerdlow et al. 2008).

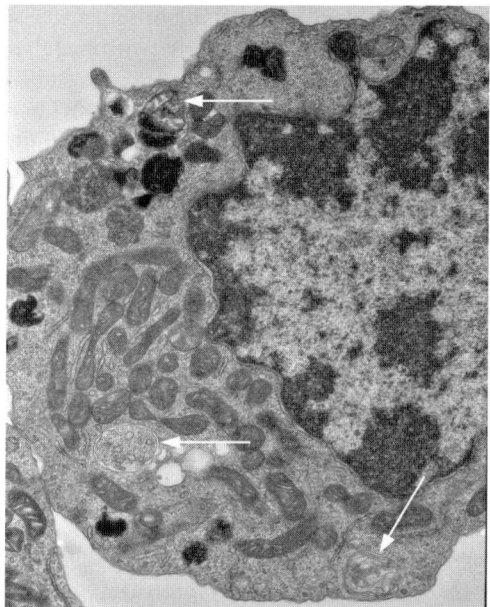

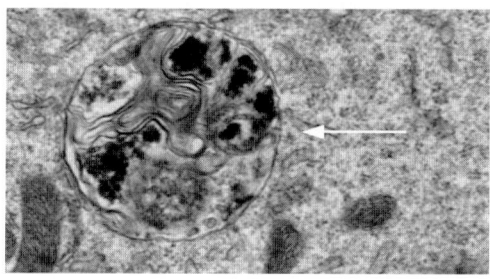

Figure 4. Electron micrograph of a malignant T cell undergoing autophagy. T cells that ectopically express Bcl-2 readily undergo autophagy in the presence of glucocorticoid hormones. Here, WEHI7.2 T cells stably expressing Bcl-2 (to inhibit apoptosis) were treated with 10^{-6} M dexamethasone for 72 hours and visualized by electron microscopy. (*A*) An electron micrograph of a single cell or (*B*) a region from within a cell. Examples of autophagosomes are shown next to the arrows.

Methods for the detection of autophagy are very well described by Klionsky and colleagues (Klionsky et al. 2008). Briefly, the most common method of analysis is the assessment of microtubule-associated protein-1 light chain-3 (LC3) by immunoblotting. LC3 is considered a marker for autophagy when it is proteolytically processed and conjugated to phosphatidylethanolamine (LC3II). GFP-tagged LC3 is commonly transfected into cells to detect LC3 aggregates or punctate GFP-LC3II, which is indicative of autophagosome formation. A second marker of autophagy is p62 degradation. The signaling adaptor p62, which is implicated in the activation of the transcription factor NFκB, is rapidly degraded by autolysosomes. Thus, p62 expression is inversely correlated with the conversion of LC3I to LC3II. Finally, as previously stated and shown in Figure 4, autophagosome formation can be readily visualized by electron microscopy, although this method is less amenable than others to quantitative interpretation.

Autophagy Induction by ER Stress Pathways

The link between calcium and autophagy was initially discovered by several groups reporting that autophagy could be provoked by ER stress (Bernales et al. 2006; Ogata et al. 2006; Yorimitsu et al. 2006). For example, both thapsigargin and tunicamycin stimulate autophagy (Ogata et al. 2006). ER stress affects Autophagy-Related Genes (*atg*), which are evolutionarily conserved and indispensable for autophagy in many cell systems. In yeast, the transcription factor Hac-1 (an ortholog of the ER-stress mammalian XBP-1) transactivates *atg8* during the unfolded protein response (Bernales et al. 2006). Other studies have shown that mutations in ER stress-related proteins such as PERK or EIF2α can inhibit autophagy (Kouroku et al. 2007). Further, knockout of several *atg* genes prevents autophagy-mediated survival in the presence of tunicamycin (Ogata et al. 2006). While these data indisputably demonstrate that ER stress induces autophagy, a direct role for calcium had not been implicated at this point in time.

Autophagy Regulation by Calcium Signaling

Direct evidence that calcium signaling stimulates autophagy was first reported by Jaattela and colleagues (Hoyer-Hansen et al. 2007). They demonstrated that ER calcium mobilization induces autophagy when stimulated by agents such as vitamin D, ionomycin, and thapsigargin. Moreover, GFP-LC3 aggregates were inhibited with BAPTA-AM, suggesting that autophagosome formation was calcium dependent. They provided further evidence that autophagy occurred by the calcium-dependent activation of AMP activated protein kinase, which required upstream activation of the calcium/calmodulin kinase kinase β. AMPK is activated during nutrient deprivation to inhibit activity of the target of rapamycin (mTOR), a negative regulator of autophagy (Hoyer-Hansen and Jaattela 2007). Further evidence supporting a direct role for calcium in the induction of autophagy was the finding that calcium phosphate precipitates could induce autophagy when transfected into HEK293 cells (Gao et al. 2008). Importantly, autophagy mediated by calcium phosphate was also Beclin-dependent. Beclin is a newly discovered BH3-only protein that mediates autophagy by forming a complex between the class III PI3 kinase Vps34 and p150, which facilitates assembly of the autophagosome (Sinha and Levine 2008).

Autophagy in the Context of T-cell Activation

As previously described, significant contributions in the calcium field have been made by investigating signal transduction pathways in activated lymphocytes. Accordingly, autophagy may also be important for regulating lymphocyte activation. For example, T cells from *atg5* knockout mice do not proliferate following ligation of the T-cell receptor, nor do they survive in the periphery (Pua et al. 2007), suggesting that autophagy is required for T-cell activation. Interestingly, it was shown that T-cell activation increases autophagy by NFkB-dependent transcription of Beclin-1 (Copetti et al. 2009a; Copetti et al. 2009b). In this study, the authors provide evidence that NFκB directly binds to the Beclin-1 promoter following activation of Jurkat T cells. Although a direct link has not been observed, it is possible that calcium-dependent activation of calcineurin stimulates this process, thus implicating a role for calcium in Beclin-1 transcription and autophagy.

In addition to producing IP_3, T-cell activation also generates reactive oxygen species (Devadas et al. 2002; Hildeman et al. 2003). Thiol groups are found on IP_3Rs and ryanodine receptors, and oxidation of both calcium channels favors their opening (Bootman et al. 1992; Bultynck et al. 2004; Joseph et al. 2006; Sun et al. 2001; Xia et al. 2000). Further, cyclic ADP ribose and NAADP govern redox reactions and are also endogenous ligands for ryanodine receptors and natural 2nd messengers produced by T-cell activation (Fliegert et al. 2007; Guse 2009). These studies have implicated a role for reactive oxygen species in regulating calcium signals. Not surprisingly, reactive oxygen species also contribute to the induction of autophagy. For example, hydrogen peroxide directly facilitates formation of the autophagosome by oxidizing Atg4 (Scherz-Shouval et al. 2007). Another study demonstrated that neurons undergo autophagy when mitochondrial fission is induced by nitric oxide (Barsoum et al. 2006). This observation is attractive in light of the fact that nitric oxide protects cardiomyocytes from apoptosis. Together, these data suggest the possibility that oxidative metabolites function as signaling molecules by activating calcium and triggering autophagy in lymphocytes, although more definitive data is necessary to support this important conclusion.

Autophagy Regulation by IP_3Rs

There is now substantial evidence that autophagy is directly regulated by IP_3 as well as by IP_3Rs. For example, inhibition of inositol monophosphate by agents such as lithium and L-690,330 induced autophagy (Sarkar et al. 2005; Sarkar and Rubinsztein 2006). Kroemer and colleagues first showed that IP_3Rs act as inhibitors of autophagy by demonstrating that knocking down IP_3Rs or inhibiting their channel activity was sufficient to induce conversion of LC3I to LC3II. (Criollo et al. 2007). Moreover, Beclin-1 is in complex with IP_3Rs along with Bcl-2 and perturbation of the Beclin-IP_3R-Bcl-2 interaction with Xestospongin B or RNA interference is sufficient for autophagy to occur (Vicencio et al. 2009). Additionally, phosphorylation of Beclin promotes its dissociation from Bcl-xL in order to induce autophagy (Zalckvar et al. 2009). Interestingly, inhibition of IP$_3$Rs or siRNA knockdown of Beclin-1 did not affect ER luminal calcium as measured in aequorin-expressing cells (Criollo et al. 2007; Vicencio et al. 2009). However, because Xestospongin significantly inhibited calcium responses following stimulation with histamine, it is possible that a lack of calcium release promotes autophagy without altering steady state levels under these conditions. Nevertheless, these findings indicate a role for calcium channels and signaling in the induction of autophagy.

CONCLUDING REMARKS

Calcium is a dynamic signal transducing ion that is absolutely required for life. Slight alterations in the frequency and/or amplitude of a calcium signal can lead to apoptosis or autophagy by numerous mechanisms. In addition, abnormal signaling not only alters calcium homeostasis in cells, but may contribute to several pathogenic states such as cancer, heart failure, diabetes, and Alzheimer's disease (Berridge 2003; Berridge 2010; Huang et al. 2010; Luciani et al. 2009; Roderick and Cook 2008). It is therefore essential to understand the pathways by which calcium regulates life and death decisions, as they will not only provide insight into normal cell physiology, but may also facilitate the development of novel targets and treatments for chronic diseases.

ACKNOWLEDGMENTS

We thank Karen McColl for editing and proofreading this manuscript as well as Hisashi Fujioka of the electron microscopy core facility at Case Western Reserve University.

REFERENCES

Altamirano F, Oyarce C, Silva P, Toyos M, Wilson C, Lavandero S, Uhlen P, Estrada M. 2009. Testosterone induces cardiomyocyte hypertrophy through mammalian target of rapamycin complex 1 pathway. *J Endocrinol* **202:** 299–307.

An J, Chen Y, Huang Z. 2004. Critical upstream signals of cytochrome C release induced by a novel Bcl-2 inhibitor. *J Biol Chem* **279:** 19133–19140.

Andreka P, Nadhazi Z, Muzes G, Szantho G, Vandor L, Konya L, Turner MS, Tulassay Z, Bishopric NH. 2004. Possible therapeutic targets in cardiac myocyte apoptosis. *Curr Pharm Des* **10:** 2445–2461.

Annis MG, Zamzami N, Zhu W, Penn LZ, Kroemer G, Leber B, Andrews DW. 2001. Endoplasmic reticulum localized Bcl-2 prevents apoptosis when redistribution of cytochrome c is a late event. *Oncogene* **20:** 1939–1952.

Aranguiz-Urroz P, Soto D, Contreras A, Troncoso R, Chiong M, Montenegro J, Venegas D, Smolic C, Ayala P, Thomas WG, et al. 2009. Differential participation of angiotensin II type 1 and 2 receptors in the regulation of cardiac cell death triggered by angiotensin II. *Am J Hypertens* **22:** 569–576.

Ashwell JD, Lu FW, Vacchio MS. 2000. Glucocorticoids in T cell development and function. *Annu Rev Immunol* **18:** 309–345.

Assefa Z, Bultynck G, Szlufcik K, Nadif Kasri N, Vermassen E, Goris J, Missiaen L, Callewaert G, Parys JB, De Smedt H. 2004. Caspase-3-induced truncation of type 1 inositol trisphosphate receptor accelerates apoptotic cell death and induces inositol trisphosphate-independent calcium release during apoptosis. *J Biol Chem* **279:** 43227–43236.

Baffy G, Miyashita T, Williamson JR, Reed JC. 1993. Apoptosis induced by withdrawal of interleukin-3 (IL-3) from an IL-3-dependent hematopoietic cell line is associated with repartitioning of intracellular calcium and is blocked by enforced Bcl-2 oncoprotein production. *J Biol Chem* **268:** 6511–6519.

Barouch LA, Harrison RW, Skaf MW, Rosas GO, Cappola TP, Kobeissi ZA, Hobai IA, Lemmon CA, Burnett AL, O'Rourke B, et al. 2002. Nitric oxide regulates the heart by spatial confinement of nitric oxide synthase isoforms. *Nature* **416:** 337–339.

Barsoum MJ, Yuan H, Gerencser AA, Liot G, Kushnareva Y, Graber S, Kovacs I, Lee WD, Waggoner J, Cui J, et al. 2006. Nitric oxide-induced mitochondrial fission is regulated by dynamin-related GTPases in neurons. *Embo J* **25:** 3900–3911.

Baus E, Andris F, Dubois PM, Urbain J, Leo O. 1996. Dexamethasone inhibits the early steps of antigen receptor signaling in activated T lymphocytes. *J Immunol* **156:** 4555–4561.

Benten WP, Lieberherr M, Sekeris CE, Wunderlich F. 1997. Testosterone induces Ca^{2+} influx via non-genomic surface receptors in activated T cells. *FEBS Lett* **407:** 211–214.

Berg I, Potter BV, Mayr GW, Guse AH. 2000. Nicotinic acid adenine dinucleotide phosphate (NAADP(+)) is an essential regulator of T-lymphocyte Ca^{2+}-signaling. *J Cell Biol* **150:** 581–588.

Bernales S, McDonald KL, Walter P. 2006. Autophagy counterbalances endoplasmic reticulum expansion during the unfolded protein response. *PLoS Biol* **4:** e423.

Berridge MJ. 1997. Lymphocyte activation in health and disease. *Crit Rev Immunol* **17:** 155–178.

Berridge MJ. 2003. Cardiac calcium signalling. *Biochem Soc Trans* **31:** 930–933.

Berridge MJ. 2010. Calcium hypothesis of Alzheimer's disease. *Pflugers Arch* **459:** 441–449.

Bian X, Hughes FM Jr, Huang Y, Cidlowski JA, Putney JW Jr. 1997. Roles of cytoplasmic Ca^{2+} and intracellular Ca^{2+} stores in induction and suppression of apoptosis in S49 cells. *Am J Physiol* **272:** C1241–1249.

Bootman MD, Taylor CW, Berridge MJ. 1992. The thiol reagent, thimerosal, evokes Ca^{2+} spikes in HeLa cells by sensitizing the inositol 1,4,5-trisphosphate receptor. *J Biol Chem* **267:** 25113–25119.

Brooks KJ, Hargreaves IP, Bates TE. 2000. Nitric-oxide-induced inhibition of mitochondrial complexes following aglycaemic hypoxia in neonatal cortical rat brain slices. *Dev Neurosci* **22:** 359–365.

Bultynck G, Szlufcik K, Kasri NN, Assefa Z, Callewaert G, Missiaen L, Parys JB, De Smedt H. 2004. Thimerosal stimulates Ca^{2+} flux through inositol 1,4,5-trisphosphate receptor type 1, but not type 3, via modulation of an isoform-specific Ca^{2+}-dependent intramolecular interaction. *Biochem J* **381:** 87–96.

Cameron AM, Steiner JP, Roskams AJ, Ali SM, Ronnett GV, Snyder SH. 1995. Calcineurin associated with the inositol 1,4,5-trisphosphate receptor-FKBP12 complex modulates Ca^{2+} flux. *Cell* **83:** 463–472.

Cante-Barrett K, Gallo EM, Winslow MM, Crabtree GR. 2006. Thymocyte negative selection is mediated by protein kinase C- and Ca^{2+}-dependent transcriptional induction of bim. *J Immunol* **176:** 2299–2306.

Chae HJ, Kim HR, Xu C, Bailly-Maitre B, Krajewska M, Krajewski S, Banares S, Cui J, Digicaylioglu M, Ke N, et al. 2004. BI-1 regulates an apoptosis pathway linked to endoplasmic reticulum stress. *Mol Cell* **15:** 355–366.

Chai J, Xiong Q, Zhang P, Zheng R, Peng J, Jiang S. 2009. Induction of Ca^{2+} signal mediated apoptosis and alteration of IP$_3$R1 and SERCA1 expression levels by stress hormone in differentiating C2C12 myoblasts. *Gen Comp Endocrinol* **166:** 241–249.

Chen R, Valencia I, Zhong F, McColl KS, Roderick HL, Bootman MD, Berridge MJ, Conway SJ, Holmes AB, Mignery GA, et al. 2004. Bcl-2 functionally interacts with inositol 1,4,5-trisphosphate receptors to regulate calcium release from the ER in response to inositol 1,4,5-trisphosphate. *J Cell Biol* **166:** 193–203.

Cheng EH, Wei MC, Weiler S, Flavell RA, Mak TW, Lindsten T, Korsmeyer SJ. 2001. BCL-2, BCL-X(L) sequester BH3 domain-only molecules preventing BAX- and BAK-mediated mitochondrial apoptosis. *Mol Cell* **8:** 705–711.

Chipuk JE, Moldoveanu T, Llambi F, Parsons MJ, Green DR. 2010. The BCL-2 family reunion. *Mol Cell* **37:** 299–310.

Cigola E, Kajstura J, Li B, Meggs LG, Anversa P. 1997. Angiotensin II activates programmed myocyte cell death in vitro. *Exp Cell Res* **231:** 363–371.

Cohen JJ, Duke RC. 1984. Glucocorticoid activation of a calcium-dependent endonuclease in thymocyte nuclei leads to cell death. *J Immunol* **132:** 38–42.

Copetti T, Bertoli C, Dalla E, Demarchi F, Schneider C. 2009a. p65/RelA modulates BECN1 transcription and autophagy. *Mol Cell Biol* **29:** 2594–2608.

Copetti T, Demarchi F, Schneider C. 2009b. p65/RelA binds and activates the beclin 1 promoter. *Autophagy* **5:** 858–859.

Cory S, Adams JM. 2002. The Bcl2 family: regulators of the cellular life-or-death switch. *Nat Rev Cancer* **2:** 647–656.

Criollo A, Maiuri MC, Tasdemir E, Vitale I, Fiebig AA, Andrews D, Molgo J, Diaz J, Lavandero S, Harper F, et al. 2007. Regulation of autophagy by the inositol trisphosphate receptor. *Cell Death Differ* **14:** 1029–1039.

Dammermann W, Zhang B, Nebel M, Cordiglieri C, Odoardi F, Kirchberger T, Kawakami N, Dowden J, Schmid F, Dornmair K, et al. 2009. NAADP-mediated Ca^{2+} signaling via type 1 ryanodine receptor in T cells revealed by a synthetic NAADP antagonist. *Proc Natl Acad Sci U S A* **106:** 10678–10683.

Davidson SM, Duchen MR. 2006. Calcium microdomains and oxidative stress. *Cell Calcium* **40:** 561–574.

Davis MC, McColl KS, Zhong F, Wang Z, Malone MH, Distelhorst CW. 2008. Dexamethasone-induced inositol 1,4,5-trisphosphate receptor elevation in murine lymphoma cells is not required for dexamethasone-mediated calcium elevation and apoptosis. *J Biol Chem* **283:** 10357–10365.

Dedkova EN, Blatter LA. 2005. Modulation of mitochondrial Ca^{2+} by nitric oxide in cultured bovine vascular endothelial cells. *Am J Physiol Cell Physiol* **289:** C836–845.

DeSouza N, Reiken S, Ondrias K, Yang YM, Matkovich S, Marks AR. 2002. Protein kinase A and two phosphatases are components of the inositol 1,4,5-trisphosphate receptor macromolecular signaling complex. *J Biol Chem* **277:** 39397–39400.

Devadas S, Zaritskaya L, Rhee SG, Oberley L, Williams MS. 2002. Discrete generation of superoxide and hydrogen peroxide by T cell receptor stimulation: selective regulation of mitogen-activated protein kinase activation and fas ligand expression. *J Exp Med* **195:** 59–70.

Diaz-Prieto N, Herrera-Peco I, de Diego AM, Ruiz-Nuno A, Gallego-Sandin S, Lopez MG, Garcia AG, Cano-Abad MF. 2008. Bcl-2 mitigates Ca^{2+} entry and mitochondrial Ca^{2+} overload through downregulation of L-type Ca^{2+} channels in PC12 cells. *Cell Calcium* **44:** 339–352.

Dimmeler S, Haendeler J, Nehls M, Zeiher AM. 1997. Suppression of apoptosis by nitric oxide via inhibition of interleukin-1beta-converting enzyme (ICE)-like and cysteine protease protein (CPP)-32-like proteases. *J Exp Med* **185:** 601–607.

Dremina ES, Sharov VS, Kumar K, Zaidi A, Michaelis EK, Schoneich C. 2004. Anti-apoptotic protein Bcl-2 interacts with and destabilizes the sarcoplasmic/endoplasmic reticulum Ca^{2+}-ATPase (SERCA). *Biochem J* **383:** 361–370.

Dremina ES, Sharov VS, Schoneich C. 2006. Displacement of SERCA from SR lipid caveolae-related domains by Bcl-2: a possible mechanism for SERCA inactivation. *Biochemistry* **45:** 175–184.

Eckenrode EF, Yang J, Velmurugan GV, Foskett JK, White C. 2010. Apoptosis protection by Mcl-1 and Bcl-2 modulation of inositol 1,4,5-trisphosphate receptor-dependent Ca^{2+} signaling. *J Biol Chem* **285:** 13678–13684.

Erin N, Bronson SK, Billingsley ML. 2003. Calcium-dependent interaction of calcineurin with Bcl-2 in neuronal tissue. *Neuroscience* **117:** 541–555.

Estrada M, Varshney A, Ehrlich BE. 2006. Elevated testosterone induces apoptosis in neuronal cells. *J Biol Chem* **281**: 25492–25501.

Fliegert R, Gasser A, Guse AH. 2007. Regulation of calcium signalling by adenine-based second messengers. *Biochem Soc Trans* **35**: 109–114.

Foyouzi-Youssefi R, Arnaudeau S, Borner C, Kelley WL, Tschopp J, Lew DP, Demaurex N, Krause KH. 2000. Bcl-2 decreases the free Ca^{2+} concentration within the endoplasmic reticulum. *Proc Natl Acad Sci U S A* **97**: 5723–5728.

Galione A, Petersen OH. 2005. The NAADP receptor: new receptors or new regulation? *Mol Interv* **5**: 73–79.

Gallo EM, Cante-Barrett K, Crabtree GR. 2006. Lymphocyte calcium signaling from membrane to nucleus. *Nat Immunol* **7**: 25–32.

Gao W, Ding WX, Stolz DB, Yin XM. 2008. Induction of macroautophagy by exogenously introduced calcium. *Autophagy* **4**: 754–761.

Germain M, Mathai JP, McBride HM, Shore GC. 2005. Endoplasmic reticulum BIK initiates DRP1-regulated remodelling of mitochondrial cristae during apoptosis. *Embo J* **24**: 1546–1556.

Goldenberg I, Grossman E, Jacobson KA, Shneyvays V, Shainberg A. 2001. Angiotensin II-induced apoptosis in rat cardiomyocyte culture: a possible role of AT1 and AT2 receptors. *J Hypertens* **19**: 1681–1689.

Gradman AH. 2009. Role of angiotensin II type 1 receptor antagonists in the treatment of hypertension in patients aged >or=65 years. *Drugs Aging* **26**: 751–767.

Green DR. 2003. Overview: apoptotic signaling pathways in the immune system. *Immunol Rev* **193**: 5–9.

Green DR. 2005. Apoptotic pathways: ten minutes to dead. *Cell* **121**: 671–674.

Guse AH. 2009. Second messenger signaling: multiple receptors for NAADP. *Curr Biol* **19**: R521–523.

Guse AH, da Silva CP, Berg I, Skapenko AL, Weber K, Heyer P, Hohenegger M, Ashamu GA, Schulze-Koops H, Potter BV, et al. 1999. Regulation of calcium signalling in T lymphocytes by the second messenger cyclic ADP-ribose. *Nature* **398**: 70–73.

Hajnoczky G, Csordas G, Das S, Garcia-Perez C, Saotome M, Sinha Roy S, Yi M. 2006. Mitochondrial calcium signalling and cell death: approaches for assessing the role of mitochondrial Ca^{2+} uptake in apoptosis. *Cell Calcium* **40**: 553–560.

Hanson CJ, Bootman MD, Distelhorst CW, Wojcikiewicz RJ, Roderick HL. 2008. Bcl-2 suppresses Ca^{2+} release through inositol 1,4,5-trisphosphate receptors and inhibits Ca^{2+} uptake by mitochondria without affecting ER calcium store content. *Cell Calcium* **44**: 324–338.

Hanson CJ, Bootman MD, Roderick HL. 2004. Cell signalling: IP_3 receptors channel calcium into cell death. *Curr Biol* **14**: R933–935.

Harr MW, Rong Y, Bootman MD, Roderick HL, Distelhorst CW. 2009. Glucocorticoid-mediated inhibition of Lck modulates the pattern of T cell receptor-induced calcium signals by down-regulating inositol 1,4,5-trisphosphate receptors. *J Biol Chem* **284**: 31860–31871.

Herold MJ, McPherson KG, Reichardt HM. 2006. Glucocorticoids in T cell apoptosis and function. *Cell Mol Life Sci* **63**: 60–72.

Hildeman DA, Mitchell T, Kappler J, Marrack P. 2003. T cell apoptosis and reactive oxygen species. *J Clin Invest* **111**: 575–581.

Hirota J, Furuichi T, Mikoshiba K. 1999. Inositol 1,4,5-trisphosphate receptor type 1 is a substrate for caspase-3 and is cleaved during apoptosis in a caspase-3-dependent manner. *J Biol Chem* **274**: 34433–34437.

Hogquist KA. 2001. Signal strength in thymic selection and lineage commitment. *Curr Opin Immunol* **13**: 225–231.

Hoyer-Hansen M, Bastholm L, Szyniarowski P, Campanella M, Szabadkai G, Farkas T, Bianchi K, Fehrenbacher N, Elling F, Rizzuto R, et al. 2007. Control of macroautophagy by calcium, calmodulin-dependent kinase kinase-beta, and Bcl-2. *Mol Cell* **25**: 193–205.

Hoyer-Hansen M, Jaattela M. 2007. AMP-activated protein kinase: a universal regulator of autophagy? *Autophagy* **3**: 381–383.

Huang CJ, Gurlo T, Haataja L, Costes S, Daval M, Ryazantsev S, Wu X, Butler AE, Butler PC. 2010. Calcium-activated calpain-2 is a mediator of beta cell dysfunction and apoptosis in type 2 diabetes. *J Biol Chem* **285**: 339–348.

Jamieson CA, Yamamoto KR. 2000. Crosstalk pathway for inhibition of glucocorticoid-induced apoptosis by T cell receptor signaling. *Proc Natl Acad Sci U S A* **97**: 7319–7324.

Jayaraman T, Marks AR. 1997. T cells deficient in inositol 1,4,5-trisphosphate receptor are resistant to apoptosis. *Mol Cell Biol* **17**: 3005–3012.

Jiao J, Huang X, Feit-Leithman RA, Neve RL, Snider W, Dartt DA, Chen DF. 2005. Bcl-2 enhances Ca^{2+} signaling to support the intrinsic regenerative capacity of CNS axons. *Embo J* **24**: 1068–1078.

Joseph SK, Nakao SK, Sukumvanich S. 2006. Reactivity of free thiol groups in type-I inositol trisphosphate receptors. *Biochem J* **393**: 575–582.

Kaiser N, Edelman IS. 1977. Calcium dependence of glucocorticoid-induced lymphocytolysis. *Proc Natl Acad Sci U S A* **74**: 638–642.

Kajstura J, Cigola E, Malhotra A, Li P, Cheng W, Meggs LG, Anversa P. 1997. Angiotensin II induces apoptosis of adult ventricular myocytes in vitro. *J Mol Cell Cardiol* **29**: 859–870.

Khan SA, Hare JM. 2003. The role of nitric oxide in the physiological regulation of Ca^{2+} cycling. *Curr Opin Drug Discov Devel* **6**: 658–666.

Kim HR, Lee GH, Ha KC, Ahn T, Moon JY, Lee BJ, Cho SG, Kim S, Seo YR, Shin YJ, et al. 2008. Bax Inhibitor-1 Is a pH-dependent regulator of Ca^{2+} channel activity in the endoplasmic reticulum. *J Biol Chem* **283**: 15946–15955.

Klee M, Pallauf K, Alcala S, Fleischer A, Pimentel-Muinos FX. 2009. Mitochondrial apoptosis induced by BH3-only molecules in the exclusive presence of endoplasmic reticular Bak. *Embo J* **28**: 1757–1768.

Klionsky DJ, Abeliovich H, Agostinis P, Agrawal DK, Aliev G, Askew DS, Baba M, Baehrecke EH, Bahr BA, Ballabio A, et al. 2008. Guidelines for the use and interpretation of assays for monitoring autophagy in higher eukaryotes. *Autophagy* **4**: 151–175.

Kockskamper J, Zima AV, Roderick HL, Pieske B, Blatter LA, Bootman MD. 2008. Emerging roles of inositol 1,4,5-trisphosphate signaling in cardiac myocytes. *J Mol Cell Cardiol* **45:** 128–147.

Kouroku Y, Fujita E, Tanida I, Ueno T, Isoai A, Kumagai H, Ogawa S, Kaufman RJ, Kominami E, Momoi T. 2007. ER stress (PERK/eIF2alpha phosphorylation) mediates the polyglutamine-induced LC3 conversion, an essential step for autophagy formation. *Cell Death Differ* **14:** 230–239.

Kruman II, Mattson MP. 1999. Pivotal role of mitochondrial calcium uptake in neural cell apoptosis and necrosis. *J Neurochem* **72:** 529–540.

Laane E, Tamm KP, Buentke E, Ito K, Kharaziha P, Oscarsson J, Corcoran M, Bjorklund AC, Hultenby K, Lundin J, et al. 2009. Cell death induced by dexamethasone in lymphoid leukemia is mediated through initiation of autophagy. *Cell Death Differ* **16:** 1018–1029.

Lam M, Dubyak G, Chen L, Nunez G, Miesfeld RL, Distelhorst CW. 1994. Evidence that BCL-2 represses apoptosis by regulating endoplasmic reticulum-associated Ca^{2+} fluxes. *Proc Natl Acad Sci U S A* **91:** 6569–6573.

Lam M, Dubyak G, Distelhorst CW. 1993. Effect of glucocorticosteroid treatment on intracellular calcium homeostasis in mouse lymphoma cells. *Mol Endocrinol* **7:** 686–693.

Latour S, Veillette A. 2001. Proximal protein tyrosine kinases in immunoreceptor signaling. *Curr Opin Immunol* **13:** 299–306.

Levine B, Klionsky DJ. 2004. Development by self-digestion: molecular mechanisms and biological functions of autophagy. *Dev Cell* **6:** 463–477.

Levine B, Yuan J. 2005. Autophagy in cell death: an innocent convict? *J Clin Invest* **115:** 2679–2688.

Lewis RS. 2001. Calcium signaling mechanisms in T lymphocytes. *Annu Rev Immunol* **19:** 497–521.

Li C, Wang X, Vais H, Thompson CB, Foskett JK, White C. 2007. Apoptosis regulation by Bcl-x(L) modulation of mammalian inositol 1,4,5-trisphosphate receptor channel isoform gating. *Proc Natl Acad Sci U S A* **104:** 12565–12570.

Lowenberg M, Tuynman J, Bilderbeek J, Gaber T, Buttgereit F, van Deventer S, Peppelenbosch M, Hommes D. 2005. Rapid immunosuppressive effects of glucocorticoids mediated through Lck and Fyn. *Blood* **106:** 1703–1710.

Lu J, Quearry B, Harada H. 2006. p38-MAP kinase activation followed by BIM induction is essential for glucocorticoid-induced apoptosis in lymphoblastic leukemia cells. *FEBS Lett* **580:** 3539–3544.

Luciani DS, Gwiazda KS, Yang TL, Kalynyak TB, Bychkivska Y, Frey MH, Jeffrey KD, Sampaio AV, Underhill TM, Johnson JD. 2009. Roles of IP_3R and RyR Ca^{2+} channels in endoplasmic reticulum stress and beta-cell death. *Diabetes* **58:** 422–432.

Ludwinski MW, Sun J, Hilliard B, Gong S, Xue F, Carmody RJ, DeVirgiliis J, Chen YH. 2009. Critical roles of Bim in T cell activation and T cell-mediated autoimmune inflammation in mice. *J Clin Invest* **119:** 1706–1713.

Magnelli L, Cinelli M, Turchetti A, Chiarugi VP. 1994. Bcl-2 overexpression abolishes early calcium waving preceding apoptosis in NIH-3T3 murine fibroblasts. *Biochem Biophys Res Commun* **204:** 84–90.

Mandic A, Hansson J, Linder S, Shoshan MC. 2003. Cisplatin induces endoplasmic reticulum stress and nucleus-independent apoptotic signaling. *J Biol Chem* **278:** 9100–9106.

Mariathasan S, Bachmann MF, Bouchard D, Ohteki T, Ohashi PS. 1998. Degree of TCR internalization and Ca^{2+} flux correlates with thymocyte selection. *J Immunol* **161:** 6030–6037.

Massion PB, Feron O, Dessy C, Balligand JL. 2003. Nitric oxide and cardiac function: ten years after, and continuing. *Circ Res* **93:** 388–398.

Mathai JP, Germain M, Shore GC. 2005. BH3-only BIK regulates BAX,BAK-dependent release of Ca^{2+} from endoplasmic reticulum stores and mitochondrial apoptosis during stress-induced cell death. *J Biol Chem* **280:** 23829–23836.

Mattiazzi A. 1997. Positive inotropic effect of angiotensin II. Increases in intracellular Ca^{2+} or changes in myofilament Ca^{2+} responsiveness? *J Pharmacol Toxicol Methods* **37:** 205–214.

McConkey DJ, Nicotera P, Hartzell P, Bellomo G, Wyllie AH, Orrenius S. 1989. Glucocorticoids activate a suicide process in thymocytes through an elevation of cytosolic Ca^{2+} concentration. *Arch Biochem Biophys* **269:** 365–370.

Minagawa N, Kruglov EA, Dranoff JA, Robert ME, Gores GJ, Nathanson MH. 2005. The anti-apoptotic protein Mcl-1 inhibits mitochondrial Ca^{2+} signals. *J Biol Chem* **280:** 33637–33644.

Mizushima N. 2007. Autophagy: process and function. *Genes Dev* **21:** 2861–2873.

Mizushima N, Levine B, Cuervo AM, Klionsky DJ. 2008. Autophagy fights disease through cellular self-digestion. *Nature* **451:** 1069–1075.

Morishima N, Nakanishi K, Takenouchi H, Shibata T, Yasuhiko Y. 2002. An endoplasmic reticulum stress-specific caspase cascade in apoptosis. Cytochrome c-independent activation of caspase-9 by caspase-12. *J Biol Chem* **277:** 34287–34294.

Morishima N, Nakanishi K, Tsuchiya K, Shibata T, Seiwa E. 2004. Translocation of Bim to the endoplasmic reticulum (ER) mediates ER stress signaling for activation of caspase-12 during ER stress-induced apoptosis. *J Biol Chem* **279:** 50375–50381.

Mustelin T, Tasken K. 2003. Positive and negative regulation of T-cell activation through kinases and phosphatases. *Biochem J* **371:** 15–27.

Nakagawa T, Yuan J. 2000. Cross-talk between two cysteine protease families. Activation of caspase-12 by calpain in apoptosis. *J Cell Biol* **150:** 887–894.

Nakagawa T, Zhu H, Morishima N, Li E, Xu J, Yankner BA, Yuan J. 2000. Caspase-12 mediates endoplasmic-reticulum-specific apoptosis and cytotoxicity by amyloid-beta. *Nature* **403:** 98–103.

Norberg E, Gogvadze V, Ott M, Horn M, Uhlen P, Orrenius S, Zhivotovsky B. 2008. An increase in intracellular Ca^{2+} is required for the activation of mitochondrial calpain to release AIF during cell death. *Cell Death Differ* **15:** 1857–1864.

Nutt LK, Chandra J, Pataer A, Fang B, Roth JA, Swisher SG, O'Neil RG, McConkey DJ. 2002a. Bax-mediated Ca^{2+} mobilization promotes cytochrome c release during apoptosis. *J Biol Chem* **277:** 20301–20308.

Nutt LK, Pataer A, Pahler J, Fang B, Roth J, McConkey DJ, Swisher SG. 2002b. Bax and Bak promote apoptosis by modulating endoplasmic reticular and mitochondrial Ca^{2+} stores. *J Biol Chem* **277:** 9219–9225.

Oakes SA, Scorrano L, Opferman JT, Bassik MC, Nishino M, Pozzan T, Korsmeyer SJ. 2005. Proapoptotic BAX and BAK regulate the type 1 inositol trisphosphate receptor and calcium leak from the endoplasmic reticulum. *Proc Natl Acad Sci U S A* **102:** 105–110.

Ogata M, Hino S, Saito A, Morikawa K, Kondo S, Kanemoto S, Murakami T, Taniguchi M, Tanii I, Yoshinaga K, et al. 2006. Autophagy is activated for cell survival after endoplasmic reticulum stress. *Mol Cell Biol* **26:** 9220–9231.

Orrenius S, McConkey DJ, Nicotera P. 1991. Role of calcium in toxic and programmed cell death. *Adv Exp Med Biol* **283:** 419–425.

Pacher P, Hajnoczky G. 2001. Propagation of the apoptotic signal by mitochondrial waves. *Embo J* **20:** 4107–4121.

Palacios EH, Weiss A. 2004. Function of the Src-family kinases, Lck and Fyn, in T-cell development and activation. *Oncogene* **23:** 7990–8000.

Palmer AE, Jin C, Reed JC, Tsien RY. 2004. Bcl-2-mediated alterations in endoplasmic reticulum Ca^{2+} analyzed with an improved genetically encoded fluorescent sensor. *Proc Natl Acad Sci U S A* **101:** 17404–17409.

Palomeque J, Delbridge L, Petroff MV. 2009. Angiotensin II: a regulator of cardiomyocyte function and survival. *Front Biosci* **14:** 5118–5133.

Pinton P, Ferrari D, Magalhaes P, Schulze-Osthoff K, Di Virgilio F, Pozzan T, Rizzuto R. 2000. Reduced loading of intracellular Ca^{2+} stores and downregulation of capacitative Ca^{2+} influx in Bcl-2-overexpressing cells. *J Cell Biol* **148:** 857–862.

Pinton P, Giorgi C, Siviero R, Zecchini E, Rizzuto R. 2008. Calcium and apoptosis: ER-mitochondria Ca^{2+} transfer in the control of apoptosis. *Oncogene* **27:** 6407–6418.

Pua HH, Dzhagalov I, Chuck M, Mizushima N, He YW. 2007. A critical role for the autophagy gene Atg5 in T cell survival and proliferation. *J Exp Med* **204:** 25–31.

Randriamampita C, Trautmann A. 2003. Ca^{2+} signals and T lymphocytes "New mechanisms and functions in Ca^{2+} signaling". *Biol of the Cell* **96:** 69–78.

Rao RV, Castro-Obregon S, Frankowski H, Schuler M, Stoka V, del Rio G, Bredesen DE, Ellerby HM. 2002. Coupling endoplasmic reticulum stress to the cell death program. An Apaf-1-independent intrinsic pathway. *J Biol Chem* **277:** 21836–21842.

Razavi HM, Hamilton JA, Feng Q. 2005. Modulation of apoptosis by nitric oxide: implications in myocardial ischemia and heart failure. *Pharmacol Ther* **106:** 147–162.

Rizzuto R, Pinton P, Ferrari D, Chami M, Szabadkai G, Magalhaes PJ, Di Virgilio F, Pozzan T. 2003. Calcium and apoptosis: facts and hypotheses. *Oncogene* **22:** 8619–8627.

Roderick HL, Cook SJ. 2008. Ca^{2+} signalling checkpoints in cancer: remodelling Ca^{2+} for cancer cell proliferation and survival. *Nat Rev Cancer* **8:** 361–375.

Rong Y, Distelhorst CW. 2007. Bcl-2 Protein Family Members: Versatile Regulators of Calcium Signaling in Cell Survival and Apoptosis. *Annu Rev Physiol* **70:** 73–91.

Rong YP, Aromolaran AS, Bultynck G, Zhong F, Li X, McColl K, Matsuyama S, Herlitze S, Roderick HL, Bootman MD, et al. 2008. Targeting Bcl-2-IP_3 receptor interaction to reverse Bcl-2's inhibition of apoptotic calcium signals. *Mol Cell* **31:** 255–265.

Rong YP, Bultynck G, Aromolaran AS, Zhong F, Parys JB, De Smedt H, Mignery GA, Roderick HL, Bootman MD, Distelhorst CW. 2009. The BH4 domain of Bcl-2 inhibits ER calcium release and apoptosis by binding the regulatory and coupling domain of the IP_3 receptor. *Proc Natl Acad Sci U S A* **106:** 14397–14402.

Sarkar S, Floto RA, Berger Z, Imarisio S, Cordenier A, Pasco M, Cook LJ, Rubinsztein DC. 2005. Lithium induces autophagy by inhibiting inositol monophosphatase. *J Cell Biol* **170:** 1101–1111.

Sarkar S, Rubinsztein DC. 2006. Inositol and IP_3 levels regulate autophagy: biology and therapeutic speculations. *Autophagy* **2:** 132–134.

Scherz-Shouval R, Shvets E, Fass E, Shorer H, Gil L, Elazar Z. 2007. Reactive oxygen species are essential for autophagy and specifically regulate the activity of Atg4. *Embo J* **26:** 1749–1760.

Schrodl K, Oelmez H, Edelmann M, Huber RM, Bergner A. 2009. Altered Ca^{2+}-homeostasis of cisplatin-treated and low level resistant non-small-cell and small-cell lung cancer cells. *Cell Oncol* **31:** 301–315.

Scorrano L, Oakes SA, Opferman JT, Cheng EH, Sorcinelli MD, Pozzan T, Korsmeyer SJ. 2003. BAX and BAK regulation of endoplasmic reticulum Ca^{2+}: a control point for apoptosis. *Science* **300:** 135–139.

Sears CE, Bryant SM, Ashley EA, Lygate CA, Rakovic S, Wallis HL, Neubauer S, Terrar DA, Casadei B. 2003. Cardiac neuronal nitric oxide synthase isoform regulates myocardial contraction and calcium handling. *Circ Res* **92:** e52–59.

Seo YW, Woo HN, Piya S, Moon AR, Oh JW, Yun CW, Kim KK, Min JY, Jeong SY, Chung S, et al. 2009. The cell death-inducing activity of the peptide containing Noxa mitochondrial-targeting domain is associated with calcium release. *Cancer Res* **69:** 8356–8365.

Shibasaki F, Kondo E, Akagi T, McKeon F. 1997. Suppression of signalling through transcription factor NF-AT by interactions between calcineurin and Bcl-2. *Nature* **386:** 728–731.

Sinha S, Levine B. 2008. The autophagy effector Beclin 1: a novel BH3-only protein. *Oncogene* **27:** S137–148

Splettstoesser F, Florea AM, Busselberg D. 2007. IP(3) receptor antagonist, 2-APB, attenuates cisplatin induced Ca^{2+}-influx in HeLa-S3 cells and prevents activation of calpain and induction of apoptosis. *Br J Pharmacol* **151:** 1176–1186.

Starr TK, Jameson SC, Hogquist KA. 2003. Positive and negative selection of T cells. *Annu Rev Immunol* **21:** 139–176.

Sun J, Xu L, Eu JP, Stamler JS, Meissner G. 2001. Classes of thiols that influence the activity of the skeletal muscle calcium release channel. *J Biol Chem* **276:** 15625–15630.

Swerdlow S, McColl K, Rong Y, Lam M, Gupta A, Distelhorst CW. 2008. Apoptosis inhibition by Bcl-2 gives way to autophagy in glucocorticoid-treated lymphocytes. *Autophagy* **4:** 612–620.

Szalai G, Krishnamurthy R, Hajnoczky G. 1999. Apoptosis driven by IP(3)-linked mitochondrial calcium signals. *Embo J* **18:** 6349–6361.

Szegezdi E, Fitzgerald U, Samali A. 2003. Caspase-12 and ER-stress-mediated apoptosis: the story so far. *Ann N Y Acad Sci* **1010:** 186–194.

Szlufcik K, Missiaen L, Parys JB, Callewaert G, De Smedt H. 2006. Uncoupled IP_3 receptor can function as a Ca^{2+}-leak channel: cell biological and pathological consequences. *Biol Cell* **98:** 1–14.

Tolosa E, Ashwell JD. 1999. Thymus-derived glucocorticoids and the regulation of antigen-specific T-cell development. *Neuroimmunomodulation* **6:** 90–96.

Tsujimoto Y, Cossman J, Jaffe E, Croce CM. 1985. Involvement of the bcl-2 gene in human follicular lymphoma. *Science* **228:** 1440–1443.

Tsunoda T, Koga H, Yokomizo A, Tatsugami K, Eto M, Inokuchi J, Hirata A, Masuda K, Okumura K, Naito S. 2005. Inositol 1,4,5-trisphosphate (IP_3) receptor type1 (IP3R1) modulates the acquisition of cisplatin resistance in bladder cancer cell lines. *Oncogene* **24:** 1396–1402.

Van Laethem F, Baus E, Smyth LA, Andris F, Bex F, Urbain J, Kioussis D, Leo O. 2001. Glucocorticoids attenuate T cell receptor signaling. *J Exp Med* **193:** 803–814.

Vanden Abeele F, Skryma R, Shuba Y, Van Coppenolle F, Slomianny C, Roudbaraki M, Mauroy B, Wuytack F, Prevarskaya N. 2002. Bcl-2-dependent modulation of Ca^{2+} homeostasis and store-operated channels in prostate cancer cells. *Cancer Cell* **1:** 169–179.

Verbert L, Lee B, Kocks SL, Assefa Z, Parys JB, Missiaen L, Callewaert G, Fissore RA, De Smedt H, Bultynck G. 2008. Caspase-3-truncated type 1 inositol 1,4,5-trisphosphate receptor enhances intracellular Ca^{2+} leak and disturbs Ca^{2+} signalling. *Biol Cell* **100:** 39–49.

Vicencio JM, Ibarra C, Estrada M, Chiong M, Soto D, Parra V, Diaz-Araya G, Jaimovich E, Lavandero S. 2006. Testosterone induces an intracellular calcium increase by a nongenomic mechanism in cultured rat cardiac myocytes. *Endocrinology* **147:** 1386–1395.

Vicencio JM, Ortiz C, Criollo A, Jones AW, Kepp O, Galluzzi L, Joza N, Vitale I, Morselli E, Tailler M, et al. 2009. The inositol 1,4,5-trisphosphate receptor regulates autophagy through its interaction with Beclin 1. *Cell Death Differ* **16:** 1006–1017.

Wang Z, Malone MH, He H, McColl KS, Distelhorst CW. 2003. Microarray analysis uncovers the induction of the proapoptotic BH3-only protein Bim in multiple models of glucocorticoid-induced apoptosis. *J Biol Chem* **278:** 23861–23867.

Wei MC, Zong WX, Cheng EH, Lindsten T, Panoutsakopoulou V, Ross AJ, Roth KA, MacGregor GR, Thompson CB, Korsmeyer SJ. 2001. Proapoptotic BAX and BAK: a requisite gateway to mitochondrial dysfunction and death. *Science* **292:** 727–730.

White C, Li C, Yang J, Petrenko NB, Madesh M, Thompson CB, Foskett JK. 2005. The endoplasmic reticulum gateway to apoptosis by Bcl-X(L) modulation of the InsP3R. *Nat Cell Biol* **7:** 1021–1028.

Willis SN, Fletcher JI, Kaufmann T, van Delft MF, Chen L, Czabotar PE, Ierino H, Lee EF, Fairlie WD, Bouillet P, et al. 2007. Apoptosis initiated when BH3 ligands engage multiple Bcl-2 homologs, not Bax or Bak. *Science* **315:** 856–859.

Winslow MM, Crabtree GR. 2005. Immunology. Decoding calcium signaling. *Science* **307:** 56–57.

Winslow MM, Neilson JR, Crabtree GR. 2003. Calcium signalling in lymphocytes. *Curr Opin Immunol* **15:** 299–307.

Xia R, Stangler T, Abramson JJ. 2000. Skeletal muscle ryanodine receptor is a redox sensor with a well defined redox potential that is sensitive to channel modulators. *J Biol Chem* **275:** 36556–36561.

Xu KY, Huso DL, Dawson TM, Bredt DS, Becker LC. 1999. Nitric oxide synthase in cardiac sarcoplasmic reticulum. *Proc Natl Acad Sci U S A* **96:** 657–662.

Xu L, Kong D, Zhu L, Zhu W, Andrews DW, Kuo TH. 2007. Suppression of IP_3-mediated calcium release and apoptosis by Bcl-2 involves the participation of protein phosphatase 1. *Mol Cell Biochem* **295:** 153–165.

Yamada T, Horiuchi M, Dzau VJ. 1996. Angiotensin II type 2 receptor mediates programmed cell death. *Proc Natl Acad Sci U S A* **93:** 156–160.

Yorimitsu T, Nair U, Yang Z, Klionsky DJ. 2006. Endoplasmic reticulum stress triggers autophagy. *J Biol Chem* **281:** 30299–30304.

Zalckvar E, Berissi H, Eisenstein M, Kimchi A. 2009. Phosphorylation of Beclin 1 by DAP-kinase promotes autophagy by weakening its interactions with Bcl-2 and Bcl-XL. *Autophagy* **5:** 720–722.

Zhang L, Li L, Liu H, Borowitz JL, Isom GE. 2009a. BNIP3 mediates cell death by different pathways following localization to endoplasmic reticulum and mitochondrion. *Faseb J* **23:** 3405–3414.

Zhang S, Hisatsune C, Matsu-Ura T, Mikoshiba K. 2009b. G-protein-coupled receptor kinase-interacting proteins inhibit apoptosis by inositol 1,4,5-triphosphate receptor-mediated Ca^{2+} signal regulation. *J Biol Chem* **284:** 29158–29169.

Zhong F, Davis MC, McColl KS, Distelhorst CW. 2006. Bcl-2 differentially regulates Ca^{2+} signals according to the strength of T cell receptor activation. *J Cell Biol* **172:** 127–137.

Zong WX, Li C, Hatzivassiliou G, Lindsten T, Yu QC, Yuan J, Thompson CB. 2003. Bax and Bak can localize to the endoplasmic reticulum to initiate apoptosis. *J Cell Biol* **162:** 59–69.

Endoplasmic-Reticulum Calcium Depletion and Disease

Djalila Mekahli[1], Geert Bultynck[1], Jan B. Parys, Humbert De Smedt, and Ludwig Missiaen

Laboratory of Molecular and Cellular Signaling, Department of Molecular Cell Biology, K.U. Leuven, Campus Gasthuisberg, 3000 Leuven, Belgium

Correspondence: Ludwig.Missiaen@med.kuleuven.be

The endoplasmic reticulum (ER) as an intracellular Ca^{2+} store not only sets up cytosolic Ca^{2+} signals, but, among other functions, also assembles and folds newly synthesized proteins. Alterations in ER homeostasis, including severe Ca^{2+} depletion, are an upstream event in the pathophysiology of many diseases. On the one hand, insufficient release of activator Ca^{2+} may no longer sustain essential cell functions. On the other hand, loss of luminal Ca^{2+} causes ER stress and activates an unfolded protein response, which, depending on the duration and severity of the stress, can reestablish normal ER function or lead to cell death. We will review these various diseases by mainly focusing on the mechanisms that cause ER Ca^{2+} depletion.

Cytosolic $[Ca^{2+}]$ ($[Ca^{2+}]_{cyt}$) is precisely regulated in time and space because Ca^{2+} controls essential cell functions like proliferation, differentiation, secretion, contraction, metabolism, trafficking, gene transcription and apoptosis, and in this way controls complex processes like development or learning behavior (Berridge et al. 2000). An abnormal $[Ca^{2+}]_{cyt}$ caused by disturbances of Ca^{2+} channels, Ca^{2+} transporters, Ca^{2+} pumps, and Ca^{2+}-binding proteins can induce multiple pathologies (Missiaen et al. 2000). Ca^{2+} channelopathies in the nervous system leading to paralysis, ataxia, or migraine can be caused by mutations in subunits of voltage-operated Ca^{2+} channels in the plasma membrane (Bidaud et al. 2006; Lorenzon and Beam 2008). Other channelopathies like malignant hyperthermia and central core disease in skeletal muscle, and some tachycardias and tachyarrhythmias in the heart are because of mutations in Ca^{2+}-release channels or Ca^{2+}-binding proteins of the sarcoplasmic reticulum (SR) (Durham et al. 2007; Lorenzon and Beam 2008; Blayney and Lai 2009; Gyorke 2009). Deafness and skin diseases can also be because of mutations in Ca^{2+} pumps (Foggia and Hovnanian 2004; Van Baelen et al. 2004; Brini and Carafoli 2009). Ca^{2+} dysregulation may also lead to more complex diseases like Alzheimer and other neurodegenerative diseases (Bezprozvanny 2009; Berridge 2010; Supnet and Bezprozvanny 2010).

Disease states associated with a decreased $[Ca^{2+}]$ in the lumen of the ER ($[Ca^{2+}]_{ER}$) have thus far received less attention. The ER controls the synthesis, modification, folding, and export of proteins. An imbalance between the demand for protein synthesis and the capacity to handle

---

[1]Both authors contributed equally to this work.

them leads to the accumulation of misfolded or unfolded proteins, which is referred to as ER stress. An unfolded protein response (UPR) is initiated to reestablish normal ER function (Schroder and Kaufman 2005; Ron and Walter 2007). If the stress is too prolonged or severe to be corrected, the adaptive response triggered by the UPR will not overcome the ER stress and a cell-death program is triggered to eliminate the damaged cell. Many diseases affect the ER environment leading to ER stress, a UPR, and apoptosis (Xu et al. 2005; Lindholm et al. 2006; Kim et al. 2008). Some of them first deplete ER Ca^{2+}, with disturbed function of luminal proteins (Michalak et al. 2002). The decreased $[Ca^{2+}]_{ER}$, rather than the increased $[Ca^{2+}]_{cyt}$, then triggers apoptosis (Nakano et al. 2006; Yoshida et al. 2006).

We will review the diseases in which a decreased $[Ca^{2+}]_{ER}$ is an upstream event in the pathophysiology and show that ER stress often plays an essential role. We will first briefly review the mechanisms controlling the $[Ca^{2+}]_{ER}$, then focus on how ER stress leads to apoptosis, and finally review the mechanisms of ER Ca^{2+} depletion in the various diseases.

Ca^{2+} HOMEOSTASIS IN THE ER/SR

To function as an intracellular Ca^{2+} store, the ER/SR needs to express at least three different types of proteins (Pozzan et al. 1994): (1) Ca^{2+} pumps for uphill transport of Ca^{2+} from the cytosol to the lumen; (2) luminal Ca^{2+}-binding proteins for storing Ca^{2+}; and (3) Ca^{2+} channels for the controlled release of Ca^{2+} to the cytosol along its electrochemical gradient. Although the ER is generally assumed to form a continuous compartment, it can be heterogeneous at the level of its Ca^{2+}-handling proteins. A heterogeneous distribution allows on the one hand localized Ca^{2+} pumping and release, and on the other hand, the setting up of Ca^{2+} signals without disturbing Ca^{2+}-dependent processes within the ER lumen (Petersen et al. 2001; Berridge 2002; Papp et al. 2003).

Ca^{2+} pumps of the SERCA type (sarco/endoplasmic-reticulum Ca^{2+}-ATPase) actively pump Ca^{2+} into the store (Fig. 1). They are encoded by three different genes, whereby each of them exists as various splice variants. SERCA2b has the highest Ca^{2+} affinity and is the most ubiquitous pump. Other isoforms have a more restricted expression pattern. Thapsigargin is a much-used specific inhibitor of the SERCA pumps. This sesquiterpene lactone irreversibly interacts with their M3-transmembrane helix. Phospholamban is the major endogenous regulator of SERCA pumps (at least for isoforms 1a, 2a, and 2b), but it is only expressed in muscle cells. This small protein decreases their Ca^{2+} affinity (Brini and Carafoli 2009; Vangheluwe et al. 2009).

Ca^{2+} in the lumen of the ER/SR is buffered by Ca^{2+}-binding proteins. Calsequestrin is the main Ca^{2+}-binding protein in skeletal and cardiac muscle (Beard et al. 2004). In other tissues Ca^{2+} binds to calreticulin (Michalak et al. 2002) and other Ca^{2+}-dependent chaperones like calnexin, 78-kDa glucose-regulated protein/immunoglobulin heavy chain binding protein (GRP78/BiP), GRP94, and various protein-disulfide isomerases (PDI) (Papp et al. 2003). All these proteins combine at least two of the following three properties: Ca^{2+} binding, regulation of Ca^{2+} pumps or Ca^{2+}-release channels, and chaperone function (Berridge 2002; Papp et al. 2003), emphasizing the close interrelation between the $[Ca^{2+}]_{ER}$ and ER function.

The main Ca^{2+}-release channels in the ER/SR belong to either the ryanodine-receptor (RyR) (Zalk et al. 2007) or the inositol 1,4,5-trisphosphate (IP_3)-receptor (IP_3R) (Foskett et al. 2007) families. In each family, three genes code for receptor subunits, which assemble to produce very large tetrameric Ca^{2+}-release channels (~2.2 MDa for the RyRs, ~1.2 MDa for the IP_3Rs). Further diversity occurs by alternative splicing and by the formation of both homo- and, at least for the IP_3R, heterotetramers. The differences in channel and regulatory properties, and in subcellular localization, allow highly specific Ca^{2+} signals propagating through the cell. RyRs are predominantly expressed in muscles and neurons although they can also be present at low levels in other cells. Skeletal muscle expresses mainly RyR1, which is activated by direct interaction with L-type voltage-operated

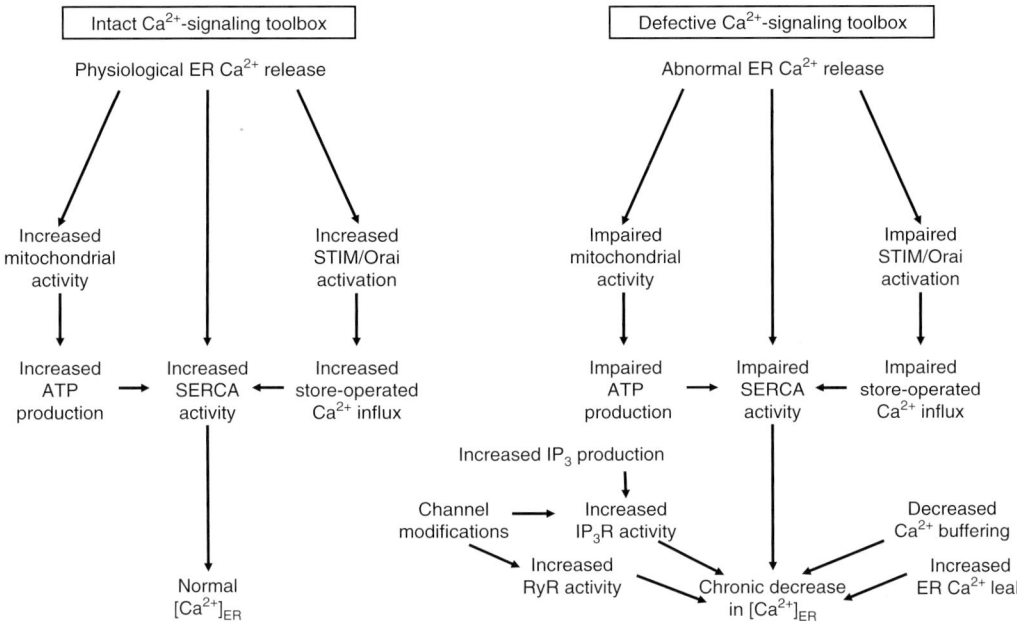

Figure 1. Normal and abnormal $[Ca^{2+}]_{ER}$. A tight coordination between ER Ca^{2+}-release and -refilling mechanisms enables proper Ca^{2+} signaling in response to physiological stimuli. Under these conditions, Ca^{2+} released from the ER stimulates mitochondrial activity and bioenergetics, leading to more ATP production. The initial decline in $[Ca^{2+}]_{ER}$ activates STIM, allowing for store-operated Ca^{2+} influx. Ca^{2+} is recycled via the SERCA pumps. Normal $[Ca^{2+}]_{ER}$ is restored and ER-related processes continue. In contrast, in pathological conditions, stress responses will occur and affect the Ca^{2+}-signaling toolbox in various ways. Impaired mitochondrial activity, store-operated Ca^{2+} influx, or SERCA activity may all cause failure in restoring normal $[Ca^{2+}]_{ER}$ in response to ER Ca^{2+}-signaling processes, and lead to a decreased $[Ca^{2+}]_{ER}$. A chronic decrease in $[Ca^{2+}]_{ER}$ may also be because of an imbalance between the Ca^{2+}-on and -off mechanisms as a result of increased IP$_3$R or RyR activity, decreased Ca^{2+} buffering, or increased ER Ca^{2+} leak.

Ca^{2+} channels, whereas the RyR2 in cardiac tissue and the RyR3 are activated by Ca^{2+} itself (Endo 2009). IP$_3$Rs on the other hand are expressed in all cell types. They generally become active when IP$_3$ is produced on cell stimulation by extracellular agonists. IP$_3$ binding at the amino terminus of the receptor induces channel opening at its carboxyl terminus (Bosanac et al. 2004). The further regulation of channel opening by cytosolic factors including Ca^{2+}, by regulatory proteins, and by phosphorylation/dephosphorylation, as well as their subcellular localization allow them to set up highly specific spatio-temporal Ca^{2+} signals (Vermassen et al. 2004; Foskett et al. 2007; Mikoshiba 2007; Vanderheyden et al. 2009).

In normal conditions, several mechanisms are operative to prevent ER Ca^{2+} depletion or overload, e.g., both Ca^{2+} channels and Ca^{2+} pumps are sensitive to luminal $[Ca^{2+}]$. The IP$_3$R becomes more sensitive to IP$_3$ when the $[Ca^{2+}]_{ER}$ increases (Irvine 1990; Missiaen et al. 1992) and also the RyR is stimulated by luminal Ca^{2+} (Nelson and Nelson 1990; Gyorke and Terentyev 2008). SERCA-mediated Ca^{2+} uptake into the ER is sensitive to $[Ca^{2+}]_{ER}$ (Takenaka et al. 1982). The release of Ca^{2+} from the ER during the generation of cytosolic Ca^{2+} signals should not decrease the $[Ca^{2+}]_{ER}$ to a level at which ER function and Ca^{2+} signaling become compromised (Sammels et al. 2010). A mechanism has evolved that couples ER Ca^{2+} depletion to an increase of Ca^{2+} entry into the cell. This phenomenon is known as "capacitative" (Putney 1986) or "store-operated" Ca^{2+} entry. STIM1 and STIM2 are ubiquitously expressed single-pass transmembrane ER and, to some extent, plasma-membrane proteins with a luminal Ca^{2+} sensor (Stathopulos et al. 2008). Depending on the extent of ER depletion, either STIM1 or STIM2 oligomerize and interact with Orai1 proteins (Brandman et al. 2007). These tetrameric Ca^{2+} channels in the plasma membrane are then responsible for an increased Ca^{2+} entry (Cahalan 2009; Deng et al. 2009; Schindl et al. 2009).

ER STRESS AND APOPTOSIS

The ER not only fulfills a crucial role in Ca^{2+} signaling, but also provides a quality-control system for the proper folding of proteins and for sensing stress (Fig. 2). A plethora of ER-resident chaperones including calreticulin, calnexin, PDI, and GRP78/BiP bind unfolded or misfolded proteins via inappropriately exposed hydrophobic or hypo-glycosylated residues (Austin 2009). Calreticulin and calnexin bind to polypeptide chains entering the ER lumen through glycosylated residues, whereas PDI mediates the correct formation of disulfide bonds. GRP78/BiP undergoes cycles of binding and release of unfolded proteins until they are properly folded and hydrophobic residues are inaccessible. ER-resident chaperones like calreticulin, GRP78/BiP, and GRP94 need a high $[Ca^{2+}]_{ER}$ for their activity (Ma and Hendershot 2004) with Ca^{2+} binding to paired anionic amino acids (Lucero and Kaminer 1999). Moreover, several of the ER chaperones also act as Ca^{2+} buffers (Lievremont et al. 1997; Papp et al. 2003). Determination of the Ca^{2+} affinities suggests up to millimolar levels in the ER (Sambrook 1990), and depletion of ER Ca^{2+} by treating cells with a Ca^{2+} ionophore or thapsigargin can lead to inappropriate secretion, aggregation, and degradation of unassembled proteins (Gaut and Hendershot 1993).

The $[Ca^{2+}]_{ER}$ must be maintained in an environment of continuous intracellular Ca^{2+} signaling. Failure of this homeostatic mechanism, for example, by inhibition of SERCA with thapsigargin, triggers a UPR to either reestablish normal ER function or to eliminate the cell (Xu et al. 2005). The adaptive mechanisms initiated by the UPR involve reduced translation of misfolded proteins, enhanced translation of ER chaperones to increase the folding capacity of the ER, and degradation of misfolded proteins through ER-assisted degradation (ERAD) (Schroder and Kaufman 2005; Malhotra and Kaufman 2007). Global mRNA translation is inhibited for a few hours to reduce the influx of new proteins into the ER, whereas alarm signals involving the activation of mitogen-activated protein kinases (MAPK) are induced (Kim et al. 2008). The UPR involves three signaling pathways: PERK (PKR-like ER kinase), Ire1 (inositol-requiring enzyme 1), and ATF6 (activating transcription factor 6).

The recognition of misfolded proteins by the Ser/Thr kinase PERK leads to phosphorylation and inactivation of the eukaryotic initiation factor 2α (eIF2α). This shuts off mRNA translation, thereby preventing the accumulation of newly synthesized proteins in the ER (Harding et al. 1999), activates the transcription factor ATF4, which increases the level of chaperones such as GRP78/BiP and GRP94, and helps to restore the cellular redox homeostasis (Harding et al. 2000, 2003).

Ire1 has endoribonuclease and Ser/Thr-kinase activity. Its endoribonuclease activity degrades many mRNAs to reduce the protein load on the ER (Hollien and Weissman 2006). Ire1 removes an intron from the mRNA of X-box-binding protein 1 (XBP1), leading to the expression of XBP1. This transcription factor is involved in the expression of several UPR and ERAD genes (Rao and Bredesen 2004). The kinase activity of Ire1 is involved in apoptotic signaling via ASK1 (apoptosis signal-regulating kinase 1) and JNK (c-Jun N-terminal kinase). JNK activates the proapoptotic BH3-only protein Bim (Lei and Davis 2003; Putcha et al. 2003), and inactivates the antiapoptotic Bcl-2 protein (Yamamoto et al. 1999). Ire1 also recruits caspase 12 (Yoneda et al. 2001), which may play a role in ER stress-induced apoptosis (Szegezdi et al. 2003). However, caspase 12 is not present in humans, and although caspase 4, its close paralogue, may perform such function, it remains uncertain whether caspase 4 is vital for ER stress-induced apoptosis (Egger et al. 2003).

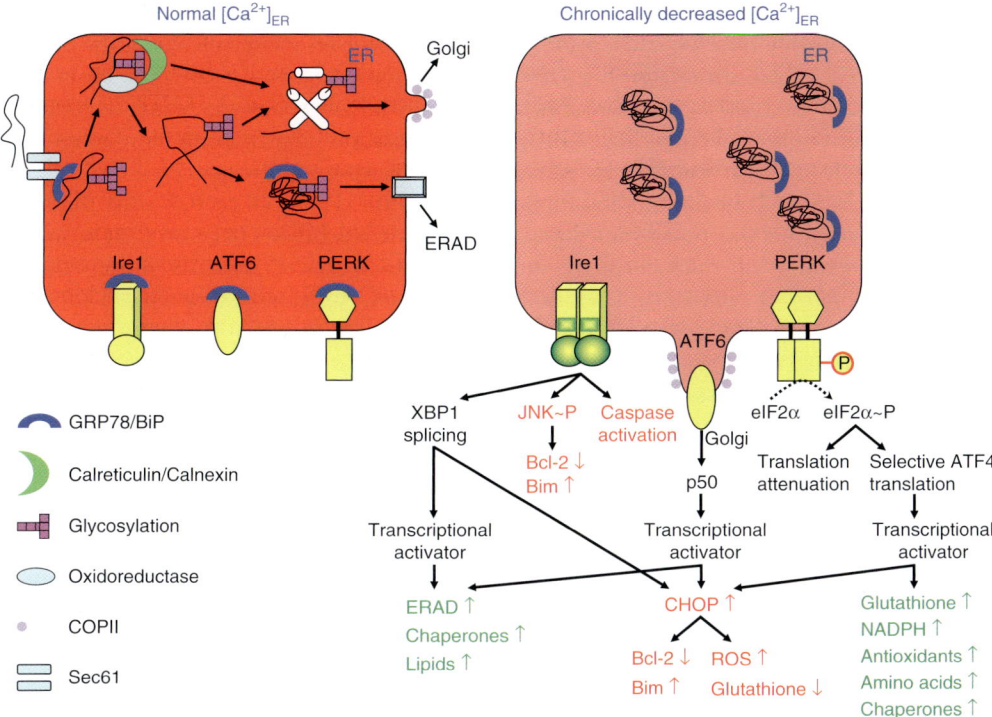

Figure 2. The UPR. At a normal $[Ca^{2+}]_{ER}$ the ER-stress sensors are scaffolded and inactivated by GRP78/BiP. Protein trafficking and quality-control mechanisms work normally. Polypeptides are translocated through Sec61 and glycosylated. This transport is facilitated by the molecular chaperone GRP78/BiP. Glucosidases then prepare the glycoprotein for binding to the ER lectins, calreticulin, and calnexin, whereas oxidoreductases catalyze disulfide-bond formation. ER-resident chaperones facilitate the proper folding of the nascent protein and prevent its aggregation. Further deglucosidation releases the ER lectins and once the protein is correctly folded and processed, the protein leaves the ER via the coat protein (COPII)-coated vesicles to the secretory pathway. Misfolded proteins, in contrast, associate with various chaperones, including GRP78/BiP, and are removed from the ER through ERAD.

In contrast, when the $[Ca^{2+}]_{ER}$ is chronically decreased, the function of chaperones becomes disturbed and unfolded proteins accumulate and act as a sponge for luminal GRP78/BiP. As a consequence, ER-stress sensors are devoid of GRP78/BiP and become activated, yielding early adaptive responses promoting survival (indicated in green) or late responses promoting apoptosis under conditions of severe or on-going ER stress (indicated in red). Ire1 undergoes dimerization and activation of its kinase and endoribonuclease activity, thereby splicing XBP1 mRNA and yielding a potent transcriptional activator that induces the expression of genes involved in ERAD, protein folding (like GRP78/BiP), and lipid synthesis. ATF6 goes to the Golgi compartment, where it is proteolytically cleaved to yield a cytosolic fragment (p50) that migrates to the nucleus and activates the transcription of UPR genes, like GRP78/BiP and CHOP. PERK dimerizes, autophosphorylates, and phosphorylates eIF2α, thereby suppressing its activity and reducing the rate of translation initiation, while increasing the rate of translation of ATF4, a potent transcription factor that augments the expression of genes involved in antioxidative stress, amino acid metabolism, and protein chaperoning. During on-going ER stress or irreparable ER damage, apoptotic pathways are activated. Ire1 phosphorylates JNK, leading to inhibition of Bcl-2 activity and activation of Bim, and recruits, releases, and activates procaspases in the cytosol. Induction of CHOP via XBP1, ATF6 or ATF4, down-regulates prosurvival Bcl-2-family members, increases prodeath proteins (like Bim) and ROS, and decreases the levels of glutathione, a ROS scavenger. In the presence of ROS, Ca^{2+} transfer to the mitochondria leads to the release of cytochrome *c*. The balance between proapoptotic and antiapoptotic Bcl-2-family members is disturbed, with activation of the intrinsic apoptotic pathway.

The transcription factor ATF6 is translocated to the Golgi during ER stress and is proteolytically activated. ATF6 stimulates ER-stress genes as a homodimer or as a heterodimer with other transcription factors like XBP1, whose transcription is also induced by ATF6 (Yoshida et al. 2001; Malhotra and Kaufman 2007). ATF6 is cytoprotective, possibly mediated by RCAN1 (regulator of calcineurin-1), an endogenous inhibitor of calcineurin (Belmont et al. 2008). This enzyme dephosphorylates the proapoptotic Bad (Bcl-2 antagonist of cell death), which then dimerizes and inhibits antiapoptotic family members such as Bcl-2 and Bcl-Xl (Wang et al. 1999).

ATF4, ATF6, and XBP1 all induce the transcription of the gene encoding CHOP (C/EBP homologous protein) (Kim et al. 2008). The Ire1-ASK1-p38-MAPK pathway enhances CHOP activity at a posttranscriptional level (Wang and Ron 1996). CHOP is involved in ER stress-induced apoptosis by down-regulating the antiapoptotic Bcl-2 (McCullough et al. 2001), and by inducing expression of the proapoptotic Bim (Puthalakath et al. 2007) and of ER oxidase 1α, thereby rendering the ER more oxidative and exacerbating ER stress (Marciniak et al. 2004). Misfolded proteins are eventually eliminated via proteins involved in the ERAD pathway, which are induced and controlled by both Ire1-XBP1 and ATF6 pathways (Yoshida et al. 2003).

No trigger for ER stress selectively elicits either adaptive responses or apoptosis. The switch between life and death is regulated by the complex interdependent UPR-signaling pathways that each may result in prosurvival or prodeath responses. The different time courses of the three main UPR branches may influence the cell fate (Lin et al. 2007a). The early termination of Ire1α activity is needed for cell death. Differential activation of PERK and Ire1α may lead to life or death (Lin et al. 2009). Cell death is induced by apoptosis and by caspase-independent necrosis. ER stress also induces autophagy (Ogata et al. 2006; Bernales et al. 2006; Hoyer-Hansen and Jaattela 2007). The PERK-ATF4 branch stimulates the expression of *ATG12*, an autophagy gene (Kouroku et al. 2007). This catabolic process removes unfolded proteins and their aggregates independently of the ubiquitin/proteasome system, thereby promoting cell survival. Ultimately, however, enhanced autophagic vacuolization may lead to non-apoptotic cell death (Levine and Kroemer 2008).

ER stress and cell death involve many Ca^{2+}-dependent processes (Kim et al. 2008) including phospholipases, scramblases, nitric-oxide (NO) synthases, calpains, calcineurin, FKBP38, fortilin, a putative modulator of Mcl-1 (myeloid cell leukemia sequence 1), death-associated protein kinase 1, mitochondrial fission, and Ca^{2+}-dependent pathways triggering autophagy. Some pathways require interplay between mitochondria and the ER in zones of close contact (Giorgi et al. 2008, 2009). These microdomains involve the close proximity of ER Ca^{2+}-release channels like the IP_3R and mitochondrial Ca^{2+}-transport mechanisms, like the voltage-dependent anion channel (VDAC) and the Ca^{2+} uniporter (Giorgi et al. 2009). Changes in ER Ca^{2+} homeostasis in this way affect mitochondrial Ca^{2+} signaling. Lowering of the $[Ca^{2+}]_{ER}$ by antiapoptotic proteins such as Bcl-2 has been described (Scorrano et al. 2003) and is expected to lower the sensitivity to apoptotic Ca^{2+} transfer from the ER to the mitochondria (Rizzuto et al. 2009). Bcl-Xl, a related antiapoptotic protein was found to induce prosurvival ER-to-mitochondria Ca^{2+} signaling by sensitizing the IP_3R to basal levels of IP_3 (White et al. 2005). ER-to-mitochondria Ca^{2+} signals can regulate cell survival by enhancing mitochondrial bioenergetics. Mitochondrial Ca^{2+} overload, on the other hand, by a larger or more persistent $[Ca^{2+}]$ rise was found to induce cell death (Rong and Distelhorst 2008). Cell death is characterized by mitochondrial outer membrane permeabilization (MOMP) and the loss of the mitochondrial transmembrane potential $\Delta\Psi_m$ (Kroemer et al. 2007). Mitochondrial Ca^{2+} overload can cause breakdown of $\Delta\Psi_m$ by activating the permeability transition pore (PTP). Loss of $\Delta\Psi_m$, however, seems to be a secondary event and not required for MOMP and the release of cytochrome *c* (Chipuk and Green 2008). Accordingly, PTP opening probably plays a role in necrosis but not apoptosis.

Deficiency of Bax and Bak confers resistance to apoptotic cell death induced by conventional anticancer therapies. SERCA inhibitors like thapsigargin, however, can efficiently kill Bax/Bak$^{-/-}$ MEFs by inducing mitochondrial Ca^{2+} overload, PTP opening and necrotic cell death (Janssen et al. 2009). In addition to PTP opening, the activation and oligomerization of the executioner proapoptotic Bcl-2-family members Bax and Bak induce MOMP in response to a variety of apoptotic triggers (Chipuk and Green 2008; Brunelle and Letai 2009). The activity of Bax/Bak is tightly controlled by proteins of the Bcl-2 family. The antiapoptotic Bcl-2-family members, including Bcl-2, Bcl-Xl and Mcl-1, neutralize and prevent oligomerization of Bax/Bak, whereas activator proapoptotic BH3-only proteins, including Bim, cleaved Bid and cytosolic p53, directly bind to Bax/Bak, causing a conformational change, membrane insertion, and oligomerization. In addition, sensitizer BH3-only proteins, including Bad, Noxa, and Puma, bind to the antiapoptotic Bcl-2-family members, neutralizing their antiapoptotic activity. Many of these proteins affect ER Ca^{2+} homeostasis by binding to the IP$_3$R and/or changing its phosphorylation, resulting in altered Ca^{2+}-flux properties of the channel (Oakes et al. 2005; White et al. 2005; Rong and Distelhorst 2008).

DIABETES MELLITUS

Intracellular Ca^{2+} signaling is perturbed in this chronic metabolic disease with hyperglycemia. Resting [Ca^{2+}]$_{cyt}$ increases, and stimulus-induced [Ca^{2+}]$_{cyt}$ increases in many tissues decrease (Levy 1999; Verkhratsky and Fernyhough 2008). The [Ca^{2+}]$_{ER}$ and SR [Ca^{2+}] ([Ca^{2+}]$_{SR}$) decrease in the pancreatic β-cell and in tissues affected by diabetic complications.

Pancreatic β-Cell

The progressive reduction in cell mass and eventually failure of the β-cell is because of apoptotic cell death. ER stress is an important mechanism of apoptosis (Eizirik et al. 2008), at least in some types of diabetes (Akerfeldt et al. 2008). The [Ca^{2+}]$_{ER}$ in the β-cell is decreased, but the mechanism involved depends on the type of diabetes. The low [Ca^{2+}]$_{ER}$ impairs proinsulin processing and transport (Guest et al. 1997). The subsequently activated UPR can lead to apoptosis resulting in insufficient insulin secretion (Oyadomari and Mori 2004; Eizirik et al. 2008). The very high secretion rate of β-cells makes them very sensitive to apoptosis induced by ER Ca^{2+} depletion (Araki et al. 2003; Cardozo et al. 2005; Tonnesen et al. 2009).

Type-1 diabetes is characterized by an autoimmune β-cell destruction caused by overproduction of NO (Gotoh and Mori 2006). Cytokines released from infiltrating T-cells and macrophages up-regulate inducible NO synthase in an NF-kB- and STAT-1-dependent manner (Eizirik et al. 2008). NO depletes ER Ca^{2+} in β-cells by acting on SERCA and on the Ca^{2+}-release channels (Oyadomari et al. 2001). NO down-regulates SERCA2b expression (Cardozo et al. 2005), perhaps through inhibition of the Sp1 transcription factor (Pirot et al. 2008). NO also reacts with superoxide anion to form peroxynitrite, which inhibits SERCA by reacting with two tyrosine residues in the channel-like domain (Viner et al. 1999; Grover et al. 2003). Peroxynitrite also activates RyR2 by poly-S-nitrosylation of the channel (Xu et al. 1998). The cytokines also up-regulate death protein 5, a BH3-only protein that contributes to Ca^{2+} depletion and ER stress (Gurzov et al. 2009). This depletion mainly occurs when IP$_3$Rs and RyRs are stimulated (Luciani et al. 2009). Type-1 diabetes was furthermore associated with the single-nucleotide polymorphism rs2296336 in the gene encoding IP$_3$R3 (Roach et al. 2006). Increased cholinergic tone with more acetylcholine-induced IP$_3$ production, and up-regulation of the IP$_3$R during hyperglycemia (Lee et al. 1999) therefore promote ER Ca^{2+} depletion. NO-induced ER stress in β-cells does not activate the ATF6 branch of the UPR (Cardozo et al. 2005; Tonnesen et al. 2009). It is still debated whether cytokine-induced ER stress is a direct cause of β-cell apoptosis or a parallel and/or downstream event (Akerfeldt et al. 2008).

Nonautoimmune type-1 diabetes in Wolfram syndrome is caused by mutations in the gene encoding the ER glycoprotein wolframin (Inoue et al. 1998; Strom et al. 1998). This genetic defect lowers $[Ca^{2+}]_{ER}$ in β-cells (Takei et al. 2006) and activates the UPR and triggers the apoptotic pathway (Yamada et al. 2006). Reconstitution of this ER-resident transmembrane protein into planar lipid bilayers induces a cation-selective ion channel (Osman et al. 2003). Wolframin also prevents ER stress via other mechanisms, e.g., by negatively regulating ATF6α through the ubiquitin-proteasome pathway (Fonseca et al. 2010).

Type-2 diabetes is characterized by insulin resistance in liver, skeletal muscle and adipose tissue, and a failure of the β-cell to compensate for the increasing demand. Insulin resistance in liver and adipose tissue may be because of ER stress (van der Kallen et al. 2009). ER Ca^{2+} depletion with thapsigargin leads to insulin resistance (Ozcan et al. 2004), but so far there are no studies linking peripheral insulin resistance to a decreased $[Ca^{2+}]_{ER}$. A high-fat diet or obesity often leads to the development of type-2 diabetes (Eizirik et al. 2008). Free fatty acids trigger β-cell loss (Leonardi et al. 2003). One model of lipotoxicity proposes that palmitate activates the UPR in β-cells (Eizirik et al. 2008). The mechanism may again involve ER Ca^{2+} depletion (Cunha et al. 2008; Gwiazda et al. 2009) by decreased expression (Roe et al. 1994; Evans-Molina et al. 2009) or activity of SERCA (Cunha et al. 2008). Peroxisome proliferator-activated receptor-γ (PPAR-γ) agonists, which improve sensitivity to insulin, also restore SERCA expression and attenuate ER stress in the β-cell (Evans-Molina et al. 2009). Despite variation in the gene encoding SERCA3 in type-2 diabetic patients (Varadi et al. 1999), insulin secretion and blood glucose levels were normal in SERCA3$^{-/-}$ mice (Arredouani et al. 2002), probably because of compensatory mechanisms.

Polymorphisms in the gene for insulin receptor substrate 1 (IRS-1) have been linked to type-2 diabetes (Almind et al. 1993). IRS-1 directly interacts with SERCA3 (Borge and Wolf 2003). Mice with deleted IRS-1 have reduced SERCA2b and 3 levels, more transient increases in $[Ca^{2+}]_{cyt}$ and less insulin secretion (Kulkarni et al. 2004).

Diabetic Cardiomyopathy

The remodeling of the SR resulting in a slower Ca^{2+} uptake, a lower $[Ca^{2+}]_{SR}$, and release of less activator Ca^{2+}, slows relaxation kinetics of the ventricle and eventually leads to systolic dysfunction, independently of vascular or valve disease (Rubler et al. 1972). Hyperglycemia causes these effects (Ren et al. 1997). The changes in SR Ca^{2+} handling depend on the type of diabetes, the experimental model, the degree of hyperglycemia, and the extent of disease progression. SR function is already abnormal at an insulin-resistant stage before the manifestation of overt type-2 diabetes (Dutta et al. 2002; Wold et al. 2005; Vasanji et al. 2006; Reuter et al. 2008).

Phospholamban, which inhibits SERCA2a, becomes up-regulated (Kim et al. 2001; Choi et al. 2002; Belke et al. 2004; Zhou et al. 2006) at an early stage of the disease (Zhong et al. 2001). Its phosphorylation by protein kinase A and Ca^{2+}/calmodulin-dependent protein kinase, which regulates the interaction with SERCA2a, and therefore stimulates Ca^{2+} uptake, decreases (Choi et al. 2002; Belke et al. 2004; Vasanji et al. 2004, 2006).

The activity and expression of SERCA2a decrease in the diabetic heart (Teshima et al. 2000; Kim et al. 2001; Trost et al. 2002; Choi et al. 2002; Belke et al. 2004; Vasanji et al. 2004; Wold et al. 2005; Zhang et al. 2008; Stolen et al. 2009; Wang et al. 2010). Reduced activity of SERCA2a is not only because of the effect on phospholamban, but also to increased formation of advanced glycation end products of SERCA2a (Bidasee et al. 2004), depressed activity of protein kinase A (Dutta et al. 2002), and sensory denervation leading to diminished production of NO and peroxynitrite, which at basal concentrations activate SERCA2a through S-nitrosylation of Cys-349 (Bencsik et al. 2008). The fact that peroxynitrite both stimulates (Adachi et al. 2004) and inhibits SERCA (Viner et al. 1999; Schmidt et al. 2003b) seems to indicate that the effect is very much dependent on the experimental conditions or perhaps on the isoform studied. SERCA2a expression decreases in a later stage of the disease (Zhong et al. 2001), perhaps by increased O-glycosylation of the transcription factor Sp1 with β-N-acetylglucosamine because of the hyperglycemia (Clark et al. 2003), or by reduced expression and activity of SIRT1, a histone deacetylase (Sulaiman et al. 2010). RyR2 function changes by formation of disulfide bonds between adjacent sulfhydryl groups (Bidasee et al. 2003a), by increased glycation (Bidasee et al. 2003b), by decreased FKBP12.6 expression and binding (Belke et al. 2004; Shao et al. 2007), by hyperphosphorylation (Shao et al. 2007; Stolen et al. 2009), and by a reduced density of T-tubules (Stolen et al. 2009). Two RyR populations appear: one with enhanced responsiveness to Ca^{2+} and another being unresponsive (Shao et al. 2007). Dysfunctional RyR2 can cause dyssynchronous and diastolic Ca^{2+} releases, sometimes with ventricular arrhythmia (Shao et al. 2007). Spontaneous Ca^{2+} sparks representing aberrant RyR2 activation increase in frequency (Yaras et al. 2005; Shao et al. 2007). SR Ca^{2+} leak increases (Belke et al. 2004; Stolen et al. 2009). Expression of RyR2 decreases (Teshima et al. 2000; Choi et al. 2002; Guner et al. 2004; Pereira et al. 2006; Zhou et al. 2006; Reuter et al. 2008; Wang et al. 2010) at later stages of the disease (Zhong et al. 2001). IP_3R1, 2, and 3 become down-regulated, but these effects may be species related (Guner et al. 2004; Zhou et al. 2006). The roles of IP_3Rs in the heart are furthermore not entirely clear.

Some treatments directly affect the Ca^{2+} signal. Overexpression of SERCA2a protects the heart from contractile dysfunction (Trost et al. 2002; Vetter et al. 2002; Sakata et al. 2007) and reverses established cardiomyopathy (Suarez et al. 2008) and the transcriptional profile induced by diabetes (Karakikes et al. 2009). SERCA2a expression and cardiac function can be normalized by PPAR-γ agonists (Shah et al. 2005), total triterpene acids from *Cornus officinalis* Sieb. (Qi et al. 2008), and the SIRT1 activator resveratrol (Sulaiman et al. 2010). Breviscapine in Chinese medicine decreases phospholamban expression and increases that of SERCA2a and RyR2 (Wang et al. 2010). Exercise training also normalizes abnormal Ca^{2+} signaling (Shao et al. 2009; Stolen et al. 2009).

Vascular Disease

Diabetes lowers $[Ca^{2+}]_{ER}$ in the smooth-muscle cells, macrophages and platelets. These changes contribute to the vascular complications including atherosclerosis (Cooper et al. 2001).

In healthy smooth-muscle cells, basal levels of NO react with superoxide anion to form peroxynitrite, which together with glutathione reacts with Cys-674 of SERCA and increases its activity (Adachi et al. 2004). The hyperglycemia of diabetes induces high levels of oxidants that irreversibly oxidize Cys-674 leading to less S-glutathionylation-induced stimulation of SERCA2 (Adachi et al. 2004) and faster degradation (Ying et al. 2008). Insulin also inhibits SERCA via enhanced nitrotyrosine formation (Kobayashi et al. 2007). SERCA2 is also redistributed to a peri-nuclear pattern (Searls et al. 2010). The subsequently decreased $[Ca^{2+}]_{ER}$ stimulates plasma-membrane Ca^{2+} influx and induces migration of the smooth-muscle cell, which contributes to neointimal hyperplasia and atherosclerosis (Tong et al. 2008). Dedifferentiation of smooth-muscle cells precedes their migration from the media to the intima. The up-regulation of the secretory-pathway Ca^{2+}-ATPase 1 (SPCA1) in diabetes (Lai and Michelangeli 2009) probably reflects the change from a contractile to a secretory cell. Sp1 and YY1, transcription factors controlling SPCA1 transcription (Kawada et al. 2005), become more active in high glucose (Han and Kudlow 1997). The expression of IP_3R and RyR decreases (Ma et al. 2008; Searls et al. 2010).

Macrophages in type-2 diabetes express more CHOP and are therefore more susceptible to ER stress-induced apoptosis. CHOP induces ER oxidase 1α, with hyperoxidation of the ER lumen and disulfide-bond formation between two cysteines in IP_3R1. This causes dissociation of the disulfide isomerase-like protein ERp44 (Kang et al. 2008) and more IP_3-induced Ca^{2+}

release (Li et al. 2009). These changes favor plaque necrosis.

Altered Ca^{2+} signaling in platelets makes them hyperreactive. Their increased adhesiveness and aggregability contribute to the development of the angiopathy (Knobler et al. 1998). The hyperglycemia causes oxidant stress in platelets (Vericel et al. 2004), which enhances tyrosine nitration of SERCA2 and in this way decreases SERCA2 function and, at least at high HbA_{1C} levels, expression in type-2 diabetic patients (Randriamboavonjy et al. 2008). PPAR-γ agonists decrease tyrosine nitration of SERCA and increase its expression. Increased levels of homocysteine in type-2 diabetic patients also release Ca^{2+} from agonist-sensitive Ca^{2+} stores (Zbidi et al. 2010). The direct stimulatory interaction of STIM1 with SERCA3 is impaired in type-2 diabetes (Lopez et al. 2008), which can explain the increased plasma-membrane Ca^{2+} entry, and the higher $[Ca^{2+}]_{cyt}$ at rest and during thrombin stimulation (Saavedra et al. 2004). SERCA3b was upregulated in type-1 diabetes (Chaabane et al. 2007). This isoform is involved in cell adhesion (Chaabane et al. 2006) and its up-regulation can thus explain the increased adhesiveness in diabetic patients.

Diabetic Nephropathy

This complication is an important cause of end-stage renal disease. Apoptosis induced by ER stress also occurs in the diabetic kidney (Liu et al. 2008). In podocytes, advanced glycation end products release ER Ca^{2+} and trigger a UPR leading to apoptosis during the early stage of the nephropathy (Chen et al. 2008). The loss of podocytes is an important determinant in the progression of the disease. Tubulointerstitial cells also show ER stress (Lindenmeyer et al. 2008), probably induced by the hyperglycemia and the massive protein reabsorption as a result of the proteinuria, but possible changes in $[Ca^{2+}]_{ER}$ were not investigated. The activated UPR selectively enhances the prosurvival pathway of the response, suggesting that diabetic damage may occur independently of any terminal UPR process (Brosius and Kaufman 2008).

Decreased IP3R1 expression in the afferent arteriole and mesangial cell leads to smaller $[Ca^{2+}]_{cyt}$ increases in response to vasoconstrictors, resulting in renal hyperfiltration and glomerular damage (Sharma et al. 1999).

Sensory Neuropathy

Diabetic neuropathy can produce prolonged changes in the nervous system, with pain, sensory loss, food ulceration, infection, gangrene and poor wound healing (Huang et al. 2002; Verkhratsky and Fernyhough 2008). The $[Ca^{2+}]_{ER}$ was decreased because of a decreased SERCA expression by a so far unidentified mechanism and by a decreased activity of the pump (Verkhratsky and Fernyhough 2008). The decreased activity of SERCA may be caused by impaired mitochondrial ATP production in diabetes because of reduced stimulation of insulin receptors (Fernyhough and Calcutt 2010). This effect on the mitochondria seems to be independent of the hyperglycemia. The decreased $[Ca^{2+}]_{ER}$ then affects protein synthesis, posttranslational modification and trafficking, which in turn diminish the supply of voltage-gated Ca^{2+} channels to the axons, thus resulting in the decrease of nerve-conductance velocity (Verkhratsky and Fernyhough 2008). Stimulus-induced $[Ca^{2+}]_{cyt}$ increases decrease (Kruglikov et al. 2004) as a result of the decreased $[Ca^{2+}]_{ER}$ and probably also as a result of decreased IP3R function. Protein glycosylation with β-N-acetylglucosamine is increased in diabetes (Hu et al. 2005). Glycosylation of IP3R1 by β-N-acetylglucosamine decreases its function (Rengifo et al. 2007).

Salivary Glands

Abnormal Ca^{2+} signaling in the salivary glands leads to dryness of the mouth, loss of taste sensation, sialosis, and other disorders of the oral cavity (Nicolau et al. 2009). SERCA is inhibited in the submandibular gland of streptozotocin-induced diabetic rats and therefore $[Ca^{2+}]_{ER}$ and IP3-induced $[Ca^{2+}]_{cyt}$ increases decrease (Fedirko et al. 2006). The decreased $[Ca^{2+}]_{ER}$ results in improper posttranslational processing, folding, and exit of ER proteins. This could explain the decreased saliva protein content and amylase activity.

NEUROLOGICAL DISEASES

Neural Ischemia

Ischemia depletes ER Ca^{2+}. Both the decreased $[Ca^{2+}]_{ER}$ (Paschen and Mengesdorf 2005) and the increased $[Ca^{2+}]_{cyt}$ (Verkhratsky 2005) contribute to cell death. The induced UPR may lead to apoptosis in the peri-infarct area (DeGracia et al. 2002). The release of Ca^{2+} amplifies the $[Ca^{2+}]_{cyt}$ increase evoked by ischemia-induced Ca^{2+} entry (Xiong et al. 2007). Inhibition of this release with dantrolene reduces cell injury (Wei and Perry 1996). The mechanism of ER Ca^{2+} depletion remains unclear. Cytosolic Ca^{2+} enhances NO synthesis, which inhibits mitochondrial electron transport, and augments the generation of reactive oxygen species (ROS) (Moncada and Erusalimsky 2002). SERCA becomes inhibited by excessive NO production (Doutheil et al. 2000), by ischemia-induced inhibition of the coupling of ATP hydrolysis to Ca^{2+} transport (Parsons et al. 1999), and by activated calpain by the increased $[Ca^{2+}]_{cyt}$ (French et al. 2006; Bevers and Neumar 2008). ER Ca^{2+}-release channels are affected during neural ischemia. RyR2 is activated by S-glutathionylation by NO and ROS (Bull et al. 2008), and by calpain-induced proteolysis (Rardon et al. 1990). Calpain causes proteolysis of the IP_3R resulting in decreased IP_3 binding, suggesting that site-specific cleavage decreases the affinity of the remaining protein species for IP_3 (Nagata et al. 1994; Dahl et al. 2000). Although this would indicate that calpain prevents Ca^{2+} release, it is also possible that the proteolysis simply removes the ligand regulation of the channel and leads to baseline Ca^{2+} release from the ER, contributing to Ca^{2+} overload (Bevers and Neumar 2008). Calpain also inhibits IP_3 metabolism by cleaving IP_3 kinase B (Pattni et al. 2003), allowing it to act longer on the IP_3R, thereby potentiating Ca^{2+} efflux from the ER (Bevers and Neumar 2008). Enhanced activity of phospholipase C and A_2 during ischemia liberates free fatty acids, which release ER Ca^{2+} (O'Neil et al. 1999).

Neurodegeneration

Ca^{2+} signaling is often abnormal in neurodegenerative diseases (Mattson 2007). Diseases with an increased $[Ca^{2+}]_{ER}$, like Alzheimer disease (Tu et al. 2006; Berridge 2010), fall outside the scope of this review. ER stress and the UPR also occur in Parkinson disease (Ryu et al. 2002), amyotrophic lateral sclerosis (Kanekura et al. 2009), and polyglutamate diseases (Lindholm et al. 2006), but the effects on $[Ca^{2+}]_{ER}$ are not well documented. We will focus on diseases with a decreased $[Ca^{2+}]_{ER}$.

Some lysosomal storage diseases lead to a decreased $[Ca^{2+}]_{ER}$. In neurons of G_{M1}-gangliosidosis, G_{M1} accumulates at the ER membrane and depletes ER Ca^{2+} stores (Tessitore et al. 2004) by interacting with the phosphorylated form of the IP_3R (Sano et al. 2009). The subsequent activation of the UPR leads to apoptosis (Sano et al. 2009). Silencing of IP_3R1 with siRNA reduces the number of apoptotic cells (Sano et al. 2009). Increased Ca^{2+} release from the ER in Gaucher disease is because of overactivation of the RyR (Korkotian et al. 1999; Pelled et al. 2005), because glucosylceramide, the lipid that accumulates in this disease, directly modulates the RyR (Lloyd-Evans et al. 2003). In Sandhoff disease, SERCA activity is inhibited by the accumulation of G_{M2}-ganglioside (Pelled et al. 2003), which depends on an exposed sialic-acid residue on G_{M2} (Ginzburg et al. 2008). The UPR is also activated by the accumulation of palmitoylated proteins in the infantile form of Batten disease, but ER Ca^{2+} handling was not investigated (Zhang et al. 2007). The decreased SERCA2 and IP_3R1 expression in Niemann-Pick A disease did not activate a UPR (Ginzburg and Futerman 2005). The suggestion that the UPR is a common mediator of apoptosis in neurodegenerative lysosomal storage diseases (Wei et al. 2008a) can therefore be questioned (Farfel-Becker et al. 2009). Increasing $[Ca^{2+}]_{ER}$ by inhibiting the RyR or by SERCA2b overexpression partially restored mutant-enzyme homeostasis in several lysosomal storage diseases (Ong et al. 2010).

Transmissible spongiform encephalopathies include Creutzfeldt-Jakob disease in humans, and bovine spongiform encephalopathy and scrapie in animals (Prusiner 1998). These diseases are associated with extracellular accumulation of a conformationally modified abnormal isoform of the prion protein, a widely expressed plasma membrane-associated glycoprotein with highest levels of expression on neurons and glia. This protein binds to the cell surface and sends a signal to the ER to release Ca^{2+} through the IP_3R and RyR (Hetz et al. 2003; Ferreiro et al. 2006, 2008). The subsequent decreased $[Ca^{2+}]_{ER}$ leads to a UPR and activates the ER-stress-induced apoptosis pathway. Dantrolene and xestospongin C, which are inhibitors of the RyR and IP_3R respectively, prevent neuronal death (Ferreiro et al. 2006, 2008).

Neuropathic Pain

Neuropathic pain is pain arising from nerve injury. The soma of sensory neurons is affected by injuring the peripheral axons. Spinal-nerve ligation depletes ER Ca^{2+} (Rigaud et al. 2009) by a loss of ER and therefore of SERCA (Gemes et al. 2009). Rigaud et al. (2009) suggested that this may trigger a UPR, but this was not directly shown. Depletion of ER Ca^{2+} stores thus contributes to the pathogenesis of neuropathic pain.

Anesthesia

General anesthesia may cause cognitive deficits after surgery (Moller et al. 1998). Inhalation anesthetics can overactivate the IP_3R, with excessive ER Ca^{2+} release leading to apoptosis (Wei et al. 2008b; Yang et al. 2008). Neurons with enhanced IP_3R activity, for example, in familial Alzheimer or Huntington disease, may be especially vulnerable.

CARDIOVASCULAR DISEASES

Atherosclerosis

Macrophages play a critical role in this chronic inflammatory disease (Fan and Watanabe 2003). They accumulate unesterified cholesterol in advanced lesions, which changes the fluidity of the ER membrane and in this way inhibits SERCA (Li et al. 2004). Depletion of ER Ca^{2+} stores induces a UPR and apoptosis (Feng et al. 2003). Excessive apoptosis plays a key role in the progression of atherosclerosis. The UPR also sets up a positive feedback loop with more Ca^{2+} release via induction of ER oxidase 1α. Hyperoxidation of the ER lumen activates Ca^{2+} release (Li et al. 2009) by disulfide-bond formation between two cysteines in IP3R1 and dissociation of the inhibitory ERp44 (Kang et al. 2008). This mechanism complements the increased ER Ca^{2+} leak through induction of a truncated variant of SERCA1 through the PERK pathway (Chami et al. 2008).

Homocysteine, a risk factor for cardiovascular disease, depletes ER Ca^{2+} in aortic smooth muscle, induces ER stress and in this way accelerates atherosclerosis (Dickhout et al. 2007). Also increased production of superoxide anion inhibits SERCA in blood vessels (Tong et al. 2009).

Endothelial dysfunction already occurs early during atherogenesis. Increased peroxynitrite formation in the endothelium inhibits SERCA, depletes ER Ca^{2+} and induces a UPR (Dickhout et al. 2005).

Chronic Heart Failure

Reduced $[Ca^{2+}]_{cyt}$ increases caused by a decreased SR Ca^{2+} content make the heart muscle too weak to pump sufficient blood through the body (Bers et al. 2003). Ca^{2+} pumping is reduced because of a decreased ratio of SERCA2a relative to phospholamban expression (Hasenfuss and Pieske 2002), or because phospholamban is either mutated with more inhibition of SERCA2a (Franz et al. 2001; Schmitt et al. 2003; Haghighi et al. 2006; Kranias and Bers 2007) or less phosphorylated (Frank et al. 2002; Bers et al. 2003; Yano et al. 2008) because of a more active protein phosphatase 1 (del Monte and Hajjar 2008). SERCA2 mutations have not been linked to heart failure (Schmidt et al. 2003a). SERCA3f, an isoform with a specific role in ER stress, becomes up-regulated (Dally et al. 2009). Enhanced Na^+-Ca^{2+} exchange leading to more extrusion of Ca^{2+} from the cell also depletes SR Ca^{2+} (O'Rourke et al. 1999).

Subconductance states of the RyR2 and decreased coupled gating of RyR2-channel clusters can increase SR Ca^{2+} leak during diastole (Reiken et al. 2003; Wehrens et al. 2003, 2005a, 2006; Lehnart et al. 2005, 2008; Huang et al. 2006; Zalk et al. 2007). Hyperactivation of RyR2 may arise from activated protein kinase A by sympathetic neurons and increased levels of catecholamines, and subsequent hyperphosphorylation of RyR2 at Ser-2809 and dissociation of FKBP12.6 (Marx et al. 2000) (but see Bers et al. 2003; Seidler et al. 2007; Yano et al. 2008). Enhanced Ca^{2+}/calmodulin-dependent protein kinase δ-dependent phosphorylation of RyR2 at Ser-2815 also increases diastolic Ca^{2+} leak and reduces SR Ca^{2+} load (Ai et al. 2005).

β-blockers prevent the hyperphosphorylation of RyR2 by protein kinase A, normalize channel function and improve cardiac function (Reiken et al. 2001; Doi et al. 2002). Heart failure can also be prevented by JTV519, which inhibits the dissociation of FKBP12.6 from RyR2; thereby stabilizing the channel, enhancing cooperativity among the subunits, and promoting coupled gating (Yano et al. 2003; Wehrens et al. 2005b). Overexpression of SERCA2 (Inesi et al. 2008; Kawase and Hajjar 2008), of pseudophosphorylated phospholamban (Hoshijima et al. 2002), or of FKBP12.6 (Huang et al. 2006), gene transfer of a phospholamban-targeted antibody (Dieterle et al. 2005), and down-regulation of phospholamban (Andino et al. 2008) can correct in vivo cardiac function. Modification of SERCA/phospholamban activity/expression is a promising target for remediation of cardiac disease. Indeed, a clinical trial of SERCA2a-gene therapy is initiated (Jaski et al. 2009).

VIRUS INFECTION

Complete virions or viral proteins can decrease the $[Ca^{2+}]_{ER}$. Some viruses stimulate Ca^{2+} release via the IP$_3$R, often by increasing the [IP$_3$] (Table 1). Other viruses release Ca^{2+} via an increased expression or function of the RyR. They may also decrease SERCA activity or expression, enhance the passive Ca^{2+} leak from the ER, or form pores in the ER membrane. Nef of human immunodeficiency virus type 1 directly interacts with the IP$_3$R and activates Ca^{2+} entry, without however inducing Ca^{2+} release (Foti et al. 1999; Manninen and Sakséla 2002).

ER Ca^{2+} depletion may be apoptotic or antiapoptotic, depending on the virus, its life cycle, and the induced pathology (Chami et al. 2006; Zhou et al. 2009). Ca^{2+} depletion by, for example, enteroviruses and human cytomegalovirus, delays apoptosis, giving the virus more time for replication. These viruses reduce ER-mitochondrial Ca^{2+} fluxes and prevent opening of the PTP with less release of cytochrome c and less caspase activation (van Kuppeveld et al. 2005; Sharon-Friling et al. 2006). ER and also Golgi Ca^{2+} depletion by e.g., enteroviruses leads to the accumulation of ER/Golgi-derived vesicles, where viral RNA replication takes place (van Kuppeveld et al. 2005), and inhibits vesicular protein trafficking and so down-regulates immune responses of the ghost (de Jong et al. 2006). In contrast, ER Ca^{2+} depletion by hepatitis C virus in liver promotes apoptosis and facilitates virion release because of translocation of Bax to the mitochondria, depolarization of the mitochondrial membrane, release of cytochrome c, and activation of caspase 3 (Benali-Furet et al. 2005). Abnormal Ca^{2+} signaling by Gp120 and Tat causes neuronal apoptosis and dysfunction and eventually AIDS dementia (Haughey and Mattson 2002). The decreased SERCA expression and increased RyR expression in Borna disease lead to ER stress, activation of the UPR and apoptotic degeneration of the cerebellum and hippocampus (Williams and Lipkin 2006). ER Ca^{2+} depletion also increases $[Ca^{2+}]_{cyt}$ and therefore activates Ca^{2+}-dependent enzymatic processes and transcription factors, promoting virus replication and the induction of a variety of responses.

Some antiviral drugs directly affect the $[Ca^{2+}]_{ER}$. Human immunodeficiency virus-protease inhibitors induce the accumulation of free cholesterol in the ER of macrophages, deplete ER Ca^{2+}, and induce ER stress and apoptosis (Zhou et al. 2005). This may explain the increased incidence of atherosclerosis and

Table 1. Effects of complete virions or viral proteins on the $[Ca^{2+}]_{ER}$.

Target	Effect	Virus or viral protein	Reference
IP$_3$R	Increased IP$_3$R activity	p12^I of human T-cell lymphotropic virus type 1	Ding et al. 2002
		glycoproteins of human herpes simplex virus type 1 and type 2	Cheshenko et al. 2003
	Increased IP$_3$R activity because of increased IP$_3$ production	influenza A virus	Hartshorn et al. 1988
		Poliovirus	Guinea et al. 1989
		gp120 and Tat of human immunodeficiency virus type 1	Dayanithi et al. 1995
			Mayne et al. 2000
			Haughey and Mattson 2002
		nonstructural protein 4 of rotavirus	Tian et al. 1995
			Dong et al. 1997
			Seo et al. 2008
		gp86 of human cytomegalovirus	Keay et al. 1995
		G-protein coupled receptor and viral macrophage inflammatory protein-I and -II of human herpes virus 8	Arvanitakis et al. 1997
			Nakano et al. 2003
RyR	Increased RyR activity	Tat of human immunodeficiency virus type 1	Norman et al. 2008
		Poliovirus	Brisac et al. 2010
	Increased RyR expression	Borna disease virus	Williams and Lipkin 2006
SERCA	Decreased SERCA activity	core protein of hepatitis C virus	Benali-Furet et al. 2005
	Decreased SERCA expression	Borna disease virus	Williams and Lipkin 2006
		latent membrane protein-1 of Epstein-Barr virus	Dellis et al. 2009
Passive Ca^{2+} leak	Enhanced passive Ca^{2+} leak	nonstructural protein 5A of hepatitis C virus	Robinson and Marchant 2008
ER membrane	Pore formation	p7 and core protein of hepatitis C virus	Griffin et al. 2003
			Bergqvist et al. 2003
		2B and 2BC proteins of entero- and rhinoviruses	Aldabe et al. 1997
			de Jong et al. 2008
		nonstructural protein 4 of rotavirus	Zhou et al. 2009
		pUL37x1 protein of human cytomegalovirus	Sharon-Friling et al. 2006
			Zhou et al. 2009
		6K protein of alphavirus	Antoine et al. 2007

cardiovascular disease in patients treated with protease inhibitors. Lopinavir and ritonavir also deplete ER Ca^{2+} and activate the UPR in intestinal epithelial cells, thus disrupting the epithelial barrier integrity with drug-induced diarrhea as a frequent side effect (Wu et al. 2010).

Bacteria can also cause ER stress. For example, Shiga toxins of *Shigella dysenteriae* serotype 1 and some serotypes of *Escherichia coli* deplete ER Ca^{2+} and trigger a UPR with apoptosis (Lee et al. 2008).

LUNG DISEASES

Asthma

SERCA2 in airway smooth muscle is downregulated in this chronic inflammatory disease with airway remodeling, leading to more sustained $[Ca^{2+}]_{cyt}$ increases and enhanced cell motility, proliferation and secretion (Mahn et al. 2009). Decreased SERCA expression might be caused by enhanced cytokine production during airway inflammation (Sathish et al. 2009).

ORMDL3 is a genetic risk factor associated with asthma (Moffatt et al. 2007). The gene encodes an ER protein (Hjelmqvist et al. 2002) that binds to and inhibits SERCA, leading to a decreased $[Ca^{2+}]_{ER}$ and an UPR (Cantero-Recasens et al. 2010).

Toxicity

Chronic exposure to cadmium in humans is associated with lung, but also bone and renal damage. Cadmium stimulates the IP_3R through IP_3 production, and inhibits SERCA (Biagioli et al. 2008). The reduced $[Ca^{2+}]_{ER}$ leads to ER stress and ER-mediated apoptosis.

LIVER DISEASES

Nonalcoholic Fatty Liver

Triglycerides and free fatty acids accumulate in the liver of obese individuals. Palmitate and stearate deplete ER Ca^{2+} stores and activate the UPR leading to cell death (Wei et al. 2009).

Cholestatic Liver Disease

Intrahepatic accumulation of bile acids induces hepatocellular injury. Glycochenodeoxycholic acid depletes ER Ca^{2+} and induces a UPR and apoptosis (Tsuchiya et al. 2006). It is unclear to what extent the decreasing $[Ca^{2+}]_{ER}$ directly contributes to the pathology.

Burn Injury

Severe burn injury impairs liver function. Thermal skin injury in rats depletes ER Ca^{2+} in the liver (Jeschke et al. 2010). This effect is because of an activation of the IP_3R by released cytochrome *c* and an increased IP_3R expression. ER Ca^{2+} depletion activates the UPR leading to apoptosis.

SKELETAL-MUSCLE DISEASES

Brody Disease

Mutations in the gene of SERCA1 (Odermatt et al. 1996) leading to reduced Ca^{2+}-pump expression or activity and hence a prolonged $[Ca^{2+}]_{cyt}$ elevation cause muscle cramping and impaired relaxation during exercise (Brody 1969). Chianina cattle congenital pseudomyotonia (Drogemuller et al. 2008) and Belgian Blue cattle congenital muscular dystony (Charlier et al. 2008) are related pathologies. Until now, no evidence for a UPR leading to apoptosis has been provided, but the ongoing contracture in cattle may induce rhabdomyolysis (Sacchetto et al. 2009).

Autosomal Centronuclear Myopathy

Centronuclear myopathies are characterized by small myofibers with centrally placed nuclei. Mutations of the muscle-specific inositol phosphatase MIP/MTMR14 cause the dominant form of the disease (Tosch et al. 2006; Shen et al. 2009). Mice deficient in this phosphatase produce less contractile force, have prolonged relaxation, and show exacerbated fatigue. $PtdIns(3,5)P_2$ and $PtdIns(3,4)P_2$ accumulate and directly activate RyR1, resulting in an increased Ca^{2+} leak, a lower $[Ca^{2+}]_{SR}$ and a higher $[Ca^{2+}]_{cyt}$. This proposed effect of $PtdIns(3,5)P_2$ and $PtdIns(3,4)P_2$ on RyR1 still needs confirmation.

Central Core Disease

Some mutations in the gene for RyR1 lead to hypotonia, proximal-muscle weakness, and central cores on muscle biopsy (Zhang et al. 1993). They can lead to a leaky channel and a reduced $[Ca^{2+}]_{SR}$, with deleterious consequences for contractions (Brini et al. 2005).

SKIN DISEASE

Darier disease is an inherited skin disorder with less adhesion between epidermal cells and abnormal keratinization. Mutations in the gene encoding SERCA2 (Sakuntabhai et al. 1999) lower the $[Ca^{2+}]_{ER}$ in keratinocytes (Foggia et al. 2006). ER stress may occur (Onozuka et al. 2006).

The fruit hull of mangosteen is used in Southeast Asia to treat skin infections and wounds (Mahabusarakam et al. 1987). α-mangostin inhibits SERCA, leading to a UPR and apoptosis (Sato et al. 2004).

CANCER

Malignant Transformation

Altered Ca^{2+} signaling may be involved in malignant transformation (Monteith et al. 2007). $[Ca^{2+}]_{ER}$ is often decreased, making the cell resistant to apoptosis. Subsequent Ca^{2+} entry increases $[Ca^{2+}]_{cyt}$ and changes gene expression, DNA repair, and cell-cycle regulation, resulting in cancer development (Korosec et al. 2006; Monteith et al. 2007; Lipskaia et al. 2009).

Human hepatitis B virus, an etiologic factor of hepatocellular carcinoma, integrates with its DNA into the gene for SERCA1 and *cis*-activates chimeric transcripts producing inactive proteins that deplete ER Ca^{2+} stores (Chami et al. 2000). Mice with a heterozygous deletion of the gene encoding SERCA2 (Liu et al. 2001) and some patients with Darier disease (Burge and Wilkinson 1992) develop squamous cell carcinomas. Neoplastic transformation has been linked to a down-regulated SERCA2 (Pacifico et al. 2003; Vanoverberghe et al. 2004; Bergner et al. 2009) or SERCA3 (Gelebart et al. 2002; Brouland et al. 2005), e.g., by somatic or germ-like mutations or epigenetic mechanisms involving promotor methylation (Endo et al. 2004; Korosec et al. 2006, 2008). ER Ca^{2+} depletion can also result from overexpression of Ca^{2+}-release channels. IP_3R3 is overexpressed in disseminated gastric cancer (Sakakura et al. 2003). The amplification of the gene for IP_3R2 increases in some tumors (Heighway et al. 1996). Increased IP_3R expression does not occur in all cancers (Bergner et al. 2009).

Anticancer Drugs

Most chemotherapeutic approaches kill tumor cells via the induction of MOMP. However, drugs that compromise the normal function and homeostasis of the ER may also induce programmed cell death or improve the therapeutic efficacy of existing anticancer drugs (Boelens et al. 2007). Some drugs primarily reduce the $[Ca^{2+}]_{ER}$ and in this way induce a UPR leading to apoptosis. Known SERCA blockers like thapsigargin and curcumin have anticancer activity (Denmeade et al. 2003; Anand et al. 2008; Bakhshi et al. 2008). Anticancer drugs like the stable analogue of the Bcl-2 antagonist HA 14-1 (Hermanson et al. 2009), artemisinin (Stockwin et al. 2009), amiloride analogues (Park et al. 2009), and 2,5-dimethyl-celecoxib (Johnson et al. 2002; Pyrko et al. 2007) also inhibit SERCA with ER stress as a result. SERCA2 expression decreases after photodynamic therapy with hypericin (Buytaert et al. 2006). ER Ca^{2+} stores are depleted by euplotin C through activated RyRs (Cervia et al. 2006), by paclitaxel through formation of Bax dimers in the ER (Liao et al. 2008), and by epigallocatechin gallate through inhibited protein processing at the level of glucosidase II (Magyar et al. 2009) and GRP78/BiP (Ermakova et al. 2006). Ca^{2+} depletion and ER stress are also induced by cisplatin (Nawrocki et al. 2005), dehydrocostuslactone (Hsu et al. 2009; Hung et al. 2010), honokiol (Chen et al. 2010), diaryl- and triarylmethanes (Abdelrahim et al. 2006), inhibitors of heat shock protein 90 (Taiyab et al. 2009), n-3 long-chain polyunsaturated fatty acids (Jakobsen et al. 2008), rhein (Lai et al. 2009), cardiotoxin III (Chien et al. 2008), homoharringtonine (Jie et al. 2007), berberine (Lin et al. 2007b), diindolylmethane (Savino et al. 2006), the multi-kinase inhibitor sorafenib (Rahmani et al. 2007), the p210 bcr-abl tyrosine-kinase inhibitor STI571 (Pattacini et al. 2004), parthenolide (Zhang et al. 2004), photodynamic therapy with tetra-*S*-glycosylated porphyrin (Thompson et al. 2008), and by many other drugs. Edelfosine leads to Bax/Bak-mediated ER Ca^{2+} depletion and apoptosis, without inducing a UPR (Nieto-Miguel et al. 2007).

Ca^{2+} depletion-induced ER stress can also lead to autophagy, for example, in response to the tyrosine-kinase inhibitor imatinib (Bellodi et al. 2009), or to necrosis, for example, in therapy-resistant tumors with down-regulated Bax or Bak (Janssen et al. 2009).

CONCLUDING REMARKS

Depletion of ER Ca^{2+} occurs in many diseases. The accompanying ER stress often triggers a UPR leading to apoptosis. The release of insufficient activator Ca^{2+} may compromise essential cell functions. We now begin to understand the molecular mechanisms that reduce the ER Ca^{2+} content. Some therapies already directly target the Ca^{2+}-signaling pathway. A better understanding of the defective Ca^{2+} signal and the development of better drugs targeting the proteins involved will eventually result in better treatments for these various diseases.

ACKNOWLEDGMENTS

Work performed in our laboratory was supported by grants from the Research Foundation—Flanders, the Concerted Actions of the K.U.Leuven, and the Interuniversity Attraction Poles Programme.

REFERENCES

Abdelrahim M, Newman K, Vanderlaag K, Samudio I, Safe S. 2006. 3,3'-diindolylmethane (DIM) and its derivatives induce apoptosis in pancreatic cancer cells through endoplasmic reticulum stress-dependent upregulation of DR5. *Carcinogenesis* **27**: 717–728.

Adachi T, Weisbrod RM, Pimentel DR, Ying J, Sharov VS, Schoneich C, Cohen RA. 2004. S-glutathiolation by peroxynitrite activates SERCA during arterial relaxation by nitric oxide. *Nat Med* **10**: 1200–1207.

Ai X, Curran JW, Shannon TR, Bers DM, Pogwizd SM. 2005. Ca^{2+}/calmodulin-dependent protein kinase modulates cardiac ryanodine receptor phosphorylation and sarcoplasmic reticulum Ca^{2+} leak in heart failure. *Circ Res* **97**: 1314–1322.

Akerfeldt MC, Howes J, Chan JY, Stevens VA, Boubenna N, McGuire HM, King C, Biden TJ, Laybutt DR. 2008. Cytokine-induced β-cell death is independent of endoplasmic reticulum stress signaling. *Diabetes* **57**: 3034–3044.

Aldabe R, Irurzun A, Carrasco L. 1997. Poliovirus protein 2BC increases cytosolic free calcium concentrations. *J Virol* **71**: 6214–6217.

Almind K, Bjorbaek C, Vestergaard H, Hansen T, Echwald S, Pedersen O. 1993. Aminoacid polymorphisms of insulin receptor substrate-1 in non-insulin-dependent diabetes mellitus. *Lancet* **342**: 828–832.

Anand P, Sundaram C, Jhurani S, Kunnumakkara AB, Aggarwal BB. 2008. Curcumin and cancer: an "old-age" disease with an "age-old" solution. *Cancer Lett* **267**: 133–164.

Andino LM, Takeda M, Kasahara H, Jakymiw A, Byrne BJ, Lewin AS. 2008. AAV-mediated knockdown of phospholamban leads to improved contractility and calcium handling in cardiomyocytes. *J Gene Med* **10**: 132–142.

Antoine AF, Montpellier C, Cailliau K, Browaeys-Poly E, Vilain JP, Dubuisson J. 2007. The αvirus 6K protein activates endogenous ionic conductances when expressed in Xenopus oocytes. *J Membr Biol* **215**: 37–48.

Araki E, Oyadomari S, Mori M. 2003. Impact of endoplasmic reticulum stress pathway on pancreatic β-cells and diabetes mellitus. *Exp Biol Med* **228**: 1213–1217.

Arredouani A, Guiot Y, Jonas JC, Liu LH, Nenquin M, Pertusa JA, Rahier J, Rolland JF, Shull GE, Stevens M, et al. 2002. SERCA3 ablation does not impair insulin secretion but suggests distinct roles of different sarcoendoplasmic reticulum Ca^{2+} pumps for Ca^{2+} homeostasis in pancreatic β-cells. *Diabetes* **51**: 3245–3253.

Arvanitakis L, Geras-Raaka E, Varma A, Gershengorn MC, Cesarman E. 1997. Human herpesvirus KSHV encodes a constitutively active G-protein-coupled receptor linked to cell proliferation. *Nature* **385**: 347–350.

Austin RC. 2009. The unfolded protein response in health and disease. *Antioxid Redox Signal* **11**: 2279–2287.

Bakhshi J, Weinstein L, Poksay KS, Nishinaga B, Bredesen DE, Rao RV. 2008. Coupling endoplasmic reticulum stress to the cell death program in mouse melanoma cells: effect of curcumin. *Apoptosis* **13**: 904–914.

Beard NA, Laver DR, Dulhunty AF. 2004. Calsequestrin and the calcium release channel of skeletal and cardiac muscle. *Prog Biophys Mol Biol* **85**: 33–69.

Belke DD, Swanson EA, Dillmann WH. 2004. Decreased sarcoplasmic reticulum activity and contractility in diabetic db/db mouse heart. *Diabetes* **53**: 3201–3208.

Bellodi C, Lidonnici MR, Hamilton A, Helgason GV, Soliera AR, Ronchetti M, Galavotti S, Young KW, Selmi T, Yacobi R, et al. 2009. Targeting autophagy potentiates tyrosine kinase inhibitor-induced cell death in Philadelphia chromosome-positive cells, including primary CML stem cells. *J Clin Invest* **119**: 1109–1123.

Belmont PJ, Tadimalla A, Chen WJ, Martindale JJ, Thuerauf DJ, Marcinko M, Gude N, Sussman MA, Glembotski CC. 2008. Coordination of growth and endoplasmic reticulum stress signaling by regulator of calcineurin 1 (RCAN1), a novel ATF6-inducible gene. *J Biol Chem* **283**: 14012–14021.

Benali-Furet NL, Chami M, Houel L, De Giorgi F, Vernejoul F, Lagorce D, Buscail L, Bartenschlager R, Ichas F, Rizzuto R, et al. 2005. Hepatitis C virus core triggers apoptosis in liver cells by inducing ER stress and ER calcium depletion. *Oncogene* **24**: 4921–4933.

Bencsik P, Kupai K, Giricz Z, Gorbe A, Huliak I, Furst S, Dux L, Csont T, Jancso G, Ferdinandy P. 2008. Cardiac capsaicin-sensitive sensory nerves regulate myocardial relaxation via S-nitrosylation of SERCA: role of peroxynitrite. *Br J Pharmacol* **153**: 488–496.

Bergner A, Kellner J, Tufman A, Huber RM. 2009. Endoplasmic reticulum Ca^{2+}-homeostasis is altered in small and non-small cell lung cancer cell lines. *J Exp Clin Cancer Res* **28:** 25. doi:101186/1756-9966-28-25.

Bergqvist A, Sundstrom S, Dimberg LY, Gylfe E, Masucci MG. 2003. The hepatitis C virus core protein modulates T cell responses by inducing spontaneous and altering T-cell receptor-triggered Ca^{2+} oscillations. *J Biol Chem* **278:** 18877–18883.

Bernales S, McDonald KL, Walter P. 2006. Autophagy counterbalances endoplasmic reticulum expansion during the unfolded protein response. *PLoS Biol* **4:** e423. doi:101371/journalpbio0040423.

Berridge MJ. 2002. The endoplasmic reticulum: a multifunctional signaling organelle. *Cell Calcium* **32:** 235–249.

Berridge MJ. 2010. Calcium hypothesis of Alzheimer's disease. *Pflugers Arch* **459:** 441–449.

Berridge MJ, Lipp P, Bootman MD. 2000. The versatility and universality of calcium signalling. *Nat Rev Mol Cell Biol* **1:** 11–21.

Bers DM, Eisner DA, Valdivia HH. 2003. Sarcoplasmic reticulum Ca^{2+} and heart failure: roles of diastolic leak and Ca^{2+} transport. *Circ Res* **93:** 487–490.

Bevers MB, Neumar RW. 2008. Mechanistic role of calpains in postischemic neurodegeneration. *J Cereb Blood Flow Metab* **28:** 655–673.

Bezprozvanny I. 2009. Calcium signaling and neurodegenerative diseases. *Trends Mol Med* **15:** 89–100.

Biagioli M, Pifferi S, Ragghianti M, Bucci S, Rizzuto R, Pinton P. 2008. Endoplasmic reticulum stress and alteration in calcium homeostasis are involved in cadmium-induced apoptosis. *Cell Calcium* **43:** 184–195.

Bidasee KR, Nallani K, Besch HR Jr, Dincer UD. 2003a. Streptozotocin-induced diabetes increases disulfide bond formation on cardiac ryanodine receptor (RyR2). *J Pharmacol Exp Ther* **305:** 989–998.

Bidasee KR, Nallani K, Yu Y, Cocklin RR, Zhang Y, Wang M, Dincer UD, Besch HR Jr. 2003b. Chronic diabetes increases advanced glycation end products on cardiac ryanodine receptors/calcium-release channels. *Diabetes* **52:** 1825–1836.

Bidasee KR, Zhang Y, Shao CH, Wang M, Patel KP, Dincer UD, Besch HR Jr. 2004. Diabetes increases formation of advanced glycation end products on sarco(endo)plasmic reticulum Ca^{2+}-ATPase. *Diabetes* **53:** 463–473.

Bidaud I, Mezghrani A, Swayne LA, Monteil A, Lory P. 2006. Voltage-gated calcium channels in genetic diseases. *Biochim Biophys Acta* **1763:** 1169–1174.

Blayney LM, Lai FA. 2009. Ryanodine receptor-mediated arrhythmias and sudden cardiac death. *Pharmacol Ther* **123:** 151–177.

Boelens J, Lust S, Offner F, Bracke ME, Vanhoecke BW. 2007. The endoplasmic reticulum: a target for new anticancer drugs. *In Vivo* **21:** 215–226.

Borge PD Jr, Wolf BA. 2003. Insulin receptor substrate 1 regulation of sarco-endoplasmic reticulum calcium ATPase 3 in insulin-secreting β-cells. *J Biol Chem* **278:** 11359–11368.

Bosanac I, Michikawa T, Mikoshiba K, Ikura M. 2004. Structural insights into the regulatory mechanism of IP_3 receptor. *Biochim Biophys Acta* **1742:** 89–102.

Brandman O, Liou J, Park WS, Meyer T. 2007. STIM2 is a feedback regulator that stabilizes basal cytosolic and endoplasmic reticulum Ca^{2+} levels. *Cell* **131:** 1327–1339.

Brini M, Carafoli E. 2009. Calcium pumps in health and disease. *Physiol Rev* **89:** 1341–1378.

Brini M, Manni S, Pierobon N, Du GG, Sharma P, MacLennan DH, Carafoli E. 2005. Ca^{2+} signaling in HEK-293 and skeletal muscle cells expressing recombinant ryanodine receptors harboring malignant hyperthermia and central core disease mutations. *J Biol Chem* **280:** 15380–15389.

Brisac C, Teoule F, Autret A, Pelletier I, Colbere-Garapin F, Brenner C, Lemaire C, Blondel B. 2010. Calcium flux between the endoplasmic reticulum and mitochondria contributes to poliovirus-induced apoptosis. *J Virol* **84:** 12226–12235.

Brody IA. 1969. Muscle contracture induced by exercise: a syndrome attributable to decreased relaxing factor. *N Engl J Med* **281:** 187–192.

Brosius FC III, Kaufman RJ. 2008. Is the ER stressed out in diabetic kidney disease? *J Am Soc Nephrol* **19:** 2040–2042.

Brouland JP, Gelebart P, Kovacs T, Enouf J, Grossmann J, Papp B. 2005. The loss of sarco/endoplasmic reticulum calcium transport ATPase 3 expression is an early event during the multistep process of colon carcinogenesis. *Am J Pathol* **167:** 233–242.

Brunelle JK, Letai A. 2009. Control of mitochondrial apoptosis by the Bcl-2 family. *J Cell Sci* **122:** 437–441.

Bull R, Finkelstein JP, Galvez J, Sanchez G, Donoso P, Behrens MI, Hidalgo C. 2008. Ischemia enhances activation by Ca^{2+} and redox modification of ryanodine receptor channels from rat brain cortex. *J Neurosci* **28:** 9463–9472.

Burge SM, Wilkinson JD. 1992. Darier-White disease: a review of the clinical features in 163 patients. *J Am Acad Dermatol* **27:** 40–50.

Buytaert E, Callewaert G, Hendrickx N, Scorrano L, Hartmann D, Missiaen L, Vandenheede JR, Heirman I, Grooten J, Agostinis P. 2006. Role of endoplasmic reticulum depletion and multidomain proapoptotic BAX and BAK proteins in shaping cell death after hypericin-mediated photodynamic therapy. *FASEB J* **20:** 756–758.

Cahalan MD. 2009. STIMulating store-operated Ca^{2+} entry. *Nat Cell Biol* **11:** 669–677.

Cantero-Recasens G, Fandos C, Rubio-Moscardo F, Valverde MA, Vicente R. 2010. The asthma-associated ORMDL3 gene product regulates endoplasmic reticulum-mediated calcium signaling and cellular stress. *Hum Mol Genet* **19:** 111–121.

Cardozo AK, Ortis F, Storling J, Feng YM, Rasschaert J, Tonnesen M, Van Eylen F, Mandrup-Poulsen T, Herchuelz A, Eizirik DL. 2005. Cytokines down-regulate the sarcoendoplasmic reticulum pump Ca^{2+} ATPase 2b and deplete endoplasmic reticulum Ca^{2+}, leading to induction of endoplasmic reticulum stress in pancreatic β-cells. *Diabetes* **54:** 452–461.

Cervia D, Martini D, Garcia-Gil M, Di Giuseppe G, Guella G, Dini F, Bagnoli P. 2006. Cytotoxic effects and apoptotic signalling mechanisms of the sesquiterpenoid euplotin C, a secondary metabolite of the marine ciliate Euplotes crassus, in tumour cells. *Apoptosis* **11:** 829–843.

Chaabane C, Corvazier E, Bredoux R, Dally S, Raies A, Villemain A, Dupuy E, Enouf J, Bobe R. 2006. Sarco/endoplasmic reticulum Ca^{2+}ATPase type 3 isoforms (SERCA3b and SERCA3f): distinct roles in cell adhesion and ER stress. *Biochem Biophys Res Commun* **345**: 1377–1385.

Chaabane C, Dally S, Corvazier E, Bredoux R, Bobe R, Ftouhi B, Raies A, Enouf J. 2007. Platelet PMCA- and SERCA-type Ca^{2+}-ATPase expression in diabetes: a novel signature of abnormal megakaryocytopoiesis. *J Thromb Haemost* **5**: 2127–2135.

Chami M, Oules B, Paterlini-Brechot P. 2006. Cytobiological consequences of calcium-signaling alterations induced by human viral proteins. *Biochim Biophys Acta* **1763**: 1344–1362.

Chami M, Gozuacik D, Saigo K, Capiod T, Falson P, Lecoeur H, Urashima T, Beckmann J, Gougeon ML, Claret M, et al. 2000. Hepatitis B virus-related insertional mutagenesis implicates SERCA1 gene in the control of apoptosis. *Oncogene* **19**: 2877–2886.

Chami M, Oules B, Szabadkai G, Tacine R, Rizzuto R, Paterlini-Brechot P. 2008. Role of SERCA1 truncated isoform in the proapoptotic calcium transfer from ER to mitochondria during ER stress. *Mol Cell* **32**: 641–651.

Charlier C, Coppieters W, Rollin F, Desmecht D, Agerholm JS, Cambisano N, Carta E, Dardano S, Dive M, Fasquelle C, et al. 2008. Highly effective SNP-based association mapping and management of recessive defects in livestock. *Nat Genet* **40**: 449–454.

Chen Y, Liu CP, Xu KF, Mao XD, Lu YB, Fang L, Yang JW, Liu C. 2008. Effect of taurine-conjugated ursodeoxycholic acid on endoplasmic reticulum stress and apoptosis induced by advanced glycation end products in cultured mouse podocytes. *Am J Nephrol* **28**: 1014–1022.

Chen YJ, Wu CL, Liu JF, Fong YC, Hsu SF, Li TM, Su YC, Liu SH, Tang CH. 2010. Honokiol induces cell apoptosis in human chondrosarcoma cells through mitochondrial dysfunction and endoplasmic reticulum stress. *Cancer Lett* **291**: 20–30.

Cheshenko N, Del Rosario B, Woda C, Marcellino D, Satlin LM, Herold BC. 2003. Herpes simplex virus triggers activation of calcium-signaling pathways. *J Cell Biol* **163**: 283–293.

Chien CM, Yang SH, Chang LS, Lin SR. 2008. Involvement of both endoplasmic reticulum- and mitochondria-dependent pathways in cardiotoxin III-induced apoptosis in HL-60 cells. *Clin Exp Pharmacol Physiol* **35**: 1059–1064.

Chipuk JE, Green DR. 2008. How do BCL-2 proteins induce mitochondrial outer membrane permeabilization? *Trends Cell Biol* **18**: 157–164.

Choi KM, Zhong Y, Hoit BD, Grupp IL, Hahn H, Dilly KW, Guatimosim S, Lederer WJ, Matlib MA. 2002. Defective intracellular Ca^{2+} signaling contributes to cardiomyopathy in type 1 diabetic rats. *Am J Physiol* **283**: H1398–H1408.

Clark RJ, McDonough PM, Swanson E, Trost SU, Suzuki M, Fukuda M, Dillmann WH. 2003. Diabetes and the accompanying hyperglycemia impairs cardiomyocyte calcium cycling through increased nuclear O-GlcNAcylation. *J Biol Chem* **278**: 44230–44237.

Cooper ME, Bonnet F, Oldfield M, Jandeleit-Dahm K. 2001. Mechanisms of diabetic vasculopathy: an overview. *Am J Hypert* **14**: 475–486.

Cunha DA, Hekerman P, Ladriere L, Bazarra-Castro A, Ortis F, Wakeham MC, Moore F, Rasschaert J, Cardozo AK, Bellomo E, et al. 2008. Initiation and execution of lipotoxic ER stress in pancreatic β-cells. *J Cell Sci* **121**: 2308–2318.

Dahl C, Haug LS, Spilsberg B, Johansen J, Ostvold AC, Diemer NH. 2000. Reduced [^{3}H]IP$_3$ binding but unchanged IP$_3$ receptor levels in the rat hippocampus CA1 region following transient global ischemia and tolerance induction. *Neurochem Int* **36**: 379–388.

Dally S, Monceau V, Corvazier E, Bredoux R, Raies A, Bobe R, del Monte F, Enouf J. 2009. Compartmentalized expression of three novel sarco/endoplasmic reticulum Ca^{2+}ATPase 3 isoforms including the switch to ER stress, SERCA3f, in non-failing and failing human heart. *Cell Calcium* **45**: 144–154.

Dayanithi G, Yahi N, Baghdiguian S, Fantini J. 1995. Intracellular calcium release induced by human immunodeficiency virus type 1 (HIV-1) surface envelope glycoprotein in human intestinal epithelial cells: a putative mechanism for HIV-1 enteropathy. *Cell Calcium* **18**: 9–18.

DeGracia DJ, Kumar R, Owen CR, Krause GS, White BC. 2002. Molecular pathways of protein synthesis inhibition during brain reperfusion: implications for neuronal survival or death. *J Cereb Blood Flow Metab* **22**: 127–141.

de Jong AS, de Mattia F, Van Dommelen MM, Lanke K, Melchers WJ, Willems PH, van Kuppeveld FJ. 2008. Functional analysis of picornavirus 2B proteins: effects on calcium homeostasis and intracellular protein trafficking. *J Virol* **82**: 3782–3790.

de Jong AS, Visch HJ, de Mattia F, van Dommelen MM, Swarts HG, Luyten T, Callewaert G, Melchers WJ, Willems PH, van Kuppeveld FJ. 2006. The coxsackievirus 2B protein increases efflux of ions from the endoplasmic reticulum and Golgi, thereby inhibiting protein trafficking through the Golgi. *J Biol Chem* **281**: 14144–14150.

Dellis O, Arbabian A, Brouland JP, Kovacs T, Rowe M, Chomienne C, Joab I, Papp B. 2009. Modulation of B-cell endoplasmic reticulum calcium homeostasis by Epstein-Barr virus latent membrane protein-1. *Mol Cancer* **8**: 59. doi:101186/1476-4598-8-59.

del Monte F, Hajjar RJ. 2008. Intracellular devastation in heart failure. *Heart Fail Rev* **13**: 151–162.

Deng X, Wang Y, Zhou Y, Soboloff J, Gill DL. 2009. STIM and Orai: dynamic intermembrane coupling to control cellular calcium signals. *J Biol Chem* **284**: 22501–22505.

Denmeade SR, Jakobsen CM, Janssen S, Khan SR, Garrett ES, Lilja H, Christensen SB, Isaacs JT. 2003. Prostate-specific antigen-activated thapsigargin prodrug as targeted therapy for prostate cancer. *J Natl Cancer Inst* **95**: 990–1000.

Dickhout JG, Sood SK, Austin RC. 2007. Role of endoplasmic reticulum calcium disequilibria in the mechanism of homocysteine-induced ER stress. *Antioxid Redox Signal* **9**: 1863–1873.

Dickhout JG, Hossain GS, Pozza LM, Zhou J, Lhotak S, Austin RC. 2005. Peroxynitrite causes endoplasmic reticulum stress and apoptosis in human vascular endothelium: implications in atherogenesis. *Arterioscler Thromb Vasc Biol* **25**: 2623–2629.

Dieterle T, Meyer M, Gu Y, Belke DD, Swanson E, Iwatate M, Hollander J, Peterson KL, Ross J Jr, Dillmann WH. 2005. Gene transfer of a phospholamban-targeted antibody improves calcium handling and cardiac function in heart failure. *Cardiovasc Res* **67**: 678–688.

Ding W, Albrecht B, Kelley RE, Muthusamy N, Kim SJ, Altschuld RA, Lairmore MD. 2002. Human T-cell lymphotropic virus type 1 p12^I expression increases cytoplasmic calcium to enhance the activation of nuclear factor of activated T cells. *J Virol* **76**: 10374–10382.

Doi M, Yano M, Kobayashi S, Kohno M, Tokuhisa T, Okuda S, Suetsugu M, Hisamatsu Y, Ohkusa T, Kohno M, et al. 2002. Propranolol prevents the development of heart failure by restoring FKBP12.6-mediated stabilization of ryanodine receptor. *Circulation* **105**: 1374–1379.

Dong Y, Zeng CQ, Ball JM, Estes MK, Morris AP. 1997. The rotavirus enterotoxin NSP4 mobilizes intracellular calcium in human intestinal cells by stimulating phospholipase C-mediated inositol 1,4,5-trisphosphate production. *Proc Natl Acad Sci* **94**: 3960–3965.

Doutheil J, Althausen S, Treiman M, Paschen W. 2000. Effect of nitric oxide on endoplasmic reticulum calcium homeostasis, protein synthesis and energy metabolism. *Cell Calcium* **27**: 107–115.

Drogemuller C, Drogemuller M, Leeb T, Mascarello F, Testoni S, Rossi M, Gentile A, Damiani E, Sacchetto R. 2008. Identification of a missense mutation in the bovine ATP2A1 gene in congenital pseudomyotonia of Chianina cattle: an animal model of human Brody disease. *Genomics* **92**: 474–477.

Durham WJ, Wehrens XH, Sood S, Hamilton SL. 2007. Diseases associated with altered ryanodine receptor activity. *Subcell Biochem* **45**: 273–321.

Dutta K, Carmody MW, Cala SE, Davidoff AJ. 2002. Depressed PKA activity contributes to impaired SERCA function and is linked to the pathogenesis of glucose-induced cardiomyopathy. *J Mol Cell Cardiol* **34**: 985–996.

Egger L, Schneider J, Rheme C, Tapernoux M, Hacki J, Borner C. 2003. Serine proteases mediate apoptosis-like cell death and phagocytosis under caspase-inhibiting conditions. *Cell Death Differ* **10**: 1188–1203.

Eizirik DL, Cardozo AK, Cnop M. 2008. The role for endoplasmic reticulum stress in diabetes mellitus. *Endocr Rev* **29**: 42–61.

Endo M. 2009. Calcium-induced calcium release in skeletal muscle. *Physiol Rev* **89**: 1153–1176.

Endo Y, Uzawa K, Mochida Y, Shiiba M, Bukawa H, Yokoe H, Tanzawa H. 2004. Sarcoendoplasmic reticulum Ca^{2+} ATPase type 2 downregulated in human oral squamous cell carcinoma. *Int J Cancer* **110**: 225–231.

Ermakova SP, Kang BS, Choi BY, Choi HS, Schuster TF, Ma WY, Bode AM, Dong Z. 2006. (-)-Epigallocatechin gallate overcomes resistance to etoposide-induced cell death by targeting the molecular chaperone glucose-regulated protein 78. *Cancer Res* **66**: 9260–9269.

Evans-Molina C, Robbins RD, Kono T, Tersey SA, Vestermark GL, Nunemaker CS, Garmey JC, Deering TG, Keller SR, Maier B, et al. 2009. Peroxisome proliferator-activated receptor γ activation restores islet function in diabetic mice through reduction of endoplasmic reticulum stress and maintenance of euchromatin structure. *Mol Cell Biol* **29**: 2053–2067.

Fan J, Watanabe T. 2003. Inflammatory reactions in the pathogenesis of atherosclerosis. *J Atheroscler Thromb* **10**: 63–71.

Farfel-Becker T, Vitner E, Dekel H, Leshem N, Enquist IB, Karlsson S, Futerman AH. 2009. No evidence for activation of the unfolded protein response in neuronopathic models of Gaucher disease. *Hum Mol Genet* **18**: 1482–1488.

Fedirko NV, Kruglikov IA, Kopach OV, Vats JA, Kostyuk PG, Voitenko NV. 2006. Changes in functioning of rat submandibular salivary gland under streptozotocin-induced diabetes are associated with alterations of Ca^{2+} signaling and Ca^{2+} transporting pumps. *Biochim Biophys Acta* **1762**: 294–303.

Feng B, Yao PM, Li Y, Devlin CM, Zhang D, Harding HP, Sweeney M, Rong JX, Kuriakose G, Fisher EA, et al. 2003. The endoplasmic reticulum is the site of cholesterol-induced cytotoxicity in macrophages. *Nat Cell Biol* **5**: 781–792.

Fernyhough P, Calcutt NA. 2010. Abnormal calcium homeostasis in peripheral neuropathies. *Cell Calcium* **47**: 130–139.

Ferreiro E, Resende R, Costa R, Oliveira CR, Pereira CM. 2006. An endoplasmic-reticulum-specific apoptotic pathway is involved in prion and amyloid-β peptides neurotoxicity. *Neurobiol Dis* **23**: 669–678.

Ferreiro E, Costa R, Marques S, Cardoso SM, Oliveira CR, Pereira CM. 2008. Involvement of mitochondria in endoplasmic reticulum stress-induced apoptotic cell death pathway triggered by the prion peptide PrP(106-126). *J Neurochem* **104**: 766–776.

Foggia L, Hovnanian A. 2004. Calcium pump disorders of the skin. *Am J Med Genet C* **131C**: 20–31.

Foggia L, Aronchik I, Aberg K, Brown B, Hovnanian A, Mauro TM. 2006. Activity of the hSPCA1 Golgi Ca^{2+} pump is essential for Ca^{2+}-mediated Ca^{2+} response and cell viability in Darier disease. *J Cell Sci* **119**: 671–679.

Fonseca SG, Ishigaki S, Oslowski CM, Lu S, Lipson KL, Ghosh R, Hayashi E, Ishihara H, Oka Y, Permutt MA, et al. 2010. Wolfram syndrome 1 gene negatively regulates ER stress signaling in rodent and human cells. *J Clin Invest* **120**: 744–755.

Foskett JK, White C, Cheung KH, Mak DO. 2007. Inositol trisphosphate receptor Ca^{2+} release channels. *Physiol Rev* **87**: 593–658.

Foti M, Cartier L, Piguet V, Lew DP, Carpentier JL, Trono D, Krause KH. 1999. The HIV Nef protein alters Ca^{2+} signaling in myelomonocytic cells through SH3-mediated protein-protein interactions. *J Biol Chem* **274**: 34765–34772.

Frank KF, Bolck B, Brixius K, Kranias EG, Schwinger RH. 2002. Modulation of SERCA: implications for the failing human heart. *Basic Res Cardiol* **97**: I72–I78.

Franz WM, Muller OJ, Katus HA. 2001. Cardiomyopathies: from genetics to the prospect of treatment. *Lancet* **358**: 1627–1637.

French JP, Quindry JC, Falk DJ, Staib JL, Lee Y, Wang KK, Powers SK. 2006. Ischemia-reperfusion-induced calpain activation and SERCA2a degradation are attenuated by exercise training and calpain inhibition. *Am J Physiol* **290:** H128–H136.

Gaut JR, Hendershot LM. 1993. The modification and assembly of proteins in the endoplasmic reticulum. *Curr Opin Cell Biol* **5:** 589–595.

Gelebart P, Kovacs T, Brouland JP, van Gorp R, Grossmann J, Rivard N, Panis Y, Martin V, Bredoux R, Enouf J, et al. 2002. Expression of endomembrane calcium pumps in colon and gastric cancer cells. Induction of SERCA3 expression during differentiation. *J Biol Chem* **277:** 26310–26320.

Gemes G, Rigaud M, Weyker PD, Abram SE, Weihrauch D, Poroli M, Zoga V, Hogan QH. 2009. Depletion of calcium stores in injured sensory neurons: anatomic and functional correlates. *Anesthesiology* **111:** 393–405.

Ginzburg L, Futerman AH. 2005. Defective calcium homeostasis in the cerebellum in a mouse model of Niemann-Pick A disease. *J Neurochem* **95:** 1619–1628.

Ginzburg L, Li SC, Li YT, Futerman AH. 2008. An exposed carboxyl group on sialic acid is essential for gangliosides to inhibit calcium uptake via the sarco/endoplasmic reticulum Ca^{2+}-ATPase: relevance to gangliosidoses. *J Neurochem* **104:** 140–146.

Giorgi C, De Stefani D, Bononi A, Rizzuto R, Pinton P. 2009. Structural and functional link between the mitochondrial network and the endoplasmic reticulum. *Int J Biochem Cell Biol* **41:** 1817–1827.

Giorgi C, Romagnoli A, Pinton P, Rizzuto R. 2008. Ca^{2+} signaling, mitochondria and cell death. *Curr Mol Med* **8:** 119–130.

Gotoh T, Mori M. 2006. Nitric oxide and endoplasmic reticulum stress. *Arterioscler Thromb Vasc Biol* **26:** 1439–1446.

Griffin SD, Beales LP, Clarke DS, Worsfold O, Evans SD, Jaeger J, Harris MP, Rowlands DJ. 2003. The p7 protein of hepatitis C virus forms an ion channel that is blocked by the antiviral drug, Amantadine. *FEBS Lett* **535:** 34–38.

Grover AK, Kwan CY, Samson SE. 2003. Effects of peroxynitrite on sarco/endoplasmic reticulum Ca^{2+} pump isoforms SERCA2b and SERCA3a. *Am J Physiol* **285:** C1537–C1543.

Guest PC, Bailyes EM, Hutton JC. 1997. Endoplasmic reticulum Ca^{2+} is important for the proteolytic processing and intracellular transport of proinsulin in the pancreatic β-cell. *Biochem J* **323:** 445–450.

Guinea R, Lopez-Rivas A, Carrasco L. 1989. Modification of phospholipase C and phospholipase A2 activities during poliovirus infection. *J Biol Chem* **264:** 21923–21927.

Guner S, Arioglu E, Tay A, Tasdelen A, Aslamaci S, Bidasee KR, Dincer UD. 2004. Diabetes decreases mRNA levels of calcium-release channels in human atrial appendage. *Mol Cell Biochem* **263:** 143–150.

Gurzov EN, Ortis F, Cunha DA, Gosset G, Li M, Cardozo AK, Eizirik DL. 2009. Signaling by IL-1β+IFN-γ and ER stress converge on DP5/Hrk activation: a novel mechanism for pancreatic β-cell apoptosis. *Cell Death Differ* **16:** 1539–1550.

Gwiazda KS, Yang TL, Lin Y, Johnson JD. 2009. Effects of palmitate on ER and cytosolic Ca^{2+} homeostasis in β-cells. *Am J Physiol* **296:** E690–E701.

Gyorke S. 2009. Molecular basis of catecholaminergic polymorphic ventricular tachycardia. *Heart Rhythm* **6:** 123–129.

Gyorke S, Terentyev D. 2008. Modulation of ryanodine receptor by luminal calcium and accessory proteins in health and cardiac disease. *Cardiovasc Res* **77:** 245–255.

Haghighi K, Kolokathis F, Gramolini AO, Waggoner JR, Pater L, Lynch RA, Fan GC, Tsiapras D, Parekh RR, Dorn GW II, et al. 2006. A mutation in the human phospholamban gene, deleting arginine 14, results in lethal, hereditary cardiomyopathy. *Proc Natl Acad Sci* **103:** 1388–1393.

Han I, Kudlow JE. 1997. Reduced O glycosylation of Sp1 is associated with increased proteasome susceptibility. *Mol Cell Biol* **17:** 2550–2558.

Harding HP, Zhang Y, Ron D. 1999. Protein translation and folding are coupled by an endoplasmic-reticulum-resident kinase. *Nature* **397:** 271–274.

Harding HP, Zhang Y, Bertolotti A, Zeng H, Ron D. 2000. Perk is essential for translational regulation and cell survival during the unfolded protein response. *Mol Cell* **5:** 897–904.

Harding HP, Zhang Y, Zeng H, Novoa I, Lu PD, Calfon M, Sadri N, Yun C, Popko B, Paules R, et al. 2003. An integrated stress response regulates amino acid metabolism and resistance to oxidative stress. *Mol Cell* **11:** 619–633.

Hartshorn KL, Collamer M, Auerbach M, Myers JB, Pavlotsky N, Tauber AI. 1988. Effects of influenza A virus on human neutrophil calcium metabolism. *J Immunol* **141:** 1295–1301.

Hasenfuss G, Pieske B. 2002. Calcium cycling in congestive heart failure. *J Mol Cell Cardiol* **34:** 951–969.

Haughey NJ, Mattson MP. 2002. Calcium dysregulation and neuronal apoptosis by the HIV-1 proteins Tat and gp120. *J Acquir Immune Defic Syndr* **31 Suppl 2:** S55–S61.

Heighway J, Betticher DC, Hoban PR, Altermatt HJ, Cowen R. 1996. Coamplification in tumors of KRAS2, type 2 inositol 1,4,5 triphosphate receptor gene, and a novel human gene, KRAG. *Genomics* **35:** 207–214.

Hermanson D, Addo SN, Bajer AA, Marchant JS, Das SG, Srinivasan B, Al-Mousa F, Michelangeli F, Thomas DD, Lebien TW, et al. 2009. Dual mechanisms of sHA 14-1 in inducing cell death through endoplasmic reticulum and mitochondria. *Mol Pharmacol* **76:** 667–678.

Hetz C, Russelakis-Carneiro M, Maundrell K, Castilla J, Soto C. 2003. Caspase-12 and endoplasmic reticulum stress mediate neurotoxicity of pathological prion protein. *EMBO J* **22:** 5435–5445.

Hjelmqvist L, Tuson M, Marfany G, Herrero E, Balcells S, Gonzalez-Duarte R. 2002. ORMDL proteins are a conserved new family of endoplasmic reticulum membrane proteins. *Genome Biol* **3:** RESEARCH0027. doi:101186/gb-2002-3-6-research0027.

Hollien J, Weissman JS. 2006. Decay of endoplasmic reticulum-localized mRNAs during the unfolded protein response. *Science* **313:** 104–107.

Hoshijima M, Ikeda Y, Iwanaga Y, Minamisawa S, Date MO, Gu Y, Iwatate M, Li M, Wang L, Wilson JM, et al. 2002. Chronic suppression of heart-failure progression by a pseudophosphorylated mutant of phospholamban via in vivo cardiac rAAV gene delivery. *Nat Med* **8:** 864–871.

Hoyer-Hansen M, Jaattela M. 2007. Connecting endoplasmic reticulum stress to autophagy by unfolded protein response and calcium. *Cell Death Differ* **14:** 1576–1582.

Hsu YL, Wu LY, Kuo PL. 2009. Dehydrocostuslactone, a medicinal plant-derived sesquiterpene lactone, induces apoptosis coupled to endoplasmic reticulum stress in liver cancer cells. *J Pharmacol Exp Ther* **329:** 808–819.

Hu Y, Belke D, Suarez J, Swanson E, Clark R, Hoshijima M, Dillmann WH. 2005. Adenovirus-mediated overexpression of O-GlcNAcase improves contractile function in the diabetic heart. *Circ Res* **96:** 1006–1013.

Huang TJ, Sayers NM, Fernyhough P, Verkhratsky A. 2002. Diabetes-induced alterations in calcium homeostasis in sensory neurones of streptozotocin-diabetic rats are restricted to lumbar ganglia and are prevented by neurotrophin-3. *Diabetologia* **45:** 560–570.

Huang F, Shan J, Reiken S, Wehrens XH, Marks AR. 2006. Analysis of calstabin2 (FKBP12.6)-ryanodine receptor interactions: rescue of heart failure by calstabin2 in mice. *Proc Natl Acad Sci* **103:** 3456–3461.

Hung JY, Hsu YL, Ni WC, Tsai YM, Yang CJ, Kuo PL, Huang MS. 2010. Oxidative and endoplasmic reticulum stress signaling are involved in dehydrocostuslactone-mediated apoptosis in human non-small cell lung cancer cells. *Lung Cancer* **68:** 355–365.

Inesi G, Prasad AM, Pilankatta R. 2008. The Ca^{2+} ATPase of cardiac sarcoplasmic reticulum: Physiological role and relevance to diseases. *Biochem Biophys Res Commun* **369:** 182–187.

Inoue H, Tanizawa Y, Wasson J, Behn P, Kalidas K, Bernal-Mizrachi E, Mueckler M, Marshall H, Donis-Keller H, Crock P, et al. 1998. A gene encoding a transmembrane protein is mutated in patients with diabetes mellitus and optic atrophy (Wolfram syndrome). *Nat Genet* **20:** 143–148.

Irvine RF. 1990. Quantal Ca^{2+} release and the control of Ca^{2+} entry by inositol phosphates–a possible mechanism. *FEBS Lett* **263:** 5–9.

Jakobsen CH, Storvold GL, Bremseth H, Follestad T, Sand K, Mack M, Olsen KS, Lundemo AG, Iversen JG, Krokan HE, et al. 2008. DHA induces ER stress and growth arrest in human colon cancer cells: associations with cholesterol and calcium homeostasis. *J Lipid Res* **49:** 2089–2100.

Janssen K, Horn S, Niemann MT, Daniel PT, Schulze-Osthoff K, Fischer U. 2009. Inhibition of the ER Ca^{2+} pump forces multidrug-resistant cells deficient in Bak and Bax into necrosis. *J Cell Sci* **122:** 4481–4491.

Jaski BE, Jessup ML, Mancini DM, Cappola TP, Pauly DF, Greenberg B, Borow K, Dittrich H, Zsebo KM, Hajjar RJ. 2009. Calcium upregulation by percutaneous administration of gene therapy in cardiac disease (CUPID Trial), a first-in-human phase 1/2 clinical trial. *J Card Fail* **15:** 171–181.

Jeschke MG, Gauglitz GG, Song J, Kulp GA, Finnerty CC, Cox RA, Barral JM, Herndon DN, Boehning D. 2010. Calcium and ER stress mediate hepatic apoptosis after burn injury. *J Cell Mol Med* **13:** 1857–1865.

Jie H, Donghua H, Xingkui X, Liang G, Wenjun W, Xiaoyan H, Zhen C. 2007. Homoharringtonine-induced apoptosis of MDS cell line MUTZ-1 cells is mediated by the endoplasmic reticulum stress pathway. *Leuk Lymphoma* **48:** 964–977.

Johnson AJ, Hsu AL, Lin HP, Song X, Chen CS. 2002. The cyclo-oxygenase-2 inhibitor celecoxib perturbs intracellular calcium by inhibiting endoplasmic reticulum Ca^{2+}-ATPases: a plausible link with its anti-tumour effect and cardiovascular risks. *Biochem J* **366:** 831–837.

Kanekura K, Suzuki H, Aiso S, Matsuoka M. 2009. ER stress and unfolded protein response in amyotrophic lateral sclerosis. *Mol Neurobiol* **39:** 81–89.

Kang S, Kang J, Kwon H, Frueh D, Yoo SH, Wagner G, Park S. 2008. Effects of redox potential and Ca^{2+} on the inositol 1,4,5-trisphosphate receptor L3-1 loop region: implications for receptor regulation. *J Biol Chem* **283:** 25567–25575.

Karakikes I, Kim M, Hadri L, Sakata S, Sun Y, Zhang W, Chemaly ER, Hajjar RJ, Lebeche D. 2009. Gene remodeling in type 2 diabetic cardiomyopathy and its phenotypic rescue with SERCA2a. *PLoS One* **4:** e6474. doi:101371/journalpone0006474.

Kawada H, Nishiyama C, Takagi A, Tokura T, Nakano N, Maeda K, Mayuzumi N, Ikeda S, Okumura K, Ogawa H. 2005. Transcriptional regulation of ATP2C1 gene by Sp1 and YY1 and reduced function of its promoter in Hailey-Hailey disease keratinocytes. *J Invest Dermatol* **124:** 1206–1214.

Kawase Y, Hajjar RJ. 2008. The cardiac sarcoplasmic/endoplasmic reticulum calcium ATPase: a potent target for cardiovascular diseases. *Nat Clin Pract Cardiovasc Med* **5:** 554–565.

Keay S, Baldwin BR, Smith MW, Wasserman SS, Goldman WF. 1995. Increases in $[Ca^{2+}]_i$ mediated by the 92.5-kDa putative cell membrane receptor for HCMV gp86. *Am J Physiol* **269:** C11–C21.

Kim I, Xu W, Reed JC. 2008. Cell death and endoplasmic reticulum stress: disease relevance and therapeutic opportunities. *Nat Rev Drug Discov* **7:** 1013–1030.

Kim HW, Ch YS, Lee HR, Park SY, Kim YH. 2001. Diabetic alterations in cardiac sarcoplasmic reticulum Ca^{2+}-ATPase and phospholamban protein expression. *Life Sci* **70:** 367–379.

Knobler H, Savion N, Shenkman B, Kotev-Emeth S, Varon D. 1998. Shear-induced platelet adhesion and aggregation on subendothelium are increased in diabetic patients. *Thromb Res* **90:** 181–190.

Kobayashi T, Taguchi K, Takenouchi Y, Matsumoto T, Kamata K. 2007. Insulin-induced impairment via peroxynitrite production of endothelium-dependent relaxation and sarco/endoplasmic reticulum Ca^{2+}-ATPase function in aortas from diabetic rats. *Free Radic Biol Med* **43:** 431–443.

Korkotian E, Schwarz A, Pelled D, Schwarzmann G, Segal M, Futerman AH. 1999. Elevation of intracellular glucosylceramide levels results in an increase in endoplasmic reticulum density and in functional calcium stores in cultured neurons. *J Biol Chem* **274:** 21673–21678.

Korosec B, Glavac D, Rott T, Ravnik-Glavac M. 2006. Alterations in the ATP2A2 gene in correlation with colon and lung cancer. *Cancer Genet Cytogenet* **171:** 105–111.

Korosec B, Glavac D, Volavsek M, Ravnik-Glavac M. 2008. Alterations in genes encoding sarcoplasmic-endoplasmic reticulum Ca^{2+} pumps in association with head and neck squamous cell carcinoma. *Cancer Genet Cytogenet* **181:** 112–118.

Kouroku Y, Fujita E, Tanida I, Ueno T, Isoai A, Kumagai H, Ogawa S, Kaufman RJ, Kominami E, Momoi T. 2007. ER stress (PERK/eIF2α phosphorylation) mediates the polyglutamine-induced LC3 conversion, an essential step for autophagy formation. *Cell Death Differ* **14:** 230–239.

Kranias EG, Bers DM. 2007. Calcium and cardiomyopathies. *Subcell Biochem* **45:** 523–537.

Kroemer G, Galluzzi L, Brenner C. 2007. Mitochondrial membrane permeabilization in cell death. *Physiol Rev* **87:** 99–163.

Kruglikov I, Gryshchenko O, Shutov L, Kostyuk E, Kostyuk P, Voitenko N. 2004. Diabetes-induced abnormalities in ER calcium mobilization in primary and secondary nociceptive neurons. *Pflugers Arch* **448:** 395–401.

Kulkarni RN, Roper MG, Dahlgren G, Shih DQ, Kauri LM, Peters JL, Stoffel M, Kennedy RT. 2004. Islet secretory defect in insulin receptor substrate 1 null mice is linked with reduced calcium signaling and expression of sarco(endo)plasmic reticulum Ca^{2+}-ATPase (SERCA)-2b and -3. *Diabetes* **53:** 1517–1525.

Lai P, Michelangeli F. 2009. Changes in expression and activity of the secretory pathway Ca^{2+} ATPase 1 (SPCA1) in A7r5 vascular smooth muscle cells cultured at different glucose concentrations. *Biosci Rep* **29:** 397–404.

Lai WW, Yang JS, Lai KC, Kuo CL, Hsu CK, Wang CK, Chang CY, Lin JJ, Tang NY, Chen PY, et al. 2009. Rhein induced apoptosis through the endoplasmic reticulum stress, caspase- and mitochondria-dependent pathways in SCC-4 human tongue squamous cancer cells. *In Vivo* **23:** 309–316.

Lee B, Jonas JC, Weir GC, Laychock SG. 1999. Glucose regulates expression of inositol 1,4,5-trisphosphate receptor isoforms in isolated rat pancreatic islets. *Endocrinology* **140:** 2173–2182.

Lee SY, Lee MS, Cherla RP, Tesh VL. 2008. Shiga toxin 1 induces apoptosis through the endoplasmic reticulum stress response in human monocytic cells. *Cell Microbiol* **10:** 770–780.

Lehnart SE, Wehrens XH, Marks AR. 2005. Defective ryanodine receptor interdomain interactions may contribute to intracellular Ca^{2+} leak: a novel therapeutic target in heart failure. *Circulation* **111:** 3342–3346.

Lehnart SE, Mongillo M, Bellinger A, Lindegger N, Chen BX, Hsueh W, Reiken S, Wronska A, Drew LJ, Ward CW, et al. 2008. Leaky Ca^{2+} release channel/ryanodine receptor 2 causes seizures and sudden cardiac death in mice. *J Clin Invest* **118:** 2230–2245.

Lei K, Davis RJ. 2003. JNK phosphorylation of Bim-related members of the Bcl2 family induces Bax-dependent apoptosis. *Proc Natl Acad Sci* **100:** 2432–2437.

Leonardi O, Mints G, Hussain MA. 2003. β-cell apoptosis in the pathogenesis of human type 2 diabetes mellitus. *Eur J Endocrinol* **149:** 99–102.

Levine B, Kroemer G. 2008. Autophagy in the pathogenesis of disease. *Cell* **132:** 27–42.

Levy J. 1999. Abnormal cell calcium homeostasis in type 2 diabetes mellitus: a new look on old disease. *Endocrine* **10:** 1–6.

Li Y, Ge M, Ciani L, Kuriakose G, Westover EJ, Dura M, Covey DF, Freed JH, Maxfield FR, Lytton J, et al. 2004. Enrichment of endoplasmic reticulum with cholesterol inhibits sarcoplasmic-endoplasmic reticulum calcium ATPase-2b activity in parallel with increased order of membrane lipids: implications for depletion of endoplasmic reticulum calcium stores and apoptosis in cholesterol-loaded macrophages. *J Biol Chem* **279:** 37030–37039.

Li G, Mongillo M, Chin KT, Harding H, Ron D, Marks AR, Tabas I. 2009. Role of ERO1-α-mediated stimulation of inositol 1,4,5-triphosphate receptor activity in endoplasmic reticulum stress-induced apoptosis. *J Cell Biol* **186:** 783–792.

Liao PC, Tan SK, Lieu CH, Jung HK. 2008. Involvement of endoplasmic reticulum in paclitaxel-induced apoptosis. *J Cell Biochem* **104:** 1509–1523.

Lievremont JP, Rizzuto R, Hendershot L, Meldolesi J. 1997. BiP, a major chaperone protein of the endoplasmic reticulum lumen, plays a direct and important role in the storage of the rapidly exchanging pool of Ca^{2+}. *J Biol Chem* **272:** 30873–30879.

Lin JH, Li H, Yasumura D, Cohen HR, Zhang C, Panning B, Shokat KM, Lavail MM, Walter P. 2007a. IRE1 signaling affects cell fate during the unfolded protein response. *Science* **318:** 944–949.

Lin JH, Li H, Zhang Y, Ron D, Walter P. 2009. Divergent effects of PERK and IRE1 signaling on cell viability. *PLoS One* **4:** e4170. doi:101371/journalpone0004170.

Lin JP, Yang JS, Chang NW, Chiu TH, Su CC, Lu KW, Ho YT, Yeh CC, Yang MD, Lin HJ, et al. 2007b. GADD153 mediates berberine-induced apoptosis in human cervical cancer Ca ski cells. *Anticancer Res* **27:** 3379–3386.

Lindenmeyer MT, Rastaldi MP, Ikehata M, Neusser MA, Kretzler M, Cohen CD, Schlondorff D. 2008. Proteinuria and hyperglycemia induce endoplasmic reticulum stress. *J Am Soc Nephrol* **19:** 2225–2236.

Lindholm D, Wootz H, Korhonen L. 2006. ER stress and neurodegenerative diseases. *Cell Death Differ* **13:** 385–392.

Lipskaia L, Hulot JS, Lompre AM. 2009. Role of sarco/endoplasmic reticulum calcium content and calcium ATPase activity in the control of cell growth and proliferation. *Pflugers Arch* **457:** 673–685.

Liu LH, Boivin GP, Prasad V, Periasamy M, Shull GE. 2001. Squamous cell tumors in mice heterozygous for a null allele of Atp2a2, encoding the sarco(endo)plasmic reticulum Ca^{2+}-ATPase isoform 2 Ca^{2+} pump. *J Biol Chem* **276:** 26737–26740.

Liu G, Sun Y, Li Z, Song T, Wang H, Zhang Y, Ge Z. 2008. Apoptosis induced by endoplasmic reticulum stress involved in diabetic kidney disease. *Biochem Biophys Res Commun* **370:** 651–656.

Lloyd-Evans E, Pelled D, Riebeling C, Bodennec J, de-Morgan A, Waller H, Schiffmann R, Futerman AH. 2003. Glucosylceramide and glucosylsphingosine modulate calcium mobilization from brain microsomes via different mechanisms. *J Biol Chem* **278:** 23594–23599.

Lopez JJ, Jardin I, Bobe R, Pariente JA, Enouf J, Salido GM, Rosado JA. 2008. STIM1 regulates acidic Ca^{2+} store refilling by interaction with SERCA3 in human platelets. *Biochem Pharmacol* **75:** 2157–2164.

Lorenzon NM, Beam KG. 2008. Disease causing mutations of calcium channels. *Channels (Austin)* **2:** 163–179.

Lucero HA, Kaminer B. 1999. The role of calcium on the activity of ERcalcistorin/protein-disulfide isomerase and the significance of the C-terminal and its calcium binding. A comparison with mammalian protein-disulfide isomerase. *J Biol Chem* **274:** 3243–3251.

Luciani DS, Gwiazda KS, Yang TL, Kalynyak TB, Bychkivska Y, Frey MH, Jeffrey KD, Sampaio AV, Underhill TM, Johnson JD. 2009. Roles of IP$_3$R and RyR Ca^{2+} channels in endoplasmic reticulum stress and β-cell death. *Diabetes* **58:** 422–432.

Ma Y, Hendershot LM. 2004. ER chaperone functions during normal and stress conditions. *J Chem Neuroanat* **28:** 51–65.

Ma L, Zhu B, Chen X, Liu J, Guan Y, Ren J. 2008. Abnormalities of sarcoplasmic reticulum Ca^{2+} mobilization in aortic smooth muscle cells from streptozotocin-induced diabetic rats. *Clin Exp Pharmacol Physiol* **35:** 568–573.

Magyar JE, Gamberucci A, Konta L, Margittai E, Mandl J, Banhegyi G, Benedetti A, Csala M. 2009. Endoplasmic reticulum stress underlying the pro-apoptotic effect of epigallocatechin gallate in mouse hepatoma cells. *Int J Biochem Cell Biol* **41:** 694–700.

Mahabusarakam W, Iriyachitra P, Taylor WC. 1987. Chemical constituents of garcinia mangostana. *J Nat Prod* **50:** 474–478.

Mahn K, Hirst SJ, Ying S, Holt MR, Lavender P, Ojo OO, Siew L, Simcock DE, McVicker CG, Kanabar V, et al. 2009. Diminished sarco/endoplasmic reticulum Ca^{2+} ATPase (SERCA) expression contributes to airway remodelling in bronchial asthma. *Proc Natl Acad Sci* **106:** 10775–10780.

Malhotra JD, Kaufman RJ. 2007. The endoplasmic reticulum and the unfolded protein response. *Semin Cell Dev Biol* **18:** 716–731.

Manninen A, Saksela K. 2002. HIV-1 Nef interacts with inositol trisphosphate receptor to activate calcium signaling in T cells. *J Exp Med* **195:** 1023–1032.

Marciniak SJ, Yun CY, Oyadomari S, Novoa I, Zhang Y, Jungreis R, Nagata K, Harding HP, Ron D. 2004. CHOP induces death by promoting protein synthesis and oxidation in the stressed endoplasmic reticulum. *Genes Dev* **18:** 3066–3077.

Marx SO, Reiken S, Hisamatsu Y, Jayaraman T, Burkhoff D, Rosemblit N, Marks AR. 2000. PKA phosphorylation dissociates FKBP12.6 from the calcium release channel (ryanodine receptor): defective regulation in failing hearts. *Cell* **101:** 365–376.

Mattson MP. 2007. Calcium and neurodegeneration. *Aging Cell* **6:** 337–350.

Mayne M, Holden CP, Nath A, Geiger JD. 2000. Release of calcium from inositol 1,4,5-trisphosphate receptor-regulated stores by HIV-1 Tat regulates TNF-α production in human macrophages. *J Immunol* **164:** 6538–6542.

McCullough KD, Martindale JL, Klotz LO, Aw TY, Holbrook NJ. 2001. Gadd153 sensitizes cells to endoplasmic reticulum stress by down-regulating Bcl2 and perturbing the cellular redox state. *Mol Cell Biol* **21:** 1249–1259.

Michalak M, Robert Parker JM, Opas M. 2002. Ca^{2+} signaling and calcium binding chaperones of the endoplasmic reticulum. *Cell Calcium* **32:** 269–278.

Mikoshiba K. 2007. IP$_3$ receptor/Ca^{2+} channel: from discovery to new signaling concepts. *J Neurochem* **102:** 1426–1446.

Missiaen L, De Smedt H, Droogmans G, Casteels R. 1992. Ca^{2+} release induced by inositol 1,4,5-trisphosphate is a steady-state phenomenon controlled by luminal Ca^{2+} in permeabilized cells. *Nature* **357:** 599–602.

Missiaen L, Robberecht W, Van Den Bosch L, Callewaert G, Parys JB, Wuytack F, Raeymaekers L, Nilius B, Eggermont J, De Smedt H. 2000. Abnormal intracellular Ca^{2+} homeostasis and disease. *Cell Calcium* **28:** 1–21.

Moffatt MF, Kabesch M, Liang L, Dixon AL, Strachan D, Heath S, Depner M, von Berg A, Bufe A, Rietschel E, et al. 2007. Genetic variants regulating ORMDL3 expression contribute to the risk of childhood asthma. *Nature* **448:** 470–473.

Moller JT, Cluitmans P, Rasmussen LS, Houx P, Rasmussen H, Canet J, Rabbitt P, Jolles J, Larsen K, Hanning CD, et al. 1998. Long-term postoperative cognitive dysfunction in the elderly ISPOCD1 study. ISPOCD investigators. International Study of Post-Operative Cognitive Dysfunction. *Lancet* **351:** 857–861.

Moncada S, Erusalimsky JD. 2002. Does nitric oxide modulate mitochondrial energy generation and apoptosis? *Nat Rev Mol Cell Biol* **3:** 214–220.

Monteith GR, McAndrew D, Faddy HM, Roberts-Thomson SJ. 2007. Calcium and cancer: targeting Ca^{2+} transport. *Nat Rev Cancer* **7:** 519–530.

Nagata E, Tanaka K, Gomi S, Mihara B, Shirai T, Nogawa S, Nozaki H, Mikoshiba K, Fukuuchi Y. 1994. Alteration of inositol 1,4,5-trisphosphate receptor after six-hour hemispheric ischemia in the gerbil brain. *Neuroscience* **61:** 983–990.

Nakano K, Isegawa Y, Zou P, Tadagaki K, Inagi R, Yamanishi K. 2003. Kaposi's sarcoma-associated herpesvirus (KSHV)-encoded vMIP-I and vMIP-II induce signal transduction and chemotaxis in monocytic cells. *Arch Virol* **148:** 871–890.

Nakano T, Watanabe H, Ozeki M, Asai M, Katoh H, Satoh H, Hayashi H. 2006. Endoplasmic reticulum Ca^{2+} depletion induces endothelial cell apoptosis independently of caspase-12. *Cardiovasc Res* **69:** 908–915.

Nawrocki ST, Carew JS, Pino MS, Highshaw RA, Dunner K Jr, Huang P, Abbruzzese JL, McConkey DJ. 2005. Bortezomib sensitizes pancreatic cancer cells to endoplasmic reticulum stress-mediated apoptosis. *Cancer Res* **65:** 11658–11666.

Nelson TE, Nelson KE. 1990. Intra- and extraluminal sarcoplasmic reticulum membrane regulatory sites for Ca^{2+}-induced Ca^{2+} release. *FEBS Lett* **263:** 292–294.

Nicolau J, De Souza DN, Simoes A. 2009. Alteration of Ca^{2+}-ATPase activity in the homogenate, plasma membrane and microsomes of the salivary glands of streptozotocin-induced diabetic rats. *Cell Biochem Funct* **27:** 128–134.

Nieto-Miguel T, Fonteriz RI, Vay L, Gajate C, Lopez-Hernandez S, Mollinedo F. 2007. Endoplasmic reticulum stress in the proapoptotic action of edelfosine in solid tumor cells. *Cancer Res* **67:** 10368–10378.

Norman JP, Perry SW, Reynolds HM, Kiebala M, De Mesy Bentley KL, Trejo M, Volsky DJ, Maggirwar SB, Dewhurst S, Masliah E, et al. 2008. HIV-1 Tat activates neuronal ryanodine receptors with rapid induction of the unfolded protein response and mitochondrial hyperpolarization. *PLoS One* **3**: e3731. doi:101371/journalpone0003731.

Oakes SA, Scorrano L, Opferman JT, Bassik MC, Nishino M, Pozzan T, Korsmeyer SJ. 2005. Proapoptotic BAX and BAK regulate the type 1 inositol trisphosphate receptor and calcium leak from the endoplasmic reticulum. *Proc Natl Acad Sci* **102**: 105–110.

Odermatt A, Taschner PE, Khanna VK, Busch HF, Karpati G, Jablecki CK, Breuning MH, MacLennan DH. 1996. Mutations in the gene-encoding SERCA1, the fast-twitch skeletal muscle sarcoplasmic reticulum Ca^{2+} ATPase, are associated with Brody disease. *Nat Genet* **14**: 191–194.

Ogata M, Hino S, Saito A, Morikawa K, Kondo S, Kanemoto S, Murakami T, Taniguchi M, Tanii I, Yoshinaga K, et al. 2006. Autophagy is activated for cell survival after endoplasmic reticulum stress. *Mol Cell Biol* **26**: 9220–9231.

O'Neil BJ, McKeown TR, DeGracia DJ, Alousi SS, Rafols JA, White BC. 1999. Cell death, calcium mobilization, and immunostaining for phosphorylated eukaryotic initiation factor 2-α (eIF2α) in neuronally differentiated NB-104 cells: arachidonate and radical-mediated injury mechanisms. *Resuscitation* **41**: 71–83.

Ong DS, Mu TW, Palmer AE, Kelly JW. 2010. Endoplasmic reticulum Ca^{2+} increases enhance mutant glucocerebrosidase proteostasis. *Nat Chem Biol* **6**: 424–432.

Onozuka T, Sawamura D, Goto M, Yokota K, Shimizu H. 2006. Possible role of endoplasmic reticulum stress in the pathogenesis of Darier's disease. *J Dermatol Sci* **41**: 217–220.

O'Rourke B, Kass DA, Tomaselli GF, Kaab S, Tunin R, Marban E. 1999. Mechanisms of altered excitation-contraction coupling in canine tachycardia-induced heart failure, I: experimental studies. *Circ Res* **84**: 562–570.

Osman AA, Saito M, Makepeace C, Permutt MA, Schlesinger P, Mueckler M. 2003. Wolframin expression induces novel ion channel activity in endoplasmic reticulum membranes and increases intracellular calcium. *J Biol Chem* **278**: 52755–52762.

Oyadomari S, Mori M. 2004. Roles of CHOP/GADD153 in endoplasmic reticulum stress. *Cell Death Differ* **11**: 381–389.

Oyadomari S, Takeda K, Takiguchi M, Gotoh T, Matsumoto M, Wada I, Akira S, Araki E, Mori M. 2001. Nitric oxide-induced apoptosis in pancreatic β cells is mediated by the endoplasmic reticulum stress pathway. *Proc Natl Acad Sci* **98**: 10845–10850.

Ozcan U, Cao Q, Yilmaz E, Lee AH, Iwakoshi NN, Ozdelen E, Tuncman G, Gorgun C, Glimcher LH, Hotamisligil GS. 2004. Endoplasmic reticulum stress links obesity, insulin action, and type 2 diabetes. *Science* **306**: 457–461.

Pacifico F, Ulianich L, De Micheli S, Treglia S, Leonardi A, Vito P, Formisano S, Consiglio E, Di Jeso B. 2003. The expression of the sarco/endoplasmic reticulum Ca^{2+}-ATPases in thyroid and its down-regulation following neoplastic transformation. *J Mol Endocrinol* **30**: 399–409.

Papp S, Dziak E, Michalak M, Opas M. 2003. Is all of the endoplasmic reticulum created equal? The effects of the heterogeneous distribution of endoplasmic reticulum Ca^{2+}-handling proteins. *J Cell Biol* **160**: 475–479.

Park KS, Poburko D, Wollheim CB, Demaurex N. 2009. Amiloride derivatives induce apoptosis by depleting ER Ca^{2+} stores in vascular endothelial cells. *Br J Pharmacol* **156**: 1296–1304.

Parsons JT, Churn SB, DeLorenzo RJ. 1999. Global ischemia-induced inhibition of the coupling ratio of calcium uptake and ATP hydrolysis by rat whole brain microsomal Mg^{2+}/Ca^{2+} ATPase. *Brain Res* **834**: 32–41.

Paschen W, Mengesdorf T. 2005. Endoplasmic reticulum stress response and neurodegeneration. *Cell Calcium* **38**: 409–415.

Pattacini L, Mancini M, Mazzacurati L, Brusa G, Benvenuti M, Martinelli G, Baccarani M, Santucci MA. 2004. Endoplasmic reticulum stress initiates apoptotic death induced by STI571 inhibition of p210 bcr-abl tyrosine kinase. *Leuk Res* **28**: 191–202.

Pattni K, Millard TH, Banting G. 2003. Calpain cleavage of the B isoform of $Ins(1,4,5)P_3$ 3-kinase separates the catalytic domain from the membrane anchoring domain. *Biochem J* **375**: 643–651.

Pelled D, Lloyd-Evans E, Riebeling C, Jeyakumar M, Platt FM, Futerman AH. 2003. Inhibition of calcium uptake via the sarco/endoplasmic reticulum Ca^{2+}-ATPase in a mouse model of Sandhoff disease and prevention by treatment with N-butyldeoxynojirimycin. *J Biol Chem* **278**: 29496–29501.

Pelled D, Trajkovic-Bodennec S, Lloyd-Evans E, Sidransky E, Schiffmann R, Futerman AH. 2005. Enhanced calcium release in the acute neuronopathic form of Gaucher disease. *Neurobiol Dis* **18**: 83–88.

Pereira L, Matthes J, Schuster I, Valdivia HH, Herzig S, Richard S, Gomez AM. 2006. Mechanisms of $[Ca^{2+}]_i$ transient decrease in cardiomyopathy of db/db type 2 diabetic mice. *Diabetes* **55**: 608–615.

Petersen OH, Tepikin A, Park MK. 2001. The endoplasmic reticulum: one continuous or several separate Ca^{2+} stores? *Trends Neurosci* **24**: 271–276.

Pirot P, Cardozo AK, Eizirik DL. 2008. Mediators and mechanisms of pancreatic β-cell death in type 1 diabetes. *Arq Bras Endocrinol Metabol* **52**: 156–165.

Pozzan T, Rizzuto R, Volpe P, Meldolesi J. 1994. Molecular and cellular physiology of intracellular calcium stores. *Physiol Rev* **74**: 595–636.

Prusiner SB. 1998. Prions. *Proc Natl Acad Sci* **95**: 13363–13383.

Putcha GV, Le S, Frank S, Besirli CG, Clark K, Chu B, Alix S, Youle RJ, LaMarche A, Maroney AC, et al. 2003. JNK-mediated BIM phosphorylation potentiates BAX-dependent apoptosis. *Neuron* **38**: 899–914.

Puthalakath H, O'Reilly LA, Gunn P, Lee L, Kelly PN, Huntington ND, Hughes PD, Michalak EM, McKimm-Breschkin J, Motoyama N, et al. 2007. ER stress triggers apoptosis by activating BH3-only protein Bim. *Cell* **129**: 1337–1349.

Putney JW Jr. 1986. A model for receptor-regulated calcium entry. *Cell Calcium* **7**: 1–12.

Pyrko P, Kardosh A, Liu YT, Soriano N, Xiong W, Chow RH, Uddin J, Petasis NA, Mircheff AK, Farley RA, et al. 2007. Calcium-activated endoplasmic reticulum stress as a major component of tumor cell death induced by

2,5-dimethyl-celecoxib, a non-coxib analogue of celecoxib. *Mol Cancer Ther* **6:** 1262–1275.

Qi MY, Liu HR, Dai DZ, Li N, Dai Y. 2008. Total triterpene acids, active ingredients from Fructus Corni, attenuate diabetic cardiomyopathy by normalizing ET pathway and expression of FKBP12.6 and SERCA2a in streptozotocin-rats. *J Pharm Pharmacol* **60:** 1687–1694.

Rahmani M, Davis EM, Crabtree TR, Habibi JR, Nguyen TK, Dent P, Grant S. 2007. The kinase inhibitor sorafenib induces cell death through a process involving induction of endoplasmic reticulum stress. *Mol Cell Biol* **27:** 5499–5513.

Randriamboavonjy V, Pistrosch F, Bolck B, Schwinger RH, Dixit N, Badenhoop K, Cohen RA, Busse R, Fleming I. 2008. Platelet sarcoplasmic endoplasmic reticulum Ca^{2+}-ATPase and mu-calpain activity are altered in type 2 diabetes mellitus and restored by rosiglitazone. *Circulation* **117:** 52–60.

Rao RV, Bredesen DE. 2004. Misfolded proteins, endoplasmic reticulum stress and neurodegeneration. *Curr Opin Cell Biol* **16:** 653–662.

Rardon DP, Cefali DC, Mitchell RD, Seiler SM, Hathaway DR, Jones LR. 1990. Digestion of cardiac and skeletal muscle junctional sarcoplasmic reticulum vesicles with calpain II. Effects on the Ca^{2+} release channel. *Circ Res* **67:** 84–96.

Reiken S, Gaburjakova M, Gaburjakova J, He KL, Prieto A, Becker E, Yi GH, Wang J, Burkhoff D, Marks AR. 2001. β-adrenergic receptor blockers restore cardiac calcium release channel (ryanodine receptor) structure and function in heart failure. *Circulation* **104:** 2843–2848.

Reiken S, Lacampagne A, Zhou H, Kherani A, Lehnart SE, Ward C, Huang F, Gaburjakova M, Gaburjakova J, Rosemblit N, et al. 2003. PKA phosphorylation activates the calcium release channel (ryanodine receptor) in skeletal muscle: defective regulation in heart failure. *J Cell Biol* **160:** 919–928.

Ren J, Gintant GA, Miller RE, Davidoff AJ. 1997. High extracellular glucose impairs cardiac E-C coupling in a glycosylation-dependent manner. *Am J Physiol* **273:** H2876–H2883.

Rengifo J, Gibson CJ, Winkler E, Collin T, Ehrlich BE. 2007. Regulation of the inositol 1,4,5-trisphosphate receptor type I by O-GlcNAc glycosylation. *J Neurosci* **27:** 13813–13821.

Reuter H, Gronke S, Adam C, Ribati M, Brabender J, Zobel C, Frank KF, Wippermann J, Schwinger RH, Brixius K, et al. 2008. Sarcoplasmic Ca^{2+} release is prolonged in nonfailing myocardium of diabetic patients. *Mol Cell Biochem* **308:** 141–149.

Rigaud M, Gemes G, Weyker PD, Cruikshank JM, Kawano T, Wu HE, Hogan QH. 2009. Axotomy depletes intracellular calcium stores in primary sensory neurons. *Anesthesiology* **111:** 381–392.

Rizzuto R, Marchi S, Bonora M, Aguiari P, Bononi A, De Stefani D, Giorgi C, Leo S, Rimessi A, Siviero R, et al. 2009. Ca^{2+} transfer from the ER to mitochondria: when, how and why. *Biochim Biophys Acta* **1787:** 1342–1351.

Roach JC, Deutsch K, Li S, Siegel AF, Bekris LM, Einhaus DC, Sheridan CM, Glusman G, Hood L, Lernmark A, et al. 2006. Genetic mapping at 3-kilobase resolution reveals inositol 1,4,5-triphosphate receptor 3 as a risk factor for type 1 diabetes in Sweden. *Am J Hum Genet* **79:** 614–627.

Robinson LC, Marchant JS. 2008. Enhanced Ca^{2+} leak from ER Ca^{2+} stores induced by hepatitis C NS5A protein. *Biochem Biophys Res Commun* **368:** 593–599.

Roe MW, Philipson LH, Frangakis CJ, Kuznetsov A, Mertz RJ, Lancaster ME, Spencer B, Worley JF III, Dukes ID. 1994. Defective glucose-dependent endoplasmic reticulum Ca^{2+} sequestration in diabetic mouse islets of Langerhans. *J Biol Chem* **269:** 18279–18282.

Ron D, Walter P. 2007. Signal integration in the endoplasmic reticulum unfolded protein response. *Nat Rev Mol Cell Biol* **8:** 519–529.

Rong Y, Distelhorst CW. 2008. Bcl-2 protein family members: versatile regulators of calcium signaling in cell survival and apoptosis. *Annu Rev Physiol* **70:** 73–91.

Rubler S, Dlugash J, Yuceoglu YZ, Kumral T, Branwood AW, Grishman A. 1972. New type of cardiomyopathy associated with diabetic glomerulosclerosis. *Am J Cardiol* **30:** 595–602.

Ryu EJ, Harding HP, Angelastro JM, Vitolo OV, Ron D, Greene LA. 2002. Endoplasmic reticulum stress and the unfolded protein response in cellular models of Parkinson's disease. *J Neurosci* **22:** 10690–10698.

Saavedra FR, Redondo PC, Hernandez-Cruz JM, Salido GM, Pariente JA, Rosado JA. 2004. Store-operated Ca^{2+} entry and tyrosine kinase pp60^{src} hyperactivity are modulated by hyperglycemia in platelets from patients with non insulin-dependent diabetes mellitus. *Arch Biochem Biophys* **432:** 261–268.

Sacchetto R, Testoni S, Gentile A, Damiani E, Rossi M, Liguori R, Drogemuller C, Mascarello F. 2009. A defective SERCA1 protein is responsible for congenital pseudomyotonia in Chianina cattle. *Am J Pathol* **174:** 565–573.

Sakakura C, Hagiwara A, Fukuda K, Shimomura K, Takagi T, Kin S, Nakase Y, Fujiyama J, Mikoshiba K, Okazaki Y, et al. 2003. Possible involvement of inositol 1,4,5-trisphosphate receptor type 3 (IP$_3$R3) in the peritoneal dissemination of gastric cancers. *Anticancer Res* **23:** 3691–3697.

Sakata S, Lebeche D, Sakata Y, Sakata N, Chemaly ER, Liang L, Nakajima-Takenaka C, Tsuji T, Konishi N, del Monte F, et al. 2007. Transcoronary gene transfer of SERCA2a increases coronary blood flow and decreases cardiomyocyte size in a type 2 diabetic rat model. *Am J Physiol* **292:** H1204–H1207.

Sakuntabhai A, Ruiz-Perez V, Carter S, Jacobsen N, Burge S, Monk S, Smith M, Munro CS, O'Donovan M, Craddock N, et al. 1999. Mutations in ATP2A2, encoding a Ca^{2+} pump, cause Darier disease. *Nat Genet* **21:** 271–277.

Sambrook JF. 1990. The involvement of calcium in transport of secretory proteins from the endoplasmic reticulum. *Cell* **61:** 197–199.

Sammels E, Parys JB, Missiaen L, De Smedt H, Bultynck G. 2010. Intracellular Ca^{2+} storage in health and disease: a dynamic equilibrium. *Cell Calcium* **47:** 297–314.

Sano R, Annunziata I, Patterson A, Moshiach S, Gomero E, Opferman J, Forte M, d'Azzo A. 2009. G$_{M1}$-ganglioside accumulation at the mitochondria-associated ER membranes links ER stress to Ca^{2+}-dependent mitochondrial apoptosis. *Mol Cell* **36:** 500–511.

Sathish V, Thompson MA, Bailey JP, Pabelick CM, Prakash YS, Sieck GC. 2009. Effect of proinflammatory cytokines on regulation of sarcoplasmic reticulum Ca^{2+} reuptake in human airway smooth muscle. *Am J Physiol* **297:** L26–L34.

Sato A, Fujiwara H, Oku H, Ishiguro K, Ohizumi Y. 2004. α-mangostin induces Ca^{2+}-ATPase-dependent apoptosis via mitochondrial pathway in PC12 cells. *J Pharmacol Sci* **95:** 33–40.

Savino JA III, Evans JF, Rabinowitz D, Auborn KJ, Carter TH. 2006. Multiple, disparate roles for calcium signaling in apoptosis of human prostate and cervical cancer cells exposed to diindolylmethane. *Mol Cancer Ther* **5:** 556–563.

Schindl R, Muik M, Fahrner M, Derler I, Fritsch R, Bergsmann J, Romanin C. 2009. Recent progress on STIM1 domains controlling Orai activation. *Cell Calcium* **46:** 227–232.

Schmidt AG, Haghighi K, Frank B, Pater L, Dorn GW, Walsh RA, Kranias EG. 2003a. Polymorphic SERCA2a variants do not account for inter-individual differences in phospholamban-SERCA2a interactions in human heart failure. *J Mol Cell Cardiol* **35:** 867–870.

Schmidt T, Zaib F, Samson SE, Kwan CY, Grover AK. 2003b. Peroxynitrite resistance of sarco/endoplasmic reticulum Ca^{2+} pump in pig coronary artery endothelium and smooth muscle. *Cell Calcium* **36:** 77–82.

Schmitt JP, Kamisago M, Asahi M, Li GH, Ahmad F, Mende U, Kranias EG, MacLennan DH, Seidman JG, Seidman CE. 2003. Dilated cardiomyopathy and heart failure caused by a mutation in phospholamban. *Science* **299:** 1410–1413.

Schroder M, Kaufman RJ. 2005. The mammalian unfolded protein response. *Annu Rev Biochem* **74:** 739–789.

Scorrano L, Oakes SA, Opferman JT, Cheng EH, Sorcinelli MD, Pozzan T, Korsmeyer SJ. 2003. BAX and BAK regulation of endoplasmic reticulum Ca^{2+}: a control point for apoptosis. *Science* **300:** 135–139.

Searls YM, Loganathan R, Smirnova IV, Stehno-Bittel L. 2010. Intracellular Ca^{2+} regulating proteins in vascular smooth muscle cells are altered with type 1 diabetes due to the direct effects of hyperglycemia. *Cardiovasc Diabetol* **9:** 8. 2010. doi: 101186/1475-2840-9-8.

Seidler T, Hasenfuss G, Maier LS. 2007. Targeting altered calcium physiology in the heart: translational approaches to excitation, contraction, and transcription. *Physiology* **22:** 328–334.

Seo NS, Zeng CQ, Hyser JM, Utama B, Crawford SE, Kim KJ, Hook M, Estes MK. 2008. Inaugural article: integrins α1β1 and α2β1 are receptors for the rotavirus enterotoxin. *Proc Natl Acad Sci* **105:** 8811–8818.

Shah RD, Gonzales F, Golez E, Augustin D, Caudillo S, Abbott A, Morello J, McDonough PM, Paolini PJ, Shubeita HE. 2005. The antidiabetic agent rosiglitazone upregulates SERCA2 and enhances TNF-α- and LPS-induced NF-kB-dependent transcription and TNF-α-induced IL-6 secretion in ventricular myocytes. *Cell Physiol Biochem* **15:** 41–50.

Shao CH, Rozanski GJ, Patel KP, Bidasee KR. 2007. Dyssynchronous (non-uniform) Ca^{2+} release in myocytes from streptozotocin-induced diabetic rats. *J Mol Cell Cardiol* **42:** 234–246.

Shao CH, Wehrens XH, Wyatt TA, Parbhu S, Rozanski GJ, Patel KP, Bidasee KR. 2009. Exercise training during diabetes attenuates cardiac ryanodine receptor dysregulation. *J Appl Physiol* **106:** 1280–1292.

Sharma K, Wang L, Zhu Y, DeGuzman A, Cao GY, Lynn RB, Joseph SK. 1999. Renal type I inositol 1,4,5-trisphosphate receptor is reduced in streptozotocin-induced diabetic rats and mice. *Am J Physiol* **276:** F54–F61.

Sharon-Friling R, Goodhouse J, Colberg-Poley AM, Shenk T. 2006. Human cytomegalovirus pUL37x1 induces the release of endoplasmic reticulum calcium stores. *Proc Natl Acad Sci* **103:** 19117–19122.

Shen J, Yu WM, Brotto M, Scherman JA, Guo C, Stoddard C, Nosek TM, Valdivia HH, Qu CK. 2009. Deficiency of MIP/MTMR14 phosphatase induces a muscle disorder by disrupting Ca^{2+} homeostasis. *Nat Cell Biol* **11:** 769–776.

Stathopulos PB, Zheng L, Li GY, Plevin MJ, Ikura M. 2008. Structural and mechanistic insights into STIM1-mediated initiation of store-operated calcium entry. *Cell* **135:** 110–122.

Stockwin LH, Han B, Yu SX, Hollingshead MG, ElSohly MA, Gul W, Slade D, Galal AM, Newton DL. 2009. Artemisinin dimer anticancer activity correlates with heme-catalyzed reactive oxygen species generation and endoplasmic reticulum stress induction. *Int J Cancer* **125:** 1266–1275.

Stolen TO, Hoydal MA, Kemi OJ, Catalucci D, Ceci M, Aasum E, Larsen T, Rolim N, Condorelli G, Smith GL, et al. 2009. Interval training normalizes cardiomyocyte function, diastolic Ca^{2+} control, and SR Ca^{2+} release synchronicity in a mouse model of diabetic cardiomyopathy. *Circ Res* **105:** 527–536.

Strom TM, Hortnagel K, Hofmann S, Gekeler F, Scharfe C, Rabl W, Gerbitz KD, Meitinger T. 1998. Diabetes insipidus, diabetes mellitus, optic atrophy and deafness (DIDMOAD) caused by mutations in a novel gene (wolframin) coding for a predicted transmembrane protein. *Hum Mol Genet* **7:** 2021–2028.

Suarez J, Scott B, Dillmann WH. 2008. Conditional increase in SERCA2a protein is able to reverse contractile dysfunction and abnormal calcium flux in established diabetic cardiomyopathy. *Am J Physiol* **295:** R1439–R1445.

Sulaiman M, Matta MJ, Sunderesan NR, Gupta MP, Periasamy M, Gupta M. 2010. Resveratrol, an activator of SIRT1, upregulates sarcoplasmic calcium ATPase and improves cardiac function in diabetic cardiomyopathy. *Am J Physiol* **298:** H833–H843.

Supnet C, Bezprozvanny I. 2010. The dysregulation of intracellular calcium in Alzheimer disease. *Cell Calcium* **47:** 183–189.

Szegezdi E, Fitzgerald U, Samali A. 2003. Caspase-12 and ER-stress-mediated apoptosis: the story so far. *Ann N Y Acad Sci* **1010:** 186–194.

Taiyab A, Sreedhar AS, Rao ChM. 2009. Hsp90 inhibitors, GA and 17AAG, lead to ER stress-induced apoptosis in rat histiocytoma. *Biochem Pharmacol* **78:** 142–152.

Takei D, Ishihara H, Yamaguchi S, Yamada T, Tamura A, Katagiri H, Maruyama Y, Oka Y. 2006. WFS1 protein modulates the free Ca^{2+} concentration in the endoplasmic reticulum. *FEBS Lett* **580:** 5635–5640.

Takenaka H, Adler PN, Katz AM. 1982. Calcium fluxes across the membrane of sarcoplasmic reticulum vesicles. *J Biol Chem* **257:** 12649–12656.

Teshima Y, Takahashi N, Saikawa T, Hara M, Yasunaga S, Hidaka S, Sakata T. 2000. Diminished expression of sarcoplasmic reticulum Ca^{2+}-ATPase and ryanodine sensitive Ca^{2+} channel mRNA in streptozotocin-induced diabetic rat heart. *J Mol Cell Cardiol* **32:** 655–664.

Tessitore A, del P Martin M, Sano R, Ma Y, Mann L, Ingrassia A, Laywell ED, Steindler DA, Hendershot LM, d'Azzo A. 2004. G_{M1}-ganglioside-mediated activation of the unfolded protein response causes neuronal death in a neurodegenerative gangliosidosis. *Mol Cell* **15:** 753–766.

Thompson S, Chen X, Hui L, Toschi A, Foster DA, Drain CM. 2008. Low concentrations of a non-hydrolysable tetra-S-glycosylated porphyrin and low light induces apoptosis in human breast cancer cells via stress of the endoplasmic reticulum. *Photochem Photobiol Sci* **7:** 1415–1421.

Tian P, Estes MK, Hu Y, Ball JM, Zeng CQ, Schilling WP. 1995. The rotavirus nonstructural glycoprotein NSP4 mobilizes Ca^{2+} from the endoplasmic reticulum. *J Virol* **69:** 5763–5772.

Tong X, Evangelista A, Cohen RA. 2009. Targeting the redox regulation of SERCA in vascular physiology and disease. *Curr Opin Pharmacol* **10:** 1–6.

Tong X, Ying J, Pimentel DR, Trucillo M, Adachi T, Cohen RA. 2008. High glucose oxidizes SERCA cysteine-674 and prevents inhibition by nitric oxide of smooth muscle cell migration. *J Mol Cell Cardiol* **44:** 361–369.

Tonnesen MF, Grunnet LG, Friberg J, Cardozo AK, Billestrup N, Eizirik DL, Storling J, Mandrup-Poulsen T. 2009. Inhibition of Nuclear Factor-kB or Bax prevents endoplasmic reticulum stress- but not nitric oxide-mediated apoptosis in INS-1E cells. *Endocrinology* **150:** 4094–4103.

Tosch V, Rohde HM, Tronchere H, Zanoteli E, Monroy N, Kretz C, Dondaine N, Payrastre B, Mandel JL, Laporte J. 2006. A novel PtdIns3P and $PtdIns(3,5)P_2$ phosphatase with an inactivating variant in centronuclear myopathy. *Hum Mol Genet* **15:** 3098–3106.

Trost SU, Belke DD, Bluhm WF, Meyer M, Swanson E, Dillmann WH. 2002. Overexpression of the sarcoplasmic reticulum Ca^{2+}-ATPase improves myocardial contractility in diabetic cardiomyopathy. *Diabetes* **51:** 1166–1171.

Tsuchiya S, Tsuji M, Morio Y, Oguchi K. 2006. Involvement of endoplasmic reticulum in glycochenodeoxycholic acid-induced apoptosis in rat hepatocytes. *Toxicol Lett* **166:** 140–149.

Tu H, Nelson O, Bezprozvanny A, Wang Z, Lee SF, Hao YH, Serneels L, De Strooper B, Yu G, Bezprozvanny I. 2006. Presenilins form ER Ca^{2+} leak channels, a function disrupted by familial Alzheimer's disease-linked mutations. *Cell* **126:** 981–993.

Van Baelen K, Dode L, Vanoevelen J, Callewaert G, De Smedt H, Missiaen L, Parys JB, Raeymaekers L, Wuytack F. 2004. The Ca^{2+}/Mn^{2+} pumps in the Golgi apparatus. *Biochim Biophys Acta* **1742:** 103–112.

Vanderheyden V, Devogelaere B, Missiaen L, De Smedt H, Bultynck G, Parys JB. 2009. Regulation of inositol 1,4,5-trisphosphate-induced Ca^{2+} release by reversible phosphorylation and dephosphorylation. *Biochim Biophys Acta* **1793:** 959–970.

van der Kallen CJ, van Greevenbroek MM, Stehouwer CD, Schalkwijk CG. 2009. Endoplasmic reticulum stress-induced apoptosis in the development of diabetes: is there a role for adipose tissue and liver? *Apoptosis* **14:** 1424–1434.

Vangheluwe P, Sepulveda MR, Missiaen L, Raeymaekers L, Wuytack F, Vanoevelen J. 2009. Intracellular Ca^{2+}- and Mn^{2+}-transport ATPases. *Chem Rev* **109:** 4733–4759.

van Kuppeveld FJ, de Jong AS, Melchers WJ, Willems PH. 2005. Enterovirus protein 2B po(u)res out the calcium: a viral strategy to survive? *Trends Microbiol* **13:** 41–44.

Vanoverberghe K, Vanden Abeele F, Mariot P, Lepage G, Roudbaraki M, Bonnal JL, Mauroy B, Shuba Y, Skryma R, Prevarskaya N. 2004. Ca^{2+} homeostasis and apoptotic resistance of neuroendocrine-differentiated prostate cancer cells. *Cell Death Differ* **11:** 321–330.

Varadi A, Lebel L, Hashim Y, Mehta Z, Ashcroft SJ, Turner R. 1999. Sequence variants of the sarco(endo)plasmic reticulum Ca^{2+}-transport ATPase 3 gene (SERCA3) in Caucasian type II diabetic patients (UK Prospective Diabetes Study 48). *Diabetologia* **42:** 1240–1243.

Vasanji Z, Cantor EJ, Juric D, Moyen M, Netticadan T. 2006. Alterations in cardiac contractile performance and sarcoplasmic reticulum function in sucrose-fed rats is associated with insulin resistance. *Am J Physiol* **291:** C772–C780.

Vasanji Z, Dhalla NS, Netticadan T. 2004. Increased inhibition of SERCA2 by phospholamban in the type I diabetic heart. *Mol Cell Biochem* **261:** 245–249.

Vericel E, Januel C, Carreras M, Moulin P, Lagarde M. 2004. Diabetic patients without vascular complications display enhanced basal platelet activation and decreased antioxidant status. *Diabetes* **53:** 1046–1051.

Verkhratsky A. 2005. Physiology and pathophysiology of the calcium store in the endoplasmic reticulum of neurons. *Physiol Rev* **85:** 201–279.

Verkhratsky A, Fernyhough P. 2008. Mitochondrial malfunction and Ca^{2+} dyshomeostasis drive neuronal pathology in diabetes. *Cell Calcium* **44:** 112–122.

Vermassen E, Parys JB, Mauger JP. 2004. Subcellular distribution of the inositol 1,4,5-trisphosphate receptors: functional relevance and molecular determinants. *Biol Cell* **96:** 3–17.

Vetter R, Rehfeld U, Reissfelder C, Weiss W, Wagner KD, Gunther J, Hammes A, Tschope C, Dillmann W, Paul M. 2002. Transgenic overexpression of the sarcoplasmic reticulum Ca^{2+}ATPase improves reticular Ca^{2+} handling in normal and diabetic rat hearts. *FASEB J* **16:** 1657–1659.

Viner RI, Ferrington DA, Williams TD, Bigelow DJ, Schoneich C. 1999. Protein modification during biological aging: selective tyrosine nitration of the SERCA2a isoform of the sarcoplasmic reticulum Ca^{2+}-ATPase in skeletal muscle. *Biochem J* **340:** 657–669.

Wang XZ, Ron D. 1996. Stress-induced phosphorylation and activation of the transcription factor CHOP (GADD153) by p38 MAP Kinase. *Science* **272:** 1347–1349.

Wang HG, Pathan N, Ethell IM, Krajewski S, Yamaguchi Y, Shibasaki F, McKeon F, Bobo T, Franke TF, Reed JC. 1999. Ca^{2+}-induced apoptosis through calcineurin dephosphorylation of BAD. *Science* **284:** 339–343.

Wang M, Zhang WB, Zhu JH, Fu GS, Zhou BQ. 2010. Breviscapine ameliorates cardiac dysfunction and regulates the myocardial Ca^{2+}-cycling proteins in streptozotocin-induced diabetic rats. *Acta Diabetol* doi:101007/s00592-009-0164-x.

Wehrens XH, Lehnart SE, Huang F, Vest JA, Reiken SR, Mohler PJ, Sun J, Guatimosim S, Song LS, Rosemblit N, et al. 2003. FKBP12.6 deficiency and defective calcium release channel (ryanodine receptor) function linked to exercise-induced sudden cardiac death. *Cell* **113:** 829–840.

Wehrens XH, Lehnart SE, Marks AR. 2005a. Intracellular calcium release and cardiac disease. *Annu Rev Physiol* **67:** 69–98.

Wehrens XH, Lehnart SE, Reiken S, van der Nagel R, Morales R, Sun J, Cheng Z, Deng SX, de Windt LJ, Landry DW, et al. 2005b. Enhancing calstabin binding to ryanodine receptors improves cardiac and skeletal muscle function in heart failure. *Proc Natl Acad Sci* **102:** 9607–9612.

Wehrens XH, Lehnart SE, Reiken S, Vest JA, Wronska A, Marks AR. 2006. Ryanodine receptor/calcium release channel PKA phosphorylation: a critical mediator of heart failure progression. *Proc Natl Acad Sci* **103:** 511–518.

Wei H, Perry DC. 1996. Dantrolene is cytoprotective in two models of neuronal cell death. *J Neurochem* **67:** 2390–2398.

Wei H, Kim SJ, Zhang Z, Tsai PC, Wisniewski KE, Mukherjee AB. 2008a. ER and oxidative stresses are common mediators of apoptosis in both neurodegenerative and non-neurodegenerative lysosomal storage disorders and are alleviated by chemical chaperones. *Hum Mol Genet* **17:** 469–477.

Wei H, Liang G, Yang H, Wang Q, Hawkins B, Madesh M, Wang S, Eckenhoff RG. 2008b. The common inhalational anesthetic isoflurane induces apoptosis via activation of inositol 1,4,5-trisphosphate receptors. *Anesthesiology* **108:** 251–260.

Wei Y, Wang D, Gentile CL, Pagliassotti MJ. 2009. Reduced endoplasmic reticulum luminal calcium links saturated fatty acid-mediated endoplasmic reticulum stress and cell death in liver cells. *Mol Cell Biochem* **331:** 31–40.

White C, Li C, Yang J, Petrenko NB, Madesh M, Thompson CB, Foskett JK. 2005. The endoplasmic reticulum gateway to apoptosis by Bcl-X$_L$ modulation of the InsP$_3$R. *Nat Cell Biol* **7:** 1021–1028.

Williams BL, Lipkin WI. 2006. Endoplasmic reticulum stress and neurodegeneration in rats neonatally infected with borna disease virus. *J Virol* **80:** 8613–8626.

Wold LE, Dutta K, Mason MM, Ren J, Cala SE, Schwanke ML, Davidoff AJ. 2005. Impaired SERCA function contributes to cardiomyocyte dysfunction in insulin resistant rats. *J Mol Cell Cardiol* **39:** 297–307.

Wu X, Sun L, Zha W, Studer E, Gurley E, Chen L, Wang X, Hylemon PB, Pandak WM Jr, Sanyal AJ, et al. 2010. HIV protease inhibitors induce ER stress and disrupt barrier integrity in intestinal epithelial cells. *Gastroenterology* **138:** 197–209.

Xiong ZG, Chu XP, Simon RP. 2007. Acid sensing ion channels–novel therapeutic targets for ischemic brain injury. *Front Biosci* **12:** 1376–1386.

Xu C, Bailly-Maitre B, Reed JC. 2005. Endoplasmic reticulum stress: cell life and death decisions. *J Clin Invest* **115:** 2656–2664.

Xu L, Eu JP, Meissner G, Stamler JS. 1998. Activation of the cardiac calcium release channel (ryanodine receptor) by poly-S-nitrosylation. *Science* **279:** 234–237.

Yamada T, Ishihara H, Tamura A, Takahashi R, Yamaguchi S, Takei D, Tokita A, Satake C, Tashiro F, Katagiri H, et al. 2006. WFS1-deficiency increases endoplasmic reticulum stress, impairs cell cycle progression and triggers the apoptotic pathway specifically in pancreatic β-cells. *Hum Mol Genet* **15:** 1600–1609.

Yamamoto K, Ichijo H, Korsmeyer SJ. 1999. BCL-2 is phosphorylated and inactivated by an ASK1/Jun N-terminal protein kinase pathway normally activated at G$_2$/M. *Mol Cell Biol* **19:** 8469–8478.

Yang H, Liang G, Hawkins BJ, Madesh M, Pierwola A, Wei H. 2008. Inhalational anesthetics induce cell damage by disruption of intracellular calcium homeostasis with different potencies. *Anesthesiology* **109:** 243–250.

Yano M, Kobayashi S, Kohno M, Doi M, Tokuhisa T, Okuda S, Suetsugu M, Hisaoka T, Obayashi M, Ohkusa T, et al. 2003. FKBP12.6-mediated stabilization of calcium-release channel (ryanodine receptor) as a novel therapeutic strategy against heart failure. *Circulation* **107:** 477–484.

Yano M, Yamamoto T, Kobayashi S, Ikeda Y, Matsuzaki M. 2008. Defective Ca^{2+} cycling as a key pathogenic mechanism of heart failure. *Circ J* **72:** A22–A30.

Yaras N, Ugur M, Ozdemir S, Gurdal H, Purali N, Lacampagne A, Vassort G, Turan B. 2005. Effects of diabetes on ryanodine receptor Ca release channel (RyR2) and Ca^{2+} homeostasis in rat heart. *Diabetes* **54:** 3082–3088.

Ying J, Sharov V, Xu S, Jiang B, Gerrity R, Schoneich C, Cohen RA. 2008. Cysteine-674 oxidation and degradation of sarcoplasmic reticulum Ca^{2+} ATPase in diabetic pig aorta. *Free Radic Biol Med* **45:** 756–762.

Yoneda T, Imaizumi K, Oono K, Yui D, Gomi F, Katayama T, Tohyama M. 2001. Activation of caspase-12, an endoplastic reticulum (ER) resident caspase, through tumor necrosis factor receptor-associated factor 2-dependent mechanism in response to the ER stress. *J Biol Chem* **276:** 13935–13940.

Yoshida H, Matsui T, Hosokawa N, Kaufman RJ, Nagata K, Mori K. 2003. A time-dependent phase shift in the mammalian unfolded protein response. *Dev Cell* **4:** 265–271.

Yoshida H, Matsui T, Yamamoto A, Okada T, Mori K. 2001. XBP1 mRNA is induced by ATF6 and spliced by IRE1 in response to ER stress to produce a highly active transcription factor. *Cell* **107:** 881–891.

Yoshida I, Monji A, Tashiro K, Nakamura K, Inoue R, Kanba S. 2006. Depletion of intracellular Ca^{2+} store itself may be a major factor in thapsigargin-induced ER stress and apoptosis in PC12 cells. *Neurochem Int* **48:** 696–702.

Zalk R, Lehnart SE, Marks AR. 2007. Modulation of the ryanodine receptor and intracellular calcium. *Annu Rev Biochem* **76:** 367–385.

Zbidi H, Redondo PC, Lopez JJ, Bartegi A, Salido GM, Rosado JA. 2010. Homocysteine induces caspase activation by endoplasmic reticulum stress in platelets from type 2 diabetics and healthy donors. *Thromb Haemost* **103:** 1022–1032.

Zhang S, Ong CN, Shen HM. 2004. Critical roles of intracellular thiols and calcium in parthenolide-induced apoptosis in human colorectal cancer cells. *Cancer Lett* **208:** 143–153.

Zhang L, Cannell MB, Phillips AR, Cooper GJ, Ward ML. 2008. Altered calcium homeostasis does not explain the contractile deficit of diabetic cardiomyopathy. *Diabetes* **57:** 2158–2166.

Zhang Y, Chen HS, Khanna VK, De Leon S, Phillips MS, Schappert K, Britt BA, Browell AK, MacLennan DH. 1993. A mutation in the human ryanodine receptor gene associated with central core disease. *Nat Genet* **5:** 46–50.

Zhang Z, Lee YC, Kim SJ, Choi MS, Tsai PC, Xu Y, Xiao YJ, Zhang P, Heffer A, Mukherjee AB. 2007. Palmitoyl-protein thioesterase-1 deficiency mediates the activation of the unfolded protein response and neuronal apoptosis in INCL. *Hum Mol Genet* **15:** 337–346.

Zhong Y, Ahmed S, Grupp IL, Matlib MA. 2001. Altered SR protein expression associated with contractile dysfunction in diabetic rat hearts. *Am J Physiol* **281:** H1137–H1147.

Zhou Y, Frey TK, Yang JJ. 2009. Viral calciomics: interplays between Ca^{2+} and virus. *Cell Calcium* **46:** 1–17.

Zhou BQ, Hu SJ, Wang GB. 2006. The analysis of ultrastructure and gene expression of sarco/endoplasmic reticulum calcium handling proteins in alloxan-induced diabetic rat myocardium. *Acta Cardiol* **61:** 21–27.

Zhou H, Pandak WM Jr, Lyall V, Natarajan R, Hylemon PB. 2005. HIV protease inhibitors activate the unfolded protein response in macrophages: implication for atherosclerosis and cardiovascular disease. *Mol Pharmacol* **68:** 690–700.

Index

A

Adenylate cyclases, phosphorylation of, 306
Adenylyl cyclases, 273–290
    activity at single cell level, 290
    overview, 273–275
    phylogenetic tree, 276
    physiological roles for Ca^{2+}-dependency of, 285–287
    regulation by Ca^{2+}-signaling pathways, 278–290
        by agonist-triggered Ca^{2+} release, 288
        by Ca^{2+} entry, 287
        compartments for Ca^{2+} and cAMP, 288–289
        direct regulation by Ca^{2+}, 278–279
        evidence for, 285, 286
        indirect regulation by PKC, 281, 282–283
        indirect regulation via CaMKII, CaMKIV, and CaN, 279–280, 282
        indirect regulation via $G\beta\gamma$, 283–285
    regulatory susceptibilities of, 275–278
    structural domains of, 275
Aequorin, 361–362
A kinase anchoring protein (AKAP), 9–11
Anesthesia, IP_3R activity and, 472
Angiotensins, apoptosis regulation by, 446, 447
Apoptosis, 443–452
    ER stress and, 464–467
    overview, 443–444
    regulation by anti-apoptotic Bcl-2 family proteins, 449–451
        IP_3 receptor interaction, 449–450
        IP_3R phosphorylation, 450
        mitochondrial cross-talk, 450
        regulation of luminal calcium, 450–451
        regulation of prosurvival calcium signals, 451
    regulation by antigen receptors, 444–445
    regulation by pharmacological agents, 448
        cisplatin, 448
        staurosporine, 448
        thapsigargin, 448
    regulation by proapoptotic Bcl-2 family proteins, 451–452
        calcium regulation by Bax and Bak, 451
        calcium regulation by BH3-only proteins, 451–452
    regulation by signaling molecules, 445–448
        angiotensins, 446, 447
        corticosteroids, 445–446
        nitric oxide, 447–448
        testosterone, 446–447
Arachidonate-regulated Ca^{2+} (ARC) channels, STIM1 roles in regulation, 74
Asthma, 474–475
Atherosclerosis, 472
ATP, ryanodine receptor (RyR) activation and, 105–106
ATPase. *See* Plasma membrane Ca^{2+} ATPase (PMCA); SERCA (sarco/endoplasmic reticulum Ca^{2+} ATPase); SPCA (secretory pathway Ca^{2+} ATPase)
Atypical periodic paralyses (APP), role of ryanodine receptors (RyRs) in, 101
Autophagy, 452–455
    induction by ER stress pathways, 453
    overview, 452–453
    regulation by calcium signaling, 453–454
    regulation by IP_3Rs, 454–455
    T-cell activation, 454
Autosomal centronuclear myopathy, 475
Axon outgrowth and pathfinding, regulation of, 345–346

B

Bak, calcium regulation by, 451
Bax, calcium regulation by, 451
Bcl-2 family proteins
    apoptosis regulation by anti-apoptotic, 449–451
    apoptosis regulation by proapoptotic, 451–452
    regulation of ER Ca^{2+} pump, 236
BH3-only proteins, calcium regulation by, 451–452
Bik, 452
Bim, 452
BiP (immunoglobulin bonding protein), 199–200
BK_{Ca} channels, in smooth muscle, 436, 437
Brody disease, 475
Buffers. *See* Calcium (Ca^{2+}) buffers, cytosolic; Calcium (Ca^{2+}) buffers, organellar
Burn injury, 475

C

Ca^{2+} current types, 2–4
Ca^{2+}-dependent inactivation (CDI), 68–69
Ca^{2+} homeostasome, 188–191

Index

Ca^{2+}-induced Ca^{2+} release (CICR), in cardiac myocytes, 404–405, 407–409
Ca^{2+} sensors, as buffers, 177
Cab45, 205
CaBPs. See Calcium-binding proteins
Cadmium toxicity, 475
Calbindin-D9K (CB-D9K), 184–185
    properties of, 178
    structure, 181
Calbindin-D28K (CB-D28K), 185–186
    functional aspects, 185–186
    properties of, 178
    structural aspects, 185
    structure, 181
Calcineurin (CaN)
    adenylyl cyclase regulation, 277, 282
    cardiac RyRs and, 409
Calcium-binding proteins
    histidine-rich and regulation of ER Ca^{2+} pump, 238
    neuronal function, regulation of, 262–266
    voltage-gated calcium channels
        Ca_V2 channels in synaptic transmission, 14–15
        excitation-contraction coupling, 11
Calcium (Ca^{2+}) buffers, cytosolic, 177–191
    Ca^{2+} homeostasome and, 188–191
    Ca^{2+} sensors as, 177
    calbindin-D9K (CB-D9K), 184–185
    calbindin-D28K (CB-D28K), 185–186
        functional aspects, 185–186
        structural aspects, 185
    calretinin, 186
    defined, 177–179
    mitochondrial Ca^{2+} buffering compared, 187–188
    parameters to characterize, 179–186
        intracellular concentration, 179
        ligand interaction, 182
        metal-binding affinities, 179–180
        metal-binding kinetics, 180, 182
        mobility, 182
    parvalbumins, 182–184
        functional aspects, 183–184
        structured aspects, 182–183
    pH buffers compared, 178
    properties of, 178
    structure, 181
Calcium (Ca^{2+}) buffers, organellar, 197–207
    endolysosomal compartment, 206
    endoplasmic reticulum, 197–201
        calreticulin, 198–199
        ERp72, 200–201
        GRP94, 200
        immunoglobulin bonding protein (BiP/GRP78), 199–200
        protein disulfide isomerase (PDI), 200
    endoplasmic reticulum Golgi intermediate complex, 206
    Golgi Ca^{2+} stores, 205–206
    mitochondria, 206
    peroxisomes, 206
    sarcoplasmic reticulum Ca^{2+} stores, 201–205
Calcium channels. See Store-operated calcium channels (SOCs); Transient receptor potential (TRP) channels; Voltage-gated calcium channels
Calcium oscillations, 159–171
    computational models for, 160–167
        classifications, 162
        $InsP_3$ oscillations and, 161–164
        perspectives, 166–167
        stochastic aspects, 164–165
    interplay between Ca^{2+} entry and Ca^{2+} release during, 167–171
    overview, 159–160
    store-operated calcium (SOC) channels, 168–171
Calcium signaling
    autophagy regulation by, 453–454
    in cardiac myocytes, 403–418
    during mammalian fertilization, 313–328
    in neuronal development, 337–348
    in smooth muscle, 423–438
    in synapse-to-nucleus communication, 371–392
    visualization during embryonic skeletal muscle formation in vertebrates, 351–366
Caldendrin, 265, 266
Calmodulin (CaM)
    Ca^{2+} channels, 11, 14–15
    IP_3 receptor regulation, 126
    neuronal functions of, 255–257
    ryanodine receptor (RyR) regulation, 106, 408
    as sensor for CDI (Ca^{2+}-dependent inactivation), 69, 407
Calneurons, 263, 265–266
Calnexin, regulation of ER Ca^{2+} pump and, 236–237
CALNUC (nucleobindin), 205
Calreticulin
    gain-of-function (overexpression), 199
    loss-of-function (deficiency), 198–199
    organellar calcium buffering, 198–199
    regulation of ER Ca^{2+} pump, 236–237
    structure, 198
Calretinin
    properties of, 178
    structural and functional aspects, 186
Calsequestrin, 201–203
    in cardiac myocytes, 408
    ryanodine receptor (RyR) regulation, 106–107, 408
Calumenin, regulation of ER Ca^{2+} pump and, 237–238
CaM. See Calmodulin (CaM)

Index

CaM kinase (CaMK)
    adenylyl cyclase regulation, 277, 279–280, 282
    CaMKII
        dendritic arborization and, 344–345
        ryanodine receptor (RyR) phosphorylation, 108
CaN. See Calcineurin (CaN)
Cancer, 476–477
    anticancer drugs, 476–477
    malignant transformation, 476
Cardiac cycle, overview of, 403–404
Cardiac myocytes
    adaptation to disease, 414–417
    calcium signaling, 403–418
        Ca^{2+} efflux, 410–412
        contraction, 409–410
        excitation-contraction coupling, 404–406, 413
        L-type VOCCs, regulation of, 406–407
        RyRs and CICR, regulation of, 407–409
    cardiac cycle, 403–404
    subcellular organization, 412–414
Cardiomyopathy, diabetic, 468–469
Cardiovascular diseases, 472–473
Catecholaminergic polymorphic ventricular tachycardia (CPVT), 408
Cation channels, nonselective
    calcium signaling in smooth muscle, 429–431
    receptor-activated channels, 430–431
CB-D9K. See Calbindin-D9K (CB-D9K)
CB-D28K. See Calbindin-D9K (CB-D28K)
CDI (Ca^{2+}-dependent inactivation), 68–69
Central core disease, 101, 475
Channels. See specific channel types
Cholestatic liver disease, 475
Chronic heart failure, 472–473
CICR (Ca^{2+}-induced Ca^{2+} release), in cardiac myocytes, 404–405, 407–409
Cisplatin, apoptosis regulation by, 448
Corticosteroids, apoptosis regulation by, 445–446
CPVT (catecholaminergic polymorphic ventricular tachycardia), 408
CRAC (Ca^{2+} release-activated Ca^{2+}) channels
    accumulation/activation at ER-PM junctions, 64–66
    overview, 58
    properties, 68–69
        Ca^{2+}-dependent inactivation (CDI), 68–69
        ion selectivity and permeation, 68
CSQ. See Calsequestrin
Cytosolic calcium buffers, 177–191

D

Dendritic arborization, 344–345
DHPRs (dihydropyridine receptors), 104
Diabetes mellitus, 467–471
    cardiomyopathy, 468–469
    nephropathy, 470
    sensory neuropathy, 470
    vascular disease, 469–470
Diacylglycerol kinases (DGK), phosphorylation of, 306
Dihydropyridine receptors (DHPRs), 104

E

EF-hand proteins
    calcium-binding proteins (CaBPs), 262–264
    cytosolic calcium buffers, 177, 181
    neuronal calcium sensor (NCS) proteins, 257, 258
Endolysosomal compartment
    calcium buffers, 206
    NAADP and physiology of, 152–153
    transient receptor potential (TRP) channels, 41–44
Endoplasmic reticulum
    Ca^{2+} homeostasis in, 462–464
    Ca^{2+} pumps of, 229–245
    calcium buffers, 197–201
        calreticulin, 198–199
        ERp72, 200–201
        GRP94, 200
        immunoglobulin bonding protein (BiP/GRP78), 199–200
        protein disulfide isomerase (PDI), 200
    calcium depletion and disease, 461–477
        asthma, 474–475
        atherosclerosis, 472
        autosomal centronuclear myopathy, 475
        Brody disease, 475
        burn injury, 475
        cadmium toxicity, 475
        cancer, 476–477
        cardiovascular diseases, 472–473
        central core disease, 475
        chronic heart failure, 472–473
        diabetes mellitus, 467–471
        liver diseases, 475
        lung diseases, 474–475
        neural ischemia, 471
        neurodegeneration, 471–472
        neurological diseases, 471–472
        neuropathic pain, 472
        skeletal muscle diseases, 475
        skin disease, 475–476
        virus infection, 473–474
    conduit from synapse to nucleus, 390–391
    ER-PM junctions
        CRAR channel accumulation/activation at, 64–66
        formation and function of, 71–72
        STIM accumulation at, 63–64
    NAADP and lysosomal-ER interactions, 150–151
    transient receptor potential (TRP) channels, 38–41

Endoplasmic reticulum Golgi intermediate complex (ERGIC), 206
Endosomes, transient receptor potential (TRP) channels and, 41–44
Epigenetic mechanisms in synaptic activity and Ca^{2+}-dependent transcriptional regulation, 385–387
ER-assisted degradation, 464, 465
ERGIC (endoplasmic reticulum Golgi intermediate complex), 206
ERp57, regulation of ER Ca^{2+} pump and, 236–237
ERp72, 200–201
ER-PM junctions
 CRAR channel accumulation/activation at, 64–66
 formation and function of, 71–72
 STIM accumulation at, 63–64
ER stress
 apoptosis and, 464–467
 autophagy induction by, 453
Excitation-contraction coupling
 in cardiac myocytes, 404–406, 413
 voltage-gated calcium channels, 8–11
Excitation-response coupling, voltage-gated calcium channels and, 8
Excitation-secretion coupling, voltage-gated calcium channels and, 12
Excitation-transcription coupling
 smooth muscle, calcium signaling in, 437
 voltage-gated calcium channels, 11–12

F

Fatty liver, nonalcoholic, 475
Fertilization, mammalian
 Ca^{2+} homeostasis in oocytes and eggs, 323–325
 $[Ca^{2+}]_i$ clearing mechanisms, 323–324
 Ca^{2+} influx mechanisms, 324–325
 Ca^{2+} oscillations, molecular players responsible for, 318–327
 calcium signaling, 313–328
 egg activation, 314–318
 IP_3R1 in maturing oocytes, 321–323
 IP_3R1 in MII eggs, 319–321
 PLCζ, 325–327
FK506-binding proteins (FKBPs)
 in cardiac myocytes, 408–409
 ryanodine receptors (RyR) regulation, 107–108, 408–409
Fluorescent Ca^{2+} reporters, 362–363

G

GBγ, adenylyl cyclase regulation by, 283–285
GCAPs (guanylyl-cyclase-activating proteins), 261
Glutamate. See Metabotropic glutamate receptor type 1 (mGluR1)

Golgi apparatus
 Ca^{2+} pumps of, 229–245
 calcium buffers, 205–206
G protein modulation, voltage-gated calcium channels and, 13–14
G_q proteins, 85
GRP78, 199–200
GRP94, 200
Guanylyl-cyclase-activating proteins (GCAPs), 261

H

Hailey-Hailey disease, 244
Hax-1, regulation of ER Ca^{2+} pump, 236
Heart failure, chronic, 472–473
Histidine-rich Ca^{2+}-binding protein, 203, 238
Hyperpolarizing-activated cyclic nucleotide-gated (HCN) channels, 404

I

Immune cells, functional roles of STIM and Orai in, 70
Immunoglobulin bonding protein (BiP), 199–200
Inositol triphosphate (IP_3), calcium oscillations and, 161–164
Inositol triphosphate receptors. See IP_3 receptors
Intracellular Ca^{2+} dynamics, 107
IP_3 receptors, 119–132
 autophagy regulation by, 454–455
 Bcl-2 interaction, 449–450
 calcium oscillation computational models, 162
 calcium signaling in smooth muscle, 433–435
 history of, 119–123
 IP_3R1 in Purkinje cells, 85–86
 phosphorylation, 450
 regulation of, 123–127
 structural determinants of activation, 127–132
 virus infections and, 474
Ischemia, neural, 471

J

Junctate, 203, 205

K

K^+ channel-interacting proteins (KChIPs), 261–262

L

Liver diseases, 475
Lung diseases, 474–475
Lysosomes
 NAADP and lysosomal-ER interactions, 150–151
 storage disease, 471
 transient receptor potential (TRP) channels, 41–44

M

Malignant hyperthermia, role of ryanodine receptors (RyRs) in, 99–101
Metabotropic glutamate receptor type 1 (mGluR1), 81–92
  mechanisms of synaptic depolarizations, 86–91
  overview, 81–83
  postsynaptic Ca^{2+} release from internal stores, 83–86
  TRPC channels and, 86–92
Metal-binding affinities, of cytosolic Ca^{2+} buffers, 179–180
Metal-binding kinetics, of cytosolic Ca^{2+} buffers, 180, 182
Mg^{2+}, ryanodine receptors (RyR) inhibition by, 105
MGluR1. See Metabotropic glutamate receptor type 1 (mGluR1)
Mitochondria
  Bcl-2 and mitochondrial cross talk, 450
  calcium buffers, 187–188, 206
Multiminicore disease (MmD), role of ryanodine receptors (RyRs) in, 101
Muscle. See Cardiac myocytes; Skeletal muscle; Smooth muscle

N

NAADP. See Nicotinic acid adenine nucleotide diphosphate (NAADP)
NAADP receptors, 141–153
  in endolysosomal physiology, 152–153
  pharmacological properties, 146
  two-pore channels, 147–148
NCS. See Neuronal calcium sensor (NCS) family
NCX. See Sodium/calcium exchanger (NCX)
Nephropathy, diabetic, 470
Neural ischemia, 471
Neurodegeneration, 471–472
Neurological diseases, 471–472
Neuronal calcium sensor (NCS) family, 257–262
  guanylyl-cyclase-activating proteins (GCAPs), 261
  K^+ channel-interacting proteins (KChIPs), 261–262
  KChIPs, 261–262
  NCS-1, 257–259
  recoverin, 259–261
Neuronal development, calcium signaling in, 337–348
  axon outgrowth and pathfinding, regulation of, 345–346
  calcium-mediated specification of neurotransmitter phenotype, 340–344
  dendritic arborization, 344–345
  frequency coding by calcium transients, 347–348
  ion channel expression, 346–347
  spatiotemporal characteristics of calcium transients, 339–340
Neuronal function, regulation of, 253–266
  calcium-binding proteins (CaBPs), 262–266
  calmodulin, 255–257
  neuronal calcium sensor (NCS) family, 257–262
  synaptotagmins and neurotransmitter release, 254–255
Neuropathic pain, 472
Neuropathy, diabetic, 470
Neurotransmitter phenotype, calcium-mediated specification of, 340
Neurotransmitter release, synaptotagmins and, 254–255
NHE (sodium/proton exchangers), phosphorylation of, 306
Nicotinic acid adenine nucleotide diphosphate (NAADP). See also NAADP receptors
  as Ca^{2+}-mobilizing messenger, 142–143
  Ca^{2+} stores targeted by, 143–144
    mammalian cells, 144–146
    sea urchin eggs, 143–144
  desensitization of NAADP-evoked Ca^{2+} release, 144–146
  discovery as Ca^{2+}-mobilizing molecule, 142
  in endolysosomal physiology, 152–153
  lysosomal-ER interactions, 150–152
  plasma membrane excitability, modulation of, 152
  structure and function, 142
Nitric oxide, apoptosis regulation by, 447–448
NMDA receptors
  antagonism of synapse-to-nucleus communication, 381–382, 383
  caldendrin and, 265
Nuclear Ca^{2+} signaling, 382, 384–385, 388–390
Nuclear pore permeability, calcium regulation of, 387

O

Oncomodulin, functional aspects of, 183
Orai
  calcium oscillations and, 168–169
  calcium signaling in smooth muscle, 429
  CDI (Ca^{2+}-dependent inactivation), participation in, 69
  functional organization of, 62
  functional roles of, 69–71
    immune cells, 70
    platelets, 70–71
    skeletal muscle, 71
    sweat glands and teeth, 71
  molecular basis for store-operated calcium entry, 59–69
  stoichiometry questions, 66–67
Organellar calcium buffers, 197–207

Index

P

P2X receptor, 430–431
P54/NEFA, 205
Parvalbumins, 182–184
    functional aspects, 183–184
    properties of, 178
    structure, 181
    structured aspects, 182–183
PDI (protein disulfide isomerase), 200
Peroxisomes, calcium buffers and, 206
Phospholamban, regulation of ER Ca^{2+} pump, 233–236
Phospholipase C (PLC)
    mGlu1-dependent slow EPSC (excitatory postsynaptic current) and, 87, 89–91
    phosphorylation of, 305–306
    PLCζ in mammalian fertilization, 325–327
Phospholipase D (PLD), phosphorylation of, 306
PKA. See Protein kinase A (PKA)
PKC. See Protein kinase C (PKC)
Plasma membrane
    ER-PM junctions
        CRAR channel accumulation/activation at, 64–66
        formation and function of, 71–72
        STIM accumulation at, 63–64
    NAADP and modulation of excitability, 152
Plasma membrane Ca^{2+} ATPase (PMCA)
    genetic manipulations, 218–219
    isoforms and tissue distribution, 216–218
    overview, 213–215
    protein kinase C (PKC) and, 304
    role in physiology and pathology, 218
    structural and regulatory characteristics, 215–216
Platelets
    functional roles of STIM and Orai in, 70–71
    SERCA3 in, 239
PLC. See Phospholipase C (PLC)
PLD (phospholipase D), phosphorylation of, 306
PMCA. See Plasma membrane Ca^{2+} ATPase (PMCA)
PMR1 mutants in yeast, 243
Programmed cell death. See Apoptosis
Protein disulfide isomerase (PDI), 200
Protein kinase A (PKA)
    ryanodine receptor (RyR) phosphorylation, 108, 409
    sodium/calcium exchange (NCX) phosphorylation, 411
    voltage-gated calcium channels, 406
Protein kinase C (PKC), 295–308
    activation of, 296, 298–304
    adenylyl cyclase regulation, 277, 281, 282–283
    imaging cPKC dynamics in living cells, 300–302
    interaction between C2 domains and phospholipids, 299–300
    overview of family, 295–296
    properties of, 297
    regulation of Ca^{2+} handling, 304–306
    regulatory loops, 306–308
    sodium/calcium exchange (NCX) phosphorylation, 411
    terminating signaling, 302–304
    voltage-gated calcium channels, 406
Protein phosphatases, phosphorylation of, 306
Purkinje cells
    cytosolic Ca^{2+} buffers, 189–191
    metabotropic glutamate receptor type 1 (mGluR1), 82–92

R

Reactive nitrogen species, ryanodine receptors (RyR) regulation, 108–109
Reactive oxygen species, ryanodine receptors (RyR) regulation, 108–109
Recoverin, 259–261
Ryanodine receptors (RyRs), 97–109
    calcium oscillation computational models, 162
    calcium signaling in cardiac myocytes, 407–409, 414, 415, 416
    calcium signaling in smooth muscle, 432–433
    calsequestrin interaction with, 201
    expression of, 99
    human diseases, roles in, 99–102
    receptor genes and isoforms, 98–99
    regulation, 104–109
        Ca^{2+}, Mg^{2+}, and ATP, 105–106
        calmodulin, 106, 408
        calsequestrin, 106–107, 408
        FK506-binding proteins, 107–108, 408–409
        phosphorylation, 108, 409
        reactive oxygen species and reactive nitrogen species, 108–109
        voltage-gated calcium channels, 104–105
    ultrastructural studies, 102–104
    virus infections and, 474
RyRs. See Ryanodine receptors (RyRs)

S

Salivary glands, abnormal Ca^{2+} signaling in, 470–471
Sarco/endoplasmic reticulum Ca^{2+} ATPase. See SERCA (sarco/endoplasmic reticulum Ca^{2+} ATPase)
Sarcolipin, regulation of ER Ca^{2+} pump, 233–236
Sarcoplasmic reticulum
    Ca^{2+} homeostasis in, 462–464
    calcium buffers, 201–205
    transient receptor potential (TRP) channels, 38–41
Sea urchin eggs
    NAADP-evoked Ca^{2+} release, 143–144
    properties of endogenous two-pore channels from, 148–149

Second messengers, in synapse-to-nucleus communication, 378–381
Secretory granules, transient receptor potential (TRP) channels and, 44
Secretory pathway Ca^{2+} ATPase. *See* SPCA (secretory pathway Ca^{2+} ATPase)
Secretory vesicles, transient receptor potential (TRP) channels and, 44
SERCA (sarco/endoplasmic reticulum Ca^{2+} ATPase)
    cardiac myocytes, 410–412, 415, 416
    pumping mechanism, 231
    regulators, 233–238
        antiapoptotic proteins Hax-1 and Bcl-2, 236
        calumenin, 237–238
        histidine-rich Ca^{2+}-binding protein, 238
        luminal proteins (calreticulin, calnexin, ERp57), 236–237
        overview, 233
        phospholamban, 233–236
        sarcolipin, 233–236
    SERCA1, 238–239
    SERCA2
        physiological role of, 230–231
        regulators, 233–238
        splicing variants, 230
        structure of SERCA2b pump, 231–233
    SERCA3, 239–240
    virus infections and, 474
Signaling pathways, regulation by adenylyl cyclases, 273–290
Signal transduction, by voltage-gated calcium channels, 2
Skeletal muscle
    diseases, 475
    functional roles of STIM and Orai in, 71
    visualization of Ca^{2+} signaling during embryonic formation in vertebrates, 351–366
        Ca^{2+} transients in mouse and rat myotubules, 355–359
        cultured *Xenopus* myocytes, 363–366
        examples of developing muscle systems, 353
        imaging methodologies, 352–355
        zebrafish embryos, 359–363
Skin disease, 475–476
Smooth muscle
    calcium signaling, 423–438
        Ca^{2+} flashes, 427–428, 429, 436
        Ca^{2+} sparklets, 428, 430
        Ca^{2+} sparks, 432–433, 436
        Ca^{2+} waves, 433–435, 436
        contractility and, 435–437
        excitation-transcription coupling, 437
        global Ca^{2+} signals, 427, 428, 435–436, 437
        imaging intracellular Ca^{2+}, 424–426
        intracellular Ca^{2+} stores, 431–435
        IP$_3$R mediated, 433–435
        nonselective cation channels, 429–431
        ryanodine receptor (RyR) mediated, 432–433, 436
        store-operated calcium channels, 429
        transcription factor activation, 437–438
        VDCCs, signals mediated by, 426–428, 435–438
        voltage-gated calcium channels, 426–428
    described, 423
SNARE proteins, 13, 255
SOCs. *See* Store-operated calcium channels (SOCs)
Sodium/calcium exchanger (NCX)
    cardiac myocytes, 404, 411, 413–414
    genetic manipulations, 223
    isoforms and tissue distribution, 221–222
    overview, 213–215
    protein kinase C (PKC) and, 305
    role in physiology and pathology, 222–223
    structural and regulatory characteristics, 219–221
Sodium/proton exchangers (NHE), phosphorylation of, 306
SPCA (secretory pathway Ca^{2+} ATPase), 240–244
    expression pattern, 240–243
        SPCA1, 240–242
        SPCA2, 242–243
    history of, 240
    role in cellular physiology, 243–244
        human disease, 244
        PMR1 mutants in yeast, 243
        studies in cell systems, 243
        studies in model organisms, 243–244
    structure, 240
Staurosporine, apoptosis regulation by, 448
STIM
    calcium oscillations and, 168–169
    calcium signaling in smooth muscle, 429
    functional roles of, 69–71
        immune cells, 70
        platelets, 70–71
        skeletal muscle, 71
        sweat glands and teeth, 71
    molecular basis for store-operated calcium entry, 59–69
        accumulation at ER-PM junctions, 63–64
        functional organization of, 61
        oligomerization, 61–63
    STIM1 roles in calcium channel regulation, 73–74
        ARC channels, 74
        TRPC channels, 73–74
        voltage-gated (Ca$_V$) channels, 74
    stoichiometry questions, 66–67
Store-operated calcium channels (SOCs), 57–75
    calcium oscillations, 168–171
    calcium signaling in smooth muscle, 429
    CRAC channel properties, 68–69
        Ca^{2+}-dependent inactivation (CDI), 68–69
        ion selectivity and permeation, 68

Store-operated calcium channels (SOCs) (*Continued*)
  described, 57–59
  ER-PM junctions
    CRAR channel accumulation/activation at, 64–66
    formation and function of, 71–72
    STIM accumulation at, 63–64
  functional roles of STIM and Orai, 69–71
    immune cells, 70
    platelets, 70–71
    skeletal muscle, 71
    sweat glands and teeth, 71
  molecular basis, 59–69
    CRAC channel accumulation/activation at ER-PM junctions, 64–66
    CRAC channel properties, 68–69
    functional organization of Orai, 62
    functional organization of STIM, 61
    molecular choreography of calcium entry, 59–61
    oligomerization of STIM, 61–63
    STIM accumulation at ER-PM junctions, 63–64
    stoichiometry questions, 66–67
  regulation of calcium entry in physiological context, 72–73
  STIM1 roles in regulation, 73–74
    ARC channels, 74
    TRPC channels, 73–74
    voltage-gated (Ca_V) channels, 74
Sweat glands, functional roles of STIM and Orai in, 71
Synapse-to-nucleus communication, calcium signaling in, 371–392
  endoplasmic reticulum as conduit, 390–391
  epigenetic mechanisms in synaptic activity and Ca^{2+}-dependent transcriptional regulation, 385–387
  extrasynaptic NMDA receptor antagonism, 381–382, 383
  global functional consequences of local signaling, 376–378
  mobile second messengers, 378–381
  nuclear Ca^{2+} signaling, 382, 384–385, 388–390
  nuclear pore permeability, 387
  overview, 371–376
Synaptic transmission, metabotropic glutamate receptor type 1 (mGluR1) and, 81–92
Synaptic vesicle, transient receptor potential (TRP) channels and, 44
Synaptotagmins, neurotransmitter release and, 254–255

T

T cells
  activation and autophagy, 454
  apoptosis and, 443–447, 449, 451–452

Teeth, functional roles of STIM and Orai in, 71
Testosterone, apoptosis regulation by, 446–447
Thapsigargin, apoptosis regulation by, 448
Total internal reflection fluorescence (TIRF)
  imaging calcium signaling in smooth muscle, 425–426
  imaging cPKC dynamics in living cells, 300–301
Transcription factor activation, by calcium signaling in smooth muscle, 437–438
Transient receptor potential (TRP) channels, 25–45
  Ca^{2+} permeable TRP pores, 27–29
  change in inwardly-driving forces for Ca^{2+} entry, 35–38
  depolarization action, 35–38
  described, 25–27
  as intracellular calcium-release channels, 38–44
    endoplasmic reticulum, 38–41
    endosomes, 41–44
    lysosomes, 41–44
    sarcoplasmic reticulum, 38–41
    secretory granules, 44
    secretory vesicles, 44
    synaptic vesicle, 44
  phylogenetic tree of superfamily, 26
  protein kinase C (PKC) and, 304–305
  in smooth muscle cells, 431
TRPA (ankyrin) family, 34–35
TRPC (canonical) family, 29–32
  metabotropic glutamate receptor type 1 (mGluR1), 86–92
  STIM1 roles in regulation, 73–74
  TRPC1, 30
  TRPC2, 30
  TRPC3, 31
  TRPC4, 31
  TRPC5, 31
  TRPC6, 31
  TRPC7, 31–32
TRPM (melastatin) family, 33–34
  TRPM1, 33
  TRPM2, 33–34
  TRPM3, 34
  TRPM4, 34
  TRPM5, 34
  TRPM6, 34
  TRPM7, 34
  TRPM8, 34
TRPM (mucolipin) family, 35
TRPP (polycystin) family, 35
TRPV (vanilloid) family, 32–33
  TRPV1, 32
  TRPV2, 32
  TRPV3, 32–33
  TRPV4, 33
  TRPV5, 33
  TRPV6, 33

Transmissible spongiform encephalopathies, 471–472
Troponins, 409–410
TRP. *See* Transient receptor potential (TRP) channels
Two-pore channels
    described, 147
    as NAADP receptors, 147–148
    phylogenetic tree, 147
    properties of endogenous from sea urchin eggs, 148–149
    single-channel properties of human, 149–150

V

Vascular disease, in diabetes mellitus, 469–470
Virus infection, 473–474
Voltage-gated calcium channels, 1–17
    Ca^{2+} current types, 2–4
    calcium oscillation computational models, 162
    calcium signaling in cardiac myocytes, 404–407, 413, 414
    calcium signaling in smooth muscle, 426–428, 435–438
    Ca_V1 channels
        excitation-contraction coupling, 8–11
        excitation-response coupling, 8
        excitation-secretion coupling, 12
        excitation-transcription coupling, 11–12
    Ca_V2 channels in synaptic transmission, 12–15
        Ca^{2+} binding proteins, 14–15
        G protein modulation, 13–14
        SNARE proteins, 13
    Ca_V3 channels, 15–16
        functional roles, 16
        molecular properties, 15–16
        regulation of, 16
    effector checkpoint model of regulation, 16–17
    excitation-contraction coupling, 8–11
        Ca^{2+} binding proteins, 11
        mechanisms, 8
        proteolytic processing and regulation via carboxy-terminal domain, 9–10
        regulation, 8–9
    molecular properties, 4–8
        channel diversity, 7
        functions of Ca^{2+} channel subunits, 6–7
        molecular basis of function, 7–8
        subunit structure, 4–5
        three-dimensional structure, 5–6
    physiological roles, 1–2
    RyR regulation, 104–105
    STIM1 roles in regulation, 74

X

Xenopus myocytes, visualization of Ca^{2+} transients in, 363–366

Z

Zebrafish embryos, Ca^{2+} signaling during muscle formation, 359–363